Wireless Communication Systems

This practically-oriented, all-inclusive guide covers all the major enabling techniques for current and next-generation cellular communications and wireless networking systems. Technologies covered include CDMA, OFDM, UWB, turbo and LDPC coding, smart antennas, wireless ad hoc and sensor networks, MIMO, and cognitive radios, providing readers with everything they need to master wireless systems design in a single volume.

Uniquely, a detailed introduction to the properties, design, and selection of RF subsystems and antennas is provided, giving readers a clear overview of the whole wireless system. It is also the first textbook to include a complete introduction to speech coders and video coders used in wireless systems.

Richly illustrated with over 400 figures, and with a unique emphasis on practical and state-of-the-art techniques in system design, rather than on the mathematical foundations, this book is ideal for graduate students and researchers in wireless communications, as well as for wireless and telecom engineers.

Ke-Lin Du is currently a researcher in the Center for Signal Processing and Communications at Concordia University, Canada. Prior to joining Concordia University in 2001, he held positions with Huawei Technologies, the China Academy of Telecommunication Technology, and the Chinese University of Hong Kong. He visited the Hong Kong University of Science and Technology in 2008. His current research interests include signal processing, wireless communications, RF systems, and neural networks. He is a Senior Member of the IEEE.

M. N. S. Swamy is currently a Director of the Center for Signal Processing and Communications in the Department of Electrical and Computer Engineering, Concordia University, where he was Dean of the Faculty of Engineering and Computer Science from 1977 to 1993. He has published extensively in the areas of circuits, systems, and signal processing, co-authoring four books. Professor Swamy is a Fellow of the IEEE, IET (UK), and EIC (Canada), and has received many IEEE-CAS awards, including the Guillemin-Cauer award in 1986, as well as the Education Award and the Golden Jubilee Medal, both in 2000.

Wireless Communication Systems

From RF Subsystems to 4G Enabling Technologies

KE-LIN DU and M. N. S. SWAMY

Concordia University, Canada

CAMBRIDGE
UNIVERSITY PRESS

CAMBRIDGE
UNIVERSITY PRESS

University Printing House, Cambridge CB2 8BS, United Kingdom

One Liberty Plaza, 20th Floor, New York, NY 10006, USA

477 Williamstown Road, Port Melbourne, VIC 3207, Australia

314-321, 3rd Floor, Plot 3, Splendor Forum, Jasola District Centre, New Delhi - 110025, India

79 Anson Road, #06-04/06, Singapore 079906

Cambridge University Press is part of the University of Cambridge.

It furthers the University's mission by disseminating knowledge in the pursuit of
education, learning and research at the highest international levels of excellence.

www.cambridge.org
Information on this title: www.cambridge.org/9780521114035

© Cambridge University Press 2010

First published 2010

A catalogue record for this publication is available from the British Library

Library of Congress Cataloging in Publication data
Du, K.-L.
Wireless communication systems / Ke-Lin Du and M.N. S. Swamy.
p.   cm.
Includes bibliographical references and index.
ISBN 978-0-521-11403-5
1. Wireless communication systems.   I. Swamy, M. N. S.   II. Title.
TK5103.2.D825   2010
621.382–dc22
2009051702

ISBN 978-0-521-11403-5 Hardback

Additional resources for this publication at www.cambridge.org/9780521114035

To My Son Cynric
K.-L. Du

and

To My Parents
M. N. S. Swamy

Contents

Preface

In the last three decades, the explosive growth of mobile and wireless communications has radically changed the life of people. Wireless services have migrated from the conventional voice-centric services to data-centric services. The circuit-switched communication network is now being replaced by the all-IP packet-switched network. Mobile communications have also evolved from the first-generation (1G) analog systems to the third-generation (3G) systems now being deployed, and the fourth-generation (4G) systems are now under development and are expected to be available by 2010. The evolution of wireless networking has also taken place rapidly during this period, from low-speed wireless local-area networks (LANs) to broadband wireless LANs, wireless metropolitan-area networks (MANs), wireless wide-area networks (WANs), and wireless personal-area networks (PANs). Also, broadband wireless data service has been expanded into broadcasting service, leading to satellite TV broadcasting and wireless regional-area networks (RANs) for digital TV. The data rate has also evolved from the 10 kbits/s voice communications to approximately 1 Gbit/s in the 4G wireless network. In addition, the 4G wireless network will provide ubiquitous communications.

Scope and Purpose

A complete wireless system involves many different areas. However, most existing textbooks on wireless communications focus only on the fundamental principles of wireless communications, while many other areas associated with a whole wireless system, such as digital signal processing, antenna design, microwave and radio frequency (RF) subsystem design, speech coding, video coding, and channel coding, are left to other books.

This book provides a broad, also in certain depth, technical view of wireless communications, covering various aspects of radio systems. Various enabling technologies for modern wireless communications are also included. Unlike the existing books in the field, this book is organized from a wireless system designer's viewpoint. We give wide coverage to the techniques that are most relevant to the design of wireless communication and networking systems. We focus ourselves on the lower layers of wireless systems, since the upper layers such as network layers and transport layers are topics of general data communication systems. Due to limited space, we do not provide lengthy mathematical details, but rather emphasize the practical aspects.

The book is divided into twenty-two chapters, including introduction, overwiew of wireless communications and networking, wireless channel and radio propagation, cellular systems and multiple access, diversity, channel equalization, modulation and detection, spread-spectrum communications, orthogonal frequency division multiplexing (OFDM), antennas, RF and microwave subsystems, A/D and D/A conversions, digital signal processing, information theory, ultra wideband (UWB) communications, speech/audio coding, image/video coding, channel coding, smart antennas, multiple input multiple output (MIMO) systems, cognitive radios, and wireless ad hoc/sensor networks. Each chapter contains some examples and problems.

Intended Audience

This book is primarily intended as a textbook for advanced undergraduate and graduate students specializing in wireless communications and telecommunication systems. It is also a good reference book for practicing engineers. The reader is supposed to have a background in electrical engineering and to be familiar with the theory of signals and systems, probabilities and stochastic processes, basic circuits, basic digital communications, linear algebra, and advanced calculus. These courses are offered in most electrical engineering undergraduate programs. The contents are useful for mobile cellular communications, satellite communication, and wireless networking.

The material in this book can be taught in two semesters. The first semester may cover Chapters 1 to 13, which deal with the principles of wireless communications, and the analog and digital designs. The second semester could cover the remaining chapters, including information theory and coding, and some advanced and emerging technologies. If only one semester is available for this course, we suggest teaching Chapters 1 to 13, 15, and selected sections from Chapters 18 to 22. Since each chapter is rather comprehensive on the topics treated and is relatively self-contained, the reader can select to read only those chapters that are of interest. MATLAB codes for the examples in the book are downloadable from the book website.

Acknowledgments

First of all, we would like to thank the anonymous reviewers, whose comments have enriched this book. Our appreciation is extended to Ayhan Altintas of Bilkent University (Turkey) for commenting on Chapters 3 and 10, Chunjiang Duanmu of Zhejiang Normal University (China) for reviewing Chapter 17, and Wai Ho Mow of Hong Kong University of Science and Technology (China) for his valuable comments on Chapter 21. The authors would like to express their special thanks to Ezio Biglieri and Giorgio Taricco from Politecnico di Torino (Italy) for helpful discussion.

K.-L. Du would like to express his gratitude to Wei Wu of Concordia University (Canada), Doru Florin Chiper of Technical University "Gh. Asachi" Iasi (Romania), Jie Zeng of Meidian Technologies (China), Yi Shen of Huazhong University of Science and Technology (China), Hong Bao and Jiabin Lu of Guangdong University of Technology (China), Qiang Ni of Brunel University (UK), Yin Yang and Daniel Gaoyang Dai from Hong Kong University of Science and Technology (China), Xiangming Li of Beijing Institute of Technology (China), Qingling Zhang of ZTE Corporation (China), Yi Zhang of Huawei Technologies (China), and Andrew Chi-Sing Leung from City University of Hong Kong (China) for their personal help during the period of preparing this book.

M. N. S. Swamy also wishes to thank his family for their support during the period of preparing this book.

We feel extremely fortunate to have worked with Cambridge University Press. We express our utmost appreciation to Philip Meyler, Publishing Director, Engineering, Mathematical and Physical Sciences, at Cambridge University Press, for his guidance. The encouragement and support provided by Philip Meyler has made the process of writing this book very joyful. Finally, special thanks go to the staff from Cambridge University Press: Sabine Koch, Sarah Matthews, Caroline Brown, Anna Marie Lovett, Richard Marston, and Sehar Tahir, without whose help the production of the book would have been impossible.

Feedback

A book of this length is certain to have some errors and omissions. While we have made significant attempts to a comprehensive description of major techniques related to modern wireless communications, there are many new emerging techniques, some of which may not have been included. Feedback is welcome via email at kldu@ieee.org or swamy@ece.concordia.ca, and we promise to reply to all the messages.

Concordia University
Montreal, Canada

1xEV-DO	1x Evolution, Data Optimized	AMR	adaptive multi-rate
		AMR-WB	adaptive multi-rate wideband
1xEV-DV	1x Evolution, Data and Voice	ANSI	American National Standards Institute
nG	nth generation		
3DES	Triple DES		
3GPP	Third-Generation Partnership Project	APS	adaptive phase-SCORE
		ARQ	automatic repeat request
3GPP2	Third-Generation Partnership Project 2	ASIC	application-specific integrated circuit
4GFSK	quaternary GFSK	ASK	amplitude shift keying
AAC	Advanced Audio Coding	AVC	Advanced Video Coding
ACAB	adaptive CAB	AWGN	additive white Guaasian noise
ACELP	algebraic codebook excited linear prediction		
		balun	balanced-to-unbalanced transformer
ACF	autocorrelation function		
ACI	adjacent channel interference	BAN	body area network
ACLR	adjacent channel leakage ratio	BCH	Bose-Chaudhuri-Hocquenghem
ACPR	adjacent channel power ratio	BCJR	Bahl-Cocke-Jelinek-Raviv
ACS	adaptive cross-SCORE	BER	bit error probability
ACTS	Advanced Communication Technology Satellite	BER	bit error rate
		BFSK	binary FSK
A/D	analog-to-digital	BICM	bit-interleaved coded modulation
ADC	A/D converter		
ADPCM	adaptive differential PCM	BJT	bipolar junction transistor
AES	Advanced Encryption Standard	BLAST	Bell Labs Layered Space-Time
AF	amplify-and-forward	BPSK	binary phase shift keying
AFC	automatic frequency control	BRAN	Broadband Radio Access Network
AGC	automatic gain control		
AM	amplitude modulation	BS	base station
AMC	adaptive modulation and coding	BSC	binary symmetric channel
		BS-CDMA	block-spreading CDMA
AMI	alternative mark inversion	CAB	cyclic adaptive beamforming
AMPS	Advanced Mobile Phone Services	CABAC	context-based adaptive binary arithmetic coding

CAVLC	context-based adaptive variable-length code	DAB	Digital Audio Broadcasting
CCF	cross-correlation function	DAC	D/A converter
CCI	co-channel interference	D-BLAST	diagonal BLAST
CCK	complementary code keying	DBPSK	differential BPSK
CCSDS	Consultative Committee for Space Data Systems	DCT	discrete cosine transform
cdf	cumulative distribution function	DDCR	decision-directed carrier recovery
CDMA	code division multiple access	DDS	direct digital synthesis
CDPD	Cellular Digital Packet Data	DEBPSK	differentially encoded BPSK
CELP	code-excited linear prediction	DECT	Digital Enhanced Cordless Telephone
CF	compress-and-forward	DEMPSK	differentially encoded MPSK
CFO	carrier frequency offset	DEQPSK	differentially encoded QPSK
CIC	cascaded integrator comb		
CIF	common intermediate format	DES	Data Encryption Standard
CIR	carrier-to-interference ratio	DF	decode-and-forward
CLS	constrained least-squares	DFE	decision-feedback equalization
CNR	carrier-to-noise ratio		
CORBA	common object request broker architecture	DFT	Discrete Fourier transform
		DiffServ	differential services
CORDIC	Coordinate Rotation Digital Computer	DM	delta modulation
		DMB	Digital Multimedia Broadcasting
CP-CDMA	cyclic prefix assisted CDMA		
CPFSK	continuous phase FSK	DMPSK	differential MPSK
CPM	continuous phase modulation	DNL	differential nonlinearity
CQF	conjugate quadrature filters	DoA	direction-of-arrival
CRC	cyclic redundancy check	DoD	Department of Defense; also direction-of-departure
CRLB	Cramer-Rao lower bound		
CRSC	circular recursive systematic convolutional	DPCM	differential PCM
		DPSK	differential phase-shift keying
CS-ACELP	Conjugate Structure ACELP		
CSI	channel state information	DQPSK	differential quarternary phase shift keying
CSMA	carrier sense multiple access		
CSMA/CA	CSMA with collision avoidance	DR	dielectric resonator
		DS	direct sequence
CSMA/CD	CSMA with collision detection	DSB	double sideband
		DSB-LC	DSB-large carrier
CT2	Second Generation Cordless Telephone	DSB-SC	DSB-small carrier
		DSL, xDSL	digital subscriber line
CVSDM	continuous variable slope DM	DS-CDMA	direct-sequence CDMA
		DSCQS	double stimulus continuous quality scale
D/A	digital-to-analog		

DSMA	digital sense multiple access	ETSI	European Telecommunications Standards Institute
DSP	digital signal processor		
DSSS	direct-sequence spread spectrum	E-UTRA	Evolved UTRA
		E-UTRAN	Evolved UTRA Network
DST	discrete sine transform	EVRC	enhanced variable rate codec
DSTTD	double-STTD	EVRC-WB	EVRC-Wideband
DTFT	discrete-time Fourier transform	EXIT	extrinsic information transfer
DVB-H	DVB-Handheld	EZW	embedded zero-tree wavelet
DVB-RCL	Digital Video Broadcasting–Return Channel for LMDS	FBSS	fast base station switching
		FCC	Federal Communications Commission
DVB-RCS	DVB-Return Channel via Satellite	FDD	frequency division duplexing
		FDE	frequency-domain equalization
DVB-S	Digital Video Broadcasting Satellite	FDMA	frequency division multiple access
DVB-S2	DVB-Satellite Second Generation	FDTD	finite difference time domain
DVB-T	Terrestrial DVB	FEC	forward error correction
DVB-T2	Terrestrial DVB Second Generation	FEM	finite element method
		FET	field-effect transistor
DWT	discrete wavelet transform	FFT	fast Fourier transform
DySPAN	Dynamic Spectrum Access Networks	FH	frequency hopping
		FH-CDMA	frequency-hopping CDMA
EBCOT	Embedded block coding with optimized truncation	FHSS	frequency-hopping spread spectrum
		FIR	finite impulse response
ECMA	European Computer Manufacturers Association	FM	frequency modultion
		FPGA	field programmable gate array
EDGE	Enhanced Data for GSM Evolution	FR	full-rate
EFR	enhanced full rate	FSK	frequency shift keying
EGC	equal gain combining	FWT	fast wavelet transform
EIA	Electronics Industry Association	GaAs	gallium arsenide
		GEO	geostationary earth orbit
EM	electromagnetic	GFSK	Gaussian FSK
ENOB	effective number of bits	GMC	generalized multi-carrier
EPC	Electronic Product Code	GMSK	Gaussian minimum shift keying
ESPAR	electronically steerable parasitic array radiator	GOB	group of blocks
ESPRIT	Estimation of Signal Parameters via Rotational Invariance Techniques	GOP	group of pictures
		GoS	grade of service
		GPRS	General Packet Radio Service

GPS	Global Positioning System	ICI	intercarrier interference
GSC	Golay Sequential Code	IDCT	inverse DCT
GSM	Global System for Mobile Communications	IDMA	interleave division multiple access
HAPS	high-altitude aeronautical platform system	IDWT	inverse DWT
HARQ	hybrid-ARQ	IEC	International Electrotechnical Commission
H-BLAST	Horizontal encoding BLAST	IETF	Internet Engineering Task Force
HBT	heterojunction bipolar transistor	IF	intermediate frequency
HDTV	high definition television	IIP3	input IP3
HEMT	high electron mobility transistor	IIR	infinite impulse response
HFET	heterostructure FET	IMD	intermodulation distortion
HiperACCESS	High-Performance Access	IMDCT	inverse MDCT
HiperLAN	High Performance Radio LAN	IMI	intermodulation interference
HiperMAN	High Performance Metropolitan Area Network	IMPATT	impact avalanche and transit time
HiSWAN	High Speed Wireless Access Network	IMT-2000	International Mobile Telecommunications 2000
HILN	harmonic and individual lines plus noise	IntServ	integrated services
HLR	home location register	INL	integral nonlinearity
HR	half-rate	IP	Internet Protocol
HSCSD	High Speed Circuit Switched Data	IP3	third-order intercept point
HSDPA	High-Speed Downlink Packet Access	IPv4/v6	Internet Protocol version 4/version 6
H-S/MRC	hybrid selection/maximum ratio combining	IS	Interim Standard
HSPA	High-Speed Packet Access	ISI	intersymbol interference
HSUPA	High-Speed Uplink Packet Access	ISM	industrial, scientific, medical
HTS	high-temperature superconductor	ISO	International Organization for Standardization
I	in-phase	ITU	International Telecommunication Union
IC	integrated circuit	ITU-R	ITU's Radiocommunication Sector
		ITU-T	ITU's Telecomunication Standarization Sector
		JPEG	Joint Photographic Experts Group
		JTRS	Joint Tactical Radio System

LAN	local area network	MANET	mobile ad hoc networking
LBG	Linde-Buzo-Gray	MAP	maximum a posteriori
LCC	lost call clearing	MASK	*M*-ary amplitude-shift
LCH	lost call hold		keying
LCMV	linearly constrained	MB-OFDM	multiband OFDM-based
	minimum variance	MCA	maximally constrained
LCR	level crossing rate		autocorrelation
LD-CELP	low-delay CELP	MC-CDMA	multi-carrier CDMA
LDPC	low density parity code	MC-DS-	multi-carrier DS-CDMA
LEACH	low-energy adaptive	CDMA	
	clustering hierarchy	MCM	multicarrier modulation
LEO	low earth orbit	MCU	microcontroller unit
LHCP	left-hand circular	MDCT	modified DCT
	polarization	MDF	magnitude difference
LINC	linear amplification using		function
	nonlinear components	MDHO	macro diversity handoff
LLC	logical link control	MDS	minimum detectable signal
LLR	log-likelihood ratio	MELP	mixed excitation linear
LMDS	Local Multipoint		prediction
	Distribution Service	MEMS	micro-electromechanical
LMS	least mean squares		system
LNA	low-noise amplifier	MESFET	metal-semiconductor field
LO	local oscillator		effect transistor
LOS	line-of-sight	MFSK	*M*-ary FSK
LOT	lapped orthogonal transform	MIC	microwave integrated circuit
LPC	linear predictive coding	MIM	metal-insulator-metal
LS	least squares	MIMO	multiple input multiple
LSB	least significant bit		output
LS-DRMTA	least squares despread	MIMO-SC	MIMO single carrier
	respread multitarget array	MIMO-SS	MIMO spread spectrum
LSF	line spectral frequency	MIPS	million instructions per
LSP	linear spectral pair		second
LTCC	low-temperature cofired	MISO	multiple-input single-output
	ceramic	ML	maximum-likelihood
LTE	Long-Term Evolution	MLSE	maximum-likelihood
LTI	linear time-invariant		sequence estimation
LTP	long-term prediction	MLSR	maximal length shift register
LUT	look-up table	MLT	modulated lapped transform
MAC	medium access control; also	MMDS	Multichannel Multipoint
	multiply-accumulate		Distribution Service
MAD	mean absolute difference	MMIC	monolithic microwave
MAHO	mobile-assisted handoff		integrated circuit
MAI	multiple-access interference	MMSE	minimum mean squared
MAN	metropolitan-area network		error

MoM	method of moments	OSIC	ordered serial (successive) interference cancellation
MOS	mean opinion score		
MOSFET	metal-oxide-semiconductor field effect transistor	OSTBC	orthogonal space-time block code
MPAM	*M*-ary pulse amplitude modulation	OVSF	orthogonal variable spreading factor
MPE	multipulse excitation	PABX	private automatic branch exchange
MPEG	Moving Pictures Experts Group	PACS	Personal Access Communication System
MPLS	multiprotol label switching		
MP-MLQ	multipulse maximum likelihood quantization	PAE	power-added efficiency
		PAL	Phase Alternation Line
MPSK	*M*-ary PSK	PAM	pulse amplitude modulation
MQAM	*M*-ary QAM	PAN	personal area network
MRC	maximum ratio combining	PAPR	peak-to-average power ratio
MS	mobile station	PCCC	parallel concatenated convolutional code
MSC	mobile switching center		
MSE	mean squared error	PCM	pulse code modulation
MSK	minimum shift keying	PCS	Personal Communications Service
MT-CDMA	multi-tone CDMA		
MUD	multiuser detection	PDC	Personal Digital Cellular
MUI	multiple-user interference	pdf	probability distribution function
MUSIC	MUltiple SIgnal Classifications		
		PDF	Portable Document Format
MVDR	minimum variance distortionless response	PDP	power delay profile
		PEAQ	perceptual evaluation of audio quality
NADC	North American Digital Cellular		
		PESQ	perceptual evaluation of speech quality
NCO	numerically controlled oscillator		
		PHS	Personal Handyphone System)
NMT	Nordic Mobile Telephone		
NRZ/-L/-M/-S	nonreturn-to-zero/-level/-mark/-space	PIC	parallel interference cancellation
NTT	Nippon Telephone and Telegraph	PLL	phase-locked loop
		PM	phase modulation
OCC	orthogonal complementary code	PN	pseudo-noise
		POCSAG	Post Office Code Standard Advisory Group
OFDM	orthogonal frequency division multiplexing		
		PPM	pulse position modulation
OFDMA	orthogonal frequency division multiple access	PSD	power spectral density
		PSI-CELP	pitch synchronous innovation CELP
OOK	on-off keying		
OQPSK	offset QPSK	PSK	phase-shift keying
OSI	Open Systems Interconnect	PSNR	peak signal-to-noise ratio

PSTN	public switched telephone network	SA-DWT	shape-adaptive DWT
PWM	pulse-width modulation	SAR	successive approximation register
Q	quadrature-phase	SAW	surface acoustic wave
Q²PSK	quadrature quadrature PSK	SB-ADPCM	subband-split ADPCM
QAM	quadrature amplitude modulation	SC	single-carrier
QCELP	Qualcomm CELP	SCCC	serially concatenated convolutional code
QCIF	quarter-CIF	SCD	spectrum cyclic density
QMF	quadrature mirror filter	SCORE	Signal Communication by Orbital Relay Equipment; also self-coherence restoral
QO-STBC	quasi-orthogonal STBC		
QoS	quality of service		
QPSK	quaternary phase shift keying	SDMA	space division multiple access
QS-CDMA	quasi-synchronous CDMA	SDR	software-defined radio
QSIF	quarter-SIF	SECAM	SEquential Couleur Avec Memoire
RAN	regional area network		
RCELP	relaxed CELP	SEGSNR	segmental SNR
RCPC	rate-compatible punctured convolutional	SEP	symbol error probability
		SER	symbol error rate
RELP	residual excited linear prediction	SFBC	space–frequency block code
RF	radio frequency	SFDR	spurious-free dynamic range
RFID	radio frequency identification	SFIR	spatial filtering for interference reduction
RHCP	right-hand circular polarization	SF-OFDM	space-frequency coded OFDM
RLE	run-length encoding		
RLS	recursive least-squares	S/H	sample-and-hold
rms	root-mean-squared	SIC	serial interference cancellation
ROC	region of convergence		
ROI	region of interest	SICM	symbol-interleaved coded modulation
RPE	regular pulse excitation		
RPE-LTP	regular pulse excitation with long-term prediction	SIF	source input format
		SiGe	silicon-germanium
RS	Reed-Solomon	SIMO	single-input multiple-output
RSSI	radio signal strength indication	SINAD	signal-to-noise-and-distortion
RTMS	Radio Telephone Mobile System	SINR	signal-to-interference-plus-noise ratio
RTP	Real-time Transport Protocol		
		SIR	signal-to-interference ratio
RZ	return-to-zero	SISO	soft-in/soft-out
SA-DCT	shape-adaptive DCT	SMV	selectable mode vocoder

SNDR	signal-to-noise-plus-distortion ratio	TDRSS	Tracking and Data Relay Satellite System
SNR	signal-to-noise radio	TEC	total electron content
SOI	signal-of-interest	TEM	transverse electromagnetic
SOVA	soft output Viterbi algorithm	TH	time hopping
SPIHT	set partitioning in hierarchical trees	THSS	time-hopping spread spectrum
SPIN	Sensor Protocols for Information via Negotiation	TIA	Telecommunications Industry Association
SQNR	signal-to-quantization-noise ratio	ToA	time-of-arrival
		TR	transmitted reference
SS7	Signaling System No. 7	TXCO	temperature-controlled crystal oscillator
SSB	single sideband		
SSMA	spread spectrum multiple access	UDP	User Datagram Protocol
		UMB	Ultra Mobile Broadband
STBC	space-time block code	UMTS	Universal Mobile Telecommunications System
STDO	space-time Doppler		
ST-MF	space-time matched filter		
ST-MUD	space-time MUD	UPE	unequal error protection
STF-OFDM	space-time-frequency coded OFDM	UQ-DZ	uniform quantizer with dead zone
ST-OFDM	space-time coded OFDM	USB	Universal Serial Bus
STP	short-term prediction	UTRA	UMTS Terrestrial Radio Access
STS	space-time spreading		
STTC	space-time trellis code	UWB	ultra wideband
STTD	space-time transmit diversity	UWC-136	Universal Wireless Communication 136
SUI	Standford University Interim	V-BLAST	vertical encoding BLAST
		VCO	voltage-controlled oscillator
TACS	Total Access Communication System	VGA	variable gain amplifier
		VLR	visitor location register
TCM	trellis-coded modulation	VMR-WB	variable multi-rate wideband
TCP	Transmission Control Protocol		
		VO	video object
TDAC	time domain aliasing cancellation	VoIP	voice over IP
		VOP	video object plane
T-DMB	Terrestrial-DMB	VQ	vector quantization
TDD	time-division duplexing	VSELP	vector-sum excited linear prediction
TDoA	time-difference-of-arrival		
TD-SCDMA	Time Division-Synchronous Code Division Multiple Access	VSWR	voltage standing-wave ratio
		VTC	Visual Texture Coding
		WAN	wide area network
TDMA	time division multiple access	WCDMA	Wideband CDMA
		WiBro	Wireless Broadband

Wi-Fi	Wireless Fidelity	XPD	cross-polarization
WiMAX	Worldwide Interoperability for		discrimination
	Microwave Access	ZCR	zero-crossing rate
WSN	wireless sensor network	ZF	zero-forcing
WSSUS	wide sense stationary,	ZMCSCG	zero-mean circularly
	uncorrelated scattering		symmetric complex Gaussian

1 Introduction

1.1 The wireless age

Subsequent to the mathematical theory of electromagnetic waves formulated by James Clerk Maxwell in 1873 [3] and the demonstration of the existence of these waves by Heinrich Hertz in 1887, Guglielmo Marconi made history by using radio waves for transatlantic wireless communications in 1901. In 1906, amplitude modulation (AM) radio was invented by Reginald Fessenden for music broadcasting. In 1913, Edwin H. Armstrong invented the superheterodyne receiver, based on which the first broadcast radio transmission took place at Pittsburgh in 1920. Land-mobile wireless communication was first used in 1921 by the Detroit Police Department. In 1929, Vladimir Zworykin performed the first experiment of TV transmission. In 1933, Edwin H. Armstrong invented frequency modulation (FM). The first public mobile telephone service was introduced in 1946 in five American cities. It was a half-duplex system that used 120 kHz of FM bandwidth [4]. In 1958, the launch of the SCORE (Signal Communication by Orbital Relay Equipment) satellite ushered in a new era of satellite communications. By the mid-1960s, the FM bandwidth was cut to 30 kHz. Automatic channel trunking was introduced in the 1950s and 1960s, with which full-duplex was introduced. The most important breakthrough for modern mobile communications was the concept of cellular mobile systems by AT&T Bell Laboratories in the 1970s [2].

The last two decades have seen an explosion in the growth of radio systems. Wireless communication systems migrated from the first-generation (1G) narrowband analog systems in the 1980s, to the second-generation (2G) narrowband digital systems in the 1990s, to the current third-generation (3G) wideband multimedia systems that are being deployed. Meanwhile, research and development in the future-generation wideband multimedia radio systems is actively being pursued worldwide.

We have experienced a cellular revolution. In 2002, mobile phones worldwide began to outnumber fixed-line phones. By November 2007, the total number of worldwide mobile phone subscriptions had reached 3.3 billion, and by 2007 over 798 million people around the world accessed the Internet or equivalent mobile Internet services at least occasionally using a mobile phone.[1] This also makes the mobile phone the most common electronic device in the world. In addition to its multimedia services such as speech, audio, video,

---

[1] http://en.wikipedia.org/wiki/Mobile_phone, retrieved on Oct 21, 2008

Table 1.1. Division of electromagnetic waves.			
Electromagnetic waves		Frequency	Wavelength
Extremely low frequency	ELF	30–300 Hz	10–1,000 km
		300–3,000 Hz	1–100 km
Very low frequency	VLF	3–30 kHz	100–10 km
Low frequency	LF	30–300 kHz	10–1 km
Medium frequency	MF	300–3,000 kHz	1,000–100 m
High frequency	HF	3–30 MHz	100–10 m
Very high frequency	VHF	30–300 MHz	10–1 m
Ultra high frequency	UHF	300–3,000 MHz	100–10 cm
Super high frequency	SHF	3–30 GHz	10–1 cm
Extreme high frequency	EHF	30–300 GHz	10–1 mm
		300–3,000 GHz	1–0.1 mm
Infrared rays		43,000–416,000 GHz	7–0.7 μm
Visible light		430,000–750,000 GHz	0.4–0.7 μm
Ultraviolet[a]		750,000–3,000,000 GHz	0.4–0.1 μm

[a] Beyond ultraviolet are X-rays and Gamma-rays.

and data, the pervasive use of wireless communications has also entered many aspect of our life, including health care, home automation, etc.

1.2 Spectrum of electromagnetic waves

The medium for wireless communications is open space, and information is transferred via electromagnetic waves. In order to separate different wireless systems, the spectrum of electromagnetic waves is divided into many frequency bands. The wavelengths and frequencies of electromagnetic waves are listed in Table 1.1.

At lower frequencies, radio waves tend to follow the earth's surface, while at higher frequencies, e.g., above about 300 MHz, they propagate in straight lines. The range from dc to SHF has been widely used for communications and other purposes such as radar, industry, heating, spectroscopy, radio astronomy, medicine, power transmission, and science. In contrast, the region of EHF waves and beyond is wide open due to technical difficulties. This is due to considerable attenuation in the atmosphere, and there are many difficulties in wave generation, amplification, detection, and modulation techniques. At above 1,000 GHz, the wave propagation turns optical. Optical communications are now restricted to optical fibers.

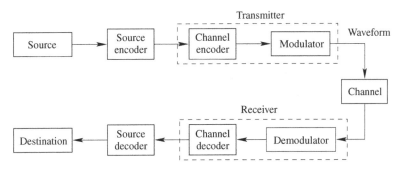

Figure 1.1

Block diagram of a general communication system.

1.3 Block diagram of a communication system

A communication system deals with information or data transmission from one point to another. The block diagram of a general digital communication system is given in Fig. 1.1. This block diagram is also applicable to remote sensing systems, such as radar and sonar, in which the transmitter and receiver may be located at the same place.

The source generates either analog signals such as speech, audio, image, and video, or digital data such as text or multimedia. The source encoder generates binary data from the source. The generated binary data is then subject to a channel encoder so that the binary data sequences can be reliably reproduced at the receiver. The channel-encoded data stream is then modulated to generate waveforms for transmission over a channel, which is a physical link such as a telephone line, a high frequency radio link, or a storage medium. The channel is subject to various types of noise. At the receiver, the above procedure is reversed so as to finally restore the original source information.

There are three types of common transmission channels: wireless channels, guided electromagnetic wave channels, and optical channels. The wireless channel can be the atmosphere or free space. Due to its open nature, there are various noise sources added to the channel. Coaxial cable line was once a major guided wave channel, and optical fiber is a special type of guided wave channel. The long-distance telephone network once used coaxial cable lines, which has now been replaced by optical fiber.

1.4 Architecture of radio transceivers

The well-known super-heterodyne receiver architecture was invented by Armstrong in 1913. Armstrong also demonstrated frequency modulation in 1933. In this section, we introduce two architectures of radio transceivers: the super-heterodyne transceiver and the direct-conversion transceiver.

1.4.1 Super-heterodyne transceivers

Conventional super-heterodyne transceivers

The 1G radio systems were analog systems using frequency division multiple access (FDMA). For each user, there is a fixed super-heterodyne transceiver. The received signal is first passed through surface acoustic wave (SAW) filters for image suppression. The filtered signal is low-noise amplified, and is then subject to one or more intermediate frequency (IF) stages (mixing and bandpass filtering) and baseband processing. For transmission, the baseband signal is first filtered, then upconverted by multiple IF conversion stages, power amplified, and finally passed to an antenna for transmission. The oscillators and filters are generally not adjustable.

The 2G systems use the same super-heterodyne transceiver architecture for conversion between radio frequency (RF) and baseband signals. Analog-to-digital (A/D) converters and digital-to-analog (D/A) converters are used for conversion between analog and digital baseband signals. The receiver typically converts the RF signal to a baseband signal after applying a few IF stages, and then separates orthogonal in-phase (I) and quadrature-phase (Q) baseband signals prior to applying A/D conversion for each of these. Due to the application of time division multiple access (TDMA) and/or code division multiple access (CDMA), each transceiver can support multiple users and this requires a wider frequency slice for each transceiver. In the digital part, many dedicated digital application-specific integrated circuits (ASICs) as well as general-purpose digital signal processors (DSPs) are used to perform various signal processing tasks such as equalization, modulation/demodulation, channel coding/decoding, and voice coding/decoding.

The super-heterodyne architecture achieves good I/Q matching, and has no problems of dc offset and LO (local oscillator) leakage. However, it suffers from the image problem, as is illustrated in Fig. 1.2. For two signals $x_1(t) = A_1 \cos \omega_1 t$ and $x_2(t) = A_2 \cos \omega_2 t$, after lowpass filtering the product $x_1(t)x_2(t)$, we get a signal of the form $\cos(\omega_1 - \omega_2)t$, which is the same as $\cos(\omega_2 - \omega_1)t$. Thus the bands that are symmetrical above or below the LO frequency will be downconverted to the same band. In order to suppress the image, an image-rejection filter has to be placed before the mixer. Since the image-rejection filter is typically realized as a passive, external component, it requires the preceding low noise-amplifier (LNA) to drive a 50-ohm load. Most RF transceivers employ two stages of downconversion to relax the Q required for each filter. In most RF applications, the overall image suppression is required to be around 60 to 70 dB [6].

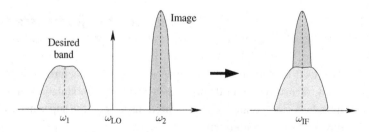

Figure 1.2   Illustration of the image problem.

Super-heterodyne transceivers for software-defined radio

In conventional super-heterodyne transceivers, the analog filters are designed for a carrier frequency and a channel bandwidth. This is not suitable for multiband systems. More recent transceivers employ low-IF sampling rather than baseband sampling. The low-IF scheme has a number of advantages. First, the dc offset problem does not occur. Second, the expensive IF SAW filter, IF phase-locked loop (PLL), and image rejection filter are not necessary. Also, the impact of near-dc flicker ($1/f$) noise on the receiver performance is significantly reduced. As to the downside, it suffers from LO pulling/leakage due to coupling or imperfect isolation between the RF components; it also requires stringent image rejection to suppress strong interferers from adjacent channels, which are images of the desired signals arising from the low IF.

The low-IF scheme is employed in the classical architecture for software-defined radio (SDR) at the base station (BS), as shown in Fig. 1.3. The wideband RF front-end replaces many narrowband transceivers used in 1G or 2G systems. The wideband front-end converts an entire band containing multiple carriers to a suitable IF signal, which is then digitalized, while conventional 2G systems shift individual carriers to baseband prior to digitalization.

Note, that in Fig. 1.3, BP3 is the anti-aliasing bandpass filter. The selection of IF_D is dependent on the frequency converters that are COTS (commercial off-the-shelf) available. Note that the RF filters can be inserted between the LNA and mixer in the receiver, and between the mixer and power amplifier in the transmitter, to reject any signal generated by the nonlinearity of the LNA or mixer.

The digital signal processing module implements an independent digital front-end, baseband processing for each carrier, and O&M (operation and management) signaling. For each carrier, a digital front-end downconverts the digital IF signal to I and Q baseband signals by a numerically controlled oscillator (NCO); this is followed by baseband processing that contains sampling rate conversion, demodulation and filtering, channel decoding, and source decoding. The transmit path is the reverse of the receive path. Note that on the transmit path the signals of all carriers are summed before D/A conversion is applied. Care must be taken to avoid numerical overflow.

1.4.2 Direct-conversion transceivers

Another common transceiver architecture is the direct-conversion or zero-IF system, also known as *homodyne system*. A direct-conversion receiver downconverts the RF signal directly to baseband by using an LO, whose frequency is exactly equal to the frequency of the RF signal. Direct conversion requires only a single frequency synthesizer and avoids an off-chip IF filter, and thus is preferred in a fully integrated design.

Since the intermediate frequency ω_{IF} is zero, the image to the RF signal is the RF signal itself. Thus, the image problem does not arise and the method eliminates the use of bulky, off-chip, front-end image rejection filters. However, since the synthesizer operates at the same frequency as the RF signal, LO leakage and frequency pulling occur. Other disadvantages include the problems of dc offset, flicker or $1/f$ noise and I/Q mismatch. These

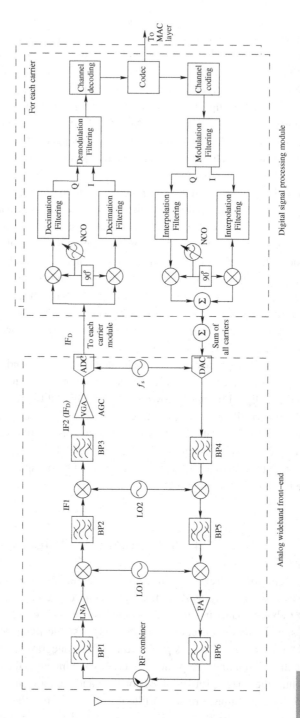

Figure 1.3 A classical SDR implementation at the BS. IF*n* stands for the *n*th IF stage, IF$_D$ for the digital IF, BP*n* for the *n*th bandpass filter, LO*n* for the *n*th LO, LNA for low noise amplifier, PA for power amplifier, VGA for variable gain adaptor, ADC for A/D converter, DAC for D/A converter, AGC for automatic gain control, and NCO for numerically controlled oscillator.

disadvantages can be overcome in the super-heterodyne architecture by using an off-chip IF filter and an extra frequency synthesizer.

LO leakage arises due to limited isolation between the LO port and the inputs of the mixer and the LNA, causing a leakage LO signal to feed through the LNA and the mixer or a large leakage interfering signal to feed through the LO input. The original RF signal as well as the LO leakage is then mixed with the LO, and a dc offset is generated in both the cases. This is known as the *self-mixing phenomenon*. Cancellation of the dc offsets is a primary concern in direct-conversion receiver design. Thus, this method requires an LO with a very high precision and stability. The LO leakage to antenna may be reradiated, creating interference to other receivers. The Federal Communications Commission (FCC) requires that the upper bounds of the in-band LO radiation is typically between −50 dBm and −80 dBm for wireless standards [6].

In direct-conversion transceivers, I/Q mismatch can be viewed as the so-called *self-image problem*, where the baseband equivalent signal is essentially interfered by its own complex conjugate [7]. I/Q mismatch at the receiver will corrupt the downconverted signal constellation, leading to a higher BER, while I/Q mismatch at the transmitter can lead to increased out-of-band emissions with nonlinear power amplifiers. Signal processing techniques may be used to correct the I/Q mismatch [1, 7].

In addition, $1/f$ noise is a severe problem in CMOS implementations, with a flicker noise corner frequency in the vicinity of 1 MHz [5]. The flicker noise in SiGe and BiCOMS technologies are much lower than that in CMOS technology. The $1/f$ noise can be reduced by incorporating large devices at the stages following the mixers, since the operating frequency is relatively low. The CMOS technology is not suitable for high-sensitivity direct-conversion receivers such as narrow-band systems, while the SiGe and BiCMOS technologies make it possible to achieve higher receiver sensitivity for wideband systems.

For the transmitter part, the power amplifier will disturb the transmit LO, and will corrupt the oscillator spectrum, despite the shielding techniques used. This is the injection-pulling or injection-locking mechanism. This influence can be reduced if the power amplifier output spectrum is sufficiently away from the LO frequency. This can be achieved by offsetting the LO frequency by mixing two voltage-controlled oscillators (VCOs). Another way to prevent LO pulling is to upconvert the baseband signal in two steps, yielding the power amplifier output spectrum that is far from the frequency of the VCOs. This is implemented in modern super-heterodyne transceivers.

1.5 Organization of the book

This book gives a comprehensive introduction to wireless communication systems. The contents are organized into four parts. Part 1 (Chapters 2 to 9) introduces the principles of wireless communications. Part 2 (Chapters 10 to 13) deals with the analog and digital implementation of wireless communication systems. Information theory and coding are treated in Part 3 (Chapters 14 to 17). Part 4 (Chapters 18 to 22) describes some advanced and emerging technologies for future-generation wireless communications.

The contents by chapters are listed below.

- In Chapter 2, we give an overview of wireless communications and its history. Circuit/packet switching and the OSI reference model are also described in this chapter.
- Electromagnetic wave propagation is subject to propagation loss. Chapter 3 introduces propagation loss models, characteristics of wireless channels, and the mechanisms of signal propagation in the channel.
- Fundamentals on multiuser communications are developed in Chapter 4. This chapter treats the cellular concept, various multiple access techniques, Erlang capacity, protocol design, quality of service (QoS), and user location.
- Wireless channels are usually in a fading state. Diversity is the common method for combating fading. Diversity ensures that the same information reaches the receiver from statistically independent channels. By combining multiple independently fading copies of the same signal, fading can be substantially reduced. Diversity is examined in Chapter 5.
- Channel estimation and equalization are necessary for signal detection. Channel estimation finds the channel information when the transmission signal propagates through the channel. Using this channel information, the equalizer can remove the influences of fading and other undesirable channel conditions, and thus restore the original transmitted signal. These topics are discussed in Chapter 6.
- Modulation is a process that incorporates the message into a carrier for transmission. The message can be embedded into the amplitude, frequency, or phase of the carrier, or a combination of these. Modulation and demodulation, which are subject to RF or microwave operations, are necessary for signal transmission. Chapter 7 introduces digital modulation and demodulation.
- Spread spectrum communications, or CDMA technology, spread each user's signal over the same wider bandwidth for transmission. At the receivers, these user signals are separated by using their specific codes. CDMA is the underlying technology for 3G cellular communications, and is introduced in Chapter 8.
- OFDM technology transmits messages simultaneously over multiple carriers in a linear band-limited channel. It is robust against multipath fading, but with a low complexity. OFDM technology has been widely implemented in high-speed wireless networking and is an enabling technique for 4G mobile communications. Chapter 9 introduces OFDM technology.
- An antenna is the interface between the RF/microwave circuits and the free space. It transmits the generated RF or microwave signals over the wireless channel, and at the same time, passes the received signal on to the RF/microwave circuits at the receiver. Antennas are described in Chapter 10.
- RF/microwave subsystems, known as the front-ends of wireless transceivers, are the analog circuits in wireless communication systems. They convert RF signals into baseband signals, and vice versa. RF/microwave subsystems are introduced in Chapter 11.
- Modern wireless communication systems are digital systems, where information processing is performed in digital form, whereas the received/transmitted signal at the

antenna is in analog form. A/D and D/A converters are used for conversion between the analog and digital signals within the wireless transceiver. A/D and D/A converters are described in Chapter 12.

- Digital signal processing is an enabling technique for digital communication systems. Chapter 13 introduces basic digital signal processing techniques that are used in wireless communications and source coding.

- Information theory was established by Shannon. It lays the theoretical foundation for source coding and channel coding, as well as the entire communication networks. Information theory is the subject of Chapter 14.

- Source coding or data compression is performed to remove redundancy in the original data so as to maximize the information storage and transmission. Speech communication is the most fundamental service provided by wireless networks. Source coding of speech and audio signals is presented in Chapter 16.

- Wireless communications are being escalated to deliver multimedia service. This involves image and video coding. Source coding for images and videos is introduced in Chapter 17.

- After the redundancy in a message is removed during source coding, the message is more vulnerable to errors. For the purpose of reliable storage or transmission over a noisy channel, error-correcting codes are used for error recovery. This is the topic of channel coding. Channel coding is described in Chapter 15.

- Use of multiple antennas is an effective solution for high-speed or high-reliability communications. Smart antennas and MIMO communications are two major multiple-antenna technologies. Smart antennas can be used for diversity combining and beamforming. Chapter 18 discusses smart antenna technology.

- Chapter 19 continues with the discussion of multiple antenna systems: MIMO technology. MIMO technology can be implemented as space-time coding or spatial multiplexing. MIMO is an enabling technique for 4G mobile communications and future-generation wireless networks.

- Ultra wideband (UWB) technology employs a spectrum in excess of 500 MHz that overlaps licensed bands in an unlicensed mode. It is an enabling technique for gigabits/s wireless networking. UWB technology is described in Chapter 20.

- Software-defined radio (SDR), or software radio, provides a solution for *one hardware platform, multiple wireless standards*. It makes possible multiband, multimode, multistandards low-power radio communications. Cognitive radio, based on the platform provided by SDR, solves the problem of crowded spectrum allocation. Both technologies are enabling techniques for 3G and 4G wireless systems. They are treated in Chapter 21.

- Wireless ad hoc networks are playing an increasing role in current and future-generation wireless and mobile networks. Wireless sensor networks, as an emerging technology, are being employed in a range of applications such as home, industry, military, public security, environment monitoring, and medical applications. Both the wireless ad hoc and sensor networks are important for ubiquitous networking. These topics are described in Chapter 22.

In each chapter, some problems are included and should be helpful to students to review the contents of the chapter.

References

[1] J. J. de Witt & G.-J. van Rooyen, A blind I/Q imbalance compensation technique for direct-conversion digital radio transceivers. *IEEE Trans. Veh. Tech.*, **58**:4 (2009), 2077–2082.

[2] V. H. MacDonald, The cellular concept. *Bell Sys. Tech. J.*, **58**:1 (1979), 15–41.

[3] J. C. Maxwell, *A Treatise on Electricity and Magnetism* (Oxford: Clarendon Press, 1873; New York: Dover, 1954).

[4] T. S. Rappaport, *Wireless Communications: Principles & Practice*, 2nd edn (Upper Saddle River, NJ: Prentice Hall PTR, 2002).

[5] B. Razavi, Design Consideration for direct-conversion receiver. *IEEE Trans. Circ. Syst. II*, **44**:6 (1997), 428–435.

[6] B. Razavi, *RF Microelectronics* (Upper Saddle River, NJ: Prentice Hall, 1998).

[7] M. Valkama, M. Renfors & V. Koivunen, Advanced methods for I/Q imbalance compensation in communication receivers. *IEEE Trans. Signal Process.*, **49**:10 (2001), 2335–2344.

2 An overview of wireless communications

2.1 Roadmap of cellular communications

2.1.1 First-generation systems

The 1G mobile cellular systems were analog speech communication systems. They were mainly deployed before 1990. They are featured by FDMA (frequency division multiple access) coupled with FDD (frequency division duplexing), analog FM (frequency modulation) for speech modulation, and FSK (frequency shift keying) for control signaling, and provide analog voice services. The 1G systems were mainly deployed at the frequency bands from 450 MHz to 1 GHz. The cell radius is between 2 km and 40 km.

The AMPS (Advanced Mobile Phone Services) technique was developed by Bell Labs in the 1970s and was first deployed in late 1983. Each channel occupies 30 kHz. The speech modulation is FM with a frequency deviation of ±12 kHz, and the control signal is modulated by FSK with a frequency deviation of ±8 kHz. The control channel transmits the data streams at 10 kbits/s. AMPS was deployed in the USA, South America, Australia, and China. In 1991, Motorola introduced the N-AMPS to support three users in a 30 kHz AMPS channel, each with a 10 kHz channel, thus increasing the capacity threefold.

The European TACS (Total Access Communication System) was first deployed in 1985. TACS is identical to AMPS, except for the channel bandwidth of 25 kHz. Speech is modulated by FM with a frequency deviation of ±12 kHz, and the control signal is modulated by FSK with a frequency deviation of ±6.4 kHz, achieving a data rate of 8 kbits/s. TACS has various versions: ETACS deployed in UK in 1987 and NTACS/JTACS that came into service in Japan in June 1991.

The NTT (Nippon Telephone and Telegraph) system, first deployed in Japan in 1979, is also based on AMPS. The NTT system has a channel spacing of 25 kHz and overlapped channels of 12.5 kHz are used to increase frequency utilization. The speech signal is FM modulated with a ±5 kHz deviation. The control signaling uses FSK with a deviation of ±8 kHz, and the data rate is 0.3 kbits/s.

The NMT (Nordic Mobile Telephone) system, developed by Ericsson, was introduced in Sweden in 1981. It uses a channel spacing of 25 kHz. The modulation for speech is FM with a frequency deviation of ±5 kHz. It transmits 1.2 kbits/s using FSK with a frequency deviation of ±3.5 kHz.

C-Netz/C-450 was deployed in Germany, Austria, Portugal, and South Africa from 1981. C-450 has a channel spacing of 20 kHz. It employs FM speech modulation with ±4 kHz deviation and FSK signaling with ± 2.5 kHz deviation, achieving a signaling data rate of 5.28 kbits/s.

During this period, several other systems were deployed, such as RadioCom in France in 1985, Comvik in Sweden in 1981, and RTMS (Radio Telephone Mobile System) in Italy in 1985. There were also some cordless telephone systems, which are low-power, low-range systems that allow a user to move in a house or a building.

For the 1G systems, the use of FDMA/FDD demands each user to occupy two slices of frequency bands, one for the uplink and the other for the downlink. For each slice of the frequency band, a transceiver must be designed, leading to a high cost. While the 1G systems aroused the market popularity internationally, they were restricted by the small coverage areas, poor speech quality, and poor battery performance. These shortcomings were overcome by introducing digital communication systems, resulting in the 2G systems.

2.1.2 Second-generation systems

The 2G systems were introduced in the early 1990s. They provide wireline-quality digital voice services based on circuit-switched data communications. These systems are featured by digital implementation. New access techniques, such as TDMA (time division multiple access) and CDMA (code division multiple access), were also introduced. In addition to 2G cellular systems, many 2G cordless phone, wireless LAN and satellite radios were also developed during this period.

The most dominant 2G cellular standards are GSM (Global System for Mobile Communications) and IS-95 (Interim Standard 95) CDMA. As of the second quarter of 2007, GSM had 2.3 billion subscribers worldwide, while CDMA had 450 million [9]. Other regional 2G cellular standards are IS-54/IS-136 TDMA and PDC (Personal Digital Cellular) TDMA in Japan. The operating frequency bands are typically between 900 MHz and 1.9 GHz.

The GSM air interface, introduced in 1990 by ETSI (European Telecommunications Standards Institute), is based on FDMA/TDMA/FDD and GMSK (Gaussian minimum shift keying) modulation. The spectrum is divided into many channels of 200 kHz bandwidth, with a channel data rate of 270.833 kbits/s. Each channel is time-divided for eight users.

IS-54, introduced in 1991 by TIA/EIA (Telecommunications Industry Association/Electronics Industry Association) of the USA and deployed in 1993, is also known as *NADC (North American Digital Cellular)* or *D-AMPS (Digital-AMPS)* system. IS-54 employs FDMA/TDMA/FDD with $\pi/4$-DQPSK (differential quaternary phase shift keying) modulation. It uses the same 30-kHz channels and frequency bands as AMPS, but has six times the capacity of AMPS. Each channel has a bit rate of 48.6 kbits/s. IS-54 uses the same 10 kbits/s FSK signaling scheme of AMPS for the forward (downlink) and reverse (uplink) control channels. IS-136, as a modification of IS-54, uses $\pi/4$-DQPSK modulation for the control channels, resulting in a higher control channel data rate for paging and short messaging. IS-136 is not compatible with IS-54. PDC, introduced in 1993, is somewhat similar to IS-54/IS-136. It is a TDMA/FDMA/FDD system with $\pi/4$-DQPSK.

Each carrier is 25 kHz wide, and supports a raw data rate of 42 kbits/s. Each carrier is time-divided into 3 full-rate (6 half-rate) channels.

The TIA/EIA IS-95 standard was introduced in 1993, and the IS-95A revision was released in 1995. IS-95A interoperates with the analog AMPS, and is the basis for the 2G CDMA deployment. The first IS-95A system was deployed in Hong Kong in 1996. IS-95 employs CDMA/FDD with OQPSK (offset quaternary phase shift keying) modulation. It uses the same bands as IS-54, but each channel is 1.2288 MHz wide, and each user has a data rate of 9.6 kbits/s. IS-95 allows variable data rates of 1.2, 2.4, 4.8, and 9.6 kbits/s. Compared to the other 2G technologies, IS-95 is significantly more complex. It also employs some techniques, such as power control, frequency and delay diversity, variable-rate coding, and soft handoff, which were later adopted in the 3G standards. The IS-95B revision uses the same physical layer as IS-95A, but further provides 64 kbits/s packet-switched data, in addition to voice services; thus, IS-95B is treated as a 2.5G technology. IS-95B was first deployed in Korea in 1999. IS-95/95A/95B are collectively known as *CDMAOne* or *IS-95*.

GSM and IS-136 migrated to 3G in two phases. In the first phase, the GPRS (General Packet Radio Service) increased the data rate to around 115 kbits/s, though the theoretical peak rate is 172.2 kbits/s if all the eight time slots in a GSM frame are used. GPRS made a transition from circuit-switched data that are used in GSM to packet-switched data. The data rate for circuit-switched data on GSM is 9.6 kbits/s. HSCSD (High Speed Circuit Switched Data), as a new implementation to transmit circuit-switched data over GSM, supports data rates up to 38.4 kbits/s, by allocating all the eight slots to one user. GPRS was also deployed over IS-136. In the second phase, the EDGE (Enhanced Data for GSM Evolution) standard further enhances the data rate to 384 kbits/s. In EDGE, a high-rate 8PSK modulation coexists with the GMSK modulation. EDGE is a convergence of the GSM and IS-136 standards.

In addition, the Cellular Digital Packet Data (CDPD) overlay network, released in 1995, provides a low-speed packet data service over the US AMPS network. It provides a data rate of 19.2 kbits/s over the 30-kHz AMPS channel.

Cordless telephone

In addition to the 2G cellular standards, there are also many 2G cordless telephone or wireless local loop (WLL) technologies such as PHS (Personal Handyphone System) in Japan, CT2 (Second Generation Cordless Telephone) in UK, CT2+ in Canada, CT3 in Sweden, DECT (Digital Enhanced Cordless Telephone) in Europe, and PACS (Personal Access Communication System) in USA.

A cordless telephone system is similar to a cellular telephone system, but provides low mobility at pedestrian speeds over a smaller area such as a building and usually as a private automatic branch exchange (PABX) or connected to the public switched telephone network (PSTN). These systems are limited to low power transmission. Cordless telephone systems are designed for microcell/indoor PCS (Personal Communications Service) use, and they typically provide coverage within a range of a few hundred meters.

CT2, introduced in 1989, is a digital version of the 1G cordless telephones. CT2 employs FDMA/TDD (time-division duplexing) with GFSK (Gaussian FSK) modulation; it uses a carrier spacing of 100 kHz, with a channel bit rate of 72 kbits/s. Both CT2+ and CT3 standards are very similar to CT2, with difference in details on available slots in the frame organization.

DECT, an ETSI standard finalized in 1992, is based on TDMA/FDMA/TDD transmission with GFSK modulation, and each channel is 1.728 MHz wide, achieving a channel bit rate of 1152 kbits/s. Each channel consists of 24 slots per 10 ms long frame, time-shared among 12 users.

PACS, introduced in 1992, is a FDMA/TDMA/FDD or /TDD technique with $\pi/4$-QPSK (quaternary phase shift keying) modulation; each carrier is 300 kHz wide, achieving a channel bit rate of 384 kbits/s, which is shared by 8 (FDD) or 4 (TDD) users. PHS employs FDMA/TDMA/TDD and $\pi/4$-DQPSK; each carrier is 300 kHz wide, achieving a channel bit rate of 384 kbits/s.

2.1.3 Third-generation systems

Currently, 3G cellular systems are being deployed worldwide. The 3G standards were developed by ITU (International Telecommunication Union) under the name of IMT-2000 (International Mobile Telecommunications 2000) or UMTS (Universal Mobile Telecommunications System) in ITU-R Rec. M.1457. The 3G cellular system is featured by wideband communications. As general requirements, it demands a data rate of 2 Mbits/s at stationary mobiles, 384 kbits/s for a user at pedestrian speed, and 144 kbits/s in a moving vehicle. It is targeted to be a global system supporting global roaming. The 3G network uses packet switching, and is typically deployed at the 2 GHz frequency band.

In June 1998, ITU-R (ITU's Radiocommunication Sector) received 11 competing proposals for terrestrial mobile systems, and approved five. Two mainstream 3G standards are WCDMA and CDMA2000, which are administered by two bodies in ITU, 3GPP (Third-Generation Partnership Project) and 3GPP2 (Third-Generation Partnership Project 2), respectively. In October 2007, ITU-R elected to include WiMAX (802.16e) in the IMT-2000 suite of wireless standards and updated ITU-R Rec. M.1457. WiMAX now is a strong contender to WCDMA and CDMA2000.

UTRA/WCDMA

WCDMA (Wideband CDMA), also known as *UTRA (UMTS Terrestrial Radio Access)*, was jointly developed by ARIB, Japan and ETSI. It is a wideband solution, with a carrier bandwidth of 5 MHz and a chip rate of 3.84 Mchips/s. WCDMA has a flexible carrier spacing with a 200 kHz carrier raster, to improve the spectrum utilization efficiency by providing flexibility to conform with that of GSM. WCDMA supports the legacy GSM at the network level, and 3GPP keeps the core network to be as close to GSM core network as possible. WCDMA employs CDMA/FDD with QPSK/BPSK (binary phase shift keying) modulation in its first release completed in 1999. WCDMA supports user data rates up to

2.3 Mbits/s both in the uplink and the downlink. The duration of a frame is 10 ms. The first commercial launch of WCDMA was in Japan in 2001.

The HSDPA (High-Speed Downlink Packet Access, 3GPP TR25.858) and HSUPA (High-Speed Uplink Packet Access, 3GPP TS25.896) standards evolved as a consequence of 3GPP to high-speed data services. HSDPA was included in 3GPP Release 5 in March 2002, and HSUPA was included in Release 6 in December 2004. They together are known as *HSPA (High-Speed Packet Access)*. They use different physical layers from WCDMA, and employ many new features. HSDPA employs orthogonal frequency division multiplexing (OFDM) technology for transmission. HSDPA supports 16QAM (quadrature amplitude modulation), achieving a data rate of up to 14.4 Mbits/s. HSUPA uses QPSK modulation only, and has a speed of up to 5.76 Mbits/s. HSDPA and HSUPA are both treated as 3.5G (3.5th generation) systems, and they have both FDD and TDD modes; they both evolved to HSPA+ (3.9G, short for 3.9th generation), which was specified in 3GPP Release 7. Downlink MIMO (multiple input multiple output) is supported in HSPA+. WCDMA was deployed in 2003, HSDPA in 2006, and HSUPA in 2007.

CDMA2000

CDMA2000, also known as *IS-2000*, was proposed by TIA/EIA. It is a narrowband multicarrier solution, with a carrier width of 1.25 MHz and a chip rate of 1.2288 Mchips/s, achieving a maximum data rate of 2.457 Mbits/s in the downlink. CDMA2000 supports the legacy IS-95 at the air interface. It adopts CDMA/FDD in the FDD mode, and TDMA/CDMA/TDD in the TDD mode. The modulation schemes are BPSK, QPSK, 8PSK and 16QAM. The use of $N = 1$ and 3 carriers has been specified, and it can be extended to 6, 9, and 12 carriers in the future, achieving an effective chip rate of $1.2288N$ Mchips/s. The frame duration is 20 ms or 5 ms.

The CDMA2000 family includes 1x (Phase 1), 1xEV-DO (Evolution, Data Optimized, CDMA2000 Rev.0), and 1xEV-DV (Evolution, Data and Voice) standards. 1xEV-DO and 1xEV-DV, together known as *IS-856* of TIA/EIA, are both backward-compatible with IS-95 and 1x. CDMA2000 1x was first deployed in Korea in October 2000, and 1xEV-DO was launched in Korea in 2002.

CDMA2000 1x is four times more efficient than TDMA networks, and has a voice capacity that is twice that of IS-95. It delivers a peak data rate of 144 kbits/s in loaded network, and delivers a peak packet data rate of 307 kbits/s in mobile environments. 3GPP2 published 1x Advanced in August 2009 for upgrading the 1x platform while sustaining backward compatibility. By taking advantage of several interference cancellation and radio link enhancements, 1x Advanced enahncements can theoretically quadruple the voice capacity of 1x systems in the same 1.25 MHz of spectrum. 1xEV-DO provides peak forward data rates of up to 2.4 Mbits/s in a 1.25 MHz channel, and achieves an average throughput of over 700 kbits/s, equivalent to cable modem speeds. The data rate on the reverse link is up to 153.6 kbits/s. 1xEV-DO offers multicast services, which enable multimedia services, such as real-time TV broadcast and movies, to an unlimited number of users. 1xEV-DV Release D (CDMA2000 Rev. D, 3GPP2 C.S0002-D) supports a peak data rate of 3.09 Mbits/s in the forward link and 1.8456 Mbits/s in the reverse link

(3.5G); it supports voice as well as data to offer smooth support for voice and legacy services.

NxEV-DO or *EV-DO Multicarrier* (3GPP2 C.S0024-B) was published in 2006. It provides a peak forward link data rate of $N \times 4.9$ Mbits/s, and a peak reverse link data rate of $N \times 1.8$ Mbits/s. It is capable of delivering a peak data rate of 73.5 Mbits/s in the forward link and 27 Mbits/s in the reverse link by using 15 carriers. This can be treated a 3.9G technology.

UTRA-TDD and TD-SCDMA

TDD avoids the uplink/downlink spectrum pair required in FDD; this is ideal for asymmetric services and is especially suitable for highly populated areas. 3GPP also has two TDD modes: UTRA-TDD and TD-SCDMA (Time Division-Synchronous Code Division Multiple Access). UTRA-TDD, developed by ETSI, is the TDD mode of UMTS. UTRA-TDD employs TDMA /CDMA/TDD with QPSK modulation. It uses the same bandwidth (5 MHz) and chip rate (3.84 Mchips/s) as UTRA-FDD. Like UTRA-FDD, UTRA-TDD supports the legacy GSM at the network level. The frame of 10 ms length is divided into 16 slots, and each slot allows up to 8 CDMA channels.

TD-SCDMA was proposed in China in 1998. It is similar to UTRA-TDD in many aspects, but uses a bandwidth of 1.6 MHz and a chip rate of 1.28 Mchips/s. TD-SCDMA is also known as the *low chiprate TDD mode* of UMTS. It employs QPSK/8PSK modulation. The maximum data rate is 2 Mbits/s. The frame duration is 5 ms, which is divided into 10 time slots. Each time slot allows up to 16 CDMA channels. The spectral efficiency of TD-SCDMA is almost twice that of UTRA-TDD. TD-SCDMA provides a cost-effective way to upgrade existing GSM networks to 3G core networks. TD-SCDMA was first deployed in China on April 1, 2008.

UWC-136/EDGE

ITU also approved UWC-136 (Universal Wireless Communication 136)/EDGE as a candidate for IMT-2000 3G standards. UWC-136/EDGE was developed by TIA/EIA to maximize commonality between IS-136 and GPRS, and to meet the ITU-R requirements for IMT-2000. UWC-136 provides backward compatibility with IS-136 and IS-136+.

UWC-136 increases the voice and data capacity of the 30 kHz channels by using enhanced modulations ($\pi/4$-DQPSK and 8PSK) with the exisiting 30 kHz IS-136+. A complementary wideband TDMA is defined to provide high data rate. By adding a 200 kHz carrier component to provide a data rate of 384 kbits/s, compatibility with GPRS and EDGE is possible. For transmission at a data rate of 2M bits/s, a carrier component of 1.6 MHz is added. EDGE also evolved to EDGE Evolution (3.5G).

DECT

DECT was also approved by ITU as a PCS solution for the IMT-2000 standard. DECT employs FDMA/TDMA/TDD. In order to increase the data rate to meet IMT-2000 requirements, in addition to its original GMSK modulation, other modulation schemes such as $\pi/2$-DBPSK, $\pi/4$-DQPSK, and $\pi/8$-D8PSK are also used.

Mobile WiMAX

Mobile WiMAX, developed on the basis of IEEE 802.16e, is a wireless metropolitan-area network (MAN) technology. IEEE 802.16e was completed in December 2005. IEEE 802.16e is based on OFDM technology. It allows OFDMA (orthogonal frequency division multiple access) with both FDD and TDD operations. MIMO technology is supported in WiMAX. It can deliver a maximum of 75 Mbits/s and cover a range of 70 miles. Mobile WiMAX can be treated as 3.9G. Mobile WiMAX is deployed in the 2 to 6 GHz licensed bands. The first commercial mobile WiMAX network was launched in Korea in June 2006.

3GPP LTE

3GPP LTE (Long-Term Evolution), also referred to as *E-UTRA (Evolved UTRA)* or *E-UTRAN (Evolved UTRA Network)*, is the project name for the evolution of UMTS, which was started in 2005. In June 2005, 3GPP approved the further study of six physical layer proposals: multicarrier WCDMA, multicarrier TD-SCDMA, and four OFDMA-based proposals. LTE is targeted at the development of a new air interface, but the evolution of UMTS via HSDPA and HSUPA is still being pursued.

LTE, publicized in 3GPP Release 8, was finalized in December 2008. LTE with OFDMA air interface is expected to be deployed in 2010 or 2011. LTE uses a number of bandwidths scalable from 1.25 MHz to 20 MHz, and both FDD and TDD can be used. Both OFDM and MIMO technologies are employed to enhance the data rate to 172.8 Mbits/s for the downlink and 86.4 Mbits/s for the uplink. LTE uses OFDM in the downlink, while in the uplink a single-carrier (SC) FDMA is used. The bandwidth of LTE is more than twice that of HSDPA. LTE has a 2 to 6 dB peak-to-average power ratio (PAPR) advantage over the OFMDA method used in mobile WiMAX.

For a 5 MHz band, HSPA+ achieves 42 Mbits/s downlink and 10 Mbits/s uplink, while LTE achieves 43.2 Mbits/s downlink and 21.6 Mbits/s uplink. But HSPA+ does not support over 5 MHz band, while LTE supports up to 20 MHz band. The modulation used in LTE is QPSK, 16QAM, or 64QAM. LTE can be treated as 3.9G.

Ultra Mobile Broadband

Ultra Mobile Broadband (UMB) is 3GPP2's project for evolution to 4G. UMB has a scalable bandwidth between 1.25 to 20 MHz with noncontiguous and dynamic channel (bandwidth) allocations, within a band between 450 MHz and 2500 MHz. It uses a combined OFDMA/OFDM/CDMA/TDMA modulation with FDD or TDD and is an all-IP network. MIMO and SDMA (space division multiple access) technologies are also used. UMB supports a forward link data rate of up to 288 Mbits/s and a reverse link data rate of up to 75 Mbits/s. UMB was published in September 2007 (3GPP2 C.S0084-0 v2.0), and is expected to be deployed in 2009. It can be treated as 3.9G.

IEEE 802.20 (Mobile-Fi)

The IEEE 802.20 standard, nicknamed *Mobile-Fi*, was started in December 2002. It was approved in June 2008. It is targeted for mobile broadband wireless access, with broadband packet-based air interface for mobile users at speeds of up to 250 km/h. It is somewhat similar to IEEE 802.16e. IEEE 802.20 is targeted at licensed bands below 3.5 GHz. It supports a peak data rate in excess of 1 Mbits/s in the downlink and 300 kbits/s in the uplink per 2.5 MHz channel bandwidth. The standard supports both TDD and FDD operations, and can be deployed by using up to 40 MHz frequency band. IEEE 802.20 can be treated as 3.9G.

The standard includes an OFDM wideband mode and a 625k-multicarrier mode. The OFDM wideband mode supports both TDD and FDD, whereas the 625k-multicarrier mode supports TDD only. Both modes are designed to support a full range of QoS attributes. The wideband mode is based on the OFDMA technique. The 625k-multicarrier mode was developed to extract maximum benefit from adaptive, multiple-antenna signal processing.

2.1.4 Fourth-generation systems

Research in 4G mobile radio systems is now underway. ITU-R defined IMT-Advanced in 2007 based on ITU-R Rec. M.1645, and it targets peak data rates of about 100 Mbits/s for highly mobile access (at speeds of up to 250 km/hr), and 1 Gbit/s for low mobility (pedestrian speeds or fixed) access. Development of IMT-Advanced standards is likely to be completed by 2010, with deployment expected around 2015. The 4G wireless systems are supposed to support the following.

- Ubiquitous, mobile, seamless communications.
- A downlink date rate of 100 Mbits/s at stationary conditions and 100 Mbits/s at 250 miles/hour in wide coverage.
- A data rate of 1 Gbit/s at stationary conditions in local area.
- Internet-based mobility with IPv6 (Internet Protocol version 6).
- QoS (quality of service)-driven.
- A high capacity with a spectral efficiency up to 10 bits/s/Hz.
- Smart spectrum with dynamic spectrum allocation within 3 to 10 GHz.
- Dynamic soft channel management.

This will enable us to watch TV and movies on a moving cellular phone. The 4G system will integrate both mobile cellular and wireless networking functions.

OFDM is the main technique to be employed for 4G radio systems. OFDM can be extended to OFDMA, or combined with CDMA or TDMA for multiple access. Some other enabling techniques for 4G mobile radio systems are the following.

- Software-defined radio enables a wide variation in adaptive RF bands, channel access modes, data rates, bit error rates (BERs), and power control that are demanded by 4G systems [8]. Software-defined radio also enables seamless adaptation of radio access technologies (RATs), which is required for 4G.

- Cognitive radio is required for dynamic spectrum management. It provides real-time band allocation for each user.
- MIMO or multiple antenna techniques permit a high spectral efficiency.
- Adaptive modulation and coding (AMC) jointly selects the most appropriate modulation and coding scheme according to channel conditions. AMC is more effective in packet networks, as used by 4G.
- Hybrid-ARQ (HARQ) increases the throughput by combining the advantages of ARQ (automatic repeat request) and channel coding.

These techniques have been incorporated into some 3G standards such as HSDPA, HSUPA, WiMAX, IEEE 802.20, CDMA2000, 3GPP HSPA+, 3GPP LTE, and 3GPP2 UMB. In addition, spectrum flexibility is identified as one main requirement for 4G systems, and multihop relaying is useful to extend the range for the high data rates [1].

Some standards that pave the way to 4G systems are mobile WiMAX, HiperMAN (High Performance Metropolitan Area Network), and WiBro (Wireless Broadband, Korean). These are point-to-multipoint broadband wireless access techniques. These techniques are wireless MANs. Mobile WiMax supports a mobility of up to 70–80 miles/hour. WiMedia, which is based on ultra wideband (UWB) modulation, is suitable for 4G at stationary conditions in local area. A framework for 4G systems was presented from Ericsson in 2005 [1]. In August 2006, Samsung demonstrated 4G services based on mobile WiMAX, which supported a data rate of 1 Gbit/s at stationary and 100 Mbits/s on the move.

The IEEE is developing extensions to both IEEE 802.11 and 802.16 to meet IMT-Advanced (4G) requirements. The IEEE 802.16 Work Group is developing its 802.16m specification, also called *WiMAX II*, which provides continuing support for legacy 802.16-OFDMA (802.16e) equipment, as a high-data-rate (100 Mbits/s) extension at 250 km/h. IEEE 802.16m also employs OFDM, AMC, HARQ, and MIMO technologies. It also supports intelligent multi-hop relays to achieve high data rate over a wide area. The transmission range is expected to be between 2 km in urban environments and 10 km in rural areas. In the meanwhile, IEEE 802.11 Working Group has launched a very high through-put (VHT) study to develop a radio capable of data rates up to 1 Gbit/s at stationary or pedestrian speeds. IEEE 802.21, a standard in its final development, defines link-layer services to enable handovers among different radio air interfaces. The combination of IEEE 802.11VHT and 802.16m via 802.21 creates a single integrated system, which is likely to be IEEE's proposal for IMT-Advanced [3]. The formal completion of this work is expected in early 2010.

Since 2008, 3GPP has also been developing the LTE Advanced as the enhancement to its LTE standard for 4G requirements. The LTE Advanced is scheduled to be published as 3GPP Release 10.

Generally, 2G systems have a spectral efficiency of less than 1 bit/s/Hz, the current 3G systems have a spectral efficiency of 1–3 bits/s/Hz, and the desired spectral efficiency for 4G systems is 10 bits/s/Hz. The evolution of mobile communication systems is demonstrated in Fig. 2.1.

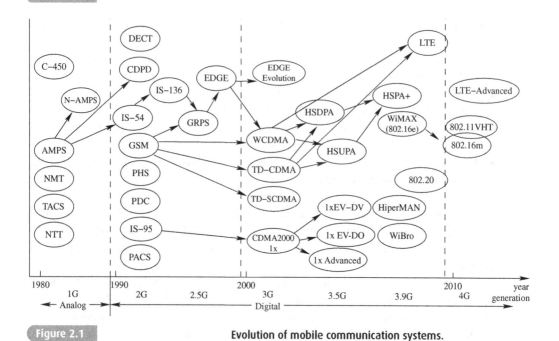

Figure 2.1
Evolution of mobile communication systems.

According to In-Stat, nearly 11% of worldwide wireless subscriptions were 3G at the end of 2008. In-Stat predicts that the percentage of 3G and 4G subscriptions will reach 30% by the end of 2013.[1]

2.1.5 Satellite communications

The concept of communications by satellite was first suggested by Arthur C. Clarke in 1945 [2]. In 1957, the USSR launched *Sputnik I*, and the USA launched *SCORE* in 1958.

Satellites are used for mobile communications. The major differences between satellite and terrestrial mobile channels are the lower elevation angle and the larger distance for the satellite case. For mobile satellite communications, low earth orbit (LEO) satellites are used so as to keep the time delays as small as possible. Mobile satellite systems not only function as standalone wireless systems, but also provide integration with terrestrial mobile systems. Mobile satellite networks have become a crucial component for future global telecommunication infrastructure, and they are standardized by several organizations such as 3GPP and Satellite UMTS. They are intended to complement and extend the existing terrestrial networks so as to provide complete global coverage. Currently, the cost and handset weight are two major obstacles for the commercial success of satellite networks.

Iridium and Globalstar are two famous first-generation LEO satellite constellations that provide hand-held telephony services, primarily for remote areas. Iridium, proposed by

---

[1] http://www.instat.com/

Motorola, consists of 66 LEO satellites in six orbital planes at an altitude of 780 km. Iridium provides global access to personal communications. It operates in the 1.5 to 1.6 GHz band and the channel data rate is 2.4 kbits/s. Iridium communication service was commercially available in 1998, but is no longer available since 2000 due to bankruptcy of its operator. Globalstar was deployed in 2001. The constellation consists of 48 satellites in orbit inclined at 52° with eight planes at an altitude of 1414 km. The data rates range from 1.2 kbits/s to 9.6 kbits/s.

High-altitude aeronautical platform systems (HAPSs) are wireless infrastructures located on unmanned aircraft to provide radio access to a relatively large area on the ground [6]. HAPSs are quasi-stationary aeronautical platforms operating at 17–22 km above the earth's surface, which is in the stratosphere layer of the atmosphere. This avoids the high cost associated with a satellite network. HAPS has been recognized as a possible high-bandwidth solution within the 3G IMT-2000 platform.

In additional to mobile communications, satellites are also used for broadcasting, such as the DVB-S (Digital Video Broadcasting Satellite) system. Satellite broadcasting uses geostationary earth orbit (GEO) satellites, since it can tolerate longer time delays and one satellite can cover one-third of the earth. GEO satellites are deployed at an altitude of 35,786 km above the equator line.

Next-generation satellite communication systems are currently being developed to support multimedia and Internet-based applications [5]. For instance, the Spaceway satellite system consists of 16 GEO and 20 medium-earth-orbit (MEO) satellites operating in the Ka band and utilizing FDMA/TDMA with QPSK modulation; it provides downlink rates of up to 100 Mbits/s, uplink rates of 16 kbits/s–6 Mbits/s, and a total capacity of up to 4.4 Gbits/s. Astrolink, Cyberstar, Skybridge and Celestri are also examples of this generation of satellite communication networks.

The Global Positioning System (GPS) is another popular application for satellite communications. It determines the position of a user, and is attractive in both military and consumer markets. GPS uses spread-spectrum code tracking circuitry to accurately track the propagation delay between the transmitter and receiver. The system has 24 GPS satellites in circular orbits 12,211 miles above the earth, eight satellites in each of three orbits that are inclined 55° with respect to the equator and are separated in longitude by 120°. The orbital period of each satellite is exactly 12 hours. Each satellite uses two carrier frequencies at 1572.42 and 1227.60 MHz, and is differentiated by using a different Gold code. Any point on the earth is illuminated by at least four satellites.

2.2 Mobile cellular networks

The mobile telecommunication network model has five basic network entities: radio systems, switching systems, location registers, processing networks, and interconnected external networks. The radio system (base station, BS) is comprised of antennas, radio transceivers, and its controller. The switching system transmits and switches the traffic from end to end. The location register is a database that stores and retrieves the location and service information of the users. The processing center provides a computing platform to enhance the capabilities of the network.

The network structure consists of three fundamental components: access network (AN), core network (CN), and intelligent network (IN). The access network connects the user equipment and the BS. The radio access network consists of BSs and radio network controller. The core network, also called *fixed network*, handles all the internal connections. The core network consists of mobile switching center (MSC), visitor location register (VLR), home location register (HLR), authentication center (AuC), gateway MSC, etc. The intelligent network is in charge of billing, location management, roaming, handover, etc.

A general mobile cellular network includes mobile stations (MSs), BSs, and MSCs. The MSCs maintain all the mobile-related information, and are in charge of mobile handoffs. The mobile cellular network relies heavily on landline connections. For general 1G and 2G mobile communication networks, as illustrated in Fig. 2.2, the MSCs are interconnected by the PSTN or data networks for carrying traffic, and by the Signaling System No. 7 (SS7) signaling network for carrying signaling information. The 3G network is highly dependent on the Internet Protocol (IP) network. Each MS has its home MSC. MSCs are usually implemented with the three modules: HLR, VLR, and AuC.

The hierarchical cell structure for future mobile communications is illustrated in Fig. 2.3.

2.2.1 Circuit/packet switching

The 2G networks are based on the circuit-switched telephone system. Every call is assigned a dedicated communication path and bandwidth. Each call is set up ahead of the transmission of the data. If there is no circuit available, the caller must try again at a later

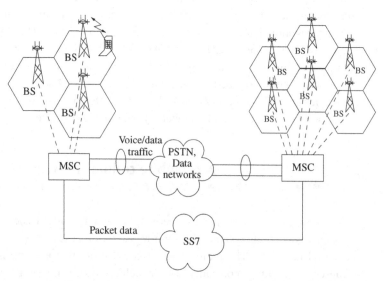

Figure 2.2   General cell network used in 1G and 2G mobile communication networks.

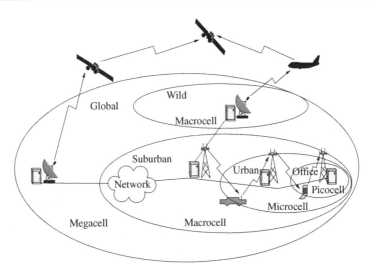

Figure 2.3
Hierarchical cell structure for future mobile communications.

time. Circuit switching is efficient for continuous data transmission such as speech communications. It is not efficient for wireless data communications, due to their short, bursty transmissions; more often the time required for setting up a circuit connection exceeds the duration of the data transmission.

Packet switching is usually used for computer networking and wireless data communications. Packet switching does not assign a physical connection for each user, but rather all users share the network bandwidth. Data are first divided into individual packets, each containing a header, the data load and also error control information. The header consists of a flag that indicates the beginning of a packet, the addresses of the source and the destination, and a control field that contains packet numbering, acknowledgments, and ARQ. These packets are then transmitted and routed by the network. The user does not have to establish a connection to the destination before sending the data. At the destination, the packets are reassembled into the original message.

Most modern WAN protocols, such as TCP/IP, X.25, and frame relay, are based on the packet switching technique. X.25 and frame relay employ the technique of virtual circuit packet switching, while the Internet implements datagram packet switching. Packet switching is also called *packet radio* in case of a wireless link.

The Internet, which uses the IP network protocol, is based on packet switching. Unlike circuit switching, Internet is never busy or never refuses packets. However, a congested network may have the problem of large packet delays or packet loss. Internet offers only best-effort QoS, on a first-come-first-served basis for packet routing.

Signaling System No. 7

Signaling constitutes the command/control infrastructure of the modern telecommunication networks. The SS7 is the most widely used one for common channel signaling between interconnected networks, being a large set of common channel signaling protocols defined by ITU-T.

SS7 is based on packet switching. It carries signaling information from end to end through multiple switching nodes. It is composed of the SS7 Network Services Part (NSP) and SS7 User Part. The SS7 NSP corresponds to the first three layers of the OSI (Open Systems Interconnect) reference model. The GSM MAP (Mobile Application Part) signaling protocol rides on top of SS7.

Internet Protocol in wireless standards

IP is a network-layer protocol that can run over any link layer and allows a variety of applications. The modularity and simplicity of IP have made it the basic networking protocol for modern communication systems, for data communications as well as voice, video, and multimedia communications. VoIP and video over IP are emerging as strong rivals to traditional circuit-switched voice and cable TV. IP-based protocols and architecture are now the basis for the 3G and future-generation mobile communications and wireless networking.

The 3G and 4G wireless networks are on top of the Internet. To deal with the address shortage of IPv4, mobile IP addressing is employed, which is based on the IP address of the MS's home agent and a temporary IP address called a *care-of address (CoA)*. The home agent redirects data from the home network to the care-of address by constructing a new IP header that contains the care-of address as the destination IP address. The IPSec protocol adds security communication that IPv4 lacks.

IPv6 (RFC 2460) increases the IP address size from 32 bits to 128 bits. Mobility is built into IPv6. The mobile IPv6 protocol does not require foreign agents in foreign subnets to configure the care-of address of the mobile nodes. Route optimization is an integral feature of IPv6. Mobile IPv6 uses IPSec for all its security requirements.

The IEEE 802.16 family of standards defines a convergence sublayer for asynchronous transfer mode (ATM) and packet services, but WiMAX is based on only IP and Ethernet (IEEE 802.3) convergence sublayer. 1xEV-DO is completely decoupled from the legacy circuit-switched wireless voice network, and this enables building all-IP 1xEV-DO networks. 3GPP LTE and WiBro support all-IP packet switching only.

2.3 Roadmap for wireless networking

The cellular wireless/mobile and satellite communication systems can be treated as wireless wide area networks (WANs), since they provide wide geographical or global coverage. Wireless WANs can also be implemented through satellites. Wireless data networks are generally classified as wireless local-area networks (LANs), wireless personal-area networks (PANs), wireless metropolitan-area networks (MANs), and wireless regional-area networks (RANs), according to the range of coverage.

Wireless LANs and wireless PANs are typically deployed in the unlicensed industrial, scientific, medical (ISM) 2.4 GHz band. Wireless networking provides complementary radio access for 3G and future mobile communications.

2.3.1 Wireless local-area networks

Wireless LANs typically cover a range of a few meters to a few hundred meters. The most well-known wireless LAN standards are the IEEE 802.11 standards. Other wireless LANs are the European HiperLAN (High Performance Radio LAN) and the Japanese HiSWAN (High Speed Wireless Access Network).

IEEE 802.11/a/b/g/n/VHT

The baseline IEEE 802.11 standard, introduced in 1997, defines 1 and 2 Mbits/s modes. It supports three different physical layers, namely frequency-hopping spread spectrum (FHSS), direct-sequence spread spectrum (DSSS), and infrared. The modulation for the FHSS scheme is GFSK for 1 Mbits/s and 4GFSK (quaternary GFSK) for 2 Mbits/s. The DSSS scheme uses DBPSK for transmission at 1 Mbits/s and DQPSK at 2 Mbits/s. Both of the spread spectrum schemes occupy the 2.4 GHz ISM band. The infrared physical layer has almost never been deployed, since it is severely limited by the short connection distance of 1 m and the line-of-sight (LOS) requirement.

IEEE 802.11b, also known as *Wi-Fi (Wireless Fidelity)*, is the most popular wireless LAN technology. It was released in October 1999. It achieves a data rate of 5.5 or 11 Mbits/s, and uses complementary code keying (CCK) modulation, which is a kind of DSSS, for backward compatibility with the DSSS-based 802.11. It uses a bandwidth of 22 MHz, and covers a range of 100 meters.

IEEE 802.11a, released in October 1999, is based on OFDM technology. IEEE 802.11g, released in June 2003, is the same as 802.11a, except that 802.11a operates in the 5 GHz band while 802.11g in the 2.4 GHz ISM band. They use BPSK, QPSK, 16QAM (quadrature amplitude modulation) and 64QAM modulations, achieving a scalable data rate of up to 54 Mbits/s. Both standards use 20 MHz bandwidth for operation, and 802.11a covers a range of 50 meters.

IEEE 802.11n, ratified on September 11, 2009, is a new generation of wireless LAN standard that achieves raw data rates of up to 300 Mbits/s by combining multiple antennas, clever encoding, and coexistence of 20 and 40 MHz bands in the 2.4 and 5 GHz bands. IEEE 802.11n supports mission-critical applications with throughput, QoS and security rivaling the Ethernet standard 100Base-T. Prior to its ratification, many draft-n products were available on the market.

It is a MIMO-OFDM system, and mandates the interoperation with the legacy 802.11a/g systems. Adaptive beamforming, space–time block code (STBC), and low density parity code (LDPC) techniques are also used as options for increasing the range and reliability of communications.

Meanwhile, since March 2007, IEEE 802.11 Working Group has been developing 802.11VHT (very high throughput) to support data rates of up to 1 Gbit/s at stationary or pedestrian speeds in the 6 GHz and 60 GHz bands, for the IMT-Advanced requirements.

IEEE 802.11 standards do not inherently support QoS due to the use of CSMA/CA (carrier sense multiple access with collision avoidance). To improve IEEE 802.11 standards, IEEE 802.11e was specified to provide QoS enhancement and 802.11i to provide security enhancements. On the basis of IEEE 802.11i, IEEE 802.11w increases the security for selected IEEE 802.11 management frames. The 802.11r task group is currently addressing secure fast roaming, and IEEE 802.11u is being proposed for interworking with non-802.11 networks.

HiperLANs

Under the project name *Broadband Radio Access Network (BRAN)*, ETSI introduced its HiperLAN/1 standard in 1996 as a competitor to IEEE 802.11b. HiperLAN/1 uses FSK and GMSK modulations, achieving a maximum bit rate of 23.5 Mbits/s, operating in the 5 GHz band. Like IEEE 802.11b, HiperLAN/1 uses a distributed MAC protocol based on CSMA/CA.

HiperLAN/2, completed in February 2000, is another ETSI standard competing with IEEE 802.11a. The HiperLAN/2 standard was derived from 802.11a. The primary difference from IEEE 802.11a arises in the medium access control (MAC) layer: HiperLAN/2 employs reservation-based TDMA/TDD protocol, instead of the CSMA/CA employed in IEEE 802.11a, for QoS support. Like 802.11a, it operates in the 5 GHz band with a maximum data rate of 54 Mbits/s.

2.3.2 Wireless personal-area networks

Wireless PANs typically cover a short range that is below ten meters. Wireless PANs are characterized by low-cost low-power implementation. Popular wireless PAN standards are the IEEE 802.15 family and ITU HomeRF. Wireless PANs typically operate in the unlicensed 2.4 GHz ISM band. They are useful in a wide variety of applications, including industrial control, public safety, automotive sensing, and home automation and networking. The concept of the wireless PAN is further extended to the wireless body area network (BAN), as wireless devices that are transported in pockets or worn on the body communicate with one another. Wireless BANs are promising in healthcare, sports and entertainment.

HomeRF

The HomeRF Shared Wireless Access Protocol (SWAP) was designed to carry voice and data within the home. HomeRF makes use of the existing PC industry infrastructure, as well as the Internet, TCP/IP and Ethernet. HomeRF achieves a maximum rate of 10 Mbits/s in its version 2.0 and a coverage range of 50 m. The HomeRF working group was disbanded in January 2003 after the IEEE 802.11b network became accessible and Microsoft began including support for Bluetooth in its Windows operating systems.

IEEE 802.15.1 (Bluetooth)

The IEEE 802.15 standards include several wireless PAN standards. IEEE 802.15.1, more widely known as *Bluetooth*, is the most widely used wireless PAN. Its first specification was published in 1999. Bluetooth, developed by Bluetooth SIG (Special Interest Group), is widely used for the interconnection of consumer electronics and computer peripherals. It is based on DS-FH (direct-sequence frequency-hopping) technology, and has a channel bandwidth of 2 MHz. Depending on its maximum transmitted power, it covers a range of up to 1 m, 10 m or 100 m. Bluetooth 1.2, ratified as IEEE 802.15.1-2005, supports a data rate of 780 kbits/s. Bluetooth 2.0, specified in November 2004, supports a data rate of up to 3 Mbits/s. Up to eight devices can be networked in an *ad hoc* piconet.

Bluetooth 3.0 was approved by Bluetooth SIG on April 21, 2009. It supports data rates up to 24 Mbits/s. It adds IEEE 802.11 as a high speed transport. UWB was antipated, but is missing in Bluetooth 3.0, because WiMedia Alliance announced in March 2009 that it was disbanding. In October 2009 Bluetooth SIG dropped development of UWB as part of the alternative MAC/PHY, Bluetooth 3.0/High Speed solution. Bluetooth 4.0 was approved on December 17, 2009. Aiming mainly at consuming less power. It can run on a standard coin-cell battery for several years. Bluetooth 4.0 supports sending small data packets at a data rate of up to 1 Mbit/s.

IEEE 802.15.3/3a (WiMedia)/3c

IEEE 802.15.3 standards are used for high-rate wireless PANs. These standards use the same MAC layer, but differ in the physical layer. IEEE 802.15.3-2003, approved in June 2003, supports a scalable data rate from 11 to 55 Mbits/s. It employs TDMA with QPSK, DQPSK, 16QAM, 32QAM, and 64QAM modulations. It is a single-carrier system operating in the 2.4 GHz band.

IEEE 802.15.3a, introduced in 2003, enhanced the data rate of 802.15.3 by using UWB technology. It uses three 528 MHz band between 3 and 5 GHz, and has a data rate of up to 480 Mbits/s at a distance of 2 m and 110 Mbits/s at 10 m. A multiband OFDM (MB-OFDM)-based UWB scheme and a direct sequence-based UWB (DS-UWB) scheme were proposed, but due to strong disputes between the two parties, the IEEE P802.15 TG3a project authorization request (PAR) was withdrawn on January 19, 2006.

WiMedia is based on the MB-OFDM UWB scheme of IEEE 802.15.3a. Wireless USB shares the same UWB physical layer with WiMedia, but it addresses very different architectural goals. Both standards have been completed. Wireless USB employs the centrally-controlled piconet architecture of the 802.15 wireless PAN, while WiMedia MAC diverges from the piconet architecture. Wireless USB has its origin in USB (Universal Serial Bus) that focuses on the host and its set of associated devices. Thus, either the host role or the device role is executed in Wireless USB. In contrast, WiMedia has its origin in consumer electronics that operates in a dynamic, even mobile environment, where every device executes all required protocol functions.

Currently, IEEE 802.15.3c wireless PAN is under development, and is expected to be approved in September 2009. IEEE 802.15.3c targets a data rate of over 2 Gbits/s and operates in millimeter wavebands including the 57–64 GHz unlicensed band. It is an OFDM implementation of UWB.

IEEE 802.15.4 (ZigBee)/4a

The IEEE 802.15.4 standards are targeted for ultralow complexity, ultralow cost, ultralow power consumption, and low-data-rate wireless connectivity. It can be used for interactive toys, wireless sensor networks (WSNs), industrial control, home (office, building, or factory) automation, and tagging and tracking. The IEEE 802.15.4 standard employs a distributed MAC protocol based CSMA/CA. ZigBee does not use all abilities of IEEE 802.15.4.

IEEE 802.15.4 employs DSSS coding and OQPSK/BPSK modulation. The basic channel access mode specified by IEEE 802.15.4-2003 is unslotted CSMA/CA. Other mechanisms are also used for beacon transmission and message acknowledgements by using a fixed timing schedule or Guaranteed Time Slots (GTS). It uses a channel of 2 MHz for operation, and achieves a data rate from 20 to 250 kbits/s.

The ZigBee 1.0 specification, ratified in December 2004, is based on IEEE 802.15.4-2003. While IEEE 802.15.4 focuses only on the lower two layers, ZigBee also provides the upper layers for the protocol stack. ZigBee operates in the ISM radio bands, which are 868 MHz in Europe, 915 MHz in USA and Australia, and 2.45 GHz in most jurisdictions worldwide. BPSK is used in the 868 and 915 MHz bands, and OQPSK in the 2.45 GHz band. The raw data rate per channel is 250 kbits/s in the 2.45 GHz band (with 16 channels), 40 kbits/s in the 915 MHz band (with 10 channels), and 20 kbits/s in the 868 MHz band (with 1 channel). At the 2450 MHz band, OQPSK modulation is employed while the 868/915 MHz bands rely on BPSK. Transmission range can be up to 75 meters, and the maximum output power of the radios is generally 0 dBm. ZigBee Pro was approved in October 2007. ZigBee Pro adds some new application profiles such as automatic meter reading.

IEEE 802.15.4a is targeted for the merging market of ZigBee and active radio frequency identification (RFID). IEEE 802.15.4a was completed in January 2007. The baseline consists of two optional physical schemes: a UWB impulse radio (operating in unlicensed UWB band) and a chirp spread spectrum (operating in unlicensed 2.4GHz band). The coverage can be beyond 100m. It provides communications and high-precision ranging/location with an accuracy of 1 m and better.

2.3.3 Wireless metropolitan-area networks

Wireless MANs cover a range of dozens of kilometers. They are targeted at filling the gap between high-data-rate wireless LANs and high mobility cellular wireless WANs.

IEEE 802.16 (WiMAX)

IEEE 802.16 standards, also known as wireless MAN or *WiMAX (Worldwide Interoperability for Microwave Access)*, provide broadband wireless access. The baseline IEEE 802.16, approved in 2001, uses the frequency band from 10 to 66 GHz. It requires LOS propagation, and uses single-carrier modulation and TDMA. The baseline 802.16 has a channel bandwidth of 20, 25, or 28 MHz. IEEE 802.16a/d/e use the frequency band from 2 to 11 GHz, and are based on OFDM technology. IEEE 802.16a was completed in January 2003, and 802.16-2004, which consists of 802.16a/d, was completed in June 2004.

WiMAX is based on 802.16-2004 and 802.16e. It is built on all-IP network architecture for plug and play network deployments. IEEE 802.16e provides support in low mobility, with speeds up to 70 km/h. WiMAX can be used to connect Wi-Fi hotspots to each other or the Internet, to act as a wireless alternative for the last mile broadband services, or to provide 3.9G mobile communications. WiMAX supports a variety of QoS requirements. Multicast and broadcast services are supported in WiMAX.

WiMAX employs TDMA/FDMA/OFDMA with both TDD and FDD. The modulation techniques used are QPSK, 16QAM and 64QAM. IEEE 802.16a/d/e (WiMAX) support a scalable channel bandwidth from 1.75 to 20 MHz, with a peak data rate is 75 Mbits/s, and covers a range of 70 miles. IEEE 802.16a/d/e define three different physical layer air interfaces: single-carrier, OFDM, and OFDMA schemes. IEEE 802.16m (WiMAX II) is targeted for completion in early 2010.

HiperMAN and WiBro

ETSI also defined the HiperACCESS (High-Performance Access) standard as an alternative to the baseline IEEE 802.16. Both standards operate in licensed bands between 10 and 66 GHz, and have channel bandwidths of 20 or 25 (for USA), and 28 MHz (for Europe), with data rates in excess of 120 Mbits/s. The baseline 802.16 employs the single-carrier modulation air interface.

The HiperMAN standard is a subset of IEEE 802.16a, due to the collaboration between ETSI and IEEE. Among the three air interfaces of IEEE 802.16a, HiperMAN uses only the OFDM scheme.

The Korean WiBro standard is based on IEEE 802.16, and provides seamless multimedia Internet services under medium or low mobility. WiBro employs OFDMA with TDD. WiBro has joined WiMAX, and promised to harmonize with the IEEE 802.16e OFDMA scheme with a FFT size of 1,024 [7]. In June 2006, WiBro was launched for commercial services in Korea.

2.3.4 Wireless regional-area networks

Wireless broadband data services were first employed in LMDS (Local Multipoint Distribution Service) and MMDS (Multichannel Multipoint Distribution Service) in the late 1990s. LMDS was deployed at 2.5 GHz, 3.5 GHz, and millimeter wave frequency bands over a range of 3–6 km. LMDS was targeted at business users and provides a data rate of up to several hundreds Mbits/s. The baseline IEEE 802.16 standard can be treated as an LMDS technology. LMDS had a rapid and short-lived success, but was phased out due to its short range and rooftop antenna installation. MMDS was deployed at 2.5 GHz to provide broadcast video services over a range of 30 to 70 km, and is now replaced by satellite TV.

IEEE 802.22, unofficially known as *Wi-TV*, was started in October 2004 and is now under development. IEEE 802.22, like WiMAX and LMDS, provides a point-to-multipoint service. It is also based on OFDMA. It is scheduled to have a cognitive radio-based air interface for use by license-exempt devices to share the underused UHF/VHF TV bands between 54 and 862 MHz in the U.S., since the band is being vacated as broadcasters move to digital systems. It can support as far as 40 km to serve as an alternative to cable or DSL, for rural fixed wireless access.

While IEEE 802.22 was targeted as a wireless RAN to complement both Wi-Fi wireless LAN and WiMAX wireless MAN, WiMAX is trying to reuse the UHF/VHF band based on cognitive radio. This introduces a great advantage to WiMAX since the coverage is much wider than at the 2.4 GHz band, leading to a significant reduction in the number of BSs.

Other broadband wireless access technologies

The high-rate wireless networks, such as IEEE 802.11, 802.16, 802.22, and 802.15.3 standards, basically provide broadband fixed wireless access. High-frequency broadband fixed wireless access provides cellular broadband two-way communications above 20 GHz. The high frequency demands LOS, and the atmospheric conditions significantly influence the signal propagation, and this limits the range to 5 km for reliable services. Due to the wide range of spectrum available, high-frequency broadband fixed wireless access is suitable for *triple play*, that is, fast Internet, multicasting/broadcasting, and telephony. This provides each user a capacity of 15–20 Mbits/s on the downlink. LMDS is such a system.

Major high-frequency broadband fixed wireless access standards are ETSI HiperACESS BRAN, IEEE 802.16, DVB-RCL (Digital Video Broadcasting–Return Channel for LMDS) on the uplink, and DVB-S on the downlink.

Next-generation high-frequency broadband fixed wireless access will be based on TDD rather than FDD, since TDD can significantly increase flexibility and spectral efficiency and cut the costs in user terminals. This technique may require adaptive power control to compensate for rain attenuation. Frequency reuse can be optimized by using the same frequencies for separate cells, separate sectors, or alternating polarization for each channel. Note that rainfall may depolarize the channel.

The evolution of wireless networking is shown in Fig. 2.4.

2.3.5 Ad hoc wireless networks

Wireless communications and networking are traditionally infrastructure-based. There are BSs, also called *access points*, that provide access (gateway) for MSs to a backbone wireless network. The BSs and backbone networks perform all networking functions, and they are usually connected for better coordination. There is no peer-to-peer communication between MSs.

Recently, ad hoc wireless networks, also known as *packet radio networks* or *multihop wireless networks*, have attracted a lot of interest. Ad hoc networks are telecommunication

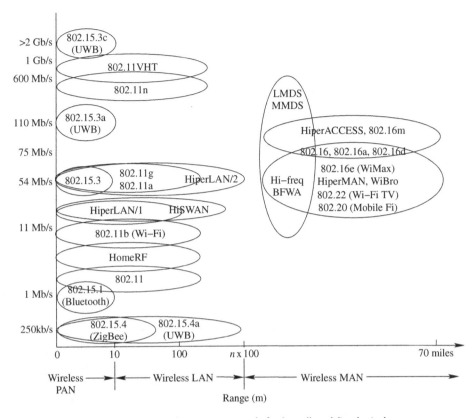

Figure 2.4 Evolution of wireless networking. BFWA stands for broadband fixed wireless access.

networks without a pre-existing infrastructure. An ad hoc wireless network is composed of many wireless mobile nodes that self-configure to form a network "on the fly". Thus, ad hoc architectures are concerned with peer-to-peer communications. Infrastructure is not used in ad hoc wireless networks. Control and networking are processed using distributed control algorithms at each node, and multihop routing is used for packet transmission. Ad hoc networking is important for future high-performance core networks, since it provides the networks with inherent flexibility and survivability, and also reduces the number of BSs required to cover a given area. In ad hoc networking, the security problem is more severe due to the distributed operation at each node.

Without the use of infrastructure ad hoc wireless networks are low cost to deploy and maintain, and also easy for network reconfiguration. Ad hoc wireless networks can form a node hierarchy, either permanently or dynamically. However, multihop and distributed control also lead to performance degradation. The need for all the nodes to collaborate in order to perform infrastructural tasks like routing and forwarding may not be solved since there is no centralized authority: Some nodes may not cooperate to keep the network running. Energy constraints are also a challenge for the design and implementation of ad hoc wireless networks.

Ad hoc wireless networks are especially useful for applications where infra-structure is difficult to deploy rapidly, such as battlefield military communications, emergency rescue, and space monitoring. They also provide a cheap alternative to their infrastructure counterparts for home networking. They can also be used for data networks, device networks, and sensor networks.

The MANET (Mobile Ad Hoc Networking) working group within IETF (Internet Engineering Task Force) is now developing IP-based protocols for ad hoc networking. Ad hoc networking is supported in IEEE 802.11, Bluetooth, 802.15.3, 802.15.4, and HiperLAN. The HiperLAN/1 is designed for ad-hoc networking that works in the 5 GHz band and is based on CSMA/CA.

2.4 Other applications

2.4.1 Paging systems

Paging is a simplex system, where brief text or voice messages are sent only from the service provider to subscribers. It is a one-way, forward direction (BS-to-MS) data communication in packet mode. Paging is the most elementary form of mobile communication, and can work as an independent network. Communication from the user to the BS is via PSTN or other public data networks. Paging is now integrated as a service into cellular systems. The paging terminal stores and forwards voice messages.

There are several propriety systems designed by Motorola, NEC, Ericsson, and the British Post Office. There is no single universal standard, and ITU-R has recommended several standards, including POCSAG (Post Office Code Standard Advisory Group) proposed by British Post Office in the late 1970s, Golay Sequential Code (GSC) paging system by Motorola, NEC by NTT, and RDS by Radio Data System.

Each packet starts with a dotting sequence (10101010...) or a preamble to establish bit timing. A word synchronization field follows the preamble, in order to mark the beginning of message words. The synchronization word is usually selected as a sequence with good correlation properties such as Barker sequences.

PSCSAG has a transmission rate of 512 bits/s using FSK modulation, and supports FSK signaling at up to 2400 bits/s with a channel bandwidth of 12.5 kHz [10]. The GSC system uses a data rate of 300 bits/s for word sync and address, 600 bits/s for preamble and data, and FSK with a deviation of ±5 kHz. The NEC system transmits at a rate of 200 bits/s using FSK with a peak deviation of ±5 kHz. The RDS system is a paging/digital broadcast system which uses a channel bandwidth of 57 kHz on FM broadcast stations. It achieves a data rate of 1187.5 bits/s with a deviation of ±2 kHz.

Newer paging systems such as FLEX in USA and ERMES in Europe, both introduced in 1993, provide up to 6400 bits/s by using FSK and 4FSK modulation. For FLEX, the BS transmitters are synchronized by GPS. The possible frequency deviations are ±1.6 KHz for FSK modulation, and ±1.6 kHz and ±4.8 kHz for 4FSK modulation.

2.4.2 Digital broadcasting systems

Broadcasting systems are now a part of a daily life. The use of AM radio, analog TV, and FM radio dates back to the first half of the twentieth century, and these technologies are based on analog communications. In order to increase the quality of reception and increase the spectrum efficiency, digital broadcasting began to replace the analog broadcasting techniques in the past few years. As of late 2009, ten countries had completely shut down analog TV broadcasts. Many countries will shut down such broadcasts within a few years. In the USA, all high-power analog TV transmissions were turned off on June 12, 2009.

The European DAB (Digital Audio Broadcasting) standard was designed in the 1980s. It is based on the OFDM technology, and the MPEG-1 Audio Layer II codec is employed. In February 2007, DAB+ was released as an upgraded version of the system. DAB+ is approximately 4 times more efficient than DAB due to the adoption of the MPEG-4 AAC+ (ISO/IEC 14496-3) audio codec.

The ESI DVB project was started in September 1993. The major DVB standards include DVB-S, DVB-S2 (DVB-Satellite Second Generation) for satellites, DVB-C for cable, DVB-T, DVB-T2 (Terrestrial DVB Second Generation) for terrestrial television, and DVB-H (DVB-Handheld) for terrestrial handhelds.[2] DVB-S and DVB-C were ratified in 1994, and DVB-T was ratified in early 1997, and DVB-S2 was ratified in March 2005. DVB-T2 has been completed and is expected to be ratified in April 2009. For DVB standards, all data is transmitted in MPEG-2 transport streams with some additional constraints (DVB-MPEG). These distribution systems differ mainly in the modulation schemes and the error correcting codes used. DVB-T/T2 are based on OFDM technology. DVB-S2/T2 use the H.264/MPEG4-AVC codec.

The ETSI DMB (Digital Multimedia Broadcasting, TS 102 427 and TS 102 428) standard is based on the DAB standard, and has some similarity with the competing mobile TV standard, DVB-H; like DVB, DMB can be either T-DMB (Terrestrial-DMB) or S-DMB (Satellite-DMB). T-DMB uses MPEG-4 Part 10 (H.264) for the video and MPEG-4 Part 3 BSAC or HE-AAC V2 for the audio. The audio and video are encapsulated in MPEG-2 TS. The first DMB service started in South Korea in May 2005. In December 2007, ITU approved T-DMB and DVB-H as the global standards.

2.4.3 RF identification

RF (radio frequency) identification systems, also called *RFIDs*, are small low-cost tags on objects in order to track their positions [4]. RFID systems are now being typically deployed at low (125 kHz), medium (13.56 MHz), and high frequency (868 MHz, 2.45 GHz) bands. At 125 kHz and 13.56 MHz, inductive coupling is used to communicate between readers and tags, whereas electromagnetic coupling is used at 868 MHz and 2.4 GHz. Systems using electromagnetic coupling often have a better reading range. The ability to propagate through solid material depends on the carrier frequency, and lower frequencies give better

---

[2] http://www.dvb.org/

propagation than higher frequencies. RFID is continuously replacing the barcoding systems, as it gets cheaper when volumes rise. RFID typically uses amplitude shift keying (ASK) and/or FSK modulation.

In USA, the American National Standards Institute (ANSI)'s X3T6 group is developing a standard for operation at the 2.45 GHz band. The International Organization for Standardization (ISO) published ISO 11784 and 11785 for animal tracking at 135 kHz, ISO 14443 for a proximity card used for RFID at 13.56 MHz, and ISO 15693 and 18092 for a vicinity card (with a maximum distance of 1 to 1.5 meters) at 13.56 MHz. ISO/IEC 18000, which is jointly developed by the ISO and the IEC (International Electrotechnical Commission), is a series of RFID standards for item management at all the RFID frequency bands. The EPCglobal Network has also developed the Electronic Product Code (EPC).

As a wireless system, the RFID tag contains a transceiver and an antenna. It can be either passive, active, or semi-passive. A passive tag contains no power source, and it responds only when a nearby reader powers it. Passive tags have a read distance ranging from about 10 cm (ISO 14443) up to a few meters (EPC and ISO 18000-6), depending on the chosen radio frequency and antenna design. Passive tags can be manufactured with a printed antenna.

The semi-passive and active tags contain a battery, which is used to power the circuit. This leads to a greater sensitivity than that of passive tags, typically 20 dB or more. Thus, it can reach a distance that is ten times longer, or provide a better reliability. An active tag also broadcasts to the reader, thus it is much more reliable even in a very adverse RF environment, or can reach a range of 500 m, but with a shorter life. Semi-passive tags use the battery only to power the circuits, but not to broadcast a signal. Like a passive tag, a semi-passive tag uses the RF energy received from the reader to respond. The battery may also store energy from the reader.

RFIDs are gaining popularity in consumer market. Both the U.S. Department of Defense (DoD) and the world's biggest retailer Wal-Mart are fully incorporating RFID into their supply chain and logistics. The U.S. DoD is using active RFID tags to reduce logistics costs and increase the visibility. RFID tags are also used in passports, as a replacement for the traditional barcodes (for example, in libraries), and for transportation payments and product tracking.

Wireless sensor networks (WSNs) are used for manufacturing and plant control that requires reliability, throughput, distributed control, and data synchronization. Sensor networks are also self-organizing networks, but with a data rate typically much lower than in a telecommunication network. RFID is potentially useful in every sector of industry and commerce that require data collection. The combination of RFID and wireless sensor networking provides a total inventory asset visibility.

2.5 Open systems interconnect (OSI) reference model

Communication network design is based on a layered protocol structure. Layering enables design modularity for network design. The OSI reference model proposed by ISO and

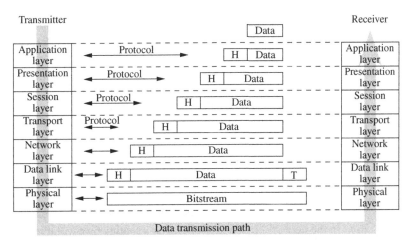

Figure 2.5

Data transmission in the OSI reference model.

ITU-T is a widely accepted framework for data network protocols. It is composed of seven layers: physical, data link, network, transport, session, presentation, and application layers. A protocol stack is a set of protocols consisting of all the layers of the network.

Each layer has its specific packet format or frame structure. The protocol structure of the OSI model is shown in Fig. 2.5. At the transmitter, the data is prefixed with a layer header at each layer, which is used to identify the layer and apply the layer functionality, and is passed downward. At the data link layer, an additional tail is added for the channel error coding. The data is then transmitted as bitstream at the physical layer. At the receiver, the process is reversed: the headers are removed at each layer, until the original data is obtained.

Physical layer

The physical layer, also called the *link layer*, is responsible for reliable transmission of raw bitstreams over a point-to-point physical link such as radio frequency, microwave, or copper wires, regardless of the types of data. The physical layer defines all the electrical and physical specifications for devices. All operations associated with real signals, such as modulation, coding, diversity, multiple-antenna techniques, equalization, and spread spectrum, are in charge of the physical layer.

Data link layer

The data link layer is the second layer. It transforms the raw transmission bitstreams, which may be corrupted by interference and noise, into a stream that appears free of transmission errors to the network layer. It is also used for traffic regulation when the transmitter is faster than the receiver.

The data link layer is typically subdivided into two sublayers: the access layer or MAC sublayer, and the logical link control (LLC) sublayer. The MAC sublayer is in charge of the allocation of bandwidth resources to multiple users. It decides which users are allowed to participate in a simultaneous transmission so as to provide service with acceptable quality to each user, taking into account their QoS requirements. The upper LLC sublayer is independent of the media, and deals with addressing and multiplexing on multiaccess media, flow control, error detection and packet retransmission by the ARQ protocol, and encryption. The most important data link control protocol is HDLC (High-level Data Link Control, ISO 3009, ISO 4335). HDLC is widely used, and it is also the basis for many other important data link control protocols.

Network layer

The network layer is in charge of the establishment and maintenance of end-to-end connections in the network. In the network layer, a routing protocol dictates the route for a packet to arrive at its destination from a source node. Dynamic resource allocation can increase the capacity, and flow control also helps to minimize the delay. These three aspects are interdependent, for QoS. IP is the network protocol for Internet.

Transport layer

The transport layer is responsible for the end-to-end functions such as error recovery, retransmission request, reordering of packages, and flow control to further strengthen error protection.

TCP (Transmission Control Protocol) and UDP (User Datagram Protocol) are the two traditional transport layer protocols in the IP network. TCP ensures reliable end-to-end transmission but has no delay bounds. UDP lacks the congestion control, and it may cause delay as well as packet loss. Neither TCP nor UDP is suitable for multimedia sessions, and Real-time Transport Protocol (RTP), defined in RFC 1889, is the popular transport protocol for this purpose. RTP typically runs over UDP but provides ordering and timing information for real-time applications. TCP also assumes that all packet losses are caused by congestion and the loss rate is small. This is not valid in a wireless network, since packet errors are very frequent due to the poor channel conditions. To improve the performance of TCP in wireless networks, the link can be made more reliable by strong error correction and link-layer ARQ. Cross-layer design for improving the interaction between the link layer and higher layers helps to improve the performance.

In Fig. 2.5, if there are intermediate nodes, the intermediate nodes perform only the functions of the first three layers, namely, physical, data link, and network layers, as the transport layer protocol functions in only the source and destination nodes to provide reliable end-to-end communications.

Session layer

The session layer controls the end-to-end dialogues/connections (sessions). It establishes, manages and terminates the connections between the local and remote applications. The session layer provides protocol negotiation, protocol configuration, and session state maintenance services. The session layer functions are performed by TCP in the TCP/IP suite.

Presentation layer

The presentation layer establishes different syntax and semantics for application layer entities. End-to-end encryption between application entities is also a presentation layer function.

In the TCP/IP suite, the MIME (Multipurpose Internet Mail Extensions) format, and the TLS (Transport Layer Security) and SSL (Secure Sockets Layer) cryptographic protocols correspond to presentation layer functions. For a video network, the presentation layer is concerned with the different encoding format of the media objects, such as MPEG, AVI, and WAV.

Application layer

The application layer provides services to user-defined application processes. It also issues requests to the presentation layer. The application layer produces data to transmit over the network and processes data received over the network. The application layer provides functions such as remote procedure call.

The upper application layer contains the user application programs. In the TCP/IP suite, FTP (File Transfer Protocol), SMTP (Simple Mail Transfer Protocol), POP3 (Post Office Protocol) , IMAP (Internet Message Access Protocol), and HTTP (Hypertext Transfer Protocol) are user application programs. Wireless Application Protocol (WAP), developed by WAP Forum, is a set of standards for accessing online services from mobile devices.

Problems

2.1 Assume the energy of a battery for a cellular phone is 2 Amp-hour. If the cellular phone draws 30 mA in idle mode and 250 mA during a call, what is the battery life if the phone is on and has 2 3-minute calls every day?

2.2 Name all the possible applications you can think of for wireless communications.

2.3 Give the reasons that FM, rather than AM, is used in the 1G systems.

2.4 Tabulate all the cellular communication standards by listing the channel bandwidth, channel spacing, peak data rate, typical data rate, modulation type, duplexing, standard body, maximum number of concurrent users.

2.5 A cellular system uses FDMA with a spectrum allocation of 12.8 MHz in each direction, a guard band of 10 kHz at the edge of the allocated spectrum, and a channel bandwidth of 30 kHz. How many channels are available?

2.6 A GEO satellite is in an equatorial orbit with orbital period $t_s = 24$ h. It appears stationary over a fixed point on the earth surface. Verify that the altitude of a GEO satellite is 35,784 km.

2.7 What are the goals of 4G wireless systems? Name the enabling techniques for achieving these goals.

References

[1] D. Astely, E. Dahlman, P. Frenger, R. Ludwig, M. Meyer, S. Parkvall, P. Skillermark, & N. Wiberg, A future radio-access framework. *IEEE J. Sel. Areas Commun.*, **24**:3 (2006), 693–706.

[2] A. C. Clark, Extraterrestrial relays. *Wireless World*, **51** (1945), 305–308.

[3] L. F. Eastwood, Jr., S. F. Migaldi, Q. Xie, & V. G. Gupta, Mobility using IEEE 802.21 in a heterogeneous IEEE 802.16/802.11-based, IMT-Advanced (4G) network. *IEEE Wireless Commun.*, **15**:2 (2008), 26–34.

[4] K. Finkenzeller, *RFID Handbook: Fundamentals and Applications in Contactless Smart Cards and Identification*, 2nd edn (Chichester, England: Wiley, 2003).

[5] M. Ibnkahla, Q. M. Rahman, A. I. Sulyman, H. A. Al-Asady, J. Yuan, & A. Safwat, High-speed satellite mobile communications: technologies and challenges. *Proc. IEEE*, **92**:2 (2004), 312–339.

[6] S. Karapantazis, F.-N. Pavlidou, Broadband communications via high-altitude platforms: a survey. *IEEE Commun. Surv. & Tutor.*, **7**:1 (2005), 2–31.

[7] W. Konhauser, Broadband wireless access solutions – progressive challenges and potential value of next generation mobile networks. *Wireless Pers. Commun.*, **37** (2006), 243–259.

[8] J. Mitola, The software radio architecture. *IEEE Commun. Mag.*, **33**:5 (1995), 26–38.

[9] S. Ortiz, Jr., 4G wireless begins to take shape. *IEEE Computer*, **40**:11 (2007), 18–21.

[10] T. S. Rappaport, *Wireless Communications: Principles & Practice*, 2nd edn (Upper Saddle River, NJ: Prentice Hall PTR, 2002).

Channel and propagation

3.1 Propagation loss

The underpinning principle for electromagnetic wave propagation is Maxwell's equations. Examples of solutions of Maxwell's equations over very large terrain profiles can be found in [73, 81]. However, due to the complex environment of wireless channels that produce reflected, diffracted, or scattered copies of the transmitted signal, analysis based on Maxwell's equations is extremely complex and also impractical since it has to be based on a lot of assumptions. This section describes a number of models, mainly empirical models, for estimating the propagation loss.

3.1.1 Free-space loss

In free space, the propagation loss from the transmit antenna to the receive antenna can be derived by the Friis power transmission equation [55]

$$P_r(d) \approx P_t G_t G_r \left(\frac{\lambda}{4\pi d} \right)^2, \tag{3.1}$$

where d is the distance between the transmit and receiver antennas, G_r and G_t are the gains of the receive and transmit antennas, respectively, P_t, P_r are the transmitted and received power, and λ is the wavelength of the carrier frequency. Thus, the power decays as d^{-2}.

In the logarithmic form, we have the free-space loss

$$L_{fs}(d) = \frac{P_t G_t G_r}{P_r} = -147.6 + 20 \log_{10} d + 20 \log_{10} f \quad \text{(dB)}, \tag{3.2}$$

where f is the transmission frequency. Thus, the free-space loss increases by 20 dB, if the distance increases by ten times. The free-space loss model is only suitable for a LOS path with no reflection multipath, such as in the case of satellite communications.

3.1.2 Plane earth loss model

The plane earth loss model, as a d^{-4} power law, is a popular empirical law in wireless communications [25, 46, 56, 62]. Measurement in macrocells typically gets a path loss exponent that is close to 4.

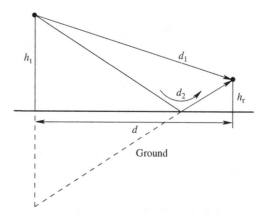

Figure 3.1   **Physical model for plane earth loss.**

The model assumes a main path accompanied by a ground-reflected path, as shown in Fig. 3.1.

The distance difference of the two paths d_2 and d_1 is given by

$$d_2 - d_1 \approx \frac{2h_t h_r}{d}, \tag{3.3}$$

where d is the ground distance between the transmit and receive antennas, and h_t, h_r are the heights of the transmit and receive antennas, respectively.

The path loss can be expressed by

$$\frac{P_r}{P_t} = \left(\frac{\lambda}{4\pi d}\right)^2 \left|1 + Re^{j\beta \frac{2h_t h_r}{d}}\right|^2, \tag{3.4}$$

where R is the reflection loss, and $\beta = \frac{2\pi}{\lambda}$ is the wave number.

Since the angle of incidence with the ground is close to 90°, the magnitude of the reflection coefficient is close to unity. For horizontally polarized antennas, $R \approx -1$, thus

$$\frac{P_r}{P_t} = 2\left(\frac{\lambda}{4\pi d}\right)^2 \left[1 - \cos\left(\frac{2\beta h_t h_r}{d}\right)\right]. \tag{3.5}$$

The cosine term can be approximated as $1 - \cos x \approx x^2/2$ for small x. This approximation can be applied when $h_t h_r \ll d$. By taking the gains G_t and G_r into consideration, we have

$$P_r(d) = P_t G_t G_r \left(\frac{h_t h_r}{d^2}\right)^2. \tag{3.6}$$

In decibels, the loss is given by

$$L_{\text{pel}} = \frac{P_t G_t G_r}{P_r} = 40\log_{10} d - 20\log_{10} h_t - 20\log_{10} h_r \text{ (dB)}. \tag{3.7}$$

This approximation is valid only when d is greater than a critical distance

$$d > d_c = \frac{4h_t h_r}{\lambda}. \tag{3.8}$$

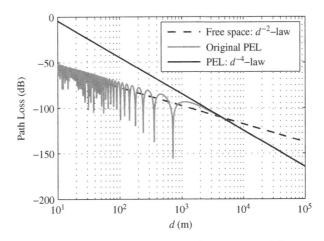

Figure 3.2 Simulation of plane earth loss for $h_t = 40$ m, $h_r = 1.5$ m, and $f = 1800$ MHz.

In this case, the received power becomes independent of the frequency and decays as the fourth power of the distance d.

Example 3.1: Given $h_t = 40$ m, $h_r = 1.5$ m, and $f = 1800$ MHz, the loss versus the distance d is plotted in Fig. 3.2, by using (3.2), (3.5), (3.7). It shows that the plane earth loss model can be approximated by the d^{-4} law when $d > d_c = 1440$ m.

The critical distance d_c can be used for cellular planning. When $d < d_c$, the propagation obeys the d^{-2} law; when $d > d_c$, it obeys the d^{-4} law. Thus, when the two-ray model is suitable, this makes d_c the natural size of a cell. If d_c is very large, the cell size can be further shrunk for microcells in order to increase the capacity and reduce the transmit power.

3.1.3 Okumura-Hata model

The Okumura-Hata model is a good, but more complex propagation model that is based on extensive empirical measurements taken in urban environments [49], and was further approximated by Hata [29]. The model includes parameters such as frequency, frequency range, heights of the transmitter and receiver, and building density. The model is the most popular model for macrocell loss prediction. The model for urban areas was standardized in ITU-R Rec. P.529.

Based on the clutter and terrain conditions, the model varies. The loss is given by [29, 62]

$$L = \left(69.55 + 26.16 \log_{10} f_c - 13.82 \log_{10} h_t\right)$$
$$+ \left(44.9 - 6.55 \log_{10} h_t\right) \log_{10} d - C \quad \text{(dB)}, \tag{3.9}$$

where d (in km) is the distance between the transmitter and receiver, h_t (in m) is the BS height, h_r (in m) is the MS height, f_c (in MHz) is the carrier frequency, and C is given by

(a)

$$C = 2\log_{10}^2\left(\frac{f_c}{28}\right) + 5.4 \tag{3.10}$$

for suburban areas,

(b)

$$C = 4.78\log_{10}^2 f_c + 18.33\log_{10} f_c + 40.94 \tag{3.11}$$

for open areas, and

(c)

$$C = \begin{cases} 3.2\log_{10}^2(11.75h_r) - 4.97, & \text{large cities}, f_c \geq 300\text{MHz} \\ 8.29\log_{10}^2(1.54h_r) - 1.1, & \text{large cities}, f_c < 300\text{MHz} \\ (1.1\log_{10}f_c - 0.7)\,h_r - (1.56\log_{10}f_c - 0.8), & \text{medium or small cities} \end{cases} \tag{3.12}$$

for urban areas.

The model was intended for macrocells, and is applicable over distances of 1–100 km, frequency range 150–1500 MHz, BS height 30–200 m, and MS height 1–10 m. The model is satisfactory in urban and suburban areas, but is not that good in rural areas. This model is suitable for 1G cellular systems, but is not applicable for current cellular systems that have smaller cell sizes and higher frequencies, and for indoor wireless systems.

3.1.4 COST-231-Hata model

The COST-231-Hata model is an extension of the Okumura-Hata model to 2 GHz. It is also an empirical model, and is suitable for microcells and small macrocells. This model is suitable when f_c is within 1.5 GHz–2 GHz, h_t is within 30–200 m, h_r is within 1–10 m, and d is within 1–20 km. It is used by the ITU-R IMT-2000 standards for the outdoor case.

The COST-231 model is given by [15, 25]

$$L_{\text{urban}} = \left(46.3 + 33.9\log_{10}f_c - 13.82\log_{10}h_t\right) + \left(44.9 - 6.55\log_{10}h_t\right)\log_{10}(d) - C + C_M \quad \text{(dB)}, \tag{3.13}$$

where C is the correction factor for mobile antenna height in urban areas, as defined in (3.12) for small to medium-sized cities, and for larger cities at frequencies $f_c > 300$ MHz, C_M is 0 dB for medium-sized cities and suburbs, and 3 dB for metropolitan areas.

Although both the Okumura-Hata and COST-231-Hata models are specified to have a BS antenna height above 30 m, they can be used when h_t is less than 30 m, as long as the surrounding buildings are well below this height. They are not suitable for microcells like urban canyons.

3.1.5 Other empirical models

The propagation loss can generally be written as a linear equation in decibels

$$P_r = P_t + K dB - 10\gamma \log_{10}\left(\frac{d}{d_0}\right) \quad \text{(dBm)}, \tag{3.14}$$

where K is a constant, d_0 is a reference distance for the antenna far field, γ is the path-loss exponent, and P_t, P_r are measured in dBm. In real environments, γ varies from 2.5 to 6 [53]. To avoid the scattering phenomenon in the antenna near field, d is required to be greater than d_0, and d_0 is typically taken as 1–10 m for indoor environments and 10–100 m for outdoor environments. This simplified model is usually used to approximate the field measurements. K is sometimes selected to be the free-space path gain at distance d_0 [25]

$$K = 20 \log_{10} \frac{\lambda}{4\pi d_0} \quad \text{(dB)}. \tag{3.15}$$

Lee's model [42] is used to predict the path loss over flat terrain. It also takes into account the influence of the heights of the transmit and receive antennas, h_t and h_r, but it is based on measurement at 900 MHz. For suburban areas, it has a $d^{3.84}$ power law. IMT-2000 also gives the outdoor-to-indoor and pedestrian models [10], both of which lead to more loss than Lee's model for certain scenarios. The IEEE 802.16 standardization group have proposed a path loss model for propagation in suburban environments [20].

The accuracy of these models typically has a difference of a few dB when compared with field measurements. In industry, it is common practice to use the field measurement and to modify the slope of the linear model over certain range of distance from the BS. These models cannot be used for microcell regions, whose distance is typically less than one mile from the BS, since other propagation phenomena dominate.

One parameter that influences the propagation loss is frequency. For millimeter waves (30–300 GHz), the attenuation is at least 20 dB more than that of a wave of 3 GHz, from the above models. Due to the small wavelength, millimeter waves cannot penetrate solid materials very well, and scattering and diffraction also occur. In addition, they suffer from foliage loss, and are significantly influenced by atmospheric gaseous losses and weather conditions such as rain fall [43]. Due to the wide range of spectrum available, they are of increasing interest.

The Olsen-Segal model [50] has been approved for terrestrial point-to-point LOS links by ITU-R. The model takes into account both the atmospheric impairments and the ground reflection. It is usually used for fading-depth predictions on satellite links, where the fading depth is defined as the ratio of the average signal power to the minimum signal power.

3.1.6 COST-231-Walfisch-Ikegami model

Although the plane earth model has a path loss exponent close to measurement, the two-path model is inapplicable since the MS typically operates without a LOS path or a ground reflection. In fact, in most cases, diffraction is a major propagation mechanism. A number of physical models based on diffraction analysis, such as the Ikegami model [31], the

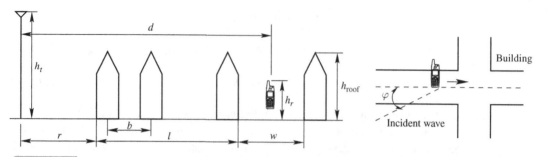

Figure 3.3 A typical propagation scenario in urban areas and definition of the parameters used in the COST-WI model.

flat-edge model, and the Walfisch-Bertoni model [74], are discussed in [62]. Theoretical analysis from these physical models can also yield a path-loss exponent close to 4, but provides more insight into the propagation mechanism.

The COST-231-Walfisch-Ikegami (COST-WI) model combines the Walfisch-Bertoni model and the Ikegami model plus some empirical correction factors to improve the agreement with the measurements in the urban environment [15, 28].[1] The model, shown in Fig. 3.3, defines more parameters, namely heights of buildings h_{roof}, widths of roads w, building separation b, and road orientation with respect to the direct radio path φ.

In the LOS case, when a mobile antenna is within a street canyon, a simple propagation loss formula based on measurement, and different from free space loss, is applied:

$$L = 42.6 + 26 \log_{10} d + 20 \log_{10} f_c \quad \text{(dB)}, \tag{3.16}$$

where d is in km and f_c is in MHz.

For the non-LOS case, the total loss is given by

$$L = L_{\text{fs}} + L_{\text{msd}} + L_{\text{rts}} \quad \text{(dB)}, \tag{3.17}$$

where L_{fs} is the free-space loss, L_{msd} is the loss of multiple-screen diffraction to the top of the final building, and L_{rts}, the rooftop-to-street loss, is for the single diffraction and scattering process down to the mobile at the street level. The expressions for L_{msd} and L_{rts} for non-LOS propagation are given in [15, 62, 69].

The COST-231-Walfisch-Ikegami model is applicable for frequency in the range 800–2000 MHz, h_t in the range of 4–50 m, h_r in the range of 1–3 m, and distance in the range from 20 m to 5 km. The model achieves the best approximation when the BS antenna height h_t is much greater than the roof height of the buildings, $h_t \gg h_{\text{roof}}$. The mean error in the estimation of the path loss is in the range of ± 3 dB and the standard deviation is 4–8 dB [15]. Large prediction errors arise when $h_t \approx h_{\text{roof}}$, while the performance is very poor when $h_t \ll h_{\text{roof}}$. This model is used in ITU-R IMT-2000 standards in the form of ITU-R Rec. P.1411, but the applicable frequency is extended to 5 GHz [32]. It is also recommended by the WiMAX Forum for system modeling in case of standard non-LOS.

---

[1] The COST-231 final report is available at http://www.lx.it.pt/cost231/final_report.htm, where the COST-WI model is introduced in Chapter 4.

3.1.7 Indoor propagation models

Indoor propagation models must be considered for PCS. Indoor radio propagation is dominated by the same mechanisms as the outdoor, but the conditions are more variable inside a building. House and office buildings have different internal and external structures. The wide variety in partitions as well as their physical and electrical characteristics makes it very difficult to find a general model to a specific indoor environment. Experimental measurements for various common building material and indoor and in-building partitions are tabulated in [56].

ITU-R Rec. P.1238 [38] gives a total path loss model for propagation within buildings

$$L = 20 \log_{10} f_c + 10\gamma \log_{10} r + L_f(n_f) - 28 \quad (\text{dB}), \tag{3.18}$$

where γ is the path loss exponent, $L_f(n_f)$ is the the floor penetration loss, which varies with the number of penetrated floors n_f. Corresponding parameters are given in [38]. COST-231 also has some models for indoor multi-wall propagation and propagation into buildings [15].

For an external signal source, each floor may introduce a loss of 1.9 dB [75]. The basement may introduce an additional 10-dB penetration loss. Inside elevators, the signal level may drop as much as 20 dB.

Radio propagation within tunnels is also important for providing communications to people in trains and mine trolleys passing through tunnels. For very short tunnels, they can be illuminated by using transmit antennas at both ends. For long tunnels, a leaky feeder radiator runs along the whole length of the tunnel to provide illumination. The major sources of attenuation are the bends and obstructions in the tunnels. Radio propagation in tunnels and mines has been studied for the frequency range 438 MHz to 24 GHz [18]. In an underground mine tunnel channel, the path amplitude distribution tends to follow a Rice distribution in the LOS case, and a Rayleigh distribution otherwise [9].

In an indoor wideband wireless body-to-body MIMO channel for wireless PANs, the measured path-loss exponent is less than 2.0, and the log-normal distributed shadowing has a standard deviation of 4.8 dB [78]. These indicate that body shadowing causes a prominent power loss in short-range body-to-body communications. The small path-loss exponent is associated with the scattering environments.

For wireless BANs, on-body channel characterization at 2.45 GHz has been investigated in [26]. Propagation measurements show that variability in the path loss due to different antenna placements and due to posture changes can be as much as 50 dB. On-body propagation channels are measured for UWB channels in [3], and at 868 MHz in [16].

A good text on propagation loss models is by Saunders and Aragon-Zavala [62], where a detailed exposition of various propagation channels is given.

Extended Saleh-Valenzuela model

The well-known Saleh-Valenzuela indoor channel model [61] was based on measurements utilizing low power ultra-short pulses (of width 10 ns and center frequency 1.5 GHz) in a medium-size, two-story office building [61].

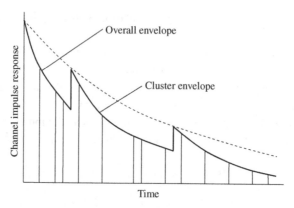

Figure 3.4 The double exponential decay of the Saleh-Valenzuela model: the mean cluster amplitude and the ray amplitude within each cluster have exponential decays.

In this model, multipath components arrive in groups (clusters), or in rays within a cluster. Cluster arrivals are Poisson-distributed with rate Λ. Within each cluster, ray arrivals are also Poisson-distributed. This model requires four parameters, namely the cluster arrival rate Λ, the ray arrival rate within a cluster λ ($\lambda > \Lambda$), the cluster decay factor Γ, and the ray decay factor γ. It is further observed that clustering also takes place in the angular domain [68]. The double-exponential decay is illustrated in Fig. 3.4.

Ray-tracing

Ray-tracing or geometrical optics and the uniform theory of diffraction (UTD) techniques are usually used to approximate the accurate solution based on Maxwell's equations. The error of geometrical optics approximation is very small when the receiver is many wavelengths away from the nearest scatter, or when all the scatters are smooth and large compared to a wavelength. The ray representation of radio propagation is especially useful at microwave and millimeter wave bands. This method is appropriate for characterizing radio wave propagation in cities, since the wavelength is much smaller compared to the dimensions of the buildings.

A number of software tools based on ray tracing are widely used for both indoor and outdoor system planning: Lucent's Wireless Systems Engineering (WiSE), Wireless Valley's Site-Planner, and Marconi's Planet EV.

3.1.8 Channel models in wireless standards

The European COST-259 directional channel model was developed as an empirical model for simulation of systems with multiple antenna elements at either the BS or the MS. The model takes into account small-scale as well as large-scale effects, and covers the macro, micro, and picocellular scenarios. The European project COST-273 extends the COST-259 model to the double-directional case.

The COST-273 model is suitable as a space-time model. The 3GPP channel model [1], which is based on COST-273, is widely used for modeling the outdoor macro- and microcell wireless environments. It is suitable for WiMAX, and can be used for other systems such as IEEE 802.11n and 802.20, with minor modifications to the parameters.

The 3GPP channel model uses too many parameters to build a fully empirical channel model. 3GPP2 provides simpler semi-empirical channel models. It defines pedestrian A and pedestrian B models for low-mobility pedestrian mobile users (at 3 km/h), and vehicular A and vehicular B models for higher-mobility vehicular mobile users (at 30 km/h). These models define the multipath profiles according to the number of multipath taps, and the power and delay of each multipath component. Each multipath component is modeled as an independent Rayleigh fading, and the correlation in the time domain is due to the Doppler effect of the specified speed.

The Erceg model [19] was obtained based on extensive measurement at 1.9 GHz in 95 macrocells across the USA. It has three modes: Erceg A for hilly terrain with moderate to heavy tree density, Erceg B for hilly terrain with light tree density or flat terrain with moderate to heavy tree density, and Erceg C for flat terrain with light tree density. The Erceg model is valid for the frequency range 1900–3500 MHz, BS height 10–80 m, MS height 2–10 m, and distance 100 m–8 km. These models correspond to the six SUI (Stanford University Interim) channel models: SUI-1 to SUI-6, with terrain type C corresponding to SUI-1 and SUI-2, terrain type B to SUI-3 and SUI-4, and terrain type A to SUI-5 and SUI-6. The Erceg model is applicable for fixed wireless deployment with MS installed under the eave or on the rooftop, and has been adopted in IEEE 802.16 for fixed broadband applications [20].

The IEEE 802.11 TGn models [21] are a set of models, which are an improved and standardized version of the extended Saleh-Valenzuela model, with overlapping delay clusters. The models are for indoor MIMO wireless LANs at both 2 and 5 GHz and for bandwidths of up to 100 MHz. Six canonical channels are modeled, namely, flat fading, residential, small office, typical office, large office, and large open spaces. For each canonical channel, the number of clusters, the values of the DoD (direction-of-departure) and DoA (direction-of-arrival) and the cluster angular spreads are fixed, and the tap-delay profiles are represented in the delay domain.

The ITU channel models, unlike the 3GPP channel model, do not model any correlation between the fading waveforms across the transmit and receive antennas. These models were developed for single-input single-output channels. They are also widely used for link-level and system-level performance simulation of 1xEV-DO and HSDPA. In case of correlation of multiple antennas, correlation matrices can be multiplied with the channel matrix **H** at both the transmit and receive ends. More details are given in Chapter 19. ITU has specified two multipath profiles (A and B) for vehicular, pedestrian, and indoor channels, respectively. Channel A is suitable for urban macro-cellular environments, while channel B is suitable for modeling rural macrocells and microcells with cell radius less than 500 m. The ITU channel models are tabulated in Table 3.1 [4].

Table 3.1. ITU channel models.		
Model	Multipath profile (μs)	Relative power profile (dB)
(<3 km/h)		
Pedestrian A	[0, 0.11, 0.19, 0.41]	[0, −9.7, −19.2, −22.8]
Pedestrian B	[0, 0.2, 0.8, 1.2, 2.3, 3.7]	[0, −0.9, −4.9, −8.0, −7.8, −23.9]
(60–120 km/h)		
Vehicular A	[0, 0.31, 0.71, 1.09, 1.73, 2.51]	[0, −1, −9, −10, −15, −20]
Vehicular B	[0, 0.3, 8.9, 12.9, 17.1, 20.0]	[−2.5, 0, −12.8, −10.0, −25.2, −16.0]
Indoor A	[0, 0.05, 0.11, 0.17, 0.29, 0.31]	[0, −3, −10, −18, −26, −32]
Indoor B	[0, 0.1, 0.2, 0.3, 0.5, 0.7]	[0, −3.6, −7.2, −10.8, −18.0, −25.2]

3.2 Channel fading

There are usually three types of channel fading for mobile communications: shadowing (slow fading), multipath Rayleigh fading, and frequency-selective fading. Reflection, diffraction, and scattering are the three major mechanisms that influence the signal propagation.

3.2.1 Log-normal shadowing

During the motion of an MS, clutter such as trees, buildings, and moving vehicles partially block and reflect the signal, thus resulting in a drop in the received power. In the frequency domain, there is a power decrease in a wide frequency range. Hence, it is called *slow fading*. The slow power variation relative to the average can be modeled by a log-normal probability distribution function (pdf).

For the log-normal distribution, the logarithm of the random variable has a normal distribution. The pdf and cumulative distribution function (cdf) are given by

$$p(r) = \frac{1}{r\sigma\sqrt{2\pi}} e^{-\frac{(\ln r - m)^2}{2\sigma^2}}, \tag{3.19}$$

$$D(r) = \Pr(x < r) = \frac{1}{2}\left[1 + \mathrm{erf}\left(\frac{\ln r - m}{\sigma\sqrt{2}}\right)\right], \tag{3.20}$$

where m and σ are the mean and deviation, $\Pr(\cdot)$ is the probability function, and $\mathrm{erf}(x)$ is the well-known error function, defined in Appendix A.

In the shadowing model, the transmit-to-receive power ratio $\psi = P_t/P_r$ is assumed to be random with a log-normal distribution [25]

$$p(\psi) = \frac{10/\ln 10}{\sqrt{2\pi}\,\sigma_{\psi_{\mathrm{dB}}}\,\psi}\exp\left(-\frac{(\psi_{\mathrm{dB}} - m_{\psi_{\mathrm{dB}}})^2}{2\sigma_{\psi_{\mathrm{dB}}}^2}\right), \quad \psi > 0, \qquad (3.21)$$

where $\psi_{\mathrm{dB}} = 10\log_{10}\psi$ in decibels, and $m_{\psi_{\mathrm{dB}}}$, $\sigma_{\psi_{\mathrm{dB}}}$ are the mean and standard deviation of ψ_{dB}. In this model, it is possible for ψ to take on a value within 0 and 1, which corresponds to $P_r > P_t$, but this is physically impossible. Nevertheless, this probability is very small when $m_{\psi_{\mathrm{dB}}}$ is a large and positive number. By a change of variable, the distribution of the dB value of ψ, $p(\psi_{\mathrm{dB}})$, is Gaussian with $m_{\psi_{\mathrm{dB}}}$ and $\sigma_{\psi_{\mathrm{dB}}}$.

Shadowing typically causes a standard deviation $\sigma_{\psi_{\mathrm{dB}}}$ of 4–13 dB [4, 25, 64], and a typical value of $\sigma_{\psi_{\mathrm{dB}}}$ is 8 dB; the mean power $m_{\psi_{\mathrm{dB}}}$ depends on the path loss and the surrounding. This fluctuation in mean power occurs on a large scale, typically dozens or hundreds of wavelengths, and thus, is also known as *large-scale fading* or *macroscopic fading*. Statistically, macroscopic fading is determined by the local mean of a fast fading signal.

Trees cause an important class of environmental clutter. A tree with heavy foliage causes shadowing. A tree with full foliage in summer has approximately a 10 dB higher loss due to shadowing than the same tree without leaves in winter as it acts as a wave diffractor [57].

Example 3.2: Given $\sigma_{\psi_{\mathrm{dB}}} = 6$ and $m_{\psi_{\mathrm{dB}}} = 10$, the pdf of shadowing is plotted in Fig. 3.5. Note that the cdf at $\psi = 1.0$ is very small, and is negligible as $m_{\psi_{\mathrm{dB}}}$ becomes larger.

Percentage of coverage area

The probability of coverage area, $U(P_0)$, is defined as the percentage of area with a received signal that is equal or greater than the threshold P_0, given a likelihood of coverage at the cell boundary. $U(P_0)$ is defined from shadowing by [39]

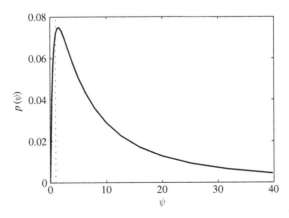

Figure 3.5 The pdf for shadowing: $\sigma_{\psi_{\mathrm{dB}}} = 6$ and $m_{\psi_{\mathrm{dB}}} = 10$.

$$U(P_0) = \frac{1}{\pi R^2} \int \Pr(P_r(r) > P_0)\, dA, \tag{3.22}$$

where dA is an incremental area, P_r is the average received power, and R is the radius of a cell.

$U(P_0)$ is derived as [39, 56]

$$U(P_0) = \frac{1}{2}\left\{1 - \mathrm{erf}(a) + e^{\frac{1-2ab}{b^2}}\left[1 - \mathrm{erf}\left(\frac{1-ab}{b}\right)\right]\right\}, \tag{3.23}$$

where $a = \frac{P_r(r)-P_0}{\sigma\sqrt{2}}$, $b = \frac{10n\log_{10}e}{\sigma\sqrt{2}}$, n the path-loss exponent, and σ the standard deviation, in decibels. In the special case when the average received level at the cell boundary $\overline{P}_r(R) = P_0$, $U(P_0)$ is given as

$$U(P_0) = \frac{1}{2}\left[1 + e^{\frac{1}{b^2}}\mathrm{erfc}\left(\frac{1}{b}\right)\right]. \tag{3.24}$$

In this case, $U(P_0)$ is independent of P_0.

Example 3.3: Given the average received level at the cell boundary $\overline{P}_r(R) = P_0$, from (3.24), the relation of $U(P_0)$ versus n/σ is plotted in Fig. 3.6.

3.2.2 Rayleigh fading

Rayleigh distribution

When both the I and Q components of the received signal, x_I and x_Q, are normally distributed, the received signal is a complex Gaussian variable. The envelope of the received signal, $r = \left(x_I^2 + x_Q^2\right)^{1/2}$, is Rayleigh distributed, and r^2 has an exponential distribution.

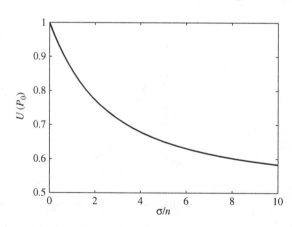

Figure 3.6   Fraction of total area with signal above the threshold P_0, $U(P_0)$, for the average received level at the cell boundary P_0.

Note that the exponential distribution is a special case of the central-χ^2 distribution with $m = 1$. The χ^2 distribution is the distribution of $Y = X^2$, where X is a Gaussian distribution.

Assuming both x_I and x_Q have a standard deviation of σ, the total power in the received signal is $E\left[r^2\right]/2 = \sigma^2$, and we have the pdfs as

$$p_r(r) = \frac{r}{\sigma^2} e^{-\frac{r^2}{2\sigma^2}}, \quad 0 \leq r < \infty, \tag{3.25}$$

$$p_{r^2}(r) = \frac{1}{2\sigma^2} e^{-\frac{r}{2\sigma^2}}. \tag{3.26}$$

The mean of r is $E[r] = \sigma\sqrt{\frac{\pi}{2}}$, and the root-mean-squared (rms) value of the distribution is $E\left[\sqrt{r^2}\right] = \sqrt{2}\sigma$.

Example 3.4: The magnitude of a complex Gaussian random variable for $\sigma = 1$ obtained for 1,000 and 10,000 samples is shown in Fig. 3.7. It is seen that as the number of samples increases, the magnitude approximates a Rayleigh distribution.

The Rayleigh distribution has a pdf defined by (3.25), where r is the amplitude of the received signal and $\overline{r^2} = 2\sigma^2$ is the predicted mean power or mean squared value of the signal amplitude. The pdf has its maximum at $r = \sigma$ with its mean value at $\bar{r} = \sigma\sqrt{\pi/2}$, and has a variance of $0.429\sigma^2$.

The cdf is given by

$$P(R) = \Pr(r < R) = 1 - e^{-\frac{R^2}{2\sigma^2}}. \tag{3.27}$$

For small r, $P(R) \approx \frac{r^2}{2\sigma^2}$.

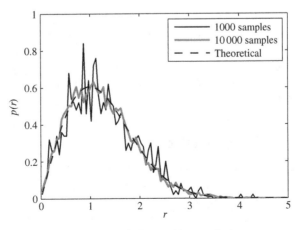

Figure 3.7   The Rayleigh pdf obtained by simulation.

Rayleigh fading in wireless environments

For wireless communications, the envelope of the received carrier signal is Rayleigh distributed; such a type of fading is thus called *Rayleigh fading*. This can be caused by multipath with or without the Doppler effect.

In the multipath case, when the dominant signal becomes weaker, such as in the non-LOS case, the received signal is the sum of many components that are reflected from the surroundings. These independent scattered signal components have different amplitudes and phases (time delays); then, the I and Q components of the received signal can be assumed to be independent zero-mean Gaussian processes. This is derived from the central limit theorem, which states that the sum of a sufficient number of random variables approaches very closely to a normal distribution. When the MS moves, the frequency shift of each reflected signal component that arises from the Doppler effect also has an influence on the fading. Successive drops in amplitudes occur at distances of $\frac{\lambda}{2}$, that is, every time period of $\frac{\lambda}{2v}$, where λ is the wavelength of the carrier frequency, and v is the speed of the MS.

Since the I and Q components of the received signal are i.i.d. zero-mean Gaussian random variables, the phase at any time instant is uniformly distributed

$$p_\phi(\phi) = \frac{1}{\pi}, -\pi \leq \phi < \pi. \tag{3.28}$$

Rayleigh fading occurs very rapidly, and hence it is known as *fast fading*. It can cause as much as 30 to 50 dB rapid power fluctuations at a scale that is comparable to one wavelength, and is thus referred to as *small-scale fading*.

The multipath model is commonly modeled as a two-ray model for illustrating Rayleigh fading. The impulse response is given by [56]

$$h(t) = \alpha_1 e^{j\theta_1(t)}\delta(t) + \alpha_2 e^{j\theta_2(t)}\delta(t - \tau), \tag{3.29}$$

where α_1 and α_2 are independent random variables with a Rayleigh pdf, θ_1 and θ_2 are two independent random variables with uniform pdf over $[0, 2\pi]$, and τ is the time delay.

The received signal is composed of multipath components. The different delays of these components lead to a multipath delay spread. If the time differences of these components are significant compared to one symbol period, intersymbol interference (ISI) occurs. The multipath delay spread has a spectrum with null magnitudes at frequency intervals of $\frac{1}{\tau}$, where τ is the time delay. Thus, this type of fading is called *frequency-selective fading*. When the delay spread is much less than a symbol period, the channel is said to exhibit *flat fading*.

Combined effect of path loss, shadowing, and Rayleigh fading

The effects of path loss, shadowing, and Rayleigh fading can be combined by adding the average path loss $m_{\psi_{dB}}$ (characterized by the path-loss model), shadowing, and Rayleigh fading. Thus, we have

$$\frac{P_r}{P_t} = C_0 - 10\gamma \log_{10} \frac{d}{d_0} - \psi_{dB} - \psi_{dB}^R \quad (dB), \tag{3.30}$$

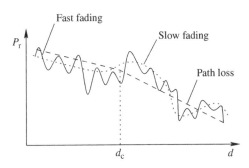

Figure 3.8 Signal power fluctuation vs. range: The combined effect of path loss (plane earth loss model), slow fading and fast fading.

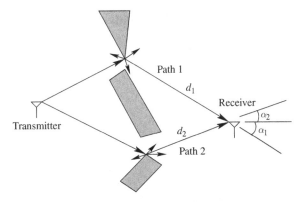

Figure 3.9   Geometry of the two-path model for Rayleigh fading.

where C_0 is a constant that depends on the antenna and channel characteristics, d_0 is a reference distance for the antenna far field, γ is the path-loss exponent, ψ_{dB} is a zero-mean Gaussian with variance $\sigma_{\psi_{dB}}^2$, which corresponds to shadowing, and ψ_{dB}^R is the power contribution associated with Rayleigh fading. The overall effect is illustrated in Fig. 3.8.

3.2.3 Two-path model of Rayleigh fading

The simple two-path model, shown in Fig. 3.9, can be used for demonstrating Rayleigh fading. The mechanisms for multipath can arise from reflection, scattering, reflection, or diffraction. The received signal is given by

$$y(t) = a_1 x(t - \tau_1) + a_2 x(t - \tau_2), \tag{3.31}$$

where $\tau_1 = d_1/c$ and $\tau_2 = d_2/c$ are the time delays corresponding to the two paths d_1 and d_2, c being the speed of light, and the coefficients a_1 and a_2 measure the reflection and propagation losses of the transmitted signal $x(t)$, and thus $a_1, a_2 \leq 1$. The two-path channel is a special case of the tapped-delay-line model, the N-tap Rayleigh fading model with $N = 2$.

Assuming the transmit signal to be a sinusoidal wave $x(t) = \cos(2\pi f_c t)$, the received signal becomes

$$y(t) = a_1 \cos(2\pi f_c t - \beta d_1) + a_2 \cos(2\pi f_c t - \beta d_2), \qquad (3.32)$$

where $\beta = \frac{2\pi}{\lambda}$ is the wave number.

If $d_1 - d_2 = m\lambda$, m being a nonzero integer, the sum of the two paths has its maximum $(a_1 + a_2) \cos(2\pi f_c t - \tau_1)$ at the receiver; if $d_1 - d_2 = \left(m + \frac{1}{2}\right)\lambda$, the sum reaches its minimum $(a_1 - a_2) \cos(2\pi f_c t - \tau_1)$. The amplitude of the received signal may be different at different receiver locations, but it does not vary at a given location.

Two-path model with Doppler

When the receiver is moving at a speed of v, the Doppler effect leads to a frequency shift of $-\frac{v}{\lambda}$ if the receiver is moving away from the transmitter, or $\frac{v}{\lambda}$ if it is moving toward the transmitter. When the receiver is moving away from the transmitter, we have

$$y(t) = a_1 \cos\left(2\pi\left(f_c - \frac{v\cos\alpha_1}{\lambda}\right)t - \beta d_1\right) + a_2 \cos\left(2\pi\left(f_c - \frac{v\cos\alpha_2}{\lambda}\right)t - \beta d_2\right), \qquad (3.33)$$

where α_1 and α_2 are the relative angles of the two paths with the moving direction. Since the speed of movement is always very small compared to the speed of light, the Doppler shift is very small. The superposition of two signals at slightly different carriers f_1 and f_2 leads to a beating envelope with a time interval of $\frac{1}{|f_1 - f_2|} = \frac{\lambda}{v|\cos\alpha_1 - \cos\alpha_2|}$. This can be more clearly seen by

$$y(t) = a_1 \cos(2\pi f_c t - \beta(vt\cos\alpha_1 + d_1)) + a_2 \cos(2\pi f_c t - \beta(vt\cos\alpha_2 + d_2)). \quad (3.34)$$

Accordingly, if $v(\cos\alpha_1 - \cos\alpha_2)t + (d_1 - d_2) = m\lambda$, we get peaks. Thus, the spacing between the peaks and that between the valleys are both $\frac{\lambda}{v|\cos\alpha_1 - \cos\alpha_2|}$. Unlike the time-invariant two-path model, the signal at the receiver will oscillate as it moves past each wavelength.

Example 3.5: Given $f_c = 1$ GHz, $a_1 = a_2 = 0.45$, $d_1 = 100$ m, $d_2 = 160$ m, $\alpha_1 = 50°$, and $\alpha_2 = 30°$, the two-path signal, given by (3.34), is illustrated in Fig. 3.10.

For multiple paths, by assuming that the total power of the multipath components does not change over the region of observation, that is, $\sum_{i=1}^{N} |a_i|^2 = C_P$, where C_P is a constant, and that the phases are uniformly distributed in $[0, 2\pi]$, the Rayleigh distribution can be derived [46].

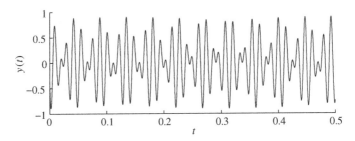

Figure 3.10   The received time-variant two-path signal for Rayleigh fading.

Delay dispersion

Multipath delay spread can be demonstrated by using a two-path model. If the time difference Δt is significant compared to one symbol period, ISI can occur. The delay spread in the time domain corresponds to frequency-selective fading in the frequency domain.

Let us assume that we have two multipaths of signal $s(t)$, with time delay τ_1 and τ_2. The channel is given by

$$h(\tau) = a_1 \delta(\tau - \tau_1) + a_2 \delta(\tau - \tau_2). \tag{3.35}$$

This leads to a spectrum of the received signal $r(t)$ to be

$$R(f) = S(f)H(f), \tag{3.36}$$

where

$$H(f) = a_1 e^{-j2\pi f\tau_1} + a_2 e^{-j2\pi f\tau_2}, \tag{3.37}$$

whose magnitude response is given by

$$|H(f)| = \sqrt{|a_1|^2 + |a_2|^2 + 2|a_1||a_2|\cos(2\pi f\Delta\tau - \Delta\phi)}, \tag{3.38}$$

$\Delta\tau = \tau_2 - \tau_1$ and $\Delta\phi = \phi_2 - \phi_1$, ϕ_1 and ϕ_2 being the phases of a_1 and a_2. This leads to dips in the transfer function, when the phase difference is $(2n + 1)180°$, n being an integer. The frequency difference between two adjacent notch frequencies is $\Delta f = 1/\Delta\tau$. These fading dips also distort the phase of the signals. For multipath components with different delays, a delay-dispersive channel is thus obtained.

Selectivity indicates that the value of the signal received is changed by the channel over time or frequency. Dispersion means that the channel is dispersed, or spread out, over time or frequency. Selectivity and dispersion are dual to each other. Selectivity in time causes dispersion in frequency, and selectivity in frequency causes dispersion in time.

3.2.4 Random frequency modulation

When an MS changes its velocity vector to the BS randomly, such a motion generates randomly varying Doppler frequency at the receiver. This is known as *random FM*.

Random FM defines the performance limit of mobile communication systems for high signal-to-noise ratio (SNR).

The joint pdf of the four random variables r, $\dot{r}$, ϕ, $\dot{\phi}$ of a Gaussian process is given by [58]

$$p\left(r, \dot{r}, \phi, \dot{\phi}\right) = \frac{r^2}{(2\pi)^2\sigma^2 v^2} e^{-\left(\frac{r^2}{2\sigma^2} + \frac{r^2\dot{\phi}^2 + \dot{r}^2}{2v^2}\right)}, \tag{3.39}$$

where σ is the same as that for Rayleigh fading, $\sigma^2 = \frac{1}{2}E\left[r^2\right]$, and $v = \frac{1}{2}E\left[\dot{r}^2 + r^2\dot{\theta}^2\right]$.

The pdf $p(\dot{\phi})$ is obtained by integrating $p\left(r, \dot{r}, \phi, \dot{\phi}\right)$ over r, $\dot{r}$, ϕ [58, 66], giving

$$p\left(\dot{\phi}\right) = \frac{1}{2}\frac{\sigma}{v}\left(1 + \frac{\sigma^2}{v^2}\dot{\phi}^2\right). \tag{3.40}$$

From this, the mean square value of random FM, $E\left[\dot{\phi}^2\right]$, is infinite. Thus, random FM power can be arbitrarily large unless the bandwidth is restricted. The probability density of random FM conditioned on the received envelope R, $p\left(\dot{\phi}|R\right)$, determines the rms bandwidth for a given R. The conditional pdf $p\left(\dot{\phi}|R\right)$ has a Gaussian distribution with a standard deviation $\mu_2 = \sqrt{v}$, and is given by [66]

$$p\left(\dot{\phi}|R\right) = \frac{p\left(\dot{\phi}, R\right)}{\sqrt{2\pi v}} = \frac{R}{\sqrt{2\pi v}} e^{-\frac{R^2\dot{\phi}^2}{2v}}. \tag{3.41}$$

Thus, for a deep fade in the signal envelope, the frequency deviation due to random FM increases proportionally. A fade depth of 20 dB produces a frequency deviation of $10\omega_{max}$, where ω_{max} is the maximum Doppler frequency in radians/s. This makes coherent digital modulation more difficult over fast fading mobile radio channels.

3.2.5 Ricean fading

When a strong stationary path such as a LOS path is introduced into the Rayleigh fading environment, the fading becomes Rice-distributed fading. Ricean fading is suitable for characterizing satellite communications or in some urban environments. Ricean fading is also a small-scale fading. In this case, the probability of deep fades is much smaller than that in the Rayleigh-fading case.

Based on the central limit theorem, the joint pdf of amplitude r and phase ϕ may be derived as [46]

$$p_{r,\phi}(r, \phi) = \frac{r}{2\pi\sigma^2} e^{-\frac{r^2 + A^2 - 2rA\cos\phi}{2\sigma^2}}, \tag{3.42}$$

where A is the amplitude of the dominant component and σ is the same as that for Rayleigh fading, $\sigma^2 = \frac{1}{2}E\left[r^2\right]$. This joint pdf is not separable, and the pdf of r or ϕ can be obtained by integrating over the other quantity. The pdf of the amplitude is a Rice distribution [59]

$$p_r(r) = \frac{r}{\sigma^2} e^{-\frac{r^2 + A^2}{2\sigma^2}} I_0\left(\frac{rA}{\sigma^2}\right), \quad 0 \leq r < \infty, \tag{3.43}$$

where $I_0(x)$ is the modified Bessel function of the first kind and zero order, and is defined as

$$I_0(x) = \frac{1}{2\pi} \int_0^{2\pi} e^{-x\cos\theta} d\theta. \tag{3.44}$$

The mean square value of r is given by

$$P_r = \overline{r^2} = 2\sigma^2 + A^2. \tag{3.45}$$

The squared-envelope r^2 has a noncentral χ-square distribution with two degrees of freedom [69].

The Rice factor K_r is defined as the ratio of the dominant component to the power in all the other components, $K_r = \frac{A^2}{2\sigma^2}$. The Rice distribution approximates the Rayleigh distribution as $K_r \ll 1$, and reduces to it at $K_r = 0$. It approximates the Gaussian distribution with mean value A as $K_r \gg 1$, and reduces to the Gaussian as $K_r \to \infty$. The factor K_r typically shows an exponential decrease with range, and varies from 20 near the BS to zero at a large distance [53].

Example 3.6: The pdfs of the Rice, Rayleigh, and Gaussian distributions are compared in Fig. 3.11. $K_r = 0$ corresponds to Rayleigh fading, $K_r = 40$ can be used to approximate the Gaussian distribution.

The dominant component changes the phase distribution from the uniformly random distribution of Rayleigh fading to clustering around the phase of the dominant component. The stronger the dominant component, the closer the resulting phase to the phase of the dominant component. This is similar to a delta function. Flat Ricean fading channel is suitable for characterizing a real satellite link.

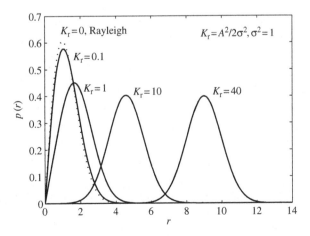

Figure 3.11   Rice distribution for different Rice factor K_r.

3.2.6 Other fading models

Nakagami fading

The Nakagami distribution is another popular empirical fading model [47, 67]

$$p(r) = \frac{2}{\Gamma(m)} \left(\frac{m}{2\sigma^2} \right)^m r^{2m-1} e^{-m \frac{r^2}{2\sigma^2}}, \quad r \geq 0, \tag{3.46}$$

where $\sigma^2 = \frac{1}{2} E[r^2]$, $\Gamma(\cdot)$ is the Gamma function, and $m \geq 1/2$ is the fading figure. The received instantaneous power r^2 satisfies a Gamma distribution. The phase of the signal is uniformly distributed in $[0, 2\pi)$, which is independent of r.

The Nakagami distribution is a general model obtained from experimental data fitting. The Nakagami distribution has a shape very similar to that of the Rice distribution. The shape parameter m controls the severity of fading. When $m = 1$, its pdf reduces to that of a Rayleigh fading. When $m \to \infty$, it becomes the additive white Gaussian noise (AWGN) channel, that is, there is no fading. When $m > 1$, the fading is close to Ricean fading, and the Nakagami and Rice distributions can approximate each other with

$$K_r = (m - 1) + \sqrt{m(m - 1)}, \quad m > 1, \tag{3.47}$$

$$m = \frac{(K_r + 1)^2}{2K_r + 1}. \tag{3.48}$$

The Nakagami distribution has a simple dependence on r, and thus is often used in tractable analysis of fading performance [67]. When the envelope r is assumed to be Nakagami distributed, the squared-envelope r^2 has a Gamma distribution [4, 69]. The Nakagami distribution is capable of modeling more severe fading than Rayleigh fading by selecting $\frac{1}{2} < m < 1$. However, due to the lack of physical basis, the Nakagami distribution is not as popular as the Rayleigh and Ricean fading models in mobile communications.

Suzuki fading

The Suzuki model [70] is a statistical model that gives the composite distribution due to log-normal shadowing and Rayleigh fading. This model is particularly useful for link performance evaluation of slow moving or stationary MSs, since the receiver has difficulty in averaging the effects of fading. It is widely accepted for the signal envelope received in macrocellular mobile channels with no LOS path.

Many other fading channel models are discussed in [67].

3.2.7 Outage probability

Fading channels lead to an oscillating SNR at different locations, and a mobile user will experience rapid variations in SNR, γ. An average SNR can be used to characterize the channel and to compute the BER. For many applications, BER is not the primary concern as long as it is below a threshold. A more meaningful measure is the outage probability,

P_{out}, which is the percentage of time that an acceptable quality of communication is not available. P_{out} can be calculated by the minimum SNR for the system to work properly. This minimum SNR, γ_{min}, can be calculated from the minimum acceptable BER, $P_{b,min}$. In this case

$$P_{out} = \Pr\left(\gamma < \gamma_{min}\right) = \int_0^{\gamma_{min}} p_\gamma(\gamma)d\gamma, \tag{3.49}$$

where $p_\gamma(\gamma)$ is the pdf of γ.

For the frequency-flat Rayleigh fading channel, $\gamma = |h|^2 \frac{E_s}{N_0}$, thus

$$P_{out} = 1 - e^{-\frac{\gamma_{min}}{E_s/N_0}}. \tag{3.50}$$

Given a target error rate P_e, the required SNR, γ_{min}, can be calculated, and then inserted into (3.49) or (3.50).

Example 3.7: For BPSK signals, the BER for coherent demodulation is given by (refer to (7.55))

$$P_b = Q\left(\sqrt{2\gamma_b}\right), \tag{3.51}$$

where the Q-function is given in Appendix A.

From this, the target γ_{min} for $P_e = 10^{-3}$, 10^{-4}, 10^{-5}, and 10^{-6}, can be calculated. The outage probability for different target error rates is plotted in Fig. 3.12. It is seen that for a given BER, a higher quality target (lower P_e) leads to a higher outage probability.

The outage probability P_{out} of a wireless channel can also be defined by

$$P_{out} = \text{cdf}(r_{min}) = \Pr\left(r < r_{min}\right), \tag{3.52}$$

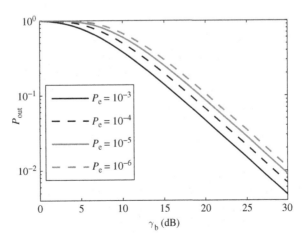

Figure 3.12   Outage probability for BPSK in the frequency-flat Rayleigh fading channel.

where r_{min} is the minimal detected signal envelope. Given an outage probability, the mean power $2\sigma^2$ can be accordingly calculated. The outage probability can also be defined by the probability that the received power at a distance is below the minimum power that is acceptable.

3.3 Doppler fading

Multipath components lead to delay dispersion, while the Doppler effect leads to frequency dispersion for a multipath propagation. Doppler spread is also known as *time-selective spread*. Frequency-dispersive channels are known as *time-selective fading* channels. Signals are distorted in both the cases. Delay dispersion is dominant at high data rates, while frequency dispersion is dominant at low data rates. The two dispersions are equivalent, since the Fourier transform can be applied to move from the time domain to the frequency domain. These distortions cannot be eliminated by just increasing the transmit power, but can be reduced or eliminated by equalization or diversity.

3.3.1 Doppler spectrum

For a moving MS, different multipath components arrive from different directions, and this gives rise to different frequency shifts v, leading to a broadening of the received spectrum. Assuming a statistical distribution of the multipath component direction θ, $p_\theta(\theta)$, and the antenna pattern $G(\theta)$, the Doppler spectrum is derived as [46, 56, 62]

$$S_D(v) = \begin{cases} \dfrac{\overline{\Omega}[p_\theta(\theta)G(\theta)+p_\theta(-\theta)G(-\theta)]}{\sqrt{v_{max}^2-v^2}}, & v \in [-v_{max}, v_{max}] \\ 0 & \text{otherwise} \end{cases} \tag{3.53}$$

where $\overline{\Omega}$ is the mean power of the arriving field, and $v_{max} = f_c v/c$ is the maximum frequency shift due to the Doppler effect, v being the speed of the MS. Note that waves from directions $-\theta$ and θ have the same Doppler shift.

According to the Clarke or Jakes model, the angle distribution of scattering is assumed to be uniform from all azimuthal directions, that is, $p_\theta(\theta) = \frac{1}{2\pi}$, for a symmetrical antenna like a dipole; this leads to

$$S_D(v) = \frac{G(\theta)\overline{\Omega}}{\pi\sqrt{v_{max}^2 - v^2}}. \tag{3.54}$$

This spectrum has a U-shape, and is known as the *classical Doppler or Jakes spectrum*. It can be derived via the Wiener-Khintchin theorem, that is, the Fourier transform of the autocorrelation of the complex envelope of $\epsilon(f; t)$ of the received signal. The autocorrelation function (ACF) is given by [22, 69]

$$\phi_c(\Delta t) = E\left[\epsilon^*(f; t)\epsilon(f; t + \Delta t)\right] = \overline{\Omega}J_0\left(2\pi v_{max}\Delta t\right), \tag{3.55}$$

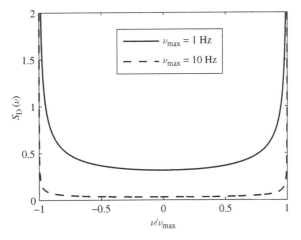

Figure 3.13

Doppler spectrum for $G(\theta)\overline{\Omega} = 1$.

where $J_0(\cdot)$ is the zero-order Bessel function of the first kind

$$J_0(x) = \frac{1}{\pi} \int_0^\pi e^{-jx\cos\theta} d\theta. \tag{3.56}$$

This leads to a nonuniform spectrum, and singularities at the maximum and minimum Doppler frequencies. The frequency dispersion can lead to transmission errors for narrowband systems and OFDM.

Note that the Jakes spectrum is derived from the 2-D isotropic scattering model, and thus is not applicable to microcells that are deployed in dense urban areas, where the streets and buildings along the streets guide the wave within a very narrow angle.

Example 3.8: From (3.54), by setting $G(\theta)\overline{\Omega} = 1$, the Doppler spectra for $v_{max} = 1$ and 10 Hz are plotted in Fig. 3.13.

Example 3.9: Assume there are 200 random paths with maximum Doppler shift $v = 200\,\text{Hz}$ with random moving direction α_n and random delay τ_n. A typical illustration of the received signal envelope is shown in Fig. 3.14. Applying the fast Fourier transform (FFT) on the received signal leads to the Doppler power spectrum density, and the result of a typical run is shown in Fig. 3.15. The spectrum is very similar to the theoretical spectrum.

The classical U-shape spectrum cannot specify the peak value and has a fixed spectral width of $2v_{max}$, whereas the measured spectrum reaches a peak near the maximum Doppler frequency and then decays to zero. The extended Clarke's model introduces fluctuating

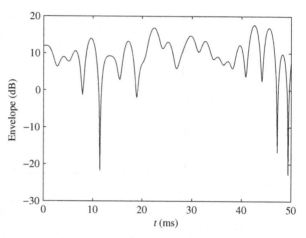

Figure 3.14   Simulation of Doppler fading with 200 scatters.

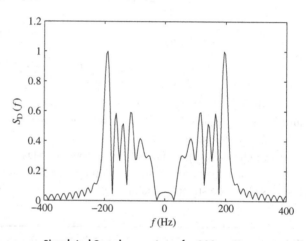

Figure 3.15   Simulated Doppler spectrum for 200 scatters.

component phases, and its statistical properties are essential for accurate spectral analysis and channel simulations. The extended Clarke's model has the ACF [22]

$$\phi_c(\Delta t) = \overline{\Omega} e^{-B|\Delta t|/2} J_0 \left(2\pi v_{max} \Delta t\right), \tag{3.57}$$

where B is a positive constant with the dimension of frequency, which determines the correlation time scale of the component phase process. As $B \to 0$, i.e., in case of absence of fluctuations in component phases, the exponential term approaches unity and the ACF of the fading approaches that for the Clarke's model. Estimation of B from the measured data can be obtained by applying statistical information geometric techniques [22].

The power spectrum of the fading process is derived as [22]

$$S_D(f) = \frac{\overline{\Omega}}{\pi} \int_{-v_{max}}^{v_{max}} \frac{B}{(B/2)^2 + [2\pi(v - \lambda)]^2} \frac{1}{\sqrt{v_{max}^2 - \lambda^2}} d\lambda. \tag{3.58}$$

3.3.2 Level crossing rates

From the Doppler spectrum, the occurrence rate of fading dips, known as the *envelope level crossing rate (LCR)*, and the average duration of fades can be derived [39, 46, 62, 69]. LCR is defined as the number of positive-going crossings of a reference level in unit time, while average duration of fades is the average time for the envelope being below a specified level. Similar to LCR, the zero-crossing rate (ZCR) is defined as the average number of positive-going zero-crossings per second for a signal. For Ricean and Rayleigh fading, these parameters can be derived in closed form [69].

The envelope LCR at level R, L_R, can be derived based on the joint pdf of the envelope and its slope, $p(r, \dot{r})$. For Ricean fading and 2-D isotropic scattering, L_R is derived as [69]

$$L_R = \sqrt{2\pi (K_r + 1)}\, v_{\max}\rho e^{-K_r-(K_r+1)\rho^2} I_0\left(2\rho\sqrt{K_r (K_r + 1)}\right), \qquad (3.59)$$

where $\rho = \dfrac{R}{r_{\mathrm{rms}}}$, r_{rms} being the rms envelope level, and $I_0(\cdot)$ is the modified Bessel function of the first kind and zero order, which is given by (3.44). For Rayleigh fading ($K_r = 0$), the expression for L_R simplifies to

$$L_R = \sqrt{2\pi}\, v_{\max}\rho e^{-\rho^2}. \qquad (3.60)$$

Example 3.10: The envelope LCR versus the level ρ is plotted in Fig. 3.16 for the 2-D isotropic scattering environment. It is seen that the fades are shallower for large K_r. It is also seen that at around $\rho = 0$ dB the envelope LCR is independent of K_r. This feature is exploited for estimation of the MS speed.

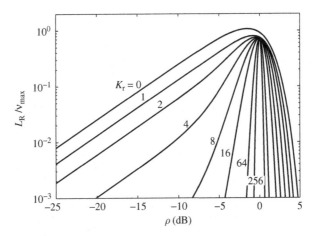

Figure 3.16   Envelope level crossing rate for Ricean fading and 2-D isotropic scattering.

Assuming the received bandpass Ricean fading signal as

$$x(t) = g_I(t) \cos 2\pi f_c t - g_Q(t) \sin 2\pi f_c t, \tag{3.61}$$

where $g_I(t)$ and $g_Q(t)$ are independent Gaussian random processes with variance σ^2 and means m_I and m_Q, the envelope $r = |g_I(t) + j g_Q(t)|$ is Ricean distributed, with $K_r = \left(m_I^2 + m_Q^2\right)/(2\sigma^2)$. Thus, the ZCR of the zero-mean Gaussian random processes $g_I(t) - m_I(t)$ and $g_Q(t) - m_Q(t)$ for 2-D isotropic scattering is given by [69]

$$L_Z = \sqrt{2} \nu_{max}. \tag{3.62}$$

The LCR and ZCR can be used to estimate the velocity of an MS. For a bandpass signal under the Ricean fading model, the velocity estimators are robust with respect to the Ricean factor K_r [5, 69]

$$\hat{\nu}_{ZCR} \approx \frac{\lambda_c \hat{L}_{ZCR}}{\sqrt{2}}, \quad \hat{\nu}_{LCR} \approx \frac{\lambda_c \hat{L}_{R_{rms}}}{\sqrt{2\pi e^{-1}}}, \tag{3.63}$$

where $\hat{}$ denotes the estimation, and $L_{R_{rms}}$ is L_R for $R = R_{rms}$. The influence of K_r and the angle of the specular component θ_0 on L_{ZCR} is given in [5]. The ZCR velocity estimator is shown to be more robust than the LCR method [5].

3.3.3 Average duration of fades

Average duration of fades is the average time that the envelope level is below level R. The probability of the envelope level being below R is given by

$$\Pr(\alpha \leq R) = \int_0^R p(\alpha) d\alpha = \frac{\sum_i t_i}{T}, \tag{3.64}$$

where t_i is the duration of the ith continuous period that is below R, and T is the total period. For the Rice distribution, $\Pr(\alpha \leq R)$ can be expressed by [69]

$$\Pr(\alpha \leq R) = 1 - Q(\sqrt{2K_r}, \rho\sqrt{2(K_r + 1)}), \tag{3.65}$$

where $Q(a, b)$ is the Marcum-Q function defined by

$$Q(a, b) = 1 - \int_0^b z e^{-\frac{z^2 + a^2}{2}} I_0(za) dz, \tag{3.66}$$

$I_0(\cdot)$ being the modified Bessel function of the first kind and zero order, which is given by (3.44).

The average duration of fades is given by

$$\bar{t} = \frac{P(\alpha \leq R)}{L_R}, \tag{3.67}$$

where L_R is given by (3.59). Thus, we have

$$\bar{t} = \frac{1 - Q\left(\sqrt{2K_r}, \rho\sqrt{2(K_r + 1)}\right)}{\sqrt{2\pi(K_r + 1)} \nu_{max} \rho e^{-K_r - (K_r + 1)\rho^2} I_0\left(2\rho\sqrt{K_r(K_r + 1)}\right)}. \tag{3.68}$$

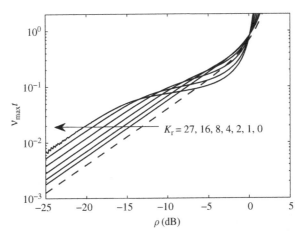

Figure 3.17
Average envelope fade duration for Ricean fading and 2-D isotropic scattering: $v_{max}\bar{t}$ versus ρ.

For Rayleigh fading ($K_r = 0$), it reduces to

$$\bar{t} = \frac{e^{\rho^2} - 1}{\rho v_{max}\sqrt{2\pi}}. \tag{3.69}$$

Example 3.11: The average duration of fades for Ricean fading in the 2-D isotropic scattering environment, which is given by (3.68), is plotted in Fig. 3.17. It is seen that for very deep fades (ρ is very small) each fade lasts a very short time ($\bar{t}$ is very small), and from Fig. 3.16, that the LCR is very small.

3.4 WSSUS model

Wireless channels are time-variant, with an impulse response $h(t, \tau)$, and can be modeled by using the theory of linear time-variant systems. Because most wireless channels are slowly time-varying, or quasi-static, this enables the use of many concepts of linear time-invariant (LTI) systems. By performing the Fourier transform to the absolute time t, or the delay τ, or both, we obtain the delay Doppler-spread function $S(v, \tau)$, the time-variant transfer function $H(t, f)$, or the Doppler-variant transfer function $B(v, f)$, respectively.

The stochastic model of wireless channels is a joint pdf of the complex amplitudes for any τ and t. The ACF is usually used to characterize the complex channel. The ACF of the received signal, $y(t) = x(t) * h(t, \tau)$, for a linear time-variant system is given by

$$R_{yy}(t, t') = \int_{-\infty}^{\infty} \int_{-\infty}^{\infty} R_{xx}(t - \tau, t' - \tau') R_h(t, t', \tau, \tau') d\tau d\tau', \tag{3.70}$$

where the ACFs are

$$R_{xx}\left(t-\tau,t'-\tau'\right)=E\left[x^*(t-\tau)x\left(t'-\tau'\right)\right],\tag{3.71}$$

$$R_h\left(t,t',\tau,\tau'\right)=E\left[h^*(t,\tau)h\left(t',\tau'\right)\right].\tag{3.72}$$

A stochastic process is said to be strictly stationary if all its statistical properties are independent of time. When only the mean is independent of time while the autocorrelation depends on the time difference $\Delta t = t' - t$, such a process is said to be a *wide sense stationary* process. The popular WSSUS (wide sense stationary, uncorrelated scattering) model is based on the dual assumptions: *wide sense stationarity* and *uncorrelated scatters*. The assumption of wide sense stationarity states that the ACF of the channel is determined by the difference Δt, that is,

$$R_h\left(t,t',\tau,\tau'\right)=R_h\left(\Delta t,\tau,\tau'\right).\tag{3.73}$$

Thus, the statistical properties of the channel do not change over time, and the signals arriving with different Doppler shifts are uncorrelated :

$$R_h\left(v,v',\tau,\tau'\right)=P_s\left(v,\tau,\tau'\right)\delta\left(v-v'\right),\tag{3.74}$$

where P_s is the scattering function, giving the Doppler power spectrum for a multipath channel with different path delays τ.

The assumption of uncorrelated scatters means that signals with different delays are uncorrelated

$$R_h\left(t,t',\tau,\tau'\right)=P_h\left(t,t',\tau\right)\delta\left(\tau-\tau'\right),\tag{3.75}$$

where P_h is the delay cross-power spectral density. In the frequency domain, we have the time-frequency correlation function R_H that depends only on the frequency difference $\Delta f = f' - f$

$$R_H\left(t,t',f,f'\right)=R_H\left(t,t',\Delta f\right).\tag{3.76}$$

Combining the two properties, the WSSUS model can be characterized by

$$R_h\left(t,t',\tau,\tau'\right)=P_h\left(\Delta t;\tau'\right)\delta\left(\tau'-\tau\right).\tag{3.77}$$

This kind of stationarity is only valid for a small geographical area. For a large scale channel behavior, the WSSUS model must be examined across consecutive time intervals. This is the quasi-WSSUS model [7], which gives satisfactory results for practical channels. The WSSUS model can be implemented using tapped-delay-line models.

The scattering function $P_s(\tau, v)$ gives the average power output of the channel as a function of the time delay τ and the Doppler shift v. It is a compact characterization of multipath-fading channel. The relationships between the autocorrelation and scattering functions can be characterized by Fig. 3.18.

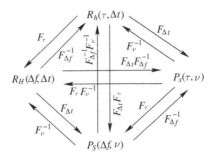

Figure 3.18 Relationships between the ACF and the scattering function. F and F^{-1} denote the Fourier and inverse Fourier transforms.

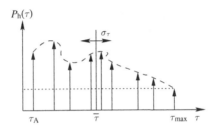

Figure 3.19   Power delay profile for a typical urban environment.

3.4.1 Delay spread

The delay power spectral density or power delay profile (PDP), $P_h(\tau)$, is obtained by integrating the scattering function $P_s(\tau, \nu)$ over the Doppler shift ν.

The PDP can be calculated by

$$P_h(\tau) = \sum_{n=1}^{N} P_n \delta(\tau - \tau_n) = \int_{-\infty}^{\infty} |h(t, \tau)|^2 dt, \tag{3.78}$$

where $P_n = a_n^2$, a_n being the amplitude of the nth delay, and the second equality holds if ergodicity holds. First arrival delay τ_A is the delay of the first arriving signal, all the other delays are known as *excess delays*, and the maximum excess delay τ_{max} is the delay corresponding to a specified power threshold. A typical PDP for an urban environment is shown in Fig. 3.19.

The average WSSUS channel delay $\overline{\tau}$ and the rms delay spread σ_τ are defined by

$$\overline{\tau} = \frac{\int_0^{\infty} \tau P_h(\tau) d\tau}{\int_0^{\infty} P_h(\tau) d\tau}, \tag{3.79}$$

$$\sigma_\tau = \sqrt{\frac{\int_0^{\infty} (\tau - \overline{\tau})^2 \tau P_h(\tau) d\tau}{\int_0^{\infty} \tau P_h(\tau) d\tau}}. \tag{3.80}$$

Table 3.2. Typical values of rms delay spreads, σ_τ [62].

Environment	$\sigma_\tau, \mu s$
Indoor cells	0.01–0.05
Mobile satellite	0.04–0.05
Open area	< 0.2
Suburban area	< 1
Urban area	1–3
Hilly area	3–10

Delay spread leads to frequency-selective fading, as the channel function resembles a tapped-delay filter. A general rule of thumb is that $\tau_{\max} \approx 5\sigma_\tau$.

The PDP has been modeled in order to understand the channel behavior and to evaluate the performance of equalizers. There are many measurements of indoor and outdoor channels [13]. The one-sided exponential profile is a suitable model for both indoor and urban channels

$$P_h(\tau) = \frac{P_T}{\sigma_\tau} e^{-\frac{\tau}{\sigma_\tau}}, \quad \tau \geq 0, \tag{3.81}$$

where P_T is the total received power.

When the excess delay spread exceeds the symbol duration by 10% to 20%, an equalizer may be required. The average delay and the delay spread of a channel diminish with decreasing cell size due to shorter propagation path.

Typical values for the rms delay spread are given in Table 3.2 [62]. Results from the COST-207 models, which were developed and standardized for the GSM system, gives typical rms delay spread as follows[14, 69]:

- for typical urban (TU) (nonhilly) area, $\sigma_\tau = 1.0\,\mu s$,
- for bad urban (BU) (hilly) area, $\sigma_\tau = 2.5\,\mu s$,
- for rural area (RA) (nonhilly) area, $\sigma_\tau = 0.1\,\mu s$, and
- for typical hilly terrain (HT) area, $\sigma_\tau = 5.0\,\mu s$.

Typical PDPs are defined for these typical channel environments in [14].

3.4.2 Correlation coefficient

The correlation coefficient of two signals is usually defined with respect to the signal envelopes x and y

$$\rho = \rho_{xy} = \frac{E[xy] - E[x]E[y]}{\sqrt{\left(E\left[x^2\right] - E[x]^2\right)\left(E\left[y^2\right] - E[y]^2\right)}}. \tag{3.82}$$

For two statistically independent signals, $\rho = 0$; when ρ is below a threshold such as 0.5, the signals are typically considered effectively as decorrelated.

For a channel with PDP of type (3.81), assuming a classical Doppler spectrum for all the components, the correlation coefficient of two signals with a temporal separation Δt and a frequency separation Δf is given by [39, 46]

$$\rho_{xy}(\Delta t, \Delta f) = \frac{J_0^2 (2\pi \nu_{\max} \Delta t)}{1 + (2\pi \sigma_\tau \Delta f)^2}, \tag{3.83}$$

where $J_0(x)$ is the zero-order Bessel function of the first kind, defined by (3.56), and $\nu_{\max}$ is the maximum Doppler frequency.

Equation (3.83) is derived based on a number of assumptions including the WSSUS model, non-LOS signal, exponential shape of the power delay profile, uniform distribution of incident power, and use of omnidirectional antennas. This equation can also be used for spatial separation, since the latter can be converted into temporal separation for MSs. For a typical urban channel model, the ρ values for 30 kHz (IS-136), 200 kHz (GSM), and 5 MHz (WCDMA) frequency separation are 0.97, 0.4, and 0.001, respectively.

3.4.3 Channel coherent bandwidth

The channel coherence bandwidth B_c is defined as the maximum frequency difference $(\Delta f)_{\max}$ that limits the correlation coefficient ρ to be smaller than a given threshold, typically 0.7. For instance, using (3.83), for $\rho = 0.5$, Δf takes on its maximum at $\Delta t = 0$, accordingly

$$B_c = (\Delta f)_{\max} = \frac{1}{2\pi \sigma_\tau}. \tag{3.84}$$

Due to the uncertainty relation between the Fourier transform pairs, there is an uncertainty relation between B_c and the rms delay spread σ_τ [46]

$$B_c \geq \frac{1}{2\pi \sigma_\tau}. \tag{3.85}$$

That is, both B_c and σ_τ can be used to characterize the channel, and they are in inverse proportion; although usually they can be related by the approximation $B_c \approx \frac{1}{\sigma_\tau}$, they cannot be exactly derived from each other.

A wireless channel is said to have *frequency coherence* if it satisfies

$$H(f) \approx \text{constant}, \quad |f_c - f| \leq B_c, \tag{3.86}$$

where f_c is the center carrier frequency.

For narrowband signals, the signal bandwidth $B \ll B_c$; then, the fading across the entire signal bandwidth is highly correlated. Thus, the fading is roughly equal across the entire signal bandwidth. This is known as *flat fading*. In this case, for linearly modulated signals, the symbol period $T_s \approx 1/B \gg 1/B_c \approx \sigma_\tau$, thus ISI is negligible. On the other hand, if the signal bandwidth $B > B_c$, the channel amplitudes at frequencies separated by more than B_c are approximately independent. Thus, the channel amplitude varies across the signal bandwidth B. The channel is thus known as a *frequency-selective fading* channel. For linearly modulated signals, $T_s < \sigma_\tau$ and thus ISI cannot be neglected. When B is close to B_c, the

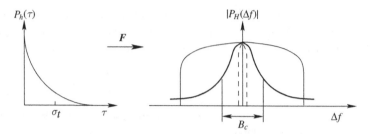

Figure 3.20 Power delay profile, frequency-flat fading, and frequency-selective fading. *F* denotes the Fourier transform.

channel has a behavior between that of flat fading and frequency-selective fading. This is depicted in Fig. 3.20

3.4.4 Doppler spread and channel coherent time

The Doppler power spectral density $P_B(\nu)$ is obtained by integrating the scattering function $P_s(\tau, \nu)$ over the time delay τ. Analogous to the derivation of the average channel delay $\bar{\tau}$ and rms delay spread σ_τ, the average Doppler shift $\bar{\nu}$ and the rms Doppler spread σ_ν can be derived as the first- and second-order moments of $P_B(\nu)$. The Doppler spread corresponds to time-selective fading.

The channel coherence time T_c is also defined according to (3.83). The coherent time measures how fast the channel changes in time: A large coherent time corresponds to a slow channel fluctuation. The coherence time is defined in a manner similar to that of the coherent bandwidth: It is defined as the time delay for which the signal autocorrelation coefficient reduces to 0.7. It also has an uncertainty relationship with the rms Doppler spread σ_ν, although usually the approximation $T_c \approx 1/\sigma_\nu$ is applied.

A channel is said to have *temporal coherence* if a narrowband (no frequency dependence), fixed (no spatial dependence) channel satisfies

$$h(t) \approx \text{constant}, \quad |t_0 - t| \le T_c, \tag{3.87}$$

where t_0 is an arbitrary time instant.

Definitions similar to that of channel coherent bandwidth can be given with respect to the channel coherent time. For example, if the coded symbol duration is much greater than the channel coherent time, the channel is a *time-selective channel*.

Both T_c and B_c are used to quantize the spread of the signal correlation coefficient $\rho(\Delta t, \Delta f)$ around the origin, and are given for the classical channel by [6]

$$T_c = \frac{9}{16\pi \nu_m}, \quad B_c = \frac{1}{2\pi \sigma_\tau} \tag{3.88}$$

for a correlation coefficient of 0.5 [51]. The coherent time decides the maximum duration for undistorted symbols.

When $T_s > T_c$, or equivalently, $B < B_D$, the channel impulse response $h(t)$ changes within a signal symbol duration, thus the channel is known as a *fast fading* channel. On the other hand, if the channel impulse response changes much slower than the symbol rate, that is, $T_s \ll T_c$ or equivalently, $B \gg B_D$, the channel is called a *slow fading* channel. Thus, a low data rate mobile moving at a high speed has a fast fading channel, while a high data rate mobile moving at a slow speed has a slow fading channel.

For mobile communications, flat slow fading and frequency-selective slow fading are the two common channel models, since fast fading can occur only when the MS is at low data rates and is moving rapidly. Various functions for characterizing stochastic channels and their relationships are given in [40].

3.4.5 Angle spread and coherent distance

The channel model can also include the directional information such as the DoA and DoD of the multipath components into its impulse response, leading to the double-directional impulse response. Analogous to the nondirectional case, a number of power spectrums such as the double directional delay power spectrum, angular delay power spectrum, angular power spectrum, and azimuthal spread can be defined [23]. Such a directional channel model is especially useful for multiple antenna systems.

Angle spread at the receiver is the spread in DoAs of the multipath components at the receive antenna array, while angle spread at the transmitter is the spread in DoDs of the multipath components that finally arrive at the receiver. Denoting the DoA by θ, the angle power spectrum or power angular profile $P_A(\theta)$ is given by

$$P_A(\theta) = \sum_{n=1}^{N} P_n \delta(\theta - \theta_n). \tag{3.89}$$

Analogous to the derivation of the channel delay and delay spread, the mean DoA $\bar{\theta}$ and the rms angle spread σ_θ can be derived as the first- and second-order moments of $P_A(\theta)$. The angle spread leads to *space-selective fading*. The coherent distance D_c is used to characterize space-selective fading, and it is calculated as the spatial separation when the correlation coefficient reduces to 0.7.

A wireless channel is said to have *spatial coherence* if for a static narrowband channel

$$h(\mathbf{r}) \approx \text{constant}, \quad |\mathbf{r}_0 - \mathbf{r}| \leq D_c, \tag{3.90}$$

where $\mathbf{r}_0$ is an arbitrary position in space. Spatial incoherence is caused by multipaths from many different directions. These waves create constructive and destructive interference, leading to spatial selectivity.

For receive antennas, the coherent distance is inversely proportional to the angle spread. An approximate rule of thumb is [4]

$$D_c \approx \frac{\lambda}{5\sigma_\theta}. \tag{3.91}$$

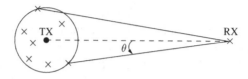

Figure 3.21   Scatters around a transmitter.

The Rayleigh fading channel assumes a uniform angle spread, and the coherent distance is given by [4]

$$D_c \approx \frac{9\lambda}{16\pi}. \tag{3.92}$$

Angle spread and coherent distance are particularly important in multiple antenna systems. D_c indicates how far apart the antennas should be spaced in order for the received signals to be statistically independent. Given a target receive antenna, D_c and σ_θ can also be defined at a transmit antenna array.

Angular spread can be derived from the scattering surrounding the transmitter. Usually the scatters are assumed to be uniformly distributed on a circle around a transmitter so that the multiple paths arrive at the receiver in a very narrow cluster [54]. This is illustrated in Fig. 3.21. The power is concentrated within a small angle spread around 0°, with the maximum power angular density at 0°.

In an indoor or a congested urban environment, all the waves are guided along the wall or the buildings along a street; extensive field measurements show that the scattering can be modeled by a Laplace distribution [54]

$$P_A(\theta) = \frac{P_T}{\sqrt{2}\sigma_\theta} e^{-\left|\frac{\sqrt{2}\theta}{\sigma_\theta}\right|}, \tag{3.93}$$

where σ_θ is the angular spread relative to the mean direction of scatter $\bar{\theta}$. In rural areas, σ is typically 1°, and in indoor environments it may be tens of degrees [62].

Example 3.12: Assume $\sigma_\theta = 20°$ and $P_T = 1$ for an indoor environment. The power angular spectrum as obtained from (3.93) is shown in Fig. 3.22.

An extension to both the PDP and PAP is the power delay-angular profile, which is a three-dimensional plot defined by

$$P(\tau, \theta) = \sum_{n=1}^{N} P_n \delta(\tau - \tau_n) \delta(\theta - \theta_n). \tag{3.94}$$

In a mobile-to-mobile communication channel, the antenna heights of both the transmitter and receiver are below the surrounding objects; it is thus likely that both the transmitter and receiver experience rich-scattering effects in the propagation paths. The double-ring scattering model is suitable for analyzing the mobile-to-mobile channel in Rayleigh fading [52] or in Ricean fading [77].

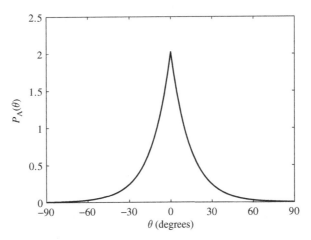

Figure 3.22   Power angular spectrum for indoor scattering.

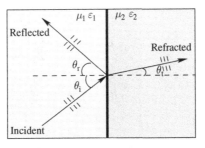

Figure 3.23   Reflection and refraction: Plane wave incident onto a plane boundary. The paper plane is the scattering plane

3.5 Propagation mechanisms

Reflection, refraction, scattering and diffraction are the four important mechanisms of radio propagation. Reflection and refraction occur when a propagating wave impinges on an object that has very large dimensions compared to the wavelength. Scattering occurs when the wave goes through a medium composed of many small objects (in terms of wavelength). Diffraction occurs when the wave path is obstructed by a surface that has sharp edges. All these mechanisms can be analyzed by using Maxwell's equations.

3.5.1 Reflection and refraction

When a plane wave is incident on a plane boundary between two media with different permeabilities μ_1, μ_2 and permittivities ϵ_1, ϵ_2, at an angle θ_i, the reflected and refracted (transmitted) waves, as shown in Fig. 3.23, can be determined by Snell's law of reflection and Snell's law of refraction, respectively

$$\theta_i = \theta_r, \tag{3.95}$$

$$\frac{\sin \theta_i}{\sin \theta_t} = \sqrt{\frac{\epsilon_2 \mu_2}{\epsilon_1 \mu_1}} = \frac{n_2}{n_1}, \tag{3.96}$$

where for media k $(k = 1, 2)$,

$$n_k = \frac{c}{v_k} = \sqrt{\mu_{rk} \epsilon_{rk}} \tag{3.97}$$

is the refractive index, $\mu_{rk} = \frac{\mu_k}{\mu_0}$ being the relative permeability, $\epsilon_{rk} = \frac{\epsilon_k}{\epsilon_0}$ the relative permittivity, v_k the wave speed, μ_0, ϵ_0 the permeability and permittivity in free space, given by $\mu_0 = 4\pi \times 10^{-7}$ henrys/m, $\epsilon_0 = 8.854 \times 10^{-12}$ farads/m.

In addition to the direction changes of the incident wave according to Snell's laws, the amplitudes of the reflected and refracted waves can be determined using electromagnetic analysis, and obtained relative to the incident wave amplitude by the Fresnel reflection and transmission coefficients R and T. The Fresnel coefficients are different, depending on whether the electric field is parallel or normal to the scattering plane. They are given by [62]

$$R_{\parallel} = \frac{E_{r\parallel}}{E_{i\parallel}} = \frac{Z_2 \cos \theta_t - Z_1 \cos \theta_i}{Z_2 \cos \theta_t + Z_1 \cos \theta_i}, \tag{3.98}$$

$$R_{\perp} = \frac{E_{r\perp}}{E_{i\perp}} = \frac{Z_2 \cos \theta_i - Z_1 \cos \theta_t}{Z_2 \cos \theta_i + Z_1 \cos \theta_t}, \tag{3.99}$$

$$T_{\parallel} = \frac{E_{t\parallel}}{E_{i\parallel}} = \frac{2Z_2 \cos \theta_i}{Z_2 \cos \theta_t + Z_1 \cos \theta_i}, \tag{3.100}$$

$$T_{\perp} = \frac{E_{t\perp}}{E_{i\perp}} = \frac{2Z_2 \cos \theta_i}{Z_2 \cos \theta_i + Z_1 \cos \theta_t}, \tag{3.101}$$

where the subscripts $\parallel$ and $\perp$ denote the cases of the electric field being parallel and normal to the scattering plane, respectively, E is the strength of the electric field, the subscripts r and i denote the reflected and incident waves, respectively, and Z_1 and Z_2 are the wave impedances of the two media, and the wave impedance is calculated by

$$Z = \sqrt{\frac{j\omega\mu}{\sigma + j\omega\epsilon}}, \tag{3.102}$$

σ being the conductivity.

The *Brewster angle* is calculated by

$$\theta_B = \tan^{-1} \frac{n_2}{n_1}. \tag{3.103}$$

At $\theta_i = \theta_B$, $R_{\parallel}$ changes its polarization state. When $\theta_i < \theta_B$, $R_{\parallel}$ and $R_{\perp}$ are both negative, and thus reflection changes the polarization of the incident wave by a 180 phase change; when $\theta_i > \theta_B$, there is no phase change for the reflected wave. In both the cases, the axis ratio (AR) changes for circularly or elliptically polarized incident waves. Note that the Brewster angle occurs only for vertical polarization.

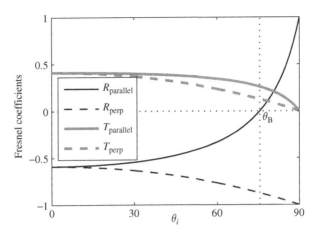

Figure 3.24 Fresnel coefficients for average ground at 1 GHz with $\sigma = 0.005$ S m^{-1} and $\epsilon_r = 15$.

Figure 3.25 Scattering on rough surface.

Example 3.13: Figure 3.24 gives an illustration of the Fresnel coefficients for average ground ($\sigma = 0.005$ S m^{-1} and $\epsilon_r = 15$) at 1 GHz. In case of reflection from the ground, $\parallel$ corresponds to vertical polarization and $\perp$ to horizontal polarization. It is seen that at $\theta_i = \theta_B = 75.5°$, $R_\parallel$ has a 180° phase change.

3.5.2 Scattering

Scattering is an important mechanism of wave propagation, due to rough terrain surface. Scattering theory assumes the roughness of a surface to be random. Different heights of the surface lead to reflection (scattering) in different directions, leading to a reduction in the power of the specularly reflected ray. When the surface turns rougher, the incident wave is reflected from many points on the surface, leading to a broadening of the scattered energy. This is illustrated in Fig. 3.25.

The degree of scattering is dependent on the incident angle as well as the roughness of the surface. The roughness can be characterized by the height of two points at the surface, and the incident waves reflected from these points have a relative phase difference

$$\Delta\phi = 4\pi \frac{\Delta h}{\lambda} \cos\theta_i, \tag{3.104}$$

where Δh is the difference in the heights at the two points, the wavelength λ characterizes the roughness of the surface, and θ_i is the angle of incidence. Thus, as θ_i becomes large, the relative phase difference decreases and the surface becomes relatively flat.

The above criterion is applicable only to a single location. The Rayleigh criterion considers a surface as smooth if this phase shift is less than 90° [62]

$$\Delta h < \frac{\lambda}{8 \cos \theta_i}. \tag{3.105}$$

For a wide area, the Rayleigh criterion gives a statistical characterization of the terrain.

The roughness parameter C is defined by

$$C = \frac{4\pi \sigma_h \theta_i}{\lambda}, \tag{3.106}$$

where σ_h is the standard deviation of the surface irregularity (height). Typically, the terrain is deemed as smooth if $C < 0.1$, irregular if $C > 10$, and quasi-smooth if $0.1 \leq C \leq 10$.

A scattering loss factor can be incorporated into the Fresnel reflection coefficients to account for the energy loss due to scattering [8, 56]

$$\rho_s = e^{-8\left(\frac{\pi}{\lambda}\sigma_h \sin \theta_i\right)^2} I_0 \left[8 \left(\frac{\pi}{\lambda}\sigma_h \sin \theta_i \right)^2 \right], \tag{3.107}$$

where $I_0(\cdot)$ is the Bessel function of the first kind and order zero.

3.5.3 Diffraction

Diffraction is caused by discontinuities in a surface on which an electromagnetic wave impinges. It allows radio waves to propagate around the curved surface of the Earth and to reach behind obstructions. Diffraction can be easily understood by using Huygen's principle. For diffraction analysis, if the wavelength is very small, geometrical optics is exact. For microwave propagation, diffraction analysis can be used to derive a diffraction coefficient that estimates the power that can be received in the shadow region behind an obstacle, such as a mountain or a building or a series of mountains or buildings.

The two canonical models for diffraction analysis of a homogeneous plane wave are diffraction by a knife edge or screen and diffraction by a wedge. In these cases, the geometrical optics solution leads to a completely incorrect field in the shadow region. Geometrical optics has also been extended to include diffraction, yielding the geometrical theory of diffraction.

Knife-edge or half-plane diffraction can be used to model the effect of a surrounding such as a hill. The received field is the sum of the direct path and the diffraction path terms. Knife-edge diffraction is illustrated in Fig. 3.26, where d_1 and d_2 are the distances from the transmitter and the receiver to the top of the edge, respectively, and h_1, h_2, and h are the heights of the transmitter, receiver, and edge, respectively. The clearance between the knife-edge and the direct path is denoted by l. If $l < 0$, the direct path is obstructed, and only the diffraction term contributes to the received field.

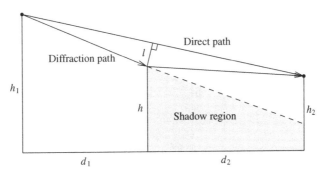

Figure 3.26   **Knife-edge diffraction.**

For single knife-edge diffraction, the propagation loss due to diffraction is given by
[56, 62]

$$L_{ke} = -20 \log_{10} \left(\left| \frac{E_d}{E_i} \right| \right) = -20 \log_{10} |F(v)| \quad \text{(dB)}, \tag{3.108}$$

where E_d is the diffracted field, E_i is the incident field, and

$$F(v) = \frac{1+j}{2} \int_v^\infty e^{-j\pi t^2/2} dt. \tag{3.109}$$

The parameter v, called the *Fresnel-Kirchoff diffraction parameter*, can be approximated by

$$v \approx l \sqrt{\frac{2(d_1 + d_2)}{\lambda d_1 d_2}}. \tag{3.110}$$

Note that $|F(v)|$ can also be represented using the Fresnel cosine and sine integrals $C(v)$ and $S(v)$

$$|F(v)| = \frac{1}{2} \left(\frac{1}{2} + C^2(v) - C(v) + S^2(v) - S(v) \right), \tag{3.111}$$

where

$$C(x) = \int_0^x \cos\left(\frac{\pi}{2} t^2 \right) dt, \quad S(x) = \int_0^x \sin\left(\frac{\pi}{2} t^2 \right) dt. \tag{3.112}$$

Example 3.14: By using the standard MATLAB routines for the Fresnel cosine and sine integrals $C(v)$ and $S(v)$, the knife-edge diffraction attenuation, given by (3.108), is plotted in Fig. 3.27. The illuminated region corresponds to negative values of v, and the shadow region corresponds to positive values of v.

Diffraction by multiple screens are especially important for wireless communications, but no general exact solution is available. The Bullington method uses a single knife-edge to replace a series of screens, but this method provides very optimistic estimates of the

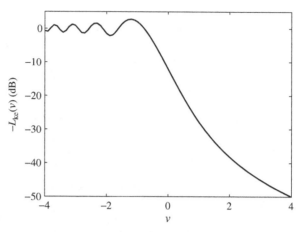

Figure 3.27 Knife-edge diffraction path loss $L_{ke}(v)$ versus Fresnel diffraction parameter v: $v > 0$ corresponds to shadow region and $v < 0$ corresponds to lit region.

received signal strength. A number of other approximate methods, such as the Epstein-Peterson method, Deygout's method, and the ITU empirical model, are available. These models are described in [46, 62].

3.6 Atmospheric effects

The atmosphere of the Earth is divided into various layers. From the ground to a height of about 90 km, it is the troposphere (0 to about 10 km), stratosphere (about 10 to 45 km), and mesosphere (about 45 to 90 km); from about 90 to 600 km, it is the ionosphere, which is within the thermosphere of the atmosphere. From about 600 to 10,000 km, it is exosphere. Above 600 km, it can be treated as free space. This is illustrated in Fig. 3.28.

The refractive index n of the Earth's atmosphere is slightly greater than unity, and at the Earth surface, it is typically 1.0003. The atmospheric refractivity N is defined as

$$N = (n - 1) \times 10^6. \tag{3.113}$$

N varies with temperature, pressure, and water vapor pressure, and these quantities vary with both location and height. According to ITU-R Rec. P.453, N varies approximately exponentially within the first few tens of kilometers of the Earth's atmosphere [33, 62]

$$N = N_s e^{-\frac{h}{H}}, \tag{3.114}$$

where h is the height above the sea level, the surface value $N_s \approx 315$, and $H = 7.35$ km.

Variation of the refractive index with height causes the ray paths to be not straight, but tends to curve slightly towards the ground. This can somewhat extend the range of signal propagation on the earth's surface.

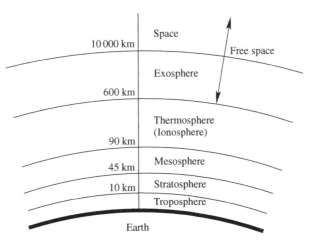

Figure 3.28   The layer model of the Earth's atmosphere.

3.6.1 Tropospheric effects

The troposphere contains particles of a wide range of sizes and characteristics, from atmospheric gases to raindrops. The resulting propagation loss arises from absorption and scattering. In the troposphere, rain is the major factor for propagation loss. The ionosphere is a region of inhomogeneous and anisotropic magnetized plasma. The ionosphere is more intense during the day than during the night. Ionosphere, rain and ice crystals also result in depolarization.

Gaseous absorption

The atmosphere of the Earth introduces losses as a result of energy absorption by the atmospheric gases such as water vapor (H_2O) or molecular oxygen (O_2). The attenuation, $\gamma(f)$, is measured in dB/km. Attenuation due to gaseous absorption can be estimated based on ITU-R Rec. P.676.

Figure 3.29 shows the specific attenuation from 1 to 350 GHz at sea-level for dry air and water vapor with a density of 7.5 g/m^3 at 15°C. This figure is plotted based on the equations given in ITU.R Rec. P.676-6. It is seen that the atmospheric gases introduce substantial absorption at high frequencies. Maximum absorption occurs at frequencies that coincide with one of the molecular resonances of water or oxygen. Water vapor (H_2O) has resonance frequencies at 22.3, 183.3, and 323.8 GHz, while oxygen (O_2) has resonance frequencies at 60.0 and 118.74 GHz. At these frequencies, significant absorption is observed.

In the frequency band of 57–64 GHz, there is a significant attenuation with a peak at 60 GHz due to the presence of oxygen. At 60 GHz, molecular oxygen can lead to an attenuation of 15 dB/km. Thus, this band is actually not suitable for microwave or satellite communications. These atmospheric effects must be considered in the link budget analysis.

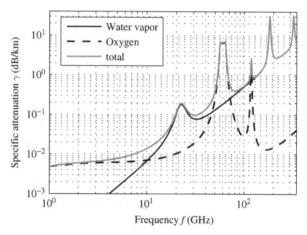

Figure 3.29 Attenuation due to gaseous absorption: 15°C, water vapor density 7.5 g/m³, pressure 1,013 hPa.

Rain fading

Attenuation arising from significant rain intensity, measured in mm/h, becomes considerable for frequencies above 5 GHz. For network deployment, local meteorological records should be inspected for the rain intensity in mm/h to calculate the link budget. Site diversity can be exploited to reduce rain fading, where an additional station at a place not covered by the rain cell provides diversity.

Propagation loss due to rain fading, when the density of the raindrops in a given region is constant, is given by

$$L = 4.343\alpha r = \gamma_R r \quad \text{(dB)}, \tag{3.115}$$

where r is the distance, $\gamma_R = 4.343\alpha$ is the loss for unit length path, and α can be obtained by using theoretical analysis, or more commonly and practically by an empirical model [37]

$$\gamma_R = aR^b \quad \text{(dB/km)}, \tag{3.116}$$

where R is the rainfall rate measured in millimeters per hour, and a and b are determined as functions of frequency (in GHz) in the range from 1 to 1000 GHz. The parameters a and b are given as power-law coefficients derived from scattering calculations in ITU-R Rec. P.838 [37].

Figure 3.30 illustrates the relation between attenuation and frequency for some discrete values of rainfall rate, which is plotted based on the equations given in ITU.R Rec. P.838. The effective rain height is determined by the latitute ϕ of the Earth station, and is given in ITU-R Rec. P.618 [35].

3.6.2 Ionospheric effects

In case of satellite-ground radio communications, the Earth's ionosphere has a significant influence on the propagation of signals. For frequencies of 1 GHz and above, if the total

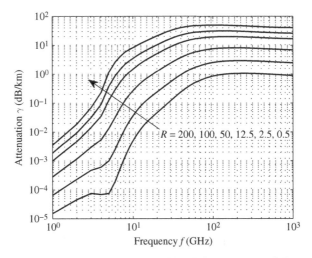

Figure 3.30   Rain attenuation γ as a function of frequency f (GHz) for various rainfall rates R (mm/h).

electron content (TEC) is as high as 10^{18} electrons per meter squared, four ionospheric effects, namely, Faraday rotation, propagation delay, dispersion, and scintillation, need to be considered [17, 62]. The ionosphere effects are strongly related to frequency, and more details are given in ITU-R Rec. P.531 [34].

TEC varies in time and space, depending on the solar disturbances and the solar cycle. When a linearly polarized wave enters an ionized medium that is influenced by an external magnetic field, the wave is split into two oppositely rotating circularly polarized waves, with slight differences in velocities and ray paths. At a receiver outside the ionized medium, the resultant linear polarization obtained from recombining the two component waves is different from the original one in the polarization angle. This phenomenon is known as *Faraday rotation*. This rotation angle is very large for frequencies of 1 GHz and below. This results in serious problems when orthogonal polarizations are used to transmit two separate signals. A simple solution is to use circular polarization, and there is no Faraday rotation in this case. This makes circularly polarized antennas essential.

Faraday rotation is the integral along the path of the product of two quantities: the component of the Earth's magnetic field along the propagation path, and the local ionospheric electron density [17]. A simple approximation for the Faraday rotation, Φ, in radians, is given by [17]

$$\Phi = 1.885 f^{-2} \text{TEC}. \tag{3.117}$$

The ionosphere also reduces the speed of a radio wave. Relative to the free-space propagation delay for the same path, the ionosphere causes an additional delay [17]

$$\Delta t = 40.3 c^{-1} f^{-2} \text{TEC} \quad \text{(s)}, \tag{3.118}$$

where c is the free-space speed. The GPS actually uses this relation for accurate ranging. From (3.118), dispersion, defined as the rate of change of the time delay with frequency, is derived as

$$\frac{dt}{df} = -80.6 f^{-3} \text{TEC} \quad \text{(s/Hz)}. \tag{3.119}$$

This introduces a phase advance relative to free space

$$\Delta\phi = 8.44 \times 10^{-7} f^{-1} \text{TEC} \quad \text{(radians)} \tag{3.120}$$

and the phase dispersion is given by

$$\frac{d\phi}{dt} = -8.44 \times 10^{-7} f^{-2} \text{TEC} \quad \text{(radians/s)}. \tag{3.121}$$

Scintillation

Scintillation is the variation in the signal amplitude, phase, DoA, and Faraday rotations, caused by spatial irregularities in the refractive indices of the troposphere and the ionosphere. The turbulence caused by the wind in the troposphere leads to rapid refractive index variations over short distances and over short time intervals. Wind present in the ionosphere causes rapid variations in the local electron density. At visible optical frequencies, scintillation is observed as the *twinkling of stars*. Rain is also a source of tropospheric scintillation.

Scintillation is not an absorptive effect as the mean power of the signal is unchanged. The intensity of the scintillation is measured by its standard deviation. The scintillation intensity can be predicted using a model described in ITU-R Rec. P.618 [35]. This effect is strong for high frequency signals.

Scintillation is easily noticeable in warm, humid climates, and is greatest in summer [62]. It is particularly severe at the time of sunset, when there is a rapid variation in the local electron density. Scintillation is most severe in the equatorial region, and in north and south high-latitude regions [17]. The influence of scintillation can be reduced by using a wide aperture antenna, because this averages the scintillation across the slightly different paths across the aperture. Spatial diversity can also be exploited to reduce the overall fade depth.

3.7 Channel sounding

There are numerous papers on channel measurements. In [2], analysis of joint statistical properties of azimuth spread, delay spread, and shadowing fading in macrocellular environments at 1.8 GHz is given, based on measurements in typical urban (TU), bad urban (BU), and suburban (SU) areas, which are according to the definitions in COST-207 [14].

The azimuth spread σ_A and the delay spread σ_D of a channel are defined as the square root of the second central moments of the respective variables θ and τ. Both σ_A and σ_D are subject to the log-normal distribution

$$\sigma_A = 10^{\epsilon_A X + \mu_A}, \quad \sigma_D = 10^{\epsilon_D Y + \mu_D}, \tag{3.122}$$

Table 3.3. Measurements of azimuth spread, delay spread, and shadowing fading in different environments ©IEEE [2].

Class	σ_s	$E[\sigma_A]$	μ_A	ϵ_A	$E[\sigma_D]$	μ_D	ϵ_D
TU-32	7.3 dB	8°	0.74	0.47	0.8 μs	−6.20	0.31
TU-21	8.5 dB	8°	0.77	0.37	0.9 μs	−6.13	0.28
TU-20	7.9 dB	13°	0.95	0.44	1.2 μs	−6.08	0.35
BU	10.0 dB	7°	0.54	0.60	1.7 μs	−5.99	0.46
SU	6.1 dB	8°	0.84	0.31	0.5 μs	−6.40	0.22

where X, Y are zero-mean Gaussian distributed random variables with unit variance, $\mu_A = E\left[\log_{10}(\sigma_A)\right]$, $\mu_D = E\left[\log_{10}(\sigma_D)\right]$ are the logarithmic means, and $\epsilon_A = \text{std}\left[\log_{10}(\sigma_A)\right]$, $\epsilon_D = \text{std}\left[\log_{10}(\sigma_D)\right]$ are the logarithmic standard deviations.

The channel gain h can be decomposed as

$$h = h_{\text{loss}}h_s, \tag{3.123}$$

where h_{loss} is the deterministic distance-dependent pass loss, and h_s is the channel's shadowing fading loss, which can again be modeled by a log-normal distributed random variable

$$h_s = 10^{\sigma_s Z/10}, \tag{3.124}$$

where Z is a zero-mean Gaussian random variable with unit variance. Thus,

$$\sigma_s = \text{std}\left[10\log_{10}(h_s)\right]. \tag{3.125}$$

According to the measurements performed in [2], the shadowing fading standard deviation σ_s is in the range of 6–10 dB, with the largest σ_s in the bad urban environment and the smallest in the suburban environment. This is in agreement with many other publications.

A summary from the measurement is given in Table 3.3, wherein TU-32, TU-21, and TU-20 are typical urban environments with BS antenna heights of 31m, 21m, and 20m, respectively.

Channel measurement and modeling campaigns have also been conducted for urban spatial radio channels in macro/microcell at 2.154 GHz for a bandwidth of 100 MHz [72], MIMO channels in microcell and picocell at 1.71/2.05 GHz [41], outdoor mobile channels at 5.3 GHz [83], a spatio-temporal channel in a suburb non-LOS microcell [71], a MIMO wireless LAN environment at 5.2 GHz [45], a MIMO outdoor-to-indoor channel at 5.2 GHz [79], a land mobile satellite channel at Ku-band (10–12 GHz), an airport surface area channel in the 5-GHz band with a 50-MHz bandwidth [44, 65], a HAPS channel in built-up areas at 2.0, 3.5, and 5.5 GHz [30], wideband underground mine tunnel channels at 2.4 GHz and 5.8 GHz [9], a double directional UWB channel in a wooden house [27], a nomadic diversity channel at 1.9 GHz with an 80-MHz bandwidth in typical indoor office and industrial environments [48], time-varying indoor and outdoor wideband 8 × 8 MIMO channels at 2.55 and 5.2 GHz with receiver movement [76], wideband MIMO mobile-to-mobile (M-to-M) channels at 2.435 GHz for vehicular communication along streets and expressways in a metropolitan area [82], indoor wireless LAN channels at 17 GHz [60] and at 60 GHz [24, 80], an UWB channel [11], an indoor wideband (100 MHz) body-to-body MIMO channel at 5.5 GHz [78], and a UWB cooperative BAN channel [12].

Problems

3.1 For what distances is the two-ray plane earth loss model valid in a macrocell ($h_t = 50$ m and $h_r = 2$ m) and a microcell ($h_t = 10$ m and $h_r = 2$ m)? Consider the frequencies 900 MHz and 1800 MHz.

3.2 Determine the path loss by the Hata model at a frequency of 1 GHz, with $h_t = 50$ m, $h_r = 2$ m, and distance $d = 2$ km.

3.3 Refer to [15, 28], and find the expressions for the COST-WI model. Consider the same data as in Problem 3.2, and determine the path loss for an MS on the street by the COST-WI model. Assume 10 buildings between the BS and the MS, a building separation of 40 m, 6 storeys per building, 5 m per storey, a street width of 20 m, and a road orientation of 30° with the direct radio path.

3.4 Determine the maximum Doppler shift for a mobile moving at 50 and 100 km/h at frequencies of 1 GHz and 2 GHz.

3.5 From the pdf of a Rayleigh channel, derive the mean and mean square values of a Rayleigh fading signal.

3.6 Calculate the LCRs and AFDs for Rayleigh fading signal at levels of 5 dB above and below the mean envelope: (a) $f = 1$ GHz, $v = 50$ km/h; (b) $f = 2$ GHz, $v = 50$ km/h; (c) $f = 1$ GHz, $v = 100$ km/h; (a) $f = 2$ GHz, $v = 100$ km/h.

3.7 The Doppler power spectrum for the I- and Q-channels of an indoor channel is typically assumed to be uniformly distributed with a maximum Doppler shift of 10 Hz. Assuming that the PDP to be $P_h(\tau) = \frac{1}{T}\left(1 - \frac{\tau}{T}\right), 0 < \tau < T$,
(a) determine the autocorrelation function and then the coherent time of the channel;
(b) calculate the rms delay spread and the coherent bandwidth of the channel at $T = 2\mu s$.

3.8 Assume that a laptop computer moves at a speed of 20 km/h in an IEEE 802.11g wireless LAN operating at 2.45 GHz band. Determine the maximum Doppler shift and the coherent time.

3.9 Calculate the Brewster angle for a wave impinging on ground of a permittivity of $\epsilon_r = 3.9$.

3.10 If the received power at 1 km is 1 dBm, determine the received power at 2 km, 5 km, and 20 km from the same transmitter for the path loss models: (a) free space; (b) $\gamma = 3$;

(c) $\gamma = 4$; (d) plane earth loss model; and (e) COST-231-Hata. Plot the models over the range of 1 km to 20 km. Assume $f = 2\,\mathrm{GHz}$, $h_t = 50\,\mathrm{m}$, $h_r = 3\,\mathrm{m}$, $G_t = G_r = 1$. Comment on the results.

3.11 The average PDP of a channel is given by

$$P(\tau) = \sum_{n=0}^{2} \frac{10^{-6}}{2n^2 + 1} \delta(\tau - 10^{-6}n).$$

What is the local average power in dBm? What is the rms delay spread of the channel? If 16QAM modulation with a bit rate of 1 Mbits/s is applied, will the signal undergo flat or frequency-selective fading?

3.12 Assume two propagation paths that are identical except that each experiences an independent log-normal shadowing. A selection diversity is used at the receiver. Derive the pdf of the receiver output.

3.13 The pdf of a random variable X is $p(x)$. Let $Y = aX^3 + b$, $a < 0$. Determine and plot the pdf of Y. Assume X is a Gaussian random variable with zero mean and unit variance. [Hint: $p_Y = \frac{\partial \Pr(Y \le y)}{\partial y}$.]

3.14 Determine the autocorrelation of the stochastic process $x(t) = A \sin(2\pi f_c t + \theta)$, where f_c is a constant and θ is the uniformly distributed phase.

3.15 In wireless communication systems, the carrier frequency in the reverse link is usually smaller than the forward link. Explain the reason.

3.16 The scattering function $P_s(\tau, \lambda)$ is nonzero and uniform for the region $0 \le \tau \le 2$ ms and $-0.2\,\mathrm{Hz} \le \lambda \le 0.2\,\mathrm{Hz}$, and is zero otherwise.
(a) Derive the PDP of the channel and the Doppler power spectrum.
(b) Calculate the multipath spread, Doppler spread, coherent time, and coherent bandwidth of the channel.
(c) For a bandwidth of 10 kHz and data transmission at 200 bits/s, design a binary communication system with frequency diversity: the modulation type, the number of subchannels and frequency separation, and the signaling interval.

3.17 Refer to [22], and implement a simulation of the extended Clarke's model by using data generated from the fundamental fading model with component phase fluctuations. Assume 200 random paths, $B = 100$ Hz, $v_{max} = 100$ Hz, and a sampling period of 1 ms.
(a) Generate the normalized ACF and compare it with the result given by (3.57).
(b) Generate the power spectrum and compare it with the result given by (3.58).
(c) Plot the power spectrum according to (3.58) for $B/v_{max} = 0, 0.2, 0.5$, and 1.

3.18 Assuming a mobile speed of 60 km/h, a carrier frequency of 920 MHz and rms delay spread of 3 μs, what are the coherent time and coherent bandwidth. For IS-95, the coded symbol rate is 19.2 kbits/s and the bandwidth is 1.2288 MHz; what type of fading is experienced by the IS-95 channel?

3.19 The channel PDP is given as $0\,\text{dB}$ at $\tau = 0$, $-6\,\text{dB}$ at $\tau = 2\,\mu s$, $-12\,\text{dB}$ at $\tau = 4\,\mu s$, and $-16\,\text{dB}$ at $\tau = 7\,\mu s$. Draw the channel PDP. Determine the mean excess delay, rms delay spread, and maximum excess delay of the channel.

3.20 Derive the Brewster angle given by (3.103).

3.21 A right-hand circularly polarized plane wave is incident on the boundary between dry air and dry earth. Describe the polarization of the reflected wave when angle of incidence θ_i and the Brewster angle θ_B are related by: (a) $\theta_i < \theta_B$; (b) $\theta_i = \theta_B$; (c) $\theta_i > \theta_B$. For dry air, $\epsilon_1 = \epsilon_0$, $\mu_1 = \mu_0$, $\sigma_1 = 0$; for dry earth, $\epsilon_2 = 2.53\epsilon_0$, $\mu_2 = \mu_0$, $\sigma_1 = 0$. Plot the Fresnel coefficients.

3.22 A microwave link operating at $5\,\text{GHz}$ with a path length of $30\,\text{km}$ has a maximum acceptable path loss of $160\,\text{dB}$. The transmit antenna is $25\,\text{m}$ above the ground level, and the receive antenna is $15\,\text{m}$ above the ground level. A hill $80\,\text{m}$ high, located $10\,\text{km}$ away from the transmitter, blocks the transmission between the transmitter and the receiver. Determine the total path loss including the free space loss and the knife-edge attenuation.

3.23 A receiver can produce acceptable BERs when the instantaneous SNR is at or above $10\,\text{dB}$. What mean SNR is required in a Rayleigh channel for acceptable BERs for 99% of the time?

3.24 Determine the minimum symbol rate to avoid the effects of Doppler spread in a mobile system operating at $900\,\text{MHz}$ with a maximum speed of $120\,\text{km/h}$.

References

[1] 3GPP, Technical Specification Group Radio Access Network. *Spatial Channel Model for Multiple Input Multiple Output (MIMO) Simulations*, 3GPP TR 25.996 V6.1.0, technical report, Sep 2003.

[2] A. Algans, K. I. Pedersen, & P. E. Mogensen, Experimental analysis of the joint statistical properties of azimuth spread, delay spread, and shadow fading. *IEEE J. Sel. Areas Commun.*, **20**:3 (2002), 523–531.

[3] A. Alomainy, Y. Hao, C. G. Parini, and P. S. Hall, Comparison between two different antennas for UWB on-body propagation measurements, *IEEE Anten. Wireless Propag. Lett.*, **4**:1 (2005), 31–34.

[4] J. G. Andrews, A. Ghosh, & R. Muhamed, *Fundamentals of WiMAX: Understanding Broadband Wireless Networking* (Upper Saddle River, NJ: Prentice Hall, 2007).

[5] M. D. Austin & G. L. Stuber, Velocity adaptive handoff algorithms for microcellular systems. *IEEE Trans. Veh. Tech.*, **43**:3 (1994), 549–561.

[6] S. Barbarossa & A. Scaglione, Time-varying fading channels. In G. B. Giannakis, Y. Hua, P. Stoica & L. Tong, eds., *Signal Processing Advances in Wireless & Mobile Communications*: **2**, ch.1 (Upper Saddle River, NJ: Prentice Hall, 2001).

[7] P. A. Bello, Characterization of randomly time-variant linear channels. *IEEE Trans. Circ. Syst.*, **11**:4 (1963), 360–393.

[8] L. Boithias, *Radio Wave Propagation* (New York: McGraw-Hill, 1987).

[9] M. Boutin, A. Benzakour, C. L. Despins & S. Affes, Radio wave characterization and modeling in underground mine tunnels. *IEEE Trans. Anten. Propagat.*, **56**:2 (2008), 540–549.

[10] P. Burns, *Software Defined Radio for 3G* (Boston: Artech House, 2003).

[11] D. Cassioli, M. Z. Win & A. F. Molisch, The ultra-wide bandwith indoor channel: from statistical model to simulations. *IEEE J. Sel. Areas Commun.*, **20**:6 (2002), 1247–1257.

[12] Y. Chen, J. Teo, J. C. Y. Lai, E. Gunawan, K. S. Low, C. B. Soh & P. B. Rapajic, Cooperative communications in ultra-wideband wireless body area networks: channel modeling and system diversity analysis. *IEEE J. Sel. Areas Commun.*, **27**:1 (2009), 5–16.

[13] J. Chuang, The effects of time delay spread on portable radio communications channels with digital modulation. *IEEE J. Sel. Areas Commun.*, **5**:5 (1987), 879–889.

[14] COST 207, *Proposal on Channel Transfer Functions to be Used in GSM Tests Late 1986*, TD(86)51-REV 3 (WG1), European Commission, Brussels, Sep 1986.

[15] COST 231, *Digital Mobile Radio Toward Future Generation Systems*, Final Report, European Commission, Brussels, 1999.

[16] S. L. Cotton & S. G. Scanlon, Characterization and modeling of the indoor radio channel at 868 MHz for a mobile bodyworn wireless personal area network, *IEEE Anten. Wireless Propagat. Lett.*, **6**:1 (2007), 51–55.

[17] K. Davies & E. K. Smith, Ionospheric effects on satellite land mobile systems. *IEEE Anten. Propagat. Mag.*, **44**:6 (2002), 24–31.

[18] Q. V. Davis & D. J. R. Martin, and R. W. Haining, Microwave radio in mines and tunnels. In *Proc. IEEE VTC*, Pittsburgh, PA, May 1984, 31–36.

[19] V. Erceg, L. J. Greenstein, S. Y. Tjandra, S. R. Parkoff, A. Gupta, B. Kulic, A. A. Julius & R. Bianchi, An empirically based pathloss model for wireless channels in suburban environments. *IEEE J. Sel. Areas Commun.*, **17**:7 (1999), 1205–1211.

[20] V. Erceg et al, *Channel Models for Fixed Wireless Applications*, Rev. 4.0, IEEE802.16.3c-01/29r4, IEEE, Jul 2001.

[21] V. Erceg et al, *IEEE P802.11 Wireless LANs: TGn Channel Models*, IEEE 802.11-03/940r4, IEEE, May 2004.

[22] S. T. Feng & T. R. Field, Statistical analysis of mobile radio reception: an extension of Clarke's model. *IEEE Trans. Commun.*, **56**:12 (2008), 2007–2012

[23] B. H. Fleury, First- and second-order characterization of direction dispersion and space selectivity in the radio channel. *IEEE Trans. Inf. Theory*, **46**:6 (2000), 2027–2044.

[24] S. Geng, J. Kivinen, X. Zhao & P. Vainikainen, Millimeter-wave propagation channel characterization for short-range wireless communications. *IEEE Trans. Veh. Tech.*, **58**:1 (2009), 3–13.

[25] A. Goldsmith, *Wireless Communications* (Cambridge, UK: Cambridge University Press, 2005).

[26] P. S. Hall et al, Antennas and propagation for on-body communication systems. *IEEE Anten. Propagat. Mag.*, **49**:3 (2007), 41–58.

[27] K. Haneda, J. Takada & T. Kobayashi, Cluster properties investigated from a series of ultrawideband double directional propagation measurements in home environments. *IEEE Trans. Anten. Propagat.*, **54**:12 (2006), 3778–3788.

[28] D. Har, A. W. Watson & A. G. Chadney, Comments on diffraction loss of rooftop-to-street in COST 231 Walfish–Ikegami model. *IEEE Trans. Veh. Tech.*, **48**:5 (1999), 1451–1452.

[29] M. Hata, Empirical formula for propagation loss in land mobile radio services. *IEEE Trans. Veh. Tech.*, **29**:3 (1980), 317–325.

[30] J. Holis & P. Pechac, Elevation dependent shadowing model for mobile communications via high altitude platforms in built-up areas. *IEEE Trans. Anten. Propagat.*, **56**:4 (2008), 1078–1084

[31] F. Ikegami, T. Takeuchi & S. Yoshida, Theoretical prediction of mean field strength for urban mobile radio. *IEEE Trans. Anten. Propagat.*, **39**:3 (1991), 299–302.

[32] ITU-R, *Propagation Data and Prediction Methods for the Planning of Short-Range Outdoor Radiocommunication Systems and Radio Local Area Networks in the Frequency Range 300 MHz to 100 GHz*, ITU-R Rec. P.1411-4, Geneva, 2007

[33] ITU-R, *The Radio Refractive Index: Its Formula and Refractivity Data*, ITU-R Rec. P.453-6, Geneva, 1997.

[34] ITU-R, *Ionospheric Propagation Data and Prediction Methods Required for the Design of Satellite Services and Systems*, ITU-R Rec. P.531-8, Geneva, 2005.

[35] ITU-R, *Propagation Data and Prediction Methods Required for the Design of Earth-Space Telecommunication Systems*, ITU-R Rec. P.618-9, Geneva, 2007.

[36] ITU-R, *Attenuation by Atmospheric Gases*, ITU-R Rec. P.676-6, Geneva, 2005.

[37] ITU-R, *Specific Attenuation Model for Rain for Use in Prediction Methods*, ITU-R Rec. P.838-3, Geneva, 2005.

[38] ITU-R, *Propagation Data and Prediction Models for the Planning of Indoor Radiocommunication Systems and Radio Local Area Networks in the Frequency Range 900 MHz to 100 GHz*, ITU-R Rec. P.1238, Geneva, 1997.

[39] W. C. Jakes, Jr. (ed.), *Microwave Mobile Communications* (New York: Wiley, 1974).

[40] R. Kattenbach, *Characterization of Time-variant Indoor Radio Channels by Means of their System and Correlation Functions*, Doctoral dissertation (in German), University of Kassel, Shaker Verlag, Aachen, 1997, ISBN 3-8265-2872-7.

[41] J. P. Kermoal, L. Schumacher, K. I. Pedersen, P. E. Mogensen & F. Frederiksen, A stochastic MIMO radio channel with experimental validation. *IEEE J. Sel. Areas Commun.*, **20**:6 (2002), 1211–1226.

[42] W. C. Y. Lee, *Mobile Cellular Telecommunications: Analog and Digital Systems*, 2nd edn (New York: McGraw-Hill, 1995).

[43] M. Marcus & B. Pattan, Millimeter wave propagation: spectrum management implications. *IEEE Microwave Mag.*, **6**:3 (2005), 54–62.

[44] D. W. Matolak, I. Sen & W. Xiong, The 5-GHz airport surface area channel – part I: measurement and modeling results for large airports. *IEEE Trans. Veh. Tech.*, **57**:4 (2008), 2014–2026.

[45] A. F. Molisch, M. Steinbauer, M. Toeltsch, E. Bonek & R. S. Thoma, Capacity of MIMO systems based on measured wireless channels. *IEEE J. Sel. Areas Commun.*, **20**:3 (2002), 561–569.

[46] A. F. Molisch, *Wireless Communications* (Chichester: Wiley-IEEE, 2005).

[47] M. Nakagami, The *m*-distribution: a general formula of intensity distribution of rapid fading. In W. C. Hoffman, ed., *Statistical Methods in Radio Wave Propagation* (Oxford: Pergamon Press, 1960), pp. 3–36.

[48] C. Oestges, D. Vanhoenacker-Janvier & B. Clerckx, Channel characterization of indoor wireless personal area networks. *IEEE Trans. Anten. Propagat.*, **54**:11 (2006), 3143–3150.

[49] Y. Okumura, E. Ohmori, T. Kawano & K. Fukuda, Field strength and its variability in VHF and UHF land mobile radio service. *Rev. Electr. Commun. Lab*, **16** (1968), 825–873.

[50] R. L. Olsen & B. Segal, New techniques for predicting the multipath fading distribution on VHF/UHF/SHF terrestrial line-of-sight links in Canada. *Canadian J. Electr. Comput. Eng.*, **17**:1 (1992), 11–23.

[51] J. F. Ossana, A model for mobile radio fading due to building reflections: Theoretical and experimental fading waveform power spectra. *Bell Syst. Tech. J.*, **43**:6 (1964), 2935–2971.

[52] C. S. Patel, G. L. Stuber & T. G. Pratt, Simulation of Rayleigh faded mobile-to-mobile communication channels. In *Proc. IEEE VTC*, Orlando, FL, Oct 2003, **1**: 163–167.

[53] A. Paulraj, R. Nabar & D. Gore, *Introduction to Space-Time Wireless Communications* (Cambridge, UK: Cambridge University Press, 2003).

[54] K. I. Pedersen, P. E. Mogensen & B. H. Fleury, A stochastic model of the temporal and azimuthal dispersion seen at the base station in outdoor propagation environments. *IEEE Trans. Veh. Tech.*, **49**:2 (2000), 437–447.

[55] D. M. Pozar, *Microwave Engineering*, 3nd edn (New York: Wiley, 2005).

[56] T. S. Rappaport, *Wireless Communications: Principles & Practice*, 2nd edn (Upper Saddle River, NJ: Prentice Hall PTR, 2002).

[57] D. O. Reudink and M. F. Wazowicz, Some propagation experiments relating foliage loss and diffraction loss at X-band and UHF frequencies. *IEEE Trans. Veh. Tech.*, **22**:4 (1973), 1198–1206.

[58] S. O. Rice, Mathematical analysis of random noise. *Bell Syst. Tech. J*, **23**:3 (1944), 282–332.

[59] S. O. Rice, Statistical properties of a sine wave plus random noise. *Bell Syst. Tech. J.*, **27**:1 (1948), 109–157.

[60] M. L. Rubio, A. Garcia-Armada, R. P. Torres & J. L. Garcia, Channel modelling and characteristic at 17 GHz for indoor broadband WLAN. *IEEE J. Sel. Areas Commun.*, **20**:3 (2002), 593–601.

[61] A. A. M. Saleh & R. A. Valenzuela, A statistical model for indoor multipath propagation. *IEEE J. Sel. Areas Commun.*, **5**:2 (1987), 128ÍC-137.

[62] S. R. Saunders & A. Aragon-Zavala, *Antennas and Propagation for Wireless Communication Systems*, 2nd edn (Chichester, England: Wiley, 2007).

[63] S. Scalise, H. Ernst & G. Harles, Measurement and modeling of the land mobile satellite channel at Ku-band. *IEEE Trans. Veh. Tech.*, **57**:2 (2008), 693–703

[64] S. Y. Seidel, T. Rapport, S. Jain, M. Lord, & R. Singh, Path loss, scattering and multipath delay statistics in four European cities for digital cellular and microcellular radiotelephone. *IEEE Trans. Veh. Tech.*, **40**:4 (1990), 721–730.

[65] I. Sen & D. W. Matolak, The 5-GHz airport surface area channel – part II: measurement and modeling results for small airports. *IEEE Trans. Veh. Tech.*, **57**:4 (2008), 2027–2035.

[66] A. U. H. Sheikh, *Wireless Communications: Theory and Techniques* (Boston, MA: Kluwer, 2004).

[67] M. K. Simon & M.-S. Alouini, *Digital Communications over Fading Channels*, 2nd edn (New York: Wiley, 2005).

[68] Q. H. Spencer, B. D. Jeffs, M. A. Jensen & L. Swindlehurst, Modeling the statistical time and angle of arrival characteristics of an indoor multipath channel. *IEEE J. Sel. Areas Commun.*, **18**:3 (2000), 347–360.

[69] G. L. Stuber, *Principles of Mobile Communication*, 2nd edn (Boston, MA: Kluwer, 2001).

[70] H. Suzuki, A statistical model for urban radio propagation. *IEEE Trans. Commun.*, **25**:7 (1977), 673–680.

[71] J. Takada, J. Fu, H. Zhu & T. Kabayashi, Spatio-temporal channel characterization in a suburban non line-of-sight microcellular environment. *IEEE J. Sel. Areas Commun.*, **20**:3 (2002), 532–538.

[72] M. Toeltsch, J. Laurila, K. Kalliola, A. F. Molische, P. Vainikainen & E. Bonek, Statistical characterization of urban spatial radio channels. *IEEE J. Sel. Areas Commun.*, **20**:3 (2002), 539–549.

[73] C. A. Tunc, A. Altintas & V. B. Erturk, Examination of existent propagation models over large inhomogeneous terrain profiles using fast integral equation solution. *IEEE Trans. Anten. Propagat.*, **53**:9 (2005), 3080–3083.

[74] J. Walfisch & H. L. Bertoni, A theoretical model of UHF propagation in urban environments. *IEEE Trans. Anten. Propagat.*, **36**:12 (1988), 1788–1796.

[75] E. H. Walker, Penetration of radio signals into buildings in the cellular radio environment. *Bell Syst. Tech. J.*, **26**:9 (1983), 2719–2734.

[76] J. W. Wallace & M. A. Jensen, Time-varying MIMO channels: measurement, analysis, and modeling. *IEEE Trans. Anten. Propagat.*, **54**:11 (2006), 3265–3273.

[77] L.-C. Wang, W.-C. Liu & Y.-H. Cheng, Statistical analysis of a mobile-to-mobile Rician fading channel model. *IEEE Trans. Veh. Tech.*, **58**:1 (2009), 32–38.

[78] Y. Wang, I. B. Bonev, J. O. Nielsen, I. Z. Kovacs & G. F. Pedersen, Characterization of the indoor multiantenna body-to-body radio channel. *IEEE Trans. Anten. Propagat.*, **57**:4 (2009), 972–979.

[79] S. Wyne, A. F. Molisch, P. Almers, G. Eriksson, J. Karedal & Fredrik Tufvesson, Outdoor-to-indoor office MIMO measurements and analysis at 5.2 GHz. *IEEE Trans. Veh. Tech.*, **57**:3 (2008), 1374–1386.

[80] H. Xu, V. Kukshya & T. S. Rappaport, Spatial and temporal characteristics of 60 GHz indoor channels. *IEEE J. Sel. Areas Commun.*, **20**:3 (2002), 620–630.

[81] A. Yagbasan, C. A. Tunc, V. B. Erturk, A. Altintas & R. Mittra, Use of characteristic basis function method for scattering from terrain profiles. *Turk. J. Electr. Eng.*, **16**:1 (2008), 33–39.

[82] A. G. Zajic, G. L. Stuber, T. G. Pratt & S. T. Nguyen, Wideband MIMO mobile-to-mobile channels: geometry-based statistical modeling with experimental verification. *IEEE Trans. Veh. Tech.*, **58**:2 (2009), 517–534.

[83] X. Zhao, J. Kivinen, P. Vainikainen & K. Skog, Propagation characteristics for wideband outdoor mobile communications at 5.3 GHz. *IEEE J. Sel. Areas Commun.*, **20**:3 (2002), 507–514.

4 Cellular and multiple-user systems

4.1 The cellular concept

The cellular concept was a major breakthrough in mobile communications, and it initiated the era of modern wireless communications. It helped to solve the problem of spectral congestion and user capacity. For wireless communications, the antennas are typically required to be omnidirectional. The cellular structure divides the geographical area into many cells. A BS equipped with an omnidirectional antenna is installed at the center of each cell. Neighboring cells use different frequency bands to avoid co-channel interference (CCI).

The same frequency bands can be used by different cells that are sufficiently far away. This leads to frequency reuse. For a distance D between two cells that use the same frequency bands and a radius R of the cell, the relative distance D/R between the two cells is the *reuse distance*. There is no CCI within a cluster of cells, where each cell uses a different frequency spectrum. The number of cells in a cluster is called the *cluster size*. The cluster size determines the capacity of the cellular system: a smaller cluster size leads to a large capacity.

The cell shape usually takes the form of hexagon, and the overall division of space is like a beehive pattern [16]. This is illustrated in Fig. 4.1 for cluster size 4 and reuse distance $D/R \approx 4$. This hexagonal cell shape is suitable when the antennas of the BSs are placed on top of buildings with a coverage radius of a few miles, such as in the 1G mobile systems. For small cells with the BS antenna placed closer to the ground, especially in typically urban street grids, diamond cells can more accurately model the contours of constant power [8]. For a given distance between reuse cells, the hexagonal geometry requires the least number of cells.

For any receiver in the cellular system, there is interference from inside as well as from outside its cell. The intracell interference is due to the nonorthogonal channelization between users, while the intercell interference is known as *CCI*. In orthogonal multiple-access techniques such as TDMA or FDMA, there is no intracell interference in ideal operating conditions. In CDMA with nonorthogonal codes, there is intracell as well as intercell interference, but all the interference is suppressed by a factor equal to the processing gain due to the code cross-correlation.

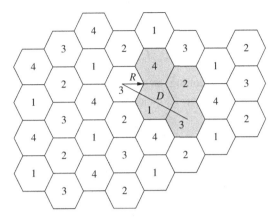

Figure 4.1   The cell structure for cluster size 4 and reuse distance $D/R \approx 4$.

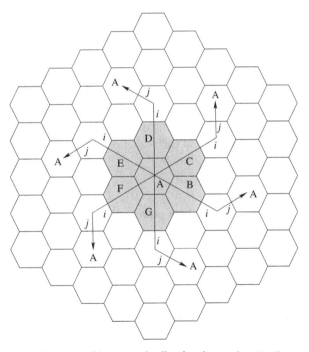

Figure 4.2   Assignment of hexagonal cells. The cluster size $N = 7$.

4.1.1 Cell planning

Cell planning in the hexagonal cellular structure is based on the i-j coordinate system, as shown in Fig. 4.2. From the current cell A, move i cells along i axis, and then further j cells along j axis, $i, j = 0, 1, 2, \ldots$

From the Pythagorian theorem, we have

$$D^2 = \left(j\sqrt{3}R + i\sqrt{3}R\sin 30°\right)^2 + \left(j\sqrt{3}R\cos 30°\right)^2 = 3R^2\left(i^2 + j^2 + ij\right). \tag{4.1}$$

The reuse distance D/R and the cluster size N have such a relation [16]

$$D/R = \sqrt{3\left(i^2 + ij + j^2\right)} = \sqrt{3N}. \tag{4.2}$$

For frequency planning, in order to maximize spectral efficiency while satisfying the minimum reuse distance, the cluster size N must satisfy such a relation [18, 19]

$$N = \frac{A_{\text{cluster}}}{A_{\text{cell}}} = i^2 + ij + j^2, \qquad i, j = 0, 1, 2, \cdots, \tag{4.3}$$

where A_{cluster} and A_{cell} are, respectively, the area of a cluster of hexagonal cells and that of one cell. Thus the cluster size can only be $N = 1, 3, 4, 7, 9, 12, 13, 16, 19, 21, 25, 27, 28, 31, 36\ldots$

Cell planning is based on (4.2), and the smallest integer out of the set for N. The distance between the desired BS and interferer, D, is calculated by link budget analysis. Typically, $N = 1$ for CDMA-based 2G and 3G standards, $N = 7$ for GSM, $N = 4$ for PDC, and $N = 21$ for analog systems such as AMPS and NMT.

Deployment of cell size

Deployment of cell size is dependent on the traffic intensity in erlangs, which can be calculated from expected total number of subscribers within an area, their mean access times and the mean duration of each call. For orthogonal systems, such as TDMA or FDMA, the received power P_r is given by

$$P_r = P_t d^{-\gamma}, \tag{4.4}$$

where P_t is the transmitted power, d is the distance between the BS to an MS, and γ is the path-loss exponent. Within the cell, $\gamma = 2$, and outside the cell, γ is selected between 2 to 4.

For a hexagonal cellular structure, the SIR received within a cell is mainly decided by the six first-tier BSs of the same frequency. For example, in Fig. 4.1, for cell 2 in the cluster, there are altogether 6 first-tier cells using the same frequency. For the worst-case carrier-to-interference power ratio (CIR) calculation, the MS is placed at the corner of a cell. Assuming that all the BSs emit the same power, from geometrical analysis, the SIR received at the MS can be approximated by [19]

$$\text{CIR} \approx \frac{1}{2\left(\frac{D}{R} - 1\right)^{-\gamma} + \left(\frac{D}{R} - \frac{1}{2}\right)^{-\gamma} + \left(\frac{D}{R}\right)^{-\gamma} + \left(\frac{D}{R} + \frac{1}{2}\right)^{-\gamma} + \left(\frac{D}{R} + 1\right)^{-\gamma}}. \tag{4.5}$$

For hexagonal cells, CIR can also be approximated by [8]

$$\text{CIR} = a_1 \left(a_2 N\right)^{\gamma/2}, \tag{4.6}$$

where $a_1 = 0.167$ and $a_2 = 3$ for hexagonal cells, and $a_1 = 0.125$ and $a_2 = 4$ for diamond cells. By specifying a target CIR value CIR_0, the minimum N can be calculated from (4.6). For example, GSM requires $\text{SIR}_0 = 7$ dB.

The user capacity in each cell C_u is given by

$$C_u = N_c = \frac{B}{NB_s},\qquad(4.7)$$

where N_c is the number of channels assigned to a cell, B is the total system bandwidth, and B_s is the bandwidth of each channel.

Similar analysis is made for nonorthogonal systems such as CDMA. For CDMA systems, the SIR that is commonly used for uplink with power control is given by [7, 8]

$$\text{SIR} = \frac{1}{\frac{\xi}{3G}(N_c - 1)(1 + \lambda)},\qquad(4.8)$$

where ξ is a constant characterizing the code cross-correlation that depends on the code properties and other system assumptions and is between 1 and 3, G is the processing gain, and λ is the ratio of the average received power from all intercell interference to that of all intracell interference under perfect intracell power control. Accordingly, by setting $C_u = N_c$ and a target SIR value in (4.8), the user capacity is derived as

$$C_u = N_c = 1 + \frac{1}{\frac{\xi}{3G}(1 + \lambda)\text{SIR}_0}.\qquad(4.9)$$

4.1.2 Increasing capacity of cellular networks

In a cellular system, CCI is a major source of interference. In populous regions, a cell may have many users, and the interferences from these omnidirectional antennas are continually increasing. The cell capacity may not support all these simultaneous users. Cell splitting and sectoring are common techniques employed for increasing the capacity of the cellular system.

The cell-splitting technique subdivides a congested cell into smaller cells, called *microcells*. Each microcell is equipped with a BS that is reassigned a frequency band. Due to the smaller size of the microcell, the antenna has a reduced height and a reduced transmission power so that the receiving power at the new and old cell boundaries are equal to each other. Cell splitting can increase the network capacity by installing more BSs. After a cell is split into microcells, changes in the frequency plan are required to retain the frequency reuse constraint. On the other hand, this also increases the number of handoffs between cells for moving users. The large amount of signaling for handoffs reduces the spectral efficiency. To reduce the interference to other cells or limit the radio coverage of the newly formed microcells, antennas are usually deliberately downtilted so that the main radiation beam is towards the ground.

Picocells and femtocells further reduce the size of a microcell. They provide better coverage in some locations and enable higher data rates at a significantly lower investment than a macrocell/microcell BS. The indoor picocell BS is a small device that can be mounted on a wall or ceiling to support users in airports, shopping malls, office buildings and other large enterprises. The femtocell BS, also called a *home BS*, provides coverage within an individual home to support several users. The femtocell BS can be deployed as a 3G access point, since 3G cellular technology suffers from inadequate indoor signal penetration. The cellular BSs are interconnected via an existing access network.

The cell-sectoring technique replaces the single omnidirectional antenna at the BS by several directional antennas, typically three sectors of 120° each. These directional antennas can be made of patched antenna arrays. Unlike omnidirectional antennas, the directional antennas interfere only with the neighboring cells in their directions. When a cell is sectorized into N sectors, interference power is reduced by roughly a factor of N under heavy loading. Handoff is necessary as MSs move between sectors, and this can be performed in the BS without the intervention of the MSC. However, cell sectoring suffers from a considerable loss in trunking inefficiency, which becomes a major concern for operation.

A more complex method to increase the system capacity is to use the smart antenna technique. Antenna arrays are used to generate beam patterns towards desired users while rejecting all interfering signals. Due to the excellent interference rejection capability, each BS can cover a wider area. The smart antenna technique achieves a significant improvement in system capacity. Smart antennas can be either a switched-beam or an adaptive array system. The switched-beam system continually chooses one pattern from many predefined patterns to steer toward the desired user. This method is incapable of enhancing the desired signal when an interfering signal is close to it, since the beam patterns are predefined. This problem can be solved by using an adaptive array, which creates a beam pattern for each user adaptively by enhancing the desired signal and suppressing all interfering signals. Adaptive antenna array techniques are much more complex for implementation, and will be treated in Chapter 18.

4.1.3 Interference in multiuser systems

Interference is the major limiting factor for the performance of cellular systems. CCI and adjacent channel interference (ACI) are the two major types of interference in cellular systems. In a multiuser system such as a CDMA system, the noise can be neglected when compared with interference from other users or sources. Such a system is said to be *interference limited*. Other interference such as intermodulation interference, intercarrier interference (ICI), and ISI may also arise.

Co-channel interference

In a multi-cell environment with frequency reuse, CCI from different cells always exists. To reduce the CCI, co-channel cells are physically separated by a minimum distance of reuse. Since CCI is from other cells, it usually has a longer delay, a larger angle spread, smaller K_r-factor, and a smaller cross-polarization discrimination. In general, the channel for CCI is close to an i.i.d. channel. In order to mitigate the CCI, multiple antennas can be used at the receiver for CCI cancellation and at the transmitter for CCI avoidance. In the IS-95 CDMA system, the frequency reuse rate is 1, and approximately 40% of the total interference is from the surrounding cells [27]. CCI is also known as *multiple-access* interference (MAI) or *multiple-user interference (MUI)* when it is from users within the same cell, since it arises from the nonorthogonality between multiple accesses, such as in the case of CDMA when nonorthogonal codes are used. In CDMA systems, CCI is typically treated as white noise. The pdf of the total CCI power as well as the signal outage probability is modeled in [23].

Adjacent channel interference

ACI arises from signals which are adjacent in frequency. Due to the imperfect receiver filters, the power in the nearby frequency channel may leak into the passband of the adjacent channel. This problem is especially severe when the receiver is far from the BS but very close to an adjacent channel user. This is referred to as the *near–far effect*. This also occurs for a BS, when an interfering MS close to the BS transmits on a channel that is close to that being used by a desired MS far from the BS. ACI can be minimized through filtering and optimization of channel assignments. Since each cell has only a fraction of all the channels, the channels within each cell can be separated in frequency.

Intermodulation interference

Intermodulation interference may arise either at the receiver or at the transmitter. Unlike ACI, intermodulation products cannot be removed by simple filtering. Intermodulation products are produced when the desired and interfering signals enter a nonlinear device, such as a mixer or a power amplifier. Intermodulation interference can be reduced by increasing the linearity of the devices and proper frequency allocation.

Intercarrier interference

Doppler spread is caused by different Doppler shifts from different incoming waves at the receiving signal, when the transmitter and/or receiver is moving. The Doppler spead f_D is given as $f_D = \frac{v}{\lambda}$, where v is the speed of the MS and λ is the wavelength of the carrier frequency f_c. f_D is typically very small compared to f_c. The received carrier frequency is $f_c + f_D$ if the mobile is moving toward the BS, or $f_c - f_D$ if moving away from the BS. In narrowband communications, the Doppler spread can account for up to a few percent of the subcarrier space. This may damage the orthogonality among the subcarriers and produce crosstalk, and thus ICI arises.

Intersymbol interference

ISI is due to multipath components of signals. When the time delay of a multipath component is significant compared to a symbol period, ISI occurs. ISI is a common form of interference in wireless communications. It can typically be mitigated by equalization or using OFDM technology.

Environmental noise

Interference may also arise due to the radiated power from different neighboring sources such as electric motors and electrical switches. This is generally known as electromagnetic pollution. The mobile radio systems have no control over such emissions. There are some noise models in the literature such as those given in [17].

One public concern about radiated electromagnetic fields is the possible hazards of radiated power to human safety. Currently, it is commonly believed that the damage due to electromagnetic power to humans is the heat effect, as in the case of a microwave oven. The heat is generated from within the body, and may cause damage to the human brain, eyes, and internal organs. For this reason, some safety radiation standards have been defined. The ANSI/IEEE standard C95.1-1992 sets the power density limits for the bands from 100 MHz to 300 MHz to $0.2 \, \text{mW/cm}^2$, for frequencies above 15 GHz to $10 \, \text{mW/cm}^2$, and in between these two bands the power density limit increases proportionally in a logarithmic fashion. The difference arises from the fact that low frequencies have strong penetration into the human body, while high frequencies cannot penetrate through the human skin.

In addition to the standards for human safety, FCC has also defined the radiation limits for individual wireless standards to minimize their interference with other wireless equipments.

4.1.4 Power control

Power control is a critical aspect for achieving the high capacity of a wireless network. Power control on the downlink leads to less intercell and intracell interference than on the uplink. This is due to the fact that on the downlink, the BS is situated at the center of the cell, and is relatively far from the other cells. In contrast, on the uplink, an MS from the cell boundary has to increase its power to ensure a sufficient received power at the BS, which is at the same level as the received power of those MSs that are close to the BS; this leads to significant interference with neighboring cells. Power control helps prolong battery life for MSs. Power control is especially important for the CDMA system, where every user in every cell shares the same frequency band.

Open-loop power control makes the loss in the forward path similar to that in the reverse path. By setting the sum of the transmit and receive powers to a preset value, a reduction in the received signal level leads to a command to increase the transmit power. Closed-loop power control allows the power from the MS to deviate from its nominal value as set by open-loop control. In case of downlink power control, the BS adjusts the transmit power for each traffic channel independently, based on the propagation and interference conditions.

4.1.5 Channel assignment

Channel assignment can be implemented in a fixed, flexible, or dynamic mode. *Fixed channel assignment* allocates the channel resources statically, and is used in the 1G macrocellular systems. In the fixed channel assignment scheme, if all the channels in a cell are occupied, a new call is blocked. To reduce the blocking probability, a cell is allowed to borrow channels from a neighboring cell, if all of its own channels are occupied, and assign a borrowed channel to a new call; this is known as *fixed channel assignment with borrowing*. The MSC supervises the borrowing.

Dynamic channel assignment is more suitable for microcellular channel assignment. It allows any cell to use any channel as long as the cochannel reuse constraint is not violated. This reduces the probability of blocking, thus increasing the trunking capacity of the system, since all the channels are accessible to all the cells. Dynamic channel assignment requires the MSC to continuously collect real-time data on channel occupancy, traffic distribution, and radio signal strength indications (RSSIs) of all the channels. Thus, it is more expensive to implement. This technique can assign the same channel to an MS that is moving from one cell to another as long as the level of CCI is tolerable, while fixed channel assignment has to implement handoff in this case. It outperforms fixed channel assignment for light nonstationary traffic, but fixed channel assignment is more efficient for heavy traffic.

Flexible channel assignment combines the advantages of both the fixed and dynamic channel assignment schemes. The total number of channels are divided into two categories: Some channels are dedicated for the cells, while others are kept in reserve with the control center which dynamically allocates the channels. Frequency borrowing, which is a variation of the flexible channel assignment, exhibits a blocking probability lower than that achieved in fixed or dynamic channel assignment.

In DECT, *dynamic channel selection* is employed to improve efficiency. This can elegantly deal with fast changing shadowing. More discussion on channel assignment is contained in [25].

4.1.6 Handoff

When an MS moves into a new cell during a conversation, the MSC automatically transfers the call to a channel of the new cell. Handoff is a fundamental function in cellular communications. Handoff, also known as *handover*, can be intercell handoff or intracell handoff. Intercell handoff is implemented when an MS traverses a cell boundary, while intracell handoff can be applied within the cell when one channel in the cell is subject to excessive interference. The handoff procedure first measures the link quality (i.e., RSSI). If it is below a certain threshold, handoff is initiated. This is followed by radio and network resource allocation. Handoffs must be performed as infrequently as possible, and be imperceptible to the users.

To perform handoff, the drop in the measured signal should not be due to instantaneous fading as the MS is moving away from the serving BS. Thus, the power level is measured as an average during a period. A threshold signal-level difference between the powers received from the two BSs is used to initiate the handoff.

In 1G systems, the BSs are in charge of the signal strength measurement, and the RSSI results are reported to the MSC. The MSC decides whether a handoff is necessary. In 2G systems that use TDMA, handoff is mobile-assisted. In mobile-assisted handoff (MAHO), every MS measures the received power from the surrounding BSs and reports the results to its serving BS. MAHO is much faster than the handoff used in the 1G systems, since the measurements are made by each mobile and the MSC does not need to continuously monitor the RSSIs.

To reduce the probability of forced termination of a call, handoff requests should be given priority. This can be achieved by reserving a few channels exclusively for handoffs, or by queuing handoff requests. The channel-reservation scheme may reduce the total traffic, but is spectrum-efficient in the dynamic channel assignment strategy. Queuing of handoff requests is possible, since there is a time interval for the received signal to drop from the handoff threshold to call termination due to insufficient signal level.

At the IEEE, a specification on media-independent handover services (IEEE 802.21 MIH) is currently being developed to enable handover and interoperability between heterogeneous network types including both IEEE 802 and non-IEEE 802 networks [10]. IEEE 802.21 provides a framework that allows higher levels to interact with lower layers to provide session continuity without dealing with the specifics of each technology.

Handoff methods

Handoff can be either hard handoff, or soft handoff, or fast base station switching (FBSS). For hard handoff, an MS is connected to only one BS at a time. There is a short interruption during the transfer from one BS to another. It is simple to implement, but it does not take advantage of the diversity of receiving signals from two or more BSs. Hard handoff is supported in most 1G and 2G standards, and is also supported in WCDMA and WiMAX.

Soft handoff is also known as *macro diversity handoff (MDHO)*, since the MS is allowed to simultaneously communicate with more than one BS, and diversity combining is applied at the MS. All the BSs involved in the MDHO of a given MS constitute the *diversity set*. DECT, IS-95 CDMA, CDMA2000, WCDMA, TD-SCDMA, and HSUPA support the MDHO. MDHO between sectors of the same BS is also supported in WCDMA.

FBSS, also known as *fast cell selection*, is similar to MDHO in that it maintains a set of BSs, called the *active set*, but the MS communicates in the uplink and downlink with only one BS at a time, referred to as the *anchor BS*. When a change of anchor BS is required, the connection is switched between BSs without explicitly performing handoff signaling. The MS simply reports the selected anchor BS to the previous anchor BS. WCDMA (including HSDPA and HSUPA) supports FBSS in its Release 7 published in September 2007. Mobile WiMAX (IEEE 802.16e) supports the mandatory hard handoff, and optionally supports FBSS and MDHO.

Both FBSS and MDHO offer superior performance to hard handoff, but they need to synchronize the BSs in the active or diversity set, and use the same carrier frequency, and share network entry-related information.

Hard handoff

To initiate a hard handoff, H, the average signal difference between the BSs and T, the temporal window length over which the signal strength is averaged must be carefully selected.

Typically, T corresponds to the time for moving 20 to 40 wavelengths, and H is of the order of the shadow standard deviation [25]. The corner effect usually occurs in urban microcellular settings, when the MS turns the corner of a street, and a building blocks the LOS signal component. The corner effect can lead to a sudden drop of the signal strength by 25–30 dB over a distance as small as 10 m [25].

In conventional hard handoff, as used in GSM, when starting a new cell, the old cell is switched off. In order to avoid frequent switching between two BSs at the borders of two cells, the hard switch occurs only when the pilot signal from the second cell is sufficiently larger (e.g., 6 dB higher) than that from the original cell. This also reduces the performance at the cell border.

Various implementations of handoff algorithms have been detailed in [25]. Hard handoff is used in all 1G and most 2G standards, such as most of the TDMA cellular systems. Some 3G standards, such as UTRA-TDD, TD-SCDMA, and mobile WiMAX, also employ hard handoff. Soft handoff is introduced in Section 8.3.6.

Mobility management

For mobile communications, the process of identifying and tracking an MS's attachment position to the network is known as *location management*. Location management and handoff management together is known as *mobility management*. Location management has two processes: location registration (update) and paging. The MS performs location registration by periodically informing the network its location so that the network authenticates the user and updates its location in the database. The network will page all the BSs within the area of the subscriber, when an incoming request for session initiation arrives at the network.

As opposed to location management, handoff management has strict real-time requirement. Handoff is performed in two steps: handoff detection and handoff execution. Handoff management should minimize handoff failures and also avoid unnecessary handoffs. There is a tradeoff between dropping probability and handoff rate. Handoffs are often given more priority than starting new sessions.

For IP-based wireless communications, IP address of the MS must remain unchanged during the handoff process. This is difficult since a user moves across two BSs that belong to different IP subnets. Mobile IP, defined by RFC 3344, is the IETF solution to this problem. Mobile IP is specifically designed as an overlay solution to Internet Protocol version 4 (IPv4), and mobility has been considered in IPv6. Mobile IP defines two addresses for each mobile node: the home address (HoA), which is issued to the mobile node by its home network, and the care-of address (CoA), which is a temporary IP address assigned to the mobile node by the visited network.

4.2 Multiple access techniques

In a multiuser system, the uplink and downlink channels are different in nature. The downlink, also called *broadcast channel* or *forward channel*, transmits signals to many receivers from one transmitter. Thus, the transmitted signal is the sum of the signals to each of the users, $s(t) = \sum_{k=1}^{K} s_k(t)$. Synchronization of all the user signals is easy to realize in the downlink. The transmitter has its constraints on total power and bandwidth. The uplink channel, also known as a *multiple access channel* or *reverse channel*, receives signals from many transmitters.

Each communication standard defines its multiple access technique that allows multiple users to efficiently share the limited physical resources, usually defined in terms of the bandwidth. No matter what multiple access technique is selected, the objective of a multiuser system is to have a high capacity with sufficient QoS. Voice transmission is delay-sensitive, but allows a maximum tolerable BER of 10^{-3}. For data transmission, the target BER is 10^{-6}.

Multiple access techniques used in cellular standards include TDMA, direct-sequence CDMA (DS-CDMA), FDMA, and their combinations. These multiple-access techniques are demand-assignment-based. Strong QoS control is achieved by using a connection-oriented MAC architecture.

In a multiuser system, orthogonal multiple access techniques are used in order to separate different users. Popular multiple access techniques are TDMA and FDMA that have orthogonal channels, and CDMA that has almost-orthogonal channels. All multiple access techniques that divide the signal space orthogonally, such as TDMA, FDMA, and orthogonal CDMA, are proved to have the same capacity in AWGN channels [8], since they are all designed to have the same number of orthogonal dimensions for a given bandwidth and duration of time. Note that this is only for the orthogonal case. In dense wireless environments, orthogonality cannot be maintained due to interference from neighboring cells. In flat and frequency-selective fading channels, different multiple access techniques lead to different channel capacities.

4.2.1 Duplexing: FDD versus TDD

Mobile radio systems can be classified as simplex, half-duplex or full-duplex. Paging systems are simplex systems, where communication is only in one direction. Half-duplex systems allow two-way communication, but use the same channel for both transmission and reception. Thus, at any given instant, a user can only transmit or receive information. An example of half-duplex systems is the walkie-talkie system. Full-duplex, simply called *duplex*, allows simultaneous transmission and reception. Duplex is the simplest multiple access technique. FDD and TDD are the two duplexing methods.

FDD

In FDD, different frequency bands are used for uplink and downlink transmissions. The frequency band with the higher frequency is used for downlink transmission, while that with the lower frequency for uplink transmission. This is in view of the fact that the lower frequency leads to a lower propagation loss and thus is suitable for MSs. FDD mode is suitable for macrocells and microcells.

In order to obtain the channel state information (CSI) at the transmitter, the receiver has to quantize and feedback the channel state to the transmitter. In FDD, the user terminals usually receive a continuous downlink signal; dummy data must be inserted when the traffic is small. The transmission of dummy data wastes spectrum and power. FDD requires two sets of RF transceivers, one for each of the uplink and downlink. FDD has inflexibility in traffic allocation, although spectrum usage can be adapted to traffic variations by adding or removing channels. The channels should not be very broad for the reason of spectrum efficiency. FDD introduces high hardware cost, since for each carrier frequency an RF circuit is required.

Current 3G mobile communications, such as IS-95, WCDMA, and CDMA2000, and other wireless techniques prefer FDD. FDD is a legacy for voice communication, which has symmetric and predictable traffic.

TDD

In TDD, uplink and downlink transmissions use the same frequency bands, but different time slots. Since the time intervals are easy to change, TDD can adapt its spectrum resources by arranging different numbers of time slots on the uplink and downlink according to the actual traffic at each TDD frame. TDD can be used for very broad channels, and this leads to a high frequency diversity. This asymmetric feature is well suited to Internet and broadcasting services. TDD can better manage dense traffic in urban areas than FDD can and is more efficient in pico- to micro-cell sized cases, while FDD works more efficiently in macro-cell cases. TDD is only suitable for pico- to micro-cells, since large propagation delays between the MS and BS may lead to overlap of transmitting and receiving time slots. Unlike FDD, dummy data transmission is made unnecessary by introducing pauses in transmission. When the available bandwidth is in noncontiguous blocks, TDD is a desirable choice. TDD leads to lower-cost RF transceivers, which do not require a high isolation for the transmission and reception of multiplexing as needed in FDD transceivers. Thus, the entire RF transceiver can be integrated into a single integrated circuit (IC).

The high flexibility and spectrum efficiency of TDD come with a price. It requires burst demodulators, and synchronization for uplink transmissions becomes more difficult since the continuous downlink signal in FDD, that can be used for frequency and clock reference, is not available. Due to the identical radio propagation on the uplink and the downlink, it is easy to perform accurate measurements and the channel impulse responses can be estimated using a Steiner estimator through a training sequence. Furthermore, the strong signals generated by all the nearby mobile transmitters fall into the receive band for TDD, yielding a relatively high PAPR, which requires high-linearity amplifiers. TDD introduces a time latency.

TDD is used for uplink and downlink separation in most 2G standards, UTRA-TDD, and TD-SCDMA. It is attractive for future-generation mobile communications and wireless networking. In 802.16, both FDD and TDD are supported. The Korean WiBro supports TDD. In WCDMA, high-speed packet data mode is introduced, and this further decreases the deployment of FDD.

4.2.2 FDMA

FDMA divides the total frequency bandwidth into some frequency channels, each being assigned to a single user. Each user accesses its own channel and interference to other users is avoided. For FDMA, frequency synchronization and stability are difficult for narrowband communications such as speech communications, and it is also sensitive to fading. Thus, FDMA is only used for analog communications, wideband systems, or a hybrid with other multiple access techniques.

Frequency guard bands are required for separating each frequency band to compensate for imperfect filter implementation, ACI, and Doppler spread. The complexity of FDMA systems is lower compared to that of TDMA systems. FDMA is a continuous transmission scheme, and this approach is efficient when the information flow to be transmitted is steady, such as voice signals, but is inefficient with data that are bursty in nature. In FDMA, it is difficult to assign each user multiple channels. In FDMA with FDD, two channels are assigned to each user. FDMA supports both analog and digital transmissions. It was the only access method for the 1G analog FM systems, and is also used in 2G and 3G systems by combining with other multiple access techniques.

It is important to mention the difference between the two terminologies: FDMA and FDD. FDMA is used to separate different users using different frequencies, whereas FDD is used to divide uplink and downlink channels of the same user by using different frequencies. Similar difference applies for TDMA and TDD.

4.2.3 TDMA

TDMA divides the time axis into periodical time frames and time slots, and each slot in a time frame is assigned to a single user to transmit data. A user can send a large amount of data by using the time slots at the same position in multiple frames, so that the receiver can easily collect and assemble the bursty packets.

The need for A/D conversion, digital modulation, and synchronization makes TDMA much more complex than FDMA. Synchronization is a major difficulty in TDMA, and high synchronization overhead is required. For the downlink, all signals originate from the same transmitter, and synchronization at the receivers may be easier. For the uplink, different users transmit from different channels with different delays, and this may destroy orthogonality in time. Multipath further destroys time division orthogonality in both the uplink and downlink channels. Temporal guard intervals are used in the time slots to combat synchronization errors and multipath, and these intervals need to be greater than the channel delay spread. Users of TDMA systems occupy a larger bandwidth compared to the FDMA case; this allows frequency diversity within the bandwidth, and the sensitivity of Doppler effect is lower. But equalizers are required for ISI.

TDMA/TDD is a discontinuous transmission scheme. This enables the MS and BS to listen during the idle time slots. This enables the CSI to be obtained at the transmitter without the need for feedback. This feature is very desirable for the precoding of MIMO systems, and MAHO. TDMA systems improve the capacity by a factor of 3 to 6 times as compared to the FDMA-based analog cellular systems [19].

As in the case of FDMA, TDMA is efficient only with steady data traffic. For voice telephony, the fixed channel or time slot allocation can guarantee real-time constant bit rate voice quality. Due to the increased number of users and types of data services, dynamic channel allocation for TDMA and FDMA will introduce delay. TDMA is dominant in wired communications and 2G wireless communications, and also is combined with other multiple access techniques in 3G standards such as UTRA-TDD and TD-SCDMA.

4.2.4 CDMA

Spread spectrum multiple access (SSMA) technology, also known as *CDMA*, spreads the narrowband user data into a much wider spectrum. A pseudo-noise (PN) sequence is used to convert a narrowband signal to a wideband noise-like signal before transmission. In SSMA, each user shares the same bandwidth and time. The receiver despreads its data by using its unique code that is orthogonal to the codes used by the other users.

SSMA takes the form of either DS-CDMA, frequency-hopping CDMA (FH-CDMA), or time-hopping CDMA (TH-CDMA).

DS-CDMA

In DS-CDMA, sometimes simply called *CDMA*, data of multiple users are spread onto the same spectrum by multiplying with a very large bandwidth PN sequence for each user signal before transmission. DS-CDMA has a soft capacity limit, and is a dynamic channel allocation multiple access technique that has no fixed number of users, as opposed to TDMA and FDMA. The system capacity is determined by the total power of the interference, and new users joining the network will gracefully degrade the signal quality of all the users.

The large bandwidth introduces inherent frequency diversity. This provides a high time resolution to distinguish the different multipath waves, yielding multipath-diversity gain by applying the rake diversity technique. Thus, DS-CDMA provides better diversity against selective fading by providing more frequency and time diversity, compared to TDMA. In addition, DS-CDMA can operate without timing coordination among users, as opposed to TDMA.

Efficient frequency reuse can be achieved in DS-CDMA, since every cell can use the same frequency and users are separated by their unique codes. The use of the same frequency band by all the cells enables soft handoff to make use of macrodiversity. Since all the users share the same channel, the near–far effect will occur, that is, the power of each user received at the BS is not equal. This requires power control at the BS.

The IS-95 and most 3G mobile cellular standards select DS-CDMA as the main multiple access scheme, where DS-CDMA is combined with FDMA or TDMA.

FH-CDMA

In FH-CDMA or FHMA, the carrier frequencies of the users are varied in a pseudorandom fashion within a wideband channel. The total wideband channel is divided into many narrowband slots. The user data is broken into uniform sized bursts, which are then transmitted on different carriers frequencies (slots) controlled by a PN sequence. FHMA often uses FSK modulation and this enables inexpensive noncoherent detection. Frequency hopping provides inherent diversity against frequency selective fading. It is effective to combat interception.

FHSS is used in the baseline IEEE 802.11. FHMA is combined with TDMA in DECT and GSM. FHMA is combined with DS-CDMA to provide DS/FHMA in Bluetooth, where the center frequency of a direct-sequence modulated signal is hopped in a pseudorandom fashion. FHSS is more widely used in military communication systems. DS-CDMA and FH-CDMA will be introduced in Chapter 8.

TH-CDMA

TH-CDMA is not as commonly used as DS-CDMA and FH-CDMA. One reason is that it suffers from serious interference if there is a continuous transmission in its coverage area, since the TH system works in an on-and-off fashion in a frame. The TH technique usually works with the FH technique, forming a hybrid TH–FH system. TH-CDMA is the traditional method for UWB communications, and will be introduced in Chapter 20.

4.2.5 OFDMA

OFDM divides the total band into a number of subcarriers. Unlike in FDMA, where each subcarrier is separated by a guard band, OFDM overlaps them. To avoid ICI, OFDM makes the subcarriers orthogonal to each other by precisely controlling their relative frequencies and timing. OFDM can be combined with any of the multiple access techniques such as TDMA, FDMA, or CDMA.

In OFDM, one user uses all the subcarriers at a given time. OFDMA, as the multiple access in OFDM, assigns each user with a distinct subset of the subcarriers dynamically. This yields MAI-free access. OFDMA can be treated as a CDMA system with complex exponential spreading codes. In WiMax, OFDM is used for fixed case (802.16d) and OFDMA for mobile case (802.16e). OFDMA is also used in 3GPP LTE. We will give more introduction to OFDM and OFDMA in Chapter 9.

4.2.6 SDMA

Space division multiple access (SDMA) separates users in a spatial way. The sectorized antenna is the simplest SDMA. SDMA is generally achieved by using directional antennas such as an adaptive antenna array or a switching antenna array. SDMA uses the angular dimension introduced by the directional antenna to channelize the signal space. The antenna array can suitably steer the beams toward the users. A system with N antenna elements can distinguish at most N users. Each user can share the same channel resources such as the frequency and time slot. The multiple access gain of SDMA is influenced by the relative locations of the users. A tracking algorithm is needed to keep a high carrier-to-interference ratio (CIR). Also, an array of N antennas can create $N-1$ nulls in its radiation pattern, and this can effectively reduce interference to the cell.

In practical systems, SDMA is usually implemented using sectorized antennas, and these antennas introduce several orthogonal sectors, in which TDMA or FDMA can be used to further channelize the channels. Due to its limited multiple access capacity, SDMA is usually used for improving the spectral efficiency by allowing channel reuse within a cell. In the future, smart antennas will likely simultaneously steer the beams in the directions of many users. Smart antenna technology will be dealt with in Chapter 18.

Some of the multiple-access techniques discussed above are illustrated in Fig. 4.3.

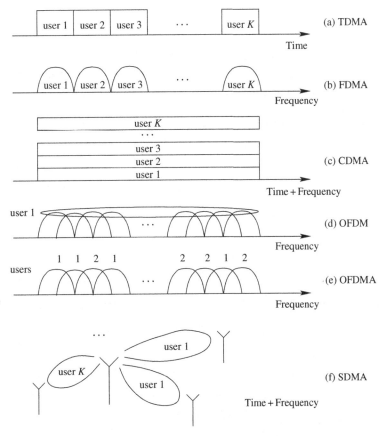

Figure 4.3   **Illustration of some multiple access techniques.**

4.3 Random multiple access

Demand-assignment-based multiple access is necessary for real-time voice and video communications. However, once a channel is assigned, it cannot be used by other users. When data generation is random and has a high PAPR, such as in the case of packet-based wireless networking, random multiple access, also called *packet radio*, is most efficient. Packet radio is easy to implement, but has low spectral efficiency and may cause delays. Although TDMA and FDMA are more efficient when all users have data to send, wasted frequency or time slots can be up to 50% [2]. That is the reason why CDMA is very successful for voice.

In order to identify the target receiver in packet radio, each packet starts with a header that contains the address or identification of the receiver. The header is followed by a data portion with or without error-correction protection. The most popular technique is ALOHA [1] and carrier-sense multiple access (CSMA). Random multiple access combined with a reservation protocol is more efficient when the data length is long.

4.3.1 ALOHA

ALOHA can be either pure or slotted ALOHA. In pure ALOHA, a packet is transmitted immediately after it is generated; after a packet is transmitted, the user waits for an acknowledgment from the receiver as to whether the packet has been received without errors. If no acknowledgment is received, the packet is assumed to be lost in a collision. In this case, the user waits for a random time and retransmits the packet. A packet is successfully received only if there is no other transmission during its transmission.

Assuming that all the packets have the same length t_p and the average transmission rate of all the transmitters is λ_p packets per second, for completely random times and different transmitters the probability that n packets are transmitted within time duration t is given by the Poisson's distribution [18]

$$\Pr(X(t) = n, t) = \frac{\left(\lambda_p t\right)^n e^{-\lambda_p t}}{n!}. \tag{4.10}$$

The possible collision time is $2t_p$, and the possibility that during this period there is zero packet transmitted is thus calculated as $\Pr(X(t) = 0) = e^{-2\lambda_p t_p}$. The effective channel throughput can be obtained by the normalized channel usage or load $R = \lambda_p t_p$

$$T = Re^{-2R}. \tag{4.11}$$

The maximum throughput for pure ALOHA is 18.4% at $R = 0.5$.

The slotted ALOHA introduces some coordination by dividing time into time slots, each of which is of the packet length t_p. All terminals are allowed to transmit their packets only at the beginning of each slot. The probability that during t_p there is zero packet transmitted

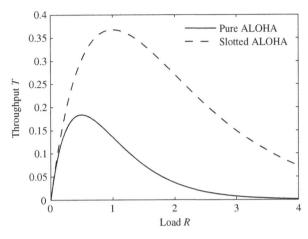

Figure 4.4
Throughput of ALOHA.

is calculated as $\Pr(X(t) = 0) = e^{-L}$, and correspondingly, the effective throughput is given by

$$T = Re^{-R}. \tag{4.12}$$

The maximum throughput for slotted ALOHA is 36.8% at $R = 1$.

Example 4.1: The throughputs of pure ALOHA and slotted ALOHA are plotted in Fig. 4.4.

ALOHA is usually used in the uplink to signal its presence and request reservations. For example, IEEE 802.15.4a uses ALOHA for channel access; in Wireless USB, the host has a number of device notification time slots to permit unassociated devices to request membership in the cluster via slotted ALOHA protocol.

4.3.2 Carrier-sense multiple access

The popular distributed contention method is CSMA. Each node must sense the channel before sending data so as to minimize the chance of collision. Thus, the channel throughput is substantially increased over that of ALOHA. The achievable throughput for CSMA can be as high as 90% under ideal conditions.

CSMA is suitable when all the users can detect the transmissions of one another and the propagation delay is small, and this is the feature of wired LAN. The IEEE 802.3 (Ethernet) standard uses CSMA with collision detection (CSMA/CD), where channel sensing is performed by sending a few bits (jamming signal) and comparing it with what is sensed on the channel.

CSMA with collision avoidance (CSMA/CA) is more suitable for the wireless case, since it is more energy-efficient for MSs. CSMA/CA attempts to avoid collisions by using explicit packet acknowledgment (ACK) from the receiving station. Collision avoidance is implemented by a four-way handshake prior to transmission. CSMA/CA uses heuristics, including random backoff, listen-before-talk, and mandated interframe delay periods, to avoid collisions among users on the shared channel. CSMA/CA is implemented in the IEEE 802.11 standards. Due to the added overhead, CSMA/CA is slower than CSMA/CD.

Hidden terminal problem

For a centralized wireless network, the BS broadcasts the status of the reverse channel via the forward channel, and thus the probability of collision is significantly reduced. For a wireless ad hoc network, carrier sensing is not very effective, as a given user may have difficulties in sensing signals transmitted from the other users. Two typical problems existing for carrier sensing are the *hidden terminal* and the *exposed terminal problems*, which arise from the fact that carrier sensing is only performed at the transmitter while its effect is determined at the receiver: the absence of complete simultaneity. Also, collision detection is not possible. A wireless transceiver cannot transmit and receive at the same time, since the transmitted signal is always much stronger than the received signal. CSMA assumes the propagation delay between two nodes to be much shorter than the duration of packet. This is usually satisfied in land-based systems, but is not the case for satellite communications.

The hidden terminal problem is illustrated in Fig. 4.5. As illustrated, nodes A and B are out of each other's range, and thus, they cannot hear each other. If node A is transmitting to node C, node B may still sense the channel as idle and starts to transmit; packets from A and B will collide at node C. Thus, nodes A and B are hidden terminals to each other. CSMA/CA can largely avoid the hidden terminal problem by using the request-to-send and clear-to-send primitives; this is implemented in the 802.11 MAC. A sending station A transmits a request-to-send and waits for C to reply with clear-to-send. Another strategy is

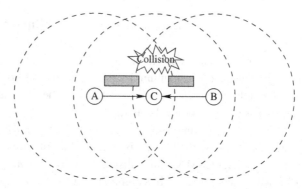

Figure 4.5   The hidden terminal problem.

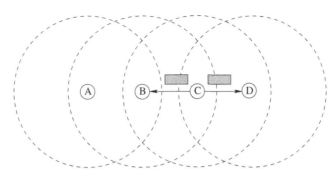

Figure 4.6   The exposed terminal problem.

the *idle signal casting multiple access (ISMA)*, wherein each terminal is informed from the BS of the other terminals' transmission.

Exposed terminal problem

The exposed terminal problem is shown in Fig. 4.6. Assume that nodes B and C intend to transmit only. When node C transmits to node D, node B can detect the transmission, as node B is within the radio coverage of node C. Thus, the transmission from nodes B and A are blocked, even if they are both idle. To alleviate this, a node must wait a random backoff time between two consecutive new packet transmission times to resolve contention.

Throughput

CSMA can be implemented in one of the three modes based on the method for rescheduling transmissions when a collision occurs, namely nonpersistent CSMA, 1-persistant CSMA, and p-persistent CSMA:

- In the nonpersistent CSMA, a user that has a packet to transmit will sense the channel. If the channel is idle, the user transmits a packet; otherwise the user waits a random time delay, and then repeats the process.
- In the 1-persistent CSMA, the user senses the channel and transmits its packet with probability 1 if it is idle; otherwise, it will wait until the channel is idle and transmit with probability 1.
- The p-persistent CSMA is a generalization of the 1-persistent CSMA to reduce collision. In p-persistent CSMA, upon sensing that the channel is idle, the user transmits its packet with probability p and delays it by τ with probability $1 - p$.

A thorough throughput analysis has been given in [13]. An optimum p-persistent CSMA can be achieved by finding the optimum value of p for each of the throughputs in terms of transmission delay. For small throughput, the 1-persistent ($p = 1$) is optimal. Slotted CSMA is also possible for fixed length packets.

The normalized throughputs for some of the CSMA protocols are given below [12, 13]

$$T = \frac{Re^{-aR}}{R(1 + 2a) + e^{-aR}} \qquad \text{(Unslotted nonpersistent CSMA),} \qquad (4.13)$$

$$T = \frac{aRe^{-aR}}{(1 + a) - e^{-aR}} \qquad \text{(Slotted nonpersistent CSMA),} \qquad (4.14)$$

$$T = \frac{R\left[1 + R + aR\left(1 + R + \frac{aR}{2}\right)\right]e^{-R(1+2a)}}{R(1 + 2a) - \left(1 - e^{-aR}\right) + (1 + aR)e^{-R(1+a)}}$$

$$\text{(Unslotted 1-persistent CSMA),} \qquad (4.15)$$

$$T = \frac{R\left(1 + a - e^{-aR}\right)e^{-R(1+a)}}{(1 + a)\left(1 - e^{-aR}\right) + ae^{-R(1+a)}} \quad \text{(Slotted 1-persistent CSMA),} \qquad (4.16)$$

$$T = \frac{Re^{-aR}}{Re^{-aR} + bR\left(1 - e^{-aR}\right) + 2aR\left(1 - e^{-aR}\right) + \left(2 - e^{-aR}\right)}$$

$$\text{(Unslotted nonpersistent CSMA/CD),} \qquad (4.17)$$

$$T = \frac{aRe^{-aR}}{aRe^{-aR} + b\left(1 - e^{-aR} - aRe^{-aR}\right) + a\left(2 - e^{-aR} - aRe^{-aR}\right)}$$

$$\text{(Slotted nonpersistent CSMA/CD),} \qquad (4.18)$$

where $a = \tau_d/t_p$. For CSMA/CD, b is the jamming length.

Example 4.2: The throughput for nonpersistent CSMA, as given by (4.13), is plotted in Fig. 4.7. Maximum throughput is achieved when $a = 0$. As $a \to 0$, $T \to R/(1 + R)$.

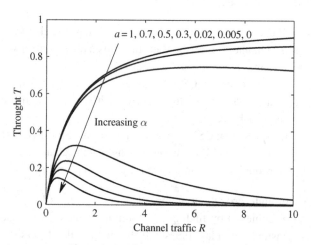

Figure 4.7

Throughput of nonpersistent-CSMA.

CSMA is generally implemented in the time domain. Although it is possible to apply the random access to frequency and code slots, the implementation is very sophisticated. For this reason, CSMA can be viewed as a type of TDMA.

Digital sense multiple access

Digital sense multiple access (DSMA) transmits channel and decode status flags on the forward channel to signify whether the reverse channel is busy and whether the data block that was just received on the reverse channel has been decoded without any errors, respectively. With the information about the reverse channel, DSMA operates in a way similar to slotted CSMA/CD. Slotted nonpersistent DSMA has been used in the CDPD network.

DSMA with delayed transmission (DSMA/DT) [15] is designed for reverse channel to improve the throughput of DSMA for long round-trip propagation and processing delay, which occurs in outdoor, high-speed environments or when the receiver requires long signal processing time.

4.3.3 Scheduling access

When multiusers share a wireless channel, scheduling must be made for continuous data transmission, since random access leads to frequent collisions. Scheduling access methods are generally distributed contention methods and polling methods. This is defined in the MAC layer of a wireless standard. For high-data-rate services, low latency operation can be achieved in a distributed contention-based MAC at low load; in heavy-load systems this can be accomplished using centralized or distributed round-robin scheduling (polling), which is static TDMA in the context of a packet-based system. Polling can achieve a throughput in excess of 90% over a channel without impairments, though it has the same drawbacks of TDMA.

In WiMAX, an MS can obtain uplink bandwidth by polling. The BS allocates dedicated or shared resources periodically to each MS, for it to request bandwidth. Polling can be either unicast or multicast. For multicast polling, a contention access and resolution mechanism for multiple MSs is applied.

ALOHA is still needed in scheduling access, since a predefined scheduling is not available at system startup. Packet-reservation multiple access (PRMA) [9] embodies this random startup by combining slotted ALOHA for bursty data with TDMA scheduling for continuous data. PRMA is used for the reverse channel, while the BS controls the forward channel. The BS broadcasts the information which the MS had successfully broadcast in the previous time slot. When there is a collision on the reverse channel, the BS transmits a negative acknowledgment (NACK).

Although CSMA/CA is the mandatory data service, the 802.11 standards support an optional channel reservation scheme based on four-way handshake between the sender and receiver nodes for providing QoS. In WiMedia, each active device is required to send a beacon frame in a selected beacon slot of the superframe. Each active device has a separate beacon slot. All the devices listen for beacons and interpret their content so as to respect declared reservations. The reservations are established via the WiMedia distributed reservation protocol.

4.4 Erlang capacity in uplink

For a given QoS, the capacity of a multiple access network is referred to as the average number of users that can be serviced. QoS can be defined as 100% minus the percentage of blocked call, minus ten times the percentage of lost calls [18]. Cellular systems rely on trunking to accommodate a large number of users in a pool of a limited number of channels. Each user is allocated a channel on a per-call basis. The trunking theory was developed by Erlang in the late 19th century. The unit of traffic density is now called *erlang*. One erlang represents the traffic intensity of a channel that is completely occupied. A channel occupied half of the time has a traffic of 0.5 erlang.

For a large system, the calls from users are random, and satisfy Poisson's distribution. Assuming the total call arrival is λ calls per second, the probability of call service time t longer than T is given by

$$\Pr(t > T) = e^{-\mu T}, \quad T > 0, \tag{4.19}$$

where the average interval for calls is $1/\mu$. Define $T_{\mathrm{tr}} = \lambda/\mu$, measured in erlang; then, T_{tr} is the average traffic offered in the unit of users (channels).

Two models are usually used to characterize the Erlang capacity: lost call clearing (LCC) and lost call hold (LCH) [27]. In the LCC model, if a new user wants to enter a network with all time-/frequency-slots occupied, it can only leave and will then re-enter after a random interval as a new user. This causes a slight increase in λ. The number of total states is thus $1 + N_c$, where N_c is the number of available channels. In the LCH model, the user that is not served will repeat its request for service and stay in the network. This leads to a slight increase in the average interval for calls $1/\mu$, and the number of states will approach infinity. In both the cases, $T_{\mathrm{tr}} = \lambda/\mu$ is slightly increased. Analysis of both the models is based on the Markov model [27].

4.4.1 Erlang B equation

The LCC model can be used to compute the call blocking probability of a time-/frequency-slot system by one BS. The probability of call blocking is given by [18, 19, 27]

$$P_{\mathrm{block}} = \frac{T_{\mathrm{tr}}^{N_c}/N_c!}{\sum_{k=0}^{N_c} T_{\mathrm{tr}}^k/k!}. \tag{4.20}$$

This is known as the *Erlang B equation*. The blocking probability during the busiest hour is defined as the *grade of service (GoS)*, which must be estimated by the system designer. The GoS for the AMPS system was designed as 2% blocking [19].

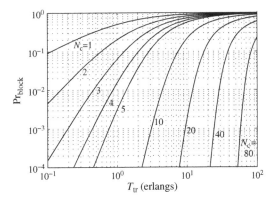

Figure 4.8

Block probability of an Erlang B system.

Example 4.3: The blocking probability for different values of N_c and T_{tr}, given by (4.20), is illustrated in Fig. 4.8. Given a blocking probability, the ratio of offered traffic to available channels, T_{tr}/N_c, can be determined using Fig. 4.8. It is seen that for small N_c, this ratio is very low, while for large N_c the ratio is slightly less than unity. For example, given $P_{block} = 0.001$, $T_{tr}/N_c = 0.066$ if $N_c = 3$, and $T_{tr}/N_c = 0.72$ for $N_c = 80$. Thus for a given P_{block}, T_{tr} increases faster than linearly with N_c, and the difference between the actual increase and the linear increase is known as the *trunking gain*. Thus, a large pool of channels is more efficient in bandwidth efficiency.

The trunking efficiency can be defined by the channel usage efficiency [25]

$$\eta_T = T_{tr}\left(1 - P_{block}\right)/N_c \tag{4.21}$$

in erlangs / channel.

4.4.2 Erlang C equation

The Erlang C system is based on the LCH model, and gives the probability of a user being on hold when there is no available channel. The blocking probability is the probability of a new call when there are N_c or more users in a system [19]

$$P_{delay} = \Pr(t_{delay} > t) = \sum_{k=N_c}^{\infty} P_k = e^{-\lambda/\mu} \sum_{k=N_c}^{\infty} (\lambda/\mu)^k/k!$$

$$= \frac{\dfrac{T_{tr}^{N_c}}{N_c!}}{\dfrac{T_{tr}^{N_c}}{N_c!} + \left(1 - \dfrac{T_{tr}}{N_c}\right)\sum_{k=0}^{N_c-1}\dfrac{T_{tr}^k}{k!}}. \tag{4.22}$$

This is the *Erlang C equation*. When N_c is very large, the results of both the Erlang B and C models are very similar.

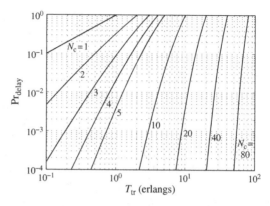

Figure 4.9

Block probability of an Erlang C system.

The GoS is defined as

$$\Pr(t_{\text{delay}} > t) = \Pr(t_{\text{delay}} > 0) \Pr(t_{\text{delay}} > t | t_{\text{delay}} > 0) \tag{4.23}$$

$$= \Pr(t_{\text{delay}} > 0) e^{-\frac{(N_c - T_{\text{tr}})t}{H}}, \tag{4.24}$$

where H is the average duration of a call. The average delay D for all calls is derived as

$$D = \Pr(t_{\text{delay}} > 0) \frac{H}{N_c - T_{\text{tr}}}. \tag{4.25}$$

Example 4.4: The blocking probability for different values of N_c and T_{tr} is illustrated in Fig. 4.9. The result is very close to that of the Erlang B system.

Erlang C model in CDMA systems

The Erlang C equation is more suitable for mobile systems and for non-frequency-/time-slot systems such as the CDMA systems. In CDMA systems, N_c is given by

$$N_c = \frac{\frac{W}{R}}{\frac{E_b}{I_0}}, \tag{4.26}$$

which is further reduced by a factor $1 - \eta$, $\eta < 1$. The quantity η is used to limit the ratio of total interference-plus-noise I_0 to noise N_0

$$\frac{I_0}{N_0} < \frac{1}{\eta}. \tag{4.27}$$

Typically, $\eta = 0.25$ to 0.1, corresponding to $I_0/N_0 = 6\,\text{dB}$ to $10\,\text{dB}$ [27]. In CDMA systems, $T_{\text{tr}} = \lambda/\mu$ is also reduced by a voice activation factor $\rho < 1$. The outage probability is given by [27]

$$P_{\text{out}} < e^{-\rho\lambda/\mu} \sum_{k=[N_c(1-\eta)]}^{\infty} (\rho\lambda/\mu)^k/k! \,. \tag{4.28}$$

In case of undesirable power control, an approximation for the outage probability is given in [27]. It is shown that even if the standard deviation of power control is $\sigma_c = 2.5\,\text{dB}$, the system capacity is reduced by 20%.

The use of ρ as well as the sector gain leads to a capacity gain over frequency-/time-slot systems. The ratio of Erlang capacity of the IS-95 CDMA system to that of the time-slot system, $(\lambda/\mu)_{\text{spread}}/(\lambda/\mu)_{\text{slotted}}$, is typically 6 : 1 (IS-95 to GSM) or 5 : 1 (IS-95 to IS-136), assuming the BS employs three-sector antennas [27].

4.5 Protocol design for wireless networks

A typical data network such as the internet has five layers, namely physical, data-link, network, transport and application layers of the OSI reference model. Wireless networks can use the same model. Very often, wireless standards have a single MAC layer that has the same functions as the data link layer. For all IEEE 802.x protocols and most other wireless standards, only the MAC and physical layers are defined. The data are delivered over existing cellular networks such as the GSM network or data communication networks such as the IP network. Unlike wired network protocols where each layer is relatively independent, many functions in wireless networks, such as power control, multiple antennas, and dynamic resource allocation, may span multiple layers of the protocol stack. For wireless networks, cross-layer design provides significant performance gain over a strictly layered structure.

4.5.1 Layered protocol design

Channel structure

The channel structure is composed of logical channels, transport channels, and physical channels. Logical channels are offered by the MAC layer to higher layers. Transport channels are offered by the physical layer to the MAC layer. Physical channels are handled in the physical layer.

A logical channel can be either a control channel or a traffic channel, depending on the type of information transferred on the channel. A transport channel is a physical channel that offers information services to the MAC layer, and it can be either a common or dedicated channel. The physical channels are defined by code, frequency, and time slot. Rate matching of data that are multiplexed to dedicated and shared channels is done at the physical layer. There are two types of channel mapping: physical to transport, and transport to logical channels.

The control architecture uses paging and access channels for the forward and reverse link control. Paging channels are used by the BS to inform MSs of the system parameters and to alert the MSs of the incoming calls. The systems parameters may include a list of BSs in the neighborhood, MS location, access parameters, and information on usable channels. Access channels are used by MSs to get access to initiate a call, by providing its identification and dialed number. The network will then decide whether to allocate a channel for the call. Access procedure is usually based on ALOHA.

Pilot channels, like paging channels, continuously transmit information to allow MSs to acquire the system parameters. It helps the MS to measure the signal strength for comparison and synchronization.

On the reverse link, transmission takes place over two types of channels: access channels and traffic channels. The reverse link is usually deemed to be less reliable than the forward link, as the transmission power of the MS is constrained. For this reason, data on the reverse channel is better protected.

Physical layer design for TDMA systems

In the TDMA system, the packet of a burst is typically of fixed length. Each packet typically contains information of the traffic and control channels. It consists of data bits, a training sequence, and tail bits. The training sequence has a fixed length and a known pattern. The received training sequence is compared with the known training sequence so as to estimate the channel and then reconstruct the rest of the original data bits; this is known as *equalization*. The training sequence is also used for synchronization. The data bits are then extracted. The packet is assigned in the form of a time slot, and the typical structure of a time slot is shown in Fig. 4.10. Typically, the first and last few symbols in a packet are subject to time-domain raised-cosine shaping to prevent abrupt time-domain signal changes.

For TDD, the guard section is used to account for the round-trip time delay, $\tau = 2d/c$, where d is the distance between the BS and the MS, and c the speed of light in free space. The guard section must be included in the transmit packet of the MS (reverse link) to avoid the overlap of the data from both the BS and the MS. In the forward link, the guard time is not assigned.

For example, in GSM, a time slot for normal burst contains 148 bits. Among the 148 bits, two blocks of 57-bit payload data are separated by a midamble (training sequence) of 26 bits, which is known and used for equalization. At either end of the time slot is 3 tail bits, which are also known and are used to control MLSE-based detection of burst data. A guard period of 8.25 bits are also appended to the end of the time slot. The symbol duration is 3.7 μs. The time slot structure is shown in Fig. 4.11. In GSM, data transmission is in a circuit-switched mode, just like voice transmission. For the GSM system, there are

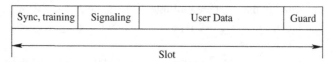

Figure 4.10   **Packet structure of a time slot.**

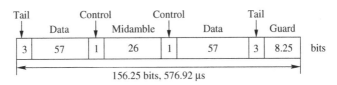

Figure 4.11

Structure of a normal burst in GSM.

also other kinds of bursts: access burst, S burst for time synchronization, and F burst for frequency synchronization.

In GSM, eight different PN-sequences are designed for midambles in the normal burst, so that different midambles can be used in different cells. These PN sequences have low cross-correlation and specific autocorrelation functions. The reason for using midamble rather than preamble is to keep the channel estimate to be sufficiently accurate at both the beginning and end of the burst, since the channel of GSM is time-varying for speeds of MSs up to 250 km/h.

A frame is a basic period in a wireless standard. In the forward link, multiple slots constitute a frame. All MSs are synchronized to the frame timing of the BS. For GSM, a frame is comprised of 8 time slots and covers 4.61 ms, 26 frames constitute a superframe covering 6.12 s, which is the period for all channel arrangement modes, and 2,048 superframes constitutes a hyperframe, which covers 3 h and 28 min.

Physical layer design for CDMA systems

In the FDD-based CDMA system, channelization is realized by using different codes. All the channels use the same frequency band. The pilot channel allows the MS to perform timing acquisition from a BS, channel estimation, and signal strength estimation of all neighboring BSs. The pilot channel uses Walsh code 0 (all-zeros code) for transmission. The pilot channel is not power controlled, since it is used by all the MSs. It is so important that typically 20% of the total BS power is allocated to the pilot channel.

In the IS-95 CDMA system, the BSs are synchronized based on the GPS. The BS uses the synchronization channel to transmit system information to the MS so that the MS synchronizes itself to the network. This channel is not scrambled. Each frame of the synchronization channel is aligned at the start of the long-code PN-sequence. The synchronization channel uses Walsh code 32. Other channels are also used to carry system information.

WCDMA is a FDD-based hybrid TDMA/CDMA scheme. Unlike IS-95, transmission in WCDMA is still based on a hierarchical time slot structure very similar to that of GSM. A frame has a duration of 10 ms, which is divided into 15 time slots.

MAC layer design

The MAC layer protocol implements admission control and scheduling. It performs request collection and request scheduling. The request scheduler is used to arbitrate the requests from the participating devices so as to achieve a high bandwidth efficiency and also a high QoS to the devices. Due to the conflicting requirements of bandwidth efficiency and QoS, a tradeoff between them must be made by a scheduling algorithm. A good scheduling algorithm must consider factors such as system throughput, channel quality, fairness, QoS, and admission control of new devices. To guarantee QoS, a central scheduling approach is the preferred solution.

For MAC layer protocol design of the TDMA system, dynamic or fixed frame structure can be used. MAC layer design of the CDMA system should consider the downlink power budget and the uplink interference limit. In MAC layer design, if the ARQ protocol is used, when a packet is received with error, instead of discarding it the MAC layer can save it and combine it with the retransmitted packet as a form of diversity. Type-I and Type-II HARQ can be used for this purpose. Both methods can substantially increase throughput when compared with the simple retransmission [11].

4.5.2 Cross-layer design

The OSI model allows multi-vendor computers to interact and communicate, but QoS is not considered as an issue in the design process. The protocol stack design is highly rigid and strict, and there is no collaboration between the different layers. This may cause problems for real-time applications such as VoIP. The challenges of mobile communication systems cannot be met with a layered design approach.

Cross-layer design is now a new hype in wireless communication systems. As energy management, security and cooperation are cross-layered in nature, cross-layer protocol interactions can improve network efficiency and QoS support. Cross-layer design is particularly important for any type of wireless network, since the state of the physical medium can significantly vary over time.

A simple cross-layer design requires adaptability in the MAC and PHY layers in response to application services. For multimedia services, the MAC layer needs to distinguish the service type and its associated QoS requirements, and map the service to the corresponding physical layer configuration. HARQ is also a commonly used cross-layer approach, as it involves the interaction of the MAC and physical layers.

Cross-layer design can take an evolutionary (such as for 4G mobile systems) or a revolutionary approach (for WSNs). The evolutionary approach is mainly used for cellular communications and wireless networking, since compatibility with existing systems and networks is extremely important. Most existing cross-layer designs are evolutionary, as they need to communicate with the rest of the world, that is, the IP network. The revolutionary approach can be applied to specific applications (such as WSNs) where backward compatibility is not important.

Layer triggers are the most basic implementation of cross-layer design. This method is cheap to implement, and it also maintains compatibility with existing layered structure. Layer triggers are predefined signals that are used to notify special events between protocols. A modern implementation of TCP has the ECN-bit in the TCP header, which can be used for transmission rate control [6].

The MobileMan reference architecture [5] implements a system-wide cross-layer design in a MANET protocol stack using 802.11. Protocols belonging to different layers cooperate by sharing network status, while still maintaining the layer separation in the protocol design. The core component is the network status repository. Whenever a protocol in the stack collects information, it will publish this to the repository, thus making it available for every other protocol.

Cross-layer scheduling can significantly enhance multiuser system capacity by using the time-varying nature of the wireless channel, and both the physical layer and the MAC layer are required to jointly adapt to the changing channel [14]. In the physical layer, adaptation can be achieved by using appropriate forward error correction (FEC) and modulation levels according to the instantaneous CSI. In the MAC layer, users with good CSI can be assigned more opportunity, and thus a higher instantaneous throughput is achieved by using the adaptive physical layer. This, however, leads to the fairness problem. Fairness can be defined as *effort fair* if the allocation of services to different users is fair, or *outcome fair* if the actual realized throughput is fair. Many scheduling models including the fairness notion are available in the literature for wireless networks [14].

Cross-layer design requires information exchange between layers. This information can be adapted at each layer. Adaptation at each layer should compensate for variations of the specific time scale at that layer. Diversity can also be exploited into each layer. Cross-layer design is especially suitable for systems that exploits multiple antennas and AMC techniques.

4.6 Quality of service

End-to-end QoS is a major concern for customers, since they pay for the service. Current 3G and future-generation cellular communications are based on IP technology to deliver data, voice, video, messaging, or multimedia. However, IP was designed for survivability and provides best-effort data, but does not provide QoS. QoS can be more easily realized by polling.

QoS requires mechanisms in both the control plane and the data plane [2]. In the control plane, users and the network are allowed to negotiate and agree on the required QoS specifications so that the network can allocate resources for each service. This is implemented by QoS policy management, signaling, and admission control. The data plane enforces the agreed-on QoS by classifying the incoming packets into several queues and allocating resources to each queue. The classification is based on the headers of the incoming packets.

QoS for IP

IP does not provide QoS. Some form of QoS can be provided by transport layer protocols that run over IP, such as TCP and RTP. These transport layer protocols cannot control the end-to-end delay or throughput that is controlled by the network. For this reason, QoS mechanisms must be placed into the network layer. For this reason, the IETF developed a number of protocols for delivering QoS over the IP network, such as the IntServ (integrated services), DiffServ (differential services), and MPLS (multiprotocol label switching).

IntServ employs the resource reservation protocol (RSVP) for signaling end-to-end QoS requirements and making resource reservations. IntServ provides a guaranteed IP QoS, but it has serious limitations such as the attendant scalability problems, and high overhead of signaling and state maintenance. This makes IntServ suitable only for small networks.

DiffServ is a multi-purpose control protocol which uses the TOS-bit in the IP header to identify and mark different types of packets, so as to provide different QoS classes for different types of data. DiffServ depends on aggregate traffic handling, rather than the per-flow traffic handling used in IntServ. DiffServ lacks the degrees of service assurance and granularity offered by IntServ, but it provides good QoS that is scalable and easy to deploy. The DiffServ mechanism is therefore suitable for providing QoS to large IP networks.

MPLS inserts a new fixed-length label between the layer 2 and IP headers of a packet. The label specifies how the packet is treated within the MPLS network. Packets are not routed using IP headers, but are switched using the information in the label. MPLS itself does not provide an end-to-end IP QoS mechanism, but it provides an infrastructure on which IntServ and DiffServ mechanisms can be implemented. However, MPLS breaks the end-to-end principle of the IP.

In WiMAX, the QoS architecture employs the concept of service flow in the MAC layer. Each service flow is associated with a set of QoS parameters, such as latency, jitter through-put, and packet error rate. In WiMAX (IEEE 802.16), four services are defined to support different types of data flows: [4]

- Unsolicited grant service (UGS) for real-time constant bit rate (CBR) traffic such as VoIP.
- Real-time polling service (rtPS) for variable-bit-rate traffic such as MPEG video.
- Non-real-time polling service (nrtPS) for delay-tolerant data service with a minimum data rate, such as FTP (File Transfer Protocol).
- Best-effort service, which does not specify any service related requirements.

IEEE 802.16e includes an additional service known as *extended rtPS (ErtPS)*, which provides a scheduling algorithm that builds on the efficiency of both UGS and rtPS. This service supports real-time service flows that generate variable size data packets on a periodic basis. 1xEV-DO supports both user- and application-level QoS.

Resource-allocation techniques

In order to support the various QoS requirements, many resource-allocation algorithms can be applied. These algorithms can take advantage of multiuser diversity and AMC. Resource allocation is usually formulated as a constrained optimization problem, which either maximizes the total data rate with a constraint on the total transmit power or mini-mizes the total transmit power subject to a constraint on user data rate. There are a number of resource-allocation techniques [2]: maximum sum-rate algorithm, maximum fairness algorithm, proportional rate constraints algorithm, and proportional fairness scheduling.

The maximum sum-rate algorithm maximizes the sum rate of all users, subject to a total transmit power [29]. A few users that are close to the BS may be allocated all the system resources, due to their good channels, while other users may never get the resources. The maximum fairness algorithm tries to maximize the minimum user data rate, and this generates equalized data rate for all users [20]. This method has an inflexible rate distribution among users, and the total throughput is limited by the user with the worst SINR. The proportional rate constraints algorithm aims to maximize the sum throughput, with the constraint that each user's data rate is proportional to a set of prespecified parameters [24]. This method is more suitable when different users require different data rates.

Unlike the other three algorithms which attempt to instantaneously achieve their respective objectives, proportional fairness scheduling achieves its objective over time [26]. In this case, in addition to throughput and fairness, latency can also be traded for additional flexibility to scheduling. Proportional fairness scheduling takes advantage of multiuser diversity while providing comparable long-term throughput for all users. Proportional fairness scheduling has been widely used in packet data systems, such as HSDPA and 1xEV-DO.

4.7 User location

User location or positioning offers important services for the users, and it also significantly increases the performance of the cellular network. Also, wireless user location is important for public safety, since a large portion of all emergency 911 (E-911) calls originates from mobile phones. For this reason, the FCC issued, in 1996, an order that mandates all wireless service providers to deliver accurate location in case of E-911 calls. Wireless E-911 services require a location accuracy to within 100 m for 67 percent of the cases. However, user location is challenging due to the adverse wireless channel. In 4G wireless networks, the location of users is required to be determined. IEEE 802.16m supports E-911. Location is also an option for IEEE 802.15.4 and 802.15.4a.

Principles of basic location techniques

Basic location techniques are RSSI positioning, DoA positioning, time-of-arrival (ToA) positioning, and time-difference-of-arrival (TDoA) positioning [21]. These techniques are illustrated in Fig. 4.12. They are all network-based techniques. For all the cases, multipath propagation may influence the location accuracy.

RSS positioning is attractive, as the implementation complexity is low; but this approach is traditionally treated as a coarse positioning method due to signal fading caused by multipath propagation. RSSI measurement can be accurate in dense sensor networks [28].

The DoA positioning technique determines the position of an MS using triangulation. This is illustrated in Fig. 4.12a. The intersection of two directional lines determines a unique position. Thus, a pair of BSs is required for location. The estimation of DoA

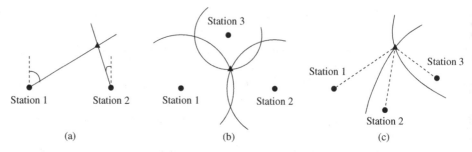

Figure 4.12   Three location techniques. (a) DoA positioning. (b) ToA positioning. (c) TDoA positioning.

requires directional antennas or antenna arrays, thus it is difficult to estimate the DoA at the MS. The method does not require synchronization at the receiver, but requires calibration of antennas. In addition, the method does not perform well in indoor environments due to multipath propagation.

The ToA positioning technique determines the position of the MS as the intersection of three circles, as illustrated in Fig. 4.12b. The ranging is based on the propagation time of the radio wave. This technique requires the synchronization of the transmitter and receiver. The Cramer-Rao lower bound (CRLB) of a ToA estimate is given by [3]

$$\sigma_\tau^2 \geq \frac{1}{8\pi^2 B^2 \gamma}, \tag{4.29}$$

where σ_τ^2 is the variance of the ToA estimate, B is the bandwidth of the received signal, and γ is the bit SNR (E_b/N_0). Since the impact of the bandwidth is quadratic, but the impact of the SNR is linear, spread-spectrum and UWB are good choices for accurate ranging. The GPS is based on spread-spectrum technology for timing accuracy.

ToA positioning is a low-complexity, accurate measurement method based on energy detection. However, it requires a precise time synchronization among the nodes. ToA estimation is made on the basis of the symbol rate samples that are obtained after a square-law device. ToA positioning is used in the GPS system, where the circles are replaced by spheres in space and a fourth sphere is required to solve the receiver-clock bias in the 3D case. This bias arises from the unsynchronized clocks between the receiver and the satellite. ToA positioning has been standardized within IEEE 802.15.4a.

The TDoA positioning technique determines the position of the MS by using trilateration. This is illustrated in Fig. 4.12c. It uses time differences rather than absolute time measurements. The difference between the time instants that the transmitted signal is received at the two BSs is converted to a difference between the distance from the transmitter to one of the two BSs and that from the transmitter to the other BS. The distance difference is used to define a hyperbolic curve, with the two BSs as the two foci. Two hyperbolas determine the location, and thereby two pairs (at least three) of BSs are required. The TDoA technique eliminates the requirement of time synchronization. Pulsed-UWB signals have a fine time resolution, and this makes pulsed-UWB technology a prominent choice for indoor positioning using ToA/TDoA estimation. Location is enabled for the IR-UWB mode of IEEE 802.15.4a.

Types of location systems

When multiple network BSs are involved in wireless user location, the method is known as *network-based wireless location*. The BSs measure the signals transmitted from an MS and relay them to a central site for estimation of the MS location. In this technique, the MS is not involved in the location-finding process and thus, no modification to the existing handset is required.

When the MS determines its location from signals received from some BSs or from the GPS, the technique is called *mobile-based location*. This requires the integration of a location estimation module such as a GPS receiver into the MS. The cell-ID solution is faster for deployment, but has a low accuracy. Mobile-based location mainly depends on satellite-based positioning techniques. The satellite-based solution is very accurate, but is not reliable in urban or in-building environments, as the receiver cannot see four satellites in its LOS. It is also expensive to deploy. Mobile and network-based techniques can be combined to provide a more robust estimate of the location in a single process. An overview of wireless location techniques and challenges can be found in [22].

The US GPS is a real-time, worldwide, satellite-based positioning system. GPS has become the standard location and navigation system. GPS positioning accuracy now exceeds that of most other positioning systems. It does not need calibration as microwave equipments do. GPS provides a precise and standard positioning service. The European Galileo, and the Russian GLONASS (Global Navigation Satellite System) are satellite-based positioning systems to rival with GPS. The Galileo system is still in development, but promises a higher accuracy than GPS. The principle of satellite positioning is the same as the network-based techniques, where a satellite corresponds to a BS for positioning.

The operating frequency of a location system determines the operating range and accuracy. In general, the higher the operating frequency, the better the positioning accuracy and the shorter the range. Microwave and UWB location systems in the super high frequency (SHF, 3–30 GHz) range are most commonly used over the range of 100 meters, with an accuracy of up to 1 cm.

A wideband system can achieve a higher location accuracy than a narrowband system. In the time-based approach, the absolute resolution of location, ϵ, is related to the signal bandwidth B by

$$\epsilon = \frac{c}{B}, \tag{4.30}$$

where c is the speed of light. For an UWB system, the bandwidth is usually in excess of 3 GHz, thus the location accuracy is of the order of 1 cm. The large bandwidth helps to resolve multipath and interference, but due to the high attenuation associated with the high-frequency portion of the signal, the range is typically limited to 100 meters. UWB-based location is especially attractive [21], and it can be used for ad hoc networking, smart homes and offices, inventory control, sensor networks, smart highways, and manufacturing automation.

In the indoor radio channel, it is difficult to accurately measure the DoA, carrier signal phase of arrival (PoA) and RSSI; thus, indoor positioning systems mainly use ToA based techniques. The UWB system is a means of measuring accurate ToA for indoor geolocation applications. Due to the high attenuation for high frequencies, the frequency band used for a UWB system is typically within 2–3 GHz. IEEE 802.15.4a is based on UWB technology, and can provide high-precision location. Indoor localization has given risen to a multitude of applications in warehousing, supply-chain management, health care, public safety, and military.

The DSSS wideband signal is also used in ranging systems. A transmitter transmits a signal spread by a known PN sequence, and the receiver cross-correlates the received signal with a locally generated PN sequence using a sliding correlator. The distance is determined from the arrival time of the first correlation peak.

Problems

4.1 Given a total bandwidth for cellular operator as 12.5 MHz. If each voice channel needs a bandwidth of 30 kHz, what is the total number of simplex channels per cluster and per cell in a 7-cell cluster?

4.2 Derive Equation (4.5). Evaluate the C/I ratio for $N = 7$ and $\gamma = 3$.

4.3 Consider a cellular system with hexagonal cells of radius $R = 2$ km. If the minimum distance between cell centers that use the same frequency is $D = 10$ km, find the reuse factor N and the number of cells per cluster. For a total of 1000 channels, determine the number of channels assigned to each cell.

4.4 If a user makes 10 calls per day with an average call duration of 6 minutes, what is the traffic due to this caller?

4.5 If there are 400 seizures (connected calls) and 20 blocked calls during the busiest hours, what is the GoS?

4.6 Consider a CDMA system with perfect power control within a cell. Assuming a target SIR of 10 dB, a processing gain of 100, the nonorthogonal spreading code with $\xi = 2$ and equal average power from inside and outside of the cell, find the user capacity of the system.

4.7 Determine the propagation delay for the case of a channel data rate of 19.2 kbits/s and packet length of 256 bits. If a LOS radio link exists for any user which is at most 10 km away from the BS, what is the best choice for the number of bits per packet if slotted ALOHA is used?

4.8 In a pure ALOHA system, the channel bit rate is 2400 bits/s. Suppose each user transmits a 60-bit message every minute on the average.
(a) Determine the maximum number of terminals that can use the channel.
(b) What is the corresponding result in the case of slotted ALOHA?

4.9 For a pure ALOHA system operating with a throughput of $T = 0.15$ and packets generated with a Poisson's distributed arrival rate, determine (a) the value of the load R and (b) the average number of attempted transmissions for sending a packet.

4.10 Consider a pure ALOHA system with a transmission rate of 1 Mbit/s. Compute the load R and throughput T for a system with 200 bit packets and an average transmission rate of $\lambda_p = 500$ packets per second. What is the effective data rate?

4.11 The maximum calls per hour in a cell is 5000 and the average call holding time is 140 seconds.
(a) Find the offered load for a GoS of 2%.
(b) How many service channels are required for handling the load?

4.12 How many users can be supported by 50 channels at a GoS of 2%? Assume the average call holding time is 120 seconds and the average busy hour call per user is 1 call per hour.

4.13 A trunk accumulated 0.5 erlang of usage while 6 calls were carried in an hour without overflow. What is the average holding time per call?

4.14 A wireless LAN operates at a data rate of 10 Mbits/s, and the packets are of length 500 bits. The maximum propagation $\tau = 0.5$ μs. Calculate the normalized throughput using (a) an unslotted nonpersistent, (b) a slotted persistent, (c) a unslotted 1-persistent, and (d) a slotted 1-persistent CSMA protocol. Remark on the results.

4.15 Calculate the capacity and spectral efficiency of a single sector DS-CDMA system if the total bandwidth is 10 MHz, the bandwidth efficiency is 0.9, the frequency reuse efficiency is 0.5, the capacity degradation factor is 0.8, the voice activity factor is 0.38, the data rate is 9.6 kbits/s, and $E_b/N_0 = 8.9$ dB.

4.16 The GSM system transmits at 270.8 kbits/s to support 8 users per frame. If each user occupies one time slot per frame, what is the raw data rate for each user? In each time slot, guard and other overheads consume a rate of 10.1 kbits/s. What is the user traffic efficiency?

4.17 A cell in a blocked call delayed cellular system has 8 channels. Assume that the probability of a delayed call is 5%, each user has a load of 0.10 erlang, and $\lambda = 0.75$ call per busy hour. Determine the maximum number of users that the cell can support. Determine the probability that a delayed call needs to wait for more than 20 seconds.

4.18 In a pure ALOHA system, the average packet arrival rate is assumed to be 10^3 packets/s. The transmission rate is 10 Mbits/s, and each packet has 1500 bits. What is the normalized throughput of the system? At what packet length is the throughput maximized?

4.19 What is the ratio of the cluster sizes needed for two systems with 9 dB and 15 dB C/I requirements?

References

[1] N. Abramson, The ALOHA system – another alternative for computer communications. In *Proc. Amer. Federation Inf. Proc. Soc. Fall Joint Comput. Conf.*, Nov 1970, 281–285.

[2] J. G. Andrews, A. Ghosh & R. Muhamed, *Fundamentals of WiMAX: Understanding Broadband Wireless Networking* (Upper Saddle River, NJ: Prentice Hall, 2007).

[3] W. C. Chung and D. S. Ha, An accurate ultra wideband (UWB) ranging for precision asset location. In *Proc. IEEE UWBST*, Reston, VA, Nov 2003, 389–393.

[4] C. Cicconetti, L. Lenzini & E. Mingozzi, Quality of service support in IEEE 802.16 networks. *IEEE Network*, **20**:2 (2006), 50–55.

[5] M. Conti, G. Maselli, G. Turi & S. Giordano, Cross-layering in mobile ad hoc network design. *IEEE Computer*, **37**:2 (2004), 48–51.

[6] S. Floyd, TCP and explicit congestion notification. *ACM Comp. Commun. Rev.*, **24**:5 (1994), 10–23.

[7] K. S. Gilhousen, I. M. Jacobs, R. Padovani, A. J. Viterbi, L. A. Weaver, Jr. & C. E. Wheatley, III, On the capacity of a cellular CDMA system. *IEEE Trans. Veh. Tech.*, **40**:2 (1991), 303–312.

[8] A. Goldsmith, *Wireless Communications* (Cambridge, UK: Cambridge University Press, 2005).

[9] D. J. Goodman, R. A. Valenzuela, K. T. Gayliard & B. Ramamurthi, Packet reservation multiple access for local wireless communications. *IEEE Trans. Commun.*, **37**:8 (1989), 885–890.

[10] IEEE 802.21 WG, *Draft IEEE Standard for Local and Metropolitan Area Networks: Media Independent Handover Services*, IEEE LAN/MAN Draft IEEE P802.21/D11.0, May 2008.

[11] S. Kallel, Analysis of memory and incremental redundancy ARQ schemes over a nonstationary channel. *IEEE Trans. Commun.*, **40**:9 (1992), 1474–1480.

[12] G. E. Keiser, *Local Area Networks* (New York: McGraw-Hill, 1989).

[13] L. Kleinroch & F. A. Tobagi, Packet switching in radio channels: part I – carrier sense multiple-access modes and their throughput-delay characteristics. *IEEE Trans. Commun.*, **23**:12 (1975), 1400–1416.

[14] V. K. N. Lau & Y.-K. R. Kwok, *Channel Adaptive Technologies and Cross Layer Designs for Wireless Systems with Multiple Antennas: Theory and Applications* (New Jersey: Wiley, 2006).

[15] K. K. Leung , J.-M. Ho & H. Chien, A new digital sense multiple access (DSMA) protocol for high-speed wireless networks. In *Proc. IEEE PIMRC*, Boston, MA, Sep 1998, **3**, 1360–1366.

[16] V. H. MacDonald, The cellular concept. *Bell Syst. Tech. J.*, **58**:1 (1979), 15–41.

[17] D. Middleton, Statistical-physical models of electromagnetic interference. *IEEE Trans. Electromagnetic Compatibility*, **19**:3 (1977), 106–127.

[18] A. F. Molisch, *Wireless Communications* (Chichester, UK: Wiley-IEEE, 2005).

[19] T. S. Rappaport, *Wireless Communications: Principles & Practice*, 2nd edn (Upper Saddle River, NJ: Prentice Hall, 2002).

[20] W. Rhee & J. M. Cioffi, Increase in capacity of multiuser OFDM system using dynamic subchannel allocation. In *Proc. IEEE VTC*, Tokyo, Japan, May 2000, 1085–1089.

[21] Z. Sahinoglu, S. Gezici & I. Guvenc, *Ultra-wideband Positioning Systems* (Cambridge, UK: Cambridge University Press, 2008).

[22] A. H. Sayed, A. Tarighat & N. Khajehnouri, Network-based wireless location. *IEEE Signal Process. Mag.*, **22**:4 (2005), 24–40.

[23] A. U. H. Sheikh, *Wireless Communications: Theory and Techniques* (Boston, MA: Kluwer, 2004).

[24] Z. Shen, J. G. Andrews & B. L. Evans, Adaptive resource allocation for multiuser OFDM with constraint fairness. *IEEE Trans. Wireless Commun.*, **4**:6 (2005), 2726–2737.

[25] G. L. Stuber, *Principles of Mobile Communication*, 2nd edn (Boston, MA: Kluwer, 2001).

[26] P. Viswanath, D. N. C. Tse & R. Laroia, Opportunistic beamforming using dumb antennas. *IEEE Trans. Inf. Theory*, **48**:6 (2002), 1277–1294.

[27] A. J. Viterbi, *Principles of Spread Spectrum Communication* (Reading, MA: Addison-Wesley, 1995).

[28] R. Zemek, D. Anzai, S. Hara, K. Yanagihara & K. Kitayama, RSSI-based localization without a prior knowledge of channel model parameters. *Int J. Wireless Inf. Networks*, **15**:3/4 (2008), 128–136.

[29] Y. J. Zhang & K. B. Letaief, Multiuser adaptive subcarrier-and-bit allocation with adaptive cell selection for OFDM systems. *IEEE Trans. Wireless Commun.*, **3**:4 (2004), 1566–1575.

Diversity

5.1 Diversity methods

Two channels with different frequencies, polarizations, or physical locations experience fading independently of each other. By combining two or more such channels, fading can be reduced. This is called *diversity*. Diversity ensures that the same information reaches the receiver from statistically independent channels. There are two types of diversity: microdiversity that mitigates the effect of multipath fading, and macrodiversity that mitigates the effect of shadowing.

For a fading channel, if we use two well-separated antennas, the probability of both the antennas being in a fading dip is low. Diversity is most efficient when multiple diversity channels carry independently fading copies of the same signal. This leads to a joint pdf being the product of the marginal pdfs for the channels. Correlation between the fading of the channels reduces the effectiveness of diversity, and correlation is characterized by the correlation coefficient, as discussed in Section 3.4.2. Note that for an AWGN channel, diversity does not improve performance.

Common diversity methods for dealing with small-scale fading are spatial diversity (multiple antennas with space separation), temporal diversity (time division), frequency diversity (frequency division), angular diversity (multiple antennas using different antenna patterns), and polarization diversity (multiple antennas with different polarizations). Macrodiveristy is usually implemented by combining signals received by multiple BSs, repeaters or access points, and the coordination between them is part of the networking protocols.

Spatial diversity

Equation (3.83) is valid based on an assumption of using omnidirectional antennas and uniform incident power. For an antenna array using omnidirectional antennas, such as in the MS case, the correlation between adjacent antenna elements can be calculated by (3.83). In this case, $f_2 - f_1 = 0$, and $v\tau = d$ is the distance between the two antennas.

Example 5.1: The correlation coefficient between two antenna elements is plotted in Fig. 5.1. At around $d = 0.2\lambda$, the correlation coefficient $\rho_{xy} = 0.5$. At $d = 0.38\lambda$, $\rho_{xy} = 0$. As a rule of thumb, for uniform incident directions, the spacing of antenna is usually taken as approximately 0.5λ.

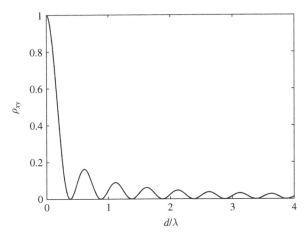

Figure 5.1

Correlation coefficient of two antennas.

This model is invalid for cellular BSs, since the interacting objects surrounding MSs make the uniform incident power assumption invalid. To obtain sufficient decorrelation, one has to increase the antenna spacing.

Temporal diversity

Signals that are received at different time instants are uncorrelated. In order to identify a time-variant channel with a band-limited Doppler spectrum, similar to the sampling theorem for identifying a signal with a band-limited spectrum, the temporal sampling rate should be at least twice the maximum Doppler frequency ν_{max}

$$f_{min} = \frac{1}{\tau_{max}} = 2\nu_{max}. \tag{5.1}$$

This criterion ensures the identifiability of the channel. For temporal diversity with sufficient decorrelation, the minimum time separation must be greater than τ_{max}. Temporal diversity can be converted into space diversity.

Temporal diversity is usually implemented by using FEC with interleaving, and ARQ. Multipath diversity is also a type of temporal diversity, collecting and combining signal multipath components to obtain a stronger signal. Transmission with adaptive modulation, which requires the knowledge of the channel, is also a kind of temporal diversity.

Frequency diversity

In this case, the same signal is transmitted at multiple frequencies, which are separated by at least the coherent bandwidth of the channel. This can again be analyzed by using (3.83), where the numerator is unity since $vt = 0$. For a correlation coefficient smaller than 0.5, $\Delta f \geq \frac{1}{2\pi\sigma_t}$. Frequency diversity is rarely implemented this way for the sake of frequency efficiency. Instead, the information is spread onto a wide frequency band to combat fading, and this is embodied in TDMA, frequency hopping, CDMA, and OFDM.

Angular diversity

Angular diversity, also known as *pattern diversity*, uses multiple antennas at the same location, each having a different pattern, so that multipath components from different DoAs are attenuated differently. These antennas should be placed by minimizing mutual coupling, which changes the beam patterns of all the antennas. Angular diversity is usually combined with spatial diversity. Smart antennas with steerable or fixed multiple narrow beams are now used in wireless systems.

Mutual coupling changes the individual antenna patterns, but it is insignificant unless the antennas are located very close to one another. Mutual coupling may result in a higher spatial correlation between the antenna signals by reradiation of the received power, leading to a reduction in the capacity.

Polarization diversity

Horizontally- and vertically-polarized multipath components have different pro-pagation over a wireless channel. Fading of signals with different polarizations is statistically independent. A scattering environment tends to depolarize a signal. Thus, receiving the signal using two different polarized antennas provides diversity. Two cross-polarized antennas with no spacing between them can provide diversity. Cross-polarized systems are of interest since they are able to double the antenna numbers using half the spacing for co-polarized antennas. Polarization diversity can achieve a diversity gain as high as that of space diversity alone in reasonable scattering areas [44], and thus it is being deployed in more and more BSs now. The sum of the multipaths is determined by their relative phases and their polarizations. This may lead to prolonged deep fades, and to avoid this problem, a fast polarization-hopping diversity scheme was proposed in [53].

Macrodiversity

Macrdiversity, or *large-scale space diversity*, also achieves high capacity by receiving and/or transmitting the same signal through multiple BSs. It is an effective method for combating shadowing. Soft handoff in CDMA systems is a well-known implementation of macrodiversity. At any instant, the BS with the best quality is chosen to serve the MS, and the local-mean power criterion is usually used since the branch selection cannot be so fast compared to the rapidly varying instantaneous signal power. In TDMA systems, hard handoff has to be used due to the nonuniversal frequency reuse; however, the use of dynamic channel assignment techniques can help TDMA benefit from the macrodiversity by implementing soft handoff.

Cooperative diversity

Cooperative diversity is an effective and promising method to combat fading in wireless channels in a mobile ad hoc network or a cellular network. Users or nodes in a wireless network share their resources and transmit cooperatively, and these users or nodes collectively function like an antenna array, thus providing diversity. For example, combining the signals transmitted from the direct and relay links provides a kind of diversity.

5.2 Combining multiple signals

Diversity is an effective method for combating fading by improving SNR. In cellular systems, diversity can also be used to reduce CCI. Diversity has a considerable impact on other aspects of wireless transmission, such as reducing the impact of random FM noise, which is produced as a result of channel fading for angle modulation.

As the receiver has multiple independent fading copies of the same signal, it has to combine these signals to improve the quality of the detected signal. Diversity can be exploited in two ways: *selection diversity* that selects and processes the best signal copy and discards all other copies, and *combining diversity* that combines all copies of the signal and processes the combined signal. Combining diversity produces a better performance, but has a much more complex receiver than selection diversity has. Combining diversity is processed at the baseband. All the diversity techniques can also be implemented in digital systems.

For multiple antennas, the gain is due to diversity gain and beamforming gain. Diversity gain is the gain due to the fact that the received copies at multiple antennas have little probability of being in a fading dip simultaneously; this corresponds to selection diversity. Beamforming gain corresponds to combining diversity which averages over noises at different antennas. For selection diversity, only one RF chain is required, while for combining diversity multiple RF chains are required.

Combining diversity exploits the information of all received signal copies. Combining can be based on maximum ratio combining (MRC) or equal gain combining (EGC). For highly time-varying channels, the phase estimation is difficult. Unlike MRC and EGC, square-law combining achieves diversity without the need for phase estimation. Square-law combining of orthogonally modulated signals, such as FSK or DS-CDMA signals that are widely used for noncoherent demodulation, is a method for exploiting diversity.

Diversity and channel coding can both improve the average probability of a detection error as well as the outage probability. Assuming maximum-likelihood (ML) detection and MRC, at high SNR, the error probability for a fading channel can be expressed by

$$P_e \propto \frac{\overline{\gamma}^{-G_d}}{G_c}, \tag{5.2}$$

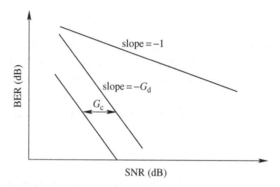

Figure 5.2

Influence of diversity and coding gains on BER.

where G_c is the coding gain, $\overline{\gamma}$ is the symbol SNR, and G_d is the diversity order. G_c and G_d influence the bit error probability (BEP), also called *bit error rate (BER)*,[1] in different ways, as shown in Fig. 5.2 on a log–log scale. In the figure, the line with slope -1 corresponds to $G_d = 1$, that is, there is no diversity gain. The coding gain G_c shifts the BER curve to the left.

From (5.2), the diversity gain or order can accordingly be defined as

$$G_d = - \lim_{\overline{\gamma} \to \infty} \frac{\log_2 P_e(\overline{\gamma})}{\log_2 \overline{\gamma}}. \tag{5.3}$$

5.2.1 Selection diversity

Selection diversity is based on the largest instantaneous power, also known as *RSSI*.

Outage probability

The pdf of the selected signal is the product of the pdfs of each diversity branch

$$p_{\gamma_{max}}(\gamma) = p\left(\max\left[\gamma_1, \gamma_2, \cdots, \gamma_{N_r}\right] < \gamma\right) = \prod_{i=1}^{N_r} p(\gamma_i < \gamma), \tag{5.4}$$

where γ_i is the instantaneous SNR on the ith diversity branch, and N_r is the number of diversity branches. For Rayleigh fading, the average SNR on the ith branch is $\overline{\gamma}_i = E[\gamma_i]$, the SNR distribution is exponential $p(\gamma_i) = \frac{1}{\overline{\gamma}_i} e^{-\gamma_i/\overline{\gamma}_i}$, and the outage probability is given by

$$P_{out}(\gamma_0) = \int_0^{\gamma_0} \prod_{i=1}^{N_r} p(\gamma_i) \, d\gamma = \prod_{i=1}^{N_r} \left(1 - e^{-\gamma_0/\overline{\gamma}_i}\right). \tag{5.5}$$

---

[1] BEP and BER are equivalent words in this book.

If the average SNR is the same for all branches $\overline{\gamma}_i = \overline{\gamma}$, the outage probability of the selection combiner is given by

$$P_{\text{out}}(\gamma_0) = \Pr(\gamma_{\max} < \gamma_0) = \left(1 - e^{-\gamma_0/\overline{\gamma}}\right)^{N_r}. \tag{5.6}$$

Differentiating $P_{\text{out}}(\gamma_0)$ with respect to γ_0 leads to the distribution for $\gamma_{\max}$

$$p_{\gamma_{\max}}(\gamma) = \frac{N_r}{\overline{\gamma}}\left(1 - e^{-\frac{\gamma}{\overline{\gamma}}}\right)^{N_r - 1} e^{-\frac{\gamma}{\overline{\gamma}}}. \tag{5.7}$$

From this, the average SNR is derived as

$$\overline{\gamma_{\max}} = \int_0^\infty \gamma p_{\gamma_{\max}}(\gamma)d\gamma = \overline{\gamma}\sum_{i=1}^{N_r}\frac{1}{i}. \tag{5.8}$$

Thus, the average SNR gain and array gain increase with N_r, the largest gain being achieved when switching from no diversity to two-branch diversity. The array gain diminishes as N_r increases.

Example 5.2: The outage probability for selection diversity, given by (5.6), is plotted in Fig. 5.3.

The BER with slow fading can be derived by averaging the BER on the AWGN channel over the pdf of the selected SNR, $\gamma_{\max}$

$$P_b = \int_0^\infty P_b(\gamma)p_{\gamma_{\max}}(\gamma)d\gamma. \tag{5.9}$$

Typically, when the average received branch symbol SNR $\overline{\gamma} \gg 1$, P_b is proportional to $\frac{1}{\overline{\gamma}^{N_r}}$.

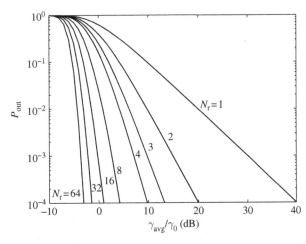

Outage probability of selection diversity in Rayleigh fading channels.

For selection diversity in Rayleigh fading channels, exact closed-form expressions for the average crossing rate of the combined signal have been given in [54] for the inphase ZCR, inphase rate of maxima, phase ZCR, and the instantaneous frequency ZCR of the output of the selection combiner.

Selection criteria

The RSSI-based selection diversity is only valid when the BER is determined by noise. When CCI is high, the high RSSI may be due to strong interference and thus the RSSI criterion is not suitable. The BER-based selection diversity is more suitable for all cases.

The BER-based selection diversity uses a training sequence, and demodulates the signal at each receive antenna and selects the antenna with the best BER. To evaluate all the N_r diversity branches, one needs N_r RF chains or needs to repeat the training sequence N_r times. In the latter case, one demodulator is not suitable for a fast changing channel. Due to the limited length of the training sequence, the calculated BER is not accurate, and a longer training sequence can improve the BER accuracy, but leads to a loss in the spectral efficiency.

Influence of correlation branches

When there is some degree of correlation between the received signals, the effectiveness of diversity will be reduced. In this case, the outage probability $P_{out}(\gamma_0)$ can be derived using the joint pdf of the varying γ_i of the N_r different branches. For two-branch diversity, the correlated multiplicative fading processes on the two branches are assumed to be jointly Gaussian, with complex cross-covariance ρ. The normalized covariance between branches is closely approximated by $|\rho^2|$.

For a threshold level γ_0, the outage probability P_{out} is derived as [34]

$$P_{out} = 1 - e^{-\frac{\gamma_0}{\overline{\gamma}_1}} Q(b, |\rho|a) - e^{-\frac{\gamma_0}{\overline{\gamma}_2}} Q(a, |\rho|b)$$
$$+ e^{-\frac{\gamma_0}{1-|\rho|^2}\left(\frac{1}{\overline{\gamma}_1}+\frac{1}{\overline{\gamma}_2}\right)} I_0\left(\frac{2|\rho|\gamma_0}{(1-|\rho|^2)\sqrt{\overline{\gamma}_1\overline{\gamma}_2}}\right), \tag{5.10}$$

where

$$a = \sqrt{\frac{2\gamma_0}{\overline{\gamma}_1(1-|\rho|^2)}}, \quad b = \sqrt{\frac{2\gamma_0}{\overline{\gamma}_2(1-|\rho|^2)}} \tag{5.11}$$

and

$$Q(x,y) = \int_y^\infty e^{-\frac{x^2+z^2}{2}} I_0(xz)z\,dz = 1 - \int_0^b e^{-\frac{x^2+z^2}{2}} I_0(xz)z\,dz. \tag{5.12}$$

Under normal operating conditions, $\overline{\gamma}_1 = \overline{\gamma}_2$. In this case, as $|\rho|$ increases, the diversity gain decreases [34]. Even at $\rho = 0.95$, a diversity gain of 4.2 dB is realized at an outage probability of 1%. As the imbalance between $\overline{\gamma}_1$ and $\overline{\gamma}_2$ increases, the diversity gain degrades as compared to the case of $\overline{\gamma}_1 = \overline{\gamma}_2$ for a given ρ.

When $\gamma_0 \ll \overline{\gamma}_1, \overline{\gamma}_2$, the outage probability can be approximated by [37]

$$P_{\text{out}} \approx \frac{\gamma_0^2}{\overline{\gamma}_1 \overline{\gamma}_2 \left(1 - |\rho|^2\right)}.$$ (5.13)

5.2.2 Maximum ratio combining

MRC is an optimum combination strategy for a slow and flat-fading channel with AWGN as the only disturbance. Each channel is a time-invariant filter with the impulse response

$$h_n(\tau) = \alpha_n \delta(\tau),$$ (5.14)

where α_n is the instantaneous attenuation of the nth diversity branch. By selecting the antenna weight $w_{\text{MRC}} = \alpha_n^*$, the signals are phase-corrected and weighted by the amplitude, and the combined output SNR is the sum of the branch SNRs

$$\gamma_{\text{MRC}} = \sum_{n=1}^{N_r} \gamma_n.$$ (5.15)

Thus, the received SNR continuously grows as the number of antennas is increased. The average combined SNR is given by

$$\overline{\gamma}_{\text{MRC}} = N_r \overline{\gamma},$$ (5.16)

where $\overline{\gamma}$ is the average SNR on each branch. Therefore, the SNR grows linearly with the diversity order.

Outage probability

Assuming each branch to have equal average branch SNR $\overline{\gamma}$ and to be uncorrelated, the distribution of γ is a χ^2 distribution with $2N_r$ degrees of freedom, mean $N_r\overline{\gamma}$, and variance $2N_r\overline{\gamma}$, given by [14, 20, 34, 38, 41]

$$p_\gamma(\gamma) = \frac{\gamma^{N_r-1} e^{-\gamma/\overline{\gamma}}}{\overline{\gamma}^{N_r} (N_r - 1)!}, \quad \gamma \geq 0.$$ (5.17)

The corresponding outage probability for a given threshold γ_0 is given by

$$P_{\text{out}} = \Pr\left(\gamma < \gamma_0\right) = \int_0^{\gamma_0} p_\gamma(\gamma) d\gamma = 1 - e^{-\gamma_0/\overline{\gamma}} \sum_{k=1}^{N_r} \frac{(\gamma_0/\overline{\gamma})^{k-1}}{(k-1)!}.$$ (5.18)

Example 5.3: The outage probability of MRC in Rayleigh fading is plotted in Fig. 5.4. Comparison with Fig. 5.3 indicates that the performance of MRC is much better than that of selection diversity. The diversity order is embodied from the slope of the BER curve or outage probability curve versus average SNR on a log-log scale. A comparison of Figs. 5.4 and 5.3 shows that MRC is 2 dB more effective than selection diversity.

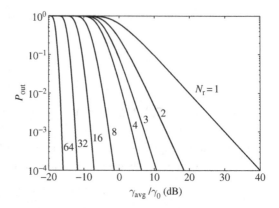

Figure 5.4 Outage probability of MRC in Rayleigh fading channels.

Ergodic error probability

The ergodic error probability can be derived in a fashion similar to that in (5.9). Since MRC is a coherent detection technique, it is only suitable for signals that can be coherently detected. The ergodic error probability $\overline{P}_s$ for Rayleigh fading is given by (7.167), when the transmit power is uniformly distributed onto all diversity paths. Equation (7.167) is reproduced below [20]

$$\overline{P}_s = a \int_0^b \left(\frac{N_r \sin^2 \theta}{N_r \sin^2 \theta + c\gamma_s} \right)^{N_r} d\theta, \tag{5.19}$$

where the values of the parameters a, b, c depend on the modulation method. For BPSK signals, the BER for equally distributed branches is given by

$$P_s = \frac{1}{\pi} \int_0^{\frac{\pi}{2}} \left(\frac{N_r \sin^2 \theta}{N_r \sin^2 \theta + \gamma_s} \right)^{N_r} d\theta. \tag{5.20}$$

Example 5.4: The ergodic BER of MRC for Rayleigh fading, as given by (5.20), is plotted in Fig. 5.5. It is seen that as $N_r \rightarrow \infty$, the performance approaches that of an AWGN channel.

Ergodic capacity

The ergodic capacity of a diversity system can also be obtained in case of MRC. Assuming that the signal is transmitted over N_r independent channels with coefficients h_m and that the signal power is equally distributed among these channels, the instantaneous capacity after MRC is

$$C(\gamma) = \log_2 \left(1 + \sum_{m=1}^{N_r} |h_m|^2 \frac{E_s}{N_r N_0} \right) = \log_2(1 + \gamma). \tag{5.21}$$

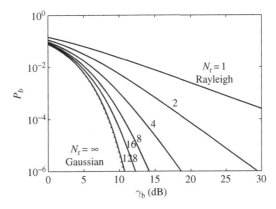

Figure 5.5

BER of BPSK over Rayleigh fading channels with diversity degree N_r using MRC.

The pdf $p_\gamma(\gamma)$ is given by (5.17) with $\overline{\gamma} = \frac{E_s/N_0}{N_r}$ for nondissipative channels with $\sigma_H^2 = 1$. The ergodic capacity is obtained by averaging $C(\gamma)$ with respect to $p_\gamma(\gamma)$

$$\overline{C} = \int_0^\infty \log_2(1+\gamma) \frac{N_r^{N_r} \gamma^{N_r-1}}{(N_r-1)! \, (E_s/N_0)^{N_r}} e^{-\frac{\gamma N_r}{E_s/N_0}} d\gamma. \tag{5.22}$$

This result can be evaluated numerically.

Example 5.5: The ergodic capacity, given by (5.22), is plotted in Fig. 5.6. It is seen that $\overline{C}$ for different values of N_r runs between the two curves for the AWGN and the Rayleigh channels, which correspond to $N_r = \infty$ and $N_r = 1$, respectively. The result for the Rayleigh channel can also be calculated by setting $N_r = 1$ or by using the closed-form solution given by (14.76), which is reproduced below [20]

$$\overline{C} = \log_2 e \cdot \exp\left(\frac{1}{\sigma_H^2 E_s/N_0}\right) \cdot \operatorname{expint}\left(\frac{1}{\sigma_H^2 E_s/N_0}\right), \tag{5.23}$$

where

$$\operatorname{expint}(x) = \int_x^\infty \frac{e^{-t}}{t} dt. \tag{5.24}$$

The result for the AWGN channel is calculated by (14.54), which is also reproduced below

$$C_{\text{AWGN}} = \log_2\left(1 + \frac{E_s}{N_0}\right). \tag{5.25}$$

The largest gains are obtained for the transition from $N_r = 1$ to $N_r = 2$. Calculation of numerical integration must be very carefully performed, or numerical instability may occur.

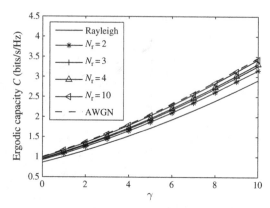

Figure 5.6 The ergodic capacity for the N_rth-order MRC diversity system over Rayleigh fading channels.

Outage capacity

The outage probability P_{out} and the outage capacity C_{out} can be obtained from each other. The outage probability P_{out} can be defined by

$$P_{out} = \Pr\left(C(\gamma) < C_{out}\right) = \int_0^{\gamma_{min}} p_\gamma(\gamma)d\gamma, \qquad (5.26)$$

where the minimum SNR γ_{min} can be calculated from C_{out}, and $p_\gamma(\gamma)$ is the pdf of γ. For the AWGN channel, $\log_2(1 + \gamma_{min}) = C_{out}$, thus $\gamma_{min} = 2^{C_{out}} - 1$.

By inserting $p_\gamma(\gamma)$ into (5.26), we have

$$P_{out} = \frac{1}{(N_r - 1)!} \int_0^{\frac{2^{C_{out}} - 1}{E_s/(N_r N_0)}} \gamma^{N_r - 1} e^{-\gamma} d\gamma. \qquad (5.27)$$

The relations of P_{out}, C_{out}, E_s/N_0 and N_r can be numerically evaluated from (5.27).

Example 5.6: Given $E_s/N_0 = 10$ dB, the relation of P_{out} versus C_{out} is plotted in Fig. 5.7. As E_s/N_0 increases, the curves shift to the right. Given P_{out}, the relation of C_{out} versus SNR can also be obtained.

For MRC scheme with independent but unbalanced branches in Nakagami-m fading, the pdf of the SNR at the combiner output, P_{out} and the ergodic error probability have been obtained in [1]. In [16], closed-form expressions for the LCR and the ACF are derived for MRC with independent but unbalanced branches, and Rayleigh fading.

Influence of correlation branches

Assuming that two branches are correlated with a correlation coefficient ρ and each branch has an equal average branch SNR $\overline{\gamma}$, the outage probability with MRC is given by [17]

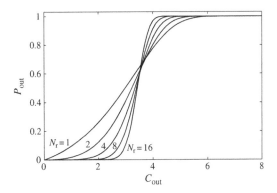

Figure 5.7 Outage probability as a function of outage capacity of flat-fading channels with Gaussian input, for different diversity degree N_r.

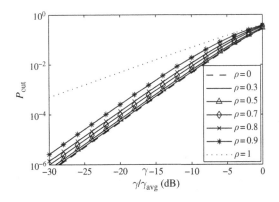

Figure 5.8 The outage probability for correlated branches with MRC.

$$P_{\text{out}} = \Pr(\gamma_1, \gamma_2 < \gamma_0) = 1 - \frac{1}{2|\rho|}\left[(1 + |\rho|)e^{-\frac{\gamma_0}{\bar{\gamma}(1+|\rho|)}} - (1 - |\rho|)e^{-\frac{\gamma_0}{\bar{\gamma}(1-|\rho|)}}\right], \quad (5.28)$$

where γ_1, γ_2 are the instantaneous SNRs on the two branches.

Example 5.7: The outage probability (5.28) is plotted in Fig. 5.8. It is seen that the diversity gain is still very high when $\rho = 0.9$. Usually it is assumed that almost full diversity gain is obtained when $\rho = 0.7$. This is the reason $\rho = 0.7$ is specified for defining the coherent time, bandwidth, and angle.

In the case when all the channels have the same average power $\sigma_H = 1$ and the correlation between any pair of diversity branches is ρ, the correlation matrix is given by

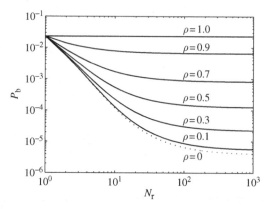

Figure 5.9 BER of BPSK over Rayleigh fading channels with diversity degree N_r using MRC, correlation coefficient ρ at $\frac{E_s}{N_0} = 10$ dB.

$$\Phi_{HH} = \sigma_H \begin{bmatrix} 1 & \rho & \cdots & \rho \\ \rho & 1 & \cdots & \rho \\ \vdots & \vdots & \ddots & \vdots \\ \rho & \rho & \cdots & 1 \end{bmatrix}. \tag{5.29}$$

The correlation matrix has $(N_r - 1)$-fold eigenvalues $\lambda_1 = \lambda_2 = \cdots = \lambda_{N_r-1} = \sigma_H^2(1-\rho)$, and $\lambda_{N_r} = \sigma_H^2[1 + \rho(N_r - 1)]$; these correspond to the average powers of the various channels [20, 38]. For BPSK signals in flat Rayleigh fading channels, from (5.20), with $\gamma_l = \lambda_l \frac{E_s}{N_0}$ and $\sigma_H = 1$, the ergodic error probability is given as

$$P_s = \frac{1}{\pi} \int_0^{\frac{\pi}{2}} \left(\frac{N_r \sin^2 \theta}{N_r \sin^2 \theta + (1-\rho)\frac{E_s}{N_0}} \right)^{N_r-1} \left(\frac{N_r \sin^2 \theta}{N_r \sin^2 \theta + [1 + \rho(N_r - 1)]\frac{E_s}{N_0}} \right) d\theta. \tag{5.30}$$

Example 5.8: Assuming $\frac{E_s}{N_0} = 10$ dB, the ergodic error probability for correlated branches with MRC, given by (5.30), is plotted in Fig. 5.9. It is seen that correlation reduces the effective diversity degree for a given diversity order N_r. For $\rho = 1$, there is no diversity gain at all.

The BER performance of MRC in Nakagami fading is analyzed for coherent and non-coherent FSK in [7], and the results are extended to include coherent PSK and DPSK. The effect of correlation is also considered for the dual diversity case [7]. In [29], performance metrics of MRC in correlated Nakagami fading, such as P_{out}, the average symbol error probability (SEP), also called *symbol error rate (SER)*[2], for coherent multichannel

---

[2] SEP and SER are equivalent words in this book.

reception, and the diversity gain, are calculated with arbitrary fading parameters and arbitrary average powers.

H-S/MRC

The tradeoff between the selection diversity and MRC is the hybrid selection scheme, called the *hybrid selection/maximum ratio combining (H-S/MRC)*, which selects the best L out of N_r signals and processes them. The main feature of the H-S/MRC scheme is the reduction of the number of RF chains.

The performance of the H-S/MRC diversity system over Rayleigh and Nakagami-*m* channels is analyzed in [12, 48]. Exact expressions for the SEP for coherent detection of MPSK and MQAM with H-S/MRC in multipath-fading channels are given in [48]. It has been shown that H-S/MRC, even with $L \ll N_r$, can achieve performance close to that of N_r-branch MRC [48]. For a H-S/MRC diversity system that has independent Nakagami fading on the diversity branches with unequal fading parameters and unequal SNRs, closed-form expressions for P_{out}, the channel capacity, and the average SEP for a general class of M-ary modulation schemes (including MPSK, MQAM, BFSK, and MSK) with coherent detection have been derived in [12].

5.2.3 Equal gain combining

MRC needs the knowledge of the channel on each branch. EGC is a simpler technique that cophases signals on all the branches and combines them with equal weights. EGC is useful for constant amplitude signals such as MPSK that has equal energy symbols.

After cophasing and combining, the envelope of the composite signal is

$$\alpha_{\text{EGC}} = \sum_{k=1}^{N_r} \alpha_k \tag{5.31}$$

and the SNR of the EGC output is given by

$$\gamma_{\text{EGC}} = \frac{1}{N_r} \left(\sum_{n=1}^{N_r} \sqrt{\gamma_n} \right)^2. \tag{5.32}$$

γ_{EGC} generally does not have a closed-form pdf nor a closed-form cdf.

With Rayleigh fading, if the branches experience uncorrelated fading, we have [41]

$$\gamma_{\text{EGC}} = \overline{\gamma} \left[1 + (N_r - 1) \frac{\pi}{4} \right]. \tag{5.33}$$

Thus, the performance of EGC is always worse than that of MRC, but is quite close to that of MRC, with a power penalty of 1.05 dB.

When applied to unequal energy constellations, EGC requires to compute the optimal bias, which depends on the measurement on each branch. Calculation of the bias for MPAM and MQAM constellations has been proposed in [26]. Similar to H-S/MRC, H-S/EGC is obtained as the hybrid of selection combing and EGC [24]. The

H-S/EGC receiver proposed in [24] allows diversity for nonconstant modulus modulation formats.

Outage probability

The outage probability for EGC is similar to, but slightly worse than, that for MRC. The distribution of γ at low values of γ is given by [10, 34]

$$p(\gamma) = \frac{2^{N_r-1}}{(2N_r-1)!} \left(\frac{1}{\prod_{k=1}^{N_r} \frac{Z_k}{Z_T}} \right) \left(\frac{\gamma^{N_r-1}}{\prod_{k=1}^{N_r} \overline{\gamma}_k} \right), \tag{5.34}$$

where Z_k is the noise power on branch k, and Z_T is the total noise power.

To maximize the performance of EGC, one can minimize $p(\gamma)$ for small γ, (5.34), with respect to γ. This yields the condition of $\prod_{k=1}^{N_r} \frac{Z_k}{Z_T} = \left(\frac{1}{N_r} \right)^{N_r}$, that is, all the branches have equal noise power level. Under this assumption, $p(\gamma)$ becomes [34]

$$p(\gamma) = \frac{2^{N_r-1} N_r^{N_r}}{(2N_r-1)!} \left(\frac{\gamma^{N_r-1}}{\prod_{k=1}^{N_r} \overline{\gamma}_k} \right) \tag{5.35}$$

and the outage probability is given by [34]

$$P_{\text{out}} = \Pr(\gamma < \gamma_0) = \int_0^{\gamma_0} p(\gamma) d\gamma = \frac{(2N_r)^{N_r}}{(2N_r)!} \left(\frac{\gamma_0^{N_r}}{\prod_{k=1}^{N_r} \overline{\gamma}_k} \right). \tag{5.36}$$

For low outage, if $N_r \gg 1$, EGC requires 1.34 dB more power than MRC does [34].

Influence of correlated branches

In the case of two branches that are correlated with a complex correlation coefficient ρ, the outage probability for EGC is approximated by [22, 37]

$$P_{\text{out}} = \Pr(\gamma_1, \gamma_2 \le \gamma_0) = 1 + \frac{\left(1 - \sqrt{|\rho|}\right) e^{-\frac{2a\gamma_0}{\overline{\gamma}}} - \left(1 + \sqrt{|\rho|}\right) e^{-\frac{2b\gamma_0}{\overline{\gamma}}}}{2\sqrt{|\rho|}}, \tag{5.37}$$

where γ_1, γ_2 are the instantaneous SNRs on the two branches, γ_0 is the specified outage SNR, $\overline{\gamma}$ is the average SNR on each of the branches,

$$a = \frac{g_2}{2\left(1 - \sqrt{|\rho|}\right)}, \quad b = \frac{g_2}{2\left(1 + \sqrt{|\rho|}\right)} \tag{5.38}$$

and $g_2 = 1.16$ for $|\rho| < 1$ and $g_2 = 1$ for $|\rho| = 1$.

Expressions have been derived in [4] for the LCR and the average duration of fades in the case of two-branch predetection selection diversity, EGC and MRC with correlated Rayleigh fading signals.

Dual-branch diversity combining over log-normal channels has been analyzed in [8, 38] for MRC, selection diversity and switch diversity, allowing for the possibility of average power unbalance and correlation between the two branches.

5.2.4 Switch diversity

Selection diversity requires a dedicated receiver on each branch to continuously monitor the branch SNR. Switch diversity, or threshold combining, is a simple selection diversity method. Selection diversity uses only one receiver to scan each of the branches in a sequential order, and outputs the first signal whose SNR is above a given threshold γ_T. Whenever the signal on the branch in use is below γ_T, switching continues until an acceptable signal level is found. This is also known as *switching diversity with feedback* [17]. To reduce the noise arising from rapid switching, the switch-and-stay strategy is usually used: the combiner switches to another branch whenever the signal on the branch in use falls below a selected threshold, regardless of the signal level on the other branches.

In FM analog mobile radio systems, switching between antennas creates noise in the signal. Thus, the switching rate must be limited, leading to a limited protection against fading. In digital radio systems, switching yields loss of phase coherence in the signal. This is not a problem for FSK and DPSK modulation: Such systems are not affected by the switching rate, achieving a higher degree of protection against fading. The BER performance of switched diversity with feedback for DPSK mobile radio has been analyzed in [49].

Switch diversity is a degenerate form of selection diversity. Switch diversity provides most of the gain in the region immediately above the threshold level. The optimum threshold should be set slightly above the lowest acceptable SNR. The outage probability for two branches with switch diversity has been given in [37]. The performance of switch diversity is between that of no diversity and selection diversity. The optimal selection of the threshold is $\gamma_T = \gamma_0$, and this leads to the same outage probability as the selection diversity [14, 38].

5.2.5 Optimum combining

The preceding combining techniques are based on maximizing the signal power at the combiner output. MRC is optimum only when there is no interference. In case of CCI, optimum combining can be derived by maximizing the output signal-to-interference-plus-noise ratio (SINR) of the combiner.

The received signal consists of the desired signal, noise, and N_I interfering signals

$$x = x_d + n + \sum_{i=1}^{N_I} x_i. \tag{5.39}$$

If the transmitted average power of the desired signal $s_d(t)$ is normalized to unity, we can write

$$x = h_d s_d(t) + n + \sum_{i=1}^{N_I} x_i, \tag{5.40}$$

where h_d is the complex channel attenuation vector of the N_r diversity branches.

The weight vector that maximizes the output SINR is derived as [50, 51]

$$w_{\mathrm{opt}} = \mathbf{R}_I h_d^*, \tag{5.41}$$

where $\mathbf{R}_I$, the correlation matrix of the received noise-plus-interference contained in the multiple received signal copies, is given by

$$\mathbf{R}_I = \sigma_n^2 \mathbf{I} + \sum_{k=1}^{N_I} E\left[x_k^* x_k^T\right], \tag{5.42}$$

σ_n^2 being the noise power. This solution is known as the *Wiener solution*.

The output SINR is given by

$$\gamma = \frac{w_{\mathrm{opt}}^H h_d h_d^H w_{\mathrm{opt}}}{w_{\mathrm{opt}}^H R_I w_{\mathrm{opt}}}. \tag{5.43}$$

The pdf of the combiner output is given in [11, 37]. More combining algorithms will be introduced in Chapter 18, when we discuss beamforming.

Performance of various diversity combiners with CCI

The performance of a diversity system in reducing interference has been derived in [17, 36] for the cases of Rayleigh or Ricean distributed desired signal and Rayleigh distributed interfering signals. The pdfs and cdfs of the SINR for both cases are given in [36], when MRC is applied. Assuming equal-power interference sources, analytical expressions are derived for the pdf of the output SINR, P_{out}, and the average BER with MRC for an arbitrary number of interference sources in [36] for BPSK modulation.

For flat-fading channels, simplified closed-form expressions for the pdf of the output SINR and the outage probability (i.e., the cdf of the output SINR) with MRC in case of CCI is given in [13]. In [5], closed-form expressions for pdf and cdf of the output SINR in case of CCI are derived, when there is a channel estimation error. For perfect channel estimation, the pdf of the output SINR reduces to that given in [13].

Closed-form expressions for P_{out} when using MRC with CCI are available for Rayleigh fading [27] and for Nakagami fading [30]. An exact closed-form expression for the average BER of coherent BPSK using MRC with correlated branches in the presence of CCI and Rayleigh fading is derived in [55]. In addition, theoretical analysis has been made for hybrid selection/MRC diversity in the presence of the Rayleigh desired signal with different CCI models in [15, 23], and expressions for the pdf of the instantaneous SINR, as well as average BER and P_{out} for some special modulation schemes given.

The performance of EGC, selection combining, switching diversity schemes with CCI have also been investigated in the literature. In [2], closed-form expressions for P_{out} in Nakagami fading with CCI are derived for EGC, selection combining, and switched diversity schemes. In [2], average BEPs of MPSK using both dual-branch EGC and dual-branch selection combining are derived in a Ricean or Nakagami fading channel with Rayleigh fading CCI. In [40], the P_{out}'s of EGC and selection combining are compared in a Rayleigh fading channel with CCI. In [39], the average BEPs of EGC and selection combining for band-limited BPSK systems in Nakagami fading with CCI are derived. The average output SINR of EGC in a correlated Nakagami fading channel with CCI is obtained in closed form in [31]. Outage and average SEP performance of EGC in a Nakagami fading channel with Rayleigh fading CCI are analyzed in [32].

Performance of optimum combining with CCI

For optimum combining, a closed-form expression is given for one interferer in [50]. In [52], a closed-form expression for the upper bound on the BER with optimum combining is given for any number of antennas N_r and interferers N_I, with coherent detection of BPSK and QAM signals, and differential detection of DPSK. Bounds on the performance gain of optimum combining over MRC are also derived, and these bounds are asymptotically tight with decreasing BER.

In [45], a performance analysis of optimum combining in the presence of N_I equal power interferers and noise has been made, subject to independent flat Rayleigh fading channels, when $N_I < N_r$. An approximate expression of the pdf of the output SINR γ has been derived analytically, and then applied to obtain the closed-form cdf of the SINR (outage probability) as

$$
P_{\text{out}} = \Pr\left(\gamma \le \gamma_0\right) = \frac{\left(1 + N_r\overline{\gamma}_I\right)^{N_I-1}}{\left(-N_r\overline{\gamma}_I\right)^{N_r-1}} \left\{ \sum_{k=0}^{N_I-1} \left(\frac{-N_r\overline{\gamma}_I}{1 + N_r\overline{\gamma}_I}\right)^k b_k \right.
$$

$$
\times \left[1 - e^{-\frac{\gamma_0(1+N_r\overline{\gamma}_I)}{\overline{\gamma}_s}} \sum_{m=0}^{k} \frac{\left(\frac{\gamma_0(1+N_r\overline{\gamma}_I)}{\overline{\gamma}_s}\right)^m}{m!} \right]
$$

$$
\left. - \left(1 + N_r\overline{\gamma}_I\right) \sum_{n=0}^{N_r-N_I-1} \left(-N_r\overline{\gamma}_I\right)^n c_n \left[1 - e^{-\frac{\gamma_0}{\overline{\gamma}_s}} \sum_{m=0}^{n} \frac{\left(\frac{\gamma_0}{\overline{\gamma}_s}\right)^m}{m!} \right] \right\}, \quad (5.44)
$$

where

$$
b_k = \frac{a_k}{\binom{k}{N_r-1}}, \quad k = 0, 1, \cdots, N_I - 1, \tag{5.45}
$$

$$
c_n = \sum_{k=0}^{N_I-1} \frac{\binom{k}{n}}{\binom{k}{N_r-1}} a_k, \quad n = 0, 1, \ldots, N_r - N_I - 1 \tag{5.46}
$$

with

$$
a_k = (-1)^{N_I-1+k} \frac{\binom{k}{N_I-1}}{(N_I-1)!} \prod_{n=1, n\neq k+1}^{N_I} (N_r - n), \quad k = 0, 1, \cdots, N_I - 1. \tag{5.47}
$$

$\overline{\gamma}_s$ is the average input SNR for the desired user and $\overline{\gamma}_I$ is the average interference-to-noise ratio (INR) for any interferer. All interferers have equal power so that the average SINR is given by

$$\overline{\gamma} = \frac{\overline{\gamma}_s}{1 + N_I \overline{\gamma}_I}. \tag{5.48}$$

In [45], closed-form expressions for the BER of coherent BPSK, DPSK, and FSK modulations have been derived for $N_I < N_r$.

Optimum combining of N_I equal-power interferers ($N_I \geq N_r$) with BPSK modulation in a flat Rayleigh-fading environment is analyzed in [35]. The pdf of the output SINR of optimum combining has a Hotelling T^2 distribution, and closed-form expressions for the outage probability and BER are derived. In [36], the results obtained for optimum combining and Rayleigh fading in [35] are extended to the case when the desired signal is subject to Ricean fading, for $N_I \geq N_r$. The performance has been studied for several channel models of the desired signal: Rayleigh, Rice, and nonfading. In all cases, the interference is assumed to be subject to independent and identically distributed (i.i.d.) Rayleigh fading. Exact analysis of optimum combining is performed in [28] when either the desired user or the interferers undergo Ricean fading.

In [42], the pdf of the output SINR of the optimum combining technique has been examined. When $N_I > N_r$, the outage probability of optimum combining with N_I equal-power interferers is upper and lower bounded by the MRC performance [13] with N_I and $N_I - N_r + 1$ interferers, respectively.

In [21], a closed-form expression of the exact BER has been derived for optimum combining with BPSK and Rayleigh fading, in the presence of N_I equal power interferences. It is shown that for large SNR, the BER of a system with N_r diversity branches and N_I interferences ($N_I < N_r$) is equivalent to a system with ($N_r - N_I$) diversity branches, but no interference [21, 51].

Optimum combining gives the best error performance only with perfect channel estimation. Optimized diversity combining is introduced in [25] that results from pilot-based ML channel estimation applied to a correlated flat Rayleigh fading channel in the presence of CCI and additive noise. Optimized diversity combining can perform significantly better than optimum combining with imperfect channel estimates.

5.3 Transmit diversity

Transmit diversity is dual to receive diversity, when the knowledge of the complex channel gain is available. For a transmit diversity system with N_t transmit antennas and one receive antenna, when the receiver CSI is available and the transmit antennas are subject to an average total energy constraint E_s, the received SNR can be combined by using an analysis similar to that for receiver MRC, and the combined SNR is given by

$$\gamma_{\text{MRC}} = \sum_{i=1}^{N_t} \gamma_i. \tag{5.49}$$

At high SNR, both transmit and receive MRC achieve full diversity order. Similar analysis can be done for EGC and selection diversity.

The transmitter CSI can be obtained by feedback from the receiver. The receiver esti-mates the channel using a pilot transmitted from the transmitter and then feeds back it to the transmitter. In TDMA, the BS can use transmission from the MS to estimate the channel, and then transmit based on this information. This is because in time division the forward and reverse links are reciprocal.

5.3.1 Open-loop transmit diversity

When there is no transmitter CSI, transmit diversity gain can be obtained by using Alam-outi's space-time diversity scheme [6]. Alamouti's scheme is for two-antenna transmit diversity. The method works over two symbol periods, during which the channel is assumed constant. During the first symbol period, two different symbols s_1 and s_2, each having an energy $E_s/2$, are transmitted from antennas 1 and 2, respectively. During the second sym-bol period, symbols $-s_2^*$ and s_1^* are, respectively, transmitted from antennas 1 and 2, again each having an energy $E_s/2$. For the two-symbol period, the received symbol is given by

$$y = \mathbf{H}_A s + n, \tag{5.50}$$

where the output $y = (y_1, y_2^*)^T$, the transmitted symbols $s = (s_1, s_2)^T$, the AWGN $n = (n_1, n_2)^T$, and the channel matrix

$$\mathbf{H}_A = \begin{bmatrix} h_1 & h_2 \\ h_2^* & -h_1 \end{bmatrix}. \tag{5.51}$$

Defining a zero-forcing receiver $z = (z_1, z_2)^T = \mathbf{H}_A^H y$, we can estimate s from

$$z = \left(\left| h_1^2 \right| + \left| h_2^2 \right| \right) s + \tilde{n}, \tag{5.52}$$

where $\tilde{n} = \mathbf{H}_A^H n$. Thus, each component of z corresponds to one transmitted symbol, and a diversity order of 2 is achieved. Alamouti's scheme can be general-ized to multiple anten-nas. The Alamouti transmit diversity scheme is known as the *Alamouti code* or *transmitter beamforming*, which is dual to receiver beamforming in a smart antenna system.

The 2×1 Alamouti code achieves the same diversity order and data rate as a 1×2 receive diversity with MRC, but with a 3 dB power penalty, due to the redundant transmission at the transmitter. For receiver MRC diversity, the received SNR linearly increases with the diversity order N_r. Due to the transmit power penalty, the received SNR does not always increase. The Alamouti code has been extended to the orthogonal space-time block code (OSTBC) for $N_t > 2$. For a single receive antenna, when an OSTBC is employed, the received SNR is given by

$$\gamma_{td} = \frac{\gamma}{N_t} \sum_{i=1}^{N_t} |h_i|^2 \rightarrow \gamma E\left[|h_1|^2 \right], \tag{5.53}$$

where the law of large numbers is applied for large N_t. This result shows that the received SNR for transmit diversity approaches the average SNR as N_t increases. More details on the Alamouti code and OSTBCs are given in Chapter 19.

5.3.2 Closed-loop transmit diversity

When the channel information is available at the transmitter by reciprocity or feedback from the receiver, closed-form transmit diversity can be implemented. Transmit selection diversity is a simple, but most effective transmit diversity [49]. Transmit selection diversity selects from N_t antennas $N' < N_t$ antennas that have the best channels for transmission. This can significantly reduce the hardware cost and complexity. The use of fewer transmit antennas for transmitting the same signal also reduces the spatial interference. Like selection diversity, transmit selection diversity achieves a full diversity order of N_t at the transmitter side.

When a single transmit antenna is selected, the power penalty relative to receive MRC diversity that is observed in the case of OSTBCs does not occur. In the case of i.i.d. Rayleigh fading, the average SNR of a single transmit antenna in an $N_t \times 1$ system is given by [9]

$$\gamma_{\text{tsd}} = \overline{\gamma} \sum_{i=1}^{N_t} \frac{1}{i},\tag{5.54}$$

which is identical to (5.8). Although full diversity is obtained, the average SNR $\overline{\gamma}$ is lower than that achieved with all the transmit antennas using beamforming.

The feedback required for transmit selection diversity is also very low, since only the indices of the required antennas are fed back. For N' active transmit antennas, only $N' \log_2 N_t$ bits per channel coherent time is required to send to the transmitter.

Another kind of closed-loop transmit diversity is linear diversity precoding, where the data rate is unchanged but is used to improve the link diversity. Linear diversity precoding is a special case of linear beamforming. Linear precoding will be introduced in Chapter 19. Compared to OSTBCs, linear precoding achieves a higher SNR than the open-loop STBCs, by a factor up to N_t [9].

5.4 Multiuser diversity

Multiuser diversity [19] takes advantage of the fact that different users have channels that fade independently. By transmitting only to users with the best channels, the system capacity as well as performance can be improved. Single-user diversity uses a point-to-point link consisting of multiple independent channels whose outputs are combined, while in multiuser diversity the multiple channels are for different users and selection diversity is used to select the user with the best channel to increases the mean SNR. Multiuser diversity leads to a significant improvement in throughput when there is a large number of users.

Transmission scheduling that employs multiuser diversity based on the channel conditions of users is known as *opportunistic scheduling*. In addition to increasing the system throughput, opportunistic scheduling also improves BER performance, since transmission is only to users with the largest SNRs. In i.i.d. Rayleigh fading channels, this maximum

SNR has a gain of $\ln K$, when the number of users, K, is large [46]. Multiuser diversity has been supported by 1xEV-DO and WiMAX to improve the overall system diversity.

However, opportunistic scheduling leads to problems of fairness and delay. The user with the best channel may occupy the system resources all the time, while other users with poor channels may never get system resources. Proportional fairness scheduling [46] helps to solve the fairness and delay problems in the downlink. Proportional fair scheduling is the baseline scheduling technique for the TDMA-based downlink of 1xEV-DO for packet data transmission.

Opportunistic scheduling requires CSI at the BS. Strategies with one-bit feedback per user can capture the double-logarithmic capacity growth of full-CSI systems, i.e., $\log \log K$, for flat Rayleigh-fading channels [33]. The one-bit feedback technique is extended for sub-channel/user selection under both correlated and uncorrelated subchannel conditions. With one-bit feedback, most of the sum-rate capacity of the full-CSI system can be achieved, and it is possible to maintain proportional fairness without any loss of throughput [33]. The results have also been extended to frequency-selective channels via a simple joint user/subchannel selection strategy.

5.4.1 Pdf and cdf

For N_u users with independent flat Rayleigh fading channels, the user SNRs $\gamma_i = |h_i|^2 E_s/N_0$ are χ^2 distributed with two degrees of freedom. The pdf $p_{\gamma_{\max}}(\gamma)$ of the strongest channel $\gamma_{\max} = \max(\gamma_1, \gamma_2, \ldots, \gamma_{N_u})$, can be derived. First, the cdf is given by

$$P_{\gamma_{\max}}(\gamma) = \Pr(\gamma_{\max} < \gamma) = \Pr(\gamma_i < \gamma | i = 1, 2, \ldots, N_{N_u}) = \prod_{i=1}^{N_u} P_{\gamma_i}(\gamma), \qquad (5.55)$$

where the last equality is obtained for independent processes γ_i. The pdf can be obtained as the derivative of the cdf with respective to γ, thus

$$p_{\gamma_{\max}}(\gamma) = \frac{dP_{\gamma_{\max}}(\gamma)}{d\gamma} = \sum_{i=1}^{N_u} p_{\gamma_i}(\gamma) \prod_{j=1, j\neq i}^{N_u} P_{\gamma_i}(\gamma). \qquad (5.56)$$

For flat-fading channels, $P_{\gamma_i}(\gamma) = 1 - e^{-\frac{\gamma}{\gamma_i}}$. Thus,

$$p_{\gamma_{\max}}(\gamma) = \sum_{i=1}^{N_u} \frac{1}{\overline{\gamma}_i} e^{-\frac{\gamma}{\overline{\gamma}_i}} \prod_{j=1, j\neq i}^{N_u} \left(1 - e^{-\frac{\gamma}{\overline{\gamma}_i}}\right). \qquad (5.57)$$

For i.i.d. channels, all $\overline{\gamma}_i = \overline{\gamma}$, we have the pdf and cdf as

$$p_{\gamma_{\max}}(\gamma) = \frac{N_u}{\overline{\gamma}} e^{-\frac{\gamma}{\overline{\gamma}}} \left(1 - e^{-\frac{\gamma}{\overline{\gamma}}}\right)^{N_u-1}, \qquad (5.58)$$

$$P_{\gamma_{\max}}(\gamma) = \left(1 - e^{-\frac{\gamma}{\overline{\gamma}}}\right)^{N_u}. \qquad (5.59)$$

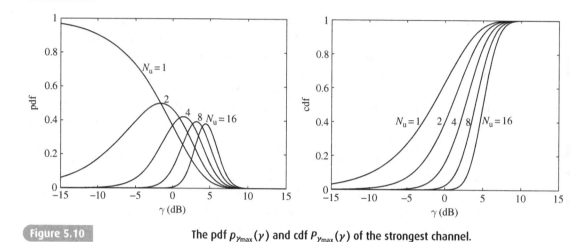

Figure 5.10
The pdf $p_{\gamma_{\max}}(\gamma)$ and cdf $P_{\gamma_{\max}}(\gamma)$ of the strongest channel.

Example 5.9: The pdf and cdf for a multiuser diversity system with $N_u = 1, 2, 4, 8,$ and 16 i.i.d. channels are plotted in Fig. 5.10. It is seen that as N_u increases, the outage probability (cdf) decreases.

The ergodic capacity $\overline{C}$ with respect to E_s/N_0 and the outage probability with respect to the sum rate can also be derived based on the pdf and cdf for the multiuser diversity case, and can be evaluated using numerical integration. $\overline{C}$ increases with N_u.

5.4.2 Multiuser diversity versus classical diversity

Like the classical diversity techniques, multiuser diversity is also used to combat the adverse fading channel by exploiting independently faded signal paths. For multiuser diversity, the independently faded signals are from different users in the network. However, the objective of classical diversity is to improve the reliability of point-point communication, while multiuser diversity aims to improve the total downlink throughput of the network.

The classical diversity techniques were designed to counteract fading. In contrast, multiuser diversity improves system performance by exploiting fading. In a fading environment, a user that has a channel strength much greater than the average level exists with a high probability; this strong channel is then fully utilized.

To exploit multiuser diversity, the BS needs to have access to channel quality measurements and the ability to schedule transmission among the users based on the instantaneous channel quality. Practical implementation of multiuser diversity may face the following problems [43, 47]:

- *Fairness and delay.* For example, a user who is close to the BS always has a better average SNR and is noise-limited; such a user benefits from multiuser diversity, but other users have high delay. There are some schedulers that exploit multiuser diversity while considering fairness [43].

- *Error and feedback delay in channel measurement.* The measurement error is usually small since the pilot signal in the downlink is shared by many users and is strong. Feedback delay has a more significant influence on the channel estimation error. A short feedback delay can be achieved by increasing the feedback frequency, which, however, increases the system overload. Some techniques for reducing the feedback delay have been studied in [43].
- *Slow and small channel fluctuations such as in the LOS path or in a little-scattering environment.* In this case, a technique called *opportunistic beamforming* that uses dumb antennas at the BS can be exploited to induce faster and larger fluctuations [46].

Opportunistic beamforming

In the opportunistic beamforming scheme, multiple antennas are deployed at the BS to induce fast and large channel fluctuations. The beamforming weights can be simply generated in a pseudorandom fashion [46]. It can substantially improve the system capacity in a slow fading environment. Opportunistic beamforming can also significantly increase the dynamic range of the fluctuations in a Ricean environment, and thus give a performance gain, particularly when the K_r-factor is large [43, 46]. Opportunistic beamforming requires no knowledge of the individual channel gains, neither at the users or at the transmitter. It relies on the multiuser diversity scheduling, which requires feedback of the overall SNR of each user. However, in an independent fast Rayleigh fading environment, the opportunistic beamforming technique does not provide any performance gain [43, 46].

In a slowly fading environment with many users in the system, opportunistic beamforming with multiuser diversity scheduling has a performance similar to that of transmit beamforming, but outperforms the space-time coding case. In a multicell environment, opportunistic beamforming has a dual benefit of opportunistic nulling in an interference-limited cellular system; this has the potential of converting a user from a low SINR, interference-limited channel to a high SINR, noise-limited channel [43, 46, 47].

Adaptive opportunistic beamforming is proposed in [18] for Ricean fading channels, wherein the beamforming weights are generated based on the estimation of the users' DoAs. With the same pilot overhead in the downlink and the same feedback overhead in the uplink, the improved scheme considerably outperforms original opportunistic beamforming [46] in both slow and fast-fading environments. This improvement becomes more pronounced when the number of users is small or the number of antennas is large.

Problems

5.1 Assume three branch diversity in a Rayleigh fading channel. Given an average SNR of 10 dB, determine the probability that the SNR is below 5 dB. What is the result for a single receiver case.

5.2 Give a high-SNR approximation to the outage probability for the parallel channel with L i.i.d. Rayleigh fading branches.

5.3 In a three-branch selection diversity, the mean SNR of each branch is randomly varying with a uniform distribution over 4 dB. Derive an expression for the outage probability.

5.4 Determine the time separation required for two signals to achieve time diversity in a Rayleigh channel at 1800 MHz with a mobile speed of 60 km/h.

5.5 Two independent branches are used to transmit a BPSK signal over an AWGN channel h, $y = \sqrt{\gamma}hx + n$, where the channel h is a random variable with pdf $p_h(h) = 0.2\delta(h - 1) + 0.8\delta(h - 4)$, the noise term n is of zero mean and a variance of $1/2$, h and x are normalized, and γ is the SNR. Assuming the CSI is available at the receiver, what is the error probability for:
(a) Equal-gain combining. (b) Maximal-ratio combining. (c) Selection combining.

5.6 Write a MATLAB program to evaluate (5.44). Plot P_{out} as a function of $\overline{\gamma}$, for $\gamma_I = 1$, $\gamma_0 = 6$ dB, $N_I = 5$ and $N_r = 8$.

References

[1] V. A. Aalo, T. Piboongungon & G. P. Efthymoglou, Another look at the performance of MRC schemes in Nakagami-m fading channels with arbitrary parameters. *IEEE Trans. Commun.*, **53**:12 (2005), 2002–2005.

[2] A. A. Abu-Dayya & N. C. Beaulieu, Outage probabilities of diversity cellular systems with cochannel interference in Nakagami fading. *IEEE Trans. Veh. Tech.*, **41**:4 (1992), 343–355.

[3] A. A. Abu-Dayya & N. C. Beaulieu, Diversity MPSK receivers in cochannel interference. *IEEE Trans. Veh. Tech.*, **48**:6 (1999), 1959–1965.

[4] F. Adachi, M. T. Feeney & J. D. Parsons, Effect of correlated fading on level crossing rates and average fade duration with pre-detection diversity reception. *IEE Proc. Pt. F*, **135**:1 (1988), 11–17.

[5] Y. Akyildiz & B. D. Rao, Maximum ratio combining performance with imperfect channel estimates. In *Proc. IEEE ICASSP*, Orlando, FL, May 2002, **3**, 2485–2488.

[6] S. M. Alamouti, A simple transmit diversity technique for wireless communications. *IEEE J. Sel. Areas Commun.*, **16**:8 (1998), 1451–1458.

[7] E. K. Al-Hussaini & A. A. M. Al-Bassiouni, Performance of MRC diversity systems for the detection of signals with Nakagami fading. *IEEE Trans. Commun.*, **33**:12 (1985), 1315–1319.

[8] M.-S. Alouini & M. K. Simon, Dual-branch diversity over correlated log-normal fading channels. *IEEE Trans. Commun.*, **50**:12 (2002), 1946–1959.

[9] J. G. Andrews, A. Ghosh & R. Muhamed, *Fundamentals of WiMAX: Understanding Broadband Wireless Networking* (Upper Saddle River, NJ: Prentice Hall, 2007).

[10] B. B. Barrow, Diversity combination of fading signals with unequal mean strengths. *IRE Trans. Commun. Syst.*, **11**:1 (1963), 73–78.

[11] V. M. Bogachev & I. G. Kiselev, Optimum combining of signals in space diversity reception. *Telecommun. Radio Eng.*, **34/35**:10 (1980), 83–85.

[12] J. Cheng & T. Berger, Capacity and performance analysis for hybrid selection/maximal-ratio combining in Nakagami fading with unequal fading parameters and branch powers. In *Proc. IEEE ICC*, Anchorage, AK, May 2003, **5**, 3031–3035.

[13] J. Cui & A. U. H. Sheih, Outage probability of cellular radio systems using maximal ratio combining in the presence of multiple interferers. *IEEE Trans. Commun.*, **47**:8 (1999), 1121–1124.

[14] A. Goldsmith, *Wireless Communications* (Cambridge, UK: Cambridge University Press, 2005).

[15] K. A. Hamdi & L. Pap, Exact BER analysis of binary and quaternary PSK with generalized selection diversity in cochannel interference. *IEEE Trans. Veh. Tech.*, **56**:4 (2007), 1849–1856.

[16] P. Ivanis, D. Drajic, & B. Vucetic, The second order statistics of maximal ratio combining with unbalanced branches. *IEEE Commun. Lett.*, **12**:7 (2008), 508–510.

[17] W. C. Jakes, Jr., ed., *Microwave Mobile Communications* (New York: Wiley, 1974).

[18] I.-M. Kim, Z. Yi, D. Kim & W. Chung, Improved opportunistic beamforming in Ricean fading channels. *IEEE Trans. Commun.*, **54**:2 (2006), 199–211.

[19] R. Knopp & P. Humblet, Information capacity and power control in single-cell multiuser communications. In *Proc. IEEE ICC*, Seattle, WA, Jun 1995, 331–335.

[20] V. Kuhn, *Wireless Communications over MIMO Channels: Applications to CDMA and Multiple Antenna Systems* (Chichester, UK: Wiley, 2006).

[21] D. Lao & A. M. Haimovich, Exact closed-form performance analysis of optimum combining with multiple cochannel interferers and Rayleigh fading. *IEEE Trans. Commun.*, **51**:6 (2003), 995–1003.

[22] W. C. Y. Lee, Mobile radio performance for two-branch equal-gain combining receiver with correlated signals at the land site. *IEEE Trans. Veh. Tech.*, **27**:4 (1978), 239–243.

[23] C. M. Lo & W. H. Lam, Performance of generalized selection combining for mobile radio communications with mixed cochannel interferers. *IEEE Trans. Veh. Tech.*, **51**:1 (2002), 114–121.

[24] Y. Ma & J. Jin, Unified performance analysis of hybrid-selection/equal-gain combining. *IEEE Trans. Veh. Tech.*, **56**:4 (2007), 1866–1873.

[25] R. K. Mallik, Optimized diversity combining with imperfect channel estimation. *IEEE Trans. Inf. Theory*, **52**:3 (2006), 1176–1184.

[26] R. K. Mallik & G. K. Karagiannidis, Equal-gain combining with unequal energy constellations. *IEEE Trans. Wireless Commun.*, **6**:3 (2007), 1125–1132.

[27] J. P. P. Martin & J. M. Romero-Jerez, Outage probability with MRC in presence of multiple interferers under Rayleigh fading channels. Electron. Lett., **40**:14 (2004), 888–889.

[28] M. R. McKay, A. Zanella, I. B. Collings & M. Chiani, Error probability and SINR analysis of optimum combining in Rician fading. *IEEE Trans. Commun.*, **57**:3 (2009), 676–687.

[29] J. Reig, Performance of maximal ratio combiners over correlated Nakagami-*m* fading channels with arbitrary fading parameters. *IEEE Trans. Wireless Commun.*, **7**:5 (2008), 1441–1444.

[30] J. M. Romero-Jerez, J. P. P. Martin & A. J. Goldsmith, Outage probability of MRC with arbitrary power cochannel interferers in Nakagami fading. *IEEE Trans. Commun.*, **55**:7 (2007), 1283–1286.

[31] N. C. Sagias, G. K. Karagiannidis, D. A. Zogas, G. S. Tombras & S. A. Kotsopoulos, Average output SINR of equal-gain diversity in correlated Nakagami-m fading with cochannel interference. *IEEE Trans. Wireless Commun.*, **4**:4 (2005), 1407–1411.

[32] N. C. Sagias, Closed-form analysis of equal-gain diversity in wireless radio networks. *IEEE Trans. Veh. Tech.*, **56**:1 (2007), 173–182.

[33] S. Sanayei & A. Nosratinia, Opportunistic downlink transmission with limited feedback. *IEEE Trans. Inf. Theory*, **53**:11 (2007), 4363–4372.

[34] M. Schwartz, W. R. Bennet & S. Stein, *Communication Systems and Techniques* (New York: McGraw-Hill, 1966).

[35] A. Shah & A. M. Haimovich, Performance analysis of optimum combining in wireless communications with Rayleigh fading and cochannel interference. *IEEE Trans. Commun.*, **46**:4 (1998), 473–479.

[36] A. Shah & A. M. Haimovich, Performance analysis of maximal ratio combining and comparison with optimum combining for mobile radio communications with cochannel interference. *IEEE Trans. Veh. Tech.*, **49**:4 (2000), 1454–1463.

[37] A. U. H. Sheikh, *Wireless Communications: Theory and Techniques* (Boston, MA: Kluwer, 2004).

[38] M. K. Simon & M.-S. Alouini, *Digital Communications over Fading Channels*, 2nd edn (New York: Wiley, 2005).

[39] K. Sivanesan & N. C. Beaulieu, Exact BER analysis of bandlimited BPSK with EGC and SC diversity in cochannel interference and Nakagami fading. *IEEE Commun. Lett.*, **8**:10 (2004), 623–625.

[40] Y. Song, S. D. Blostein & J. Cheng, Outage probability comparisons for diversity systems with cochannel interference in Rayleigh fading. *IEEE Trans. Wireless Commun.*, **4**:4 (2005), 1279–1284.

[41] G. L. Stuber, *Principles of Mobile Communication*, 2nd edn (Boston, MA: Kluwer Academic Publishers, 2001).

[42] Y. Tokgoz, B. D. Rao, M. Wengler, & B. Judson, Performance analysis of optimum combining in antenna array systems with multiple interferers in flat Rayleigh fading. *IEEE Trans. Commun.*, **52**:7 (2004), 1047–1050.

[43] D. Tse & P. Viswanath, *Fundamentals of Wireless Communications* (Cambridge, UK: Cambridge University Press, 2005).

[44] A. M. D. Turkmani, A. A. Arowojolu, P. A. Jefford & C. J. Kellett, An experimental evaluation of the performance of two-branch space and polarization diversity schemes at 1800 MHz. *IEEE Trans. Veh. Tech.*, **44**:2 (1995), 318–326.

[45] E. Villier, Performance analysis of optimum combining with multiple interferers in flat Rayleigh fading. *IEEE Trans. Commun.*, **47**:10 (1999), 1503–1510.

[46] P. Viswanath, D. N. C. Tse, & R. Laroia, Opportunistic beamforming using dumb antennas. *IEEE Trans. Inf. Theory*, **48**:6 (2002), 1277–1294.

[47] P. Viswanath, Opportunistic communication: a system view. In H. Bolcskei, D. Gesbert, C. B. Papadias & A.-J. van der Veen, eds., *Space-Time Wireless Systems: From Array Processing to MIMO Communications* (Cambridge, UK: Cambridge University Press, 2006), pp. 426–442.

[48] M. Z. Win & J. H. Winters, Virtual branch analysis of symbol error probability for hybrid selection/maximal-ratio combining. *IEEE Trans. Commun.*, **49**:11 (2001), 1926–1934.

[49] J. H. Winters, Switched diversity with feedback for DPSK mobile radio systems. *IEEE Trans. Veh. Tech.*, **32**:1 (1983), 134–150.

[50] J. H. Winters, Optimum combining in digital mobile radio with cochannel interference. *IEEE J. Sel. Areas Commun.*, **2**:4 (1984), 528–539.

[51] J. H. Winters, J. Salz, and R. D. Gitlin, The impact of antenna diversity on the capacity of wireless communication systems. *IEEE Trans. Commun.*, **42**:2–4 (1994), 1740–1750.

[52] J. H. Winters & J. Salz, Upper bounds on the bit-error rate of optimum combining in wireless systems. *IEEE Trans. Commun.*, **46**:12 (1998), 1619–1624.

[53] K. T. Wong, S. L. A Chan & R. P. Torres, Fast-polarization-hopping transmission diversity to mitigate prolonged deep fades in indoor wireless communications. *IEEE Anten. Propagat. Mag.*, **48**:3 (2006), 20–27.

[54] H. Zhang & A. Abdi, On the average crossing rates in selection diversity. *IEEE Trans. Wireless Commun.*, **6**:2 (2007), 448–451

[55] X. Zhang & N. C. Beaulieu, A closed-form BER expression for BPSK using MRC in correlated CCI and Rayleigh fading. *IEEE Trans. Commun.*, **55**:12 (2007), 2249–2252.

6 Channel estimation and equalization

6.1 Channel estimation

Traditionally, receivers rely on the transmitter-assisted training sequences to extract the desired reference signal for multiuser channel estimation and equalization. There has also been substantial research done in blind channel estimation and equalization, where no training sequence is used.

One of the most popular parameter estimation methods is the maximum-likelihood (ML) method. Consider a slowly time-varying (frequency-selective) channel model

$$r(m) = \sum_k h(k)s(m-k) + n(m), \tag{6.1}$$

where $r(m)$ is a vector obtained by stacking p consecutive received samples in the mth symbol duration, $w(m)$ is the corresponding p-dimensional AWGN vector at the receiver, and $s(m-k)$ is the input at time $m-k$. Assume that the channel has a finite impulse response of order L. For N symbols, we have $r = \left(r^T(N-1), \ldots, r^T(0)\right)^T$ and

$$r = Sh + n \tag{6.2}$$

where S is a block Hankel matrix of size $Np \times (L+1)p$, h is the vector of channel parameters, and w is the stacked AWGN vector:

$$S = \begin{pmatrix} s(N-1)\mathbf{I}_p & s(N-2)\mathbf{I}_p & \cdots & s(N-L-1)\mathbf{I}_p \\ \vdots & \vdots & \vdots & \vdots \\ s(0)\mathbf{I}_p & s(-1)\mathbf{I}_p & \cdots & s(-L)\mathbf{I}_p \end{pmatrix}, \tag{6.3}$$

$$h = \begin{pmatrix} h(0) \\ \vdots \\ h(L) \end{pmatrix}, \quad n = \begin{pmatrix} n(N-1) \\ \vdots \\ n(0) \end{pmatrix}, \tag{6.4}$$

with $\mathbf{I}_p$ being the $p \times p$ identity matrix.

For training-based channel estimation, the input vector s is the known training symbols, and the corresponding observation r is also known at the receiver. The ML estimation solution is given by

$$\hat{h} = \arg \min_h \|r - Sh\|^2 = S^\dagger r, \tag{6.5}$$

where $\mathbf{S}^\dagger$ is the pseudo-inverse of $\mathbf{S}$. This is also the classical linear least squares estimator. This method realizes the minimum mean squared error (MMSE) among all unbiased estimators and achieves the CRLB.

The frequency of channel estimation is determined by the Doppler spread of the channel $\nu_{\max}$. The pilot spacing M should satisfy the Nyquist sampling rate

$$M \le \frac{1}{2\nu_{\max}T}, \tag{6.6}$$

where T is the symbol period.

6.1.1 Adaptive channel estimation

The least mean squares (LMS) algorithm, proposed by Widrow et al [36], is the most popular adaptation technique. The LMS algorithm is a gradient-based search algorithm. It minimizes the mean square error between the desired signal $d(t)$ and the model filter output $y(t) = \mathbf{w}^T(t)\mathbf{x}(t)$, that is

$$\mathbf{w} = \arg\min E\left[\|d(t) - y(t)\|^2\right], \tag{6.7}$$

where $\mathbf{w}$ is the weighting vector, and $\mathbf{x}$ is the input delay line. This yields the LMS algorithm [7]

$$\mathbf{w}(t+1) = \mathbf{w}(t) + 2\mu e(t)\mathbf{x}(t), \tag{6.8}$$

where the step size μ should be selected in a range that guarantees convergence, and e is the error defined by

$$e(t) = d(t) - \mathbf{w}^T(t)\mathbf{x}(t). \tag{6.9}$$

The algorithm can be initialized by setting $\mathbf{x}(0) = \mathbf{w}(0) = \mathbf{0}$.

The LMS algorithm is the most widely used one in adaptive filtering. It has low computational complexity, unbiased convergence to the Wiener solution, and stable behavior when implemented with finite-precision arithmetic, but the convergence speed is relatively low. For rapidly time-varying channels, a fast-convergent algorithm such as the recursive least-squares (RLS) or Kalman filtering is needed to avoid numerical instability. The RLS algorithm solves the least squares problem in a recursive form. It has a rate of convergence that is typically an order of magnitude faster than the LMS algorithm. This is, however, at the price of a large computational complexity. More details of these adaptive algorithms are given in Section 13.5. Various adaptation techniques are discussed in [7, 16].

Channel estimation using a training sequence reduces the spectrum efficiency. For example, the GSM uses a training sequence of 26 bits in every 148-bit frame. Training-sequence-based channel estimation is used practically in all the wireless standards, since the blind methods have many complexity and stability problems.

6.1.2 Blind channel estimation

Blind techniques for channel estimation utilize qualitative information about the channel and the input. No explicit training signal is used, and the receiver estimates the channel from the signals received during the normal data transmission. Blind techniques typically employ the signal properties such as cyclostationarity, finite alphabet, and constant modulus. But, they usually suffer from convergence problems. Semiblind techniques combine the merits of both the training-based and blind techniques, and are more promising.

The blind estimation of a channel as well as its input signal is an unidentifiable problem. This makes the conventional CRLB inapplicable. By incorporating some forms of constraint to make the problem identifiable, some CRLBs have been derived. Various blind channel estimators can be compared based on tight performance bounds.

The blind channel identification problem consists of estimating the channel $\mathbf{H}$ and the input signal $s(t)$ from N samples of $\mathbf{y}(t)$. For a single-input multiple-output (SIMO) channel

$$\mathbf{y}(t) = \sum_{k=0}^{L-1} \mathbf{h}_k s(t-k) + \mathbf{n}(t) = \mathbf{H}\mathbf{s}(t) + \mathbf{n}(t), \quad t = 1, 2, \ldots, \tag{6.10}$$

where $\mathbf{y}(t)$ is the $N_r \times 1$ output signal, $\mathbf{h}_k$ is the channel impulse response vector at the kth time tap, L is the channel length, $s(t)$ is the input signal and $\mathbf{n}(t)$ is AWGN. $\mathbf{H} = [\mathbf{h}_0, \mathbf{h}_1, \cdots, \mathbf{h}_{L-1}]$, and $\mathbf{s}(t) = (s(t), s(t-1), \ldots, s(t-L-1))^T$. By stacking all the N samples of $\mathbf{y}(t)$ into an $MN \times 1$ vector $\mathbf{y}$, a Toepliz form is obtained. From this equation, it is clear that the identification of $\mathbf{H}$ and s is not possible unless another constraint is exercised.

Blind channel estimation has to rely on the known statistics of the transmitted signals. The signal properties that can be used for blind equalization include constant modulus (envelope), statistical properties like cyclostationarity, and finite symbol alphabet. There are also some training-based semiblind channel estimation algorithms that exploit the advantages of both the training-based and the blind approaches [34, 35].

6.2 Channel equalization

Channel equalizers are used to reduce the ISI arising from the delay spread or band-limitation of the channel by adjusting the pulse shape so that it does not interfere with the neighboring pulses. Equalization is a MIPS (million instructions per second)-intensive function in cellular phone receivers. Major equalization techniques are linear equalization [21], decision-feedback equalization (DFE), and Viterbi maximum-likelihood sequence estimation (MLSE) [31]. Adaptive equalization is especially useful in TDMA systems. These receivers for frequency-selective channels can be easily extended to multiple antenna cases [26]. Channel equalization was first used in echo cancellation in telephone networks.

For all these algorithms, a training sequence is transmitted over the channel to aid the receiver to calculate the channel impulse response **H**, which is used for equalization. Classical channel equalization methods such as the linear equalizer and the DFE are based on training sequences. There are some blind algorithms, which avoid the use of training sequences.

Adaptive equalizers are typically implemented by using a transversal filter with a tap-spacing of the symbol interval T. The symbol-spaced equalizers are more sensitive to the sampling instant. The fractionally-spaced equalizer uses tap-spacing less than T [13]. The fractionally-spaced equalizer is equivalent to a symbol-spaced equalizer preceded by a matched filter [30]. Since an exact matched filter is usually difficult to obtain, the fractional-spaced equalizer is very attractive. In real systems, the output of a matched filter is usually oversampled so as to extract the timing information and to mitigate the influence of timing errors. Thus a $T/2$-spaced transversal filter can be easily implemented [30], and this provides a better accuracy than the symbol-spaced equalizers.

6.2.1 Optimum sequence detection

For frequency-selective channels, the optimum sequence detection is to maximize the *a-posteriori* probability $\Pr(x|r)$, that is, the probability that x is transmitted given the received r. Applying Bayes' rule, we have

$$\Pr(x|r) = p_{r|x}(r)\frac{\Pr(x)}{p_r(r)}. \qquad (6.11)$$

MAP sequence detection

The maximum *a-posteriori* (MAP) sequence detector is obtained as

$$\hat{x} = \arg\max_{x\in\mathcal{X}^L} \Pr(x|r) = \arg\max_{x\in\mathcal{X}^L} p_{r|x}(r)\Pr(x), \qquad (6.12)$$

where L is the sequence length, and $\mathcal{X} = \{X_i\}$ is a finite alphabet of symbols.

Maximum-likelihood sequence estimation

The a priori probability $\Pr(x)$ is typically known for communications systems, since the symbols are usually uniformly distributed. In this case, the MAP detector is equivalent to the maximum likelihood sequence detector

$$\hat{x} = \arg\max_{x\in\mathcal{X}^L} \Pr(x|r) = \arg\max_{x\in\mathcal{X}^L} p_{r|x}(r). \qquad (6.13)$$

The Viterbi algorithm is an efficient algorithm for this purpose [31]. The Viterbi algorithm minimizes the error probability when detecting sequences.

Symbol-by-symbol MAP detector

The optimum symbol-by-symbol MAP detector [1] minimizes the symbol error probability (SEP)

$$\hat{x}(k) = \arg\max_{X_j \in \mathcal{X}} \Pr\left(x(k) = X_j | \boldsymbol{r}\right) = \arg\max_{X_j \in \mathcal{X}} \sum_{\boldsymbol{x} \in \mathcal{X}^L, x(k) = X_j} \Pr(\boldsymbol{x}|\boldsymbol{r})$$

$$= \arg\max_{X_j \in \mathcal{X}} \sum_{\boldsymbol{x} \in \mathcal{X}^L, x(k) = X_j} p_{\boldsymbol{r}|\boldsymbol{x}}(\boldsymbol{r}) Pr(\boldsymbol{x}). \tag{6.14}$$

The above equation differs from (6.12) in that all sequences $\boldsymbol{x}$ with $x(k) = X_j$ contribute to the decision.

Symbol-by-symbol maximum-likelihood detection

A symbol-by-symbol ML detector can also be obtained as

$$\hat{x}(k) = \arg\max_{X_j \in \mathcal{X}} p_{\boldsymbol{r}|x(k)=X_j}(\boldsymbol{r}) = \arg\max_{X_j \in \mathcal{X}} \sum_{\boldsymbol{x} \in \mathcal{X}^K, x(k) = X_j} p_{\boldsymbol{r}|\boldsymbol{x}}(\boldsymbol{r}). \tag{6.15}$$

For memoryless channels like AWGN or flat-fading channels, the detection can be based on the current symbol rather than the whole sequence; then, the ML detector reduces to

$$\hat{x}(k) = \arg\max_{X_j \in \mathcal{X}} p_{\boldsymbol{r}|X_j}(r(k)). \tag{6.16}$$

The time index k can be dropped.

In addition, there are also multiple-symbol ML [6] and multiple-symbol MAP [18] detection techniques.

6.2.2 Linear equalizers

A linear equalizer is usually implemented with a finite impulse response (FIR) transversal filter that consists of a delay line of $2K + 1$ τ-second taps, as shown in Fig. 6.1. The tap weights are adapted to improve the shape of the system impulse response so as to match the unequalized channel.

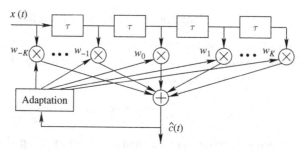

Figure 6.1   The transversal FIR filter for linear equalization.

A complex transmit symbol sequence $\{c_i\}$ is sent over a dispersive, noisy channel, and a sequence $\{x_i\}$ is received. The sequence $\{x_i\}$ is used as input to the equalizer. The output of the equalizer is given as

$$\hat{c}(t) = \sum_{i=-K}^{K} w_i x(t - i\tau). \tag{6.17}$$

The estimate $\{\hat{c}_i\}$ should be as close to $\{c_i\}$ as possible. Thus, the linear equalizer has a transfer function of the form

$$H_{eq}(z) = \sum_{i=-K}^{K} w_i z^{-i}. \tag{6.18}$$

Usually, the ISI caused by the channel distortion is limited to a finite number of symbols on both sides of the desired symbol. The time delay τ between adjacent taps may be selected as the symbol interval T, yielding a *symbol-spaced equalizer*. When τ is selected to be less than T, typically $\tau = T/2$, we get a fractionally spaced equalizer.

Linear equalizers result in noise enhancement. They perform well in the wired line case, but perform very poorly in the wireless case. This is due to the deep spectrum nulls in the passband in the wireless channel.

Zero-forcing equalizer

If the channel is ISI distorted by $H_c(f)$, the linear equalizer removes the ISI distortion by applying channel inversion, $H_c^{-1}(f)$. Thus, the combination of channel and equalizer leads to a completely flat transfer function. This is an ideal equalizer, also know as *zero-forcing (ZF) equalizer*, since it forces the ISI to zero at the sampling instants $t = kT, k = 0, 1, \cdots$. As a result, the output of the ZF equalizer is simply

$$z_k = c_k + n_k, \quad k = 0, 1, \ldots \tag{6.19}$$

where c_k is the desired symbol and n_k is the additive noise. The ZF equalizer enhances the noise of the channel, at frequencies where the transfer function of the channel attains small values, since it performs channel inversion.

For the $(2K + 1)$-tap equalizer, the value of K should satisfy $2K + 1 \geq L$ so that the equalizer spans the length of the ISI, where L is the number of signal samples spanned by the ISI. From (6.17), by applying the ZF condition to each sample at $t = mT$

$$c(mT) = \sum_{i=-K}^{K} w_i x(mT - i\tau) = \begin{cases} 1, & m = 0 \\ 0, & m = \pm 1, \cdots, \pm K. \end{cases} \tag{6.20}$$

This is a set of $2K+1$ linear equations for the coefficients w_i of the ZF equalizer. In matrix form

$$\mathbf{Xw} = \mathbf{c}, \tag{6.21}$$

where $X_{ij} = x(iT - j\tau)$, $i, j = -K, -K + 1, \cdots, K$, $\mathbf{w} = (w_{-K}, \cdots, w_K)^T$, $\mathbf{c} = (0, \cdots, 0, 1, 0, \cdots, 0)^T$. Thus, we have the ZF solution

$$\mathbf{w} = \mathbf{X}^{-1}\mathbf{c}. \tag{6.22}$$

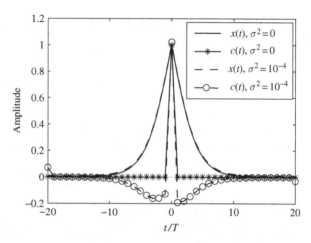

Figure 6.2   The distorted pulse and the equalized pulse.

Example 6.1: Assume the input to the ZF equalizer is given by

$$x(t) = \frac{2}{1 + e^{|t|/(2T)}},$$

where T is the symbol period. The pulse is sampled at the rate $2/T$. Assume that the noise at the receiver, $n(t)$, is zero-mean AWGN with variance σ^2. The results of a 5-tap ZF equalizer ($K = 2$) are shown in Fig. 6.2 for $\sigma^2 = 0$ and $\sigma^2 = 10^{-4}$. The result for $\sigma^2 = 10^{-4}$ is from a random run. It is seen that the ISI is substantially eliminated by equalization for $\sigma^2 = 0$. Note that the performance of the ZF equalizer degrades rapidly if we introduce noise into the model.

MMSE equalizer

Although the ZF equalizer eliminates the ISI, noise enhancement makes the BER performance undesirable. A better choice is the MMSE equalizer. In the MMSE equalizer, the equalizer filter parameters are updated by minimizing the mean squared error (MSE) criterion,

$$\min J = E\left[|c_k - \hat{c}_k|^2\right], \tag{6.23}$$

where $\hat{c}_k = z_k$ is the equalizer output. By this means, the ISI is removed as much as possible.

The optimum MMSE solution is given by

$$w_{\text{opt}} = \left(\mathbf{R}_{xx}^T\right)^{-1} p^T, \tag{6.24}$$

where $w = (w_{-L}, \cdots, w_L)^T$, the correlation matrix of the received signal $\mathbf{R}_{xx} = E\left[u_m u_m^H\right], p = E\left[u_m c_m^*\right]$, and

$$u_m = (x(mT + K\tau), x(mT - (-K + 1)\tau), \cdots, x(mT - K\tau))^T. \tag{6.25}$$

Thus

$$\mathbf{R}_{xx} = \frac{1}{2K+1}\mathbf{X}^T\mathbf{X} + \sigma_0^2\mathbf{I}, \qquad (6.26)$$

with σ_0^2 being the variance of the noise at the receiver.

The components of $\mathbf{R}_{xx}$ and p are given by

$$R_{xx}(i,j) = R_{xx}(i-j) = E\left[x(mT - i\tau)x^*(mT - j\tau)\right] + \delta_{ij}\sigma_0^2, \qquad (6.27)$$

$$p(i) = E\left[x(mT - i\tau)c_m^*\right] \qquad (6.28)$$

for $i,j = -K, -K+1, \cdots, 0, \cdots, K$, where $E[\cdot]$ is over all values of m, and δ_{ij} is the Krocknecker delta function. Usually, $E[\cdot]$ is implemented by averaging over samples 1 to K.

This solution is known as the Wiener solution. The MMSE is given by

$$J_{\min} = 1 - p\mathbf{R}_{xx}^{-1}p^T. \qquad (6.29)$$

Adaptive equalizers

Solving (6.22) or (6.24) requires a complexity of the order of $(2K+1)^3$ complex operations on each iteration (each symbol period T) due to the matrix inversion operation. This is the reason why the LMS, RLS, or Kalman filtering algorithm is usually used for adaptation [27].

In either the ZF or the MMSE equalizer case, we need to solve a set of linear equations for each sample. In practice, a simple iterative procedure can be applied so that for each input sample, only one iteration is performed. This can be derived by defining an MSE from the set of linear equations, and then applying steepest descent to the MSE. One can also apply RLS or the conjugate-gradient method.

The adaptive implementation of the MMSE equalizer using the LMS method is given by

$$w_{k+1} = w_k + \mu e_k x_k^*, \qquad (6.30)$$
$$e_k = c_k - \hat{c}_k, \qquad (6.31)$$

where the step size μ should be selected to ensure stability and a fast convergence [7].

Example 6.2: Consider a channel

$$h = [0.1811, -0.2716, 0.0453, 0.9054, 0.2263, 0.0453, -0.1358].$$

For a bitstream of 1000 bits, and $\sigma = 0.1$, simulation of an adaptive 7-tap MMSE equalizer using the LMS algorithm is illustrated in Fig. 6.3. The MSE obtained is the average of 1000 independent runs. It is seen that $\mu = 0.02$ achieves a good tradeoff of stability and convergence speed.

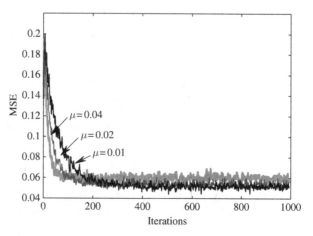

Figure 6.3

Simulation of the adaptive MMSE equalizer using different step sizes.

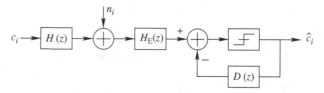

Figure 6.4

Structure of a decision feedback equalizer.

Given an input of sampled signal sequences, $s_{in}(kT_0)$, the equalizer is a time-varying, N-tap FIR filter, and the output of the equalizer $\hat{x}(kT_0)$ is given by convolution

$$\hat{x}(kT_0) = \sum_{i=0}^{N-1} w_i(kT_0)\, s_{in}((k-i)T_0). \tag{6.32}$$

The output $\hat{x}(kT_0)$ is compared with the training signal $x(kT_0)$, and the error signal is used to adjust the filter weight values for the next instant, $w_i((k+1)T_0)$.

The adaptive algorithm is switched off at the end of the training sequences. The trained equalizer can then work for the transmitted data. For mobile systems, the channel is time-varying, and the decision-directed operation can continue adapting to the channel for the transmitted data [8]. The adaptive algorithm is still working, but the training sequence is replaced by the output of the decision circuits.

6.2.3 Decision-feedback equalizers

The DFE is a nonlinear equalizer that uses a feedforward filter and a feedback filter. The DFE performs better than linear equalizers in channels with severe amplitude distortion, and thus is widely used in wireless communications. The feedback filter subtracts the ISI due to the earlier pulses from the current pulse (postcursor ISI). The structure of a DFE is shown in Fig. 6.4.

The DFE consists of a feedforward filter $H_E(z)$, which is a conventional linear equalizer, and a feedback filter $D(z)$. The impact of the received symbol to future samples can be computed and then subtracted from the received signal. Since the ISI is computed using the signal after hard decision, the additive noise from the feedback signal is eliminated, leading to a smaller error probability than that in the case of a linear equalizer.

Due to the use of the feedback filter, the DFE performs much better than the linear equalizer for the same number of taps. However, it may suffer from error propagation, and the difficulty in analyzing the stochastic nonlinear dynamic system that governs the error propagation causes the DFE to be not as widely used as the linear equalizer. Error propagation is negligible for small BER, which can often be achieved via coding. Error propagation makes DFEs more sensitive to channel estimation errors.

DFEs can be implemented as an MMSE-DFE or a ZF-DFE. The SNR of the ZF-DFE is simply $\text{SNR}^{\text{ZF}-\text{DFE}} = \frac{\sigma_c^2}{\sigma_n^2}$, where σ_c is the deviation of $c(k)$. The MMSE-DFE targets at minimizing the MSE by achieving a balance between the noise enhancement and the residual ISI. Since the postcursor ISI does not contribute to noise enhancement, the objective is to minimize the sum of the noise and the average precursor ISI. For the ZF-DFE, all the precursor ISI is eliminated, and the postcursor ISI is subtracted by the feedback branch.

DFE with Gaussian distributed input symbols is considered as the canonical receiver structure due to its simple structure and optimality. The ZF-DFE achieves a capacity very close to the Shannon capacity of the channel, and asymptotically approximates the channel capacity for large SNR [10]. The unbiased MMSE-DFE is optimum since it achieves the channel capacity for all SNRs [3, 10]. However, these results are based on the assumption of correct decisions. Unfortunately, DFE suffers from error propagation, which degrades the performance, especially at low SNR. Channel coding is the common method for approaching the channel capacity, but it cannot be applied in DFE in a straight forward manner. This is because DFE requires zero-delay decisions, while channel decoding is typically based on a block or a sequence of symbols.

6.2.4 MLSE equalizer

The MLSE equalizer achieves a much better performance than DFEs do, but with a much higher complexity. The linear equalizers and DFEs provide hard decision as to which symbol has been transmitted. Unlike the linear equalizers and DFEs that adjust the received signal via filtering to remove the distortion, the MLSE determines the sequence of symbols that has most likely been transmitted. This is very similar to the decoding of convolutional codes.

The output signal of the time-discrete channel, or the received signal at the receiver, can be written as

$$r_i = \sum_{n=0}^{L} h_n c_{i-n} + n_i, \tag{6.33}$$

where L is the length of the channel h, and n_i is AWGN with variance σ_n^2. For a sequence of N received symbols, the joint pdf of the vector of received symbols r, conditioned on the data vector c and impulse response vector h, is given by

$$p(r|c,h) = \frac{1}{\left(2\pi\sigma_n^2\right)^{N/2}} \exp\left(-\frac{1}{2\sigma_n^2}\sum_{i=1}^{N}\left|r_i - \sum_{n=0}^{L}h_n c_{i-n}\right|^2\right). \qquad (6.34)$$

For a given h, the MLSE of c maximizes $p(r|c,h)$; this corresponds to the minimization of

$$\sum_{i=1}^{N}\left|r_i - \sum_{n=0}^{L}h_n c_{i-n}\right|^2. \qquad (6.35)$$

An exhaustive search for c is the most straightforward but also the most computationally complex method. The Viterbi algorithm is the one that is used in practice [31]. It was originally proposed for ML decoding of convolutional codes. The similarity between an ISI channel and a convolutional encoder was recognized by Forney, and hence he applied the Viterbi algorithm to channel equalization [12].

MLSE estimators offer the best BER performance among all the equalizers. However, MLSE has a computational complexity that increases exponentially with the length of the impulse response of the channel, and the size of signal constellation. This makes it impractical for systems with a large constellation and/or a long channel impulse response. A sequential decoding algorithm for decoding convolutional codes can be use instead, especially for a long encoder constraint length and a moderate-to-high SNR [19]. This sequential sequence estimation has been applied to signal detection in ISI channels [20].

The Viterbi equalizer estimates the data without changing the distorted sequence. The lattice sequence estimator [24] and the sphere decoding algorithm [9, 15] were developed for the near-ML detection of a lattice-type modulation, such as MPAM and MQAM, transmitted over a linear channel, with a significantly reduced complexity. The method limits the search among possible candidates to those located within a sphere centered on the received vector. ML detection can be formulated as an integer LS problem, and can be solved at considerable complexity reduction by using fast algorithms [15] of sphere decoding [9]. Turbo equalization is an iterative equalization and decoding technique for coded data transmission over ISI channels.

There is a fundamental link between orthogonality deficiency of the channel matrix, $od(\mathbf{H})$, and the performance of the linear equalizers [22]. When $od(\mathbf{H})$ is strictly upper-bounded by unity, linear equalizers collect the same diversity order as ML equalizers do, and the outage capacity loss relative to ML equalizers is constant with respect to SNR. Based on these results, hybrid equalizers can be designed. Hybrid equalizers can trade off between performance and complexity by tuning $od(\mathbf{H})$.

6.2.5 Viterbi algorithm

MLSE operates on the received signal $r(t)$, $0 \leq t \leq \tau$ to produce an estimate sequence $\mathbf{b}^* = \left(b_1^*, \ldots, b_N^*\right)$ of the transmitted symbol sequence $\mathbf{b} = (b_1, b_2, \ldots, b_N)$ with the minimum error probability. This can be represented by the maximization of a pdf in the form

$$\max p\left[r(t), 0 \le t \le \tau | b_1 = \hat{b}_1, b_2 = \hat{b}_2, \cdots, b_N = \hat{b}_N\right]. \tag{6.36}$$

This is the definition for MAP detection, and it is equivalent to MLSE detection for equiprobable sequences. MLSE detection determines the total distances of all possible paths through the trellis, for all possible input sequences, and selects the path with the smallest metric. For a symbol of m possible values, the brute force method uses m^N calculations.

The Viterbi algorithm [31, 32] solves the MLSE problem optimally but using a complexity that linearly increases with N rather than exponentially. In the case of equalization, the Viterbi algorithm is often referred to as a *Viterbi equalizer*.

The ML decoder chooses an estimate sequence b^* of b, which maximizes the log-likelihood function $\log P(r|b)$. For a discrete memoryless channel,

$$\log P(r|b) = \sum_{i=1}^{N} \log P(r_i|b_i). \tag{6.37}$$

The log-likelihood $\log P(r|b)$ is called the *metric* associated with path b, and is denoted $M(r|b)$. The terms $\log P(r_i|b_i)$ are called *branch metrics*, denoted $M(r_i|b_i)$. Thus,

$$M(r|b) = \sum_{i=1}^{N} M(r_i|b_i). \tag{6.38}$$

The partial path metric for the first j branches of a path is defined as

$$M^j(r|b) = \sum_{i=1}^{j} M(r_i|b_i). \tag{6.39}$$

From the trellis diagram, at a node of two routes, we can exclude the one with the smaller sum of measurements. By defining the state measurement of node i at time j, $M_i(j)$, as the maximum among all the sums of all branch measurements to this node (state), we have

$$M_i(j+1) = \max\left[M_{i'}(j) + m_{i',i}, M_{i''}(j) + m_{i'',i}\right], \tag{6.40}$$

where i' and i'' are the nodes allowed to transfer to node i in one symbol period, and $m_{i',i}$ and $m_{i'',i}$ are the branch measurements of the two transfer branches. Equation (6.40) is usually called the *Viterbi algorithm*.

For discrete memoryless channel, the Viterbi algorithm is implemented as follows [19]:

- Step 1. Set the time instant $j = K$, K being the constraint length. Calculate the partial metric M^j for the single path entering each state. Store the path with the largest metric, called the *survivor*, and its metric for each state.
- Step 2. Set $j = j + 1$. Calculate M^j for all the paths entering each state i by adding the branch metric entering state i to the metric of the connecting survivor at time $j-1$,

$M_{i'}(j - 1)$. For each state i, store the survivor and its metric, and eliminate all the other paths.
- Step 3. Repeat Step 2, until $j \geq N + K$, where N is the length of the input sequence.

The Viterbi algorithm always finds the maximum likelihood path through the trellis. The algorithm starts with the all-zero state of the trellis at time $j = 0$, with the initial cumulative path metrics M for all states being zero.

The basic operation reduces the maximum number of routes from 2^N to only $4N$ computations, for binary sequences. At each state in each layer of the trellis diagram, an add-compare-select (ACS) operation is required. To implement the Viterbi algorithm, two sets of registers are needed: registers for storing the states of the routes $M_i(j)$ and for storing the selection at each node. The Viterbi algorithm must keep records of $2^{N(K-1)}$ surviving paths and their corresponding metrics. For this reason, when the Viterbi algorithm is used for decoding of convolutional codes, the number of computations as well as the memory requirement makes convolutional codes feasible only for small N and K.

Commercially available Viterbi equalizer/decoder chips are typically limited to 64 states. These can be used for equalization for pulses that each smear over $K = 7$ bit intervals. In IS-95, the constraint length $K = 9$ is selected. Hardware implementation of the Viterbi algorithm has been discussed in many publications [14, 17].

The Viterbi algorithm was originally proposed for decoding convolutional codes using MLSE [31]. The channel memory which introduces ISI is analogous to the encoder memory in a convolutional code [12], thus it is also used for the equalization and detection of digital signals. It is generally applicable to detection of a discrete-time finite-state Markov process immersed in additive memoryless noise. The trellis diagram is used for characterizing a finite state machine. The Viterbi algorithm is well suited for estimating the most likely input sequence to the finite state machine, thus implementing equalization and detection simultaneously in case of ISI distortion. The Viterbi algorithm is a technique for finding the shortest route through a graph. More description of the Viterbi algorithm is given in Chapter 15.

6.2.6 Frequency-domain equalizers

Compared with the conventional time-domain equalization, frequency-domain equalization (FDE) exhibits a substantially low complexity that grows with the number of symbols of dispersion [33]. Also, adaptive algorithms generally converge faster and are more stable in the frequency domain [28]. FDE is an effective technique for high-data-rate wireless communication systems suffering from very long ISI.

FDE for single-carrier systems (SC-FDE) shares some common elements with OFDM. Compared to OFDM, FDE offers similar performance and complexity for broadband wireless systems [5], but it is more robust without heavy interleaving and error-correction coding and is less sensitive to nonlinear distortion and carrier synchronization difficulties [5], thereby allowing for the use of low-cost power amplifiers.

The SC-FDE system is shown in Fig. 6.5. At the transmitter, a cyclic prefix is periodically added to the baseband data sequence $\{x(n)\}$ and modulated onto a single-carrier

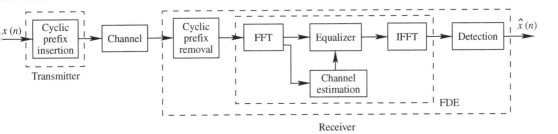

Figure 6.5   Simplified block diagram of an SC-FDE system.

frequency for transmission across the time-varying and frequency-selective fading channel. At the receiver, the cyclic prefix is removed. The FFT is utilized to convert the
time-domain data signal into frequency-domain signal. Then, frequency-domain channel
estimation and equalization are applied to mitigate ISI. The IFFT is applied to convert the
equalized frequency-domain signal into a time-domain signal for detection and estimation.
Like OFDM, FDE applies the processing in the frequency domain and uses cyclic prefix.
OFDM will be introduced in Chapter 9.

Single-carrier FDMA (SC-FDMA) [25] is a multiple-access scheme based on SC-FDE.
It has similar performance and similar overall complexity to OFDM. SC-FDMA has the
inherent characteristics of a low PAPR. This is the reason why SC-FDMA is preferred
over OFDM in the uplink communications in 3GPP LTE, where lower PAPR greatly
enhances the transmit power efficiency of MSs. For subchannel allocation, SC-FDMA can
be implemented by sub-band [25] and interleaved approaches [29] for allocating resources
to users.

Single-carrier systems generally require knowledge of the channel and the variance of
the additive noise process for channel equalization. Training blocks are utilized to estimate
the channel transfer function for the blocks. The channel transfer functions for the data
blocks are estimated by interpolating the channel transfer functions of the training blocks
at the current and the next frames. Adaptive SC-FDE systems with explicit channel and
noise-power estimation have been considered in [4, 23, 37].

A single-carrier frequency-domain DFE, which employs a frequency-domain feedforward filter and a short time-domain feedback filter, is described in [2]. It significantly
outperforms a linear FDE, and yields a capacity very close to that of OFDM, with a
computational complexity similar to that of OFDM.

6.2.7 Blind equalizers

As in the case of blind channel estimation discussed earlier, blind adaptive equalization
exploits known statistical properties of the transmitted signal for estimation of both the
channel and the data. The constant modulus algorithm is the well-known blind adaptation
algorithm. This principle is applicable for blind channel estimation, blind equalization,
blind source separation and blind beamforming in the wireless communication field. For

constant envelope signals such as FM and M-ary PSK (MPSK) signals, the amplitude of the signal is the same, and thus such a constant-modulus cost function can be defined by

$$J_{CM} = E\left[(1 - |y_n|)^2\right] \tag{6.41}$$

or by a similar function. The approximation of J_{CM} is formed from the equalizer output y_n, and no training signal is required for estimation.

Blind equalization algorithms can also be derived from high-order statistics of the signals. For example, the entropy of the received symbols is smaller than that in case of no ISI, since ISI introduces dependence; thus, maximizing the joint entropy of the equalizer outputs removes the ISI. Blind adaptive equalization algorithms are typically based on implicit high-order statistics criteria or the constant modulus criteria. Generally, blind equalization algorithms have convergence and complexity problems, making them rarely used in industry.

6.2.8 Precoding

Channel equalization can also be implemented at the transmitter side. Linear preequalization at the transmitter has exactly the same performance as linear equalization at the receiver does. Precoding is done in a nonlinear way. Similar to the evolution from the ZF linear equalizer to the DFE, precoding replaces the linear filter by a nonlinear feedback structure to avoid power enhancement. Since equalization is implemented at the transmitter, error propagation that occurs in the case of the DFE is avoided. Precoding is applicable only if the channel $H(z)$ is known at the transmitter. Tomlinson-Harashima precoding and flexible precoding are the two fundamental channel precoding techniques. More exposition of precoding is given in [11]. Flexible precoding, also known as *distribution-preserving precoding*, is employed in ITU V.34 voiced-band modem standard. Flexible precoding can be viewed as being derived from linear equalization at the receiver.

The nonlinear device used in precoding is a modulo operation. There is a precoding loss as the transmit power penalty. Tomlinson-Harashima precoding converts the ISI channel $H(z)$ into a memoryless AWGN channel. Flexible precoding may cause error propagation, and for this reason Tomlinson-Harashima precoding is preferable for high rate transmission such as DSL. Implementation of Tomlinson-Harashima precoding is also very simple.

Signal shaping is intended to decrease average transmit power. The constellation expansion also causes an increase in the peak power of the transmit signal [11]. Precoding and signal shaping are combined in V.34 and V.90.

6.3 Pulse shaping

Filters for transmitters and receivers have different requirements. The transmitter filter band-limits the signal spectrum to the Nyquist bandwidth B, while the receiver filter must have high stopband attenuation to reject the out-of-band noise. The cascade of the transmitter and receiver filters must satisfy the Nyquist criterion so as to avoid ISI. This is called *Nyquist filtering*. Pulse shaping reduces sidelobe energy associated with a rectangular pulse. It is targeted at shaping the transmitted pulses to allow transmission at a rate close to $2B$ symbols/s (Nyquist rate).

An ISI-free signal is any signal that passes through zero at all but one of sampling instants. This is known as the *Nyquist criterion*

$$p(kT) = \begin{cases} 1, & k = 0 \\ 0, & k \neq 0. \end{cases} \tag{6.42}$$

That is, $p(t)$ has a zero crossing at every T seconds. Note that in this section T is the sampling period.

In the frequency domain, the folded spectrum $P_\Sigma(f)$ of $p_k(t)$ must satisfy [27]

$$P_\Sigma(f) = \frac{1}{T} \sum_{n=-\infty}^{\infty} P\left(f + \frac{n}{T}\right) = 1, \tag{6.43}$$

from which one can design pulses in the frequency domain that yield zero ISI.

The sinc function is an ISI-free signal, and it has a desired rectangular frequency response called the *desired Nyquist filter*. The sinc pulse is, however, physically unrealizable. Also, in the time domain, the signal decays slowly, at a rate of $1/t$.

6.3.1 Raised-cosine filtering

The raised-cosine pulse shaping function is a popular subset of the Nyquist filter family that satisfies the Nyquist criterion for an ISI-free baseband impulse response. In the frequency domain, the raised-cosine filter is defined by

$$R_\alpha(f) = \begin{cases} \frac{1}{2}\left[1 + \cos\frac{\pi}{a}\left(fT - \frac{1-\alpha}{2}\right)\right], & -f_2 < f < -f_1 \\ 1, & |f| \leq f_1 \\ \frac{1}{2}\left[1 + \cos\frac{\pi}{a}\left(fT + \frac{1-\alpha}{2}\right)\right], & f_1 < f \leq f_2 \\ 0, & \text{otherwise} \end{cases}, \tag{6.44}$$

where $\alpha \in (0, 1]$ is the roll-off factor, $f_1 = \frac{1-\alpha}{2T}$ and $f_2 = \frac{1+\alpha}{2T}$.

The pulse $p(t)$ in the time domain corresponds to $R_\alpha(f)$ is given by

$$p(t) = \frac{\sin(\pi t/T)}{\pi t/T} \frac{\cos(\alpha \pi t/T)}{1 - 4(\alpha t/T)^2}. \tag{6.45}$$

The raised-cosine pulse decays as a function of $1/t^3$.

Example 6.3: The frequency- and time-domain representations of the raised-cosine function are shown in Figs. 6.6 and 6.7, respectively. The impulse response of a raised-cosine filter is similar to the sinc pulse but with sidelobe that decreases with increasing α.

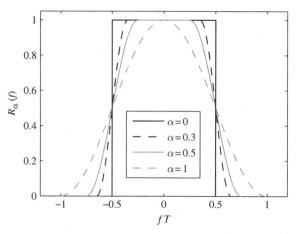

Figure 6.6   The raised-cosine filter in the frequency domain.

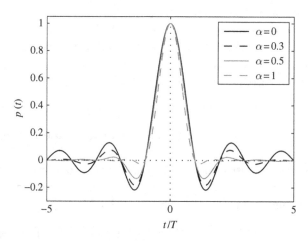

Figure 6.7   The raised-cosine filter in the time domain.

In practice, the raised-cosine filter is usually applied to shape rectangular pulses to generate ISI-free signals. The pulse shaping filter is designed as

$$H(f) = R_\alpha(f)/\mathrm{sinc}(Tf), \tag{6.46}$$

where $\mathrm{sinc}(Tf)$ is the frequency response of a rectangular pulses with width T.

Example 6.4: The frequency response of the pulse shaping filter, given by (6.46), is plotted in Fig. 6.8.

When a Nyquist filter is raised-cosine pulse shaped, the combined filter is an anti-ISI filter for a rectangular pulse shape, which has a sinc spectrum. This effectively increases the

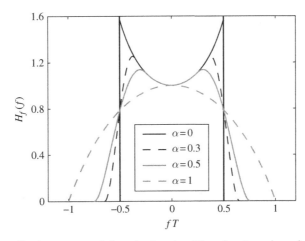

Figure 6.8

Amplitude response of the raised-cosine filtered rectangular pulse.

system bandwidth from that of the Nyquist filter, $\frac{1}{T}$, to $\frac{1+\alpha}{T}$. This requires some additional bandwidth for realizable anti-ISI filtering. This is suitable for a large channel space. An increase in α can reduce the PAPR of the transmitted signal in a SC-FDMA system [25], leading to reduced requirement on the linearity of the transmitter power amplifier.

6.3.2 Root-raised-cosine filtering

Commonly, the transmitter and receiver filters have the form of $H_T(f) = R_\alpha^\beta(f)$ and $H_R(f) = R_\alpha^{1-\beta}(f)$ to satisfy the Nyquist criterion, where β is selected to optimize the performance. This is given by the relation

$$H_N(f) = H_T(f)H_{\text{ch}}(f)H_R(f), \tag{6.47}$$

where H_N and H_{ch} are the Nyquist and channel frequency responses, respectively.

From (6.47), if $H_T(f)$ and $H_R(f)$ are related by

$$H_T^*(f) = H_R(f), \tag{6.48}$$

Equation (6.48) satisfies the requirement for matched filtering, except for the linear phase factor. By a careful selection of H_T and H_R, the system can satisfy the Nyquist and matched filtering criteria simultaneously. A popular selection is

$$H_T(f) = H_R(f) = \sqrt{R_\alpha(f)}. \tag{6.49}$$

This function is known as the *root-raised-cosine function*. It has better spectral properties than the raised-cosine function, but decays less rapidly in the time domain. In practical implementation, the transmitter and receiver filters also include a term to compensate the channel distortion.

The impulse response for the root-raised-cosine filter is given as

$$p(t) = \frac{\sin\left(\frac{(1-\alpha)\pi t}{T}\right) + \frac{4\alpha t}{T}\cos\left(\frac{(1+\alpha)\pi t}{T}\right)}{\frac{\pi t}{T}\left[1 - \left(\frac{4\alpha t}{T}\right)^2\right]}. \qquad (6.50)$$

Raised-cosine pulse shaping and root-raised-cosine filtering are widely used in wireless communications for avoiding ISI. In most practical communication systems, the matched filter must be used in the receiver since it is the optimum. Thus root-raised-cosine filtering is typically used at both the transmitter and the receiver. Root-raised-cosine pulse filtering is employed in IS-95, WCDMA, and TD-SCDMA with $\alpha = 0.22$, in IS-54/136 with $\alpha = 0.35$, in IEEE 802.16 with $\alpha = 0.25$, in ZigBee with $\alpha = 1$, in Bluetooth with $\alpha = 0.4$, and in PACS with $\alpha = 0.5$.

Problems

6.1 Consider a wireless channel with three multipath components

$$H(f) = 0.8 - 0.5e^{-j2\pi f T_s} - 0.3e^{-j4\pi f T_s}.$$

Neglecting the thermal noise, the nth received sample is given by

$$r(n) = 0.8c(n) - 0.5c(n-1) - 0.3c(n-2),$$

where $c(n)$ is the nth transmitted symbol. Design a ZF equalizer to compensate for the linear distortion caused by the channel.

6.2 Suppose that the sample values of the channel response are

$$[x(-5), x(-4), x(-3), x(-2), x(-1), x(0), x(1), x(2), x(3), x(4), x(5)]$$

$$= [0.01, -0.02, 0.03, -0.15, 0.2, 1, -0.1, 0.12, -0.06, 0.04, 0.005].$$

(a) Calculate the ZF equalizer coefficients.
(b) Set $K = 2$. Verify whether the values of $[c(-2), c(-1), c(0), c(1), c(2)] = [0, 0, 1, 0, 0]$.
(c) What are the values for $c(-3)$ and $c(3)$?

6.3 Let the pulse shaping function $g(t) = \text{sinc}(t/T_s)$. Find the matched filter for $g(t)$. [Hint: $g^*(-t)$.]

6.4 Show that the impulse response of a root raised cosine filter is given as

$$x(t) = \frac{1}{\sqrt{T}}\frac{(4\alpha t/T)\cos[\pi(1+\alpha)t/T] + \sin[\pi(1-\alpha)t/T]}{(\pi t/T)\left[1 - (4\alpha t/T)^2\right]}.$$

Plot the impulse response.

6.5 Show that a pulse having the raised-cosine spectrum satisfies the Nyquist criterion for any value of the roll-off factor α.

6.6 4PAM is used for transmitting at a bit rate of 9600 bits/s on a channel with frequency response

$$C(f) = \frac{1}{(1 + j(f/2400))^2}, \quad |f| \le 2400$$

and $C(f) = 0$ otherwise. The noise is zero-mean AWGN with PSD $N_0/2$ W/Hz. Determine the magnitude responses of the optimum transmitting and receiver filters.

6.7 For a telephone channel in the frequency range 300 Hz to 3000 Hz, (a) select a power-efficient constellation to achieve a data rate of 9600 bits/s.
(b) Assuming an ideal channel, if root-raised-cosine pulse is used as the transmitter pulse, select the roll-off factor.

6.8 A channel has an impulse response $h(\tau) = 1 - 0.2\tau$. Derive an expression for the coefficients of a T-space ZF equalizer.

6.9 Write a computer program to design: (a) the ZF equalizer and (b) the MMSE equalizer. Test the program using some specific channel conditions.

6.10 Write a computer program to plot the impulse responses of the raised-cosine and root-raised-cosine filter. Plot for the cases of $\alpha = 0, 0.22, 0.5$, and 1.

References

[1] K. Abend & B. D. Fritchman, Statistical detection for communication channels with intersymbol interference. *Proc. IEEE*, **58**:5 (1970), 779–785.
[2] N. Benvenuto & S. Tomasin, On the comparison between OFDM and single carrier modulation with a DFE using a frequency-domain feedforward filter. *IEEE Trans. Commun.*, **50**:6 (2002), 947–955.
[3] J. M. Cioffi, G. P. Dudevoir, M. V. Eyuboglu & G. D. Forney, Jr., MMSE decision-feedback equalizers and coding – part I: equalization results. *IEEE Trans. Commun.*, **43**:10 (1995), 2582–2594.
[4] J. Coon, M. Sandell, M. Beach & J. McGeehan, Channel and noise variance estimation and tracking algorithms for unique-word based singlecarrier systems. *IEEE Trans. Wireless Commun.*, **5**:6 (2006), 1488–1496.
[5] A. Czylwik, Comparison between adaptive OFDM and single carrier modulation with frequency domain equalization. In *Proc. IEEE VTC*, Phoenix, AZ, May 1997, **2**, 865–869.
[6] D. Divsalar & M. K. Simon, Multiple-symbol differential detection of MPSK. *IEEE Trans. Commun.*, **38**:3 (1990), 300–308.
[7] K.-L. Du & M. N. S. Swamy, *Neural Networks in a Softcomputing Framework* (London: Springer, 2006).
[8] K.-L. Du & M. N. S. Swamy, An adaptive space-time multiuser detection algorithm for CDMA systems. In *Proc. IEEE WiCOM*, Wuhan, China, Sep 2006, 1–5.

[9] U. Fincke & M. Pohst, Improved methods for calculating vectors of short length in a lattice, including a complexity analysis. *Math. of Computation*, **44** (1985), 463–471.

[10] R. F. H. Fischer, C. Windpassinger, A. Lampe & J. B. Huber, MIMO precoding for decentralized receivers. In *Proc. IEEE ISIT*, Lausanne, Switzerland, June-Jul 2002, 496.

[11] R. F. H. Fischer, *Precoding and Signal Shaping for Digital Transmission* (New York: Wiley-IEEE Press, 2002).

[12] G. D. Forney, Jr., Maximum likelihood sequence estimation of digital sequence in the presence of intersymbol interference. *IEEE Trans. Inf. Theory*, **18**:3 (1972), 363–378.

[13] R. D. Gitlin and S. B. Weinstein, Fractionally-spaced equalization: an improved digital transversal equalizer. *Bell Syst. Tech. J.*, **60**:2 (1981), 275–296.

[14] M. Guo, M. O. Ahmad, M. N. S. Swamy & C. Wang, FPGA design and implementation of a low-power systolic array-based adaptive Viterbi decoder. *IEEE Trans. Circ. Syst. I*, **52**:2 (2005), 350–365.

[15] B. Hassibi & H. Vikalo, On the expected complexity of sphere decoding. In *Proc. Asilomar Conf. Signals Systems Computers*, Pacific Grove, CA, Nov 2001, **2**, 1051–1055.

[16] S. Haykin, *Adaptive Filter Theory*, 4th edn (Upper Saddle River, NJ: Prentice Hall, 2002).

[17] Y. Jiang, Y. Tang, Y. Wang & M. N. S. Swamy, A trace-back-free Viterbi decoder using a new survival path management algorithm. In *Proc. IEEE ISCAS*, Phoenix, AZ, May 2002, **1**, 261–264.

[18] B. Li, W. Tong & P. Ho, Multiple-symbol detection for orthogonal modulation in CDMA system. *IEEE Trans. Veh. Tech.*, **50**:1 (2001), 321–325.

[19] J. Lin & D. J. Costello, Jr., *Error Control Coding: Fundamentals and Applications* (Englewood Cliffs, NJ: Prentice Hall, 1983).

[20] E. M. Long & A. M. Bush, Decision-aided sequential sequence estimation for intersymbol interference channels. In *Proc. IEEE ICC*, Boston, MA, Jun 1989, 26.1.1–26.1.5.

[21] R. W. Lucky, Techniques for adaptive equalization of digital communication systems. *Bell Syst. Tech. J.*, **45**:2 (1966), 255–286.

[22] X. Ma & W. Zhang, Fundamental limits of linear equalizers: diversity, capacity, and complexity. *IEEE Trans. Inf. Theory*, **54**:8 (2008), 3442–3456.

[23] M. Morelli, L. Sanguinetti & U.Mengali, Channel estimation for adaptive frequency-domain equalization. *IEEE Trans. Wireless Commun.*, **4**:5 (2005), 2508–2518.

[24] W. H. Mow, Maximum likelihood sequence estimation from the lattice viewpoint. *IEEE Trans. Inf. Theory*, **40**:5 (1994), 1591–1600.

[25] H. G. Myung, J. Lim & D. J. Goodman, Peak-to-average power ratio of single carrier FDMA signals with pulse shaping. In *Proc. IEEE PIMRC*, Helsinki, Finland, Sep 2006, 1–5.

[26] A. Paulraj, R. Nabar & D. Gore, *Introduction to Space–Time Wireless Communications* (Cambridge, UK: Cambridge University Press, 2003).

[27] J. G. Proakis & M. Salehi, *Digital Communications*, 5th edn (New York: McGraw-Hill, 2008).

[28] J. J. Shynk, Frequency-domain and multirate adaptive filtering. *IEEE Signal Process. Mag.*, **9**:1 (1992), 14–35.

[29] U. Sorger, I. De Broeck & M. Schnell, Interleaved FDMA – a new spreading-spectrum multiple-access scheme. In *Proc. IEEE ICC*, Atlanta, GA, Jun 1998, 1013–1017.

[30] G. L. Stuber, *Principles of Mobile Communication*, 2nd edn (Boston, MA: Kluwer, 2001).

[31] A. J. Viterbi, Error bounds for convolutional codes and an asymptotically optimum decoding algorithm. *IEEE Trans. Inf. Theory*, **13**:2 (1967), 260–269.

[32] A. J. Viterbi, *Principles of Spread Spectrum Communication* (Reading, MA: Addison-Wesley, 1995).

[33] T. Walzman & M. Schwartz, Automatic equalization using the discrete frequency domain. *IEEE. Trans. Inf. Theory*, **19**:1 (1973), 59–68.

[34] F. Wan, W.-P. Zhu & M. N. S. Swamy, A semiblind channel estimation approach for MIMO-OFDM systems. *IEEE Trans. Signal Process.*, **56**:7 (2008), 2821–2834.

[35] F. Wan, W.-P. Zhu & M. N. S. Swamy, Frequency-domain semi-blind channel estimation for MIMO-OFDM Systems. Submitted to *IEEE Trans. Wireless Commun.*

[36] B. Widrow, J. M. McCool, M. G. Larimore, and J. C. R. Johnson, Adaptive switching circuits. In *Proc. IRE WESCON Conv.*, Los Angeles, CA, 1960, 96–104.

[37] Y. R. Zheng & C. Xiao, Channel estimation for frequency-domain equalization of single-carrier broadband wireless communications. *IEEE Trans. Veh. Tech.*, **58**:2 (2009), 815–823.

7.1 Analog modulation

Before we start to deal with digital modulation, we first briefly introduce analog modulation. Given a carrier waveform

$$x(t) = a\cos(\omega t + \phi), \tag{7.1}$$

a message can be modulated on the amplitude, frequency, phase, or a combination of these, leading to various modulation techniques. Analog modulation, such as amplitude modulation (AM), phase modulation (PM), or frequency modulation (FM), is only used in analog wireless communication systems.

7.1.1 Amplitude modulation

For a baseband signal $x_{BB}(t)$, the conventional AM generates a waveform of the type

$$x_{AM}(t) = A\left[1 + mx_{BB}(t)\right]\cos \omega_c t, \tag{7.2}$$

where m is the modulation index, and $\omega_c = 2\pi f_c$ is the angle frequency of the carrier.

The spectrum of the modulated signal is obtained by taking the Fourier transform of the signal $x_{AM}(t)$

$$X_{AM}(f) = \frac{A}{2}\left[m\left(X_{BB}\left(f - f_c\right) + X_{BB}\left(f + f_c\right)\right) + \delta\left(f - f_c\right) + \delta\left(f + f_c\right)\right]. \tag{7.3}$$

Double-sideband (DSB) AM defines the modulated waveform as

$$x_{DSB\text{-}AM}(t) = Ax_{BB}(t)\cos \omega_c t. \tag{7.4}$$

The spectrum of the modulated signal is given by

$$X_{DSB\text{-}AM}(f) = \frac{A}{2}X_{BB}\left(f - f_c\right) + \frac{A}{2}X_{BB}\left(f + f_c\right). \tag{7.5}$$

The spectra of conventional AM and DSB-AM are illustrated in Fig. 7.1. Both techniques have the same bandwidth of the modulated signal, which is twice that of the baseband bandwidth

$$B_{AM} = 2B_{BB}. \tag{7.6}$$

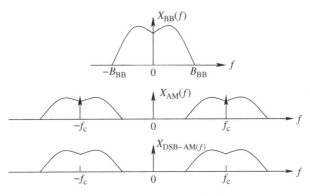

Figure 7.1   Spectra of the conventional AM and the DSB-AM.

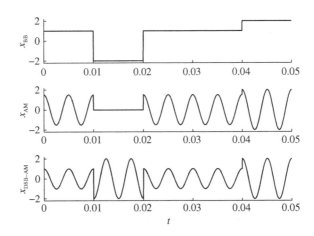

Figure 7.2   Modulated waveforms of conventional AM and DSB-AM.

There is a spectral line at f_c for the conventional AM, and this corresponds to a sinu-soidal component at f_c. This component does not transmit any information, leading to a lower power efficiency, compared to DSB-AM. For this reason, conventional AM is some-times termed *DSB-large carrier (DSB-LC)*, while DSB-AM is termed *DSB-small carrier (DSB-SC)*. However, the presence of this component makes $1 + mx_{BB}(t)$ always positive, resulting in easy demodulation when applying cheap envelope detection. Single-sideband (SSB) AM is obtained by filtering out one of the sidebands.

Example 7.1: Given a digital-like waveform and the AM parameters $m = 0.5, A = 1$, and $f_c = 200$, the amplitude modulated waveforms of the conventional AM and the DSB-AM are shown in Fig. 7.2.

AM signals can be demodulated using a coherent demodulator. Demodulation that uses an LO identical in frequency and phase to the modulator LO is referred to as *synchronous* or *coherent demodulation.* Coherent SSB- and DSB-AM demodulators have the same SNR performance, while coherent conventional AM has a smaller output SNR performance. For signals using conventional AM, demodulation can also be done by using an envelope detector, and such a method is known as *noncoherent demodulation.* The noncoherent demodulator is comparable to the coherent demodulator in terms of the output SNR for large input SNR, but introduces significant degradation for small input SNR.

AM is mainly used in broadcast radios and sound modulation in television, and finds limited application in other wireless systems, since the modulated amplitude is more susceptible to noise and a highly linear power amplifier is required in the transmitter.

7.1.2 Phase modulation and frequency modulation

Both PM and FM methods are angle modulation schemes, and they are nonlinear modulation schemes. The phase-modulated and frequency-modulated signals are, respectively, defined by

$$x_{PM}(t) = A \cos \left[\omega_c t + m x_{BB}(t) \right], \tag{7.7}$$

$$x_{FM}(t) = A \cos \left[\omega_c t + m \int_{-\infty}^{t} x_{BB}(t) dt \right], \tag{7.8}$$

where m is the phase or frequency modulation index.

Example 7.2: For the same message given in Example 7.1 and the modulation parameters: $A = 1, f_c = 200$, and $m_{FM} = 1000$ or $m_{PM} = 10$, the waveforms of the FM and PM signals are plotted in Fig. 7.3.

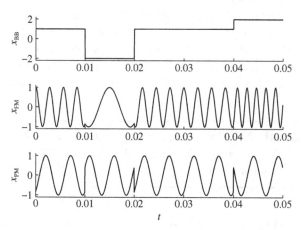

Figure 7.3   Modulated waveforms of FM and PM.

FM is more widely used than PM. This is due to the fact that frequency modulation and demodulation are simple to implement. In reality, any circuit having a transfer function that is sensitive to frequency can be used as a frequency demodulator. For example, a highpass filter can actually convert an FM signal to an AM signal, whose amplitude corresponds to the frequency of the FM signal.

For implementation of FM, the baseband signal that bears information is passed through a pre-emphasis filter to emphasize weaker higher frequency signal components before it is passed to either a direct or indirect FM modulator. This is because the discriminator's output noise PSD, which is proportional to the square of the frequency [15], has a much more adverse influence on the higher frequency components of the information-bearing signal. At the receiver, the emphasized components are de-emphasized. More details on analog FM may be found in [32].

The nonlinear dependence of PM or FM signals upon the baseband signal makes the bandwidth of the modulated signal infinite. Thus, the effective bandwidth of the signal, B_{FM} or B_{PM}, can be defined as that containing 98% of the signal power, and this bandwidth is given by *Carson's rule*

$$B_{FM}, B_{PM} \approx 2(\beta + 1)B_{BB}, \tag{7.9}$$

where $\beta = mA_m/\omega_m$, when $x_{BB}(t) = A_m \cos \omega_m t$. The parameter β is also called *modulation index*. Accordingly,

$$x_{FM}(t) = A \cos(\omega_c t + \beta \sin \omega_m t). \tag{7.10}$$

The SNR of the FM radio can be much better than the AM radio, but needs more channel bandwidth. The output SNRs of FM and AM are related by [27]

$$SNR_{FM} = \frac{3(2 + m^2)}{2m^2}\beta^2 SNR_{AM}, \tag{7.11}$$

where m is the AM modulation index. Thus, the output SNR of FM increases in square law, when its bandwidth increases linearly.

PM and FM are more important than AM, since they are also used in the analysis of oscillators and frequency synthesizers. FM finds applications in two-way voice radio, and is used in most 1G mobile communication systems. Demodulation of FM signals can be performed by using a discriminator and an envelope detector.

7.2 Introduction to digital modulation

Digital modulation techniques are more important for modern wireless systems. Compared to analog systems, digital systems achieve better spectral efficiency, have better noise and fade-rejecting capability, and require lower transmit power. In addition, error correction and encryption can be easy to implement in digital systems.

As in analog systems, the amplitude, frequency, and phase of a carrier $A \cos(\omega t + \phi)$ are used to modulate digital symbols, and accordingly the three basic classes of digital modulation techniques are amplitude-shift keying (ASK), frequency-shift keying (FSK), and phase-shift keying (PSK). Another important class of digital modulation technique is quadrature amplitude modulation (QAM), which employs both the amplitude and phase for modulation. These modulation techniques are widely used in various standards for wireless communications.

In recent wireless standards, multiple modulation techniques are used, in an adaptive manner. The modulation parameters such as the modulation level, symbol rate, or coding rate of the channel coding are adaptively adjusted according to the channel and traffic. Adaptive modulation can be preassigned or be dependent on the measurement of the channel and traffic. Adaptive modulation is embodied in all 3G/4G and some recent wireless network standards.

7.2.1 Signal space diagram

Digital signal symbols are represented by different states in the signal space diagram, known as the *constellation diagram*. The shape of the constellation is related to the modulation method. Typically, the constellation takes form of squares or circles, for simple demodulation. It is interesting to note that rectangular and hexagonal constellations have a better power efficiency than the corresponding square and circular constellations associated with M-ary QAM (MQAM) and M-ary PSK (MPSK), with a saving of 1.3 dB in power [11]. This gain in power, however, leads to increased complexity in constellation. This method is usually not used, since coding provides much better performance.

The carrier modulated waveform can be expressed in the complex envelope form

$$s_{\mathrm{BP}}(t) = \Re \left[s_{\mathrm{LP}}(t) e^{j 2\pi f_c t} \right], \tag{7.12}$$

where

$$s_{\mathrm{LP}}(t) = s_{\mathrm{LP},I}(t) + j s_{\mathrm{LP},Q}(t) \tag{7.13}$$

is the complex baseband signal, and $s_{\mathrm{BP}}(t)$ is the upconverted bandpass signal, which is the physically existing signal. In quadrature form, the bandpass waveform is expressed by

$$s_{\mathrm{BP}}(t) = s_{\mathrm{LP},I}(t) \cos 2\pi f_c t - s_{\mathrm{LP},Q}(t) \sin 2\pi f_c t. \tag{7.14}$$

In digital modulation, the bandpass signal is transmitted at each baud interval as a member of a finite set of finite energy waveforms, $\{s_{\mathrm{BP},0}(t), s_{\mathrm{BP},1}(t), \cdots, s_{\mathrm{BP},M-1}(t)\}$. The baseband counterpart is denoted by $\{s_{\mathrm{LP},0}(t), s_{\mathrm{LP},1}(t), \cdots, s_{\mathrm{LP},M-1}(t)\}$. Each complex waveform $s_{\mathrm{LP},m}$ can be projected onto a set of N orthogonal basis functions, yielding a signal vector representation

$$s_{\mathrm{LP},m} = \left(s_{\mathrm{LP},m,0}, s_{\mathrm{LP},m,1}, \cdots, s_{\mathrm{LP},m,N-1} \right). \tag{7.15}$$

The N orthogonal basis functions can be obtained from the M signals by using the Gram-Schmidt orthogonalization [9].

The energy in the waveform $s_{\text{BP},m}(t)$ is given by

$$E_m = \langle s_{\text{BP},m}, s_{\text{BP},m}\rangle = \|s_{\text{BP},m}\|^2 = \int_{-\infty}^{\infty} s_{\text{BP},m}^2(t)dt$$

$$\approx \frac{1}{2}\int_{-\infty}^{\infty} s_{\text{LP},m}^2(t)dt = \frac{1}{2}\|s_{\text{LP},m}\|^2. \tag{7.16}$$

The approximation is applicable when the bandwidth of the baseband signal B_{BB} is much less than f_c.

The squared Euclidean distance between two symbols $s_{\text{BP},i}(t)$ and $s_{\text{BP},j}(t)$ is given by

$$d_{ij}^2 = \|s_{\text{BP},i} - s_{\text{BP},j}\|^2 = \frac{1}{2}\|s_{\text{LP},i} - s_{\text{LP},j}\|^2$$

$$= \int_{-\infty}^{\infty}\left[s_i(t) - s_j(t)\right]^2 dt$$

$$= E_{s_i} + E_{s_j} - 2\sqrt{E_{s_i}E_{s_j}}\,\Re(\rho_{ij}), \tag{7.17}$$

where ρ_{ij} is the correlation coefficient between the two symbols

$$\Re(\rho_{ij}) = \frac{\langle s_{\text{BP},i}, s_{\text{BP},j}\rangle}{\|s_{\text{BP},i}\|\cdot\|s_{\text{BP},j}\|} = \frac{s_{\text{BP},i}s_{\text{BP},j}}{|s_{\text{BP},i}|\cdot|s_{\text{BP},j}|}$$

$$= \Re\left(\frac{\langle s_{\text{LP},i}, s_{\text{LP},j}\rangle}{\|s_{\text{LP},m}\|\,\|s_{\text{LP},j}\|}\right). \tag{7.18}$$

For antipodal binary signals, $\rho_{12} = -1$ and d_{ij} reaches its maximum. For orthogonal signals $\rho_{12} = 0$.

7.2.2 Demodulation and detection

The optimum receiver of any modulation alphabet represented in a signal space diagram is a correlator, or a matched filter, matched to all the possible transmit waveforms. The coherent receiver compensates for phase rotation due to the channel by means of carrier recovery. By adjusting the magnitude of the channel attenuation, in the absence of noise, the received signal r is equal to the transmitted signal s. The ideal detector is known as the *MAP detector*. For the assumption of equiprobable symbols, the MAP detector is identical to the ML detector.

When carrier recovery is difficult, differential detection becomes an attractive alternative. The receiver needs just to compare the phases and/or amplitudes of two subsequent symbols. In order to implement differential detection, the transmitted signals need to be differentially encoded. When the carrier phase is completely unknown, noncoherent detection such as envelope detection can be used. Generally, coherent detection provides the best bit error probability (BEP) performance, and is discussed here. The block diagrams of MPSK, M-ary pulse amplitude modulation (MPAM) and QAM demodulators can be found in [28].

For M-ary signaling, it is typically assumed that the symbol energy is equally divided among all bits and that Gray encoding is used. Based on these assumptions, when coherent ML detection is employed, the bit SNR and the symbol SNR are related by

$$\gamma_b \approx \frac{\gamma_s}{\log_2 M} \tag{7.19}$$

and the BEP and the symbol error probability (SEP) are related by

$$P_b \approx \frac{P_s}{\log_2 M}. \tag{7.20}$$

The bit SNR and symbol SNR are defined by

$$\gamma_b = \frac{E_b}{N_0}, \quad \gamma_s = \frac{E_s}{N_0}, \tag{7.21}$$

where the average bit energy E_b is calculated by $E_b = P_R T_b$, P_R being the average received power and T_b the bit period. The noise power spectrum N_0 is constant, measured in watts/Hz.

Coherent maximum-likelihood detection

Consider a channel of the form

$$y = \mathbf{H}x + e \tag{7.22}$$

where the vector e has components that are zero-mean complex AWGN variables with variance σ^2, and $\mathbf{H}$ is known at the receiver. Thus, y is a complex Gaussian random vector with mean $\mathbf{H}x$ and covariance matrix $\sigma^2 \mathbf{I}$.

The conditional pdf (or likelihood function) of y, conditioned on x, can be written as

$$p(y|x) = \frac{1}{\pi^n \sigma^{2n}} e^{-\frac{1}{\sigma^2} \|y - \mathbf{H}x\|^2}, \tag{7.23}$$

where n is the dimension of y. Given y, x can be estimated by maximizing the likelihood function. This is equivalent to the minimization of the ML metric

$$\hat{x} = \arg\min_x \|y - \mathbf{H}x\|^2. \tag{7.24}$$

7.2.3 Error probability in the Gaussian channel

The error probability is determined by the decision zone of each symbol in the constellation diagram, and can be analyzed by using the ML receiver structure.

From the fundamental theory of detection and estimation of signals in noise, for a Gaussian channel, the probability of symbol s_i being misjudged as symbol s_j is given by [28, 39]

$$\Pr\left(s_i, s_j\right) = Q\left(\sqrt{\frac{d_{ij}^2}{2N_0}}\right) = \frac{1}{2}\mathrm{erfc}\left(\frac{1}{2}\sqrt{\frac{d_{ij}^2}{N_0}}\right), \tag{7.25}$$

where d_{ij} is the Euclidean distance between the two symbols, and N_0 is the noise power. The Q-function is related to the complementary error function, erfc(x), by (refer to Appendix A)

$$\text{erfc}(x) = 2Q\left(\sqrt{2}x\right). \tag{7.26}$$

The Q-function is upper-bounded by the Chernoff bound

$$Q(t) \le e^{-t^2/2}. \tag{7.27}$$

For binary signals, all symbols have the same energy E_s, and we have

$$\Pr\left(s_i, s_j\right) = Q\left(\sqrt{\gamma_s\left(1 - \Re\left(\rho_{ij}\right)\right)}\right). \tag{7.28}$$

For binary orthogonal signals such as the binary FSK (BFSK), minimum shift keying (MSK) and binary pulse position modulation (PPM), we have $\Pr\left(s_i, s_j\right) = Q\left(\sqrt{\gamma_s}\right)$. For antipodal signaling, $\Pr\left(s_i, s_j\right) = Q\left(\sqrt{2\gamma_s}\right)$.

The union bound

Given M equally likely symbols such as MPSK, the union bound on the SEP P_s is given by

$$P_s \le \frac{1}{M}\sum_{i=1}^{M}\sum_{k=1, k\neq i}^{M} Q\left(\frac{d_{ik}}{\sqrt{2N_0}}\right). \tag{7.29}$$

Defining the minimum distance of the constellation as $d_{\min} = \min\left(d_{ik}\right)$, the above bound can be simplified with a looser bound

$$P_s \le (M-1)Q\left(\frac{d_{\min}}{\sqrt{2N_0}}\right). \tag{7.30}$$

The nearest-neighbor approximation to P_s is given by the probability of error associated with constellations at the minimum distance $d_{\min}$ scaled by $M_{d_{\min}}$, the number of neighbors at this distance [13]

$$P_s \approx M_{d_{\min}}Q\left(\frac{d_{\min}}{\sqrt{2N_0}}\right). \tag{7.31}$$

The nearest-neighbor approximation is less than the loose bound, and also slightly less than the union bound. At high SNR, it is quite close to the exact probability of symbol error.

For an M-ary modulation, a symbol contains $\log_2 M$ bits. The $\log_2 M$ possible bits should be designed to map to symbol s_i, $i = 1,\ldots,M$, in such a way that a decoding error is associated with an adjacent decision region, leading to only one bit error. This mapping scheme is known as the Gray coding. In this case, the error probability of a symbol being its nearest neighbors is very low. Thus, we have the BEP as

$$P_b \approx \frac{P_s}{\log_2 M}. \tag{7.32}$$

7.3 Baseband modulation

Baseband modulation represents digital sequences by pulse waveforms that are suitable for baseband transmission. Baseband modulation waveforms are also called *line codes* or *PCM codes*.

7.3.1 Line codes

Basic classes of line codes are nonreturn-to-zero (NRZ), return-to-zero (RZ), pseudoternary, and biphase. NRZ and RZ codes can be subdivided into unipolar and bipolar. Many more complex substitution codes and block codes are also available, and they are described in [39].

The different line codes are used for different applications. Selection of line code is based on the following requirements on performance [39]:

- *Adequate timing information in the received data sequence.* For this purpose, line codes with higher transition density are preferable.
- *A spectrum suitable for the channel.* The PSD of the line code should have small bandwidth compared with the channel bandwidth to avoid the occurrence of ISI.
- *Narrow bandwidth.*
- *Low error probability.*
- *Error detection capability.*
- *Bit sequence independence (transparency).* Attributes of the code are independent of the source statistics.
- *Differential coding.* Differential coded sequences are immune from polarity inversion. Some line codes have differential coding inherent in them.

Among NRZ codes, the NRZ-L (NRZ-level) waveform is the most common waveform in digital logic. It uses a level of A for 1, and a level of $-A$ or 0 for 0 (depending on bipolar or unipolar). NRZ-M (NRZ-mark) and NRZ-S (NRZ-space) are two differentially encoded NRZ codes. In telecommunications, NRZ codes are limited to short-haul links due to the lack of the timing information. RZ codes have more transitions in the waveform that render easy timing, but this introduces a wider bandwidth.

Pseudoternary codes use three levels $\pm A$ and 0. The AMI (alternative mark inversion) codes in this class are often called *bipolar codes* in telecommunications. They are implemented as AMI-RZ and AMI-NRZ. Other codes in this class are the dicode NRZ and dicode RZ. Dicodes and AMI codes are related by differential coding.

Biphase codes use half-period pulses with differential phases. The Bi-Φ-L (biphase-level) format is better known as the *Manchester* code. Some biphase codes, such as Bi-Φ-L, Bi-Φ-M (biphase-mark), Bi-Φ-S (biphase-space), typically ensure at least one transition in a bit duration, thus providing adequate timing information to the demodulator. The conditioned Bi-Φ-L code, as a differentially encoded Bi-Φ-L, is immune from polarity inversion in the circuit. The delay modulation code, also called the *Miller code*, is also a biphase code. The Manchester code is specified for the IEEE 802.3 standard (Ethernet) for baseband coaxial cable, and differential Manchester has been specified in the IEEE 802.5 standard for token ring, using baseband coaxial cable or twisted pair.

PSDs

Most of the digital modulated baseband signals can be written in the form

$$s(t) = \sum_{k=-\infty}^{\infty} a_k g(t - kT), \tag{7.33}$$

where a_k's are random data bits, $g(t)$ is the pulse shape in $[0, T]$, and T is the bit duration. For NRZ codes, $g(t) = A, 0 \leq t \leq T$, and 0 otherwise, A being the signal amplitude; for RZ codes, $g(t) = A, 0 \leq t \leq T/2$, and 0 otherwise. In the unipolar case, $a_k = 1$ for binary 1, and 0 for binary 0; in the bipolar case, $a_k = 1$ for binary 1, and -1 for binary 0.

For wide sense stationary signals, the PSD can be calculated by taking the Fourier transform of the autocorrelation $R(\tau)$, using the Wiener-Khinchine theorem. For cyclostationary signals, $R(\tau)$ is the time average of the time-dependent $R(t, \tau)$ in a period. The derivation of the PSD is based on the stochastic mode of the bit sequences. For uncorrelated data sequences $\{a_k\}$, the PSD is derived as [39]

$$\Phi_s = \frac{|G(f)|^2}{T} \left[\sigma_a^2 + m_a^2 R_b \sum_{n=-\infty}^{\infty} \delta (f - nR_b) \right], \tag{7.34}$$

where $R_b = 1/T$ is the bit rate, $G(f)$ is the Fourier transform of $g(t)$, and σ_a^2 and m_a are the variance and mean of $\{a_k\}$.

The fractional out-of-band power is defined as

$$P_{ob}(B) = 1 - \frac{\int_{-B}^{B} \Phi_s(f)df}{\int_{-\infty}^{\infty} \Phi_s(f)df}. \tag{7.35}$$

This can be obtained by numerical integration.

The PSD of $s(t)$ is derived, for NRZ codes, as [39]

$$\Phi_s(f) = \frac{A^2 T}{4} \left(\frac{\sin \pi f T}{\pi f T} \right)^2 + \frac{A^2}{4} \delta(f) \quad \text{(unipolar NRZ)}, \tag{7.36}$$

$$\Phi_s(f) = A^2 T \left(\frac{\sin \pi f T}{\pi f T} \right)^2 \quad \text{(bipolar NRZ)}. \tag{7.37}$$

These PSD expressions are valid for NRZ-L, NRZ-M, and NRZ-S codes, since the statistics properties of their sequences are the same.

For unitary average symbol energy, $A = \sqrt{2}$ in both the cases. The two PSDs have the same shape, but the unipolar case has an impulse at dc.

The PSDs of all the basic line codes are given in [39]. The bandwidths of some of the line codes are also given in [39], and they are listed in Table 7.1, where B_{null} stands for the null bandwidth, $B_{90\%}$ for the 90% energy bandwidth, and $B_{99\%}$ for the 99% energy

Table 7.1. Bandwidths of basic line codes.

Line codes	B_{null}	$B_{90\%}$	$B_{99\%}$
bipolar NRZ(-L,-M,-S)	$1.0R_b$	$0.85R_b$	$10R_b$
unipolar NRZ(-L,-M,-S)	$1.0R_b$	$0.54R_b$	$5R_b$
bipolar RZ	$2.0R_b$	$1.7R_b$	$22R_b$
unipolar RZ	$2.0R_b$	$1.6R_b$	$22R_b$
AMI-RZ, dicode RZ	$1.0R_b$	$1.71R_b$	$20R_b$
AMI-NRZ, dicode NRZ	$1.0R_b$	$1.53R_b$	$15R_b$
Bi-Φ(-L,-M, -S)	$2.0R_b$	$3.05R_b$	$29R_b$

bandwidth. The NRZ code has a narrow bandwidth ($B_{\text{null}} = R_b$). The RZ code increases the density of transitions for ease of timing recovery, but leads to a doubling of the bandwidth.

BEPs

The BEPs of line codes transmitted through an AWGN channel can be obtained for optimum detection. The BEPs for the basic line codes can be derived from (7.28). The results for some line codes are given as [39]

$$P_b = Q\left(\sqrt{2\gamma_b}\right) \quad \text{(bipolar NRZ-L, bipolar-RZ, Bi-Φ-L)}, \tag{7.38}$$

$$P_b \approx 2Q\left(\sqrt{2\gamma_b}\right) \quad \text{(bipolar NRZ-M/-S, conditioned Bi-Φ-L)}, \tag{7.39}$$

$$P_b = Q\left(\sqrt{\gamma_b}\right) \quad \text{(unipolar NRZ-L, unipolar-RZ, Bi-Φ-M/-S)}, \tag{7.40}$$

where $\gamma_b = \frac{E_b}{N_0}$.

Example 7.3: The BEPs for different line codes, given by (7.38)–(7.40), are plotted in Fig. 7.4.

7.3.2 Pulse time modulation

Pulse time modulation is a type of baseband modulation scheme that is in between being pure analog or pure digital. It uses a constant amplitude. Pulse time modulation generates a sequence of pulses with a spectrum around the sampling frequency and its harmonics. Popular pulse time modulation schemes are pulse-width modulation (PWM) and PPM. They have fixed symbol intervals, and this enables much easier multiplexing and demultiplexing in the time domain. A PWM or PPM signal can be generated by uniform sampling or natural sampling.

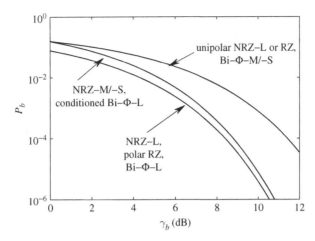

Figure 7.4

BEPs of some line codes.

PCM code, as will be discussed in Chapter 12, is also a baseband coding scheme. It is widely used in source coding. PCM codes can be either NRZ or RZ. The digital PPM denotes a PCM code of k bits by a narrow pulse at one of the 2^k positions for a symbol period T in the time domain, leading to a bandwidth expansion of $2^k/k$. Compared to PCM, digital PPM can achieve a 5–11 dB improvement in sensitivity, but requires more transmission bandwidth [39]. PPM is usually implemented differentially as differential PPM to avoid the difficulties in synchronization. PPM is an M-ary modulation technique that can be detected noncoherently.

PPM is inherently sensitive to multipath interference, and for this reason, it is mainly used in optical or UWB wireless communications, where there is no or very little multipath interference. The 16PPM and 4PPM are used in the infrared physical mode of the baseline IEEE 802.11 standard. They are also used in the ISO 15693 standard for contactless Smart card, and the EPC standard for RFID.

7.4 Pulse amplitude modulation

Pulse amplitude modulation (PAM) can be in the baseband or in the passband. M-ary amplitude-shift keying (MASK) is MPAM in the passband case (with a carrier). NRZ and RZ line codes are baseband binary PAM. All information is encoded into the signal amplitude.

MPAM is one-dimensional linear modulation. For passband MPAM, that is, for MASK, the transmitted signal over one symbol period is given by

$$s_i(t) = A_i g(t) \cos(2\pi f_c t), \quad 0 \leq t \leq T, \tag{7.41}$$

Figure 7.5   The constellation of 4PAM. The information bits are Gray-encoded.

where $A_i = (2i - 1 - M)d$, $i = 1, 2, \cdots, M$ (for bipolar MPAM) or $A_i = (i - 1)d$ (for unipolar MPAM), and $g(t)$ is a real-valued pulse shaping function. The bandpass MPAM signal is a DSB-SC AM signal.

The MPASK signals have energies

$$E_i = \int_0^T s_i^2(t)dt = \frac{1}{2}A_i^2 \int_0^T g^2(t)dt = \frac{1}{2}A_m^2 E_g, \tag{7.42}$$

where E_g denotes the energy in the pulse $g(t)$. On a signal constellation diagram, $s_i = A_i\sqrt{\frac{1}{2}E_g}$, $i = 1, 2, \ldots, M$. For the rectanglular pulse $g(t) = \sqrt{\frac{2}{T}}$, $E_g = 2$, we have $s_i = A_i$. The signal constellation $\{s_i\}$ is one-dimensional, and the case of 4PAM is illustrated in Fig. 7.5.

For baseband MPAM, the cosine factor in (7.41) is dropped, that is, the transmitted signal over one symbol period is given by

$$s_i(t) = A_i g(t), \quad 0 \le t \le T. \tag{7.43}$$

On-off keying (OOK) is a unipolar MASK with $M = 2$. Like AM, ASK can be demodulated coherently or by using an envelope detector. For binary digital modulation, given the same probability of error, coherent FSK and coherent ASK requires the same power, which is 3 dB higher than that required for coherent PSK. Noncoherent detection techniques such as envelope detection lead to worse BEP performance than their respective coherent version. ASK has a high sensitivity to amplitude noise, and thus is less popular than PSK and FSK in wireless communications.

PSD

The PSD of MPAM contains two terms, the first being a continuous term and the second a discrete term with spectral lines at uniform frequency intervals of $1/T$. When the amplitudes are symmetrically distributed around zero, their mean m_A is zero, resulting in the second term being zero, and the discrete spectral lines disappear. Thus, for bipolar MASK there are no spectral lines, while for unipolar MASK there are spectral lines.

The PSD for the symmetrical bipolar MPAM (not necessarily uniform spaced) is given by

$$\Phi_{\tilde{s}}(f) = \frac{\sigma_A^2|G(f)|^2}{T}, \tag{7.44}$$

where $G(f)$ is the Fourier transform of $g(t)$. Note that $\Phi_{\tilde{s}}(f)$ is the PSD of the complex envelope of the signal.

When $g(t)$ is a rectangular pulse of unit amplitude, the PSD of the symmetrical bipolar MASK is given by

$$\Phi_{\tilde{s}}(f) = \sigma_A^2 T \left(\frac{\sin(\pi f T)}{\pi f T} \right)^2, \tag{7.45}$$

which has the same shape as that of MPSK. σ_A^2 is the variance of A_i.

If $m_A \neq 0$, as in unipolar MPAM, and $g(t)$ is a rectangular pulse, the PSD is given by

$$\Phi_{\tilde{s}}(f) = \sigma_A^2 T \left(\frac{\sin(\pi f T)}{\pi f T} \right)^2 + m_A^2 \delta(f). \tag{7.46}$$

There is a spectral line at dc. For the unipolar uniform MASK, $m_A = \frac{(M-1)A}{2}$. For MASK, the corresponding passband PSD is given by

$$\Phi_s(f) = \frac{1}{2} \left[\Phi_{\tilde{s}} (f - f_c) + \Phi_{\tilde{s}} (-f - f_c) \right], \tag{7.47}$$

which has a spectral line at f_c and $-f_c$.

Error probability

A bipolar MASK can only be demodulated coherently, since the sign of the signal cannot be differentiated by noncoherent demodulation. A unipolar MASK can be demodulated either coherently or noncoherently. The optimum receiver is a matched filter followed by an amplitude detector. To differentiate the demodulated amplitudes, a threshold detector with $M - 1$ thresholds is used. For coherent detection of bipolar MASK signals, carrier recovery can be based on a squaring loop or Costas loop. Noncoherent detection uses an envelope detector, followed by a threshold detector.

The probability of symbol error for MPAM with equal amplitude spacings, when subject to coherent detection, is given by [39]

$$P_s = \frac{2(M - 1)}{M} Q \left(\sqrt{\frac{\Delta^2}{2N_0}} \right), \tag{7.48}$$

where Δ is the amplitude spacing, $\Delta = |A_i - A_{i-1}| = 2d$.

Each symbol conveys $\log_2 M$ bits. Assuming equally likely symbols, the average energy is given by

$$\overline{E_s} = \frac{1}{M} \sum_{i=1}^{M} A_i^2. \tag{7.49}$$

From (7.48) and (7.49), the probabilities of symbol error for MASK with symmetrical uniformly-spaced bipolar and unipolar amplitude distribution, when subject to coherent detection, are derived as

$$P_s = \frac{2(M-1)}{M} Q\left(\sqrt{\frac{6\gamma_s}{M^2-1}}\right) \qquad \text{(coherent bipolar MASK)}, \qquad (7.50)$$

$$P_s = \frac{2(M-1)}{M} Q\left(\sqrt{\frac{3\gamma_s}{2M^2-3M+1}}\right) \qquad \text{(coherent unipolar MASK)}, \qquad (7.51)$$

where $\gamma_s = \frac{\overline{E_s}}{N_0} \approx \gamma_b \log_2 M$. The bipolar case is has an SNR advantage over the unipolar case, by 3 to 6 dB, with an asymptotic ratio of 6 dB. Thus, for coherent detection, bipolar MASK is preferred.

Noncoherent demodulation of unipolar MASK introduces a small performance degradation, which is less than 1 dB for a BEP of 10^{-5}. For large M, this degradation is negligible. The error rate is derived as [39]

$$P_s = \frac{1}{M}\left[e^{-\frac{1}{2}B_M\gamma_s} + (2M-3)Q\left(\sqrt{B_M\gamma_s}\right)\right] \qquad \text{(noncoherent unipolar MASK)}, \qquad (7.52)$$

where

$$B_M = \frac{3}{2M^2-3M+1}. \qquad (7.53)$$

MASK has the same PSD as MPSK for the same symbol rate and pulse shape. This implies that they have the same bandwidth. The uniform bipolar MASK is inferior to MPSK in terms of error probability by 3.98 to 5.17 dB for $M \geq 4$.

Example 7.4: The SEPs of the uniformly spaced MASK, for the cases of bipolar, coherent demodulation, given by (7.50); unipolar, coherent demodulation, (7.51); and unipolar, noncoherent demodulation (7.52), are plotted in Fig. 7.6.

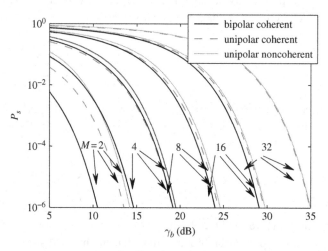

Figure 7.6   SEPs of uniformly spaced MASK.

7.5 Phase shift keying

PSK is a modulation technique that conveys data by changing or modulating the phase of a reference signal called the carrier wave. The demodulator determines the phase of the received signal and maps it back to the symbol it represents. The receiver compares the phase of the received signal to a reference signal, hence this system is termed *coherent*. PSK has a constant envelope, and this relaxes the requirements on the transmitter power amplifier. PSK is more bandwidth efficient than FSK, and more power efficient than ASK and FSK.

When the phase is differentially encoded, the demodulator then determines the changes in the phase of the received signal rather than the phase itself. This is differential phase-shift keying (DPSK). DPSK can be significantly simpler to implement than PSK since the demodulator does not need a copy of the reference signal to determine the exact phase of the received signal, thus it is a non-coherent scheme. In return, it produces a higher BEP at demodulation. When the communication channel introduces an arbitrary phase shift, the demodulator is unable to discriminate the constellation points; in this case, the data is often differentially encoded prior to modulation. PSK and DPSK are illustrated in Fig. 7.7, by QPSK and DQPSK. Note that their phase or differential phase is Gray coded.

7.5.1 Binary phase shift keying

In BPSK, the carrier signal has a constant amplitude but its phase is switched between two values, which are separated by π, to represent 0 and 1, respectively. Typically, the two phases are 0 and π, and the signals are represented by

$$s_1, s_2(t) = \pm A \cos 2\pi f_c t, \quad kT \leq t \leq (k+1)T, \tag{7.54}$$

for 1 and 0, respectively. The signals are called *antipodal*, and they have a correlation coefficient $\rho_{12} = -1$, leading to the minimum BEP for a given $\gamma_b = E_b/N_0$. BPSK is a binary antipodal ASK.

Demodulation of BPSK can be coherent. The coherent detector can be in the form of either a correlator or a matched filter. A reference signal must be used in the receiver, which

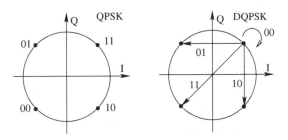

Signal constellation diagrams of QPSK and DQPSK.

Figure 7.7

must be synchronous to the received signal in frequency and phase. A carrier recovery circuit can be used. At passband, a correlator is typically used, since a matched filter with $h(t) = \cos 2\pi f_c(T - t)$ is difficult to implement.

BEP

In an uncoded AWGN channel, for coherent demodulation, BPSK has the BEP as a function of γ_b

$$P_b = Q\left(\sqrt{2\gamma_b}\right) \quad \text{(coherent BPSK)}. \tag{7.55}$$

PSD

The PSD of the baseband BPSK signal is given as

$$\Phi_{\tilde{s}}(f) = A^2 T \left(\frac{\sin \pi f T}{\pi f T}\right)^2 \quad \text{(BPSK)}. \tag{7.56}$$

The null bandwidth $B_{\text{null}} = 2R_b$, $B_{90\%} \approx 1.7R_b$, and $B_{99\%} \approx 20R_b$, R_b being the data bit rate.

BPSK signals have a mainlobe at the carrier frequency and many sidelobes in the power spectrum. The use of raised cosine filtering can increase the spectral roll-off, and thus the spectral efficiency. This also makes the signal to no longer have a constant envelope. The raised cosine filtering method is applicable to QPSK.

Differential BPSK

Differentially encoded BPSK (DEBPSK) signals can be coherently demodulated or differentially demodulated. The PSD of the DEBPSK signal is the same as that of BPSK. The modulator and demodulator structures for various BPSK and DEBPSK schemes are given in [39].

For DEBPSK signals that use differential demodulation, the scheme is called *differential BPSK (DBPSK)*. DBPSK does not use any coherent reference signal for demodulation, but uses the previous symbol as the reference for demodulating the current symbol. This leads to a suboptimum receiver.

When DBPSK uses differential coherent demodulation, which requires a reference signal but does not require phase synchronization, we get the optimum DBPSK. The BEP of the optimum DBPSK is given by [39]

$$P_b = \frac{1}{2}e^{-\gamma_b} \quad \text{(optimum DBPSK)}. \tag{7.57}$$

The suboptimum receiver is used in practice as the DBPSK receiver. The performance of the suboptimum receiver is given in [25]. Its error performance is slightly inferior to that of the optimum case, with a loss smaller than 2 dB. When an ideal narrow-band IF filter with a bandwidth of $W = 0.57/T$ is placed before the correlator, the best performance is achieved with a loss of 1 dB [25, 39]

$$P_b = \frac{1}{2}e^{-0.8\gamma_b} \quad \text{(suboptimum DBPSK)}. \tag{7.58}$$

The DEBPSK signal can also be demodulated coherently, and this scheme is known as *DEBPSK*. It is the same as coherent BPSK, but uses differential encoding to eliminate phase ambiguity in the carrier recovery circuit for coherent PSK. The corresponding BEP is given as [39]

$$P_b = 2Q\left(\sqrt{2\gamma_b}\right)\left[1 - Q\left(\sqrt{2\gamma_b}\right)\right] \quad \text{(DEBPSK)}. \tag{7.59}$$

For large SNR, $P_b \approx 2Q\left(\sqrt{2\gamma_b}\right)$, which is two times the BEP of coherent BPSK without differential encoding.

Example 7.5: The BEPs of the coherent BPSK, coherent DEBPSK, optimum DPSK and suboptimum DBPSK schemes are plotted in Fig. 7.8.

7.5.2 M-ary phase shift keying

The reason for using MPSK is to increase the bandwidth efficiency of PSK. For MPSK, the modulated signal is defined as

$$
\begin{aligned}
s_i(t) &= Ag(t)\cos\left(2\pi f_c t + \phi_i\right), \quad 0 \le t \le T, \quad i = 0, 1, \cdots, M-1 \\
&= A\cos\phi_i g(t)\cos 2\pi f_c t + A\sin\phi_i g(t)\sin 2\pi f_c t \\
&= s_i^I g(t)\cos 2\pi f_c t + s_i^Q g(t)\sin 2\pi f_c t, \tag{7.60}
\end{aligned}
$$

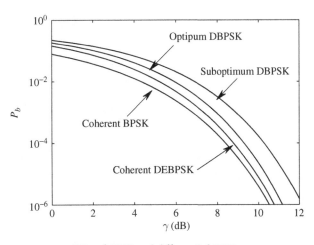

Figure 7.8   BEPs of BPSK and differential BPSKs.

where $\phi_i = \frac{2\pi i}{M}$, $A = \sqrt{E_s}$, $s_i^I = A \cos \phi_i$, $s_i^Q = A \sin \phi_i$, and $g(t)$ is a pulse shaping function given by $g(t) = \sqrt{\frac{2}{T}}$, $0 \le t \le T$, or by another function, which satisfies the orthonormal conditions

$$\int_0^T g^2(t) \cos^2 (2\pi f_c t)\, dt = 1, \tag{7.61}$$

$$\int_0^T g^2(t) \cos (2\pi f_c t) \sin (2\pi f_c t)\, dt = 0. \tag{7.62}$$

The carrier frequency f_c is selected as an integer multiple of the symbol rate, thus the signal initial phase in any symbol interval is one of the M phases, ϕ_i.

On the signal constellation diagram, the minimum distance between two constellation points is given by

$$d_{\min} = 2A \sin \left(\frac{\pi}{M}\right). \tag{7.63}$$

The constellations of Gray coded QPSK and 8PSK are shown in Fig. 7.9. For MPSK, there are M points uniformly distributed on a unit circle, each point representing a possible value of the symbol. The symbol uses Gray code such that adjacent signal points differ by only one bit. Using Gray coding, a symbol error is most likely to cause a single bit error. The Gray encoder, as baseband signal generator, outputs $\cos \phi$ and $\sin \phi$, where ϕ is the phase of the signal.

When $M \ge 4$, the MPSK signals are two-dimensional and can be written as

$$s_i(t) = s_i^I \cos 2\pi f_c t - s_i^Q \sin 2\pi f_c t. \tag{7.64}$$

The quadrature modulator can be implemented as shown in Fig. 7.10.

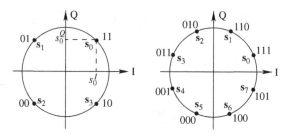

Figure 7.9   MPSK constellations. (a) QPSK. (b) 8PSK.

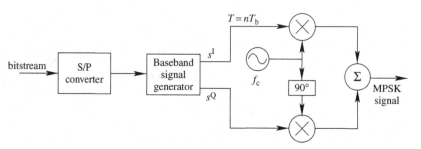

Figure 7.10   MPSK modulator.

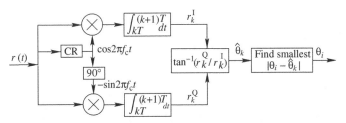

Coherent MPSK demodulator. CR denotes coherent reference.

Coherent MPSK demodulation uses two correlators, and a carrier recovery circuit is used to get the reference signal. At the receiver, the received signal is

$$r(t) = s(t) + n(t), \tag{7.65}$$

where $n(t)$ is the AWGN channel noise. The structure of coherent MPSK demodulator is shown in Fig. 7.11.

SEP

For coherent demodulation, the SEP is the probability of $\hat{\phi}_i$ being outside its decision region, which is an angle of $2\pi/M$. The formulas cannot be given in a closed form for $M > 4$, and numerical integration has to be used for obtaining the SEP. The result in integration form is given in [39].

For large SNR, the SEP can be approximated by

$$P_s \approx 2Q\left(\sqrt{2\gamma_s}\sin\frac{\pi}{M}\right) \quad \text{(coherent MPSK)}. \tag{7.66}$$

This corresponds to the nearest-neighbor approximation to P_s. For MPSK, each constellation point has two nearest neighbors at distance $d_{\min}$.

Example 7.6: The SEP of MPSK, given by (7.66), is plotted in Fig. 7.12 for $M = 2, 4, 8,$ 16, and 32.

Gray coding is usually used for signal assignment in MPSK. For two adjacent signals in the constellation, the n-tuple symbols have only one differing bit. For Gray-encoded MPSK, the BEP under the Gaussian condition is approximately

$$P_b \approx \frac{P_s}{n} = \frac{P_s}{\log_2 M} \tag{7.67}$$

for $M > 2$ and $P_s < 10^{-3}$. For low SNR, a more accurate BEP expression for Gray-coded coherent MPSK is given in [19]. Since the exact BEP expression for $M = 2$ and 4 are available, here we give its form for $M \geq 8$

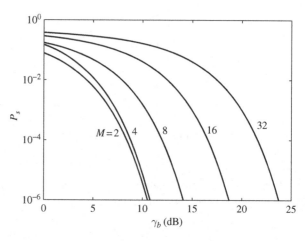

Figure 7.12

SEP of MPSK.

$$P_b \approx \frac{2}{\log_2 M} \sum_{i=1}^{M/4} Q\left(\sqrt{\frac{2E_b \log_2 M}{N_0}} \sin \frac{(2i-1)\pi}{M}\right). \tag{7.68}$$

The approximation given by (7.66) and (7.67) corresponds to a lower bound, while (7.68) is an upper bound. The expression using the first two terms in (7.68) gives the best approximation of MPSK for $M \geq 8$.

PSD

It is well-known that the PSD of a bandpass signal is just the shift version of the PSD of the baseband signal or complex envelope. The PSD of the complex envelope of the MPSK signal is derived as [39]

$$\Phi_{\tilde{s}}(f) = A^2 n T_b \left(\frac{\sin \pi f n T_b}{\pi f n T_b}\right)^2 \quad \text{(MPSK)}, \tag{7.69}$$

where $n = \log_2 M$ and the bit duration $T_b = T/n$. For unit bit energy $E_b = 1$, we have $A = \sqrt{2}$ and $T_b = 1$. The PSD of MPSK is the same as that of BPSK in terms of symbol rate, but in terms of bit rate it is n times narrower than that of BPSK.

Since the minimum passband required for transmission of the symbols is $1/T$, the maximum bandwidth efficiency is $R_b/T = \log_2 M$. Hence, the bandwidth efficiency increases with M.

Example 7.7: The PSD of MPSK is plotted in Fig. 7.13 for $M = 2, 4, 8, 16,$ and 32. The null bandwidth $B_{\text{null}} = 2R_s/n$, $B_{90\%} \approx 1.7R_s/n$, $B_{99\%} \approx 20R_s/n$. MPSK signals have a mainlobe at the carrier frequency and many sidelobes in the power spectrum. The use of raised cosine filtering can increase the spectral rolloff, and thus the spectral efficiency. This also makes the envelope of the signal no longer constant.

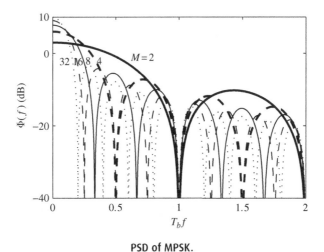

Figure 7.13

PSD of MPSK.

Differential MPSK

Differential MPSK (DMPSK) is referred to as differentially encoded and differentially coherently demodulated MPSK. The differentially coherent demodulation is a noncoherent scheme, since phase coherent reference signals are not needed. This scheme is used in case of random phase in the received signal. Like DEBPSK, when the differentially encoded MPSK signal is coherently demodulated, we get differentially encoded MPSK (DEMPSK), which is used to eliminate phase ambiguity in the carrier recovery process. The modulator uses the MPSK modulator, preceded by two differential encoders, one on each channel, before the carrier multiplier. The DMPSK and DEMPSK demodulators are given in [39]. The DEMPSK demodulator is in fact the coherent MPSK, followed by a differential decoder. The PSD for differentially encoded MPSK signals is the same as that of MPSK for an equally likely original bitstream.

The SEP for DMPSK is in an integral form [39]

$$P_s = \frac{\sin \frac{\pi}{M}}{2\pi} \int_{-\pi/2}^{\pi/2} \frac{e^{-\gamma_s \left(1 - \cos \frac{\pi}{M} \cos x\right)}}{1 - \cos \frac{\pi}{M} \cos x} dx. \tag{7.70}$$

For $M = 2$, it reduces to (7.57). For other M values, it has to be numerically evaluated. For large SNR, it can be approximated by

$$P_s \approx 2Q\left(\sqrt{2\gamma_s} \sin \frac{\pi}{\sqrt{2M}}\right) \quad \text{(optimum DMPSK).} \tag{7.71}$$

This result is asymptotically 3 dB inferior in SNR to the SEP of the MPSK, which is given by (7.66) for $M \geq 4$.

The SEP for DEMPSK is also given in integral form [39]. For $M = 2$, it reduces to (7.59). For $M = 4$, it reduces to [39]

$$P_s = 4Q\left(\sqrt{\gamma_s}\right) - 8\left[Q\left(\sqrt{\gamma_s}\right)\right]^2 + 8\left[Q\left(\sqrt{\gamma_s}\right)\right]^3 - 4\left[Q\left(\sqrt{\gamma_s}\right)\right]^4 \quad \text{(DEQPSK).} \tag{7.72}$$

The last three terms can be ignored for large SNR, and the SEP is about twice that of coherent QPSK. In fact, for large SNR

$$P_s \approx 2P_s^{\text{MPSK}} \quad \text{(DEMPSK)},\tag{7.73}$$

where P_s^{MPSK} is the SEP for MPSK. This corresponds to 0.5 dB or less degradation in SNR.

Applications

MPSK has a maximum spectral efficiency of $\log_2 M$ bits/s/Hz. The popular MPSK schemes uses $M = 2$, 4, and 8. 8PSK is usually the highest-order PSK constellation deployed. For more than 8 phases, the BEP becomes too high, and in this case, better, though more complex, QAM modulation can be used. 8PSK is used by EDGE. Since 8PSK has a BEP performance close to (0.5 dB better than) that of 16QAM but its data rate is only 3/4 that of 16QAM, 16QAM is more frequently used than 8PSK. 8PSK is used in DVB-S/S2/T.

MPSK and its variants are widely used in various wireless communication standards. For example, QPSK and its variants are used in IS-95, CDMA2000, WCDMA, TD-SCDMA, WiMedia, satellite communication systems, and DBS. In IEEE 802.11b, DBPSK, DQPSK, and QPSK are used, depending on the data rate required. Likewise, in IEEE 802.11g, BPSK, QPSK, 16QAM, and 64QAM are used. Bluetooth uses $\pi/4$-DQPSK and 8DPSK in its version 2, and uses GMSK in version 1. ZigBee uses DBPSK at 868/915 MHz, but uses OQPSK and 16QAM at 2.4 GHz. 8PSK is used in EDGE, UWC-136, TD-SCDMA. In EDGE, 8PSK is used in a way analogous to $\pi/4$-DQPSK, by inserting a $\pi/8$ phase rotation at each symbol transmission to eliminate zero crossings. 3GPP LTE uses QPSK, 16QAM and 64QAM. In 802.16 standards family, QPSK and 16QAM are mandatory, and 64QAM is optional. DQPSK is used in DAB.

High-order modulations like MPSK and MQAM require more power for a given error performance. MQAM also has a strict requirement on the power amplifier, thus it is typically not used in MSs.

7.5.3 Quaternary phase shift keying

QPSK is an MPSK scheme with $M = 4$, and it is the most popular MPSK scheme. QPSK increases the bandwidth efficiency by a factor of two when compared to BPSK, while having the same BEP performance. Other MPSK schemes increase the bandwidth efficiency, but yield BEP degradation. QPSK modulation is used in CDMA-based systems, IS-54/136, UWC-136, PWT, PDC, PHS, DVB-S/S2/T, LMDS, and satellite systems.

The QPSK signal can be defined by

$$s(t) = \frac{A}{\sqrt{2}} I_k g(t) \cos 2\pi f_c t - \frac{A}{\sqrt{2}} Q_k g(t) \sin 2\pi f_c t, \quad kT \le t \le (k+1)T,\tag{7.74}$$

where $I_k = \pm 1$ and $Q_k = \pm 1$ corresponds to the odd- and even-numbered bits, with mapping $1 \to +1$ and $0 \to -1$.

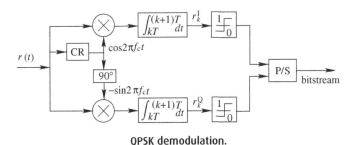

Figure 7.14   **QPSK demodulation.**

For QPSK, the bitstream is first reshuffled into the symbol stream, each symbol having two bits. The symbol rate is only half the bit rate. Each symbol can have one of four values, which are modulated to one of the four positions in the constellation diagram.

The modulator and demodulator for MPSK are applicable to QPSK. Due to the special property of the QPSK constellation, a simple demodulator can be used. Coherent detection requires a pilot-signal-assisted fading compensation for demodulation. The QPSK correlation receiver is illustrated in Fig. 7.14. The I- and Q-channel signals can be demodulated separately as two individual BPSK signals. A parallel-to-serial (P/S) converter is used to combine the two sequences into one. The received symbol is decided by the QPSK decision table.

The QPSK signal can be treated as two independent BPSK signals along the in-phase and quadature directions. For the AWGN channel, from Fig. 7.14, the two channels are independent and the average BEP for each channel is the same. The exact BEP is derived from the detection theory as

$$P_b = Q\left(\sqrt{2\gamma_b}\right) \quad \text{(coherent QPSK)}. \tag{7.75}$$

It is the same as that of BPSK, given by (7.55) and (7.139), under both the AWGN and flat Rayleigh conditions.

A symbol represents two bits from the I- and Q-channels. The SEP is given as

$$
\begin{aligned}
P_s &= 1 - \Pr\,(\text{both bits are correct}) \\
&= 1 - (1 - P_b)^2 \\
&= 2Q\left(\sqrt{\gamma_s}\right) - \left[Q\left(\sqrt{\gamma_s}\right)\right]^2.
\end{aligned} \tag{7.76}
$$

By using $\gamma_s = 2\gamma_b$ and $P_b \approx P_s/2$ for Gray encoding, and the union bound and the nearest-neighbor approximation, an approximate formula for P_b is also derived, $P_b \approx Q(\sqrt{2\gamma_b})$, which is coincidently the same as the exact form (7.75). Thus, QPSK achieves twice the data rate as BPSK, with the same bandwidth and BEP performance.

Like all DEMPSK schemes, differentially encoded QPSK (DEQPSK) has a BEP that is twice as large as that for coherent QPSK under both the AWGN and flat Rayleigh fading conditions. For the AWGN channel

$$P_b = 2P_b^{\text{QPSK}} = 2Q\left(\sqrt{2\gamma_b}\right) \quad \text{(DEQPSK)}, \tag{7.77}$$

because a symbol error leads to errors of two consecutive phase transitions. In DEQPSK, information dibits are represented by the phase difference $\Delta\phi_i$ from symbol to symbol. A possible phase assignment is given by $00 \rightarrow 0, 01 \rightarrow \pi/2, 10 \rightarrow -\pi/2$, and $11 \rightarrow \pi$.

For DQPSK, the BEP can be derived from (7.71) as

$$P_b \approx Q\left(\sqrt{4\gamma_b}\sin\frac{\pi}{4\sqrt{2}}\right) \quad \text{(optimum DQPSK)}, \tag{7.78}$$

which is 2 to 3 dB inferior to coherent QPSK. Like DBPSK, a suboptimum demodulator using the previous symbol as reference is available, and the BEP is given as [39]

$$P_b \approx e^{-0.59\gamma_b} \quad \text{(suboptimum DQPSK)}. \tag{7.79}$$

At high SNR it is inferior to the optimum DQPSK by less than 1 dB. DQPSK is used in IEEE 802.11 and 802.11b.

QPSK is nominally a constant envelope format, but it has deep amplitude dips at bit transitions. This is due to the 180° phase shift between symbols, which causes the trajectories in the I-Q diagram to pass through the origin for these transitions. A number of variants of QPSK are described below.

Quadrature quadrature PSK (Q^2PSK) [30] has a nonconstant envelope. It uses four basis signals, as opposed to two for QPSK. Q^2PSK has the same BEP as QPSK and MSK in an AWGN channel, but with a doubled bandwidth efficiency. The structure of the demodulator is very similar to that of MSK. A constant-envelope version of Q^2PSK can be obtained by a simple coding scheme that adds a fourth bit for every three information bits, leading to a code rate of 3/4. There is still an increase of 50% in bandwidth efficiency over MSK.

Offset QPSK

OQPSK delays the Q-channel signal (even bitstream) by half a symbol period, prior to carrier multiplication.

$$s(t) = \frac{A}{\sqrt{2}}I(t)g(t)\cos 2\pi f_c t - \frac{A}{\sqrt{2}}Q\left(t - \frac{T}{2}\right)g\left(t - \frac{T}{2}\right)\sin 2\pi f_c t, \quad -\infty < t < \infty, \tag{7.80}$$

The OQPSK signal has a symbol period of $T/2$. At the symbol boundary, only one of the two bits in an (I_k, Q_k) pair may change sign, but not both. Thus, it avoids 180° phase transitions by doubling the phase switching rate. Phase changes can only be 0 and $\pm 90°$. This leads to less out-of-band interference due to band-limiting and amplifier nonlinearity, as compared to QPSK. Also, the reduced amplitude variation allows the use of a more power-efficient, less linear RF power amplifier. The PSD and BEP of OQPSK are the same as those of QPSK.

Consequently, if a bandpass filter is used to control spillover (as used in satellite transmitters), a much smaller envelope variation (due to the FM-to-AM conversion effect) is achieved. IS-95 uses QPSK for the forward link and OQPSK for the reverse link. OQPSK is also used for satellite communications.

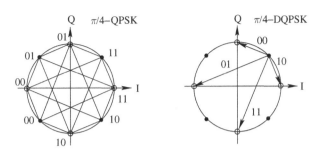

Constellations of $\pi/4$-QPSK and $\pi/4$-DQPSK.

$\pi/4$-QPSK

The $\pi/4$-QPSK modulation alternatively selects the modulated signal points from two QPSK constellations that have a phase difference of $\pi/4$, as shown in Fig. 7.15, where the dotted constellations are used for $t = 2mT$, $m = 0, 1, \cdots$, and the circled constellations for $t = (2m + 1)T$. This method leads to a phase transition for every symbol, and thus enables a receiver to perform timing recovery and synchronization. The phase changes $\Delta\phi_k$ are confined to odd multiples of $\pi/4$, and information is carried by $\Delta\phi_k$.

Since information is carried by $\Delta\phi_k$, $\pi/4$-QPSK is a form of differentially encoded QPSK, but differs from the DEQPSK in the differential coding rules. It can be differentially demodulated. Differential coherent demodulation can reduce the adverse effect of the fading channel. $\pi/4$-QPSK can also be demodulated by coherent demodulation and FM-discriminator detection. Coherent demodulation provides a BEP performance of 2 to 3 dB better than the other schemes do [18, 39]. The coherent detector can be the same as that for QPSK, but it is followed by differential decoding. The coherently demodulated $\pi/4$-QPSK has the same BEP as that of DEQPSK

$$P_b \approx Q\left(\sqrt{2\gamma_b}\right) \quad \text{(coherent } \pi/4\text{-QPSK)}, \tag{7.81}$$

and noncoherently demodulated $\pi/4$-QPSK has the same BEP as that of DQPSK.

The $3\pi/8$-8PSK modulation scheme used in EDGE exploits an idea similar to that in $\pi/4$-QPSK. It has a 16-point constellation, which is formed from two 8-point mappings, offset by a phase of $3\pi/8$ radians. This avoids symbol-to-symbol transitions through the origin, and thus lowers the PAPR compared to the normal 8PSK.

$\pi/4$-DQPSK

The $\pi/4$-DQPSK modulation combines differential encoding with $\pi/4$-QPSK. The constellation of $\pi/4$-DQPSK modulation is also shown in Fig. 7.15. For each symbol, the modulated signal has a phase shift which is that of the previous symbol plus a nonzero phase shift. The nonzero phase shift has four states, corresponding to the four possible values of a symbol. The transitions between subsequent signal constellations never pass through the origin, leading to much smaller fluctuations of the envelope than that of QPSK.

$\pi/4$-DQPSK can be differentially demodulated. For $\pi/4$-DQPSK with Gray coding, the BEP is quite complicated [29, 35]

$$P_b = Q(a, b) - \frac{1}{2} I_0(ab) e^{-\frac{1}{2}(a^2 + b^2)}, \tag{7.82}$$

where I_0 is the zero-order modified Bessel function of the first kind, which is given by (3.44),

$$a = \sqrt{2\gamma_b \left(1 - \frac{1}{\sqrt{2}}\right)}, \qquad b = \sqrt{2\gamma_b \left(1 + \frac{1}{\sqrt{2}}\right)}, \tag{7.83}$$

and $Q(a, b)$ is the Marcum Q function, defined by (3.66).

With coherent demodulation, $\pi/4$-DQPSK has the same BEP performance as DQPSK. It produces better results than OQPSK in the presence of multipath fading. $\pi/4$-DQPSK modulation, like QPSK, is used in CDMA-based systems, IS-54/136, UWC-136, PWT, PDC, PHS, LMDS, and satellite systems.

Example 7.8: The modulated signals of the BPSK, QPSK, OQPSK, and 8PSK modulation schemes are plotted in Fig. 7.16. Here, $A = 1$ and $f_c = 2$.

Example 7.9: The BEPs of all the popular PSK schemes discussed above are plotted in Fig. 7.17.

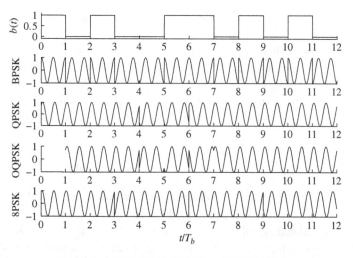

Figure 7.16   Modulated BPSK, QPSK, OQPSK, and 8PSK signals.

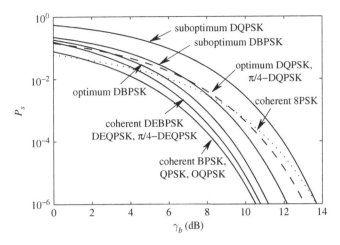

Figure 7.17

BEPs of some popular PSK schemes.

7.6 Frequency shift keying

FSK is another constant-envelope modulation technique. FSK modulates the digital signal using predetermined output frequency. FSK is appropriate for channels that lack phase stability, while phase estimation is necessary for coherent detection used in linear modulation methods such as PAM, PSK, and QAM. Since the frequency modulation technique is nonlinear, it tends to occupy a higher bandwidth than the amplitude and phase modulations do.

7.6.1 Binary frequency shift keying

For BFSK, the instantaneous frequency is usually shifted between two discrete values, $f_1 = f_c - \Delta f$ and $f_2 = f_c + \Delta f$. A positive frequency deviation from the operating frequency is represented by 1, and a negative frequency deviation is represented by 0. That is, the modulated waveform is $\cos 2\pi f_1 t$ for 0 and $\cos 2\pi f_2 t$ for 1. The BFSK signal is represented by

$$s_1(t) = Ag(t)\cos\left(2\pi f_1 t + \phi_1\right), \quad kT \le t < (k+1)T, \tag{7.84}$$

$$s_2(t) = Ag(t)\cos\left(2\pi f_2 t + \phi_2\right), \quad kT \le t < (k+1)T, \tag{7.85}$$

where the rectangular window $g(t) = \sqrt{2/T}$ is defined in $kT \le t < (k+1)T$, and $E_b = A^2 T/2$.

In order to implement coherent FSK that uses coherent detection, it is required that $\phi_1 = \phi_2$ at $t = 0$ and that f_1 and f_2 are suitably selected so as to make $s_1(t)$ and $s_2(t)$ orthogonal

$$\int_{kT}^{(k+1)T} s_1(t)s_2(t)dt = 0. \tag{7.86}$$

This corresponds to [39]

$$f_1 = \frac{n-m}{4T}, \quad f_2 = \frac{n+m}{4T}, \tag{7.87}$$

where n and m are integers. Thus we have

$$f_2 - f_1 = 2\Delta f = \frac{m}{2T}, \quad f_c = \frac{n}{2T}. \tag{7.88}$$

That is, the nominal carrier frequency f_c must be an integer multiple of $1/2T$ for orthogonality. A coherent FSK signal may have discontinuous phase at bit boundaries.

For $\phi_1 = \phi_2$, when $f_2 - f_1 = \frac{k}{T}$, the phase continuity is maintained at bit transitions. For $k = 1$, the FSK is called *Sunde's FSK*. MSK is a particular form of FSK that has the minimum separation for orthogonality, namely $1/2T$, as well as continuous phase.

Example 7.10: The modulated signals of Sunde's FSK and coherent FSK with discontinuous phase are plotted in Fig. 7.18. For Sunde's FSK, bit 1 corresponds to $f_2 = 2/T$, while bit 0 corresponds to $f_1 = 1/T$. For coherent FSK with discontinuous phase, $f_2 = 6/4T$, $f_1 = 3/4T$.

PSD

Unlike QPSK, which has a continuous power spectrum with main lobe at f_c and many sidelobes around it, the power spectrum of FSK, in addition to the continuous power spectrum, has also spectral lines added at some frequencies.

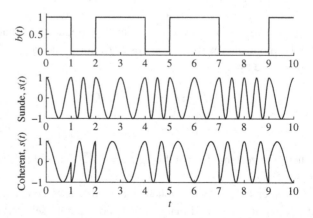

Figure 7.18   FSK signals: Sunde's FSK and coherent FSK with discontinuous phase.

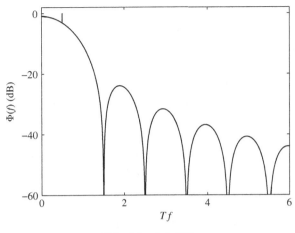

Figure 7.19   **PSD of Sunde's FSK.**

The PSD of a passband signal can be derived from the baseband signal based on (7.47). For Sunde's FSK, the I- and Q-channel are independent of each other, and thus the PSD of the complex envelope is the PSDs of the two components

$$\Phi_{\tilde{s}}(f) = \Phi_I(f) + \Phi_Q(f). \tag{7.89}$$

The complete baseband PSD of the BFSK is given as [39]

$$\Phi_{\tilde{s}}(f) = \frac{A^2}{4}\left[\delta\left(f - \frac{1}{2T}\right) + \delta\left(f + \frac{1}{2T}\right)\right] + T\left(\frac{2A\cos(\pi Tf)}{\pi\left[1 - (2Tf)^2\right]}\right)^2, \tag{7.90}$$

where $A = \sqrt{2}$ for a unity signal energy in one side.

Example 7.11: The one-sided PSD is illustrated in Fig. 7.19. A spectrum line occurs at $fT = 0.5$. For passband spectrum, the spectrum lines occur at $f = f_c \pm \frac{1}{2T}$, which are the two frequencies of the BFSK. From the figure, the null bandwidth is $B_{\text{null}} = 1.5R_b$, thus the null-to-null bandwidth at f_c is $3R_b$. The two-sided bandwidths at f_c are $B_{90\%} \approx 1.23R_b$ and $B_{99\%} \approx 2.12R_b$. The transmission bandwidth is usually set as $B = 2R_b$.

BEP

A coherent FSK signal can be coherently or noncoherently detected. For coherent demodulation, two LOs that operate at f_1 and f_2 are required. For Sunde's FSK, the BEP performance for any two equally likely binary signals can be derived from (7.28), where $\rho_{12} = 0$ (orthogonal) and $E_s = E_b$. The BEP for coherent detection is given by

$$P_b = Q(\sqrt{\gamma_b}). \tag{7.91}$$

The result is 3 dB inferior to that of coherent BPSK.

A noncoherent FSK signal can only be demodulated by noncoherent demodulation. Information of an FSK signal is conveyed at zero-crossing points during each bit period, and a limiter is used for demodulation. The output of a limiter contains only two states, zero and one. Noncoherent demodulation can be implemented with the correlator-squarer structure. For each f_i, $i = 1, 2$, the I and Q signals at the correlator output are squared and summed, and the maximum sum corresponds to the detected frequency. Noncoherent demodulation using envelope detection is more common, since the LOs are not needed and it requires only about 1 dB more power than coherent FSK demodulation under the Gaussian condition for $P_b \leq 10^{-4}$. For this reason, practical FSK systems never use coherent demodulation. The method uses two bandpass filters to convert the FSK signal into two ASK signals, which are then demodulated with an envelope or rectifier detector.

For orthogonal, equiprobable, equal-energy, noncoherent BFSK signals, the BEP is given by [31, 39]

$$P_b = \frac{1}{2} e^{-\frac{\gamma_b}{2}}. \tag{7.92}$$

For orthogonality between noncoherent FSK signals, the minimum frequency separation is $1/T$, in contrast to $1/2T$ for coherent FSK [39]. Thus, more bandwidth is required for noncoherent FSK at the same symbol rate.

Example 7.12: The BEPs for coherently and noncoherently demodulated FSK are plotted in Fig. 7.20. Noncoherent demodulation typically requires 1 dB more power than coherent modulation.

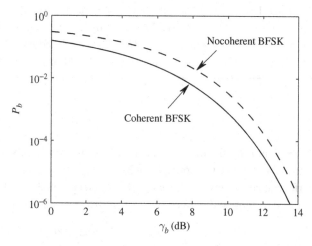

Figure 7.20   **BEPs of FSK using coherent and noncoherent demodulation.**

7.6.2 *M*-ary frequency shift keying

For *M*-ary FSK (MFSK) modulation, *M* different frequencies, usually equally spaced, are used. The modulated MFSK signal is defined by

$$s_i(t) = Ag(t)\cos\left[2\pi\left(f_c + \alpha_i\Delta f_c\right)t + \phi_i\right], \quad kT \leq t < (k+1)T \qquad (7.93)$$

for $i = 1, 2, \ldots, M$, where $\alpha_i = 2i - 1 - M$ and $g(t) = \sqrt{2/T}$.

When all ϕ_i's are the same, the signal is coherent and the demodulation can be coherent or noncoherent. Otherwise, the signal is noncoherent and demodulation must be noncoherent.

Like BFSK, for orthongonality of the signals, the frequency separations between any two of them must be $m/2T$ for the coherent case and m/T for the noncoherent case, $m \geq 1$. Typically, a uniform frequency separation between adjacent frequencies is used for MFSK. The PSD of MFSK is very complex, and is plotted in [1, 39]. GFSK and 4-level GFSK, which are obtained by prefiltering the BFSK and 4FSK signals by a lowpass Gaussian filter, are used in the baseline IEEE 802.11 standard. GFSK are also used in DECT, CT2, Bluetooth, and paging. FSK is used in AMPS as well as most 1G mobile systems.

A simple way to generate MFSK signals is to use *M* oscillators operating at the *M* frequencies, and switch between them at each symbol period. This usually leads to phase discontinuity at the switching times due to the phase offsets between oscillators. This discontinuity in phase leads to a broadening of spectrum. To eliminate the phase discontinuity, the continuous phase FSK (CPFSK) modulation technique is usually used.

SEP

The principles for demodulation of BFSK are applicable to MFSK. The modulators and demodulators for BFSK can be easily extended to MFSK, and their structures are given in [39]. The coherent MFSK demodulator can use the general form of detector for *M*-ary equiprobable, equal-energy signals with known phases. It consists of a bank of *M* correlators or matched filters.

The orthogonal MFSK waveforms have a constellation of *M* *M*-dimensional orthogonal vectors s_i, $i = 1, \cdots, M$, whose *i*th element is $\sqrt{E_s}$ and all other elements are zero. The basis function for s_i is

$$\psi_i = \sqrt{2/T}\cos 2\pi\left(f_c + (i-1)\Delta f\right)t. \qquad (7.94)$$

For coherent detection, the SEP derivation for MFSK is more complex for $M > 2$. The SEP can be estimated by (7.30). If the signal set is equal-energy and orthogonal, the distances between any two signals are the same, $d = \sqrt{2E_s}$, thus we have

$$P_s \leq (M-1)Q\left(\sqrt{\gamma_s}\right). \qquad (7.95)$$

For $M = 2$, it is an exact representation, while for $P_s \leq 10^{-3}$ it is a good approximation.

For MFSK, noncoherent receivers are easier and cheaper to implement than coherent receivers. The square-law detector is a noncoherent detector for *M*-ary orthogonal signals. It takes the correlator-squarer or matched-filter-squarer form. The matched-filter envelope detector is also used. The conventional frequency discriminator is even simpler for MFSK demodulation. For matched-filter envelope-demodulated orthogonal MFSK, the SEP is given by [33, 39]

$$P_s = \sum_{m=1}^{M-1}(-1)^{m+1}\binom{M-1}{M}\frac{1}{m+1}e^{\frac{-m\gamma_s}{m+1}}, \tag{7.96}$$

where $\binom{M-1}{M}$ is the binomial coefficient. For BFSK ($M = 2$), the expression reduces to (7.92). The first term in the summation provides an upper bound as

$$P_s \leq \frac{M-1}{2}e^{-\gamma_s/2}. \tag{7.97}$$

For equal-energy, equiprobable, and orthogonal MFSK, the probability of bit error is derived as [35, 39]

$$P_b = \frac{M}{2(M-1)}P_s. \tag{7.98}$$

For large M, $P_b \approx \frac{1}{2}P_s$.

Example 7.13: The SEPs for noncoherent demodulation are plotted in Fig. 7.21. From the figure, it is seen that for given BEP an increase in M leads to a reduction in the bit SNR. This is different from the cases of MPSK and MQAM. Because all the M signals are orthogonal, the power efficiency increases, but the bandwidth efficiency decreases, as M increases. The orthogonal property of MFSK motivates the OFDM scheme as a means of providing power-efficient modulation.

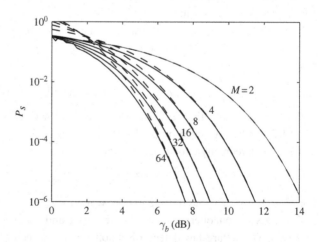

Figure 7.21 SEP of MFSK using noncoherent demodulation. The dotted lines correspond to upper bound.

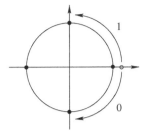

Figure 7.22
State diagram of MSK modulation.

7.6.3 Minimum shift keying

MSK or *minimum frequency-shift keying* can be viewed as OQPSK plus half-sinusoidal pulse shaping. MSK encodes each bit as a half sinusoid. The MSK signal can be defined as

$$s(t) = AI(t + T)\cos\left(\frac{\pi t}{2T}\right)\cos 2\pi f_c t + AQ(t)\sin\left(\frac{\pi t}{2T}\right)\ \sin 2\pi f_c t, \quad -\infty < t \le \infty,$$

$$(7.99)$$

where T, as in OQPSK, is the bit period, as composed to the symbol period for QPSK. The symbol duration for MSK and OQPSK is a one-bit period, while that for QPSK is a two-bit period. The state diagram of MSK is the same as that of OQPSK, and is shown in Fig. 7.22.

The MSK signal is shown to be a special FSK signal with two frequencies $f_-, f_+ = f_c \pm \frac{1}{4T}$ [39]. The frequency separation is $\Delta f = \frac{1}{2T}$, which is the minimum separation for two FSK signals to be orthogonal; hence the name *minimum shift keying*. MSK carrier phase is continuous at bit transitions.

MSK is a particularly spectrally efficient form of coherent CPFSK. It is a CPFSK scheme with a modulation index $h = 0.5$. When the MSK signal is realized in this manner, it is called *fast frequency shift keying (FFSK)*.

MSK can also be implemented in a serial fashion. In this case, the precise synchronization and balancing for the Q-channel is no longer needed, and this is especially suitable for high bit rates. Many MSK-type schemes have been proposed to improve the bandwidth efficiency of MSK. They can be continuous phase modulation with constant envelope, or based on pulse shaping in the Q-channel such as the sinusoidal FSK and many other symbol-shaping pulses. These are given in detail in [39]. These shaping schemes can generally have better spectral sidelobe roll-offs, but have a wider main lobe than the MSK spectrum and that of conventional PSK.

MSK has the advantage of not introducing any ISI. As a binary modulation scheme, the spectral efficiency of MSK is still very low. MSK is used in NASA's Advanced Communication Technology Satellite (ACTS) system.

Example 7.14: The modulated signal of MSK modulation is plotted in Fig. 7.23. Here, $A = 1$ and $f_c = 2$.

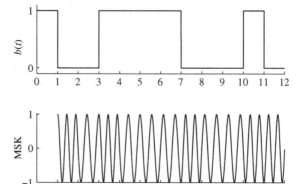

Figure 7.23

Modulated MSK signal.

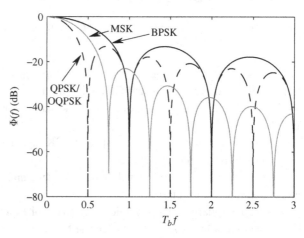

Figure 7.24

PSDs of MSK, BPSK, QPSK, and OQPSK.

PSD

The PSD for MSK is given as [39]

$$\Phi_{\tilde{s}}(f) = \Phi_I(f) + \Phi_Q(f) = \frac{16A^2T}{\pi^2}\left[\frac{\cos(2\pi Tf)}{1-(4Tf)^2}\right]^2. \tag{7.100}$$

Example 7.15: The PSDs of MSK, QPSK or OQPSK, and BPSK are plotted in Fig. 7.24. For MSK, the null bandwidth B_{null} is $0.75R_b$, as compared to $1.0R_b$ for BPSK and $0.5R_b$ for QPSK or OQPSK. The bandwidths containing 90% of the power, $B_{90\%}$, are $0.76R_b$ for MSK, $0.8R_b$ for QPSK or OQPSK, and $1.7R_b$ for BPSK, respectively. For MSK, $B_{99\%} \approx 1.2R_b$, while the corresponding values are $10R_b$ for QPSK or OQPSK, and $20R_b$ for BPSK.

Unlike FSK, which has spectral lines at certain frequencies, MSK does not have. The power spectrum decreases faster than that of OQPSK, QPSK, and FSK, leading to less out-of-band energy. Thus, MSK provides an advantage over other schemes in case of a more stringent in-band power specification.

Since MSK can be viewed as a form of OQPSK, it has the same BEP as OQPSK, QPSK and BPSK for coherent detection. This is only valid for the infinite bandwidth case. However, from the PSDs, the BEP performance of QPSK or OQPSK would be better than that of MSK, when the system bandwidth is below $0.75R_b$, since the in-band power for MSK is lower.

Demodulation

The MSK signal can be demodulated by using the same techniques as for FSK (since MSK is a type of FSK). It can also be demodulated as a continuous phase modulation (CPM) scheme with trellis demodulation using the Viterbi algorithm (since MSK is a CPFSK scheme) [39].

Coherent demodulation of MSK is very similar to that of QPSK. The BEP is derived as

$$P_b = Q\left(\sqrt{2\gamma_b}\right) \quad \text{(coherent MSK)}, \tag{7.101}$$

which is the same as that of BPSK, QPSK, and OQPSK. Since MSK is a type of FSK, it can also be demodulated noncoherently with about 1 dB loss in power efficiency.

To ensure the orthogonality between the two channels, it is necessary to have $f_c = \frac{n}{4T}$, $n = 2, 3, \cdots$ [39]. However, for the typical case $f_c \gg 1/T$, and even if it is not a multiple of $\frac{1}{4T}$, orthogonality is almost retained.

The MSK signal has no discrete spectral line that can be used for synchronization. By passing through a squarer, strong discrete spectral lines are generated at $2f_-$ and $2f_+$. There is a 180° phase ambiguity in carrier recovery due to the squaring operation. One solution is to differentially encode the data stream before modulation.

7.6.4 Gaussian minimum shift keying

The power spectrum of MSK has a wide mainlobe. GMSK is obtained by narrowing the mainlobe of MSK signals using a premodulation Gaussian lowpass filter, that is, using Gaussian instead of sinusoidal pulse shaping [22]. The transfer function of the Gaussian filter is given by

$$H(f) = e^{-\frac{2\ln 2 f^2}{B_b^2}}, \tag{7.102}$$

where B_b is the 3-dB bandwidth of the baseband shaping filter. Thus, smaller B_b corresponds to a higher frequency efficiency. The impulse response of the Gaussian filter is given by [28]

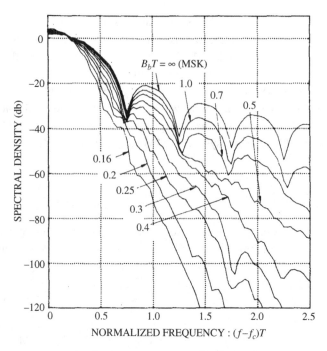

Figure 7.25

PSD of GMSK. ©IEEE, 1981. ([22], Fig. 2).

$$g(t) = Q\left(\frac{2\pi B_b \left(t - \frac{T}{2}\right)}{\sqrt{\ln 2}}\right) - Q\left(\frac{2\pi B_b \left(t + \frac{T}{2}\right)}{\sqrt{\ln 2}}\right), \tag{7.103}$$

which can be approximated by a Gaussian response.

Since the Gaussian filter does not satisfy the Nyquist criterion for ISI cancellation, reducing the spectral occupancy introduces ISI. ISI increases as B_b decreases. This reduces the BEP performance of MSK. The filter parameter is specified by the normalized bandwidth $B_b T$. A decrease in $B_b T$ narrows the power spectrum, but leads to an increase in ISI and also a degradation in power efficiency. At 99.99% bandwidth, the power degradation factor β with respect to MSK due to premodulation filter is 0.76 for $B_b T = 0.20$, 0.84 for $B_b T = 0.25$, 0.89 for $B_b T = 0.30$, and 0.97 for $B_b T = 0.5$. GMSK signals have a phase that changes linearly with time.

The PSD of GMSK is simulated in [22]. For $B_b T = \infty$, it reduces to MSK. A smaller $B_b T$ leads to a tighter PSD. The spectrum with $B_b T = 0.25$ is quite tight, and is usually used. The PSD of GMSK is plotted in Fig. 7.25. We have $B_{90\%} = 0.52 R_b$, $B_{99\%} = 0.79 R_b$ for $B_b T = 0.2$, $B_{90\%} = 0.57 R_b$, $B_{99\%} = 0.86 R_b$ for $B_b T = 0.25$, and $B_{90\%} = 0.69 R_b$, $B_{99\%} = 1.04 R_b$ for $B_b T = 0.5$.

A Costas loop demodulator is implemented in [22]. For GMSK, the measured BER can be approximated by [22]

$$P_b \approx Q\left(\sqrt{2\alpha \gamma_b}\right) \quad \text{(GMSK)}, \tag{7.104}$$

where α is the degradation factor due to the Gaussian filter, $\alpha = 0.68$ for GMSK with $B_b T = 0.25$, $\alpha = 0.85$ for simple MSK ($B_b T \to \infty$). With coherent detection in the AWGN channel, theoretically $\alpha = 1$ for MSK.

GMSK increases the spectral efficiency of MSK. Like MSK, GMSK signals can also be demodulated by using coherent detection, differential detection, and frequency discriminator techniques. Coherent detection generally gives the best result. Differential detection does not suffer from the threshold effect and cancels the phase distortion between adjacent symbols. Like MSK, GMSK is a constant-envelope modulation scheme that achieves a high power-efficiency MS using a class C amplifier.

The GMSK modulator consists of a bit stuffing system, a differential encoder, a Gaussian lowpass filter and an FM modulator. The bit stuffing system repeats each bit once to eliminate small and ambiguous phase change sequence patterns and to provide a symmetric detection. The differential encoder encodes information bits using the carrier phase differences.

GMSK is used in the GSM, GRPS, EDGE, and CDPD systems. In the GSM system, $B_b T$ is selected as 0.3. In CDPD, $B_b T = 0.5$. GFSK is similar to GMSK, but it utilizes a Gaussian filter to smooth positive/negative frequency deviations of FSK. GFSK is used in DECT and CT2. In DECT, $B_b T = 0.5$, and in CT2 $B_b T = 0.3$. Although GMSK is a good choice for voice modulation, it is not desirable for data modulation. This is because a much lower BER is required for data, which limits the value α and consequently reduces the spectral efficiency of GMSK for data.

7.6.5 Continuous phase modulation

CPM is a constant amplitude modulation scheme that is jointly power and bandwidth efficient. With suitable design, CPM may achieve higher bandwidth efficiency than QPSK and higher-order MPSK. Although high-order QAM can outperform MPSK in terms of power or bandwidth efficiency, their nonconstant envelope imposes strict restrictions on power amplifiers. The constant-amplitude property of CPM makes it desirable in certain cases. The MSK and GMSK are two practical CPM schemes.

The CPM signal is defined as [3]

$$s(t) = A \cos \left(2\pi f_c t + \Phi(t, \boldsymbol{a}) \right), \quad -\infty < t < \infty, \tag{7.105}$$

where A is a constant, $\boldsymbol{a} = \{a_k\}$ is the transmitted M-ary symbol sequence, which takes values $\pm 1, \pm 3, \ldots, \pm(M-1)$, and

$$\Phi(t, \boldsymbol{a}) = 2\pi h \sum_{k=-\infty}^{\infty} a_k q(t - kT) \tag{7.106}$$

with

$$q(t) = \int_{-\infty}^{t} g(\tau) d\tau \tag{7.107}$$

The term $g(t)$ is the frequency shape pulse, which has a smooth pulse shape over $0 \leq t \leq LT$, but is zero outside, and h is the modulation index. If $L \leq 1$, $g(t)$ is a *full-response pulse shape*, and it is a *partial-response pulse shape* if $L > 1$.

In order to develop a practical maximum-likelihood CPM detector, h should be chosen as a rational number to make a finite number of phase states. There is memory in the received signal for $L > 1$, and this enables the use of the Viterbi algorithm for partial response CPM [4]. When the CPM signal state at $t = kT$ is defined by $s_k = (\theta_k, a_{k-1}, a_{k-2}, \ldots, a_{k-L+1})$, where $\Phi(t, \boldsymbol{a})$ is decomposed into the sum of the cumulate phase θ_k and the instant phase $\theta(t, \boldsymbol{a}_k)$, the total number of signal states is pM^{L-1}, for p phase states. With the Viterbi algorithm, the number of paths to be searched is only the number of states, pM^{L-1}.

When $g(t)$ is a rectangular pulse of a length of L symbols [3, 39]

$$g(t) = \begin{cases} \frac{1}{2LT}, & 0 \leq t \leq LT \\ 0, & \text{otherwise} \end{cases}. \tag{7.108}$$

We get CPFSK for $L = 1$. If further $M = 2$ and $h = 1/2$, we get MSK.

A CPM signal with integer h has discrete frequency components in PSD, enabling carrier and symbol timing in CPM receivers [39]. Many modulation schemes are given in [39]. Demodulation is mainly based on the optimum ML detection, even though other demodulators are also used. MSK-type demodulators work well for binary CPM wth $h = 1/2$. Differential receivers and discriminator receivers are also used for binary partial response CPM.

Synchronization is a difficult problem for the CPM technique. Today, synchronization methods such as the MSK-type synchronizer, square loop and fourth-power loop synchronizers are in practical use for some binary CPM with $h = 1/2$ such as GMSK.

The multi-h CPM is a special CPM where h is cyclically changed for successive symbol intervals. This leads to an increase in the minimum Euclidean distance and an improvement in the error performance, when compared to CPM. Its modulator and demodulator are very similar to that of CPM but with slight modifications. More detail on CPM and multi-h CPM is addressed in [39].

7.7 Quadrature amplitude modulation

QAM is a modulation scheme which communicates data by changing (modulating) the amplitude of two sinusoidal carrier waves, which are out of phase with each other by 90°. Unlike MPAM or MPSK, which has one degree of freedom for encoding the information, MQAM encodes information in both the amplitude and the phase of the transmitted signal. Thus, MQAM is more spectrally efficient by encoding more bits per symbol for a given average energy.

Unlike MPSK, where all the M signals are distributed on a unit circle with equidistance, the output signals generated by the rectangular MQAM modulator are distributed on an orthogonal grid on the constellation diagram. The M symbols are gray-encoded.

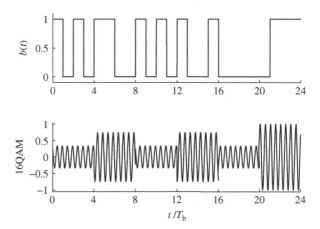

Figure 7.26

Modulated 16QAM signal.

The modulated signal is given by

$$s_i(t) = \Re\left\{A_i e^{j\phi_i} g(t) e^{j2\pi f_c t}\right\}$$

$$= \left[A_i \cos\phi_i g(t)\right]\cos(2\pi f_c t) - \left[A_i \sin\phi_i g(t)\right]\sin(2\pi f_c t) \qquad (7.109)$$

for $kT \le t \le (k+1)T$, $i = 1, 2, \cdots M$. For the rectangular window, the energy in symbol $s_i(t)$ is $E_{s_i} = A_i^2 T/2$. The modulated signal can be written as

$$s_i(t) = I_i g(t)\cos(2\pi f_c t) - Q_i g(t)\sin(2\pi f_c t), \qquad (7.110)$$

where

$$I_i = s_{i,1} = A_i \cos\phi_i, \quad Q_i = s_{i,2} = A_i \sin\phi_i. \qquad (7.111)$$

Example 7.16: The modulated signal of 16QAM modulation is plotted in Fig. 7.26. Here, the bit period $T_b = 1$ and $f_c = 2$, and the waveform is normalized.

MQAM constellations

The constellation diagram of MQAM can be specified in any combination of A_i and θ_i. Three types of QAM constellations are popular, namely type-I, type-II and type-III constellations. The type-I (star) constellation places a fixed number of signal points uniformly on each of the N concentrated rings, where N is the number of amplitude levels. This is also known as *star-QAM*. The points on the inner ring are very close in distance, and are thus most affected by errors. The type-II constellation improves the error performance of type-I, by decreasing the number of points on the inner circles, and making the distance between two adjacent points on the outer circles to be approximately equal to that on the inner circles. The type-III (square) constellation has a very small performance improvement over

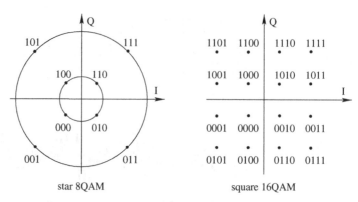

Figure 7.27

Constellations of Gray coded star 8QAM and square 16QAM.

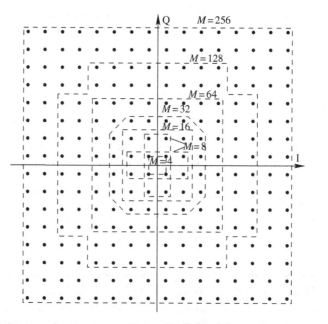

Figure 7.28

Square QAM constellations.

type-II constellation, but its implementation is much simpler. For this reason, the type-III constellation is most widely used. A type-I 8QAM and a type-III 16QAM are shown in Fig. 7.27.

Square MQAM is the most popular QAM scheme. It usually takes the form of a square constellation such as 4QAM, 16QAM, 64QAM, and 256QAM. 4QAM is the same as QPSK. These constellations are more often used, due to the relative ease of their circuit implementation. Square MQAM implementation for 32QAM, 128QAM, and 512QAM is also possible. The square QAM has a maximum possible minimum Euclidean distance $d_{\min}$ among its phasors (constellation states), for a given average symbol power. It is most appropriate for the AWGN channel. The square QAM constellations are shown in Fig. 7.28

for $M = 4, 16, 32, 64, 256, 512$. Two rectangular constellations for $M = 8$ are also shown in the figure.

Star MQAM may also be used due to its relatively simple detector and lower PAPR [14]. Star MQAM can be treated as multi-level MPSK, each level having a different amplitude. Although it is not optimum in terms of d_{min} under the constraint of average phasor power, it allows the use of efficient differential encoding and decoding methods. This makes it suitable for fading channels. The method of differential coding for star QAM is determined depending on the purpose: based on either avoiding carrier recovery or enabling differential detection of signals.

Good constellation mappings may be hard to find for MQAM signals, especially for irregular constellation shapes. It may be also hard to find a Gray code mapping where all the symbols differ from their adjacent symbols by exactly one bit.

Application

An MQAM system requires a smaller minimum carrier-to-noise power ratio than an MPSK system does. A higher level of encoding requires a higher minimum carrier-to-noise power ratio (CNR or C/N). Gray codes are used to map binary symbols to phasor states in the constellation.

MQAM modulation is most widely used in various wireless systems. Lower-order QAM schemes have better cell overlap control and good tolerance to distortion, but lower spectral efficiency. Higher-order QAM schemes provide higher data rates at the cost of stricter C/N requirements, smaller coverage radii for the same availability, and hardware complexity, and more severe cell-to-cell interference. For satellite communication systems, QPSK is usually used in the uplink direction at a lower frequency channel, and MQAM is used in the downlink direction at a higher frequency band.

Due to the high spectral efficiency, MQAM are widely used in broadband wireless and satellite multimedia communication systems, such as DVB. In IEEE 802.11n, the 256QAM modulation is used in order to achieve a data rate in excess of 600 Mbits/s. In WiMax, adaptive modulation is applied, higher-order modulation being used for MSs that are closer to the BS. In LTE, QPSK, 16QAM, and 64QAM are used in the downlink. MQAM is also used in DVB-S/C/T.

PSD

For symmetrical constellations, the means of I_k and Q_k are zero, but the variances depend on the constellation shape

$$\sigma_I^2 = E\left[I_k^2\right], \quad \sigma_Q^2 = E\left[Q_k^2\right]. \tag{7.112}$$

The PSD for QAM is derived by [39]

$$\Phi_{\tilde{s}}(f) = \frac{|P(f)|^2}{T}\left(\sigma_I^2 + \sigma_Q^2\right) = \frac{2P_{avg}}{E_p}|P(f)|^2, \tag{7.113}$$

where $P(f)$ is the spectrum of the signal pulse shape $g(t)$, P_{avg} is the average power of the signal, and E_p is the pulse energy of $g(t)$ in $[0, T]$. From this, we can see that the PSD shape of QAM is independent of its constellation, but decided by $g(t)$.

For the rectangular pulse, the PSD is derived as

$$\Phi_{\tilde{s}}(f) = A_{avg}^2 T \left(\frac{\sin \pi fT}{\pi fT} \right)^2$$

$$= A_{avg}^2 nT_b \left(\frac{\sin \pi fnT_b}{\pi fnT_b} \right)^2. \tag{7.114}$$

It has the same shape as the PSD of MPSK. For square MQAM ($M = 4^n$, $n = 1, 2, \ldots$)

$$\Phi_{\tilde{s}}(f) = \frac{2}{3} E_0 (M - 1) \left(\frac{\sin \pi fnT_b}{\pi fnT_b} \right)^2, \tag{7.115}$$

where E_0 is the energy of the signal with the lowest amplitude.

SEP

The QAM modulator and demodulator are based on the same quadrature modulator and demodulator used for MPSK. Coherent detection is used, and a pilot-signal-aided fading compensation technique is required. For square MQAM, the r_{1k} and r_{2k} can be detected separately by two multi-threshold detectors to generate I_k and Q_k.

The received MQAM signal can be demodulated by first correlating with two phase-shifted basis functions $\cos 2\pi f_c t$ and $\sin 2\pi f_c t$. The correlator outputs are then sampled and passed to the detector. Detection is based on the minimum distance between the received amplitudes of I and Q and the set of standard amplitudes of I and Q for modulation.

Clock synchronization for MQAM is the same as that for MPSK. Carrier synchronization is required for square QAM constellation, but it is not required for star QAM if differential encoding is used. Differential coding is needed for square QAM to resolve the phase ambiguity in carrier recovery. This is because all square QAM schemes have 4-fold symmetry, and the four-time loop for carrier recovery may introduce a phase ambiguity of $\pi/2$. This is similar to the M-fold symmetry for MPSK. Differential coding can be designed to remove the symmetrical ambiguity, but it violates the Gray code rule, leading to an increase in BER. For K-bit square QAM, only the first two bits are differentially encoded, while other bits are Gray coded in each quadrant. This same rule applies for other constellations. For star QAM, the use of differential coding can eliminate carrier recovery.

MQAM modulation with a square constellation of size $M = 2^k$ with k even, such as $M = 4, 16, 64, 256$, can be treated as two independent MPAM systems with constellation size $\sqrt{M}$ over the I and Q signal components, each having half the total MQAM system energy. Thus, the probability of symbol error for MQAM can be derived as

$$P_s = 1 - \left(1 - P_{\sqrt{M}} \right)^2 = 2 P_{\sqrt{M}} - P_{\sqrt{M}}^2, \tag{7.116}$$

where $P_{\sqrt{M}}$ is the SEP for each branch, that is, an $\sqrt{M}$-ary PAM with one-half of the average power of the QAM signal. At high SNR

$$P_s \approx 2P_{\sqrt{M}} = \frac{4(\sqrt{M}-1)}{\sqrt{M}} Q\left(\sqrt{\frac{3\gamma_s}{M-1}}\right), \qquad (7.117)$$

where γ_s is the average symbol SNR.

For square QAM with $M = 2^k$ for k odd, it has a rectangular constellation, and there is no equivalent $\sqrt{M}$-ary PAM system. Each constellation point may have two, three, or four nearest neighbors. The inner points have four nearest neighbors, while the outer points have two or three. The distance $d_{\min} = 2d$. The nearest-neighbor approximation to P_s can be derived, using four nearest neighbors

$$P_s \approx 4Q\left(\sqrt{\frac{3\gamma_s}{M-1}}\right). \qquad (7.118)$$

For nonrectangular constellations, the solution (7.118) is an upper bound on P_s. The nearest-neighbor approximation for nonrectangular constellation is given by [13]

$$P_s \approx M_{d_{\min}} Q\left(\frac{d_{\min}}{\sqrt{2N_0}}\right), \qquad (7.119)$$

where $M_{d_{\min}}$ is the largest number of nearest neighbors for any constellation point.

For Gray coded QAM, the BEP is calculated by

$$P_b \approx \frac{P_s}{\log_2 M}. \qquad (7.120)$$

The BEP of Gray coded coherent square QAM can be more accurately computed by [19]

$$P_b \approx \frac{4}{\log_2 M}\left(1 - \frac{1}{\sqrt{M}}\right)\sum_{i=1}^{\frac{\sqrt{M}}{2}} Q\left((2i-1)\sqrt{\frac{3\left(\log_2 M\right)\gamma_b}{M-1}}\right) \quad \text{(square QAM).} \quad (7.121)$$

The first two terms in the summation give a very accurate result. If all terms are included, then it is an upper bound, while the first term corresponds the lower bound. The approximation becomes exact for $M = 4$. A comparison between P_s of MPSK and that of MQAM clearly shows that when $M \geq 8$, every doubling of M leads to an asymptotic 3 dB increase in power saving over MPSK.

Example 7.17: The BEPs of QAM using the accurate expression for $M = 4$ and the two-term approximation for $M \geq 8$ are plotted in Fig. 7.29.

In [6], exact closed-form expressions for the BEP of the coherent demodulation of Gray coded MPAM, square MQAM, rectangular $I \times J$-ary QAM have been derived, under the AWGN channel. A general, numerically efficient algorithmic method for the

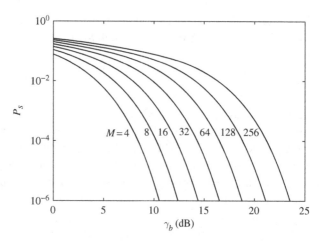

Figure 7.29

BEP of square QAM.

exact evaluation of the BEP/SEP in arbitrary 2-D constellations with arbitrary labeling (bits-to-symbol mapping) and arbitrary signaling has been proposed in [36].

For MQAM demodulation, the channel amplitude is used to scale the decision region so as to obtain the correct transmitted symbol, and this scaling is known as *automatic gain control (AGC)*. The channel gain is typically estimated at the receiver using pilot symbols.

For both MPSK and MQAM, the maximum (ISI free) spectral efficiency is $\eta_s = \log_2 M$ bits/s/Hz. MPSK leads to constant-envelope signals. MQAM has a large PAPR in the resulting signal, and a linear amplifier is required.

More details on various modulation techniques and their PSDs are given in [35, 39].

7.8 Bandwidth efficiencies of *M*-ary modulation

For *M*-ary modulation, the bit rate is related to the symbol rate by

$$R_b = (\log_2 M)R_s. \tag{7.122}$$

For MPSK modulation, the modulated signal has a null-to-null bandwidth of $2R_s$ Hz, that is

$$B_{\text{null-to-null}} = 2R_s = \frac{2R_b}{\log_2 M} \quad \text{(Hz)}. \tag{7.123}$$

The same results are obtained for MQAM and DMPSK. Thus, the bandwidth efficiencies (bits/s/Hz) for MPSK, DMPSK, and MQAM are given by

$$\eta = \frac{R_b}{B_{\text{null-to-null}}} = \frac{1}{2}\log_2 M \quad \text{(bits/s/Hz)} \quad \text{(MPSK, DMPSK, MQAM).} \tag{7.124}$$

For coherent MFSK, the minimum spacing between two tones is $\Delta f = 1/(2T_s) = R_s/2$. For *M* tones, the bandwidth is $(M-1)R_s/2$. There is an allowance of R_s Hz at either end of the band to get the null. Thus, the total bandwidth is $(M+3)R_s/2$. The bandwidth efficiency for coherent MFSK is given by

$$\eta = \frac{2\log_2 M}{M+3} \quad \text{(bits/s/Hz)} \quad \text{(coherent MFSK)}. \tag{7.125}$$

The bandwidth efficiency for noncoherent MFSK is given by

$$\eta = \frac{\log_2 M}{2M} \quad \text{(bits/s/Hz)} \quad \text{(noncoherent MFSK)}. \tag{7.126}$$

7.9 Matched filtering

Matched filtering is an optimum filtering technique that results in the maximum SNR at the receiver decision instants. For the AWGN channel, the matched filtering criterion can be represented by [21, 28]

$$H(f) = kX^*(f)e^{-j\omega T_0}, \tag{7.127}$$

where $H(f)$ is frequency response of the pulse shaping filter, k is a scaling factor, $X(f)$ is the voltage spectrum of the pulse, $*$ is the conjugation operator, and T_0 is the duration of the pulse. The delay term is used for causality. As $|H(f)| = k|X(f)|$, the magnitude response of the matched filter is identical to the transmitted signal spectrum. On the other hand, the phase of $H(f)$ is the negative of the phase of $X(f)$ shifted by ωT_0.

Matched filtering makes use of the coherent property of the pulse frequency components, while the noise components are incoherent.

The matched filtering criterion (7.127) can be transformed into the time domain by using the inverse Fourier transform

$$h(t) = kx^* (T_0 - t). \tag{7.128}$$

Since the impulse response of the matched filter is a time-reversed copy of the expected pulse, the output is the autocorrelation of either the input pulse or impulse response of the matched filter

$$y(t) = kR_{xx}(\tau) = kR_{hh}(\tau), \tag{7.129}$$

where $\tau = T_0 - t$, and $x(t) = s(t) + n(t)$, $s(t)$ being the transmitted pulse and $n(t)$ white noise with double-sided PSD $N_0/2$. The matched filter is thus also called a *correlation receiver*. The matched filter output at the sampling instant $t = T_0$ is identical to the output of the signal correlator.

For digital communications, the variable delay τ is unnecessary, since the timing is usually known and also only the peak value of the correlation function, $R_{xs}(0)$, is important. The output of the correlator reaches the maximum at the end of the pulse of period T_0. The matched filter output is given by $\int_0^{T_0} s^2(t)dt = E_s$, E_s being the normalized symbol energy. The decision instant SNR is given by

$$\frac{S}{N} = \frac{2E_s}{N_0}. \tag{7.130}$$

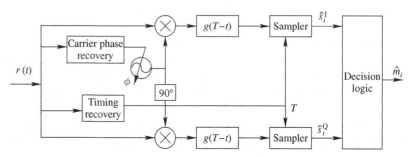

Figure 7.30

Signal correlator in digital communications.

Figure 7.31

Receiver with carrier phase and timing recovery. T is the sampling period.

This SNR value depends only on pulse energy and the input noise power spectral density, but independent of pulse shape. The signal correlator is shown in Fig. 7.30.

A single matched filter can only detect one state of digital symbol. For reception of an M-ary signal, M parallel correlators are used, each having a reference pulse corresponding to one of the M symbol states. The outputs of all the M correlators are compared, and the one with the maximum value corresponds to the detected symbols.

7.10 Synchronization

For digital demodulation and detection, synchronization of symbol timing and estimation of the carrier frequency and phase are very challenging tasks. Symbol timing is necessary for estimating the received symbol and is used for driving the sampling devices at the demodulators. Carrier frequency and phase are used in all coherent demodulators. Due to the complex channel, synchronization of timing and carrier phase is most difficult. In case of multipath components, the receiver is usually synchronized to the strongest multipath component or the first arriving multipath component that exceeds a power threshold.

The carrier phase and timing recovery structure for the amplitude and phase demodulator is shown in Fig. 7.31. Carrier phase and symbol timing can be jointly estimated by using the ML method. They can also be detected separately by assuming the other to be known.

Synchronization at the receiver has several layers: carrier synchronization; symbol timing recovery; slot, frame, and superframe synchronization. Synchronization is typically accomplished through initial acquisition, tracking, and reacquisition. Acquisition is a hypothesis-testing problem that determines the interval in which the optimum estimation lies, while tracking fine-tunes the estimation to the exact value.

For cellular communication systems, BSs use precise rubidium clocks or GPS signals as frequency references, while MSs use quartz oscillators. The BS periodically transmits a high-precision frequency reference, and the MS can adjust its local oscillator using this reference. Time synchronization information is also transmitted from the BS to the MS.

7.10.1 Carrier synchronization

Carrier frequency offsets are caused by drifts of LOs at the transmitters and receivers, and/or the Doppler shift. Carrier synchronization is necessary for coherent demodulation, where the phase of the received signal must be known so that the LO is exactly synchronized to the carrier, both in frequency and phase. This is practically very difficult.

Carrier synchronization can be achieved by sending a pilot tone, which has a strong spectral line at the carrier frequency. This spectral line enables the receiver to easily lock onto it and generate a local coherent carrier. Carrier synchronization can also be achieved by using a carrier recovery circuit which extracts the phase and frequency of the received signal to generate a clean reference signal. Carrier synchronization is implemented by some variant of PLL such as a Costas loop.

Carrier phase recovery

Assuming a known timing, the ML phase estimation $\hat{\phi}$ for an unmodulated carrier plus noise, $r(t) = A \cos(2\pi f_c t + \phi) + n(t)$, is given by [13]

$$\int_{T_0} r(t) \sin\left(2\pi f_c t + \hat{\phi}\right) dt = 0, \tag{7.131}$$

where T_0 is a finite time interval $T_0 \geq T$. This solution is accomplished by using a PLL that satisfies (7.131), as shown in Fig. 7.32. In the figure, if $z(t) = 0$, then $\hat{\phi}$ is the ML estimate of ϕ. If $z(t) \neq 0$, then the VCO adjusts its phase $\hat{\phi}$.

For a modulated carrier, there are decision-directed and non-decision-directed parameter estimation methods. The data decision structure can remove the data decision from the modulation of the received signal, and the resulting unmodulated carrier is then passed to the above PLL for phase recovery. A non-decision-directed loop structure can use a squaring or Mth-power loop, or a Costas loop. The Mth-power loop takes the Mth-power of the signal, filters out the frequency Mf_c by a bandpass filter, and the output is sent to a PLL. The Costas loop is very similar to a PLL. The decision-directed PLL is superior

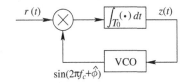

Figure 7.32

Phase-locked loop for carrier phase recovery.

in performance to both the Costas loop [8] and the squaring (or Mth-power) loop, when the demodulation error rates are below 10^{-2}. More discussion on carrier phase recovery is given in [28].

Carrier frequency synchronization

The LO can be synchronized with an incoming carrier wave by either using a pilot carrier transmitted along with the incoming carrier wave, or using a carrier-recovery circuit. The pilot carrier has the same phase and frequency as the modulated signal, and can be used to phase-lock the LO. Carrier-recovery circuits are typically based on PLLs.

The carrier-recovery loop and the timing synchronizer are ideally locked to the carrier of the LOS or the minimum-delay multipath component, which has the largest amplitude. In practice, the synchronizer typically locks to the first found component above a threshold to avoid the complex search process.

For PSK signals such as BPSK or QPSK signals, there is no spectral line at the carrier frequency. A nonlinear device is used to generate such a spectral line. A Mth-power loop or a Costas loop can be used for this purpose. For MPSK, the Mth-power loop can produce a spectral line at Mf_c. A PLL tracks and locks onto the frequency and phase of this Mf_c component. A divided-by-M device produces the desired carrier at f_c and almost the same phase of the received signal. For a BPSK signal $m(t)\cos(2\pi f_c t + \phi)$, $m(t) = \pm 1$, a simple squarer, which uses a diode, bandpass filtering, and a divide-by-two circuit, generates an LO signal synchronized to the input signal, $\cos(2\pi f_c t + \phi)$. Mixing the LO with the input signal yields the demodulated $m(t)$. Thus, for BPSK signals, the PLL is not necessary. The Mth-power device is difficult in circuit implementation, especially at high frequencies. The Costas loop avoids this device, but it requires a good balance between the I- and Q-channels. The performance of the Mth-power loop and the Costas loop implementation are the same [12].

For QAM, carrier recovery is typically realized by using the fourth-power loop or by decision-directed carrier recovery (DDCR). DDCR is suitable for fixed-link QAM systems, while the fourth-power loop is more suitable for digital radio in fading channels, since it does not depend on data decisions. The fourth-power device generates a spectral line at $4f_c$. DDCR can be used for any constellation, but it has a BER threshold: below this threshold, its performance is extremely good; otherwise, the decision is not reliable.

Frequency synchronization is typically realized using automatic frequency control (AFC). It is often feedback-based and the algorithm requires the PLL operation suitable for DSP implementation. The received signal constellation rotation is corrected as phase error. By monitoring the phase error in the signal constellation, phase correction can be conducted by an analog component with a digital control signal or by using a DSP.

7.10.2 Symbol timing recovery

Symbol timing recovery or clock synchronization ensures the received signal is sampled as close as possible to the optimal sampling point for detection. The clock signal is recovered

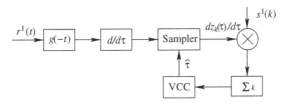

Figure 7.33 **Decision-directed timing estimator.**

from the demodulated waveform. Digital interpolation is usually used to find the value of the signal at the optimal sampling point from two or more neighboring samples.

Usually the spectrum of the received waveform does not contain a discrete component at the clock frequency, due to distortion of the transmission link and noise. But a periodic component at the clock frequency exists in the squared waveform for digital signals. Timing estimation can be performed on either the in-phase or quadrature branch. The timing estimator can be either a decision-directed estimator or a non-decision-directed estimator.

ML estimation

Assuming a known carrier phase ϕ, the ML estimation of the timing delay τ can be derived. The ML estimation of τ is required to satisfy [13, 28]

$$\sum_k s_I(k)\frac{d}{d\tau}z_k(\tau) = 0, \tag{7.132}$$

where $s_I(k)$ is the amplitude of the in-phase component of the symbol transmitted over the kth symbol period, and

$$z_k(\tau) = \int_{T_0} r(t)g(t - kT - \tau)dt. \tag{7.133}$$

For decision-directed estimation, the estimator is as shown in Fig. 7.33. The voltage-controlled clock (VCC) has an input as the LHS of (7.132) and an output of timing estimator $\hat{\tau}$. The feedback loop ensures that the input will become zero and in this case the output is the ML estimation of τ. In the figure, the in-phase equivalent lowpass received signal is $r_I(t) = s_I(t; \tau) + n_I(t)$, and $s_I(t; \tau) = \sum_k s_I(k)g(t - kT - \tau)$.

Early-late gate synchronizer

A well-known non-decision-directed timing estimation method is the early-late gate synchronizer [13, 28]. It exploits the symmetry of autocorrelation and the fact that the maximum of the autocorrelation is at $\tau = 0$. It is applied to the output of the matched filter or correlator. Bit timing is established by adaptive timing based on an error signal generated by, for example, early-late gate integrators. The architecture of the early-late gate synchronizer is shown in Fig. 7.34. The loop adjusts the timing of the VCO square wave so that it is in perfect synchronization with the demodulated signal $m(t)$. Timing

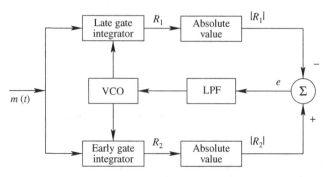

Figure 7.34

Early-late gate timing estimator.

recovery estimates the timing offsets before sampling. The early-late gate is used in IS-95 [38].

Timing recovery is conventionally performed using a decision-directed PLL. Conventionally timing recovery is an independent step relative to equalization and channel decoding. This timing recovery approach fails at low SNR. Iterative timing recovery performs timing recovery multiple times while interacting with the decoder, to enable reliable operation at very low SNR, with a marginal increase in complexity when compared to the conventional timing recovery [5]. The eye diagram is a useful tool for accessing the distortion of a digital signal. The vertical eye opening is an indication of the amount of noise that can be tolerated without causing an incorrect decision. The eye diagram can be obtained by triggering the oscilloscope with the recovered symbol timing clock signal and displaying the symbol stream using a sweep time for 2 or 3 symbol periods.

The filter and square timing recovery technique [24] is a simple and well-known one. It exploits the cyclostationary property for timing estimation. It is based on an explicit spectral line regeneration using a nonlinear device. In [41], this symbol timing recovery technique is refined by implementing interpolation.

Frame timing

Frame timing can be accomplished by using training symbols at the beginning of a data frame, and frame synchronization is performed by repetitively correlating the received training sequence with its replica known at the receiver until a peak is identified. The receiver can then adjust its timing based on the time at the peak. Frame timing can be tracked by the same operation. In case of frequency offset, the received signal should be correlated with the frequency-shifted original training sequence. For estimation of slot timing, a primary synchronization sequence is included at the beginning of the transmitted data. The signal at the ADC output is filtered by a correlator matched to the primary synchronization sequence.

Some BSs provide the synchronization channel for MSs to achieve slot and frame synchronization. Synchronization is searched by performing correlations for at least a frame boundary. Also the location of each time slot corresponds to a strong peak.

7.11 Differential modulation

Differential modulation is a noncoherent technique. The differential scheme does not require channel knowledge, but incurs 3 dB loss in the required SNR when compared to its coherent counterpart.

The differential scheme is traditionally used for PSK-modulated symbols $s(n)$. But the idea is applicable to other modulation schemes. The transmitted symbol $x(n)$ is obtained by

$$x(n) = x(n-1)s(n), \quad n \geq 0 \tag{7.134}$$

and $x(-1) = 1$.

After transmission over a flat-fading channel h, the received signal at the receiver is given by

$$y(n) = \sqrt{\gamma}\, hx(n) + w(n), \tag{7.135}$$

where γ is the average symbol SNR, and $w(n)$ is white Gaussian noise with mean 0 and variance 1.

Assuming the channel h is time-invariant for at least two symbol periods, we have

$$
\begin{aligned}
y(n) &= \sqrt{\gamma}\, hx(n-1)s(n) + w(n) \\
&= [y(n-1) - w(n-1)]s(n) + w(n) \\
&= y(n-1)s(n) + v(n),
\end{aligned}
\tag{7.136}
$$

where $v(n) = w(n) - w(n-1)s(n)$.

Assuming $|s(n)| = 1$, we obtain that $v(n)$ is Gaussian with mean zero and variance 2. The ML demodulation of $s(n)$ from (7.136) gives

$$\hat{s}(n) = \arg \min_{s(n) \in \mathcal{S}} \|y(n) - y(n-1)s(n)\|^2. \tag{7.137}$$

This does not require any information of the channel h. From (7.136), the SNR at the output of the differential decoder is derived as

$$\gamma_{\text{diff}} = \frac{|h|^2 \gamma}{2}. \tag{7.138}$$

The 3 dB loss in SNR compared to the coherent ML decoding is due to the fact that the noise variance is increased from 1 for $w(n)$ to 2 for $v(n)$.

7.12 Error probability in fading channels

Thus far, the error probabilities given for different modulation schemes are for the AWGN channel. This section gives the corresponding error probabilities in fading channels.

7.12.1 Flat Rayleigh fading channel

Wireless channels with relatively small bandwidth are typically characterized as the Rayleigh-fading channel. The Rayleigh fading leads to a significant degradation in BEP performance compared to the Gaussian channel, since the Rayleigh fading decreases the SNR. In the Gaussian channel, the BEP decreases exponentially with the SNR, while in the Rayleigh fading channel it decreases linearly with the average SNR. Diversity, spread spectrum, and OFDM are the three techniques to overcome fading.

The BEP of Rayleigh fading can be derived as

$$P_b^{\text{Rayleigh}} = \int_0^\infty P_b\left(\overline{\gamma_b}|r\right) p_r(r) dr, \tag{7.139}$$

where $p_r(r)$ is the Rayleigh pdf, given by (3.25), and $\overline{\gamma_b} = \frac{\overline{E_b}}{N_0} = \frac{2\sigma^2 E_b}{N_0}$ is the mean bit SNR, σ being a parameter of the Rayleigh pdf. The SEP of the Rayleigh fading channel can be derived similarly.

The SEP for Rayleigh fading can also be derived from (7.25). First, at each instant, the amplitude of the received signal, r, satisfies the Rayleigh distribution. Define the received power as $P_{\text{inst}} = r^2$, and then the mean power is $\overline{P} = 2\sigma^2$. Since $dP_{\text{inst}} = 2rdr$, the pdf of the received power $p_{P_{\text{inst}}}(P_{\text{inst}})$ can be derived. By scaling P_{inst} by $1/N_0$, the pdf of the symbol SNR γ_s, $p_{\gamma_s}(\gamma)$, is derived as

$$p_{\gamma_s}(\gamma) = \frac{1}{\overline{\gamma}_s} e^{-\frac{\gamma}{\overline{\gamma}_s}}, \tag{7.140}$$

where $\overline{\gamma}_s = \frac{\overline{E_s}}{N_0}$ is the mean symbol SNR. The SEP of the Rayleigh fading channel is finally obtained by averaging over the distribution of the symbol SNR

$$P_s^{\text{Rayleigh}} = \int_0^\infty p_{\gamma_s}(\gamma) P_s(\gamma) d\gamma, \tag{7.141}$$

where $P_s(\gamma)$ is the SEP in the AWGN channel. This procedure is also valid for Ricean fading and other amplitude distributions.

In the AWGN channel, many coherent demodulation methods have an approximate or exact value for P_s of the form

$$P_s(\gamma_b) \approx c_1 Q\left(\sqrt{c_2 \gamma_b}\right). \tag{7.142}$$

Another general form for P_s in the AWGN channel is

$$P_s(\gamma_b) \approx c_1 e^{-c_2 \gamma_b}, \tag{7.143}$$

where c_1 and c_2 are constants.

Corresponding to (7.142), the SEP in the Rayleigh fading channel can be derived by substituting (7.142) in (7.141):

$$P_s^{\text{Rayleigh}} = \frac{c_1}{2}\left(1 - \sqrt{\frac{c_2\overline{\gamma}_b}{2 + c_2\overline{\gamma}_b}}\right). \tag{7.144}$$

At high $\overline{\gamma}_b$, $P_s^{\text{Rayleigh}} \approx \frac{c_1}{2c_2\overline{\gamma}_b}$.

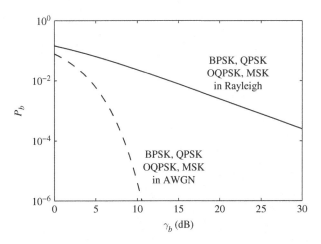

Figure 7.35

BEPs of BPSK, QPSK, OQPSK, and MSK in the AWGN and flat Rayleigh fading channels.

Also, corresponding to (7.143), the SEP in the Rayleigh fading channel can be obtained by substituting (7.143) in (7.141):

$$P_s^{\text{Rayleigh}} = \frac{c_1}{1 + c_2 \overline{\gamma}_b}. \tag{7.145}$$

At high $\overline{\gamma}_b$, $P_s^{\text{Rayleigh}} \approx \frac{c_1}{c_2 \overline{\gamma}_b}$.

Example 7.18: The BEPs for BPSK (also QPSK, OQPSK, and MSK) in the AWGN and Rayleigh fading channels are plotted in Fig. 7.35. It is shown that at the BEP of 10^{-5} the AWGN channel requires a bit SNR γ_b of 10 dB, while the flat Rayleigh fading channel requires 44 dB; this means a loss of 34 dB. To increase the BEP performance, one can increase the transmitting power by 34 dB. Proper coding techniques can have a coding gain of 34 dB, and thus are power efficient; this is achieved at the price of reduced data rate.

Based on the two error probability representations given by (7.144) and (7.145), BEPs in the Rayleigh fading channel can be obtained. Some values of c_1 and c_2 are listed in Table 7.2.

Closed-form results for MPSK, DMPSK, and MQAM

For MPSK, a closed-form expression for the flat Rayleigh fading channel is given by [10, 33]

$$P_s^{\text{Rayleigh}} = \frac{M-1}{M} - \alpha \left[\frac{1}{2} + \frac{1}{\pi} \tan^{-1} \left(\alpha \cot \frac{\pi}{M} \right) \right] \quad \text{(MPSK, Rayleigh)}, \tag{7.146}$$

Table 7.2. List of c_1 and c_2 for different modulation schemes under a flat Rayleigh fading channel.

Modulation	c_1	c_2	BEP (AWGN)	BEP (Rayleigh)
coherent BPSK, QPSK, OOPSK, MSK	1	2	(7.142)	(7.144)
optimum DBPSK	$\frac{1}{2}$	1	(7.143)	(7.145)
DEBPSK, DEQPSK, $\pi/4$-QPSK	2	2	(7.142)	(7.144)
suboptimum DBPSK	$\frac{1}{2}$	0.8	(7.143)	(7.145)
optimum DQPSK, $\pi/4$-DQPSK	1	1.112	(7.142)	(7.144)
MPSK ($M > 4$)	$\frac{2}{\log_2 M}$	$2\log_2 M \sin^2 \frac{\pi}{M}$	(7.142)	(7.144)
DEMPSK	$\frac{4}{\log_2 M}$	$2\log_2 M \sin^2 \frac{\pi}{M}$	(7.142)	(7.144)
optimum DMPSK	$\frac{2}{\log_2 M}$	$2\log_2 M \sin^2 \frac{\pi}{\sqrt{2}M}$	(7.142)	(7.144)
coherent BFSK	1	1	(7.142)	(7.144)
noncoherent orthogonal BFSK	$\frac{1}{2}$	$\frac{1}{2}$	(7.143)	(7.145)
coherent GMSK	1	2α	(7.142)	(7.144)[a] [22]
MQAM ($M = 2^k$ for k even)	$\frac{4(\sqrt{M}-1)}{\sqrt{M}\log_2 M}$	$\frac{3\log_2 M}{M-1}$	(7.142)	(7.144)

[a] $\alpha = 0.68$ for $BT_s = 0.25$

where

$$\alpha = \sqrt{\frac{\overline{\gamma}_s \sin^2\left(\frac{\pi}{M}\right)}{1 + \overline{\gamma}_s \sin^2\left(\frac{\pi}{M}\right)}}. \tag{7.147}$$

For DMPSK, a closed-form SEP formula for fast Rayleigh channels is derived as [40]

$$P_s = \frac{M-1}{M} + \frac{|\rho_z|\tan(\pi/M)}{\xi(\rho_z)}\left[\frac{1}{\pi}\arctan\left(\frac{\xi(\rho_z)}{|\rho_z|}\right) - 1\right] \quad \text{(DMPSK, fast Rayleigh)}, \tag{7.148}$$

where

$$\rho_z = \frac{\overline{\gamma}}{1+\overline{\gamma}}\rho_\alpha, \tag{7.149}$$

$$\xi(\rho_z) = \sqrt{1 - |\rho_z|^2 + \tan^2\left(\frac{\pi}{M}\right)}. \tag{7.150}$$

ρ_α denotes the correlation between α_k and α_{k-1}, α_k being the received signal phase difference. When $M = 2$, it reduces to

$$P_s = \frac{1}{2}\left(1 - \frac{\overline{\gamma}}{1+\overline{\gamma}}|\rho_\alpha|\right), \tag{7.151}$$

which is the same as that give in [33]

As $\overline{\gamma} \to \infty$, $|\rho_z| \to |\rho_\alpha|$, P_s reveals the presence of an error floor for DMPSK on fast fading channels. On slow fading channels, $\rho_\alpha = 1$, and thus $P_s \to 0$, indicating the error floor vanishes.

For slow fading, $\rho_\alpha = 1$, we have from (7.148)

$$P_s = \frac{M-1}{M} - \frac{1}{\pi}\sqrt{\frac{\overline{\gamma}^2\sin^2\left(\frac{\pi}{M}\right)}{1+2\overline{\gamma}+\overline{\gamma}^2\sin^2\left(\frac{\pi}{M}\right)}} \times \left[\pi - \arccos\left(\frac{\overline{\gamma}}{1+\overline{\gamma}}\cos\left(\frac{\pi}{M}\right)\right)\right], \tag{7.152}$$

which is identical to the result given in [10].

Based on the exact instantaneous BEP expression given in [6], an exact closed-form BEP formula for coherently detected MQAM with Gray code bit mapping in the Rayleigh fading channel is derived in [7]. To simplify the formula, tight invertible lower and upper bounds on the BEP are derived. Using these bounds, tight lower and upper bounds on the BEP-based outage probability are also obtained in a log-normal shadowing environment.

In addition, the SEP performance of coherent MPSK in a situation where the phase error, quadrature error, and I/Q mismatch problems take place all concurrently over an AWGN and arbitrary fading channel has been investigated in [26].

7.12.2 Flat Ricean fading channel

The SEP for the Ricean fading channel can also be computed from (7.141), but $p_{\gamma_s}(\gamma)$ for Ricean fading is used. Closed-form BEP expressions are derived for optimum DBPSK and noncoherent BFSK [39]

$$P_b = \frac{K_r + 1}{2(K_r + 1 + \overline{\gamma}_b)}e^{-\frac{K_r\overline{\gamma}_b}{K_r+1+\overline{\gamma}_b}} \quad \text{(optimum DBPSK)}, \tag{7.153}$$

$$P_b = \frac{K_r + 1}{2(K_r + 1) + \overline{\gamma}_b}e^{-\frac{K_r\overline{\gamma}_b}{2(K_r+1)+\overline{\gamma}_b}} \quad \text{(noncoherent BFSK)}. \tag{7.154}$$

For other modulation schemes, numerical integration has to be used for calculating the BEP performance in the Ricean channel.

The BEP performance in the Ricean fading channel lies between that in the AWGN and Rayleigh fading channels. For $K_r \to 1$, it reduces to Rayleigh fading, while for $K_r \to \infty$, we obtain the AWGN channel.

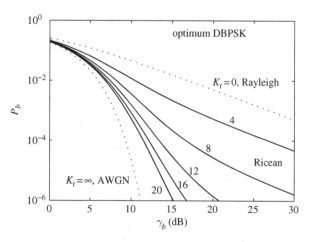

Figure 7.36   BEPs of the optimum DBPSK in the AWGN, Rayleigh fading, and Ricean fading channels.

Example 7.19: The BEPs for the optimum DBPSK are plotted in Fig. 7.36 in case of the AWGN, Rayleigh fading, and Ricean fading channels.

7.12.3 Alternative form of the Q-function

The Q-function is most fundamental for BEP calculation. Its integration limit that runs to infinity makes its calculation very difficult. Recently, the following representation of the Q-function has been introduced and it makes the computation much more efficient [13, 21, 33]:

$$Q(x) = \frac{1}{\pi} \int_0^{\pi/2} e^{-\frac{x^2}{2\sin^2\theta}} d\theta, \quad x > 0. \tag{7.155}$$

By using this new Q-function, one can very easily calculate the BEP or SEP related to the Q-function. The alternative Q-function makes the derivation of closed-form BEP or SEP, or its numerical calculation much easier.

The SEP in fading channels can be written in a generic form [21]

$$P_s(\gamma) = \int_{\theta_1}^{\theta_2} f_1(\theta) e^{-\gamma f_2(\theta)} d\theta. \tag{7.156}$$

The alternative form of Q-function is more useful for calculating the BEP or SEP in fading channels.

The SEPs of MPSK, MASK, and MQAM, given in terms of Q-function can be calculated by using the alternative Q-function representation [17]

$$P_s = a \int_0^b e^{-\frac{c\gamma_s}{\sin^2\theta}} d\theta, \tag{7.157}$$

where

$$a = \frac{2}{\pi} \frac{M-1}{M}, \qquad b = \frac{\pi}{2}, \qquad c = \frac{3}{M^2 - 1} \qquad \text{(MASK)}, \qquad (7.158)$$

$$a = \frac{4}{\pi} \frac{\sqrt{M}-1}{\sqrt{M}}, \qquad b = \frac{\pi}{2}, \qquad c = \frac{3}{2(M-1)} \qquad \text{(MQAM)}, \qquad (7.159)$$

$$a = \frac{1}{\pi}, \qquad b = \pi \frac{M-1}{M}, \qquad c = \sin^2(\pi/M) \quad \text{(MPSK)}. \qquad (7.160)$$

7.12.4 Error probability using moment-generating functions

The average SEP can be expressed as

$$\overline{P}_s = \int_{\theta_1}^{\theta_2} f_1(\theta) M_\gamma \left(-f_2(\theta)\right) d\theta, \qquad (7.161)$$

where the moment-genrating function $M_\gamma(s)$ is the Laplace transform of $p_\gamma(\gamma)$, that is, $M_\gamma = \int_0^\infty p_\gamma(\gamma) e^{s\gamma} d\gamma$.

For the Rayleigh distribution

$$M_\gamma(s) = \frac{1}{1 - s\overline{\gamma}}. \qquad (7.162)$$

For the Rice distribution

$$M_\gamma(s) = \frac{1 + K_r}{1 + K_r - s\overline{\gamma}} e^{\frac{K_r s\overline{\gamma}}{1+K_r-s\overline{\gamma}}}. \qquad (7.163)$$

The average (ergodic) error probability in fading for modulation with P_s in the AWGN channel, which is given by (7.142), can be calculated by

$$\overline{P}_s = \frac{\alpha_M}{\pi} \int_0^{\pi/2} M_{\gamma_s} \left(\frac{-\beta_M}{2 \sin^2 \theta}\right) d\theta. \qquad (7.164)$$

From (7.157), the exact average probability of error for M-ary linear modulation in fading is given by

$$\overline{P}_s = a \int_0^b M_{\gamma_s} \left(\frac{-c}{\sin^2 \theta}\right) d\theta. \qquad (7.165)$$

For statistically independent diversity branches, we have [17]

$$\overline{P}_s = a \int_0^b \prod_{l=1}^{N_r} M_{\gamma_s,l} \left(\frac{-c}{\sin^2 \theta}\right) d\theta = a \int_0^b \left[M_{\gamma_s}\left(\frac{-c}{\sin^2 \theta}\right)\right]^{N_r} d\theta, \qquad (7.166)$$

where N_r is the diversity order.

The ergodic error probability in the Rayleigh fading channel can thus be obtained by inserting the MGF (7.162) into (7.166)

$$\overline{P}_s = a \int_0^b \left(\frac{N_r \sin^2 \theta}{N_r \sin^2 \theta + c\gamma_s}\right)^{N_r} d\theta \qquad (7.167)$$

when the transmit power is uniformly distributed onto all diversity paths. For Ricean fading, the MGF (7.163) is inserted into (7.166).

The MGFs for many distributions are given in [33].

7.13 Error probabilities due to delay spread and frequency dispersion

Frequency-selective fading leads to ISI, while Doppler gives rise to spectrum broadening, which leads to ACI. Doppler spread has a significant influence on the BEP performance of modulation techniques that use differential detection. BEPs due to delay spread (frequency-selected fading) or frequency dispersion (Doppler spread) cannot be reduced by increasing the transmit power, thus they are called *error floors*. For low-data-rate systems such as sensor networks or paging, the Doppler shift can lead to an error floor of an order of up to 10^{-2}. Generally, for high-data-rate wireless communications such as mobile communications and wireless LAN, error due to frequency dispersion is of the order of up to 10^{-4}, which is very small compared with that arising from noise.

Error probability due to frequency dispersion

For DPSK in fast Ricean fading, where the channel decorrelates over a bit period, the BEP is given by [13, 33]

$$\overline{P}_b = \frac{1}{2} \left(\frac{1 + K_r + \overline{\gamma}_b(1 - \rho_c)}{1 + K_r + \overline{\gamma}_b} \right) e^{-\frac{K_r \overline{\gamma}_b}{1 + K_r + \overline{\gamma}_b}}, \tag{7.168}$$

where K_r is the fading parameter of Ricean distribution, and ρ_c is the channel correlation coefficient after a bit period T_b. For the uniform scattering Doppler spectrum model, from Section 3.3.1, the Doppler power spectrum is

$$S(f) = \frac{P_0}{\pi \sqrt{v_{\max}^2 - f^2}}, \quad |f| < v_{\max}, \tag{7.169}$$

then $\rho_c = \phi_c(T)/\phi_c(0) = J_0(2\pi v_{\max}T)$. For the rectangular Doppler power spectrum model, $S(f) = P_0/2v_{\max}$, $|f| < v_{\max}$, then $\rho_c = \mathrm{sinc}(2v_{\max}T)$.

For Rayleigh fading, $K_r = 0$, from (7.168), we have

$$\overline{P}_b = \frac{1}{2} \left(\frac{1 + \overline{\gamma}_b(1 - \rho_c)}{1 + \overline{\gamma}_b} \right). \tag{7.170}$$

The irreducible error floor for DPSK is obtained by limiting $\overline{\gamma}_b \to \infty$ in (7.168)

$$\overline{P}_{b,\mathrm{floor}} = \frac{(1 - \rho_c)e^{-K_r}}{2}. \tag{7.171}$$

The irreducible bit error floor for DQPSK in fast Ricean fading is derived as [13, 16]

$$\overline{P}_{b,\text{floor}} = \frac{1}{2} \left(1 - \sqrt{\frac{(\rho_c/\sqrt{2})^2}{1 - (\rho_c/\sqrt{2})^2}} \right) e^{-\frac{(\sqrt{2}-1)K_r}{\sqrt{2}-\rho_c}}, \qquad (7.172)$$

where ρ_c is the channel correlation coefficient after a symbol time T.

For the uniform scattering model and Rayleigh fading, the irreducible error for DPSK is obtained from (7.171) as

$$\overline{P}_{b,\text{floor}} = \frac{1}{2} [1 - J_0 (2\pi \nu_{\max} T_b)] \approx 0.5 (\pi \nu_{\max} T_b)^2. \qquad (7.173)$$

For small $\nu_{\max} T_b$, the BEP is generally given by

$$\overline{P}_b^{\text{Doppler}} = K (\nu_{\max} T_b)^2, \qquad (7.174)$$

where K is a constant. Note that the error floor decreases as the data rate $R = 1/T_b$ increases. For MSK, $K = \pi^2/2$.

Error probability due to delay dispersion

For delay dispersion, ISI influences a significant percentage of each symbol. For Rayleigh fading, when the maximum excess delay of the channel is much smaller than the symbol duration, the error floor due to delay dispersion is given by [21]

$$\overline{P}_{b,\text{floor}} = K \left(\frac{\sigma_\tau}{T_b} \right)^2, \qquad (7.175)$$

where σ_τ is the rms delay spread of the channel, given by (3.80), and K is a constant, which depends on the system implementation. For differentially detected MSK, $K = \frac{4}{9}$ [21]. The error floor due to delay dispersion is typically more significant than that due to frequency dispersion for high data rates.

7.14 Error probability in fading channels with diversity reception

The average SEP in a flat-fading channel is derived by the distribution of the SNR

$$\overline{P}_s = \int_0^\infty p_\gamma(\gamma) P_s(\gamma) d\gamma. \qquad (7.176)$$

For BPSK signals in the Rayleigh fading channel, the SEP for N_r diversity branches with MRC is given by [28]

$$\overline{P}_s = \left(\frac{1-\mu}{2} \right)^{N_r} \sum_{k=0}^{N_r-1} \binom{N_r - 1 + k}{k} \left(\frac{1+\mu}{2} \right)^k, \qquad (7.177)$$

where

$$\mu = \sqrt{\frac{\overline{\gamma}_s}{1+\overline{\gamma}_s}}. \tag{7.178}$$

For large $\overline{\gamma}_s$, the SEP is well approximated by [28]

$$\overline{P}_s = \left(\frac{1}{4\overline{\gamma}_s}\right)^{N_r} \binom{2N_r - 1}{N_r}. \tag{7.179}$$

Thus, the BEP decreases with the N_rth power of the average SNR. For a MIMO system of N_t transmitting antennas and N_r receiving antennas, there are altogether N_tN_r diversity branches, and thus N_r in the above equation is replaced by N_tN_r.

For the AWGN channel, consider the error probability of the form

$$P_s = \frac{1}{2}e^{-\alpha\gamma_s}, \tag{7.180}$$

where $\alpha = \frac{1}{2}$ for noncoherent FSK and $\alpha = 1$ for differential coherent PSK. For N_r diversity branches, we have the error probability for both cases [31, 32]

$$P_b = \frac{1}{2} \prod_{k=1}^{N_r} \frac{1}{1+\alpha\overline{\gamma}_{s,k}} \quad \text{(MRC)}, \tag{7.181}$$

$$P_b = \frac{1}{2} \frac{N_r!}{\prod_{k=1}^{N_r}\left(k+\alpha\overline{\gamma}_s\right)} \quad \text{(Selection diversity)}, \tag{7.182}$$

$$P_b = \frac{\left(\frac{N_r}{2}\right)^{N_r}\sqrt{\pi}}{2\left(N_r - \frac{1}{2}\right)} \prod_{k=1}^{N_r} \frac{1}{1+\alpha\overline{\gamma}_{s,k}} \quad \text{(EGC)}. \tag{7.183}$$

For coherent FSK and coherent PSK in the AWGN channel, we have

$$P_s = Q\left(\sqrt{\frac{\alpha\gamma_s}{2}}\right), \tag{7.184}$$

where $\alpha = \frac{1}{2}$ for coherent FSK and $\alpha = 1$ for coherent PSK. For N_r diversity branches, the average error probability is derived from (7.176), by taking $p(\gamma)$ obtained from (5.17). There is no convenient closed-form solution. In the case of all branches being equal in strength $\overline{\gamma}_s$, we have [31]

$$P_s = \frac{1}{2\sqrt{\pi}} \frac{1}{(\alpha\overline{\gamma}_s)^{N_r}} \frac{\left(N_r - \frac{1}{2}\right)!}{N_r!}. \tag{7.185}$$

The error probability of discriminator detected FSK that uses two branch diversity is also given in [2, 32] for MRC and selection diversity.

By using the moment-generating function and the trigonomic form of the Q-function, the average SEP of an MRC diversity system can be derived as [21]

$$\overline{P}_s = \int_{\theta_1}^{\theta_2} d\theta f_1(\theta)\left[M_{\gamma_s}\left(-f_2(\theta)\right)\right]^{N_r}. \tag{7.186}$$

For BPSK in Rayleigh fading, the SEP is alternatively given by [21]

$$\overline{P}_s = \frac{1}{\pi} \int_0^{\pi/2} \left(\frac{\sin^2 \theta}{\sin^2 \theta + \overline{\gamma}_s} \right)^{N_r} d\theta. \tag{7.187}$$

Diversity can also improve the SEP performance of frequency-selective and time-selective fading channels. Generally, the SEP with diversity is approximately the N_rth power of the BEP without diversity.

The BEP of BPSK for MRC in a Rayleigh fading channel can be extended to the SEP of MPSK in a Rayleigh fading channel in a straightforward manner [37]. MRC receivers using BPSK and MPSK modulation in spatially correlated Rayleigh fading channels have been analyzed in [34], and closed-form expressions for error probabilities of these modulations with MRC derived. Exact average SEP expressions for arbitrary M-ary rectangular QAM, when used along with MRC receive diversity over independent but not necessarily identically distributed Nakagami fading channels, are derived in [20].

Asymptotic BEP and SEP expressions are derived in [23] for coherent MRC with MPAM, MPSK and MQAM modulation, differential EGC with DMPSK, and square law combining with BFSK, in correlated Ricean fading and nonCGaussian noise (including interference). Compared to BPSK with MRC, BFSK with square law combining has an asymptotic performance loss of 6 dB independent of the type of noise and N_r. The asymptotic performance loss of differential EGC compared to MRC is always 3 dB for AWGN, but depending on N_r it may be larger or smaller than 3 dB for non-Gaussian noise.

Problems

7.1 For baseband data transmission, if $N_0 = 10^{-7}$ W/Hz and the signal amplitude $A = 20$ mV, what is the maximum data rate for $P_s = 10^{-5}$?

7.2 A sine wave is used for the PSK and QPSK signaling schemes, respectively. The duration of a symbol is 10^{-5} s. If the received signal is $s(t) = 0.01 \sin(2\pi 10^6 t + \theta)$ volts and the measured noise power at the receiver is 10^{-7} watts, calculate E_b/N_0.

7.3 Consider the sequence 1001 1110 0001.
(a) Differentially encode it and obtain the differentially encoded sequence **b**.
(b) Biphase-modulate **b** by a sinusoidal carrier of arbitrary phase, and plot the modulated signal.
(c) Draw the block diagram of the demodulator.
(d) Show how the demodulator generates the original sequence.

7.4 Generate the message 0010011001 through MSK transmitter and receiver. Sketch the waveforms at the input and output.

7.5 Given MPSK modulation with $M = 2, 4, 8$, and 16, calculate the bit SNR E_b/N_0 required for BEPs of $10^{-5}, 10^{-6}, 10^{-7}$.

7.6 Calculate the bit rate that can be supported by a 4 kHz channel for modulation: (a) BPSK, (b) QPSK, (c) 8PSK, (d) 16QAM, (e) coherent BFSK, (f) coherent 4FSK, (g) noncoherent BFSK, (h) noncoherent 8FSK. If $N_0 = 10^{-7}$ W/Hz, find the carrier power required for each modulation to support $P_b = 10^{-5}$. (Hint: The carrier power $P_c = \frac{A^2}{2} = E_b R_b$.)

7.7 For BPSK and MSK, compare their bandwidths that contain 90% power with their null-to-null bandwidths.

7.8 Consider PSK modulation with a carrier component

$$x_c(t) = A \sin\left[2\pi f_0 t + d(t) \cos^{-1} a\right],$$

where $d(t) = +1$ or -1 with bit intervals of T_b. Assuming the total power is P_T, show that the powers in the modulation and carrier components are, respectively, given as

$$P_c = a^2 P_T, \quad P_m = (1 - a^2)P_T.$$

7.9 Compare binary ASK, BPSK, and BFSK in terms of the transmission bandwidth for a constant data rate. The required bandwidth can be taken as the null-to-null bandwidth.

7.10 For $P_b = 10^{-5}$, what is the dB increase in E_b/N_0 required for DPSK, 16QAM in slow Rayleigh fading over the nonfading cases.

7.11 Plot the BEP for M-ary noncoherent FSK, DMPSK, MPSK, and MQAM in Rayleigh fading.

7.12 Use numerical integration to evaluate the BEP for MQAM in Ricean fading.

7.13 Plot $g(t)$ given by (7.103), for different values of $B_b T$ between 0 and 1. The horizontal axis is t/T and the vertical axis is $g(t)T$.

7.14 A matched filter has the frequency response $H(f) = \frac{1 - e^{-j2\pi fT}}{j2\pi f}$. Determine the corresponding impulse response $h(t)$. What is the signal waveform to which this filter is matched?

7.15 Assume BPSK is used for transmitting information over an AWGN channel with a PSD of $\frac{N_0}{2} = 10^{-9}$ W/Hz. The transmitted signal energy $E_b = \frac{1}{2}A^2 T_b$, where T_b is the bit interval and A is the signal amplitude. Determine A to achieve an error probability of 10^{-6} when the data rate is 100 kbits/s.

7.16 A PCM encoded speech signal uses 8 bits per sample, sampling at 8 kHz. The PCM data is then transmitted over an AWGN channel via 8PAM. Determine the bandwidth required for transmission.

7.17 Consider the three 8QAM constellations, shown in Fig. 7.37. The minimum distance between adjacent points is $2A$, and all the signal points are equally probable. Determine the average transmitted power for each constellation. Which constellation is more power efficient?

7.18 Show that a $\pi/4$-DQPSK signal can be demodulated by using an FM discriminator.

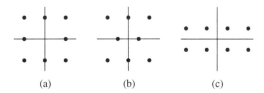

(a)   (b)   (c)

Figure 7.37

Figure for Problem 7.17.

7.19 Given a binary sequence 110010100101, plot the MSK modulated waveform as well as its I and Q components.

7.20 Plot the GMSK modulated waveform for bit sequence 11010001, assuming $B_b T_b = 0.3$.

7.21 The BEPs for BPSK, MSK, and QPSK are the same. What are the differences in terms of bandwidth efficiency?

7.22 Gray encoding ensures that only one bit changes when the decimal number changes by one unit. For a binary representation of a decimal number $b_1 b_2 \ldots b_n$, b_1 being the MSB, write a program to convert it into a Gray code $g_1 g_2 \ldots g_n$, according to such a relation [9]: $g_1 = b_1$, and $g_n = b_n \oplus b_{n-1}$, where $\oplus$ denotes modulo-2 addition. Tabulate the conversion for decimal numbers 0 through 31.

7.23 Refer to [7], find the closed-form BEP expressions for MQAM with Gray code bit mapping in the Rayleigh fading channel in equation (6), and the upper bound (10) and lower bound (12). Plot the BEP formulas as a function of average SNR for $M = 4, 16, 64, 256$.

References

[1] R. R. Anderson & J. Salz, Spectra of digital FM. *Bell Syst. Tech. J.*, **44**:6 (1965), 1165–1189.

[2] H. W. Arnold & W. F. Bodtmann, Switched-diversity FSK in frequency selective Rayleigh fading. *IEEE J. Sel. Areas Commun.*, **2**:4 (1984), 540–547.

[3] T. Aulin & C.-E. Sundberg, Continuous phase modulation – part I: full response signaling. *IEEE Trans. Commun.*, **29**:3 (1981), 196–209.

[4] T. Aulin, N. Rydbeck & C.-E. Sundberg, Continuous phase modulation – part II: partial response signaling. *IEEE Trans. Commun.*, **29**:3 (1981), 210–225.

[5] J. R. Barry, A. Kavcic, S. W. McLaughlin, A. Nayak & W. Zeng, Iterative timing recovery: methods for implementing timing recovery in cooperation with iterative error-control decoding. *IEEE Signal Process. Mag.*, **21**:1 (2004), 89–102.

[6] K. Cho & D. Yoon, On the general BER expression of one- and two-dimensional amplitude modulations. *IEEE Trans. Commun.*, **50**:7 (2002), 1074–1080.

[7] A. Conti, M. Z. Win & M. Chiani, Invertible bounds for M-QAM in Rayleigh fading. *IEEE Trans. Wireless Commun.*, **4**:5 (2005), 1994–2000.

[8] J. P. Costas, Synchronous communications. *Proc. IRE*, **44**:12 (1956), 1713–1718.

[9] K.-L. Du & M. N. S. Swamy, *Neural Networks in a Softcomputing Framework* (London: Springer, 2006).

[10] N. Ekanayake, Performance of M-ary PSK signals in slow Rayleigh fading channels. *Electron. Lett.*, **26**:10 (1990), 618–619.

[11] G. D. Forney, Jr. & L. F. Wei, Multidimensional constellations – part I: introduction, figures of merit, and generalized cross constellations. *IEEE J. Sel. Areas Commun.*, **7**:6 (1989), 877–892.

[12] F. M. Gardner, *Phaselock Techniques*, 2nd edn (New York: Wiley, 1979).

[13] A. Goldsmith, *Wireless Communications* (Cambridge, UK: Cambridge University Press, 2005).

[14] L. Hanzo, M. Munster, B. J. Choi & T. Keller, *OFDM and MC-CDMA for Broadband Multi-User Communications, WLANs and Broadcasting* (New York: Wiley-IEEE, 2003).

[15] W. C. Jakes, Jr., ed., *Microwave Mobile Communications* (New York: Wiley, 1974).

[16] P. Y. Kam, Tight bounds on the bit-error probabilities of 2DPSK and 4DPSK in nonselective Rician fading. *IEEE Trans. Commun.*, **46**:7 (1998), 860–862.

[17] V. Kuhn, *Wireless Communications over MIMO Channels: Applications to CDMA and Multiple Antenna Systems* (Chichester, UK: Wiley, 2006).

[18] C.-L. Liu & K. Feher, $\pi/4$-QPSK modems for satellite sound/data broadcast systems. *IEEE Trans. Broadcast.*, **37**:1 (1991), 1–8.

[19] J. Lu, K. B. Letaief, J. C.-I. Chuang & M. L. Liou, M-PSK and M-QAM BER computation using signal-space concepts. *IEEE Trans. Commun.*, **47**:2 (1999), 181–184.

[20] A. Maaref & S. Aissa, Exact error probability analysis of rectangular QAM for single- and multichannel reception in Nakagami-m fading channels. *IEEE Trans. Commun.*, **57**:1 (2009), 214–221.

[21] A. F. Molisch, *Wireless Communications* (Chichester, UK: Wiley-IEEE, 2005).

[22] K. Murota & K. Hirade, GMSK modulation for digital mobile telephony. *IEEE Trans. Commun.*, **29**:7 (1981), 1044–1050.

[23] A. Nezampour, A. Nasri, R. Schober & Y. Ma, Asymptotic BEP and SEP of quadratic diversity combining receivers in correlated Ricean fading, non-Gaussian noise, and interference. *IEEE Trans. Commun.*, **57**:4 (2009), 1039–1049.

[24] M. Oerder & H. Meyr, Digital filter and square timing recovery. *IEEE Trans. Commun.*, **36**:5 (1988), 605–612.

[25] J. H. Park, Jr., On binary DPSK detection. *IEEE Trans. Commun.*, **26**:4 (1978), 484–486.

[26] S. Park & S. H. Cho, SEP performance of coherent MPSK over fading channels in the presence of phase/quadrature error and I-Q gain mismatch. *IEEE Trans. Commun.*, **53**:7 (2005), 1088–1091.

[27] D. M. Pozar, *Microwave and RF Wireless Systems* (New York: Wiley, 2001).

[28] J. G. Proakis & M. Salehi, *Digital Communications*, 5th edn (New York: McGraw-Hill, 2008).

[29] J. G. Proakis & D. G. Manolakis, *Digital Signal Processing: Principle, Algorithms, and Applications*, 4th edn (Upper Saddle River, NJ: Pearson Prentice Hall, 2007).

[30] D. Saha & T. G. Birdsall, Quadature-quadrature phase-shift keying. *IEEE Trans. Commun.*, **37**:5 (1989), 437–448.

[31] M. Schwartz, W. R. Bennet, & S. Stein, *Communication Systems and Techniques* (New York: McGraw-Hill, 1966).

[32] A. U. H. Sheikh, *Wireless Communications: Theory and Techniques* (Boston, MA: Kluwer, 2004).

[33] M. K. Simon & M.-S. Alouini, *Digital Communications over Fading Channels*, 2nd edn (New York: Wiley, 2005).

[34] D. B. Smith & T. D. Abhayapala, Maximal ratio combining performance analysis in practical Rayleigh fading channels. *IEE Proc.-Commun.*, **153**:5 (2006), 755–761.

[35] G. L. Stuber, *Principles of Mobile Communication*, 2nd edn (Boston, MA: Kluwer, 2001).

[36] L. Szczecinski, S. Aissa, C. Gonzalez, & M. Bacic, Exact evaluation of bit- and symbol-error rates for arbitrary 2-D modulation and nonuniform signaling in AWGN channel. *IEEE Trans. Commun.*, **54**:6 (2006), 1049–1056.

[37] V. V. Veeravalli, On performance analysis for signaling on correlated fading channels. *IEEE Trans. Commun.*, **49**:11 (2001), 1879-C1883.

[38] A. J. Viterbi, *Principles of Spread Spectrum Communication* (Reading, MA: Addison-Wesley, 1995).

[39] F. Xiong, *Digital Modulation Techniques*, 2nd edn (Boston, MA: Artech House, 2006).

[40] Q. T. Zhang & X. W. Cui, A closed-form expression for the symbol-error rate of M-ary DPSK in fast Rayleigh fading. *IEEE Trans. Commun.*, **53**:7 (2005), 1085–1087.

[41] W.-P. Zhu, Y. Yan, M. O. Ahmad & M. N. S. Swamy, Feedforward symbol timing recovery technique using two samples per symbol. *IEEE Trans. Circ. Syst. I*, **52**:11 (2005), 2490–2500.

Spread spectrum communications

8.1 Introduction

Spread spectrum communications was originally used in the military for the purpose of interference rejection and enciphering. In digital cellular communications, spread spectrum modulation is used as a multiple-access technique. Spectrum spreading is mainly performed by one of the following three schemes.

- *Direct sequence (DS)*: Data is spread and the carrier frequency is fixed.
- *Frequency hopping (FH)*: Data is directly modulated and the carrier frequency is spread by channel hopping.
- *Time hopping (TH)*: Signal transmission is randomized in time.

The first two schemes are known as *spectral spreading*, and are introduced in this chapter. Time hopping is known as *temporal spreading*, and will be introduced in Chapter 20. Spectrum spreading provides frequency diversity, low PSD of the transmitted signal, and reduced band-limited interference, while temporal spreading has the advantage of time diversity, low instantaneous power of the transmitted signals, and reduced impulse interference.

CDMA is a spread spectrum modulation technology in which all users occupy the same time and frequency, and they can be separated by their specific codes. For DS-CDMA systems, at the BS, the baseband bitstream for each MS is first mapped onto M-ary symbols such as QPSK symbols; each of the I and Q signals is then spread by multiplying a spreading code and then a scrambling code. The spread signals for all MSs are then amplified to their respective power, summed, modulated to the specified band, and then transmitted. Each MS receives the summed signal, but extracts its own bitstream by demodulating, descrambling, and despreading the sum signal using its unique scrambling and spreading codes. In a spread spectrum system, bandwidth spreading and user separation are done using different codes: spreading codes and scrambling codes.

CDMA technology uses a set of binary CDMA codes that are orthogonal or almost orthogonal to one another. Each single pulse of the code waveform is called a *chip*, and it has a duration of T_c. Assuming that each chip has a rectangular pulse shape, the corresponding frequency response is a sinc function; thus, the bandwidth between the two zero points of the main lobe is $1/T_c$, and this can be defined as the bandwidth of the transmit signal. The number of chips in one symbol is often referred to as the *processing gain* or *spreading factor* $G_p = T/T_c$, T being the symbol period; thus, the processing gain is the ratio of the spread bandwidth of the transmitted data to the data rate, and spreading increases the bandwidth by a factor G_p. In addition, the processing gain allows the removal of up to $G_p - 1$ interfering signals of other users.

Scrambling multiplies the bitstream of a physical channel by a user-specific PN sequence. This effectively prevents the data from being eavesdropped. Scrambling also changes the dc components of the bitstreams to zero, since the PN sequence is of zero mean. At the receive end, the received scrambled bitstream can be descrambled by using the same PN sequence. Scrambling is also used for wireless systems other than spread spectrum systems.

The use of spread spectrum modulation has a number of advantages such as resistance to eavesdropping, resistance to multipath fading, interference rejection and multiple access capabilities. As opposed to TDMA and FDMA, in CDMA all the cells can use the same frequency. The capacity can be increased by adding some cells without modifying the frequency scheme in existing cells.

CDMA cellular systems employ universal frequency reuse. Any technique that reduces multiple-access interference leads to a capacity gain. For voice communications, voice activity detection and speech coding can reduce MAI, and this leads to a significant capacity increase over TDMA systems.

PSD of the DS-CDMA signal

The power spectrum of an infinite random sequence of data $d(t)$ with rate $R = 1/T$ bits/s is given by

$$S_D(f) = T \left(\frac{\sin \pi fT}{\pi fT} \right)^2. \tag{8.1}$$

The power spectrum of the spreading sequence $p(t)$ is the same as that of the spread sequence $d(t)p(t)$, and is given by

$$S_{ss}(f) = \frac{1}{f_c} \left(\frac{\sin \pi f/f_c}{\pi f/f_c} \right)^2, \tag{8.2}$$

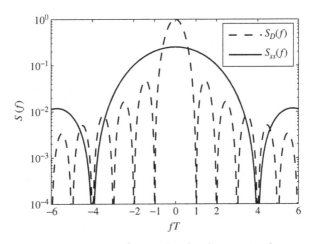

Figure 8.1   Power spectrums of narrowband and CDMA signals.

where $f_c = G_p/T = G_pR$. Thus, a narrowband waveform is spread over a wider bandwidth.

Example 8.1: Given $G_p = 4$, the PSDs of a narrowband signal and the corresponding DS-CDMA signal can be calculated by (8.1) and (8.2). They are plotted in Fig. 8.1. Clearly, the bandwidth of the DS-CDMA signal is four times that of the narrowband signal.

8.2 Spreading sequences

8.2.1 Properties of spreading sequences

A spreading sequence, $c = \{c_k\}$, is a periodic deterministic sequence with a period N. The duration for each c_k is the chip period T_c. The spreading sequence is used to generate a spreading waveform by applying a real chip amplitude shaping function that has a peak amplitude of unity. Given a symbol duration T, the chip duration T_c should be selected so that T is an integer multiple of T_c, and the processing gain or spreading factor $G_p = T/T_c$.

When $G_p = N$, the code is a *short code*, and each data symbol is spread by a full period of the spreading sequence. When $G_p \ll N$, the code is called a *long code*, and each data symbol is spread by a portion of the spreading sequence. In the following, the use of short code is assumed, and for this reason G_p is usually replaced by N.

The length of a code is the period of the code, G_pT_c. Putting it another way, there are G_p chips in a code. Two codes, c_i and c_j, are said to be *orthogonal* if their cross-correlation function (CCF) over the period is equal to zero:

$$\text{CCF}(i,j) = \int_0^{G_pT_c} c_i(t)c_j(t)dt = 0, \quad \text{for all } i \neq j. \tag{8.3}$$

Orthogonality allows multiple information streams to be multiplexed for transmission.

De-spreading is a process of data recovery from the composite spread spectrum signal by applying the same CDMA code that is used for spreading. In order to obtain the original data by correlation, the autocorrelation function (ACF) of the spreading sequence should be a Dirac delta function. For a spreading sequence $p(t)$, the desirable ACF at time iT_c is given by

$$\text{ACF}(i) = \begin{cases} G_p, & i = 0 \\ 0, & \text{otherwise} \end{cases}. \tag{8.4}$$

For nonorthogonal spreading sequences, the receiver achieves a finite interference suppression by a factor ACF/CCF.

For spread spectrum systems that use long codes, accordingly, partial period ACF and CCF can be defined over a portion (G_p chips) of the long codes; partial period ACF and CCF are dependent on the delay and the starting point of the subsequence. The partial period correlations are difficult to analyze, and are usually treated statistically.

Selection of the spreading code is based on three factors: autocorrelation, cross-correlation, and the number of codes.

- Good autocorrelation property helps to despread the original code perfectly, and to mitigate the ISI. It is also useful for synchronization and reduction of interchip interference in a rake receiver.
- Desirable cross-correlation property is that all the codes are orthogonal to one another so that all the other users' information is demodulated as noise at the receiver. For unsynchronized systems, orthogonality must be guaranteed between the codes with arbitrary delays.
- The number of codes is also important since all the users in a cell and its surrounding cells must have different codes to identify them. A large number of codes can support more users.

For the uplink of CDMA systems, the popular spreading sequences are the m-sequence, Gold sequence, and Kasami sequence. These sequences have a good cross-correlation property between their time-shifted versions. The m-sequence has a good ACF property, and also has $2^m - 1$ codes. The Gold sequence is obtained by combining a subset of m-sequence, and has $2^m + 1$ codes. The Kasami sequence can have a great number of codes. For the downlink, the Walsh-Hadamard codes can be used, since signals of different users can be synchronized before transmission.

Among popular spreading sequences, the m-sequence, Gold code, and Kasami code have nonzero off-peak autocorrelations and crosscorrelations; the Walsh code and the orthogonal Gold code are orthogonal only in the case of perfect synchronization, but have nonzero off-peak autocorrelations and cross-correlations in the asynchronous case. The Barker sequences are specially designed sequences that have almost ideal aperiodic ACFs. Complementary codes are a kind of orthogonal codes.

There are also some codes with good autocorrelation properties [11]. These codes are useful as reference signals for synchronization, channel estimation, or channel sounding. For example, the Zadoff-Chu sequence is used as uplink reference signal in LTE.

8.2.2 Pseudo-noise sequences

PN sequences can be used as codes. Although different PN sequences of a given length are not completely orthogonal to one another, they can be easily generated by a PN code generator. PN sequences are typically generated using a maximal length shift register (MLSR), hence the generated PN sequence is also called *MLSR sequence* or *m-sequence*. The MLSR sequence generator is shown in Fig. 8.2. At each time instant, the register right-shifts its content by one bit. The output sequence a_n is given by

$$a_n = c_1 a_{n-1} + c_2 a_{n-2} + \cdots + c_r a_{n-r} = \sum_{i=1}^{r} c_i a_{n-i}, \tag{8.5}$$

where $c_1, \ldots, c_r$ are connection weights with 1 for connection and 0 for no connection, and all a_i's are the bits in the register. The addition is a modulo-2 addition. For an m-stage shift register, the period of the generated sequence is at most $2^m - 1$.

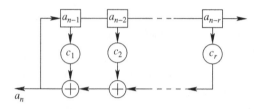

Figure 8.2   MLSR sequence generator.

The m-sequence always has an odd length $N = 2^m - 1$. It satisfies the three properties of a random sequence or a large PN-sequence [22, 63]:

- *Balance of 1 and 0*: Out of the length $2^m - 1$, 2^{m-1} bits are 1 and $2^{m-1} - 1$ bits are 0, since the case of all zero bits is eliminated;
- *Run-length property*: The probability of n continuous 0 or 1 is $1/2^n$, for $n \leq m - 1$, and $1/2^{m-1}$ for $n = m$;
- *Shift property*: If the sequence is shifted by any nonzero number of elements, the resulting sequence has half of its elements the same as in the original sequence and half different from the original sequence.

In addition, the m-sequence has the *shift-and-add property*: If we shift a sequence of length $2^m - 1$ by r bits, $r < 2^m - 1$, we obtain an m-sequence with a different initial vector. The modulo-2 addition of the original and the shifted m-sequences leads to a new m-sequence.

The m-sequence is a type of cyclic code, and it is generated and characterized by a generator polynomial, and can be analyzed by using the algebraic coding theory. The m-sequence of period N has an autocorrelation property that is very close to the desirable case

$$\text{ACF}(i) = \begin{cases} N, & i = kN \\ -1, & i \neq kN \end{cases}, \tag{8.6}$$

where k is any integer. The normalized $\overline{\text{ACF}}(i) = \text{ACF}(i)/\text{ACF}(0)$ is very close to the delta function $\delta(i)$ for large N. Thus m-sequences are ideal for minimizing the ISI effect. Their cross-correlation is very small. For the m-sequence, the suppression factor $\text{ACF}/\text{CCF} = N$.

The power spectrum of the m-sequence is derived by taking the Fourier transform of the ACF of the waveform $a(t)$. It is derived as

$$S_a(f) = \begin{cases} \frac{N+1}{N^2} \text{sinc}^2 (fT_c), & f = \frac{i}{NT_c}, i \neq 0 \\ \frac{1}{N^2}, & f = 0 \end{cases}. \tag{8.7}$$

The ACF and PSD for the m-sequence are plotted in Fig. 8.3.

Usually, the length of the m-sequence is selected as $N = T/T_c$, so that the spreading code is a short spreading code since the repetition of the autocorrelation leads to a peak every symbol period. These short-periodic peaks yield significant ISI from multipath components in the first few symbols after the desired symbol. This can be avoided by using a long spreading code, where $N \gg T/T_c$. When a long spreading code is used, the demodulation

Table 8.1. Properties of m-sequences. Adapted from [53] ©1980, IEEE.

m	$N = 2^m - 1$	Number of m-sequences	ACF	$t(n)$
3	7	2	5	5
4	15	2	9	9
5	31	6	11	9
6	63	6	23	17
7	127	18	41	17
8	255	16	95	33
9	511	48	113	33
10	1023	60	383	65
11	2047	176	287	65
12	4095	144	1407	129
13	8191	630	≥ 703	129
14	16383	756	≥ 5631	257
15	32767	1800	≥ 2047	257
16	65535	2048	≥ 4095	513

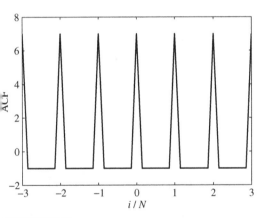
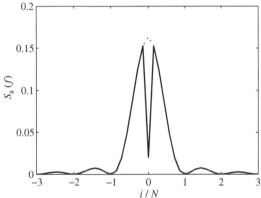

Figure 8.3   ACF and PSD for an m-sequence waveform, $N = 7$.

is taken over a partial period $T = G_p T_c \ll N T_c$, where G_p is the processing gain. This partial-period autocorrelation can roughly attenuate ISI by a factor of G_p [20].

For each m, there exist a pair of m-sequences with three-valued CCFs, -1, $-t(m)$, $t(m) - 2$, where

$$t(m) = \begin{cases} 2^{(m+1)/2} + 1, & \text{for } m \text{ odd} \\ 2^{(m+2)/2} + 1, & \text{for } m \text{ even} \end{cases} . \tag{8.8}$$

Such a pair of m-sequences is called a *preferred pair of m-sequences*. In other cases, the CCF peak is higher.

Table 8.1 lists some properties of m-sequences. The number of m-sequences for an MLSR of length m is given in [56]. This number is much smaller compared to its sequence length, and cannot provide a sufficient number of codes for CDMA systems. Also, the maximum cross-correlation levels are too large. Thus, m-sequences are not used alone in CDMA applications. The shift-register connections for generating m-sequences with $m = 2$ to 34 are listed in [15, 49].

8.2.3 Gold sequences

The Gold code [19] is generated by using the preferred pairs of m-sequences by a process of all possible cyclically shifted modulo-2 additions of the preferred pair. The m-sequence family has one such unique preferred pair for each sequence length; the preferred pairs have good correlation properties. Due to the use of the preferred pair, both the autocorrelations and cross-correlations of Gold codes take on the values $\{-1, -t(m), t(m) - 2\}$, where $t(m)$ is given by (8.8) [24]. Gold codes have lower peak cross-correlations than m-sequences, but have worse autocorrelation properties than m-sequences.

The family of the $2^m - 1$ derived sequences plus the preferred pair are collectively known as Gold codes; there are all together $2^m + 1$ Gold codes of code length $2^m - 1$, for two m-sequences of order m. Like the m-sequence, all the $2^m + 1$ Gold codes are balanced, with 2^{m-1} ones and $2^{m-1} - 1$ zeros.

The autocorrelation property of the spreading codes can effectively eliminate ISI. In [2], minimum autocorrelation spreading codes were designed for this purpose. The Gold code has been extensively used in spread-spectrum communication systems, such as in IS-95, WCDMA, and UTRA-TDD standards as the scrambling code, and in satellite systems such as the GPS and NASA's Tracking and Data Relay Satellite System (TDRSS).

Orthogonal Gold codes

The cross-correlation of the Gold codes is -1 for many code offsets. By attaching an additional 0 to the original Gold codes, the cross-correlation can turn to 0. A total of 2^r orthogonal Gold codes can be obtained by zero-padding from a preferred pair of two r-stage MLSRs.

Orthogonal Gold codes have the same cross-correlation property as the Walsh codes of the same length, but have better characteristics in terms of autocorrelation. The orthogonal Gold code is desirable when the code's auto-correlations must be low to avoid falsely registering the main peak of the autocorrelation function [24]. Orthogonal Gold codes of length 256 chips have been used in WCDMA for fast cell search.

8.2.4 Kasami sequences

The Kasami sequences [29] have properties similar to the preferred pairs of m-sequences and are also derived from m-sequences in a fashion similar to the generation of the Gold sequences. Consider two m-sequences of periods $2^m - 1$ and $2^{m/2} - 1$ that are generated from two preferred polynomials $p_1(x)$ and $p_2(x)$, which are of order m and $m/2$, respectively. The set of Kasami sequences is generated by using the long sequence and the sum of the long sequence with all $2^{m/2} - 1$ cyclic shifts of the short sequence. Thus, the number of Kasami sequences is $2^{m/2}$, each with a period of $2^m - 1$. This is the small set of Kasami sequences.

The small set has $2^{m/2}$ binary sequences of $2^m - 1$ for m even. For the small set, Kasami codes achieve the Welch lower bound for autocorrelation and cross-correlation for any set of $2^{m/2}$ sequences of length $2^m - 1$ [20, 24, 56]. The off-peak autocorrelation and cross-correlation of the Kasami sequences are also three-valued $\{-1, -t(m), t(m) - 2\}$, with

$$t(m) = 2^{m/2} + 1. \tag{8.9}$$

The large set of Kasami sequences includes both the small set of Kasami sequences and a set of Gold sequences as its subsets. The autocorrelation and cross-correlation properties of the large Kasami set are inferior to those of the small Kasami set, but the large set has a large number of sequences.

In WCDMA, the uplink short scrambling code is a complex code $c = c_I + jc_Q$, where c_I and c_Q are two different codes from the extended very large Kasami set of length 256. The uplink long scrambling code is based on the Gold sequence, and the downlink scrambling code uses the Gold sequence.

8.2.5 Walsh sequences

The Walsh codes, also known as the *Walsh-Hadamard codes*, are generated by rearranging the Hadamard codes; they are orthogonal to one another. The K-bit Walsh codes can be generated recursively by using

$$\mathbf{B}_{n+1} = \begin{bmatrix} \mathbf{B}_n & \vdots & \mathbf{B}_n \\ \cdots & \cdots & \cdots \\ \mathbf{B}_n & \vdots & \overline{\mathbf{B}_n} \end{bmatrix}. \tag{8.10}$$

The recursion starts from $\mathbf{B}_0 = 0$ or 1. Each row of the matrix can be used as a code. Thus, the length as well as the number of codes can be $2, 4, \cdots, 2^K$. The Walsh code has a length of 2^K. The correlation of two Walsh codes is nonzero only if the two codes are the same. The Walsh-Hadamard codes can be generated by using a Walsh-Hadamard sequence generator [63]. DS-CDMA with Walsh codes can support at most $N = T/T_c$ users.

Walsh codes have perfect orthogonality. Unlike the aforementioned codes, which still have good orthogonality properties between their delay versions, the orthogonality of Walsh codes are destroyed by delay dispersion. In a multipath environment, even though the users are synchronous, multipath destroys the synchronism of the channel. Thus, when Walsh codes are used in a multipath channel, there will be interference between users. Equalization of the spreading codes can mitigate this interference.

A drawback of the Walsh code is that the codewords have poor autocorrelation properties. By multiplying the Walsh codes with a scrambling code, better autocorrelation properties are achieved and the mutual orthogonal property is retained.

In the downlink, Walsh codes are usually used so that different users can be separated completely. However, the number of Walsh codes is so limited that it is not enough for assignment to all the users in the cell as well as in the neighboring cells. This problem can be mitigated by multiplying the Walsh codes by a scrambling code, since the resulting code sequences are also orthogonal. In cellular networks, each BS uses a different scrambling code $c^s(k)$ for cell identification. Each user has a code that is obtained as a Walsh code $c_u^w(k)$ multiplied by the scrambling code in its cell

$$c_u(k) = c_u^w(k)c^s(k). \tag{8.11}$$

In this case, the new code maintains orthogonality for synchronous signals and suppresses asynchronous signals like a random code. Walsh codes multiplied by different scrambling codes appear to be noise to each other. This method is used in IS-95, WCDMA, and CDMA2000.

The first sequence in the Walsh code set is the all-zeros sequence. It gives the highest possible autocorrelation $ACF(n) = N - n$, where N is the length of the sequence and n is the number of bit shifts. For this reason, the all-zero sequence is usually used as the pilot sequence.

8.2.6 Orthogonal variable spreading factor sequences

The orthogonal variable spreading factor (OVSF) codes are derived from the Walsh codes. These codes are of different lengths, and they are used for spreading when different users require different data rates.

The procedure for generating OVSF codes is shown by the graph of Fig. 8.4. At each node of the tree, a code $w_k(n)$ of length n generates two new codes by the rule

$$w_k(n) \rightarrow \begin{cases} w_{2k-1}(2n) & = & [w_k(n), w_k(n)] \\ w_{2k}(2n) & = & [w_k(n), -w_k(n)] \end{cases} . \tag{8.12}$$

The variable spreading factor is used for rate-matching.

The OVSF code is used in WCDMA with spreading factors of 4–512, and in TD-SCDMA with spreading factors of 1–16 (1,2,4,8,16). In WCDMA, downlink transmission

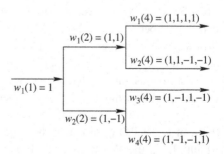

Figure 8.4   **Graph for OVSF code generation.**

at the peak data rate requires three channelization codes with a spreading factor of 4; that is, three-fourths of the code resources is allocated to a single user. In the uplink, an MS transmitting with the peak data rate is seen as a large interference, but does not affect the code resource availability of other intracell users. The OVSF code is also used in HSUPA.

The multi-rate orthogonal Gold code that is generated by combining a Walsh code generator and an orthogonal Gold code generator can also be used [24].

8.2.7 Barker sequences

A Barker sequence is a binary sequence whose off-peak values of its ACF are 1, 0, or -1. All the known Barker sequences (excluding their negations and reversals) are:

$+ + -$	length 3
$+ - ++$ and $+ - --$	length 4
$+ + + - +$	length 5
$+ + + - - + -$	length 7
$+ + + - - - + - - + -$	length 11
$+ + + + + - - + + - + - +$	length 13

where a $+$ sign stands for the positive pulse and a $-$ sign for the negative pulse. The length-11 and length-13 Barker sequences have the best autocorrelation properties among all the binary codes [21]. Barker sequences are commonly used for word synchronization in paging systems, and the length-11 Barker sequence is also used in IEEE 802.11.

In general, any sequence whose aperiodic autocorrelation reaches the lower bound, $|ACF| = 1$, is called a *Barker sequence*. The longest known quadriphase Barker sequence is of length 15 [21]. Polyphase Barker sequences up to length 36 have been reported in [16, 42, 43].

Quasi-Barker sequences achieve the peak aperiodic autocorrelation sidelobe level $|ACF| = 1$ only within a certain window centered at the mainlobe. Quasi-Barker sequences exist for all lengths. The best quasi-Barker sequences of a given length are those that achieve the largest window and can thus tolerate the highest degree of mis-synchronization. By exhaustive search, all the best biphase and quadriphase quasi-Barker sequences, with maximum window sizes, of lengths up to 36 and 21, respectively, are listed in [26].

8.2.8 Complementary codes

The Golay complementary sequences are sequence pairs for which the sum of the autocorrelation functions is zero for all nonzero delay shifts [17, 18]. The complementary codes used for CCK also have the property that the sum of their aperiodic ACFs are zero for all nonzero delays [56]. The orthogonal complementary codes (OCCs) [57] are orthogonal in both the synchronous and asynchronous cases, thus offering MAI-free operation. The OCCs can hardly be used in real systems due to the very small set size: Only $G_p^{1/3}$ users can be supported for a processing gain of G_p. In comparison, the Walsh codes and OVSF codes are not truly orthogonal codes, since they possess very high out-of-phase ACFs and CCFs.

Unlike traditional CDMA codes that use a single code, the orthogonality of the OCC is based on a flock of element codes. That is, each user is assigned a flock of element codes as its signature code in an OCC-based CDMA system. The OCC is a *truly perfect orthogonal code*, since it provides zero out-of-phase ACFs and zero CCFs. These ACF and CCF properties make the rake receiver and power control unnecessary.

The OCC was considered in [7] for MC-CDMA (multi-carrier CDMA) based communication systems. It has also found application in the TD-LAS (Large-Area-Synchronous) CDMA system [9, 34], which has been approved by 3GPP2 as an enhanced standard. In the TD-LAS CDMA system, pair-wise OCCs, called *loosely synchronized (LS) codes*, have been used as spreading codes of the users. The LAS code family is obtained by combining the LS code with a large area (LA) code. The ACFs of all LAS codes are ideal, and there exists an interference-free window (IFW) or a zero-correlation zone (ZCZ) in the CCFs of the access codes around the origin. The use of LAS code family achieves an intracell-interference-free cellular quasi-synchronous (QS)-CDMA system with low intercell interference. The TD-LAS CDMA system is still at the developing stage. A systematic construction of many families of generalized LS codes has been proposed in [58].

CDMA systems implemented based on the OCCs have a number of advantages over the current 3G CDMA systems [8]:

- OCC-based CDMA is suitable for bursty traffic. For detecting bits at the edges of a packet or frame, OCC-based DS-CDMA can yield zero partial CCFs.
- In WCDMA, UTRA-TDD, and TD-SCDMA, the rate change can be made only in powers of two due to the application of the OVSF code. In OCC-based CDMA, the transmission rates can be continuously adjusted by simply shifting more than one chip (at most N chips) between two consecutive off-stacking (OS) bits. If N chips are shifted, OS spreading reduces to DS spreading, yielding the lowest data rate.
- OS spreading also helps support asymmetrical transmission in the uplink and downlink by simply adjusting relative offset chips between two neighboring spreading modulated bits.
- The same processing gain is applicable to all different transmission rates. For CDMA-based 3G systems that use the OVSF code, a slower transmission rate corresponds to a higher processing gain.

8.3 Direct-sequence spread spectrum

The spread spectrum scheme described thus far is DS-CDMA. It is widely used in cellular standards. In DS-CDMA, each user u is assigned a unique spreading code c_u. The *load of the system* is defined as the ratio between the number of active users K and the spreading factor N

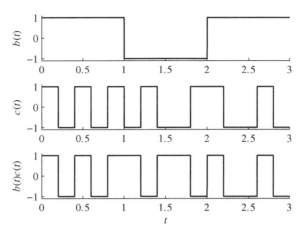

Spread waveform of a bitstream for a DS-CDMA system with BPSK modulation. The top figure is the bitstream, the middle figure is the spreading sequence, and the bottom figure is the spread sequence.

$$\beta = \frac{K}{N}. \tag{8.13}$$

When $\beta = 1$ the system is said to be *fully loaded*.

Example 8.2: For the data stream of each user, each symbol is first spread by a spreading sequence of length G_p. Given a bitstream with BPSK modulation and $G_p = 5$, the spread waveform is plotted in Fig. 8.5.

8.3.1 DS-CDMA model

Uplink

When K active users transmit their information asynchronously over a common AWGN channel, the received signal at a single antenna of the BS can be modeled by

$$r(t) = \sum_{i=1}^{K} \sum_{j=-\infty}^{\infty} A_i b_i(j) c_i (t - jT - \tau_i) + n(t), \tag{8.14}$$

where A_i is the received amplitude of the ith user, $b_i(j)$ is the transmitted symbol of the ith user in $jT \leq t \leq (j+1)T$, T is the symbol period, $c_i(t)$ is the signature waveform of the ith user, τ_i is the delay of the ith user, and $n(t)$ is AWGN with zero mean and a double-sided PSD of σ^2 W/Hz.

Equation (8.14) corresponds to the case of baseband transmission. In cellular systems, the first term on the RHS needs to be upconverted to the carrier frequency f_c by multiplying it by $\cos(2\pi f_c t)$ before transmission. At the receiver, the received signal is first

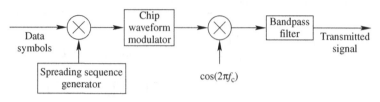

Figure 8.6   **Functional block diagram of the DS-CDMA transmitter at the MS.**

downconverted by multiplying it by $\cos(2\pi f_c t + \phi)$, where ϕ is matched to the carrier phase arising from channel delay. The functional block diagram of the DS-CDMA transmitter at the MS is shown in Fig. 8.6. After the sum signal is received at the BS, it is first downconverted to the baseband, and each user's bitstream is then extracted from the sum baseband signal by correlating with the user-specific scrambling and spreading codes and demodulation.

When all τ_i, $i = 1, \ldots, K$, are the same, that is, all users' signals arrive at the BS at the same time, we get a synchronous CDMA system. In this case, we can take $\tau_i = 0$.

The signature waveforms have the properties

$$c_i(t) = 0, \quad \text{for } t \notin [0, T], \tag{8.15}$$

$$\int_0^T c_i^2 dt = 1 \tag{8.16}$$

for all i.

Assuming that short codes ($N = G_p$) are used, the signature waveform can be represented by

$$c_i(t) = \sum_{k=0}^{N-1} c_{i,k} P_c(t - kT_c), \tag{8.17}$$

where $P_c(t)$ is a rectangular chip waveform of duration $T_c = T/N$, that is, $P_c(t) = 1/\sqrt{T_c}$, for $t \in [0, T_c]$, and 0 otherwise. Thus, $c_i = [c_{i,0}, \ldots, c_{i,N-1}]$ is the normalized spreading sequence assigned to the ith user.

Pulse shaping is usually applied to the spread sequence. The pulse shaping function is denoted by $g(t)$. For rectangular pulses, $g(t) = \sqrt{2/T}$, $0 \le t \le T$, T being the symbol period.

Downlink

The functional block diagram of the DS-CDMA receiver at the MS is depicted in Fig. 8.7. The received signal at the MS is the sum signal transmitted by the BS, which has the same form as in (8.14). The sum signal is first downconverted to the baseband, then descrambled and despread using the scrambling and spreading codes of that specific MS, and is finally demodulated to generate the bitstream for the MS.

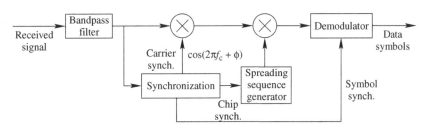

Figure 8.7   **Functional block diagram of the DS-CDMA receiver at the MS.**

By using a chip matched filter and sampling at the chip rate, the received signal during the jth symbol interval, $jT \leq t \leq (j+1)T$, is given by the discrete-time N-dimensional vector

$$r(j) = \sum_{i=1}^{K} A_i b_i(j) c_i + n(j), \tag{8.18}$$

where $r(j)$ and $n(j)$ are N-vectors whose kth entries are obtained by integrating $r(t)$ over the kth chip period.

So far, we have assumed that the short spreading sequence ($G_p = N$) is used in the DS-CDMA model, that is, the period of each periodic spreading sequence is the same as the symbol interval. In IS-95 and CDMA2000, the period of the spreading sequence is larger than the symbol interval ($N > G_p$), and such long spreading sequences are aperiodic. In this case, modifications to (8.14) and (8.17) are necessary.

The matched filter is applied to the baseband signal obtained by despreading, $\hat{x}(t)$, and the symbols are obtained as

$$\hat{b}_l = \int_0^T \hat{x}(t) * g^*(-t) dt. \tag{8.19}$$

For single-user cellular systems with flat-fading channels, the SNR is the same as that for the narrowband case

$$\text{SNR} = |h(l)|^2 \frac{E_s}{N_0}, \tag{8.20}$$

where l corresponds to the symbol index. Spread spectrum gives no advantage in this case [32]. The spread bandwidth usually leads to a frequency-selective behavior of the mobile channel. The rake receiver is used to overcome the frequency-selective fading.

8.3.2 Conventional receiver

The conventional single-user CDMA receiver at the BS uses a bank of matched filters that are matched to the user spreading waveforms. For each user, the received signal is descrambled using a different PN code, then correlated with its spreading code, and finally sampled at the symbol rate.

The received signal for a symbol period can be expressed in the matrix form

$$r = \mathbf{S}\mathbf{A}b + n, \tag{8.21}$$

where r is an N-dimensional vector, $\mathbf{S}$ contains all signatures, $\mathbf{S} = [c_1 \, c_2 \, \cdots \, c_K]$, $\mathbf{A} =$ diag $(A_1, A_2, \ldots, A_K)$, $b = (b_1, b_2, \ldots, b_K)^T$, and n is the noise vector.

For the synchronous CDMA channel, the output of the matched filter is given by

$$y = \mathbf{S}^H r = \mathbf{S}^H \mathbf{S}\mathbf{A}b + \mathbf{S}^H n, \tag{8.22}$$

where y is a K-dimensional vector, $y = (y_1, y_2, \ldots, y_K)^T$. The matched filter output for user u is given by

$$y_u = c_u^H r = \int_0^T r(t)c_u(t)dt, \quad u = 1, \ldots, K. \tag{8.23}$$

For orthogonal codes, $\mathbf{S}^H \mathbf{S} = \mathbf{I}_{K \times K}$, thus (8.22) reduces to

$$y = \mathbf{S}^H r = \mathbf{A}b + \mathbf{S}^H n. \tag{8.24}$$

A decision of the symbols can be made based on this equation. For BPSK modulation, decision can be made by using the signum function. For an M-ary constellation, due to the application of power control, $\mathbf{A} = \sqrt{P}\mathbf{I}_{K \times K}$, and $y \approx \sqrt{P}b$, where P is the power of all the received signals. The decision is just the same as the demodulation of the constellation.

For nonorthogonal codes, (8.22) can be expressed by

$$y = \mathbf{R}\mathbf{A}b + n_1, \tag{8.25}$$

where the crosscorrelation matrix $\mathbf{R} = \mathbf{S}^H \mathbf{S} = \left[R_{ij} \right]$ is given by

$$R_{ij} = \int_0^T c_i(t)c_j(t)dt, \tag{8.26}$$

and $n_1 = \mathbf{S}^H n$. A decision on the user data can then be made based on y_k.

This matched filter structure is simple, but the performance is highly dependent on the power of each user and the cross-correlations between the code waveforms with random delays. In frequency-selective fading channels, the conventional receiver is simply a rake receiver.

8.3.3 Rake receiver

Multipath signals cause significant ISI in TDMA systems, and a time-domain filter called an *equalizer* is usually used to minimize the ISI. Unlike in other systems, in DS-CDMA systems the multipath signals can be employed to advantage by using the rake receiver [48].

Only when the multipaths can be differentiated can they be made use of by the rake receiver. Spectrum spreading facilitates the situation by making the multipath copies resolvable. As long as the relative time delay between two multipaths is greater than T_c, each multipath is separable. According to the delay and addition property of PN-sequences, the different delayed versions of the signal are almost noncoherent. Multipath delays can be obtained by searching the pilot sequences.

With coherent demodulation, an increase in the number of independent Rayleigh multipaths L leads to a better performance. For multipaths of fixed amplitudes and phases, the desirable performance bound is the same as that for the case of a single component, assuming that the energy of the single component and the total energy of all the multipaths are the same [63]. In this case, the rake receiver, like the matched filter, combines the transmitter filter and the multipath channel.

Analysis indicates that as the number of Rayleigh multipaths increases, the performance of the system approaches that of nonfading propagation [63]. This is the advantage of diversity.

Structure of rake receiver

A rake receiver consists of a bank of correlators (matched filters), each synchronized with a different multipath of time delay τ. The outputs of these correlators are then weighted and combined. Thus, a rake receiver is actually a tapped delay line filter, and is also a diversity receiver. A BS typically deploys a four-finger rake receiver: According to the ITU Pedestrian-A channel model, only four multipath components carry energy. When a path has a delay τ_i relative to the delay the rake finger is tuned to, the rake receiver with nonideal spreading sequences suffers from interpath or interchip interference.

The MRC strategy can theoretically obtain the optimum weights for the rake receiver. This requires that each resolvable multipath component is assigned a rake finger. Given τ_{max} to be the maximum delay of the channel, τ_{max}/T_c taps are required, where each multipath component is assumed to be at an integer multiple of a chip period. In outdoor environments, the number of taps may be very large, and for the reason of implementation, we have to select a subset of the resolvable multipaths before applying MRC.

The principle of the rake receiver is shown in Fig. 8.8. In the figure, $c(t)$ is the spreading code, b_l is the symbol transmitted over the symbol period $[lT, (l+1)T]$, and b_l^m is the symbol transmitted over the symbol time $[lT - mT_c, (l+1)T - mT_c]$, which is assumed to be constant.

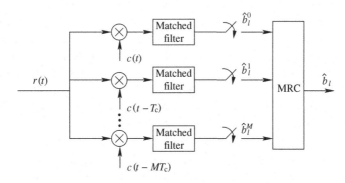

Principle of the rake receiver.

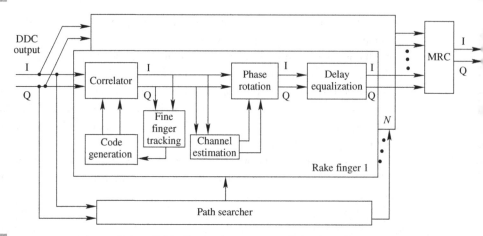

Figure 8.9 **Implementation of the rake receiver. DDC stands for digital down-converter.**

The architecture for implementing a rake receiver is shown in Fig. 8.9. Oversampled baseband I and Q signals are fed to each finger from a digital down-converter output, and also fed to the path searcher. The path searcher despreads the received data by using correlations with different time delays of the code. The range of time delays is decided by the radius of the cell plus allowed delay spread. The peaks in the correlation correspond to strong multipath signals. A peak detection threshold is set for selecting the suitable peaks. For each peak, its associated time delay is forwarded to a finger for further processing. This timing information is used for code generation and code synchronization at the rake and also allows the rake to lock on a multipath. The path searcher is usually computationally more expensive than the rake.

The incoming data is multiplied by the despreading code, and this result is integrated to produce symbols. The data is processed at the oversampling rate, and this leads to phase rotation in the symbol data. Phase rotation can be removed by using phase data from the channel estimation that is based on the pilot information. The finger then tracks any small change in the time delay of the path with the feedback loop controlled by the fine finger tracking function. I and Q signals are generated from the correlator periodically on 1/4 chip earlier or later of the code assigned by the path searcher. The early and late correlations are examined so as to correct the timing of the path. Finally, a rake combiner, usually using the MRC strategy, is used to combine the active outputs so as to maximize the SNR.

The rake receiver achieves the optimum performance when there is no ISI. That is, the maximum delay τ_{max} must be smaller than T. When there is ISI, an equalizer can be applied to the rake receiver output, which is sampled at intervals T. The combination of rake receiver and equalizer can also be replaced by a *chip-based equalizer*, which is optimum but most complex for implementation. When smart antenna technology is combined with the rake receiver for a CDMA system, we get a space-time processing system [10, 36]. Two CDMA beamforming rake receiver schemes are given and compared in [5, 36].

IS-95, WCDMA and CDMA2000 use the rake receiver, but not multiuser detection (MUD). This is because the spreading factor is too large and the computational load grows exponentially with the spreading factor for MUD. MUD is applied in TD-SCDMA, since the maximum spreading factor is 16 and thus the computational complexity is tractable. A typical TD-SCDMA system uses a circular array with eight antennas.

8.3.4 Synchronization in CDMA

For CDMA systems such as IS-95, it is reasonable to achieve time synchronization before accurate phase and frequency are obtained. In principle, matched filtering or cross-correlation is the optimum method for initial synchronization. The timing acquisition process is based on the autocorrelation of the spreading sequence. Since all periodic spreading sequences have an autocorrelation that peaks at zero delay, the synchronizer adjusts its estimate of the delay to maximize the autocorrelation output of the integrator. A sharp autocorrelation function facilitates fine synchronization.

The sliding correlator is usually implemented in discrete time intervals of $\frac{1}{2}T_c$; it correlates the received signal with the known synchronization sequence, over the time interval of NT_c, until the correlator output exceeds a threshold.

When timing acquisition is achieved with $\tau = 0$, to account for frequency error Δf, the signal error is scaled by a factor, which is approximated by [63]

$$D(\Delta f) = \left[\frac{\sin(\pi N \Delta f T_c)}{\pi N \Delta f T_c} \right]^2, \tag{8.27}$$

where N is the spreading factor. This signal deteriorating factor is small when $N\Delta f T_c$ is small.

After timing acquisition achieves a timing error of a small fraction of T_c, timing tracking must be applied to further reduce the error to zero. This is achieved by using the delay-locked loop (DLL), which is very similar to the early-late gate circuitry. Once timing is accurately obtained, phase and frequency can be accurately estimated by using the PLL technique.

The pilot sequence is very important for CDMA systems. It is a nonmodulated signal. The pilot signal is shared by all the users, and it has a strong power.

8.3.5 Power control

Power control is necessary to reduce CCI, and is implemented in most cellular standards in some form. The DS-CDMA system has the problem of the near–far effect, which mandates implementation of power control to limit the amplitudes of user signals so as to maintain orthogonality and system capacity. Power control can also reduce the interference to other systems, and reduce the power consumption at mobile terminals. Power control is a critical aspect of CDMA-based cellular systems.

Before an MS establishes a connection with the BS, it cannot be power-controlled by the BS. The initial transmission power of the MS is estimated from its nominal power, its current power, and the received power from the BS. Power control can take an open-loop or closed-loop form. Open-loop power control can compensate for slow-varying and log-normal shadowing effects, whereas closed-loop power control can compensate for power fluctuations arising from fast Rayleigh fading, which is frequency-dependent and occurs over every half-wavelength.

In open-loop power control, the sum of the downlink and uplink powers (in dB) is maintained constant. Open-loop control is based on AGC, but the powers of different MSs can differ by several dB at the BS. Power control is based on the SIR E_b/I_0 measured at the BS. For IS-95, E_b/I_0 measured at the BS should be in the range of 3 to 7 dB [63]. The BS transmits power control instructions to the MS to increase or decrease its power. E_b/I_0 in dB is normally distributed with a large standard deviation in case of a fading channel. A closed-loop power control is necessary to help reduce the standard deviation to typically within 1.5 to 2.5 dB [63].

Power control is performed via the uplink and downlink transmissions of power control instructions over the traffic channel. These power control bits are usually not error-protected so as to avoid delays due to decoding. Power control is more stringent on the uplink than on the downlink. The MS has to report the quality of the downlink to the BS. In the case of closed-loop power control during soft handoff, the MS receives from two or more BSs power-control instructions that may be conflicting. The rule for handling this is to power down the MS if any of the BSs commands to power down and to power up the mobile only when all the BSs command to power up.

For downlink power control, open-loop control is sufficient, as power control in the downlink does not affect the functioning of the system. Uplink power control is rather crude, usually ±6 dB around its nominal value for IS-95.

In IS-95, fast closed-loop power control is applied to the reverse link at a rate of one instruction per frame. Open and fast closed-loop power control schemes are supported in both the reverse and forward links in CDMA2000, WCDMA, UTRA-TDD, and TD-SCDMA, with a speed of up to 800 Hz, 1,500 Hz, 100 Hz, and 200 Hz, respectively. In GSM, power control loops run at 2 Hz. WiMAX also supports closed-loop power control for the uplink, but leaves the downlink power control to the manufacturer.

8.3.6 Soft handoff

In CDMA systems, all the BSs use the same frequency. Thus, an MS can transmit to and receive from two or more BSs. The received signals from multiple BSs constitute multipaths, and can be used to improve the performance.

Soft handoff is an important feature of CDMA systems, and it also has significant impact on power control. Unlike hard handoff used in GSM, an MS receives and transmits signals from and to two or more BSs. Signals coming from different BSs have different delays, and these signals can be combined by using a rake receiver. Due to the use of diversity, soft handoff significantly increases the performance while the MS is on the border of two cells. Soft handoff can improve coverage by a factor of 2 to 2.5 in cell area, thus leading to a reduction in the number of BSs by this factor [62]. The reverse channel capacity is also increased by a factor greater than 2 [62]. However, the MS must use the Walsh codes in the multiple BSs at the same time. It also requires a lot of signaling during handoff.

To implement soft handoff, each cell or sector transmits in the downlink simultaneously its specific pilot signal and its user signals. For synchronous CDMA, all BSs use the same pilot code and they are distinguished by using different phase shifts of the same pilot, while for asynchronous CDMA, each BS is assigned a distinct scrambling code. The pilot PN-sequence signal can be added to the user signals, or it is multiplied with each of the user signals. All the pilot PN sequences can use the same m-sequence but with different initial vectors. The pilot signals are used by MSs for handoff implementation.

When the searcher finds a new, sufficiently strong pilot signal, the MS reports to its original BS. The BS then reports this to the switching center, and the switching center instructs the new BS to transmit and receive signals from the MS. The MS communicates with two BSs at the same time, with the same information. The rake receiver at the MS combines the signals from the two BSs, just as in the case of processing multipaths with different time delays.

Soft handoff improves the system performance, but at a cost of consuming more spectrum resources. Soft handoff can help increase the capacity of a heavily loaded cell by making use of the neighboring cells. For a given outage probability P_{out}, the transmit power should be above a threshold γ dB; for soft handoff, the required threshold is 6 to 8 dB, which is several dB less than that needed for hard handoff [63]. This leads to an increase in the cell size by a factor of two or more. Thus, soft handoff supports a larger capacity than hard handoff does.

During soft handoff, the second BS must transmit the same information as the first BS does. When the power of the second BS received at the MS is 6 dB less than that from the first BS, the second BS is activated. This first BS continues to communicate with the MS, until its power received at the MS is 6 dB less than that from the second BS. Thus, during soft handoff, there is more interference caused to the other users.

In IS-95 or CDMA2000, each BS transmits the same PN-sequence as a pilot signal, but with an offset of 64 chips from each other. The pilot signal is a PN sequence of 32,768 chips, or 26.7 ms. This can be used to estimate the signal strengths of all the surrounding BSs. WCDMA supports soft, interfrequency, and inter-RAT handoffs, where RAT stands for radio access technology, such as GSM, CDMA2000, or UTRA-TDD.

TD-SCDMA employs baton handover, which takes advantage of both hard and soft handoffs. Baton handover, which is similar to the procedure of handing over a baton in a relay, depends on the user positioning capability provided by TD-SCDMA BSs using smart antenna technology.

8.4 Multiuser detection

8.4.1 Introduction

In the conventional CDMA system, MAI and multipath fading (ISI) are mitigated by rigorous power control in conjunction with single-user rake receivers. The rake detector [48] effectively combats multipath fading by coherently combining resolvable multipath replicas of the desired signal; but it is based on the assumption of path resolvability, which is not always true. The rake structure is not flexible for multiuser solution. The rake receiver also does not make use of the inherent diversity of the spread spectrum. MAI is caused by cochannel signals that are not orthogonal to the desired signal, and is reduced by strict power control. Power control instructions also consume frequency spectrum and the performance is unsatisfactory.

In a multiple-access channel such as a CDMA channel, MUD, also known as *multiuser equalization* or *joint detection*, can be applied to detect the data of the users. MUD improves the data detection process by exploiting the cross-correlation among the signals to be demodulated. By using MUD, receivers can effectively reduce the influence of the near–far effect, i.e., MAI or CCI, as well as ISI due to multipath delays [10]. When using MUD in CDMA systems, interference signals are not treated as noise, instead their spreading codes are used to mitigate MAI. MUD has become a key technique to overcome the effects of MAI and multipath fading, thus substantially increasing the capacity of CDMA systems.

MUD is typically not used on downlink channels, since downlink channels are synchronous and all interference is typically eliminated by using orthogonal codes. Also, the power consumption and complexity of MUD make it impractical for its implementation in MSs.

MUD achieves a BER performance that greatly exceeds that of the conventional correlator-based receiver. The optimum MUD receiver [59] is an MLSE receiver, which is based on the ML or MAP criterion. The Viterbi algorithm is a well-known, efficient MLSE algorithm. The complexity of such algorithms grows exponentially with the number of users.

Many low complexity MUD algorithms, such as the linear MUD and iterative MUD algorithms, are available. The linear MUD algorithm applies a linear transformation to the matched filter outputs to reduce the MAI effects on between them. The decorrelating and MMSE detectors are two popular MUD algorithms. The matched filter detector is a simple example of the linear MUD algorithm. The serial interference cancellation (SIC) and parallel interference cancellation (PIC) algorithms are two simple, nonlinear MUD techniques. The turbo MUD technique is a nonlinear, iterative MUD algorithm, and it is used in the context of the convolutionally encoded multiple-access channel. It uses soft decision strategies to reduce the probability of error propagation. The turbo MUD algorithm can approach the performance of the MLSE MUD algorithm.

For IS-95A networks, the voice capacity per cell is about 20 to 25 users. CDMA2000 1x increases it to 35 to 40 users. In 3G standards, such as CDMA2000 and WCMDA, MUD has been specified as an option. This option will not be activated by most mobile communications operators due to its complexity. In UTRA-TDD, each slot allows up to 8 codes for multiple access, while in TD-SCDMA each slot allows up to 16 codes; this requires a relatively small spreading factor. Since the computational complexity increases at least linearly with the spreading factor, MUD is supported easily in the two TDD modes. Only TD-SCDMA applies MUD as a mandatory part, and in the TD-SCDMA system the maximum spreading factor is 16. In UTRA-TDD and TD-SCDMA, smart antenna technology with MUD was developed based on the algorithm given in [4].

8.4.2 Optimum multiuser detector

The optimum multiuser detector performs joint MAP detection for all users based on the received sequence r. Mathematically,

$$\hat{b}^{MAP} = \arg \min_{b \in \mathcal{B}^K} \Pr(b|S, r) = \arg \max_{b \in \mathcal{B}^K} p_{r|b,S}(r) \Pr(b), \qquad (8.28)$$

where $\mathcal{B}$ is the symbol set, and the second equality is obtained by applying Bayes's rule.

The hypothesis $\hat{b}$ is generally uniformly distributed, thus $\Pr(b)$ is known at the receiver. In this case, joint MAP detection is equivalent to joint ML detection

$$\hat{b}^{ML} = \arg \max_{b \in \mathcal{B}^K} p_{r|b,S}(r). \qquad (8.29)$$

The conditional pdf in (8.29) is a multivariable Gaussian function. By taking its natural logarithm, the optimum MAP or ML detection algorithm is finally obtained as

$$\hat{b} = \arg \min_{b \in \mathcal{B}^K} \|r - SAb\|^2$$

$$= \arg \min_{b \in \mathcal{B}^K} \int_0^T \left[r(t) - \sum_{k=1}^{K} b_k A_k c_k(t) \right]^2 dt$$

$$= \arg \min_{b \in \mathcal{B}^K} \|r\|^2 - 2r^H SAb + b^H (SA)^H (SA)b$$

$$= \arg \max_{b \in \mathcal{B}^K} 2r^H SAb - b^H ARAb. \qquad (8.30)$$

Note that the search space is the entire K-dimensional complex space, rather than a finite alphabet. Since b consists of discrete values, an exhaustive search is necessary. The complexity grows exponentially with the number of users and the length of the sequence.

In general, the optimum MUD receiver for bit sequence consists of a bank of single-user correlators followed by a Viterbi algorithm, and the algorithm has a complexity per binary decision of $O\left(2^K\right)$, K being the number of users. The optimum detector is based on the conventional matched filter detector, but the discrete decision device for each of the users is replaced by the joint Viterbi algorithm. This is illustrated in Fig. 8.10.

Suboptimal MUD algorithms based on the genetic algorithm have also been given [12, 66]. The genetic algorithm-based MUD algorithm will be more practical when the

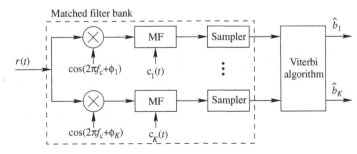

Figure 8.10   Optimum MUD detector based on MLSE.

parallel hardware for the genetic algorithm is available. Nevertheless, the genetic algorithm provides a benchmark when the exact solution is computationally intractable. The optimum MUD can be resolved using the collective computational behavior of such neural networks as the Hopfield network [30] or the cellular neural network [6]. Many suboptimal algorithms have been proposed based on the criterion (8.30) using various optimization techniques such as the gradient descent and the genetic algorithm [25].

8.4.3 Linear multiuser detection

Linear MUD is very similar to linear equalization, and concepts like ZF and Wiener filtering are usually used.

Decorrelation receiver

The decorrelation receiver [37] is the ZF version of the multiuser detector. It searches the unconstrained vector $b \in \mathcal{B}^K$ that minimizes the ML objective function [32]

$$\tilde{b} = \arg \min_{b \in \mathcal{B}^K} \| r - \mathbf{S}\mathbf{A}b \|^2. \tag{8.31}$$

By setting the derivative of the squared Euclidean distance with respect to $\tilde{b}^H$ to zero and applying the Wirtinger calculus $\partial \tilde{b}/\partial \tilde{b}^H = \mathbf{0}$ (see Appendix B), we have

$$-\mathbf{S}^H r + \mathbf{S}^H \mathbf{S}\mathbf{A}\tilde{b} = \mathbf{0}. \tag{8.32}$$

From this, the ZF solution $\tilde{b}_{\mathrm{ZF}}$ is obtained. For $K \leq N$, $\mathbf{S}$ generally consists of linear independent columns. The correlation $\mathbf{R}$ has a full rank and its inverse exists. Thus, the symbols can be estimated by

$$\tilde{b}_{\mathrm{ZF}} = \mathbf{A}^{-1}\mathbf{R}^{-1}y = b + \mathbf{A}^{-1}\mathbf{R}^{-1}\mathbf{S}^H n, \tag{8.33}$$

where

$$y = \mathbf{S}^H r \tag{8.34}$$

is the output of the receiver after despreading. For $K > N$, $\mathbf{R}^{-1}$ has to be replaced by the pseudoinverse $\mathbf{R}^\dagger$.

The final output is obtained by a hard decision,

$$\hat{b} = \mathcal{Q}\left(\tilde{b}\right). \tag{8.35}$$

where $\mathcal{Q}$ is the quantization function.

Like the ZF equalizer, the decorrelation receiver leads to noise enhancement, which is determined by the conditioning of the correlation matrix, especially for high load β. In order to compute the optimal cross-correlation matrix, knowledge of all the user parameters is required. The noise covariance matrix is given by

$$\Phi_{\mathrm{ZF}} = E\left[\left(\tilde{b}_{\mathrm{ZF}} - b\right)\left(\tilde{b}_{\mathrm{ZF}} - b\right)^H\right] = \sigma_n^2 \mathbf{R}^{-1}, \tag{8.36}$$

where σ_n^2 is the variance of n. For $K > N$, $\mathbf{R}^{-1}$ is replaced by pseudoinverse $\mathbf{R}^\dagger$.

For BPSK or QPSK modulation, the BER performance for the kth user is given by [59]

$$P_{b,k} = Q\left(\frac{A_k}{\sigma_n\sqrt{\left(\mathbf{R}^{-1}\right)_{kk}}}\right). \tag{8.37}$$

MMSE receiver

Similarly to the MMSE equalizer, the MMSE MUD receiver targets at a balance between interference suppression and noise enhancement. The objective function to minimize is the MSE $E\left[|\boldsymbol{b} - \hat{\boldsymbol{b}}|^2\right]$.

For the linear MMSE MUD receiver, a linear transform $\mathbf{W}$ is applied such that the MSE is minimized [38]

$$\mathbf{W} = \arg\min_{\mathbf{W}\in\mathcal{C}^{K\times N}} E\left[\|\boldsymbol{b} - \mathbf{W}\boldsymbol{y}\|^2\right], \tag{8.38}$$

where $\mathcal{C}$ denotes the complex space and $\mathbf{W}$ is a $N \times K$ matrix.

Again, by setting the partial derivative of the squared Euclidean distance with respect to $\mathbf{W}$ to zero and considering the Wirtinger calculus $\partial\mathbf{W}^H/\partial\mathbf{W} = 0$, we have the well-known Wiener solution. The final result is given as [1]

$$\mathbf{W}_{\text{MMSE}} = \mathbf{A}^{-1}\left(\mathbf{R} + \sigma_n^2\mathbf{A}^{-2}\right)^{-1}. \tag{8.39}$$

From this, the MMSE receiver can be treated as a tradeoff between the matched filter and decorrelator: If $\sigma_n^2 \to 0$, the second term in the inverse is reduced, and a decorrelator is obtained; at the other extreme, for $\sigma_n^2 \to \infty$, $\mathbf{R}$ is negligible, and a matched filter is obtained.

The derived MMSE solution is given by

$$\tilde{\boldsymbol{b}} = \mathbf{A}^{-1}\left[\mathbf{R} + \sigma_n^2\mathbf{A}^{-2}\right]^{-1}\boldsymbol{y}. \tag{8.40}$$

The estimate symbol vector $\hat{\boldsymbol{b}}$ is decided by quantizing $\tilde{\boldsymbol{b}}$.

The MMSE is given by

$$J_{\text{MMSE}} = \min_{\mathbf{W}\in\mathcal{C}^{K\times N}} E\left[\|\boldsymbol{b} - \mathbf{W}\boldsymbol{y}\|^2\right] = \text{trace}\left(\left(\mathbf{I} + \sigma_n^2\mathbf{A}\mathbf{R}\mathbf{A}\right)^{-1}\right). \tag{8.41}$$

The total signal distortion for the MMSE receiver is much lower than that for the decorrelation receiver. The MMSE receiver exhibits an excellent performance, but it requires training data. Both the decorrelating detector and the linear MMSE detector achieve the optimal near-far resistance.

8.4.4 Serial/parallel interference cancellation

The linear ZF and MMSE MUD receivers have a complexity that grows cubically with the system size in view of the need for the calculation of the matrix inverse of $\mathbf{M}$,

$$M = \begin{cases} R & \text{(ZF)} \\ R + \sigma_n^2 A^{-2} & \text{(MMSE)} \end{cases}. \qquad (8.42)$$

The calculation of M^{-1} is basically via the solution of a set of linear equations. The set of linear equations can be solved iteratively, and this yields the linear SIC and PIC algorithms. Linear SIC and PIC are decision-driven MUD algorithms.

Linear SIC

Application of the Gauss-Seidel algorithm to the set of linear equations leads to the SIC technique. The algorithm always converges for the Hermitian positive definite matrix M [23]. This is satisfied for CDMA systems.

The SIC receiver is a suboptimum but practical multiuser receiver [31]. SIC just detects users in the order of their signal strength. The signal of each user is subtracted from the total signal, based on which the next user is detected. Thus, the SIC receiver is a decision-feedback receiver. The SIC receiver is illustrated in Fig. 8.11.

In each stage of the SIC receiver, a conventional matched filter detector and an interference canceller are included. The first stage estimates and removes the strongest user signal. The user signals should be detected from the strongest to weakest power to reduce error propagation. Sorting also leads to a faster convergence.

The SIC receiver simply performs a hard decision upon the kth user's correlator as the initial data estimate $\hat{b}_{k,n}$, which is the estimate of $b_{k,n}$, the kth user's original transmitted data at time nT_b. The estimate $\hat{b}_{k,n}$ is multiplied by its respective synchronized spreading sequence and carrier signal to produce a replica of the original signal, which is passed through a transversal filter that emulates channel $h_k(t)$. The CCI replica of each user is subtracted from the original received signal $r^{(1)}(t) = r(t)$ to produce the desired received signal of the user with the second largest power, denoted user m, and we obtain $r^{(2)}(t)$. The signal $r^{(2)}(t)$ is passed to a second set of correlators to estimate the data bits $b_{m,n}$, of user m. A cascade of such CCI canceller and data decision stages will lead to a better BER performance.

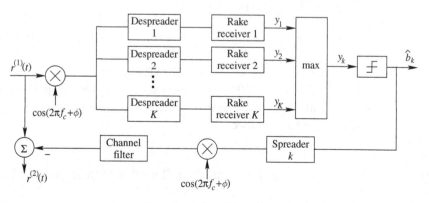

Figure 8.11   One stage of an SIC receiver.

The direct implementation of matrix inversion leads to a complexity of $O\left(K^3\right)$, while the iterative implementation has a complexity of $O(nK^2)$, where n is the number of iterations. As long as $n < K$, the iterative implementation always leads to a reduction in the computational cost.

At each iteration, interference cancellation is implemented for all the K users. Because of the cascade structure, to reduce the bit delay, we need to limit the number of stages. Only a few of the strongest users are removed at these stages, and the final output is input to a conventional matched filter detector to extract the remaining signals. The order of cancellation is based on the ranking of the power of the matched filter outputs.

The SIC receiver provides a BER performance that is much better than that of the conventional matched filter detector, especially for the near-far problem [33]. The SIC receiver suffers from error propagation, and this degradation can be improved by power control. In the adaptive SIC receiver, the interference canceller is adjusted by the LMS algorithm. This yields a more accurate amplitude estimate than the SIC receiver yields, and hence a much lower BER. The SIC technique can theoretically achieve the Shannon capacity, and practical SIC implementations can approach this capacity [61].

Linear PIC

Application of the Jacobi algorithm for solving the set of linear equations [23] leads to the linear PIC algorithm. The PIC receiver cancels all the users simultaneously. The method first makes decisions for all the users based on the received total signal, and obtains $\hat{b}_k^0$, $k = 1, 2, \ldots, K$. These decoded user signals are then respread. For a given user k, all the other $K - 1$ users' respread signals are treated as interferers, and are subtracted from the total signal. The remnant signal is then subject to the same procedure, until the decision no longer changes or the prespecified number of iterations is reached.

The PIC receiver has a lower latency, but has a performance degradation due to the near-far effect. The convergence of the PIC depends highly on the eigenvalue distribution of $\mathbf{M}$ [32]. The convergence property is generally poor.

8.4.5 Combination of linear MUD and nonlinear SIC

The Bell Laboratories layered space-time (BLAST) scheme [13, 14], originally proposed for multiple-antenna systems, can be directly applied to CDMA systems, since both systems have a similar architecture

$$r = Sb + n. \tag{8.43}$$

For CDMA systems, this equation is obtained when strict power control is applied. The method first uses a linear MUD stage to suppress the interference prior to the nonlinear SIC stage.

At the first step, the linear filter w_1 is applied to suppress the interference of user 1. w_1 is drawn as the first column of $\mathbf{W}^{(1)} = \mathbf{W}_{\text{ZF}}$ or $\mathbf{W}_{\text{MMSE}}$, and

$$\tilde{b}_1 = w_1^H r. \tag{8.44}$$

The symbol is then obtained as $\hat{b}_1 = \mathscr{Q}\left(\tilde{b}_1\right)$.

By implementing interference cancellation, we have

$$\tilde{r}_2 = r - s_1 \hat{b}_1, \tag{8.45}$$

where s_1 is the first column of $\mathbf{S}$.

The residual $\tilde{r}_2$ is then processed by a second filter w_2, which is the first column of $\mathbf{W}^{(2)}$, $\mathbf{W}^{(2)}$ being the ZF or MMSE filter obtained by removing s_1 from $\mathbf{S}$. To suppress the interference of the second user, we have

$$\tilde{b}_2 = w_2^H \tilde{r}_2. \tag{8.46}$$

This procedure is repeated until all the users have been detected. To avoid matrix inversions, QL decomposition can be applied [32].

The ML space-time multiuser detectors for flat-fading channels and multipath fading channels are presented in [40] and [44]. Space-time MMSE multiuser detectors have been proposed in [10, 45]. MUD for CDMA systems has been extended to antenna arrays, and several space-time MUD algorithms are described in [64].

8.5 Bit error probability and system capacity

8.5.1 BER performance

The DS-CDMA system is an interference-limited system, where the noise is negligible compared with the interference from all the other users. By using the standard Gaussian approximation, the MAI is assumed to be a Gaussian random variable. For a synchronous K-user system with N chips in each symbol, in the downlink, the SIR at an MS receiver is given by [56]

$$\text{SIR} = \frac{N}{K-1}. \tag{8.47}$$

If noise is considered, the SINR is accordingly given by

$$\text{SINR} = \left(\frac{N_0}{E_s} + \frac{K-1}{N}\right)^{-1}. \tag{8.48}$$

In the uplink, signals from different users travel through different channels, and this leads to the near-far effect, where different users have different powers at the BS receiver.

Assuming random spreading codes with N chips per symbol, random start time, and random carrier, the average SINR for asynchronous users is given by [20, 50, 56]

$$\text{SINR} = \left(\frac{N_0}{E_s} + \frac{K-1}{3N} \right)^{-1}. \tag{8.49}$$

For the interference-limited system, the SIR is given by

$$\text{SIR} = \frac{3N}{K-1}. \tag{8.50}$$

The above equations for both the uplink and the downlink are known as the *standard Gaussian approximations* to SINR and SIR.

The calculated SIR can be used to replace γ_b in the error probability equations. For example, the BER for asynchronous DS-CDMA with BPSK modulation is given by

$$P_b = Q\left(\sqrt{\gamma_b}\right) = Q\left(\sqrt{\frac{3N}{K-1}}\right). \tag{8.51}$$

The standard Gaussian approximation underestimates the BER performance when the number of simultaneous users K is small or the processing gain N is large. For practical systems, the SIR for nonrandom spreading codes can be approximated by

$$\text{SIR} = \frac{3N}{\psi(K-1)}, \tag{8.52}$$

where ψ is a constant, whose value depends on the system assumptions. For PN sequences, $\psi = 2$ or 3, depending on the system assumptions [20].

The improved Gaussian approximation proposed in [41] is much more accurate than the standard Gaussian approximation, but its complexity is too high. A simple but approximate Gaussian approximation is given in [28], and the result is very close to that of the improved Gaussian approximation[41]. For BPSK signals, the BER performance for the AWGN channel is given by [28]

$$P_b \approx \frac{2}{3} Q\left(\sqrt{\frac{3N}{K-1}}\right) + \frac{1}{6} Q\left(\frac{N}{\sqrt{(K-1)N/3 + \sqrt{3}\sigma}}\right)$$
$$+ \frac{1}{6} Q\left(\frac{N}{\sqrt{(K-1)N/3 - \sqrt{3}\sigma}}\right), \tag{8.53}$$

where

$$\sigma^2 = (K-1)\left[\frac{23}{360}N^2 + \left(\frac{1}{20} + \frac{K-2}{36}\right)(N-1)\right]. \tag{8.54}$$

This result is very accurate for all values of K and N.

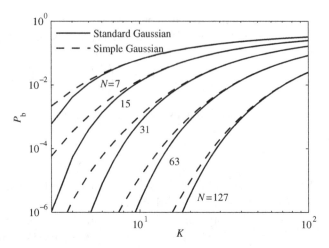

Figure 8.12 BER performance for different K and N: Standard Gaussian approximation vs. simple improved Gaussian approximation.

Example 8.3: Based on the standard Gaussian approximation and the simple improved Gaussian approximation to the SIR, the BERs for the asynchronous CDMA system with BPSK modulation are given by (8.52) and (8.53), respectively. These results are plotted in Fig. 8.12. It is seen than the standard Gaussian approximation is accurate for large K, but underestimates the BER performance for small K.

From the above equations, we can solve for the total number of active users for a given P_b or SINR. For speech communications, the voice activity occupies only about 3/8 of the call duration. Cell sectorization and smart antenna technology can be used to reduce the interference. All these can be used to increase the cell capacity. In a cellular network, interference from other cells should also be considered into the capacity analysis.

8.5.2 Uplink capacity

The uplink capacity in a cell depends on the user SNRs and the system matrix $\mathbf{S}$. For an asynchronous CDMA system with strict power control, signals from other $K - 1$ users appear as interference at the kth user receiver. The sum capacity for K active users is derived as [1]

$$R_{\text{CDMA}} = KR_k \approx 1.44B \quad \text{(bits/s)}, \tag{8.55}$$

where B is the signal bandwidth and R_k is the Shannon capacity of the kth user. Thus, the sum capacity does not increase with K.

For the synchronous link with the individual codes being orthogonal to one another, when strict power control is implemented, the sum capacity of the K users is given by [1, 49]

$$\sum_{k=1}^{K} R_k \leq B \log_2 \left(1 + \frac{KP}{BN_0} \right) \quad \text{(bits/s)}, \tag{8.56}$$

where P is the average power of each user's signal. Thus, the sum capacity of the CDMA system increases with K, as is the case for FDMA and TDMA.

In a cell, the average capacity per user is given by

$$C_u = \frac{\overline{C}}{K} \quad \text{(bits/s)}, \tag{8.57}$$

where $\overline{C}$ is the average capacity of the cell.

The spectral efficiency is defined as the average number of information bits transmitted per chip

$$\eta = \frac{\overline{C}}{N} = \beta C_u \quad \text{(bits/s/Hz)}, \tag{8.58}$$

where β is defined by (8.13). Note that as the bandwidth of a DS-CDMA signal is approximately equal to the reciprocal of the chip duration, 1 bit/chip is equivalent to 1 bit/s/Hz. If the code rate supported by each user is R_c bits/symbol, the spectral efficiency is given by

$$\eta = \frac{KR_c}{N} = \beta R_c \quad \text{(bits/s/Hz)}. \tag{8.59}$$

For an M-ary modulation scheme, $R_c = \log_2 M$ bits/symbol.

In the case of orthogonal spreading codes that are used for synchronous CDMA transmission in flat-fading channels, no MAI exists; thus, there are K independent, parallel data streams. Each stream has a capacity C_u determined by its SNR only. The spectral efficiency η linearly increases with β up to $\beta = 1$ for a fixed SNR. However, for $\beta > 1$, the so-called *Welch-bounded sequence* has to be used to keep η constant at its value at $\beta = 1$ [60].

The spectral efficiencies for the cases of random spreading codes with different receivers in AWGN multiple-access channels, such as the single-user matched filter, the decorrelator (ZF), the MMSE receiver, and the optimum MUD receiver, are given in [60]. The results for spectral efficiency are shown in Fig. 8.13 for $E_b/N_0 = 10$ dB and $K \to \infty$. The spectral efficiencies achievable by an optimum joint decoder with no spreading and with orthogonal codes for $K \leq N$ are also plotted for comparison. It is seen that the capacity of the MMSE receiver is always higher than that of the single-user matched filter and ZF receivers, but is lower than that of the optimum MUD receiver and the receiver with orthogonal codes. The ZF receiver typically approximates the performance of the MMSE and the optimum MUD receiver for small β. As load increases, η reaches an optimum, and then decreases dramatically and drops to zero below $\beta = 1$.

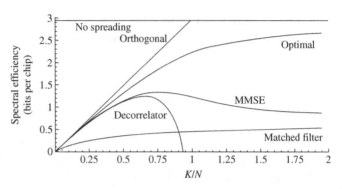

Figure 8.13 The uplink spectral efficiencies of DS-CDMA for different receivers for $E_b/N_0 = 10$ dB and $K \to \infty$. ©IEEE 1999. ([60], Fig. 1).

8.6 Other DSSS techniques

In addition to DS-CDMA, there are some other DSSS techniques, such as single-carrier cyclic prefix assisted CDMA (CP-CDMA) [3], multicarrier CDMA (MC-CDMA) [65], block-spreading CDMA (BS-CDMA) [46], and interleave division multiple access (IDMA) [39, 47].

Single-carrier CP-CDMA is used for broadband cellular systems [3]. It is a block-based transmission scheme, where a cyclic prefix is inserted to each data block. This alleviates the inter-block interference if the cyclic prefix length is larger than the maximum delay spread of the channel. The use of a cyclic prefix transforms the linear convolution into circular convolution, so that FFT-based linear equalizers can be used to recover the transmitted symbols for each user in frequency-selective fading channels.

MC-CDMA, which will be introduced in Section 9.13, combines DS-CDMA with OFDM [65]. Unlike the single-carrier CP-CDMA system that transmits the data block directly, the MC-CDMA system transmits the IFFT version of the data block. Due to the insertion of a cyclic prefix, FFT-based linear receivers are also applicable for MC-CDMA systems.

A block-spreading CDMA (BS-CDMA) system produces a significantly improved multiuser performance without using complex MUD techniques for uplink transmission [46]. Code orthogonality is easily maintained when channel variation across the consecutive blocks is negligible in a block-based high-speed transmission, leading to MAI-free transmission in a slow fading channel. The system uses FDE at the receiver to combat ISI over frequency-selective fading channels. Despreading is implemented prior to equalization, reducing the frequency domain process to a symbol-wise operation.

Interleave division multiple access

IDMA [39, 47] may be considered as a special DS-CDMA scheme. In a DS-CDMA system, the bitstream is first channel-encoded, interleaved, and then spread for each user, and the spreader is user-specific. In contrast, in IDMA, spreading is placed prior to interleaving, and the interleaver is user-specific. The different users are distinguished by their unique chip interleavers. In IDMA, FEC and spreading can be combined in a single encoder, which is the same for all users; as a consequence, very low-rate encoding is used, and the spreader may be used to simplify the overall encoder. The combined use of low-rate channel coding and chip-level interleaving allows IDMA for simple soft-input soft-output (SISO) iterative decoding techniques with low-complexity MUD. Like multicode CDMA, the data rate can be controlled by the number of signature sequences assigned given a predefined target BER.

IDMA with a sufficiently large number of superimposed codewords, with optimum power allocation, and in conjunction with a suitable receiver is capacity-achieving [27]. Single-carrier IDMA realizes time diversity, where chip interleaving is capable of increasing the achievable time-diversity in time-selective channels, and joint coding and spreading design.

The MMSE-filtered PIC strategy is used as a low-complexity SISO MUD [47]. Due to low-rate coding and chip-level processing, rather than spreading and bit-level processing in DS-CDMA, for each transmitted symbol, the interference for one chip is independent from that for another chip due to the chip-interleaver, and thus correlation between users is virtually eliminated. Hence, the optimal filtering after interference cancellation is simply the summation of the log-likelihood-ratio (LLR) values for all the chips, and matrix inversion is avoided, resulting in a low-complexity implementation that grows linearly with the number of users. This is achieved at a price of more memory for interleaver sizes.

IDMA has been generalized to the multicarrier case. In the multicarrier interleave-division-multiplexing (IDM)-aided IDMA (MC-IDM-IDMA) [67], each user transmits multiple streams differentiated by stream-specific chip interleavers. This concept is similar to the multicode scheme employed in the HSPA system.

Possible applications include 4G cellular standards, wireless LANs, ad-hoc networks [55], and UWB systems [35]. Channel coding is introduced in Chapter 15.

8.7 DSSS and DS-CDMA in wireless standards

DSSS, as well as its multiuser version DS-CDMA, is the most popular spread spectrum technology. DS-CDMA is widely used in CDMA-based 2G and 3G mobile communications standards including IS-95, CDMA2000, WCDMA, UTRA-TDD, and TD-SCDMA. Both IS-95 and CDMA2000 use synchronous DS-CDMA architecture. Each BS has a code clock obtained from combining the PN short code and long code generators by spreading. A BS code clock is synchronized by loading the shift registers with the start bits.

DSSS is also widely used in wireless networking. The baseline IEEE 802.11 supports both FHSS and DSSS. In the DSSS mode of the baseline IEEE 802.11, each data bit is spread by an 11-chip Barker sequence, and the data stream is spread over an 11 MHz band. DBPSK and DQPSK are used for the data rates of 1 Mbit/s and 2 Mbits/s, respectively.

IEEE 802.11b uses eight-chip CCK, which is an extension of DSSS modulation. IEEE 802.15.4 (ZigBee) also uses DSSS modulation using PN chip sequences.

DS-CDMA is usually combined with FDMA or TDMA. In the combined DS-CDMA/FDMA system, the total bandwidth is divided into multiple sub-bands, and DS-CDMA is the multiple access method employed in each sub-band. This approach is used in IS-95 and WCDMA. When DS-CDMA is combined with TDMA, each user is assigned one timeslot. Users in different cells are distinguished by different spreading codes. This eliminates the near-far effect. Such an idea is embodied in UTRA-TDD.

IS-95

In the downlink, IS-95 uses 64 Walsh codes for each band, each code being designated as a channel. Each channel in the same band must use a unique code, while the same code can be used in different bands. Among the 64 codes: W0, W1, $\cdots$, W63, some are used for forward channel broadcasting, such as W0 for the pilot channel, W1 to W7 for the paging channels and W32 for the synchronization channel, while all the remaining codes plus some unused paging codes are assigned to users as their unique identifications and also for traffic.

The pilot channel sends all zeros (W0 code). It serves as the beacon signal that defines the radius of the cell, and the signal level is 4 to 6 dB higher than all other channels. The pilot channel is also used as a timing or demodulation reference at the MS receiver, and also for signal measurement during handoffs. The sync channel is used by the MS to acquire initial time synchronization. The paging channels are used by the BS to page an idle MS in the case of an incoming call, or carry control messages for call setup. The MS is listening on a paging channel when it is idle.

The uplink does not have pilot and synchronization channels for synchronization, thus the Walsh code cannot be used on the uplink and instead a long PN code of a length of $2^{42} - 1$ chips are used for channelization. In order to provide isolation among BSs or sectors, each BS or sector is assigned a unique short PN code with a length of $2^{15} - 1 = 32767$ chips, which is superimposed on top of a Walsh code; that is, each downlink channel including the pilot channel is first spread by its Walsh code, and then spread by a quadrature pair of short PN sequences, PN_I and PN_Q. The same pair of short PN sequences for the pilot channel are used in all BSs, but each BS in the network is assigned a signature generated by phase offset in steps of 64 chips of the pilot PN sequence pair.

CDMA2000

CDMA2000 reuses several IS-95 common channels and adds several new forward common channels. The pilot channel must be receivable at all locations in a cell, since it is used for synchronization, multipath channel estimation, coherent detection, frequency correction, as well as handover decisions. IS-95 uses BPSK prespreading, and fixed 64-bit Walsh codes. In CDMA2000, Walsh codes from 4 chips to 256 chips are used for variable information rates.

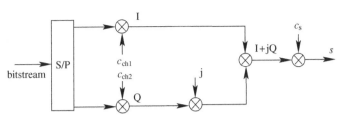

Figure 8.14
Spreading for WCDMA downlink channels.

WCDMA

WCDMA also uses the Walsh codes. They are used to multiplex users on the downlink and to identify MSs on the uplink. Each complex-valued downlink physical channel except the synchronization channel is split into I and Q parts which are spread by using different codes. This is shown in Fig. 8.14.

WCDMA uses two types of codes: channelization codes for spreading and scrambling codes for multiple access. The channelization codes are OVSF codes, whose code length is 4–256 for the uplink and 4–512 for the downlink. After channelization using a code, scrambling is applied to the result. The scrambling code c_s is only for identification purposes (for separating users on the uplink and for identifying BSs on the downlink), while the channelization code is used in the downlink to separate different intracell users. Separation of intercell users is implemented by using the scrambling code in the downlink.

The scrambling code is a complex-valued code, which is derived from two real-valued codes. It can be a long or a short code. The short scrambling code is derived from the very large Kasami code of length 256, while the long code is derived from the Gold code of length $2^{25} - 1$. The long code is truncated to a length of one frame (10 ms). For the uplink, both the short and long scrambling codes can be used. In case MUD is implemented in the BS, the short code is recommended for complexity reasons; otherwise, the long code should be used, since it provides a better whitening of interference. For the downlink, only the long code is used.

Multirate CDMA systems

In multirate CDMA systems, such as WCDMA, UTRA-TDD, and TD-SCDMA, multirate is realized by changing the spreading factor N, which is implemented by using the OVSF code. Since the chip rate $1/T_c$ is constant and $T = NT_c$, decreasing N leads to a higher data rate. However, a large spreading factor corresponds to a good capability for interference suppression. Thus, low spreading users require either a higher power level, or very few interferers in a cell, or a sophisticated detection technique. The use of the OVSF code is inflexible for rate-matching for multimedia applications, since the spreading factor must be made a power of two.

The multicode technique is another solution for multirate transmission. This method assigns more spreading codes to a subscriber that requires a high data rate. This method consumes more spreading codes, but can perform like a conventional single-rate CDMA system. The multicode technique is used in CDMA2000 and also in WCDMA. Multicode transmission is also allowed in UTRA-TDD. CDMA2000 also uses repetition and puncturing to achieve multirate. In HSDPA, AMC is used instead, and the method fixes the spreading factor to $N = 16$.

The principal drawback of the multicode technique is that the transmitted signals may have a high PAPR. In [54], constant-amplitude codes are designed to reduce the PAPR in multicode systems to the favorable value 1.

Remarks on spread spectrum technologies

DS-CDMA is now the mainstream technology for 3G mobile communications. It is effective for multiplexing a large number of variable-rate users in a cellular environment. Although DSSS is used for WCDMA and CDMA2000, CDMA is not an appropriate technology for high data rates. As 3G evolves toward higher rates, notably in HSDPA/HSUPA, 3GPP LTE, and 1xEV-DO, CDMA technology is actually losing its position. In HSDPA and 1xEV-DO, very small spreading factors are used and dynamical TDMA scheduling is employed based on channel conditions and latency. In 3GPP LTE, OFDM technology is applied.

CDMA technology is suitable for low-data-rate communications, such as voice, where many users are multiplexed to yield a high overall system capacity. For high-data-rate systems, each user must be assigned multiple codes, introducing considerable self-interference. For this reason, OFDM technology is widely used in high-data-rate wireless communications. We will introduce OFDM technology in Chapter 9.

8.8 Frequency-hopping spread spectrum

FHSS, or *code-controlled multifrequency-FSK modulation*, is another type of spectrum spreading technique. Unlike DS-CDMA which occupies the whole frequency band, FHSS uses only one among a set of narrow channels and hops through all of them in a predetermined spreading sequence at each predefined time interval. In the FHSS system, as in the DSSS system, the processing gain G_p is also defined as the ratio of the spread bandwidth of the transmitted signal to the data rate. The principle of FHSS is shown in Fig. 8.15, where B is the spread bandwidth, and $N = G_p$ is the period of the code.

At the transmitter, the modulated signal is mixed with the synthesizer output pattern to produce the FH signal. For an angle-modulation scheme, the transmitted signal for the kth hop is given by

$$s(t) = \sqrt{2P_{\text{avg}}} \cos\left(2\pi f_i t + \phi(t) + \phi_k\right), \quad (k-1)T_h \le t \le kT_h, \tag{8.60}$$

where P_{avg} is the average power, f_i is the carrier frequency for the kth hop, $\phi(t)$ is a continuous phase term, ϕ_k is a random phase angle for the kth hop, and T_h is the interval of each hop.

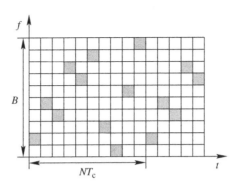

Figure 8.15   **Principle of FHSS.**

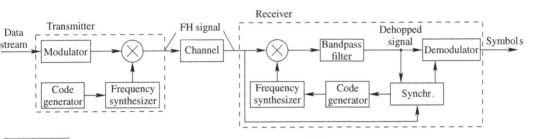

Figure 8.16   **The transceiver of the FHSS system.**

At the receiver, the mixing operation removes the frequency-hopping pattern from the received signal, and the dehopped signal is obtained. The architecture of an FHSS system is depicted in Fig. 8.16.

FH can be either slow FH or fast FH. The time spent in one frequency is called *dwell time* or *hop interval*, T_h. Fast FH changes the carrier frequency several times during one symbol period, the chip period $T_c = T_s/k$ for some integer k. Thus, each symbol is spread over a large bandwidth, and there is frequency diversity on each symbol that combats fading or interference. Fast FH is not widely used in commercial wireless systems, and its market share is supplanted by the DS-CDMA technology.

Slow FH transmits one or multiple symbols over each frequency, $T_c = kT_s$ for some integer k. Slow FH is usually used with TDMA so that each time slot is transmitted on a carrier frequency according to a hopping scheme. Slow FH is used in GSM, DECT, HomeRF, Bluetooth, and the baseline IEEE 802.11. In Bluetooth, the full-duplex signal implements FH at up to 1,600 hops per second amongst 79 frequencies at 1-MHz intervals. FH is more commonly used in military systems to avoid jammers. 3GPP LTE achieves frequency diversity also by FH.

For FHSS, MFSK is the commonly used modulation method. FH/MFSK is often detected using a noncoherent technique, due to the difficulty in rapid carrier synchronization as the carrier frequency is hopped. Beacon frames that contain the pattern and time can be used for synchronization between the receiver and the transmitter. In the FH scheme of the baseline 802.11, two-level GFSK and four-level GFSK modulation are used for the data rates of 1 Mbit/s and 2 Mbits/s, respectively.

8.8.1 Error performance of FHSS

Slow frequency hopping

For FHSS systems using BFSK modulation, if no two users utilize the same frequency band at the same time, the error probability for BFSK is given by

$$P_0 = \frac{1}{2} e^{-\frac{\gamma_b}{2}}. \tag{8.61}$$

In the case of two users transmitting on the same frequency band, a collision occurs. The error probability of the collided band should be 0.5, and thus the overall error probability is given by

$$P_e = P_0 (1 - P_c) + \frac{1}{2} P_c, \tag{8.62}$$

where P_c is the probability of collision.

Given M frequency slots and K users, when all the users are assumed to hop synchronously, the probability of collision is

$$P_c = 1 - \left(1 - \frac{1}{M}\right)^{K-1} \approx \frac{K-1}{M}, \tag{8.63}$$

where the approximation is made for large M. Inserting P_c into (8.62), we have

$$P_e = \frac{1}{2} e^{-\frac{\gamma_b}{2}} \left(1 - \frac{K-1}{M}\right) + \frac{K-1}{2M}. \tag{8.64}$$

For $\gamma_b \to \infty$, $P_e = \frac{K-1}{2M}$, which demonstrates an irreducible error rate due to MAI.

In the case of asynchronous hopping, the probability of collision is given by [52]

$$P_c = 1 - \left[1 - \frac{1}{M}\left(1 + \frac{1}{N_b}\right)\right]^{K-1}, \tag{8.65}$$

where N_b is the number of bits per hop. Obviously, P_c in the asynchronous case is higher than that in the synchronous case. The error probability P_e can be correspondingly obtained by (8.62).

The error probability of slow FH/MFSK on the AWGN channel with non-coherent square-law detection is given by (7.96).

Fast frequency hopping

For BFSK modulation, if fast FH is used, that is, each bit corresponds to N_h hops, the error rate probability for an FH system is approximated by [8]

$$P_e = \sum_{n=r}^{N_h} \binom{N_h}{r} P_{e0}^n (1 - P_{e0})^{N_h - n}, \tag{8.66}$$

where P_{e0} is the error probability of a single jamming trial, which is J/N, J being the number of jammers and N the number of channels, and r is the number of wrong chip decisions necessary to cause a bit error. P_{e0} can be calculated from (8.62).

A fast FH system offers a performance much better than that of a slow FH system, but at the cost of complexity. For a three-hops-per-bit FH system, if at least $r = 2$ frequencies are correct, the decision is a correct bit per symbol, thus $P_e \approx 3P_{e0}^2 (1 - P_{e0}) \approx 3P_{e0}^2$, which is much better than that in the case of one hop per bit, namely, $P_e = P_{e0}$.

8.8.2 FHSS versus DSSS

Both DSSS and FHSS have no impact on the performance in the AWGN channel, but they can improve the performance in the Rayleigh fading channel and in the case of narrowband interference. Between FHSS and DSSS, the nature of interference reduction differs: DSSS results in a reduced-power interference on the entire band all the time, while FHSS has a full power interferer on a narrowband channel over a chip period. Thus, coding with interleaving is needed for FHSS, and a collision in one or two frequencies can be recovered by coding.

Implementation of FHSS is inexpensive. FHSS is superior to DSSS in terms of immunity to interference. In case of a strong interference in a segment of the band, the FHSS transmitter will use the remaining band effectively with a decrease in the throughput. FH leads to a whitening of the received signal and an averaging over all the frequencies. When different users use different FH sequences, the FH technique can also be used as a multiple-access technique. Some spectral regions with steady interference can be omitted from the hopset, and this is a process called *spectral notching*. In contrast, a single interferer with sufficient power can stop communication of a whole DSSS system. Interference from a narrowband jammer in DSSS systems can be alleviated by adaptive filtering [51]. FHSS has the technical constraints of producing fast frequency synthesizers, especially in the case of fast hopping.

Implementation of DSSS requires synchronization and tracking of the phase of the received signal, which is typically difficult in the presence of fast fading. Due to their complexity, DSSS systems consume more power than FSSS systems. Transmission time in DSSS is shorter than that in FHSS, since no frequency changing is necessary. DSSS achieves a higher processing gain and is more widely used in high rate systems.

Multiuser FHSS, also referred to as *FH-CDMA* or *FH-SSMA*, is implemented by assigning each user a unique spreading code sequence for hop pattern generation. FH-CDMA is typically used in uplinks, and for military purposes. FH-CDMA typically supports a smaller number of users than DS-CDMA. The near-far effect is also an issue for FSK-based FHSS systems: A near transmitter has sidelobes that may extend to the band of the desired signal. Choosing CPFSK can generate an inherently narrow signal spectrum. Thus, the advantage of FH-CDMA over DS-CDMA is its robustness to the near-far effect: hopping can mitigate performance degradation caused by strong interferers [20]. FHSS and FH-CDMA are used in some low-rate systems. For example, FH-CDMA is the core multiple-access technology in Bluetooth, and FHSS is employed in GSM.

Problems

8.1 An FHSS system using 8FSK employs 500 different frequencies. Determine the processing gain.

8.2 In IS-95, assume $K = 30$ users share a channel of 1.25 Mhz. Each user has a chip rate of 1.2288 Mchips/s and a baseband data rate of 13 kbits/s, and the PN code lengths are 32,678 chips. If each user is provided with a maximum E_b/N_0 of 7.8 dB, determine the bit error probability for a user, and the processing gain of IS-95.

8.3 A CDMA system has a bandwidth of $W = 1.6$ Mhz, supports a data rate of $R = 12.5$ kbits/s, and the minimum acceptable $E_b/N_b = 8.9$ dB. Determine the maximum number of users that can be supported by a single cell:
(a) with omnidirectional BS antennas and no voice activity detection.
(b) with three-sectors at the BS and voice activity is detected as $\alpha = 3/8$.

8.4 Given a code length (processing gain) of 255 and the desired bit error probability of 10^{-4} for a conventional receiver, how many equal-power users can be supported at $E_b/N_0 = 20$ dB?

8.5 Plot the power spectrum of an m-sequence with chip duration T_c and period NT_c.

8.6 A preferred pair of m-sequences of length 31 is given as

$$b_1 = 1010111011000111110011010010000$$
$$b_2 = 1011010100011101111100100110000$$

Verify that the pair satisfies the three-valued cross-correlation for preferred pairs of m-sequences. Demonstrate that a Gold code of length 31 achieves the permitted values of cross correlations. Verify that it is true for $b_1 + Db_2$ and $b_1 + D^3b_2$, where D is the delay operator.

8.7 Consider the m-sequence generated using the primitive polynomial $g(D) = 1 + D^2 + D^5$. Show that the properties for m-sequences are satisfied.

8.8 Plot the ACF and the power spectrum for the m-sequence specified by $g(D) = 1 + D^2 + D^3 + D^4 + D^5$. The shift-register clock rate is 10 kHz.

8.9 A rate-1/3 convolutional code with $d_{\text{free}} = 8$ is used to encode a data sequence at a rate of 1000 bits/s. BPSK modulation is employed. The DSSS has a chip rate of 10 Mchips/s. (a) Determine the coding gain. (b) Determine the processing gain. (c) Determine the jamming margin for $E_b/J_0 = 10$ dB, where J_0 is the spectral density of the combined interference.

8.10 A DSSS BPSK signal has a processing gain of 100. Determine the jamming margin against a continuous-tone jammer for an error probability of 10^{-4}.

8.11 Write a MATLAB program for constructing Gold code sequences for length $n = 2^m - 1$, $m = 3$ to 9. Verify their cross-correlation peak values by searching all the Gold codes. For example, the parity polynomials for a pair of preferred sequence for constructing Gold code of length $n = 7$ are given as $h_1(D) = D^3 + D + 1$ and $h_2(D) = D^3 + D^2 + 1$. Find from the literature all the pairs of preferred sequence for $m = 3$ to 9.

8.12 A DS/BPSK system employs an LFSR of length 15 to generate PN sequences. What is the processing gain? If the required bit error probability is 10^5, what is the maximum interference?

8.13 In the IS-95 system, the conventional single-user CDMA receiver is used. For a bandwidth of 1.25 MHz and a user data rate of 9.6 kbits/s, what is the maximum number of voice users that can be accommodated?

8.14 An IS-95 mobile transmits with a power of 24 dBm. If the BS wishes the MS to transmit with a power of 18 dBm, how long does it take for the change? [Hint: Power control frequency is 800 Hz and the step is 1 dB.]

8.15 For the DS-CDMA system in the AWGN multiple-access channel, the spectral efficiencies for the different receivers are given in [60]. Write a MATLAB program to reproduce Fig. 8.13.

8.16 A slow FH BPSK system with noncoherent detection operates at $E_b/J_0 = 15$, the hopping bandwidth is 2 GHz, and the bit rate is 100 bits/s. What is the processing gain of the system?

8.17 For a slow FH, 16FSK system, if each hop lasts over 6 symbols, what is the processing gain?

8.18 For a fast FH, 16FSK system, if each 16FSK symbol experiences 5 hops, what is the processing gain?

8.19 For an FH, MFSK wireless LAN system, an m-sequence is generated by a 20 stage LFSR. Each state of the register corresponds to a hopping frequency. The minimum hopping step is 200 Hz. The register clock rate is 2 kHz. 4FSK modulation is applied and the data rate is 2.4 kbits/s. Determine the hopping bandwidth, the chip rate, the processing gain, and chips per symbol.

References

[1] M. A. Abu-Rgheff, *Introduction to CDMA Wireless Communications* (Oxford, UK: Academic Press, 2007).

[2] A. I. Amayreh & A. K. Farraj, Minimum autocorrelation spreading codes. *Wireless Pers. Commun.*, **40**:1 (2007), 107–115.

[3] K. L. Baum, T. A. Thomas, F. W. Vook & V. Nangia, Cyclic-prefix CDMA: an improved transmission method for broadband DS-CDMA cellular systems. In *Proc. IEEE WCNC*, Orlando, FL, Mar 2002, **1**, 183–188.

[4] J. J. Blanz, A. Papathanassiou, M. Haardt, I. Furio & P. W. Baier, Smart antennas for combined DOA and joint channel estimation in time-slotted CDMA mobile radio systems with joint detection. *IEEE Trans. Veh. Tech.*, **49**:2 (2000), 293–306.

[5] P. Burns, *Software Defined Radio for 3G* (Boston: Artech House, 2003).

[6] D. C.-H. Chen, B. J. Sheu & W. C. Young, A CDMA communication detector with robust near-far resistance using paralleled array processors. *IEEE Trans. Circ. Syst. Video Tech.*, **7**:4 (1997), 654–662.

[7] H.-H. Chen, J. F. Yeh & N. Seuhiro, A multi-carrier CDMA architecture based on orthogonal complementary codes for new generations of wideband wireless communications. *IEEE Commun. Mag.*, **39**:10 (2001), 126–135.

[8] H.-H. Chen & M. Guizani, *Next Generation Wireless Systems and Networks* (Chichester, UK: Wiley, 2006).

[9] CWTS-SWG2 LAS-CDMA, *Physical layer aspects of TD-LAS high speed packet technology*, LAS TR 25.951, v1.0.0, Jul 2001.

[10] K.-L. Du & M. N. S. Swamy, An adaptive space-time multiuser detection algorithm for CDMA systems. In *Proc. IEEE WiCOM*, Wuhan, China, Sep 2006, 1–5.

[11] K.-L. Du & W. H. Mow, Search for long low-autocorrelation binary sequences by evolutionary algorithms. Submitted to *IEEE Trans. Aerospace Electron. Syst.*

[12] C. Ergun & K. Hacioglu, Multiuser detection using a genetic algorithm in CDMA communications systems. *IEEE Trans. Commun.*, **48**:8 (2000), 1374–1383.

[13] G. J. Foschini & M. J. Gans, Layered space-time architecture for wireless communication in a fading environment when using multiple antennas. *Bell Labs Tech. J.*, **1**:2 (1996), 41–59.

[14] G. J. Foschini & M. J. Gans, On limits of wireless communications in a fading environment when using multiple antennas. *Wireless Pers. Commun.*, **6**:3 (1998), 311–335.

[15] G. D. Forney, Jr., Coding and its application in space communications. *IEEE Spectrum*, **7**:6 (1970), 47–58.

[16] M. Friese, Polyphase Barker sequences up to length 36. *IEEE Trans. Inf. Theory*, **42**:4 (1996), 1248–1250.

[17] R. L. Frank, Polyphase complementary codes. *IEEE Trans. Inf. Theory*, **26**:6 (1980), 641–647.

[18] M. J. E. Golay, Complementary series. *IRE Trans. Inf. Theory*, **7**:2 (1961), 82–87.

[19] R. Gold, Optimum binary sequences for spread-spectrum multiplexing. *IEEE Trans. Inf. Theory*, **13**:4 (1967), 619–621.

[20] A. Goldsmith, *Wireless Communications* (Cambridge, UK: Cambridge University Press, 2005).

[21] S. W. Golomb and R. A. Scholtz, Generalized Barker sequences. *IEEE Trans. Inf. Theory*, **ll**:4 (1965), 533–537.

[22] S. W. Golomb, *Shift Register Sequences* (San Francisco, CA: Holden-Day, 1967).

[23] G. Golub & C. van Loan, *Matrix Computations*, 3rd edn (Baltimore, MD: John Hopkins University Press, 1996).

[24] L. Hanzo, M. Munster, B. J. Choi & T. Keller, *OFDM and MC-CDMA for Broadband Multi-User Communications, WLANs and Broadcasting* (New York: Wiley-IEEE, 2003).

[25] F. Hasegawa, J. Luo, K. R. Pattipati, P. Willett & D. Pham, Speed and accuracy comparison of techniques for multiuser detection in synchronous CDMA. *IEEE Trans. Commun.*, **52**:4 (2004), 540–545.

[26] K. M. Ho & W. H. Mow, Searching for the best biphase and quadriphase quasi-Barker sequences. In *Proc. IEEE ICCCAS*, Chengdu, China, Jun 2004, **1**, 43–47.

[27] P. A. Hoeher & H. Schoeneich, Interleave-division multiple access from a multiuser theory point of view. In *Proc. Int. Symp. Turbo Codes Appl.*, Munich, Germany, Apr 2006.

[28] J. M. Holtzman, A simple, accurate method to calculate spread-spectrum multiple-access error probabilities. *IEEE Trans. Commun.*, **40**:3 (1992), 461–464.

[29] T. Kasami, Weight distribution of Bose-Chaudhuri-Hocquenghem codes. In R. C. Bose and T. A. Dowling, eds., *Combinatorial Mathematics and its Applications* (Chapel Hill, NC: University of North Carolina Press, 1967), pp. 335–357.

[30] G. I. Kechriotis & E. S. Manolako, Hopfield neural network implementation of the optimal CDMA multiuser detector. *IEEE Trans. Neural Netw.*, **7**:1 (1996), 131–141.

[31] R. Kohno, Interference cancellation and multiuser detection. In M. Shafi, S. Ogose & T. Hattori, eds., *Wireless Communications in the 21st Century* (New York: IEEE Press, 2002).

[32] V. Kuhn, *Wireless Communications over MIMO Channels: Applications to CDMA and Multiple Antenna Systems* (Chichester, UK: Wiley, 2006).

[33] K.-C. Lai, R. Chandrasekaran, R. E. Cagley & J. J. Shynk, Multistage interference cancellation algorithms for DS/CDMA signals. In G. B. Giannakis, Y. Hua, P. Stoica, L. Tong, eds., *Signal Processing in Wireless & Mobile Communications: Trends in Single- and Multi-user Systems*, **2** (Upper Saddle River, NJ: Prentice Hall, 2001), pp. 267–314.

[34] D. B. Li, The perspectives of large area synchronous CDMA technology for the fourth-generation mobile radio. *IEEE Commun. Mag.*, **41**:3 (2003), 114ÍC-118.

[35] K. Li, X. Wang, G. Yue & L. Ping, A low-rate code-spread and chip-interleaved time-hopping UWB system. *IEEE J. Sel. Areas Commun.*, **24**:4 (2006), 864–870.

[36] J. Liberti & T. Rappaport, *Smart Antenna for Wireless Communications* (Englewood Cliffs, NJ: Prentice Hall, 1999).

[37] R. Lupas & S. Verdu, Linear multi-user detectors for synchronous code-division multiple-access channels. *IEEE Trans. Inf. Theory*, **35**:1 (1989), 123–136.

[38] U. Madhow & M. L. Honig, MMSE interference suppression for direct sequence spread-spectrum CDMA. *IEEE Trans. Commun.*, **42**:12 (1994), 3178–3188.

[39] R. Mahadevappa & J. G. Proakis, Mitigating multiple access interference and inter-symbol interference in uncoded CDMA systems with chip-level interleaving. *IEEE Trans. Wireless Commun.*, **1**:4 (2002), 781–792.

[40] S. Y. Miller & S. C. Schwartz, Integrated spatial-temporal detectors for asynchronous Gaussian multiple-access channels. *IEEE Trans. Commun.*, **43**:2–4 (1995), 396–411.

[41] R. K. Morrow, Jr. & J. S. Lehnert, Bit-to-bit error dependence in slotted DS/SSMA packet systems with random signature sequences. *IEEE Trans. Commun.*, **37**:10 (1989), 1052–1061.

[42] W. H. Mow, Best quadriphase codes up to length 24. *Electron. Lett.*, **29**:10 (1993), 923–925.

[43] W. H. Mow & S.-Y. R. Li, Aperiodic autocorrelation and crosscorrelation of polyphase sequences. *IEEE Trans. Inf. Theory*, **43**:3 (1997), 1000–1007.

[44] M. Nagatsuka & R. Kohno, A spatially and temporally optimal multiuser receiver using an antenna array for DS/CDMA. *IEICE Trans. Commun.*, **E78-B**:11 (1995), 1489–1497.

[45] C. B. Papadias & H. Huang, Linear space-time multiuser detection for multipath CDMA channels. *IEEE J. Sel. Areas Commun.*, **19**:2 (2001), 254–265.

[46] X. Peng, A. S. Madhukumar, F. Chin & T. T. Tjhung, Block spreading CDMA system: a simplified scheme using despreading before equalisation for broadband uplink transmission. *IET commun.*, **3**:4 (2009), 666–676.

[47] L. Ping, L. Liu, K. Wu & W. K. Leung, Interleave division multiple access. *IEEE Trans. Wireless Commun.*, **5**:4 (2006), 938–947.

[48] R. Price & P. E. Green, Jr, A communication technique for multipath channels. *Proc. IRE*, **46**:3 (1958), 555–570.

[49] J. G. Proakis & M. Salehi, *Digital Communications*, 5th edn (New York: McGraw-Hill, 2008).

[50] M. B. Pursley, Performance evaluation for phase-coded spread-spectrum multiple-access communication – part I: system analysis. *IEEE Trans. Commun.*, **25**:8 (1977), 795–759.

[51] K. D. Rao, M. N. S. Swamy & E. I. Plotkin, A nonlinear adaptive filter for narrowband interference mitigation in spread spectrum systems. *Signal Process.*, **85**:3 (2005), 625–635.

[52] T. S. Rappaport, *Wireless Communications: Principles & Practice*, 2nd edn (Upper Saddle River, NJ: Prentice Hall, 2002).

[53] S. V. Sarwate & M. B. Pursley, Crosscorrelation properties of pseudrandom and related sequences. *Proc. IEEE*, **68**:5 (1980), 593–613.

[54] K.-U. Schmidt, Quaternary constant-amplitude codes for multicode CDMA. *IEEE Trans. Inf. Theory*, **55**:4 (2009), 1824–1832.

[55] H. Schoeneich & P. A. Hoeher, Adaptive interleave-division multiple access – a potential air interference for 4G bearer services and wireless LANs. In *Proc. IEEE/IFIP WOCN*, Muscat, Oman, Jun 2004, 179–182.

[56] G. L. Stuber, *Principles of Mobile Communication*, 2nd edn (Boston, MA: Kluwer, 2001).

[57] N. Suehiro & M. Hatori, N-Shift cross-orthogonal sequence. *IEEE Trans. Inf. Theory*, **34**:1 (1988), 143–46.

[58] X. Tang & W. H. Mow, Design of spreading codes for quasi-synchronous CDMA with intercell interference. *IEEE J. Sel. Areas Commun.*, **24**:1 (2006), 84–93.

[59] S. Verdu, *Multiuser Detection* (Cambridge, UK: Cambridge University Press, 1998).

[60] S. Verdu & S. Shamai, Spectral efficiency of CDMA with random spreading. *IEEE Trans. Inf. Theory*, **45**:2 (1999), 622–640.

[61] A. J. Viterbi, Very low rate convolutional codes for maximum theoretical performance of spread-spectrum multiple-access channels. *IEEE J. Sel. Areas Commun.*, **8**:4 (1990), 641–649.

[62] A. J. Viterbi, A. M. Viterbi, K. Gilhousen & E. Zehavi, Soft handoff extends CDMA cell coverage and increases reverse channel capacity. *IEEE J. Sel. Areas Commun.*, **12**:8 (1994), 1281–1288.

[63] A. J. Viterbi1995, *Principles of Spread Spectrum Communication* (Reading, MA: Addison-Wesley, 1995).

[64] X. Wang & H. V. Poor, Space-time multiuser detection in multipath CDMA channels. *IEEE Trans. Signal Process.*, **47**:9 (1999), 2356–2374.

[65] N. Yee, J.-P. Linnartz & G. Fettweis, Multicarrier CDMA in indoor wireless radio networks. In *Proc. IEEE PIMRC*, Yokohama, Japan, Sep 1993, 109–113.

[66] K. Yen & L. Hanzo, Genetic-algorithm-assisted multiuser detection in asynchronous CDMA communications. *IEEE Trans. Veh. Tech.*, **53**:5 (2004), 1413–1422.

[67] R. Zhang & L. Hanzo, Three Design Aspects of Multicarrier Interleave Division Multiple Access. *IEEE Trans. Veh. Tech.*, **57**:6 (2008), 3607–3617.

Orthogonal frequency division multiplexing

9.1 Introduction

OFDM, also known as *simultaneous MFSK*, has been widely implemented in high-speed digital communications in delay-dispersive environments. It is a multicarrier modulation (MCM) technique. OFDM was first proposed by Chang in 1966 [11]. Chang proposed the principle of transmitting messages simultaneously over multiple carriers in a linear band-limited channel without ISI and ICI. The initial version of OFDM employed a large number of oscillators and coherent demodulators. In 1971, DFT was applied to the modulation and demodulation process by Weinstein and Ebert [81]. In 1980, Peled and Ruiz introduced the notion of cyclic prefix to maintain frequency orthogonality over the dispersive channel [60]. The first commercial OFDM-based wireless system is the ETSI DAB standard proposed in 1995.

A wide variety of wired and wireless communication standards are based on the OFDM or MCM technology. Examples are

- *digital broadcasting systems* such as DAB, DVB-T (Terrestrial DVB), and DVB-H;
- *home networking* such as digital subscriber line (xDSL) technologies;
- *wireless LAN standards* such as HyperLAN/2 and IEEE 802.11a/g/n;
- *wireless MANs* such as IEEE 802.16a/e (WiMAX), ETSI HiperACESS, IEEE 802.20 (mobile-Fi), WiBro, and HiperMAN2;
- *wireless WANs* such as 3GPP LTE and 3GPP2 UMB;
- *powerline communications* such as HomePlug;
- *wireless PANs* such as UWB radios (IEEE 802.15.3a/3c/4a).

It is commonly deemed that OFDM is a major modulation technique for beyond-3G wireless multimedia communications.

Features of OFDM technology

In OFDM technology, the multiple carriers are called *subcarriers*, and the frequency band occupied by the signal carried by a subcarrier is called a *sub-band*. OFDM achieves orthogonality in both the time and frequency domains.

The most attractive feature of OFDM is its robustness against multipath fading, but with a low complexity; that is, it can reliably transmit data over time-dispersive or frequency-selective channels without the need of a complex time-domain channel equalizer. OFDM can be efficiently implemented using FFT.

Other advantages of OFDM are listed as follows.

- OFDM achieves robustness against narrowband interference, since narrowband interference only affects a small fraction of the subcarriers.
- Frequency diversity can be exploited by coding and interleaving across subcarriers in the frequency domain.
- Different modulation formats and data rates can be implemented on different subcarriers depending on the channel quality of individual sub-bands.
- OFDM enables bandwidth-on-demand technology and higher spectral efficiency.
- Contiguous bandwidth for operation is not required.

However, the high PAPR of the OFDM signal places a strict requirement on power amplifiers. For this reason, 3GPP LTE employs SC-FDMA for transmission to reduce the strict requirement on power amplifiers in MSs. Also, multicarrier systems are inherently more susceptible to frequency offset and phase noise; thus, frequency jitter and Doppler shift between the transmitter and receiver cause inter-carrier interference (ICI), which becomes a challenge in case of medium or high mobility.

9.2 Principle of OFDM

For high-speed communications, the multipath characteristics of the environment cause the channel to be frequency-selective. In the OFDM system, a frequency-selective channel is equally divided into N frequency-flat subchannels. Specifically, the N subcarriers are assumed to be at the frequencies $f_n = nW/N$, $n = 0, 1, \ldots, N - 1$, where W is the total bandwidth that is available, and usually $W = N/T$.

These subcarriers are orthogonal: Their inter-carrier spacing is equivalent to the inverse of the symbol duration, $1/T$, and the spectral peak of each subcarrier must coincide with the zero-crossings of all the other subcarriers. For PAM modulation with rectangular pulse shape, mutual orthogonality of the subcarriers is easily seen from the relation

$$\int_{iT}^{(i+1)T} e^{j2\pi f_k t} e^{-j2\pi f_n t} dt = \delta_{nk}. \tag{9.1}$$

Example 9.1: Given an OFDM signal using 7 subcarriers, the spectrum of each modulated subcarrier is a sinc function due to the rectangular pulse shape in the time domain. The frequency spectrum of the subcarriers is shown in Fig. 9.1a. The power spectrum of the OFDM signal is shown in Fig. 9.1b. The first spectrum sidelobes are only about 17.3 dB below the passband level. By using other windows for pulse shaping, the sidelobes can be suppressed, but this introduces ACI.

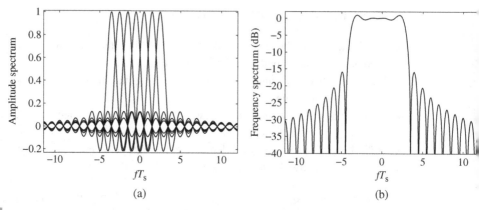

Spectrum of OFDM. (a) Amplitude spectrum of each carrier. (b) Power spectrum of the OFDM signal.

OFDM leads to an equation of the same form as that of IFFT operation. Assuming that the complex symbol to be transmitted on the ith subcarrier at instant k is $d_{i,k}$, the transmitted signal is given by

$$s(t) = \sum_{k=-\infty}^{\infty} s_k(t) = \sum_{k=-\infty}^{\infty} \sum_{i=0}^{N-1} d_{i,k} g_i(t - kT), \qquad (9.2)$$

where $g_i(t)$ is a normailzed, frequency-shifted rectangular pulse

$$g_i(t) = \frac{1}{\sqrt{T}} e^{j2\pi f_i t}, \quad 0 \le t \le T, \qquad (9.3)$$

and 0 otherwise. The complex symbol $d_{i,k}$ can be obtained by ASK, PSK, or QAM.

Without loss of generality, for symbol $k = 0$, if the symbol is sampled at $t_u = uT/N$, $u = 0, 1, \ldots, N - 1$, we have

$$s_u = \frac{1}{\sqrt{T}} \sum_{i=0}^{N-1} d_{i,0} e^{j2\pi iu/N}, \quad u = 0, \ldots, N - 1. \qquad (9.4)$$

The is exactly the IFFT of the transmitted symbols. Thus, these modulations can be performed by using IFFT.

At the receiver, demodulation can be implemented by multiplying the received signal on each carrier f_i, that is, $d_{i,k} g_i(t)$, by $e^{-j2\pi f_i t}$ and integrating over a symbol duration. FFT can be used for this purpose. Spectral overlapping among subcarriers can be separated at the receiver due to the orthogonality property of the subcarriers. This makes the implementation much simpler than the traditional implementation that uses multiple oscillators.

For the baseband OFDM signal, the zero-to-null bandwidth is N/T, thus the sampling frequency should be at least $2N/T$ to avoid aliasing. When implemented using IFFT, only N samples are available for a symbol period T. Thus, N zeros can be padded to increase the number of samples to $2N$, and $2N$-point IFFT can then be used to generate the OFDM signal. Those subcarriers with zero data are known as *dummy or virtual subcarriers*. This oversampling strategy can be implemented by padding L (even) zeros, half of them before and half after the data sequence. For example, in IEEE 802.11a, the signal samples are generated by 64-point IFFT, among which 12 points correspond to virtual subcarriers. The virtual subcarriers create separation between the repetitions of the PSD. At the receiver, N-point FFT is used to demodulate the signal.

9.3 OFDM transceivers

OFDM splits the bitstream into multiple parallel bitstreams of reduced bit rate, modulates them using an M-ary modulation, and then transmits each of them on a separate subcarrier. The amplitude spectrum of each modulated subcarrier using PSK or QAM has a sinc shape. At the peak spectrum response of each subcarrier, the spectral responses of all other subcarriers are identically zero. This is shown in Fig. 9.1a. This improves the spectral efficiency. The use of narrowband flat-fading subchannels can effectively resist frequency selective fading.

Each modulated subcarrier in the OFDM signal can be an MPSK, MASK, or MQAM signal. Thus, the OFDM signal can be obtained first by 1:N serial-to-parallel (S/P) conversion and then by converting each of the N bitstreams to one of the N subcarriers f_i. Their sum is then upconverted to the RF band. For each subcarrier, there is a complete modulator, which typically consists of a subcarrier oscillator, one (for MASK) or two (for MQAM and MPSK) multipliers, and an adder. A phase shifter is also required for MQAM and MPSK. For passband modulation, a mixer and a bandpass filter are required for upconversion.

The receiver consists of a downconverter to translate the signal to the baseband. The baseband OFDM signal is then demodulated by a bank of N demodulators at N subcarrier frequencies, one for each of the subcarriers. The subcarrier demodulator can be a standard MPSK, MASK, or MQAM demodulator, which consists of oscillators, multipliers, integrators, and threshold detectors.

The basic implementation of the OFDM transceiver is rarely used in practical systems, since for large N the implementation is too complex and thus is impractical. Almost all modern OFDM systems use the DFT-based digital implementation. The block diagram of the OFDM system is shown in Fig. 9.2.

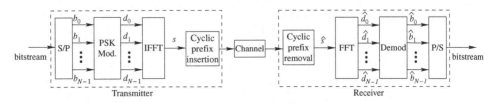

Figure 9.2   **The OFDM system.**

The input high-speed bitstream is first converted into N parallel low-speed bitstreams. N is selected as a power of two, since it is used as the FFT block size. Each low-speed bitstream is then converted into symbols, and M-ary modulated to its subchannel with the modulated data sequence $d_n = d_{I_n} + jd_{Q_n}$, $n = 0, 1, \ldots, N - 1$. All the N modulated subchannel data are fed to an IFFT algorithm or circuit to generate an OFDM signal. MPSK modulation is usually used, since it produces a constant amplitude signal, thus reducing problems with amplitude fluctuations due to fading. IFFT is used to find the time waveform for the spectrum. A cyclic prefix is then appended to the OFDM signal, and serves as guard time.

OFDM demodulation reverses the process of OFDM modulation. At the receiver, the received signal is filtered using a wide bandpass filter. An orthogonal detector is applied to downconvert the signal to the IF band. The cyclic prefix is first removed. An FFT circuit is then used to obtain the Fourier coefficients of the signal $\hat{d}_n$, $n = 0, 1, \ldots, N - 1$, in the observation period $[kT_s, kT_s + T]$, where T is the symbol duration, and $T_s = T_g + T$ is the extended OFDM symbol duration, T_g being the guard interval.

By comparing $b_i(k)$ and $\hat{b}_i(k)$, $i = 0, 1, \ldots, N - 1$, the BER performance can be calculated. OFDM transmission preserves the orthogonality among the subchannels, and the BER performance is thus decided by the modulation scheme in each subchannel.

9.4 Cyclic prefix

The frequency-selective channel is divided into frequency-flat fading channels, and this substantially eliminates ISI. However, delay dispersion can lead to appreciable errors even when $\sigma_\tau/T < 1$, and also leads to a loss of orthogonality between the subcarriers and thus to ICI. The cyclic prefix is a kind of guard interval that effectively eliminates the effects of delay dispersion. The OFDM data frame is cyclically extended with the last L symbols of the frame, which corresponds to the maximum channel delay τ_{max}. This can effectively eliminate ISI and interframe interference. An empty guard interval can avoid ISI, but not ICI. When the cyclic prefix is used in the guard interval, ICI can also be avoided.

Due to the orthogonality of the subchannels, individual subchannels can be separated by using an FFT circuit at the receiver if there is neither ISI nor ICI. Channel equalization is easily accomplished by using a complex scale factor for each subchannel. Multipath fading may cause each subchannel to spread the power into the adjacent channels. Distortion arising from ISI and ICI can be reduced by increasing the number of subcarriers. This, however, leads to carrier instability against Doppler and a large FFT size. OFDM inserts a guard time, usually selected as two to four times the expected delay spread, to eliminate ISI, but the guard time reduces the data throughput. To reduce ICI, OFDM symbols are cyclically extended into the guard interval to ensure that an OFDM symbol has an integer number of cycles in the DFT intervals.

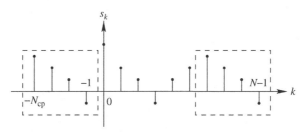

Figure 9.3   **Cyclic prefix is the repetition of the last part of the symbol to be transmitted.**

Assume that a cyclic prefix is prepended to the normal OFDM symbol, $\tau_{\text{max}} \leq T_{\text{cp}}$. The cyclic prefix spans the duration from $-T_{\text{cp}} < t < 0$, and the normal OFDM symbol from $0 < t < T$. The cyclic prefix is a copy of the last part of the transmitted symbol. For a carrier spacing W/N, the symbol period $T = N/W$. The number of samples of the symbol is N, and the number of the samples in the cyclic prefix is $N_{\text{cp}} = NT_{\text{cp}}/T$. This is shown in Fig. 9.3. Cyclic prefix can recover the orthogonality of the subcarriers. Usually, cyclic prefix occupies 10% of symbol duration. At the receiver side, for each received block, the cyclic prefix is first removed, and the remainder is fed to FFT.

With the cyclic prefix, the transmitted OFDM signal can be represented by

$$s(k) = \begin{cases} \text{ifft}\,(d_0, d_1, \ldots, d_{N-1}), & k = 0, 1, \ldots, N-1 \\ s(N+k), & k = -N_{\text{cp}}, \ldots, -1 \end{cases}, \tag{9.5}$$

$$x(t) = \sum_{k=-N_{\text{cp}}}^{N-1} s(k)w\left(t - \frac{k}{N}T\right), \quad -\frac{N_{\text{cp}}}{N}T \leq t \leq T, \tag{9.6}$$

where $w(t)$ is the time-domain window function, such as a rectangular window or the raised cosine function.

Example 9.2: Given a randomly generated bitstream, we use OFDM modulation with $N = 128$ and $N_{\text{cp}} = 13$, and the modulation on each subcarrier is BPSK with amplitude $A = 1$. The modulated OFDM signal is plotted in Fig. 9.4, for a period of $T_s = T + T_{\text{cp}}$, where the rectangular window is used.

The use of cyclic prefix, although elegant and simple, reduces the spectral efficiency and introduces a power penalty. Both the rate loss and the power loss are a fraction $\frac{N_{\text{cp}}}{N+N_{\text{cp}}}$. The length of the cyclic prefix N_{cp} is determined by the channel length τ_{max}. The efficiency can be improved by increasing the number of subcarriers N. This, however, increases the total symbol duration $N + N_g$, which may be critical for time varying channels, since the channel must remain constant during one symbol period, otherwise the orthogonality

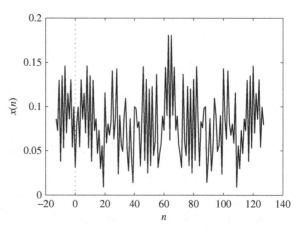

Modulated OFDM signal $x(n)$ for a BPSK bitstream, $N = 128$ and $N_{cp} = 13$.

between subcarriers will be lost. The wasted power in the cyclic prefix causes interference to neighboring users.

Zero-padded OFDM

Most OFDM systems are based on the traditional cyclic-prefix OFDM (CP-OFDM) modulation scheme. An alternative is zero-padded OFDM, which replaces this time-domain redundancy by null samples [55]. This solution relies on a larger FFT demodulator, and has the merit of guaranteeing symbol recovery irrespective of the channel null locations for coherent modulation. It enables semiblind pilot-based channel estimation with improved tracking capability of channel variations. The price paid is somewhat increased receiver complexity. In terms of power amplifier-induced clipping effects, zero-padded OFDM introduces slightly more nonlinear distortions, and therefore, needs slightly increased power backoff than CP-OFDM.

In the MB-OFDM technology, which was standardized by the WiMedia Alliance for UWB operation, the cyclic prefix is just a zero prefix of null data to avoid the transmitter power penalty. At the receiver, the tail of the transmitted symbols is duplicated to the position of the cyclic prefix to recreate the same effect as a cyclic prefix. Although the zero-prefix scheme avoids the power penalty, additional noise from the received tail symbols is added into the signal, raising the noise power to $\frac{N+N_{cp}}{N}\sigma_n^2$, whereas in the cyclic prefix scheme, the tail is ignored. For this reason, most OFDM-based standards employ the cyclic prefix scheme.

By capitalizing on the advantages of zero-padded OFDM, the pseudorandom-postfix OFDM replaces the null samples inserted between each OFDM modulated block by a known vector weighted by a pseudorandom scalar sequence to replace the guard interval contents of CP-OFDM [52]. Unlike cyclic-prefix and zero-padded OFDM modulators, the receiver can exploit pseudorandomly weighted postfix sequences for channel estimation, and thus, the pilot overhead is avoided. The method has the ability to estimate and track the channel variations semi-blindly using the first-order statistics of the received signal.

Case study: IEEE 802.11a and WiMAX

IEEE 802.11a uses a 20-MHz channel spacing in the 5-GHz unlicensed band. IEEE 802.11g is virtually identical to IEEE 802.11a, but operates on the unlicensed 2.4-GHz ISM band. IEEE 802.11a uses OFDM with 64 subcarriers for each 20-MHz channel, thus 64-point IFFT is used to create an OFDM symbol. Among the 64 subcarriers, the inner 52 subcarriers are user-modulated and transmitted, and the outer 12 subcarriers are null-carriers for ACI reduction. The 52 useful tones are indexed from -26 to 26, without a DC component. Among the 52 tones, the 4 tones indexed $-21, -7, 7, 21$ are used for pilot tones, and they are BPSK-modulated by a PN sequence; the other 48 tones are used for data transmission. The power in the pilot subcarriers is higher so as to allow reliable channel tracking even at low SNR.

The cyclic prefix consists of $N_{cp} = 16$ samples, thus the total number of samples in an OFDM symbol is 80, corresponding to a duration of $T_s = 80 T_{sample} = 4\mu s$, where $T_{sample} = 1/(20\,\text{MHz}) = 0.05\ \mu s$ is the sampling period. The symbol rate is $R_s = 1/T_s = 0.25$ Msymbols/s. Depending the channel state, rate adaptation is achieved by selecting the modulation as BPSK, QPSK, 16QAM, or 64QAM, and/or choosing the rate of convolutional codes to be $1/2, 2/3$, or $3/4$. This leads to different data rates ranging from 6 Mbits/s to 54 Mbits/s. For example, for the combination of 64QAM and $R_c = 3/4$, the data rate is $\log_2 64$ bits/symbol $\times 0.25$ Msymbols/s $\times 3/4 \times 48 = 54$ Mbits/s.

For fixed WiMAX (IEEE 802.16d), 256 subcarriers are used, which are composed of 192 data subcarriers for data, 8 pilot subcarriers for various estimation purposes, and 56 null subcarriers for guard bands. The DC subcarrier is nulled to avoid DC offset. The null subcarrier is the center of the signal frequency band. For mobile WiMAX (IEEE 802.16e), OFDMA technology is used; it has a scalability supported by four different FFT sizes, 128, 512, 1024, and 2048. In OFDM, one user can use all the subcarriers at any given time, while in OFDMA multiple users are assigned subsets of subcarriers.

9.5 Spectrum of OFDM

For the baseband OFDM signal, f_i must be integer multiples of $1/2T$, and the minimum frequency separation between subcarriers is $1/T$ (refer to FSK). The frequency subcarriers are $f_0 + iR_s$, $i = 0, 1, \cdots, N-1$, where $R_s = 1/T$ is the symbol rate. For ease of analysis, f_0 is usually set to zero; but in practical use, the zero frequency subcarrier is unused to avoid the dc component, as in IEEE 802.11a.

The PSD of the OFDM signal is derived from the assumption of random data. It is just the superposition of the PSDs of all the sub-band signals. For the OFDM signals that use MASK, QAM with symmetrical constellation, and MPSK modulation, their PSDs have the same shape. The normalized PSD at the positive frequency part is given by

$$\Phi_{\tilde{s}}(f) = \sum_{i=0}^{N-1} \Phi_{\tilde{s}_i}(f) = \sum_{i=0}^{N-1} \left(\frac{\sin \pi\,(f - f_i)\,T}{\pi\,(f - f_i)\,T} \right)^2, \quad f \geq 0. \tag{9.7}$$

For unipolar MASK, a discrete part is added to the above spectrum. The null-bandwidth of the total PSD is $N/T = NR_s$, and the null-to-null bandwidth is $2NR_s$. As $N \to \infty$, we get a steep spectrum cut-off at $f/(NR_s) \to 1$.

Bandpass OFDM is typically scheduled as

$$f_i = f_c - \frac{N-1}{2T} + \frac{i}{T}, \quad i = 0, 1, \ldots, N - 1. \tag{9.8}$$

That is, the subcarriers are uniformly distributed in the range $f_c - \frac{N-1}{2T}$ to $f_c + \frac{N-1}{2T}$, with a symmetry about the nominal carrier frequency f_c. This can be obtained by upconverting the baseband OFDM signal with a frequency shift of $f_c - \frac{N-1}{2T}$. The total PSD for the positive frequency part is also given by (9.7), but with a frequency shift of $f_c - \frac{N-1}{2T}$. The null-to-null bandwidth is $(N + 1)R_s$. As $N \to \infty$, we get a steep cut-off in the spectrum.

For both the baseband and passband OFDM, the spectrums exhibit a steep cut-off as $N \to \infty$. For this reason, bandpass OFDM achieves the Nyquist rate, since the normalized bandpass frequency $(f - f_c)/(NR_s) \to 1$, as $N \to \infty$. The same rule applies for the zero-to-null band of the baseband OFDM. This corresponds to a spectral efficiency of 2 bauds/Hz, which is the theoretically highest Shannon bandwidth efficiency.

Under a Rayleigh-fading channel, the overall capacity for each OFDM block is a random variable; in case of large N and finite power, the distribution of the instantaneous capacity is shown to be approximately Gaussian [15]. In the limit as $N \to \infty$, the capacity approaches a constant value equal to the capacity of the infinite-bandwidth Gaussian channel [15].

For baseband OFDM, the subcarriers are required to be orthogonal. The subcarriers at the RF band are not required to be orthogonal to one another, because demodulation is performed at the baseband after downconverting from the RF band. Thus, f_c need not to be an integer multiple of $1/(2T)$, as required for f_i in baseband OFDM.

Spectrum shaping

In practical wireless standards, the spectrum of the transmitted OFDM signal must be subject to a spectral mask to avoid interference to adjacent bands. This requires the reduction of the sidelobes of the spectrum of the OFDM signal. Reducing the spectrum sidelobes can be implemented by windowing or pulse shaping.

The raised cosine function is commonly used for pulse shaping. It can eliminate ISI by introducing a roll-off segment at each side of the window, leading to a total pulse length of $(1 + \alpha)T_s$, where α is the roll-off factor and T_s is the extended symbol period. This introduces overlapping between adjacent symbols.

In OFDM, a cyclic extension is employed. The last length-T_g portion of the symbol is duplicated, and is then added to the symbol as a prefix. Thus, the extended symbol period is given by

$$T_s = T + T_g, \tag{9.9}$$

where T_g is the guard time.

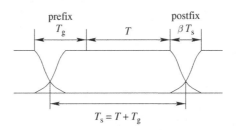

Figure 9.5   Pulse shaping of OFDM symbols. Cyclic extension is also shown.

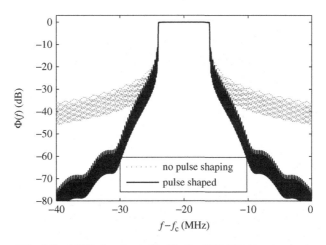

Figure 9.6   PSD of the OFDM signal specified in the IEEE 802.11a standard.

Similarly, since the total pulse length $(1 + \alpha)T_s$ is longer than the extended symbol duration T_s, the first length-αT_s portion of the symbol is duplicated and added as a postfix to the extended symbol to deal with overlapping. This is shown in Fig. 9.5. This accounts for the overlapping period by αT_s.

In OFDM, pulse shaping is performed on each carrier. Unlike the case in Section 6.3.1, where raised cosine filtering is applied in the frequency domain, here the raised cosine function is applied along the time domain. The raised-cosine time-window and its spectrum are dual to (6.44) and (6.45), respectively; they can be obtained by applying the transformation $t \rightarrow f$ and $1/T \rightarrow T$. Thus, the spectrum (Fourier transform) of the raised cosine time-window function can be derived as [89]

$$W(f) = \frac{T_s \cos(\pi f \alpha T_s)}{1 - (2f\alpha T_s)^2} \frac{\sin(\pi f T_s)}{\pi f T_s}. \tag{9.10}$$

The overall PSD of the pulse-shaped OFDM signal is thus given by

$$\Phi_w(f) = \sum_{i=0}^{N-1} |W(f - f_i)|^2. \tag{9.11}$$

Example 9.3: In IEEE 802.11a, a raised-cosine filter with $\alpha = 0.025$ is used to shape the PSD of the 64-channel QAM-OFDM to a 40-dB attenuation at 30 MHz from the nominal carrier frequency f_c. The channel separation is $1/(3.2\mu s) = 0.3125$ MHz. With a 0.8 μs guard interval, $T_s = 4\mu s$. The PSD of the IEEE 802.11a OFDM signal is plotted in Fig. 9.6, where the PSD with no pulse shaping is also plotted for comparison. With the raised cosine pulse shaping, the specification of 40-dB attenuation at 30 MHz away from the carrier frequency can be realized.

9.6 Fading mitigation in OFDM

For the OFDM system in a frequency-flat slow fading channel, if the rms excess delay σ_τ is much less than the symbol period, the multipath signals received are not resolvable and a single-path signal with a fading envelope and random phase is formed. This type of fading does not introduce any ISI and ICI, and all the subcarriers experience the same complex fading factor h.

The performance of OFDM will degrade in a fading channel other than the frequency-flat slow fading channel. Frequency-selective slow fading is characterized by multiple resolvable signal paths in the channel. Fading and multipath will incur both ISI and ICI. Frequency-flat fast fading arises from Doppler shift, and the multiplicative fading factor varies in a symbol period. Doppler shift introduces ICI, leading to an error floor in the BEP performance. For frequency-selective fast fading, both ISI and ICI occur, and the error floor in the BEP performance is also present.

The most important measure for fading mitigation in OFDM is through channel estimation and equalization. Other techniques for fading mitigation include differential modulation, and diversity schemes such as the widely used coded OFDM and MIMO-OFDM.

As in single-carrier systems, differential demodulation and detection can be used for OFDM in fading channels where carrier synchronization for coherent demodulation is difficult to accomplish. For OFDM, differential encoding and detection can be performed in the time domain or in the frequency domain. Time-domain differential encoding and detection are performed on consecutive symbols for each subcarrier, and in the frequency-domain counterparts to consecutive subcarriers for each symbol period. To prevent fast fading that generates large phase noise and an increase in BEP, in the time-domain case, the symbol rate should be greater than the Doppler frequency spread ν_{max}; likewise, in the frequency-domain case, the subcarrier spacing should be much smaller than the channel coherent bandwidth $1/\sigma_\tau$. The SNR in differential detection is 3 dB inferior to that with ideal coherent detection, but practical coherent detection also introduces some loss in SNR due to channel estimation accuracy and power loss in pilots. Thus, differential detection is typically 1 to 2 dB inferior in SNR to coherent detection [89].

ISI and ICI in the OFDM system can be completely eliminated by using cyclic prefix. However, an uncoded OFDM system does not have any frequency diversity, and each subcarrier can be in a fading dip. To reduce BEP, some form of diversity, such as coding, must be used to improve the BEP. When coding is applied, the bitstream can be first coded by using turbo/convolutional code or Reed-Solomon (RS) code, followed by interleaving. In the MIMO-OFDM system, space-time trellis coding (STTC) or space-time block coding (STBC) can be further applied. MIMO and channel coding are, respectively, described in Chapters 19 and 15.

To mitigate fading in OFDM, conventional equalization can be used on each subcarrier, but it also enhances the noise power. Precoding [68] uses the same idea as equalization, but the fading is inverted at the transmitter rather than at the receiver, thus noise power remains $N_0 B_N$, where B_N is the bandwidth of each subchannel. Precoding is common in wireline multicarrier systems such as DSL. However, in the wireless case, due to a Rayleigh fading channel, the power for channel inversion will approach infinity; also, precoding requires subchannel flat-fading gains at the transmitter, which is difficult to obtain in a rapidly fading channel.

Adaptive loading, where power and data rate on each subchannel are adapted to maximize the total rate of the system by using adaptive modulation, can also be used to mitigate fading. Like precoding, the subchannel fading at the transmitter is required.

9.7 Channel estimation

Channel estimation and equalization are necessary for combating the performance degradation caused by fading. This requires estimation of the channel. Channel estimation is based on the assumption of perfect synchronization, thus it is a part of the coherent detection. Channel estimation is usually performed jointly with carrier and timing synchronization. It can be pilot-assisted or decision-directed.

The pilot-assisted method uses the pilot symbols sparsely distributed across the time slots as well as the subcarriers. Estimation of the channel for each subcarrier at each symbol period is performed by two-dimensional interpolation. At the beginning, preambles that contain one or two preamble OFDM symbols are sent prior to the user data symbols for the purpose of synchronization and initial channel estimation. Pilot symbols are then inserted according to a known pattern among the subcarriers in order to track the time-varying channel to maintain accurate channel estimates.

Decision-directed channel estimation uses the previously demodulated symbols. The decision-directed method is not as reliable as the pilot-assisted method and it suffers from error propagation, but it uses less bandwidth. It is capable of tracking fast fading, while the pilot-assisted method with sparsely inserted pilots is not able to do so. The decision-directed method can use the same least squares (LS) or linear MMSE algorithm as the pilot-assisted method does, but the input to the algorithm is the previously demodulated symbols [27].

The pilot-assisted and decision-directed methods can be combined to combat the shortcomings of the two methods. After the channel is estimated, the symbols can be detected by using a ZF or MMSE detector.

Due to shaping of the transmit spectrum, practical OFDM systems are usually not fully loaded. The subcarriers that are set to null are referred to as *virtual carriers*. Cyclic prefix and/or virtual carriers have been employed in subspace-based blind channel estimation techniques [41, 54], and semiblind techniques that also take advantage of the training data [54]. In the following, we introduce pilot-based channel estimation.

9.7.1 Pilot arrangement for channel estimation

The pilot-assisted method requires an arrangement of the pilots. Figure 9.7 illustrates three types of pilot arrangement, namely block-type, comb-type and scattered-type. In the block-type pilot arrangement, one specific symbol full of pilot subcarriers is transmitted periodically. The block-type pilot arrangement is suitable for estimation of slow fading channels, and it has been used in the IEEE 802.16e-2005 OFDM mode standard. In the comb-type arrangement, a number of subcarriers are reserved for pilot signals, which are transmitted continuously. The comb-type pilot arrangement can be used for equalization when the channel changes in one OFDM block. N_p pilots are uniformly inserted into $x(k)$ on each carrier. The comb-type arrangement requires interpolation of the channel. For fast fading channels, the comb-type arrangement leads to better channel estimation. The scattered-type is not periodical, and the number of pilots and their positions can be changed dynamically according to the channel.

Based on the Nyquist theorem, the maximum sampling distance along the time axis is derived as [57]

$$N_t \leq \frac{1}{2v_{\max}T_s} = \frac{1}{2\left(1 + \frac{N_{cp}}{N}\right)v_{\max}}, \tag{9.12}$$

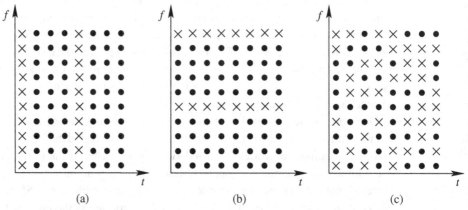

Figure 9.7   Pilot Arrangement. (a) Block-type. (b) Comb-type. (c) Scattered-type. A "×" represents a pilot.

where $v_{\max}$ is the maximum Doppler spread, and T_s is the extended OFDM symbol duration, $T_s = \left(1 + \frac{N_{cp}}{N}\right) T$.

By duality between the Fourier transform and the inverse Fourier transform (see Chapter 13), in the frequency domain the sampling theorem requires that [51, 57]

$$N_f \leq \frac{1}{\sigma_\tau \Delta f} = \frac{N}{N_{cp}}, \tag{9.13}$$

where N_f is the sampling distance along the frequency axis, Δf is the subcarrier spacing, $\Delta f = 1/T$, and σ_τ is the multipath spread of the channel.

9.7.2 Pilot-assisted channel estimation

Operation of OFDM requires knowledge of the channel. For frequency-selective flat fading, the channel impulse response for the nth subchannel at time k is $h_{n,k}$. Also, since channel estimation is based on the pilot symbols, the data on each subchannel is known. Based on this, an LS estimate of the channel is given by

$$\hat{h}_{n,k}^{LS} = \frac{r_{n,k}}{d_{n,k}}, \tag{9.14}$$

where $d_{n,k}$ is the known symbol on subchannel n at time k, and $r_{n,k}$ is the received value on subchannel n at time k. The LS estimator has a coarse estimate, but it is simple.

Assuming AWGN with variance σ_n^2 on each subchannel, the linear MMSE or Wiener estimate is given by [51]

$$\hat{h}_k^{MMSE} = \mathbf{R}_{hh} \left(\mathbf{R}_{hh} + \sigma^2 \mathbf{I}_{N \times N}\right)^{-1} \hat{h}_k^{LS}, \tag{9.15}$$

where $\hat{h}_k^{LS}$ is obtained by stacking all the N LS estimates at time k, $\hat{h}_k^{LS} = \left(\hat{h}_{1,k}^{LS}, \hat{h}_{2,k}^{LS}, \ldots, \hat{h}_{N,k}^{LS}\right)^T$, and $\mathbf{R}_{hh}$ is the autocovariance matrix of the channel gains

$$\mathbf{R}_{hh} = \mathrm{E}\left[\boldsymbol{h}_k \boldsymbol{h}_k^H\right], \tag{9.16}$$

$\boldsymbol{h}_k = \left(h_{1,k}, h_{2,k}, \cdots, h_{N,k}\right)^T$. For the wide sense stationary channel, $\mathbf{R}_{hh}$ is independent of time k.

The method is accurate, but the complexity is $O\left(N^2\right)$ complex multiplications, which is very high for large N. This method is used for initial acquisition of the channel. To reduce the complexity, the structure of OFDM can be used and $\mathbf{R}_{hh}$ can be eigen-decomposed,

$$\mathbf{R}_{hh} = \mathbf{U} \boldsymbol{\Lambda} \mathbf{U}^H, \tag{9.17}$$

where the dimension of the space is approximately $N_{cp} + 1$, and the transforms $\mathbf{U}$ and $\mathbf{U}^H$ are, respectively, DFT and IDFT.

After the channel is estimated, the final estimate of the data symbol at time k is given by a one-tap equalizer

$$\hat{d}_{n,k} = \frac{r_{n,k}}{\hat{h}_{n,k}}, \tag{9.18}$$

where the subscript n runs for the subcarriers, and k corresponds to the time instant. The channel is assumed to be constant for a period of one symbol.

After the channel is acquired, it also needs to be tracked by using pilot symbols scattered into the OFDM time-frequency grid, spaced by N_f subcarriers and N_t OFDM symbols. At the pilot positions, the channel is estimated by the LS or MMSE estimation, while at other positions the channel can be estimated by interpolation.

Case study: LTE

In LTE, data symbols are mapped to resource blocks based on CSI at the transmitter. The uplink frame is the same for both FDD and TDD. Each frame is 10 ms long and is divided into 10 subframes. In TDD mode, each subframe is allocated to either the downlink or uplink. Each subframe is divided into two time slots. A slot consists of seven DFT-spread OFDM symbols with cyclic prefix of a normal length or six symbols with cyclic prefix of an extended length.

A resource block is defined for one slot, and contains 12 adjacent subcarriers spaced by 15 kHz, or 180 kHz. A resource block in uplink TDD mode is illustrated in Fig. 9.8. Every fourth symbol is a pilot symbol for channel estimation at the receiver, while the others are data symbols. The symbols are time multiplexed, occupying the entire transmission bandwidth of 180 kHz. The pilot symbol is chosen as a Zadoff-Chu sequence, and different Zadoff-Chu sequences are assigned to resource blocks of different cells.

Channel sounding is also required in the uplink. For uplink, channel sounding is achieved by reserving blocks within the same subframe for this purpose. The blocks containing sounding signals occupy a bandwidth of multiple resource blocks, and these blocks of different users share either the time, frequency, or code domain. Different Zadoff-Chu sequences are assigned to different users.

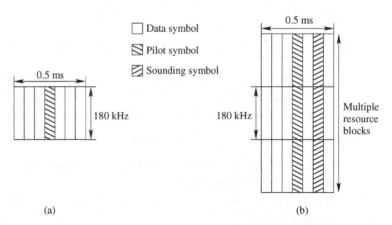

(a)   (b)

Figure 9.8 Symbols within a resource block of LTE in uplink TDD mode. (a) A regular resource block. (b) A resource block with a sounding symbol.

9.8 Peak-to-average power ratio

The OFDM signal consists of many independently modulated subcarriers, which may lead to a large PAPR for some OFDM symbols. When N signals have the same phase, they produce a peak power, which is N times the average power.

9.8.1 Peak factor: definition and impact

The PAPR or peak factor is defined by

$$\text{PAPR} = \frac{\max_t |x(t)|^2}{E\left[|x(t)|^2\right]}. \tag{9.19}$$

For constant amplitude signals, PAPR=0 dB, for a sine wave PAPR=3 dB.

For OFDM systems, in order to reduce the overhead of the cyclic prefix, the number of subcarriers N should be as large as possible. However, for N subcarriers the maximum possible PAPR is N, although the possibility of PAPR being N is very low.

The maximum PAPRs for MQAM-OFDM and MPSK-OFDM signals are derived in [89]. For the square MQAM with $M = 2^k$ for k even, the PAPR is given as

$$\text{PAPR} = \frac{3N(\sqrt{M} - 1)}{\sqrt{M} + 1} \quad \text{(MQAM-OFDM).} \tag{9.20}$$

For MPSK,

$$\text{PAPR} = N \quad \text{(MPSK-OFDM).} \tag{9.21}$$

The probability of maximum PAPR happening is very low. For system design, the statistical distribution of the instantaneous PAPR is more important. For large N, the IFFT output $x(n)$, $0 \le n \le N - 1$, is zero-mean complex Gaussian, since the central limit theorem is applicable to the IFFT expression. In this case, the OFDM signal has an envelope that is Rayleigh distributed with variance σ^2, and a phase of uniform distribution. The probability of the PAPR exceeding a threshold $P_0 = \sigma_0^2/\sigma^2$ is given by [77, 89]

$$\Pr\left(\text{PAPR} > P_0\right) = 1 - \left(1 - e^{-P_0}\right)^{\alpha N}, \tag{9.22}$$

where $\alpha = 1$ for $N \ge 512$, and $\alpha = 2.8$ for $N \le 256$. The result is quite accurate for $N \ge 128$.

Example 9.4: The probability of the PAPR exceeding P_0 is plotted in Fig. 9.9, for N=64, 128, ..., 2048. It is seen that the probability of PAPR being greater than a moderate P_0 decreases rapidly.

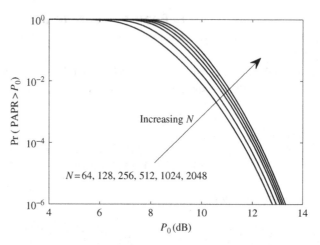

Figure 9.9 The probability of the PAPR being greater than P_0.

The high PAPR becomes a critical challenge to RF power amplifier design. In order to operate in the linear region to avoid distortion, the peak must be within this region. This can be realized by reducing the average power of the input signal, known as *input back-off*, which also results in a proportional output back-off. Back-off reduces the power efficiency of the power amplifier, which may limit the battery life for MSs. Back-off also reduces the coverage range. For a given average power requirement, high PAPR requires power amplifiers with a large linear region, and this increases the cost of the power amplifiers. In addition, high PAPR corresponds to a large dynamic range of the signal, and thus requires high resolution for both the DACs at the transmitter and the ADCs at the receiver.

Thus, OFDM requires high-power amplifiers, with a wide linear range, and hence, class A amplifiers are required. If a cheaper nonlinear amplifier is used, the distortions introduced destroy orthogonality between the subcarriers, and also lead to more out-of-band emissions.

9.8.2 Peak factor reduction techniques

The PAPR can be reduced by amplitude clipping or companding, by using a PAPR reduction code [82], by phase adjustments [6, 53], or by precoding [70]. PAPR reduction techniques are reviewed in [31], and criteria for selecting a PAPR reduction technique are also given. Here, we only mention some basic ideas.

From Fig. 9.9, it is seen that large peaks are rare to happen. Thus, clipping the high peaks would not introduce too much performance degradation. Clipping is the simplest and most widely used PAPR reduction technique. Amplitude clipping introduces out-of-band emission (spectral growth) and in-band distortion, and hence multiplicative and additive time-domain modifications of the OFDM signal have been proposed to alter the amplitude of the time-domain OFDM samples rather than clipping [33]. The companding technique for PAPR reduction uses the μ-law to compress the signal amplitude [78]. The spectral regrowth and in-band distortion arising from clipping can also be improved by an iterative process of clipping and filtering [4].

Channel codes are employed not only as a means of error control but also simultaneously as a sequence restrictor so as to select only low peak OFDM symbols from a set of block codes for transmission. The Golay complementary sequences [30] are particularly suitable for PAPR reduction [82]. The polyphase Golay complementary sequences [26] are suitable for MPSK-OFDM, since all the elements in the sequence have the same amplitude. The peak factor of $x(t)$ that employs any Golay complementary sequence for the complex symbols on all the subcarriers, $\{d_{n,k}, n = 0, 1, \ldots, N - 1\}$, is bounded by 3 dB, that is, it achieves a maximum PAPR of 2 [63]. A systematic method for encoding an arbitrary data sequence into a Golay complementary sequence is given in [20]. Coding for PAPR reduction has a large overhead, thus reducing the data throughput. Iterative channel codes such as turbo codes [1] and LDPC codes have also been used for this purpose [18].

The phase adjustment method multiplies the complex symbol, $d_{n,k}$, to be transmitted at the nth subcarrier by $\exp\left(j\phi_{k,l}\right)$, where $\phi_{k,l}$ is a predefined phase selected as the lth value of L possible discrete phase values for symbol k. The ensemble of the phase adjustments is known to both the transmitter and the receiver. By minimizing the PAPR, the value of l can be derived. This approach cannot guarantee a given maximum peak factor. In the partial transmit sequence (PTS) technique [53], an input data block of N symbols is partitioned into disjoint sub-blocks. The subcarriers in each sub-block are weighted by a phase factor for that sub-block. The phase factors are selected such that the PAPR of the combined signal is minimized. In the selected mapping (SLM) technique [6], the transmitter generates a set of modified data blocks, all representing the same original data block, and selects the most favorable ones for transmission. Specifically, each data block is multiplied component-wise by different phase sequences. Exploiting the fact that the modulation symbols belong to a given constellation and that the multiple signals generated by the PTS or SLM processes are widely different in a Hamming distance sense, the simplified ML decoder for SLM and PTS proposed in [36] operates without side information. Compared to the QAM constellation, the hexagonal constellation has more regularly-spaced signal points in a given area. These extra degrees of freedom can be exploited to eliminate data rate loss due to the side information in the PTS and SLM techniques [32]. The techniques proposed in [32] achieve almost identical PAPR reduction and SEP performance as those of the techniques in [36], but with significantly reduced complexity at the receiver.

There are a number of special sequences that can be used to encode the information sequences $\{d_{n,k}, n = 0, 1, \ldots, N - 1\}$ or to adjust the phase $\{\phi_{k,l}, l = 0, 1, \ldots, L - 1\}$ so as to reduce amplitude variations. Such codes include the Shapiro-Rudon sequences, Golay codes, m-sequences, Newman phases, and Schroeder phases [33]. According to the comparison made in [33], the Newman phases show the lowest peak factor among a number of techniques.

Precoding-based PAPR reduction is implemented by multiplying each data block by the same precoding matrix prior to OFDM modulation and transmission [70]. Precoding avoids block-based optimization, and this method is data-independent. A properly selected precoding matrix distributes the power of each modulated symbol over the OFDM block, and this makes the PAPR of OFDM modulated signals very close to that of single-carrier signals. This method works with an arbitrary number of subcarriers and any type of baseband modulation used.

Constant envelope OFDM (CE-OFDM) [74] transforms the OFDM signal, by way of phase modulation, to a 0-dB constant-envelope signal, thus eliminating the PAPR problem. At the receiver, phase demodulation is applied ahead of the conventional OFDM demodulator. The BEP performance of CE-OFDM is analyzed in AWGN and fading channels in [74]. CE-OFDM achieves good performance in dense multipath with the use of cyclic prefix transmission in conjunction with a frequency domain equalizer.

9.8.3 Amplitude clipping or companding

Peak factor can be reduced by modifying the time-domain signal when its amplitude exceeds a limit. The transfer function for amplitude clipping for a baseband multi-carrier signal $x(t)$ is given by

$$y(t) = \begin{cases} -l, & x \le -l \\ x(t), & |x| < l \\ l, & x \ge l \end{cases}, \tag{9.23}$$

where l is the absolute level of clipping.

Error probability due to in-band distortion

The in-band distortion has been modeled based on the Bussgang theorem in [19, 58]. The clipped output can be modeled as

$$\tilde{x}(n) = \alpha x(n) + d(n), \quad n = 0, 1, \ldots, N - 1, \tag{9.24}$$

where the distortion $d(n)$ is uncorrelated with $x(n)$, and the attenuation factor α is given by

$$\alpha = 1 - e^{-\beta^2} + \frac{\sqrt{\pi}\beta}{2}\text{erfc}(\beta), \tag{9.25}$$

with the clipping level β being defined as

$$\beta = \frac{l}{\sqrt{E[|x(n)|^2]}} = \frac{l}{\sqrt{E_x}}, \tag{9.26}$$

with E_x being the average input power of the OFDM signal before clipping. The attenuation factor $\alpha < 1$ and is very close to unity as $\beta > 8$ dB. In this case, the uncorrelated distortion can be approximated by the clipped-off signal $c(n) = \tilde{x}(n) - x(n)$.

The variance of the distortion is given by [19]

$$\sigma_d^2 = E_x \left(1 - e^{-\beta^2} - \alpha^2\right). \tag{9.27}$$

Assuming that the distortion is Gaussian, the signal-to-noise-plus-distortion ratio (SNDR) of an OFDM symbol is given by

$$\text{SNDR} = \frac{\alpha^2 E_x}{\sigma_d^2 + N\frac{N_0}{2}}, \tag{9.28}$$

where $\frac{N_0}{2}$ is the variance of the channel noise on a single carrier. Substituting this SNDR into the BEP equation of an M-ary modulation, we get the BEP for the modulation. The BEP exhibits an irreducible floor as $N_0 \to 0$ or $E_b/N_0 \to \infty$. The BEP for an OFDM system can be calculated as the average BEP on a single carrier.

Example 9.5: For MQAM with average power $E_x = E_b N \log_2 M$, the SNDR can be calculated using (9.28). From (7.117), the BEP can be approximated by replacing γ by SNDR

$$P_b \approx \frac{4}{\log_2 M} \left(1 - \frac{1}{\sqrt{M}}\right) Q\left(\sqrt{\frac{3\text{SNDR}}{M-1}}\right). \tag{9.29}$$

For 64QAM modulation, the BEP is plotted in Fig. 9.10, for $\beta = 1.4$, 1.6, 1.8, 2.0, 2.2, 2.6, and ∞. The irreducible floor is clearly demonstrated for large E_b/N_0.

Error probability due to clipping

The probabilities of error due to clipping vary across subcarriers, and the overall probability is dominated by that of the lower subcarriers. A bound on the BEP floor caused

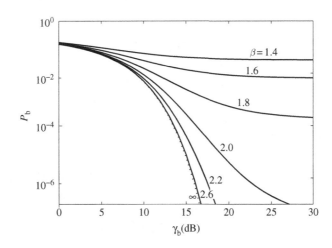

Figure 9.10 The BEP for a clipped OFDM signal in the AWGN channel for a single carrier and 64QAM modulation.

by clipping in a multi-carrier system is given in [5] under the assumption of the absence of noise and other channel impairments. Assuming that each subcarrier carries a square-QAM constellation of L^2 points, with the same power for all the subcarriers, the overall probability of symbol error is upper bounded for a clipping by [5]

$$P_s = \frac{8N(L-1)}{\sqrt{3}L} e^{-\frac{\beta^2}{2}} Q\left(\left[\frac{3\pi\beta^2}{\sqrt{8\left(L^2-1\right)}}\right]^{1/3}\right). \tag{9.30}$$

This bound is obtained for continuous-time signals, where the clipping events satisfy the Poison probability. For practical transceivers operating with discrete-time values, the clipping probability has a corresponding discrete-time value. Based on this, a more tight error bound is obtained as [5]

$$P_s = \frac{8N(L-1)}{L} Q(\beta) Q\left(\left[\frac{3\pi\beta^2}{\sqrt{8\left(L^2-1\right)}}\right]^{1/3}\right). \tag{9.31}$$

Example 9.6: The SEP of the OFDM system caused by clipping, given by (9.31), is plotted in Fig. 9.11 for QPSK (4QAM), 16QAM, 64QAM and 256QAM.

There is also an additive Gaussian noise model for SEP analysis [49]. In [5], it was pointed out that the additive Gaussian model achieves an SEP that is substantially lower than that from the above model, and the result of the above model is very close to that from simulation [5]. Thus, the OFDM signal is relatively insensitive to clipping at the transmitter.

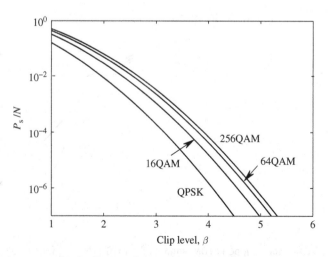

Figure 9.11   The SEP due to amplitude clipping, for QPSK (4QAM), 16QAM, 64QAM and 256QAM.

The SEP (9.31) is derived when there exists clipping only. In practice, we are more interested in the SEP at the OFDM receiver, where there exist noise and fading. In case of channel noise and fading together with clipping at the receiver, the SEP performance degrades significantly. Noise can reduce the effective distance between the constellation points, thus increasing the SEP.

A noiseless, slowly varying Rayleigh fading channel can be modeled by the pdf [5]

$$p_z(z) = \frac{\pi}{2} z e^{-\pi z^2/4}, \quad z > 0, \tag{9.32}$$

where the mean is normalized to unity. Assuming that the channel is known and is constant over one OFDM symbol duration, an upper bound on the error probability is derived as [5]

$$P_s = 4\pi \frac{N(L-1)}{L} \int_0^\infty z e^{-\pi z^2/4} Q(\beta z) Q\left(\left[\frac{3\pi \beta^2 z^2}{\sqrt{8\left(L^2-1\right)}}\right]^{1/3}\right) dz. \tag{9.33}$$

This bound is tight for lower subcarriers of the multi-carrier signal.

Example 9.7: The SEP for a Rayleigh fading channel with clipping at the OFDM receiver, as given by (9.33), is plotted in Fig. 9.12 numerically, for QPSK (4QAM), 16QAM, 64QAM and 256QAM.

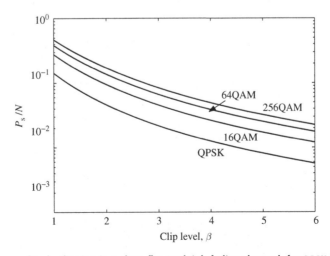

Figure 9.12   The SEP due to amplitude clipping in a slow, flat Rayleigh fading channel, for QPSK (4QAM), 16QAM, 64QAM and 256QAM.

9.9 Intercarrier interference

Cyclic prefix ensures orthogonality between the subcarriers in a delay-dispersive (frequency-selective) environment, thus eliminating ICI due to frequency selectivity. However, the frequency-dispersive (time selective) channel due to the Doppler effect also creates ICI. It also leads to random frequency modulation, leading to errors. For a large symbol duration, a small Doppler shift can cause a considerable ICI. Another source that introduces ICI is due to errors in the local oscillators.

To eliminate ICI, we need to optimally select the carrier spacing and OFDM symbol length. A short symbol duration T causes a smaller Doppler-induced ICI, but it must be above a minimum value for the reason of spectral efficiency: The cyclic prefix, determined by the $\tau_{\max}$, should occupy 10% of the symbol duration.

In OFDM, the rectangular time-domain signal has a sinc shape in the frequency domain, which decays slowly in the frequency domain. The slope of the sinc function is large when its value is close to zero, and as a result, a small Doppler shift causes a large ICI. By suitably selecting a pulse shape whose spectrum has a large mainlobe and small sidelobes, ICI due to the Doppler effect can be decreased; this, however, leads to a slower decay in the time domain, and thus to errors induced by the delay spread. FDE techniques such as ZF and MMSE equalizers can also be used to reduce ICI.

For large N, the central limit theorem can be applied and the ICI can be treated as a Gaussian variable. Based on this, the autocorrelation of the ICI for 2-D isotropic scattering can be derived [72]. The variance of the ICI caused by the Doppler shift is derived as [72, 89]

$$\phi_{cc}(0) = \frac{E_{av}}{T} - \frac{E_{av}}{TN^2}\left(N + 2\sum_{i=1}^{N-1}(N-i)J_0\left(2\pi\nu_{\max}Ti\right)\right), \qquad (9.34)$$

where E_{av} is the average symbol energy, T is the symbol period, J_0 is the Bessel function of the first kind and zero order, and $\nu_{\max}$ is the maximum Doppler frequency. It is seen that the variance of the ICI is independent of the signal constellation. The SIR of the OFDM signal is given by

$$\text{SIR} = \frac{E_{av}/T}{\phi_{cc}(0)}. \qquad (9.35)$$

Example 9.8: The SIR as a function of $\nu_{\max}T$, as given by (9.35), is plotted in Fig. 9.13 for $N = 64$ to 2048. It is seen that SIR decreases with the increase of $\nu_{\max}$ and N.

From Fig. 9.13, it is seen that there is an SIR floor, as $\nu_{\max}T$ increases. Thus, ICI results in an error floor, which can be estimated by substituting SIR given by (9.35) for the average SNR γ in the BEP expression for the data modulation technique on each subcarrier. The theoretical BEP result is very close to that obtained from random simulations.

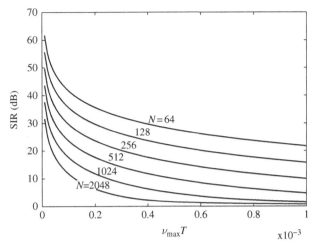

Figure 9.13

Influence of ICI on SIR.

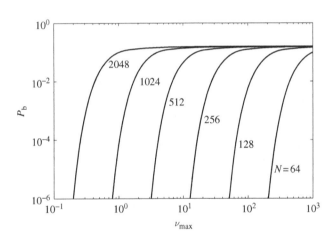

Figure 9.14

The error floor of QPSK due to ICI.

Example 9.9: For QPSK, the BEP is given by $P_b = Q\left(\sqrt{\gamma}\right)$. By substituting SIR given by (9.35) for γ, the corresponding BEP due to ICI is obtained. Given a bit rate of $R_b = 10$ Mbits/s, the BEP against the Doppler frequency ν_{max} is plotted in Fig. 9.14 for $N = 64$, 128, 256, 512, 1028, and 2048. Bit error floor is clearly demonstrated in the figure.

For a particular fading channel such as the Rayleigh fading channel, we have $\gamma_b = \alpha^2 \gamma_b$, α being the fading amplitude, and the average SEP is derived by (7.141) with γ_s being replaced by SIR. The same expression for the error floor in the BEP performance is applicable for both the frequency-flat fast fading and the frequency-selective fast fading channels.

Standard OFDM systems using a guard-time interval or cyclic prefix combat ISI, but provide no protection against ICI. This drawback has led to the introduction of pulse-shaping OFDM systems [71].

A universal bound on the ICI power, P_{ICI}, is given in [42] based on the Doppler spectrum

$$P_{ICI} \leq \frac{(2\pi f_D T)^2}{12}, \tag{9.36}$$

where f_D is the maximum Doppler spread and T is the OFDM symbol duration. Upper and lower bounds on P_{ICI} of OFDM in a Gaussian scattering channel (e.g., with a Gaussian DoA pdf $p(\theta)$) are also given in [40] is a function of $f_D T$, the frequency offset, and the mobile moving direction. It is found in [40] that the Gaussian scattering channel can be best described as the classical Doppler spectrum model.

9.10 Synchronization

Time and frequency synchronization between the transmitter and receiver is of critical importance to OFDM systems. Multicarriers make OFDM susceptible to time and frequency synchronization errors, i.e., phase noise and frequency offset. Frequency jitter and Doppler shift cause ICI and time-varying phase shift. Sample timing errors could destroy the orthogonality among the subcarriers, leading to an increase in the BEP, or at least introducing to the signal symbol a phase shift, which is linearly proportional to the subcarrier frequency.

9.10.1 Influence of frequency offset

In the OFDM system, frequency offset is introduced by mismatch between the transmit and receive sampling clocks and misalignment between the reference frequencies of the transmitter and receiver. Frequency offset causes ICI due to the rotation of the signal constellation on each subcarrier, in turn causing a loss in SNR. The SNR loss is due to the amplitude loss, since the desired subcarrier is no longer sampled at the peak of the equivalent sinc function of the DFT. As a consequence, the adjacent subcarriers cause interference as they are not sampled at their zero crossings. This can be seen from Fig. 9.1a.

The degradation caused by the sampling frequency offset can be expressed in terms of the SNR loss as [61]

$$L_{SNR} \approx 10 \log_{10} \left(1 + \frac{\pi^2}{3} \gamma \left(k \frac{\delta f}{f_s} \right)^2 \right) \tag{9.37}$$

$$= 10 \log_{10} \left(1 + \frac{\pi^2}{3} \gamma (k t_\Delta)^2 \right) \quad \text{(dB)}, \tag{9.38}$$

where k is the subcarrier index, $\gamma = \frac{E_s}{N_0}$ is the real channel SNR, $f_s = N/T$, T being the symbol period, δf is the sampling frequency offset between the receiver and the transmitter, and t_Δ is the normalized sampling error given by

$$t_\Delta = \frac{T'' - T'}{T'}, \tag{9.39}$$

T'' and T' being the sampling periods at the transmitter and receiver, respectively. This approximation is quite accurate except for the smallest and highest carrier indexes. Thus, when the number of subcarriers and the sampling error t_Δ are small so that $kt_\Delta \ll 1$, the degradation can be ignored.

The overall effect of the carrier frequency offset (CFO) on SNR is analyzed in [61], and is given by (9.37); for small frequency offset δf, the SNR loss is approximated by [62]

$$L_{\mathrm{SNR}} \approx \frac{10}{3 \ln 10} (\pi T' k \delta f)^2 \gamma \quad (\mathrm{dB}), \tag{9.40}$$

where T' is the sampling period at the receiver. The maximum L_{SNR} is at $k = N$.

Frequency error reduces the useful signal amplitude at the frequency-domain sampling point by a factor of $\mathrm{sinc}(\delta f / \Delta f)$, where $\delta f / \Delta f$ is the normalized frequency synchronization error, and Δf is the frequency difference between two adjacent subcarriers [33]. Assuming that the effects of the frequency error can be approximated by white Gaussian noise with variance σ_{ICI}^2, the equivalent SNR is given by [33]

$$\gamma' = \frac{\mathrm{sinc}\left(\frac{\delta f}{\Delta f}\right) \sigma_a^2}{\sigma_{\mathrm{ICI}}^2 + \sigma_a^2 / \gamma}, \tag{9.41}$$

where σ_a^2 is the average symbol power. The ICI variance can be expressed by [33]

$$\sigma_{\mathrm{ICI}}^2 = \sigma_a^2 \sum_{i=-N/2-1, i \neq 0}^{N/2} \left\| \mathrm{sinc}\left(i + \frac{\delta f}{\Delta f}\right) \right\|^2. \tag{9.42}$$

Example 9.10: The ICI variance versus the normalized frequency synchronization error is plotted in Fig. 9.15, for $N = 16, 64, 2048$. It is shown that the ICI noise variance virtually has no change for $N > 64$.

Example 9.11: Based on the equivalent SNR given by (9.41), the BEP performance for QPSK modulation of subcarriers is plotted in Fig. 9.16, for different values of $\delta f / \Delta f$. As $\delta f / \Delta f$ increases, a bit error floor is clearly demonstrated.

In the presence of the frequency offset that considers the Doppler effect, the effective SNR for each subcarrier p is derived as [66]

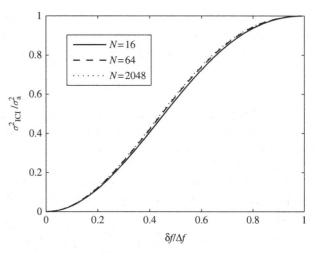

Figure 9.15   ICI variance σ_{ICI}^2 as a function of the normalized frequency shift $\delta f/\Delta f$.

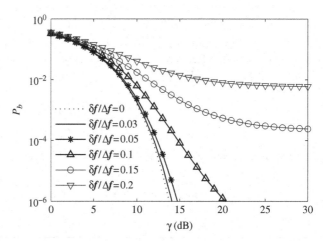

Figure 9.16   Influence of frequency synchronization error $\delta f/\Delta f$ on the BEP over the AWGN channel for QPSK signals.

$$\gamma(p) = \frac{\sigma_s^2(p)\sigma_H^2(p)\left|\mu_N\left(\frac{\delta f}{\Delta f}+p\beta\right)\right|^2}{\sum_{n\in\mathcal{N}_a,n\neq p}\sigma_s^2(p)\sigma_H^2(p)\left|\mu_N\left(\frac{\delta f}{\Delta f}+n(1+\beta)-p\right)\right|^2+\sigma_w^2}, \tag{9.43}$$

where $\sigma_H^2(p)$ is the variance of the subchannel H_p, $\sigma_s^2(p)$ is the variance of the symbols on subchannel p, σ_w^2 is the variance of the noise at subchannel p,

$$\mu_N(\theta) = \frac{1}{N}\sum_{n=0}^{N-1}e^{j2\pi n\theta/N} = \frac{1}{N}\frac{\sin(\pi\theta)}{\sin(\pi\theta/N)}e^{j\pi\theta(N-1)/N}, \tag{9.44}$$

$\beta = v/c$ denotes the relative Doppler shift, v being the velocity component along the radial direction, and the frequency offset

$$\frac{\delta f}{\Delta f} = \frac{f_{\text{off}}(1 + \beta)}{\Delta f} = \frac{(1 + \beta)f_c - f_{\text{LO}}}{\Delta f} \tag{9.45}$$

denotes the normalized shift of the carrier, f_{off} being the frequency offset at the transmitter, f_{LO} the frequency of the LO at the receiver. $\mathcal{N}_a$ is the set of active subcarriers $\mathcal{N}_a \subset \{-N/2 + 1, \ldots, N/2\}$. The parameters $\frac{\delta f}{\Delta f}$, β, and symbols can be estimated by using the ML approach [66].

Error performance of OFDM systems with CFO has also been analyzed in the literature [7, 21]. In [21], exact closed-form BEP/SEP expressions have been derived for an OFDM system with BPSK/QPSK modulation and CFO in AWGN, frequency-flat and frequency-selective Rayleigh fading channels. An exact BEP expression for an OFDM system with windowing reception has been derived in [7].

9.10.2 Phase noise effects on OFDM

Phase noise effects introduced by the LO in any receiver can only be mitigated by improving the performance of the oscillator itself. Phase noise can be interpreted as a parasitic phase modulation in the oscillator's signal. Two different kinds of effects that are introduced by phase noise into the OFDM signal are common phase error and ICI [3]. When there is a timing error or a carrier phase offset, phase rotation will occur. The phase noise is the sum of the common phase error and the carrier phase noise. The common phase error is the same for all subcarriers, and can be corrected by some kind of phase rotation. Carrier phase noise is caused by the imperfection of the transmitter and receiver oscillators. The ICI resembles Gaussian noise.

When the ratio of the phase noise bandwidth B_ϕ to the OFDM inter-carrier spacing Δf, $B_\phi/\Delta f$ is small, the common phase error dominates over the ICI. This error can be corrected together with the channel effects by channel estimation, and the SEP is substantially reduced after the correction. As this ratio approaches unity, the ICI increases and the correction capabilities decrease. When this ratio is greater than unity, it cannot be corrected since the ICI dominates the phase noise.

SNR degradation caused by the phase noise is the same for both the OFDM and single-carrier systems. For small phase noise variance ($\sigma^2 \ll 1$), the SNR degradation is given by [3, 50]

$$\Delta\gamma = 10 \log_{10}\left(1 + \sigma^2 \frac{E_s}{N_0}\right) \quad \text{(dB)}, \tag{9.46}$$

where the variance σ^2 can be calculated based on the phase-noise pdf and is given by

$$\sigma^2 = \int_0^b \frac{2N_{0p}}{C} df, \tag{9.47}$$

with N_{0p} being the PSD of the phase noise and C the related carrier power. The frequency b is due to the band-limited phase noise or filtering in the receiver.

The influence of carrier phase noise on system performance has also been analyzed in [62]. The carrier phase noise is modeled as the Wiener process $\theta(t)$ with $E[\theta(t)] = 0$ and $E\left[(\theta(t + t_0) - \theta(t_0))^2\right] = 4\pi\beta|t|$, where β, in hertz, is the single-sided 3-dB line width of the Lorentzian power spectral density of the free-running carrier generator. For small impairments, the degradation in SNR can be approximated by [62]

$$D \approx \frac{11}{6\ln 10}\left(4\pi N \frac{\beta}{W}\right)\gamma \quad \text{(dB)}, \tag{9.48}$$

where $\gamma = E_s/N_0$, and W is the bandwidth of the OFDM signal.

Although the SNR degradation caused by the phase noise is independent of the number of subcarriers in the OFDM signal and the phase noise bandwidth, the situation is different for correcting the common phase error by any phase correction scheme that takes advantage of the pilot-based correction mechanism.

9.10.3 Influence of timing offset

Since the bandwidth of the phase noise spectrum is wider than the subcarrier spacing, this leads to energy spillage into other subchannels and thus ICI. When there is a timing error, the FFT window at the receiver spans samples from two consecutive OFDM symbols, leading to an inter-OFDM symbol interference.

Assuming that the Fourier transform of $f(t)$ is $F(\omega)$, the time-shift version of $f(t)$, $f(t-\tau)$, has the Fourier transform $e^{-j\omega\tau}F(\omega)$. Thus, a misalignment τ of the FFT window at the receiver leads to a phase error of $2\pi\Delta f\tau$ between two adjacent subcarriers. The phase error introduced has a considerable influence on the BEP performance of the OFDM system, especially in case of coherent detection. In order to use coherent detection, a phase correction mechanism has to be employed.

As long as the timing offset τ satisfies $0 \leq \tau \leq \tau_{\max} - T_g$, where $\tau_{\max}$ is the maximum channel delay spread and $T_g = T_{cp}$ is the guard time, it still introduces ICI due to the loss of orthogonality and thus a degradation in performance. In this case, the timing offset can be included into the channel estimation. If τ is not within this window, ISI occurs even if the phase shift is accounted for. The ICI loss can be approximated by [2, 59]

$$\text{ICI} \approx 10\log_{10}\left(2\left(\frac{\tau}{T_s}\right)^2\right) \quad \text{(dB)}. \tag{9.49}$$

From this, we see that timing synchronization is not a critical problem as long as the induced phase change is corrected, since typically $\tau \ll T_s$.

9.10.4 Implementation of synchronization

Synchronization is usually implemented in two steps: acquisition (coarse synchronization) phase and tracking (fine synchronization) phase, to achieve a minimal computational load and a minimal information overhead. Coarse timing and frequency acquisition relies on pilot symbols or pilot tones embedded into the OFDM symbols, while fine timing and frequency tracking exploits the cyclic extension of the OFDM signal. For TDMA-based OFDM systems, frame synchronization between a BS and an MS also needs to be maintained. This added redundancy can be further used for frequency synchronization as well as FFT window alignment.

Initially a frame synchronization circuit is used to detect the frame starting point, which is usually achieved by correlating the incoming signal with a known preamble. This circuit can also be used for initial gain control. Since timing offset only introduces a phase rotation, which linearly increases with the order of the subcarrier and does not violate orthogonality of the symbols, it can be compensated after FFT. Frequency offset influences the orthogonality, and must be corrected before applying FFT. This requires a fast coarse acquisition, and the PLL technique is not viable for this purpose. In stead, a pilot sequence is usually transmitted. Synchronization of the OFDM system can be performed in the following steps:

- Coarse timing recovery/frame synchronization;
- Coarse frequency offset correction;
- Fine frequency correction (after FFT);
- Fine timing correction (after FFT).

Timing offset and CFO detections can be implemented by using the correlation, ML, and MMSE approaches.

Synchronization using pilot symbols

Synchronization using pilot symbols is reliable and widely used in communications, although it has a small bandwidth overhead. This method can be used to synchronize frame (packet or slot) timing, symbol timing, and to estimate CFO at the initial stage of synchronization.

In [69], Schmidl and Cox proposed a synchronization scheme that uses a two-symbol training sequence. The first training symbol consists of two identical halves in the time domain. Correlation between the two halves is used for frame and symbol timing. The second training symbol is composed of two PN sequences, one on the odd frequencies for estimating the subchannels and the other on the even frequencies to determine the frequency offset. The second training symbol is used for fine frequency estimation. The selection of a particular PN sequence does not have much influence on the performance of the synchronization, but the PN sequence should be selected so as to make it easy to implement or to have a low PAPR.

In [16], the two pilot symbols modulated by two PN sequences in [69] are replaced by a single pilot symbol modulated by an m-sequence. This improved version reduces the overhead and maximizes the pilot symbol SNR at the receiver. The method can also estimate the subsample timing offset. However, its use of a matched filter leads to higher complexity. The Schmidl-Cox Scheme has been used for synchronization in IEEE 802.11a.

Timing synchronization

A positive offset in timing synchronization causes ISI between two successive OFDM symbols. Thus, the estimated start sample should be shifted several samples, although the shift decreases tolerance to multipath fading.

For the first training symbol $u(t)$, at time t in the first half, we have $u(t) = u(t + T/2)$, thus,

$$r^*(t)r(t + T/2) = |u(t)|^2 e^{\pi T \Delta f + \theta(t+T/2) - \theta(t)} \approx |u(t)|^2 e^{\pi T \Delta f}, \qquad (9.50)$$

where $r(t)$ is the received signal, $\theta(t)$ is the phase shift caused by the channel to all the samples, T is the symbol period, and Δf is the CFO; for a slow-fading channel, $\theta(t)$ is slowly changing. It is seen that the influence of the channel is negligible, but the frequency offset introduces a phase of $\pi T \Delta f$.

Timing synchronization can be achieved by cross-correlation. Frame/symbol timing synchronization can be based on (9.50). Assuming that there are $L = N/2$ samples in one-half of the first training symbol, the correlation between the samples in the two halves is defined by

$$R(d) = \sum_{m=0}^{L-1} r^*_{d+m} r_{d+m+L}, \qquad (9.51)$$

where d is the time index of the first sample in a window of $2L$ samples. At the start of a frame $R(d)$ has a peak.

Sampling clock offset is the misalignment between the DAC clock at the transmitter and the ADC clock at the receiver. The distortion arising from sampling frequency offset linearly increases with the index of the subcarrier. Sampling clock offset also introduces ISI. A sampling clock offset estimation method is given in [87].

Carrier synchronization

In the Schmidl-Cox scheme [69], at the receiver, the two identical halves in the first training symbol remain identical but there is a phase difference between them, which is caused by the CFO. Thus, the first training symbol is also used for coarse frequency offset estimation.

In (9.50), the CFO introduces a phase of $\phi = \pi T \Delta f$ in the correlation between two samples $T/2$ apart in the first training symbol. Since $R(d)$ is a sum of these correlations, the phase shift of $R(d)$ introduced by the CFO is also ϕ. If $|\phi| \leq \pi$, the CFO can be estimated by

$$\widehat{\Delta f} = \frac{\angle R(d)}{\pi T}. \qquad (9.52)$$

Otherwise, an additional frequency offset $2i/T$, for integer i, exists due to the periodic nature of the phase:

$$\widehat{\Delta f} = \frac{\angle R(d)}{\pi T} + \frac{2i}{T}. \qquad (9.53)$$

This additional frequency offset can be further determined by using the second training symbol. The estimated CFO with respect to the subcarrier spacing is $\hat{\epsilon} = \widehat{\Delta f} T$, where $1/T$ is the subcarrier spacing.

Carrier frequency offset estimation using one symbol with two identical halves can be modified into using two consecutive symbols carrying identical information [87]. Accordingly, (9.52) can be written as

$$\hat{\epsilon} = \widehat{\Delta f} T = \frac{1}{2\pi} \angle \left(\sum_{m=0}^{N-1} r_{d+m}^* r_{d+m+N} \right), \tag{9.54}$$

where r_{d+m} and r_{d+m+N} are the two consecutive OFDM symbols, and d is the time index of the first sample in a window of $2N$ samples.

Compensation of the frequency offset is performed by multiplying each distorted sample by a complex factor, α^m, where m is the index of the sample. In the case of two identical half symbols, $\alpha = e^{-j\pi \widehat{\Delta f} T/L} = e^{-j2\pi \epsilon/N}$. To save computation time, α^m can be recursively computed by using $\alpha^m = \alpha \alpha^{m-1}$ or using a lookup table.

Synchronization by correlating the cyclic extension

Both frequency- and time-domain synchronization can be simultaneously obtained by using the cyclic prefix by means of correlation [33, 34, 75]. In fact, most timing recovery methods for OFDM rely on correlation. The scheme is used for fine frequency and timing corrections after FFT.

Assuming that the CFO is $\Delta f = \epsilon/T$, the downconverted output will be $s(t)e^{j2\pi \epsilon t/T}$. If the signal is sampled at N times the symbol rate, the received signal samples in the AWGN channel are given by

$$r(k) = s(k - \tau_0) e^{j2\pi \epsilon (k - \tau_0)/N} + n(k), \tag{9.55}$$

where τ_0 is the timing offset in the samples.

For $2N + N_{cp}$ consecutive samples, where $N + N_{cp}$ samples correspond to a complete symbol with N_{cp} samples for the cyclic prefix. Assuming that the samples in the cyclic prefix are indexed as

$$\mathcal{I} = \{\tau_0, \tau_0 + 1, \cdots, \tau_0 + N_{cp} - 1\}, \tag{9.56}$$

the samples with indexes

$$\mathcal{I}' = \{\tau_0 + N, \cdots, \tau_0 + N + N_{cp} - 1\} \tag{9.57}$$

are replicas of the cyclic prefix. All received samples in the observation interval are identified as $r(1), \ldots, r(2N + N_{cp})$.

By correlating the samples in the cyclic prefix and their replicas, we have [75]

$$E\left[r(k)r^*(k+m)\right] = \begin{cases} \sigma_s^2 + \sigma_n^2, & m = 0 \\ \sigma_s^2 e^{-j2\pi \epsilon}, & m = N \\ 0, & \text{otherwise} \end{cases} \tag{9.58}$$

for $k \in \mathcal{I} \cup \mathcal{I}'$, where σ_s^2 and σ_n^2 are the signal power and the noise power, respectively. For $k \notin \mathcal{I} \cup \mathcal{I}'$, the correlation is zero. The correlation given in (9.58) can be used for estimation of the timing offset and the frequency offset.

The peak of the correlation (9.58) can be used to estimate the timing offset, and the correlation for $m = N$ can be used for estimating the frequency offset. More precise estimation of the timing offset and the frequency offset is given by the joint ML estimation of τ_0 and ϵ [39, 75].

Another approach for synchronizing the frame position and the carrier frequency using the cyclic prefix correlation has been proposed in [34]. The implementation can be simplified by only using the sign bits of the in-phase and the quadrature components of the received OFDM signal for frame synchronization and frequency offset compensation.

The cyclic prefix correlation technique does not introduce any spectrum overhead, and it is actually a blind technique. However, it has some disadvantages. The peak of the correlation output varies from symbol to symbol, due to the randomness of the data. For small N, the sidelobes of the correlation output are comparable to the peaks. The method is only valid for finding symbol timing, but cannot detect the start of a frame, such as the start of a data packet.

Other synchronization schemes

Other schemes for OFDM synchronization are also available in the literature. In the DVB standard, a null symbol is used as the first OFDM symbol in the time frame and the receiver detects it by monitoring the symbol energy [23]. This method is only valid for continuous transmission systems, since in other cases there is no difference between the null symbol and the idle period between signal bursts. Pilot designs for consistent frequency-offset estimation of OFDM systems in frequency-selective fading channels are treated in [43].

Virtual carriers have been exploited for blind CFO estimation [45]. The optimal virtual carrier placement that minimizes the CRLB of the CFO estimation is achieved by placing the virtual carriers with even spacing across the whole OFDM symbol [28]. This placement also maximizes the SNR and minimizes the theoretical MSE of the CFO estimation [86].

9.11 OFDM-based multiple access

OFDM is usually used as a modulation format for a single user, where all the subcarriers are used by a single user. For the purpose of multiple access, different users can be assigned different subcarriers, and each user can have multiple subcarriers. This strategy makes subcarrier administration very difficult, and orthogonality between users does not obtain.

More often, OFDM is combined with TDMA or packet radio. Each user occupies the full band during a time slot or a packet transmission. This strategy is used in IEEE 802.11a. OFDM can also be combined with FDMA, where each user is assigned a subset of adjacent subcarriers, and bands of different users are separated by frequency guard.

In OFDMA, users share subcarriers and time slots. Thus, OFDMA is actually a hybrid of FDMA and TDMA. Different users within the same cell use different tones, and thus do not interfere other users. OFDM and CDMA can be combined into MC-CDMA, where spreading can be either in the time or the frequency domain.

OFDMA

OFDMA has the advantages of the single-user OFDM that we discussed earlier. In addition, it is a flexible multiple-access technique for many users with wide variation in data rates and QoS requirements. OFDMA has a reduced PAPR as compared to OFDM, since each MS uses only a small fraction of all the subcarriers. Lower data rates and bursty data can be transmitted more efficiently than single-user OFDM, TDMA, or CSMA.

OFDMA systems are sensitive to multipath fading. Different subchannels experience different fading. For those strongly attenuated subchannels, error correction has to be introduced, and this leads to a reduced spectral efficiency. MC-CDMA can better solve this problem. OFDMA requires overheads in both directions: the transmitter requires CSI from the receivers, and the receivers also need the information of their subcarrier assignment.

The OFDMA approach allows for increased multiuser diversity and adaptive modulation, which can boost the capacity. Multiuser diversity is obtained by adaptive subcarrier allocation so that data are sent to one or a subset of users with good channel conditions. Adaptive modulation transmits as high a data rate as possible for a good channel and at a lower rate for poor channels.

OFDMA can be implemented using fast hopping across all tones in a predetermined pseudorandom pattern, making it an SS technology. This adds the advantages of FH systems in cellular deployment. OFDMA is at least three times more efficient than CDMA at the physical layer [12]. Compared to CDMA and TDMA, OFDMA takes advantage of the granularity of OFDM in its control layer design, resulting in efficient MAC and link layers for data. OFDMA eliminates some of the disadvantages of CDMA, such as the difficulty of adding capacity via microcells and the use of fixed bandwidth size. OFDMA is used in the mobile WiMAX (IEEE 802.16e).

In [84], an algorithm that combines subcarrier, bit, and power allocations for OFDMA has been proposed to minimize the overall transmit power for given transmission rates and QoS requirements of the users by using instantaneous CSI. This approach allows all the subcarriers to be used more effectively, because a subcarrier will be left unused only if it is in deep fade to all users and all users transmit in all the time slots.

When the number of subcarriers is high, OFDMA has the disadvantage of high PAPR, sensitivity to CFO, and high complexity at MSs. Based on a layered FFT structure, quadrature OFDMA (Q-OFDMA) [92] achieves the same guard-interval overhead and same bandwidth occupation as OFDMA systems, but with reduced PAPR and improved CFO robustness and frequency diversity. It also achieves low-complexity downlink receivers.

9.12 Performance of OFDM systems

The performance of the OFDM system in the AWGN channel is identical to that of a serial modem, since AWGN in the time domain corresponds to AWGN of the same average power in the frequency domain and the OFDM subcarriers are orthogonal. The optimum receiver for OFDM consists of a bank of correlators, one for each subcarrier. The error probability for each subcarrier is independent of the other subcarriers due to the orthogonal nature. The overall error probability can be obtained from the error probability of each subcarrier. For OFDM systems using a fixed signal constellation in all the subcarriers, the total BEP is dominated by the subchannels that have the worst performance. The overall capacity of the OFDM system is the sum of the individual subchannel capacities.

The error performance of OFDM systems in various fading channels has been investigated in the literature. In [80], the SEP performance of MQAM and BPSK OFDM systems over frequency-selective Rayleigh-fading channels with Doppler spread is analyzed. The SEP performance for the MQAM OFDM systems over frequency-selective fast Ricean fading with Doppler spread is analyzed in [91]. A closed-form expression for the BEP of DMPSK OFDM systems in frequency-selective slow/fast Rayleigh/Ricean fading channels with diversity reception has been derived in [48]. In [22], the SEP of OFDM systems in slow Nakagami-m fading channels is given in closed form, and the analysis extended to a system using MRC diversity reception.

Water filling or bit loading

Multipath incurs frequency-selective fading. The distribution of the SNR at any frequency follows a Rayleigh distribution. The performance of the OFDM system over a stationary channel can be optimized by maximizing the overall bit rate for a given error probability, or by minimizing the error probability for a given bit rate. When the total power is constrained and the symbol rate is set to a target value, each subchannel is usually assigned the same error probability to achieve the minimum overall error probability. This approach has been shown to be suboptimum [83]. The optimum solution is derived by applying constraint on the SEP of each subchannel to the target value, rather than the BEP. This finally leads to the well-known water-filling solution. For the more common case of minimization of the BEP for a fixed bit rate under a transmitter-power constraint, the BEP must be below some target value. There are some algorithms to solve this problem [14].

In OFDM, bit loading harnesses the channel knowledge. At frequencies with high SNR, higher-order modulation is used, while at frequencies with low SNR, lower-order modulation is employed or no information is sent at all. Bit loading can be based on maximizing the total rate subject to total-transmitted-power constraint, or minimizing the total transmitted power subject to the total-rate constraint. Bit loading approaches the water-filling solution for transmit-spectrum allocation in order to achieve the channel capacity. It is noted in [15] that the use of OFDM with equal power allocated on all subcarriers does not lead to any capacity reduction for large N and finite power.

Bit-loading algorithms can be classified into three categories: incremental allocation [35], channel-capacity approximation-based allocation [13, 56], and BEP expression-based allocation [25]. Incremental allocation starts from an all-zero allocation, and an additional bit is allocated to the subcarrier requiring the smallest incremental energy until either the total power or the aggregate BEP constraints are violated [35]. In [13, 56], the channel-capacity approximation is used for computing the bit allocation across all subcarriers. In [25], closed-form error-probability expressions are used for bit and power allocation on the subcarriers with a specified error-rate constraint.

Comparison with single-carrier techniques

Since OFDM is implemented with FFT/IFFT, it has a complexity of $O\left(T_m B \log B\right)$, as opposed to $O\left(B^2 T_m\right)$ for a standard time-domain equalizer, where $B \approx 1/T$ is the bandwidth or the data rate, T being the symbol period, and T_m the delay spread. A typical wireless environment experiences delay spread of up to 800 ns. For a data rate of 10 Mbits/s, this covers several symbols if a single carrier is used. An equalizer such as a DFE can be used, but there must be a sufficient number of taps. The equalizer coefficients must be trained for each packet, leading to a large overhead. OFDM is most suitable for high-speed transmission at say, 60 GHz.

Recently, SC-FDE have been shown to have similar performance and complexity compared to the OFDM technique, but have a number of advantages over OFDM [17] (refer to Section 6.2.6). It may become an alternative to OFDM in future-generation wireless systems.

Spread spectrum and OFDM represent two opposite approaches to combating frequency-selective channels. The DS-CDMA system spreads the bandwidth, and then uses the rake receiver for multipath separation. Due to imperfect separation, path crosstalk remains. In contrast, OFDM converts a frequency-selective channel into many flat-fading channels, enabling very simple receivers. Compared with DS-CDMA, OFDM has the advantages of computational efficiency, no near-far problem since no correlation is used, and fast synchronization. These are achieved at the expense of an SNR loss due to the guard interval. The SNR is changed by a factor

$$\delta = 1 - \frac{N_{\text{cp}}}{N + N_{\text{cp}}} = \frac{1}{1 + \frac{N_{\text{cp}}}{N}}, \tag{9.59}$$

where N_{cp} is the number of samples in the cyclic prefix and N is the number of subchannels. This SNR loss can be reduced by decreasing the ratio N_{cp}/N. In addition, the narrowband subcarriers generate flat-fading conditions, thus no frequency diversity can be exploited. The combination of the two approaches has been explored for future mobile radio systems [33, 38, 90].

9.13 Multi-carrier CDMA

Recently, the MC-CDMA system, as a combination of OFDM and CDMA, has received attention for feasibility of an air interface for 4G mobile communications [33]. MC-CDMA effectively mitigates multipath interference and also provides multiple-access capability.

Unlike DS-CDMA that applies spreading sequences in the time domain, MC-CDMA applies spreading sequences in the frequency domain, mapping a different chip to a different OFDM subcarrier. MC-CDMA transmits a data symbol on all the subcarriers simultaneously. The overall MC-CDMA transmitter can be implemented by a DS-CDMA spreader followed by an OFDM transmitter. A code symbol is first mapped to a vector by using an N-dimensional spreading sequence, N being the number of subcarriers. By performing symbol spreading on the ith symbol for user j, $d_{j,i}$, we have the spread symbols

$$\tilde{d}_{j,i} = d_{j,i}c_j, \quad j = 1, \ldots, K, \tag{9.60}$$

where K is the number of users, and $c_j = (c_{j,0}, \ldots, c_{j,N-1})$ is the code vector of the jth user, $c_{j,k}$ being the kth chip of the spreading sequence of the jth user. All the code vectors c_j are orthogonal, and the Walsh codes are especially suitable as the spreading codes. Instead of $(d_{1,i}, \cdots, d_{K,i})$, the sequence $(\tilde{d}_{1,i}, \ldots, \tilde{d}_{K,i})$ is then OFDM-modulated.

The transmitted signal of the ith symbol of the jth user, $s_{j,i}(t)$, is denoted by [33, 90]

$$s_{j,i}(t) = \sum_{k=0}^{N-1} d_{j,i}c_{j,k}e^{2\pi(f_0+kf_d)t}p(t-iT), \tag{9.61}$$

where f_0 is the frequency of the lowest subcarrier, f_d is the subcarrier separation, and $p(t)$ is the rectangular pulse with value 1 for $t \in [0, T]$ and 0 otherwise. When $f_d = 1/T$, the transmitted signal can be generated using IFFT.

When each user transmits M bits during a signaling interval, this leads to a spreading factor or processing gain $G = N/M$. When the number of users $K = G$, we get a fully loaded system, and the MAI dominates the system performance, which becomes poor. The fully loaded MC-CDMA system usually uses Walsh codes.

At the receiver, the signal is first OFDM-demodulated; the output is the convolution of the channel and the spread symbol sequence $(\tilde{d}_{1,i}, \ldots, \tilde{d}_{K,i})$. By using the jth user's spreading sequence, the ith symbol of the jth user, $d'_{j,i}$, can be obtained. By using a ZF or MMSE equalizer, the unspread symbols $(d_{1,i}, \ldots, d_{K,i})$ can be obtained.

The ith received symbol at the kth carrier is given as

$$r_{k,i} = \sum_{j=0}^{K-1} H_k d_{j,i}c_{j,k} + n_{k,i}, \tag{9.62}$$

where H_k is the frequency response of the kth subcarrier, and $n_{k,i}$ is the corresponding noise sample.

For the jth user, the final decision variable $d_{j,i}$ is given by

$$d_{j,i} = \sum_{k=0}^{N-1} c_{j,k} g_k r_{k,i}, \qquad (9.63)$$

where the gain g_k is given by the reciprocal of the estimated channel transfer factor of subcarrier k. To maintain orthogonality between different users, FDE must be performed on the received subcarrier symbols. BEP analysis has been performed over both the Rayleigh and Ricean fading channels for various equalization methods in [90].

MC-CDMA also has the PAPR problem. Spreading usually has no significant impact on PAPR [51]. The average transmitted power is proportional to the number of simultaneously used spreading codes [33]. MUD can also be implemented in MC-CDMA. Various multicarrier-based CDMA systems are described in [79].

Multicode MC-CDMA using MMSE combining provides a performance superior to that of OFDM for the same data rate and the same bandwidth [37]. The best performance is achieved by using the largest possible spreading factor, that is, the number of subcarriers, since by spreading the same data symbol over all subcarriers, frequency diversity can be maximally exploited by using MMSE combining [37].

Spreading codes for MC-CDMA

For MC-CDMA, the Walsh code and the orthogonal Gold code can be used. Complex orthogonal sequences such as the family of Frank codes and Zadoff-Chu codes have also been applied for MC-CDMA [64]. Compared to the orthogonal Gold code and the Walsh code, the Frank code and the Zadoff-Chu code have lower off-peak autocorrelation, which helps peak factor reduction and reliable code-acquisition for synchronization [33]. The family of Zadoff-Chu codes is shown to be similar to the set of Newman phases and Schroeder phases [33]. Both the Frank code and the Zadoff-Chu code have perfect periodic autocorrelations. The Zadoff-Chu code achieves the lowest peak factor among the four spreading sequences. The all-zero sequence of the Walsh code yields the highest possible autocorrelation, and hence yields the highest possible peak factor of N.

These spreading codes are used when all users occupy the same set of subcarriers. When different MSs are assigned nonoverlapping sets of subcarriers, we need only one spreading sequence, and any of the sequence that can produce a low power factor, such as the Newman phases, can be used.

Many results for CDMA systems such as the SIC receiver and the MMSE receiver can be modified and applied to MC-CDMA [29].

Fourier code

The columns of a Fourier matrix can be considered as orthogonal spreading codes. The spreading code for user k is $c_k = (c_{k,0}, c_{k,1}, \ldots, c_{k,L})^T$, and the chip $c_{k,l} = e^{-j2\pi lk/L}$. If Fourier spreading is applied in MC-CDMA systems, the FFT operation is canceled by the

IFFT for the OFDM operation if they are of the same size [8]. This yields an SC-FDE system with cyclic prefix. The Fourier code leads to an equal or lower PAPR, compared with the Walsh code [9]. The Fourier code is applied in DFT-spread OFDM, which is used for the uplink of LTE.

Zadoff-Chu code

The Zadoff-Chu code is defined as

$$
c_{k,l} = \begin{cases} e^{j2\pi k(ql+l^2/2)/L} & \text{(for } L \text{ even)} \\ e^{j2\pi k(ql+l(l+1)/2)/L} & \text{(for } L \text{ odd)} \end{cases} \tag{9.64}
$$

where q is any integer and k is prime with L. For L prime, a set of $L-1$ sequences is obtained. The Zadoff-Chu code has an optimum periodic ACF and a low constant-magnitude periodic CCF. When used for MC-CDMA, the PAPR bounds for MC-CDMA uplink signals are 2 for the Zadoff-Chu code, ≤ 4 for the Golay code, $\leq 2L$ for the Walsh code, and $\leq 2(t(m)-1-\frac{t(m)+2}{L})$ for the Gold code, where $t(m)$ is given by (8.8) [24].

Comparison with other techniques

The main advantage of MC-CDMA over other OFDM-based multiple access techniques is the inherent frequency diversity provided by MC-CDMA. This advantage is at the cost of higher MAI.

In DS-CDMA, the rake receiver is used to exploit the multipath diversity, but the number of fingers (diversity order) is limited by the complexity. In comparison, MC-CDMA can easily achieve a high diversity order by transmitting the same information on several subcarriers.

The maximum achieved order of frequency diversity is limited by [33, 65]

$$
L \approx \frac{W}{B_c} \tag{9.65}
$$

where W is the total bandwidth of the channel and B_c is its coherent bandwidth. Spreading over more than L subcarriers leads to no more increase in terms of diversity gain, but a rapid increase in MAI, which causes a decrease in the overall performance of the system.

In a single-cell scenario, MC-CDMA achieves frequency diversity when there is no CSI at the transmitter, while OFDMA achieves a higher capacity by exploiting multiuser diversity over the frequency domain with CSI at the transmitter [47]. In a multi-cell multiuser downlink scenario, OFDMA and MC-CDMA have been compared in terms of system ergodic capacity and system goodput in [10]. For high-speed data communications, OFDMA outperforms MC-CDMA; the achievable goodput of OFDMA can sometimes be even higher than the ergodic capacity of MC-CDMA. For voice users, MC-CDMA shows noticeably higher goodput than the OFDMA. These results are also applicable for the MIMO case [10, 47].

9.14 Other OFDM-associated schemes

Multi-carrier DS-CDMA and multi-tone CDMA

In addition to MC-CDMA, there are some other schemes for combining OFDM and CDMA, such as multi-carrier DS-CDMA (MC-DS-CDMA) and multi-tone CDMA (MT-CDMA) [33]. In MC-CDMA, at the transmitter the bitstream having a bit rate of $1/T$ is first spread, yielding a bit rate of N/T; then by assigning it to the N subcarriers, the bitstream on each subcarrier is reduced to $1/T$ again.

MC-DS-CDMA is very similar to MC-CDMA. In MC-DS-CDMA, the bit rate R_b is first reduced by a factor of N before being assigned to N carriers, each of the bitstreams is then spread by a factor of G, before applying MCM. To maintain orthogonality between subcarriers, MC-DS-CDMA has the subcarrier separation $f_d = R_b G/N$.

MT-CDMA [76] is similar to MC-DS-CDMA in data mapping and spreading, but it uses spreading codes which are approximately N_c times longer than that used in MC-DS-CDMA; thus, the processing gain of MT-CDMA is N_c times that of MC-DS-CDMA. Orthogonality between subcarriers is not maintained in MT-CDMA.

Non-FFT-based OFDM

The widely adopted OFDM schemes are all FFT-based. Other non-FFT-based schemes such as the wavelet-transform-based OFDM (wavelet OFDM) and the DCT-based OFDM have also been proposed recently. MCM is very similar to sub-band coding. Like sub-band coding, many MCM schemes that use this idea, such as wavelet-based fractal modulation [85], wavelet packet modulation [44], wavelet PAM [46], and overlapped discrete wavelet multitone [67], have been proposed. In addition, the MASK-OFDM can be efficiently implemented by DCT [88].

Transform-domain communication system

The transform-domain communication system (TDCS) [73] is a recent technique that synthesizes a smart adaptive waveform to avoid interference at the transmitter rather than at the receiver. TDCS avoids existing users and jammers by notching or removing their bands prior to generating the time-domain fundamental modulation waveform by using IFFT. Like MC-CDMA, TDCS also achieves multiple access using PN sequences. Like OFDM and MC-CDMA, TDCS is also FFT-based.

Generalized multi-carrier System

In order to retain high orthogonality between subcarriers as well as high spectral efficiency, the OFDM technology is usually used at low mobility (low Doppler spread) and a small time delay spread. In case of high mobility of the MS, where the Doppler shift and time delay spread are high, the generalized multi-carrier (GMC) scheme can be applied, since no assumption of orthogonality between subcarriers is made and each carrier signal can be transmitted on a non-flat-fading channel. GMC is a nonothogonal multicarrier transmission technology with a relatively high complexity.

Problems

9.1 IEEE 802.16e is designed to allow a delay spread of up to $\sigma_\tau = 20$ μs and the maximum speed of $v = 125$ km/h. In the OFDMA mode, the subcarrier spacing is designed to be $\Delta f = 11.16$ kHz, and the OFDM symbol duration is 100 μs. Determine the maximum frequency-domain spacing and time-domain spacing for the pilots.

9.2 For an OFDM system with $N = 4$ subcarriers, if the CP has a length of $N_g = 2$ and the wireless channel has a length of $L = 3$, the received signal is given by

$$r(n) = 0.8c(n) - 0.5c(n-1) - 0.3c(n-2).$$

Two blocks of binary data symbols are fed to the IDFT unit, $c_0 = [1, -1, -1, 1]^T$ and $c_1 = [-1, -1, -1, 1]^T$. Show the estimated symbols at the receiver. Assume the channel is known to the receiver. Compare this result with that of Problem 6.1.

9.3 Consider a signal with a bandwidth of 1 MHz and a data rate of 1 Mbit/s. The wireless channel has a delay spread of 10 μs.
(a) Design an OFDM system for this purpose. Give the number of subchannels, the modulation, and the data rate on each subchannel.
(b) If raised cosine pulse with the roll-off factor $\alpha = 1$ is used, and the subcarriers are separated by the minimum bandwidth necessary for retaining orthogonality, redesign the OFDM system.
(c) For case (b), assuming the SNR on each subchannel is 15 dB, find the maximum constellation size for MQAM modulation that can be transmitted over each subchannel with a target BEP of 10^{-4}, as well as the corresponding data rate of the system.

9.4 Find the data rate of an IEEE 802.11a system, if 16 out of the available 48 subchannels are BPSK-modulated with a rate-1/2 channel code and the remaining are 16QAM-modulated with a rate-3/4 channel code.

9.5 Evaluate the cost of the cyclic prefix in OFDM in terms of (a) extra channel bandwidth, and (b) extra signal energy.

9.6 Given $\beta = 1/150$, $N = 256$, $\delta = 0$, $\sigma_H^2(n) = 1$, $\sigma_s^2 = 1$, $\sigma_w^2 = 10^{-3}$. Using (9.43), plot the SNR degradation versus the subcarrier index n, for various estimated values of β, that is, for different values of $\hat{\beta}/\beta$. Compare your plot with Fig. 3 in [66].

References

[1] M. Al-Akaidi & O. Daoud, Reducing the peak-to-average power ratio using turbo coding. *IEE Proc. Commun.*, **153**:6 (2006), 818–821.

[2] J. G. Andrews, A. Ghosh, & R. Muhamed, *Fundamentals of WiMAX: Understanding Broadband Wireless Networking* (Upper Saddle River, NJ: Prentice Hall, 2007).

[3] A. G. Armada, Understanding the effects of phase noise in orthogonal frequency division multiplexing (OFDM). *IEEE Trans. Broadcast.*, **47**:2 (2001), 153–159.

[4] J. Armstrong, Peak-to-average power reduction for OFDM by repeated clipping and frequency domain filtering. *Electron. Lett.*, **38**:8 (2002), 246–247.

[5] A. R. S. Bahai, M. Singh, A. J. Goldsmith & B. R. Saltzberg, A new approach for evaluating clipping distortion in multicarrier systems. *IEEE J. Sel. Areas Commun.*, **20**:5 (2002), 1037–1046.

[6] R. W. Bauml, R. F. H. Fisher & J. B. Huber, Reducing the peak-to-average power ratio of multicarrier modulation by selected mapping. *Electron. Lett.*, **32**:22 (1996), 2056–2057.

[7] N. C. Beaulieu and P. Tan, On the effects of receiver windowing on OFDM performance in the presence of carrier frequency offset. *IEEE Trans. Wireless Commun.*, **6**:1 (2007), 202–209.

[8] K. Bruninghaus & H. Rohling, On the duality of multi-carrier spread spectrum and single-carrier transmission. In *Proc. Int. Workshop Multi-Carrier Spread Spectrum (MC-SS)*, Oberpfaffenhofen, Germany, Apr 1997, 187–194.

[9] A. Bury & J. Lindner, Comparison of amplitude distributions for Hadamard spreading and Fourier spreading in multi-carrier code division multiplexing. In *Proc. IEEE GLOBECOM*, San Francisco, CA, Nov/Dec 2000, 857–860.

[10] P. W. C. Chan, E. S. Lo, V. K. N. Lau, R. S. Cheng, K. B. Letaief, R. D. Murch & W. H. Mow, Performance comparison of downlink multiuser MIMO-OFDMA and MIMO-MC-CDMA with transmit side information – multi-cell analysis. *IEEE Trans. Wireless Commun.*, **6**:6 (2007), 2193–2203.

[11] R. W. Chang, Synthesis of band-limited orthogonal signals for multichannel data transmission. *Bell Syst. Tech. J.*, **45** (1966), 1775–1796.

[12] H.-H. Chen & M. Guizani, *Next Generation Wireless Systems and Networks* (Chichester, UK: Wiley, 2006).

[13] P. S. Chow, J. M. Cioffi & J. A. C. Bingham, A practical discrete multitone transceiver loading algorithm for data transmission over spectrally shaped channels. *IEEE Trans. Commun.*, **43**:2–4 (1995), 773–775.

[14] J. M. Cioffi, G. P. Dudevoir, M. V. Eyuboglu & G. D. Forney, Jr., MMSE decision-feedback equalizers and coding – part I: equalization results. *IEEE Trans. Commun.*, **43**:10 (1995), 2582–2594.

[15] A. Clark, P. J. Smith & D. P. Taylor, Instantaneous capacity of OFDM on Rayleigh-fading channels. *IEEE Trans. Inf. Theory*, **53**:1 (2007), 355–361.

[16] A. Coulson, Maximum likelihood synchronization for OFDM using a pilot symbol:
     algorithms; analysis. *IEEE J. Sel. Areas Commun.*, **19**:12 (2001), 2486–2494; 2495–
     2503.

[17] A. Czylwik, Comparison between adaptive OFDM and single carrier modulation with
     frequency domain equalization. In *Proc. IEEE VTC*, Phoenix, AZ, May 1997, **2**, 865–
     869.

[18] O. Daoud & O. Alani, Reducing the PAPR by utilisation of the LDPC code. *IET*
     Commun., **3**:4 (2009), 520–529.

[19] D. Dardari, V. Tralli & A. Vaccari, A theoretical characterization of nonlinear distor-
     tion effects in OFDM systems. *IEEE Trans. Commun.*, **48**:10 (2000), 1755–1764.

[20] J. Davis & J. Jedwab, Peak-to-mean power control in OFDM, Golay complementary
     sequences, and Reed-Muller codes. *IEEE Trans. Inf. Theory*, **45**:7 (1999),
     2397–2417.

[21] P. Dharmawansa, N. Rajatheva & H. Minn, An exact error probability analysis of
     OFDM systems with frequency offset. *IEEE Trans. Commun.*, **57**:1 (2009), 26–31.

[22] Z. Du, J. Cheng & N. C. Beaulieu, Accurate error-rate performance analysis of OFDM
     on frequency-selective Nakagami-m fading channels. *IEEE Trans. Commun.*, **54**:2
     (2006), 319–328.

[23] K. Fazel, S. Kaiser, P. Robertson & M. J. Ruf, A concept of digital terrestrial television
     broadcasting. *Wireless Pers. Commun.*, **2**:1/2 (1995), 9–27.

[24] K. Fazel & S. Kaiser, *Multi-Carrier and Spread Spectrum Systems: From OFDM and
     MC-CDMA to LTE and WiMAX*, 2nd edn (Chichester, UK: Wiley, 2008).

[25] R. Fischer, A new loading algorithm for discrete multitone modulation. In *Proc. IEEE
     GLOBECOM*, London, UK, Mar 1996, 724–728.

[26] R. L. Frank, Polyphase complementary codes. *IEEE Trans. Inf. Theory*, **26**:6 (1980),
     641–647.

[27] P. K. Frenger & N. A. B. Svensson, Decision-directed coherent detection in multi-
     carrier systems on Rayleigh fading channels. *IEEE Trans. Veh. Tech.*, **49**:2 (1999),
     490–498.

[28] M. Ghogho, A. Swami & G. Giannakis, Optimized null-subcarrier selection for
     CFO estimation in OFDM over frequency-selective fading channels. In *Proc. IEEE
     GLOBECOM*, San Antonio, TX, Nov. 2001, **1**, 202–206.

[29] S. Glisic, *Advanced Wireless Communications: 4G Technologies*, 2nd edn (Chich-
     ester, UK: Wiley-IEEE, 2007).

[30] M. J. E. Golay, Complementary series. *IRE Trans. Inf. Theory*, **7**:2 (1961), 82–87.

[31] S. H. Han & J. H. Lee, An overview of peak-to-average power ratio reduction
     techniques for multicarrier transmission. *IEEE Wireless Commun.*, **12**:2 (2005),
     56–65.

[32] S. H. Han, J. M. Cioffi & J. H. Lee, On the use of hexagonal constellation for peak-to-
     average power ratio reduction of an OFDM signal. *IEEE Trans. Wireless Commun.*,
     7:3 (2008), 781–786.

[33] L. Hanzo, M. Munster, B. J. Choi & T. Keller, *OFDM and MC-CDMA for Broadband
     Multi-User Communications, WLANs and Broadcasting* (New York: Wiley-IEEE,
     2003).

[34] M.-H. Hsieh & C.-H. Wei, A low-complexity frame synchronization and frequency offset compensation scheme for OFDM systems over fading channels. *IEEE Trans. Veh. Tech.*, **48**:5 (1999), 1596–1609.

[35] D. Hughes-Hartogs, *Ensemble Modem Structure for Imperfect Transmission Media*, U.S. Patents Nos. 4,679,227, Jul 1987; 4,731,816, Mar 1988; and 4,833,796, May 1989.

[36] A. D. S. Jayalath & C. Tellambura, SLM and PTS peak-power reduction of OFDM signals without side information. *IEEE Trans. Wireless Commun.*, **4**:5 (2005), 2006–2013.

[37] R. Kimura & E. Adachi, Comparison of OFDM and multicode MC-CDMA in frequency selective fading channel. *Electron. Lett.*, **39**:3 (2003), 317–318.

[38] V. Kuhn, *Wireless Communications over MIMO Channels: Applications to CDMA and Multiple Antenna Systems* (Chichester, UK: Wiley, 2006).

[39] N. Lashkarian & S. Kiaei, Class of cyclic-based estimators for frequency-offset estimation of OFDM systems. *IEEE Trans. Commun.*, **48**:12 (2000), 2139–2149.

[40] K. N. Le, Bounds on inter-carrier interference power of OFDM in a Gaussian scattering channel. *Wireless Pers. Commun.*, **47**:3 (2008), 355-C362.

[41] C. Li & S. Roy, Subspace-based blind channel estimation for OFDM by exploiting virtual carriers. *IEEE Trans. Wireless Commun.*, **2**:1 (2003), 141–150.

[42] Y. Li & L. J. Cimini, Bounds on the interchannel interference of OFDM in time-varying impairments. *IEEE Trans. Commun.*, **49**:3 (2001), 401–404.

[43] Y. Li, H. Minn, N. Al-Dhahir & A. R. Calderbank, Pilot designs for consistent frequency-offset estimation in OFDM systems. *IEEE Trans. Commun.*, **55**:5 (2007), 864–877.

[44] A. R. Lindsey, Wavelet packet modulation for orthogonally transmultiplexed communication. *IEEE Trans. Signal Process.*, **45**:5 (1997), 1336–1339.

[45] H. Liu & U. Tureli, A high-efficiency carrier estimator for OFDM communications. *IEEE Commun. Lett.*, **2**:4 (1998), 104–106.

[46] J. N. Livingston & C. Tung, Bandwidth efficient PAM signaling using wavelets. *IEEE Trans. Commun.*, **44**:12 (1996), 1629–1631.

[47] E. S. Lo, P. W. C. Chan, V. K. N. Lau, R. S. Cheng, K. B. Letaief, R. D. Murch & W. H. Mow, Adaptive resource allocation and capacity comparison of downlink multiuser MIMO-MC-CDMA and MIMO-OFDMA. *IEEE Trans. Wireless Commun.*, **6**:3 (2007), 1083–1093.

[48] J. Lu, T. T. Tihung, F. Adachi & C. L. Huang, BER performance of OFDMCMDPSK system in frequency-selective Rician fading with diversity reception. *IEEE Trans. Veh. Tech.*, **49**:4 (2000), 1216–1225.

[49] D. J. G. Mestdagh, P. Spruyt & B. Biran, Analysis of clipping effect in DMT-based ADSL systems. In *Proc. IEEE ICC*, New Orleans, LA, May 1994, 293–300.

[50] M. Moeneclaey, The effect of synchronization errors on the performance of orthogonal frequency-division multiplexed (OFDM) systems. In *Proc. COST 254 (Emergent Techniques for Communication Terminals)*, Toulouse, France, Jul 1997, Paper 5.1.

[51] A. F. Molisch, *Wireless Communications* (Chichester, UK: Wiley-IEEE, 2005).

[52] M. Muck, M. de Courville, and P. Duhamel, A pseudorandom postfix OFDM modu-lator – semi-blind channel estimation and equalization. *IEEE Trans. Signal Process.*, **54**:3 (2006), 1005–1017.

[53] S. H. Muller & J. B. Huber, OFDM with reduce peak-to-average power ratio by optimum combination of partial transmit sequences. *Electron. Lett.*, **33**:5 (1997), 368–369.

[54] B. Muquet, M. de Courville & P. Duhamel, Subspace-based blind and semi-blind channel estimation for OFDM systems. *IEEE Trans. Signal Process.*, **50**:7 (2002), 1699–1712.

[55] B. Muquet, Z. Wang, G. B. Giannakis, M. de Courville & P. Duhamel, Cyclic prefix-ing or zero padding for wireless multicarrier transmissions? *IEEE Trans. Commun.*, **50**:12 (2002), 2136–2148.

[56] S. Nader-Esfahani & M. Afrasiabi, Simple bit loading algorithm for OFDM-based systems. *IET Commun.*, **1**:3 (2007), 312–316.

[57] R. Nilsson, O. Edfors, M. Sandell & P. O. Borjesson, An Analysis of two-dimensional pilot-symbol assisted modulation for OFDM. In *Proc. IEEE ICPWC*, Bombay, India, Dec 1997, 71–74.

[58] H. Ochiai & H. Imai, Performance analysis of deliberately clipped OFDM signals. *IEEE Trans. Commun.*, **50**:1 (2002), 89–101.

[59] J. M. Paez-Borrallo, Multicarrier vs. monocarrier modulation techniques: An intro-duction to OFDM. In *Proc. Berkeley Wireless Research Center Retreat*, Jan 2000.

[60] A. Peled & A. Ruiz, Frequency domain data transmission using reduced compu-tational complexity algorithms. In *Proc. IEEE ICASSP*, Denver, CO, Jun 1980, 964–967.

[61] T. Pollet, P. Spruyt & M. Moeneclaey, The BER performance of OFDM systems using non-synchronized sampling. In *Proc. IEEE Globecom*, San Francisco, CA, Nov–Dec 1994, **1**, 253–257.

[62] T. Pollet, M. van Bladel & M. Moeneclaey, BER sensitivity of OFDM systems to car-rier frequency offset and Wiener phase noise. *IEEE Trans. Commun.*, **43**:2–4 (1995), 191–193.

[63] B. M. Popovic, Synthesis of power efficient multitone signals with flat amplitude spectrum. *IEEE Trans. Commun.*, **39**:7 (1991), 1031–1033.

[64] B. M. Popovic, Spreading sequences for multicarrier CDMA systems. *IEEE Trans. Commun.*, **47**:6 (1999), 918–926.

[65] J. G. Proakis & D. G. Manolakis, *Digital Signal Processing: Principle, Algorithms, and Applications*, 4th edn (Upper Saddle River, NJ: Pearson Prentice Hall, 2007).

[66] A.-B. Salberg & A. Swami, Doppler and frequency-offset synchronization in wide-band OFDM. *IEEE Trans. Wireless Commun.*, **4**:6 (2005), 2870–2881.

[67] S. Sandberg & M. Tzannes, Overlapped discrete multitone modulation for high speed copper wire communications. *IEEE J. Sel. Areas Commun.*, **13**:9 (1995), 1571–1585.

[68] A. Scaglione, G. B. Giannakis & S. Barbarossa, Redundant filterbank precoders and equalizers – I: unification and optimal designs; II: blind channel estimation, synchronization, and direct equalization. *IEEE Trans. Signal Process.*, **47**:7 (1999), 1988–2006; 2007–2022.

[69] T. M. Schmidl & D. C. Cox, Robust frequency and timing synchronization for OFDM. *IEEE Trans. Commun.*, **45**:12 (1997), 1613–1621.

[70] S. B. Slimane, Reducing the peak-to-average power ratio of OFDM signals through precoding. *IEEE Trans. Veh. Tech.*, **56**:2 (2007), 686–695.

[71] T. Strohmer & S. Beaver, Optimal OFDM design for time-frequency dispersive channels. *IEEE Trans. Commun.*, **51**:7 (2003), 1111–1122.

[72] G. L. Stuber, *Principles of Mobile Communication*, 2nd edn (Boston, MA: Kluwer, 2001).

[73] P. J. Swackhammer, M. A. Temple & R. A. Raines, Performance simulation of a transform domain communication system for multiple access applications. In *Proc. IEEE MILCOM*, Atlantic City, NJ, Oct–Nov 1999, **2**, 1055–1059.

[74] S. C. Thompson, A. U. Ahmed, J. G. Proakis, J. R. Zeidler & M. J. Geile, Constant envelope OFDM. *IEEE Trans. Commun.* **56**:8 (2008), 1300–1312.

[75] J.-J. van de Beek, M. Sandell & P. Borjesson, ML estimation of time and frequency offset in OFDM systems. *IEEE Trans. Signal Process.*, **45**:7 (1997), 1800–1805.

[76] L. Vandendorpe, Multitone direct sequence CDMA system in an indoor wireless environment. In *Proc. IEEE Symp. Commun. Veh. Technol.*, Delft, Netherlands, Oct 1993, 4.1.1–4.1.8.

[77] R. van Nee & A. de Wild, Reducing the peak-to-average power ratio of OFDM. In *Proc. IEEE VTC*, Ottawa, Canada, May 1996, **3**, 2072–2076.

[78] X. Wang, T. T. Tjhung & C. S. Ng, Reduction of peak-to-average power ratio of OFDM system using a companding technique. *IEEE Trans. Broadcast.*, **45**:3 (1999), 303–307.

[79] Z. Wang & G. B. Giannakis, Block spreading for multipath-resilient generalized multi-carrier CDMA. In G. B. Giannakis, Y. Hua, P. Stoica & L. Tong, eds., *Signal Processing in Wireless & Mobile Communications: Trends in Single- and Multi-user Systems*, **2** (Upper Saddle River, NJ: Prentice Hall, 2001), pp. 223–266.

[80] T. R. Wang, J. G. Proakis, E. Masry & J. R. Zeidler, Performance degradation of OFDM systems due to Doppler spreading. IEEE Trans. Wireless Commun., **5**:6 (2006), 1422–1432.

[81] S. Weinstein & P. Ebert, Data transmission by frequency division multiplexing using the discrete Fourier transform. *IEEE Trans. Commun.*, **19**:10 (1971), 628–634.

[82] T. A. Wilkinson & A. E. Jones, Minimisation of the peak-to-mean envelope power ratio of multicarrier transmission schemes by block coding. In *Proc. IEEE VTC*, Chicago, IL, Jul 1995, **2**, 825–29

[83] T. J. Willink & P. H. Wittke, Optimization and performance evaluation of multicarrier transmission. *IEEE Trans. Inf. Theory*, **43**:2 (1997), 426–440.

[84] C. Y. Wong, R. S. Cheng, K. B. Letaief & R. D. Murch, Multiuser OFDM with adaptive subcarrier, bit, and power allocation. *IEEE J. Sel. Areas Commun.*, **17**:10 (1999), 1747–1758.

[85] G. W. Wornell & A. V. Oppenheim, Wavelet based representations for a class of self-similar signals with application to fractal modulation. *IEEE Trans. Inf. Theory*, **38**:2 (1992), 785–800.

[86] Y. Wu, S. Attallah & J. W. M. Bergmans, On the optimality of the null subcarrier placement for blind carrier offset estimation in OFDM systems. *IEEE Trans. Veh. Tech.*, **58**:4 (2009), 2109–2115.

[87] W. Xiang, T. Pratt & X. Wang, A software radio testbed for two-transmitter two-receiver space-time coding OFDM wireless LAN. *IEEE Commun. Mag.*, **42**:6 (2004), S20–S28.

[88] F. Xiong, M-ary amplitude shift keying OFDM system. *IEEE Trans. Commun.*, **51**:10 (2003), 1638–1642.

[89] F. Xiong, *Digital Modulation Techniques*, 2nd edn (Boston, MA: Artech House, 2006).

[90] N. Yee, J.-P. Linnartz & G. Fettweis, Multicarrier CDMA in indoor wireless radio networks. In *Proc. IEEE PIMRC*, Yokohama, Japan, Sep 1993, 109–113.

[91] R. Y. Yen, H.-Y. Liu & W. K. Tsai, QAM symbol error rate in OFDM systems over frequency-selective fast Ricean-fading channels. *IEEE Trans. Veh. Tech.*, **57**:2 (2008), 1322–1325.

[92] J. A. Zhang, L. Luo & Z. Shi, Quadrature OFDMA systems based on layered FFT structure. *IEEE Trans. Commun.*, **57**:3 (2009), 850–860.

Antennas

10.1 Maxwell's equations

Maxwell's equations are a set of four fundamental equations that govern all electromagnetic (EM) phenomena [28]. Maxwell's equations can be in differential or integral form. In differential form, the four basic laws are given as

$$\nabla \times \boldsymbol{E} = -\frac{\partial \boldsymbol{B}}{\partial t} \qquad \text{(Faraday's law)} \tag{10.1}$$

$$\nabla \times \boldsymbol{H} = \frac{\partial \boldsymbol{D}}{\partial t} + \boldsymbol{J} \qquad \text{(Maxwell-Ampere law)} \tag{10.2}$$

$$\nabla \cdot \boldsymbol{D} = \rho \qquad \text{(Gauss's electric law)} \tag{10.3}$$

$$\nabla \cdot \boldsymbol{B} = 0 \qquad \text{(Gauss's magnetic law)} \tag{10.4}$$

where $\boldsymbol{E}$ is electric field intensity (volts/m^2), $\boldsymbol{D}$ electric flux density (coulombs/m^2), $\boldsymbol{H}$ magnetic field intensity (amperes/m), $\boldsymbol{B}$ magnetic flux density (webers/m^2), $\boldsymbol{J}$ electric current density (amperes/m^2), and ρ electric charge density (coulombs/m^3).

Among the four Maxwell's equations, any one can be derived from the other three by using the equation of continuity

$$\nabla \cdot \boldsymbol{J} = -\frac{\partial \rho}{\partial t} \qquad \text{(Equation of continuity)} \tag{10.5}$$

There are three constitutive relations

$$\boldsymbol{D} = \epsilon \boldsymbol{E} \tag{10.6}$$

$$\boldsymbol{B} = \mu \boldsymbol{H} \tag{10.7}$$

$$\boldsymbol{J} = \sigma \boldsymbol{E} \tag{10.8}$$

where ϵ, μ, and σ are, respectively, the permittivity (farads/m), permeability (henrys/m), and conductivity (siemens/m) of the medium. These parameters are tensors in anisotropic media and scalars in isotropic media.

The three independent equations in Maxwell's equations and the three constitutive relations make all the six variables in Maxwell's equations solvable.

Maxwell's equations in time-harmonic form are obtained by setting the field quantities as harmonically oscillating functions with a single frequency ω. Using the complex phasor notation, $e^{j\omega t}$, (10.1), (10.2), and (10.5) can be written in a simplified form as

$$\nabla \times \boldsymbol{E} = -j\omega\boldsymbol{B} \qquad (10.9)$$

$$\nabla \times \boldsymbol{H} = j\omega\boldsymbol{D} + \boldsymbol{J} \qquad (10.10)$$

$$\nabla \cdot \boldsymbol{J} = -j\omega\rho \qquad (10.11)$$

The time-domain solution can be derived by using the Fourier transform.

In order to solve Maxwell's equations, a number of boundary conditions must be specified to make the solution unique. For a second-order partial differential equation (PDE), the boundary conditions can be either the Dirichlet boundary condition (a value of the function ϕ is specified), a Neumann boundary condition (a value of the normal derivative, $\frac{\partial\phi}{\partial n}$, is specified), or a mixed boundary condition (a linear combination of a Dirichlet and a Neumann boundary condition).

10.2 Introduction to computational electromagnetics

Computational electromagnetics solves Maxwell's equations subject to a set of boundary conditions associated with the domains by using numerical techniques. They are now fundamental for antenna and microwave/RF design.

Three methods are widely used in computational electromagnetics, namely, the method of moments (MoM), the finite difference time domain (FDTD) method, and the finite element method (FEM). These methods are known as *full-wave* computational electromagnetics methods. The full-wave methods approximate Maxwell's equations numerically, without making any physical approximations. There are powerful commercial computer codes for all these methods.

The first step for all the three methods is to discretize the domain and this process is known as *meshing*; this subdivides the domain into many elements. As a rule of thumb, each wavelength is segmented into ten segments in each geometrical dimension to maintain sufficient accuracy. For the FDTD method, Maxwell's equations are then converted into difference equations within each element, iteration goes on in each element for each time instant and the result is passed to neighboring elements. No linear algebra is required. For the MoM and the FEM, the second step is to select the interpolation function, known as the *shape function*, which relates the spatial variation of the unknowns within each element. This is followed by constructing the equation for each element and assembling them to form a system of equations. Finally, the system of linear equations is solved by using a direct solver such as Gaussian elimination or forward and backward substitution based on LU-factorization, or by using a recursive solver such as the conjugate-gradient method [12].

A good introductory text for the three methods is by Davidson [10]. In addition to full-wave methods, asymptotic techniques, such as physical optics, geometrical optics, and the uniform theory of diffraction (UTD), are based on high-frequency approximation to Maxwell's equations, and thus the validity increases with frequency.

10.2.1 Method of moments

The MoM was first introduced by Harrington into computational electromagnetics [18]. This method is also known as the *boundary element method (BEM)*, where only the surface is meshed. The MoM is the most popular computational electromagnetics method for antenna engineering. It is also widely used in RF computational electromagnetics. The MoM is based on the integral form of Maxwell's equations, and has been traditionally applied in the frequency domain.

The Numerical Electromagnetic Code – Method of Moments, widely known as *NEC-2*, is a powerful program for antenna modeling, that is available in the public domain. NEC-2 has no graphic abilities. There are some commercial codes incorporating all the functionality of NEC-2, but with better graphical functions and other enhancements. Some examples of the MoM codes are FEKO, Advanced Design System (ADS), Momentum, Zeland's IE3D, SuperNEC, Ansoft's Ensemble, and Sonnet Software's EM; these codes can be used by inputting the geometry of the antenna.

The MoM uses an integral equation method to solve for the current density on the surface of the conductors. Once the current density on the surface of the antenna is known, the radiation pattern, directivity and gain can be computed using the far-field model. Input impedance, voltage standing-wave ratio (VSWR) and scattering parameters can also be calculated. It is suitable for simulation of wire antennas and printed antennas.

The MoM is very efficient for modeling structures that consist entirely of perfectly or highly conducting radiators or scatterers, when compared to the FEM or the FDTD method. Since it is a boundary element method, the boundary conditions are built into the MoM formulation, while for the FEM and the FDTD methods the boundary conditions must be explicitly imposed. The MoM includes the radiation condition automatically, and this makes it especially suitable for radiation or scattering problems.

The MoM is difficult to implement for arbitrary geometry and three-dimensional (3-D) structures that use nonhomogeneous, magnetodielectric, or anisotropic materials. These complex cases and more general Maxwell's equations can be solved using the FDTD method or the FEM, which is computationally more demanding.

10.2.2 Finite difference time-domain method

The FDTD method is derived from the differential form of Maxwell's equations. In difference methods, the derivatives are approximated by finite differences. The FDTD method directly approximates the differential operators in Maxwell's curl equations on a grid on time and space by using Yee's finite difference scheme [50]. The FDTD method is currently the most popular method for electrodynamic problems in computational electromagnetics. A comprehensive study of the FDTD method is given in [40].

The FDTD method may introduce numerical dispersion, that is, the phase velocity v_p and the wave (group) speed v_g are different. This can be minimized by selecting a time-step at the *Courant limit*, i.e., $\Delta t = \Delta z/c$, where c is the speed of the wave propagation, and the spatial step size Δz is no more than one-tenth of the minimum wavelength (corresponding to the maximum frequency). The frequency dispersion is frequency dependent, and worsens rapidly when the frequency is above a certain value.

Implementation of the FDTD method does not lead to complex matrix computation. In contrast, the MoM and the FEM lead to a large system of linear equations. For FEM, the coefficient matrix is sparse, while the MoM generates a full matrix. The FDTD method provides a very simple implementation of a full-wave solver, and much less work is required as compared to an MoM or FEM implementation.

For 3D FDTD, halving the mesh size leads to an increase in run time by a factor of 16, and doubling the frequency leads to an increase in run time by a factor of 32 to 45 [10]. Memory also becomes a serious issue in 3D FDTD.

Unlike the MoM and the FEM that operate in the frequency domain, the FDTD method operates in the time domain; as a result, by using a wideband source the FDTD method can compute a wideband response in one run, while the MoM and the FEM have to recompute the system response for each frequency. The result from the FDTD method is more suitable for computer visualization of the field dynamics. The Fourier transform is used to yield a frequency response. Thus, the FDTD method is preferable for wideband systems.

When a suitable absorbing boundary condition (ABC) is available, the FDTD method can also be applied to antennas. The perfectly matched layer (PML)-based ABC is a very accurate mesh truncation technique [6], and is integrated into the software package, Microwave Studio.

Commercial software packages for the FDTD method include Remcon's XFDTD, and Sonnet Software and CST's joint product, Microwave Studio. XFDTD is an implementation of the standard FDTD. Microwave Studio is a transient solver that uses the finite integration technique; The finite integration technique can be rewritten as a standard FDTD method for Cartesian grids. Zeland (now a part of Mentor Graphics) also provides Fidelity as a FDTD simulator.

10.2.3 Finite element method

The FEM is a general method for solving partial differential equations, subject to certain boundary conditions, and is very useful in providing solutions to engineering and physical problems. The FEM can be formulated via the conventional Ritz variational method or Galerkin's weighted residual method. Both methods start from the partial differential form of Maxwell's equations. The former finds a variational functional whose minimum corresponds to the solution, while the latter introduces a weighted residual and solves the weighting functions. In most applications, both methods lead to identical equations.

The FEM is traditionally formulated in the frequency domain. The FDTD method typically employs a rectangular, staggered grid, while the FEM allows elements of arbitrary geometry. This provides the FEM with more geometrical modeling flexibility by using triangular elements for two dimensions and tetrahedral elements for three dimensions. As with the FDTD, the FEM does not include the radiation condition. For open regions, as in radiation or scattering problems, one has to employ an artificial absorbing region within the mesh or a hybridization with the MoM to terminate the mesh.

The FEM is especially suitable for volumetric configuration with complex geometries and inhomogeneous material or material with frequency-dependent properties. The FEM can also couple electromagnetic solutions with mechanical or thermal solutions. The FEM is much more complex to implement than the FDTD method, but is more accurate than the FDTD. However, for electromagnetic eigenvalue problems, the standard FEM introduces spurious modes, while this does not occur in the FDTD.

Both the FEM and FDTD methods are theoretically capable of solving any problem, but they are computationally too expensive for problems that are more suitable for the MoM. For all the three methods, the mesh has to become finer for an increasing frequency, and all these methods scale badly with frequency. Some fast methods try to replace the traditional direct matrix solution algorithms with iterative solvers.

Several commercial software packages are available for finite element analysis of electromagnetic problems. Ansoft's High-Frequency Structure Simulators (HFSS) is the most popular. Ansys provides FEMLAB, and Agilent also provides its HFSS. There are a number of excellent texts on the FEM, including the one by Jin [20].

10.3 Antenna fundamentals

An antenna is a device that converts a guided wave into a free-space wave, or vice versa. It connects the energy of electrons and the energy of photons. Radiation is produced by accelerated or decelerated charge. Maxwell's equations constitute the theoretical foundation for antenna and microwave design.

The transmission between a transmit and a receive antenna is illustrated in Fig. 10.1. The link can be characterized by the Friis transmission formula [13]

$$\frac{P_r}{P_t} = \frac{A_{et}A_{er}}{r^2\lambda^2},\tag{10.12}$$

where P_r (in watts) is the received power, P_t (in watts) is the transmitted power, A_{et} (in m^2) is the effective aperture of the transmit antenna, A_{er} (in m^2) is the effective aperture of the receive antenna, r (in m) is the distance between the antennas, and λ (in m) is the wavelength.

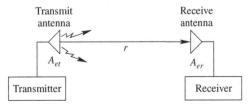

Transmission between two antennas.

10.3.1 Radiation patterns

The radiation pattern of an antenna is a three-dimensional quantity associated with its spherical coordinates θ and ϕ. The radiation pattern is also a reception pattern due to antenna reciprocity. The radiation pattern can be characterized by using its normalized field pattern or its normalized power pattern.

The normalized power pattern is defined by

$$P_n(\theta,\phi) = \frac{S(\theta,\phi)}{S(\theta,\phi)_{\max}}, \tag{10.13}$$

where $S(\theta,\phi)$ is the power per unit area, measured in watts/m^2, or the amplitude of the Poynting vector,

$$S(\theta,\phi) = \frac{E_\theta^2(\theta,\phi) + E_\phi^2(\theta,\phi)}{Z_0} \tag{10.14}$$

and $Z_0 = E/H = 376.7$ ohms is the intrinsic impedance of free space.

A radiation pattern has its main lobe and minor lobes. The half-power beamwidth (HPBW), or 3-dB beamwidth, and the beamwidth between the first nulls (FNBW) are important pattern parameters. A typical beampattern and its parameters are illustrated in Fig. 10.2. According to the principle of reciprocity, the pattern of an antenna is the same for both reception and transmission.

Beam area

The beam solid angle or beam area, Ω_A, of an antenna is defined by

$$\Omega_A = \int_0^{2\pi} \int_0^{\pi} |P_n(\theta,\phi)|^2 \sin(\theta)d\theta d\phi. \tag{10.15}$$

The pattern can usually be separated as $P_n(\theta,\phi) = P_e(\theta)P_a(\phi)$, where $P_e(\theta)$ and $P_a(\phi)$ are elevation and azimuth patterns, respectively. Ω_A can be approximated by using

$$\Omega_A \approx \theta_{HP}\phi_{HP}, \tag{10.16}$$

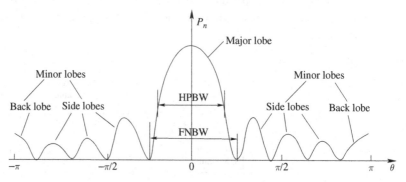

Figure 10.2

Antenna radiation pattern.

where θ_{HP} and ϕ_{HP} are the HPBWs in the two principal planes, and minor lobes are neglected.

Radiation intensity

The power radiated from an antenna per unit solid angle is referred to as the *radiation intensity U*, measured in watts/radian^2. Accordingly, the normalized power pattern can be defined by

$$P_n(\theta, \phi) = \frac{U(\theta, \phi)}{U(\theta, \phi)_{max}} \tag{10.17}$$

Unlike the definition given in (10.13), where the amplitude of the Poynting vector varies with the distance from the antenna, the radiation intensity is independent of the distance, assuming the case of far field of the antenna.

10.3.2 Antenna field zones

The field of an antenna is usually divided into two regions, the near field or Fresnel zone, and the far field or Fraunhofer zone. The boundary between the two regions is selected as

$$R_0 = 2\frac{L^2}{\lambda} \tag{10.18}$$

where L is the largest dimension of the antenna. The antenna field zones are shown in Fig. 10.3.

In the far or Fraunhofer field, the radiation is close to a transverse electromagnetic (TEM) wave radiating radially outward, and the energy drops at a rate of $1/r^2$, where r is the distance from the antenna. The power pattern is independent of the distance. In the near or Fresnel region, the power flow is not radial and the power pattern is dependent on the distance; the energy drops as $1/r^3$. There is oscillating energy flow as well as radially outward energy flow. The outflow is the radiated power, while the oscillating flow is reactive energy that is trapped in the vicinity of the antenna. The near-field or far-field patterns can be measured in an anechoic chamber.

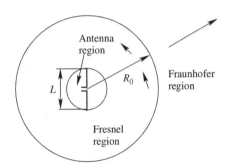

Antenna field zones.

10.3.3 Antenna gain and directivity

At the far field, the field E and the power density W, measured in watts/m^2, are related by (10.14)

$$W = S(\theta, \phi) = \frac{E^2}{Z_0}.$$ (10.19)

For an isotropic antenna, the power density at a distance R in the far-field is given by

$$W_{iso} = \frac{P_{rad}}{4\pi R^2},$$ (10.20)

where P_{rad} is the radiated power. Correspondingly, the radiation intensity U is given by

$$U_{iso} = \frac{P_{rad}}{4\pi}.$$ (10.21)

The gain of a transmit antenna in the direction (θ, ϕ) can be defined as the ratio of the radiation intensity to that for a lossless isotropic antenna

$$G(\theta, \phi) = \frac{U(\theta, \phi)}{P_T/(4\pi)},$$ (10.22)

where P_T is the output power when the antenna is matched to its feed.

Directivity D corresponds to the case when there is loss in the hypothetical isotropic antenna, and is defined by

$$D = \frac{P(\theta, \phi)_{max}}{P(\theta, \phi)_{av}} = \frac{U(\theta, \phi)}{P_{rad}/(4\pi)}.$$ (10.23)

The unit of D is usually dBi, denoting decibels over isotropic. Directivity can also be defined by

$$D = \frac{4\pi}{\Omega_A}.$$ (10.24)

A specified directivity parameter is a primary goal for many antenna structures.

Directivity and gain are thus related by

$$G(\theta, \phi) = \eta_\Omega D(\theta, \phi),$$ (10.25)

where $\eta_\Omega = P_{rad}/P_T \leq 1$ is the ohmic or radiation efficiency of the antenna, and is given by

$$\eta_\Omega = \frac{R_{rad}}{R_{rad} + R_L},$$ (10.26)

R_L being a loss resistance and R_{rad} the radiation resistance of the antenna. R_L always exists due to conductance loss, and is a function of frequency.

The spatial resolution of an antenna can be defined as half the beamwidth between the first nulls, FNBW/2. Usually, this value is approximately equal to HPBW. The number of point sources of radiation uniformly distributed over the sky that can be resolved by an antenna is given by

$$N = \frac{4\pi}{\Omega_A} = D.$$ (10.27)

The impedance presented by an antenna to a transmission line can be represented by a 2-terminal network with an equivalent impedance Z. This impedance is called the *terminal or driving point impedance*. If the antenna is isolated and lossless, the terminal impedance is the same as the self-impedance of the antenna, which contains a self-resistance (radiation resistance) and a self-reactance. The self-impedance is the same for both reception and transmission. According to the reciprocity theorem for antennas, the antenna impedance for both transmitting and receiving operation modes is the same.

10.3.4 Effective area and effective height

For a receive antenna, its effective area or aperture A_e, measured in m^2, is defined as the ratio of the received power P to the power density W of the planar wave

$$A_e = \frac{P}{W},\tag{10.28}$$

where the plane wave is assumed to have the same polarization as that of the antenna. D and A_e are related by

$$D = 4\pi \frac{A_e}{\lambda^2}.\tag{10.29}$$

All antennas have an effective aperture, which can be calculated using this equation.

For an aperture antenna such as a horn antenna or a parabolic reflector antenna, the effective aperture A_e is less than its physical aperture A_p due to the nonuniform field response in the aperture antenna. The aperture efficiency, $\epsilon = A_e/A_p$, is usually 50 to 80% for horn and parabolic reflector antennas. Large dipole and patch arrays can achieve a higher aperture efficiency.

Another parameter related to the aperture is the effective height h_e, measured in meters. It is defined as the ratio of the induced voltage to an incident field E of the same polarization

$$h_e = \frac{V}{E}.\tag{10.30}$$

The effective height is useful for characterizing transmitting tower-type antennas. A_e and h_e are related by

$$A_e = \frac{h_e^2 Z_0}{4R_{\text{rad}}}.\tag{10.31}$$

The quality factor, Q, of an antenna is defined by [25]

$$Q = \frac{\text{energy stored per second}}{\text{energy lost per second}} = \frac{X}{R_{\text{rad}} + R_L} = \frac{f_c}{B},\tag{10.32}$$

where f_c is the center frequency and B is the bandwidth. Increasing R_{rad} or R_L reduces Q, and thus increases the bandwidth.

10.3.5 Antenna temperature

The noise received as electromagnetic radiation by an antenna as well as the thermal noise in the antenna itself can be modeled by a noise temperature or temperature of the antenna's radiation resistance. For a receive antenna, the equivalent noise temperature, measured in kelvins (K), is defined by

$$T_{ant} = T_A + T_{th} = \frac{G_N(f)}{k_B}, \tag{10.33}$$

where the antenna temperature T_A is due to the electromagnetic radiation and T_{th} is the thermal noise in the antenna. The equivalent noise is assumed to be white with the one-sided noise power spectral density (PSD) $G_N(f)$, measured in watts/Hz, and k_B is Boltzmann's constant $k_B = 1.38 \times 10^{-23}$ J/K. T_A is determined by T_s, the temperature or brightness temperature of the sky or source that the antenna points toward, and can be calculated by

$$T_A = \frac{1}{\Omega_A} \int_0^{2\pi} \int_0^{2\pi} T_s(\theta, \phi) P_n(\theta, \phi) d\Omega, \tag{10.34}$$

where $d\Omega = \sin\theta d\theta d\phi$. Note that the cosmic background noise is independent of frequency and is 2.7 K everywhere in the sky. Radio noise is treated in ITU-R Rec. P.372.

The overall noise temperature of a radio receiver is given by

$$T_{sys} = T_{ant} + T_e, \tag{10.35}$$

where T_e is the equivalent input noise temperature of the receiver.

10.3.6 Polarization

Polarization of an electromagnetic wave is referred to as the orientation of the electric field vector, $\boldsymbol{E}$. For a wave traveling along the z-direction, if $\boldsymbol{E}$ is along the x- or y-direction only, the wave is linearly polarized in the x- or y-direction. In general, the electric field has both x- and y-components, and the wave is referred to as *elliptically polarized*:

$$\boldsymbol{E} = \hat{x}E_x \sin(\omega t - \beta z) + \hat{y}E_y \sin(\omega t - \beta z + \phi), \tag{10.36}$$

where $\beta = \frac{2\pi}{\lambda}$ is the wave number and ω is the angular frequency. The elliptical polarization is characterized by axial ratio, the ratio of the major to minor axes of the polarization ellipse. When $E_x = E_y$, the wave is *circularly polarized*; it is left-hand circular polarization (LHCP) if $\phi = 90°$, or right-hand circular polarization (RHCP) if $\phi = -90°$.

An antenna is blind to a wave of opposite polarization. Most cosmic radio sources are unpolarized, and any polarized antennas can receive only half of the available power [25]. Dipoles generate a linearly polarized wave, and end-fire antennas such as axial-mode helical antennas generate circular polarization. Circular polarization can also be generated by a pair of crossed $\lambda/2$ dipoles that have equal currents with quadrature phases; transmission is along the direction that is perpendicular to the antenna plane. The circular polarizations are opposite in the two axial directions.

The receive voltage is proportional to the dot-product of the electric field and the polarization vector of the receive antenna. The received power has a polarization loss factor

$$\epsilon_{\text{pol}} = |e_i \cdot e_r|^2, \tag{10.37}$$

where e_i is the unit electric field vector of the incident wave and e_r is the unit polarization vector of the received antenna. For this reason, a linearly polarized antenna can only receive half the power of a circularly polarized signal reaching it, since $e_i = (\hat{x} + \hat{y})/\sqrt{2}$ and $e_r = \hat{y}$, making $\epsilon_{\text{pol}} = 0.5$. Similarly, an LHCP antenna is completely mismatched to an RHCP wave, and a horizontally polarized antenna is completely mismatched to a vertically polarized wave.

Cross-polarization discrimination

A radiation pattern can be expressed as

$$E = E_\theta(\theta, \phi)\hat{\theta} + E_\phi(\theta, \phi)\hat{\phi}, \tag{10.38}$$

where $E_\theta(\theta, \phi)$ and $E_\phi(\theta, \phi)$ are the amplitudes of the electric field vector in the elevation and azimuth directions, respectively.

In wireless communications, the interactions of environment not only attenuate the transmitted signals, but also depolarize them. Cross-polarization discrimination (XPD) is defined as the power ratio of the copolarization and cross-polarization components of the mean incident wave. A higher XPD corresponds to less energy coupled in the cross-polarized channel. XPD can be defined for the two cases: azimuth transmission (χ_θ) and elevation transmission (χ_ϕ)

$$\chi_\theta = \frac{E\left[|E_{\theta\theta}|^2\right]}{E\left[|E_{\theta\phi}|^2\right]}, \quad \chi_\phi = \frac{E\left[|E_{\phi\phi}|^2\right]}{E\left[|E_{\phi\theta}|^2\right]}, \tag{10.39}$$

where $E_{\theta\phi}$ denotes the received ϕ-polarized electric field of the transmitted θ-polarized electric field. Similar explanations apply for $E_{\theta\theta}$, $E_{\phi\phi}$, and $E_{\phi\theta}$. XPD can be approximated by a Gaussian distribution [44].

10.3.7 Receiving and transmitting power efficiency

Transmit and receive antennas can be characterized by their equivalent circuits, as shown in Fig. 10.4. The antenna is modeled by a network $Z_A = R_A + jX_A$. The impedance at the generator or receiver side is assumed to be purely resistant. The antenna impedance Z_A can be impedance matched to a lossless transmission line of the characteristic impedance $Z_0 = R_A$ by using a stub tuner (refer to Section 11.5). The input impedance of antennas can be analyzed by using a network analyzer.

In the transmitting case, the radiation power via the antenna is given by

$$P_{\text{rad}} = \frac{R_{\text{rad}}}{R_{\text{rad}} + R_L + R_t} P_t, \tag{10.40}$$

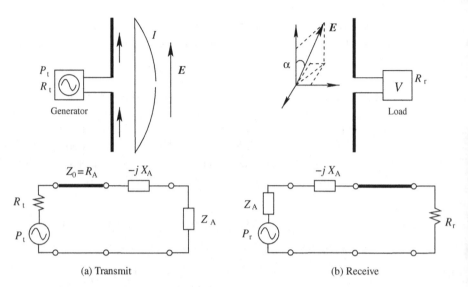

Figure 10.4 The equivalent circuits for (a) a transmit and (b) a receive antenna. Note that the transmission line segment is composed of two parts. One part is used for impedance matching, and it can be treated as an impedance of $-jX_A$. The other part is a lossless transmission line of characteristic impedance $Z_0 = R_A$.

where R_{rad} is radiation resistance of the antenna, R_L is the loss resistance of the antenna, $R_A = R_{rad} + R_L$, P_t is the transmitted power, and R_t is the resistance of the transmitter (generator). When R_L and R_t are negligible, P_{rad} is close to P_t and the radiating efficiency approaches 100%.

The maximum power delivered to the antenna occurs when the conjugate matching occurs

$$R_A = R_{rad} + R_L = R_t,$$ (10.41)

where R_t is the resistance of the transmitter. The power supplied by the transmitter during conjugate matching is given by [5]

$$P_t = \frac{V_g^2}{4\,(R_{rad} + R_L)},$$ (10.42)

where V_g is the voltage amplitude of the generator. Half of this power is dissipated as heat in the internal resistance R_t and the other half is delivered to the antenna.

In the receiving case, in case of perfect match, the receiver impedance R_r should be equal to the antenna radiation resistance R_A, that is, $R_r = R_A$. In this case, the power received by the receiver is

$$P_r = \frac{R_r}{R_r + R_A} P_R = 0.5 P_R,$$ (10.43)

where P_R is the power collected by the antenna. Thus, at best, 50% of the total power collected by the antenna is transferred to the receiver, while the other half is scattered [25].

10.4 Antennas for wireless communications

Antennas are used to radiate electromagnetic energy by using an oscillating current distribution. An antenna can be either a resonant or a nonresonant antenna. A resonant antenna uses resonance to increase the radiating current. A resonant antenna radiates almost all the radio signal fed to it, if working at a resonant frequency; otherwise, a large portion of the fed signal is not radiated. This makes resonant frequency only suitable for narrowband use. Conventional wired antennas and patch antennas are resonant antennas.

A nonresonant antenna, also called *frequency-independent antenna*, has approximately constant input impedance and radiation characteristics over a wide range of frequency. The dimensions of a nonresonant antenna are decided by its lower and higher frequency limits. A nonresonant antenna can be designed as a wideband antenna such as a log-periodic dipole antenna or a spiral antenna, but its size is usually very large.

Wired antennas, such as dipole and monopole antennas, and patch antennas are most widely used in wireless communications. The dipole has a favorable radiation pattern, but has a relatively large size and needs a differential feed. The microstrip-patch antenna is inexpensive, lightweight, and easy to manufacture. The performance of microstrip printed-dipole arrays are compared to that of arrays of free-standing dipoles in a MIMO channel in [30, 43]. The effects of parameters such as the dielectric thickness and the permittivity on the MIMO performance of the printed-dipole array are analyzed as well in [43]. The rectangular patch has attractive radiation characteristics and has a good polarization, thus it is most commonly used. These antennas typically have small bandwidth.

Aperture antennas are also used for wireless communications. The parabolic reflector is usually employed for satellite communications, and the horn antenna and the corner reflector can be used as sector antennas in a BS. The helical antenna is also widely used for satellite and space communications.

For antenna design, the important parameters are the directivity and gain, bandwidth, and input impedance. Impedance matching between the antenna and the coaxial cables with a 50-ohms impedance is necessary to improve the VSWR or power reflection. Most commercial antennas have a VSWR of 1.5:1 or less, and the corresponding reflected power is 4% of the incoming power.

10.4.1 Antennas for base stations

For cellular BSs, the coverage area must be large. This requires the radiation pattern to be omnidirectional in the horizontal (azimuthal) plane. The directivity can be increased by decreasing the beamwidth in the vertical (elevation) plane. Dipoles, monopoles, or folded dipoles can be used for this purpose. For cellular and point-to-multipoint communications using BSs with multiple sectors, the beamwidth of each antenna can be adjusted to its sector, and directional antennas such as Yagi or corner reflectors can be applied. Dead spots arise due to the presence of nulls in the radiation pattern. Dead spots lead to call drops, and should be avoided. The nulls in the antenna pattern may increase the number of handoffs, resulting in increased management overhead of the system.

The vertical beamwidth may vary depending on coverage. This beam may be tilted downward so that the main beam is toward its coverage area to reduce interference to neighboring cells. Depending on the radiation characteristics, bandwidth as well as installation, the type of antennas used for BSs may be dipoles, monopoles, patch arrays, corner reflectors, or aperture antennas such as horn antennas.

For indoor use such as wireless LAN, a leaky line, which is a coaxial cable with a leaky outer conductor, can be used as the radiation antenna. The leaky-wave antenna is a kind of traveling-wave antenna. Parabolic reflectors are widely used in satellite communications and broadcasting systems. The parabolic reflector is very effective for enhancing the gain of the antenna, and the antenna is installed at the focus of the reflector.

Diversity can effectively increase the system capacity. Space diversity uses multiple antennas, and polarization diversity makes use of the low correlation between oppositely polarized electric field components. Diversity based on multiple antennas can be easily implemented in BSs, since there is no size restriction as in the MS case. Cross-polarized antennas are also widely used in cellular BS installations as it reduces spacing needs and tower loads. With the use of MIMO technology in the next-generation wireless communications, cross-polarization will become an approach to exploit diversity in both BSs and MSs. 3GPP/3GPP2 have defined a cross-polarized channel model for MIMO systems [1].

10.4.2 Antennas for mobile stations

The antennas for MSs should have generally an omnidirectional antenna pattern, because the signal arrives in a continuously changing random direction and the MS is changing its orientation.

For MSs, a vertical polarized antenna such as a vertical dipole or monopole antenna made of a flexible antenna element, called *whip antenna*, is usually desirable. Such an antenna is subject to size restriction, and it is required to have adequate bandwidth and high efficiency. A retractable whip antenna is usually used in mobile handsets. Such a whip antenna can be treated as a monopole with the phone chassis as the ground. The length of the whip can be either $\lambda/4$ or $3\lambda/8$. The use of $3\lambda/8$ is to shift the current maximum away from the user and also to reduce the current on the chassis. In order to reduce the size, a dipole or monopole antenna can be embedded in a dielectric medium. In this case, the size can be reduced by a factor of $1/\sqrt{\epsilon_r}$, where ϵ_r is the relative permittivity of the dielectric.

A normal-mode helical antenna can be used in combination with a whip antenna. When the whip antenna is retracted, the helical antenna becomes the major receive antenna. Dual band can be easily achieved in one helical antenna by using two pitch angles. A helical antenna can be modeled as an array of loop antennas. It has an elliptical polarization, which approximates a circular polarization when the number of turns is large.

The sleeve antenna [21] avoids the big ground plane in the monopole by using a coax feed line that passes through a metallic sleeve without direct contact. The sleeve is used as a choke to prevent the antenna current from leaking into the outer surface of the coaxial cable. The sleeve moves the virtual antenna feed up the monopole. The bandwidth increases because the current at the feed point remains nearly constant over a wide frequency range. The overall length of the sleeve and the monopole is $\lambda/4$. It is inexpensive, compact, and widely used in MSs such as vehicles.

Antennas can also be installed inside MSs. Planar antennas using length $\lambda/4$ microstrip are usually mounted on the chassis. The microstrip can meander to adapt to the size restriction. Such internal antennas may have reduced sensitivity and allow reduced transmission power. Planar antennas of inverted-F shape can be used when dual-band is required [31]. By electrically tuning the frequency using a PIN diode, a dual-band planar inverted-F antenna is capable of operating in several frequency bands, covering the 850, 900, 1800, 1900 and 2000 MHz frequency bands that are used for GSM, PCS, and UMTS systems, with a total efficiency over 40% [24]. Design of the various antennas mentioned in this section is given in [25].

Electro-textiles are conductive fabrics that interpolate conductive metal/polymer threads with normal fabric threads. These fabrics can be integrated into clothing and they are washable, durable and flexible. Antennas made of electro-textile materials are now available for distributed body-worn electronics in wireless BANs [23, 32].

Minimization of antennas

MSs are restricted by size. Thus the antennas attached to them must be very small, but with acceptable performance. Reducing the size of an antenna will influence its bandwidth, gain, efficiency, and polarization purity, and also influence its feeding, since the transmission depends on the wavelength.

In order to use multiple antennas in MSs, measures must be taken to mitigate the influence of mutual coupling. Fractal geometry can be used for miniaturizing antennas. Fractals have no characteristic size, and they are composed of many copies of themselves at different scales. The fractal contours can add electrical length in less volume [16]. Increasing the fractal dimension of the antenna leads to a higher degree of miniaturization. Fractal antennas are more effective in decreasing mutual coupling for antenna arrays, since for the same center-to-center spacing, the fractal antennas have a larger edge-to-edge separation. Fractals have been used for miniaturization of antennas and for designing multiband, wideband or even ultra wideband (UWB) antennas [45]. Fractal elements and arrays are also ideal candidates for use in reconfigurable systems.

10.5 Dipole antennas

The Hertzian electric dipole, often called *short dipole*, is a pair of electric charges, which vary sinusoidally with time but have equal magnitude with opposite signs, at the two ends of an electrically short length. On the short length of the dipole, the current is uniformly distributed.

The Hertzian dipole cannot efficiently radiate energy due to the need of a high voltage to generate large current. A small horizontal loop antenna may be treated as the magnetic counterpart of a short vertical dipole: They have the same far-field patterns, but with opposite polarization [25].

For a Hertzian dipole of length Δl along the z-axis, the E- and H-fields are given by [25]

$$E_r = Z_0 \frac{I_0 \Delta l \cos\theta}{2\pi r^2} \left(1 + \frac{1}{j\beta r}\right) e^{-\beta r}, \tag{10.44}$$

$$E_\theta = jZ_0 \frac{\beta I_0 \Delta l \sin\theta}{4\pi r} \left(1 + \frac{1}{j\beta r} - \frac{1}{(\beta r)^2}\right) e^{-\beta r}, \tag{10.45}$$

and $E_\phi = 0$, $H_\phi = E_\theta / Z_0$, $H_r = H_\theta = 0$, where $Z_0 = 120\pi$ is the space impedance and I_0 is the constant current on the wire.

The radiation resistance R_{rad}, in ohms, is given by

$$R_{\text{rad}} = \frac{2\pi}{3} \left(\frac{\Delta l}{\lambda}\right)^2. \tag{10.46}$$

In free space $\eta = 120\pi$ ohm, and thus

$$R_r = 80\pi^2 \left(\frac{\Delta l}{\lambda}\right)^2 = 789 \left(\frac{\Delta l}{\lambda}\right)^2. \tag{10.47}$$

For the radiator to be efficient, its length must be comparable to the wavelength. This explains why radiation is not considered in low-frequency circuit design.

The directivity is given by

$$D = \frac{3}{2} \sin^2 \theta. \tag{10.48}$$

The commonly used antennas are the wire, or thin-wire, antennas. The $\lambda/2$ dipole antenna, as shown in Fig. 10.5, is a centrally fed element with a sinusoidally varying current.

A slot antenna is complementary to a dipole or a strip, if the dipole or the strip is cut from a metal sheet, leaving the slot. They have the same pattern, but with E and H interchanged. This is shown in Fig. 10.6. Radiation occurs from both sides of the sheet.

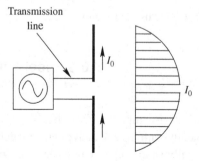

Figure 10.5   The $\lambda/2$ dipole antenna.

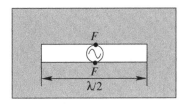 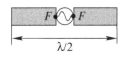

Figure 10.6 A $\lambda/2$ slot antenna which is a slot cut from an infinite flat sheet, and a complementary $\lambda/2$ dipole antenna.

10.5.1 Wire dipole antennas

For the thin-wire linear dipole, the current distribution on the wire is close to sinusoidal, if the wire diameter is less than $\lambda/100$. The current distribution is a standing wave that is generated from the forward traveling wave and reflected wave of equal amplitude from the distal end of the wire. When the total length is equal to one half of a wavelength, the antenna is at resonance.

Radiation pattern

For the thin-wire linear dipole, the current on the wire can be approximated by

$$I(z') = I_{\max} \sin\left[\frac{2\pi}{\lambda}\left(\frac{l}{2} - |z'|\right)\right]. \tag{10.49}$$

The current is zero at the ends, $z' = \pm l/2$.

The far field is given by [25]

$$H_\phi = \frac{j[I_0]}{2\pi r}\left[\frac{\cos[(\beta L \cos\theta)/2] - \cos(\beta L/2)}{\sin\theta}\right], \tag{10.50}$$

$$E_\theta = Z_0 H_\phi, \tag{10.51}$$

where L is the length of the dipole, r is the distance from the center of the dipole to the far-field point, $[I_0] = I_0 e^{j\omega[t-(r/c)]}$, I_0 being the amplitude of the sinusoidal current on the wire, $\beta = 2\pi/\lambda$ is the wave number, and $Z_0 = 120\pi$ is the space impedance. The radiation pattern is a function of θ only, and the radiation pattern is omnidirectional in the azimuthal plane.

If half the length of a dipole is $\lambda/4$, that is, $L = \lambda/2$, we get the basic $\lambda/2$-dipole:

$$P(\theta) = \frac{\cos^2\left(\frac{\pi}{2}\cos\theta\right)}{\sin^2\theta}. \tag{10.52}$$

The 3-dB beamwidth is $78°$, as compared to $90°$ for the short dipole. The directivity (gain) is 1.64 or 2.15 dB.

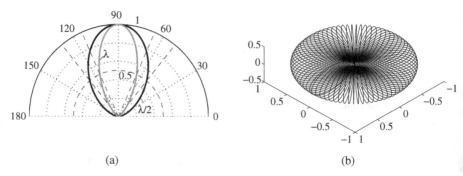

(a)   (b)

Figure 10.7 (a) Dipole patterns for the half-wave ($\lambda/2$) and full-wave (λ) dipoles. (b) The 3-D pattern for the $\lambda/2$ dipole.

Example 10.1: The normalized patterns of the dipole antennas for $L = \lambda/2$ (half-wave) and $L = \lambda$ (full-wave) are plotted in Fig. 10.7a, at any azimuthal angle ϕ. Note that the elevation angle θ is defined over $[0, \pi]$. The beamwidths between half-power points are $78°$ and $47.8°$ for the $\lambda/2$ and λ antennas, respectively. A 3-D view of the pattern for the $\lambda/2$ dipole is plotted in Fig. 10.7b.

Radiation resistance

The radiation resistance can be calculated by equating the radiated power, which is calculated by integrating the Poynting vector over a large sphere, to a power delivered to the radiation resistance at the maximum current

$$P = \oiint \mathbf{S} \cdot \mathbf{ds} = \frac{1}{2}\sqrt{\frac{\mu}{\epsilon}} \oiint |H_\phi|^2 \, ds = \left(I_0/\sqrt{2}\right)^2 R_{\text{rad}}, \tag{10.53}$$

where R_{rad} is the radiation resistance at the current maximum point I_0, which is the center of the antenna or at the terminals of the transmission line for the $\lambda/2$-dipole.

The radiation resistance is equal to the self-resistance. For the $\lambda/2$-dipole, the radiation resistance R_{rad} is 73.2 ohms. A general result for the self-resistance of dipole of any length, $R_{11} = R_{\text{rad}}$, is given in [25]. The terminal impedance also has an inductive reactance and the impedance is given by

$$Z = 73.2 + j42.5 \quad \text{(ohms)}. \tag{10.54}$$

Since the phase velocity of the radio wave along the wire is slightly less than the free-space velocity, the wavelength is slightly less than that in free space. By shortening the antenna by a few percent to 0.475λ, the reactance is reduced to zero, and thus the antenna is resonant. In this case, the radiation or terminal resistance is about 67 ohms.

The $\lambda/2$-dipole has a terminal resistance of 73.2 ohms, while the connecting transmission lines have a different impedance. For example, the coaxial cable has an impedance of 50 ohms, while a 2-wire line has a characteristic impedance of 300 to 600 ohms. An impedance transformer is required to match the impedance.

The dipole so arranged has a vertical polarization. A vertical electric field will induce a maximum voltage at the antenna output $V = V_{max}$, while a horizontal electric field will induce zero voltage at the antenna output. At an angle of α, the induced voltage is given by

$$V = V_{max} \cos \alpha. \tag{10.55}$$

10.5.2 Baluns

When a dipole is fed at the center by a coaxial line, as illustrated in Fig. 10.8 a net current (I_3) is induced on the outside of the coaxial cable. This current is not shielded and will radiate.

For feeding the dipole antenna, a two-wire line can be used to support the dipole so as to prevent the current from flowing to the surface of the outer conductor of the coaxial cable. This prevents the entire coaxial cable from becoming a radiating antenna. Such a structure is called a *balun*, which is an acronym for *balanced-to-unbalanced transformer*. A balun changes a single-ended signal (signal referenced to ground), such as an unbalanced coaxial line, to a balanced signal with equal potentials but of opposite polarity. The balanced output can be treated as two separated ports with anti-phase signals. The balun connects the dipole (with a balanced structure) and the coaxial cable (with an unbalanced structure).

In addition to application in antenna feeding networks, baluns are also used in mixers and frequency multipliers. The balun functions as an impedance transformer. It can also be used as an antiphase power splitter. There are many balun structures in the literature. The planar version of the Marchand balun is very attractive for its planar structure and wideband performance [35]. A balun may have a limited bandwidth.

The feed network is used for power distribution between the transmitter and the antenna as well as impedance matching. Proper impedance matching makes the reflection coefficients and VSWR to be within the specified levels. In the receiver case, a balun is not necessary if the SNR is adequate.

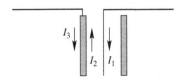

Figure 10.8   The currents in the coaxial line when feeding a dipole antenna.

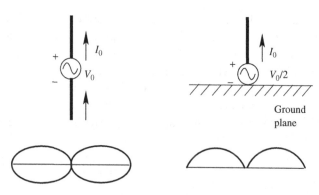

Figure 10.9   **Monopole vs Dipole.**

10.5.3 Wire monopoles

The monopole is a variant of the dipole that uses the ground plane to provide the other half of the antenna. It has half the architecture of the dipole and as such, its self-impedance is half that of the dipole. For the monopole of length $\lambda/4$, its self-impedance is $Z = 36.6 + j21.2$ ohms. Comparison of a dipole and a monopole is shown in Fig. 10.9.

A monopole has twice the directivity of its dipole counterpart. A popular monopole antenna is the quarter-wave ($\lambda/4$) monopole. For the $\lambda/4$ monopole, the directivity is 3.28 or 5.15 dBi. To produce the same field strength as a dipole, a monopole above the ground needs only half the power. While at the receiver side, the gain of a monopole is only half that of the dipole.

The monopole antenna is unbalanced, and thus avoids the use of a balun. Both the dipole and monopole antennas are widely used in wireless communications.

10.6 Patch antennas

Patch antennas are also widely used in wireless communications. They can be used in MSs for cellular communications, in MSs and BSs for wireless networking, and also for beamforming at the BSs for satellite communications. Printed microstrip antennas have very little manufacturing cost, and they can be made as resonant-slot antennas to better control polarization.

Patch antennas are commonly a rectangular metal patch on a thin layer of dielectric above a ground plane. A patch antenna of length L, width W, and thickness t is shown in Fig. 10.10. Feeding is at the center of one side of the patch. This produces linearly polarized radiation with a maximum broadside to the patch. Feeding typically uses either a coaxial probe or a microstrip line. When fed by a microstrip line, a quarterwave transformer can be used for impedance matching between the patch and the microstrip line.

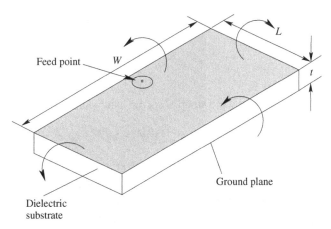

Figure 10.10

Patch antenna. The feed line can be coax feed or transmission line.

10.6.1 Microstrip antennas

Analysis and design of microstrip patch antennas can be based on the simple but approximate transmission-line model [7], or on the accurate but computationally complex cavity model [5, 27]. Both methods provide physical insight. In the transmission-line model, the microstrip radiator element is viewed as a transmission-line resonator with no transverse field variations, and the radiation occurs mainly from the fringing fields; the patch can be modeled as two slots that are separated by the length of the resonator. In the cavity model, the region between the patch and the ground plane is treated as a cavity that is enclosed by magnetic walls around the periphery and electric walls at the top and bottom. The far-field radiation is calculated from the equivalent magnetic current around the periphery. More complex numerical methods include the MoM [29] and FDTD methods. The result based on the transmission-line model is introduced here.

The patch can be modeled by $n = W/t$ parallel-plate microstrip transmission lines of length L. The thickness t is usually a few percent of wavelength in the dielectric. The microstrip characteristic impedance Z_c is given by [25]

$$Z_c = \frac{Z_0}{(n+2)\sqrt{\epsilon_r}} = \frac{Z_0}{\left(\frac{W}{t} + 2\right)\sqrt{\epsilon_r}}, \tag{10.56}$$

where two microstrip transmission lines are added to account for the fringing effect, and $Z_0 = 377$ ohms is the free-space impedance. The resonant length L is critical, and is usually a few percent less than $\lambda/2$, where λ is the wavelength in the dielectric, $\lambda = \lambda_0/\sqrt{\epsilon_r}$.

Radiation from the patch can be modeled by two radiating slots, one at the left and the other at the right of the pad [25]. The slots are of length W and width of a few $\lambda/100$. If $W = \lambda$, the input impedance is calculated as $R_{\text{in}} \approx 50$ ohms. When W is smaller, R_{in} increases in proportion to the reduction.

The radiation pattern can be obtained by combining that of two in-phase slot dipoles. The radiation pattern of the patch is broad, with a typical directivity $D \approx 4$ or 6 dBi.

The radiation resistance of a resonant $\lambda/2$ patch is given by [25]

$$R_r = 90 \frac{\epsilon_r^2}{\epsilon_r - 1} \left(\frac{L}{W}\right)^2 \quad \text{(ohms)}, \tag{10.57}$$

and $X_r = 0$.

For VSWR < 2 and small t/λ_0, the bandwidth is empirically given by [25]

$$\text{BW} = 3.77 \frac{(\epsilon_r - 1)}{\epsilon_r^2} \frac{W}{L} \frac{t}{\lambda_0}, \tag{10.58}$$

which is a linear function of t. However, t must be small, since a large t will lead to larger surface waves, more fringing leakage, and smaller directivity. Thus, the microstrip antenna has an inherently narrow bandwidth of only a couple of percent. An increase in W leads to an increase in bandwidth. However, W should be less than λ to prevent excitation of higher-order modes. A commonly used substrate for microstrip antennas is fiberglass-reinforced synthetic substrate, with ϵ_r typically between 2.1 and 2.6, and low dielectric loss (tan δ within 0.0006 and 0.002) [26].

10.6.2 Broadband microstrip antennas

Regularly shaped microstrip antennas are featured by narrow bandwidth, lower gain, and lower power-handling capability. By use of thick substrates with a low dielectric constant, the bandwidth can be improved to 5% to 10%. Bandwidth can be enhanced by modifying the shape of the patches, by using stacked (multilayer) or co-planar parasitic patches, or by using impedance-matching networks. Many bandwidth-enhancement techniques for microstrip antennas have been reviewed in [26].

Among various shape-modifying techniques, the technique of adding a U-shaped slot or using an L-shaped probe can provide bandwidths in excess of 30% [37, 47]. The U-slot patch antenna consists of a probe-fed rectangular patch with a U-shaped slot, which is cut symmetrically around the center of the patch; this antenna has an average gain of 7 dBi, and the L-probe-fed rectangular patch achieves an average gain of 7.5 dBi [37]. With the use of L-slot, L-shaped feed line and truncated corner, a wideband circularly-polarized slot antenna achieves a gain of around 6–7.5 dBi [47].

The bandwidth is usually defined as the frequency range over which VSWR is less than 2 or 1.5. The bandwidth of a microstrip antenna is related to its quality factor Q by [26]

$$\text{BW} = \frac{\text{VSWR} - 1}{Q\sqrt{\text{VSWR}}}. \tag{10.59}$$

The planar monopole antenna provides extremely wideband impedance characteristics. An array of planar monopoles can be used for wideband beamforming.

A reconfigurable microstrip antenna can adjust its operating frequency, radiation patterns, and polarization. It has a compact size. The patch antenna with switchable slot (PASS) is a reconfigurable slot-loaded patch antenna that inserts a switch in the center of the slot to control its configuration. The switch can be a PIN diode or a micro-electromechanical system (MEMS)-based switch. PASS has been design for dual-band wireless LAN or GPS applications [46]. Frequency-reconfigurable antenna structures that incorporate switches, such as RF MEMS and PIN diodes, for frequency selection are desirable for mobile multiradio platforms [48].

10.7 Polarization-agile antennas

Polarization is a key parameter in antenna design. Polarization-agile antennas [14] are those antennas whose polarization can be dynamically changed at different times. Polarization-agile antennas can make use of polarization diversity in wireless communications to mitigate multipath fading. Different polarizations can be used for frequency-reuse radio transceivers, where the two orthogonal polarizations are used, one for transmitting and the other for receiving. Polarization-agile antennas can also be used in MIMO systems.

A circularly-polarized microstrip antenna can be implemented by exciting two orthogonal modes with equal magnitude. A simple polarization-agile antenna can be implemented by using a switchable phase shifter, as shown in Fig. 10.11. By controlling the phase difference between ϕ_1 and ϕ_2, four types of polarization, i.e., two opposite linear polarizations and two opposite circular polarizations, can be achieved. A square microstrip antenna, which is fed at the four middle points of the edges with equal amplitude but phases 0, 90°, 180°, and 270°, results in an excellent axis ratio (AR) in the entire VSWR bandwidth [26]. A similar result is obtained for circular microstrip antennas. A square-patch antenna with a slot along the diagonal direction can realize circular polarization with a single feed [26, 46]. A square-patch antenna embedded with a pair of L-type slits can switch its polarization between linear and circular polarization by using PIN diodes [39]. By using switches such as PIN diodes or RF MEMS, the polarization can be configured. Some RF MEMS-switch controlled pixel-patch antennas are capable of both polarization and frequency reconfiguration.

Wave polarizers can be used to convert between a linearly polarized wave and an elliptically or circularly polarized wave. A polarizer can be designed by using several conductive-strip or dielectric slab gratings. By suitably selecting grating separation and rotation angles or the depth of the slabs, a very small pattern distortion and insertion

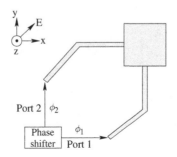

Figure 10.11   Polarization-agile antenna using switchable phase shifter.

loss can be achieved. A polarization cascade matrix gives the reflection and transmission properties between the incident wave and the output linearly polarized wave. For each grating, the elliptically polarized wave is represented by two orthogonal linearly polarized waves. Each grating slab will retard the phase of an electric component by an angle, and thus change the polarization. The theory underlying this is given in [19]. A parasitic axial-mode helix can also be used as a polarizer to transform linearly polarized radiation to a circularly polarized wave [25].

10.8 Antenna arrays

Antenna arrays are usually used to provide desired radiation patterns. In a phased array, the phase of each element can be varied so as to adjust the radiation pattern. The elements can be antennas of the same type. This is illustrated in Fig. 10.12. All the branches are fed by a common feed. The feed cables for all the branches are made to be of equal length by using the corporate structure. Otherwise, the cables introduce different phase delays, which, in addition to the desired phase changes, must be compensated by phase shifters. Each antenna element is usually assigned with equal feed power by using power splitters, and phase shifters and attenuators are then connected to each element to adjust its phase and amplitude independently.

Three types of antenna arrays are usually used: uniform linear arrays, uniform planar arrays and uniform circular arrays, where all the antenna elements are identical. An important parameter for antenna arrays is the array factor, also known as the *beamforming gain*.

The application of antenna arrays in smart antenna or MIMO systems is different from that of the conventional phased arrays. In a phased array, all the elements may have a common feed, whereas each element in a smart antenna or MIMO system has a separate feed whose current and phase can be adjusted independently. For application in smart antenna systems, uniform linear arrays or uniform planar arrays are conventionally used.

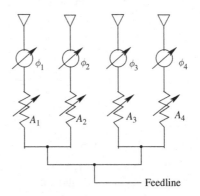

Figure 10.12   Phased array with phase shifters and attenuators. The array has a corporate structure.

Uniform circular arrays are presently gaining more popularity due to their compact size and symmetry [11]. Uniform linear arrays have ambiguity in a differentiating the elevation angles θ and $-\theta$. In practice, the field of view of a uniform linear array is restricted to $120°$, or a loss in spatial resolution will occur. For BSs that require omnidirectional antenna coverage, multiple uniform linear arrays along different sectors can be used. However, uniform circular arrays provide a better solution. Uniform circular arrays are usually composed of monopoles. Uniform planar arrays are more complex for implementation in smart antenna or MIMO systems.

For use in BSs, multiple patch antennas can be used as a phased array by suitably designing the feeding system so that it has a desired beam pattern. These patch arrays can be placed on a planar board or on a cylinder. Printed patch array antennas are popular due to their ease of manufacture using microstrip lines, which are etched on one side of a PCB. A multilayer PCB is usually used so as to accommodate the connections and circuitry. Patch arrays are also used to replace the parabolic dishes for satellite and space communications. Although the microstrip element is inherently narrowband, log-periodic patch arrays achieve broadband ratios of up to 4 to 1 with moderate gain.

Parasitic array antennas

In conventional array antennas or antenna arrays, all antenna elements are active. For digital beamforming, each antenna element is connected to its own RF circuit branch and ADC. This leads to a significant increase in the dc power and cost. A switched parasitic antenna offers characteristics similar to an array antenna with several fixed beams, but uses only one RF branch. Due to the compact size and low cost, it is suitable for small wireless devices [33].

A switched parasitic antenna can be designed as an array of $\lambda/2$-dipoles, composed of one central active element and two co-linear parasitic elements. The two parasitic elements may be either open- or short-circuited by using PIN diodes and digital control signals [33].

The electronically steerable parasitic array radiator (ESPAR) antenna is a low-power-consumption, compact, smart antenna [38]. The inter-element spacing can be as small as 0.05λ [41]. The ESPAR antenna is a reactively controlled directive-array antenna, in which only the central element is actively fed and connected to the RF port while all the surrounding elements are parasitic. By tuning the load reactance of parasitic elements, beam steering is achieved. The load reactances are implemented using reversely biased varactor diodes. ESPAR is especially suitable for low-power-consumption, low-cost MSs.

10.8.1 Array factor

For the uniform linear array, when the elements have identical amplitude, but progressive phase shift α across the array, the array factor is given by

$$\text{AF} = \sum_{n=1}^{N_a} e^{j(n-1)\psi} = \frac{\sin(N_a\psi/2)}{N_a \sin(\psi/2)}, \tag{10.60}$$

where N_a is the number of array elements, d is the inter-element spacing, and

$$\psi = \frac{2\pi}{\lambda} d \cos\theta + \alpha, \tag{10.61}$$

with θ being the elevation angle and λ the wavelength. The array factor can be designed by suitably selecting α. In fact, both the amplitude and phase of the antenna elements can be adjusted to control the array factor. The array factors for the uniform planar array and the uniform circular array for identical amplitude and progressive phase shift of antenna elements are derived in [5].

A distributed antenna is a phased array that is composed of multiple identical antenna elements with currents of the same amplitude and phase. A distributed antenna provides a better power efficiency than a single antenna for the same total delivered power [11]. This also introduces reliability.

Beampattern design for antenna arrays requires the pattern to satisfy some specifications, and some optimization techniques are used to select the optimum amplitude and phase at each antenna element. For example, a uniform linear array with amplitudes of Dolph-Tchebysheff distribution and uniform phase difference between adjacent elements can generate optimum patterns for a specified sidelobe level with all sidelobes of the same level [25].

10.8.2 Mutual coupling and spatial correlation

For smart antenna or MIMO systems, the capacity offered by multiple antennas may not be realized when the antenna elements are highly correlated. Physical limitations on the MSs force multiple antennas to be spaced closely. The spatial correlation coefficient ρ is defined to measure the relationship between the signals of two antennas as [42]

$$\rho = \frac{\left| \int_0^{2\pi} \int_0^{\pi} E_1^*(\theta,\phi) E_2(\theta,\phi) P(\theta,\phi) \sin\theta \, d\theta \, d\phi \right|^2}{\iint E_1^*(\theta,\phi) E_1(\theta,\phi) P(\theta,\phi) \sin\theta \, d\theta \, d\phi \iint E_2^*(\theta,\phi) E_2(\theta,\phi) P(\theta,\phi) \sin\theta \, d\theta \, d\phi}, \tag{10.62}$$

where E_i, $i = 1, 2$, is the electric field of antenna i when all other antennas are matched, $P(\cdot)$ is the distribution of DoA (θ, ϕ), and superscript $*$ denotes conjugation. Mutual coupling and antenna spacing can change E_i, and hence, the spatial correlation. Mutual coupling from other antenna elements or nearby conducting objects modify the induced current on an antenna element; thus, many parameters of the antennas including spatial correlation are modified. Mutual coupling as well as spatial correlation is mostly influenced by the spacing between antennas. To reduce the influence of mutual coupling, the distance between antenna elements needs to be at least half a wavelength, and a spacing of one half-wavelength is usually recommended.

Mutual coupling introduces mutual impedance that changes the radiation property of each antenna. For two identical thin-wire linear dipoles that are placed side by side, the mutual impedance is given in [25]. For the special case of the antenna length $L = n\frac{\lambda}{2}$ for n odd, the mutual resistance and mutual reactance, measured in ohms, are given by [25]

$$R_{21} = 30 \left\{ 2\text{Ci}(\beta d) - \text{Ci}\left[\beta \left(\sqrt{d^2 + L^2} + L \right) \right] - \text{Ci}\left[\beta \left(\sqrt{d^2 + L^2} - L \right) \right] \right\}, \quad (10.63)$$

$$X_{21} = -30 \left\{ 2\text{Si}(\beta d) - \text{Si}\left[\beta \left(\sqrt{d^2 + L^2} + L \right) \right] - \text{Si}\left[\beta \left(\sqrt{d^2 + L^2} - L \right) \right] \right\}, \quad (10.64)$$

where d is the distance between the two parallel antennas, Ci and Si are the cosine and sine integrals,

$$\text{Ci}(x) = \gamma + \ln(x) + \int_0^x \frac{\cos t - 1}{t} dt, \quad \text{Si}(x) = \int_0^x \frac{\sin t}{t} dt, \quad (10.65)$$

with $\gamma = 0.577$ being Euler's constant, R_{21} and X_{21} the mutual resistance and mutual reactance on antenna 2 caused by antenna 1, and

$$Z_{21} = R_{21} + jX_{21} = Z_{12} = R_{12} + jX_{12}. \quad (10.66)$$

Mutual impedances between two thin-wire dipoles of various configurations are also given in [25].

Example 10.2: The mutual resistance and reactance as a function of the distance between the two antennas, given by (10.63) and (10.64) are plotted in Fig. 10.13 for $\lambda/2$ dipoles. It is seen that the total resistance for each antenna, $R_{11} - R_{21}$, is significantly lower than the self resistance R_{11} when the distance is below 0.4λ. This leads to considerable loss in radiation efficiency, since the antenna current must be large for radiating the power, and the antenna loss resistance consumes a considerable amount of energy. When the spacing is $\lambda/2$, the mutual resistance R_{21} is -12.7 ohms, and this value oscillates around 0, and takes small values when the spacing is close to a multiple of $\lambda/2$.

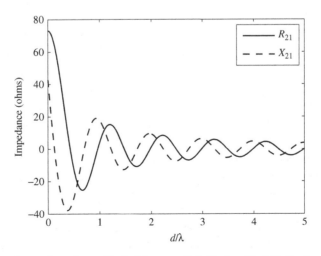

Figure 10.13 Mutual resistance R_{21} and reactance (X_{21}) of two parallel side-by-side $\lambda/2$-dipoles versus the distance between the two antennas, d.

When the antenna patterns are dissimilar, the correlation is small. Orthogonal patterns can be obtained by a suitable selection of angle, space, or polarization diversity. In this case, small spacings can be used. Dissimilar patterns can also be obtained using different types of antennas. Spatial correlation reduces the number of independent channels in a MIMO system, and thus significantly diminishes the capacity of a MIMO system.

10.9 Wideband antennas

The useful bandwidth of an antenna is decided by both its pattern and impedance characteristics. Assuming the VSWR is within the desired range, the fractional bandwidth (FBW) of an antenna can be defined as

$$\text{FBW} = \frac{f_2 - f_1}{f_0} \tag{10.67}$$

where f_1 and f_2, $f_1 < f_2$, are the frequency limits, and $f_0 = (f_1 + f_2)/2$ is the center frequency. For narrowband communications, FBW is usually less than 5%; for wideband antennas, FBW can be up to 25%; for UWB antennas, it is between 25 and 200%. The bandwidth can also be defined as $\frac{f_2}{f_1}$:1. When the impedance varies more rapidly than the pattern, the bandwidth is usually small and can be defined as Q at f_0.

In contrast to the dipole, which is a resonant, high-Q antenna with its input impedance changing rapidly with frequency, broadband antennas are nonresonant low-Q radiators whose input impedance remains essentially constant over a wide frequency range. Wideband antennas require the phase center and VSWR to be constant across the whole bandwidth of operation.

10.9.1 Implementation of wideband antennas

Rumsey's principle indicates that the impedance and pattern of an antenna are frequency-independent if the antenna shape is specified in terms of the angle only. This requires an architecture of infinite size. By suitable truncation of the architecture, these antennas can provide the broadband property. Examples are the biconical dipole antenna, the biconical vee antenna, the flat-plane bow-tie dipole antenna, the axial-mode helical antenna, the planar log-spiral antenna, the self-complementary toothed log-periodic antenna, and the log-periodic dipole array. Little energy is reflected from the open end of the antenna, thus the input impedance remains essentially constant over a wide bandwidth. The biconical dipole, the log-spiral antenna, and the bow-tie dipole provide omnidirectional radiations.

Conical monopoles can achieve impedance bandwidth, which is dependent on the radius of the cylindrical stub and increases with increased radius. Tapering the transition for the feed probe is often employed in wideband elements, such as biconical dipoles and conical monopoles. A cheap alternative to the cones is to use a planar element to replace the cylindrical stub, leading to a planar monopole. Square planar monopoles of different size typically have an impedance bandwidth of 80% [4]. Such planar monopole antennas are especially suitable for multiband and UWB systems. A significant increase in impedance bandwidth can be achieved by trimming the square edge near the ground plane, yielding an asymmetrical or symmetrical pentagonal monopole. An increase in the bevel angle can increase the upper-edge frequency and thus, the impedance bandwidth. The use of a shorting post can reduce the lower-edge frequency by introducing an extra mode [4]. The planar antenna can be replaced by a wire grid to reduce the influence of wind.

For multiband SDR, multiband antennas must be used. This requires the use of a wideband antenna. Tunable microstrip antennas provide an alternative to large bandwidth antennas. Tunable microstrip antennas can be designed by changing the length of the small stub attached to the regularly shaped microstrip antennas [26]. For a rectangular microstrip antenna, a stub can be placed along one of its edges. When the length of the stub is small, tunability is achieved. However, when the length is comparable to $\lambda/4$, dual-band properties are exhibited. For a rectangular microstrip antenna, the TM_{10} and TM_{30} modes have the radiation patterns in the broadside direction with the same polarization at both the resonant frequencies, and thus these modes can be used for dual-band operation.

The aperture-coupled microstrip-patch antenna is a low-cost solution for BSs. This design provides dual linear polarization, thus allowing polarization diversity. The planar structure can be simulated accurately by using the MoM. Spiral antennas have the characteristics of broad bandwidth and circular polarization. They can be implemented as wires, microstrips, or slots.

Broadband reconfigurable antennas

Broadband reconfigurable antennas are especially desirable for SDRs. The bandwidth can be adjusted over a wide range of frequency, when its environment or application is changed, by reconfiguring the system structure using switches. The switches can be made using RF MEMS technology, since it allows for lower resistance and excellent efficiency as compared to electrical switches [9, 34].

The self-structuring antenna [9] consists of a number of wires or patches interconnected by controllable switches. These switches are used to alter the electrical shape of the antenna. The self-structuring antenna is capable of selecting its state from one of 2^n discrete electrical configurations for n switches. The self-structuring antenna is composed of an antenna template, controllable switches, a sensor system for measuring important quantities such as signal strength, VSWR and input impedance, an efficient binary search algorithm, and a microprocessor controller. The binary search algorithm can be based on the genetic algorithm or simulated annealing [12] so that it can find a suboptimal configuration in a short time. The self-structuring antenna can be used as a wideband antenna or as a no-design antenna. In [34], a design that is able to electronically change the operating frequency of a rectangular ring slot antenna by adjusting the MEMS switches to change the circumference of the ring is described.

10.9.2 Ultra wideband antennas

UWB antennas are quite different from conventional narrowband antennas. FCC specifies the 3.1 to 10.6 GHz band for UWB communications. Due to the intended applications in consumer electronics, UWB antennas should be of small size, efficient, omnidirectional in pattern, and have low distortion, and large bandwidth. The phase center and VSWR of UWB antennas are required to be constant across the whole bandwidth of operation.

Conventional resonant antennas are not suitable for UWB applications, since they can only radiate waves at the resonant frequency. A nonresonant antenna is usually electrically large. It is dispersive for UWB pulses by radiating different frequency components from different parts of the antenna, leading to extended and distorted waveforms. Also, special care must be taken to make it efficient. Thus it is also not suitable for UWB systems. UWB systems transmit very low power pulses, and antennas are significant pulse-shaping filters; for UWB systems, any distortion in the frequency domain increases the complexity for detection at the receiver.

For UWB pulse signals, also called *digital waves*, the radiation from an antenna is significantly different from that of continuous wave narrowband signals. Since UWB systems are typically used for indoor environments, the near-field radiation must be considered. For fast transient radiation, the frequency-domain model fails due to the near-field dispersion.

The biconical and bowtie monopoles/dipoles are usually used for UWB applications [25]. Broadband planar monopole antennas are particularly attractive for use as UWB antennas due to ease of fabrication.

A number of antennas can be used as impulse antennas under certain conditions [15]. The conical antenna, when used as a receive antenna, generates its output as the integral of the incident electric field. A monopole antenna is used as a simpler version of the conical antenna, and thus also has an integral effect for the voltage wave from the driving point or for the received electric field. The D-dot probe antenna is an extremely short monopole antenna, and its output is the derivative of the incident electric field. A TEM horn antenna, when used as a transmit antenna, generates the radiated electric field as the derivative of the input driving-point voltage.

Planar UWB antennas

Planar monopole and dipole antennas are well-suited for UWB systems, since they are compact and easy to fabricate. They can generate a very large impedance bandwidth and near constant group delays if designed appropriately [8]. They can be viewed as a microstrip antenna on a very thick substrate (very large t) with $\epsilon_r = 1$ (suspended in air). For a rectangular planar monopole, as shown in Fig. 10.14, the impedance bandwidth depends mainly on the width W of the plate, the diameter d of the feeding probe, and the length p of the probe. When the frequency is above 1 GHz, the SMA connector is generally used for feeding the antenna, and thus $d = 0.12$ cm. The lower edge frequency of the antenna is then given by [26]

$$f_L = \frac{7.2}{L + r + p} \quad \text{(GHz)}, \qquad (10.68)$$

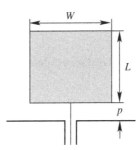

Figure 10.14   Planar rectangular monopole.

where the height L of the monopole antenna, the equivalent radius r corresponding to an equivalent cylindrical monopole antenna, $r = \frac{W}{2\pi}$, and the feed probe length p are all in centimeters.

A planar circular or elliptical monopole antenna can generate a larger bandwidth than other planar monopole antennas can [3]. It can be designed to include the whole UWB band very easily [3, 17]. For example, when the radius of a circular monopole is 2.5 cm and the feed probe length $p = 0.1$ cm, the bandwidth for VSWR≤ 2 is from 1.17 GHz to 12 GHz, corresponding to a bandwidth ratio of 1:10.2 [26].

The antenna impedance of a narrowband antenna is an intrinsic property of the antenna. In contrast, the impedance of a UWB antenna is not an intrinsic property, but a design choice. For instance, a planar elliptical dipole yields an excellent match to 50 ohms [36], but a planar elliptical monopole also offers a good match to 50 ohms [3]. Impedance of these antennas can be adjusted by varying the gap between the dipole elements and by varying the gap between a monopole element and the ground plane.

When UWB systems share the frequency bands with existing wireless standards, band-stop filtering is required to prevent interference. This is normally done by a conventional filter in the RF receiver front end. UWB antennas can also be designed with a band-notch characteristic to aid narrowband signal rejection. This can be realized by modifying the planar monopoles by adding internal or external slot structures [22, 51].

Many UWB antennas that cover the entire UWB band of 3.1–10.6 GHz are available in the literature [2, 23, 49]. In [2], planar monopole UWB antennas of elliptical and circular shape have been fabricated using simple design formulas. These antennas have near-omnidirectional characteristics and a radiation efficiency of more than 90%. A planar dipole UWB antenna of bow-tie shape has been designed in [49]. Planar UWB textile antennas have been designed for wireless BAN applications in [23].

Problems

10.1 From Maxell's equations, derive the wave equations for E and H as

$$\nabla^2 E + \omega^2 \epsilon \mu \left(1 - j\frac{\sigma}{\omega\epsilon}\right) E = 0, \quad \nabla^2 H + \omega^2 \epsilon \mu \left(1 - j\frac{\sigma}{\omega\epsilon}\right) H = 0.$$

In space, the conductivity is very small, $\sigma \ll \omega\epsilon$. [Hint: Use the vector identity $\nabla \times \nabla \times E = \nabla(\nabla \cdot E) - \nabla^2 E$.]

10.2 Find the far-field distance for an antenna with maximum dimension of 1 m, operating at 2 GHz.

10.3 If the BS transmits with a power of 50 W via a unity-gain antenna at 2 GHz, find the received power in dBm at a free space distance of 100 m. Assume the receive antenna has a unity gain. What is the received power at a free-space distance of 10 km?

10.4 Prove that the directivity of a half-wave dipole is equal to 1.64.

10.5 A half-wave dipole radiates a power of 100 W in free space at a frequency of 900 MHz. Calculate the electric and magnetic field strengths at a point $r = 100$ m, $\theta = 60°$, and $\phi = 30°$.

10.6 Eight half-wave dipoles are arranged side-by-side along a circle of radius $a = \lambda/2$. Calculate the open-circuit voltage on any antenna element, when all the antennas are fed with current $I_0 \cos \omega t$.

10.7 A radio transmitter on the Moon is operated at $f = 2.8$ GHz with a transmit power of $P_T = 2$ W. The gain of the antenna is $G_T = 1000$. A receiver is located on the Earth, and the propagation delay from the Moon to the Earth is 1.27 s. For a detection threshold $P_0 = 10^{-14}$ W, what is the gain of the receive antenna G_R? What is the area of the receive antenna?

10.8 Derive the directivity of the half-wave dipole to be

$$D(\theta) = 1.64 \left| \frac{\cos\left(\frac{\pi}{2}\cos\theta\right)}{\sin\theta} \right|^2.$$

10.9 An elliptically polarized wave is traveling along the z direction in free space. It has components in volts/m as

$$E_x = 2\sin(\omega t - \beta z), \quad E_y = 5\sin(\omega t - \beta z + 45°).$$

What is the average power density conveyed by the wave? What is the axial ratio of the wave? Is it a left-circularly polarized or right-circularly polarized wave?

10.10 Find the maximum effective aperture of a microwave antenna with a directivity of 50.

10.11 Two isotropic sources have equal amplitude and opposite phase. They are separated by a distance of 2λ. Plot the far-field power pattern. If the two sources are two half-wave dipoles configured side by side, plot the far-field power pattern.

10.12 An antenna has a uniform field $E = 1$ V/m (rms) at a distance of 100 m for $30° \le \theta \le 45°$ and $0° \le \phi \le 60°$, and $E = 0$ elsewhere. The antenna terminal current is 2 A (rms). Calculate the directivity, effective aperture, and radiation resistance.

10.13 An antenna system is composed of two quarter-wave monopoles above a perfect ground. The two quarter-wave monopoles are separated by a quarter-wavelength. The two antennas are fed currents of equal amplitudes but 90° out of phase. The antenna system radiates a power of 1 kW. Determine the magnitude and phase of the required driving-point voltages for the two antennas.

10.14 An omnidirectional antenna is 10 m above the Earth surface. What is the distance to the horizon it can cover? Assume a smooth Earth and a medium refractivity. [Hint: $d = \sqrt{2Rh}$.]

10.15 An isotropic antenna is radiating in free space. At a distance of 100 m, the total electric field is measured to be 2 V/m. Determine the power density and the radiated power.

10.16 The radiation intensity of an antenna is given by

$$U(\theta, \phi) = \cos^4 \theta \sin^2 \phi$$

for $0 \leq \theta \leq \pi/2$ and $0 \leq \phi \leq 2\pi$, and is zero elsewhere. Find (a) the directivity and (b) the HPBW in the elevation plane.

10.17 A half-wave dipole is connected to a generator with impedance of $50 + j25$. The dipole has a loss resistance of 2 ohms. The peak voltage of the generator is 5 V, and the radiation resistance of the dipole is $73 + j43.5$ ohms. Find (a) the power supplied by the source, (b) the power radiated by the antenna, and (c) the radiation efficiency.

10.18 A 4-cm long dipole carries a phasor current $I_0 = 10e^{j45°}$ A. If the wavelength is 10 cm, find the E- and H-fields at a distance of $r = 1$ m and $\theta = 30°$.

10.19 Verify that the directivity of a Hertzian dipole is given by (10.48).

10.20 Design a microstrip patch antenna operating at 2.4 GHz. The substrate is Duroid 5880 ($\epsilon_r = 2.2$) with a thickness of 0.04 in.

References

[1] 3GPP, *Spatial Channel Model for Multiple Input Multiple Output MIMO Simulations*, TR 25.996, ver. 6.1.0, Sep 2003.

[2] A. M. Abbosh & M. E. Bialkowski, Design of ultrawideband planar monopole antennas of circular and elliptical shape. *IEEE Trans. Anten. Propagat.*, **56**:1 (2008), 17–23.

[3] N. P. Agarwall, G. Kumar & K. P. Ray, Wide-band planar monopole antennas. *IEEE Trans. Anten. Propagat.*, **46**:2 (1998), 294–295.

[4] M. J. Ammann & Z. N. Chen, Wideband monopole antennas for multi-band wireless systems. *IEEE Anten. Propagat. Mag.*, **45**:2 (2003), 146–150.

[5] C. A. Balanis, *Antenna Theory: Analysis and Design*, 2nd edn (New York: Wiley, 1997).

[6] J.-P. Berenger, A perfectly matched layer for the absorption of electromagnetic waves. *J. Comput. Physics*, **114**:2 (1994), 185–200.

[7] K. R. Carver & J. W. Mink, Microstrip antenna technology. *IEEE Trans. Anten. Propagat.*, **29**:1 (1981), 2–24.

[8] Z. N. Chen, M. J. Ammann, X. Qing, X. H. Wu, T. S. P. See & A. Cai, Planar antennas. *IEEE Microwave Mag.*, **7**:6 (2006), 63–73.

[9] C. M. Coleman, E. J. Rothwell, J. E. Ross & L. L. Nagy, Self-structuring antennas. *IEEE Anten. Propagat. Mag.*, **44**:3 (2002), 11–22.

[10] D. B. Davidson, *Computational Electromagnetics for RF and Microwave Engineering* (Cambridge, UK: Cambridge University Press, 2005).

[11] K.-L. Du, Pattern analysis of uniform circular array. *IEEE Trans. Anten. Propagat.*, **52**:4 (2004), 1125–1129.

[12] K.-L. Du and M. N. S. Swamy, *Neural Networks in a Softcomputing Framework* (London: Springer, 2006).

[13] H. T. Friis, A note on a simple transmission formula. *Proc. IRE*, **34**:5 (1946), 254–256.

[14] S. Gao, A. Sambell & S. S. Zhong, Polarization-agile antennas. *IEEE Anten. Propagat. Mag.*, **48**:3 (2006), 28–37.

[15] M. Ghavami, L.B. Michael & R. Kohno, *Ultra Wideband: Signals and Systems in Communication Engineering*, 2nd edn (Chichester, UK: Wiley, 2007).

[16] J. P. Gianvittorio & Y. Rahmat-Samii, Fractal antennas: a novel antenna miniaturization technique, and applications. *IEEE Anten. Propagat. Mag.*, **44**:1 (2002), 20–36.

[17] M. Hammoud, P. Poey & F. Colombel, Matching the input impedance of a broadband disc monopole. *Electron. Lett.*, **29**:4 (1993), 406–407.

[18] R. F. Harrington, *Field Computation by Moment Methods* (New York: Macmillan, 1968).

[19] N. Hill & S. Cornbleet, Microwave transmission through a series of inclined gratings. *Proc. IEE*, **120**:4 (1973), 407–412.

[20] J. Jin, *The Finite Element Method in Electromagnetics*, 2nd edn (New York: Wiley, 2002).

[21] H. Jasik, ed., *Antenna Engineering Handbook* (New York: McGraw-Hill, 1961).

[22] A. Kerkhoff and H. Ling, A parametric study of band-notched UWB planar monopole antennas. In *Proc. IEEE AP-S Int. Symp.*, Monterey, CA, Jun 2004, **2**, 1768–1771.

[23] M. Klemm & G. Troester, Textile UWB antennas for wireless body area networks. *IEEE Trans. Anten. Propagat.*, **54**:11 (2006), 3192–3197.

[24] M. Komulainen, M. Berg, H. Jantunen, E. T. Salonen & C. Free, A frequency tuning method for a planar inverted-F antenna. *IEEE Trans. Anten. Propagat.*, 56:4 (2008), 944–950.

[25] J. D. Kraus & R. J. Marhefka, *Antennas: For All Applications*, 3rd edn (New York: McGraw-Hill, 2002).

[26] G. Kumar & K. P. Ray, *Broadband Microstrip Antennas* (Norwood, MA: Artech House, 2003).

[27] Y. T. Lo, D. Solomon, & W. F. Richards, Theory and experiment on microstrip antennas. *IEEE Trans. Anten. Propagat.*, **27**:2 (1979), 137–145.

[28] J. C. Maxwell, *A Treatise on Electricity and Magnetism* (Oxford, UK: Clarendon Press, 1873; New York: Dover, 1954).

[29] E. H. Newman & P. Tulyathan, Analysis of microstrip antennas using method of moments. *IEEE Trans. Anten. Propagat.*, **29**:1 (1981), 47–53.

[30] U. Olgun, C. A. Tunc, D. Aktas, V. B. Erturk & A. Altintas, Particle swarm optimization of dipole arrays for superior MIMO capacity. *Microwave Opt. Tech. Lett.*, **51**:2 (2009), 333–337.

[31] J. Ollikainen, O. Kivekas, A Toropainen, and P. Vainikainen, Internal dual-band patch antenna for mobile phones. In *Proc. Millenium Conf. Anten. Propagat. (AP2000)*, Davos, Switzerland, Apr 2000, paper no. 1111 .

[32] Y. Ouyang & W. J. Chappell, High frequency properties of electro-textiles for wearable antenna applications. *IEEE Trans. Anten. Propagat.*, **56**:2 (2008), 381–389.

[33] S. L. Preston, D. V. Thiel, J. W. Lu, S. G. O'Keefe & T. S. Bird, Electronic beam steering using switched parasitic patch elements. *Electron. Lett.*, **33**:1 (1997), 7–8.

[34] R. J. Richards & H. J. De Los Santos, MEMS for RF/microwave wireless applications: the next wave. *Microwave J.*, **44**:3 (2001), 20–41

[35] I. D. Robertson & S. Lucyszyn, ed., *RFIC and MMIC Design and Technology* (London, UK: IEE Press, 2001).

[36] H. Schantz, Planar elliptical element ultra-wideband dipole antennas. In *Proc. IEEE A-S Int. Symp.*, San Antonio, TX, Jun 2002, **3**, 44–47.

[37] A. K. Shackelford, K.-F. Lee & K. M. Luk, Design of small-size wide-bandwidth microstrip-patch antennas. *IEEE Anten. Propagat. Mag.*, **45**:1 (2003), 75–83.

[38] C. Sun, A. Hirata, T. Ohira & N. C. Karmakar, Fast Beamforming of electronically steerable parasitic array radiator antennas: theory and experiment. *IEEE Trans. Anten. Propagat.*, **52**:7 (2004), 1819–1832.

[39] Y. J. Sung, Reconfigurable patch antenna for polarization diversity. *IEEE Trans. Anten. Propagat.*, **56**:9 (2008), 3053–3054.

[40] A. Taflove & S. C. Hagness, *Computational Electrodynamics: the Finite Difference Time Domain Method*, 3rd edn (Boston, MA: Artech House, 2005).

[41] M. Taromaru & T. Ohira, Electronically steerable parasitic array radiator antenna: principle, control theory and its applications. In *Proc. General Assembly of URSI*, New Delhi, India, Oct 2005, Paper no. C02.1.

[42] K. Tsunekawa & K. Kagoshima, Analysis of a correlation coefficient of built-in diversity antennas for a portable telephone. In *Proc. IEEE AP-S Int. Symp.*, Dallas, TX, May 1990, vol 1, 543–46.

[43] C. A. Tunc, D. Aktas, V. B. Erturk & A. Altintas, Capacity of printed dipole arrays in MIMO channel. *IEEE Ant. Propagat. Mag.*, **50**:5 (2008), 190–198.

[44] R. G. Vaughan, Polarization diversity in mobile communications. *IEEE Trans. Veh. Tech.*, **39**:3 (1990), 177–186.

[45] D. H. Werner & S. Ganguly, An overview of fractal antenna engineering research. *IEEE Anten. Propagat. Mag.*, **45**:1 (2003), 38–57.

[46] F. Yang & Y. Rahmat-Samii, Patch antennas with switchable slots (PASS) in wireless communications: concepts, designs, and applications. *IEEE Anten. Propagat. Mag.*, **47**:2 (2005), 13–29.

[47] S.-L. S. Yang, A. A. Kishk & K.-F. Lee, Wideband circularly polarized antenna with
     L-shaped slot. *IEEE Trans. Anten. Propagat.*, **56**:6 (2008), 1780–1783.
[48] S. Yang, C. Zhang, H. K. Pan, A. E. Fathy & V. K. Nair, Frequency-reconfigurable
     antennas for multiradio wireless platforms. *IEEE Microwave Mag.*, **10**:1 (2009),
     66–83.
[49] K. Y. Yazdandoost & R. Kohno, Ultra wideband antenna. *IEEE Commun. Mag.*, **42**:6
     (2004), S29–S32.
[50] K. Yee, Numerical solution of initial boundary value problems involving Maxwell's
     equations in isotropic media. *IEEE Trans. Anten. Propagat.*, **14**:3 (1966), 302–307.
[51] H. Yoon, H. Kim, K. Chang, Y. J. Yoon & Y. H. Kim, A study on the UWB antenna
     with band-rejection characteristic. In *Proc. IEEE AP-S Int. Symp.*, Monterey, CA, Jun
     2004, **2**, 1784–1787.

11 RF and microwave subsystems

11.1 Introduction

The term *microwaves* is used to describe electromagnetic waves with frequencies from 300 MHz to 300 GHz, corresponding to wavelengths in free space from 1 m to 1mm. Within the microwave range, from 30 GHz to 300 GHz the wavelengths are between 1 mm and 10 mm, and hence these waves are known as *millimeter waves*. Below 300 MHz the spectrum of electromagnetic waves is known as the *radio frequency (RF)* spectrum, while above the microwave spectrum are the infrared, visible optical, ultraviolet, and x-ray spectrums. Wireless communications uses only the electromagnetic waves in the range of the microwave and RF spectrums. In the wireless communications literature, the term RF is often used to represent the entire RF and microwave spectrums.

11.1.1 Receiver performance requirements

The requirements on RF receivers are typically more demanding than those on transmitters. In addition to the requirements on gain and noise figure, the receiver must have:

- *A good sensitivity* to the minimum power at the antenna for a given BER requirement. For example, the GSM standard requires a reception dynamic range from -102 dBm to -15 dBm, IEEE 802.11g requires a reception range of -92 dBm to -20 dBm, for WCDMA it is -117 to -25 dBm (before spreading), for CDMA2000 it is -117 dBm to -30 dBm, and for WideMedia it is -80.8 dBm/MHz (or -72.4 dBm/MHz at highest speed) to -41.25 dBm/MHz. For multiple data rates, a higher data rate requires a higher sensitivity, since it requires a larger SNR. The AWGN sensitivity is computed by

$$\text{Sensitivity} = \text{NF}_R + \text{SNR} + \text{margin}$$

where NF_R denotes the noise figure of the receiver. The high sensitivity demands a high gain, as high as 100 to 120 dB, to restore the received signal to its original baseband level. The total gain should be distributed over the RF, IF, and baseband stages for reasons of stability and cost. The receiver sensitivity of WCDMA is -106 dBm with a nominal receiver noise figure of 9 dB, therefore the required signal power before despreading is -117 dBm.

Table 11.1. Radio specifications for GSM/WCDMA/Wi-Fi/WiMAX [45].				
	WiMAX (802.16)	Wi-Fi (802.11b)	WCDMA	GSM
Frequency band	2–11 GHz	2.412–2.484 GHz	2.010–2.025 GHz	890–915 MHz
	10–66 GHz		1.900–1.910 GHz	
Channel spacing	1.25–28 MHz	25 MHz	5 MHz	200 kHz
Channel bandwidth	1.25–28 MHz	20 MHz	3.84 MHz	200 kHz
Sensitivity	(−93.2 to −80)	−76 dBm	−117 dBm	−102 dBm
	+10log(BW)			
Maximum input signal	−20 dBm	−10 dBm	−25 dBm	−15 dBm
Input noise	−116 to −103 dBm	−104 dBm	−111 dBm	−124 dBm
Required SNR	9.8–23 dB	14 dB	7.2 dB	9 dB

- *A wide dynamic range* for handling interference in the channel.
- *A good adjacent channel selectivity* for receiving the wanted signal while reject-ing interference present on an adjacent channel. The minimum adjacent chan-nel leakage power ratio for WCDMA is greater than 33 and 43 dB for adjacent channel frequency relative to assigned channel frequency at ±5 and ±10 MHz, respectively.
- *A good blocking ability* to maintain performance of a wanted signal in the presence of an interferer.
- *A desirable intermodulation rejection ability* to reject interfering signals arising from intermodulation with the wanted signal.
- *A good isolation* from the transmitter to prevent saturation of the receiver in a duplex system. The isolation is required to be not less than 100 dB. For a half-duplex system, the transmitter and receiver are not operating simultaneously and a transmit/receive switch can be used. For a full-duplex system, both transmitter and receiver operate simultaneously but on different frequency bands; duplexing is achieved using bandpass filters.

The radio specifications of GSM/WCDMA/Wi-Fi/WiMAX are summarized in Table 11.1 for GSM power class II, WCDMA TDD mode, IEEE 802.11b, and IEEE 802.16 [45].

For RF transmitters, it is generally required to have a tight occupied bandwidth, low out-of-band emissions, low spurious emissions, and the ability to inhibit intermodulation.

11.1.2 Architecture of RF subsystems

The architecture of radio tranceivers has been discussed in Section 1.4. The RF front-end or subsystem is critical to the performance of the whole radio transceiver. Also, the emission of the RF front-end must strictly comply with the electromagnetic interference and electromagnetic compatibility requirements that are specified by FCC for a wireless standard.

A typical RF subsystem architecture for SDRs has been shown in Fig. 1.3. On the receive path, the received RF signal is first subject to a bandpass filter and LNA before it is transformed to the IF signal. Conversely, on the transmit path, the IF signal is transformed to RF signal and then amplified with the power amplifier. Depending on implementation, there may be more than one IF stage. For a wideband system, all the RF components, such as the mixers, preamplifiers, and LNAs, are required to be of wideband nature, and these components can be used for multiband RF front-ends. The filters have to accommodate for the widest bandwidth among the multiple radio standards.

Oscillators are necessary for frequency conversion, and PLLs are used for generating stable oscillators. For phased-array antennas, phase shifters are essential components. Phase shifters can be fabricated by using PIN diodes, field-effect transistor (FET) switches or MEMS switches. Attenuators and amplifiers may be used in phased arrays and automatic gain control (AGC). Baluns are required in many microwave components, such as balanced mixers, push-pull amplifiers, multipliers, and phase shifters, to improve the performance and reduce the cost of the RF module. The transmit/receive switch, which is used to connect the antenna to form a transmit or receive train, can be implemented using a circulator. We will introduce all the important RF or microwave components that constitute a RF subsystem in this chapter.

Discrete gallium arsenide (GaAs) transistors traditionally dominate gigahertz (GHz)-band RF circuits. These require significant area and have large power consumption. More and more RF designs are now based on CMOS technologies. Some wireless transceivers implement system-on-chip (SOC) technology. The system-on-package (SOP) scheme allows combination of CMOS monolithic microwave integrated circuits (MMICs) with low-temperature cofired ceramic (LTCC) multilayer passive components, which are then integrated in an LTCC substrate. Integrating high-quality passive components on ceramic substrates significantly increases the output power.

RF Globalnet[1] is a leading online trade publication for the RF/microwave design industry. Its vast product directory contains detailed information on all kinds of RF/microwave and antenna products, services, and test equipment.

11.2 RF system analysis

Before we proceed, we first give the definition of power spectral density (PSD), also called *spectral density* or *spectrum*, $S_x(f)$, of a random signal $x(t)$. PSD is a measure of how much power a signal carries in a unit bandwidth (Hz), centered around frequency f. The value of $S_x(f)$ can be measured by applying a bandpass filter of 1-Hz bandwidth to the frequency f and then averaging the output power over a sufficiently long time

$$S_x(f) = \lim_{T \to \infty} \frac{\overline{|X_T(f)|^2}}{T}, \tag{11.1}$$

---

[1] http://www.rfglobalnet.com

where $X_T(f)$ is the Fourier transform of $x(t)$ in a period $[0, T]$,

$$X_T(f) = \int_0^T x(t)e^{-j2\pi ft}dt. \tag{11.2}$$

11.2.1 Noise

Noise is one of the most important considerations in communication system design. For analog systems, thermal noise is generated by the thermal agitation of the electrons inside a conductor in equilibrium. Noise is additive for cascaded analog systems, thus each processing block of an RF subsystem must be allocated a reasonable gain and noise figure to reduce the noise of the whole subsystem.

For analog systems, the most common noises are thermal noise, shot noise, and flicker noise. For digital systems, the major noise is the quantization noise arising from A/D conversion. Overflow and rounding errors can also be treated as noise.

Thermal noise

Thermal noise, also called *Johnson noise*, is a consequence of Brownian motion of free charge carriers, usually electrons, in a resistive medium. For a resistor R at temperature T, the random motion in the resistor leads to small random voltage fluctuations at the terminals of the resistor. The rms voltage is given by the Rayleigh-Jean approximation [36, 37]

$$V_n = \sqrt{4kTBR}. \tag{11.3}$$

A Thevenin equivalent circuit for a noisy resistor can be used, and the noise power of thermal noise of the channel, measured in watts, is given by Nyquist's formula

$$N = k_B TB, \tag{11.4}$$

where k_B is Boltzmann's constant 1.381×10^{-23} watts/Hz/K, T is the noise temperature in kelvins (K), and B is the bandwidth of the channel. In decibels referencing to milliwatts (dBm), $k_B = -198.6$ dBm/Hz/K. The thermal noise power at room temperature ($T = 290$ K) is given by

$$N = -174 + 10 \log_{10}(B) \quad \text{(dBm)}. \tag{11.5}$$

Thermal noise is independent of frequency, thus it is referred to as *white noise*.

The two-sided PSD of thermal noise is defined as

$$S_n(f) = \frac{N}{2B} = \frac{k_B T}{2} = \frac{N_0}{2}, \tag{11.6}$$

where $N_0 = k_B T$ is in watts/Hz. This is the conventional definition used for communication systems, over the frequency range from $-B$ to B Hz. The one-sided PSD $S_n(f) = N_0$, since noise is only defined over 0 to B Hz. The probability density function (pdf) of white noise is zero-mean Gaussian. In reality, $S_n(f)$ is flat for only $|f| < 100$ GHz, and dropping beyond this frequency, leading to a finite total noise power of the resistor, N [37].

Shot noise

Shot noise occurs only when there is a direct current flow and a potential barrier over which the charge carriers hop. Some potential barriers are the junction of a PN diode, the emitter-base junction of a bipolar transistor, and the cathode of a vacuum tube. Shot noise is also a Gaussian white process, with a PSD

$$\overline{I_n^2} = 2qI, \tag{11.7}$$

where q is the charge of an electron and I the average current. Due to the requirement for a potential barrier, shot noise only exists in nonlinear devices. In a bipolar transistor, both the base and collector currents are sources of shot noise, while in FETs only the dc gate leakage current produces shot noise.

Flicker noise

Flicker, or $1/f$ or *pink*, noise is the most mysterious type of noise. Flicker noise is ubiquitous, but no universal mechanism is identified yet for the flicker noise. Thus, it can only be characterized by empirical parameters. Flicker noise occurs in most active devices. Flicker noise power is concentrated, with a PSD of approximately $1/f$ characteristic at low frequencies, usually below a few kilohertz, and has a flat, weak spectrum above a few kilohertz.

Flicker noise may arise from random trapping of charge in bulk material, at junction interfaces and at the surface of the substrate as in the FET. Note that the flicker noise also has a Gaussian pdf, but its PSD is not white but proportional to $1/f$. Due to the nonlinarity of active devices, the $1/f$-shaped spectrum may be translated into the RF range. Flicker noise is also exhibited in forward-biased junctions, as in the case of bipolar transistors.

For a three-terminal active device, the broadband noise performance can be split into three zones, as illustrated in Fig. 11.1. The flicker noise typically occurs up to about 1 kHz for bipolar technologies, but above 10 MHz for FET technologies. When frequency is very large, the device capacitance leads to an increase in the noise, known as the *capacitance-coupled noise*.

Flicker noise also shows up in resistors, where it is usually called *excess noise*. A resistor is found to exhibit flicker noise only when dc current follows through it, with the noise

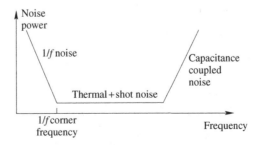

Figure 11.1   The noise power density spectrum of a transistor.

increasing with the current [27]. The common carbon composition resistors exhibit the most significant flicker noise, while metal film and wirewound resistors have the least flicker noise.

In a practical system such as in a low noise system, thermal noise is dominant, and thus physical cooling can effectively improve its noise performance. Noise power can be represented by using an equivalent thermal noise temperature $T_e = \frac{N}{k_B B}$ kelvins (K); shot and flicker noises can be treated in a similar way.

11.2.2 Noise figure

Noise figure measures the amount of noise added by a device. It is defined by the ratio of the input SNR to the output SNR of a functional block, or more often by the ratio of the output noise to the input noise

$$F = \frac{\mathrm{SNR}_i}{\mathrm{SNR}_o} = \frac{N_o}{N_i}. \tag{11.8}$$

For practical devices, $F > 1$, i.e., $F > 0$ dB. The two definitions are equivalent, since the output signal $S_o = G S_i$, and $N_o = G(N_i + N_e)$, where N_e is the device noise as an equivalent input noise to the device. Thus N_o in the right-hand side of the equation is replaced by $N_i + N_e$. This definition is usually called *noise factor*, and noise figure is given by $10 \log_{10} F$, in decibels.

Noise figure is typically specified for all devices and systems at an input noise temperature of $T = 290$ K for fair comparison. In this case, N_o is calculated by $N_o = 290 k_B B$. An equivalent input noise temperature T_e can be used to characterize the system noise, and thus the noise figure can be calculated by

$$F = \frac{N_o}{N_i} = \frac{k_B(T + T_e)B}{k_B T B} = 1 + \frac{T_e}{290}. \tag{11.9}$$

Or an equivalent noise temperature can be calculated by

$$T_e = 290(F - 1) \quad (K). \tag{11.10}$$

For a typical LNA, F is 3.0 dB, corresponding to $T_e = 290$ K.

For passive, lossy devices such as transmission lines, attenuators, or mixers, their gains G_l are less than unity. The lossy device has an equivalent noise temperature $\left(\frac{1}{G_l} - 1\right) T$, and the noise figure is given by [36]

$$F = 1 + \left(\frac{1}{G_l} - 1\right)\frac{T}{290}. \tag{11.11}$$

When the transmission line is at room temperature 290 K, $F = 1/G_l$ and thus a 3-dB attenuator has a noise figure of 3 dB.

The total noise figure of a cascaded system is calculated by using the Friis noise formula [36, 37]

$$F_{\mathrm{tot}} = F_1 + \frac{F_2 - 1}{G_1} + \frac{F_3 - 1}{G_1 G_2} + \cdots + \frac{F_n - 1}{G_1 G_2 \cdots G_{n-1}}, \tag{11.12}$$

where F_i and G_i are the noise figure and gain of the ith element in the cascaded chain. Alternatively, the overall equivalent noise T_e can be obtained by [36]

$$T_e = T_{e1} + \frac{T_{e2}}{G_1} + \frac{T_{e3}}{G_1 G_2} + \cdots + \frac{T_{en}}{G_1 G_2 \cdots G_{n-1}}, \qquad (11.13)$$

where T_{ei} is the equivalent noise temperature of the ith element. Thus the first few devices have a more significant effect on the noise power. The first element in the chain should therefore have a low noise factor and a high gain. For example, if the bandpass filter preceding an LNA has a loss of 2 dB, it corresponds to a noise figure $F_1 = 2$ dB. If the LNA has a noise figure $F_2 = 3$ dB, the overall noise figure becomes $F_1 F_2 = 5$ dB. Thus, the in-band loss for the bandpass filter is most critical for the system performance.

11.2.3 Link budget analysis

The carrier-to-noise ratio (CNR) defines the ratio of the signal power to the noise power over a channel. CNR can be regarded as a figure of merit of a communication system. For a wireless communication system, CNR can be calculated by using the link equation

$$\text{CNR} = \frac{P_{\text{ERP}} L_P G_R}{N}, \qquad (11.14)$$

where P_{ERP} is the effective radiated power from the transmit antenna, L_P is the propagation loss, G_R is the gain of the receive antenna, and N is the noise power. P_{ERP} is calculated by

$$P_{\text{ERP}} = P_t L_{\text{CT}} G_T \qquad (11.15)$$

where P_t is the output power at the transmitter power amplifier, L_{CT} is the loss of the connecting cable between the power amplifier and the transmit antenna, and G_T is the gain of the transmit antenna. N is attributed to thermal noise and is given by (11.4). The carrier-to-interference ratio (CIR) is defined as the ratio of the signal power to the noise power plus interfering power.

For an RF subsystem, each analog/RF functional block along the link leads to a loss or a gain to the overall power. For wireless system design, it is important to give a system link budget from the point after DAC in the transmit subsystem to the point before ADC in the receive subsystem. This is illustrated by Fig. 11.2. In the figure, the gain or loss at each functional block is given, and the noise figure of each block at the receiver is also given.

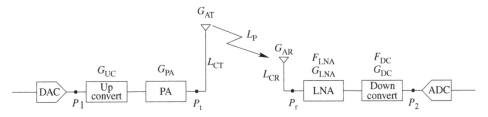

Figure 11.2   RF system link budget analysis.

In decibels, we can get the following three equations for

$$P_t = P_1 + G_{UC} + G_{PA}, \tag{11.16}$$

$$P_r = P_t - L_{CT} + G_{AT} - L_P + G_{AR} - L_{CR}, \tag{11.17}$$

$$P_2 = P_r + G_{LNA} + G_{DC}. \tag{11.18}$$

The cumulative noise effect must also be considered in the receive subsystem, and this can be based on the noise figure. In cellular RF design, one usually performs noise figure analysis for the receive subsystem only. This is because the thermal noise from the transmit subsystem is significantly attenuated by the propagation, and the ambient thermal noise surrounding the receive antenna is much higher than the attenuated transmit noise. Many standards such as GSM and 802.11g require an overall noise figure of around 10 dB, and WideMedia requires a noise figure 6.6 dB. For 3G standards, a noise figure of 5 dB is recommended.

11.3 Transmission lines

11.3.1 Fundamental theory

A transmission line is a distributed parameter network, as shown in Fig. 11.3. It is usually represented by a two-wire line.

The voltage $V(z)$ and current $I(z)$ at any point z on the line are related by the transmission line equations, which under sinusoidal steady state conditions, are given in the phasor form

$$\frac{dV(z)}{dz} = -(R + j\omega L)I(z), \tag{11.19}$$

$$\frac{dI(z)}{dz} = -(G + j\omega C)V(z), \tag{11.20}$$

where R (ohms/m) and L (henrys/m) are, respectively, the series resistance and inductance per unit length for both conductors, and G (siemens/m) and C (farads/m) are shunt conductance and capacitance per unit length, respectively.

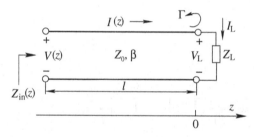

Figure 11.3   Transmission line terminated with load.

Traveling wave solutions to the wave equations (11.19) and (11.20) can be found as

$$V(z) = V_o^+ e^{-\gamma z} + V_o^- e^{\gamma z}, \tag{11.21}$$

$$I(z) = I_o^+ e^{-\gamma z} + I_o^- e^{\gamma z}, \tag{11.22}$$

where

$$\gamma = \alpha + j\beta = \sqrt{(R + j\omega L)(G + j\omega C)} \tag{11.23}$$

is the complex propagation constant, with α being the attenuation coefficient and $\beta = 2\pi/\lambda$ the wave number, V_o^+ and I_o^+ are the voltage and current amplitudes of the incident wave along $+z$ direction, V_o^- and I_o^- are the voltage amplitude and current amplitudes of the reflected wave along $-z$ direction, and Z_0 is the characteristic impedance of the transmission line and is defined by

$$Z_0 = \sqrt{\frac{R + j\omega L}{G + j\omega C}}. \tag{11.24}$$

It can be shown that

$$Z_0 = \frac{V_o^+}{I_o^+} = \frac{-V_o^-}{I_o^-}. \tag{11.25}$$

It should be pointed that the distributed parameters R, L and G are not constant, but are frequency-dependent due to the skin effect – the current mainly flows near the surface of the conductor. Usually, the transmission line is assumed to be lossless or with low loss, that is, $R = G = 0$. Accordingly, we have $Z_0 = \sqrt{\frac{L}{C}}$, $\gamma = j\beta = j\omega\sqrt{LC}$, and the phase velocity $v_p = \frac{\omega}{\beta} = \frac{1}{\sqrt{LC}}$.

Reflection coefficient

Consider a lossless transmission line terminated by a load Z_L, as shown in Fig. 11.3. Since Z_L is at $z = 0$, we have $V(0) = Z_L I(0)$. The voltage reflection coefficient Γ is defined as the ratio of the amplitude of the reflected voltage wave to that of the incident voltage wave, and it can be derived as

$$\Gamma = \frac{V_o^-}{V_o^+} = \frac{Z_L - Z_0}{Z_L + Z_0}. \tag{11.26}$$

In order to obtain $\Gamma = 0$, $Z_L = Z_0$. In this case, the load is said to be *matched to the line*, and there is no reflection of the incident wave. For a short circuit ($Z_L = 0$), $\Gamma = -1$, while for an open circuit ($Z_L = \infty$), $\Gamma = 1$.

Γ is the reflection coefficient at the load ($z = 0$), i.e., $\Gamma = \Gamma(0)$. It can be generalized to any point z:

$$\Gamma(z) = \frac{V_o^- e^{j\beta z}}{V_o^+ e^{-j\beta z}} = \Gamma(0)e^{2j\beta z}. \tag{11.27}$$

Return loss

The average power flow is constant at any point on the line and is given by

$$P_{\text{av}} = \frac{1}{2} \frac{|V_o^+|^2}{Z_0} \left(1 - |\Gamma|^2\right). \tag{11.28}$$

This result is obtained by assuming that the generator is matched so that there is no rere-flection of the reflected wave from the generator side. This is also the power delivered to the load.

The return loss (RL) is defined as the power loss for reflected power due to the load

$$\text{RL} = -20 \log_{10} |\Gamma| \quad (\text{dB}). \tag{11.29}$$

For a matched load ($\Gamma = 0$), the return loss is ∞. When $|\Gamma| = 1$, all incident power is reflected, and thus the return loss is $0\,\text{dB}$.

Voltage standing wave ratio

When the load is matched to the line, the amplitude of the voltage at any point z on the line is a constant $|V(z)| = |V_o^+|$. In case of an unmatched load, reflection leads to a standing wave on the line; the voltage amplitude is given by

$$|V(z)| = |V_o^+| \cdot \left|1 + |\Gamma| e^{j(\theta + 2\beta z)}\right|, \tag{11.30}$$

where θ is the phase of Γ, $\Gamma = |\Gamma| e^{j\theta}$. The amplitude $|V(z)|$ takes its maximum and minimum values, $|V(z)|_{\text{max}}$ and $|V(z)|_{\text{min}}$, at $e^{j(\theta + 2\beta z)} = 1$ and -1, respectively. The mismatch of the line can be measured by the standing wave ratio, more frequently known as *voltage standing wave ratio (VSWR)*, defined by

$$\text{VSWR} = \frac{|V(z)|_{\text{max}}}{|V(z)|_{\text{min}}} = \frac{1 + |\Gamma|}{1 - |\Gamma|}. \tag{11.31}$$

Thus, VSWR≥ 1. For matched load, VSWR$= 1$. For a short or open circuit, $\Gamma = -1$ or 1, thus VSWR$=\infty$.

Input impedance

The input impedance seen in the direction of the load is derived by

$$Z_{\text{in}} = \frac{V(-l)}{I(-l)} = \frac{1 + \Gamma(-l)}{1 - \Gamma(-l)} Z_0 = \frac{1 + \Gamma e^{-2j\beta l}}{1 - \Gamma e^{-2j\beta l}} Z_0. \tag{11.32}$$

Inserting (11.26) into (11.32), we have

$$Z_{\text{in}} = Z_0 \frac{Z_L + jZ_0 \tan \beta l}{Z_0 + jZ_L \tan \beta l}. \tag{11.33}$$

From (11.33), the following points can be reached:

- For a short circuit ($Z_L = 0$),

$$Z_{in} = jZ_0 \tan \beta l \tag{11.34}$$

which is purely imaginary, periodic in z with a period of $\lambda/2$, and can take any value between $-j\infty$ and $j\infty$.

- For an open circuit ($Z_L = \infty$),

$$Z_{in} = -jZ_0 \cot \beta l \tag{11.35}$$

which is also purely imaginary, periodic in z with a period of $\lambda/2$.

- For $l = \lambda/2$ or $l = n\lambda/2$, $n = 1, 2, \ldots$

$$Z_{in} = Z_L. \tag{11.36}$$

This indicates that a $\lambda/2$ line does not change the load impedance.

- For $l = \lambda/4$ or $l = \lambda/4 + n\lambda/2$, $n = 0, 1, 2, \ldots$

$$Z_{in} = \frac{Z_0^2}{Z_L}. \tag{11.37}$$

This result is very important for impedance transformation and matching. The quarter-wavelength line is known as a *quarter-wave transformer*.

Insertion loss

Consider the case of a lossless transmission line of characteristic impedance Z_0 feeding a line of characteristic impedance Z_1. If the latter is terminated in a load of impedance Z_1 or is of infinite length, there is no reflection from its load, and Z_{in} seen from the feedline is Z_1. The reflection coefficient is thus given by

$$\Gamma = \frac{Z_1 - Z_0}{Z_1 + Z_0}. \tag{11.38}$$

At the intersection $z = 0$, the incident wave is transmitted to the load line, and also reflected to the feedline. The wave on the load line is outgoing with an amplitude of $TV_o^+ e^{-j\beta z}$, $z > 0$, where T is the transmission coefficient. At $z = 0$,

$$T = 1 + \Gamma. \tag{11.39}$$

Insertion loss (IL) is defined by

$$IL = -20 \log_{10} |T| \quad (dB). \tag{11.40}$$

Smith chart

The Smith chart is the most widely used graphical aid for solving transmission line problems. The Smith chart is a polar plot of the voltage reflection coefficient Γ [36]. Any passively realizable reflection coefficient is a unique point on the Smith chart. The Smith chart can be used to convert between the reflection coefficients and normalized impedances (or admittances), using the impedance (or admittance) circles.

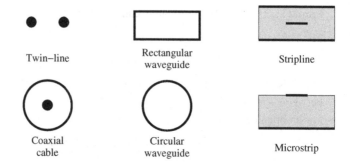

Figure 11.4   **Structures of some transmission lines and waveguides.**

11.3.2 Types of transmission line

Practical transmission lines and waveguides are twin-line, coaxial cable, stripline, microstrip, rectangular waveguide, and circular waveguide, and they are shown in Fig. 11.4. They typically operate in TEM mode, that is, $E_z = H_z = 0$. For transmission lines in TEM mode, there is a one-to-one correspondence between electric field and voltage, and between magnetic field and current.

Coaxial cables have very high bandwidth and power-handling capacity, while planar transmission lines are compact and easily integrated with active devices. The characteristic impedances of these transmission lines can be found in [36]. For example, the characteristic impedance of a coaxial cable is given by

$$Z_0 = \frac{\eta}{2\pi} \ln\left(\frac{b}{a}\right) \tag{11.41}$$

where a is the radius of the inner conductor, b is the inner radius of the outer conductor, and $\eta = \sqrt{\mu/\epsilon}$ is the intrinsic impedance of the medium. μ and ϵ are, respectively, the permittivity and permeability of the material filling the space between the conductors.

The microstrip line is the most important type of transmission line. The grounded metalized surface covers only one side of the dielectric substrate, and this causes the electromagnetic wave along the microstrip line to be not a pure TEM; but the TEM approximation is sufficiently accurate as long as the height of the dielectric substrate is very small relative to the wavelength. The analytical result for the effective dielectric constant ϵ_r^e and the characteristic impedance Z_0 of a lossless microstrip line is given as [17, 19, 36]:
For $W/h \leq 1$:

$$\epsilon_r^e = \frac{\epsilon_r + 1}{2} + \frac{\epsilon_r - 1}{2}\left[\left(1 + 12\frac{h}{W}\right)^{-0.5} + 0.04\left(1 - \frac{W}{h}\right)^2\right], \tag{11.42}$$

$$Z_0 = \frac{60}{\sqrt{\epsilon_r^e}} \ln\left(\frac{8h}{W} + 0.25\frac{W}{h}\right). \tag{11.43}$$

For $W/h \geq 1$:

$$\epsilon_r^e = \frac{\epsilon_r + 1}{2} + \frac{\epsilon_r - 1}{2} \left(1 + 12\frac{h}{W}\right)^{-0.5}, \tag{11.44}$$

$$Z_0 = \frac{120\pi}{\sqrt{\epsilon_r^e}} \left[\frac{W}{h} + 1.393 + 0.677 \ln\left(\frac{W}{h} + 1.444\right)\right]^{-1}. \tag{11.45}$$

In the above, W is the width of the microstrip line. For a very thin conductor, this solution provides an accuracy better than one percent [17].

Given Z_0 and ϵ_r, an expression for determining $\frac{W}{h}$ is given as [17, 19, 36]

$$\frac{W}{h} = \begin{cases} \frac{8e^A}{e^{2A} - 2}, & \text{if } \frac{W}{h} < 2 \\ \frac{2}{\pi}\left[B - 1 - \ln(2B - 1) + \frac{\epsilon_r - 1}{2\epsilon_r}\left(\ln(B - 1) + 0.39 - \frac{0.61}{\epsilon_r}\right)\right], & \text{if } \frac{W}{h} > 2 \end{cases}, \tag{11.46}$$

where

$$A = \frac{Z_0}{60}\sqrt{\frac{\epsilon_r + 1}{2}} + \frac{\epsilon_r - 1}{\epsilon_r + 1}\left(0.23 + \frac{0.11}{\epsilon_r}\right), \quad B = \frac{377\pi}{2Z_0\sqrt{\epsilon_r}}. \tag{11.47}$$

The strip thickness is usually very small, and its effect is usually neglected. The effect of strip thickness on the characteristic impedance Z_0 and the effective dielectric constant ϵ_r^e is given in [3, 19].

11.4 Microwave network analysis

A waveguide is characterized by electric field and magnetic field, and their relation by wave impedance. The measurement of voltage and current at microwave frequencies is difficult. Equivalent voltage, current, and impedance for waveguides can be defined. Once such voltages and currents are defined at various points in a microwave network, the microwave system can be analyzed in a way similar to the conventional circuit theory.

Consider an N-port microwave network, illustrated in Fig. 11.5, where each port corresponds to a transmission line or a waveguide. At a phase reference plane of voltage and current phasors, called *terminal plane*, of port n, the equivalent voltage V_n and current I_n are given by their components for incident and reflected waves:

$$V_n = V_n^+ + V_n^-, \quad I_n = I_n^+ - I_n^-, \tag{11.48}$$

which is the result at $z = 0$ from the transmission line theory.

Impedance and admittance matrices

The impedance matrix $\mathbf{Z} = [Z_{ij}]$ relates the voltages and currents by

$$V = \mathbf{Z}I, \tag{11.49}$$

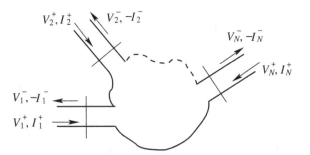

Figure 11.5   An *N*-port microwave network.

where $V = (V_1, V_2, \ldots, V_N)^T$ and $I = (I_1, I_2, \ldots, I_N)^T$. The entries Z_{ij} are defined by

$$Z_{ij} = \left. \frac{V_i}{I_j} \right|_{I_k=0, k \neq j}. \tag{11.50}$$

That is, Z_{ij} can be obtained by driving port j with current I_j, open-circuiting all other ports, and measuring the open-circuit voltage at port i.

Similarly, the admittance matrix $\mathbf{Y} = [Y_{ij}]$ relates the voltages and currents by

$$I = \mathbf{Y}V. \tag{11.51}$$

Obviously, $\mathbf{Z}$ and $\mathbf{Y}$ are inverse of each other. The entries Y_{ij} are given by

$$Y_{ij} = \left. \frac{I_i}{V_j} \right|_{V_k=0, k \neq j}. \tag{11.52}$$

That is, Y_{ij} is obtained by driving port j with voltage V_j, short-circuiting all other ports, and measuring the short-circuit current at port i.

For a reciprocal network, which contains no active devices, ferrites or plasmas, $Z_{ij} = Z_{ji}$ and $Y_{ij} = Y_{ji}$. For a lossless network, all Z_{ij}'s and Y_{ij}'s are purely imaginary. A two-port reciprocal network can be represented by a T (for Z_{ij}) or π (for Y_{ij}) equivalent circuit.

Scattering matrix

The scattering matrix $\mathbf{S} = [S_{ij}]$ relates the incident voltage waves to the reflected voltage waves for an N-port microwave network

$$V^- = \mathbf{S}V^+ \tag{11.53}$$

where $V^- = (V_1^-, V_2^-, \ldots, V_N^-)^T$ and $V^+ = (V_1^+, V_2^+, \ldots, V_N^+)^T$. The component S_{ij} is defined by

$$S_{ij} = \left. \frac{V_i^-}{V_j^+} \right|_{V_k^+=0, k \neq j} \tag{11.54}$$

That is, S_{ij} is calculated by driving port j with an incident wave of voltage V_j^+ and no incident wave on all other ports, and measuring the reflected wave V_i^- at port i. This requires all other ports be terminated in matched loads to prevent reflection. As a result, S_{ii} is the reflection coefficient of port i when all other ports are impedance matched, S_{ij} is a *transmission coefficient* from port j to port i when all other ports are impedance matched. Note $\Gamma_i = S_{ii}$ only when all other ports are matched.

S can be derived from **Z**. For reciprocal networks, **S** is symmetric, $S_{ij} = S_{ji}$. For lossless networks, **S** is a unitary matrix, $\mathbf{S}^* = [\mathbf{S}^T]^{-1}$, where the superscript $*$ indicates conjugation. In many practical cases, the characteristic impedances of all the ports are identical, often 50 ohms.

If port n has a real characteristic impedance Z_{0n}, a generalized scattering matrix is defined by

$$\widetilde{V}^- = \mathbf{S}\widetilde{V}^+, \tag{11.55}$$

where

$$\widetilde{V}^- = \left(\frac{V_1^-}{\sqrt{Z_{01}}}, \frac{V_2^-}{\sqrt{Z_{02}}}, \cdots, \frac{V_N^-}{\sqrt{Z_{0N}}} \right)^T, \quad \widetilde{V}^+ = \left(\frac{V_1^+}{\sqrt{Z_{01}}}, \frac{V_2^+}{\sqrt{Z_{02}}}, \cdots, \frac{V_N^+}{\sqrt{Z_{0N}}} \right)^T. \tag{11.56}$$

Accordingly

$$S_{ij} = \frac{V_i^-/\sqrt{Z_{0i}}}{V_j^+/\sqrt{Z_{0j}}}\bigg|_{V_k^+=0, k \neq j}. \tag{11.57}$$

A signal flow graph is very useful for analysis of microwave networks in terms of transmitted and reflected waves [36]. An unknown impedance at microwave frequencies is mainly measured by using a vector network analyzer. The S-parameters of a microwave network can also be measured using a vector network analyzer.

Transmission matrix

In practical applications, two-port or cascaded two-port networks are usually used. A 2-by-2 transmission matrix, or *ABCD matrix*, can be defined for each two-port network

$$\begin{bmatrix} V_1 \\ I_1 \end{bmatrix} = \begin{bmatrix} A & B \\ C & D \end{bmatrix} \begin{bmatrix} V_2 \\ I_2 \end{bmatrix}, \tag{11.58}$$

where V_1 and I_1 are the input voltage and current to port 1, and V_2 and I_2 are the output voltage and current from port 2. All the currents are specified to be along one direction.

For multiple two-ports in cascade, one can get the overall transmission matrix by multiplying the individual transmission matrices. This method is useful since a library of transmission matrices can be built up for many elementary two-port networks. The

$ABCD$-parameters can be derived from the Y- or Z-parameters of a network. For a reciprocal network, $AD - BC = 1$.

11.5 Impedance matching

Impedance matching is important, for example, for reducing reflection and improving the sensitivity of a receiver. A matching network for connecting a load to a transmission line can always be found, if the load impedance $Z_L = R_L + jX_L$ has nonzero real part, i.e., $R_L \neq 0$. Although theoretical results can be obtained by using computer analysis, the Smith chart provides a fast, sufficiently accurate solution and is widely used in practice. There are many lossless matching networks. The Bode-Fano criterion gives a theoretical limit on the minimum reflection coefficient magnitude for any matching network [36].

11.5.1 Stub tuners

Impedance matching can be implemented by connecting a single open-circuited or shorted-circuited length of transmission line, known as a *stub*, in parallel (shunt) or in series with the feedline at a certain distance from the load. A proper length of open or shorted transmission line can result in any desired value of reactance. Given a reactance, its open-circuit and shorted-circuit implementations have a difference of $\lambda/4$ in stub length.

The shunt stub, as shown in Fig. 11.6, is especially suitable for microstrip or stripline realization. For microstrip or stripline implementation, open-circuited stubs are easier to manufacture since via holes on the circuit board are avoided. For coaxial lines or waveguides, short-circuited stubs can effectively prevent radiation.

The single-stub tuners are fixed. Double-stub tuners provide better adjustability. Double-stub tuners are often fabricated in coaxial line, and the adjustable stubs are connected in parallel to the feed coaxial line. However, the double-stub tuner cannot match all load impedances.

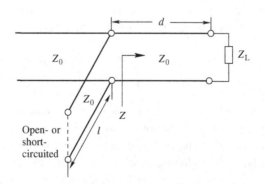

Figure 11.6   The single, shunt stub tuner.

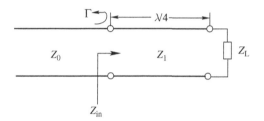

Figure 11.7
Quarter-wave transformer for impedance matching.

11.5.2 Quarter-wave transformer

Assume that we use a feedline of characteristic impedance Z_0 to feed a load resistance $Z_L = R_L$. A quarter-wave transformer, characterized by (11.37), can be used for impedance matching. This is shown in Fig. 11.7. For impedance matching,

$$Z_0 = Z_{\text{in}} = \frac{Z_1^2}{Z_L}. \tag{11.59}$$

This yields

$$Z_1 = \sqrt{Z_0 Z_L}. \tag{11.60}$$

Here Z_L is assumed to be real.

The quarter-wave transformer is used to match a real load impedance to a transmission line. A complex load impedance can be first transformed to a real impedance by using a transmission line or a reactive stub of suitable length. The quarter-wave transformer is then applied.

The quarter-wave transformer matches the load at the design frequency f_0. At f_0, the matching section has a length of $\lambda_0/4$. But at other frequencies f, the length is not a quarter wavelength, and hence mismatch occurs. Given a maximum tolerable reflection coefficient Γ_m and assuming TEM waves, the fractional bandwidth is given by [36]

$$\frac{\Delta f}{f_0} = 2 - \frac{4}{\pi} \cos^{-1} \left[\frac{\Gamma_m}{\sqrt{1 - \Gamma_m^2}} \frac{2\sqrt{Z_0 Z_1}}{|Z_L - Z_0|} \right]. \tag{11.61}$$

It is seen that the bandwidth increases when Z_L is close to Z_0. In microwave design, Γ_m can be determined by the VSWR specification.

When a bandwidth requirement cannot be satisfied by using a single section quarter-wave transformer, a multi-section transformer can be used.

11.5.3 Multisection matching transformers

Multisection matching transformers can be either a binomial transformer or a Chebyshev transformer. Each section is a quarter-wave transformer. The binomial transformer is maximally flat near the design frequency. As opposed to the binomial transformer, the Chebyshev transformer achieves a maximal bandwidth, but with a Chebyshev equal-ripple passband. These transformers are actually passband filters.

For an N-section binomial transformer, assuming the reflection coefficient Γ_n between any two adjacent sections is small, the design equation is given by [36]

$$\ln \frac{Z_{n+1}}{Z_n} \approx 2^{-N} C_n^N \ln \frac{Z_L}{Z_0}, \quad n = 0, 1, \ldots, N, \tag{11.62}$$

where the binomial coefficients $C_n^N = \frac{N!}{(N-n)!n!}$.

The fractional bandwidth is given by [36]

$$\frac{\Delta f}{f_0} = 2 - \frac{4}{\pi} \cos^{-1} \left[\frac{1}{2} \left(\frac{\Gamma_m}{|A|} \right)^{1/N} \right], \tag{11.63}$$

where

$$A = 2^{-N} \frac{Z_L - Z_0}{Z_L + Z_0}. \tag{11.64}$$

A transformer with more sections achieves wider bandwidth. A thorough exposition on the design of binomial as well as Chebyshev multisection matching transformers is given in [31].

11.6 Microwave resonators

A resonator is any structure that is able to contain at least one oscillating electromagnetic field. Resonators can be classified as lumped-element (or quasilumped-element) and distributed-element resonators. Microwave resonators are useful for many applications including the realization of filters, oscillators, frequency meters, and tuned amplifiers. They can be implemented using transmission lines or waveguides.

11.6.1 RLC resonant circuits

Near resonance, resonators can be modeled by a series or parallel RLC equivalent circuit, as shown in Fig. 11.8. The input impedance is

$$Z_{in} = R + j\omega L - j\frac{1}{\omega C} \qquad \text{(series RLC circuit)}, \tag{11.65}$$

$$Z_{in} = \left(\frac{1}{R} + \frac{1}{j\omega L} + j\omega C \right)^{-1} \qquad \text{(parallel RLC circuit)}. \tag{11.66}$$

The complex power delivered to the resonator is given by

$$P_{in} = \frac{1}{2} V I^* = \frac{1}{2} Z_{in} |I|^2 = P_l + 2j\omega \left(W_m - W_e \right), \tag{11.67}$$

where P_l is power dissipated by the resistor R, $W_m = \frac{1}{4} |I_L|^2 L$ is the average magnetic energy stored in the inductor L, and $W_e = \frac{1}{4} |V_C|^2 C$ is the average electric energy stored in the capacitor C. Here, I_L is the current flowing through the inductor and V_C is the voltage across the capacitor.

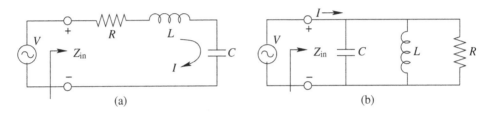

Figure 11.8   RLC resonator circuits. (a) The series RLC circuit. (b) The parallel RLC circuit.

At resonance, $W_m = W_e$, we have the resonance frequency ω_0

$$\omega_0 = \frac{1}{\sqrt{LC}} \tag{11.68}$$

and a purely resistive input impedance

$$Z_{\text{in}} = R. \tag{11.69}$$

The quality factor, Q, of a resonant circuit is defined as

$$Q = \omega \frac{\text{average energy stored}}{\text{power dissipated}} = \omega \frac{W_m + W_e}{P_l}. \tag{11.70}$$

At resonance, Q is given by

$$Q = \frac{\omega_0 L}{R} = \frac{1}{\omega_0 RC} \quad \text{(series RLC circuit)}, \tag{11.71}$$

$$Q = \frac{R}{\omega_0 L} = \omega_0 RC \quad \text{(parallel RLC circuit)}. \tag{11.72}$$

For both the circuits, the half-power fractional bandwidth of the resonator is given by

$$\text{BW} = \frac{2\Delta\omega}{\omega_0} = \frac{1}{Q}. \tag{11.73}$$

When a resonant circuit is coupled to other circuitry, the loaded Q, Q_L, is different from the unloaded Q. Since the resonant frequency is decided by L and C, the resonant circuit should be connected to an external resistor R_L. For a series resonator, R_L should be added in series with R, and the resulting external Q, Q_e, can be defined by $Q_e = \frac{1}{\omega_0 R_L C}$. For parallel resonators, R_L should be added in parallel with R, and Q_e can be defined by $Q_e = \omega_0 R_L C$. In both cases, the loaded Q can be expressed by

$$\frac{1}{Q_L} = \frac{1}{Q_e} + \frac{1}{Q}. \tag{11.74}$$

11.6.2 Transmission line resonators

The RLC resonator can be implemented in microstrip line by using microstrip capacitor and inductor components. Such implementations may resonate at some higher frequencies at which their sizes are no longer very small compared to a wavelength. In this sense, they are not lumped or quasilumped implementations any more. Here we are more interested in distributed-element implementations.

In order to implement a resonator and evaluate its Q, we consider lossy transmission lines. For a low-loss transmission line, the incident and reflected waves should have a form of $e^{-\gamma z}$ and $e^{\gamma z}$, where $\gamma = \alpha + j\beta$. Accordingly, Z_{in} in (11.33) can be recalculated by replacing $j\beta z$ with $(\alpha + j\beta)z$. By comparing Z_{in} of the transmission line with that of a series or parallel RLC resonator circuit and making some approximation, we can obtain a number of transmission line resonators [36]:

- *Short-circuited $\lambda/2$ or $n\lambda/2$ line resonator ($l = n\lambda/2$, $n = 1, 2, \ldots$).* The equivalent series RLC parameters corresponding to Z_{in} are given by

$$R = Z_0\alpha l, \quad L = \frac{Z_0\pi}{2\omega_0}, \quad C = \frac{1}{\omega_0^2 L}. \tag{11.75}$$

At resonance, $Z_{in} = R$, and $Q = \frac{\omega_0 L}{R} = \frac{\pi}{2\alpha l}$. $Q = \frac{\beta}{2\alpha}$ for the first resonance at $l = \lambda/2$ and decreases as l increases.
- *Short-circuited $\lambda/4$ line resonator ($l = \lambda/4$).* It is a parallel-style RLC resonator with

$$R = \frac{Z_0}{\alpha l}, \quad C = \frac{\pi}{4\omega_0 Z_0}, \quad L = \frac{1}{\omega_0^2 C}. \tag{11.76}$$

At resonance, $Z_{in} = R$, and $Q = \omega_0 RC = \frac{\pi}{4\alpha l} = \frac{\beta}{2\alpha}$.
- *Open-circuited $\lambda/2$ or $n\lambda/2$ line resonator, ($l = n\lambda/2$, $n = 1, 2, \ldots$).* It is also a parallel resonant circuit, with

$$R = \frac{Z_0}{\alpha l}, \quad C = \frac{\pi}{2\omega_0 Z_0}, \quad L = \frac{1}{\omega_0^2 C}. \tag{11.77}$$

At resonance, $Z_{in} = R$, and $Q = \omega_0 RC = \frac{\pi}{2\alpha l}$. Again $Q = \frac{\beta}{2\alpha}$ for $l = \lambda/2$. This type of resonator is more suitable for microstrip implementation.

In order to achieve maximum power transfer between a resonator and a feedline, the resonator must be matched at the resonant frequency. In this case, $Z_{in} = R = Z_0$. Thus the unloaded Q equals the external Q, $Q = Q_e$, and the resonator is said to be *critically coupled* to the feed.

A $\lambda/2$ open-circuited microstrip resonator can be fed by gap-coupling to a microstrip feedline. The gap in the microstrip line can be modeled as a series capacitor C. This is shown in Fig. 11.9. In this case, resonance occurs when $Z = 0$. Coupling to the feedline lowers the resonant frequency. When the normalized susceptance of C, $b_c = Z_0\omega C \ll 1$,

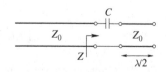

Figure 11.9   **A microstrip resonator coupled to a microstrip feedline and its equivalent circuit.**

the resonant frequency is closed to the unloaded resonator. At $b_c = \sqrt{\frac{\pi}{2Q}}$, critical coupling occurs.

11.6.3 Waveguide cavities

Resonators can also be implemented using a closed section of waveguide. Both the ends are closed in order to prevent radiation loss. This corresponds to the short-circuited case of transmission line. This closed section becomes a cavity. Electric and magnetic energies are stored within the cavity, while dissipation occurs in the metallic wall and the dielectric. Coupling with other components is done via a small aperture or probe. For a rectangular or circular waveguide cavity, resonance occurs when the cavity has a length of $n\lambda_g/2$, $n = 1, 2, \ldots$, where λ_g is the guided wavelength.

For metallic cavities, the Q is decided by the Q due to conductor loss, Q_c, and the Q due to dielectric loss, Q_d,

$$\frac{1}{Q} = \frac{1}{Q_c} + \frac{1}{Q_d}. \tag{11.78}$$

Circular cavities are often used to make microwave frequency meters. The top wall is movable for tuning the resonant frequency. High Q is easier to obtain in the TE_{011} mode, and this corresponds to a higher frequency resolution.

Dielectric resonators are made of low-loss high dielectric material; they are similar to metallic cavities, but have some field leakage from their sides and ends. Dielectric resonators generally have a smaller size and a lower cost. They can be easily integrated into microwave integrated circuits or coupled to a microstrip line.

Due to the conductor loss, Q decreases as $1/\sqrt{f}$ for a cavity or transmission line resonator [36]. At very high frequency, Q will be too small for use. The Fabry-Perot resonator reduces the side walls of a cavity resonator to reduce the conductor loss and the possible number of resonant modes, and becomes an open resonator with two parallel metal plates. Spherical or parabolic reflecting mirrors can be used to confine the energy to a region. Thus, quasi-optical resonators are especially useful for millimeter or submillimeter waves.

11.7 Power dividers and directional couplers

Power dividers and directional couplers are used for power division or combining. Power dividers take the form of three-port networks (T-junctions), while directional couplers take the form of four-port networks.

11.7.1 Three-port networks

Analysis based on the scattering matrix shows that a three-port network cannot be lossless, reciprocal, and matched at all ports, at the same time. If nonreciprocity is assumed, input

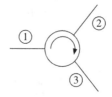

Figure 11.10   Symbol of a circulator.

matching at all ports and lossless property can be satisfied. Such a device is called a *circulator*, and can be made by using anisotropic materials such as ferrite. If only two ports are matched, a lossless and reciprocal three-port network can be realized. On the other hand, if it is lossy, it can be reciprocal and matched at all ports.

Circulators

A circulator can have two possible sets of S parameters

$$\mathbf{S}_1 = \begin{bmatrix} 0 & 0 & 1 \\ 1 & 0 & 0 \\ 0 & 1 & 0 \end{bmatrix}, \quad \mathbf{S}_2 = \begin{bmatrix} 0 & 1 & 0 \\ 0 & 0 & 1 \\ 1 & 0 & 0 \end{bmatrix}. \tag{11.79}$$

For $\mathbf{S}_1$, power flow occurs from ports 1 to 2, 2 to 3, and 3 to 1. The circulator is represented by the symbol in Fig. 11.10. For $\mathbf{S}_2$, the opposite circularity holds; this can be obtained by changing the polarity of the ferrite bias field.

A circulator can be used as an isolator when port 3 is terminated with a matched load. The S-parameter matrix of an isolator is $\mathbf{S} = \begin{bmatrix} 0 & 0 \\ 1 & 0 \end{bmatrix}$; this indicates that both port 1 and port 2 are matched, but power transmission occurs only from port 1 to port 2. Since $\mathbf{S}$ is not unitary, the isolator is lossy.

Isolators are used to isolate a high-power source from a load to prevent damage or interference. There are a number of ferrite isolators such as the resonance isolator and the field displacement isolator. Circulators can be used as diplexers or used in reflection-type phase shifters/attenuators.

Power dividers

The T-junction power divider is used for power division or combining. It takes the form of waveguide or microstrip. It can be designed as a lossless or resistive power divider. The output power ratio can be specified. Both the lossless and resistive power dividers cannot achieve isolation between output ports.

The Wilkinson power divider, as shown in Fig. 11.11, is a lossy three-port network with all ports matched and isolation between output ports, that is, $S_{32} = S_{23} = 0$. The Wilkinson power divider can divide the input power according to any ratio. It uses a resistor connecting the two output ports. When the two output ports are matched, no power is

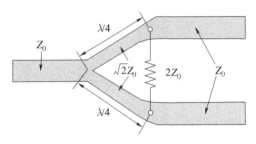

Figure 11.11

The Wilkinson power divider in microstrip form. The power is equally split.

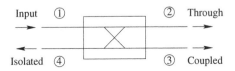

Figure 11.12

Symbol of directional coupler.

dissipated in the resistor. Only reflected power from port 2 or 3 is dissipated in the resistor. It is suitable for microstrip or stripline implementation. The Wilkinson divider has also been generalized to an N-way divider or combiner.

11.7.2 Four-port networks

A directional coupler is any reciprocal, lossless, matched four-port network. Two particular choices of directional couplers, namely the symmetrical and antisymmetrical couplers, are commonly used in practice. Their scattering matrices are, respectively, derived as

$$\mathbf{S} = \begin{bmatrix} 0 & \alpha & j\beta & 0 \\ \alpha & 0 & 0 & j\beta \\ j\beta & 0 & 0 & \alpha \\ 0 & j\beta & \alpha & 0 \end{bmatrix}, \quad \mathbf{S} = \begin{bmatrix} 0 & \alpha & \beta & 0 \\ \alpha & 0 & 0 & -\beta \\ \beta & 0 & 0 & \alpha \\ 0 & -\beta & \alpha & 0 \end{bmatrix}. \tag{11.80}$$

In both cases, $\alpha^2 + \beta^2 = 1$.

A directional coupler uses port 1 as input port, there are outputs from the other three ports, as shown in Fig. 11.12. Power outputs from ports 2, 3, 4 are respectively known as through power P_2, coupled power P_3, and isolated power P_4. A directional coupler is characterized by three quantities, namely coupling C, directivity D, and isolation I

$$C = 10\log_{10}\frac{P_1}{P_3} = -20\log_{10}\beta \quad \text{(dB)}, \tag{11.81}$$

$$D = 10\log_{10}\frac{P_3}{P_4} = 20\log_{10}\frac{\beta}{|S_{14}|} \quad \text{(dB)}, \tag{11.82}$$

$$I = 10\log_{10}\frac{P_1}{P_4} = -20\log_{10}|S_{14}| \quad \text{(dB)}. \tag{11.83}$$

It is seen that $I = D + C$ dB. For an ideal coupler ($S_{14} = 0$), both D and I are infinite.

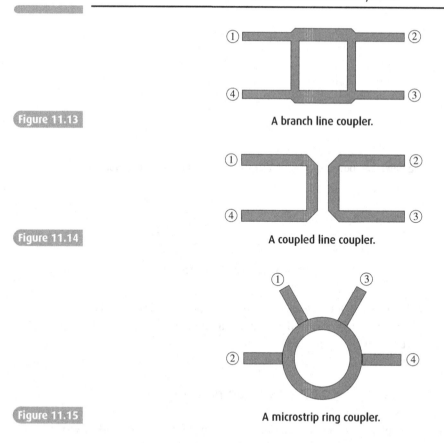

Figure 11.13   A branch line coupler.

Figure 11.14   A coupled line coupler.

Figure 11.15   A microstrip ring coupler.

Hybrid couplers are directional couplers with $C = 3$ dB, that is, $\alpha = \beta = 1/\sqrt{2}$. The outputs of the through and coupled arms have equal power. The phase difference between the output ports is $90°$ (quadrature hybrid), which is an example of a symmetrical coupler), or $180°$ (magic-T hybrid or rate-race hybrid), which is an example of an antisymmetrical coupler.

The quadrature hybrid can be designed as a branch-line hybrid (shown in Fig. 11.13), coupled line coupler (shown in Fig. 11.14), or Lange coupler. The bandwidth of coupling response C of the coupled line coupler can be improved by using multiple $\lambda/4$ sections, as in the case of multisection matching transformer. Coupling in a coupled line coupler is generally too loose to achieve C of 3 dB or 6 dB. The Lange coupler uses an interdigital geometry to increase the coupling. The $180°$ hybrid can be fabricated as a ring hybrid in microstrip or stripline form, shown in Fig. 11.15, tapered coupled line hybrid, or as a waveguide hybrid junction called *magic-T*.

The Bethe hole coupler is a single-hole waveguide directional coupler. In order to improve the bandwidth of the directivity response D, multiple hole waveguide couplers can be designed. Like the multisection matching transformer, the multihole waveguide coupler can be designed with a binomial response or Chebyshev response for its directivity. The spacing between the holes is $\lambda_g/4$.

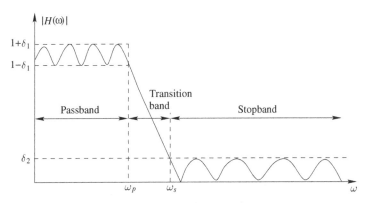

Figure 11.16 **Typical magnitude characteristics of physically realizable lowpass filters.**

11.8 RF/microwave filters

RF/microwave filters typically have lowpass, highpass, bandpass, or band-reject frequency response. Bandpass filters are particularly important in wireless communication systems. For filter design, the insertion loss and passband characteristics are important.

The insertion loss method flexibly controls the passband and stopband amplitude and phase responses in a systematic way, and hence is widely used in analog filter design. The method generates lumped-element circuits. Distributed implementation, especially microstrip implemention, which is of critical importance in modern wireless design, can then be obtained from the lumped-element circuits.

Filters can also be implemented using coupled resonators. In addition to microstrip filters fabricated by using conventional materials, more and more recent researches use advanced materials such as high-temperature superconductors (HTS), ferroelectric materials, MEMS technology, MMIC technology, or LTCC technology [19].

Ideal filters are noncausal and thus physically unrealizable. In practical realizations, we have to realize causal filters that approximate the ideal filters as closely as possible. For a practical filter, a small amount of ripple in the passband and stopband is tolerable. A typical filter design specification is illustrated in Fig. 11.16, where δ_1, δ_2 denote the amplitudes of passband and stopband ripples, and ω_p, ω_s are the passband and stopband edge frequencies.

11.8.1 Insertion loss method

In the insertion loss method, a filter response is characterized by its insertion loss, or power loss ratio, P_{LR}, which is the ratio of the incident power and the power delivered to load

$$P_{LR} = \frac{1}{1 - |\Gamma(\omega)|^2}. \tag{11.84}$$

Note that $P_{LR} = 1/|S_{12}|^2$ if both load and source are matched. Since $|\Gamma(\omega)|^2$ is an even function of ω, it can be expressed as a polynomial of ω^2. Thus, we have

$$P_{LR} = 1 + \frac{M(\omega^2)}{L(\omega^2)}, \tag{11.85}$$

where $M(\omega^2)$ and $L(\omega^2)$ are polynomials of ω^2. In the following, we introduce several methods for designing lowpass filters.

Maximally flat filters

The binomial or Butterworth filter has the maximally flat passband response. For an Nth-order lowpass filter, it is specified as

$$P_{LR} = \frac{1}{|S_{21}(j\omega)|^2} = 1 + \epsilon^2 \left(\frac{\omega}{\omega_c}\right)^{2N}. \tag{11.86}$$

The passband is from $\omega = 0$ to $\omega = \omega_c$, and at the edge of the passband ω_c, $P_{LR} = 1 + \epsilon^2$. Its first $2N - 1$ derivatives are zero at $\omega = 0$. For $\omega \gg \omega_c$, P_{LR} increases at a rate of $20N$ dB/decade. It is an all-pole filter.

Example 11.1: For $\epsilon = 0.4$, the transfer function $|S_{21}(j\omega)|^2$ corresponding to different orders is illustrated in Fig. 11.17.

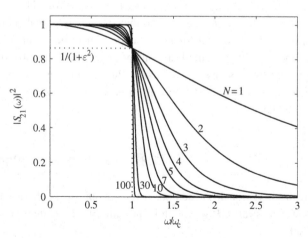

Amplitude response of the Butterworth filter: $\epsilon = 0.4$.

Equal ripple filters

The type-I Chebyshev filter exhibits equal-ripple passband and maximally flat stopband. It is also an all-pole filter. For an Nth-order lowpass filter, it specifies P_{LR} as

$$P_{\mathrm{LR}} = 1 + \epsilon^2 T_N^2 \left(\frac{\omega}{\omega_c} \right), \tag{11.87}$$

where ω_c is the passband edge frequency, and $T_N(x)$ is the Nth-order Chebyshev polynomial

$$T_1(x) = x, \quad T_2(x) = 2x^2 - 1, \quad T_3(x) = 4x^3 - 3x, \quad T_4(x) = 8x^4 - 8x^2 + 1,$$
$$T_5 = 16x^5 - 20x^3 + 5x, \quad T_6(x) = 32x^6 - 48x^4 + 18x^2 - 1, \quad \cdots \tag{11.88}$$

The recurrence is given by

$$T_N(x) = 2xT_{N-1}(x) - T_{N-2}(x). \tag{11.89}$$

Alternatively, $T_N(x)$ can be written as

$$T_N(x) = \begin{cases} \cos(N \arccos(x)), & |x| \le 1 \\ \cosh(N \operatorname{arccosh}(x)), & |x| > 1 \end{cases} . \tag{11.90}$$

The type-I Chebyshev filter has a sharp cutoff, but has ripples of amplitude $\epsilon^2 + 1$. For $\omega \gg \omega_c$, P_{LR} increases at 20 dB/decade, but still $6(N-1)$ dB greater than the binomial response.

The type-II Chebyshev filter has both poles and zeros. It exhibits a maximally flat passband but an equiripple behavior in the stopband. P_{LR} is defined as

$$P_{\mathrm{LR}} = 1 + \epsilon^2 \frac{T_N^2 \left(\frac{\omega_s}{\omega_c} \right)}{T_N^2 \left(\frac{\omega_s}{\omega} \right)}, \tag{11.91}$$

where ω_s is the stopband frequency.

Elliptic function filters

The elliptic (or Cauer) function filter has equal-ripple passband and stopband. It has both poles and zeros. P_{LR} is given as

$$P_{\mathrm{LR}} = 1 + \epsilon^2 U_N^2 \left(\frac{\omega}{\omega_c} \right), \tag{11.92}$$

where ω_c is the passband frequency, U_N is the Jacobian elliptic function of order N, and ϵ controls the passband ripple. As compared to the Butterworth and Chebyshev filters, it has better cutoff rate.

Linear phase filters

A linear phase response in the passband is very important for communication systems to prevent distortion. In addition to the amplitude requirement for a lowpass filter, a linear phase response of the voltage transfer function can be defined as

$$\phi(\omega) = A\omega \left[1 + p \left(\frac{\omega}{\omega_c} \right)^{2N} \right], \tag{11.93}$$

which has maximally flat group-delay $\tau_d = \frac{d\phi}{d\omega}$. Design for linear phase filters is more complicated.

The Bessel filter has a transfer function derived from a Bessel polynomial. It has a maximally flat group-delay in the passband; this is in a sense complementary to the Butterworth response, which has a maximally flat amplitude. The Bessel filter is an all-pole filter with the transfer function

$$H(s) = \frac{1}{B_N(s)}, \tag{11.94}$$

where $B_N(s)$ is the Nth-order Bessel polynomial.

The Gaussian filter, whose transfer function is derived from a Gaussian function, has zero overshoot on the step and impulse response. The response is very close to that of the Bessel filter, giving poor selectivity and high sensitivity in exchange for superior delay and phase linearity.

The amplitude and delay responses of the five types of filters, i.e., the Butterworth, Chebyshev, elliptic, Bessel, and Gaussian lowpass filters, are illustrated in Fig. 11.18.

11.8.2 Prototyping

Filter design starts from lowpass filter prototypes, which are normalized in impedance and frequency. The maximally flat filters, the equal-ripple filters, the linear phase filters and the Gaussian filters are all-pole filters. They all can be implemented by using a doubly-terminated lossless ladder network or its dual, as shown in Fig. 11.19.

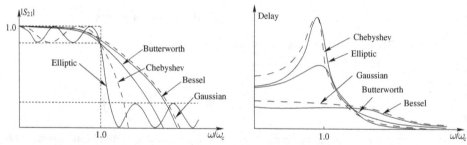

Figure 11.18 The amplitude and phase responses of Butterworth, Chebyshev, elliptic, Bessel, and Gaussian lowpass filters, for the same order ($N = 2$).

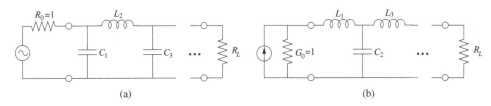

Figure 11.19   Two dual LC ladder circuits for lowpass filter prototypes.

For Nth-order maximally flat and equal-ripple lowpass filter prototypes, the values for circuit elements for $N = 1$ to 10 are tabulated, and the attenuation characteristics for various N are also given in [19, 31, 36]. The element values for Gaussian filters of order 2 to 10 are tabulated in [19].

In many wireless communication systems, flat group delay as well as selectivity is required for a bandpass filter. For the prototype lowpass linear phase filter, design values are also tabulated for $N = 1$ to 10 for the ladder circuits [31, 36]. Another lowpass prototype for linear phase filters was introduced in [19, 40]. The optimized element values are tabulated in [19] for filters of orders 4/6/8/10 and a return loss S_{11} of -20 dB/-30 dB at the passband. There is a tradeoff between linear phase response and selectivity in the filter design.

The prototype circuits and values for the elliptic function lowpass filter are also given in [19]. The prototype circuits are very similar to the LC ladder circuit, but using series parallel-resonant branches, or its dual using shunt series-resonant branches.

For designs of the same order, the equal-ripple response has the sharpest cutoff but the worst group delay. The maximally flat response has a flat amplitude attenuation and also a flat group delay in the passband, but with a slightly lower cutoff. The linear phase filter has constant group delay, but the worst cutoff.

Transformations and distributed implementation

These lowpass filter prototypes are then scaled for arbitrary frequency and impedance, and can also be transformed into highpass, bandpass or bandstop types. These transformations are given in [19, 36]. The design in lumped-element circuit can then be implemented in distributed form. For microstrip filters, the desired source impedance is normally 50 ohms.

In the following, we describe some conventional methods for microstrip filter design. Some advanced RF/microwave filters are introduced in [19]. A most comprehensive text on microstrip filter design is by Hong [19].

11.8.3 Stub filters

The lumped-element design needs to be transformed into a distributed circuit such as a microstrip circuit. Richard's transformation and Kuroda's identities are used to transform the lumped-element circuits into distributed ones. Richard's transformation converts

Figure 11.20   A lowpass filter using stubs. Each section has a length of λ/8.

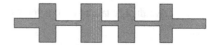

Figure 11.21 A Hi-Z-low-Z lowpass filter. Note that the thick sections correspond to Z_l, the narrow sections to Z_h, and the sections at the two ends correspond to Z_0.

lumped elements to transmission line sections, and four Kuroda's identities are used to separate the filter elements by using transmission line sections. Richard's transformation is used to transform inductors and capacitors into short-circuited and open-circuited stubs of λ/8 at ω_c, respectively. All the stubs are of the same length, and are called *commensurated lines*.

Kuroda's identities employ redundant transmission line sections to achieve a practical microwave filter implementation: They physically separate transmission line stubs, convert between series and shunt stubs, and change impractical characteristic impedances into practical ones. The additional transmission line sections are also λ/8 long at ω_c, and are called *unit elements*.

This generates filters using stubs. A lowpass filter using stubs is illustrated in Fig. 11.20. A simple design procedure for using stubs λ/8 long is detailed in [36].

The stub filter implementation is suitable for all kinds of filters, and the open-circuited stub is more suitable for microstrip filters. The design equations for using $\lambda_g/4$ short-circuited stubs or $\lambda_g/2$ open-circuited stubs are given in [19, 31].

11.8.4 Stepped-impedance lowpass filters

Stepped-impedance, or *Hi-Z-low-Z*, lowpass filter is a popular microstrip or stripline design that uses alternating transmission line sections of very high and very low characteristic impedances. Such filters are easier to design and cover less space than filters implemented using stubs.

Due to approximation introduced, they are usually used when a sharp cutoff is not required, such as for rejecting unwanted mixer products. The designed lowpass filter is illustrated in Fig. 11.21, and has a one-to-one correspondence with the LC-ladder circuit shown in Fig. 11.19b. The method converts an inductor to a length of transmission line of a fixed high impedance Z_h, and a capacitor to a length of transmission line of a fixed low impedance Z_l. The values of Z_h and Z_l should be set to the highest and lowest characteristic impedances that can be fabricated. The value of each inductor or capacitor is determined by the length of the transmission line [36]

$$\beta l \approx \frac{LZ_0}{Z_h} \quad \text{(for inductors)}, \tag{11.95}$$

$$\beta l \approx \frac{CZ_l}{Z_0} \quad \text{(for capacitors)}, \tag{11.96}$$

where Z_0 is the filter impedance, L and C are the normalized values of the lowpass prototypes, and $\beta = 2\pi/\lambda$.

11.8.5 Coupled line bandpass filters

Like Kuroda's identities, the impedance (K) or admittance (J) inverter can be used for conversion between series-connected elements and shunt-connected ones. The J-inverter converts a shunt LC resonator into a series LC resonator, while the K-inverter performs the inverse operation. They can be constructed by using a quarter-wave transformer. The K and J inverters are especially useful for bandpass or bandstop filters of narrow bandwidths [36].

Coupled resonator circuits are particularly useful for designing RF/microwave narrow-band bandpass or bandstop filters. There is a general technique for designing coupled resonator filters for any type of physical resonator including waveguide filters, dielectric resonator (DR) filters, ceramic combline filters, and microstrip filters [19]. The method depends on the coupling coefficients of the coupled resonators and the external quality factors of the input and output resonators.

Parallel-coupled half-wavelength resonator bandpass filters

Parallel-coupled transmission lines can be used for filter design. The microstrip or stripline implementation is easy to fabricate, and is suitable for fractional bandwidths less than 20%. Narrowband bandpass filters can be implemented using cascaded coupled line sections. A bandpass filter is illustrated in Fig. 11.22.

These coupled-line sections are identical; each coupled-line section is $\lambda/2$ long in the vicinity of the bandpass region of the filter, and can be modeled by a shunt-parallel LC resonator. The odd- and even-mode impedances of the coupled part are determined by the spacing between two coupled-line sections. Each coupled-line section has a $\lambda/4$ line coupled with the previous one, and a $\lambda/4$ line coupled with the following one. The coupled part of two coupled-line sections can be modeled by a J-inverter. Cascaded coupled-line sections thus have an equivalent circuit in the same form as that of the bandpass or bandstop filter. The design equations for parallel-coupled-line bandpass filters are given in [19, 36].

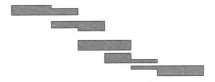

Figure 11.22   Coupled line bandpass filter.

Figure 11.23   End-coupled resonator bandpass filter.

Figure 11.24   Hairpin-line bandpass filter.

End-coupled half-wavelength resonator bandpass filters

The end-coupled microstrip bandpass filter, as shown in Fig. 11.23, uses many open-end microstrip resonators, each of length approximately half a guided wavelength at the midband frequency f_0 of the bandpass filter. The gap between two adjacent open ends can be modeled by a capacitor. All the microstrip-line sections have impedance Z_0. Design is based on the length of each section and the gaps between these sections. Design equations are given in [19, 36].

Hairpin-line bandpass filters

The hairpin-line bandpass filter, as shown in Fig. 11.24, is very compact. It can be viewed as folding the resonators of the parallel-coupled bandpass filter into a U shape. Thus, the same design equations as for parallel-coupled half-wavelength resonator filters can be used, but must take the reduction of the coupled-line lengths due to folding into consideration [19]. The two arms of each hairpin resonator must also be well separated to reduce coupling. Due to the large coupling arising from the compact architecture, full-wave EM simulation is usually used to verify the design, that is, to calculate S_{21} and S_{11}.

Interdigital bandpass filters

The interdigital bandpass filter, as shown in Fig. 11.25, consists of an array of TEM-mode or quasi-TEM-mode transmission line resonators, each of length $\lambda/4$ at the midband frequency and short-circuited at one end and open-circuited at the other end, arranged in an alternating orientation. Grounding is implemented by using via holes. Each resonator can have different length and width, and coupling is adjusted by the spacing between adjacent resonators.

The second passband of the filter is centered at about three times the midband frequency of the desired first passband, and there is no spurious response in between. But for the parallel-coupled half-wavelength filters, a spurious passband at around twice the midband frequency almost always occurs [19]. The design equations are given in [19]. Use of full-wave EM analysis for design verification is also desirable.

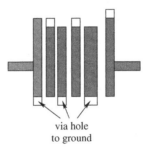

via hole
to ground

Figure 11.25

Interdigital bandpass filter

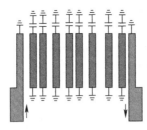

Figure 11.26

Combline bandpass filter.

Combline bandpass filters

The combline filter has a structure very similar to the interdigital filter. This is shown in Fig. 11.26. It is comprised of an array of coupled resonators. These resonators are short-circuited at one end, and at the other end they are connected to the ground with a lumped capacitance in between. A second passband at about three times the midband frequency occurs. The design equations are given in [19].

The larger the value of the lumped capacitance, the shorter the resonator line. This leads to a more compact structure with a wider stopband between the first passband and the unwanted second passband. This provides a convenient method for filter tuning. The pseudocombline filter is a variant of the combline filter, wherein the shorted-circuited ends are replaced by $\lambda_g/4$ open-circuited stubs.

11.8.6 Computer-aided design for RF/microwave filter design

Computer-aided design (CAD) has been extensively used for RF/microwave filter simulation, design and validation. For filter design, the most important performance parameters are S_{21} and S_{11}. These are conventionally obtained by using the linear model obtained from cascaded ABCD matrices. Linear simulation based on network analysis is simple and fast. However, linear analysis is based on analytical circuit models, and the models are valid only for certain frequencies and physical parameters. For complex RF/microwave structures or for a more accurate solution, full-wave EM solvers are preferred.

In Chapter 10, we have introduced a number of full-wave EM methods and the popular commercial EM solvers. There are also some other integrated packages such as Agilent's

Advanced Design System (ADS), and Applied Wave Research (AWR)'s Microwave Office. These integrated packages incorporate both full-wave EM solvers and linear simulators that are based on transmission line theory, and also provide optimizers. Ansoft's Harmonica is a linear and nonlinear simulator.

When full-wave EM simulators are used, the RF/microwave filter structure is divided into cells with 2D or 3D meshing. The smaller the meshing size, the smaller the simulation error but the longer the simulation time. By repeating the EM simulation using different mesh sizes, one can find a very good compromise between the simulation error and time. As a rule of thumb, EM solvers are best suited to problems with physical dimensions of the order of $0.1\lambda_g$ to $10\lambda_g$. Problems with smaller electrical lengths can be solved using quasi-static solvers, while large problems require more computing resources.

Neural networks [12] have also been used for filter modeling [33]. After training from existing examples or the results from full-wave EM simulators, a neural network extracts the nonlinear relation of existing devices. The method can provide the same degree of accuracy as that afforded by EM simulators, but with less computation. For example, an EM simulator is used to obtain the S parameters for all components to be modeled in a range of parameters and frequencies, and a neural network is then trained using these results. The trained neural network can be generalized over this range.

Computer-aided analysis is only suitable for validation of a filter. Given filter design specifications, an optimization design procedure can be applied. The designer can provide an initial design for a specific design method, and then adjust the parameters in a systematic manner. The performance of each new design is evaluated by the computer-aided analysis, and is also compared to the filter specifications. This procedure is repeated until the optimum design is achieved. Many optimization techniques including the conjugate gradient method and evolutionary algorithms are described in [12].

For filter synthesis by optimization, an objective function must be defined. For a two-port filter topology, the following error function can be used as the objective function [19]

$$E(f, \mathbf{\Phi}) = \sum_{i=1}^{I} \left| S_{21}(f_i, \mathbf{\Phi}) - S_{21}^d(f_i) \right|^2 + \sum_{i=1}^{J} \left| S_{11}(f_j, \mathbf{\Phi}) - S_{11}^d(f_j) \right|^2, \qquad (11.97)$$

where S_{21} and S_{11} are obtained from computer analysis, S_{21}^d and S_{11}^d are the desired frequency response from the specifications, f_i's are frequencies at specified points, I and J are the numbers of evaluation points, and $\mathbf{\Phi}$ represents all designable parameters for a given filter topology. Many commercial computer-aided design packages can perform filter synthesis by optimization directly.

11.8.7 Filters for wireless communications

A filter is used to reject the out-of-band interference prior to further processing. The insertion loss of the filter is an important factor. In wireless systems, the receive and transmit filters are combined by using a diplexer to allow them to share the same antenna.

Filters can be designed by using circuits of electrically connected resonators (such as ladder filters) or as wave propagation coupled resonators. These filters can be manufactured using the thin-film resonator technology, which uses thin films of piezoelectric materials to manufacture resonators and filters over a range of 500 MHz and 20 GHz. Bandpass filters are more common in wireless communications.

Since bandpass and bandstop filters require LC segments that behave as either series or parallel resonant circuits, as seen from coupled line bandpass or bandstop filters, coupled resonators are also used for the design of such filters. Compared to the coupled line filter, the stub filter is more compact and easier to design. A quarter-wave open-circuited or short-circuited transmission line stub is a series or parallel resonant circuit, respectively. However, a filter using stub resonators often uses characteristic impedances that are difficult to fabricate. The capacitive-gap coupled resonator filter is a kind of bandpass filter suitable for fabrication in microstrip or stripline form. The direct-coupled cavity filter is also popular in waveguide form. Design details for various filters can be found in references [19, 31, 36].

At RF frequencies from 800 MHz to about 4 GHz, DR bandpass filters are generally used, and they provide small size and high Q. At IF frequencies below 100 MHz, bandpass filters typically use quartz crystal or SAW devices. Crystal resonators have high-Q resonance. Resonators are also used in SAW devices. SAW filters have very sharp cutoff. Above 4 GHz, waveguide resonator-based bandpass filters are usually used.

Filters for base stations

Filters used in wireless BSs can be either coaxial-cavity resonator filters or DR filters [30]. The coaxial-cavity filter offers low-cost design, but has limited Q. Coaxial TEM filters are widely used in BSs. They typically use combline or interdigital structure. Combline filters can be designed with a bandwidth of 1–50%, and are commonly deployed. Interdigital filters are used when wider bandwidth is needed. Coaxial TEM filters and diplexers can be machined from a single block of metal.

The DR filter is emerging as a more popular solution in view of its high performance. A DR filter typically consists of a number of dielectric resonators mounted inside cavities within a metal. It has a high unloaded Q, high dielectric constant ϵ_r, and small temperature drift. DR filters can operate in different modes such as transverse electric (TE) or transverse magnetic (TM) modes. Each mode has a different filter size, unloaded Q and spurious performance. The unloaded Q is typically the reciprocal of the loss tangent $\left(\frac{1}{\tan \delta}\right)$ of the dielectric material. The loaded Q is typically 70–80% of the unloaded Q, and the difference is due to the size of the enclosure, support structure of dielectric resonators inside the cavity, and tuning screws. Dielectric resonators usually have poor spurious performance. There are also filters that combine the small-size advantages of coaxial-resonator filters and the high Q values of DR filters. These devices are typically machined from metals.

HTS filters realize small-size, high-order filters with low insertion loss. Most superconducting filters are simply microstrip filters using HTS thin films instead of conventional conductor films. Receiver sensitivity can be significantly improved by designing the receiver to operate in a cryogenic environment. This eliminates thermal noise from the LNA and improves the filter loss performance. Due to the low resistance of HTS materials, it is possible to use planar thin-film technology to provide HTS filters that are two orders of magnitude smaller in size than conventional DR filters. Yttrium barium copper oxide (YBCO) and thallium barium calcium copper oxide (TBCCO) are the two most popular HTS materials, with critical temperature T_c of 92 K and 105 K, respectively [19]. HTS microstrip filters are desirable for future wireless systems.

Filters for mobile stations

MSs are restricted by size, and bandpass microstrip filters may not be useful. SAW filters are unique passive microwave components that are suitable for MSs. The use of SAW filters in modern MSs is decreasing due to the high integration of RF front ends. Meanwhile, multiband MSs require more SAW filters. SAW-based filters or resonators provide high-Q for frequencies up to 2 GHz. For use in MSs, SAW devices must withstand powers as high as 34 dBm for 2G and 3G mobile communications. The size of SAW filters has shrunk significantly in recent years. SAW devices have superior performance such as high stability, low insertion loss, high stopband rejection, linear phase characteristics, narrow transition width between passband and stopband, small influence of the shape factor on frequency response, and temperature stability [44, 52].

SAW devices are based on the propagation of standing microacoustic waves. Transduction and reflection are the basic mechanisms in most SAW devices. A SAW device consists of a piezoelectric substrate with metallic structures such as interdigital transducers, and reflection or coupling gratings on its polished surface [44]. Due to the piezoelectric effect, a microwave input signal at the transmitting interdigital transducer stimulates a microacoustic wave that propagates along the surface of the elastic solid, and bounces into the cavity formed by the gratings, thus generating resonance. A SAW generates an electric charge distribution at the receiving interdigital transducer, and thus outputs a microwave electrical signal. Design of interdigital transducers is of critical importance in terms of frequency response characteristics.

Many passive components such as inductances and capacitors for matching and building LC filters can be integrated into SAW devices by using LTCC technologies. The balun functionality and impedance transformations can also be integrated into SAW modules. SAW technology now typically covers the frequency range from 10 MHz up to 3 GHz, since the frequency range is bounded by their size. SAW devices require proper packaging.

11.9 Phase shifters

Phase shifters can be easily implemented using transmission lines. A switched-line phase shifter is shown in Fig. 11.27. By changing the length of a transmission line by Δl, a phase shift φ is realized:

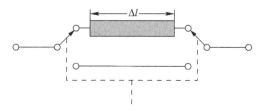

Figure 11.27   A switched-line phase shifter.

$$\varphi = \beta \Delta l = \frac{2\pi f \Delta l}{c} \sqrt{\epsilon_r}, \tag{11.98}$$

where $\beta = 2\pi/\lambda$, λ being the wavelength of the signal passing along the line.

Phase shifts can also be achieved by changing the wave number β, which can be modified by changing the relative permittivity ϵ_r of a dielectric. Switches are required for switching between transmission lines of different electrical lengths or different dielectrics. Popular RF switches are PIN diodes, FETs, and MEMS switches.

The ferrite phase shifter is based on the interaction between the electromagnetic waves and the spinning electrons in a magnetized ferrite. The latching ferrite phase shifter requires only a pulsed current to change its state. It has low insertion loss and can handle significantly higher power, but is complex and expensive. Compared with ferrite phase shifters, PIN diode or FET phase shifters have smaller size, integrability with planar circuitry, higher speed, and lower cost; but they have higher insertion loss, because they require continuous bias current.

MEMS phase shifters have a very low insertion loss, less than 2 dB, over a full phase shift of 360° at microwave frequency ranges. They have high isolation and low drive power. The combination of MEMS with new dielectric tunable materials can achieve a lightweight, low cost as well as continuous phase change [52]. The MEMS bridge that is made of polymers having elastic modulus of one-tenth that of metals can effectively reduce the actuation voltage, which is around 100 V with conventional metal bridges.

Digital phase shifters are very similar to digital attenuators. They use switches to set the different states of a cascade of discrete phase shifters. They are usually designed for binary phase shifts of 180°, 90°, 45°, 22.5°, 11.25°, etc arranged in cascade to realize all possible combination of the phase. The switched-line phase shifter is most fundamental, but its size is dependent on the lowest frequency of operation and the number of bits. A satisfactory single-pole two-throw (SP2T) switch is essential. When the off line has a length of a multiple of $\lambda/2$, resonance may occur; this problem must be considered when selecting the lengths of the two transmission lines.

There are several types of digital phase shifter. The loaded-line phase shifter can be used for small phase shift (typically less than 45°), and it is realized by loading a transmission line with a shunt susceptance, which introduces a phase shift in the transmitted wave. The reflection-type phase shifter is based on the total path length change between the two reflected waves. The switched-filter phase shifter is similar to the switched-line phase shifter, but the transmission lines are replaced by a lowpass filter (for phase lag) and a highpass filter (for phase advance). The switched-filter phase shifter is much smaller than an equivalent switched-line, reflection-type, or loaded-line phase shifter, and is most popular for a digital MMIC phase shifter [41].

An analog phase shifter is one where continuous phase variation is possible. Analog phase shifters are desirable for adaptive phase arrays. An analog phase shifter can typically be implemented as a reflection-type phase shifter [41]. A line stretcher is also a phase shifter.

11.10 Basic concepts in active RF circuits

From this section on, we discuss various active RF circuits and components. We first give some basic concepts in this section.

The use of nonlinear devices leads to undesired harmonics and intermodulation products. These spurious signals increase the conversion loss and signal distortion of active RF components such as mixers and amplifiers. Amplitude nonlinearity between input and output power can be expanded into a power series

$$v_o = k_1 v_i + k_2 v_i^2 + k_3 v_i^3 + \cdots . \tag{11.99}$$

For a single-carrier input signal, this yields harmonic products, which can be eliminated by filtering in case of narrowband communications. For wideband communications with an octave or greater bandwidth, when more than one carrier frequency (or tone) is used, intermodulation products will occur. These intermodulation products are very close to the desired signals and cannot be easily removed by filtering.

Minimum detectable signal

For a given system noise power, the minimum detectable signal (MDS) determines the minimum SNR at the demodulator. The minimum SNR at the demodulator, $\text{SNR}_{o,\min} = \left(\frac{S_o}{N_o}\right)_{\min}$, is associated with the MDS, $S_{i,\min}$, by

$$S_{i,\min} = \frac{S_{o,\min}}{G} = k_B B \left[T_A + (F - 1)T_0 \right] \text{SNR}_{o,\min}, \tag{11.100}$$

where $T_0 = 290$ K, T_A is the antenna temperature, F is the receiver noise figure, G is the overall power gain, B is the channel bandwidth measured in Hz, $S_{o,\min}$ is the output signal, and k_B is Boltzmann's constant. In decibels

$$S_{i,\min} = -174 + 10 \log_{10} B + 10 \log_{10} \left(\frac{T_A - T_0}{T_0} + F \right) + \text{SNR}_{o,\min} \quad \text{(dBm)}. \tag{11.101}$$

The sum of the first three terms is the total noise of the system, and is known as the *noise floor*, N_{floor}. In view of this, the MDS is very weak for a narrowband channel. Assuming $T_A = T_0$, the third term reduces to F in dB. For example, GSM requires an MDS of -102 dBm with an SNR of 9 to 12 dB at a BER of 10^{-3}. Thus, the maximum noise figure is between 7 to 10 dB.

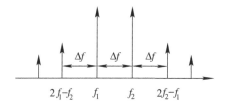

Figure 11.28

Third-order intermodulation products.

For digital demodulation, we have $\frac{S_o}{N_o} = \frac{E_b}{n_0} \frac{R_b}{B}$, where R_b is the bit rate and B is bandwidth. The MDS power $S_{i,\min}$ can be converted to the receiver voltage sensitivity (rms) by

$$V_{i,\min} = \sqrt{2Z_0 S_{i,\min}}, \qquad (11.102)$$

where Z_0 is the impedance of the receive antenna.

The dynamic range of a receiver is defined as the ratio of the maximum allowable signal power to MDS power. The maximum allowable signal power can be selected as the third-order intercept point (IP3).

Intermodulation products

For two carrier frequencies (two tones) f_1 and f_2, $\Delta f = f_2 - f_1 > 0$, the output spectrum contains harmonics of the form

$$m\omega_1 + n\omega_2, \quad m, n = 0, \pm 1, \pm 2, \ldots, \qquad (11.103)$$

where $\omega_1 = 2\pi f_1$ and $\omega_2 = 2\pi f_2$. When the harmonic is a combination of both the frequencies, it is referred to as an *intermodulation product*, with an order of $|m| + |n|$.

The third-order term in (11.99), $k_3 v_i^3$, as well as the higher-order odd-power terms, lead to intermodulation products, and the third-order term is usually more significant. The third-order term generates frequencies $2f_2 - f_1 = f_2 + \Delta f$ and $2f_1 - f_2 = f_1 - \Delta f$. These terms set the dynamic range of the system. They are neighboring carrier frequencies, as shown in Fig. 11.28.

The spectrum resulting from a nonlinear device has many high-order distortion terms. This is known as *spectral regrowth* or *intermodulation distortion (IMD)*, which forms the ACI. Some frequency components of the intermodulation products may fall into the IF band; and they are called *spurious responses*. The order must be above 6 to 10 so that the power included in these components is sufficiently small. A spectrum analyzer generates a frequency-domain representation of an input signal, and is a valid tool for measuring harmonics and intermodulation products.

One-dB compression point and third-order intercept point

The input-output power relationship for the nonlinear device is given by (11.99). The third-order IMD of two nearby interferers is of critical importance. IP3 is defined for the

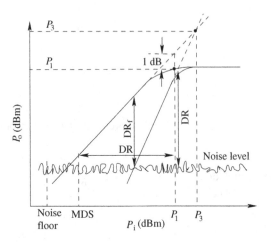

Figure 11.29   **Input vs. output power for nonlinear devices.**

third-order IMD that intercepts with the linear small signal line. IP3 is a unique quantity that can be used for comparing the linearity of different circuits. The third-order IMD is due to $k_i v_i^3$ in (11.99), so it has a slope of 3 in the input/output power diagram (dB), as shown in Fig. 11.29.

There is a saturation zone in the figure. This is because k_3 is typically negative. The 1-dB compression point, whose power level is denoted as P_1, is used to characterize the linear operating range of the device. The second-order term has a slope of 2; it is omitted in the figure, since it can be easily removed by filtering. IP3 is denoted by P_3. For amplifiers, P_1 and P_3 are usually denoted in terms of their output power, while for mixers they are in terms of the input power.

When two signals $A \cos \omega_1 t$ and $A \cos \omega_2 t$ are used as input to a nonlinear device, the 1-dB compression point and the input IP3 (IIP3) in amplitude are related by [9, 37]

$$\left(\frac{A_{1dB}}{A_{IIP3}} \right)^2 \approx -9.6 \quad \text{(dB)}. \tag{11.104}$$

The 1-dB compression and IP3 are the two important parameters for characterizing a circuit's linearity. The Nth-order intercept point is defined as the point where the fundamental signal and the Nth-order harmonic intercept with their linear extrapolations. IP3 is used to describe a receiver's ability to handle large signals. That is, a large IP3 corresponds to a small intermodulation signal.

Nonlinear Effects of Cascaded Systems

For cascaded RF systems, each component has its own linearity problem. Given the 1-dB compression point and IP3 of each component, the 1-dB compression point and IP3 of the entire system can be determined.

Assuming that the magnitudes of the multiple tones are the same, the overall 1-dB compression point and IP3 are given by [29, 37]

$$\frac{1}{P_{1dB}} = \frac{1}{P_{1dB,1}} + \frac{G_1}{P_{1dB,2}} + \frac{G_1 G_2}{P_{1dB,3}} + \frac{G_1 G_2 G_3}{P_{1dB,4}} + \cdots, \tag{11.105}$$

$$\frac{1}{P_{IP3}} \approx \frac{1}{P_{IP3,1}} + \frac{G_1}{P_{IP3,2}} + \frac{G_1 G_2}{P_{IP3,3}} + \frac{G_1 G_2 G_3}{P_{IP3,4}} + \cdots, \tag{11.106}$$

where the subscript 1dB,k denotes the 1-dB compression point of the kth stage, and the subscript IP3,k denotes the input IP3 of the kth stage. Thus, the later stages influence the overall 1-dB compression point and overall IP3 more significantly. To increase the overall 1-dB compression point or overall IP3, the gains of the components should be reduced, and the gain of the first component has the most significant influence. This is in contrast to the case of the noise figure, given by (11.12), and becomes a tradeoff for RF system design.

Dynamic range

The dynamic range of a circuit is determined by its noise figure and linearity. The noise figure determines the MDS, and the IIP3 specifies the maximum input power with a tolerable distortion. In an RF front-end, the noise figure is usually dominated by the LNA and the linearity is dominated by the mixer.

For different applications, the definition of dynamic range may be different. For mixers, the dynamic range is defined in accordance with the input power. It is the range between the MDS and the 1-dB compression point, DR $= P_1 - $ MDS. For amplifiers, it is in terms of the output power, and is defined as the range from the noise floor to the 1 dB compression point, DR $= P_1 - N_o$.

For LNAs and some mixers, operation is limited to lower power level so that power contribution from the third-order intermodulation is lower than the output noise level N_o, thus leading to a minimal spurious response. This range is referred to as the *spurious-free dynamic range (SFDR, DR_f)*, and is given by [35]

$$DR_f = \frac{2}{3}(P_3 - N_o) \quad \text{(dB)}. \tag{11.107}$$

In [37], the SFDR is defined as the upper end of the dynamic range on the intermodulation behavior and the lower-end on the sensitivity, and is given by [37]

$$DR_f = \frac{2}{3}(P_3 - N_{\text{floor}}) - SNR_{o,\text{min}} \quad \text{(dB)}. \tag{11.108}$$

Adjacent channel power ratio

For power amplifiers, ideally the phase shift, or time delay, should be constant for all power levels. Phase in practical amplifiers changes with the power level, thus with amplitude. Phase nonlinearity introduces intermodulation products in a fashion similar to amplitude nonlinearity.

Distortion is usually defined as CIR power ratio, where the interference is composed of the intermodulation products. In order to make the contribution of the intermodulation products negligible to the system CNR, they should be at least 10 dB lower than the carrier level.

The sum of all intermodulation products in an adjacent channel is the adjacent-channel power level. The adjacent channel power ratio (ACPR), also known as *adjacent channel leakage ratio (ACLR)*, is defined as the ratio of the adjacent-channel power to the carrier power. For personal communications, ACPR is usually lower than −35 dB. For example, the WCDMA standard requires ACPRs at 5 MHz frequency offset to be −33 dB or better, and ACPR at 10 MHz frequency offset to be −43 dB or better.

11.11 Modeling of RF components

Semiconductor-based RF components such as various diodes and transistors are fundamental to active RF circuits. The physical modeling of various diodes, transistors, and FETs is given in detail in [48]. In this section, we briefly describe the physical characteristics of these components.

11.11.1 Diodes

PN diode

A pn junction is obtained by connecting a p-type and an n-type semiconductor, with similar impurity concentrations. The PN diode has a dc V-I characteristic

$$I(V) = I_s \left(e^{\alpha V} - 1\right), \tag{11.109}$$

where $\alpha = \frac{q}{nk_BT}$, q being the charge of an electron, k_B Boltzmann's constant, T the temperature, n the ideality factor, and I_s is the saturation current, which is typically less than 10^{-6} A. The I-V characteristic is illustrated Fig. 11.30.

Small-signal approximation gives

$$I(V) = I_0 + \frac{v}{R_j} + \frac{\alpha}{2R_j}v^2 + \cdots, \tag{11.110}$$

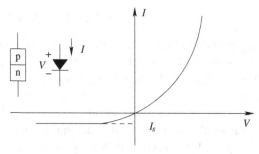

Figure 11.30   V-I characteristic of a PN diode.

where $I_0 = I(V_0)$ is the dc bias current corresponding to the dc bias voltage V_0, v is a small ac signal voltage, and $R_j = \frac{1}{\alpha(I_0 + I_s)}$ is the junction resistance of the diode.

Diodes have a typical nonlinear response, and are usually used for mixer design. A diode can be used as rectifier to convert a fraction of RF signal to dc power. This is useful for power monitoring, automatic gain control, and signal strength indication. An input small-voltage RF (ac) signal can lead to a dc-rectified current plus ac signals. The ac signals are usually filtered. The nonlinearity of the diode can also be used for demodulating an amplitude modulated RF signal, that is, for signal detection.

The PN diode can be modeled by a current source in parallel with two capacitors: C_j for the junction capacitance, and C_d for the diffusion capacitance. Most microwave applications rely on C_j, which has a nonlinear dependence on voltage. Diodes in these applications are known as *varactors*. Varactor diodes can be used in designing amplitude modulators, phase shifters, and frequency multipliers.

PIN diode

The PIN diode is similar to the PN diode but has a very minimally doped region, called *intrinsic (I) region*, sandwiched between the p-type (p$^+$) and n-type (n$^+$) regions. The I region has high resistivity, and yields very small junction capacitance, due to the wider depletion region. The I region increases the breakdown voltage of the device, allowing a high reverse voltage. The PIN diode has higher impedance under reverse bias and thus has more isolation than the PN diode. The PIN diode can handle medium to large RF power levels.

For a forward biased PIN diode, holes and electrons are injected into the I region. These charges do not annihilate each other immediately, but stay alive for an average carrier lifetime τ, yielding an average stored charge, Q, which lowers the effective resistance R_S of the I region [47]

$$R_S = \frac{W^2}{(\mu_n + \mu_p)Q}, \tag{11.111}$$

where $Q = I_F\tau$ (in coulombs), I_F being the forward bias current, W is the I region width, μ_n is the electron mobility, and μ_p is the hole mobility. The equation is valid at frequencies higher than the I region transmit time frequency, $f > \frac{1300}{W^2}$, f being in MHz and W in μm.

For a zero or reverse biased PIN diode, there is no stored charge in the I region and the diode can be modeled as a capacitor, C_T, shunted by a parallel resistance R_P. C_T is given by

$$C_T = \frac{\epsilon A}{W}, \tag{11.112}$$

where ϵ is the dielectric constant of silicon and A is the area of the diode junction. The equation is valid when the frequency is above the dielectric relaxation frequency of the I region, $f > \frac{1}{2\pi\rho\epsilon}$, with ρ being the resistivity of the I region [47]. At lower frequencies, the PIN diode behaves like a varactor. R_P is proportional to voltage and inversely proportional to frequency. Its value is typically higher than the reactance of C_T, and is less significant.

The PIN diode is mainly used as a switch, between the reverse biased and forward biased states, since it has V-I characteristic that is particularly suitable for this purpose.

The intrinsic layer exhibits a lower, variable resistance as a function of the forward bias, so that the PIN diode can be used as a variable resistor or attenuator. PIN switches and attenuators may be used as RF amplitude modulators. Square wave or pulse modulation uses PIN diode switch designs, whereas linear modulators use attenuator designs.

Schottky diode

The Schottky junction is a metal-semiconductor hetero-junction. It can be modeled using the same model as that of the PN diode, but without C_d. It works like a pn junction, but without the charge accumulation problem of the pn junction. This enables its use for forward bias applications such as rectification, mixing, and detection.

Like the PN diode, the Schottky diode can be used as a varactor, but a smaller capacitance variation is generally obtained due to the smaller breakdown voltage of the Schottky diode. Schottky diodes can be manufactured in Si or GaAs.

Gunn diode

The Gunn diode is manufactured using III-V compound semiconductors, such as GaAs or PIn (Phosphorus-Indium). It is manufactured as an n-doped layer embedded between two thinner n^+-doped layers. The Gunn diode exhibits negative resistance for suitably biased voltage. Gunn diodes can produce continuous power output of several hundred milliwatts from 1 to 100 GHz. They are mainly used in the design of negative resistance amplifiers and oscillators. The tunnel diode is a pn diode with a high-impurity density in between, and it also has negative resistance when suitably biased.

IMPATT diode

The avalanche diode is a reverse-biased pn junction diode. It has negative resistance when fed with suitable current. The dc V-I characteristic is depicted in Fig. 11.31. The IMPATT (impact avalanche and transit time) diode is a special kind of avalanche diode. It is a very powerful microwave source, which provides the highest output power in the

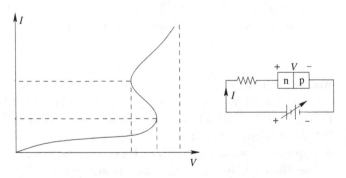

Figure 11.31   V-I characteristic of an avalanche diode.

millimeter-wave frequency range among solid-state devices. The IMPATT diode consists of a reverse-biased pn junction and a drift zone, or a reverse-biased PIN diode.

The IMPATT diode exhibits a dynamic negative resistance and is often used in the design of negative resistance oscillators and amplifiers, when high output power is required. It may be operated at frequencies from 10 up to 350 GHz, at relatively high powers. It is noisier than the Gunn diode, and thus is rarely used for local oscillators in receivers. The IMPATT diode can be manufactured using Si, GaAs, or InP. GaAs technology achieves a higher efficiency than Si technology.

The BARITT (barrier injection transit time) diode is similar to the IMPATT diode. It generally has a lower power output, but it generates lower AM noise. This makes it useful for local oscillators at frequencies up to 94 Ghz.

Step-recovery diode

The step-recovery diode, or *snap diode*, has a PIN structure. The doping level is gradually decreased as the junction is approached. This reduces the switching time, since the smaller amount of stored charge near the junction can be released more rapidly when changing from forward to reverse bias and the conduction abruptly halts. The abrupt current change is utilized to generate step-like voltage waveforms, and the step recovery diode can be used to produce sampling devices. The step-recovery diode is commonly used in high-order frequency multipliers, since it provides an efficiency much higher than that provided by the varactor diode.

11.11.2 Transistors

Transistors are the key active elements in RF and microwave circuits. They are most widely used for signal generation (oscillators), signal amplification, and many other functions such as switching and signal processing.

Bipolar transistors

The bipolar transistor was invented in 1948, and it established the foundation of modern electronics. It is a three-terminal device. The emitter, base, and collector regions have a p-n-p or n-p-n doping strategy, yielding a device consisting of two back-to-back pn junctions. This is shown in Fig. 11.32.

For the npn transistor, the emitter and collector currents are given by [41]

$$I_e \approx I_c \approx \frac{q D_n n_{b0} A e^{\frac{q V_{be}}{k_B T}}}{W}, \tag{11.113}$$

where D_n is the electron diffusion coefficient, A is the cross-section area of the device, q is the electron charge, n_{b0} is the equilibrium concentration of electrons in the base, V_{be} is the base-emitter voltage, and W is the width of the base. The base current is modeled by [41]

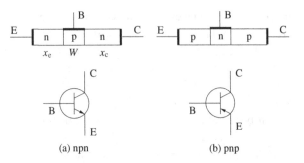

Figure 11.32

Schematics of the npn and pnp bipolar transistors.

$$I_b = \frac{qD_p p_{e0} A e^{\frac{qV_{be}}{k_B T}}}{x_e}, \tag{11.114}$$

where D_p is the hole diffusion coefficient and p_{e0} is the equilibrium concentration of holes in the emitter.

Since the emitter is much more heavily doped than the base, $p_{e0} \ll n_{b0}$. From (11.113) and (11.114), we have $I_b \ll I_e \approx I_c$. This feature can be used for voltage and current amplification, based on the common-base and common-emitter configurations, respectively, and current-voltage (I-V) curves are constructed accordingly.

In the common-emitter configuration, the input is V_{be} and the output is I_c. The transconductance g_m is defined by

$$g_m = \frac{\partial I_c}{\partial V_{be}} = \frac{I_c q}{k_B T}. \tag{11.115}$$

This value is obtained by differentiating (11.113).

Bipolar transistor technology has used heterojunction strategies to achieve higher operating frequencies. Three mature bipolar transistor technologies are the traditional silicon bipolar junction transistor (BJT), silicon-germanium (SiGe) heterojunction bipolar transistor (HBT), and GaAs-based HBT processes. The HBT technologies have higher transconductance and output resistance, higher power handling capacity, and higher breakdown voltage than the BJT technology. BJTs are only suitable for lower gigahertz frequencies, while HBTs can be used for very high operating frequencies and high power applications. In addition, the InP-based HBT process was predicted to provide the highest levels of RF performance, when compared with other bipolar processes, and has received much attention in recent years [41]. Silicon bipolar transistors usually have a much lower flicker noise than GaAs devices do.

Field effect transistors

The GaAs metal-semiconductor field effect transistor (MESFET) was the first three-terminal active RF device. The MESFET has source, drain, and gate terminals, as shown in Fig. 11.33. The MESFET is one of the two types of FET. The FET, sometimes called a *unipolar transistor*, uses either electrons (in n-channel FET) or holes (in p-channel FET)

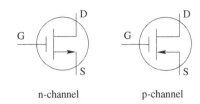

The symbols of p-channel and n-channel MESFETs.

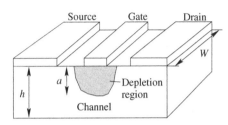

The schematic of a MESFET.

for conduction. The GaAs MESFET has three metal contacts at the gate, drain, and source terminals. The source and drain contacts are ohmic, while the gate contact constitutes a Schottky diode junction.

In general, the n-type MESFET is used, where the device cap and channel are doped n-type and thus current is carried by electrons rather than the slow holes. The common-source configuration is usually used. The GaAs MESFET is widely used for realizing oscillators, amplifiers, mixers, and switches. The MESFET can also be manufactured in Si technology, but limited to the operating frequency of 1 GHz. The schematic of a MESFET is shown in Fig. 11.34.

The transconductance g_m, which characterizes the gain mechanism of an FET, is defined by

$$g_m = \frac{\partial I_{ds}}{\partial V_{gs}} \tag{11.116}$$

at a given V_{ds}. It is derived as [41]

$$g_m = \frac{\epsilon_0 \epsilon_r v_{eff} W}{a}, \tag{11.117}$$

where ϵ_0 and ϵ_r are the free space and semiconductor permittivities, respectively, v_{eff} is the velocity of carriers in the device channel, W is the Schottky contact width, and a is the thickness of the depletion region.

For a GaAs MESFET, given V_{gs}, the I-V curves are illustrated in Fig. 11.35. The curves have linear and saturation regions. When V_{gs} turns more negative, the size of the depletion region is increased, leading to a decreasing drain-source current. At a certain negative V_{gs}, the depletion region thickness extends over the full channel depth and no drain-source current can pass through. This is referred to as the *pinch-off voltage*.

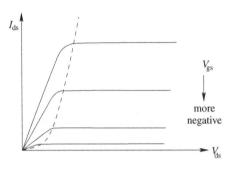

Figure 11.35   I-V curves of a GaAs MESFET.

The high electron mobility transistor (HEMT) is an FET that operates in very similar a way to a GaAs MESFET. Unlike GaAs MESFETs, HEMT devices have heterostructures, which allow operation at higher frequencies. The HEMT has larger electron mobility, and this leads to improved high-frequency noise and gain characteristics. Modern HEMT technologies are pseudomorphic GaAs HEMTs, lattice-matched and pseudomorphic InP HEMTs, and metamorphic GaAs HEMTs. For power-amplifier applications in 3G wireless communications, linearity is a key requirement. Both HBT and HEMT technologies deliver impressive power and linearity performance.

MOSFET

The metal-oxide-semiconductor field effect transistor (MOSFET or MOS FET) is another type of FET. The MOSFET is by far the most common FET in both digital and analog integrated circuits (ICs), and is the core component in CMOS ICs. According to the p- and n-channel doping, it can be of nMOSFET or pMOSFET type.

For the nMOSFET, the source and drain are n+ regions and the body is a p-region, whereas for the pMOSFET the source and drain are p+ regions and the body is an n-region. The symbol of the MOSFET is very similar to that of the MESFET.

Like the MESFET, the MOSFET is often used to convert an input voltage into an output current, and the transconductance is given by $g_m = I_o/V_{in}$. By using an appropriate impedance at the output or input of the MOSFET, a voltage or current gain can be provided.

A MOSFET has two operating modes. When the gate-to-source voltage V_{gs} is smaller than the threshold voltage V_{th}, the channel is in *off or accumulation mode*; when $V_{gs} > V_{th}$, the channel is in *on or inversion mode*. There is no distinct line between on and off. When V_{gs} is equal to or slightly smaller than V_{th}, there is a small current flowing from drain to source; this current is known as the *subthreshold current*. This intermediate state is known as the *subthreshold region* or *weak inversion region*.

In the subthreshold region, there is an exponential relation between the drain current I_d and V_{gs}

$$I_d = I_{d0} \left(\frac{W}{L} \right) e^{\frac{V_{gsq}}{nk_BT}}, \qquad (11.118)$$

where the constant I_{d0} is close to 20 nA, $n \approx 1.5$, W and L are the gate width and length, $k_B = 1.38 \times 10^{-23}$ J/K is Boltzmann's constant, and T is the temperature in kelvins. This mode is generally not of interest in RF design.

When a MOSFET is in on mode, it has two operating regions: the triode and saturation regions. In the triode region, the drain-source voltage V_{ds} is smaller than the over-drive voltage V_{od}

$$V_{ds} < V_{od} = V_{gs} - V_{th}. \tag{11.119}$$

In this case, the drain current is given by [11, 27]

$$I_d = \mu_n C_{ox} \frac{W}{L} \left[(V_{gs} - V_{th}) V_{ds} - \frac{V_{ds}^2}{2} \right], \tag{11.120}$$

where μ_n is the mobility and C_{ox} is the gate capacitance per unit area. If $V_{ds} \ll V_{od}$, the V_{ds}^2 term is negligible and I_d is proportional to V_{ds}. In this case, the MOSFET can be used as an effective resistor.

When $V_{ds} > V_{od}$, the MOSFET is in the saturation region. A general relation of I_d and V_{od} of a MOSFET in saturation is given in [11, 27]. For large L, the drain current and the small-signal transconductance can be approximated by

$$I_d = \frac{\mu_n C_{ox}}{2} \frac{W}{L} V_{od}^2, \tag{11.121}$$

$$g_m = \mu_n C_{ox} \frac{W}{L} V_{od}. \tag{11.122}$$

For the MOSFET, three basic sources of noise occur, namely, the thermal noise due to the resistive channel, the shot noise due to the junction diodes, and the flicker noise arising from the surface traps between the silicon and the gate oxide. Another important noise in the MOSFET is induced gate noise, caused by some of the channel thermal noise being coupled into the gate. The induced gate noise increases with frequency, and is also called *blue noise*. The induced gate noise is correlated to the thermal noise, with a correlation of $j0.395$ for long channel devices [27]. When operating at radio frequency, the flicker noise is negligible, and the thermal noise and the induced gate noise are the dominant contributors to the noise figure. In a mixer, where both low-frequency and high-frequency signals appear, the flicker noise is also important. Although a MOSFET creates more noise than a bipolar device, a well designed CMOS circuit provides a noise figure comparable to those of bipolar and HBT circuits [11]. More detail on MOSFET physics is given in [27].

The success of the MOSFET has been mainly due to the development of digital CMOS logic, which uses as building blocks pMOSFET-nMOSFET pairs that are connected together at both gates and both drains. The CMOS structure produces very small power consumption. Silicon CMOS technology is now a viable technology for low-GHz RF applications. RF CMOS is now receiving wide attention since it can provide a very inexpensive solution. RF CMOS technology is starting to challenge GaAs and bipolar technologies for

RF applications. However, CMOS is still inferior in terms of the noise and linearity performance. An advantage CMOS offers over bipolar and MESFET technologies is the ability to integrate a high degree of functionality on a single chip.

A number of nonlinear models for modeling MOSFET, MESFET, HEMT, and BJT devices including HBT are given in [14, 41].

11.12 Switches

Microwave switches are used for directing signals. They are also used to construct other circuits such as phase shifters and attenuators. The selection of a switch is dependent on the signal level and speed of operation. Mechanical switches can deliver high power, and are widely used in TV, AM, FM and other broadcasting systems. High-power mechanical switches can be in the form of waveguide or coaxial connector ports, but they have low speed of operation. Solid-state switches are preferred when operation speed is more important than power handling capacity. PIN diodes and FETs are used for this purpose. MEMS switches are also receiving more attention.

PIN switches

The PIN switch is obtained by biasing the PIN diode to either a high or low impedance state. A single-pole single-throw (SPST) RF switch can be formed by using either a single series or shunt connected PIN diode. A series connected PIN switch has a low insertion loss over a wide range of frequency, and a shunt connected PIN switch provides high isolation. For a series switch, the insertion loss and power dissipation are functions of the diode resistance, while the isolation depends primarily on the capacitance of the PIN diode. In shunt switch designs, the isolation and power dissipation rely on the diode's forward resistance, whereas the insertion loss is primarily determined by the capacitance.

It is difficult for a single PIN diode, either in shunt or series to achieve an isolation of more that 40 dB at radio and microwave frequencies. Compound switches and tuned switches are used for this purpose. A compound switch combines series and shunt diodes, while a tuned switch employs a resonant structure. These PIN switches can be used as duplexers (transmit/receive switches). Analyses of various PIN switches are given in [47].

FET switches

GaAs FET devices, which have a very high mobility of the carriers in GaAs, switch at nanosecond scale, much faster than the PIN diode. The GaAs FET switch has other inherent advantages over the PIN diode, such as simplified bias network, negligibly small dc power consumption, and simplified driver circuit design. Also it is compatible for MMICs. PIN switches are increasingly being replaced by GaAs FET switches, especially for low to medium power applications.

The gate voltage V_{gs} acts as the control signal for the FET switch. From Fig. 11.35, when the gate voltage $V_{gs} = 0$, the GaAs MESFET is in low impedance or on state; when the gate voltage is greater than the pinch-off voltage, it is in high impedance or off state. Silicon FETs can handle high power at low frequencies, but the performance drops rapidly as frequency increases.

GaAs FET switches can be combined with passive components into an MMIC. MMIC switches operate over a broad bandwidth and have a much lower power consumption, compared with that of PIN switches. However, MMIC switches introduce a higher insertion loss than PIN diodes. For frequencies above 1 GHz, solid-state switches have a typical insertion loss of 1–2 dB at on state, and isolation of −20 dB to −25 dB at off state [52]. The PIN diode and the GaAs MESFET have good high-frequency performance, but only for signals with small power.

MEMS switches

RF MEMS is recently gaining wide interest. A classical review on RF MEMS is given in [4]. Most MEMS devices employ silicon as the basic material. RF MEMS switches [10] implement short circuit or open circuit operations in the RF transmission line by using mechanical movement. RF MEMS switches are extensively used in phase shifters and switching networks up to 20 GHz, and are targeted for operating at 0.1 to 100 GHz.

Switching is achieved via a micro-actuator generated by various actuation mechanisms, such as electrostatic, electrothermal, magnetic, piezoelectric or electromagnetic actuation. The current mature RF MEMS technique uses electrostatic force for the mechanical movement. The electrostatic mechanism has low power consumption. Unlike ICs, a MEMS device contains movable, fragile parts, and thus its packaging is expensive.

RF MEMS switches offer a substantially higher performance than PIN diodes or FET switches do [38]: near-zero power consumption, very high isolation, very low insertion loss of 0.1 dB up to 40 GHz, very low intermodulation products due to high linearity that is around 30 dB better than that of PIN or FET switches, higher power handling capability, and very low manufacturing cost.

In addition to high packaging cost, the RF MEMS switch also has some drawbacks such as relative low speed (in the order of 10^{-6} second versus 10^{-9} second for solid-state switches), a high-drive voltage of 20–100 V for the electrostatic RF MEMS, and relatively low reliability [38, 52].

In wireless communications, RF MEMS can be used in the switch network, reconfigurable Butler matrix, switched filter banks, switched phase shifters, transmit/receive switches, SP2T (single-pole two-throw) to SP4T (single-pole four-throw) switches, and antenna diversity SP2T. RF MEMS is a state-of-the-art technology for enabling ubiquitous wireless connectivity, since it is being widely used in the reconfigurable hardware of SDRs. RF MEMS is an efficient way to miniaturize RF radio: Since passive components account for 80% the total components, RF MEMS switches enable a reconfigurable passive network by incorporating MEMS ohmic switches and variable capacitors.

More description on RF MEMS is given in [52]. Almost all RF devices, including switches, inductors, capacitors, filters, phase shifters, transmission lines and RF components, and antennas, have their micromachined versions, and these are reviewed in [52].

11.13 Attenuators

An attenuator is a device that reduces the amplitude or power of a signal without distorting its waveform. As opposed to an amplifier, an attenuator provides loss instead of gain. Attenuators are frequently used for gain control. For example, in a multichannel communication system, the gains through different channels need to be equalized. Attenuators are also used in phased array antennas and smart antenna systems.

Attenuators are usually passive devices made from resistors. The resistive characteristics of the PIN diode and the transistor can be used for this purpose. Attenuators are typically implemented in digital form. A digital attenuator consists of a cascade of attenuation units, each having an on or off state. When all the units are switched on, the maximum attenuation is achieved; when all the units are off, the system has the minimum attenuation, which is known as the *zero-state or reference insertion loss* of the attenuator. Switches are the most important components in a digital attenuator.

Attenuators are required to provide the specified attenuation over a certain frequency band. In applications such as phased arrays, the insertion phase is required to be invariant for all possible attenuation states. The GaAs dual-gate FET can be used to achieve equiphase but different gain paths [41], where gate 1 is used for gain level control and gate 2 for on-/off-state control.

Analog attenuators can provide a continuous amplitude control. This is desirable for correction of any degradation arising from fabrication. Analog attenuators are usually used in AGC control loops and adaptive beamforming systems, and as variable taps in direct implementation of digital filters. Various types of switched attenuators and analog reflection-type attenuators are described in [41].

The resistance characteristic of the forward biased PIN diode, as given by (11.111) can be used in an attenuator. The PIN diode attenuator is extensively used in AGC and RF leveling applications. It can take the form of a simple series or shunt mounted diode acting as a lossy reflective switch, or a more complex structure that maintains a constant matched input impedance across the full dynamic range of the attenuator [47].

Example 11.2: For a typical PIN diode, $W = 250 \ \mu$m, $\tau = 4 \ \mu$s, $\mu_n = 0.13$ m^2/v· s, and $\mu_p = 0.05$ m^2/v· s. Figure 11.36 gives the resistance of the forward biased PIN diode as function of forward current I_F.

In an extreme case, the attenuator becomes a microwave termination, which terminates a microwave transmission line without much reflection. The resistive, power-absorbing material is used for this purpose. The impedance of the termination should be equal to the waveguide impedance for minimum reflection.

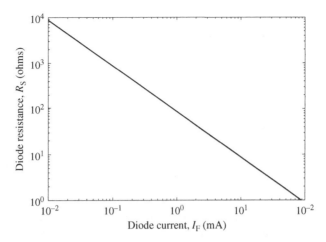

Figure 11.36 The resistance of the PIN diode, R_S, as a function of forward current I_F.

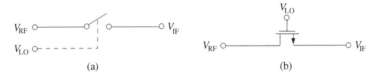

Figure 11.37 (a) A simple mixer. (b) Realization using switch.

11.14 Mixers

Mixers are essential for modulators, demodulators, frequency translators, and phase-locked loops. Mixers perform frequency translation by multiplying two input signals. Mixers can be made from nonlinear devices, such as diodes or transistors.

A simple switch can be used as a mixer, since the output is the RF input when the switch is on, or zero when the switch is off. This operation can be viewed as multiplying the RF signal by a rectangular waveform. This switch can be implemented by using a transistor, as shown in Fig. 11.37. This is a passive mixer. For an ideal switch, there is no distortion created by the mixer. For a practical switch, its resistance depends on the LO drive voltage as well as the input voltage, thus distortion is introduced. The simple switch mixer is a passive mixer, since it provides no gain. Active mixers generally provide gain, which is the IF power delivered to the load divided by input RF power.

11.14.1 Operation of mixers

The square-law term of the nonlinearity is used for mixers. When the RF signal of frequency ω_{RF} and LO signal of frequency ω_{LO} are combined by using a T-junction or directional coupler, the square law generates dc, $2\omega_{RF}$, $2\omega_{LO}$, and $\omega_{RF} \pm \omega_{LO}$ terms. Only the $\omega_{RF} \pm \omega_{LO}$ terms are used for mixers, while all other terms are ignored. The dc term can be biased by using chokes, while the unwanted frequency terms can be removed by filtering.

For mixing, the IF frequency is related to the RF frequency and LO frequency by

$$\omega_{IF} = \omega_{RF} - \omega_{LO}. \tag{11.123}$$

The mixer produces the RF frequency $\omega_{LO} + \omega_{IF} = \omega_{RF}$ and a spurious image RF frequency at $\omega_{LO} - \omega_{IF} = \omega_{RF} - 2\omega_{IF}$. For a down-converter, if $\omega_{RF} = \omega_{LO} \pm \omega_{IF}$, the filtered mixer output is the ω_{IF} or $-\omega_{IF}$ term. These two terms are indistinguishable. The image response can be eliminated by using RF filtering before mixing, but it is very difficult since the desired RF frequency $\omega_{LO} + \omega_{IF}$ is very close to the spurious image frequency $\omega_{LO} - \omega_{IF}$, as $\omega_{IF} \ll \omega_{LO}$. Thus, this image RF frequency should be outside the RF bandwidth B_{RF} so that it can be filtered, that is,

$$f_{IF} > \frac{B_{RF}}{2}. \tag{11.124}$$

Image reject filters are usually placed after the RF amplifier, so that they have less influence on the noise figure of the receiver and also remove harmonics due to the RF amplifier. The image reject filters are typically ceramic DR filters.

For upconversion or modulation, an LO signal and an IF signal are combined to generate $\omega_{LO} \pm \omega_{IF}$ terms. The upper frequency is called the *upper sideband*, the lower frequency is called the *lower sideband*. When filtering out one of the sidebands, we get single sideband (SSB) modulation, while double sideband (DSB) modulation retains both sidebands.

Mixers require impedance matching at all the three ports (RF, LO and IF). Undesirable harmonic power can be dissipated in resistive terminations or blocked by reactive terminations. The conversion loss of a mixer is defined as the ratio of available input power to output power. Practical diode mixers typically have a conversion loss of 4 to 7 dB, and transistor mixers have a lower conversion loss or even have a conversion gain of a few dB [35].

For heterodyne receivers that use image-reject filters, the mixer at the first downconversion stage should employ conjugate matching at the input. The first bandpass filter is the preselect filter, which is placed ahead of the first amplifier (LNA) and must have low insertion loss to retain a low noise figure. A down-conversion mixer often follows the LNA. Due to the high gain of the LNA, the mixer can have a higher noise figure but it requires a higher IIP3. Because the input signal to the mixer is higher than that of the LNA, the mixer is required to have higher linearity than the LNA, and in most cases the mixer's IIP3 dominates the IIP3 of the front-end. The IF frequency is usually selected to be less than 100 MHz, due to the limitation on ADC/DAC components.

The noise figure of a mixer is very important for a system. The noise figure in the SSB case is twice that of the DSB case, that is, 3 dB more than that of DSB, if the signal and image bands are subject to equal gains at the RF port of a mixer. This is because the noise of the image band is folded into the IF band. Noise figure for a practical mixer varies from 1 dB to 5 dB, and diode mixers generally achieve lower noise figure than transistor mixers do.

Impedance matching at RF and LO inputs ensures good signal sensitivity and noise figure. Isolation between RF and LO ports is also important to prevent LO-power leakage. As
discussed in Chapter 1, the isolation between any two ports of a mixer is critical. Otherwise,
LO leakage, dc offset or even-order distortion problems may occur.

11.14.2 Types of mixers

The most fundamental diode mixer is the single-ended diode mixer. The double-balanced
diode mixer is popular since it provides complete port-to-port isolation between all the
three ports and has excellent rejection of spurious mixing products, but it may require a
large chip area in MMIC implementation. The Schottky diode is the most commonly used
diode in mixer circuits. The GaAs Schottky diode can be used up to 1000 GHz, and is
relatively cheap.

Compared with diode mixers, bipolar or FET mixers have a higher dynamic range, and
require less LO power. FET mixers can be integrated with other FET-based RF circuitry.
There are a number of FET mixers such as the single-ended FET mixer, the dual-gate
FET mixer, the differential FET mixer, and the Gilbert-cell mixer. The MOSFET and
the Si MESFET can be used for frequencies up to several GHz, the GaAs MESFET can
be used up to Ku band, and the HEMT is used mainly at Ka band and above. A conversion gain can be achieved by using active FET mixers. FET mixers provide noise
performance comparable to that of diode mixers. The dual-gate FET mixer is also widely
used.

The single-ended mixer often has poor RF input matching and RF/LO isolation. There
are several types of mixers: the single-ended mixer, balanced mixer, double-balanced
mixer, and image rejection mixer [36]. They all generate good conversion loss.

- The single-ended mixer outputs all harmonic combinations of the RF and LO signals.
- The balanced mixer combines two identical single-ended mixers with a 3 dB hybrid
  junction (90° or 180°) to produce either better input VSWR or better RF/LO isolation.
  The balanced mixer using a 90° hybrid generates good RF VSWR, but poor RF/LO
  isolation, while that using a 180° hybrid suppresses all even harmonics of the LO.
- The double-balanced mixer uses two 180° hybrid balanced mixers. Like the 180° hybrid
  balanced mixer, it has good RF/LO isolation, but poor input VSWR. It suppresses all
  even harmonics of both the LO and RF signals, thus yielding a very low conversion loss.
- The image rejection mixer is the most complex, but exhibits the best performance. It can
  generate good RF VSWR and also good RF/LO isolation.

Diode and FET mixer design is described in [41]. Conventionally, mixing exploits the
nonlinearity in diodes or FETs, but this generates spurious mixing products and intermodulation. The resistive FET mixer is based on an unbiased FET, which has a linear channel
conductance, thus eliminating spurious responses. Resistive FET mixers are thus desirable
[41]. The Gilbert-cell mixer is usually used in RF IC design; it provides better isolation
between the LO and RF port than a passive mixer. It can also have positive conversion gain.
It provides a smaller noise than a passive mixer, and requires a small LO drive amplitude,

but consumes more dc power. Bipolar and CMOS mixer IC design is discussed in [37]. CMOS mixer design is also described in [11].

11.15 Amplifiers

An amplifier has a dynamic range within which it can keep a linear relation between its input power and output power. Below the lower end of the range, the input power is too low and the output is dominated by the noise of the amplifier. The level at the lower end is known as the noise floor of the system, typically varying from -60 to -100 dBm over the bandwidth of the system for transistors. Above the dynamic range is the saturation area. If the input power is too large, the amplifier can be damaged.

Three types of amplifier are used in wireless systems, namely LNAs for receivers, power amplifiers for transmitters, and IF amplifiers for both receivers and transmitters. Important parameters for amplifiers are the power gain, noise figure, and intercept points. The intercept points determine the linearity of amplifiers, which is a critical parameter.

11.15.1 Requirements in wireless systems

For BSs, power amplifiers may be required to provide output power of 10–100 W. For MSs, the output power may be up to 1W. Since most components cannot provide such a power, various power combining techniques are used. For power amplifier design, the input power is usually high, and the transistor does not perform linearly. This makes power amplifier design much more complex than LNA design. For MSs, efficiency of the power amplifier is a critical concern for the lifetime of the battery.

For SDRs, a suitable wideband LNA should have a large gain, acceptable input impedance matching, and low noise figure across the entire band. This broadband nature precludes the use of resonant circuits. A common-gate amplifier is an attractive CMOS solution, where the input parasitic capacitance can be embedded in an LC ladder filter for impedance matching to get a maximally flat frequency response over a wide bandwidth [21].

Quasi-optical amplifiers [5] are competitive with other solid-state and vacuum-based power amplifier technologies in the upper microwave and millimeter-wave bands. Although still lagging in power-added efficiency (PAE), quasi-optical amplifiers have advantages of superior linearity, low noise, high SFDR, and added functionality such as electronic beam steering. This makes quasi-optical amplifiers attractive as robust LNAs.

For satellite communications, cryogenic cooling of receivers can reduce their noise temperature, since antenna noise is determined by celestial sources and the atmosphere. In the absence of strong celestial source, the antenna is toward a very cold sky with a cosmic microwave background radiation of 2.725 K, which is modified by the atmosphere. InP heterostructure FET (HFET) technology is mature and is suitable for application in extremely low-noise amplifiers [34].

Most wireless standards employ quadrature modulation and/or multiple carriers. In this case, the power of a signal varies or fluctuates significantly over time. The signals transmitted from BSs typically have a high PAPR, since they each serve multiple users by transmitting signals at multiple carrier frequencies with random phases at the same time. For CDMA systems, the superposition of multiple signals leads to a sum term similar to white Gaussian noise, which has a high PAPR. Signals at BSs are similar for all standards, and have an output power spectrum similar to that of AWGN, with a PAPR of, typically, 10 dB. At MSs, the PAPR values are different for different formats: For example, the PAPR is 0 dB for constant envelope modulation such as FM used in AMPS and GMSK in GSM, 5.1 dB for OQPSK in CDMA IS-95, and it would be greater than 13 dB for OFDM.

High linearity is necessary for efficient use of spectrum. However, an increase in amplifier linearity generally leads to a decreasing power efficiency. Nonlinear amplifiers usually have a high power efficiency, and their linearity can be improved via a linearization technique. Linearization of amplifiers can be realized by employing the feedforward linearization technique, linear amplification using nonlinear components (LINC) [7], or an analog or digital predistorter followed by an amplifier.

11.15.2 Structure of amplifiers

The basic topology of an amplifier includes an active device such as a FET or a BJT, dc feed, input matching network, and output matching network. An RF choke is used to filter the dc drain. The amplifier is a two-port network that consists of a source for feeding the input and a load connected to the output. This is shown in Fig. 11.38.

The power delivered to the load is given by

$$P_L = \frac{1}{2}\Re\left(V_L I_L^*\right) = \frac{1}{2}|I_L|^2 R_L,$$

(11.125)

where V_L and I_L are the voltage and current of the load, respectively. The output power P_{out} is related to P_L by

$$P_L = P_{\text{out}}\left(1 - |\Gamma_L|^2\right),$$

(11.126)

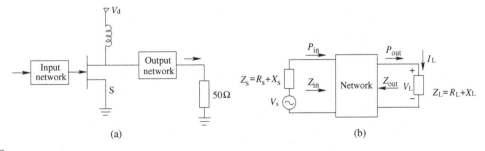

(a)   (b)

Figure 11.38   Architecture of an amplifier. (a) Circuit with amplifier. (b) Schematic diagram.

where Γ_L is the reflection coefficient at the load. When $Z_L = Z_{\text{out}}^*$, $\Gamma_L = 0$ and the maximum RF power delivery occurs.

The efficiency of an amplifier can be characterized by two metrics: the drain efficiency η and the maximum PAE η_{PAE}, defined by

$$\eta = \frac{P_L}{P_{\text{dc}}} \times 100\%, \tag{11.127}$$

$$\eta_{\text{PAE}} = \frac{P_L - P_{\text{in}}}{P_{\text{dc}}} \times 100\% = P_{\text{in}} \frac{G - 1}{P_{\text{dc}}} \times 100\% = \left(1 - \frac{1}{G}\right)\eta, \tag{11.128}$$

where P_L is the power delivered to the load, P_{dc} is the power drawn from the supply, and the network (power) gain $G = P_L/P_{\text{in}}$.

Thus a high PAE requires a high η as well as a sufficient power gain. For a large power gain, we have $\eta \approx \eta_{\text{PAE}}$. The typical amplifier performance for MSs is $P_{\text{out}} = 20$ to $30\,\text{dBm}$, $\eta = 30$ to 60%, $G = 20$ to $30\,\text{dB}$, IMD$= -30$ dBc, power control at 1-dB step, and output spurs and harmonics at -50 to 70 dBc [37].

Amplifiers must be band-specific, due to the strict requirement on linearity. For wideband or SDR systems, several amplifiers may be needed, each for a specific band or standard. Amplifiers are required to work at low supply voltages and high operating frequencies. Some amplifier circuits use a switch that is on and off periodically at the input frequency. The switch is usually implemented using active devices such as BJTs or FETs with a large gate width. Due to the prevalence of complex modulation techniques that result in a high PAPR, there is a more strict linearity requirement on amplifiers. SDR systems dictate the use of wideband amplifiers with good linearity and good power efficiency.

11.15.3 Classification of amplifiers

According to the method of operation, efficiency, linearity, and power-output capability, amplifiers are commonly grouped into class A, B, AB, C, D, E, F, F^{-1} (dual of class F), and S. Analysis of these classical amplifiers is based on the assumption that both the input and output waveforms are considered sinusoidal.

Class A amplifiers

The class A amplifier operates linearly across the full input and output range. For the class A amplifier, the transistor is biased with the input signal. A single-stage class A amplifier typically provides a power gain of more than $10\,\text{dB}$. It is inherently a linear circuit, and is used mostly as LNAs or power amplifiers at MSs. The maximum efficiency is equal to 50%, and is less than 40% in practical designs [9, 11, 14, 37]. The device always dissipates the same amount of power, even at zero input. The dc power consumption is relatively high over the entire range of operation, leading to relatively low PAE. Typically, the first low-power stage of multistage amplification is the class A amplifier.

Class B amplifiers

The class B amplifier uses two transistors in a push-pull mode, and is implemented using a low-loss high-frequency transformer. Each transistor conducts half of the carrier period. The class B amplifier is able to self-adjust its dc power consumption according to the input power level, leading to a higher PAE than that of the class A amplifier. Class B provides a better efficiency than Class A, achieving a maximum efficiency of $\eta = \pi/4 \approx 79\%$ [14, 37]. The gain of the class B amplifier changes with the input signal level, resulting in a higher distortion than that of the class A amplifier. For a long-channel CMOS class B amplifier, the maximum efficiency reaches $\eta \approx 8/(3\pi) \approx 85\%$, and the PAE is usually between 30% and 40% [11]. Class B has essentially no power dissipation with zero input signal. It is typically used as an audio amplifier, and is usually replaced by class AB for RF applications.

Class AB amplifiers

The class AB amplifier has a drain current duty cycle that lies between those of class A and class B. For a very small input signal, the class AB amplifier essentially operates as the class A amplifier. The efficiency is between 50% and 78.5% at the 1-dB compression point, depending on the angle of conduction. Class AB has higher output power and better PAE, typically 40% at the maximum power level. However, the dc power consumption is relatively constant over the entire range of operation, and its PAE is much smaller when the output power is lower than P_{1dB} [11]. For large dynamic range systems such as OFDM-based 802.11a, the power amplifier needs to be backed off by 8 to 10 dB from its output P_{1dB} point, and this may reduce the PAE to as low as 5%.

The Class AB amplifier has a higher efficiency and a cooler heatsink than the class A amplifier, but causes some tolerable nonlinear effects. The class AB amplifier is biased to a quiescent point, which is somewhere in the region between the cutoff point and the class A bias point ($I_{max}/2$). A harmonic short is placed across the output port to prevent any harmonic voltage from being generated at the output. Such a harmonic short can be realized by using a parallel shunt resonator that has a resonant frequency at the fundamental. The class B amplifier corresponds to a zero level of quiescent bias (zero-bias); the current waveform is a perfectly halfwave-rectified sinusoid, containing only even harmonics [8].

Most RF amplifier ICs are based on the class AB mode. Final stages of a multistage amplifier mainly use class AB. A linearization technique such as predistortion can be applied to Class AB to achieve high linearity.

Class C amplifiers

For the class C amplifier, the output transistor is on for less than a half cycle to improve the power efficiency. Class C is a truly nonlinear mode of operation. It achieves a maximum efficiency of 100%, and typically between 85 and 90%. This efficiency is achieved at the cost of reduced power-handling capacity, gain, and linearity. The class C amplifier exhibits a high efficiency only when operating near cutoff, and is only suitable for constant envelope signals such as FM signals; it also needs a resonant circuit to filter out the unwanted distortion. Since it only reproduces the peaks of the signal, it has a lower gain.

Classes D, E, F, F^{-1} amplifiers

Classes D, E, F, F^{-1} are switch-mode amplifiers that use a transistor switch to pump a resonant tank circuit. They achieve a very high efficiency, but have relatively poor linearity. These devices generate a flatter, squarer periodic waveform, rather than a sinusoidal waveform. The existence of higher harmonics helps to improve the power and efficiency of these amplifiers. The switched-mode amplifiers are inherently constant-envelope amplifiers, and thus linear modulation cannot be obtained. These switch-mode amplifiers are rejected by the wireless communications industry, but they were at one time commonly used in AM and SSB broadcasting.

The class D amplifier uses two transistors, and achieves a theoretical efficiency of 100%, and very high power-handling capacity. However, it is limited by the requirement for a perfect switch, since otherwise large crowbar dissipation may occur.

Classes E and F use only one transistor. They achieve PAE efficiency above 80% at frequency above 5 GHz, and a maximum theoretical efficiency $\eta = 100\%$ while delivering full power. There is a tradeoff between efficiency and output harmonic content. Class F usually achieves an efficiency superior to that of class E, and better power-handling capacity than class E, although it has half the power-handling capacity of class D. As a result, the class F amplifier is more widely used in practice. Analysis and design of the class E amplifier in CMOS is given in detail in [39].

Class F^{-1} (inverted class F) was introduced mostly for low-voltage amplification used for monolithic application. Class F^{-1} is dual to class F mode with the collector voltage and current waveforms being mutually interchanged. Like class F, it also achieves a maximum theoretical efficiency of 100%. The Class E/F amplifier combines the advantages of both class E and class F^{-1} operations [23].

Class S amplifiers

Class S uses an appropriate PWM of the input switch control voltage to a switch-mode amplifier such as class D, then uses a low-loss lowpass filter to generate output signal with excellent linearity. It achieves a maximum theoretical efficiency of 100%. Class S is well suited for CMOS integration, except for the lowpass filter. Since the switch has to be n times faster than that in a non-PWM amplifier, where n determines the desired dynamic range, it is difficult to use the PWM over carrier frequencies above 10 MHz. This makes the class S amplifier useless for wireless communications.

Remarks

Nonlinear amplifiers typically have a much higher power efficiency than linear power amplifiers. This is because they do not need a large amount of dc power for maintaining a constant gain over the entire range of operation. For power amplifier design, heat sink or good thermal dissipation is essential for the transistor packages.

Classes A, AB, and B amplifiers are most widely used in wireless transmitters due to their high linearity. They are so categorized according to the duty cycle of their drain currents. A parallel class A&B amplifier is developed in [11] as a parallel combination of a class A and a class B amplifier, and it provides a better performance compromise for both large and small input levels in comparison to class A, B, and AB amplifiers. With proper ratio between the sizes for MOSFETs of the class A and B amplifiers, a parallel class A&B amplifier can have a larger linear range, a better power efficiency, and much lower dc power consumption. The PAE of the parallel class A&B amplifier is almost the same as that of the class A amplifier at low input and approximates that of the class B at high input.

In practical wireless systems, PAE is usually very low, typically at 10%. An overdriven class B is obtained by increasing the amplitudes of both the current and voltage waveforms but keeping the truncated peak values the same as in the conventional class B; this leads to a maximum efficiency $\eta = 88.6\%$ [14]. Efficiency typically drops rapidly as the amplifier output is below the saturation power level. The efficiency of the class A amplifier increases linearly with the output power. The PAE can be improved by varying the dc current and/or dc voltage supplied to the amplifier as output power changes, as in the case of the dynamic supply voltage (DSV) amplifier [2]. More discussion on amplifiers is given in [37]. Waveform analysis based on reduced conduction angle and design on different classes of amplifiers are given in [9, 14, 27].

Power combining

In order to increase the output power of amplifiers made of solid-state devices, power-combining and phased-array techniques can be employed. The conventional power-combining method uses the binary Wilkinson power splitter/combiner. An increase in output power causes an increase in the length of transmission line as well as the number of combining nodes. The recent phased-array technique employs spatial or quasi-optical power combining by coupling the components to beams or modes in free space [18]. Spatially combined power sources can be implemented using an amplifier array or an oscillator array.

11.15.4 Linearization techniques

Classes E and F are becoming popular since they have high efficiency and output power [2]. Band-efficient modulation techniques typically require linear amplifiers to minimize spectrum regrowth, and multicarrier modulation techniques need linear amplifiers to avoid cross-modulation. For MSs, most linear amplifiers employ the class A technique, but the efficiency is too low for battery usage.

In order to achieve high linearity while providing the efficiency of the switching-mode circuits, linearization techniques such as LINC [7], feedforward linearization, and predistortion linearization [8] can be applied to nonlinear but power-efficient amplifiers. Switch-mode amplifiers rely on the high-frequency characteristics of the transistor due to high-speed switching transitions.

LINC combines the outputs of two nonlinear amplifiers, which are represented by two constant envelope vectors and the result is obtained by changing the phase difference between the two vectors. This technique requires an efficient power combiner at the output, which is hard to integrate in CMOS.

Predistortion

Predistortion linearization places a small device on the amplifier input, which consumes little power, and provides linearization comparable to the more complex feedforward linearization. The input signal is split into a distortion path and a linear path. The pre-distorion technique can take analog or digital form. The analog predistortion technique was extensively used to correct the nonlinear characteristics of traveling wave tubes, which are still used today for high power applications in the upper GHz bands. A pre-distorter/amplifier combination typically introduces a spectrum of distortion products that significantly exceeds the spectral bandwidth of the uncorrected amplifier output due to the high-order distortion products introduced by the predistortion process [9]. This can be troublesome for meeting regulatory spectral distortion masks for alternative channel power.

The predistortion technique is mainly used as a complement to a system using feedback or feedforward linearization technique. An RF predistortion amplifier typically provides a distortion correction of 5 to 10 dB. A simple analog predistorter can be made by using one or more diodes. The predistorter has a gain, but the use of passive components makes the gain negative.

Digital predistortion is now gaining more research interest [22]. Based on DSP technology, it uses a look-up table (LUT) that is obtained from *a priori* characterization of the modulator. DSP predistortion is replacing analog predistortion in most applications. A hybrid analog and DSP implementation uses DSP to control the various amplitude and phase scaling adjustments.

Feedforward linearization

The feedforward linearized amplifier is an open-loop correction system. It makes use of a main amplifier and an error amplifier. A correction signal with equal amplitude but opposite phase is added to the main amplifier output by using a coupler to generate a linearization process. Compared to a feedback linearized amplifier, the inherent delay for a closed-loop system is avoided and thus the correction is performed at a much faster speed. Also the stability problem facing a feedback system is avoided.

Feedforward linearized amplifiers need to maintain accurate gain and phase tracking. But drift due to temperature and aging is still a problem, and there are now dozens of patents for feedforward adaptation and drift cancellation. A feedforward amplifier typically provides 25 dB distortion correction for a single forward loop. Multiple feedforward loops can be used for greater linearity. Feedforward amplifiers are in mass production and have become key components in the infrastructure of mobile communications.

The efficiency of feedforward linearized amplifiers is low, typically 10–15% [9]. Feedforward solutions have thus far dominated the power amplifier market of multicarrier communications, since for multicarrier applications the reduction in intermodulation products is very desirable. However, the extra cost arising from couplers and other components makes such techniques much more expensive than the digital predistortion technique.

11.15.5 Microwave transistors for amplifiers

There are several types of power amplifiers: solid-state power amplifiers, traveling-wave-tube amplifiers, and klystron power amplifiers. At high power (<100W), traveling-wave-tube amplifiers and klystron power amplifiers provide good performance in terms of size, cost, and efficiency, but solid-state power amplifiers are superior in linearity. At low power, solid-state power amplifiers such as microwave transistor amplifiers are more desirable, since they are low-cost, reliable, and can be easily integrated into microwave integrated circuits (MICs) and MMICs, and can presently be used at frequencies up to 100 GHz. Microwave tubes are still necessary for very high power and/or very high frequency today, due to their better noise figures, tuning characteristics, temperature responses, and cost.

Microwave transistors are the crucial components for amplifier design. They are also widely used for oscillators, switches, phase shifters, mixers, and active filters. The silicon (Si) BJTs and GaAs MESFETs are most widely used. Silicon BJTs are cheap, capable of high gain and power capacity at low frequencies, while GaAs MESFETs have better noise figure and can operate at high frequencies. This is because the GaAs MESFET has a higher electron mobility than silicon BJTs and has no shot noise. The silicon BJT can be used up to 10 GHz, the GaAs MESFET up to 100 GHz, and the GaAs HBT allows higher frequencies. BJTs are preferred over GaAs MESFETs at frequencies 2 to 4 GHz due to their higher power gain and lower cost. The GaAs MESFET provides good noise figure performance up to 18 GHz. The HEMT provides the best noise figure and is most suitable for low noise and millimeter-wave applications. For typical bias currents in RF applications, MOSFET devices have better linearity than BJTs.

For power amplifiers, the power output is also a concern. Conventionally, for BSs the large transmit power had to be obtained using microwave tubes. Recent progress in wide bandgap semiconductor materials such as SiC and the GaN-based alloys has enabled the fabrication of microwave transistors to replace microwave tubes [50]. The 4H-SiC MESFET and AlGaN/GAN HFET provide RF output power typically of the order of 4–6 W/mm and 10–12 W/mm with PAE approaching the ideal values for class A and B operations, compared to 1 W/mm for the GaAs MESFET. The 4H-SiC MESFET can operate over a frequency range of 3 to 30 GHz with good PAE, and AlGaN/GaN HFETs can operate in a wider range from 3 GHz to in excess of 30 GHz. The high power throughput in these power amplifiers is difficult to completely dissipate through the substrate and thus, these power amplifiers suffer from self-heating effects. Recent AlGaN/GaN HFETs achieve an RF output power density as high as 30 W/mm, and microwave amplifier output power approaching

100 W for single-chip operation [51]. The GaAs HBT also demonstrates excellent output power capability.

11.15.6 Stability

Passive circuits are unconditionally stable. For active circuits such as amplifiers, stability analysis must be made, or the designed amplifier may become an oscillator. To ensure stability, the input and output ports of the amplifier must have $|\Gamma_{\text{in}}| < 1$ and $|\Gamma_{\text{out}}| < 1$, where the input and output reflection coefficients Γ_{in} and Γ_{out} are, respectively, determined by the reflection coefficients of the source and load matching networks, Γ_S and Γ_L. The amplifier is unconditionally stable subject to the necessary and sufficient conditions [36, 37, 43]

$$K = \frac{1 - |S_{11}|^2 - |S_{22}|^2 + |\Delta|^2}{2\,|S_{12}S_{21}|} > 1, \tag{11.129}$$

$$|\Delta| < 1, \tag{11.130}$$

where

$$\Delta = S_{11}S_{22} - S_{12}S_{21}, \tag{11.131}$$

S_{ij} are the S-parameters of the amplifier, and K is known as *Rollett's stability factor*. As S_{12} decreases, that is, as the reverse isolation of the circuit increases, the stability is increased.

Another criterion for unconditional stability, which also gives relative stability, is given by [15, 36]

$$\mu = \frac{1 - |S_{11}|^2}{\left|S_{22} - S_{11}^*\Delta\right| + |S_{21}S_{12}|} > 1. \tag{11.132}$$

The greater the value of μ, the better the stability.

When the device is potentially unstable, the load and source stability circles can be plotted on the Smith chart to find the range of values of Γ_S and Γ_L, and the stable and unstable regions can be identified [36].

11.15.7 Transistor amplifier design

For amplifier design, we need first to conduct stability analysis. The maximum gain of an amplifier is achieved when the source and load matching sections provide a conjugate match between the source or load impedance and the transistor, that is, $\Gamma_{\text{in}} = \Gamma_S^*$ and $\Gamma_{\text{out}} = \Gamma_L^*$. For a single-stage transistor amplifier, the transducer power gain is given by

$$G_T = G_S G_0 G_L, \tag{11.133}$$

where G_S, G_0, and G_L are, respectively, the gain contributions of the source-matching section, the transistor, and the load-matching section. $G_0 = |S_{21}|^2$ is a fixed value determined by the transistor, while G_S and G_L are decided by the source and load matching sections.

This leads to narrowband frequency response, because most transistors have a significant impedance mismatch (large $|S_{11}|$ and $|S_{22}|$). In many practical cases S_{12} can be ignored, and the transistor is then unilateral. This can significantly simplify the design procedure. Design can be made using the Smith chart.

The design procedure is detailed in [36], and exploits the constant gain circles for G_S and G_L on the Smith chart, which represents the loci of Γ_S and Γ_L.

Broadband amplifiers can be designed at the expense of gain and complexity. A fairly flat gain can be obtained if the amplifier has a gain smaller than the maximum gain, but the source and load matching is poor. The balanced amplifier solves the input and output matching problem by using 90° couplers to cancel the input and output reflections from two identical amplifiers [36]. The distributed amplifier uses multiple identical FETs, whose gates and drains are, respectively, connected to different transmission lines at equal spacing. Such an amplifier is known as a *traveling wave amplifier*.

Conjugate matching between the source and the input impedance of the transistor achieves the maximum gain, but generally not the minimum noise figure, $F_{\min}$. For LNA design, a compromise has to be made. Design can be done by plotting constant gain circles and constant noise figure circles and selecting a suitable tradeoff between them. The design makes use of the noise parameters of the particular transistor being used, the minimum noise figure $F_{\min}$ of the transistor, the optimum reflection coefficient Γ_{opt} that results in minimum noise figure, and R_N equivalent noise resistance of the transistor. More discussion is given in [37, 41]. For multistage LNA design, from the Friis noise formula (11.12), the first amplifier stage should achieve the optimum noise figure rather than the optimum gain, while the later stages can be designed with higher gain, since their noise figures are less important.

11.16 Oscillators

Oscillators are nonlinear circuits that convert dc power to an RF waveform, typically a sinusoidal waveform. Local oscillators (LOs) are used with mixers for frequency synthesis. The frequency of the LO in a radio transceiver is typically adjustable over the signal band. This is implemented by using a PLL that contains a VCO. The VCO is an oscillator whose frequency is set by a voltage signal. For the VCO, the tuning range is measured in MHz/V.

11.16.1 Analysis methods

An oscillator can be viewed as a transistor amplifier with linear positive feedback. When the denominator of the transfer function is zero at a particular frequency, a zero input can lead to a nonzero output, and thus the circuit produces an oscillating waveform at that frequency. Unlike amplifier design that aims at a high stability, an oscillator works only when the circuit is not stable.

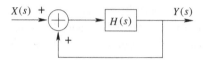

Figure 11.39   Feedback circuit of an oscillator.

Feedback model

For a simple linear feedback system, as shown in Fig. 11.39, the system transfer function is given by

$$H_{\text{tot}}(s) = \frac{Y(s)}{X(s)} = \frac{H(s)}{1 - H(s)}. \tag{11.134}$$

For self-sustaining oscillation, two conditions must be satisfied at ω_0:

$$|H(j\omega_0)| = 1, \quad \angle H(j\omega_0) = 0. \tag{11.135}$$

These are known as the *Barkhausen criteria*. In order to start up oscillation, an input signal containing a spectral component at the desired frequency must be available. This signal is already available as the white noise.

In most RF oscillators, a frequency-selective network called a *resonator* is included in the feedback loop to stabilize the frequency. The feedback model is a two-port model, and is valuable for low-frequency oscillators.

Negative-resistance model

Another approach to describing the operating principle of an oscillator circuit is through the one-port negative-resistance model. This model is more useful for microwave oscillator analysis. Losses in the resonator circuit are compensated by a negative resistance demonstrated by an active device. Oscillation occurs at ω_0 when [37]

$$\Re[Z_T(s)] + \Re[Z_A(s)]|_{s=j\omega_0} = 0, \tag{11.136}$$

where Z_T and Z_A are the resonator impedance and the impedance of the active circuit. The power produced by the active circuit is delivered to the load.

The Gunn and IMPATT diodes have a negative-resistance behavior in a certain frequency range. The Gunn oscillator is a voltage-controlled oscillator, while an IMPATT oscillator is a current-controlled oscillator. A Gunn oscillator is of relative low power and can be used as an LO or transmitter oscillator. An IMPATT oscillator is noisier but provides higher power. Negative-resistance diode oscillators are the more powerful solid-state sources at very high frequencies, and are now the only viable solid-state source in the frequency range of millimeter waves. They are one-port networks designed using these diodes, which are biased to generate an impedance with a negative real part, or created by terminating a potentially unstable transistor with an impedance that drives the device in an unstable region. A positive resistance implies energy dissipation, and similarly a negative resistance implies an energy source.

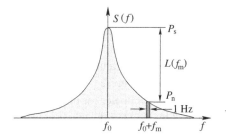

Figure 11.40   Output spectrum $S(f)$ of a typical oscillator.

11.16.2 Phase noise

Consider a generated oscillating signal

$$x(t) = A \cos (\omega_c t + \phi_n) \tag{11.137}$$

where ϕ_n is the phase noise. Phase noise leads to spectrum spread from the ideal spectrum line at ω_c, as shown in Fig. 11.40. A low-noise RF VCO requires a high-Q device, since the phase noise of the oscillator is proportional to $\frac{1}{Q_T^2}$, where Q_T is the overall quality factor.

The stability of frequency f_0 is specified in ppm/°C, and it ranges from 0.5 to 2 ppm/°C for typical wireless systems. To quantify the phase noise, we need to consider the noise power in a unit bandwidth at an offset f_m with respect to f_0. This quantity, denoted by $\mathscr{L}(f_m)$, measured in dBc/Hz, is defined as the ratio of the noise power in a single side-band per hertz bandwidth to the carrier power, at an offset frequency, f_m, from the carrier frequency f_0, where dBc means dB with respect to carrier. The phase noise is typically required to be within the range -80 to -110 dBc/Hz at a 10-kHz offset from the carrier. Phase noise spills the spectrum of a carrier signal into adjacent carriers, leading to ICI. Phase noise of an LO also corrupts the information in phase modulation. By increasing the Q of the LC oscillator, phase noise can be reduced; for this reason, passive resonators are usually incorporated into oscillators.

Modern wireless systems require stable, low-phase noise VCOs with wide tuning range. Electronically tunable capacitors are key components in such VCOs. Variable capacitors (varactors) with high Q are not available in standard silicon or GaAs processes, thus they are still external to the chip, while MEMS tunable capacitors have lower loss and greater tuning range [52].

Oscillator phase noise arises from thermal and flicker ($1/f$) noise sources. Small changes in frequency can also be modeled as phase noise. The two-sided PSD associated with phase noise is $S_\theta (f_m) = 2\mathscr{L} (f_m)$. The PSD of oscillator phase noise that uses a transistor amplifier can be modeled by Leeson's model. The Leeson phase noise model is given by [26, 28]

$$\mathscr{L} (f_m) = \frac{P_n(f_m)}{P_s} = \frac{F k_B T}{2 P_s} \left(\frac{f_0}{Q f_m} \right)^2, \tag{11.138}$$

where Q is the quality factor of the resonator, f_0 is the oscillator frequency, $P_n(f_m)$ is the noise power, P_s is the signal power measured at the input of the transistor, f_m is the frequency offset, F is the noise factor for active devices in saturation, k_B is Boltzmann's constant, and T is the temperature. From this equation, it is seen that Q, F and P_s have significant effects on phase noise. A low-noise VCO requires small F, and large Q and P_s. Q is the most effective parameter for improving the phase noise, but the Q of the integrated LC VCO is limited by the Q of the spiral inductor used.

Phase noise degrades receiver selectivity, since unwanted noise signal is also down-converted to the IF band. The maximum phase noise required for achieving an adjacent channel rejection or selectivity S dB is given by [35]

$$\mathscr{L}(f_m) = 10\log_{10}\left(\frac{P_n}{P_s}\right) = C - S - I - 10\log_{10}(B) \quad \text{(dBc/Hz)}, \qquad (11.139)$$

where C is the desired signal level measured in dBm, I is the interference signal level measured in dBm, and B is the bandwidth of the IF band in Hz. IS-54 requires phase noise to be about -115 dBc/Hz at 60-kHz offset.

11.16.3 Classification of RF oscillators

Diode and transistor microwave sources

Solid-state microwave sources can be grouped into diode or transistor oscillators. The most common diode sources are the Gunn and IMPATT diode oscillators, both of which convert a dc bias to RF power in the frequency of 2 to 100 GHz. Most microwave diode oscillators are built in coaxial resonator or waveguide cavity structure for high quality (Q) factor. The efficiency of negative-resistance diodes is very low, and a heatsink must be included to reduce the diode temperature to a reliable operating value.

Transistor oscillators have lower operating frequency or power capacity, but have some advantages over diode oscillators. They can be easily integrated with MIC or MMIC circuitry, and can also be much more flexible than a diode source. The well-known Hartley, Clapp and Colpitts oscillators are LC transistor oscillators. These oscillators are very similar, but differ in the system employed to feed back from the resonator and the implementation of the resonator itself. The schematic of the Colpitts oscillator is shown in Fig. 11.41.

For the Colpitts oscillator, a gross estimation of the oscillation frequency can be obtained from the resonator elements as

$$\omega_0 = \frac{1}{\sqrt{L\frac{C_1 C_2}{C_1 + C_2}}}. \qquad (11.140)$$

The loaded Q can be increased by increasing C_2 and decreasing C_1 in the capacitive divider, or by increasing both C_1 and C_2 but decreasing L. A more accurate oscillation frequency can be obtained by using the equivalent circuit of the transistor and by applying the oscillation condition.

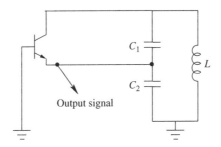

Figure 11.41   **Schematic of the Colpitts oscillator.**

Most discrete RF oscillators incorporate only one active device to minimize the noise as well as the cost. In RF oscillator ICs, the cost for using multiple active devices is not a problem. In this case, two major sources of phase noise are thermal and flicker noise. The BJT provides lower flicker noise than GaAs devices, and is more popular in low phase-noise microwave oscillators. HEMT and HBT devices are also suitable for low-noise applications, and they can work at higher frequencies.

In RF applications, the oscillator frequency is typically required to be adjustable, so as to be able to select the frequency-divided channels. The resonant frequency of an oscillator can be adjusted by changing the L and C values of the network. This can usually be more easily implemented using voltage-controlled capacitors, such as varactors. The resulting oscillator is known as *VCO*. A varactor is a diode whose junction capacitance varies with the applied dc reverse bias; dc blocking capacitors and/or RF chokes must also be used with a varactor to prevent detuning or shortening the RF circuit. A varactor can be used as a capacitor in a transistor oscillator circuit, and thus forming a VCO.

Resonator-based oscillators

Oscillator stability can be enhanced by using a high-Q tuning network [36]. Resonator-based oscillators, also known as *harmonic oscillators*, are characterized by an equivalence of two energy storage elements, operating in resonance, to produce a periodic output signal. The Hartley, Clapp and Colpitts oscillators are all harmonic oscillators

The Q of a resonant network using lumped elements or microstrip lines is typically limited to a few hundred, and the Q of dielectric resonators (DRs) can be several thousand. For waveguide cavity resonators, a Q of 10^4 or more is achieved. The quartz crystal resonator achieves an unloaded Q as high as 100,000 and a temperature drift of a few ppm/°C.

The quartz crystal resonator is composed of a slab of crystal between two metallic plates, and oscillates on the piezoelectric effect. It is a stable frequency source for wireless systems at frequencies up to a few hundred MHz. At its usual operating point, a crystal has an inductance, and for this reason a crystal is used in place of the inductor in a Colpitts or Pierre transistor oscillator. The quartz crystal resonator can be typically used up to 30 MHz. For a higher frequency, SAW devices that use a piezoelectric material allow high-volume, low-cost applications such as front-end filters for MSs. Unfortunately, neither of these resonators is suitable for IC implementation.

The transistor DR oscillator is now gaining popularity over the entire microwave and millimeter wave frequency range [36]. A small, high-Q DR is used for the load of an FET oscillator circuit. This achieves a very good temperature stability and a compatibility with MIC design. In order to generate RF power, the device must have a negative resistance, which can be realized by using a negative-resistance diode or a transistor. A DR is usually coupled to an oscillator circuit by placing it in close proximity to a microstrip line.

Off-chip quarter-wave resonators, in which piezoelectric material is used as the dielectric, are widely used in MSs. This high dielectric constant allows realizing physically small resonators with Q as high as 20,000.

Resonator-based oscillators have excellent jitter performance. The use of an off-chip LC tank or crystal makes MMIC integration difficult. Bipolar and CMOS LC oscillator IC design has been discussed in [37].

Ring oscillators

The VCO is conventionally implemented in either BJT or GaAs MESFET technology, while most of the low-frequency digital circuits are implemented in CMOS technology. Recently, more and more oscillators are being designed based on the analog CMOS technology [32, 49]. The ring oscillator is usually used in bipolar or CMOS IC implementation.

The ring oscillator consists of a ring of N delay stages with a wire inversion, for N odd, and the ring oscillates with a period of $2N$ times the state delay. The oscillation frequency is given by

$$f_{\text{osc}} = \frac{1}{2NT_{\text{pd}}}, \tag{11.141}$$

where T_{pd} is the propagation delay, which can be adjusted by changing the load or the current drive of the inverters. The current-starved ring oscillator is especially suitable for IC implementation [27]. The delay element is typically an inverter or a differential pair, corresponding to either single-ended or differential ring oscillator. The element delay is controlled by a bias voltage. The differential configuration provides better common-mode rejection. The ring oscillator is shown in Fig. 11.42.

The ring oscillator is used as the VCO in jitter-sensitive applications. It has high speed and ease of integration. The ring oscillator is extremely popular, since it is derived from digital-like building blocks. Compared to resonator-based oscillators, ring oscillators have relatively large tuning range and are simple, but have substantially inferior phase noise performance since all the energy stored during one cycle is dissipated.

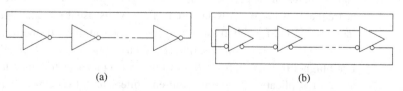

(a)   (b)

Figure 11.42   Schematic of the ring oscillator. (a) Single-ended. (b) Differential.

Monolithic integrated ring oscillators require little silicon due to the absence of inductors. Voltage controlled ring oscillators are especially useful for designing as fully integrated, low-jitter clock-recovery PLLs. Jitter performance of a ring depends primarily on the individual gate and not on the number of gates in the ring or the ring operating frequency [32]. In addition, quadrature outputs can be extracted directly from an approximately selected pair of nodes in the ring, if the ring oscillator consists of four or more inverters.

Relaxation oscillators

The relaxation oscillator operates by alternately charging and discharging the energy stored in a capacitor or inductor by using a nonlinear device such as a transistor, causing abrupt changes in the output waveform. The relaxation oscillator is often used to produce a nonsinusoidal output, such as a square wave or sawtooth.

11.17 Frequency synthesis

The frequency synthesizer derives many frequencies from a stable oscillator, thus avoiding the use of multiple independent oscillators in a multichannel system. The stable oscillator is usually a crystal-controlled oscillator. Frequency synthesizers are used in almost all modern wireless communication systems. Direct synthesis, PLL, and direct digital synthesis (DDS) are three common frequency synthesis methods.

Practical frequency synthesizers are usually a combination of the three methods. PLLs with a programmable divider, or direct digital synthesizers, are often combined with mixers to generate the desired frequencies. In this section, we introduce PLL and direct synthesis. DDS achieves the same goal as PLL, but in a different way. DDS generates the signal in digital domain, and then applies D/A conversion and lowpass filtering to obtain the analog waveform. DDS is introduced in Section 13.6.2.

11.17.1 Composition of phase-locked loops

A PLL is a closed-loop feedback control system that automatically adjusts the frequency of a controlled oscillator such as VCO until it matches the reference (input) signal in both frequency and phase. It is used to synchronize electronic tunable oscillators. A PLL has very good frequency accuracy and phase noise characteristics, but may have a long settling time. It can be implemented in either analog or digital form. The PLL is more widely used in modern wireless systems, especially where certain synchronization is required.

A PLL is composed of a phase detector, a lowpass filter, a variable oscillator such as VCO, and a negative feedback path with optional frequency divider. The architecture of a

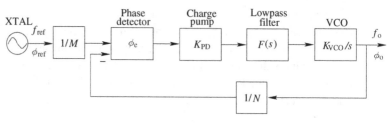

Basic PLL model.

basic PLL is shown in Fig. 11.43. There may also be a divider in the reference path. The divider makes the output clock of the PLL a rational multiple of the reference, so that both input frequencies at the phase detector are the same and only the phases are compared at the phase detector. The phase detector compares the phase of the reference and that from the oscillator, and adjusts the charge pump to change the control voltage. The lowpass filter smoothes out the signal from the charge pump.

Phase detector, phase-frequency detector/charge-pump

A phase detector generates an output voltage corresponding to the phase difference between two input signals. For an ideal phase detector, the output signal contains a dc component, which is proportional to the phase difference of two input signals

$$\bar{v}_o = K_{PD}\phi_e, \tag{11.142}$$

where K_{PD} is the gain of the phase detector.

Design of an analog phase detector is very similar to that of a balanced mixer [35]. A mixer can be used as a phase detector to transform the phase difference into the amplitude of a signal, after lowpass filtering. The phase detector can be simply an exclusive-OR gate, JK-flipflop, or S-R latch. It can maintain a phase difference of 90°, but cannot lock the signal unless the two frequencies are close.

A phase-frequency discriminator operates as a frequency discriminator for large initial errors and then as a coherent phase detector once the system falls within the lock range. It defines a simple finite-state machine to determine which signal has an earlier or more frequent zero-crossing. This two-step procedure ensures that the frequency is locked at the designed value. Typically, the phase-frequency discriminator is used with a charge pump to supply an amount of charge in proportion to the phase error. In a PLL, the charge pump can be used to amplify the phase error obtained by the phase detector, and to provide infinite gain for a static phase difference [37]. A charge-pump PLL is also called a *digital PLL*. The charge pump is a current source that is turned on or off by the difference of the UP and DN signals of the phase-frequency discriminator, which is proportional to the phase/frequency errors ϕ_e. The charge-pump output is $V_{con} = K\phi_e$, which is used for VCO control.

VCO

The VCO can be made by using LC circuits, whose oscillation arises from charging and discharging a capacitor through an inductor. The voltage-controlled capacitor can be used to change the frequency of the LC oscillator in response to a control voltage. A reverse-biased semiconductor diode has voltage-dependent capacitance, and varactors are widely used in VCOs. LC-tuned oscillators, such as the Colpitts, Clapp or Hartley topology, achieve the low-noise requirements for wireless applications, and are suitable for bipolar and CMOS IC technologies [46]. The appropriate tuning range and phase noise are the two major design specifications. YIG (Yttrium Iron Garnet or ferrite sphere) resonators or varactor diodes can be used for tuning. YIG is a high-Q resonator, while varactors provide lower Q but a cheaper and more compact solution, dominating the commercial applications.

An ideal VCO is characterized by

$$\omega_o = \omega_0 + K_{VCO} V_{con}, \tag{11.143}$$

where ω_0 is the reference frequency. Thus the excess phase in the generated output signal is given by

$$\phi_o(t) = K_{VCO} \int_{-\infty}^{t} V_{con} dt. \tag{11.144}$$

This corresponds to a transfer function

$$\frac{\Phi_o(s)}{V_{con}(s)} = \frac{K_{VCO}}{s}. \tag{11.145}$$

That is, changing phase is realized by changing frequency and then performing integration.

The VCO phase noise can be modeled as an additive component. Analysis shows that the VCO phase noise experiences a highpass transfer function, and thus its contribution to the output phase can be lowered by increasing the bandwidth of the PLL [37].

Reference frequency

The reference frequency is generated by using a crystal oscillator. Crystals have very high Q, but are tunable within a very small frequency range. When an ac signal is applied to the crystal at a frequency near the mechanical resonance, it excites a mechanical vibration, which in turn generates a current flow of an electrical resonance at the electrical terminals. For reference of very accurate high frequencies, SAW devices can be used.

Frequency multiplier

Frequency multipliers and frequency dividers are essential algebraic functions used in most synthesizer architectures. Diode multipliers can utilize either varactors, step recovery diodes, or Schottky diodes. Transistor multipliers are based on the device nonlinearity under large-signal conditions.

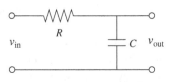

Figure 11.44

One-pole RC lowpass filter.

Frequency divider

Frequency dividers can use either varactors or step-recovery diodes. The division ratio is usually 2. Digital dividers are more commonly used, and the division ratio can be fixed or programmable. Digital dividers can be made of silicon or GaAs.

The dividers may be programmable to produce many frequencies from a single stable, accurate reference oscillator. The fractional-N PLL can have a VCO output frequency at an integer-plus-fractional multiple of the reference frequency. This can be achieved by also including a divider, $1/M$, between the reference crystal and the reference input to the phase detector. In this case, the output frequency of the VCO is multiplied by N/M. A fractional-N PLL can also be achieved by using an accumulator [42].

Loop filter

A lowpass filter is used to remove high-order harmonics. It can be a simple loop filter, such as a one-pole RC circuit. For the one-pole RC circuit, as shown in Fig. 11.44, the loop filter transfer function $F(s) = 1/(1 + sRC)$.

11.17.2 Dynamics of phase-locked loops

The loop transfer function between the output phase ϕ_o and input reference phase ϕ_{ref} is derived as

$$H(s) = \frac{\phi_o(s)}{\phi_{\text{ref}}(s)} = \frac{\frac{1}{M} K_{\text{PD}} \frac{K_{\text{VCO}}}{s} F(s)}{1 + \frac{1}{N} K_{\text{PD}} \frac{K_{\text{VCO}}}{s} F(s)} = \frac{\frac{K_{\text{PD}} K_{\text{VCO}}}{MRC}}{s^2 + \frac{s}{RC} + \frac{K_{\text{PD}} K_{\text{VCO}}}{NRC}}. \tag{11.146}$$

The last equality is for the one-pole RC loop filter. This is a second-order PLL, since the denominator is of the second degree in s. In this case, we can rewrite as

$$H(s) = \frac{\frac{N}{M} \omega_n^2}{s^2 + 2s\zeta \omega_n + \omega_n^2}, \tag{11.147}$$

where the natural frequency of the loop ω_n and the damping factor ζ are given by

$$\omega_n = \sqrt{\frac{K_{\text{PD}} K_{\text{VCO}}}{NRC}}, \quad \zeta = \frac{1}{2} \sqrt{\frac{N}{K_{\text{PD}} K_{\text{VCO}} RC}} \tag{11.148}$$

A more complex design enables the adjustment of ω_n and ζ independent of each other.

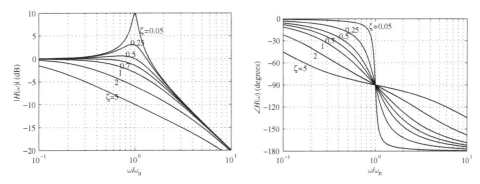

Figure 11.45

Amplitude and Phase response of $H(j\omega)$.

All second-order loops are inherently stable. Note that a high-order loop filter can be used to replace the one-pole filter. However, each additional pole introduces phase shift, and stability is more difficult to maintain [16]. To ensure stability the poles of $H(s)$ must be in the left half of the s plane. Most practical PLLs are second-order PLLs.

Example 11.3: The frequency and phase responses of the transfer function (11.147) are plotted in Fig. 11.45, for $N/M = 1$. When $\zeta = 0.707$, we get critical damping, and the maximally flat frequency response for $|H(\omega)|$. An ideal PLL should have a high ω_n and ζ close to 0.707.

Accordingly, the loop error response can be derived as

$$H_e(s) = \frac{\phi_e(s)}{\phi_{\text{ref}}(s)} = \frac{1}{M} - \frac{1}{N}H(s) = \frac{1}{M}\frac{s^2 + 2s\zeta\omega_n}{s^2 + 2s\zeta\omega_n + \omega_n^2}. \tag{11.149}$$

The step error time response of the PLL is the phase error $\phi_e(t)$ in response to a unit step at the input, and can be obtained by the inverse Laplace transform of $H_e(s)/s$

$$\phi_e(t) = \frac{1}{2M}\left[\left(\frac{\zeta}{\sqrt{\zeta^2 - 1}} + 1\right)e^{-\omega_n\left(\zeta - \sqrt{\zeta^2 - 1}\right)t} - \left(\frac{\zeta}{\sqrt{\zeta^2 - 1}} - 1\right)e^{-\omega_n\left(\zeta + \sqrt{\zeta^2 - 1}\right)t}\right]$$

$$\tag{11.150}$$

for $t > 0$.

Example 11.4: The time-domain response of the phase error, given by (11.150), is plotted in Fig. 11.46. It is seen that at $\zeta = 0.707$ the maximally flat frequency response is achieved. The typical transient response of a PLL is enveloped by $e^{-\zeta\omega_n t}$, until it is stabilized at the final frequency.

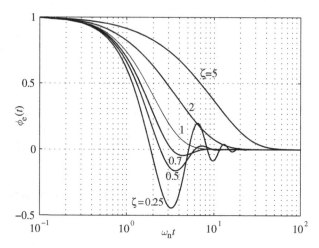

Figure 11.46   Time-domain response of ϕ_e for a step input, $M = 1$.

The 3-dB bandwidth of the loop is given by [35]

$$B = \omega_n \left(1 - 2\zeta^2 + \sqrt{2 - 4\zeta^2 + 4\zeta^4} \right)^{1/2} \tag{11.151}$$

and the acquisition time of the PLL is given by

$$T_a = \frac{2.2}{B}. \tag{11.152}$$

Tracking and acquisition characteristics are important aspects of a PLL. The lock time is typically defined in terms of the phase difference between the input and output. For a wide tracking and acquisition frequency range, most PLLs also incorporate frequency comparison. When the VCO is far from the input frequency, frequency detection is started so as to drive the VCO frequency closer to the input frequency. When the two frequencies are sufficiently close, phase detection is started for final phase-locking. A phase/frequency detector can be used for this purpose [37]. National Semiconductor provides a number of PLLs from 50 MHz to 5 GHz and integrated PLL plus VCO from 0.8 GHz to 2.4 GHz.

11.17.3 Direct frequency synthesis

In order to realize an output frequency that is a multiple of the input frequency, in the feedback path of the PLL (Fig. 11.43), a division ratio $1/N$ is applied, and the division is the modulus operation. This can achieve $\omega_{\text{out}} = N\omega_{\text{in}}$. This is very important for RF synthesizers.

Usually, we need to generate an output frequency given by

$$f_{\text{out}} = f_0 + kf_{\text{ch}} \tag{11.153}$$

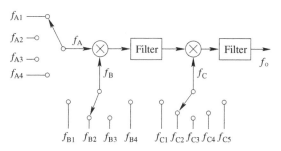

Figure 11.47

An example of direct frequency synthesis.

where f_0 is the lower end of the range, the integer k is used to select the channel, and f_{ch} is the bandwidth of each channel. PLLs must be used to maintain the exact relationship between the input and output frequencies, in locked state.

A direct frequency synthesizer creates its output frequency by mixing two or more signals to produce sum or difference frequencies, using frequency multiplication, frequency division, or their combination. The reference source is often a temperature-controlled crystal oscillator (TXCO). Selection of different output frequencies is performed by using switches. This approach achieves fast frequency switching and low phase noise, but is more hardware intensive and expensive, and generates a large number of spurious signals. It is not very flexible for generating all desirable frequencies. It requires a large number of reference oscillators. A direct frequency synthesizer is illustrated in Fig. 11.47. The output frequency is given as $f_o = f_A \pm f_B \pm f_C$. All together 13 reference oscillators are used, and $4 \times 4 \times 5 = 80$ different combinations of frequency are obtained. This approach is now mainly used for RF test and measurement.

A number of RF synthesizer architectures such as integer-N, fractional-N, and dual-loop architectures are discussed in [37]. These architectures can be used to generate many different types of input to output frequency relations.

11.18 Automatic gain control

For wireless receivers, the dynamic range of the received signal may be up to 100 dB or more, due to power fluctuations arising from shadowing and fading. In order to track the fast fluctuations, an AGC with good tracking capability is used. An AGC is absolutely necessary to raise the MDS to a level of a few milliwatts (dBm) so that the ADC can correctly quantize the signal. The total gain should be well distributed among the RF, IF and baseband stages. A moderate gain at RF is necessary to prevent too high a 1-dB compression point or IP3 and to set a good noise figure for the receiver.

While most of the gain can be fixed at different stages, an AGC with a range of 20 to 60 dB may usually be used to adapt the receiver to different levels of the received signal. It is common to implement AGC at the IF stage, as shown in Fig. 11.48, so that the AGC will attenuate the signal and any noise inserted by the IF amplifier in the same proportion to

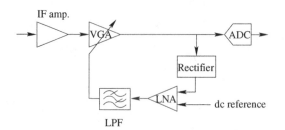

IF amp.

LPF

Figure 11.48   Architecture of an AGC circuit.

maintain a constant noise figure while the overall receiver gain is variable. The AGC loop is in general a nonlinear system with a gain acquisition time depending on input signal level. With a logarithm function applied to the RSSI or rectifier output, the AGC system can operate linearly in decibel. The rectifier is a rms-to-dc converter.

The AGC processes information in the digital domain, and sends control information to the analog component. The analog component then adjusts its power level prior to A/D conversion. Depending on the sum of the fixed gains at different stages, the AGC circuit may compensate as a variable gain amplifier (VGA) or attenuate as a variable voltage-controlled attenuator. The final output signal at the ADC or demodulator is detected and rectified to a dc signal, which is compared with a reference level. The result is amplified, lowpass filtered, and then used as the control to the AGC circuit.

The VGA can be implemented by varying the gain of an RF transistor amplifier, and the attenuator can be realized by using a PIN diode attenuator. The PIN diode attenuator generally results in lower power drain, less frequency pulling, and lower RF signal distortion, especially when using diodes with thick I regions and long carrier lifetimes. Using the PIN diode, wide dynamic range attenuation with low signal distortion can be achieved at frequencies ranging from below 1 MHz up to well over 1 GHz [47].

At the RF stage, the input signal is strictly restricted to the linear section to avoid saturation. This requires the maximum received signal power $S_{i,\max}$ to satisfy

$$S_{i,\max} + G_{RF} < \min\left(P_1, P_3\right),\qquad(11.154)$$

where G_{RF} is the gain of the RF stage, and P_1 and P_3 are, respectively, the 1-dB compression point and IP3. An AGC attenuator at the RF stage helps to reduce saturation, but it also degrades the noise figure.

Feedforward AGCs have a more rapid tracking capability than feedback AGCs, but feedback AGCs are more robust to misadjustment or errors. Feedback AGCs use the RSSI value and a gain control amplifier. Feedback AGCs are based on the signal power error.

Amplitude acquisition using AGC circuits usually occurs during a preamble duration. The preamble duration must exceed the acquisition time of the AGC loop but should be as small as possible to reduce channel bandwidth. A generalized design of AGC circuits that have constant settling time is available [24]. The settling time is independent of the absolute gain, and the method is applicable for an arbitrary monotonic nonlinear function in the gain control characteristic of the VGA. The method is also fully applicable to digital AGC loops. A digital programmable dB-linear CMOS AGC is described in [13], and a cascode 6-bit digitally controlled AGC implemented. In CMOS technology, most VGAs are implemented by using binary weighted arrays of resistors or capacitors, and these arrays are digitally controlled, leading to discrete gain steps.

In the superheterodyne structure, the gain control is continuous. In the direct-conversion receiver, discrete step gain control is implemented for both the I and Q channels in the baseband in order to minimize the mismatch between the two channels. For the low-IF architecture, the receiver AGC is very similar to that of the direct-conversion receiver.

11.19 MICs and MMICs

MICs and MMICs are used when the volume is high. In an MIC, a number of discrete active and passive components are externally attached to an etched circuit on a common substrate. An MMIC is a microwave circuit in which all active and passive components are made on one piece of semiconductor substrate. The frequency of operation can range from 1 GHz to above 100 GHz.

MMICs typically have a performance inferior to standard hybrid MICs. This is because of the low Q of the passive components in MMIC implementation. But MMICs are more compact, more reliable, more cost-effective for mass production than hybrid MICs, and have very good reproducibility. Hybrid MICs are cheaper for simple circuits. Hybrid MICs may be preferable for LNAs and power amplifiers, since the best transistors can be selected.

11.19.1 Major MMIC technologies

Today, GaAs, silicon (Si) bipolar, and BiCMOS techniques constitute the major portion of RF IC market [37]. The GaAs FET and heterojunction devices offer high power tolerance, but are usually a low-yield, high-cost solution. Most power amplifiers and front-end switches employ GaAs technologies. On the contrary, silicon technologies are more suitable for a higher level of integration in VLSI technology, and thus provide a cheap solution. A BiCMOS technology can incorporate SiGe HBTs into the CMOS technology. This achieves a tradeoff between performance and cost.

GaAs has been more widely used than silicon for MMICs, since it is suitable for both high-frequency transistors and low-loss passive components. The vast majority of MMICs above a few GHz are based on GaAs MESFET. Improvements on the silicon bipolar devices make them suitable for over 25 GHz, and most applications up to 5 GHz are dominated by silicon [41]. In recent years, heterojunction devices with both materials have been produced.

The GaAs MESFET technology is easily fabricated using ion-implantation, and has good noise figure and output power performance. It can be used for circuits operating up to 30 GHz. The GaAs-based pseudomorphic HEMT has a considerably larger transconductance than the MESFET, and can operate at over 100 GHz. The GaAs HBT has a very high gain, and offers exceedingly high power density and efficiency. Due to its large parasitic resistance and capacitance, its noise figure is considerably higher than that of HEMTs [41].

Silicon bipolar technology has advanced rapidly, and SiGe HBTs can operate at over 25 GHz. Silicon bipolar technology can be readily integrated with conventional CMOS digital IC technology, leading to the BiCMOS process that achieves digital/analog circuit integration.

CMOS technology has also been applied to RFIC/MMIC design for up to a few GHz. Recently, a number of millimeter-wave single-chip transmitters and receivers have been fabricated in CMOS technology [20, 25]. The application of silicon technology to MMIC is desirable, since it is a mature technology in the digital market and is cheap. CMOS technology is receiving active research in RF IC design.

11.19.2 Approach to MMIC design

MMIC design can employ the all-transistor technique, the lumped-element technique, or the distributed-element technique [41]. The all-transistor technique is suitable when the operating frequency is below 5 GHz, benefiting from the low capacitance of the microwave transistor. It has high packing density, and has been popular in wireless communications. The RF CMOS, silicon bipolar, and GaAs HBT technologies are more suitable for the all-transistor technique.

When the frequency is above 5 GHz, the capacitances at the input and output of the transistors introduce much higher impedances, and the all-transistor technique is not applicable. The lumped-element method is suitable for the frequency range below 20 GHz, and the classical type of MMIC employs the lumped-element method. Figure 11.49 shows an MMIC amplifier operating at 1 GHz to 2 GHz. This method is featured by the spiral inductors. It offers higher performance in terms of noise figure and power handling than the all-transistor technique. A large number of passive components are used for matching networks.

When the frequency is above 20 GHz, the spiral inductors will generate self-resonance, and the distributed design approach is suitable. The distributed elements are typically

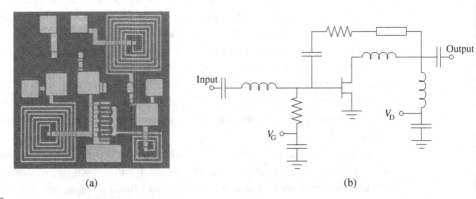

(a)   (b)

Figure 11.49 A 1 to 2 GHz MMIC amplifier using the lumped-element technique. (a) Photograph. (b) Circuit diagram. From [41] Fig.1.9. ©IEE.

designed by using microstrip or coplane waveguide. Microstrip is more widely used in MMIC design. Coplane waveguide is extensively used for millimeter-wave circuits, since it provides low dispersion and low inductance grounding. Using the coplane waveguide or other coplanar techniques, through-substrate via-holes are not required for grounding, thus the packing density can be increased; moreover, lumped elements exhibit less parasitic capacitance [41]. The lowest frequency of operation is determined by the chip size.

The majority of MMIC designs is based on the traditional microstrip design developed for hybrid MICs, where passive components take more area than active devices. Multi-layer techniques and micromachined passive components are used to make more compact packaging or to improve the performance of the circuits.

Computer-aided design techniques are an integral part of MMIC design. For RF IC design, computer-aided design tools are still in their infancy, and designers have to rely on experience and approximate simulation techniques. The standard SPICE software is used for ac analysis and is based on linear, time-invariant models, and thus is not suitable for RF circuits due to nonlinearity, time variance, and noise. Major computer-aided design packages for MMIC design are Ansoft's Serenade Desktop, Agilent's Advanced Design System (ADS), and Applied Wave Research (AWR)'s Microwave Office. These software packages typically contain or interface with some linear simulators as well as EM solvers for design simulation. Cadence's Analog Artist is ideally suited for low RF frequency MMIC designs. A good monograph that addresses all major aspects of MMIC design is [41].

11.19.3 Passive lumped components

The inductor and the capacitor are two critical passive devices used in RF ICs. The inductor is used for filtering, isolation, and compensation for parasitic capacitors, and the capacitor is used for ac coupling, filtering, and decoupling of supply voltages. The resistor is also used in RF ICs. The via-hole is an essential component for all but the most simple microwave circuits. It provides a low inductance grounding within the circuit.

Planar microwave lumped inductors and capacitors have physical dimensions that are much smaller than the free space wavelength of the highest operating frequency. Due to the small size, they are particularly useful for MMICs. But they have lower Q and lower power handling capability than distributed elements. Lumped components are suitable only when the total line length of a lumped inductor or overall size of a lumped capacitor is a small fraction of a wavelength. However, this is usually not satisfied, and other parasitics also degrade the accuracy of the model. In this case, the basic design equations can only be used for initial design, and a full-wave EM solver is used to validate and optimize the design.

Resistors

In RF IC design, on-chip resistors can be realized by many methods. The deposited thin-metal resistor has the lowest resistance. The doped semiconductor layer, such as mesa resistors, implanted planar resistors, well resistors and polysilicon resistors can also be

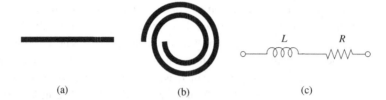

Figure 11.50 Lumped-inductors. (a) Straight-line inductor. (b) Circular spiral inductor. (c) Their circuit representation.

implemented. Thin-film resistors have better linearity and lower temperature coefficients than GaAs mesa resistors.

Lumped inductors

A high-impedance, straight-line section is a form of inductor. It can be used for low inductance values up to 3 nH. A spiral inductor can provide larger inductance, typically 10 nH. These models can be characterized by an inductor in series with a resistor. The inductance can be calculated from the physical dimensions, and the approximate equations for inductance L and the associated resistance R are given for the straight-line inductor and the circular spiral inductor in [6, 19], and other structures in [6]. The inductance of one single turn is smaller than that of a straight line with the same length and width, due to the proximity effect.

The unloaded Q of an inductor is defined by

$$Q = \frac{\omega L}{R}. \tag{11.155}$$

Straight-line and circular spiral inductors are shown in Fig. 11.50.

Spiral inductors of square and octagonal shapes are also used. The circular spiral inductor is most area-efficient, but it requires infinite angle step. Octagonal inductors are the most popular inductors on chip. The realization of a spiral inductor requires two layers of metal, and an airbridge or a dielectric via hole for the connection to the center of the inductor. The single-layer air core inductor is also practical, and it consists of a single layer of turns and uses air as the dielectric.

Lumped capacitors

The interdigital capacitor is usually used as a lumped-element capacitor for capacitance less than 1.0 pF, above which its size leads to considerable distributed effects. The metal-insulator-metal (MIM) capacitor can achieve higher capacitance such as 30 pF. These capacitors can be modeled by a capacitor C in series with a resistor R. There are also capacitors with nonlinear characteristic with respect to voltage, such as MOS caps and varactors. A varactor is a variable capacitor and is usually made of an active device such as a PN diode or a MOSFET. It is mainly used to tune the resonant frequency of an LC tank such as in a VCO.

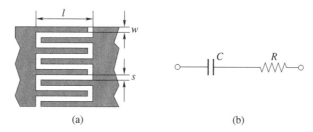

Figure 11.51

The interdigital capacitor and its circuit representation.

The MIM capacitor is constructed as a thin layer of low-loss dielectric between two metal plates. The capacitance is given by

$$C = \frac{A\epsilon_0\epsilon_r}{d}, \tag{11.156}$$

where A is the area of each plate, ϵ_r is the relative permittivity, d is the dielectric thickness, and ϵ_0 is the free space permittivity.

For the interdigital capacitor, as shown in Fig. 11.51, when the finger width W equals the finger spacing s, the maximum capacitance density is achieved; in this case, if the substrate thickness $h \gg W$, the capacitance is given by a very simple closed-form expression [1, 19]

$$C = 3.937 \times 10^{-5} l (\epsilon_r + 1) [0.11(n - 3) + 0.252] \quad \text{(pF)}, \tag{11.157}$$

where l is in μm, n is the number of fingers, and ϵ_r is the relative dielectric constant of the substrate. The resistance is given by

$$R = \frac{4}{3}\frac{R_s l}{Wn}, \tag{11.158}$$

where R_s is the surface resistance rate of the conductor in ohms/m^2.

The quality factor Q for the capacitor is defined by

$$Q = \frac{1}{\omega CR}. \tag{11.159}$$

The MIM capacitor has a high Q in general.

11.19.4 RF CMOS

The entire RF front-end can be integrated into an MMIC. The GaAs process is more popular than Si CMOS technology. However, in recent years, more attention is being paid to the low-cost RF CMOS technology. RF CMOS technology is preferable for low cost and better integration with DSP chips, but it has limitations in terms of noise and linearity properties when compared with GaAs and SiGe processes.

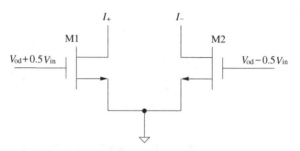

Figure 11.52   **A common-source grounded MOSFET pair used as a linear transconductor cell.**

Linearity in RF CMOS

High linearity is required for a system with a large dynamic range. This requires the transconductance of a linear MOSFET to be constant over the input voltage. From (11.122), the transconductance depends on the input voltage, which indicates a nonlinear property of the MOSFET. For the long-channel MOSFET, only the second-order distortion can be cancelled at the output by using a differential design. A common-source grounded MOSFET pair, as shown in Fig. 11.52, is usually used as a linear transconductor cell. Most CMOS circuits are based on the use of linear transconductors.

For long-channel MOSFET pair, the output current is derived as [11]

$$I_{out} = I_+ - I_- = \mu_n C_{ox} \frac{W}{L} V_{od} V_{in}. \tag{11.160}$$

For short-channel MOSFETs, odd-order distortion is generated at the output even with a differential design. This is characterized by the input referred voltage IP3, which increases with V_{od}. The dc power consumption also increases with V_{od}, and thus a compromise between linearity and power consumption must be made. A larger channel length L leads to a higher linearity, but a slower circuit response.

The MOSFET has a lower linearity than the GaAs FET and BJT under the same conditions. The linearity of the differential pair is not sufficient for most RF circuits, and linearization techniques are often used to improve the basic differential pair. A linearization technique using harmonic cancellation is given in [11]. It is a feedforward technique in which the unwanted harmonics are cancelled. This technique has been applied to the common-gate LNA, common-source LNA, and a double-balanced Gilbert mixer in CMOS.

A MOSFET has a lower IP3 than a GaAs device. In the CMOS implementation, the Gilbert multiplier cell is very widely used in mixers and AGC amplifiers. Designs of LNAs, mixers, and power amplifiers using CMOS technology are given in detail in [27].

11.19.5 Impedance matching

In order to match an RF IC to a standard external impedance such as 50 ohms, a matching network is implemented in the RF IC. The quarter-wave transformer, which uses a transmission line of a quarter wavelength, is a popular solution to impedance matching.

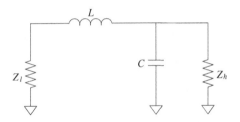

Figure 11.53   LC network for impedance matching.

However, due to the size restriction, it is seldom used in RF ICs. Instead, an LC network, known as *L-match*, as shown in Fig. 11.53, is usually used in RF ICs [11, 27, 39].

The L-match circuit realizes a transform

$$Z_l = \frac{Z_h}{Q_C^2}, \tag{11.161}$$

where Q_C is the quality factor of the capacitor C, and Z_l and Z_h are the low and high impedance at the resonant frequency $\omega_0 = 1/\sqrt{LC}$. There is relation

$$Q_C = \frac{1}{\omega_0 C Z_h} \approx Q_L = \frac{\omega_0 L}{Z_l}. \tag{11.162}$$

The lattice-type LC balun can be derived from an L-match network. It is used for power-combining in RF ICs [39].

Problems

11.1 Consider an IS-95 system. Assume the path loss of 140 dB in a rural environment, a frequency of 860 MHz, a transmit antenna gain of 10 dB, a receive antenna gain of 0 dB, a data rate of 9.6 kbits/s, antenna feed line loss of 8 dB, and all other loss of 15 dB, a fade margin of 8 dB, and a required E_b/N_0 of 9.6 dB, the receive amplifier gain of 18 dB, the total noise figure of 6 dB, and a noise temperature of 290 K, and a link margin of 8 dB. Determine the total transmit power required.

11.2 An rms voltmeter (assumed noiseless) having an effective noise bandwidth of 20 MHz is used to measure the noise voltage of a 10-kΩ resistor at room temperature. What is the meter reading?

11.3 A pager operating at a center frequency of 100 MHz has a noise bandwidth of 10 kHz. If the antenna efficiency is 40 percent and the noise figure is 10 dB, what is the minimum signal power into the receiver for a SNR of 5 dB?

11.4 For a lossless microstrip line with width a and dielectric strip thickness d, the capacitance per unit length is $C = \epsilon \frac{a}{d}$ and the inductance per unit length is $L = \mu \frac{d}{a}$. What is the characteristic impedance?

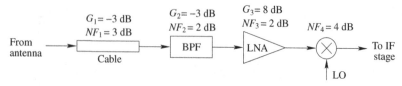

Figure 11.54

Figure for Problem 11.6.

11.5 Explain why the single-wire transmission line is rarely used compared to waveguide or coaxial lines.

11.6 The block diagram of an RF stage of a receiver is shown in Fig. 11.54. The signal from the antenna is connected by the cable to the BPF, then to the LNA, and the output of the mixer is passed to the IF stage. Calculate the noise figure. If the BPF and the LNA are placed before the cable, what is the noise figure?

11.7 Given a QPSK signal $x(t) = A \cos \omega t + B \sin \omega t$, where $A, B = \pm 1$. Due to nonideal LOs, the I/Q mismatch at LO is modeled as

$$I_{LO}(t) = (1 + \epsilon)A \cos(\omega t - \theta), \quad Q_{LO}(t) = (1 - \epsilon)B \sin(\omega t + \theta).$$

Plot the constellation diagram of the downconverted baseband signal for $\theta = 10°$ and $\epsilon = 0.2$. Compare it with that of ideally demodulated QPSK.

11.8 Given a signal level of -100 dBm, the signal bandwidth of 1 MHz, $C/N = -2$ dB for demodulation, and an interfering tone of 71 dB above the signal at 900 kHz offset from the desired signal, for a superheterodyne receiver, what is the required phase noise of the LO in dBc/Hz? Assume the phase noise is constant.

11.9 Determine the available power in dBm at room temperature in a 10 MHz bandwidth for a resistance $R = 100$ kΩ. (Hint: the noise power $P = kTB$.)

11.10 A GEO satellite is used as a regenerative repeater. For the satellite-to-earth link, the satellite antenna has a gain of 8 dB, the earth station antenna has a gain of 50 dB. The center frequency for the downlink is 3.5 GHz, and the signal bandwidth is 1 MHz. For required communication the E_b/N_0 is required to be 10 dB. What is the transmitted power for the satellite? Assume $N_0 = 4 \times 10^{-20}$ W/Hz.

11.11 A transmission line has the parameters: $L = 0.2$ μH/m, $C = 400$ pF/m, $R = 4$ Ω/m, and $G = 0.01$ S/m. Calculate the propagation constant and the characteristic impedance of the line at 1 GHz. What are results in the absence of loss ($R = G = 0$)?

11.12 The transmission line parameters of a coaxial line are given by [36]

$$L = \frac{\mu}{2\pi} \ln \frac{b}{a} \text{ (H/m)}, \quad C = \frac{2\pi \epsilon'}{\ln(b/a)} \text{ (F/m)},$$

$$R = \frac{R_s}{2\pi} \left(\frac{1}{a} + \frac{1}{b} \right) \text{ (}\Omega\text{/m)}, \quad G = \frac{2\pi \omega \epsilon''}{\ln(b/a)} \text{ (S/m)},$$

where a, b are the radii of the inner and outer conductors, R_s is the surface resistivity of the inner and outer conductors, the material filling the space between the conductors has a complex permittivity $\epsilon = \epsilon' - j\epsilon'' = \epsilon'(1 - j\tan\delta)$ and a permeability $\mu = \mu_0\mu_r$. The RG-402/U semi-rigid coaxial cable has an inner conductor diameter of 0.91 mm, and the inner diameter of the outer conductor of 3.02 mm. The conductors are copper, and the dielectric material is Teflon. Calculate R, L, G, C at 1 GHz. Compare your results to the manufacture's specifications of 50 Ω and 0.43 dB/m, and explain the reason for the difference. The dielectric constant of Teflon $\epsilon_r = 2.08$, $\tan\delta = 0.0004$ at $25°$. For copper, $\sigma = 5.813 \times 10^7$ S/m, the skin depth $\delta_s = \sqrt{\frac{2}{\omega\mu\sigma}}$, and $R_s = \frac{1}{\sigma\delta_s}$.

11.13 Design a quarter-wave transformer to match a 50-Ω to a 75-Ω cable. Plot the VSWR for $0.5 \leq f/f_0 \leq 2.0$, where f_0 is the frequency at which the line is $\lambda/4$ long.

11.14 Design a microstrip line with $Z_0 = 100$ Ω. The substrate thickness is 0.128 cm, with $\epsilon = 2.20$. What is the guided wavelength on this line at 2.0 GHz?

11.15 Given a four-port network with the scattering matrix

$$S = \begin{bmatrix} 0.2e^{j90°} & 0.8e^{j-45°} & 0.2e^{j45°} & 0 \\ 0.8e^{j-45°} & 0 & 0 & 0.3e^{j45°} \\ 0.2e^{j45°} & 0 & 0 & 0.7e^{j-45°} \\ 0 & 0.3e^{j45°} & 0.7e^{j-45°} & 0 \end{bmatrix},$$

(a) Is the network lossless?
(b) Is the network reciprocal?
(c) Find the return loss at port 1 when all other ports are matched with loads.
(d) Find the insertion loss and phase delay between ports 2 and 4, when all other ports are matched with loads.

11.16 A nonlinear device is governed by the law

$$v_2(t) = a_1 v_1(t) + a_2 v_1^2(t),$$

where $v_1(t)$ is input, $v_2(t)$ is output, and a_1 and a_2 are constants. Assume the input is an AM signal

$$v_1(t) = A_c[a + k_a m(t)]\cos(2\pi f_c t).$$

Determine the output $v_2(t)$. What is the condition for restoring the message signal $m(t)$ from $v_2(t)$?

11.17 A satellite system has a downlink C/N_0 of 80 dB·Hz. The parameters of the links are: The EIRP of the satellite is 55 dBW, downlink carrier frequency is 12.2 GHz, the data rate is 10 Mbits/s, the C/N_0 required at the earth receiver is 10 dB, the distance between the satellite and the earth receiver is 41,000 km. Assume that the efficiency of the dish antenna is 50%, and the ambient temperature is 310 K. Determine the diameter of the dish antenna at the receiver.

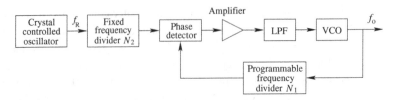

Figure 11.55

Figure for Problem 11.25.

11.18 Given the link: The transmitter power is 30 dBm, transmitter antenna gain is 5 dB, receiver antenna gain is 1.5 dB, line loss at the transmitter is 2 dB, line loss at the receiver is 1.5 dB, fade margin is 15 dB, noise PSD is -173 dBm/Hz, a bit rate of 384 kbits/s, and the required E_b/N_0 is 24 dB. Determine the maximum allowable path loss.

11.19 Calculate the power reflected to the source by an antenna with a VSWR of 2.0 for an input power of 1 W.

11.20 A coaxial three-way power divider has an input VSWR of 1.6 over a frequency range of 2.4 to 3.0 GHz. What is the return loss? If the insertion loss is 0.5 dB, what are the percentages of power reflection and transmission?

11.21 Given two linearly polarized antennas in free space oriented at $10°$ to each other. Calculate the polarization mismatch loss.

11.22 A BS transmits a power of 20 W into a feeder cable with a loss of 8 dB. The transmit antenna has a gain of 10 dB in the direction of the receiver. The receiver has a antenna gain of 0 dB, a feeder loss of 3 dB, and a sensitivity of -104 dBm. Calculate the EIRP and the maximum acceptable path loss.

11.23 A receiver has a noise bandwidth of 200 kHz and requires a input SNR of 9 dB for an input signal of -98 dBm. Determine the maximum allowable value of the receiver noise figure, and the equivalent input noise temperature of such a receiver.

11.24 A receiver operating at room temperature has a noise figure of 6 dB and a bandwidth of 1 GHz. The input 1-dB compression point is 1 dBm. Find the minimum detectable signal and dynamic range.

11.25 Determine the output frequencies of the frequency synthesizer shown in Fig. 11.55 for $N_1 = 20$ and $N_1 = 30$. Assume that $f_R = 1$ MHz and $N_2 = 100$.

11.26 Compare the advantages and disadvantages of the MMIC, HMIC, and hybrid printed circuit approaches.

11.27 For Problem 10.20, the designed antenna is fed by a 50-Ω microstrip line and a quarter-wave transformer is used for impedance matching. Find the dimensions of the feed line and the quarter-wave transformer. Draw the design.

11.28 A 10-dB directional coupler has a directivity of 40 dB. If the input power $P_1 = 1$ W, calculate the power output at the other three ports.

References

[1] G. D. Alley, Interdigital capacitors and their application to lumped-element microwave integrated circuits. *IEEE Trans Microwave Theory & Techniques*, **18**:12 (1970), 1028–1033.

[2] P. M. Asbeck, L. Larson, Z. Popovic & T. Itoh, Power amplifier approaches for high efficiency and linearity. In T. Itoh, G. Haddad & J. Harvey, eds., *RF Technologies for Low Power Wireless Communications* (New York: Wiley-IEEE, 2001), pp. 189–227.

[3] I. J. Bahl & R. Garg, Simple and accurate formulas for microstrip with finite strip thickness. *Proc. IEEE*, **65** (1977), 1611–1612.

[4] E. R. Brown, RF MEMS switches for reconfigurable integrated circuits. *IEEE Trans. Microwave Theory Tech.*, **46**:11 (1998), 1868–1880.

[5] E. R. Brown & J. F. Harvey, System characteristics of quasi-optical power amplifiers. *IEEE Circ. Syst. Mag.*, **1**:4 (2001), 22–36.

[6] M. Caulton, S. P. Knight & D. A. Daly, Hybrid integrated lumped-element microwave amplifiers. *IEEE Trans. Microwave Theory Tech.*, **16**:7 (1968), 397–404.

[7] D. C. Cox, Linear amplification with nonlinear components. *IEEE Trans. Commun.*, **22**:12 (1974), 1942–1945.

[8] S. C. Cripps, *Advanced Techniques in RF Power Amplifier Design* (Boston: Artech House, 2002).

[9] S. C. Cripps, *RF Power Amplifiers for Wireless Communications*, 2nd edn (Boston: Artech House, 2006).

[10] H. J. De Los Santos, *RF MEMS Circuit Design for Wireless Communications* (Boston: Artech House, 2002).

[11] Y. Ding & R. Harjani, *High-Linearity CMOS RF Front-End Circuits* (New York: Springer, 2005).

[12] K.-L. Du & M. N. S. Swamy, *Neural Networks in a Softcomputing Framework* (London: Springer, 2006).

[13] H. O. Elwan, T. B. Tarim & M. Ismail, Digitally programmable dB-linear CMOS AGC for mixed-signal applications. *IEEE Circ. Devices Mag.*, **14**:4 (1998), 8–11.

[14] A. Grebennikov, *RF and Microwave Power Amplifier Design* (New York: McGraw-Hill, 2005).

[15] M. L. Edwards & J. H. Sinsky, A new criterion for linear 2-port stability using a single geometrically derived parameter. *IEEE Trans. Microwave Theory Tech.*, **40**:12 (1992), 2303–2311.

[16] W. E. Egan, *Phase-Lock Basics* (New York: Wiley, 1998).

[17] E. O. Hammerstard, Equations for microstrip circuit design. In *Proc. 5th European Microwave Conf.*, Hamburg, Germany, Sep 1975, 268–272.

[18] J. Harvey, E. R. Brown, D. B. Rutledge & R. A. York, Spatial power combining for high-power transmitters. *IEEE Microwave Mag.*, **1**:4 (2000), 48–59.

[19] J.-S. Hong & M. J. Lancaster, *Microstrip Filters for RF/Microwave Applications* (New York: Wiley, 2001).

[20] D. Huang, R. Wong, G. Qun, N. Y. Wang, T. W. Ku, C. Chien & M.-C. F. Chang, A 60 GHz CMOS differential receiver front-end using on-chip transformer for 1.2 volt operation with enhanced gain and linearity. In *IEEE Symp. VLSI Circuits Dig.*, Honolulu, HI, Jun 2006, 144–145.

[21] A. Ismail & A. Abidi, A 3–10-GHz low-noise amplifier with wideband LC-ladder matching network. *IEEE J. Solid-State Circ.*, **39**:12 (2004), 2269–2277.

[22] P. Jardin & G. Baudoin, Filter lookup table method for power amplifier linearization. *IEEE Trans. Veh. Tech.*, **56**:3 (2007), 1076–1087

[23] D. Kee, I. Aoki, A. Hajimiri & D. B. Rutledge, The class-E/F family of ZVS switching amplifiers. *IEEE Trans. Microwave Theory Tech.*, **51**:6 (2003), 1677–1690.

[24] J. M. Khoury, On the design of constant settling time AGC circuits. *IEEE Trans. Circ. Syst. II*, **45**:3 (1998), 283–294.

[25] E. Laskin, M. Khanpour, R. Aroca, K. W. Tang, P. Garcia & S. P. Voinigescu, A 95 GHz receiver with fundamental-frequency VCO and static frequency divider in 65 nm digital CMOS. In *Proc. ISSCC*, San Francisco, CA, Feb 2008, 180–181.

[26] T. H. Lee & A. Hajimiri, Oscillator Phase Noise: A Tutorial. *IEEE J. Solid State Circ.*, **35**:3 (2000), 326–336.

[27] T. H. Lee, *The Design of CMOS Radio-Frequency Integrated Circuits*, 2nd edn (Cambridge, UK: Cambridge University Press, 2004).

[28] D. B. Leeson, A simple model of feedback oscillator noise spectrum. *Proc. IEEE*, **54**:2 (1966), 329–30.

[29] J. G. Ma, Design of CMOS RF ICs for wireless applications: system-level compromised considerations. In J. G. Ma, ed., *Third Generation Communication Systems: Future Developments and Advanced Topics* (Berlin: Springer, 2004), pp. 199–236.

[30] R. R. Mansour, Filter technologies for wireless base stations. *IEEE Microwave Mag.*, **5**:1 (2004), 68–74.

[31] G. L. Matthaei, L. Young & E. M. T. Jones, *Microwave Filters, Impedance-Matching Networks, and Coupling Structures*, (Dedham, MA: Artech House, 1980).

[32] J. A. McNeill, Jitter in Ring Oscillators. *IEEE J. Solid-State Circ.*, **32**:6 (1997), 870–879.

[33] A. Patnaik & R. K. Mishra, ANN techniques in microwave engineering. *IEEE Microwave Mag.*, **1**:1 (2000), 55–60.

[34] M. W. Pospieszalski, Extremely low-noise amplification with cryogenic FETs and HFETs: 1970–2004. *IEEE Microwave Mag.*, **6**:3 (2005), 62–75.

[35] D. M. Pozar, *Microwave and RF Wireless Systems* (New York: Wiley, 2001).

[36] D. M. Pozar, *Microwave Engineering*, 3rd edn (New York: Wiley, 2005).

[37] B. Razavi, *RF Microelectronics* (Upper Saddle River, NJ: Prentice Hall, 1998).

[38] G. M. Rebeiz & J. B. Muldavin, RF MEMS switches and switch circuits. *IEEE Microwave Mag.*, **2**:4 (2001), 59–71.

[39] P. Reynaert & M. Steyaert, *RF Power Amplifiers for Mobile Communications* (Dordrecht, The Netherlands: Springer, 2006).

[40] J. D. Rhodes, A lowpass prototype network for microwave linear phase filters. *IEEE Trans. Microwave Theory Techn.*, **18**:6 (1970), 290–301.

[41] I. D. Robertson & S. Lucyszyn (ed), *RFIC and MMIC Design and Technology* (London: IEE Press, 2001).

[42] U. L. Rodhe, *Microwave and Wireless Synthesizers: Theory and Design* (New York: Wiley, 1997).

[43] J. Rollett, Stability and power gain invariants of linear two-ports. *IRE Trans. Circ. Theory*, **9**:1 (1962), 29–32.

[44] C. C. W. Ruppel, L. Reindl & R. Weigel, SAW devices and their wireless communications applications. *IEEE Microwave Mag.*, **3**:2 (2002), 65–71.

[45] A. Rusu, D. R. D. L. Gonzalez & M. Ismail, Reconfigurable ADCs enable smart radios for 4G wireless connectivity. *IEEE Circ. & Devices Mag.*, **22**:3 (2006), 6–11.

[46] C. Samori, S. Levantino & A. L. Lacaita, Integrated LC oscillators for frequency synthesis in wireless applications. *IEEE Commun. Mag.*, **40**:5 (2002), 166–171.

[47] Skyworks, *Design With PIN Diodes*, Application Note APN1002, 200312 Rev. A, Jul 2005.

[48] A. Suarez & T. Fernandez, RF devices: characteristics and modelling. In I. A. Glover, S. R. Pennock & P. R. Shepherd, eds., *Microwave Devices, Circuits and Subsystems for Communications Engineering* (Chichester, UK: Wiley, 2005).

[49] M. Thamsirianunt & T. A. Kwasniewski, CMOS VCO's for PLL frequency synthesis in GHz digital mobile radio communications. *IEEE J. Solid-State Circ.*, **32**:10 (1997), 1511–1524.

[50] R. J. Trew, Wide bandgap semiconductor transistors for microwave power amplifiers. *IEEE Microwave Mag.*, **1**:1 (2000), 46–54.

[51] R. J. Trew, G. L. Bilbro, W. Kuang, Y. Liu, and H. Yin, Microwave AlGaN/GaN HFETs. *IEEE Microwave Mag.*, **6**:1 (2005), 56–66.

[52] V. K. Varadan, K. J. Vinoy & K. A. Jose, *RF MEMS and Their Applications* (Chichester, UK: Wiley, 2003).

A/D and D/A conversions

12.1 Introduction

Analog input signals are converted into digital signals for digital processing and transmission. The analog-to-digital (A/D) converter (ADC) performs this functionality using two steps: the sample-and-hold (S/H) operation, followed by digital quantization. The ADC is primarily characterized by the sampling rate and resolution. A sampling rate of above twice the Nyquist frequency is a must; otherwise, aliasing occurs and the result is not usable. A higher sample rate leads to a more accurate result, but a more complex system. The successive-approximation ADC successively increases the digital code by digitizing the difference until a match is found. The successive-approximation ADC is the most popular type of ADC. The sigma-delta (Σ-Δ) ADC uses oversampling and noise shaping to significantly attenuate the power of quantization noise in the band of interest.

The digital-to-analog (D/A) converter (DAC) is used to convert the processed digital signal back to an analog signal by comparing it to the input voltage. This chapter introduces ADCs and DACs that are used in wireless communication systems.

12.2 Sampling

12.2.1 Ideal and natural sampling

An analog signal $x(t)$, bandlimited to $f_{\max}$, can be transformed into digital form by periodically sampling the signal at time nT, where T is the sampling period.

Let the sampling function be a periodic impulse train given by

$$p(t) = \sum_{m=-\infty}^{\infty} \delta(t - mT), \tag{12.1}$$

where $\delta(t)$ is the Dirac δ function, which takes on the value 1 at $t = 0$ and is 0 otherwise.

The coefficient in the Fourier series of $p(t)$ is given by $\frac{1}{T} \int_{-T/2}^{T/2} \delta(t) e^{-j2\pi n f_s t}$, where $f_s = 1/T$ is the sampling frequency (see Chapter 13). Hence,

$$p(t) = \sum_{m=-\infty}^{\infty} \delta(t - mT) = \frac{1}{T} \sum_{n=-\infty}^{\infty} e^{j2\pi n f_s t}. \tag{12.2}$$

The sampled version of $x(t)$, $\hat{x}(t)$, is the product of the continuous-time signal $x(t)$ and the sampling function $p(t)$:

$$\hat{x}(t) = x(t)p(t) = \sum_{k=-\infty}^{\infty} x(kT)\delta(t - kT). \tag{12.3}$$

The sampled signal is held until it is quantized into a b-bit digit.

In the frequency domain, the spectrum of the sampling function $p(t)$ is the Fourier transform of the impulse train and is given by (see Chapter 13)

$$P(f) = \frac{1}{T} \sum_{k=-\infty}^{\infty} \delta\left(f - kf_s\right). \tag{12.4}$$

It is seen that $P(f)$ is also a periodic impulse train.

The spectrum of the sampled digital signal is then given by the convolution of $X(f)$ and $P(f)$. Thus,

$$\hat{X}(f) = X(f) * P(f) = X(f) * \frac{1}{T} \sum_{k=-\infty}^{\infty} \delta\left(f - kf_s\right) = \frac{1}{T} \sum_{k=-\infty}^{\infty} X\left(f - kf_s\right). \tag{12.5}$$

This is known as the *Poisson summation formula*. Thus, it is seen that sampling of a signal $x(t)$ produces a periodic repetition of its spectrum.

When the sampling function is a periodic train of rectangular pulses rather than a periodic train of impulses, as shown in Fig. 12.1, we call it *natural sampling*, as opposed to *ideal sampling*. The rectangular pulse train has a Fourier series given by

$$p(t) = \sum_{m=-\infty}^{\infty} p_{\text{rec}}(t - mT) = \sum_{k=-\infty}^{\infty} c_k e^{j2\pi f_s kt}, \tag{12.6}$$

where

$$c_k = \frac{1}{T} \frac{\sin(k\pi\tau/T)}{k\pi\tau/T} \tag{12.7}$$

with τ being the pulse width. The sampled signal is thus given by

$$\hat{x}_{\text{rec}}(t) = \sum_{n=-\infty}^{\infty} x(nT)p_{\text{rec}}(t - nT). \tag{12.8}$$

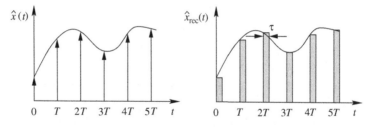

Figure 12.1   Ideal and natural sampling.

For natural sampling, the spectrum of the sampled signal is accordingly

$$\hat{X}_{\text{rec}}(f) = \sum_{k=-\infty}^{\infty} c_k X\left(f - kf_s\right).$$ (12.9)

Comparing (12.9) to (12.5), it is seen that (12.9) denotes the same spectrum as (12.5) does, except that the displaced side-bands vary with c_k.

12.2.2 Sampling theorem

Theorem 12.1 (Sampling Theorem): A bandlimited signal with the maximum frequency f_{max} is specified uniquely by its values at uniformly spaced intervals of $T_s = 1/f_s$ apart, if the sampling frequency satisfies

$$f_s \geq 2f_{\text{max}}.$$ (12.10)

The minimum allowable sampling rate $f_s = 2f_{\text{max}}$ is called the *Nyquist rate*, $f_s/2 = f_{\text{max}}$ is the Nyquist frequency, and the interval $\left[-f_s/2, f_s/2\right]$ is known as the *Nyquist frequency interval*. The sampling is known as *lowpass sampling*, since all the frequency components from 0 Hz to $f_s/2$ are recoverable. Thus, for a baseband signal with a frequency band limited from $-B/2$ to $B/2$ Hz, its digitalization requires a sampling frequency $f_s \geq B$.

According to (12.5), if the signal is bandlimited to f_{max} and $f_s \geq 2f_{\text{max}}$, the replicas do not overlap; then, in the Nyquist interval, we have

$$T\hat{X}(f) = X(f), \quad -\frac{f_s}{2} \leq f \leq \frac{f_s}{2}.$$ (12.11)

That is, the sampled signal spectrum $\hat{X}(f)$ is identical to the original spectrum $X(f)$ within the Nyquist interval.

There are also several generalizations of the sampling theorem [7, 19]. One class of uniform sampling theorem is given as: If $x(t)$ and its first $M - 1$ derivatives are available, then each of these can be uniformly sampled at the rate of f_s/M without losing any information. The second generalization is a class of nonuniform sampling theorems. There is, however, no closed-form expression for reconstruction of $x(t)$ from these samples. The problem of reconstruction of signals from nonuniform samples has been discussed in [14, 16].

12.2.3 Aliasing and antialiasing

From (12.5), it is seen that sampling causes spectrum replication. When spectrum replication leads to confusion of the original frequency f with $f + mf_s$, $m = \pm 1, \pm 2, \cdots$, *aliasing* occurs. Analog reconstruction from aliased sampling will map all the out-of-band signals onto the Nyquist interval by using the aliased frequency $f_a = f$

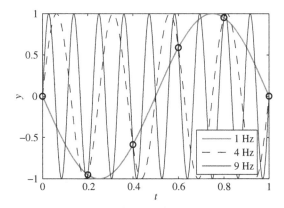

Figure 12.2 A 1 Hz, a 4 Hz, and a 9 Hz signal yield the same output, when sampled at 5 Hz.

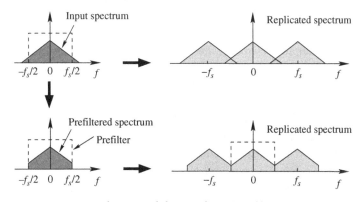

Figure 12.3 Aliasing and the antialiasing prefilter.

mod (f_s). Aliasing distorts irreversibly the frequency response within the Nyquist interval.

Example 12.1: Aliasing is illustrated in Fig. 12.2. For three signals, $y(t) = -\sin(2\pi t)$, $y(t) = -\sin(8\pi t)$, and $y(t) = -\sin(18\pi t)$, when sampled at 5 Hz, the outputs for the three signals coincide.

To avoid aliasing, an antialiasing prefilter is used to limit the cutoff frequency to $f_{max} \leq f_s/2$. Thus, the filter band is limited to $\left[-f_s/2, f_s/2\right]$. In this case, spectrum replication does not overlap. This is illustrated in Fig. 12.3.

An antialiasing prefilter must precede the ADC to bandlimit the input spectrum to within the Nyquist interval. SAW filters are antialiasing filters with excellent roll-off in the transition band but with a high insertion loss. A high quality LNA can precede the SAW filter to achieve adequate noise figure.

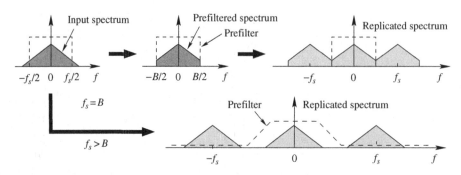

Figure 12.4

Oversampling and the antialiasing prefilter.

12.2.4 Oversampling and decimation

When f_s is greater than twice the required Nyquist frequency, the signal is oversampled. Oversampling increases the periodicity of the spectral replicas, and allows a less sharp cutoff for the prefilter. Oversampling is usually used to relax the attenuation requirements on the lowpass antialiasing filter that has a cutoff frequency at the Nyquist frequency, thus reducing the order of the filter [13].

After oversampling, the original baseband $[-B/2, B/2]$ is replicated into many spectral replicas that are separated by a wider spectrum spacing of $f_s > B$, as shown illustrated in Fig. 12.4. This makes prefilter design much easier.

If the sampling rate is reduced to B, some frequency components in $[-f_s/2, f_s/2]$ will be outside the Nyquist interval $[-B/2, B/2]$, and aliasing is introduced inside the interval. To prevent aliasing, those out-of-band frequencies must be removed by a lowpass digital filter before resampling at the lower rate. Such a filter is known as a *digital decimation filter*. Such filters just throw away some samples at fixed time intervals, achieving lowpass filtering and sampling.

12.2.5 Bandpass sampling theorem

For a bandwidth B that is centered on f_c, the lowpass sampling scheme requires a sampling frequency of $2(f_c + B/2)$. Bandpass sampling allows sampling rates consistent with the bandwidth B, rather than the highest frequency $f_c + B/2$ of the signal spectrum. Bandpass sampling is of significant importance for wireless communications.

When f_c is very large, f_s can be selected as $2B \leq f_s \leq 2f_c + B$, by using the frequency-domain repetition caused by the sampling process. This is called *bandpass-limited subsampling or undersampling*. Bandpass-limited subsampling also achieves frequency downconversion at the same time. A bandpass filter is used after bandpass sampling. The bandpass sampling criterion is stated as follows [15].

Theorem 12.2 (Bandpass Sampling): A bandpass signal is specified uniquely by its values at uniformly spaced intervals of $T_s = 1/f_s$ apart, if

$$2\frac{Q}{k} \le \frac{f_s}{B} \le 2\frac{Q-1}{k-1}, \tag{12.12}$$

where $B = f_H - f_L$, $Q = f_H/B$, k is a positive integer and $k \le \lfloor Q \rfloor$, with $\lfloor \cdot \rfloor$ denoting the integer part of the quantity within. When $k = Q$, there is zero tolerance in the sampling rate. Thus, k is usually selected to be less than Q.

Example 12.2: The condition (12.12) is plotted in Fig. 12.5. Only the wedge regions satisfy the bandpass sampling condition, while the forbidden region generates aliasing. From the plot, we can see that the waveform may be reproduced from sample values if the sampling rate is $f_s \ge 2B$.

Reconstruction of the signal is done by using the formula

$$x(t) = \sum_{n=-\infty}^{\infty} x(nT)g(t - nT), \tag{12.13}$$

where

$$g(t) = \frac{\sin \pi Bt}{\pi Bt} \cos 2\pi f_c t \tag{12.14}$$

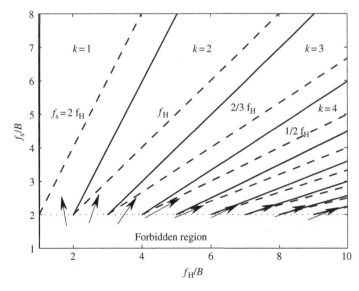

Figure 12.5   Condition for bandpass sampling: The wedges correspond to alias-free sub-band sampling.

is the inverse Fourier transform of the bandpass frequency gating (rectangular) function, and it is the ideal interpolation function for lowpass sampling, modulated by a carrier with frequency f_c. More detailed analysis on bandpass sampling is given in [15, 20]. Bandpass sampling is extremely popular in software radio systems. To avoid aliasing, bandpass filtering is a prerequisite for implementing bandpass sampling.

12.3 Quantization

Quantization is the mapping of continuous amplitude values into a finite number of bits. After sampling, each sample $x(mT)$ must be held constant for a period of at most T seconds by using a sample/hold circuit; during this time the ADC must convert it to a quantized sample, $x_Q(mT)$, represented in a number, n, of bits.

12.3.1 Uniform quantization

Usually, the ADC equally divides the full-scale range R into 2^n quantization levels, and the quantization step size or resolution is given by

$$Q = \frac{R}{2^n}. \tag{12.15}$$

The denominator can be changed to $2^n - 1$, since the up end R is not realizable as a level. Quantization is performed by rounding. Assuming that the quantization error $e = x_Q - x$ is uniformly distributed in the interval of $[-Q/2, +Q/2]$ with zero mean, the worst-case quantization error is $\pm Q/2$. The mean-square value of e or the quantization noise power is given by

$$\sigma_e^2 = E[e^2] = \frac{1}{Q} \int_{-Q/2}^{Q/2} e^2 de = \frac{Q^2}{12}. \tag{12.16}$$

The rms error is thus $e_{\text{rms}} = Q/\sqrt{12}$.

For binary coding, one needs to consider the polarity of the input signal. For unipolar analog input with range $[0, R)$, the natural binary coding is used. For bipolar analog input with range $[-R/2, R/2)$, the natural binary coding can lead to the coding of the level $x_Q = 0$ as $10 \cdots 0$, and two's complement code is usually used instead. Two's complement code is obtained by complementing the MSB (most significant bit), replacing b_{n-1} by $1 - b_{n-1}$.

Example 12.3: If the reference voltage is $+10$ V, and the input signal is $+2$ V, for a 10-bit coding, then the sampled digital value is

$$x_Q = \frac{v_i}{v_{\text{ref}}} \times max\_value = \frac{2}{10} \times 1023 = 205 = 11011101_b.$$

After digital processing, a digital signal may be converted to an analog signal. The DAC performs an operation which is the reverse of that of the ADC. The analog output for sampled value 205 is calculated by

$$v_o = x_Q \times v_{\text{ref}}/max\_value = 205 \times 10/1023 = 2.0 \quad \text{(V)}.$$

The above quantization technique is known as the *uniform pulse code modulation (PCM)* technique. It is a uniform quantization technique followed by a process of representing the quantized samples into bits of fixed length. The signal-to-quantization-noise ratio (SQNR) is obtained from (12.15) and (12.16) as

$$\text{SQNR} = \frac{\overline{x^2}}{\sigma_e^2} = 3 \times 2^{2n} \frac{\overline{x^2}}{(R/2)^2}$$

$$= 4.8 + 6n + 10 \log_{10} \frac{\overline{x^2}}{(R/2)^2} \quad \text{(dB)}, \tag{12.17}$$

where $\overline{x^2}$ is the mean-square value of x. Nonuniform PCM first applies a nonlinear element to the input signal to reduce its dynamic range, followed by a uniform PCM. This will be introduced in Chapter 16.

Midtread and midrise quantizers

For uniform quantization, the quantizer is typically symmetric in the sense that there are equal numbers of positive and negative levels. The midrise and midtread are two popular quantizers and are shown in Fig. 12.6. The midtread quantizer has a zero output, and there are an odd number of output levels. The midrise quantizer does not have a zero level, and thus has an even number of output levels. For an n-bit quantizer, the midtread allows for $2^n - 1$ different codes, as opposed to 2^n codes allowed by the midrise quantizer. The midtread quantizer is more popular for quantization, since it results in more zero levels, leading to a larger compression ratio in data compression, such as in speech and video coding.

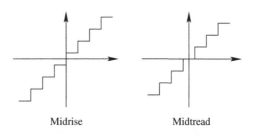

Midrise   Midtread

The midtread and midrise quantizers.

Other quantization techniques

General ADC techniques use PCM, differential PCM, scalar quantization, and vector quantization (VQ) for quantization of the held samples. The performance of these lossy data compression techniques is limited by the *rate-distortion bound*. After the samples are quantized into digital form, entropy coding can be applied to represent them using as few bits as possible. Entropy coding will be introduced in Chapter 14.

Uniform and nonuniform PCM are two scalar quantization techniques, where each dimension of the source is quantized separately. Nonuniform quantization provides better performance than uniform quantization due to its variable length of the quantization regions. The Lloyd-Max conditions give the optimal solution to nonuniform quantization [11]: the quantization levels are the centroids of the quantization regions and the boundaries are the midpoints between the quantization levels.

VQ quantizes several dimensions of the source together, thus achieving much higher compression. More detail on VQ is given in Subsection 19.2.6 and [4].

12.3.2 Improving resolution by oversampling

When the ADC is limited by the resolution, oversampling can be used to achieve higher accuracy. In the case of oversampling, each sample may have a low accuracy, but many samples can be averaged to remove noise [13]. This is similar to multiple measurements of a quantity x. Assuming that the mean-square error is σ_x^2 in a single measurement, L independent measurements of the quantity can reduce it to σ_x^2/L.

Consider two sampling rates f_s and $f_s' > f_s$, with n bits and n' bits, respectively. To maintain the same level of quality in the two cases, the PSDs of the quantization noise in the two cases are required to be the same, that is,

$$\frac{\sigma_e^2}{f_s} = \frac{\sigma_e'^2}{f_s'} = \eta. \tag{12.18}$$

The PSDs of quantization noise for both the cases are illustrated in Fig. 12.7a.

By defining the oversampling ratio $L = f_s'/f_s$, we have

$$L = \frac{\sigma_e'^2}{\sigma_e^2}. \tag{12.19}$$

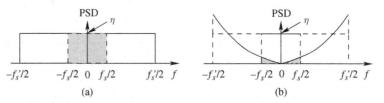

Figure 12.7 Spectrum of oversampled quantization noise power. (a) Without noise shaping. (b) With noise shaping.

Substituting (12.16) in (12.19), we have [13]

$$\Delta n = n - n' = 0.5 \log_2 L. \tag{12.20}$$

From this, we can see that a saving of half a bit is achieved when L is doubled.

A noise shaping quantizer operating at an oversampling rate f'_s can reshape the flat noise spectrum to squeeze most of the power out of the f_s Nyquist interval. The spectrum of the oversampling noise shaping quantizer is shown in Fig. 12.7b. The pth-order noise shaping filter at f'_s has a magnitude response

$$|H_{NS}(f)| = \left| 2 \sin\left(\frac{\pi f}{f'_s}\right) \right|^p, \quad -f'_s/2 \le f \le f'_s/2. \tag{12.21}$$

For small f, we have

$$|H_{NS}(f)| = \left| \left(\frac{2\pi f}{f'_s}\right) \right|^p, \quad |f| \ll f'_s/2. \tag{12.22}$$

After noise shaping, the total quantization noise power within the f_s Nyquist interval is calculated by

$$\sigma_e^2 = \frac{\sigma_e'^2}{f'_s} \int_{-f_s/2}^{f_s/2} |H_{NS}(f)|^2 \, df. \tag{12.23}$$

For a large oversampling rate L, (12.22) can be used in the integrand, and this gives

$$\sigma_e^2 = \sigma_e'^2 \frac{\pi^{2p}}{2p+1} \frac{1}{L^{2p+1}}. \tag{12.24}$$

Inserting (12.19) and (12.24) into (12.20), the gain in bits is derived as [13]

$$\Delta n = (p + 0.5) \log_2 L - 0.5 \log_2 \left(\frac{\pi^{2p}}{2p+1}\right). \tag{12.25}$$

This leads to a saving of more bits when L is doubled. For example, for $L = 8$, without noise shaping a gain of 1.5 bits is achieved, but with noise shaping of order $p = 2$ a gain of 5.4 bits is achieved. Practical values for the order p are typically 1, 2, 3.

Sigma-delta modulation is ideal for ADCs and DACs. Such an ADC employs high sampling rate and spreads the quantization noise across the band up to $f_s/2$. The DAC uses sigma-delta modulation in a loop that is the reverse of the loop for the ADC. For example, the Analog Devices AD9772 is a $2\times$ oversampled interpolating 14-bit DAC, and it handles 14-bit input word rates up to 150 Msamples/s, and the output word rate is 300 Msamples/s at maximum [8].

12.4 Analog reconstruction

Analog reconstruction transforms the sampled digital signal into an analog one. Filling the gaps between the samples results in a smoother signal, and this operation can be treated as lowpass filtering.

Ideal reconstructor

The ideal reconstruction filter $H(f)$ is an ideal lowpass filter with

$$H(f) = T = \frac{1}{f_s}, \quad -\frac{f_s}{2} \leq f \leq \frac{f_s}{2} \tag{12.26}$$

and zero otherwise. For a bandlimited spectrum $Y(f)$ with nonoverlapped replicas, then the sampled spectrum satisfies

$$\hat{Y}(f) = \frac{1}{T} Y(f), \quad -\frac{f_s}{2} \leq f \leq \frac{f_s}{2}. \tag{12.27}$$

The analog signal obtained has a spectrum

$$Y_a(f) = H(f)\hat{Y}(f) = T \cdot \frac{1}{T} Y(f) = Y(f). \tag{12.28}$$

Thus, the ideal reconstructor can completely remove the replicated spectrum images.

The corresponding impulse response of $H(f)$ is given by

$$h(t) = \text{sinc}(\pi t/T) = \frac{\sin(\pi t/T)}{\pi t/T}. \tag{12.29}$$

The sinc function has infinite anticausal part, and thus is noncausal and not realizable. An approximation to the ideal reconstructor can be obtained by truncating it to finite length.

Staircase reconstructor

A simple and widely used implementation is to hold the current sample constant for a period $T = 1/f_s$ to fill the gap until the next sample appears. The reconstructed signal thus has a staircase amplitude. The nonsmooth characteristic introduced by the staircase introduces spurious high-frequency components, which are outside the Nyquist interval. The staircase DAC has an impulse response

$$h(t) = 1, \quad 0 \leq t \leq T \tag{12.30}$$

and 0 otherwise. The corresponding frequency response is

$$H(f) = T\frac{\sin(\pi f/f_s)}{\pi f/f_s} e^{-j\pi f/f_s}. \tag{12.31}$$

The staircase reconstructor is a filter with a sinc-like shape spectrum. It is seen that $h(t)$ and $H(f)$ for the ideal as well as the staircase reconstructor are dual to each other.

The staircase reconstructor filter shapes the replicated spectrum images and leaves some remnant image components, as illustrated in Fig. 12.8. These image components in the vicinity of the Nyquist interval may be significant, and may introduce aliasing into the Nyquist interval. Thus, an anti-image lowpass postfilter must be applied to reject the spectral replicated images as much as possible.

By increasing the clock frequency to f_s' through interpolation, the first image of the spectrum can be centered on $\pm f_s'$ rather than on $\pm f_s$. D/A conversion is then applied. This permits a significant relaxation of the specifications for the anti-image lowpass postfilter. If the oversampling rate is increased by a factor of K, at the cutoff frequency $f_c = f_s/2$,

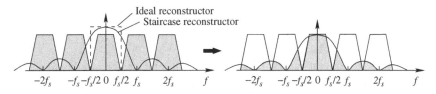

Figure 12.8

Spectrum of the staircase reconstructor.

we have $H_{DAC}(f_c) = \frac{\sin(\pi/2K)}{\pi/2K}$. For large K, this attenuation approaches 0 dB, and thus the aperture effect of the DAC is eliminated. For small oversampling rates, an inverse shape of the DAC frequency response can be applied over the passband, that is, an equalizer, which can be designed using the frequency sampling design method.

It is worth mentioning the fundamental difference between aliasing and imaging: aliasing causes loss of information, while imaging does not. We will give more description on sampling rate conversion in Section 13.7.

12.5 Parameters for A/D and D/A converters

The static nonlinearity of an ADC or DAC, caused by its physical imperfection, can be characterized by its dc transfer function, and is presented in terms of the number of bits of integral nonlinearity (INL) and differential nonlinearity (DNL) errors. These errors are measured in units of least significant bits (LSBs). The INL error is obtained by drawing a line that best fits the converter's dc transfer function and calculating the difference between the line and the actual transfer function. The DNL error is calculated as the difference between the actual step width and that for 1 LSB. These static parameters are very important in high-resolution imaging. For example, ADI provides AD9821 (12-bit, 40-Msamples/s) and AD9854B (12-bit, 30-Msamples/s) ADCs for high-performance digital still cameras and camcorders.

For evaluating ADCs or DACs for SDRs, the most important parameters are the sampling rate, bits of resolution, and SFDR. The signal-to-noise-and-distortion (SINAD) ratio is defined by the ratio of the rms signal power to the rms of all the unwanted spectral components but excluding the dc component. Sometimes SINAD is referred to as signal-plus-noise-and-distortion-to-noise-and-distortion ratio, since the measured signal may include noise and distortion, in this case, $SINAD = (S + N)/N = 1 + S/N$, N being the noise and distortion. In either case, the SINAD of a high-performance ADC should be very close to its SNR. SFDR is the ratio of the rms signal amplitude to the rms value of the peak spurious spectral component. SFDR is more important than SNR or SINAD for SDRs, since it combines linearity and quantization noise performance.

Whereas jitters may happen in ADCs, glitches may occur in DACs. The decoder in a DAC attempts to simultaneously adjust several switches to change the output voltage level so as to match the input digital word. When there are small timing discrepancies between the switches, glitches occur. This may cause significant distortion to the output signal. Glitches can be reduced by double buffering the input or adding a deglitching circuit to the output of the decoder in the DAC.

A straight binary DAC with one current switch per bit (e.g., R-2R DAC) produces code-dependent glitches. A DAC with one current source per code level does not have code-dependent glitches, but is not practical for high resolutions. A practical approach is to decode the first few MSBs into a code and have one current switch per level. This substantially reduces the code-dependent glitch. This process is called *segmentation* and is quite common in low distortion DACs, e.g., in the Analog Devices' AD977x-family DACs and the AD985x-family DDS devices [8].

12.5.1 SNR of A/D and D/A converters

The noise of an ADC contains thermal noise that includes the capacitance noise of the S/H stage, quantization noise, and aperture-jitter noise.

The variance of the quantization error is the quantization noise power, given by $\sigma_e^2 = Q^2/12$. For an analog sinewave of amplitude A, the average signal power is $A^2/2$, the quantization step size is $Q = \frac{2A}{2^n}$, where n is the ADC resolution in bits. The SQNR can be calculated by

$$\text{SQNR} = 10 \log_{10} \left(\frac{A^2/2}{Q^2/12} \right) = 1.76 + 6.02n \quad \text{(dB)}. \tag{12.32}$$

For the more general case [9]

$$\text{SQNR} = 6.02n + 4.77 - 10 \log_{10} \eta \quad \text{(dB)}, \tag{12.33}$$

where η is the PAPR of the signal. For sinusoidal signals, $\eta = 2$. The SNR of a practical ADC cannot reach the theoretical figure, especially when the number of bits is increased.

If the signal is treated as a zero-mean random process and its power is σ_x^2, then the SQNR is given by

$$\text{SQNR} = 10 \log_{10} \frac{\sigma_x^2}{\sigma_e^2} = 10 \log_{10} \frac{\sigma_x^2}{A^2} + 4.77 + 6.02n \quad \text{(dB)}, \tag{12.34}$$

where A takes half the full-scale range, $A = R/2$.

For a signal with a bandwidth of B and a sampling rate f_s, taking (12.20) into account, for the sinewave of amplitude A, the signal SNR and the conversion resolution are related by [6]

$$\text{SQNR} = 1.76 + 6.02n + 10 \log_{10} L \quad \text{(dB)}, \tag{12.35}$$

where $L = \frac{f_s}{2B}$ is the oversampling ratio. Thus, the SNR rises with 6.02 dB per bit of quantizer resolution, and 3.01 dB per doubling of the oversampling ratio, which is equivalent to 0.5 bit quantizer resolution.

When the signal has a normal distribution of amplitude, with zero mean and variance σ_x^2, based on (12.34) and taking oversampling into account, the corresponding result is derived as [5]

$$\text{SQNR} = 10.8 + 6.02n + 10\log_{10} L + 10\log_{10} \frac{\sigma_x^2}{V_{\text{pp}}^2} \quad \text{(dB)}, \tag{12.36}$$

where the peak-to-peak voltage $V_{\text{pp}} = 2A = R$.

Clock jitter is referred to as varying or jittering of the time interval between successive clock edges. This variation in time interval is random with a mean and a variance. Clock jitter increases the system noise, causes uncertainty in the sampled phases, and introduces ISI. Traditional nonspread narrowband modulation is more sensitive to clock jitter and a small amount of clock jitter can lead to a considerable SNR reduction of the ADC output. For the ADC, if the rms of the clock jitter is σ_τ, the signal-to-jitter-noise ratio (SJNR) is limited to

$$\text{SJNR} = \frac{1}{(2\pi f_s \sigma_\tau)^2} = -20\log_{10}(2\pi f_s \sigma_\tau) \quad \text{(dB)}. \tag{12.37}$$

The maximal jitter is determined by the frequency of the input signal and the resolution of the ADC [9]

$$\sigma_{\tau,\text{max}} = \frac{1}{2^n \pi f_{\text{max}}}. \tag{12.38}$$

For a well referenced clock, the SJNR is much higher than the SQNR, and the SNR is determined by the SQNR. For IF sampling in SDRs, an rms sampling clock jitter of 1 picosecond is recommended.

When the noise plus distortion is high, an ADC is characterized by the effective number of bits (ENOB) rather than the actual number of bits, since the conversion is prone to error and there are mistakes in the bits. The ENOB is calculated by [9]

$$\text{ENOB} = \frac{\text{SINAD} - 1.763}{6.02}, \tag{12.39}$$

where SINAD, in decibels, corresponds to that for the case when all the three noise sources are included.

In case of bandpass sampling, the SNR is given by [10]

$$\text{SNR} = \frac{1}{32 f_c^2 \sigma_\tau^2} \frac{f_s}{2B}, \tag{12.40}$$

where f_c is the carrier frequency and B is the signal bandwidth. Thus, the SNR increases with the oversampling ratio.

12.5.2 SFDR and dithering

Dithering can effectively increase the SFDR. It is performed by adding pseudorandom noise to the input analog signal prior to sampling and digitalization. Dithering smears out the energy in the spurious signals over a wide frequency band.

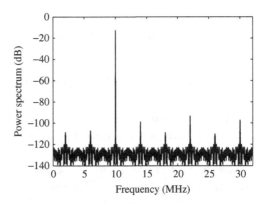

Figure 12.9

Sampling without dithering.

Example 12.4: The spectrum of a sine wave of 10 MHz, when sampled by using a 12-bit ADC at a sample rate of 32 MHz without dithering, is plotted in Fig. 12.9. It is shown that there are many spurious components in the Nyquist band.

Example 12.5: Corresponding to the case of Example 12.4, when the signal is sampled with dithering, the spectrums are shown in Fig. 12.10 for different dithering schemes. In Fig. 12.10a, the added noise signal is AWGN with $\sigma = 2 \times 10^{-4}A$, where A is the amplitude of the sine wave. This can effectively reduce the level of the spurious components by approximately 10 dB. By removing the dither from the reconstructed sine wave, the power of the spurious components is further reduced by approximately 5 dB and is shown in Fig. 12.10b. In Fig. 12.10c, the added noise is AWGN with $\sigma = 2 \times 10^{-3}A$. From Fig. 12.10c, it is seen that the added noise is so large that the SFDR is actually lower than that without dithering. In Fig. 12.10d, the spurious components in Fig. 12.10c are reduced to approximately -110 dBFS by removing the dither. Thus, the addition and removal of dithering can increase the SFDR of ADCs.

Three common types of dithers are noises with Gaussian, rectangular, and triangular pdfs. For the zero-mean Gaussian dither, the variance is recommended to be $\sigma_v^2 = \frac{1}{4}Q^2$, corresponding to $v_{\text{rms}} = Q/2$, or half-LSB. For the rectangular-type dither, the pdf has a width of Q, or 1-LSB, and a height of $1/Q$. This is equivalent to $\sigma_v^2 = \frac{1}{12}Q^2$. This leads to a decrease in SNR of 3 dB for the rectangular dither, and a decrease in SNR of 6 dB for the Gaussian dither [13].

To prevent aliasing arising from the wideband noise signal, the noise signal should be lowpass filtered before being added to the input sine wave. When the power of the added noise is orders of magnitude lower than that of the signal, as shown in Example 12.5, the added noise signal need not be lowpass filtered and this does not incur a problem.

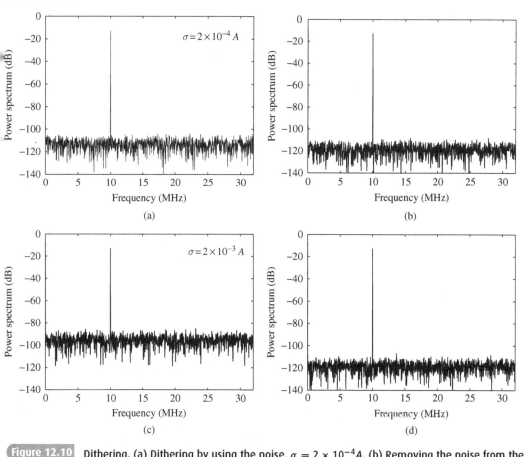

Figure 12.10 Dithering. (a) Dithering by using the noise, $\sigma = 2 \times 10^{-4}A$. (b) Removing the noise from the reconstructed signal in (a). (c) Dithering by using the noise, $\sigma = 2 \times 10^{-3}A$. (d) Removing the noise from the reconstructed signal in (c).

Dithering increases the system noise energy, thus raising the noise floor and leading to a decrease in SNR; this is acceptable if the SFDR is the prime requirement of the system. One method to reduce the noise floor is to subtract the digitalized dithering signal following digital conversion. Although the dithering addition and removal scheme can improve the SFDR, it is not practical since a copy of the dither is required at the receiving end. In many applications, dithering is not necessary, since the inherent analog noise of the mixers and other devices provides some dithering.

12.6 A/D converter circuits

Three common types of ADC circuits are the flash ADC, the successive approximation register (SAR) ADC, and the sigma-delta ADC. The SAR ADC is most widely used today, and the sigma-delta ADC is gaining more attention.

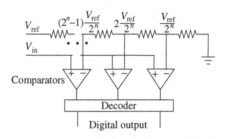

Figure 12.11   **Structure of the flash ADC.**

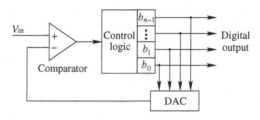

Figure 12.12   **Structure of the successive-approximation ADC.**

12.6.1 Flash A/D converters

The structure of a flash (or parallel) ADC is shown in Fig. 12.11, where all the resistors have the same resistance and n is the wordlength of the ADC. The decoder is used to transform the output of the comparators into parallel PCM output.

Flash ADCs have the fastest conversion time, and thus are suitable for applications with high sampling rates, such as in SDRs and UWB applications. These ADCs are constrained by the high complexity, and hence are typically used in the case of a wordlength of 10 bits or less. However, they are expensive, and have a high power consumption. The resistors in the voltage divider chain must have high accuracy. The number of comparators and complexity of the decoder are of the order $O(2^n)$.

12.6.2 Successive-approximation register A/D converters

The SAR ADC consists of only one analog comparator, and utilizes a control logic, a register of n bits, and a DAC of n bits, as illustrated in Fig. 12.12. The method uses the register $b_{n-1} \cdots b_2 b_1$ to store the converted digit output; the output is further converted into an analog value by the DAC and fed back to the analog comparator. The comparator compares the analog input and the DAC output, and the error signal is used to tune the register. Initially, all the b bits are set to zero in a SAR, then starting from the MSB b_{n-1}, the bit is turned on sequentially and the difference between the analog input x and the analog signal converted from the DAC, x_Q, is tested to decide whether it is left on $(x \geq x_Q)$

or turned off ($x < x_Q$). The process is repeated for the second MSB and so on, until all the n bits in the register are tested. This method requires n tests, and the result in the SAR $b_{n-1} \ldots b_1 b_0$ is the digital output.

SAR ADCs are very fast, although not as fast as flash ADCs. They have low complexity and are inexpensive. They are the most widely used ADCs for sampling rates up to some MHz.

12.6.3 Sigma-delta A/D converters

Unlike the Flash and SAR ADCs, the sigma-delta (Σ-Δ) ADC uses a 1-bit or multiple-bit ADC with noise shaping to achieve high resolutions. The single bit codes whether the current sample has a higher or lower voltage than the previous sample. High resolution is achieved by oversampling. Conversion from this oversampled 1- or more-bit stream into a slower, high-resolution stream is performed by using digital decimation filters. A sigma-delta ADC is usually split into a sigma-delta modulator and a decimator. The sigma-delta modulator is used as the A/D interface, while the latter, sometimes called a *serial-to-parallel (S/P) converter*, is a filter followed by a down-sampler.

Sigma-delta modulator

The sigma-delta modulator is a combination of delta modulator and demodulator. A generic sigma-delta modulator is shown in Fig. 12.13, where $A(z)$ is the transfer function of the delta demodulator and $B(z)$ the transfer function of the loop-filter of the delta modulator. Usually, $A(z) = B(z)$ for DM, and this is also widely used for the sigma-delta modulator, although choosing $A(z) \neq B(z)$ can lead to better performance [5].

The noise transfer function $H_e(z)$ and the signal transfer function $H_x(z)$ are, respectively, given by

$$H_e(z) = \frac{1}{1 - B(z)}, \quad H_x(z) = \frac{A(z)}{1 - B(z)}. \tag{12.41}$$

For a classical second-order sigma-delta modulator using double-integration, as shown in Fig. 12.14, it can be shown that [1, 5]

$$H_e(z) = \left(1 - z^{-1}\right)^2, \quad H_x(z) = z^{-1} \tag{12.42}$$

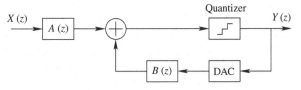

Figure 12.13   Architecture of the generic sigma-delta modulator.

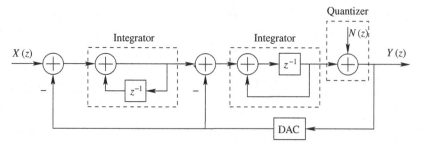

Figure 12.14 The classical second-order sigma-delta modulator. DAC is assumed to be ideal with unit gain.

and

$$Y(z) = X(z)z^{-1} + E(z)(1 - z^{-1})^2. \tag{12.43}$$

That is, the input signal $X(z)$ is filtered by an all-pass filter, while the quantization noise $E(z)$ by a highpass filter.

The SNR is derived as [5]

$$\text{SNR} = 10\log_{10}\frac{\sigma_x^2}{Q^2/12} + 10\log_{10}\frac{\pi}{\int_0^{\pi/L} 16\sin^4\frac{\omega}{2}d\omega}, \tag{12.44}$$

where the first term stands for the SQNR of the quantizer without noise shaping and the second term gives the contribution to SNR within the Lth fraction of the total frequency band caused by noise shaping, L being the oversampling ratio. An approximation is obtained by applying the 3rd-order Taylor series expansion to the sine function [5]

$$\text{SNR} \approx 20\log_{10}\frac{\sigma_x}{V_{\text{pp}}} + 6.02n + 50\log_{10}L - 2.1 \quad (\text{dB}), \quad \text{for } L \geq 4. \tag{12.45}$$

An order-M sigma-delta modulator is a straightforward extension of the first-order sigma-delta modulator by setting $H_x(z) = z^{-1}$ and $H_e(z) = (1 - z^{-1})^M$. The result given by (12.45) can be generalized to the Mth-order with noise shaping [1, 5]

$$\text{SNR} \approx 20\log_{10}\frac{\sigma_x}{V_{\text{pp}}} + 6.02n + 10(2M + 1)\log_{10}L + 10.8$$

$$+ 10\log_{10}\frac{2M + 1}{\pi^{2M}} \quad (\text{dB}), \quad \text{for } L \geq 4. \tag{12.46}$$

Thus, for every doubling of the oversampling ratio L, an extra $(6M + 3)$ dB of SNR, or an extra $(M + 1/2)$ bits of resolution, is obtained. The SNR also increases by 6.02 dB for every additional bit of quantization resolution.

Comparison of (12.36) and (12.45) shows that a higher resolution is achieved with sigma-delta modulators, leading to more efficient and lower-power implementations. This makes the sigma-delta ADC most desirable for mobile communications. The sigma-delta ADC has the highest resolution, and is the only choice when a resolution of more than 16 bits is desired [9]. Sigma-delta ADCs are widely used for audio ADCs. A reconfigurable sigma-delta ADC for a GSM/WCDMA/Wi-Fi/WiMAX zero-IF receiver has been presented in [17].

Discussion

The above result is for lowpass sigma-delta modulators. The result can be transformed to a bandpass sigma-delta modulator with equivalent noise shaping characteristic, and thus a bandpass sigma-delta ADC can be obtained from a lowpass prototype by transformation [1, 12]. The bandpass sigma-delta ADC operates in a manner similar to that of the corresponding lowpass one. In a communication system, A/D conversion at either the IF or RF stage yields a more robust system with improved IF strip testability, and is necessary for SDRs.

From (12.46), the effectiveness of increasing the order of noise shaping, M, is significantly diminished as the oversampling ratio L is decreased. However, the effectiveness of the quantization resolution n is independent of L. Multi-bit modulators are particularly attractive. While multi-bit quantization is tolerant of nonlinearity in the quantizer's ADC due to noise shaping, it imposes stringent linearity requirements on the quantizer's DAC [12]. The error due to nonlinearity in the DAC enters the modulator at its input. Thus, the linearity and resolution of the demodulator are limited by the multi-bit DAC. Several methods for reducing the dependence of multi-bit modulators on the linearity of the DAC are described in [12]. The multi-bit modulator has a more complex analog circuitry than that of the 1-bit modulator, and has to use matched components. Noise-shaping modulators using multi-bit quantizers can be used for both ADCs and DACs.

Instead of the multi-bit quantizer, the 1-bit quantizer can be used. Although the 1-bit quantizer is inherently linear, its large quantization noise can cause instability for $M > 2$. To maintain stability for $M > 2$, modifications in the architecture are necessary; this, however, generally reduces the SNR. The full dynamic range given by (12.46) can be achieved for $M > 2$ by cascading multiple first-order modulators with 1-bit quantizers. This requires precise quantization noise shaping in the first stage, especially for a low oversampling ratio. In practice, the number of cascaded first-order stages is limited to $M = 3$. The 1-bit noise-shaping modulator is popular for use in CMOS integrated circuit data converters, since the use of a 1-bit DAC does not require precision component matching. At a given oversampling ratio, the resolution that a 1-bit noise-shaping modulator can achieve is limited. Also, due to the substantial quantization noise, the design of analog output filters for oversampled ADCs is difficult [12].

Although sigma-delta modulation is inferior to PCM in terms of distortion for a given bit budget n, it is robust to analog implementation. Using an imprecise quantizer, the error made by a PCM scheme is bounded below by a positive constant, whereas the error made by a kth-order sigma-delta modulation scheme still decays as n^{-k}, and thus can be made arbitrarily small irrespective of the quantizer precision [2]. Due to the relaxed requirements in terms of the precision of analog circuitry and digital decimation filtering, the sigma-delta ADC has become a good candidate for fabrication using CMOS technology. The task of decimation can be handled entirely by a DSP. Most IC implementations of the sigma-delta ADC use switched-capacitor circuits.

As an alternative to PCM for A/D conversion, beta-encoders with error correction have been introduced in [2, 3]. An n-bit beta-encoder quantizes a real number by computing one of its n-bit truncated β-expansions. These encoders have almost optimal rate-distortion properties similar to that of PCM; also, like sigma-delta modulators, they exploit the redundancy of beta-expansions and thus they are robust with respect to quantizer imperfections.

Since high SNR is achieved only at the center frequency of noise shaping filters, digital-ization of multiple narrowband channels is not possible for the conventional sigma-delta modulators. This restricts their application in BSs where multiple channels are processed at the same time. One solution is to design $B(z)$ as a parameterizable multi-notch function [12]. For more on the theory and design of the delta-sigma data converter, the reader is referred to [12].

12.7 D/A converter circuits

The DAC is commonly implemented in the form of a multiplying DAC. The output of such a converter is the product of an array of reference currents or voltages and the input digital code. The input digital code is the PCM code in parallel format. The multiplying DAC has a fast settling time. A generic current source multiplying DAC is shown in Fig. 12.15.

The different current sources can be realized by using a common input reference voltage V_{ref} but different resistors so that

$$I_i = 2^i I_0 = \frac{V_{\text{ref}}}{R_i}, \tag{12.47}$$

where R_i is the resistance of the resistor in the ith branch. This leads to $R_i = R_0/2^i$. Such a circuit has some problems for large word length. For SDR applications, typically $n = 14$. This will lead to $R_{13} = R_0/8192$. This makes the manufacture of the resistors very difficult, in view of the precision required.

The R-$2R$ ladder structure uses a common reference voltage, but uses current division. All the resistors are either R or $2R$. It is more suitable for manufacturing and integration on a single chip. The charge redistribution structure is quite common in CMOS implemen-tation of multiplying DACs, in the same way as in switched-capacitor filters. The R-$2R$ ladder and charge redistribution structures are discussed in more detail in [18].

Counting DACs and bitstream DACs are two other classes of DACs. Like in the ADC, a S/H device is required to counteract problems that occur during the transition from one

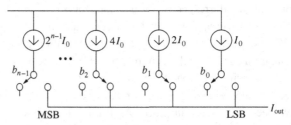

Figure 12.15   A generic current source multiplying DAC. The digital code is in the form $b_{n-1} \cdots b_1 b_0$.

sample to another, such as glitches and ringing. The zero-order hold, which holds the output constant until the DAC receives the next sample, is commonly used. Sample-and-hold is typically integrated into the DAC.

12.8 A/D and D/A converters for software-defined radios

For SDRs, the dynamic linearity of an ADC is an important parameter. This parameter is defined by harmonic performance and level of intermodulation products. The SNR is dependent on the number of bits that a sample is represented by. The SNR is improved by the ratio between the bandwidth of a single modulated carrier and that of the Nyquist zone. This ratio is improved by digital filtering following A/D conversion. SFDR is a parameter that is more important than SNR or SINAD for evaluation of the performance of ADCs or DACs in SDRs.

For mobile communication systems, the hostile RF environment prevents sophisticated modulation schemes such as 512QAM. Thus mobile systems cannot benefit much from a sampling resolution above 16 bits. This is a reason for many DSPs to perform 16-bit fixed-point arithmetic.

The successive-approximation method is a very common A/D conversion technique. Due to the sequential nature, the successive approximation ADC is usually implemented for low sampling rates. This is due to the fact that an n-bit ADC uses n successive iterations and the clock rate must be n-times the sampling rate. This limits its application for ADCs used in wideband SDRs. For higher rates, the flash ADC, which determines all bits simultaneously, can be used. For n-bit conversion, $2^n - 1$ comparators are required, thus the flash ADC is limited to low-resolution applications.

Higher resolution can be achieved by cascading two or more lower-resolution successive approximation or flash ADCs. Such ADCs are known as *pipeline ADCs*. For example, ADI provides the AD6644 ADC, which is a 14-bit, 65-Msamples/s device using the multipass conversion technique. The multipass technique is based on the successive-approximation technique, but is much more complex. It uses three cascaded successive-approximation ADCs, and is actually a divide-and-conquer implementation.

Statistics drawn from 914 commercially available ADCs are given in [9]. These ADCs are grouped into seven types of structures: flash, half-flash, folding, SAR, pipelined, sigma-delta, and unknown. The numbers of bits and the sampling rates are plotted in Fig. 12.16. The pipelined and unknown structures have the best overall performance. The SAR ADC has widely ranging sampling rates, and is popular for its range of speeds and resolutions, as well as low cost and power dissipation. Sigma-delta ADCs have the highest resolution with relatively low sampling rates from K samples/s to M samples/s, while flash ADCs have the highest sampling rates up to G samples/s but with a resolution limited to no more than 8 bits. Between these two structures are the unknown structures compromising speed or resolution.

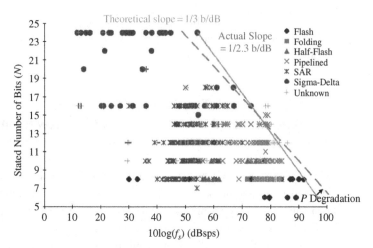

Figure 12.16 Statistics of 914 commercially available ADCs: The number of bits versus sampling rate. $P = 2^n f_s$ is a figure of merit. From [9], Fig.1. ©IEEE 2005.

Today, ADCs available for SDR span from 12–16 bits with SFDR around 100 dB, and sampling frequency ranging from 60–100 MHz. DACs with 12–14 bits, SFDR of 95 dB, sampling rate up to 400 MHz are now available. Major ADC and DAC vendors for SDR are ADI, National Semiconductor, TI, and Intersil. The UWB technique continues pushing the boundaries of ADCs. The Maxim MAX19693 is a 12-bit, 4.0-Gsamples/s DAC, and consumes 1180-mW at 4.0 Gsamples/s.[1] Atmel's AT84AS008 is a 10-bit, 2.2-Gsamples/s ADC.[2] The LTC2209 of the Linear Technology Corp is a 16-bit, 160-Msamples/s ADC having an SFDR of 100 dBc at baseband, consuming 1450 mW.[3] TI's ADS548x is a family of 16-bit, 80- to 200-Msamples/s ADCs, with an SFDR of 87 to 98 dBc at baseband and a power consumption of up to 2160 mW.[4]

Problems

12.1 Calculate the SQNR for a signal quantized using 10-bit PCM (uniformly distributed quantization levels). The input signal has the following distribution: (a) $p_x = \frac{1}{V_{fs}}$, for $-\frac{V_{fs}}{2} \le x \le \frac{V_{fs}}{2}$, and 0 elsewhere; (b) $p_x = \frac{1}{V_{fs}}$, for $-\frac{V_{fs}}{3} \le x \le \frac{2V_{fs}}{3}$, and 0 elsewhere.

12.2 For ADC selection, if the full-scale range of the converter is 10 volts and the rms quantization error is less than 1 millivolt, how many bits are required for the converter? What is the dynamic range of the converter?

[1] http://www.maxim-ic.com/
[2] http://www.atmel.com/
[3] http://www.linear.com/
[4] http://www.ti.com/

12.3 Verify that the mean and variance of the random variable $v = 0.5R(u - 0.5)$, where R is the full-scale range and u is a uniform random number in $[0, 1]$, are $m_v = 0$ and $\sigma_v^2 = R^2/48$.

12.4 The triangular dither pdf is given by

$$p(v) = \begin{cases} \frac{Q - |v|}{Q^2}, & \text{if } -Q \leq v \leq Q \\ 0, & \text{otherwise} \end{cases}$$

and $\sigma_v^2 = \frac{1}{6}Q^2$, where Q is the quantization resolution. The triangular pdf is the convolution of two rectangular ones, and thus v, a random variable with the triangular pdf, can be obtained as the sum of two independent rectangular random numbers, $v_i = Q(u_i - 0.5)$, $u_i = [0, 1]$, $i = 1, 2$. Consider the following dither sinusoid:

$$y(n) = x(n) + v(n) = A\cos(2\pi f_0 n) + v(n).$$

Let $A = 0.5Q$, $f_0 = 0.0025$ cycles/sample, $f_s = 40$ kHz. If the signal is quantized using a 4-bit ADC, and the full range $R = 16$ volts, plot the spectrums of the quantized undithered sinusoid $x_Q(n)$ and the quantized dithered sinusoid $y_Q(n)$.

12.5 Generate an array of 10,000 random numbers with a symmetrical triangular pdf defined on [-1, 1].

References

[1] P. M. Azizi, H. V. Sorensen & J. Van der Spiegel, An overview of sigma-delta converters. *IEEE Signal Process. Mag.*, **13**:1 (1996), 61–84.

[2] I. Daubechies, R. A. DeVore, C. S. Gunturk & V. A. Vaishampayan, A/D conversion with imperfect quantizers. *IEEE Trans. Inf. Theory*, **52**:3 (2006), 874–885.

[3] I. Daubechies & O. Yilmaz, Robust and practical analog-to-digital conversion with exponential precision. *IEEE Trans. Inf. Theory*, **52**:8 (2006), 3533–3545.

[4] K.-L. Du & M. N. S. Swamy, *Neural Networks in a Softcomputing Framework* (London: Springer, 2006).

[5] T. Hentschel & G. Fettweis, Software radio receivers. In F. Swarts, P. van Rooyen, I. Oppermann & M. P. Lotter, eds., *CDMA Techniques for Third Generation Mobile Systems* (Norwell, MA: Kluwer, 1999), pp. 257–283.

[6] A. Jamin, P. Mahonen & Z. Shelby, Software radio implementability of wireless LANs. In E. Del Re, ed., *Software Radio: Technologies and Services* (London: Springer, 2001).

[7] A. J. Jerri, The Shannon sampling theorem – its various extensions and applications: a tutorial review. *Proc. IEEE*, **65**:11 (1977), 1565–1596.

[8] W. Kester, ed., *Mixed-signal and DSP Design Techniques (Analog Devices Inc.)* (Oxford, UK: Newnes/Elsevier, 2003).

[9] B. Le, T. W. Rondeau, J. H. Reed & C. W. Bostian, Analog-to-digital converters. *IEEE Signal Process. Mag.*, **22**:6 (2005), 69–77.

[10] J.-F. Luy, T. Mueller, T. Mack & A. Terzis, Configurable RF receiver architectures. *IEEE Microwave Mag.*, **5**:1 (2004), 75–82.

[11] J. Max, Quantization for minimum distortion. *IRE Trans. Inf. Theory.*, **6**:2 (1960), 7–12.

[12] S. R. Norsworthy, R. Schreier & G. C. Temes, eds., *Delta-Sigma Data Converters: Theory, Design, and Simulation* (New York: IEEE Press, 1997).

[13] S. J. Orfanidis, *Introduction to Signal Processing* (Englewood Cliffs, NJ: Prentice Hall, 1995).

[14] E. I. Plotkin, M. N. S. Swamy & Y. Yoganadam, A novel iterative method for the reconstruction of signals from nonuniformly spaced samples. *Signal Process.*, **37**:2 (1994), 203–213.

[15] J. G. Proakis & D. G. Manolakis, *Digital Signal Processing: Principle, Algorithms, and Applications*, 4th edn (Upper Saddle River, NJ: Pearson Prentice Hall, 2007).

[16] J. Romero, E. I. Plotkin & M. N. S. Swamy, Reproducing kernels and the use of root loci of specific functions in the recovery of signals from nonuniform samples. *Signal Process.*, **49**:1 (1996), 11–23.

[17] A. Rusu, D. R. D. L. Gonzalez & M. Ismail, Reconfigurable ADCs enable smart radios for 4G wireless connectivity. *IEEE Circ. & Devices Mag.*, **22**:3 (2006), 6–11.

[18] D. Stranneby & W. Walker, *Digital Signal Processing and Applications*, 2nd edn (London: Elsevier, 2004).

[19] P. P. Vaidyanathan, Multirate digital filters, filter banks, polyphase networks, and applications: a tutorial. *Proc. IEEE*, **78**:10 (1990), 56–93.

[20] R. G. Vaughan, N. L. Scott & D. R. White, The theory of bandpass sampling. *IEEE Trans. Signal Process.*, **39**:9 (1991), 1973–1984.

13 Signals and signal processing

13.1 Basic transforms

We introduce in this section the Fourier, Laplace, and z-transforms, which are closely related. The Fourier transform transfers a time-domain signal into its frequency-domain version, and it is a fundamental transform used in signal processing. The Laplace transform is useful for characterizing an analog system, and the z-transform is useful for digital signal processing.

13.1.1 Fourier transform

It is well-known that a periodic analog piece-wise continuous signal $x(t)$ can be represented by a Fourier series

$$x(t) = \sum_{n=-\infty}^{\infty} d_n e^{jn\omega t}, \tag{13.1}$$

where

$$d_n = \frac{1}{T} \int_{-T/2}^{T/2} x(t) e^{-jn\omega t} dt, \tag{13.2}$$

T being the period of the signal, and $\omega = 2\pi/T$ is the fundamental (first harmonic) angular frequency.

On the other hand, in communication systems, we often have to deal with non-periodic signals, and such signals can be characterized by the Fourier transform. The Fourier transform $X(f)$ of a time-domain signal $x(t)$ is defined by

$$X(f) = \int_{-\infty}^{\infty} x(t) e^{-j2\pi ft} dt \tag{13.3}$$

and the inverse Fourier transform by

$$x(t) = \int_{-\infty}^{\infty} X(f) e^{j2\pi ft} df. \tag{13.4}$$

Often the pair can be written in terms of the angular frequency $\omega = 2\pi f$ as

$$X(\omega) = \int_{-\infty}^{\infty} x(t) e^{-j\omega t} dt, \tag{13.5}$$

$$x(t) = \frac{1}{2\pi} \int_{-\infty}^{\infty} X(\omega)e^{j\omega t}d\omega. \qquad (13.6)$$

A sufficient condition for $x(t)$ to have a Fourier transform is that $x(t)$ is absolutely integrable over (∞, ∞). It should be emphasized that this is only a sufficient, but not a necessary condition. An example is $x(t) = \sin \omega_0 t$ for which $X(\omega) = j\pi[\delta(\omega + \omega_0) - \delta(\omega - \omega_0)]$.

For an LTI system, the input and output are related by

$$y(t) = h(t) * x(t), \qquad (13.7)$$

where $h(t)$ is the impulse response of the LTI system, and $*$ represents the convolution operator. Applying the Fourier transform yields

$$Y(f) = H(f)X(f), \qquad (13.8)$$

where $H(f)$, $Y(f)$, and $X(f)$ are the Fourier transforms of $h(t)$, $y(t)$, and $x(t)$, respectively.

13.1.2 Laplace transform

The unilateral Laplace transform of a time-domain signal $x(t)$ is defined by

$$X(s) = \int_{0}^{\infty} x(t)e^{-st}dt, \qquad (13.9)$$

where $s = \sigma + j\omega$ is a complex value. The unilateral Laplace transform is typically used for causal signals. When the lower integration limit is $-\infty$, we get the bilateral Laplace transform. The Laplace transform reduces to the Fourier transform for $\sigma = 0$; it can transform signals that have no Fourier transform.

The inverse Laplace transform is defined as

$$x(t) = \frac{1}{2\pi} e^{\sigma t} \int_{-\infty}^{\infty} X(\sigma + j\omega)e^{j\omega t}d\omega. \qquad (13.10)$$

Computation using this definition is very complex. Laplace transforms for many functions are given in most textbooks on signals and systems.

For the filter given in time-domain form (13.7), the Laplace transform gives

$$Y(s) = H(s)X(s). \qquad (13.11)$$

The s-domain is usually used to characterize continuous-time systems.

The Laplace transform is one of the most powerful tools for analysis and design of linear systems and networks. The Laplace transform unravels the structure of a signal in terms of its poles and zeros. For a stable system, the transfer function $H(s)$ has all its poles located in the left half of the s-plane.

Performing the inverse Laplace transform

Given a real rational function of the form

$$F(s) = \frac{N(s)}{D(s)} = P(s) + \frac{N_1(s)}{D(s)}, \qquad (13.12)$$

where $N(s)$, $D(s)$, $P(s)$, and $N_1(s)$ are polynomials with real coefficients, and the second term in the last equation is a proper fraction, with the degree of the numerator smaller than the degree of the denominator. The inverse Laplace transform can be easily obtained: the inverse Laplace transform of $P(s)$ can be obtained by using

$$\mathcal{L}^{-1}[1] = \delta(t), \quad \mathcal{L}^{-1}[s^n] = \frac{d^n\delta(t)}{dt^n}. \tag{13.13}$$

The proper fraction $\frac{N_1(s)}{D(s)}$ can be expressed in partial fractions

$$\frac{N_1(s)}{D(s)} = \sum_i \frac{A_i}{s - p_i} + \sum_{j,j\neq i} \sum_{k=1}^{K_j} \frac{B_{j,k}}{(s - p_j)^k}, \tag{13.14}$$

where A_i's and $B_{j,k}$'s are constants, $K_j \geq 2$ for all j. Thus, the inverse transform of $\frac{N_1(s)}{D(s)}$ can be obtained by using

$$\mathcal{L}^{-1}\left(\frac{1}{s - p_i}\right) = e^{p_i t}, \quad \mathcal{L}^{-1}\left(\frac{1}{(s - p_i)^k}\right) = \frac{t^{k-1}}{(k - 1)!}e^{p_i t}. \tag{13.15}$$

For signal design, we usually assume there are no multiple-order poles, and thus the second sum-term in (13.14) does not occur.

In order to implement an analog signal processing system, many element building blocks, such as adder, multiplier, differentiator, and integrator, must be used. These building blocks can be active circuits employing the basic operational amplifier. Filters implemented using active circuits are known as *active filters*. For filter design, there are also many passive analog lowpass prototypes making use of LC ladder circuits, as introduced in Chapter 11.

13.1.3 z-transform

The z-transform is the discrete-time counterpart of the Laplace transform, and it is a generalization of the Fourier transform of a sampled signal. The two-sided z-transform is defined as

$$X(z) = \sum_{n=-\infty}^{\infty} x(n)z^{-n}, \tag{13.16}$$

where $z = e^s = re^{j\omega}$, $r = e^\sigma$, is a complex variable. Similarly, the one-sided z-transform of the sequence is defined by

$$X(z) = \sum_{n=0}^{\infty} x(n)z^{-n}. \tag{13.17}$$

For causal signals, $x(n) = 0$ when $n < 0$, the two- and one-sided z-transforms are identical.

When $z = e^{-st}$, the z-transform and the Laplace transform of the causal sampled signal are identical. The mapping between the s-plane and the z-plane is shown in Fig. 13.1. The

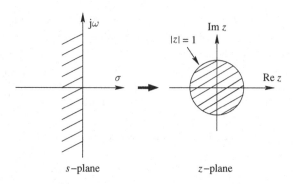

Figure 13.1   **Mapping between the s-plane and the z-plane.**

left and right halves of the complex s-plane are mapped onto the inside and outside of the unit circle $|z| = 1$ in the z-plane, respectively. The $j\omega$-axis in the s-plane maps onto the unit circle in the z-plane, since $\sigma = 0$ corresponds to the unit circle in the z-plane: $z|_{\sigma=0} = e^{j\omega t}$.

The z-transform of a signal exists under certain conditions. The extension of the Fourier transform from the frequency axis to the complex space enables the Laplace transform and the z-transform for transient signals analysis, while the Fourier transform is suitable only for steady-state or time-invariant signal analysis. The z-transforms of some common sequences are given in many textbooks, e.g., [39].

The z-transform is a linear transform just as the Fourier and Laplace transforms are. Hence, all the properties of a linear transform are applicable to the z-transform including the convolution property. For an LTI system, the input and output sequences are related by

$$y(n) = \sum_{k=-\infty}^{\infty} x(k)h(n-k) = x(n) * h(n), \tag{13.18}$$

where $h(n)$ is the impulse response of the system. The convolution property leads to the relation

$$Y(z) = H(z)Z(z). \tag{13.19}$$

$H(z)$ is known as the *transfer function* of the system.

Causality and convergence

When the input $x(n)$ exists only for $n \geq 0$, the signal is causal (physically realizable). A causal system produces an output response at time t_1 for an input at t_0, where $t_0 \leq t_1$. When $x(n)$ exists for both $n \geq 0$ and $n < 0$, the signal is double-sided, and thus non-causal. LTI systems can be so classified according to the causality of $h(n)$. Noncausal systems are counter-intuitive, they violate the sense of causality, and thus are physically not implementable.

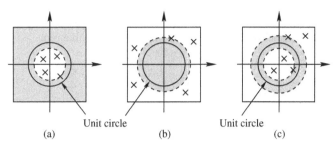

Figure 13.2 Stable ROCs for (a) causal, (b) anticausal, and (c) mixed signals. The x's represent poles. Note that the ROCs cannot contain any poles.

The z-transform has a region of convergence (ROC) so that the series in (13.16) converges. When the z-transform is a rational function, it is broken into partial fraction expansion

$$X(z) = \frac{N(z^{-1})}{D(z^{-1})} = \sum_m A_{-m} z^{-m} + \sum_i \frac{A_i}{1 - p_i z^{-1}} + \sum_{j,j \neq i} \sum_{k=1}^{K_j} \frac{B_{j,k}}{(1 - p_j z^{-1})^k}, \quad (13.20)$$

where A_{-m}, A_i, and B_j are expansion coefficients, $K_j \geq 2$ for all j. The first sum-term corresponds to time delays, and the index m can be a positive integer, a negative integer, or zero.

Causal signals have ROCs that are outside the maximum pole circle, $|z| > \max_i |p_i|$, while anticausal signals have an ROC that is inside the minimum pole circles, $|z| < \min_i |p_i|$. Mixed signals have an ROC that is an annular region between two circles: the poles that lie inside the inner circle contributing causally and the poles that lie outside the outer circles contributing anticausally.

A necessary and sufficient condition for the stability of a signal is that the ROC of the corresponding z-transform contains the unit circle [54]. This is illustrated in Fig. 13.2. For a stable causal system all its poles must lie strictly inside the unit circle in the z-plane, that is, $1 > \max_i |p_i|$. For a stable mixed signal, those poles inside the unit circle contribute causally and those outside the unit circle contribute anticausally. When the poles are on the unit circle, these signals are marginally stable: They neither diverge nor converge to zero for large n, but rather remain bounded.

Inverse z-transform

The inverse z-transform is defined as

$$x(n) = \frac{1}{2\pi j} \oint_\Gamma X(z) z^{n-1} dz, \quad (13.21)$$

where Γ is a closed contour lying within the region of convergence.

When $X(z)$ is a ratio of polynomials, $N\left(z^{-1}\right)/D\left(z^{-1}\right)$, the inverse z-transform can be very efficiently computed by using the residue theorem

$$x(n) = \frac{1}{2\pi j}\oint_{\Gamma} X(z)z^{n-1}dz = \sum_{k=1}^{K}\text{res}_{z=p_k}\left(X(z)z^{n-1}\right), \qquad (13.22)$$

where

$$X(z)z^{n-1} = \frac{N(z)}{\prod_{k=1}^{K}(z-p_k)^{m_k}}, \qquad (13.23)$$

m_k being the multiplicity of p_k, and the residue of $X(z)$ with respect to p_k is given by

$$\text{res}_{z=p_k}(X(z)) = \frac{1}{(m_k-1)!}\frac{d^{m_k-1}}{dz^{m_k-1}}\left[(z-p_k)^{m_k}X(z)\right]\Bigg|_{z=p_k}. \qquad (13.24)$$

The inverse z-transform of $X(z)$ may be not unique. It can be unique only by specifying the corresponding ROC.

When $X(z)$ is a ratio of two polynomials in z^{-1}, we usually do not use the above residue theorem to solve the inverse z-transform. Instead, it is usually broken into partial fraction expansion of the form (13.20). The inverse transform can be performed according to

$$z^{-m} \longrightarrow \delta(n-m), \quad \text{for } m \in Z, \qquad (13.25)$$

$$\frac{1}{1-p_iz^{-1}} \longrightarrow \begin{cases} p_i^n u(n), & |z| > |p_i| \\ -p_i^n u(-n-1), & |z| < |p_i| \end{cases}. \qquad (13.26)$$

If there are multiple-order poles, one has to resort to some tables for finding the transforms. The MATLAB function `residuez` performs the partial fraction expansion.

13.2 Discrete-time Fourier transform

The discrete-time Fourier transform (DTFT) is obtained by evaluating the z-transform on the unit circle, $z = e^{j\omega}$. For a discrete signal $x(n)$, its frequency response is given by

$$X(\omega) = \sum_{n=-\infty}^{\infty} x(n)e^{-j\omega n}. \qquad (13.27)$$

Note that here $\omega = 2\pi f/f_s$, in radians/sample, where f_s is the sampling frequency. The spectrum $X(\omega)$ of a signal only exists for a stable signal, that is, a signal whose ROC of z-transform contains the unit circle.

The inverse DTFT recovers the discrete signal $x(n)$ from its spectrum $X(\omega)$ over the Nyquist interval

$$x(n) = \frac{1}{2\pi}\int_{-\pi}^{\pi} X(\omega)e^{j\omega n}d\omega. \qquad (13.28)$$

Parseval's theorem relates the total energy of a sequence to its spectrum

$$\sum_{n=-\infty}^{\infty} |x(n)|^2 = \frac{1}{2\pi} \int_{-\pi}^{\pi} |X(\omega)|^2 d\omega. \tag{13.29}$$

The shape of a spectrum $X(\omega)$ is influenced by the pole/zero pattern of the corresponding z-transform $X(z)$. When a pole is close to the unit circle, the denominator will have a small distance at the frequency corresponding to the pole, leading a peak there. Similarly, when a zero is close to the unit circle, it causes a dip in the frequency spectrum at a certain frequency. By suitably placing the poles and zeros, one can design some simple filters such as resonator, notch, or comb filters. This has been discussed in [54]. Notch or comb filters are used for cancellation of periodic interference such as harmonics, while peaking comb filters are used for enhancement of periodic signals in noise.

The discrete Fourier transform (DFT) and its fast implementation, the fast Fourier transform (FFT) are most important for digital signal processing. They are mainly used for calculation of the frequency spectrum of a signal, efficient implementation of convolution, and speech or image coding. The discrete cosine transform (DCT), as a variant of DFT, is widely used for image coding. Note that the DTFT is defined for aperiodic signals.

13.2.1 Windowing

Implementation of DTFT must employ time-windowing, since the number of samples may be infinite, $x(nT)$, $-\infty < n < \infty$. We need to keep only a finite number of samples, $x(nT)$, $0 \le n \le L - 1$. Thus, DTFT is actually performed on the windowed signal

$$x_L(n) = x(n)w(n). \tag{13.30}$$

A rectangular window of length L is defined by

$$w(n) = \begin{cases} 1, & 0 \le n \le L - 1 \\ 0, & \text{otherwise} \end{cases}. \tag{13.31}$$

The DTFT of the length-L signal at a single frequency is thus given by

$$X(\omega) = \sum_{n=0}^{L-1} x(n)e^{-j\omega n} = \sum_{n=-\infty}^{\infty} x_L(n)e^{-j\omega n} = X_L(\omega). \tag{13.32}$$

Windowing reduces the frequency resolution of the computed spectrum, and the resolvable frequency resolution is given by $\Delta f = \frac{1}{LT}$. Also, spurious high-frequency components are introduced into the spectrum due to the hard clipping of the signal $x(n)$ by the window, leading to frequency leakage. This is because the rectangular window has the spectrum of a sinc function, which has many undesirable sidelobes. The height of the sidelobes is 13 dB lower than the mainlobe. The mainlobe width determines the frequency resolution limits of the windowed signal spectrum

$$\Delta f \ge \Delta f_w = \frac{1}{LT}. \tag{13.33}$$

From this a desirable data length L can be calculated according to the frequency resolution.

The rectangular window cuts the data to zero sharply, and this leads to significant frequency leakage. By using windows that cut off to zero gradually, the sidelobes can be effectively suppressed. The widely used Hamming window can suppress the sidelobe to at least 40 dB lower than the mainlobe. The Hamming window is given by

$$w(n) = \begin{cases} 0.54 - 0.46 \cos\left(\frac{2\pi n}{L-1}\right), & 0 \le n \le L-1 \\ 0, & \text{otherwise} \end{cases}. \tag{13.34}$$

The Kaiser window has a controllable suppression of the sidelobes by adjusting the shape parameter α. The Kaiser window is defined by [40]

$$w(n) = \frac{I_0\left(\alpha\sqrt{1-(n-M)^2/M^2}\right)}{I_0(\alpha)}, \quad 0 \le n \le L-1, \tag{13.35}$$

where $L = 2M + 1$, and $I_0(x)$ is the modified Bessel function of the first kind and zeroth order. It is also symmetric about its middle $n = M$. When $\alpha = 0$, it reduces to the rectangular window.

Example 13.1: The rectangular, Hamming, and Kaiser windows are plotted in Fig. 13.3. It is seen that the Kaiser window approximates the Hamming window for $\alpha = 5$ and reduces to the rectangular window for $\alpha = 0$.

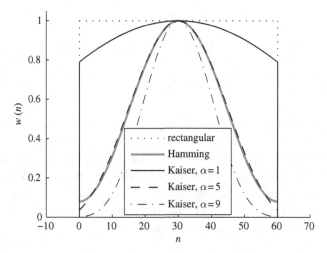

Figure 13.3   **Illustration of different windows.**

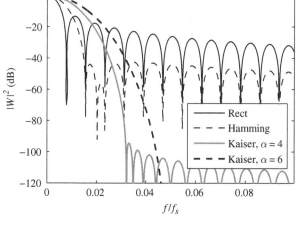

Figure 13.4 Illustration of different windows in the frequency domain, for $L = 128$.

The Fourier transform of these windows are given as

$$W_{\text{rec}}(f) = e^{-j\pi fLT} \frac{\sin(\pi fLT)}{\pi f}, \qquad (13.36)$$

$$W_{\text{Hamming}}(f) = \frac{0.04}{j\pi f} \left(1 - e^{-j2\pi ft}\right) + 0.46e^{-j\pi ft} \frac{\sin(\pi fLT)}{\pi f} \frac{1}{1 - (fLT)^2}. \qquad (13.37)$$

The closed-form expression for the Kaiser window does not exist, and an approximation is given by [35]

$$W_{\text{Kaiser}}(f) = \frac{LT}{I_0(\pi\alpha)} \frac{\sinh\left(\sqrt{\pi^2\alpha^2 - (LT\pi f)^2}\right)}{\sqrt{\pi^2\alpha^2 - (LT\pi f)^2}}. \qquad (13.38)$$

Example 13.2: The frequency spectrums of the rectangular, Hamming, and Kaiser windows, for $L = 128$, are plotted in Fig. 13.4. It is observed that the rectangular window has the narrowest mainlobe (best frequency resolution) but the sidelobes are only 13 dB lower than the mainlobe. The Hamming window has a wider mainlobe with the sidelobe 40 dB lower. The Kaiser windows for $\alpha = 4$ and 6 are wider than that of the Hamming window, but the sidelobes are more than 90 dB lower than the mainlobes.

Nonrectangular windows lead to a lower and wider mainlobe, thus reducing the frequency resolution. The effective width of the mainlobe is given by [54]

$$\Delta f_w = c \frac{1}{LT}, \qquad (13.39)$$

where $c = 1$ for the rectangular window, $c \approx 2$ for the Hamming window, and $c = \frac{6(R+12)}{155}$ for the Kaiser window, R being the variable sidelobe suppression in decibels and from which α can be calculated [40]. A longer window achieves a better frequency resolution Δf. By using the Kaiser window with adjustable sidelobe level R, very weak sinusoids can be pulled out of the DTFT spectrum, while for the Hamming window, sinusoid amplitude 40 dB less than the strongest sinusoid in the signal cannot be detected in the spectrum.

13.2.2 DFT

The N-point DFT of a length-L signal is the DTFT evaluated at N equally spaced frequencies over the Nyquist interval, $0 \leq \omega \leq 2\pi$, that is, $\omega_k = \frac{2\pi k}{N}$, $k = 0, 1, \cdots, N - 1$. The N-point DFT is given by

$$X_k = X(\omega_k) = \sum_{n=0}^{L-1} x(n) e^{-j\omega_k n} = \sum_{n=0}^{L-1} x_n W^{kn}, \quad k = 0, 1, \cdots, N - 1, \qquad (13.40)$$

where $W = e^{-j2\pi/N}$.

In matrix form

$$X = \mathrm{DFT}(x) = \mathbf{A}x, \qquad (13.41)$$

where $x = (x_0, x_1, \ldots, x_{L-1})^T$, $X = (X_0, X_1, \ldots, X_{N-1})^T$, and $\mathbf{A} = [A_{kn}]_{N \times L}$. The DFT requires NL complex multiplications.

Usually, N and L are selected to be equal. In this case, the inverse DFT can be derived from (13.41) by using matrix inversion. If $L < N$, $N - L$ zeros are padded at the end of the data record; zero padding has no effect on the DTFT. If $L > N$, the data record can be reduced to N by modulo-N wrapping, since the length-N wrapped signal $\tilde{x}$ has the same N-point DFT as the original unwrapped signal x

$$X = \mathbf{A}x = \tilde{X} = \tilde{\mathbf{A}}\tilde{x}. \qquad (13.42)$$

For $L > N$, the required number of complex multiplications for DFT after wrapping is N^2 and wrapping requires $L - N$ complex multiplications, thus the total cost for wrapped DFT is $N^2 + L - N$ MACs (multiply-accumulates); if the Cooley-Tukey version of FFT [21] is used for the implementation of DFT, the required number of MACs reduces to $\frac{1}{2}N \log_2 N + L - N$.

The inverse DFT of X gives an unique wrapped version of x, i.e., $\tilde{x}$ [54]

$$\tilde{x} = \mathrm{IDFT}(X) = \tilde{\mathbf{A}}^{-1}X = \frac{1}{N}\tilde{\mathbf{A}}^*X = \frac{1}{N}\left[\mathrm{DFT}\left(X^*\right)\right]^*, \qquad (13.43)$$

where $*$ stands for complex conjugate. Note that $\tilde{x} = x$ only if $N \geq L$. Similarly, IFFT can be performed by

$$\mathrm{IFFT}(X) = \frac{1}{N}\left[\mathrm{FFT}\left(X^*\right)\right]^*. \qquad (13.44)$$

The discrete Hartley transform can be viewed as the real counterpart of DFT, when it is applied to real signals. It has efficient fast algorithms [15, 45]. The discrete Hartley transform can be used as a tool for performing fast convolutions and for performing FFT and fast DCT.

13.2.3 FFT

FFT is based on a divide-and-conquer approach to the computation of DFT. The simplest Cooley-Tukey version of FFT [21], known as the *radix-2 FFT*, requires N to be a power of two, so that the dimension of the DFT is successively divided in half until it reduces to unity. FFT can be implemented as decimation-in-time FFT or decimation-in-frequency FFT. Both FFT algorithms have a computational cost of $\frac{N}{2} \log_2 \left(\frac{N}{2}\right)$ complex multiplications and $N \log_2 N$ complex additions. Note that a complex multiplication requires four real multiplications and two real additions, and a complex addition corresponds to two real additions. When the number of samples is not a power of two, in order to use FFT, zero padding can be used. Zero padding does not increase the true resolution in frequency, but interpolates the spectrum at more points. An efficient bit-reversal algorithm for FFT implementation is given in [64].

When N is a power of four, the radix-4 FFT algorithm can be used. The radix-4 FFT further reduces the computational complexity to $\frac{3}{4} N \log_4 \left(\frac{N}{4}\right)$ complex multiplications, but has the same number of complex additions, compared to the radix-2 FFT. This leads to a reduction in the number of multiplications by at least 25%. These algorithms are described in most signal processing texts such as [57].

In the radix-2 decimation-in-frequency FFT algorithm, the even-numbered points can be computed independent of the odd-numbered points. The split-radix FFT, known as *radix 2/4 FFT*, exploits the idea by using the radix-2 algorithm for computing the even-numbered samples and radix-4 for computing the odd-numbered samples. The split-radix FFT requires the lowest number of multiplications and additions, amongst the radix-2, radix-4, radix-8, and split-radix FFTs [31]. A best implementation of the split-radix FFT can reduce the total number of multiplications to less than half that of the radix-2 FFT, while the number of additions remains almost the same [61]. Radix-2/4 and 2/8 FFT algorithms have also been proposed for computing the DFT of length $N = q \times 2^m$ [5, 8].

In [65], DFT algorithms for real signals have been derived from the radix-2 FFT for complex data. The number of required real multiplications for the proposed algorithms is one half that of the radix-2 FFT, and the number of required real additions is also less than that for the radix-2 FFT. In [66], FFT algorithms were designed by operating on vectorized data. This leads to a significant reduction in the overhead operations associated with the radix-2 FFT while improving the simplicity and regularity of their implementations.

FFT pruning algorithms increase the efficiency of the FFT by removing operations on input values which are zero, and on output values which are not required. In [3], a new pruning FFT algorithm was proposed, using a simple modification in the decimation-in-time or decimation-in-frequency FFT algorithm. Also, the number of inputs or desired outputs can be arbitrary and can be located in any position of the input/output vector. Efficient pruning algorithms for computing the DFT for a subset of output samples have been proposed based on the radix-2 decimation-in-time and decimation-in-frequency FFTs in [7].

The radix-2 and split-radix FFT algorithms have been generalized to 2-D, 3-D, or an arbitrary dimension by several authors [10, 11, 14]. These algorithms, known as *vector radix FFT algorithms*, substantially reduce the number of arithmetic operations over the method that implements FFT dimension by dimension. The radix-2/8 approach is the best among all the existing radix-based approaches for 1-, 2-, and 3-D cases [11]. These results have been extended to the discrete Hartley transform [9, 12, 13].

FFT based on logarithmic number system (log-FFT) has also been proposed [67]. Log-FFT provides better SNR than that FFT implemented with a fixed-point or floating-point number system. Use of finite-precision log-FFT substantially reduces the complexity, compared to fixed-point FFT [74]. C codes for the DTFT and the decimation-in-time radix-2 FFT algorithm are given in [39, 54].

Convolution using FFT

Convolution of two sequences h and x corresponds to the frequency domain relation

$$y = h * x \longleftrightarrow Y(\omega) = H(\omega)X(\omega). \tag{13.45}$$

Thus, y can be derived as the inverse DTFT of $Y(\omega)$. By using the modulo-N circular convolution, we have

$$\tilde{y} = \text{IFFT}\big(\text{FFT}(h) \cdot \text{FFT}(x)\big), \tag{13.46}$$

which has a complexity of $\frac{3}{2}N \log_2 N + N$ complex multiplications.

To make the circular convolution $\tilde{y}$ equal to the ordinary convolution y, it is necessary to have $N \geq L_y$, where L_y is the length of the sequence of y. Assuming the signal x having a length of L and the filter h having an order of M, we have $L_y = L + M$. The sequences h and x are zero-padded to length N before applying the N-point FFT.

Power spectrum density

A spectral power estimate can be obtained as

$$S_{xx}(\omega) = |X(\omega)|^2, \tag{13.47}$$

where $X(\omega)$ is the Fourier transform of the signal $x(n)$. From Parseval's theorem, given by (13.29), the energy of the signal $x(n)$ can be expressed by

$$\sum_{n=-\infty}^{\infty} |x(n)|^2 = \frac{1}{2\pi} \int_{-\pi}^{\pi} S_{xx}(\omega) d\omega. \tag{13.48}$$

For a signal with infinite energy, the PSD is defined by

$$\Phi_{xx}(\omega) = \lim_{N \to \infty} \frac{1}{2N+1} \left| \sum_{n=-N}^{N} x(n) e^{-j\omega n} \right|^2 = \sum_{k=-\infty}^{\infty} R_{xx}(k) e^{-j\omega n}, \tag{13.49}$$

where $R_{xx}(k) = E\big[x(n)x^*(n+k)\big]$ is the ACF. Thus, the PSD can be obtained as the DFT of the ACF.

13.3 Digital filters

13.3.1 FIR and IIR filters

An LTI system can be either a finite impulse response (FIR) or an infinite impulse response (IIR) system, depending on whether its impulse response has finite or infinite duration. An LTI system is characterized uniquely by its impulse response sequence $h(n)$, which is the output of the system when the input is unit impulse $\delta(n)$. For an input sequence $x(n)$, the output sequence is the convolution of the input with the impulse response

$$y(n) = h(n) * x(n) = \sum_{j} x(j)h(n - j) = \sum_{m} h(m)x(n - m). \qquad (13.50)$$

The impulse response coefficients $h(0), h(1), \cdots, h(M)$ are known as the *filter coefficients* or *filter taps*.

IIR filters

For the IIR filter,

$$y(n) = \sum_{m=0}^{\infty} h(m)x(n - m), \quad 0 \le n < \infty. \qquad (13.51)$$

The z-transform of $h(n)$ is known as the transfer function of the filter

$$H(z) = \sum_{n=-\infty}^{\infty} h(n)z^{-n}, \qquad (13.52)$$

which can be represented in a recursive form

$$H(z) = \frac{B(z)}{A(z)} = \frac{b_0 + \sum_{i=1}^{M} b_i z^{-i}}{1 + \sum_{i=1}^{N} a_i z^{-i}}, \qquad (13.53)$$

where a_i and b_i are the filter coefficients. The IIR filter has L zeros and N poles, an order of N. The output of the filter is given by

$$y(n) = -\sum_{i=1}^{N} a_i y(n - i) + \sum_{i=0}^{M} b_i x(n - i). \qquad (13.54)$$

For stable and causal IIR filters the speed of response is controlled by the poles nearest to the unit circle. The closer the poles to the unit circle, the sharper the peak of the time response, but the slower the filter reaches its steady state response.

FIR filters

For the FIR filter, $h(n)$ extends only over a finite time interval, from $0 \leq n \leq M$. The length M is referred to as the *filter order*

$$y(n) = \sum_{m=0}^{M} h(m)x(n-m), \quad 0 \leq n \leq M. \tag{13.55}$$

The FIR filter corresponds to the case when $A(z)$ in (13.53) reduces to 1.

Linear-Phase FIR Filters

An FIR filter has an exact linear phase property only when it has symmetric impulse response $h(n) = h(M-n)$ or antisymmetric impulse response $h(n) = -h(M-n)$. In the latter case, a constant phase shift of $\pi/2$ is introduced. An FIR filter with nonlinear phase delay causes ISI.

The one-sample delay, z^{-1}, can be implemented using memory or by a shift register. Various block- and sample-based DSP implementations of FIR and IIR filtering in C language are given in [54].

FIR filter versus IIR filter

The FIR filter is often used in communication system design. It is always stable, and is free of limit cycles that arise from finite wordlength representation. By selecting the FIR coefficients as being symmetric or antisymmetric about the center of the filter, linear phase response is achieved, and this effectively avoids distortion. This is important since both the amplitude and the phase carry information.

The IIR filter contains feedback from the past output values. It can generally achieve the same magnitude response of an FIR filter by using a lower-order design. Thus, for a given frequency response, the FIR filter usually has more coefficients, resulting in more computation and longer processing delay. However, IIR filters are much more difficult to design. An IIR implementation is not unconditionally stable: it is stable only when the poles are all inside the unit circle. IIR implementation may be used in consumer electronics, where price and power are primary concerns. The low computation load of IIR enables the use of lower-cost DSP, and consumes less power.

13.3.2 Stability

According to the definition of causality, some filters, such as FIR smoothing filters, FIR interpolation filters, and inverse filters, are double-sided (noncausal) filters. For FIR smoothing and interpolation filters, their negative-side or anticausal part has finite duration. The filtering equation is given by

$$y(n) = \sum_{m=-D}^{M} h(m)x(n-m). \tag{13.56}$$

Such filters can be made causal by the following transformation so that they are implementable in real time

$$y_D(n) = \sum_{m=0}^{M+D} h_D(m)x(n-m), \tag{13.57}$$

where $y_D(n) = y(n-D)$ and $h_D(m) = h(m-D)$ are the delayed versions of $y(n)$ and $h(m)$ by time D.

For LTI systems, stability is more important than causality. A stable LTI system has an impulse response tending to zero in a sufficiently fast manner as $n \rightarrow \pm\infty$, so that a bounded input yields a bounded output. A necessary and sufficient condition for stability in this sense is given by

$$\sum_{n=-\infty}^{\infty} |h(n)| < \infty. \tag{13.58}$$

A stable doubled-sided system can be implemented by taking a sufficiently large negative integer $n = -D$ and setting $h(n) = 0$ for $n < -D$. This is because $h(n)$ tends to zero for large negative n. Such treatment is useful for the design of the inverse filter, $H_{inv}(z) = 1/H(z)$, which is used for equalization.

The stability of an IIR filter is determined by the coefficients in the denominator $A(z)$, namely a_i. To maintain the stability of the filter, the ROC must contain the unit circle. To further make the stable filter causal, all the poles must lie inside the unit circle.

13.3.3 Inverse filters

Given input $x(n)$ and filter $h(n)$, the output $y(n)$ is obtained by convolution

$$y(n) = h(n) * x(n). \tag{13.59}$$

If one needs to calculate $x(n)$ from $y(n)$, the inverse filter can be used

$$x(n) = h_{inv}(n) * y(n). \tag{13.60}$$

In the z-domain, we have

$$H_{inv}(z) = \frac{1}{H(z)}. \tag{13.61}$$

Inverse filtering is typically used for channel equalization. In digital communications, the channel distorts the original transmitted signal, and an inverse filter is needed to restore the transmitted signal, which is distorted by the channel. Such a filter is called a *channel equalizer*. For inverse filter design, we need to consider the stability and also the effect of noise on the restored result.

Assuming $H(z)$ to be both stable and causal, the poles of $H(z)$ must be strictly inside the unit circle but the zeros can be anywhere on the z-plane. Since $H_{inv}(z) = \frac{1}{H(z)}$, those zeros of $H(z)$ becomes poles of $H_{inv}(z)$, which may be outside the unit circle. In this case, the stable inverse z-transform $h_{inv}(n)$ is anticausal. This stable and anticausal filter can be implemented by clipping off at some large negative time $n = -D$. For multiple poles outside the unit circle, the one closest to the circle controls the decay time-constant of the negative-time tail of $h_{inv}(n)$, and D is actually decided by it.

13.3.4 Minimum-, maximum-, and mixed-phase systems

For the FIR system

An Mth-order FIR system has M zeros

$$H(z) = b_0 \left(1 - z_1 z^{-1}\right) \left(1 - z_2 z^{-1}\right) \cdots \left(1 - z_M z^{-1}\right), \tag{13.62}$$

where z_i's are zeros and b_0 is a gain constant.

The frequency response is given by $H(\omega) = H\left(e^{j\omega}\right)$. When all the zeros of the filter are inside the unit circle, for each real-valued zero or each pair of complex-conjugate zeros, the corresponding term or the pair of complex-conjugate factors in $H(\omega)$ have a net phase change of zero between $\omega = 0$ and $\omega = \pi$, leading to a total system phase relation

$$\angle H(\pi) - \angle H(0) = 0. \tag{13.63}$$

Such a system is called a *minimum-phase* system.

On the contrary, when all the zeros are outside of the unit circle, each real-value zero will contribute a net phase change of π over the frequencies $\omega = 0$ and $\omega = \pi$, and each complex-conjugate pair contributes 2π over the same frequency range. The resulting system phase change is given by

$$\angle H(\pi) - \angle H(0) = M\pi. \tag{13.64}$$

Such a system is a *maximum-phase* system. When some of the zeros are distributed inside as well as outside the unit circle, the system is called a *mixed-phase system*.

The derivative of the phase characteristic is a measure of the time delay of the signal frequency components. The minimum-phase systems correspond to minimum delay, and are thus especially important.

For the IIR system

An IIR system with a transfer function rational form

$$H(z) = \frac{B(z)}{A(z)} \tag{13.65}$$

is a minimum-phase system if all its poles and zeros are inside the unit circle. When all the poles are inside the unit circle, we get a stable and causal system; such a system can be further classified as maximum-phase if all the zeros are outside the unit circle and mixed-phase if some of the zeros are outside the circle. As a result, a stable pole-zero system that is minimum-phase has a stable inverse $H^{-1}(z) = \frac{A(z)}{Z(z)}$.

A mixed-phase pole-zero system can always be implemented as a cascade of a minimum-phase system and a stable, all-pass, maximum-phase system

$$H(z) = H_{\min}(z)H_{ap}(z),$$ (13.66)

where the all-pass filter $H_{ap}(z)$ is an IIR filter whose magnitude response is unity for all ω, i.e., $|H(\omega)|^2 = 1$.

13.3.5 Notch and comb filters

A comb filter is a narrowband peaking filter with several equally spaced passbands starting at $f = 0$, whereas a dual notch filter has narrowband nulls at these frequency points. Peaking (comb) and notch filters can be designed using pole/zero placement. They are complementary with transfer functions $H(z)$ and $H(s)$, frequency responses $H(\omega)$, and the magnitude responses squared $|H(\omega)|^2$ [54]

$$H_{comb}(z) + H_{notch}(z) = 1;$$ (13.67)

$$H_{comb}(\omega) + H_{notch}(\omega) = 1;$$ (13.68)

$$|H_{comb}(\omega)|^2 + |H_{notch}(\omega)|^2 = 1.$$ (13.69)

The pole/zero placement-based method is sufficient for narrow-width filters. For wider peak or notch widths Δf at peak/notch frequency f_0 with sampling rate f_s, one can use second-order analog peak or notch filter prototypes and transform them into digital filters by using the bilinear transformation [54].

The notch or peaking filter can be designed by using a second-order transfer function of the form

$$H(z) = \frac{B(r, z)}{A(\rho, z)},$$ (13.70)

where

$$A(\rho, z) = \left(1 - \rho e^{j\omega_0}z^{-1}\right)\left(1 - \rho e^{-j\omega_0}z^{-1}\right),$$ (13.71)

$$B(r, z) = \left(1 - r e^{j\omega_0}z^{-1}\right)\left(1 - r e^{-j\omega_0}z^{-1}\right)$$ (13.72)

for $0 < r \leq 1$ and $0 < \rho \leq 1$. It is clear that the larger the distances from the zeros to the unit circle and the smaller the distances from the poles to the unit circle, the larger the magnitude of $H(\omega)$. Hence, by properly placing the poles and zeros we can obtain various magnitude response. If $\rho \gtrsim r$, the pole is closer to the unit circle than the zero, leading to a peak at $\omega = \omega_0$; if $\rho \lesssim r$ there is a dip at the frequency. If $r = 0$, it becomes a resonator at $\omega = \omega_0$. Finally, if $r = 1$, we gets an exact zero, or a notch, at $\omega = \omega_0$. High-order resonators, comb and notch filters can be obtained by cascading second-order ones.

The case of $r = 1$ for a notch filter can be extended to multiple frequencies. In this case, the numerator becomes

$$B(z) = \prod_{i=1}^{M} \left(1 - e^{j\omega_i}z^{-1}\right)\left(1 - e^{-j\omega_i}z^{-1}\right) \qquad (13.73)$$

and the denominator is obtained by replacing z^{-1} with ρz^{-1}. All the zeros are on the unit circle at the desired notch frequencies. If ρ is close to but less than unity, we get notch filters. Note ρ should be less than unity, since this helps to ensure the poles are inside the unit circle to ensure the stability of the system.

If in the numerator z in (13.73) is replaced by $r^{-1}z$, and r is close to but less than ρ so that the notch zeros are moved inside the unit circle and behind the poles, this leads to sharp peaks at $\omega = \omega_i$.

Notch and comb filters are used for signal enhancement/noise reduction. Notch filters are used to remove these harmonics of a noise signal as shown in Fig. 13.5, while comb filters are used to enhance harmonics of the desired signal as shown in Fig. 13.6. Parametric equalizer filters can be designed as a linear combination of notch and peaking filters.

For a comb filter that enhances all harmonics of a fundamental frequency f_1 at a sampling rate of $f_s = Df_1$, one can use a transfer function of the form

$$H_{\text{comb}}(z) = b\frac{1 + z^{-D}}{1 - az^{-D}}, \qquad b = \frac{1 - a}{2}. \qquad (13.74)$$

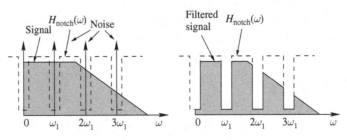

Figure 13.5   **Notch filter for noise reduction.**

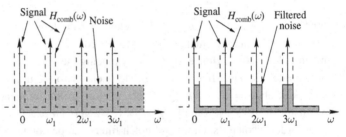

Figure 13.6   **Comb filter for signal enhancement.**

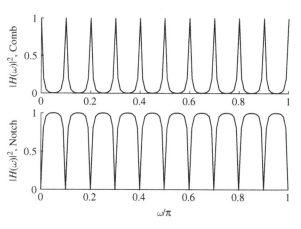

Figure 13.7

Illustration of a comb filter and its dual multi-notch filter.

The filter parameters can be determined by the desired 3-dB width of the peaks, $\Delta\omega$, and can be obtained using the bilinear transformation method [54]

$$\beta = \tan\left(\frac{D\Delta\omega}{4}\right), \quad a = \frac{1-\beta}{1+\beta}, \quad b = \frac{\beta}{1+\beta}. \tag{13.75}$$

The frequency response magnitude squared is given by

$$|H_{\text{comb}}(\omega)|^2 = \frac{\beta^2}{\tan^2(\omega D/2) + \beta^2}. \tag{13.76}$$

Example 13.3: The magnitude responses squared of a comb filter and its dual multi-notch filter with $D = 20$ and 3-dB width $\Delta\omega = 0.01\pi$ are shown in Fig. 13.7.

Based on the complementary properties, the equations for notch filters can be easily obtained from (13.67) to (13.69). Comb filters with variable gain and peak width can also be designed using IIR filters. Retrieval of a signal corrupted by a strong FM interference has been studied by applying constrained notch filtering in [76–78].

13.4 Digital filter design

Digital filter design starts with a set of design specifications and produces a set of impulse response coefficients $\mathbf{h} = (h_0, h_1, \cdots, h_{M-1})^T$ for FIR filters, or the numerator and denominator coefficients for IIR filters.

Popular filters are lowpass, highpass, bandpass, and bandstop filters. These filters have symmetric frequency response in the Nyquist interval, and have real-valued and symmetric impulse responses. Ideal differentiator and Hilbert transformer filters have antisymmetric frequency responses, and real-valued and antisymmetric impulse responses.

For the same cutoff frequencies, the lowpass/highpass and bandpass/bandstop filters are complementary. That is,

$$d_{LP}(n) + d_{HP}(n) = \delta(n) \longleftrightarrow D_{LP}(\omega) + D_{HP}(\omega) = 1, \tag{13.77}$$

$$d_{BP}(n) + d_{BS}(n) = \delta(n) \longleftrightarrow D_{BP}(\omega) + D_{BS}(\omega) = 1. \tag{13.78}$$

The frequency response of an ideal lowpass filter over the Nyquist interval is defined by

$$D(\omega) = \begin{cases} 1, & -\omega_c \leq \omega \leq \omega_c \\ 0, & -\pi \leq \omega < -\omega_c, \text{ or } \omega_c < \omega \leq \pi \end{cases}. \tag{13.79}$$

The impulse response is given by

$$d(n) = \frac{\sin(\omega_c n)}{\pi n}, \quad -\infty < n < \infty, \tag{13.80}$$

which is double-sided and infinite, and $d(0) = \omega_c/\pi$.

Similarly, the impulse response for the bandpass filter with cutoff frequencies $\omega_a < \omega_b$ is given by

$$d(n) = \frac{\sin(\omega_b n) - \sin(\omega_a n)}{\pi n}, \quad -\infty < n < \infty. \tag{13.81}$$

13.4.1 FIR digital filter design

Window method

In order to implement these filters, $d(n)$ must be clipped at a large negative value of n, say $-M$, by using a rectangular or other window, and then making it causal by introducing an appropriate delay:

$$h(n) = d(n - M)w(n), \quad -\infty < n < \infty, \tag{13.82}$$

where $w(n)$ is nonzero only when $n = 0, 1, \ldots, L - 1$, and $L = 2M + 1$. The rectangular, Hamming, or Kaiser window can be used. For the rectangular window, we have

$$h(n) = d(n - M), \quad n = 0, 1, \ldots, L - 1. \tag{13.83}$$

The filter can be implemented by

$$y(n) = \sum_{m=0}^{2M} h(m)x(n - m). \tag{13.84}$$

The transfer function is given by

$$H(z) = z^{-M} \sum_{k=-M}^{M} d(k)z^{-k} = z^{-M}\hat{D}(z), \tag{13.85}$$

where $\hat{D}(z)$ is the truncated z-transform. The window design leads to a linear phase property, arising from the delay factor $z^{-M} = e^{-j\omega M}$, since $\hat{D}(z)$ has the same symmetry or antisymmetry property as $D(z)$.

The rectangular window has the Gibbs phenomenon of Fourier series: it has largest ripples of approximately 8.9% near the passband-to-stopband discontinuity on both sides, and this is independent of the window size L. The Hamming window is a popular choice. The ripples in the passband/stopband have maximum overshoots of about 0.2%, but a wider transition width is obtained compared to the rectangular window. The Kaiser window design provides a good control over the filter design specifications, and the designer can arbitrarily specify the passband and stopband overshoot δ_{pass}, δ_{stop} and the transition width Δf. The M and α parameters of the Kaiser window (13.35) are computed from the filter specifications.

The Kaiser window method is a simple and popular filter design method for most filter specifications [4]. However, it may result in a long filter length that is not satisfactory for a stringent applications.

An FIR filter $H(z)$ can be sharpened by cascading the same filter, leading to a new, sharpened transfer function $H^2(z)$ [34].

Parks-McClellan method

The window method is a straightforward design procedure for approximating the desired magnitude response. The method is, however, not efficient for designing FIR filters with specific specifications. A number of optimization-based FIR design methods such as the weighted least squares method and the Chebyshev method are given in the literature [29].

The Parks-McClellan method [42] is a Chebyshev optimal method, which is based on the optimum equiripple approximation in both the passband and the stopband. It is used for linear-phase FIR filter design. The method uses the minimax criteria, and thus the designed filter is an equiripple filter. It generally leads to a shorter filter than the Kaiser window method. The Parks-McClellan method employs the Remez exchange algorithm, and is a common method for designing optimal linear-phase FIR filters. The method approximates the passband specification while requiring as few taps as possible. A computer program is available for designing all types of linear-phase FIR filters [43].

Frequency sampling method

For an arbitrary frequency response $D(\omega)$, the frequency integral may be very difficult to obtain. In this case, the frequency integral can be obtained by using the frequency sampling method

$$\tilde{d}(k) = \frac{1}{N} \sum_{i=-M}^{M} D(\omega_i) e^{j\omega_i k}, \quad -M \leq k \leq M, \tag{13.86}$$

where $N = 2M + 1$, and $\omega_i = 2\pi i/N$, $-M \leq i \leq M$. After the frequency response is obtained, a suitable window method can be applied.

FIR filter design using MATLAB

MATLAB provides the Filter Design and Analysis Tool (FDATool) as the interface to its Signal Processing Toolbox and Filter Design Toolbox. FDATool provides easy and optimal filter design.

For the window method, given the design specifications, if the Kaiser window is used, the MATLAB function `kaiserord` is used to estimate the order of the FIR filter. After the order of the FIR filter is obtained, the `fir1` function can be used to design the filter. MATLAB provides the `fir1` function for linear-phase FIR filter design using the window method, `fir2` for FIR filter design using the frequency sampling method, and `firls` for FIR filter design using least-squares error minimization.

For using the Parks-McClellan method, the `remezord` function is used to estimate the FIR order, given the FIR design specifications. Its output can be used to design the FIR using the `remez` function. The `remez` function gives the Parks–McClellan optimal equiripple FIR filter design.

Example 13.4: Design an FIR lowpass filter with specifications: $\omega_{s1} = 0.28$, $\omega_{p1} = 0.30$, $\omega_{p2} = 0.40$, and $\omega_{s2} = 0.42$, $\delta_p = 0.1$, $\delta_s = 0.001$. The results obtained by using the Kaiser window method, the Hamming window method, and the Parks-McClellan method are shown in Fig. 13.8. For the Hamming window method, the filter order is estimated by using `remezord`. The order obtained by `remezord` is very conservative. Thus, the corresponding filter obtained does not meet the passband specifications. This can be solved by extending the passband range and increasing the filter order N by trial and error until the specifications are satisfied.

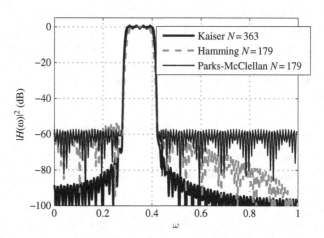

Figure 13.8 Magnitude responses of FIR bandpass filters. The normalized frequency $\omega = 1$ represents the Nyquist frequency.

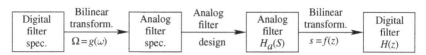

Figure 13.9   **IIR filter design using the bilinear transformation method**

13.4.2 IIR filter design

By far the most popular IIR design method is the bilinear transformation method. The digital filter specifications are first mapped onto equivalent analog filter specifications, based on which an analog filter is designed. The designed analog filter is then mapped back to obtain the corresponding digital filter. This procedure is shown in Fig. 13.9. This method cannot transform all the properties of an analog filter to a digital one. Many other IIR filter design methods are available in [4, 29].

The digital filter is based on the z-plane design, while the analog filter is based on the s-plane design. The relation between the two planes is given by a transformation of the type

$$s = f(z). \tag{13.87}$$

In the frequency domain, $s = j\Omega$ and $z = e^{j\omega}$, where Ω and ω are the analog and digital frequencies, respectively. From (13.87), we have

$$\Omega = g(\omega). \tag{13.88}$$

For a digital lowpass filter design, we select

$$s = f(z) = \frac{1 - z^{-1}}{1 + z^{-1}} \tag{13.89}$$

and thus

$$\Omega = g(\omega) = \tan\left(\frac{\omega}{2}\right). \tag{13.90}$$

Similarly, for bandpass filter design, we use

$$s = \frac{1 - 2cz^{-1} + z^{-2}}{1 - z^{-2}} \tag{13.91}$$

and accordingly

$$\Omega = \frac{c - \cos\omega}{\sin\omega}. \tag{13.92}$$

When $c = 1$, the bandpass filter reduces to the lowpass filter, and $c = -1$ leads to a highpass one. It requires $|c| \leq 1$ to ensure that the left half of the s-plane is mapped into the inside of the unit circle in the z-plane to guarantee the stability and causality of the designed digital filter. The bilinear transformations for the highpass and bandstop cases can be similarly defined [54].

A lowpass filter $H(z)$ can be transformed into a highpass filter by substituting $z \to -z$. This changes the impulse response $h_n \to (-1)^n h_n$. A bandpass or bandstop digital filter can also be designed by using a lowpass filter prototype with a bandpass or bandstop version of the bilinear transformation.

Higher-order filters achieve sharp cutoffs. The lowpass filter prototype can be designed using the analog Butterworth, Chebyshev, or elliptic filter design methods. Compared to the Butterworth filter, the Chebyshev and elliptic filters have a steeper transition width and also a smaller filter order, for a given set of design specifications. The Chebyshev filter can take the form of either equiripple passband and monotonic stopband (type-I), or equiripple stopband and monotonic passband (type-II).

IIR filter design using MATLAB

In MATLAB Signal Processing Toolbox, there are several functions for analog filter design, `buttap`, `cheb1ap`, `cheb2ap`, and `ellipap`, and several functions for analog frequency transformations `lp2bp`, `lp2hp`, `lp2lp`, `lp2bs` etc.

MATLAB provides functions for IIR filter design: `buttord` and `butter` (Butterworth), `cheb1ord` and `cheb1` (Chebyshev type-I), `cheb2ord` and `cheb2` (Chebyshev type-II), and `ellipord` and `ellip` (elliptic filters).

Example 13.5: Consider the design of an IIR bandpass filter with $\omega_{s1} = 0.2$, $\omega_{p1} = 0.3$, $\omega_{p2} = 0.5$, $\omega_{s2} = 0.6$, the passband ripple $R_p = 1 \, \text{dB}$, and stopband attenuation $R_s = 60 \, \text{dB}$. The magnitude responses of the designed filters using the Butterworth, Chebyshev type-I, Chebyshev type-II, and elliptic filters are plotted in Fig. 13.10. Among the four designs, the Butterworth filter has the largest order $N = 11$, while the elliptic filter has the lowest order $N = 5$.

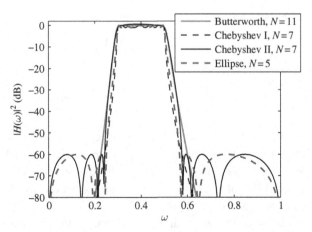

Figure 13.10 Magnitude responses of IIR bandpass filters. The normalized frequency $\omega = 1$ represents the Nyquist frequency.

Many signal processing programs, including FIR and IIR filter design programs, are available in [39] in the C and MATLAB languages.

Yule-Walker approximation

In addition to the maximally flat approximation, the Chebyshev approximation, and the equiripple approximation we have discussed so far, the least-squares approximation is also widely used in filter design. The least-squares approximation minimizes the error function over a frequency region

$$J(\omega) = \int W\left(e^{j\omega}\right) \left| H\left(e^{j\omega}\right) - D\left(e^{j\omega}\right) \right|^2 d\omega, \qquad (13.93)$$

where $D\left(e^{j\omega}\right)$ is the desired frequency response, $H\left(e^{j\omega}\right)$ is the response of the filter designed, and $W\left(e^{j\omega}\right)$ is a weighting function.

The Yule-Walker approximation is an IIR filter design technique that approximates arbitrary magnitude response. The method minimizes the error between the desired magnitude and the magnitude of the IIR filter $H\left(e^{j\omega}\right)$ in the least squares sense. In MATLAB, the `yulewalk` function is given for this purpose.

13.4.3 Hardware implementation of digital filters

For the fixed-point implementation of digital filters, round-off errors and coefficient quantization errors must be considered. These errors arise from rounding the filter coefficients, or the internal computation result, to a finite number of digits. These errors may lead to inaccurate frequency response for FIR filters and may lead to instability for IIR filters.

For IIR filters, conventional direct or canonical filter realizations are extremely sensitive to these errors. IIR filters can be efficiently implemented in cascade of second-order sections. The cascade second-order realizations may be made more robust to such round-off errors by suitably ordering/pairing the conjugate pairs. After the design of a hardware implementation of a digital filter is completed, one must reexamine the stability and the specifications of the filter.

For the implementation of high-order filters with closely clustered poles in the z-plane, the cascaded form can lead to less stability and inaccurate frequency responses. Modern DSP chips are reducing these errors by using long wordlength for coefficient storage and double precision accumulators for calculation. The use of floating-point DSPs can further reduce these errors.

13.5 Adaptive filters

Estimation theory started from Kolmogorov and Wiener's work in the 1940s for (wide sense) stationary processes. When the MMSE criterion is used for the solution, the resulting estimation is known as the *Wiener estimation*. The Wiener filter requires the knowledge of the auto- and cross-correlations, and it also involves complex matrix inversion operation. To adjust the filter coefficients as new data arrive in real time, the LMS algorithm [75] modifies the MSE along the steepest gradient direction. For a time-varying channel, a forgetting factor is introduced for calculating the autocorrelation matrix, and the RLS algorithm is derived by using the matrix inversion lemma.

For a nonstationary signal plus noise, Kalman proposed an optimum solution using the method of least squares in the 1960s. The method tries to find the estimate that minimizes the conditional-error covariance matrix, and the solution is known as the *Kalman estimation*. The criterion is the minimum variance, which can be viewed as a stochastic version of the least squares criterion. The RLS algorithm is a simplified version of the Kalman filter. The Kalman filter is the optimum filter in the Gaussian noise case, and is the best linear filter in the non-Gaussian case, in the sense of the minimum variance criterion. In the frequency domain, the Kalman filter can be regarded as a lowpass filter with a varying cutoff frequency [63]. The Kalman filter is introduced in most texts for adaptive signal processing such as [63, 71]. The 2-D Kalman filter is an extension to the Kalman filter [81].

In addition to these adaptive implementations of linear filters, many nonlinear signal processing applications based on neural networks and fuzzy logic [30] are available in the literature.

13.5.1 Wiener solution

The most common processor model is the adaptive linear combiner, which is widely used for adaptive interference cancellation, predictive coding, equalization, and beamforming. It has M inputs, each being scaled by a different coefficient or weight

$$y(k) = x^T(k)w(k) = w^T(k)x(k), \tag{13.94}$$

where the inputs $x(k) = (x_0(k), x_1(k), \cdots, x_{M-1}(k))^T$, and the weights $w(k) = (w_0(k), w_1(k), \cdots, w_{M-1}(k))^T$.

These weights are adjustable. The filter output is compared with the desired signal, and the error is used to adjust the weights so that the error is further reduced. The error at time k is defined by

$$e(k) = d(k) - y(k) = d(k) - w^T x(k), \tag{13.95}$$

where $d(k)$ is the desired signal at time k. Accordingly

$$\begin{aligned} e^2(k) &= \left(d(k) - w^T x(k)\right)\left(d(k) - x^T(k)w\right) \\ &= d^2(k) + w^T x(k)x^T(k)w - 2d(k)x^T(k)w. \end{aligned} \tag{13.96}$$

Assuming that $e(k)$, $d(k)$ and $x(k)$ are statistically stationary, the MSE $E\left[e^2(k)\right]$ is given by

$$\text{MSE} = E\left[e^2(k)\right] = E\left[d^2(k)\right] + w^T \mathbf{R}_{xx}w - 2r_{dx}^T w, \tag{13.97}$$

where

$$\mathbf{R}_{xx} = E\left[x(k)x^T(k)\right], \quad r_{dx} = E\left[d(k)x^T(k)\right]. \tag{13.98}$$

The Wiener filter is obtained by differentiating $E\left[e^2(k)\right]$ with respect to w and setting the derivative to zero

$$w = \mathbf{R}_{xx}^{-1} r_{dx}. \tag{13.99}$$

The MSE obtained by the Wiener solution is minimum

$$\mathrm{MSE}_{\min} = E\left[e_k^2\right] = E\left[\sigma_d^2\right] - r_{dx}^T w. \tag{13.100}$$

The Wiener filter can also be formulated in the frequency domain by minimizing $E\left[E^2(f)\right]$, where $E(f)$ is the frequency-domain representation of $e(t)$ with respect to $W(f)$ [71].

Digital filters are typically comprised of a tapped delay line with unit delays and coefficients. For the FIR or transversal filter, $x_n = \left(x_n, x_{n-1}, \ldots, x_{n-M+1}\right)^T$ is the input vector at time n, and w is the channel impulse response vector of length M. The Wiener solution can also be applied to an IIR filter, but results in a set of nonlinear equations.

13.5.2 LMS algorithm

The Wiener solution is computationally expensive due to the matrix inversion operation. Conventional gradient-descent and Newton's methods can be applied to the MSE to find the optimum solution. These methods can find the optimum solution in a few steps. However, since the performance surface is noisy, in order to estimate the gradient at each step, many samples are required.

The LMS algorithm is widely used in adaptive signal processing. For the adaptive linear combiner, the LMS algorithm uses a special estimate of the gradient. It uses $e^2(k)$ as an estimate of $E\left[e^2(k)\right]$ for the purpose of gradient estimation, thus

$$\hat{\nabla}(k) = \frac{\partial e^2(k)}{\partial w} = -2e(k)x(k). \tag{13.101}$$

The LMS algorithm is a gradient-descent algorithm of the form

$$w(k+1) = w(k) - \mu \hat{\nabla}(k) = w(k) + 2\mu e(k)x(k), \tag{13.102}$$

where the choice of the step size μ must be $0 < \mu < \frac{2}{\lambda_{\max}}$, $\lambda_{\max}$ being the maximum eigenvalue of $\mathbf{R}_{xx}$, to ensure convergence [30, 71]. The spread in the convergence speed of weights is proportional to the spread in eigenvalues of $\mathbf{R}_{xx}$, that is, $\lambda_{\max}/\lambda_{\min}$.

The LMS algorithm is simple and efficient, since the gradient estimate is obtained from a single data input. It converges to a solution much faster than the gradient-descent algorithm, particularly for a large number of weights.

13.5.3 RLS algorithm

The RLS algorithm is a recursive formulation of the Wiener solution. It can also be derived by applying the matrix inversion lemma to $\mathbf{R}_{xx}^{-1}$. Like the LMS algorithm, the RLS is also a most widely used adaptation algorithm. The RLS algorithm can be derived by exponentially weighting the data to remove gradually the effects of old data on $w(k)$ and to allow the tracking of slowly varying signal characteristics. It is given as [39]

$$w(k) = w(k-1) + \mathbf{G}(k)e(k), \tag{13.103}$$

$$\mathbf{P}(k) = \frac{1}{\gamma} \left[\mathbf{P}(k-1) - \mathbf{G}(k)\mathbf{x}^T(k)\mathbf{P}(k-1) \right], \tag{13.104}$$

where

$$\mathbf{G}(k) = \frac{1}{\alpha(k)} \mathbf{P}(k-1)\mathbf{x}(k), \tag{13.105}$$

$$\alpha(k) = \gamma + \mathbf{x}^T(k)\mathbf{P}(k-1)\mathbf{x}(k). \tag{13.106}$$

$\mathbf{P}(k)$ is essentially a recursive way of computing the inverse matrix $\mathbf{R}_{xx}^{-1}$. The forgetting factor γ is typically elected between 0.98 and 1. Smaller values add weight to the more recent data, leading to fluctuations in the estimation. When $\gamma = 1$, the RLS reduces to the LS.

The RLS algorithm is one-order of magnitude faster in convergence than the LMS algorithm. However, the RLS method has two main problems [39]. The blow-up problem occurs if the signal $x_{k,i}$ is zero for a long time. This causes $\mathbf{P}(k)$ to grow exponentially as a result of division by the forgetting factor γ. The second problem is its sensitivity to round-off errors, and this may lead to a negative definite $\mathbf{P}$ matrix and eventually to instability. This latter problem can be solved by using the UD factorization of $\mathbf{P}$ [39].

13.6 Digital up-conversion and digital down-conversion

The digital front-end of a wireless transceiver performs digital up-conversion in the transmit path and digital down-conversion in the receive path. Digital up-conversion is used to transform baseband signal to digital bandpass signal, whereas digital down-conversion performs the reverse operation. The idea is to modulate the signal using an NCO. For the classical SDR structure, as shown in Fig. 1.3, both digital down- and up-conversion contain a digital frequency-conversion part and a sampling-rate conversion part.

For numerical implementation, an LUT is conventionally employed for calculating the sine and cosine of the angle. The carrier frequency is adjustable. To avoid a large LUT, the CORDIC (Coordinate Rotation Digital Computer) processor [72] can be used to calculate the sine and cosine functions. CORDIC is an iterative method that computes the rotation of a two-dimensional vector (rotation mode) or converts between cartesian and polar coordinates (vectoring mode) by using only additions and shift-operations.

Typically, for each carrier the digital down- or up-conversion module requires several thousand MIPS (million instructions per second). This is equivalent to the processing power of one or more high-end DSPs such as Texas Instruments (TI)'s TMS320C6x or Analog Devices Inc (ADI)'s TigerSHARC [38]. Some spare capacity is also required. When the budget for processing power is tight, hand-crafted assembly code may have to be used to provide a processing capacity margin. In order to reduce cost and power consumption as well as space, digital up-/down-conversion can be conveniently implemented using ASICs or field programmable gate arrays (FPGAs). There are many commercially available digital up- and down-conversion chips that are designed for popular wireless communication standards. Major vendors of digital down- and up-conversion chips are Intersil, ADI and TI. Their chips are software programmable ASICs, also known as *application-specific standard parts (ASSPs)*, and they are programmable within a specified range. These devices have internal microcode processors, and can be programmed via microprocessor interfaces.

Digital down- and up-conversion chips have on-chip NCOs. The NCO generates digital sine waveforms for digital mixers. Sine and cosine NCOs are used for quadrature I and Q mixing, respectively. A digital mixer multiplies the input digital signal with a digital sine (or cosine) waveform, and produces sum and difference frequencies. The SFDR of the NCO must be considerably better than the SFDR of the digital down-/up-conversion output, since the spurious response in the NCO also mixes with unwanted out-of-band signals.

13.6.1 Numerically controlled oscillators

A simple numerical oscillator

From (13.70), we see that the filter

$$H(z) = \frac{1}{1 + a_1 z^{-1} + a_2 z^{-2}} = \frac{1}{\left(1 - re^{j\omega_0} z^{-1}\right)\left(1 - re^{-j\omega_0} z^{-1}\right)} \tag{13.107}$$

is a resonator at $\omega = \omega_0$. We have $a_1 = -2r\cos\omega_0$ and $a_2 = r^2$. The unit sample response is given by

$$h(n) = \frac{r^n}{\sin\omega_0}\sin(n+1)\omega_0 u(n), \tag{13.108}$$

where $u(n)$ is the Heaviside step function, $u(n) = 1$ for $n \geq 0$ and 0 otherwise. When the poles are placed on the unit circle ($r = 1$), we obtain a digital sinusoidal oscillator

$$h(n) = \frac{1}{\sin\omega_0}\sin(n+1)\omega_0 u(n). \tag{13.109}$$

Coupled-form oscillator

An oscillator can be modeled as an amplifier with its output fed back to its input via a phase-shifting network [68]. As mentioned earlier in Chapter 11, two necessary conditions for oscillation, namely, Barkhausen's criteria, are: the total loop gain is one, and the total loop phase shift is a multiple of 2π radians.

In order to generate two sinusoidals at the same time, an oscillator iteration can be written as [68]

$$\begin{pmatrix} x_1(n) \\ x_2(n) \end{pmatrix} = \begin{bmatrix} a & b \\ c & d \end{bmatrix} \begin{pmatrix} x_1(n-1) \\ x_2(n-1) \end{pmatrix}. \tag{13.110}$$

This is known as the *coupled-form oscillator*. It is not driven by any input, but it requires initial conditions in order to start self-sustaining oscillation.

The discrete-time equivalents of Barkhausen's criteria are given as

$$ad - bc = 1, \quad |a + d| < 2. \tag{13.111}$$

After some manipulation, the generated sine waves are given by

$$\begin{bmatrix} x_1(n) \\ x_2(n) \end{bmatrix} = \begin{bmatrix} \cos(n\theta) \\ \psi \cos(n\theta + \phi) \end{bmatrix}, \tag{13.112}$$

where

$$\theta = \cos^{-1}\left(\frac{a+d}{2}\right), \quad \psi = \sqrt{\frac{-c}{b}}, \quad \phi = \arg\left(\frac{(d-a) + j\sqrt{4 - (a+d)^2}}{2b}\right) \tag{13.113}$$

and

$$\begin{bmatrix} x_1(0) \\ x_2(0) \end{bmatrix} = \begin{bmatrix} 1 \\ \psi \cos(\phi) \end{bmatrix}. \tag{13.114}$$

The angle θ is the step angle for iteration, and ϕ is the phase shift between the two state variables. To obtain a quadrature oscillator, $\phi = \pm 90°$, accordingly $a = d$.

By selecting different parameters for ψ, ϕ, and θ, different types of oscillators are designed. Accordingly, a, b, c, and d can be decided. DSP implementation can be easily realized by using (13.110). The generated sine waves are always of the same frequency. A number of oscillators in this framework are also introduced in [68]. To combat accumulated errors, a dynamic amplitude control using an AGC is also given. For NCOs, the oscillator frequency can be changed via θ. A procedure for designing a dynamic amplitude, frequency controlled oscillator is also given in [68].

When Barkhausen's criteria are satisfied, (13.110) corresponds to vector rotation in the two-dimensional coordinate system, and the coupled-form oscillator can be implemented using the CORDIC algorithm.

13.6.2 Direct digital frequency synthesis

DDS is a technique for generating waveforms. DDS generates a digital sine wave from an LUT, and uses a DAC and a lowpass filter to generate the analog waveform. This method is extremely accurate. A cheap solution is to use the CORDIC algorithm [72] to generate a digital waveform. For analog modulation, the signal to be modulated is fed as input to the CORDIC processor, and the result is then transformed to analog form using a DAC [70]. The maximum frequency that DDS can generate is limited by the sampling rate of existing DACs, typically a few hundred megahertz.

ICs are available for generating waveforms up to hundreds of MHz. The DDS technique has a number of advantages: an arbitrary tuning frequency resolution, the phase and frequency of the waveform controllable in one sample period, implementation using integer arithmetic, guaranteed stability and non-requirement of AGC, and phase continuity in case

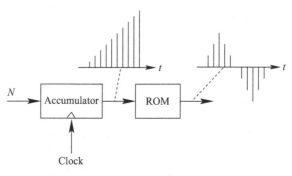

Figure 13.11   DDS principle.

of frequency change. These features make DDS desirable for realizing PM and tunable waveform generators.

DDS is used to generate any periodic, discrete-time waveform of known frequency f_o. It is implemented in two steps. An accumulator is used to generate a discrete-time phase value, followed by a phase-to-waveform converter to generate the desired signal. This is shown in Fig. 13.11. In the figure, N is a programmable step. The accumulator is a register of word length M bits. When the ROM contains a table for generating a sine function, the generated waveform is a digital representation of a sinusoid. In this case, the DDS system is actually an NCO. By changing N, we can synthesize waveforms of different frequency. The generated digital waveform can be converted into an analog waveform after being D/A-converted and filtered.

The output frequency f_o is given by [22]

$$f_o = \frac{f_s}{2^M} \Delta_{ACC}, \qquad (13.115)$$

where Δ_{ACC} is the phase increment during one sample period, coded as an integer of $M-1$ bits and f_s is the sampling frequency. Thus the maximum and minimum output frequencies are $f_s/2$ and $f_s/2^M$, respectively.

For phase-to-waveform conversion, the phase is converted into amplitude using $P \leq M$ bits. The output waveform is stored in an LUT with 2^P entries. To generate a sine wave, the corresponding sine function is stored in the LUT

$$LUT(i) = a \sin\left(\frac{2\pi i}{2^P}\right) \qquad (13.116)$$

where a is a peak amplitude. To generate two quadrature signals, both $LUT(i)$ and $LUT\left(i + \frac{2^P}{4}\right)$ are read to generate the two waveforms. The LUT memory size is of $2^P(P-1)$ bits.

One can increase P to improve the SFDR of generated waveform. Techniques for improving SFDR can be implemented through waveform compression, through the CORDIC algorithm, by using odd-number Δ_{ACC}, or by adding a dither signal to the ACC phase values, followed by noise shaping [22]. The CORDIC algorithm configured in the rotation mode can behave as a quadrature phase-to-amplitude converter that directly generates sine and cosine waveforms [70].

A digital waveform generator can be implemented as a circular buffer, which can be accessed from a table. This is the so-called *wavetable synthesis*. Various wavetable generators and oscillators are implemented in C and hardware DSP in [54]. The inputs to the generator are amplitude, frequency and the wavetable, and the output is the generated oscillating signal.

DDS offers a number of advantages over PLL-based frequency synthesis. First, it avoids analog VCO devices, leading to a low phase noise, which is approximately that of the clock. Since there is no analog feedback loop, DDS is much faster for channel switching than the PLL technique. It also allows direct signal modulation in the digital domain. The frequency steps can be made very small by increasing the word length. However, due to Nyquist's sampling theorem, the sampling rate would be very high for the RF band, exceeding today's DAC capability. This is the major restriction that prevents its wide use in the RF band.

An all-digital PLL can be realized by feeding back the generated digital waveform to a register that performs phase comparison so as to stabilize the generated waveform [33], just like an analog loop with a sinusoidal phase detector in the analog domain. Stability analysis of first-order and second-order all-digital PLLs is given in [33].

13.7 Sampling-rate conversion

In SDR applications, the wideband ADC uses a fixed sampling rate, while different air-interfaces use different chip or symbol rates. This requires multirate processing.

The sampling rate needs to be an integer multiple of the modulation symbol rate to achieve synchronization between the receiver and the transmitter. For multiple carrier and/or symbol rate systems, it is not uncommon that the sampling rate is not an integer multiple of the symbol rate. For example, if the sampling rate is 2.4 times the symbol rate, the oversampling rate can be selected as 3 or a large integer multiple. Sample-rate conversion between them is necessary. This can be achieved by first applying interpolation by 5, filtering, and then decimation by 4, that is, $2.4 \times 5/4 = 3$.

Decimation periodically removes all samples except every R_dth sample. This leads to a reduction in sample rate by a factor of R_d, hence a reduction in bandwidth by a factor of R_d. Interpolation increases the sampling rate by R_i. Interpolation is achieved by first stuffing $R_i - 1$ zero-valued samples, then using a lowpass filter to remove images, and using a gain stage to restore the amplitude of the original signal.

A large oversampling rate requires more computation for signal processing. Since synchronization is performed at the receiver side, the transmitter side does not need sampling rate conversion, but interpolation is still required to raise the digital sampling rate to be an integer multiple of the symbol rate. The resampling processing must be followed by filtering to eliminate aliasing. Interpolation can be very efficiently implemented as polyphase filters.

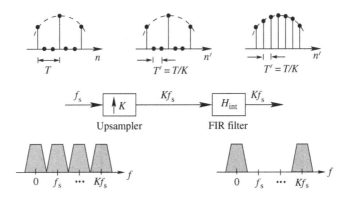

Increasing sampling rate by interpolation.

13.7.1 Interpolation

Interpolation is used to increase the sampling rate by resampling the original low-rate samples. For an upsampling factor of K, $K - 1$ zeros are inserted between any two adjacent low-rate samples. The upsampled signal is given by

$$x_{up}\left(n'\right) = \begin{cases} x(n), & n' = nK \\ 0, & \text{otherwise} \end{cases}. \tag{13.117}$$

An FIR interpolation filter is a lowpass filter that removes the high frequencies introduced by the inserted zeros, and as a result, in the time domain, the $K-1$ zeros are replaced by the calculated interpolated values. The interpolation filter operates at a sampling rate of Kf_s, and is also called the *oversampling digital filter*, as an analogy to the oversampling analog filter. Use of digital oversampling helps to alleviate the need for high-quality analog anti-image postfilters after the DAC. The interpolation procedure is shown in Fig. 13.12.

The ideal K-fold interpolation filter is a lowpass filter, operating at the fast rate $f'_s = Kf_s$, and having a cutoff frequency at the low-rate Nyquist frequency $f_c = f_s/2$. Digital oversampling keeps one of every Kth image of those of the low-rate images, thus making the post-image filter much cheaper. The FIR interpolation filter can be designed by truncating the impulse response $d\left(k'\right)$ to finite length $N = 2KM + 1$:

$$d\left(k'\right) = \frac{\sin\left(\pi k'/K\right)}{\pi k'/K}, \quad -KM \le k' \le KM. \tag{13.118}$$

The impulse response $d\left(k'\right)$ is first delayed by KM samples to make it causal, then a window is applied

$$h\left(n'\right) = w\left(n'\right) d\left(n' - KM\right), \quad n' = 0, 1, \dots, N-1. \tag{13.119}$$

The ideal interpolation is obtained by the convolution of the unsampled input with the windowed impulse response; for the rectangular window

$$y_{up}\left(n'\right) = \sum_{k'=-KM}^{KM} d\left(k'\right) x_{up}\left(n' - k'\right), \quad n' = 0, 1, \dots, N-1. \tag{13.120}$$

The function of an ideal interpolation filter is to reproduce the interpolated samples correctly. In the frequency domain, it reshapes the low-rate sampled spectrum $X(f)$ by using the ideal interpolation filter $D(f)$, and the output spectrum is given by

$$Y_{up} = D(f)X(f). \tag{13.121}$$

For the ideal interpolator, we have

$$D(f) = \begin{cases} K, & |f| \leq f_s/2 \\ 0, & f_s/2 < |f| \leq f_s'/2 \end{cases}. \tag{13.122}$$

Thus, the filter removes all replicas at multiples of the input sampling rate f_s, except those that are multiples of the output rate Kf_s. Many interpolation techniques are discussed in [71].

Polyphase form

The high-rate FIR interpolating filter can be implemented by K FIR subfilters, called *polyphase filters*, each operating at a low rate of f_s and having $1/K$ length of the original filter. This leads to considerable saving in complexity and is also suitable for parallel processing. The polyphase form works on the low-rate sample rather than the upsampler output. The ith polyphase subfilter is defined by [54]

$$d_i(k) = d(kK + i), \quad -M \leq k \leq M - 1 \tag{13.123}$$

and

$$y_i(n) = \sum_{k=-M}^{M-1} d_i(k)x(n - k), \quad i = 0, 1, \ldots, K - 1. \tag{13.124}$$

The polyphase form requires a complexity of only $2KM$ multiplications, while the direct form requires $2K^2M$ multiplications.

Multistage form

Interpolation filters can be implemented in the multistage form, by gradually increasing the sampling rate until the final rate is reached. Each stage contains an upsampler with a factor K_i and an interpolator H_i, $i = 1, \ldots, N_{stage}$. The overall interpolation factor is $K = \prod K_i$ and the overall frequency response is $H(f) = \prod H_i(f)$. Such multistage realizations allow additional savings in complexity over the multiphase form. Each interpolator removes all replicas at multiples of its input rate, except those at multiples of its output rate. Finally $K - 1$ out of K replicas are removed and only those replicas at multiples of Kf_s remain. Interpolators at each stage can be designed using the Kaiser window method.

The effective passband ripple of the whole filter $H(f)$ is worse than the ripples of the individual factors. For a multistage filter, when the oversampling ratios are placed in an ascending order from the first to the last stage, it can lead to reduced passband ripples [54]. The first filter has typically a very narrow desired transition width Δf, and thus has the most stringent specifications. Following stages have wider transition widths, and thus can be implemented with shorter filter lengths, and the simple hold or linear interpolators can be used. These simple hold or linear interpolators do not have flat passband. The first filter can be designed by equalizing all the passband attenuation of the subsequent stages. The general multistage design procedure is given in [23].

Hold and linear interpolators

The ideal interpolator is an ideal lowpass filter. It can be treated as the sampled version of the ideal analog reconstructor, sampled at the high rate f_s'. Like the analog reconstructor, two simple interpolators are the hold and linear interpolators. Unlike the ideal interpolator that is approximated by using long filter length, these simple interpolators can be used in the case of later stages of multistage implementation of interpolators, where the specifications are not stringent. These filters can also implemented in the polyphase form.

Interpolation between low-rate samples can be performed by holding the previous sample constant until the next low-rate sample. This is the simplest hold interpolator resulting from the S/H analog reconstructor. The linear interpolator between two low-rate samples, which is based on the linear analog reconstructor, is also used. An interpolation filter that uses more adjacent samples for interpolation usually yields more accurate values.

The impulse responses of the hold and linear interpolators are given, respectively, by

$$d\left(k'\right) = \begin{cases} 1, & 0 \le k' < K - 1 \\ 0, & \text{otherwise} \end{cases} \quad \text{(hold)}, \tag{13.125}$$

$$d\left(k'\right) = \begin{cases} 1 - \frac{|k'|}{K}, & |k'| \le K - 1 \\ 0, & \text{otherwise} \end{cases} \quad \text{(linear)}. \tag{13.126}$$

The corresponding filter implementations in the polyphase form are given by

$$y_{\text{up}}(nK + i) = x(n) \quad \text{(hold)}, \tag{13.127}$$

$$y_{\text{up}}(nK + i) = \left(1 - \frac{i}{K}\right) x(n) + \frac{i}{K} x(n + 1) \quad \text{(linear)}, \tag{13.128}$$

for $i = 0, 1, \ldots, K - 1$. Their frequency responses are given by [54]

$$D(f) = \frac{\sin(\pi f / f_s)}{\sin\left(\pi f / L f_s\right)} e^{-j\pi (K-1) f / K f_s} \quad \text{(hold)}, \tag{13.129}$$

$$D(f) = \frac{1}{L} \left| \frac{\sin\left(\pi f / f_s\right)}{\sin\left(\pi f / K f_s\right)} \right|^2 \quad \text{(linear)}. \tag{13.130}$$

Both frequency responses are periodic in f with period $f_s' = K f_s$, and are zero at $f_s' = n f_s$, $n = 1, 2, \ldots, K - 1$.

Example 13.6: The frequency responses of the hold and linear interpolators are compared with the ideal interpolator in Fig. 13.13, for $K = 10$. The linear interpolator has much better frequency response than the hold interpolator.

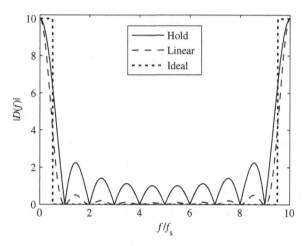

Figure 13.13   Frequency responses of ideal, hold, and linear interpolators, for $K = 10$.

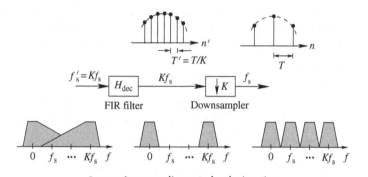

Figure 13.14   Decreasing sampling rate by decimation.

13.7.2 Decimation

Decimation is the reverse of interpolation: It reduces the sampling rate from f_s' to f_s'/K. This downsampling process keeps one out of every K high-rate samples and discards all the remaining $K - 1$ samples. The downsampled signal is defined by

$$x_{\text{down}}(n) = x'(nK). \tag{13.131}$$

In the ideal case, the high-rate signal $x'\left(n'\right)$ has no frequency components outside the low-rate Nyquist interval $\left[-f_s/2, f_s/2\right]$. The downsampled signal $x_{\text{down}}(n)$ is equal to the low-rate signal $x(n)$ which corresponds to the resampled signal from the original analog signal at the lower rate f_s. The decimation process is shown in Fig. 13.14. Note the downsampling process leads to replicas of the original spectrum at multiples of the low rate f_s.

In general, a digital lowpass filter called the *decimation filter* is located in front of the downsampler to remove those frequency components of the high-rate signal that are outside the low-rate Nyquist interval $\left[-f_s/2, f_s/2\right]$, and in this case, the downsampler output gives the correct decimation result. Without this decimation prefilter, the out-of-band frequency components will lead to aliasing, and the downsampler output will not be the correct decimation output. The combination of the decimation prefilter and downsampler is called a decimator.

In multistage implementation of decimators, the most stringent decimator is placed as the last stage, as opposed to the case of the interpolation filter. The earlier decimators can be implemented using the decimation version of the hold interpolator, which is a simple FIR averaging filter that averages K successive high-rate samples. Conversely to the interpolator, the passbands of the earlier decimators can be equalized by the last decimator. The use of decimators can also alleviate the need for high quality analog prefilters prior to the ADC.

13.7.3 Sample rate converters

In order to achieve a sampling rate by any rational factor, K/M, so that the input and output rates are related by $f_s' = (K/M)f_s$, a common oversampling rate is required

$$f_s'' = Mf_s' = Kf_s \tag{13.132}$$

This can be achieved by first interpolating from f_s by a factor of K and then decimating f_s' by a factor of M. The interpolating filter and decimating filter both operate at f_s' and are lowpass filters, thus they can be combined into a single lowpass filter by selecting a cutoff frequency $f_c = \frac{1}{2} \min\left(f_s, f_s'\right)$. This sampling rate conversion process is shown in Fig. 13.15.

The combined filter has a complexity that is M times smaller than the polyphase rate Kf_s for full interpolation, since the downsampler selects only every Mth filter output from the interpolator output.

The sequence of the cascade decimator and interpolator in a sample rate converter is very important. Decimation by a factor M followed by interpolation by K is not equivalent to interpolation by K followed by decimation by M. They are identical if and only if K and M are relatively prime [69]. This is important for efficient implementation of filter banks. More discussion on multirate digital signal processing is given in [23, 39, 54].

13.7.4 Cascaded integrator comb (CIC) filters

CIC filters efficiently perform decimation and interpolation without using multiplication [37]. This multiplierless architecture is especially attractive for hardware implementation. The basic building blocks are integrator and comb. An integrator is a single pole IIR filter, $H_I(z) = 1/\left(1 - z^{-1}\right)$, or $y(n) = y(n-1) + x(n)$ in the time domain. A comb is an

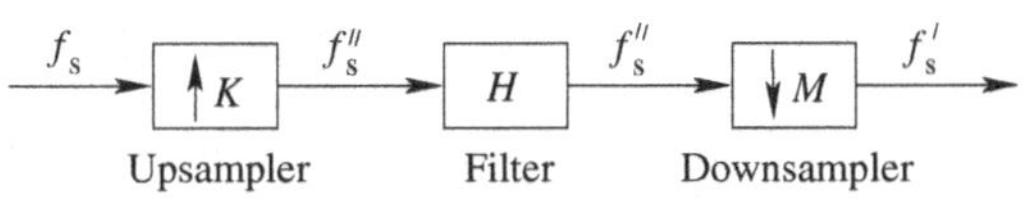

Figure 13.15 Sampling rate converter from f_s to $\frac{K}{M}f_s$. The filter is a combined interpolation/decimation filter.

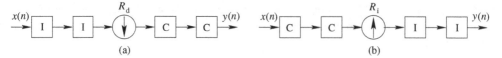

Figure 13.16
Two-stage CIC filters. (a) Decimation by a factor of R_d. (b) Interpolation by padding R_i zeros.

odd-symmetric FIR filter, $H_C(z) = 1 - z^{-M}$, or $y(n) = x(n) - x(n-M)$ in the time domain, M usually being one or two. A CIC filter is realized by cascading the integrator with the comb filter or vice versa.

A decimating CIC can be formed by cascading N integrators, followed by a decimator that reduces the rate by a factor of R_d, and N combs. An interpolating CIC can be built by cascading N combs, followed by an interpolator that increases the rate by a factor of R_i, and N integrators. Two-stage decimating and interpolating CIC filters are shown in Fig. 13.16.

A single integrator by itself is unstable. For CIC filters, the integrators and comb filters work at different sampling frequencies: one at a high frequency and the other at a low frequency. The Noble identities are used to generate an equivalent frequency response of their cascade at the high frequency, in which all the integrator and comb filters use the same sampling frequency. The transfer function for a CIC filter is [37]

$$H(z) = H_I^N(z)H_C^N(z^R) = \frac{(1 - z^{-RM})^N}{(1 - z^{-1})^N} = \left(\sum_{k=0}^{RM-1} z^{-k} \right)^N, \qquad (13.133)$$

where R can be R_i or R_d. Thus, a CIC filter is equivalent to N FIR filters, and it is unconditionally stable. Since these FIR filters are symmetric, the CIC filter has a linear phase and hence constant group delay.

The magnitude response of an N-stage CIC filter at high frequency is given by [37]

$$|H(f)| = \left| \frac{\sin(\pi Mf)}{\sin(\pi f/R)} \right|^N. \qquad (13.134)$$

However, the passband of a CIC filter is not flat, and this problem can be alleviated by a compensation filter. Design of the compensation filter is described in [2].

For example, the quadrature digital up-conversion chip from ADI, AD9856, consists of two interpolating filters, a quadrature mixer, a DDS circuit, and a 12-bit DAC with a sampling frequency up to 200 MHz. Each interpolating filter includes three half-band filters with an upsampling rate of 2 or 4, and a CIC filter, which can integrate the baseband signal by an integer factor between 2 and 63 inclusive. The total upsampling rate is in the range of 8 to 504. The DDS circuitry is employed to generate a complex local clock of frequency ranging from DC to 80 MHz with 32-bit resolution.

Half-band filters

For an FIR filter with an odd number of coefficients $M = 2K + 1$, as given by (13.55), in most applications, the impulse response $h(n)$ is assumed to be Hermitian-symmetric with respect to $n = K$, thus we get the linear-phase filter, which is in fact a zero-phase filter.

If every other coefficient except h_K is selected as zero, that is [46, 69]

$$h(2n) = \begin{cases} c, & n = K \\ 0, & \text{otherwise} \end{cases} \qquad (13.135)$$

for $n = 0, 1, \ldots, N - 1$, where c is a constant, typically taken as 0.5, we get a zero-phase filter

$$H(z) + H(-z) = 2c = 1 \qquad (13.136)$$

and

$$H(e^{j\omega}) + H(e^{j(\pi - \omega)}) = 1. \qquad (13.137)$$

Thus, the filter has a symmetry with respect to the half-band frequency $\pi/2$. It also has symmetric passband and stopband edges and peak errors (see Fig. 11.16). Thus, this filter is referred to as the *half-band filter*. It is a lowpass filter whose passband extends from zero to one-fourth of the Nyquist band. The half-band filter is useful for digital frequency convertion by ignoring every second sample, since every other $h(n)$ is zero. This achieves decimation by two.

The half-band filter can be easily extended to the Mth band filter [46, 69]

$$h(Mn) = \begin{cases} c, & n = K \\ 0, & \text{otherwise} \end{cases} \qquad (13.138)$$

for $n = 0, 1, \ldots, N - 1$, where c typically takes the value $1/M$. Accordingly

$$\sum_{k=0}^{M-1} H\left(zW^{k-1}\right) = Mc = 1. \qquad (13.139)$$

That is, the M uniformly shifted frequency responses add up to a constant for all ω, for the Mth band filter $H(z)$.

The Mth band filter has applications in the exact reconstruction of a signal $x(n)$ after it has been split into M sub-bands. In most applications, the Mth band filter is selected to be a linear-phase lowpass filter with cutoff π/M. Standard techniques for designing the FIR Mth band filter are given in [46].

13.8 Discrete cosine transform

From this section on, we will introduce some transforms that are fundamental to source coding.

The length-N DCT of a signal $s(n)$ is defined by [1]

$$S(k) = \sqrt{\frac{2}{N}} c(k) \sum_{n=0}^{N-1} s(n) \cos\left(\frac{\pi\left(n + \frac{1}{2}\right)k}{N}\right) \qquad (13.140)$$

for $k = 0, 1, \ldots, N - 1$, where

$$
c(t) = \begin{cases} \frac{1}{\sqrt{2}}, & t = 0 \\ 1, & t > 0 \end{cases}. \tag{13.141}
$$

The inverse DCT (IDCT) is accordingly given by

$$
s(n) = \sum_{k=0}^{N-1} \sqrt{\frac{2}{N}} c(k) S(k) \cos\left(\frac{\pi \left(n + \frac{1}{2} \right) k}{N} \right) \tag{13.142}
$$

for $n = 0, 1, \ldots, N - 1$.

The DCT and IDCT are linear transforms that map between a real signal and real DCT coefficients. In matrix form

$$
\mathbf{S} = \mathbf{C}s, \tag{13.143}
$$

$$
s = \mathbf{C}^T \mathbf{S}, \tag{13.144}
$$

where $\mathbf{C} = [C_{kn}]_{N \times N}$,

$$
C_{kn} = \sqrt{\frac{2}{N}} c(k) \cos\left(\frac{\pi \left(n + \frac{1}{2} \right) k}{N} \right), \tag{13.145}
$$

which is a unitary matrix, $\mathbf{C}^{-1} = \mathbf{C}^T$.

It is easily shown that the Parseval's relation holds for DCT

$$
\sum_{k=0}^{N-1} S^2(k) = \sum_{n=0}^{N-1} s^2(n). \tag{13.146}
$$

When the DCT is applied to voice or video signals, the energy is typically concentrated in the first few transform coefficients, thus the coefficients with very low energy can be discarded without introducing significant distortion [29].

The DCT consists essentially of the real part of the DFT. The DCT can be implemented based on the FFT, and a fast DCT is more efficient than the FFT-based implementation. In terms of compression, the DCT outperforms the DFT. This is because the DFT introduces sharp discontinuities at the beginning and end of the sequence for a periodic sequence of length N, which introduces high frequencies as well as adjustment in low frequency components. The DCT can be obtained from the DFT by mirroring the N-point sequence to obtain a $2N$-point sequence, and then taking the first N points of the resulting $2N$-point DFT. In this way, the discontinuity at the edges is completely eliminated. This is illustrated in Fig. 13.17.

In the worst case, the DCT can be computed by using a length-$2N$ DFT [1]. A popular fast DCT algorithm is given in [17], which requires $N \log_2 N - \frac{3N}{2} + 4$ real multiplications and $\frac{3N}{2} \left(\log_2 N - 1 \right) + 2$ real additions. This algorithm is six times faster compared to the length-$2N$ FFT [17].

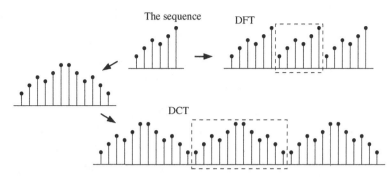

Figure 13.17 A comparison of the DFT and the DCT: The DFT introduces sharp edges while the DCT introduces continuous edges.

The DCT also substantially outperforms the DFT in terms of energy compaction for correlated sources [59]. For Markov sources with high correlation coefficient,

$$\rho = \frac{E\left[x_n x_{n+1}\right]}{E\left[x_n^2\right]}. \tag{13.147}$$

The DCT approximates the Karhunen-Loève transform in terms of the compaction ability [59]. This makes the DCT the most popular transform for information compression.

Two-dimensional DCT

The two-dimensional (2-D) DCT can be implemented using two one-dimensional (1-D) DCTs: one in the horizontal direction followed by one in the vertical direction. For $N \times N$ input data $\mathbf{s}$, the 2-D DCT can be implemented as

$$\mathbf{S} = \mathbf{CsC}^T. \tag{13.148}$$

The 2-D DCT is defined for image processing. The 2-D DCT extracts all coefficients for an $N \times N$ block as

$$S(v, u) = \frac{1}{\sqrt{2N}} c(v)c(u) \sum_{x=0}^{N-1} \sum_{y=0}^{N-1} s(y, x) \cos \frac{(2x+1)u\pi}{2N} \cos \frac{(2y+1)v\pi}{2N}, \tag{13.149}$$

where $S(v, u)$ corresponds to the amplitudes of the frequency components, and $s(y, x)$ is a pixel value in the $N \times N$ block.

The 2-D IDCT for image reconstruction is given by

$$s(y, x) = \frac{1}{\sqrt{2N}} \sum_{u=0}^{N-1} \sum_{v=0}^{N-1} c(u)c(v)S(v, u) \cos \frac{(2x+1)u\pi}{2N} \cos \frac{(2y+1)v\pi}{2N}. \tag{13.150}$$

Like fast DCT, fast 2-D DCT is also based on simple lookup tables. The product of the two cosine terms can be tabulated as a table at the beginning and multiplications are avoided. In addition, efficient 2-D IDCT implementations based on an on-line CORDIC and distributed arithmetic have been given in [79, 80].

Sinusoidal family of unitary transforms

The discrete sine transform (DST) is complementary to the DCT. Unlike the DCT, the compaction ability of the DST is very close to that of the statistically optimal Karhunen-Loève transform when the correlation coefficient ρ, defined by (13.147), is very small. This feature leads to its often being used as the complementary transform to the DCT for image and audio coding [59]. For signals with high correlation, the DCT yields better results; however, for signals with a low correlation coefficient, the DST yields lower BERs.

The DCT and DST belong to a sinusoidal family of unitary transforms. Some other members in this family are the even discrete cosine transform II (EDCT-II) [47], the DCT-II, the DST-II, the DCT-IV, and the DST-IV [48, 52]. The DCT-II is known as the standard DCT. These transforms can generally be implemented based on the DFT [47, 52] or the standard DCT [50].

Systolic architectures are suitable for implementation in VLSI form for real-time computation of the DFT and the DCT, and extension of the architectures to the 2-D transform computation is given in [51]. Systolic array VLSI architectures for the DST and prime length DST are given in [18, 44], and a unified design framework for prime length DCT/IDCT with a high throughput has been presented in [20]. A memory-based unified systolic array implementation for DCT, DST, IDCT, and inverse DST has been proposed in [19].

13.9 Wavelet transform

The wavelet transform is similar to the Fourier transform. The Fourier transform decomposes a signal into a weighted sum of sinusoids. The wavelet transform decomposes a signal into a weighted sum of wavelet functions by simply correlating a set of wavelet functions with the signal. In analogy to the frequency spectrum obtained by using the FFT, a wavelet spectrum can be obtained by using the wavelet transform. Using the wavelet transform, localized temporal and frequency information can be obtained. Wavelet analysis is very useful for processing nonstationary signals.

The set of wavelet functions is generated from a single mother wavelet by using translations and dilations. Given a mother wavelet $\psi(t)$, the wavelets can be generated by

$$\psi_{a,b}(t) = \frac{1}{\sqrt{a}} \psi\left(\frac{t-b}{a}\right), \tag{13.151}$$

where $a > 0$, $-\infty < b < \infty$, and a is referred to as the scale or dilation variable.

An infinite number of real or complex functions can be used as the mother wavelet. To be used as a mother wavelet, a function must be zero mean over the time axis, and be square-integrable. Some well-known wavelets are the Morlet or modified Gaussian wavelet, and the Mexican hat wavelet. The Mexican hat wavelet is the normalized second derivative of a Gaussian function.

The wavelet transform of a real signal $x(t)$ is defined as

$$X(a, b) = \int_{-\infty}^{\infty} \psi_{a,b}^*(t)x(t)dt, \tag{13.152}$$

where * is the complex conjugate. The inverse transform is defined by

$$x(t) = \frac{1}{c_\psi} \int_{b=-\infty}^{\infty} \int_{a=0}^{\infty} X(a, b)\psi_{a,b}(t)\frac{dadb}{a^2}, \tag{13.153}$$

where

$$c_\psi = 2\pi \int_{-\infty}^{\infty} \frac{|\psi(\omega)|^2}{\omega}d\omega < \infty, \tag{13.154}$$

$\psi(\omega)$ being the Fourier transform of $\psi(t)$.

Example 13.7: The Morlet wavelet is given by

$$\psi = e^{j\omega_0 t}e^{-t^2/2}, \tag{13.155}$$

which has a frequency spectrum (using the Fourier transform) given by

$$H(\omega) = \sqrt{2\pi}e^{-\frac{(\omega-\omega_0)^2}{2}}. \tag{13.156}$$

The Morlet wavelet and its frequency spectrum are shown in Fig. 13.18, for $\omega_0 = 4$.

Example 13.8: For a function $x(t) = \cos\left(\sin t^2\right)$, its wavelet transform, $X(a, b)$, based on the Morlet wavelet is demonstrated in Fig. 13.19.

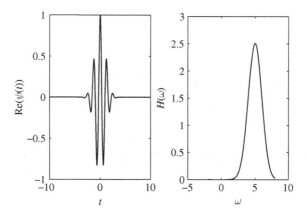

The Morlet wavelet and its Fourier transform, $\omega_0 = 5$.

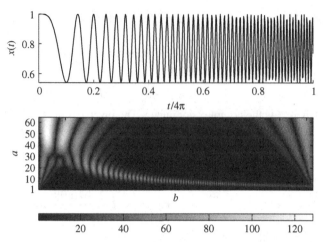

Figure 13.19 The continuous wavelet transform of $x(t) = \cos(\sin t^2)$, by using the Morlet wavelet with $\omega_0 = 5$.

13.9.1 Discrete wavelet transform

A popular approach to discretization of a and b is given by

$$a = a_0^m, \quad b = nb_0 a_0^m, \qquad (13.157)$$

where $a_0 > 1$, m and n are integers. Thus, the wavelet set is given by

$$\psi_{m,n}(t) = a_0^{-m/2} \psi \left(a_0^{-m} t - nb_0 \right), \quad m, n \in Z. \qquad (13.158)$$

From (13.152), the wavelet coefficients are accordingly obtained as

$$X(m, n) = a_0^{-m/2} \int_{-\infty}^{\infty} \psi^* \left(a_0^{-m} t - nb_0 \right) x(t) dt. \qquad (13.159)$$

This is called the *discrete-time wavelet transform (DTWT)* or *discrete wavelet transform (DWT)*. The inverse DWT (IDWT) is obtained by

$$x(t) = \sum_{m=-\infty}^{\infty} \sum_{n=-\infty}^{\infty} X(m, n) \psi_{m,n}(t). \qquad (13.160)$$

The DWT and IDWT processes are known as analysis (decomposition) and synthesis (reconstruction). These operations are in analogy to filter banks.

The fast wavelet transform (FWT) [41] is derived based on dyadic wavelets, which are obtained by scaling the mother wavelet by powers of two, that is, $a_0 = 2$, $b_0 = 1$. The FWT is essentially the classical two-channel sub-band coder using conjugate quadrature filters (CQFs) or quadrature mirror filters (QMFs). The dyadic wavelet transform corresponds to signal analysis with octave band decomposition, since every increment of m doubles the width in the time domain and halves the width in the frequency domain. The FWT is asymptotically faster than the FFT, requiring a complexity of $O(N)$ as opposed to $O(N \log N)$ for the FFT [62].

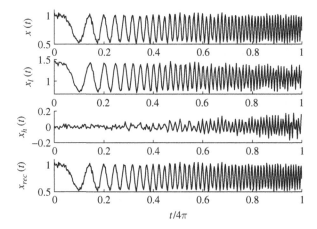

Figure 13.20

Analysis and synthesis of $x(t)$, by using the db2 wavelet.

Example 13.9: For a function $x(t) = \cos\left(\sin t^2\right) + 0.02\mathcal{N}(0, 1)$, $\mathcal{N}(0, 1)$ being zero-mean unit-variance Gaussian noise, the one-stage discrete wavelet transform generates a low frequency $x_l(t)$ (called an approximation), and a high frequency signal $x_h(t)$ (called a detail), based on the db2 wavelet of the Daubechies family wavelets; this is demonstrated in Fig. 13.20. The reconstructed signal x_{rec} is shown to be exactly $x(t)$. The analysis process can repeatedly decompose the approximation signal $x_l(t)$ into next-level approximation and detail. This is known as *wavelet analysis*. When the details as well as the approximations can be split, we get *wavelet packet analysis*.

The DWT is especially useful for signal de-noising and compression. MATLAB provides the *wavemenu* graphic interface for 1-D and 2-D wavelet analysis. The 2-D DWT is especially useful for image de-noising and compression, and is introduced in Chapter 17. The Gabor filter is an example of a wavelet filter, and is widely used in image processing for texture analysis, segmentation and classification. Lifting-based DWT provides advantages over the convolution-based DWT [28].

13.9.2 Multiresolution analysis

Multiresolution analysis is one of the most important applications of the wavelet transform. The complete representation of $f(t)$ requires a discrete wavelet family of an infinite number of orthogonal functions, $\{\psi_{mn}(t)\}$. By introducing a lowpass scaling function $\phi(t)$, with the set $\{\phi(t-n), n \in Z\}$ being orthonormal, $f(t)$ can be represented by multiresolution analysis of the signal [26]

$$f(t) = \sum_{n=-\infty}^{\infty} c_{M,n} 2^{-M/2} \phi\left(2^{-M}t - n\right) + \sum_{m=1}^{M} \sum_{n=-\infty}^{\infty} d_{mn} 2^{-m/2} \psi\left(2^{-m}t - n\right), \quad (13.161)$$

where M is a finite integer, and

$$c_{M,n} = \langle f(t), \phi_{M,n}(t) \rangle = 2^{-M/2} \int_{-\infty}^{\infty} f(t)\phi\left(2^{-M}t - n\right) dt, \qquad (13.162)$$

with $\phi_{M,n}(t) = 2^{-M/2}\phi\left(2^{-M}t - n\right)$. Note that the wavelet and scaling functions are orthogonal to each other. All the scaling functions are of the same scale, since they are shifted versions of $\phi\left(2^{-M}t\right)$, while the wavelet functions have M scales. The simplest wavelets are Haar wavelets.

Wavelets can be built from a lowpass filter $H_0(z)$ and a highpass filter $H_1(z)$. The multiresolution analysis equation is derived by [59]

$$\phi(t) = \sum_{k=-\infty}^{\infty} h_0(k)\phi_{1,k}(t) = \sum_{k=-\infty}^{\infty} h_0(k)\sqrt{2}\phi(2t - k), \qquad (13.163)$$

where $\phi_{1,k}(t) = \sqrt{2}\phi(2t - k)$, $\phi(t)$ is a scaling function, and

$$h_0(k) = \int_{-\infty}^{\infty} \phi(t)\sqrt{2}\phi(2t - k)dt. \qquad (13.164)$$

The scaling function $\phi(t)$ is closely related to the mother wavelet

$$\psi(t) = \sum_{k=-\infty}^{\infty} h_1(k)\phi_{1,k}(t) = \sum_k h_1(k)\sqrt{2}\phi(2t - k). \qquad (13.165)$$

Example 13.10: The Haar scaling function $\phi(t)$ is a unit rectangular pulse of length 1, $\phi(t) = u(t) - u(t - 1)$, where $u(t) = 1$ for $t \geq 0$ and $u(t) = 0$ for $t < 0$; thus, $\phi(t)$ satisfies $\phi(t) = \phi(2t) + \phi(2t - 1)$. The Haar wavelet is obtained by $\psi(t) = \phi(2t) - \phi(2t - 1)$. They correspond to the sum (lowpass filtering) and the difference (highpass filtering) of $\phi(2t)$. The Haar scaling and wavelet functions are plotted in Fig. 13.21. The spectrum of $\phi(t)$

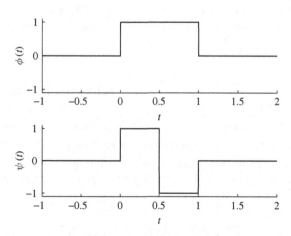

Figure 13.21   The Haar scaling function and wavelet.

is a sinc function that has a lowpass property, while the spectrum of $\psi(t)$ has a bandpass property with its energy concentration moved to higher frequency. The PSD of $\psi(2t)$ shifts its energy concentration further to the higher frequency.

Multiresolution analysis can be implemented using a hierarchical filter structure similar to that used in sub-band coding. By integrating both sides of the multiresolution analysis equation and after manipulation, three equations are obtained for $h_0(k)$ [59]

$$\sum_{k=-\infty}^{\infty} h_0(k) = \sqrt{2}, \quad \sum_{k=-\infty}^{\infty} h_0^2(k) = 1, \quad \sum_{k=-\infty}^{\infty} h_0(k)h_0(k - 2m) = \delta_m. \tag{13.166}$$

When the wavelet function is orthogonal to the scaling function at the same scale, that is,

$$\int_{-\infty}^{\infty} \phi(t - k)\psi(t - m)dt = 0, \tag{13.167}$$

we have [59]

$$h_1(k) \doteq \pm(-1)^k h_0(N - k), \tag{13.168}$$

$$\sum_{k=-\infty}^{\infty} h_0(k)h_1(n - 2k) = 0, \tag{13.169}$$

where N is the length or the filter. Further

$$\sum_{k=-\infty}^{\infty} h_1(k) = 0. \tag{13.170}$$

In [59], $h_0(k)$'s for 4-/12-/20-tap Daubechies and 6-/12-/18-tap Coiflet, lowpass filters are tabulated.

The wavelet transform is suitable for multiresolution analysis. Natural signals such as image, speech/audio signals are well suited for this type of analysis, since it divides frequency into octaves, and the human visual cortex uses a multifrequency channel decomposition to process images. The practical implementation of wavelet compression is very similar to sub-band coding. When the DWT is applied to image compression, some coefficients are discarded to increase the compression rate.

13.10 Filter banks

We are now in a position to discuss filter banks. When the spectrum of an input signal is divided into several nonoverlapping frequency bands that cover the Nyquist interval, we get a parallel filter bank. Filter banks are typically used for multirate sample conversion and sub-band coding of speech or picture signals.

The basic building blocks in a multirate DSP system are decimators and interpolators. Multirate systems can be implemented in the polyphase form. The decimator is the cascade

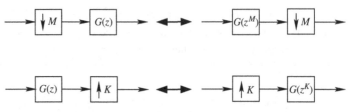

Figure 13.22   Two identity transforms for multirate systems.

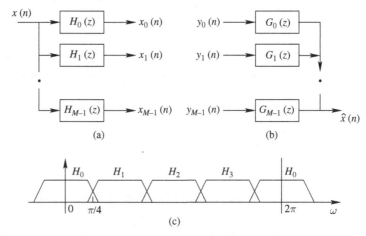

(a)   (b)

(c)

Figure 13.23 Analysis and synthesis filter banks. (a) Analysis bank. (b) Synthesis bank. (c) Typical frequency response of uniform DFT filter bank, for $M = 4$. (Fig.10 [69])©1990, IEEE.

of a lowpass pre-aliasing filter and a downsampler, and the interpolator is the cascade of an upsampler and a lowpass post-image filter. These filters are the canonical components used in filter banks. In filter banks, if a downsampler (decimator) is followed by a transfer function $G(z)$, or a downsampler (interpolator) is preceded by $G(z)$, they can be transformed by the identity transforms for multirate systems provided that $G(z)$ is rational [69]. Two identity transforms known as the Noble identities are shown in Fig. 13.22. They are most valuable for efficient implementation of filters and filter banks.

Sub-band coding can effectively reduce the number of bits for quantization of each filter output and sampling rates of each filter, leading to an overall reduction in the total bits for coding. The DWT is an example of a filter bank. More discussion on filter banks is given in [29].

Analysis and synthesis banks

Two basic types of filter banks are the analysis and synthesis banks. The analysis bank uses a set of analysis filters $H_k(z), k = 1, 2, \ldots, M$, to split a signal $x(n)$ into M sub-band signals $x_k(n)$, while the synthesis bank combines the M sub-band signals $y_k(n)$ into a reconstructed signal $\hat{x}(n)$ by using M synthesis filters $G_k(z)$. These are illustrated by Fig. 13.23. Perfect reconstruction is an important topic for filter banks.

When all $H_k(z)$'s are derived from the prototype filter $H_0(z)$ by $H_k(z) = H_0(zW^k)$, $k = 0, 1, \cdots, M - 1$, where $W = e^{-j2\pi/M}$, such an analysis filter is referred to as a *uniform DFT filter bank*. The frequency responses of $H_k(z)$ are uniformly shifted versions of $H_0(e^{j\omega})$, and this is shown in Fig. 13.23c.

Common transforms as filter banks

The DCT does a job similar to that of a filter bank. The DCT divides a signal into several frequency components, while a filter bank divides the frequency band into several channels. The block DCT transform used for data compression is shown to be equivalent to a perfect reconstruction M-band filter bank [29]. Both the analysis and synthesis filter banks have linear phase.

Example 13.11: As given in (13.145), the elements of the DCT matrix C_{kn} are defined by

$$C_{kn} = C_k(n) = \sqrt{\frac{2}{N}} c(k) \cos\left(\frac{\pi\left(n + \frac{1}{2}\right)k}{N}\right) \qquad (13.171)$$

for $k = 0, 1, \cdots, N - 1$. The DCT and IDCT together constitute an analysis-and-synthesis filter bank, where $h_i(n) = C_i(-n)$ and $g_i(n) = C_i(n)$, for $i = 0, 1, \ldots, N - 1$. Since the impulse response is antisymmetric, $C_k(n) = (-1)^k C_k(N - 1 - n)$, $k = 0, 1, \ldots, N - 1$, the filter bank has linear phase. The analysis and synthesis filters have the same magnitude response, and they are shown in Fig. 13.24 for $N = 8$.

The cosine-modulated filter bank is suitable for design and implementation of filter banks with a large number of sub-bands, since it has a simple design procedure and has low

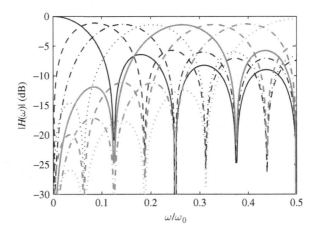

Figure 13.24

Magnitude responses of the DCT filters $h_i(n)$, for $N = 8$.

complexity due to its fast DCT-type implementation. The cosine-modulated filter bank has a nonlinear phase for the analysis filters, and thus is not desirable for image coding. The lapped orthogonal transform (LOT) is a block transform with overlapped basis functions so as to reduce blocking effects, which occur in the case of the DCT. The blocking effect is seen as discontinuities that appear across the block boundaries. The LOT-based design has linear phase analysis filters and has fast implementation. The LOT is also based on the DCT, which enables fast implementation [49]. Design and implementation of these filter banks are addressed in [29].

In [36], the design of general biorthogonal modulated filter banks (cosine-modulated and modified DFT) has been developed, using the polyphase representation. Working with the design parameters of the number of channels M, the prototype filter length N, and the system delay D, necessary and sufficient conditions for perfect reconstruction are derived using the polyphase representation. These perfect reconstruction conditions are shown to be identical for all types of modulation: modulation based on the DCT, both DCT-III/DCT-IV and DCT-I/DCT-II, and modulation based on the modified DFT. A quadratic-constrained least-squares technique has been derived for the design of prototype filters for perfect reconstruction modulated filter banks, with full generality in setting the parameters M, N, and D. The best subchannelization (greatest stopband attenuation) is obtained for $D = N - 1$.

For digital signals, the wavelet transform is shown to be a special case of critically decimated filter banks [29]. Nonuniform perfect reconstruction filter banks with integer decimation factors are also useful for linear multirate systems.

13.11 Sub-band coding

Sub-band coding, also known as *frequency-domain coding*, was first introduced in [24]. For sub-band coding, the signal is split into M sub-band signals using M analysis filters $H_k(z)$. Each sub-band signal is decimated by M, and is then quantized using a number of bits that depend on the energy content. This strategy is a generalization of the decimation process. The reconstruction of the signal is performed by first interpolating each sub-band signal by M to restore the original sampling rate and then combining all the interpolated sub-band signals using synthesis bank filters $G_k(z)$, which also eliminate the images. For $M = 2$, $\left|H_0\left(e^{j\omega}\right)\right|$ is usually an image of $\left|H_1\left(e^{j\omega}\right)\right|$ with respect to the quadrature frequency $\pi/2$, thus the name *QMF*. The parallel structure is shown in Fig. 13.25.

For $M > 2$, such an analysis/synthesis system for sub-band coding is called the *M-band maximally decimated analysis/synthesis system*, but is sometimes also called a *pseudo-QMF filter bank* for simplicity. For sub-band coding, the M analysis filters $H_k(z)$ divide the band into overlapping lowpass, bandpass and highpass portions. The overlaps allow that no portion of the frequency response is severely attenuated, which, however, introduces aliasing when the corresponding subband signals are decimated. The synthesis filters $G_k(z)$ are used to cancel this aliasing as well as the images introduced by interpolation at the reconstruction stage.

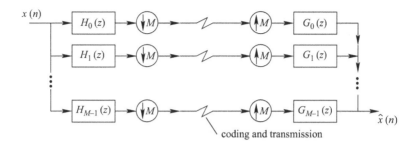

Figure 13.25 The M-band maximally decimated analysis/synthesis system for sub-band coding. For $M = 2$, it is the two-band QMF bank.

The magnitude response of the M sub-bands can be made to have different amplitudes β_k; this leads to FIR filters with adjustable multilevel response

$$G(z) = \sum_{k=0}^{M-1} \beta_k H\left(zW^k\right). \tag{13.172}$$

By selecting a large M and different β_k, one can adjust the magnitude response of the filter.

The pseudo-QMF bank is very important for modern audio coding. Frequency-domain coding has an advantage over time-domain coding, in that each frequency component can be allocated different numbers of bits. This allows us to control the level of quantization noise in each component. Filter banks are also used in the multirate generalized multi-carrier (GMC) CDMA model [73].

13.11.1 Two-channel perfect reconstruction filter banks

Quadrature mirror filter bank

For the two-band QMF bank, we need to define filters $h_0(n)$, $h_1(n)$, $g_0(n)$, and $g_1(n)$ that allow perfect reconstruction of the input $x(n)$. Let $y_0(n)$ and $y_1(n)$ be the transmitted intermediate data streams. We have, from the synthesis filter bank

$$\hat{X}(z) = Y_0\left(z^2\right) G_0(z) + Y_1\left(z^2\right) G_1(z) \tag{13.173}$$

and from the analysis filter bank

$$Y_i(z) = \frac{1}{2}\left(H_i\left(z^{1/2}\right) X\left(z^{1/2}\right) + H_i\left(-z^{1/2}\right) X\left(-z^{1/2}\right)\right), \quad i = 0, 1. \tag{13.174}$$

Combining (13.173) and (13.174) yields

$$\hat{X}(z) = \frac{1}{2}\left(H_0(z)G_0(z) + H_1(z)G_1(z)\right) X(z)$$

$$+ \frac{1}{2}\left(H_0(-z)G_0(z) + H_1(-z)G_1(z)\right) X(-z). \tag{13.175}$$

In order to remove the aliasing term, the synthesis filters can be chosen as

$$G_0(z) = H_1(-z), \quad G_1(z) = -H_0(-z) \tag{13.176}$$

and only the first term in (13.175) remains:

$$\hat{X}(z) = \frac{1}{2}\left(H_0(z)H_1(-z) - H_1(z)H_0(-z)\right)X(z). \tag{13.177}$$

To achieve perfect reconstruction, we can set

$$P(z) - P(-z) = H_0(z)H_1(-z) - H_1(z)H_0(-z) = z^{-D}, \tag{13.178}$$

where

$$P(z) = H_0(z)H_1(-z). \tag{13.179}$$

That is, the reconstructed signal is a D-samples-delayed replica of the input signal.

There are a number of methods to achieve perfect reconstruction. The FIR QMF filter order must be even for perfect reconstruction. The QMF solution [25] is the most popular one, which takes the form

$$H_1(z) = H_0(-z). \tag{13.180}$$

This implies that if H_0 is a lowpass filter then H_1 is highpass. The four filters are related by

$$H_1(z) = H_0(-z) \longleftrightarrow h_1(n) = (-1)^n h_0(n), \tag{13.181}$$

$$G_0(z) = H_0(z) \longleftrightarrow g_0(n) = h_0(n), \tag{13.182}$$

$$G_1(z) = -H_0(-z) \longleftrightarrow g_1(n) = -(-1)^n h_0(n). \tag{13.183}$$

Finally, we have

$$\hat{X}(z) = \frac{1}{2}\left(H_0^2(z) - H_0^2(-z)\right)X(z). \tag{13.184}$$

In case of perfect reconstruction, the output signal is a time-delayed version of the input signal, thus the filter should be designed so that

$$H_0^2(z) - H_0^2(-z) = z^{-D}. \tag{13.185}$$

No exact solution for FIR filters with more than 2 taps has been found to satisfy the QMF perfect reconstruction condition, but it can be approximated very well by using long FIR filters. The Haar filters, given by $H_0(z) = \frac{1}{2}(1+z^{-1})$ and $H_1(z) = \frac{1}{2}(1-z^{-1})$ are examples of a QMF filter bank. For a QMF bank, the QMF symmetry constraint $H_1(z) = H_0(-z)$ makes the implementation of linear-phase FIR filters only exist in the two-tap case.

Conjugate quadrature filter bank

Another solution that is better adapted for FIR implementation is the CQF solution [60]. The synthesis filters $g_0(n)$ and $g_1(n)$ are just time reversals of the analysis filters of length N

$$g_0(n) = h_0(N - 1 - n), \tag{13.186}$$

$$g_1(n) = (-1)^n h_0(n) = h_1(N - 1 - n) \tag{13.187}$$

and

$$h_1(n) = -(-1)^n h_0(N - 1 - n). \tag{13.188}$$

These filters can be represented in the z-domain as [6]

$$G_0(z) = z^{-(N-1)} H_0(z^{-1}), \tag{13.189}$$

$$G_1(z) = z^{-(N-1)} H_1(z^{-1}), \tag{13.190}$$

$$H_1(z) = z^{-(N-1)} H_0(-z^{-1}). \tag{13.191}$$

These filters satisfy the alias cancellation conditions for even N. Note that the filters $H_0(z)$ and $H_1(z)$ do not have linear phase. The perfect reconstruction condition leads to

$$H_0(z)H_0\left(z^{-1}\right) + H_0(-z)H_0\left(-z^{-1}\right) = 1 \tag{13.192}$$

when $D = N - 1$. This condition corresponds to the *power complementary condition* in the frequency domain,

$$|H_0(\omega)|^2 + |H_1(\omega)|^2 = 1. \tag{13.193}$$

A filter bank satisfying the power complementary condition gives perfect reconstruction, but the reverse does not necessarily hold.

13.11.2 Pseudo-QMF filter bank

The two-channel CQF is generalized into the pseudo-QMF filter bank. The pseudo-QMF filter bank is a near-perfect reconstruction filter bank. It is used in the layer-I and layer-II audio coders in the MPEG-1 and MPEG-2 standards for time to frequency mapping.

The pseudo-QMF bank consists of M channels. The analysis and synthesis filters are given by

$$h_k(n) = h(n) \cos\left(\frac{(k + \frac{1}{2})\pi}{M}\left(n - \frac{N-1}{2}\right) + \phi_k\right), \tag{13.194}$$

$$g_k(n) = h_k(N - 1 - n) \tag{13.195}$$

for $k = 0, 1, \ldots, M - 1$, where N is the length of the lowpass prototype window $h(n)$, and the phase ϕ_k satisfies

$$\phi_k - \phi_{k-1} = m\pi + \frac{\pi}{2}, \quad m \in Z. \tag{13.196}$$

This corresponds to the anti-aliasing conditions between adjacent bands.

The perfect reconstruction condition requires to that the power complementary equation be satisfied within $|f| \leq \frac{f_s}{2M}$:

$$|H(e^{j\omega})|^2 + \left|H\left(e^{j(\pi/M - \omega)}\right)\right|^2 = 1, \quad \text{for } |\omega| \leq \frac{\pi}{2M}. \tag{13.197}$$

13.11.3 Modified DCT (MDCT)

Like the pseudo-QMF, the MDCT is also used for time-to-frequency mapping of signals. The MDCT is a block transform method that transforms a block of time-domain samples into its frequency-domain representation, and the method is called *transform coding*. Mathematically, both sub-band coding and transform coding are essentially using the same technology. Usually, coders with a small number of frequency bands are referred to as sub-band coders, while coders with many frequency channels are called transform coders.

The MDCT corresponds to a perfect reconstruction filter bank. It can be used in the case of a very high number of frequency bands. The MDCT/inverse MDCT (IMDCT) is used in the MPEG AAC (ISO/IEC 13818-7) and Dolby AC-2 and AC-3 [6] audio standards. A hybrid filter bank that cascades the pseudo-QMF and the MDCT is used in MPEG Layer III Audio.

The MDCT is a special case of a class of transforms called *time domain aliasing cancellation (TDAC)* transform, and is also known as an *oddly-stacked TDAC (OTDAC)* transform [55, 56]. In the TDAC transform, an overlap-and-add procedure is used to exactly cancel the time-domain aliasing by using an overlap between blocks. The TDAC can lead to perfect reconstruction filter banks without increasing the data rate. The MDCT allows us to have 50% overlap between successive data blocks without increasing the overall data rate. It is a special case of the $N = 2M$ pseudo-QMF filter bank that produces exact perfect reconstruction but allows a wider set of prototype filters, where M is the number of frequency channels [6]. The MDCT phase term cancels frequency aliasing between all pairs of frequency bands, not merely adjacent bands. The MDCT allows lower-cost filters than the pseudo-QMF due to a lesser requirement on stopband attenuation, and has become the transform of choice for most of the newer audio coders. The implementation of the MDCT using the FFT reduces the number of operations to $O\left(N \log_2 N\right)$.

The MDCT transforms N inputs from the ith and $(i-1)$th sets of $N/2$ inputs into $N/2$ frequency-domain outputs

$$X(k) = \sum_{n=0}^{N-1} x(n) w_a(n) \cos\left(\frac{2\pi}{N}(n+n_0)\left(k+\frac{1}{2}\right)\right), \quad k = 0, 1, \ldots, \frac{N}{2} - 1, \quad (13.198)$$

where $n_0 = \frac{2+N}{4}$, and w_a is the analysis window.

At synthesis stage the $N/2$ frequency-domain outputs are transformed back into N time-domain samples, via the IMDCT

$$\hat{x}(n) = w_s(n) \frac{2}{N} \sum_{k=0}^{N/2-1} X(k) \cos\left(\frac{2\pi}{N}(n+n_0)\left(k+\frac{1}{2}\right)\right), \quad n = 0, 1, \ldots, N-1, \quad (13.199)$$

where w_s is the synthesis window.

The windows w_a and w_s should satisfy the condition that their overlapping portions between adjacent blocks should be time-reversals of each other, and the perfect reconstruction condition

$$w_a^i(n)w_s^i(n) + w_a^{i-1}\left(\frac{N}{2} + n\right) w_s^{i-1}\left(\frac{N}{2} + n\right) = 1, \quad n = 0, 1, \ldots, \frac{N}{2} - 1. \quad (13.200)$$

The windows can be a Kaiser window or a sine window. The resolution of the filter bank can be altered by changing the window length N. The blocking effect requires the use of smooth windows for the input signals and overlap-and-add of the output data for reconstruction of the signal [6].

There are some algorithms available for efficient implementations of MDCT /IMDCT [16, 32]. In [53], a parallel/pipelined algorithm for the computation of both the MDCT and IMDCT has been proposed based on the FFT.

Problems

13.1 A signal $x(t) = \sin(\pi t) + 4\sin(2\pi t)\cos(5\pi t)$, where t is in msec. The signal is sampled at a rate of 5 kHz. Determine the signal $x_a(t)$ aliased with $x(t)$. Then, give two other signals $x_1(t)$ and $x_2(t)$ that are aliased with $x_a(t)$, that is, $x_1(nT) = x_2(nT) = x_a(nT)$. [Hint: Expand $x(t)$ into a sum of sines and cosines.]

13.2 Implement in MATLAB a CIC filter with six stages of integrate and six stages of comb to reduce the sampling rate by a factor of four. Filter a sinusoidal signal with a frequency of 0.3 Hz to illustrate the effect of aliasing. Assume the normalized sampling rate is 1 Hz.

13.3 Determine the impulse response h of the following FIR filters: (a) $y(n) = 2x(n) + 4x(n-1) + 5x(n-2) + 2x(n-4)$; (b) $y(n) = x(n) - x(n-5)$.

13.4 A causal IIR filter has impulse response $h(n) = 5\delta(n) + 3(0.5)^{n-1}u(n)$. Working with the convolutional equation $y(n) = \sum_m h(m)x(n-m)$, derive the difference equation for $y(n)$.

13.5 Calculate the output y of the following filter with input: $h = [1, 2 - 1, 1]$, $x = [1, 1, 2, 1, 2, 2, 1, 1]$. (a) Use $y = h * x$. (b) Use z-transform $Y(z) = H(z)X(z)$.

13.6 Using the unit-step identity $u(n) - u(n-1) = \delta(n)$, determine the z-transform of: (a) $x(n) = u(n)$ and (b) $x(n) = -u(-n-1)$. Verify that their z-transforms are the same. One of the signals is causal and the other is anticausal. What are their ROCs?

13.7 Compute all possible inverse z-transforms of

$$X(z) = \frac{5 + z^{-1}}{1 - 0.25z^{-2}}.$$

13.8 Prove the modulation property of z-transforms. If the z-transform of $x(n)$ is $X(z)$, show the z-transform of $a^n x(n)$ is $X(z/a)$. In the frequency domain, derive the spectrum of $e^{j\omega_0 n}x(n)$ is $X(\omega - \omega_0)$, by using $a = e^{j\omega_0}$.

13.9 For a filter with frequency response

$$H(\omega) = \frac{-0.5 + e^{-j\omega 4}}{1 - 0.5e^{-j\omega 4}},$$

where ω is the digital frequency in radians/sample, determine the causal impulse response $h(n)$ for all $n \geq 0$.

13.10 For a digital filter,

$$H(z) = \frac{1 - z^{-16}}{1 - az^{-16}},$$

where $0 < a < 1$, (a) What are the poles and zeros? Plot them on the z-plane. (b) Draw the magnitude response $|H(\omega)|$ over $0 \leq \omega \leq 2\pi$. (c) Determine the causal/stable impulse response $h(n)$.

13.11 Design a multi-notch filter with $D = 8$ and the 3-dB width in the range of 0 and 0.05π.

13.12 Determine the length-9, rectangularly windowed impulse response that approximates: (a) an ideal lowpass filter of cutoff frequency $\omega_c = \pi/5$, (b) an ideal bandpass filter with cutoff frequencies $\omega_a = \pi/5$ and $\omega_b = \pi/3$.

13.13 Using the Kaiser window, design a bandpass digital filter with $f_s = 20$ kHz, $f_{sa} = 3$ kHz, $f_{pa} = 4$ kHz, $f_{pb} = 7$ kHz, $f_{sb} = 8$ kHz, $A_{pass} = 0.1$ dB, $A_{stop} = 80$ dB.

13.14 Determine whether the following systems are causal and/or stable: (a) $h(n) = 0.5^n u(n)$, (b) $h(n) = u(n + 2) - u(n - 1)$, (c) $h(n) = \sin(\pi n/4)u(n)$, (d) $h(n) = 0.5^n u(n) + 2^n u(-n - 1)$.

13.15 Consider a discrete-time LTI system with impulse response $h(n)$. If the input $x(n)$ is a periodic sequence with period N, show that the output $y(n)$ also has a periodic of N.

13.16 A discrete causal LTI system has the system function

$$H(z) = \frac{(1 + 0.4z^{-1})(1 - 9z^{-2})}{1 + 0.64z^{-2}}.$$

Is the system stable? Express $H(z)$ as the product of a minimum-phase system and an all-pass system.

13.17 Design an FIR lowpass filter satisfying the specifications: $0.95 < H(e^{j\omega}) < 1.05$ for $0 \leq |\omega| \leq 0.26\pi$, $-0.1 < H(e^{j\omega}) < 0.1$ for $0.4\pi \leq |\omega| \leq \pi$. The cutoff $\omega_c = 0.3\pi$. Which windows can be used for this specification? What is the filter length?

References

[1] N. Ahmed, T. Natarajan & K. R. Rao, Discrete cosine transform. *IEEE Trans. Computers*, **23**:1 (1974), 90–93.

[2] Alterra, *Understanding CIC Compensation Filters*, Application Note 455, v1.0, Apr 2007.

[3] R. G. Alves, P. L. 0sorio & M. N. S. Swamy, General FFT pruning algorithm. In *Proc. IEEE Midwest Symp. Circ. Syst.*, Lansing MI, Aug 2000, 1192–1195.

[4] A. Antoniou, *Digital Signal Processing: Signals, Systems, and Filters* (New York: McGraw-Hill, 2006).

[5] G. Bi & Y. Q. Chen, Fast DFT algorithms for length $N = q * 2^m$. *IEEE Trans. Circ. Syst. II*, **45**:6 (1998), 685–690.

[6] M. Bosi & R. E. Goldberg, *Introduction to Digital Audio Coding and Standards* (Boston, MA: Kluwer, 2003).

[7] S. Bouguezel, M. O. Ahmad & M. N. S. Swamy, Efficient pruning algorithms for the DFT computation for a subset of output samples. In *Proc. IEEE ISCAS*, Bangkok, Thailand, May 2003, **4**, 97–100.

[8] S. Bouguezel, M. O. Ahmad & M. N. S. Swamy, A new radix-2/8 FFT algorithm for length-$q \times 2^m$ DFTs. *IEEE Trans. Circ. Syst. I*, **51**:9 (2004), 1723–1732.

[9] S. Bouguezel, M. O. Ahmad & M. N. S. Swamy, A new split-radix FHT algorithm for length-$q * 2^m$ DHTs. *IEEE Trans. Circ. Syst. I*, **51**:10 (2004), 2031–2043.

[10] S. Bouguezel, M. O. Ahmad & M. N. S. Swamy, A note on "Split vector-radix-2/8 2-D fast Fourier transform". *IEEE Signal Process. Lett.*, **12**:3 (2005), 185.

[11] S. Bouguezel, M. O. Ahmad & M. N. S. Swamy, New radix-$(2 \times 2 \times 2)/(4 \times 4 \times 4)$ and radix-$(2 \times 2 \times 2)/(8 \times 8 \times 8)$ DIF FFT algorithms for 3-D DFT. *IEEE Trans. Circ. Syst. I*, **53**:2 (2006), 306–315.

[12] S. Bouguezel, M. O. Ahmad & M. N. S. Swamy, Multidimensional vector radix FHT algorithms. *IEEE Trans. Circ. Syst. I*, **53**:4 (2006), 905–917.

[13] S. Bouguezel, M. O. Ahmad & M. N. S. Swamy, A split vector-radix algorithms for the 3-D discrete Hartley transform. *IEEE Trans. Circ. Syst. I*, **53**:9 (2006), 1966–1976.

[14] S. Bouguezel, M. O. Ahmad & M. N. S. Swamy, A general class of split-radix FFT algorithms for the computation of the DFT of length-2^m. *IEEE Trans. Signal Process.*, **55**:8 (2007), 4127–4138.

[15] R. N. Bracewell, The fast Hartley transform. *Proc. IEEE*, **72**:8 (1984), 1010–1018.

[16] V. Britanak & K. R. Rao, An efficient implementation of the forward and inverse MDCT in MPEG audio coding. *IEEE Signal Process. Lett.*, **8**:2 (2001), 48–51.

[17] W.-H. Chen, C. H. Smith & S. C. Fralick, A fast computational algorithm for the discrete cosine transform. *IEEE Trans. Commun.*, **25**:9 (1977), 1004–1009.

[18] D. F. Chiper, M.N.S. Swamy, M.O. Ahmad & T. Stouraitis, A systolic array architecture for the discrete sine transform. *IEEE Trans. Signal Process.*, **50**:9 (2002), 2347–2354.

[19] D. F. Chiper, M. N. S. Swamy, M. O. Ahmad & T. Stouraitis, Systolic algorithms and a memory-based design approach for a unified architecture for the

computation of DCT/DST/IDCT/IDST. *IEEE Trans. Circ. Syst. I*, **52**:6 (2005), 1125–1137.

[20] D. F. Chiper, M. N. S. Swamy & M. O. Ahmad, An efficient unified framework for implementation of a prime-length DCT/IDCT with high throughput. *IEEE Trans. Signal Process.*, **55**:6 (2007), 2925–2936.

[21] J. W. Cooley & J. W. Tukey, An alorithm for the machine computation of complex Fourier series. *Math. Comput.*, **19**:2 (1965), 297–301.

[22] L. Cordesses, Direct digital synthesis: a tool for periodic wave generation (Part 1; Part 2). *IEEE Signal Process. Mag.*, **21**:4 (2004), 50–54; **21**:5 (2004), 110–112, 117.

[23] R. E. Crochiere & L. R. Rabiner, *Multirate Digital Signal Processing* (Englewood Cliffs, NJ: Prentice Hall, 1983).

[24] R. E. Crochiere, S. A. Weber & J. L. Flanagan, Digital coding of speech in sub-bands. *Bell Syst. Tech. J.*, **55**:8 (1976), 1069–1085.

[25] A. Croisier, D. Esteban & C. Galand, Perfect channel splitting by use of interpolation, decimation, and tree decomposition techniques. In *Proc. Int. Conf. Inf. Sci. Syst.*, Patras, Greece, Aug 1976, 443–446.

[26] I. Daubechies, Orthonormal bases of compactly supported wavelets. *Commun. Pure Appl. Math.*, **41**:7 (1988), 909–996.

[27] I. Daubechies, The wavelet transform, time frequency localization and signal analysis. *IEEE Trans. Inf. Theory*, **36**:5 (1990), 961–1005.

[28] I. Daubechies & W. Sweldens, Factoring wavelet transformations into lifting schemes. *J. Fourier Analysis & Applic.*, **4** (1998), 247–269.

[29] P. S. R. Diniz, E. A. B. da Silva & S. L. Netto, *Digital Signal Processing: System Analysis and Design* (Cambridge, UK: Cambridge University Press, 2002).

[30] K.-L. Du & M. N. S. Swamy, *Neural Networks in a Softcomputing Framework* (London: Springer, 2006).

[31] P. Duhamel. Implementation of split-radix FFT algorithms for complex, real, and real-symmetric data. *IEEE Trans. Acoustics Speech Signal Process.*, **34**:2 (1986), 285–295.

[32] P. Duhamel, Y. Mahieux & J. P. Petit, A fast algorithm for the implementation of filter banks based on time-domain aliasing cancellation. In *Proc. IEEE ICASSP*, Toronto, Canada, May 1991, 2209–2212.

[33] W. E. Egan, *Phase-Lock Basics* (New York: Wiley, 1998).

[34] R. W. Hamming, *Digital Filters*, 3rd edn (Mineola, NY: Dover, 1998).

[35] F. J. Harris, On the use of windows for Harmonic analysis with the discrete Fourier transform. *Proc. IEEE*, **66**:1 (1978), 51–84.

[36] P. N. Heller, T. Karp & T. Q. Nguyen, A general formulation of modulated filter banks. *IEEE Trans. Signal Process.*, **47**:4 (1999), 986–1002.

[37] E. B. Hogenauer, An economical class of digital filters for decimation and interpolation. *IEEE Trans. Acoust. Speech Signal Process.*, **29**:2 (1981), 155–162.

[38] X. H. Huang, K.-L. Du, A. K. Y. Lai & K. K. M. Cheng, A unified software radio architecture. In *Proc. IEEE SPAWC*, Taoyuan, Taiwan, Mar 2001, 330–333.

[39] E. C. Ifeachor & B. W. Jervis, *Digital Signal Processing: A Practical Approach*, 2nd edn (Harlow, UK: Prentice Hall, 2002).

[40] J. F. Kaiser & R. W. Schafer, On the use of the I_0-sinh window for spectrum analysis. *IEEE Trans. Acoust. Speech Signal Process.*, **28**:1 (1980), 105–107.

[41] S. Mallat, A theory for multiresolution signal decomposition: the wavelet representation. *IEEE Trans. Pat. Anal. Mach. Intell.*, **11**:7 (1989), 674–693.

[42] J. H. McClellan & T. W. Parks, A unified approach to the design of optimum FIR linear-phase digital filters. *IEEE Trans. Circ. Theory*, **20**:6 (1973), 697–701.

[43] J. H. McClellan, T. W. Parks & L. R. Rabiner, 'A computer program for designing optimum FIR linear-phase filters. *IEEE Trans. Audio Electroacoustics*, **21**:6 (1973), 506–526.

[44] P. K. Meher & M. N. S. Swamy, New systolic algorithm and array architecture for prime-length discrete sine transform. *IEEE Trans. Circ. Syst. II*, **54**:3 (2007), 262–266.

[45] P. K. Meher & M. N. S. Swamy, High throughput memory-based architecture for DHT using a new convolutional formulation. *IEEE Trans. Circ. Syst. II*, **54**:7 (2007), 606–610.

[46] F. Mintzer, On half-band, third-band and Nth band FIR filters and their design. *IEEE Trans. Acoust. Speech Signal Process.*, **30**:5 (1982), 734–738.

[47] N. R. Murthy & M. N. S. Swamy, On the algorithms for the computation of even discrete cosine transform-2 (EDCT-2) of real sequences. *IEEE Trans. Circ. Syst.*, **37**:5 (1990), 625–627.

[48] N. R. Murthy & M. N. S. Swamy, On the computation of running discrete cosine and sine transforms. *IEEE Trans. Signal Process.*, **40**:6 (1992), 1430–1437.

[49] N. R. Murthy & M. N. S. Swamy, On the hardware implementation of the lapped orthogonal transform and the modulated lapped transform. In *Proc. IEEE ISCAS*, San Diego, CA, May 1992, **1**, 145–148.

[50] N. R. Murthy & M. N. S. Swamy, On a novel decomposition of the DCT and its application. *IEEE Trans. Signal Process.*, **41**:1 (1993), 480–485.

[51] N. R. Murthy & M. N. S. Swamy, On the real-time computation of DFT and DCT through systolic architectures. *IEEE Trans. Signal Process.*, **42**:4 (1994), 988–991.

[52] N. R. Murthy & M. N. S. Swamy, On the on-line computation of DCT-IV and DST-IV transforms. *IEEE Trans. Signal Process.*, **43**:5 (1995), 1249–1251.

[53] N. R. Murthy & M. N. S. Swamy, A parallel/pipelined algorithm for the computation of MDCT and IMDCT. In *Proc. IEEE ISCAS*, Bangkok, Thailand, May 2003, **4**, 540–543.

[54] S. J. Orfanidis, *Introduction to Signal Processing* (Englewood Cliffs, NJ: Prentice Hall, 1995).

[55] J. P. Princen & A. B. Bradley, Analysis/synthesis filter bank design based on time domain aliasing cancellation. *IEEE Trans. Acoust. Speech Signal Process.*, **34**:5 (1986), 1153–1161.

[56] J. P. Princen, A. W. Johnson & A. B. Bradley, Subband/transform coding using filter bank designs based on time domain aliasing cancellation. In *Proc. IEEE ICASSP*, Dallas, TX, Apr 1987, **12**, 2161–2164.

[57] J. G. Proakis & D. G. Manolakis, *Digital Signal Processing: Principle, Algorithms, and Applications*, 4th edn (Upper Saddle River, NJ: Pearson Prentice Hall, 2007).

[58] S. Samadi, M. O. Ahmad & M. N. S. Swamy, Characterization of nonuniform perfect-reconstruction filter banks using unit step signal. *IEEE Trans. Signal Process.*, **52**:9 (2004), 2490–2499. – Also see, Correction to "Characterization of nonuniform perfect-reconstruction filter banks using unit step signal", *ibid.*, **52**:10 (2004), 2946.

[59] K. Sayood, *Introduction to Data Compression*, 2nd edn (San Mateo, CA: Morgan Kaufmann, 2000).

[60] M. J. T. Smith & T. P. Barnwell III, Exact reconstruction techniques for tree-structured sub-band coders. *IEEE Trans. Acoust. Speech Signal Process.*, **34**:3 (1986), 431–441.

[61] H. V. Sorensen & C. S. Burrus, Efficient computation of the DFT with only a subset of input or output points. *IEEE Trans. Signal Process.*, **41**:3 (1993), 1184–1200.

[62] G. Strang & T. Nguyen, *Wavelets and Filter Banks* (Wellesley, MA: Wellesley-Cambridge Press, 1997).

[63] D. Stranneby & W. Walker, *Digital Signal Processing and Applications*, 2nd edn (London: Elsevier, 2004).

[64] D. Sundararajan, M. O. Ahmad & M. N. S. Swamy, A fast FFT bit-reversal algorithm. *IEEE Trans. Circ. Syst. II*, **41**:10 (1994), 701–703.

[65] D. Sundararajan, M. O. Ahmad & M. N. S. Swamy, Fast computation of the discrete Fourier transform of real data. *IEEE Trans. Signal Process.*, **45**:8 (1997), 2010–2022.

[66] D. Sundararajan, M. O. Ahmad & M. N. S. Swamy, Vector computation of the discrete Fourier transform. *IEEE Trans. Circ. Syst. II*, **45**:4 (1998), 449–461.

[67] E. E. Swartzlander, Jr., D. V. S. Chandra, H. T. Nagle, Jr. & S. A. Starks, Sign/logarithm arithmetic for FFT implementation. *IEEE Trans. Comput.*, **32**:6 (1983), 526–534.

[68] C. S. Turner, Recursive discrete-time sinusoidal oscillators. *IEEE Signal Process. Mag.*, **20**:3 (2003), 103–111.

[69] P. P. Vaidyanathan, Multirate digital filters, filter banks, polyphase networks, and applications: a tutorial. *Proc. IEEE*, **78**:10 (1990), 56–93.

[70] J. Valls, T. Sansaloni, A. Perez-Pascual, V. Torres, & V. Almenar, The use of CORDIC in software defined radios: a tutorial. *IEEE Commun. Mag.*, **44**:9 (2006), 46–50.

[71] S. V. Vaseghi, *Advanced Signal Processing and Digital Noise Reduction* (Chichester, UK: Wiley & Teubner, 1996).

[72] J. E. Volder, The CORDIC trigonometric computing technique. *IRE Trans. Electron. Comput.*, **8**:3 (1959), 330–334

[73] Z. Wang & G. B. Giannakis, Block spreading for multipath-resilient generalized multi-carrier CDMA. In G. B. Giannakis, Y. Hua, P. Stoica & L. Tong, eds., *Signal Processing in Wireless & Mobile Communications: Trends in Single- and Multi-user Systems 2* (Upper Saddle River, NJ: Prentice Hall, 2001), pp. 223–266.

[74] Y. Wang, H. M. Lam, C.-Y. Tsui, R. S. Cheng & W. H. Mow, Low complexity OFDM receiver using log-FFT for coded OFDM system. In *Proc. IEEE ISCAS*, Scottsdale, AZ, May 2002, **3**, 445–448.

[75] B. Widrow, J. M. McCool, M. G. Larimore & J. C. R. Johnson, Adaptive switching circuits. In *Proc. IRE WESCON Conv.*, Los Angeles, CA, 1960, 96–104.

[76] D. Wulich, E. I. Plotkin & M. N. S. Swamy, Discrete time-varying filter and PLL for synchronous estimation of parameters of a sine signal corrupted by a closely spaced FM interference. *Signal Process.*, **21**:2 (1990), 183–194.

[77] D. Wulich, E. I. Plotkin & M. N. S. Swamy, Constrained notch filtering of nonuniformly spaced samples for enhancement of an arbitrary signal corrupted by a strong FM interference. *IEEE Trans. Signal Process.*, **39**:10 (1991), 2359–2363.

[78] D. Wulich, E. I. Plotkin, M. N. S. Swamy & W. Tong, PLL Synchronized time-varying constrained notch filter for retrieving a weak multiple sine signal jammed by FM interference. *IEEE Trans. Signal Process.*, **40**:11 (1992), 2866–2870.

[79] Y. Yang, C. Wang, M. O. Ahmad & M. N. S. Swamy, An on-line CORDIC based 2-D IDCT implementation using distributed arithmetic. In *Proc. IEEE ISSPA*, Kuala Lumpur, Malaysia, Aug 2001, 296–299.

[80] Y. Yang, C. Wang, M. O. Ahmad & M. N. S. Swamy, An FPGA implementation of an on-line radix-4 CORDIC 2-D IDCT core. In *Proc. IEEE ISCAS*, Scottsdale, AZ, May 2002, **4**, 763–766.

[81] C. R. Zou, E. I. Plotkin & M. N. S. Swamy, 2-D fast Kalman algorithms for adaptive parameter estimation of nonhomogeneous Gaussian Markov random field model. *IEEE Trans. Circ. Syst. II*, **41**:10 (1994), 678–692.

Fundamentals of information theory

14.1 Basic definitions

The foundation of information theory was established by Shannon, who formulated three important fundamental theorems: source-coding, channel-coding, and rate-distortion theorems [21, 22]. Source coding or data compression is used to remove redundancy in a message so as to maximize the information storage or spectral efficiency for transmission, and the source-coding theorem answers the question as to how much information is left after the removal of the redundancy. In contrast, channel coding adds redundancy to a message to combat errors caused by the channel. The rate-distortion theorem relates the minimum average bit rate for representing a discrete or continuous random variable with a specified distortion, and this is very useful for quantization. Channel capacity of a communication network is another important topic of information theory.

Information

Consider a random process X that takes values from a finite alphabet $\mathcal{X}$ consisting of elements X_i with probability $\Pr(X_i)$, $i = 1, \ldots, M$. Information of an event or a symbol X_i is defined as

$$I(X_i) = \log_2 \frac{1}{\Pr(X_i)} = -\log_2 \Pr(X_i) \quad \text{(bits)}. \tag{14.1}$$

Base 2 is generally used, since binary representation for symbols is used in digital communications.

Entropy

For a signal with statistically independent discrete sources (symbols) $\mathcal{X} = \{X_i | i = 1, \ldots, M\}$, by defining A_i as the event of $X = X_i$, the entropy of variable X is obtained by taking the average of the information for all the events

$$H(X) = E[I(A_i)] = \sum_{m=1}^{M} \Pr(X_m) \log_2 \frac{1}{\Pr(X_m)} \quad \text{(bits/symbol)}. \tag{14.2}$$

The entropy is treated as the information inherent in the variable X, and it is always a positive number. In communication systems, typically all symbols are assumed to have equal possibility $\Pr(X_m) = 1/M$; this leads to the maximum possible entropy $H = \log_2 M$ bits/symbol. The entropy function is concave with respect to $\Pr(X_i)$.

Example 14.1: For an alphabet of only two symbols with probabilities $\Pr(X_1) = P$ and $\Pr(X_2) = 1 - P$, the entropy function is

$$H(P) = -P\log_2(P) - (1 - P)\log_2(1 - P). \tag{14.3}$$

This is plotted in Fig. 14.1. The entropy achieves the maximum 1 bit at the highest uncertainty $\Pr(X_1) = \Pr(X_2) = 0.5$. At $P = 0$ or 1, the symbols are known *a priori*, thus no information is contained.

Mutual information

The concept of mutual information is to obtain information about A (as an event of the process X) by observing B (as an event of the process Y). It describes the average information common to X and Y. The mutual information of two events A and B is defined as

$$I(A;B) = \log_2\frac{1}{\Pr(A;B)} = -\log_2 \Pr(A;B) = \log_2\left(\frac{\Pr(A|B)}{\Pr(A)}\right) \quad \text{(bits)}, \tag{14.4}$$

where $\Pr(A)$ and $\Pr(B)$ are assumed to be nonzero, and $\Pr(A|B)$ is the conditional probability, that is, the probability of A for given B.

Using Bayes' rule, $I(A;B) = I(B;A)$. When A and B are completely independent, $\Pr(A|B) = \Pr(A)$, and $I(A;B) = 0$. When A and B are completely dependent, then $\Pr(A|B) = 1$, and $I(A;B) = -\log_2(\Pr(A)) = -\log_2(\Pr(B))$. The maximum information inherent in A is given by $I(A;A) = I(A)$; for this reason, $I(A)$ is also called the *self-information* of the event A.

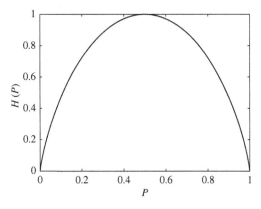

Figure 14.1 Binary entropy function.

For two signals X and Y with statistically independent symbols, $\mathcal{X} = \{X_i | i = 1, \ldots, M\}$ and $\mathcal{Y} = \{Y_i | i = 1, \ldots, N\}$, the mutual information of X and Y is defined by taking the average of the mutual information for all the events

$$I(X; Y) = E[I(A; B)] = \sum_{i=1}^{M} \sum_{j=1}^{N} \Pr(X_i, Y_j) I(X_i; Y_j)$$

$$= \sum_{i=1}^{M} \sum_{j=1}^{N} \Pr(X_i, Y_j) \log_2 \left(\frac{\Pr(X_i | Y_j)}{\Pr(X_i)} \right) \text{ (bits)}, \qquad (14.5)$$

whose maximum that is achievable is called the *channel capacity*. Given the mutual information of event A and event B, $I(A, B)$, in the channel context, event A stands for the transmitted data symbol and event B for the received symbol.

Conditional self-information

The mutual information is the most important entropy. In addition to mutual information and self-information, conditional self-information of two events A and B is defined as

$$I(A|B) = \log_2 \frac{1}{\Pr(A|B)} = -\log_2 \Pr(A|B) \quad \text{(bits)}. \qquad (14.6)$$

Joint information and entropy

Likewise, the joint information of A and B, as two events of process X and of process Y, is defined as

$$I(A, B) = -\Pr(A, B) \log_2 \Pr(A, B) \quad \text{(bits)}. \qquad (14.7)$$

Let us assume the input symbol $X_i \in \mathcal{X}$ and the output symbol $Y_j \in \mathcal{Y}$. A stands for event $X = X_i$, and B stands for event $Y = Y_j$. The joint entropy of the two processes is defined by

$$H(X, Y) = E[I(A, B)] = -\sum_{i} \sum_{j} \Pr\left(X_i, Y_j\right) \log_2 \Pr\left(X_i, Y_j\right) \quad \text{(bits)}. \qquad (14.8)$$

Conditional entropy

The conditional entropy, $H(X|Y)$, is defined by

$$H(X|Y) = -\sum_{i=1}^{M} \sum_{j=1}^{N} \Pr\left(X_i, Y_j\right) \log_2 \left(\Pr(X_i|Y_j)\right) \quad \text{(bits)}, \qquad (14.9)$$

where M and N are the numbers of elements in $\mathcal{X}$ and $\mathcal{Y}$. The conditional entropy characterizes the uncertainty caused by the channel. Note that

$$H(X, Y) = H(Y) + H(X|Y) = H(X) + H(Y|X). \qquad (14.10)$$

Relationships

The mutual information satisfies the chain rule

$$I(X;Y) = E[I(A;B)] = \sum_{i=1}^{M}\sum_{j=1}^{N} \Pr(X_i, Y_j) \log_2 \left(\frac{\Pr(X_i|Y_j)}{\Pr(X_i)} \right)$$

$$= \sum_{i=1}^{M}\sum_{j=1}^{N} \Pr(X_i, Y_j) \log_2 \left(\Pr(X_i|Y_j) \right) - \sum_{i=1}^{M} \Pr(X_i) \log_2(\Pr(X_i))$$

$$= -H(X|Y) + H(X). \tag{14.11}$$

Thus, we have the following relationships

$$I(X;Y) = H(X) - H(X|Y) = H(Y) - H(Y|X) = I(Y;X), \tag{14.12}$$

$$I(X;Y) = H(X) + H(Y) - H(X,Y), \tag{14.13}$$

$$I(X;X) = H(X). \tag{14.14}$$

These relationships can be illustrated by a Venn Diagram, as shown in Fig. 14.2 [3]. The mutual information $I(X;Y)$ is shown as the intersection of the information in X with the information in Y.

Since the conditional probabilities $\Pr(Y_j|X_i)$ are determined by the channel, the only parameter that can be optimized for a given channel for a maximum channel capacity is the statistics of the input alphabet.

Differential entropy

For a random process X that consists of continuously distributed variable x, the entropy definition for discrete symbols is not applicable. Instead, the differential entropy is defined by

$$h(X) = E\left[-\log_2 p(x) \right] = - \int_{-\infty}^{\infty} p(x) \log_2 p(x) dx \quad \text{(bits)}, \tag{14.15}$$

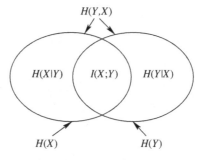

Figure 14.2   Relationships between mutual information and entropies.

where $p(x)$ is the pdf of X. Unlike the entropy, the differential entropy may be a negative value.

Example 14.2: Consider a random variable with uniform distribution between 0 and a,

$$p(x) = \begin{cases} \frac{1}{a}, & 0 \le x \le a \\ 0, & \text{elsewhere} \end{cases}. \tag{14.16}$$

The differential entropy is

$$h(X) = -\int_0^a \frac{1}{a} \log_2 \frac{1}{a} dx = \log_2 a \quad \text{(bits)}. \tag{14.17}$$

For $a < 1$, we have $h(X) < 0$.

Given a real Gaussian process with mean μ_X and variance σ_X^2, we have

$$h_g(X) = \frac{1}{2} \log_2 \left(2\pi e \sigma_X^2 \right) \quad \text{(bits)}. \tag{14.18}$$

For a circularly symmetric complex random Gaussian process with variances σ_X^2, the real and imaginary parts are independent, the pdf is the product of the two independent Gaussian pdfs, and the entropy is derived as

$$h_g(X) = \log_2 \left(\pi e \sigma_X^2 \right) \quad \text{(bits)}. \tag{14.19}$$

Thus, the differential entropy of a complex Gaussian random variable is the joint entropy of two independent real Gaussian variables with half the variance. The differential entropy of any continuous random process X, $h(X)$, is upper bounded by the differential entropy of the memoryless Gaussian source $h_g(X)$ [21].

For a multivariate normal distribution with mean $\boldsymbol{\mu}$ and covariance matrix $\boldsymbol{\Sigma}$, the pdf is given by

$$p(\boldsymbol{x}) = \frac{1}{\left(\sqrt{2\pi} \right)^n |\boldsymbol{\Sigma}|^{1/2}} e^{\frac{1}{2}(\boldsymbol{x}-\boldsymbol{\mu})^T \boldsymbol{\Sigma}^{-1}(\boldsymbol{x}-\boldsymbol{\mu})}. \tag{14.20}$$

We have

$$h(x_1, x_2, \ldots, x_n) = h(\mathcal{N}_n(\boldsymbol{\mu}, \boldsymbol{\Sigma})) = \frac{1}{2} \log_2 (2\pi e)^n \det(\boldsymbol{\Sigma}) \quad \text{(bits)}. \tag{14.21}$$

Joint Entropy

Similar to the discrete case, if two random processes X and Y have a joint pdf $p(x, y)$, the joint entropy $h(X, Y)$ is defined by

$$h(X, Y) = -\int_{-\infty}^{\infty} \int_{-\infty}^{\infty} p(x, y) \log_2 p(x, y) dx dy \quad \text{(bits)}. \tag{14.22}$$

Conditional Joint Entropy

Likewise, if X and Y have a joint pdf $p(x, y) = p(x)p(y|x)$, the conditional joint entropy $h(X|Y)$ can be defined by

$$h(X|Y) = -\int_{-\infty}^{\infty} \int_{-\infty}^{\infty} p(x, y) \log_2 p(x|y) dx dy \quad \text{(bits)}. \tag{14.23}$$

Thus

$$h(X|Y) = h(X, Y) - h(Y). \tag{14.24}$$

Mutual Information

The mutual information between two random variables with joint density $p(x, y)$ is defined by

$$I(X; Y) = \int_{-\infty}^{\infty} \int_{-\infty}^{\infty} p(x, y) \log_2 \frac{p(x, y)}{p(x)p(y)} dx dy \quad \text{(bits)}. \tag{14.25}$$

From this definition, we can derive

$$I(X; Y) = h(X) - h(X|Y) = h(Y) - h(Y|X) = I(Y; X). \tag{14.26}$$

From (14.24) and (14.26), we have

$$I(X; Y) = h(X) + h(Y) - h(X, Y). \tag{14.27}$$

Equations (14.24), (14.26), and (14.27) resemble their discrete random variable counterparts.

14.2 Lossless data compression

Data compression can be either lossless or lossy. Lossless compression is absolutely necessary for many types of data, including executable code and computer documents. This requires that the restored data file is identical to the original. In contrast, data files of images, audio, and other acquired signals allow some variations for storage or transmission.

The simplest lossless data compression method is run-length encoding (RLE). RLE is most efficient for data files that frequently contain the same character repeated many times in a row. This may be a number of spaces in a text file, or a long run of zeros between two songs in a music, or runs of same values in a sampled digital signal. For RLE, the other parts of the data do not need to be encoded, but only the long runs of the same values are encoded. For example, a run of 100 zeros can be simply encoded as (0,100). RLE followed by either Huffman or arithmetic encoding is a common strategy for lossless data compression.

14.2.1 Source coding theorem

For real coding systems, the average codeword length $\overline{R}$ is calculated as the sum of the probability P_m times its binary coding length l_m:

$$\overline{R} = \sum_{m=1}^{M} P_m l_m, \tag{14.28}$$

where M is the number of symbols. The source-coding theorem [22] states that the code efficiency of any real code must satisfy

$$\eta = \frac{H(X)}{\overline{R}} \leq 1. \tag{14.29}$$

The source coding or noiseless coding theorem can also be described as follows [3, 18]:

Theorem 14.1 (Source coding): *Consider a discrete memoryless source (DMS) with finite entropy $H(X)$, $\mathcal{X}$ being the ensemble of symbols X_i, $i = 1, \ldots, M$, with the corresponding probability of occurrence P_k, $k = 1, \ldots, M$. It is possible to construct a code that has an average length $\overline{R}$ satisfying*

$$H(X) \leq \overline{R} < H(X) + 1 \tag{14.30}$$

subject to the prefix condition.

The prefix condition ensures the received sequence to be uniquely and instantaneously decodable. The prefix condition states: given a codeword having k members $\{b_1, b_2, \ldots, b_k\}$, there is no other codeword of length l, for $1 \leq l \leq k - 1$, with elements $\{b_1, b_2, \ldots, b_k\}$. For example, a code consisting of symbols $\{1, 00, 01\}$ is prefix free, while a code like $\{1, 10, 00, 01\}$ is not prefix free. Given a received stream 11001, it can be decoded as 1,1,00,1 with the former code, but when the latter code is used, the decoder does not know whether it is 1,1,00,1 or 1,10,01.

A necessary and sufficient condition for the existence of a binary code with code words having lengths l_k, $k = 1, \ldots, K$ that satisfies the prefix condition, known as the *Kraft–McMillan inequality*, is [3]

$$\sum_{k=1}^{K} 2^{-l_k} \leq 1. \tag{14.31}$$

Shannon-Fano coding is the first well-known method that is based on the entropy of the symbols. Symbols with low probabilities are encoded into more bits, and those with high probabilities with fewer bits. The code satisfies the prefix condition, so that a coded message can be uniquely decoded. The Huffman coding shares most of the characteristics of the Shannon-Fano coding. Both the Shannon-Fano coding and the Huffman coding use a binary tree data structure, but the former is a top-down technique that works from the root to the leaves, while the latter is a bottom-up technique that starts from the leaves to the root. The two techniques are close in performance and complexity, but the Huffman coding is optimal [3].

14.2.2 Huffman coding

The Huffman code [13] is a well-known *variable-length code*, which minimizes the average codeword length, subject to the constraint that the codewords satisfy the prefix condition. Huffman code may not be unique. Huffman coding is based on the probabilities $P(X_i)$, $i = 1, \ldots, M$, of the alphabets X_i.

The idea of Huffman coding is to assign longer codewords to less probable symbols and shorter codewords to more probable ones. This idea was first used in the Morse code invented in 1840s. The method starts from merging the two least probable sources into one new source with a probability of the sum of the two probabilities. This process is repeated until there is only one merged source, and a tree is generated. Then, starting from the root of the tree, assign 0 and 1 to the two branches at each node, until all the branches of the tree are reached. This generates the code tree and code is generated for each symbol of the alphabet.

Example 14.3: Design a Huffman code for a source with ten symbols in the alphabet $\mathcal{X} = \{X_0, X_1, \ldots, X_9\}$ with corresponding probabilities 0.08, 0.06, 0.20, 0.14, 0.09, 0.07, 0.02, 0.19, 0.11, 0.04. The procedure for deriving the code is illustrated in Fig. 14.3. From the figure, we calculate the average codeword length $\overline{R} = 3.13$ bits/symbol, while the calculated entropy for the source $H(X) = 3.0819$ bits/symbol. The histogram of the alphabet is shown in Fig. 14.4. When the histogram demonstrates rapid peaks, the compression is high.

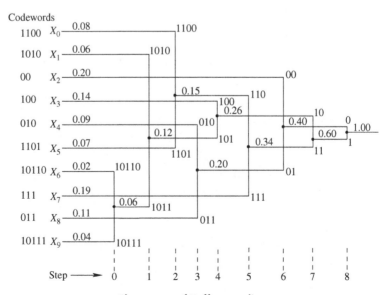

Figure 14.3   The process of Huffman coding.

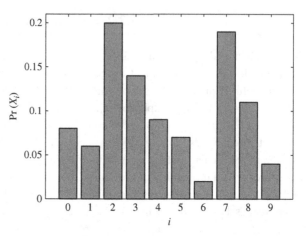

Figure 14.4 The histogram of the alphabets.

Huffman coding requires the knowledge of the source statistics, which is usually not available. For this reason, Huffman coding is usually applied by reading the original data twice. The first pass is used for frequency calculation and Huffman tree construction, and the data is compressed during the second pass. Such a method is a semiadaptive method and is normally too slow for practical implementation.

To design a Huffman code for a DMS, the probabilities of occurrence of all the source alphabets must be obtained. These probabilities can be empirically obtained by observing a long information sequence emitted from the source. For a discrete source with memory, the joint probability of the blocks of length $L \geq 2$ must also be known. Estimation of joint probabilities is extremely expensive, and thus the Huffman coding is not suitable for real sources with memory.

Reversible variable-length codes

Huffman coding is very sensitive to transmission errors. An error in a sequence of variable-length codes may cause a decoder to lose synchronization, leading to propagation of the error into the decoded sequence. Reversible variable-length codes that can be decoded in either the forward or the backward direction can reduce errors. A decoder first decodes each packet in the forward direction. If an error is detected, the packet is decoded in the reverse direction from the next resynchronization marker. This can limit the error to just one macroblock, in case of image coding.

Higher-order Huffman encoding

Huffman coding can be used in a more efficient way by encoding a block of L symbols into one codeword at a time. In this case, the average number of bits per source symbol $\bar{R}$ will be more close to the entropy $H(X)$ [18]

$$H(X) \le \overline{R} < H(X) + \frac{1}{L}. \tag{14.32}$$

Thus, this method always provides a performance as good as or better than that of Huffman coding, although the computational burden is significantly greater. This is known as a *Lth-order Huffman encoding*.

Adaptive Huffman coding

Adaptive or dynamic Huffman coding implements coding in a one-pass procedure. It starts with an empty Huffman tree and the tree is modified as symbols are being read and processed (compressed or decompressed) [20]. The performance of the adaptive Huffman coder is very close to that of the Huffman coder. For Huffman coding, a copy of the probability table is required to be transmitted with the compressed data; otherwise, the decoder has no way to decode the data. For a large alphabet, the overhead arising from this table reduces the compression ratio significantly. Adaptive Huffman coding adjusts the Huffman tree on the fly, by using data previously seen. The encoder and the decoder use identical models for compression and decompression. As a result, adaptive Huffman coding leads to significantly improved compression.

The adaptive Huffman code is a universal code (i.e., a code that does not depend on the distribution of the source). It is an asymptotically optimal universal data compression algorithm. The adaptive Huffman encoder is suitable as a real-time coder, but it is more vulnerable to errors and is more difficult to implement. The Unix *compact* program uses adaptive Huffman coding.

Detailed C programs for Huffman coding and adaptive Huffman coding are given in [17].

14.2.3 Exponential-Golomb variable-length codes

Huffman coding needs to store a predefined code table at both the encoder and the decoder. Also, when the size of the alphabet is very large, the Huffman table is very large. The Golomb code [12] and the exponential-Golomb (Exp-Golomb) code [23] are other types of variable-length codes, and they can be generated on the fly for an input symbol. Golomb code divides the alphabet into groups of equal size, while the group size in the Exp-Golomb code increases exponentially. The Exp-Golomb code is a universal code for integers, and it is simpler and faster to encode and decode than the Huffman code.

Exp-Golomb codes are constructed to a format of the type

$$\left[(M \text{ zeros}|1|(M\text{-bit INFO})\right], \tag{14.33}$$

which corresponds to a total of $(2M + 1)$ bits. Given the index of a codeword *code_num*, the codeword can be constructed as

$$M = \text{floor}\left(\log_2(code\_num + 1)\right), \tag{14.34}$$

$$\text{INFO} = code\_num + 1 - 2^M. \tag{14.35}$$

Table 14.1. Exp-Golomb codewords.	
code_num	codeword
0	1
1	010
2	011
3	00100
4	00101
5	00110
6	00111
7	0001000
8	0001001
9	0001010
...	...

The decoding process is given as follows: count the number of zeros that are followed by one as M; read the M-bit INFO field; then

$$code\_num = \text{INFO} - 1 + 2^M. \qquad (14.36)$$

The parameter to be encoded is first mapped to *code_num*, by producing short codewords for frequently-occurring values and longer codewords for less common values. The coding result is given in Table 14.1. Exp-Golomb coding is used in the JPEG-LS image and H.264/MPEG-4 AVC video compression standards.

14.2.4 Arithmetic coding

Huffman coding achieves the entropy of the source only when the symbol probability is an integer power of $\frac{1}{2}$. Arithmetic coding [19] is a direct extension of the Shannon-Fano coding scheme [3]. The idea of arithmetic coding is to efficiently calculate the probability mass function $\Pr(X_i)$ and the cumulative probability function $F(X_i)$ for the source symbol X_i. Using the ideas of Shannon-Fano coding, we can use a number in the cumulative probability interval $[F(X_i) - \Pr(X_i), F(X_i)) = [L_i, U_i)$ as the code for X_i.

Methods of arithmetic coding vary, but they all have certain common features. Each source symbol X_i, $i = 1, \ldots, M$, is assigned a disjoint subinterval of the unit interval $[0, 1)$. In a pure arithmetic coding method, the subintervals for the M symbols partition $[0, 1)$. This allows the use of a fractional number of bits for each symbol. This method replaces a stream of input symbols with a single floating-point output number in the interval $[0, 1)$, and more digits (bits) are assigned for a longer, complex message [15, 17]. The encoding process narrows the range of a floating number with every new symbol, and decoding is the inverse procedure that expands the range in proportion to the probability of each extracted symbol.

Table 14.2. Symbol probabilities of the arithmetic coding example.

Symbol	Probability	$[L_i, U_i)$ in $[0, 1)$
a	0.2	$[0, 0.2)$
b	0.1	$[0.2, 0.3)$
c	0.1	$[0.3, 0.4)$
d	0.3	$[0.4, 0.7)$
e	0.1	$[0.7, 0.8)$
f	0.05	$[0.8, 0.85)$
g	0.15	$[0.85, 1.0)$

Table 14.3. Coding process of the message *cbf gbadd*.

Message	Increase in the Interval (upper − lower)×(new portion)	Encoded Interval
c	–	$[0.3, 0.4)$
cb	$[0.02, 0.03)$	$[0.32, 0.33)$
cbf	$[0.008, 0.0085)$	$[0.328, 0.3285)$
cbfg	$[0.000425, 0.0005)$	$[0.328425, 0.3285)$
cbfgb	$[0.000015, 0.0000225)$	$[0.328440, 0.3284475)$
cbfgba	$[0.0, 0.0000015)$	$[0.328440, 0.3284415)$
cbfgbad	$[0.0000006, 0.00000105)$	$[0.3284406, 0.32844105)$
cbfgbadd	$[0.000000180, 0.000000315)$	$[0.328440780, 0.328440915)$

Example 14.4: Given the alphabet $\{a, b, c, d, e, f, g\}$, with the corresponding probabilities of $0.2, 0.1, 0.1, 0.3, 0.1, 0.05, 0.15$, code the message *cbfgbadd* by using arithmetic coding.

The first step is to assign each symbol a portion in the $[0, 1)$ interval, which corresponds to the cumulative probability function. The result is shown in Table 14.2. To code the first symbol c, we have its range is $[0.3, 0.4)$. The next symbol b is in the range $[0.2, 0.3)$ of the new interval $[0.3, 0.4)$, leading to an updated range of $0.3 + 0.1 \times [0.2, 0.3) = [0.32, 0.33)$. The coding process is shown in Table 14.3. The final interval $[0.328440780, 0.328440915)$ represents the message, and any number within this range represents the whole message.

Given a floating-point number, the decoding process extracts the symbols one-by-one, until a specified number of digits is reached. The decoding process can be formulated as

$$R_{n+1} = \frac{R_n - L_n}{U_n - L_n} \tag{14.37}$$

Table 14.4. Decoding process of the message *cbf gbadd.*

Encoded number (encoded − lower)/(upper − lower)	Interval	Decoded symbol
0.3284408	[0.3, 0.4)	*c*
0.284408	[0.2, 0.3)	*b*
0.84408	[0.8, 0.85)	*f*
0.8816	[0.85, 1.0)	*g*
0.21067	[0.2, 0.3)	*b*
0.1067	[0, 0.2)	*a*
0.5335	[0.4, 0.7)	*d*
0.445	[0.4, 0.7)	*d*

where R_n is a code within the cumulative probability range $[L_n, U_n)$ for the nth symbol. Therefore, the cumulative probabilities have to be transmitted with the code for decoding.

Example 14.5: Given a number that represents the encoded message, 0.3284408, by using the same probability intervals as the encoder given by Example 14.4, demonstrate the decoding process.

At the first step, we find that the message 0.3284408 is within [0.3, 0.4). From Table 14.2, we decode the first symbol as c. By applying (14.37), the message for the sequence removing the first symbol is obtained as 0.284408. The decoding process is shown in Table 14.4. For the given number, the decoding process can be continued after the message *cbfgbadd* is obtained, so we have to truncate the message to the specified digits. Only when the number is very close to 0.328440780 can we get the exact message.

Arithmetic coding yields a number 0.328440780 that can be represented as a fixed-point number of 23 bits, $0.01010100000101001011001_2$, corresponding to $0.32844078540\ldots$ The sequence of data symbols can be represented by $\log_2 \frac{1}{P_c} + \log_2 \frac{1}{P_b} + \log_2 \frac{1}{P_f} + \log_2 \frac{1}{P_g} + \log_2 \frac{1}{P_b} + \log_2 \frac{1}{P_a} + \log_2 \frac{1}{P_d} + \log_2 \frac{1}{P_d} = 22.8205$ bits, which is very close to the optimum coding.

Unlike Huffman coding that has a one-to-one mapping between the symbols and the codewords, arithmetic coding uses a single variable-length codeword to code a variable-length block of symbols. In arithmetic coding that uses floating-point representation, as the length of the message increases, the range of the message becomes smaller, which may be too small for computer representation given a large message length. Binary arithmetic coding is used to solve this problem. The binary arithmetic coding is relatively complex, and is well expounded in [7]. Detailed C programs for arithmetic coding are given in [7, 17].

Context-based arithmetic coding

Like Huffman coding, arithmetic coding can be implemented by a fixed or adaptive model. In the fixed model, the algorithm needs two passes of the data, the first pass for symbol statistics, and the second pass for coding. The probability table needs to be included in the compressed file. The adaptive model can dynamically estimate the symbol probabilities at the time of coding. Like Huffman coded data, arithmetic coded data is prone to channel errors, due to the cumulative impact on decoding of the subsequent data.

Context-based arithmetic coding is a popular adaptive arithmetic coding method that adapts the assigned probability to a symbol, according to the context of its spatial and/or temporal neighbors. Use of arithmetic coding can improve the efficiency over Huffman coding by approximately 5–10% [7]. Arithmetic coding plays an important role in advanced image/video coding standards, and context-based arithmetic coding is employed in the JBIG, JPEG-LS, JPEG2000, H.263, H.264, MPEG-4 Visual, and MPEG-4 Audio standards. A similar idea is also used in the context-based adaptive variable-length code (CAVLC).

14.2.5 Dictionary-based coding

Dictionary-based coding was first proposed by Lempel and Ziv in their 1977 and 1978 papers [28]. The proposed methods are now known as *LZ77* and *LZ78*, respectively. The idea is to encode by decomposing the source bitstream into many short, previously unseen substreams. A variable-length string of symbols is encoded by a single token, which is an index to a phrase dictionary. Dictionary-based coding can be applied without knowledge of the source statistics, and can be used for general-purpose data compression due to its simplicity and versatility. Compression and decompression are simply a table-lookup and string replacement operation.

Dictionary-based compression requires low system resources, and this has made such compression algorithms more popular in recent years. It has become the standard algorithm for file compression on computers. Both the LZ77 and LZ78 algorithms have an asymptotic rate approaching the entropy of the source [3]. The Lempel–Ziv codes are asymptotically optimal universal codes. LZ78 typically compresses ASCII text files by a factor of 2.

LZ77 and LZSS

LZ77 uses a dictionary based on a sliding fixed-size window that moves through the text, and the implementation is relatively simple. The LZ77 algorithm has a number of performance problems, which made it not a very practical algorithm to implement, until some of these problems were overcome in one of its variants, LZSS. LZSS improves the way the text window is maintained and the way the output tokens are represented.

It has been shown in [29] that the LZ77 universal compression is *essentially optimal* for finite N-blocks, whereas the adaptive universal Huffmann coding is not essentially optimal for finite blocks. Many general-purpose file compression and archive programs, such as *arj* and *LHarc*, are based on LZSS. Other examples are popular programs such as all the *zip* variants (e.g. GNU *gzip*, *winzip*, *pkzip*) and the PNG (Portable Network Graphics) image format. The output from LZSS is a stream of tokens referring to individual characters or strings. In *arj* and *LHarc*, Huffman coding is further performed on the output of LZSS.

LZ78 and LZW

LZ78 builds its dictionary out of all the previously seen symbols. LZ78 scans the input sequence and matches it with a dictionary it maintains. When a failure of match occurs, the resulting word is then added to the dictionary. LZ78 requires two passes over the string: The first pass parses the string, and the second pass generates the coded string. Unlike LZ77, LZ78 has to maintain a dictionary tree. LZ78 can potentially have an unlimited list of previously seen phrases. For LZ78, a string can be extremely long, leading to a high compression ratio. When implementing LZ78, both the encoder and decoder can start with a nearly empty dictionary. *Arc* and *pkarc* are general-purpose file compression and archival programs based on LZ78.

Welch [25] proposed a variant of LZ78, which is known as the *Lempel-Ziv-Welch (LZW) coding algorithm*. LZW improves LZ78 by outputting only phrases and never single characters. LZW gained immediate popularity after its appearance. A dictionary keeps a list, or a codebook of frequently occurring patterns. When these patterns appear in the source output, they are referred to the dictionary, while infrequently occurring patterns can be encoded by some other methods. The LZW algorithm employs a dynamic dictionary.

LZW is a variable-to-fixed-length algorithm, and is widely used in the compression of computer files. It typically achieves a compression ratio of 2:1 for text and executable code, and 5:1 for extremely redundant data files such as computer source code and acquired signals. If an ergodic source generates the input string, the LZW ratio can be arbitrarily close to the entropy of the source.

The GIF (Graphics Interchange Format) image file is a lossless image compression technique that employs LZW coding. It is also used in TIFF (Tagged Image File Format) and PostScript file formats. The UNIX *compress* program is also based on LZW. *Compress* provides a significant by better compression than *compact*, and it is also faster. LZW is also the basis of many personal computer utilities for doubling the capacity of the hard disk. LZW is a patent of Unisys Corporation, but the patient expired in 2004. Detailed C programs for LZSS and LZW coding are given in [17].

Example 14.6: We now demonstrate the encoding of the LZ78 algorithm for the binary input sequence: 0010010111011001011 . . . Assume that 0 and 1 are stored in the codebook.

As 0 and 1 have been stored in the codebook, the first previously unseen shortest subsequence is 00, thus Stored subsequences: 0, 1, 00

Sequence to be parsed: 10010111011001011 . . . The next unseen shortest subsequence is 10. This procedure is given in Table 14.5.

In the third line of Table 14.5, since 00 is composed of two 0's in the codebook and 0 is at position 1 of the codebook, so it is represented as 11. This process is also applied to other

Table 14.5. Procedure for Lempel–Ziv encoding.

Stored subsequences	0	1	00	10	01	011	101	100	1011
Position in codebook	1	2	3	4	5	6	7	8	9
Digital representation			11	21	12	52	42	41	71
Binary coding			0010	0100	0011	1011	1001	1000	1111

stored subsequences. The binary representation of the subsequence is given in the fourth line. The last symbol of each subsequence in the codebook is an innovation symbol, which makes the subsequence different from a previously stored subsequence. In the fourth line, each binary code has four bits; the first three denote the binary representation of pointer of its root subsequence and the last bit is the innovation symbol.

For decoding, for example, the encoded binary representation for the subsequence at position 9 in the codebook is 1111, thus the last bit is 1 and first three bits 111 points to position 7 in the codebook whose stored subsequence 101, thus the decoded subsequence is 1011.

From this example, we see that, unlike Huffman coding, the Lempel-Ziv algorithm employs fixed-length codewords to represent source symbols of different lengths. This makes Lempel-Ziv coding especially suitable for synchronous transmission. In practice, a 12-bit fixed block size is employed, corresponding to a codebook size of 4096. At present, the Lempel-Ziv algorithm has become the standard algorithm for file compression and achieves a better compression ratio than Huffman coding.

14.3 Rate-distortion theorem

The rate-distortion theorem quantifies the minimum average bit rate R_D, in bits/sample, for representing a discrete or continuous random variable X with a distortion less than D. The lossless data compression we have discussed earlier in this chapter is a special case of rate-distortion theory applied to a discrete source with zero distortion. Shannon was the first to establish the rate-distortion theorem in connection with the quantization of signals [21, 22].

The rate-distortion function $R(D)$ for a source X with distortion measure $d(x, \hat{x})$ is defined as

$$R(D) = \min_{p(\hat{x}|x)} I\left(X; \hat{X}\right), \tag{14.38}$$

subject to

$$\sum_{(x,\hat{x})} p(x)p\left(\hat{x}|x\right) d\left(x, \hat{x}\right) \leq D, \tag{14.39}$$

where the minimization runs over all $p\left(\hat{x}|x\right)$ for which $p\left(x,\hat{x}\right) = p(x)p\left(\hat{x}|x\right)$ satisfies the expected distortion constraint. The two common distortion measures are the Hamming distance and the MSE.

The rate-distortion theorem with a distortion measure states that for any given distortion D, there exists a coding scheme that maps the source output into codewords such that the minimum rate $R(D)$ is sufficient to reconstruct the source output with an average distortion arbitrarily close to D. The rate distortion theorem can be stated as follows [3, 22]:

Theorem 14.2 (Rate distortion): *If $R \geq R(D)$, there exists a sequence of codes $\hat{X}^n (X^n)$ with number of codewords $\left|\hat{X}^n(\cdot)\right| \leq 2^{nR}$ and $E\left[d\left(X^n, \hat{X}^n (X^n)\right)\right] \to D$. If $R < R(D)$, no such codes exist.*

Note that the sequence $X^n = X_1, X_2, \ldots, X_n$, and $\hat{X}^n$ is a quantized representation of X^n.

The rate-distortion function for a memoryless Gaussian source X, $\mathcal{N}\left(0, \sigma_X^2\right)$, with the MSE distortion measure, is given by [3, 22]

$$R_g(D) = \begin{cases} \frac{1}{2} \log_2\left(\frac{\sigma_X^2}{D}\right), & 0 \leq D \leq \sigma_X^2 \\ 0, & D > \sigma_X^2 \end{cases}, \tag{14.40}$$

where the MSE distortion D is given by

$$D = E\left[(x - \hat{x})^2\right] = E\left[e^2\right]. \tag{14.41}$$

This rate-distortion function for the memoryless Gaussian source is the upper bound of any memoryless, discrete-time, non-Gaussian source with zero mean and finite variance σ_X^2 [5]. The rate-distortion theorem implies that no information needs to be transmitted if the distortion $D \geq \sigma_X^2$. The rate-distortion theorem is most fundamental for quantization.

Example 14.7: The distortion function for a memoryless Gaussian source, (14.40), is plotted in Fig. 14.5.

The lower bound on the rate-distortion function is given by the Shannon lower bound for an MSE distortion measure. For a continuous-amplitude, memoryless source X, the lower bound is given as

$$R^*(D) = h(X) - \frac{1}{2} \log_2(2\pi e D), \tag{14.42}$$

where $h(X)$ is the differential entropy of a continuous source X. The rate-distortion theorem states that $R(D)$ is the minimum information rate necessary to represent the output of a discrete-time, continuous-amplitude, memoryless stationary Gaussian source, based on the MSE distortion measure per symbol.

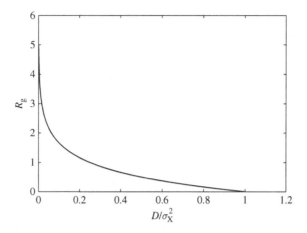

Figure 14.5

Rate distortion function of a Gaussian source.

For uniform signal pdf, as in the case of a uniform quantizer, we can derive, from (14.40), that the SNR is given by

$$\text{SNR} = \frac{\sigma_X^2}{\sigma_q^2} = 2^{2R(D)} \approx 6.02R(D) \quad \text{(dB)}, \tag{14.43}$$

where the MSE of the quantization $\sigma_q^2 = D$.

Also from (14.40), the distortion-rate function is derived as

$$D(R) = 2^{-2R}\sigma_X^2 \approx -6.02R + 10\log_{10}\sigma_X^2 \quad \text{(dB)}, \tag{14.44}$$

where R is the number of bits per sample.

The rate distortion function $R(D)$ for a multivariate normal vector source with Euclidean MSE distortion gives the allocation of R bits to the different components to minimize the total distortion. This can be realized by reverse water-filling [3].

14.4 Channel capacity

Channel capacity is the maximum rate for reliable transmission of information over the channel. Nyquist showed that a channel of bandwidth W Hz is capable of transmission of PAM signals at a rate of $2W$ samples per second without causing ISI.

Spectral or bandwidth efficiency is referred to as the rate of information transmission per unit of occupied bandwidth. It can be defined as

$$\eta_s = \frac{\text{Transmission rate}}{\text{Channel bandwidth}} = \frac{R_s H}{B} \quad \text{(bits/s/Hz)}, \tag{14.45}$$

where R_s is the symbol rate, H is the entropy, that is, the average amount of information, measured in bits, conveyed per symbol, and B is the occupied bandwidth. For an

M-ary modulation, $H = \log_2 M$ bits/symbol. Nyquist's sampling theorem limits $R_s \leq 2B$ symbols/s.

The channel capacity C, also known as *Shannon capacity*, is defined as the maximum average mutual information for a channel with power constraint S

$$C = \max_{p_X(x):E[X^2]\leq S} I(X;Y). \tag{14.46}$$

This capacity is determined by the transition probability, i.e., the conditional probability $\Pr(Y|X)$, which can be expressed by the error probability.

The rate R of an (M,n) code is defined as

$$R = \frac{\log_2 M}{n} \quad \text{(bits per transmission)}, \tag{14.47}$$

where M is the constellation size of the symbols, and n is the length of the code. A rate R is said to be *achievable* if there exists a sequence of $(\lceil 2^{nR} \rceil, n)$ codes such that the maximum probability of error $\lambda_n \to 0$ as $n \to \infty$. There is duality between channel capacity and rate distortion for arbitrary pairs of i.i.d. (independent and identically distributed) correlated state information available at the sender and at the receiver, respectively [4]. The channel coding theorem was also given by Shannon [3, 21].

Theorem 14.3 (Channel coding): *All rates below capacity are achievable. Specifically, for any $R < C$, there exists a sequence of $(\lceil 2^{nR} \rceil, n)$ codes with maximum probability of error $\lambda_n \to 0$. Conversely, any sequence of $(\lceil 2^{nR} \rceil, n)$ codes with $\lambda_n \to 0$ must have $R \leq C$.*

According to the channel coding theorem, it is possible to transmit data without errors as long as the bit rate is below the channel capacity. Channel coding is targeted at achieving this absence of errors. Channel coding is most important for improving the BER of digital communication systems that are susceptible to noise and interference.

For discrete memoryless channels, Shannon defined the capacity as the maximum mutual information between the channel input X and the channel output Y for all input probability distribution

$$C = \max_{\Pr(X_i)} I(X;Y) = \max_{\Pr(X_i)} \sum_{X_i} \sum_{Y_j} \Pr(X_i) \Pr(Y_j|X_i)$$

$$\times \log_2 \frac{\Pr(Y_j|X_i)}{\sum_{X_k} \Pr(X_k) \Pr(Y_j|X_k)}, \tag{14.48}$$

where $\Pr(Y_j|X_i)$ is the channel transition probability.

14.4.1 Capacity of the AWGN channel for Gaussian distributed input

For an AWGN channel, the output Y and the input X are related by

$$Y = X + N, \tag{14.49}$$

where the output Y and input X are continuous, and N is a white Gaussian noise with $\mathcal{N}\left(0, \sigma_N^2\right)$ and a two-sided PSD $N_0/2$.

The mutual information $I(X; Y)$ is given from (14.26) as

$$
\begin{aligned}
I(X; Y) &= h(Y) - h(Y|X) \\
&= h(Y) - h(X + N|X) \\
&= h(Y) - h(N).
\end{aligned}
\tag{14.50}
$$

In the second equality, the information in Y from known X is determined by the noise N. The last equality is due to the independence of N from X.

For independent real Gaussian processes X and N, the powers satisfy $\sigma_Y^2 = S + \sigma_N^2$, where $S = \sigma_X^2$. The entropy of Y is bounded by $\frac{1}{2} \log 2\pi e \left(S + \sigma_N^2\right)$, and $h(N) = \frac{1}{2} \log 2\pi e \sigma_N^2$; thus, we have the one-dimensional (1-D) capacity

$$
\begin{aligned}
C_{1d} = I(X; Y) &= h(Y) - h(N) \\
&= \frac{1}{2} \log_2 \left(\pi e \sigma_Y^2\right) - \frac{1}{2} \log_2 \left(\pi e \sigma_N^2\right) \\
&= \frac{1}{2} \log_2 \left(1 + \frac{S}{\sigma_N^2}\right) \quad \text{(bits per transmission)} \tag{14.51} \\
&= \frac{1}{2} \log_2 \left(1 + 2\frac{E_s}{N_0}\right) \quad \text{(bits/s/Hz)}. \tag{14.52}
\end{aligned}
$$

The last equality is obtained by inserting $S = BE_s$ and $\sigma_N^2 = B\frac{N_0}{2}$. The dimensional capacity is thus equal to η_{max}. Note that *bits per transmission* is the same as *bits/s/Hz*.

Likewise, for independent complex Gaussian processes X and N, we can derive the two-dimensional (2-D) capacity as

$$
C_{2d} = \log_2 \left(1 + \frac{S}{\sigma_N^2}\right) \tag{14.53}
$$

$$
= \log_2 \left(1 + \frac{E_s}{N_0}\right) \quad \text{(bits/s/Hz)}. \tag{14.54}
$$

In order to obtain the above result, we exploit that $h(Y)$ is bounded by $\log_2 \pi e \left(S + \sigma_N^2\right)$, $h(N) = \log_2 \pi e \sigma_n^2$, and $\sigma_N^2 = BN_0$.

At the highest spectral efficiency, $E_s = C_{2d}E_b$ for complex-valued signal alphabet; thus, we have from the capacity for the AWGN channel with Gaussian input

$$
\frac{E_b}{N_0} = \frac{2^{C_{2d}} - 1}{C_{2d}}. \tag{14.55}
$$

From this, we have $\gamma_b = \frac{E_b}{N_0} \rightarrow \ln 2 = -1.59$ dB as $C_{2d} \rightarrow 0$ (for infinite bandwidth), and increases exponentially as $C_{2d} \rightarrow \infty$. $\gamma_b = -1.59$ dB is the lower bound of γ_b, and is called the *Shannon bound*. The same lower bound is obtained for a real-valued signal alphabet.

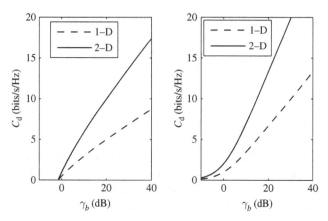

Figure 14.6 The dimensional capacity for the AWGN channel with Gaussian distributed input.

Example 14.8: The channel capacities of the AWGN channel versus $\gamma_b = \frac{E_b}{N_0}$ and $\gamma_s = \frac{E_s}{N_0}$ are plotted in Fig. 14.6, for both the 1-D and 2-D cases.

Channel capacity for band-limited channels

The channel capacity C for a band-limited channel is given by the well-known formula [21]

$$C = B \log_2 \left(1 + \frac{S}{BN_0} \right) \quad \text{(bits/s)}, \tag{14.56}$$

where B is the channel bandwidth in Hz, S is the average signal power, and BN_0 is the noise power. This result is valid for both real and complex band-limited Gaussian channels. The maximum spectral efficiency is given by the normalized capacity C/B bits/s/Hz.

Equation (14.56) can be derived as follows. For a period of T, the power per sample is $ST/2BT = S/2B$, where $2BT$ is the number of samples, based on the Nyquist theorem; the noise has a power of $\frac{N_0}{2} 2B = N_0 B$, and the noise variance per sample is thus $(N_0 B) T/2BT = N_0/2$. As a result, the capacity per sample is derived from (14.51) as

$$C = \frac{1}{2} \log_2 \left(1 + \frac{\frac{S}{2B}}{\frac{N_0}{2}} \right) = \frac{1}{2} \log_2 \left(1 + \frac{S}{N_0 B} \right) \quad \text{(bits/sample)}. \tag{14.57}$$

Since there are $2B$ samples each second, we obtain the capacity given by (14.56).

As $B \to \infty$, we get from (14.56),

$$C_\infty = \lim_{B \to \infty} \frac{S}{N_0} \log_2 \left(1 + \frac{S}{N_0 B} \right)^{\frac{N_0 B}{S}} = \frac{S}{N_0} \log_2 e = 1.44 \frac{S}{N_0}. \tag{14.58}$$

This is the maximum channel capacity. A practical system usually increases SNR and reduces B to improve the spectrum efficiency. The relation of C versus B is illustrated in Fig. 14.7.

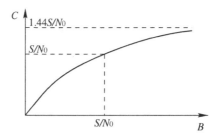

Figure 14.7 Shannon capacity C versus B for a fixed S/N_0 ratio.

For parallel Gaussian channels with a common power constraint, an optimization process, known as *water-filling*, is used to maximize the channel capacity [3].

Feedback

It is proved in [3] that feedback does not increase the capacity for discrete memoryless channels. But it helps to significantly reduce the complexity of encoding or decoding. The same is true for the AWGN channel. For channels with memory, feedback does increase the capacity. The capacity without feedback can be calculated using water-filling, but the capacity with feedback is more difficult to characterize. Feedback increases the capacity of a non-AWGN channel by at most half a bit per transmission [3].

14.4.2 Capacity of the AWGN channel for discrete input alphabets

There is no closed-form capacity bound for discrete input alphabets. They can only be evaluated numerically according to (14.48). An example is given here for this purpose.

Example 14.9: The binary symmetric channel (BSC) is a mathematical model for binary transmission over a Gaussian channel with hard decisions at the output. The capacity of the BSC is given by

$$C = 1 - H_b(P), \qquad (14.59)$$

where $H_b(P)$ is the binary entropy function,

$$H_b(P) = -P\log_2 P - (1 - P)\log_2(1 - P). \qquad (14.60)$$

P is the crossover probability, that is, the probability of input 0 becoming output 1 or vice versa, or the error probability. For the Gaussian channel using BPSK modulation, the error probability for optimal detection is $P_b = Q\left(\sqrt{2\gamma_b}\right)$ from (7.55). The channel capacity C versus γ_b is plotted in Fig. 14.8.

In case of soft decoding, the channel input is discrete, while the channel output may be continuous. Assuming the input constellation $\{c_m\}$ of size M, (14.48) can be written as

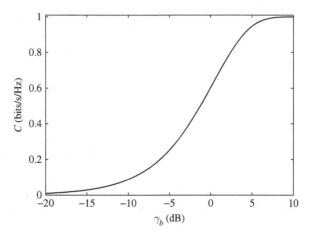

Figure 14.8 The channel capacity of the BSC.

$$C = \max_{P(c_i),i=0,\dots,M-1} \sum_{m=0}^{M-1} P(c_m) \int_{-\infty}^{\infty} p(y|c_m) \log_2 \frac{p(y|c_m)}{\sum_j P(c_j) p(y|c_j)} dy, \qquad (14.61)$$

where $P(c_m)$ is the probability of the constellation point c_m, and $p(y|c_m)$ is the conditional pdf. The maximization is subject to the average power constraint

$$\sum_{m=0}^{M-1} P(c_m) c_m^2 \le S. \qquad (14.62)$$

For M-ary modulation, it is common to assume that the channel input probabilities are equal with $P(c_m) = \frac{1}{M}$.

One-dimensional case

For a real-valued random variable y, the conditional pdf is given by

$$p(y|c_m) = \frac{1}{\sqrt{2\pi}\sigma} e^{-\frac{(y-c_m)^2}{2\sigma^2}}. \qquad (14.63)$$

Thus, for a one-dimensional constellation with equal-distance points, from (14.61), we have

$$C = \frac{1}{M} \sum_{m=0}^{M-1} \int_{-\infty}^{\infty} \frac{1}{\sigma\sqrt{2\pi}} e^{-\frac{(y-c_m)^2}{2\sigma^2}} \log_2 \frac{e^{-\frac{(y-c_m)^2}{2\sigma^2}}}{\frac{1}{M}\sum_j e^{-\frac{(y-c_j)^2}{2\sigma^2}}} dy. \qquad (14.64)$$

Substituting $y = z\sigma + c_m = (z + \bar{c}_m)\sigma$, we have

$$C = \log_2 M - \frac{1}{M} \sum_{m=0}^{M-1} \int_{-\infty}^{\infty} \frac{1}{\sqrt{2\pi}} e^{-\frac{z^2}{2}} \log_2 \sum_{i=0}^{M-1} e^{-\frac{(\bar{c}_m - \bar{c}_i)^2}{2}} e^{-z(\bar{c}_m - \bar{c}_i)} dz. \qquad (14.65)$$

Note here $\bar{c}_m$ is c_m normalized by σ. The corresponding SNR is given by

$$\gamma = \frac{1}{M} \sum_{m=0}^{M-1} \bar{c}_m^2. \tag{14.66}$$

For a given constellation, the relation between C and γ can be evaluated by numerical integration.

For MASK, we have $\bar{c}_m = (2m + 1 - M)d_0, i = 0, 1, \ldots, M - 1$. Thus,

$$\gamma = \frac{1}{M} \sum_{m=0}^{M-1} (2m + 1 - M)^2 d_0^2 = K_M d_0^2. \tag{14.67}$$

Thus, $d_0 = \sqrt{\frac{\gamma}{K_M}}$. Substituting $z = d_0 u$, after manipulation, we finally get

$$C = \log_2 M - \frac{1}{M\sqrt{2\pi}} \sqrt{\frac{\gamma}{K_M}} \sum_{m=0}^{M-1} \int_{-\infty}^{\infty} e^{-\frac{\gamma}{2K_M} u^2}$$

$$\times \log_2 \left(\sum_{i=0}^{M-1} e^{-\frac{2\gamma}{K_M}(m-i)(m-i+u)} \right) du \quad \text{(MASK)}. \tag{14.68}$$

Example 14.10: The capacity for MASK signaling in the AWGN channel is plotted in Fig. 14.9. It is seen that at low SNR, the capacity of MASK signaling is close to that of the AWGN channel.

Two-dimensional case

For the complex AWGN channel, the constellation is in the complex plane. MPSK and MQAM have signaling constellations in the complex plane. The channel input c_m and output y are both complex numbers. The conditional pdf $p(y|c_m)$ is given by

$$p(y|c_m) = \frac{1}{2\pi\sigma^2} e^{-\frac{|y-c_m|^2}{2\sigma^2}}. \tag{14.69}$$

Substituting $y = z\sigma + c_m = (z + \bar{c}_m)\sigma$, where $z = u + jv$, and using a procedure similar to that used for the real AWGN channel, the channel capacity can be derived. When all the symbols have equal probability, the capacity is derived as

$$C = \log_2 M - \frac{1}{M} \sum_{m=0}^{M-1} \int_{-\infty}^{\infty} \int_{-\infty}^{\infty} \frac{1}{2\pi} e^{-\frac{|z|^2}{2}}$$

$$\times \log_2 \sum_{i=0}^{M-1} e^{-\frac{|\bar{c}_m - \bar{c}_i|^2}{2}} e^{-\text{Re}[z^*(\bar{c}_m - \bar{c}_i)]} du dv. \tag{14.70}$$

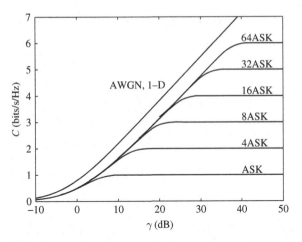

Figure 14.9   The channel capacity of the MASK modulation in the AWGN channel.

The passband signal has an SNR

$$\gamma = \frac{1}{2M} \sum_{m=0}^{M-1} |\bar{c}_m|^2 . \tag{14.71}$$

For MPSK, let $\bar{c}_m = Ae^{j\frac{2\pi m}{M}}$, $m = 0, 1, \ldots, M-1$. From (14.71), we have $\gamma = \frac{A^2}{2}$. Thus, $A = \sqrt{2\gamma}$. Substituting $u = u_1 A$ and $v = v_1 A$, after manipulation we finally obtain

$$C = \log_2 M - \frac{\gamma}{\pi M} \sum_{m=0}^{M-1} \int_{-\infty}^{\infty} \int_{-\infty}^{\infty} e^{-(u_1^2 + v_1^2)\gamma} \log_2 \sum_{i=0}^{M-1} e^{-2\gamma \left(1 - \cos \frac{2\pi(m-i)}{M}\right)}$$

$$\times e^{-2\gamma \left[u_1\left(\cos \frac{2\pi m}{M} - \cos \frac{2\pi i}{M}\right) + v_1\left(\sin \frac{2\pi m}{M} - \sin \frac{2\pi i}{M}\right)\right]} du_1 dv_1 \quad \text{(MPSK)}. \tag{14.72}$$

Similarly, C as a function of γ can be derived for any MQAM signaling. This is left as Problem 14.23.

Example 14.11: The capacity for MPSK signaling in the AWGN channel is plotted in Fig. 14.10. At low SNR the capacity of MPSK signaling is close to that of the AWGN channel.

14.4.3 Area spectral efficiency

Area spectral efficiency (ASE) is a capacity measure that can better characterize the capacity of a cellular network than the Shannon capacity. In cellular systems, the channels are reused at a distance D, and the area covered by each channel is roughly a circle with area $A = \pi D^2/4$. ASE is defined as the maximum throughput per unit bandwidth per unit area that is achieved in a cell

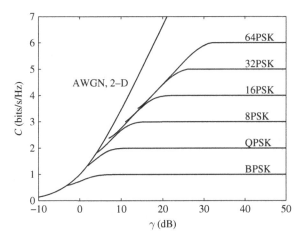

Figure 14.10 The channel capacity of the MPSK modulation in the AWGN channel.

$$\eta_{ASE} = \frac{C_{SR}/B}{\pi D^2/4} \quad (\text{bits/s/Hz/m}^2), \tag{14.73}$$

where C_{SR} is the system throughput or sum-rate capacity, i.e., the maximum of the sum of data rates of all users in a cell,

$$C_{SR} = \max \sum_{k=1}^{K} R_k, \tag{14.74}$$

with each user's capacity being calculated as Shannon limit by using SIR obtained from the analysis of the intercell and intracell interferences.

An overall investigation of ASE for the uplink of a cellular system with orthogonal channelization has been conducted in [1]. The investigation is based on the assumption of variable-rate transmission as well as different intercell interference conditions, different fading models, cell sizes, and system loads. For best-case and average interference, the optimal reuse factor is unity, no matter whether spreading is used for interference reduction or not. ASE also decreases as an exponential of a fourth-order polynomial of the cell radius.

14.5 Source-channel coding theorem

Shannon's source-channel separation theorem states that source coding and channel coding can be performed separately and sequentially without loss of optimality. However, the two-stage encoding method is optimal only for infinite block length codes, which induces infinite complexity and delay. This makes separate source and channel coding undesirable for most real-time applications. Moreover, separate source and channel coding is designed for the worst case scenario, and it is not applicable for multi-user and nonergodic channel environments [3, 11].

The redundancy in the source is usually suited to the channel, and joint source-channel coding provides a performance superior to the two-stage method. The joint source-channel coding approach shares information between the source coder and the channel coder, and utilizes the soft information from the physical layer [11]. It allows the coder to better exploit the changes in the channel conditions or variations of the source contents.

The source coding theorem requires $R > H$ for data compression, and the channel coding theorem requires $R < C$ for data transmission [3]. The source-channel coding theorem addresses that $H < C$ guarantees reliable transmission of source over a channel. To send a sequence of symbols $V^n = V_1, V_2, \ldots, V_n$ over the channel, one needs to map it onto a codeword $X^n(V^n)$ and send the codeword over the channel. The receiver makes an estimate of the sequences, $\hat{V}^n$, from its received sequence Y^n.

Theorem 14.4 (Source-Channel coding): *All sources with entropy H below the channel capacity C can be reliably transmitted. Specifically, if $V^n = V_1, V_2, \ldots, V_n$ is a finite alphabet stochastic process that satisfies the asymptotic equipartition property (AEP) and $H(V) < C$, there exists a source-channel code with probability of error $\Pr(\hat{V}^n \neq V^n) \to 0$. Conversely, for any stationary stochastic process V, if $H(V) > C$, it is not possible to send the process over the channel with arbitrarily low probability of error, $\Pr(\hat{V}^n \neq V^n) \geq \epsilon > 0$, where ϵ is a constant.*

The AEP is the direct consequence of the weak law of large numbers. The AEP states that $-\frac{1}{n} \log_2 \Pr(V^n) \to H$ with probability 1, as $n \to \infty$, where V^n can be a sequence of i.i.d. random variables or a sequence of states of any stationary ergodic source and $\Pr(V^n)$ is the probability of observing the sequence $V^n = V_1, V_2, \ldots, V_n$. Thus, $\Pr(V^n) \to 2^{-nH}$.

Similarly, AEP for continuous random variables drawn from the density $p_V(v)$ is given by $-\frac{1}{n} \log_2 p_V(V^n) \to h(V)$ with probability 1, as $n \to \infty$.

The LZ77 encoder does not completely eliminate the redundancy present in the input sequence. For example, when an LZ77 phrase has multiple matches, the LZ77 encoder can issue a pointer to any of those matches and a particular choice carries some additional bits of information. A RS channel coder exploits the inherent redundancy left by the LZ77 encoder to detect and correct a limited number of errors. The resulting LZRS'77 algorithm is backward-compatible with LZ77 in [16], that is, a file compressed with the LZRS'77 can still be decompressed by a generic LZ77 decoder.

14.6 Capacity of fading channels

The ergodic capacity $\overline{C}$ represents the average capacity among all channel states, and is mainly used for fast fading channels when coding is performed over many channel states. The outage capacity C_{out} is particularly used for slowly-fading channels.

14.6.1 Capacity with CSI at receiver only

Ergodic capacity

The ergodic capacity of a fading channel with receiver CSI (channel state information) for an average power constraint $\overline{S}$ is given by

$$\overline{C} = \int_0^\infty C(\gamma) p_\gamma(\gamma) d\gamma = \int_0^\infty \log_2(1+\gamma) p_\gamma(\gamma) d\gamma \quad \text{(bits/s/Hz)}, \tag{14.75}$$

where the received SNR $\gamma = |h|^2 E_s/N_0$, h being the instantaneous channel coefficient, and $p_\gamma(\gamma)$ is the pdf of γ. In a fading environment with receiver CSI, the Shannon capacity is less than that of an AWGN channel with the same average SNR.

For the Rayleigh channel, $p_\gamma(\gamma) = \frac{1}{\overline{\gamma}} e^{-\gamma/\overline{\gamma}}$ with $\overline{\gamma} = \sigma_h^2 \frac{E_s}{N_0}$. Inserting it into (14.75), it can be shown that [14]

$$\overline{C} = \log_2 e \cdot \exp\left(\frac{1}{\sigma_h^2 E_s/N_0}\right) \cdot \operatorname{expint}\left(\frac{1}{\sigma_h^2 E_s/N_0}\right), \tag{14.76}$$

where

$$\operatorname{expint}(x) = \int_x^\infty \frac{e^{-t}}{t} dt. \tag{14.77}$$

Example 14.12: The ergodic capacity of the Rayleigh fading channel is plotted in Fig. 14.11. For large SNR, a loss of roughly 2.5 dB is observed due to fading. This is very small compared to the BER loss of approximately 17 dB in the uncoded case. This is because the channel coding theorem assumes infinitely long codewords so that a high diversity gain can be achieved at the decoder. It is observed that both AWGN and Raleigh fading channels achieve the lower limit of −1.59 dB.

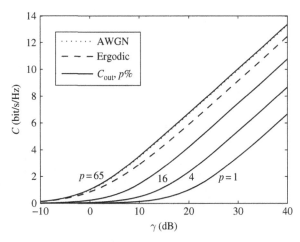

Figure 14.11 Ergodic and outage capacities of a flat Rayleigh fading channel for Gaussian input, where $p\%$ denotes the outage probability.

Outage capacity

The relation between an outage probability P_{out} and the corresponding outage capacity C_{out} is given by (5.26). We reproduce it here

$$P_{\text{out}} = \text{Pr}\,(C(\gamma) < C_{\text{out}}) = \int_0^{\gamma_{\min}} p_\gamma(\gamma)d\gamma, \qquad (14.78)$$

where $\gamma_{\min}$ can be obtained from C_{out}. For the AWGN channel, $C_{\text{out}} = \log_2(1 + \gamma_{\min})$, thus $\gamma_{\min} = 2^{C_{\text{out}}} - 1$.

For a Rayleigh fading channel with the same SNR as that of the AWGN channel, that is, $\bar{\gamma} = E_s/N_0$, we have, from (3.50)

$$P_{\text{out}} = 1 - \exp\left(\frac{1 - 2^{C_{\text{out}}}}{E_s/N_0}\right). \qquad (14.79)$$

From (14.79), C_{out} is obtained as a function of P_{out} and E_s/N_0

$$C_{\text{out}} = \log_2\left(1 - \frac{E_s}{N_0}\ln\,(1 - P_{\text{out}})\right). \qquad (14.80)$$

Example 14.13: The outage capacity, given by (14.80), is also plotted in Fig. 14.11, for different values of P_{out}. When $P_{\text{out}} = 64\%$, C_{out} is greater than the capacity of the AWGN channel. As expected, for a small P_{out} the outage capacity C_{out} drops dramatically for a given SNR.

Example 14.14: The relation between P_{out} and C_{out}, given by (14.79), is plotted in Fig. 14.12 for different SNRs.

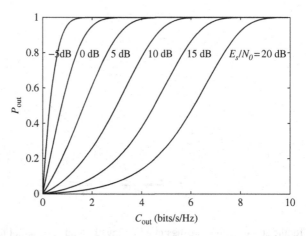

Figure 14.12 P_{out} as a function of C_{out} for the flat Rayleigh fading channel with Gaussian input.

The outage probability is an effective tool for characterizing nonergodic channels, since we cannot compute the ergodic capacity or the average error probability. In such cases, the error probability $P_e(h)$ and the capacity $C(h)$ are random variables. The outage probability can be defined as $P_{\text{out}} = \Pr\{P_e(h) > P_0\}$ or $P_{\text{out}} = \Pr\{C(h) < C_0\}$.

14.6.2 Capacity with CSI at transmitter and receiver

If both the transmitter and the receiver have CSI, the transmitter can adapt its transmission strategy. Assuming the transmitted power $S(\gamma)$ to vary with the received SNR γ subject to an average power constraint $\overline{S}$, the fading channel capacity is defined by maximizing [9]

$$C = \int_0^\infty B \log_2 \left(1 + \frac{S(\gamma)\gamma}{\overline{S}}\right) p_\gamma(\gamma) d\gamma \qquad (14.81)$$

subject to

$$\int_0^\infty S(\gamma) p_\gamma(\gamma) d\gamma = \overline{S}. \qquad (14.82)$$

By using the Lagrange multiplier method, the optimal power adaptation is derived as [9]

$$\frac{S(\gamma)}{\overline{S}} = \begin{cases} \frac{1}{\gamma_0} - \frac{1}{\gamma}, & \gamma \geq \gamma_0 \\ 0, & \gamma < \gamma_0 \end{cases}, \qquad (14.83)$$

where γ_0 is the cutoff SNR. Thus, only when $\gamma > \gamma_0$, is there data transmission. The cutoff SNR γ_0 can be derived from the power constraint

$$\int_{\gamma_0}^\infty \left(\frac{1}{\gamma_0} - \frac{1}{\gamma}\right) p(\gamma) d\gamma = 1. \qquad (14.84)$$

Finally, the capacity formula is given by [9]

$$C = \int_{\gamma_0}^\infty B \log_2 \left(\frac{\gamma}{\gamma_0}\right) p(\gamma) d\gamma. \qquad (14.85)$$

For Rayleigh fading, this capacity exceeds that of the AWGN channel with the same average SNR, which is in contrast to the case of CSI at the receiver end only.

Zero-outage capacity

When the transmitter uses the CSI to maintain a constant received power, the channel appears to be a time-invariant AWGN channel to the encoder and decoder. This power adaptation scheme, called *channel inversion*, is given by [9]

$$\frac{S(\gamma)}{\overline{S}} = \frac{\sigma}{\gamma}, \qquad (14.86)$$

where σ is the constant received SNR that satisfies the transmit power constraint (14.82), from which σ is given by $\sigma = \frac{1}{E\left[\frac{1}{\gamma}\right]}$.

Finally the capacity of the fading channel is given by

$$C = B \log_2(1 + \sigma). \tag{14.87}$$

This capacity can be achieved using a fixed-rate encoder and decoder designed for the AWGN channel with SNR σ. Since the receiver gets constant power and there is no channel outage, this capacity is also called *zero-outage capacity*. Zero-outage capacity may be significantly smaller than Shannon capacity in fading environments, since $E\left[\frac{1}{\gamma}\right]$ may be very large. The channel-inversion strategy is used in CDMA systems to combat the near-far effect.

Outage capacity

The non-outage capacity can be significantly improved by modifying the power adaptation strategy in the following way: suspend transmission in outage channel states (when $\gamma < \gamma_0$) and use a constant data rate in the other states. Such a capacity is referred to as the *outage capacity*. The power adaptation scheme compensates for fading only when $\gamma \geq \gamma_0$

$$\frac{S(\gamma)}{\overline{S}} = \begin{cases} \sigma/\gamma, & \gamma \geq \gamma_0 \\ 0, & \gamma < \gamma_0 \end{cases}, \tag{14.88}$$

where γ_0 is calculated from the outage probability: $P_{\text{out}} = \Pr(\gamma < \gamma_0)$. Accordingly,

$$\sigma = \frac{1}{E_{\gamma_0}[1/\gamma]}, \quad E_{\gamma_0}[1/\gamma] = \int_{\gamma_0}^{\infty} \frac{1}{\gamma} p_\gamma(\gamma) d\gamma. \tag{14.89}$$

This scheme is known as *truncated channel inversion*.

The outage capacity is thus given by

$$C = B \log_2 \left(1 + \frac{1}{E_{\gamma_0}[1/\gamma]}\right) \Pr(\gamma \geq \gamma_0). \tag{14.90}$$

By searching over all γ_0, a maximum outage capacity can be achieved. The maximum outage capacity is still less than the Shannon capacity (14.85), but is easier to implement.

Dirty paper coding

The notion of dirty paper coding (DPC) [2] can be used to improve channel capacity. If the transmitter has perfect, noncausal knowledge of interference to a given user, then the capacity of the channel is the same as that in the case of no interference to the user, or is equivalent to the case where the receiver has perfect knowledge of the interference so that it can subtract it. Based on this idea, the interference can be presubtracted at the transmitter without increasing the transmit power [6].

Comparisons between different transmission schemes

The AWGN channel has a larger capacity than that of all fading channels. However, at low SNR, fading channels with transmitter and receiver CSI have the same capacity as the

AWGN channel. When the SNR is lower than $0\,\mathrm{dB}$, the fading channel with transmitter and receiver CSI has a larger capacity than the AWGN channel. This is because the fading channel will occasionally have higher SNR due to its distribution over an infinite range and high data rates are transmitted during these periods.

Simulation in [9] shows that transmitter adaptation yields a negligible capacity gain compared to using only receiver CSI. Nevertheless, the power adaptation scheme with transmitter and receiver CSI requires a more complex transmitter and a feedback path for sending the CSI from the receiver to transmitter, but the decoder in the receiver is simpler. On the other hand, the nonadaptive policy leads to a simple transmission scheme, but its code design must make use of the channel correlation statistics and the decoder complexity is increased. Channel inversion and truncated channel inversion use codes designed for the AWGN channel and are easy to implement, but there is a large capacity loss in severe fading conditions.

Channel inversion is much less power-efficient in a fading channel, compared to water-filling, since the majority of the power is used to invert the bad channel condition. However, channel inversion achieves a fixed rate in all fading states. Thus, channel inversion eliminates the delay associated with the time-scale of channel variations. Without diversity, the capacity is typically very small. With increase of the diversity order, the probability of occurrence of a bad channel is reduced, and accordingly the capacity increased.

14.6.3 Capacity of frequency-selective fading channels

A frequency-selective fading channel with frequency response $H(f)$ can be divided into many subchannels. For time-invariant channels, one can select each subchannel to be of bandwidth B and $H(f) = H_j$ constant over the subchannel. Each subchannel is a flat-fading channel. Such a channel $H(f)$ is known as *block fading*.

The total power of the subchannels is subject to the power constraint $\sum_j S_j \leq S$, and the capacity is the sum of the data rates on each subchannel [9]

$$C = \sum_j B \log_2 \left(1 + \frac{|H_j|^2 S_j}{N_0 B}\right). \tag{14.91}$$

The optimum power allocation can be derived using the Lagrange multiplier method, leading to the water-filling power allocation on each subchannel

$$\frac{S_j(\gamma_j)}{S} = \begin{cases} \frac{1}{\gamma_0} - \frac{1}{\gamma_j}, & \gamma_j \geq \gamma_0 \\ 0, & \gamma_j < \gamma_0 \end{cases}, \tag{14.92}$$

where $\gamma_j = \frac{|H_j|^2 S}{N_0 B}$ is the SNR of the jth subchannel assuming that the entire power budget is allocated to it, and the cutoff value γ_0 is derived from the power constraint as

$$\sum_j \left(\frac{1}{\gamma_0} - \frac{1}{\gamma_j}\right) = 1. \tag{14.93}$$

Finally, the capacity is given by [9]

$$C = \sum_{j:\gamma_j \geq \gamma_0} B \log_2 \left(\frac{\gamma_j}{\gamma_0} \right). \tag{14.94}$$

Note that this optimum capacity is achieved by transmitting at different rates and powers over the subchannels. This idea is used in multicarrier modulation. For time-invariant frequency-selective fading channels, given continuous $H(f)$, similar results can be obtained, wherein the capacity is given in integral form.

The capacity of time-varying frequency-selective fading channels is very difficult to calculate, due to the effect of ISI. This capacity can be approximated by dividing the channel into many subchannels, each having a bandwidth of the channel-coherent bandwidth B_c. Each of these subchannels is assumed to be independent, time-varying, and flat fading with $H(f_j, t_i) = H_j(i)$ on the subchannel. The derivation for the optimum capacity can be made using the Lagrange multiplier method subject to power constraints on each subchannel and on the whole channel.

Approximate water-filling schemes can greatly simplify transmitter and receiver design. In a constant-power allocation strategy, the transmitter allocates zero power to subchannels that would receive zero power in exact water-filling, but allocates constant power in subchannels that would receive positive power in exact water-filling. A constant-power allocation strategy was observed to be close to the optimal in the adaptive modulation setting [10]. In [27], the performance of constant-power water-filling algorithms was investigated for the ISI channel and for the i.i.d. fading channel. Upper bounds on the maximum difference between the achievable rate under constant-power water-filling and that under true water-filling is given. In particular, for the Rayleigh fading channel, the spectral efficiency loss due to constant-power water-filling is at most 0.266 bit/s/Hz. The performance bound allows the development of a low-complexity, logarithm-free, power-adaptation algorithm.

14.7 Channel capacity for multiuser communications

For multiuser communications, each user has its own QoS constraints such as target data and error rates. In this case, we need to consider not only the maximum cell throughput or sum rate, but also the set of achievable individual data rates for all users, called *capacity region*. In the following, we discuss the case when the BS and all the users have a single antenna. This can be extended to the multiuser MIMO case.

14.7.1 AWGN channel

Given a common BS and N_u mobile users, for the uplink the optimum strategy is to transmit all user signals x_u simultaneously using the entire bandwidth. The received signal at time k is given as

$$r(k) = \sum_{u=1}^{N_u} x_u(k) + n(k). \tag{14.95}$$

The achievable rate R_u of user u is limited by the individual capacity in the absence of all the other users (using (14.54))

$$R_u \leq C_u = \log_2\left(1 + \frac{E_{s,u}}{N_0}\right). \tag{14.96}$$

By summing up the rates of all the N_u users, we may obtain the sum rate. However, this sum rate is too optimistic, since mutual interference is not considered. The sum rate can be obtained as the capacity achieved by a single user that transmits with the power of all N_u users

$$\sum_{u=1}^{N_u} R_u \leq C_{\max} = \log_2\left(1 + \frac{\sum_{u=1}^{N_u} E_{s,u}}{N_0}\right). \tag{14.97}$$

Equations (14.96) and (14.97) constitute the multiple access channel capacity [3, 14]. More generally, we have the following theorem [3]:

Theorem 14.5 (Capacity of multiple access channel): *The capacity region of the m-user multiple access channel $(\mathcal{X}_1 \times \mathcal{X}_2 \times \ldots \mathcal{X}_m, p(y|x_1, x_2, \ldots, x_m), \mathcal{Y})$ is the closure of the convex hull of the rate vectors satisfying*

$$R(M) \leq I\left(X(M); Y|X\left(M^c\right)\right), \quad \text{for all } M \subseteq \{1, 2, \ldots, m\}, \tag{14.98}$$

for some product distribution $p_1(x_1)\, p_2(x_2)\ldots p_m(x_m)$ on $\mathcal{X}_1 \times \mathcal{X}_2 \times \ldots \mathcal{X}_m$, where M^c is the complement of M, and $R(M) = \sum_{i \in M} R_i$, $X(M) = \{X_i : i \in M\}$.

Two-user case

For the two-user case, (14.97) can be written as

$$\begin{aligned}
C_{\max} &= \log_2\left(1 + \frac{E_{s,1} + E_{s,2}}{N_0}\right) \\
&= \log_2\left(1 + \frac{E_{s,1}}{N_0}\right) + \log_2\left(1 + \frac{E_{s,2}}{E_{s,1} + N_0}\right).
\end{aligned} \tag{14.99}$$

In the last equality, $E_{s,1}$ and $E_{s,2}$ can be interchanged. This maximum sum rate can be realized by using successive decoding with interference cancellation, which is discussed for CDMA systems in Chapter 8. If detection begins with user two, the second term corresponds to the maximum rate of user two, C_2^{MUI}. Thus, if the rate of user two, $R_2 \leq C_2^{\mathrm{MUI}}$, it can be perfectly cancelled, and user one can thus achieve its maximum rate C_1 given as the first term, $R_1 = C_1 = I(x_1(k); r(k)|x_2(k))$, which is exactly the maximum mutual information between $x_1(k)$ and $r(k)$ when $x_2(k)$ is known. Thus, we obtain a rate pair $\left(C_1, C_2^{\mathrm{MUI}}\right)$. Alternatively, a rate pair $\left(C_1^{\mathrm{MUI}}, C_2\right)$ can also be obtained if we start with the detection of user one. The capacity region is the area inside the pentagon, as shown in Fig. 14.13.

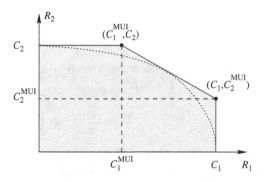

Figure 14.13 The capacity region for uplink transmission in the AWGN channel and two users. The dotted line corresponds to orthogonal multiple access schemes.

The maximum sum rate is achieved at $\left(C_1, C_2^{\text{MUI}}\right)$ and $\left(C_1^{\text{MUI}}, C_2\right)$ by using successive interference cancellation. The line between these two points can be achieved by approximately switching between the two transmission strategies. It is noted that $C_{\max} \geq \max(C_1, C_2)$. The optimum strategy with successive decoding can be generalized for N_u users in a straightforward manner. For each of the $N_u!$ possible detection orders, a point consisting of N_u individual rates is obtained. Usually, the N_u users have an ascending order of detection.

Some orthogonal multiple access schemes

For orthogonal multiple access schemes such as TDMA and FDMA, the time or frequency is shared among the users, and the capacity is inferior to that discussed above. Consider the two-user case with the received symbol energies $E_{s,1}$ and $E_{s,2}$, and assume that the two users obtain the portions α $(0 < \alpha < 1)$ and $1 - \alpha$ of the resource, respectively. For TDMA, the received powers for the two users are, respectively, $E_{s,1}/(\alpha T)$ and $E_{s,2}/((1 - \alpha)T)$, and the noise power is $N_0 B = N_0/T$ for both the users, where T is the symbol period. For FDMA, the noise powers for the two users are, respectively, $\alpha B N_0$ and $(1 - \alpha)B N_0$. Thus, the corresponding capacities are

$$C_1 = \alpha \log_2 \left(1 + \frac{E_{s,1}}{\alpha N_0}\right), \quad C_2 = (1 - \alpha) \log_2 \left(1 + \frac{E_{s,2}}{(1 - \alpha)N_0}\right). \tag{14.100}$$

The maximum capacity is given by

$$C_{\max} = C_1 + C_2. \tag{14.101}$$

This capacity reaches the maximum sum rate only at a few points, as shown in Fig. 14.13.

In the downlink, a common BS transmits to many mobile users. Each user receives all the transmitted signals and extracts its desired one. Unlike the uplink case, where each user has its own power constraint, the total transmitted power is shared among the users. The individual rate R_u also needs to satisfy the constraint $R_u \leq C_u$ or $R_u \leq C_u^{\text{MUI}}$. The same strategy of successive detection of the uplink can be applied to the downlink.

14.7.2 Flat-fading channels

In the flat-fading uplink channel, simultaneous transmission with successive decoding still achieves capacity. Without CSI at the transmitter, the best strategy is that all users transmit with the same average signal power E_s/T, and then the ergodic maximum sum rate is given by [14]

$$\overline{C}_{max} = E\left[\log_2\left(1 + \frac{E_s}{N_0}\sum_{u=1}^{N_u}|H_u|^2\right)\right] \leq \log_2\left(1 + N_u\frac{E_s}{N_0}\right), \tag{14.102}$$

where the upper bound is based on the assumption of $E[|H_u|^2] = 1$ for $u = 1, 2, \ldots, N_u$. For Rayleigh fading, the sum of the squares of the channel coefficient magnitudes is χ^2-distributed with $2N_u$ degrees of freedom. The upper bound corresponds to the maximum sum rate for the AWGN uplink channel.

We now consider the flat-fading downlink channel. Without CSI at the transmitter, capacity can be achieved by superposition coding with successive decoding at the receiver, as in the case of the AWGN downlink case. If the common BS has CSI, the maximum throughput can be significantly increased by using the *multiuser diversity* technique, which has been discussed in Section 5.4. The basic idea of multiuser diversity is that only the user with the largest instantaneous channel gain $|h_u(k)|^2$ transmits. The optimum solution is that the system distributes the power of each user onto its active time instances based on the water-filling principle. For i.i.d. channels, each user can transmit the same average data rate and fulfill its power constraint.

The capacity regions of the MIMO broadcast channel and the MIMO MAC channel are discussed in [24]. The problem of finding the optimal transmitter covariance matrices that achieve the sum capacity in a Gaussian vector multiple-access channel is addressed in [26] by using an iterative water-filling algorithm. The capacities of MIMO channels for both single-user and multiuser systems are summarized in [8].

14.8 Estimation theory

Estimation theory is concerned with the problem of finding the optimum value for an unknown continuous variable, given a collection of measurements. The quality of estimation is determined by the accuracy of the measurements. The Cramer-Rao inequality is the most important information-theoretic inequality that pertains to estimation theory. It is a lower bound on the performance of an estimator, known as the *Cramer-Rao lower bound (CRLB)*.

Assume $\mathbf{c} = (c_1, \ldots, c_m)$ to be an unknown vector that is to be estimated, and the measurement $\mathbf{x}$ to be a random variable vector. The vector estimator is given by its mean vector and covariance matrix

$$E\left[\hat{\mathbf{c}}\right] = E\left[\hat{\mathbf{c}}(\mathbf{x})\right], \tag{14.103}$$

$$\mathbf{\Sigma}\left(\hat{c}\right) = E\left[\left(\hat{c}(x) - E\left(\hat{c}\right)\right)\left(\hat{c}(x) - E\left(\hat{c}\right)\right)^T\right].\tag{14.104}$$

For estimation problems, the variance of the estimates of the unknown parameters is subject to the CRLB for any estimator [3].

Theorem 14.6 (Cramer-Rao Lower Bound): *For any unbiased estimator $\hat{c}$, $E[\hat{c}] = c_0$, its covariance matrix $\mathbf{\Sigma}(c)$ satisfies*

$$\mathbf{\Sigma}(c) \geq \mathbf{J}^{-1}(c),\tag{14.105}$$

i.e.,

$$w^T \mathbf{\Sigma} w \geq w^T \mathbf{J}^{-1} w, \quad \text{for any } w,\tag{14.106}$$

where $\mathbf{J}$ is the Fisher information matrix, defined below.

Let $p(x|c)$ denote the joint pdf of x conditioned on the unknown vector c

$$p(x|c) = p\left(x_1, \ldots, x_n | c_1, \ldots, c_m\right).\tag{14.107}$$

The Fisher information matrix $\mathbf{J}(c)$ is an $m \times m$ matrix, and is defined by

$$\mathbf{J}(c) = E\left[\left(\frac{\partial l}{\partial c}\right)\left(\frac{\partial l}{\partial c}\right)^T\right]\Bigg|_{c_0},\tag{14.108}$$

where $l = \ln p(x|c)$ is the log-likelihood function. The elements are given by

$$J_{ij}(c) = \int_{-\infty}^{\infty} p(x|c)\left[\frac{\partial l}{\partial c_i}\right]\left[\frac{\partial l}{\partial c_j}\right] dx.\tag{14.109}$$

For the parameter c_i, we may write

$$\text{var}\left[\hat{c}_i\right] \geq \left[\mathbf{J}^{-1}\right]_{ii}, \quad i = 1, 2, \ldots, m.\tag{14.110}$$

For a zero-mean Gaussian vector $x(t)$, the measurement error $e = x - E[x]$ is also Gaussian with zero mean, and the Fisher information matrix can be derived as

$$J_{ij} = \frac{1}{2}\text{tr}\left(\mathbf{R}^{-1}\frac{\partial \mathbf{R}}{\partial c_i}\mathbf{R}^{-1}\frac{\partial \mathbf{R}}{\partial c_j}\right),\tag{14.111}$$

where tr is the trace operator for a matrix and $\mathbf{R}$ is the covariance matrix of x. The CRLB is generally evaluated numerically, since analytical inversion of $\mathbf{R}$ is intractable, except for the case of each signal being stationary.

In wireless communications, CRLB sets the lower bounds for the estimation of parameters such as the channel, position, delay of multipath, and signal direction.

Problems

14.1 A random variable has the distribution $P(X = k) = p(1 - p)^{k-1}$, $k = 1, 2, 3, \ldots$. What is the entropy of X?

14.2 Prove that $I(X; Y) = H(X) + H(Y) - H(X, Y)$.

14.3 Show that the entropy of an n-dimensional Gaussian vector $X = (x_1, \ldots, x_n)^T$ with zero mean and covariance matrix $\mathbf{M}$ is $H(X) = \frac{1}{2} \log_2(2\pi e)^n |\mathbf{M}|$.

14.4 Which of the following are uniquely decodeable? (a) $\{0, 10, 11\}$. (b) $\{0, 01, 11\}$. (c) $\{0, 01, 10\}$. (d) $\{110, 11, 10\}$. (e) $\{0, 01\}$. (g)$\{110, 100, 00, 10\}$. Apply the Kraft-McMillan inequality to these codes.

14.5 Is it possible for the following codes to be Huffman codes? (a) $\{00, 01, 10, 110\}$. (b) $\{10, 01\}$.

14.6 A DMS has an alphabet of eight letters, X_i, $i = 1, 2, \ldots, 8$, with probabilities of $0.15, 0.10, 0.35, 0.10, 0.14, 0.06, 0.03,$ and 0.07. Encode the source using Huffman coding. Compare the average number of bits per letter with the entropy of the source.

14.7 For a random variable X with an alphabet of seven letters, with probabilities of $\frac{1}{28}$, $\frac{2}{28}, \frac{3}{28}, \frac{4}{28}, \frac{5}{28}, \frac{6}{28}$, and $\frac{7}{28}$, find (a) the binary Huffman codes and their average length, (b) the ternary Huffman codes and their average length.

14.8 Given the binary source $\{0, 1\}$ with $p_0 = 0.125$ and $p_1 = 0.875$, encode the sequence 11110111011110 using arithmetic coding.

14.9 Prove that the capacity C of a discrete memoryless channel with input alphabet $\mathcal{X} = \{X_1, X_2, \ldots, X_N\}$ and output alphabet $\mathcal{Y} = \{Y_1, Y_2, \ldots, Y_M\}$ satisfies $C \leq \min\{\log_2 M, \log_2 N\}$.

14.10 Use the sliding window version of the Lempel-Ziv algorithm with a window size of 10 to parse and encode the binary sequence 1011001110 0111010110 0100110011 11001101011 10000010000. What is the encoder output as a binary sequence? Decode the encoded sequence.

14.11 Use the LZ78 algorithm to solve the above problem.

14.12 Determine the capacity of the channel illustrated in Fig. 14.14.

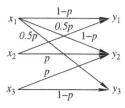

Figure for Problem 14.12.

14.13 If the SNR of a wireless channel is 10 dB and the RF bandwidth is 30 kHz, determine the theoretical maximum data rate that can be supported. If the RF bandwidth is 200 kHz, what is the result?

14.14 Compare the channel spectral efficiencies of IS-54, GSM, PDC, and IS-95. What are their theoretical spectral efficiencies at a SNR of 20 dB?

14.15 A receiver has a received signal power of -110 dBm, a received noise power spectral density of -160 dBm/Hz, and a bandwidth of 2000 Hz. Determine the maximum rate of error-free information transfer for this system.

14.16 Plot on the same figure the capacities of an AWGN channel with binary antipodal signaling and binary orthogonal signaling as a function of E_b/N_0. Optimal bit-by-bit detection is assumed at the receiver.

14.17 For any concave function f and a random variable x, establish Jensen's inequality

$$E[f(x)] \leq f(E[x]).$$

[Hint: geometrical analysis can be applied.]

14.18 For an additive noise channel, $Y = X + Z$, where the noise Z is independent of X, $\Pr(Z = 0) = \Pr(Z = a) = \frac{1}{2}$, and X takes 0 or 1, derive the channel capacity.

14.19 Consider a zero-mean complex random vector x with fixed covariance $E[xx^*] = \mathbf{K}$. Show that the differential entropy satisfies

$$h(x) \leq \log_2 \det(\pi e \mathbf{K})$$

with equality when x is a complex Gaussian distribution with zero mean and covariance $\mathbf{K}$.

14.20 Evaluate the differential entropy $h(X)$ for the exponential density $p_X(x) = \lambda e^{-\lambda x}$, $x \geq 0$.

14.21 Consider a pair of parallel Gaussian channels

$$Y_i = X_i + Z_i, \quad i = 1, 2,$$

where Z_i's are zero-mean AWGNs with variance σ_i^2, and $\sigma_1^2 > \sigma_2^2$. There is a power constraint $E\left(X_1^2 + X_2^2\right) \leq 2P$. According to water-filling, at what power does the channel transfer from one channel mode with noise variance σ_1^2 to two-channel mode?

14.22 Derive the rate-distortion function given by (14.40).

14.23 Derive and plot the capacity for square-MQAM signaling with $M = L^2$, $L = 2, 4, 8, 16$, in the AWGN channel. The reader can use the MATLAB code for Example 14.11, which is available at the website for this book. Like MPSK, the capacity of MQAM at high SNR is $\log_2 M$ bits/s/Hz. Note that MQAM signaling approaches the channel capacity $\log_2 M$ at a lower SNR compared to MPSK signaling.

14.24 Plot the capacity with BPSK, QPSK, and 8PSK modulation schemes, over an ergodic Rayleigh flat-fading channel, as a function of the average SNR.

14.25 Consider a sequence of i.i.d. measurements $x_1, x_2, \ldots, x_n$ of a Rayleigh-distributed random variable with parameter f.
(a) Determine the CRLB.
(b) Consider the estimator $\hat{f} = \frac{1}{2N} \sum_{k=1}^{N} x_k^2$. Determine its variance and check if it achieves the CRLB.

References

[1] M.-S. Alouini & A. J. Goldsmith, Area spectral efficiency of cellular mobile radio systems. *IEEE Trans. Veh. Tech.*, **48**:4 (1999), 1047–1066.

[2] M. H. N. Costa, Writing on dirty paper. *IEEE Trans. Inf. Theory*, **29**:3 (1983), 439–441.

[3] T. M. Cover & J. A. Thomas, *Elements of Information Theory*, 2nd edn (Hoboken, NJ: Wiley, 2006).

[4] T. M. Cover & M. Chiang, Duality between channel capacity and rate distortion with two-sided state information. *IEEE Trans. Inf. Theory*, **48**:6 (2002), 1629–1638.

[5] J. R. Deller, Jr., J. H. L. Hansen & J. G. Proakis, *Discrete-Time Processing of Speech Signals* (New York: Wiley-IEEE, 2000).

[6] U. Erez, S. Shamai & R. Zamir, Capacity and lattice strategies for cancelling known interference. In *Proc. IEEE ISITA*, Honolulu, HI, Nov 2000, 681–684.

[7] M. Ghanbari, Standard Codecs: Image Compression to Advanced Video Coding (London: IEE Press, 2003).

[8] A. Goldsmith, S. A. Jafar, N. Jindal & S. Vishwanath, Capacity limits of MIMO channels. *IEEE J. Sel. Areas Commun.*, **21**:5 (2003), 684–702.

[9] A. Goldsmith, *Wireless Communications* (Cambridge, UK: Cambridge University Press, 2005).

[10] A. J. Goldsmith & S.-G. Chua, Variable-rate variable-power MQAM for fading channels. *IEEE Trans. Commun.*, **45**:10 (1997), 1218–1230.

[11] A. J. Goldsmith & M. Effros, Joint design of fixed-rate source codes and multiresolution channel codes. *IEEE Trans. Commun.*, **46**:10 (1998), 1301–1312.

[12] S. W. Golomb, Run-length encoding. *IEEE Trans. Inf. Theory*, **12**:3 (1966), 399–401.

[13] D. A. Huffman, A method for the construction of minimum redundancy codes. *Proc. IRE*, **40**:9 (1952), 1098–1101.

[14] V. Kuhn, *Wireless Communications over MIMO Channels: Applications to CDMA and Multiple Antenna Systems* (Chichester, UK: Wiley, 2006).

[15] G. G. Langdon, An introduction to arithmetic coding. *IBM J. Res. Dev.*, **28**:2 (1984), 135–149.

[16] S. Lonardi, W. Szpankowski & M. D. Ward, Error resilient LZ'77 data compression: algorithms, analysis, and experiments. *IEEE Trans. Inf. Theory*, **53**:5 (2007), 1799–1813.

[17] M. Nelson, *The Data Compression Book* (New York: M&T Books, 1992).

[18] J. G. Proakis & M. Salehi, *Digital Communications*, 5th edn (New York: McGraw-Hill, 2008).

[19] J. J. Rissanen, Generalized Kraft inequality and arithmetic coding. *IBM J. Res. Dev.*, **20**:3 (1976), 198–203.

[20] K. Sayood, *Introduction to Data Compression*, 2nd edn (San Mateo, CA: Morgan Kaufmann, 2000).

[21] C. E. Shannon, A mathematical theory of communication. *Bell Syst. Tech. J.*, **27** (1948), 379–423, 623–656.

[22] C. E. Shannon, Coding theorems for a discrete source with a fidelity criterion. In *Proc. IRE National Convention Record*, New York, Mar 1959, part 4, 142–163.

[23] J. Teuhola, A compression method for clustered bitvectors. *Inf. Process. Lett.*, **7**:6 (1978), 308–311.

[24] S. Viswanath, N. Jindal & A. Goldsmith, Duality, achievable rates, and sum-rate capacity of Gaussian MIMO broadcast channels. *IEEE Trans. Inf. Theory*, **49**:10 (2003), 2658–2668.

[25] T. A. Welch, A technique for high-performance data compression. *IEEE Computer*, **17**:6 (1984), 8–19.

[26] W. Yu, W. Rhee, S. Boyd & J. M. Cioffi, Iterative water-filling for Gaussian vector multiple-access channels. *IEEE Trans. Inf. Theory*, **52**:1 (2004), 145–152.

[27] W. Yu & J. M. Cioffi, Constant-power waterfilling: performance bound and low-complexity implementation. *IEEE Trans. Commun.*, **54**:1 (2006), 23–28.

[28] J. Ziv & A. Lempel, Compression of individual sequences via variable-rate coding. *IEEE Trans. Inf. Theory*, **24**:5 (1978), 530–536.

[29] J. Ziv, The universal LZ77 compression algorithm is essentially optimal for individual finite-length N-blocks. *IEEE Trans. Inf. Theory*, **55**:5 (2009), 1941–1944.

Channel coding

15.1 Preliminaries

A channel is an abstract model describing how the received (or retrieved) data is associated with the transmitted (or stored) data. Channel coding starts with Claude Shannon's mathematical theory of communication [63].

Error detection/correction coding

Channel coding can be either error detection coding or error correction coding. When only error detection coding is employed, the receiver can request a transmission repeat, and this technique is known as *automatic repeat request (ARQ)*. This requires two-way communications. An ARQ system requires a code with good error-detecting capability so that the probability of an undetected error is very small.

Forward error correction (FEC) coding allows errors to be corrected based on the received information, and it is more important for achieving highly reliable communications at rates approaching channel capacity. For example, by turbo coding, an uncoded BER of 10^{-3} corresponds to a coded BER of 10^{-6} after turbo decoding. For applications that use simplex (one-way) channels, FEC coding must be supported since the receiver must detect and correct errors, and no reverse channel is available for retransmission requests.

Another method using error detection coding is error concealment. This method processes data in such a way that the effect of errors is minimized. Error concealment is especially useful for applications that carry data for subjective appreciation, such as speech, music, image, and video. Loss of a part of the data is acceptable, since there is still some inherent redundancy in the data. Interpolation or extrapolation can be used to generate values to replace the values with errors. Error concealment is usually used when ARQ is difficult or impossible; for example, in speech application, the vocoder only generates output speech, but no feedback is available. Hybrid with FEC is also possible.

From the viewpoint of implementation, ARQ is simple and provides higher system reliability than FEC. However, its throughput drops rapidly with increasing channel error rate. FEC systems have a constant throughput, but the probability of a decoding error is much higher than the probability of an undetected error in ARQ systems. To achieve high reliability, a long powerful code has to be used, leading to expensive decoding. For this reason, ARQ is preferred for error control in data communication systems, while FEC is the only choice in voice communication or data storage systems, where the return channel is not available, or the delay is intolerable, or retransmission is not possible. Hybrids of ARQ and FEC combine the advantages of both methods, and are known as *HARQ*, which will be introduced in Section 15.12.

Block and convolutional codes

Channel codes are traditionally classified into block codes and convolutional codes. For block codes, redundancy is added to blocks of data, while for convolutional codes redundancy is added continuously.

For an (n, k) or (n, k, d) block code, each codeword has a length of n bits, where the first k bits are the information bits and the following $n - k$ bits are redundant parity bits generated from the k information bits by an algebraic relation, and d is the minimum Hamming distance of the code. The ratio $R_c = k/n$ is known as the *code rate*, and $0 < R_c \leq 1$. The encoder is memoryless, and can be implemented with a combinatorial logic circuit.

Convolutional codes are generated by the discrete-time convolution of the information bit sequence with the impulse response of the encoder. The duration of the impulse response is characterized by the memory of the encoder. Thus, the convolutional encoder generates a continuous sequence of coded bits from a continuous sequence of input bits. The encoder accepts k-bit blocks of input information sequence and produces an encoded sequence of n-bit blocks. Each encoded block depends not only on the k-bit input block but also on m previous message blocks. The encoder thus has a memory order of m. The code obtained is an (n, k, m) convolutional code. The code rate is given by $R_c = k/n$. Since the encoder has memory, it must be implemented by using a sequential logic circuit.

Finite fields

A field F is a set of elements that is closed under two binary operations, namely, addition and multiplication. A field having a finite number of elements is called a *finite field* or *Galois field*. A finite field of q elements is denoted by GF(q), with the number of elements defined as $q = p^m$, where p is a prime number and m is a positive integer. The most common finite field is the set $\{0, 1\}$ under the operations of standard modulo-2 addition and multiplication.

15.2 Linear block codes

Linear block codes have the property that the modulo-2 addition of any two codewords must be a codeword and the all-zeros word is also a permitted codeword. Block coding is widely used for error detection/correction in data storage systems and communication systems.

Linear block coding has a general coding equation

$$c = m\mathbf{G},$$

(15.1)

where the coded message $c = [c_1 c_2 \ldots c_n]$, the message word $m = [m_1 m_2 \ldots m_k]$, and G is the $k \times n$ generator matrix. Entries of c, m, and G are elements belonging to GF(p), where GF(p) is the finite field of the integers modulo a prime p. For the common binary code, GF(2) $= \{0, 1\}$.

The (n, k) block code is generated by encoding k information digits and outputting an n-digits codeword, which is composed of the k information digits and $(n - k)$ redundant parity check digits. The rate, or efficiency, of the code is $R_c = k/n$. Normally R_c is between $1/2$ and 1.

In many cases, the first k bits of the coded message c remain the same as the message word, and the remaining $n - k$ bits are used as parity bits. Such codes are called *systematic codes*. This can be expressed by setting

$$G = [I_k | P], \tag{15.2}$$

where I_k is a $k \times k$ identity matrix and P is a $k \times (n - k)$ matrix.

To decide whether the received codeword $\hat{c}$ is valid, it is multiplied by a parity check matrix H. This yields an $(n - k)$-dimensional syndrome vector s_{syn}

$$s_{\text{syn}} = \hat{c} H^T. \tag{15.3}$$

The entries of s_{syn} correspond to the $n - k$ parity bits. If s_{syn} is a zero vector, the received codeword is valid.

The selection of H must retain the relationship

$$HG^T = O, \tag{15.4}$$

where H is an $(n - k) \times k$ matrix. It is equivalent to $cH^T = 0$. One simple solution is

$$H = \left[-P^T \; I_{n-k} \right]. \tag{15.5}$$

For binary codes, the minus sign can be dropped since modulo-2 subtraction is identical to modulo-2 addition.

Since $cH^T = 0$, the syndrome only depends on the error word e. This is given by

$$s_{\text{syn}} = \hat{c} H^T = (e + c) H^T = e H^T \tag{15.6}$$

A nonzero syndrome identifies an error. The syndrome digits are simply linear combinations of the error digits.

The popularity of linear block codes may be due to the algebraic decoding methods. They can be decoded at a complexity of $O\left(d_{\text{min}}^2\right)$, with d_{min} being the minimum distance of the code, by using finite-field arithmetic which is efficient in VLSI implementation.

15.2:1 Error detection/correction

The *Hamming weight* of a block codeword c of length n, denoted $w(c)$, is defined as the number of nonzero elements of the code.

The *Hamming distance* between two codewords c_1 and c_2, denoted $d(c_1, c_2)$, is defined as the number of places where they differ.

The *minimum distance* of a block code C, denoted d_{min}, is defined as

$$d_{min} = \min\{d(c_1, c_2) \,|\, c_1, c_2 \in C, c_1 \neq c_2\}. \tag{15.7}$$

The Hamming distance is related to the Hamming weight by

$$d(c_1, c_2) = w(c_1 + c_2). \tag{15.8}$$

For example, the Hamming distance between $c_1 = (11001010101)$ and $c_2 = (01111001010)$ is 8, which is equal to the weight of $c_1 + c_2 = (10110011111)$, that is 8.

For linear block codes, the sum of two codewords is also a codeword. Thus, it follows from (15.7) that

$$d_{min} = \min\{d(c) \,|\, c \in C, c \neq 0\} = w_{min}. \tag{15.9}$$

Thus, the minimum distance of a linear block code equals the minimum weight of its nonzero code words. For linear block codes, there are a number of bounds on d_{min}, such as the Hamming or sphere packing bound, and the Plotkin bound. These bounds are described in [56].

A block code with minimum distance d_{min} is capable of detecting all the error patterns of $d_{min} - 1$ or fewer errors, or guaranteeing the correction of all the error patterns of $\lfloor (d_{min} - 1)/2 \rfloor$ or fewer errors. That is, the error-detection capability t_{ed} and error-correction capability t_{ec} are determined by [47, 66]

$$t_{ed} = d_{min} - 1, \tag{15.10}$$

$$t_{ec} = \left\lfloor \frac{d_{min} - 1}{2} \right\rfloor, \tag{15.11}$$

where $\lfloor \cdot \rfloor$ rounds down the number to an integer. For all Hamming codes, $d_{min} = 3$ and $t_{ec} = 1$. Thus one error in each codeword can be corrected.

A block code with random error correcting capability t_{ec} is usually capable of correcting many error patterns of $t_{ec} + 1$ or more errors. For an (n, k) linear code, it is capable of correcting a total of 2^{n-k} error patterns, including those with t_{ec} or fewer errors; however, it is capable of detecting $2^n - 2^k$ error patterns [47]. For large n, $2^{n-k} \ll 2^n - 2^k$.

The upper bound on the average probability of an undetected error for an (n, k) linear systematic code over a BSC is given as [47]

$$P_{u,E} \leq 2^{-(n-k)} \left[1 - (1-p)^n\right] \leq 2^{-(n-k)}, \tag{15.12}$$

where p is the transition probability of the BSC. Thus, $P_{u,E}$ decreases exponentially with the number of parity-check digits, $n - k$.

If a binary linear block code (n, k, d_{min}) is able to correct to t_{ec} errors, each possible error pattern needs to be assigned a unique syndrome. The following Hamming bound is obtained:

$$2^{n-k} \geq \sum_{r=0}^{t_{ec}} \binom{n}{r}. \tag{15.13}$$

The left-hand side is the number of syndromes, and the right-hand side is the number of error patterns with $w(e) \leq t_{ec}$. When equality holds, the code is called a *perfect code*.

15.2.2 Simple parity check and Hamming codes

Simple parity check codes

The most simple channel coding is parity checking. The simple parity check code is a linear block code. It performs mod-2 addition of all bits in a block, and appends the result as parity bit to the end of the frame. For a length n codeword, there is $k = n-1$ information bits, and thus $R_c = (n-1)/n$. In case of even parity, when the code word contains an even number of ones, the parity bit is zero; in case of odd parity, when the code word contains an odd number of ones, the parity bit is zero.

For even-parity condition, the parity bit p is formed as $p = c_1 \oplus c_2 \oplus \cdots \oplus c_n$. At the receiver, the parity bit is recalculated to decide whether the correct data is received. This method can only detect an odd number of bit errors, but cannot detect the errors in case of an even number of bit errors. The minimum Hamming weight is 2, which is also the code's minimum distance, d_{min}. Thus, the code is a $(n, n-1, 2)$ code.

The single parity bit can only be used for error detection, most notably in the 8-bit ASCII codes. By using more parity bits, one can also estimate the position of the bit errors. This can be performed by arranging the codeword into a two-dimensional array, and then applying parity for each row and for each column. The resulting parity bits are then appended to the end of the codeword. The received codeword will be mapped to the same two-dimensional array, and the parity bits are examined at each column and each row. For a single bit error, the intersection of the row and column corresponding to the parity failure is the error position, and it thus can be corrected.

Hamming codes

One special and well-known class of linear codes is the *Hamming codes* [36]. It was invented shortly after the channel-coding theorem was published in 1948. For such codes, the columns of **H** contain all 2^{n-k} possible bit combinations of length k (except the all-zero word). The resulting Hamming code has a size of $(2^m - 1, 2^m - 1 - m)$, for $m = 2, 3, \cdots$. There are m redundant bits in the code. Its minimum distance is 3. The Hamming codes are a class of single-error-correcting perfect code, since they correct all single errors but no patterns of more than one error.

An (n, k) linear code can be extended by adding an overall parity-check bit. This is known as an *extended* $(n + 1, k)$ *code*. The parity-check bit is added so as to give even parity codewords, as this preserves the linearity of the code. The extended Hamming code has a minimum distance of 4.

Example 15.1: Given the parity check matrix

$$\mathbf{H} = \begin{bmatrix} 1 & 1 & 1 & 0 & 1 & 0 & 0 \\ 1 & 1 & 0 & 1 & 0 & 1 & 0 \\ 1 & 0 & 1 & 1 & 0 & 0 & 1 \end{bmatrix}, \tag{15.14}$$

verify that the corresponding block code is a Hamming code.

The code can be easily verified to be a $(7, 4)$ Hamming code. This can be solved by first obtaining the generator matrix $\mathbf{G}$ from $\mathbf{H}$, from (15.2) and (15.5), as

$$\mathbf{G} = \begin{bmatrix} 1 & 0 & 0 & 0 & 1 & 1 & 1 \\ 0 & 1 & 0 & 0 & 1 & 1 & 0 \\ 0 & 0 & 1 & 0 & 1 & 0 & 1 \\ 0 & 0 & 0 & 1 & 0 & 1 & 1 \end{bmatrix}. \tag{15.15}$$

We can then obtain altogether 16 codewords from 16 messages of length 4, starting from all-zeros message to all-ones message, by using $c = m\mathbf{G}$. The minimum distance of the 16 codewords in c is calculated as 3, by using the MATLAB code *Example_Hamming_Code.m*.

A more powerful linear block coding scheme is cyclic redundancy check (CRC) coding, which will be introduced in Section 15.4.

15.2.3 Syndrome decoding

The decoding of the block code is to find a code that is closest to the transmitted codeword. Syndrome decoding, also called *table-lookup decoding*, can be used for any linear block code. This leads to minimum decoding delay and minimum error probability. The method calculates the syndrome $s_{\mathrm{syn}} = \hat{c}\mathbf{H}^T$, then finding the error pattern that has the same syndrome. The decoded result is given by $c = \hat{c} + e$. For large $n - k$, this decoding scheme is too complex to implement, since it needs either a large storage for the table or a complex combinatorial logic.

From (15.6), given a syndrome, there are 2^{n-k} distinct error patterns. Only the error pattern that has the smallest number of nonzero components is the most probable one. The lookup table is obtained as the truth table of n switching functions

$$e_i = f_i (s_0, s_1, \ldots, s_{n-k-1}), \quad i = 0, 1, \ldots, n - 1, \tag{15.16}$$

where $s_0, s_1, \ldots, s_{n-k-1}$ are the syndrome digits. This table can be implemented using a combinatorial logic with $n - k$ syndrome digits as inputs and n outputs as the estimated error pattern.

Example 15.2: For the $(7,4)$ Hamming code, given message $m = (1100)$, the generated codeword is $c = (1100001)$. The received codeword is $\hat{c} = (1100011)$, show the procedure of error correction.

First, the syndrome is computed as $s = \hat{c}\mathbf{H}^T = (010)$, where $\mathbf{H}$ is given by (15.14). Then, from $s = e\mathbf{H}^T$, we have

$$\begin{cases} e_0 + e_1 + e_2 + e_4 = 0 \\ e_0 + e_1 + e_3 + e_5 = 1 \\ e_0 + e_2 + e_3 + e_6 = 0 \end{cases}. \tag{15.17}$$

Table 15.1. The lookup table for syndromes and correctable error patterns.

s_0	s_1	s_2	e_0	e_1	e_2	e_3	e_4	e_5	e_6
0	0	0	0	0	0	0	0	0	0
1	0	0	0	0	0	0	0	0	1
0	1	0	0	0	0	0	0	1	0
0	0	1	0	0	0	0	1	0	0
1	1	0	0	0	0	1	0	0	0
0	1	1	0	0	1	0	0	0	0
1	1	1	0	1	0	0	0	0	0
1	0	1	1	0	0	0	0	0	0

There are a total of $2^4 = 16$ error patterns satisfying the above equations:

$$(0000010), (0001001), (0010111), (0011100)$$
$$(0100100), (0101111), (0110001), (0111010)$$
$$(1000101), (1001110), (1010010), (1011011)$$
$$(1100011), (1101000), (1110110), (1111101).$$

The error pattern $\mathbf{e} = (0000010)$ has the smallest number of nonzero digits, and is the most possible error pattern. It is selected as the lookup table item for syndrome (010). The decoded codeword is obtained as

$$c^* = \hat{c} + e = (1100011) + (0000010) = (1100001), \qquad (15.18)$$

which is exactly the transmitted codeword.

A lookup table (LUT) for all the correctable error patterns for the (7,4) Hamming code can be obtained for all the syndromes. The result is listed in Table 15.1. From this truth table, the error pattern can be obtained from the syndrome by a simple logic circuit, and a decoding circuit can be accordingly designed. From this table, we can see that Hamming codes can correct only the error patterns of single errors and no others. For a Hamming code, all the columns in **H** are linearly independent; it can correct exactly one error and the location of the error is given by the syndrome.

15.3 Hard/soft decision decoding

Decoding takes place by either hard decision or soft decision. In case of soft decision, the channel output is used directly without quantization, while in hard decision an optimal decision is based on each input level. As a hard decision, the decoder makes a decision on a rigid condition. Soft decoding can achieve better performance, but at a much greater complexity. For binary coding, if the demodulator output uses binary quantization, $Q = 2$, the decoder is said to make *hard decisions*; if the quantization level $Q > 2$ or the output is unquantized, the demodulator will make *soft decisions*.

For a given power, the encoder produces more bits compared to the uncoded case, this results in a reduction in E_b, and thus the coded bitstream has a larger BER. This is compensated by the coding gain. When $P_e = 10^{-5}$ and hard decision decoding is employed, the coding gain is 3 to 5 dB for block codes, 4 to 5.5 dB for convolutional codes with Viterbi decoding, and 6.5 to 7.5 dB for concatenated codes of Reed-Solomon block codes and convolutional code using Viterbi decoding [70].

Hard decision decoding

Hard decision decoding uses the minimum distance decoding that detects or corrects errors based on the Hamming distance. For an (n, k) block code, if the error correction capability is t_{ec} errors, under hard decision decoding, a received codeword may be decoded in error if it contains more than t_{ec} errors.

Assuming the bit errors occur independently on an AWGN channel, for a linear block code, the probability of codeword error is upper bounded by

$$P_e \leq \sum_{i=t_{ec}+1}^{n} \binom{n}{i} p^i (1 - p)^{n-i}, \tag{15.19}$$

where p is the error probability of the bit transmission in the codeword. For coherent modulation, typically $p = \alpha_M Q\left(\sqrt{\beta_M E_c / N_0}\right)$, where E_c is the energy per codeword symbol.

For (n, k) block codes, there are k information bits, thus $E_c = kE_b/n$. Thus block codes with a large number of parity bits (k/n is small) have a small E_c, leading to an increase in the error probability. This loss in the error probability is, however, compensated by the error correction capability of the codes. At high SNR, coding yields a coding gain. At low SNR, coding yields a higher error probability than uncoded modulation does, leading to a negative coding gain. This is due to reduced energy per symbol.

At high SNR, the codeword error probability can be approximated by the nearest-neighbor errors. For a given nearest neighbor at distance d_{min}, a simple upper bound is given by [31, 56]

$$P_e \leq (M - 1)[4p(1 - p)]^{d_{min}/2}, \tag{15.20}$$

where $M = 2^k$ is the number of codewords.

The BEP after decoding depends on the specific code, decoder, and the bit-to-codeword mapping. The BEP for block codes using hard decision decoding in the AWGN channel is often approximated by [31]

$$P_e \approx \frac{1}{n} \sum_{i=t_{ec}+1}^{n} i \binom{n}{i} p^i (1 - p)^{n-i}. \tag{15.21}$$

The probability of correct decoding, P_c, is given by the probability of no errors occurring, that is, $P_c = (1 - p)^n$. The probability of decoding failure is thus

$$P_f = 1 - P_e - P_c. \tag{15.22}$$

Soft decision decoding

Soft decisions may be made depending on the actual conditions such as SNR. A soft decision can be regarded as a hard decision plus a measure of confidence. The coding gain provided by an error-correcting code to a system can be defined as the reduction, which can be realized due to the code, in the required E_b/N_0 for a given BEP.

Coding gain obtained using soft decision decoding depends on the code rate, R_c, the number of information bits per codeword, the minimum distance of the code, $d_{\min}$, and the channel SNR, E_b/N_0. In soft decision decoding, the decoding is based on the minimum Euclidean distance, d_E, between the received signal and the signal corresponding to the codeword. The codeword-error is upper-bounded by

$$P_e \leq (M - 1)Q\left(\frac{d_E}{\sqrt{2N_0}}\right), \tag{15.23}$$

where N_0 is the one-side noise-power spectral density, and

$$d_E = \sqrt{c_0 d_{\min} E}, \tag{15.24}$$

with $c_0 = 1$ for orthogonal signaling and $c_0 = 2$ for antipodal signaling, and E the energy per bit of the codewords. The energy per information bit, E_b, is given by

$$E_b = \frac{nE}{k} = \frac{E}{R_c}. \tag{15.25}$$

Thus,

$$P_e \leq (M - 1)Q\left(\sqrt{\frac{c_0 d_{\min} R_c E_b}{2N_0}}\right). \tag{15.26}$$

The Viterbi algorithm, as a decoding algorithm for convolutional codes, implements soft decision decoding using the trellis diagram [76]. The trellis diagram as well as the Viterbi algorithm can also be applied to the decoding of block codes [78]. For an AWGN channel, decoders using soft decisions can typically improve coding gain by about 2 dB compared to hard-decision decoders [47, 56, 66].

Example 15.3: The codeword error probability for hard and soft decoding of a $(7, 4)$ Hamming code is shown in Fig. 15.1, which is based on (15.20) and (15.26). As expected, the bipolar scheme has a 3-dB SNR advantage over the orthogonal binary modulated scheme. The upper bound for soft decoding is slightly lower than that for hard decoding.

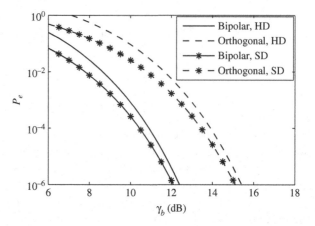

Figure 15.1 The codeword error probability for binary modulated (7,4) Hamming codes, hard and soft decoding.

15.4 Cyclic codes

Cyclic codes, also called *CRC codes*, are a special class of linear block codes. Cyclic codes are based on a cyclic algebraic structure. An (n, k) linear block code C is called a *cyclic code* if every cyclic shift of a codeword in C is also a codeword in C. All codewords can be obtained from a single codeword by shifting and addition. For cyclic codes, data encoding and error detection can be efficiently performed by means of a shift register. The location of errors can also be easily found.

Extra bits are appended to each frame. At the receiver, the CRC bits are recalculated and compared with that in the codeword. The cyclic code is popular due to its effectiveness in error protection, low overhead and ease of implementation due to the efficient encoding and decoding algorithms.

15.4.1 Encoder and decoder

Encoder

The codeword can be generated using a code polynomial in a dummy variable x. This representation has the advantage that cyclic shifts may be expressed as multiplication by x. A k-bit message, $[m_{k-1} \cdots m_1 m_0]$, can be represented by a polynomial of order $k - 1$

$$\boldsymbol{m}(x) = m_{k-1}x^{k-1} + \cdots + m_1 x + m_0. \tag{15.27}$$

$\boldsymbol{m}(x)$ is then multiplied, or left bit shifted, by the order of the generator polynomial $\boldsymbol{g}(x)$, namely $n - k$. The extended version of $\boldsymbol{m}(x)$ is divided by $\boldsymbol{g}(x)$,

$$x^{n-k}\boldsymbol{m}(x) \mod \boldsymbol{g}(x) \tag{15.28}$$

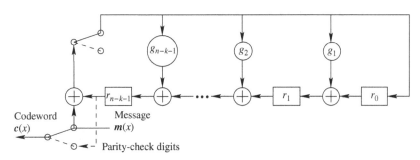

Figure 15.2   **The encoding circuit for an (n, k) cyclic code.**

and the remainder $r(x)$ appended to the original $m(x)$ to obtain the codeword as

$$c(x) = x^{n-k}m(x) + r(x). \qquad (15.29)$$

There exist circuits for performing polynomial multiplication and division [47, 56].

The encoded codeword is in a systematic form, where the leftmost k digits are the original information digits and the rightmost $n - k$ digits are the parity-check digits. The encoding circuit is shown in Fig. 15.2. The circuit is operated as follows.

- First, turn on the switches at the position in the figure and shift the k information bits into the circuit and simultaneously into the channel.
- Then turn off the feedback connection and shift the parity-check bits $r_{n-k-1}, \cdots, r_1, r_0$ out and send them into the channel.

Thus a complete codeword of n digits is formed. An encoding circuit can also be implemented using its parity polynomial $h(x)$ [47].

Encoding is traditionally implemented using shift registers and logic circuits. It can also be very easily implemented by using DSP, where the shift operations can be implemented as FIR and IIR filters.

Example 15.4: Given the generator polynomial $g(x) = x^3 + x^2 + x + 1$, the message $m(x) = x^3 + x + 1$, then the extended version $m(x)$ is divided by $g(x)$

$$x^3 \left(x^3 + x + 1 \right) \quad \bmod \left(x^3 + x^2 + x + 1 \right) \qquad (15.30)$$

By long division of polynomials, we have the remainder x and the quotient $x^3 + x^2 + x$. Note that for modulo-2 operation $-x \bmod 2 = x \bmod 2$. Thus the encoded message is $x + x^3 \left(x^3 + x + 1 \right) = x^6 + x^4 + x^3 + x$, or $1011|010$.

Generator polynomial

An (n, k) cyclic code is completely specified by the generator polynomial $g(x)$. The degree of $g(x)$ is equal to the number of parity-check digits of the code, $n - k$. The generator polynomial $g(x)$ is unique, and takes the form of

$$g(x) = x^{n-k} + g_{n-k-1}x^{n-k-1} + \cdots + g_1x + 1. \tag{15.31}$$

Every code polynomial is a multiple of $g(x)$, and every binary polynomial of degree $n - 1$ or less that is a multiple of $g(x)$ is a code polynomial [47]. $g(x)$ is a factor of $x^n + 1$.

On the other hand, if $g(x)$ is a factor of $x^n + 1$, and has a degree of $n - k$, then it can generates an (n, k) cyclic code.

Decoder

Decoding of cyclic codes consists of the same three steps as for decoding linear codes. The linear property as well as the cyclic property also simplifies the decoding of these codes.

On reception the received data is again divided by $g(x)$. If there is no error, the remainder is zero. If there is an error, the decoder can identify the position of the error digit via the remainder and the syndrome table.

As in the linear block codes, the syndrome depends only on the error, which can, in turn, be obtained from the syndrome (used as index) by using an LUT. For a cyclic code in systematic form, the syndrome can be computed easily by using a division circuit composed of a shift circuit with feedback [47]. The division circuit has a complexity that is linearly proportional to the number of parity-check digits, $n - k$.

The general-purpose cyclic code decoder is known as the *Meggitt decoder* [54]. For practical implementation of the Meggitt decoder, some restrictions are put on the error patterns to be corrected. Error-trapping decoding is a practical variant of the Meggitt decoder, and it can be implemented using a very simple combinatorial logic circuit for error detection and correction. Error-trapping decoding is most effective for decoding single-error-correcting codes and burst-error-correcting codes. The majority-logic decoding is also effective for certain classes of cyclic codes [51]. For more details, refer to [47].

An (n, k) cyclic code is capable of detecting any error burst of length $n - k$ or less, including the end-around bursts. For bursts of length $l = n - k + 1$, the undetectable probability is $2^{-(n-k-1)}$, while for $l > n - k + 1$, the undetectable probability of $2^{-(n-k)}$ [47].

A careful selection of the generator polynomial is necessary to ensure that all errors can be detected. A generator of order k allows detection of all burst errors affecting up to k consecutive bits. Two popular generator polynomials are

$$g(x) = x^{16} + x^{12} + x^5 + 1 \tag{15.32}$$

for wide area networks, and

$$g(x) = x^{32} + x^{26} + x^{23} + x^{22} + x^{16} + x^{12} + x^{11} + x^{10}$$
$$+ x^8 + x^7 + x^5 + x^4 + x^2 + x + 1 \tag{15.33}$$

for local area networks.

15.4.2 Types of cyclic codes

The simple parity check code is a cyclic code, since any cyclic shift of a codeword still leads to an even weight word, and thus is a codeword. The generator matrix is $[\mathbf{I}_{n-1}|\mathbf{1}_{n-1,1}]$, and the generator polynomial is $1 + x$. The Hamming code can be put in cyclic form: The columns of the parity check matrix $\mathbf{H}$ can be arranged so that the resulting code is a cyclic code.

The Bose-Chaudhuri-Hocquenghem (BCH) code is a binary cyclic code. The nonbinary BCH code is a generalization of the binary BCH code to a code with p^m symbols, where p is a prime number. The Reed-Solomon (RS) code is the most important subclass of the nonbinary cyclic code. Both the binary and nonbinary BCH codes are based on the theory of Galois fields.

Golay code

The (23, 12) Golay code [30] is another binary perfect code. It is capable of correcting any combination of three or fewer errors in a block of 23 digits. It has $d_{\min} = 7$. The (23, 12) Golay code is used in the GSC paging system and deep space missions.

The (23, 12) Golay code is generated by either of the two generator polynomials

$$g_1(x) = x^{11} + x^{10} + x^6 + x^5 + x^4 + x^2 + 1 \tag{15.34}$$

or

$$g_2(x) = x^{11} + x^9 + x^7 + x^6 + x^5 + x + 1 \tag{15.35}$$

which are factors of $x^{23} + 1$,

$$x^{23} + 1 = (x + 1)g_1(x)g_2(x). \tag{15.36}$$

The encoding can be implemented by an 11-stage shift register with feedback connections. To achieve its error-correcting capability, $t_{ec} = 3$, a refined error-trapping scheme such as the Kasami decoder can be used [47].

The (23, 12) Golay code is often modified by adding a parity bit, making the extended (24, 12) Golay code. The extended code has $d_{\min} = 8$, but it is not a perfect code. The extended code has a rate of $1/2$ and has a performance slightly better than the original code. The performance of the Golay code is also upper-bounded by (15.19).

BCH code

The BCH code can be treated as a generalization of the Hamming code for multiple-error correction, and the Hamming code is a special case of the BCH code, with $t_{ec} = 1$. It typically outperforms all other block codes of the same size (n, k) at moderate to high SNR, and can be designed to correct almost any required number of errors. The BCH code is used in the POCSAG, NEC, FLEX, and RDS paging standards, DVB-S2, AMPS, and PDC. For example, PDC selects rate-1/2 BCH as its channel coding technique, and FLEX uses BCH (32, 21) code with interleaving.

For the BCH code, there is a large selection of n, R_c, and t_{ec}. For any positive integer m, $m \geq 3$, and t_{ec}, $t_{ec} < 2^{m-1}$, there exists a BCH code with $n = 2^m - 1$, $n - k \leq mt_{ec}$, and $d_{min} \geq 2t_{ec} + 1$. The BCH code is capable of correcting any combination of t_{ec} or fewer errors in a block of $n = 2^m - 1$ digits.

The generator polynomial is specified in terms of its roots from the Galois field GF (2^m). Let α be a primitive element in the Galois field GF (2^m). The generator polynomial $g(x)$ is the lowest-degree polynomial over GF(2) for which α, $\alpha^2, \ldots, \alpha^{2t_{ec}}$ are the roots. Let $\phi_i(x)$ be the minimum polynomial of α^i. Then $g(x)$ is given as the least common multiple of $\phi_i(x)$, $i = 1, 2, \ldots, 2t_{ec}$

$$g(x) = \text{LCM} \left\{ \phi_1(x), \phi_2(x), \ldots, \phi_{2t_{ec}}(x) \right\}. \tag{15.37}$$

Comprehensive tables for all the BCH codes of length $2^m - 1$ with $m \leq 10$ are given in [47], listing n, k, d_{min}, and their generator polynomials. The coding gain of the BCH code is larger than that of the Hamming code, and it continues to increase with the code length, while the coding gain of the Hamming code rapidly tends to a limit as the code length increases. Decoding of BCH codes can be performed by using Berlekamp's iterative algorithm [9] and Chien's search algorithm [14].

Reed-Solomon code

The binary BCH code can be generalized into the nonbinary BCH code in a straightforward manner. For any s and t_{ec}, there exists a q-ary (q being any power of a prime number p) BCH code of length $n = q^s - 1$ with no more than $2st_{ec}$ parity-check digits, which is capable of correcting any combination of t_{ec} or fewer errors. For $q = 2$, we get the binary BCH codes.

The RS codes correspond to the nonbinary BCH codes in the case of $s = 1$. An (n, k, d_{min}) RS code maps k q-ary information symbols into n q-ary encoded symbols, and d_{min} is the minimum number of symbols that differ between codewords. Typically $q = 2^m$, and thus the RS code is made of n symbols, each symbol being m bits long. For a t_{ec}-error-correcting RS code with symbols from GF(q),

$$n = q - 1, \quad n - k = 2t_{ec}, \quad d_{min} = 2t_{ec} + 1. \tag{15.38}$$

That is, the length of the code is one less than the size of code symbols, and the minimum distance is one greater than the number of parity-check digits. Given an RS code, its error correcting power is given by

$$t_{ec} = \frac{n - k}{2}. \tag{15.39}$$

The RS codes have the best error-correcting capability among codes of the same length and dimension. RS codes are very effective for correcting burst errors, which typically occur at a fading dip.

The RS code has a minimum distance of $d_{min} = n - k + 1$, which is the largest possible d_{min} for any linear code (n, k). The probability of codeword error is also upper bounded by

(15.19), with p replaced by P_M, the error probability associated with M-ary modulation. The probability of information symbol error can also be approximated by (15.21), with P_b substituted by P_s and p by P_M.

An RS code with code symbol from GF (2^m) can correct t_{ec} or fewer m-bit bytes. Binary codes derived from the RS code are more effective against clustered errors than random errors, since clustered errors usually cause several errors per byte and accordingly relatively few byte errors. Such codes are capable of correcting any combination of [47]

$$\lambda = \frac{t_{ec}}{1 + \lfloor (l + m - 2)/m \rfloor} \tag{15.40}$$

or fewer bursts of length l, or correcting any single burst of length $(t_{ec} - 1) m + 1$ or less. Simultaneously, it corrects any combination of t_{ec} or fewer random errors. Decoding of nonbinary BCH and RS codes can be performed by using Berlekamp's iterative algorithm [9].

The RS code is used for data storage and retrieval on CDs (compact discs). It works at a symbol level, and is very effective against burst errors. Concatenated and cross-interleaved RS coding is used to handle random and burst errors. The RS code is also used in CDPD, DAB+, and DVB-C/H.

Example 15.5: The BERs of a number of linear block codes, namely, Hamming (7, 4), (15, 11), (31, 26), (63, 57), BCH (127, 64) with $t_{ec} = 10$, and RS (127, 106) with $t_{ec} = 10$, are calculated for coherent BPSK modulation, when hard decision is employed. The result is shown in Fig. 15.3. The simulation is based on (15.21), where $p = Q\left(\sqrt{2E_c/N_0}\right) = Q\left(\sqrt{2R_c\gamma_b}\right)$. Note that the BER performance of the uncoded cases is better than that of most of the coded cases at low SNR.

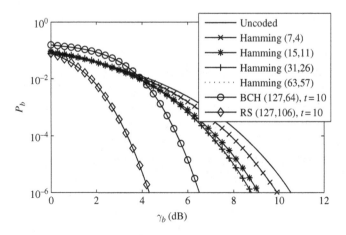

Figure 15.3   **BER for different linear block codes.**

15.5 Interleaving

On many communication channels, errors occur as random errors as well as burst errors. Codes for random error correction are generally not efficient for burst error correction. Interleaving is a most effective method for correcting random errors as well as single or multiple bursts.

Interleaving is a channel coding method that achieves time diversity. It effectively resists a deep fading or a noisy burst, and the successive digits are not corrupted at the same time. Block, convolutional, and coded modulation codes are designed for the AWGN channel. These codes are typically interleaved to mitigate the effect of error bursts. The interleaver reshuffles N continuous digits into N independent digits, and the respective BER performance can be easily explained by using the BER equation (7.141). For an uncoded system, the interleaver breaks up the error bursts, but does not really lead to a reduction in the mean BER. The interleaver reduces the mean BER only for coded bitstreams that are then decoded.

The simplest interleaver is the block interleaver. By interleaving an (n, k) code with t_{ec} random error correcting capability to degree λ, a $(\lambda n, \lambda k)$ code is obtained, which is capable of any combination of t_{ec} bursts of length λ or less. A sequence is divided into blocks of length L_π. The symbols within each block are then permuted by writing them column-wise into a rectangular array of m rows and n columns, $L_\pi = m \times n$. If the bitstream is read into the block row-wise, it will be read out and transmitted column-wise, or vice versa. As a result, two adjacent bit errors are spaced by n (or m) symbols. The value of n (or m) should be selected to be sufficiently large so that the errors affect different codewords, and are corrected by their respective CRC bits independently. At the receiver, the de-interleaver stores the received data by sequentially increasing the row number of each successive digit, and then clocks out the data row-wise, one row at a time.

The interleaver should have a size that is large enough so that fading is independent across each received codeword. During transmission, symbols in the same codeword are separated by $n - 1$ symbols. If the separation in time is greater than the channel coherent time, T_c, that is, $nT > T_c \approx 1/B_D$, where T is the symbol duration and B_D is the channel Doppler spread, the interleaver is called *deep interleaver* and the symbols in the same codeword are subject to independent fading.

For an (n, k) block code with soft decision decoding, the minimum Hamming distance of the code, $d_{\min}$, provides the diversity order. For hard decoding, the diversity order is reduced by a factor of 2 relative to soft decision decoding. Thus, designs for block coding and interleaving should maximize the Hamming distance of the code. Random interleaving permutes the row and column indices randomly by modulo arithmetic, and it can be used to replace rectangular interleaving.

Both interleavers and deinterleavers have inherent delays since the operations have to be performed after all the L_π symbols are received, and the interleaving delay is $L_\pi T$. For voice communications, a tolerable maximum delay is 40 ms, and this must be considered when designing wireless communications systems. However, for MSs, the typical fading dips are separated by a distance $\lambda/4$, which corresponds to a duration of up to 100 ms. If the maximum latency is smaller than the duration of fading dips, such as in the case of voice communications, the effectiveness of the interleaver is greatly reduced.

Interleaving has no influence in the case of nonfading or fixed amplitude and phase. In contrast, FEC not only improves the performance of channels of fixed amplitude and phase, but also that of fading channels. FEC also leads to a significantly smaller time delay. Convolutional interleaving provides the same interleaving depth as block interleaving, but with lower delay and less memory.

15.6 Convolutional codes

Unlike block codes that are limited to codeword blocks, convolutional codes have a structure that effectively extends over the entire transmitted bitstream. Convolutional coding requires a simple encoder consisting of a shift register and adders, and a sophisticated decoder. Convolutional codes are also linear codes, that is, the sum of any two codewords is another codeword and the all-zeros sequence is also a codeword. The convolutional encoding process can be viewed as delay-line binary FIRs.

For an (n, k, K) convolutional code, the convolutional encoder consists of a shift register with K stages of k bits and n linear modulo-2 addition functions for output, as shown in Fig. 15.4. For encoding, k bits are shifted into the encoder at a time, and an output of n bits is generated $(k < n)$, leading to a code rate $R_c = k/n$. The parameter K is known as the *constraint length*; it is the number of shifts through a finite state machine over which a single input data bit can influence the output bit. Typically, n and k are small integers, but K must be sufficiently large to achieve low error probability. When $k = 1$, the information sequence is processed continuously.

The best known code, prior to the turbo code, is the RS-Viterbi code, which is a combination of the RS block code and the Viterbi-decoded short constraint-length convolutional code. Convolutional coding is widely used in 2G and 3G mobile communications standards.

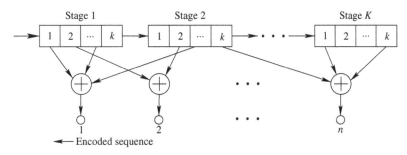

Figure 15.4   Schematic of the convolutional encoder.

15.6.1 Encoding of convolutional codes

The structure of the convolutional encoder is shown in Fig. 15.4. The convolutional encoder is characterized by K-stage shift registers together with n modulo-2 adders and a multiplexer for serializing the encoder outputs. The encoder is a linear feedforward shift register.

The convolutional encoder is a linear system. The n-bit encoder output can be obtained as the convolution of the input sequence a with the impulse responses of the n branches of the encoder. The impulse responses of the ith encoder branch for the jth input bit $a_j = (0, 0, 1, \cdots)$ is denoted $g_j^{(i)}$. The impulse responses have a duration of at most K

$$g_j^{(i)} = \left(g_{j,0}^{(i)}, g_{j,1}^{(i)}, \ldots, g_{j,K}^{(i)} \right), \tag{15.41}$$

which are also known as *generator sequences*.

The encoding equations are given by

$$b^{(l)} = a * g^{(l)} = \sum_{p=1}^{k} a^{(p)} * g_p^{(l)}, \quad l = 1, 2, \ldots, n, \tag{15.42}$$

where $*$ denotes discrete convolution and all operations are modulo-2,

$$a^{(p)} = \left(a_p, a_{p+k}, a_{p+2k}, \ldots \right), \quad p = 1, 2, \ldots, k, \tag{15.43}$$

$$b^{(l)} = \left(b_0^{(l)}, b_1^{(l)}, b_2^{(l)}, \ldots \right). \tag{15.44}$$

The encoding equation can be written in matrix form as

$$b = aG, \tag{15.45}$$

where G is called the *generator matrix* of the convolutional code, and all operations are modulo-2. For an (n, k, K) code, the generator matrix is given by

$$G = \begin{bmatrix} G_0 & G_1 & G_2 & \cdots & G_K & & \\ & G_0 & G_1 & G_2 & \cdots & G_K & \\ & & G_0 & G_1 & G_2 & \cdots & G_K \\ & & & \ddots & & & \ddots \end{bmatrix}, \tag{15.46}$$

where G_l, $l = 0, 1, \ldots, K$, is a $k \times n$ submatrix

$$G_l = \begin{bmatrix} g_{1,l}^{(1)} & g_{1,l}^{(2)} & \cdots & g_{1,l}^{(n)} \\ g_{2,l}^{(1)} & g_{2,l}^{(2)} & \cdots & g_{2,l}^{(n)} \\ \vdots & \vdots & \ddots & \vdots \\ g_{k,l}^{(1)} & g_{k,l}^{(2)} & \cdots & g_{k,l}^{(n)} \end{bmatrix}. \tag{15.47}$$

For a message a of length kL, G has kL rows and $n(K + L)$ columns and b has a length of $n(K + L)$.

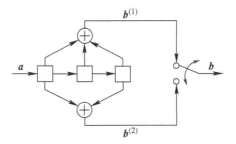

Figure 15.5   A simple convolutional encoder, $R_c = 1/2$ and $K = 3$.

The codeword is a single sequence obtained as the multiplexing of the n output sequences

$$b = \left(b_0^{(1)}, b_0^{(2)}, \ldots, b_0^{(n)}, b_1^{(1)}, b_1^{(2)}, \ldots, b_1^{(n)}, \ldots\right). \tag{15.48}$$

Example 15.6: For a simple convolutional encoder with $R_c = 1/2$ and $K = 3$, as shown in Fig. 15.5, derive the encoded sequence for an input data sequence $a = (1000)$.
    For this encoder, $k = 1$, $K = 3$, $n = 2$, and

$$g^{(1)} = (0, 1, 1, 1), \quad g^{(2)} = (0, 1, 0, 1). \tag{15.49}$$

For input sequence $a = (1000)$, we have

$$b^{(1)} = (1000) * (0111) = (0111000), \tag{15.50}$$

$$b^{(2)} = (1000) * (0101) = (0101000). \tag{15.51}$$

By multiplexing the two subsequences, we can easily derive that the encoded sequence is (11101100).

For convolutional coding, it is assumed that the encoder is initialized with the zero state and information bit sequences are padded with $(K - 1)k$ zeros to bring back the convolutional encoder to the all-zeros state. If the input sequence has a length that is not a multiple of k, it is padded with zeros so as to get an input length of mk, m being a positive integer. The rate of the code is thus given as

$$R_c = \frac{mk}{(m + K - 1)n}. \tag{15.52}$$

Typically, the sequence is long and $m \gg K$, and in this case $R_c \approx \frac{k}{n}$.

Alternative representation

The encoding equations can be represented by polynomials. Let $a^{(i)}(D)$ be the ith input sequence and $b^{(j)}(D)$ be the jth output sequence, where D is the delay operator. The generator polynomial $g_i^{(j)}(D)$ corresponds to the encoder transfer function relating input i to output j. For a k-input, n-output linear system, there are a total of kn transfer functions, which can be packed into a matrix

$$\mathbf{G}(D) = \begin{bmatrix} g_1^{(1)}(D) & g_1^{(2)}(D) & \cdots & g_1^{(n)}(D) \\ g_2^{(1)}(D) & g_2^{(2)}(D) & \cdots & g_2^{(n)}(D) \\ \vdots & \vdots & \ddots & \vdots \\ g_k^{(1)}(D) & g_k^{(2)}(D) & \cdots & g_k^{(n)}(D) \end{bmatrix}. \tag{15.53}$$

By denoting

$$\mathbf{A}(D) = \left[a^{(1)}(D), a^{(2)}(D), \cdots, a^{(k)}(D) \right] \tag{15.54}$$

as the k-tuple of input sequences and

$$\mathbf{B}(D) = \left[b^{(1)}(D), b^{(2)}(D), \cdots, b^{(n)}(D) \right] \tag{15.55}$$

as the n-tuple of output sequences, the encoding equations can be represented by

$$\mathbf{B}(D) = \mathbf{A}(D)\mathbf{G}(D). \tag{15.56}$$

After multiplexing, the codeword can be represented by

$$b(D) = b^{(1)}\left(D^n\right) + Db^{(2)}\left(D^n\right) + \cdots + D^{n-1}b^{(n)}\left(D^n\right). \tag{15.57}$$

From the above equations, the codeword can be expressed by

$$b(D) = \sum_{i=1}^{k} a^{(i)}\left(D^n\right) g_i(D), \tag{15.58}$$

where

$$g_i(D) = g_i^{(1)}\left(D^n\right) + Dg_i^{(2)}\left(D^n\right) + \cdots + D^{n-1}g_i^{(n)}\left(D^n\right), \quad i = 1, 2, \ldots, k. \tag{15.59}$$

Example 15.7: For the $(2, 1, 3)$ encoder given by Fig. 15.5, and $a = (1000)$, we have

$$g(D) = g^{(1)}\left(D^2\right) + Dg^{(2)}\left(D^2\right) = D^2 + D^3 + D^4 + D^6 + D^7, \tag{15.60}$$

$$b(D) = a\left(D^2\right)g(D) = D^2 + D^3 + D^4 + D^6 + D^7, \tag{15.61}$$

which corresponds to the codeword (0011101100). This result is the same as that given in Example 15.6.

Systematic codes

Like block codes, convolutional codes may be systematic codes if the first k output sequences are exactly the same as the k input sequences, i.e.,

$$b^{(i)} = a^{(i)}, \quad i = 1, 2, \ldots, k. \tag{15.62}$$

Thus, $g_i^{(j)} = 1$ for $j = i$ and 0 otherwise, where $i, j = 1, 2, \ldots, k$. The corresponding transfer function matrix is given by

$$\mathbf{G}(D) = \left[\mathbf{I}_{k \times k} \big| \mathbf{G}_{k \times (n-k)}(D) \right], \tag{15.63}$$

where $\mathbf{G}_{k \times (n-k)}(D)$ is the rightmost $n - k$ columns of $\mathbf{G}(D)$. In this case, the first k output sequences are called *information sequences*, and the other $n - k$ output sequences are called *parity sequences*. Systematic codes require less hardware for decoding than nonsystematic codes. Also no inverting circuit is required for recovering the information sequences from the codeword.

For nonsystematic codes, an inverter is required for recovering the information sequences such that

$$\mathbf{G}(D)\mathbf{G}^{-1}(D) = D^q \mathbf{I}_{k \times k} \tag{15.64}$$

for a delay $q \geq 0$, where the right inverse $\mathbf{G}^{-1}(D)$ is an $n \times k$ matrix. In this case, we can get the information sequences by

$$b(D)\mathbf{G}^{-1}(D) = a(D)\mathbf{G}(D)\mathbf{G}^{-1}(D) = a(D)\mathbf{D}^q. \tag{15.65}$$

If the inverse $\mathbf{G}^{-1}(D)$ does not exist, the code will become a catastrophic code, for which a finite number of channel errors will cause an infinite number of decoding errors [47].

Any code with $\mathbf{G}^{-1}$ existing is noncatastrophic, and systematic codes are always noncatastrophic [47, 62]. When using nonsystematic codes, catastrophic codes must be avoided. A nonsystematic code can be identified as a catastrophic code if all the generator polynomials have a common factor, or if the finite state diagram contains a closed loop with zero weight, other than the loop in the all-zero state.

15.6.2 Encoder state and trellis diagrams

Encoder state diagram

A convolutional encoder is a sequential circuit, thus it can be represented by a state diagram, or *finite state machine*, whose state is defined as the contents of the shift register. For an (n, k, K) encoder, there is a total of kK-bit memory. Since the first stage is controlled by the input, there are $2^{k(K-1)}$ different possible states. The output depends only on the input and the current state but not on the previous states, and thus it can be interpreted as a Markovian process of first order. For a rate-k/n code, each new block of k input bits leads to a transition to a new state on the state diagram, and there are 2^k branches entering and leaving each state.

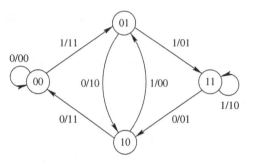

Figure 15.6
State diagram for the simple convolutional encoder with $R_c = 1/2$ and $K = 3$.

From the state diagram, the transfer function, $T(D)$, for the code can be defined. The transfer function gives us the distance properties of the convolutional code. The minimum distance of the code, d_{free}, is obtained.

Example 15.8: For the simple convolutional encoder shown in Fig. 15.5, the state diagram is shown in Fig. 15.6. The arcs or lines with arrows and labels indicate allowed transitions from the associated input to output. The states are labelled as the contents of the encoder memory. From the state diagram, we can see that $k = 1$ and that there are four states in the state diagram, thus $K = 3$. The memory constraint length, K, is easily determined from the number of states. Assuming that the encoder starts with the all-zero state, for an input sequence (1000), we have

$$
\begin{array}{lcccccc}
\text{input:} & 1 & \longrightarrow & 0 & \longrightarrow & 0 & \longrightarrow & 0 \\
\text{state:} & 00 & \longrightarrow & 01 & \longrightarrow & 10 & \longrightarrow & 00 \\
\text{output:} & 11 & \longrightarrow & 10 & \longrightarrow & 11 & \longrightarrow & 00.
\end{array}
$$

The encoded sequence is 11101100, and this result is the same as given in Example 15.6.

Trellis diagram

Although the state diagram fully describes a convolutional encoder, it does not contain temporal information that is necessary for decoding. The trellis diagram, created by Forney [24], overcomes this disadvantage. It expands the encoder state diagram by adding a time axis, and can be reduced to a state diagram. The trellis diagram simplifies the tree representation by merging nodes in the tree corresponding to the same encoder state. It arranges the finite states vertically, and the horizontal axis represents time. A state transition is represented by a branch with a label of the corresponding input and output bits. The trellis is fully developed after K steps.

The performance of the convolutional coder is dependent on the Hamming distance between the valid paths through the trellis. Since the code stream of a convolutional code cannot be divided into distinct codewords, as is the case for block codes, the distance

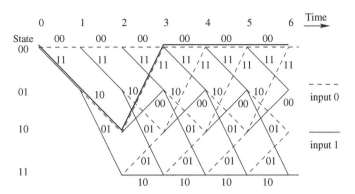

Figure 15.7 Trellis diagram for the simple convolutional encoder with $R_c = 1/2$ and $K = 3$. The encoded sequence 11101100 is shown as the bold trace.

between complete code sequences is considered. The *free Hamming distance* of a convolutional code is defined as the minimum distance between any pair of code sequences that starts from and ends at the same state. This distance has a significant impact on the BER performance, just as the minimum Hamming distance has on the performance of a block code.

The convolutional codes are linear codes. The free Hamming distance between the all-zero sequences and a sequence that starts and ends at the zero state becomes the *free Hamming weight*, that is, the number of nonzero symbols.

Example 15.9: For the simple convolutional encoder shown in Fig. 15.5, the trellis diagram is shown in Fig. 15.7. Each node corresponds to the state of the encoder after a transition from the previous node for an input bit. The labels on the branches are the encoder output bits corresponding to a state transition, and the input is represented by a dashed line (for 0) and a solid line (for 1). The shortest path from the zero state to the all-ones state gives the memory order $m = K - 1 = 2$. A path corresponds to a code sequence. For input (1000), the encoded sequence 11101100 is shown as the bold trace in the diagram.

15.6.3 Sequence decoders

Decoding of convolutional codes is much more complex than decoding of block codes, since there are no distinct codewords, only potentially infinite code sequences. Thus, in principle, the decoder may have to wait an unlimited time to give a possible code sequence.

Decoding of convolutional codes can be performed in a number of ways. Convolutional code decoders are categorized into two classes, namely the *tree decoders* and the *trellis decoders*, which make use of the code tree and the trellis, respectively. The depth-first (sequential) algorithms and limited-size breadth-first algorithms such as the M-algorithm are well-known tree decoders. The Viterbi algorithm is the major trellis decoder.

Viterbi decoding

The Viterbi algorithm implements MLSE, which searches for the path through the code trellis most closely resembling the received signal sequence. The principle is to determine the path to each node that is closest to the received sequence, and this path is known as the *survivor path*. Since there are 2^m states in the encoder state diagram for a memory order $m = K - 1$, the decoder must reserve 2^m words for storage of the survivors. Each word must be capable of storing the surviving path as well as its metric. The Viterbi algorithm is relatively easy to implement for codes with small memory orders. This also offers possibility for joint decoding and equalization using the same Viterbi decoder.

Viterbi decoding is dominant in the decoding of convolutional codes with short constraint lengths ($K \le 10$). This limits the available free distance, and the error probability cannot be made arbitrarily small. The complexity of the Viterbi algorithm increases linearly with the sequence length, N. The Viterbi algorithm implements a nearest-neighbor decoding strategy, and it is given in Section 6.2.5.

Example 15.10: From Example 15.6, we know for an input sequence 1000, the output sequence from the simple encoder is 11101100. Now given a received code sequence 11101100 and the simple encoder, decode the input sequence.

Decoding starts from the zero state. The decoding process is shown in Fig. 15.8. In the figure, on each branch there is a two-bit number representing the output bits and another number representing the branch metric, and the digit at each node denotes the cumulative metric. The branch metric is obtained as the Hamming distance between the two-bit input

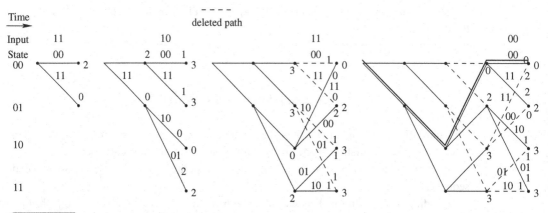

Figure 15.8 Trellis diagram for decoding the received code sequence 11101100, corresponding to the simple convolutional encoder with $R_c = 1/2$ and $K = 3$.

and the output bits of the state. At each stage of decoding, one of the two paths leaving a node is deleted, since it generates a larger cumulative metric. When there is a tie between the metrics of two branches that enter one node, an arbitrary decision is made. After all the received code sequence is processed, the path that leads to the shortest cumulative metric corresponds to the decoded sequence. The shortest path is shown as the bold path. From the trellis diagram of the encoder, as shown in Fig. 15.5, we conclude that the decoded sequence is 1000.

In Example 15.10, the Hamming distance is used as the metric, but any appropriate metric can be used. If soft information is available, the Euclidean distance can be used, and this results in soft decision decoding.

When a convolutional code is used for encoding short packets of data, it is common practice to add $K - 1$ zeros to the end of the data before encoding so as to make the encoder return to the zero state. These added zeros are known as *tail bits*, and this addition leads to a slight increase in redundancy.

The convolutional code was first used in satellite and deep space communications before 1980. The Viterbi algorithm was implemented as the major decoding technique. The Linkabit Corporation and the Harris Corporation were among the first that designed high-speed hardware Viterbi decoders [47]. Qualcomm Inc. provides a number of ASIC decoders for standard codes.

Sequential decoding

The Viterbi algorithm is only suitable for small constraint length K, since its complexity increases exponentially with K. For long-constraint-length codes, sequential decoding can be applied. The most popular sequential decoding algorithm is the Fano algorithm [21]. Unlike the Viterbi algorithm, the computation required for each decoded bit may be highly variable.

Both the stack algorithm and the Fano algorithm are sequential decoding algorithms of code trees. The complexity of the algorithm is very much dependent on the quality of the channel, and the reordering of the stack leads to an increasing complexity. The stack-bucket algorithm introduces a significant increase in speed with a very slight degradation in performance. The Fano algorithm is a depth-first tree search procedure in its purest form. The algorithm stores only one path. Compared to the stack-bucket algorithm, the Fano algorithm decodes at a slower speed for high rates, but at a faster speed for moderate rates [47]. It almost always finds the same path as the stack-bucket algorithm. Some good codes for use with sequential decoding are also listed in [47].

In the Viterbi algorithm, the fixed number of 2^K computations must be performed per decoded information block. For sequential decoding, the computational complexity is essentially independent of the constraint length, K; hence, a large K can be used, generating arbitrarily low achievable error probability. For noisy frames, the decoding time may exceed some upper limit, causing some information to be erased. Sequential decoding has a computational advantage over the Viterbi algorithm when the received sequence is not too noisy. The number of computations is a random variable. The average number of computations for a sequential decoder is normally much less than that of the Viterbi algorithm.

A sequential decoder almost always produces the maximum-likelihood path, achieving the same error probability as the Viterbi decoder. For high code rates, sequential decoding is optimum and has the same error probability as Viterbi decoding; for low rates, it is suboptimum. Since Viterbi decoding is only practical for small K, the suboptimum performance of sequential decoding at low rates can be compensated by using larger K.

Sequential decoding may introduce a typical erasure probability of 10^{-3} due to input buffer overflow. The high erasure probability, which usually occurs for very noisy received sequences, can be exploited. In Viterbi decoding, the noisy frame will always be decoded and the decoding error is likely to be very high. In sequential decoding, these noisy frames will be erased, and an ARQ retransmission is then activated.

M-algorithm

The M-algorithm is a purely breadth-first synchronous algorithm that moves forward in the time dimension. Only M candidate paths are kept at each iteration. Unlike depth-first algorithms, the M-algorithm is independent of the channel quality, since there are always M paths retained. It has a simpler metric than that of sequential decoding. Its implementation is straightforward. The complexity of the M-algorithm is independent of the code size and constraint length. For this reason, very long constraint-length codes are usually selected to ensure appropriately large d_{free}. The M-algorithm is not a viable choice for decoding binary convolutional codes, but it may work well with nonbinary modulation [62].

Remarks

In addition to the above algorithms, majority-logic or threshold decoding for complete orthogonalizable codes leads to an inferior performance when compared to Viterbi and sequential decoding, but the implementation is much simpler [47, 51].

Sequential decoding is popular for relatively low transmission speeds. However, the metric used may contain the bias term accounting for the different path lengths, and the method is difficult to parallelize. Breadth-first algorithms, such as the Viterbi algorithm and the M-algorithm, do not have the metric bias term, and are more suitable for VLSI implementations. They are popular for high-speed transmissions.

15.6.4 Trellis representation of block codes

Block codes can also be represented by a terminated trellis. This trellis view of block codes enables soft decoding. A code trellis is a visual method of keeping track of the p^k codewords of a block code. This representation is based on the parity check matrix $\mathbf{H}$, and the number of states is 2^{n-k}. Trellis representation of linear block codes was first introduced in [5], and was later explored by several other authors [52, 78].

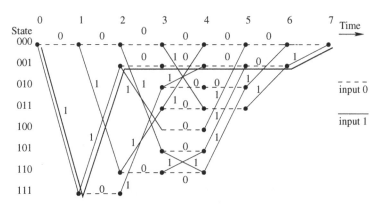

Figure 15.9   Trellis diagram of the (7, 4) Hamming code.

An $(n - k) \times n$ parity check matrix $\mathbf{H}$ can be written as

$$\mathbf{H} = \begin{bmatrix} \mathbf{h}_0 \ \mathbf{h}_1 \ \cdots \mathbf{h}_{n-1} \end{bmatrix}, \tag{15.66}$$

where $\mathbf{h}_i$'s are the column vectors of $\mathbf{H}$. A valid linear block code $\mathbf{c} = (c_0 c_1 \cdots c_{n-1})$ has a syndrome

$$s_{\text{syn}} = \mathbf{c}\mathbf{H}^T = \sum_{i=0}^{n-1} c_i \mathbf{h}_i = \mathbf{0}. \tag{15.67}$$

This relationship is used for construction of the trellis of a linear block code. The states at time i, s_i, and at time $i + 1$, s_{i+1}, are related by

$$s_{i+1} = s_i + c_i \mathbf{h}_i \tag{15.68}$$

and at time n, from (15.66), we have $s_n = \mathbf{0}$.

Example 15.11: Given a (7, 4) Hamming code with the parity check matrix

$$\mathbf{H} = \begin{bmatrix} \mathbf{h}_0 \ \mathbf{h}_1 \ \cdots \mathbf{h}_{n-1} \end{bmatrix} = \begin{bmatrix} 1 & 1 & 1 & 0 & 1 & 0 & 0 \\ 1 & 1 & 0 & 1 & 0 & 1 & 0 \\ 1 & 0 & 1 & 1 & 0 & 0 & 1 \end{bmatrix}, \tag{15.69}$$

the trellis diagram of the code is shown in Fig. 15.9. The codeword 1100001 is represented by a bold line starting and terminating at the zero state.

Viterbi decoding, also called *trellis decoding*, of block codes has also attracted many investigations, but is restricted mainly by its complexity. Based on the trellis, binary block codes can be decoded by using the BCJR (Bahl-Cocke-Jelinek-Raviv) algorithm [5].

15.6.5 Coding gain and error probability

The performance of a convolutional code is determined by its decoding algorithm and its distance properties. The minimum free distance d_{free} is a most important measure.

Minimum free distance and weight

The minimum free distance of a convolutional code is defined as the minimum distance between any two codewords in the code

$$d_{\text{free}} = \min\left\{ d\left(\boldsymbol{b}_1, \boldsymbol{b}_2\right) \big| \boldsymbol{a}_1 \neq \boldsymbol{a}_2 \right\}. \qquad (15.70)$$

When $\boldsymbol{a}_1$ and $\boldsymbol{a}_2$ have different length, the shorter sequence is padded with zeros.

The definition of the weight of a codeword is similar to that of a block code, but for a codeword, it is produced by a nonzero input information sequence. As a linear code, we also have

$$d_{\text{free}} = \min\{w(\boldsymbol{b}) | \boldsymbol{a} \neq \boldsymbol{0}\}. \qquad (15.71)$$

That is, d_{free} is the minimum weight of the codeword produced by any nonzero information sequence. For a given rate and encoder memory, the free distance of a nonsystematic code is larger than that of a systematic code [47].

Coding gain

The coding gain for a convolutional code over an uncoded BPSK or QPSK system is upper-bounded by [47]

$$G \leq G_c = 10 \log_{10} \left(\frac{R_c d_{\text{free}}}{2} \right) \quad (\text{dB}) \qquad (15.72)$$

in the hard-decision case. The corresponding gain in the soft-decision case is given by [47, 56]

$$G \leq G_c = 10 \log_{10} \left(R_c d_{\text{free}} \right) \quad (\text{dB}). \qquad (15.73)$$

That is, soft decision introduces an additional 3 dB in the asymptotic coding gain.

For hard-decision Viterbi decoding, the coding gain, G, reduces by approximately 2 dB for small E_b/N_0 when compared to soft-decision decoding, but over the entire range of E_b/N_0 ratio, it introduces a loss of between 2 and 3 dB for the AWGN channel [47]. The asymptotical coding gain, G_c, is reduced by 3 dB for large E_b/N_0 [47]. It is found that a soft quantization of $Q = 8$ achieves a performance within about 0.25 dB of the optimum performance that an unquantized demodulator achieves [47].

The minimum free distance d_{free}, the corresponding G_c, and the generators for a number of binary, short-constraint-length K convolutional codes at several code rates R_c including $R_c = 1/n, n = 2, \ldots, 8, 2/3$, and 3/4 have been tabulated in [47, 56, 62]. The listed d_{free} are the largest possible values for the given R_c and K. These codes are nonsystematic codes, since for a given R_c and encoder memory, d_{free} is larger for nonsystematic codes than for systematic codes. The value of d_{free} can be increased by decreasing R_c or by increasing K.

Error probability

The BER performance of a finite-length, N, convolutional code with ML decoding and binary antipodal signaling on the AWGN channel is upper bounded by [62]

$$P_b \leq \sum_{d=d_{\text{free}}}^{\infty} \frac{N_d \tilde{w}_d}{N} Q\left(\sqrt{d\frac{2R_c E_b}{N_0}}\right), \tag{15.74}$$

where $\tilde{w}_d$ is the average information weight per codeword

$$\tilde{w}_d = \frac{w_d}{N_d}, \tag{15.75}$$

with w_d being the total information weight of all codewords of weight d and N_d being the number of codewords of weight d.

For the BSC (i.e. hard-decision decoding), the minimum truncation length $\tau_{\text{min}} = 4K$ in most cases, K being the memory order or total constraint length of the encoder $K = \sum K_i$, and the error contribution due to truncation is negligible. For small SNR, $\tau_{\text{min}} = 5.8\,K$ is sufficient so as not to introduce much error probability due to truncation [25]. A truncation length of $4K$ to $5K$ is usually employed [47]. In practice, decoding may not always start with the first branch transmitted after the encoder is set to the all-zero state, and the effect of lack of initial branch synchronization is negligible after $5\,K$ branches are decoded. Thus, the decoding decision over the first $5K$ branches is usually discarded and the succeeding decision is reliable [25].

For hard-decision decoding, the ensemble average error-rate performance of a convolutional code on a discrete memoryless channel, similar to block coding, can be upper-bounded by [56, 76]

$$\overline{P}_b < \frac{(q-1)q^{-KR_0/R_c}}{\left(1 - q^{-(R_0 - R_c)/R_c}\right)^2}, \quad R_c \leq R_0, \tag{15.76}$$

where q is the number of channel input symbols, K is the constraint length of the code, and R_0 is the cutoff rate, which is defined as

$$R_0 = 1 - \log_2\left(1 + e^{-R_c \gamma_b}\right) \quad \text{(bits/dimension)} \tag{15.77}$$

for M-ary binary-coded signals with antipodal signaling [56]. For block coding corresponding to M-ary binary-coded signals, the average probability of error [56]

$$\overline{P}_e < 2^{-n(R_0 - R_c)} \tag{15.78}$$

for $R_c < R_0$; $\overline{P}_e \to 0$ as the code block length $n \to \infty$.

15.6.6 Convolutional coding with interleaving

Convolutional interleavers are used with convolutional codes. The data are interleaved in a continuous stream. This leads to a smaller latency when compared with the block interleaver. The minimum length of two convolutional sequences is known as the effective length of the code, and should be as large as possible.

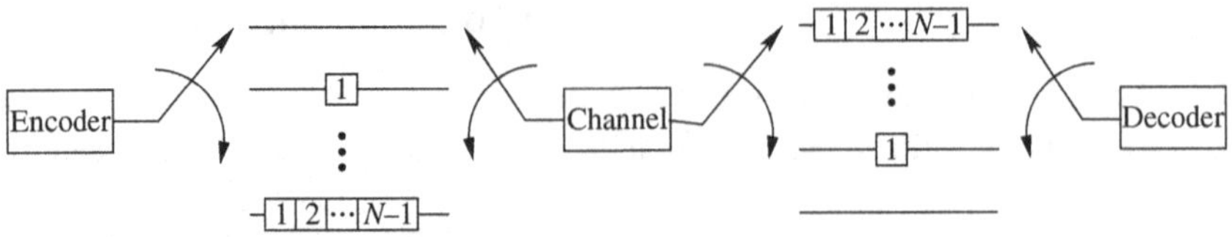

Figure 15.10   **Block diagram of the convolutional interleaver.**

The block diagram of a convolutional interleaver is shown in Fig. 15.10. The encoder shifts its output bits into a buffer of increasing size, from no buffer to buffer size $N - 1$. These buffered output bits are transmitted to the channel in the same order. At the decoder, reverse operation is performed. Thus, the delay at the encoder output increases progressively. Each encoder output symbol is separated by $N - 1$ other symbols by the interleaver. The total buffer size required is $N(N - 1)/2$, and the delay is $N(N - 1)T$, where T is the symbol time. The error probability of a convolutional interleaver is given in [56].

Interleavers can be realized as the bit-interleaved coded modulation (BICM) or symbol-interleaved coded modulation (SICM). In BICM, the bits are interleaved before being mapped to modulated symbols, while in SICM the modulated symbols are interleaved prior to transmission. BICM achieves a much better performance than SICM [12], and BICM is now dominant for coded modulation for fading channels; for example, it is used in many OFDM-based standards like IEEE 802.11, 802.16 and 802.22. In BICM, the code diversity is the smallest number of distinct bits in case of errors, as opposed to the number of channel symbols in SICM.

15.6.7 Punctured convolutional codes

When a high-rate convolutional code such as $(n-1)/n$ is used, the trellis has 2^{n-1} branches that enter each state, and thus 2^{n-1} metric computations per state must be performed in the Viterbi algorithm. The same number of comparisons are also required for selection of the best path at each state. As a result, the complexity for decoding is very high.

The rate $1/n$ codes are most widely employed because this reduces the decoding effort, and higher rates can also be obtained by puncturing. Puncturing is a popular method for adapting the code rate. After encoding, only a subset of the code bits is transmitted, and thus the code rate is increased. This puncturing scheme does not affect the decoder; thus, a single decoder can be used for a number of code rates. Puncturing must be used carefully since it may generate catastrophic codes.

Punctured convolutional codes are high-rate codes obtained from puncturing rate-$1/n$ codes, so that the decoder maintains the low complexity of the rate $1/n$ code. This also reduces the free distance of the rate $1/n$ code. Puncturing can be applied in a periodical process by using a puncturing matrix $\mathbf{P}$. High-rate codes with good distance properties can be obtained by puncturing rate-1/2 maximum free distance codes. A list of puncturing codes with rate $(n - 1)/n$, $n = 2, \ldots, 8$, obtained by puncturing rate-1/2 codes with constraint lengths $K = 3, \ldots, 9$, are tabulated in [56].

Decoding of punctured convolutional codes is performed in the same way as decoding the $1/n$ parent code. Error events in a punctured code are generally longer than that in the low-rate $1/n$ parent codes. Rate-compatible punctured convolutional (RCPC) codes [33] are desirable for adaptive channel coding. They are obtained by puncturing the same low rate $1/n$ convolutional code by different amounts. The puncturing matrices should satisfy a rate-compatibility criterion: The lower-rate codes should contain the same coded bits as all higher-rate codes.

Given the code rate of the parent code $R = 1/n$ with constraint length L_c, the parent code is completely determined by n generator polynomials $G^j(D)$, each of length L_c. Puncturing is performed periodically with a period of L_p codewords. The puncturing pattern matrix defines the transmitted and punctured bits during one period

$$\mathbf{P} = \begin{bmatrix} \boldsymbol{p}_1 & \boldsymbol{p}_2 \cdots \boldsymbol{p}_{L_p} \end{bmatrix}, \tag{15.79}$$

which is an $n \times L_p$ matrix. The columns $\boldsymbol{p}_i$ are periodically assigned to successive whole codewords $\boldsymbol{b}[j] = \begin{bmatrix} b_1(j), \ldots, b_n(j) \end{bmatrix}^T$. Generally, $\mathbf{P}$ contains $l + L_p$ ones, with $1 \leq l \leq (n-1)L_p$. Thus the code rate is obtained as [29]

$$\frac{L_p}{nL_p} = \frac{1}{n} \leq R_c = \frac{L_p}{L_p + l} \leq \frac{L_p}{L_p + 1}. \tag{15.80}$$

That is, a family of $(n-1)L_p$ different codes is obtained. The largest achievable code rate is decided by L_p.

For decoding, the positions of punctured bits have to be filled with dummy bits such as zeros, since zeros do not affect the incremental metric for Viterbi decoding. Since puncturing yields a reduced Hamming distance between code sequences, the truncation length should be increased to make a reliable decision.

15.6.8 Trellis-coded modulation

Channel coding, such as block coding and convolutional coding, leads to a coding gain at the cost of spectral efficiency or data rate. Although this is attractive for power-limited applications, it is not desirable for band-limited applications. A natural idea is to map the coded data into high-order modulation symbols. The result of this method is usually not satisfactory. The trellis-coded modulation (TCM) [74] uses a joint modulation and encoding process, leading to a significant coding gain without bandwidth expansion.

TCM employs a rate $m/(m + r)$ convolutional encoder, and maps the coded bits onto signal points by a technique called *mapping by set partitioning*. The critical step for the design of TCM codes is the mapping of the convolutional encoder outputs to points in the expanded signal constellation. Redundancy is added to the code by increasing the dimension of the signal space, while some symbol sequences are not allowed to be in the enlarged signal space. TCM uses M-ary modulation and simple convolutional coding with mapping by set partitioning. Decoding is performed by using a soft-decision Viterbi decoder. More detail on TCM code design is given in [67].

TCM can be viewed as a generalization of convolutional coding. Convolutional codes attempt to maximize the minimum Hamming distance between code symbol sequences, whereas TCM codes try to maximize the Euclidean distance. For the AWGN channel, a coding gain of 3–6 dB can be obtained with respect to an uncoded system by using TCM codes with 4–128 encoder states, without a loss of bandwidth or data rate [74]. For this reason, the TCM code is especially attractive for power-efficient or bandwidth-efficient communications, such as cellular communications and MSs.

TCM was adopted in the ITU-T Rec. V.32/33/34 for data communications over the standard telephone network; these standards are used for analog full-duplex modems, and they use QAM constellations. V.34, also called *V.fast* or *V.last*, achieves a rate of 33.6 kbits/s, which is very close to the information-theoretic capacity. Note that V.90, the 56 kbits/s modem in common use today, relies on a digital channel rather than a noisy voice-band channel. TCM is also used in IEEE 802.15.3.

Similarly, the turbo code can also be employed in coded modulation schemes, to improve spectral efficiency. Turbo-coded modulation can be based on TCM or on multilevel coded modulation. The former, called *turbo trellis coded modulation (turbo-TCM)*, uses one turbo code, while the later uses multiple separate turbo codes. Both approaches have quite similar performance, but the turbo-TCM approach is a good compromise between complexity and performance.

15.7 Conventional concatenated codes

Concatenated coding was first proposed by Forney in 1966 [23]. The conventional concatenated code is a simple cascade of the RS code and the convolutional code. The turbo code is a parallel concatenated convolutional code (PCCC) [10]. The serially concatenated convolutional code (SCCC) is another concatenated code. This section describes the conventional concatenated code, and the turbo code and the SCCC are described in the following two sections.

In Forney's paper on convolutional codes, he proposed two forms of convolutional codes: nonrecursive nonsystematic and recursive systematic. Based on the first form, Forney proposed the conventional nonrecursive convolutional codes. The recursive systematic convolutional codes are now used in the turbo code or PCCCs. A recursive systematic code can be obtained from a nonrecursive nonsystematic code by choosing one of the generator polynomials, $g_i^{(j)}(D)$, as the denominator, as shown in Fig. 15.11 for $R_c = 1/2$ and $K = 3$. The recursive code and its nonrecursive counterpart have the same distance spectra, but different IOWEF (input-output weight enumeration function).

Simple concatenated codes

The RS code is inefficient in error correction, where the errors are randomly distributed. The simple concatenated code combines codes designed for random error correction and

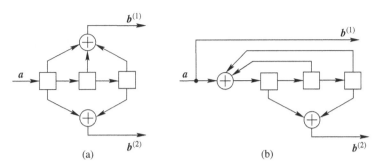

Figure 15.11 Convolutional encoders with $R_c = 1/2$ and $K = 3$. (a) Nonrecursive nonsystematic encoder: $g_1(D) = 1 + D + D^2$ and $g_2(D) = 1 + D^2$. (b) Recursive systematic encoder: $g_1(D) = 1$ and $g_2(D) = (1 + D^2)/(1 + D + D^2)$.

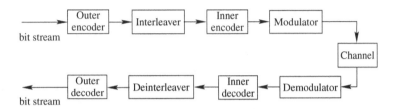

Figure 15.12   **Block diagram of concatenated coding.**

codes for burst error correction. The method uses two smaller, cascaded codes to construct one long code.

The block diagram of simple concatenated coding is shown in Fig. 15.12. The inner (n_1, k_1) binary code, C_1, protects the data in the usual way by correcting random errors, and the outer (n_2, k_2) nonbinary code, C_2, with symbols from GF (2^{k_1}) further eliminates the remaining, typically burst errors. The critical issue is to find a good combination of the inner and outer codes. If the inner code is a block code, then the length of the burst is k_1 bits. The RS code is suitable for the outer code, C_2.

Encoding is implemented in two steps [47]: The k_1k_2 information bits are divided into k_2 bytes of k_1 information bits, and the k_2 bytes are encoded into an n_2-byte codeword according to the rules for C_2. The second step is to encode each k_1-digit byte into a n_1-digit codeword in C_1, yielding a string of n_2 codewords of C_1, a total of n_2n_1 digits. The resulting digits are transmitted one C_1 codeword at a time, resulting in an (n_1n_2, k_1k_2) binary linear code. If $d_{\min}(C_1) = d_1$ and $d_{\min}(C_2) = d_2$, the minimum distance of the concatenated code is at least d_1d_2.

Decoding is also performed in two steps. It is a straightforward combination of the decoding implementations for codes C_1 and C_2. The hardware required is roughly the total of that of both the codes.

The inner code, C_1, can also be selected as a convolutional code. The concatenation of the inner convolutional code with the outer RS code is an important class of codes for fading channels. The data is first convolutionally encoded, and the output bitstream is interleaved. This is because the convolutional code may not see error bursts, but

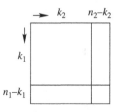

Figure 15.13   The product code.

decoding of a convolutional code using the Viterbi decoder will result in error bursts at low SNR. The RS code is then used to effectively remove burst errors arising from deep fading.

Product codes

The product code [20] is obtained by using an (n_1, k_1) inner block code, symbol-interleaving it to a degree k_2, and then applying an (n_2, k_2) outer block code with the same symbol size. This yields an $n_1 \times n_2$ block code. Block interleaving is applied between the two block codes. Data is written row by row from left to right and from top to bottom, and is read out column by column from top to bottom and from left to right. This is illustrated in Fig. 15.13.

Decoding is performed by decoding the block code in one direction, followed by decoding in the other direction. In general, if the minimum distances of the row and column codes are d_1 and d_2, correcting t_1 and t_2 errors, respectively, then the minimum distance of the product code is $d_1 d_2$, correcting $2t_1 t_2 + t_1 + t_2$ errors. The code rate is the product of the rates of the two codes, R_1 and R_2.

The serial concatenated block codes including the product code achieve capacity-approaching performance when being decoded with the iterative decoding technique. In this case, they are also termed *block turbo codes*. The Chase-II algorithm can be used to avoid the complexity associated with turbo decoding, but it only provides a near-ML decoding of the product code [57]. The trellis-based MAP algorithm is described in [35].

Applications

Concatenated codes achieve a very low error probability at a complexity which is much lower than that of a single code for the same error probability performance. The decoding process is in the reverse order to that of the coding process. Concatenation of the RS code with an inner convolutional code is a common configuration. It is widely used as wireless communication channel coding. Such codes are used in IEEE 802.16, 802.16a/d/e, DVB, and as an optional scheme in WCDMA.

The classical concatenated code is also used in the Consultative Committee for Space Data Systems (CCSDS) standard. Concatenation of an RS code with another RS code is also used for CD-ROM and DVD storage. For DVD, the code is actually an RS product code.

The block turbo code is also defined as an optional channel coding scheme in IEEE 802.16e, and consists of two binary extended Hamming codes that are applied on the original and interleaved information bit sequences, respectively.

Prior to 1995, concatenated codes were widely used when extremely high coding gains were required. Now, the turbo code and the LDPC code are gaining more popularity. The turbo and LDPC codes are special variants of the convolutional code, and they are used as integral or optional channel coding techniques in 3G and some other recent wireless standards.

15.8 Turbo codes

The turbo code is a PCCC [10]. It is created by a concatenation of several parallel, simple component codes. These parallel component codes are obtained from the first code by interleaving. By using an iterative decoding scheme, the turbo code was the first code capable of approaching Shannon's capacity limit to within some hundredths of a decibel on both the AWGN and the interleaved flat fading channels, whilst the complexity of the turbo decoder is much smaller than that of the Viterbi decoder. In the turbo code, the component codes can be either convolutional codes or block codes.

The turbo code is now used in DVB, all CDMA-based 3G standards and beyond, and IEEE 802.20. It will be replacing the RS code and the RS-Viterbi code in future systems. In CDMA-based 3G systems, the convolutional code and the turbo code are adopted for speech and data traffic, respectively. The turbo code is also used in the CCSDC standard to replace its old traditional concatenated coding system. The convolutional turbo code is also defined as an optional channel coding scheme in IEEE 802.16e (WiMAX). WiMAX employs the duo-binary turbo code, which is obtained by encoding two consecutive bits from the uncoded bitstream [11].

Just as TCM is an extension of binary convolutional coding to large signal constellations, so is turbo-coded modulation an extension of turbo coding to larger signal constellations. Turbo-TCM is a parallel concatenation of the TCM code. Serial concatenation of binary error control codes with TCM is receiving more interest in the context of space-time coding [62]. Block turbo-coded modulation is based on the traditional product code.

15.8.1 Turbo encoder

The turbo code can be generated by parallel concatenation of circular recursive systematic convolutional (CRSC) codes M times. Each parallel path, except the first one, encodes the k information bits by permutation (interleaving), drawn at random. For a recursive encoder with v memory units, the probability that any given sequence is an RZ sequence is $\frac{1}{2^v}$. The probability that the sequence remains RZ for all the M encoders is $\frac{1}{2^{Mv}}$, which is very low. This multiconcatenated code is similar to a random code. The classical turbo code is a concatenation of two parallel CRSC codes, as shown in Fig. 15.14. The substreams

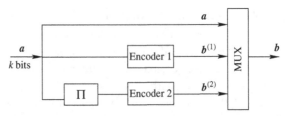

Figure 15.14   The classical turbo encoder.

generated by individual encoders are finally multiplexed. Unlike convolutional coding, which codes on a continuous stream, the turbo encoder codes data block by block, whose size is determined by the size of the interleaver.

In Fig. 15.14, each RSC component may contain a parity sequence. If higher code rates are desired, the parity section of the RSC component can be punctured. A punctured turbo code is obtained by puncturing the specified parity-check symbols from a rate-$1/n$ turbo code. b_1 and b_2 correspond to the parity-check symbols generated by the interleaved input bits. The interleavers are used to permute the input bits. The input bitstream is grouped into blocks of length N, where N is the size of the interleaver. In order to adjust the overall code rate, appropriate puncturing of a, and the output parity check bits $b^{(1)}$, $b^{(2)}$ can be performed.

For the turbo code, each encoder processes the same information bits, but in a different order due to interleaving. The total code rate is given by

$$R_c = \frac{k}{n_1 + \cdots + n_M} = \frac{1}{\frac{1}{R_{c,1}} + \frac{1}{R_{c,2}} + \cdots \frac{1}{R_{c,M}}}, \tag{15.81}$$

where $R_{c,i}$ is the rate of the ith constituent encoder.

For the classical turbo encoder, if the two component encoders have parity-check polynomials $h_0(D)$ and $h_1(D)$, respectively, and the interleaver of encoder 1 has a length of N, it is identified as an (h_0, h_1, N) turbo code, where h_0 and h_1 are the octal representations of $h_0(D)$ and $h_1(D)$. For example, if $h_0(D) = D^2 + 1$, $h_1(D) = D$ and $N = 32$, we obtain a $(5, 2, 32)$ turbo code.

The interleaver used is a pseudorandom block interleaver: the bitstream is read into the buffer by rows and columns, but is read out in a pseudorandom order. The size of the interleaver should be as large as possible. The pseudorandom interleaving patterns can be generated in many ways. The m-sequence can serve this purpose. A simple deterministic algorithm for generating pseudorandom interleaving patterns is based on the quadratic congruence [47, 68]

$$c_m = \frac{km(m+1)}{2} \quad \text{mod } K, \quad 0 \leq m \leq K \tag{15.82}$$

to generate an index mapping function $c'_m \rightarrow c_{m+1} \mod K$, where K is the interleaver size and k is an odd integer. For $K = 8$ and $k = 1$, we have $(c_0, c_1, \ldots, c_7) = (0, 1, 3, 6, 2, 7, 5, 4)$. This implies that index 0 in the interleaved sequence is mapped to index 1 in the original sequence, and so forth. The resulting permutation is $\Pi_8 = \left(c'_0, c'_1, \ldots, c'_7\right) = (1, 3, 7, 6, 0, 4, 2, 5)$.

The s-random interleaver [19] ensures the generation of sequences with reasonable Hamming weight. The interleaver has a fundamental influence on the overall minimum distance of the code, and it largely determines the error floor of the BER performance of the code. The large interleavers also result in a significant latency, which rules it out for some delay-limited applications such as speech communications.

WCDMA uses the classical turbo encoder, which consists of two RSC encoders of constraint length $K = 4$. The overall code rate is approximately $1/3$. Prior to encoding, both constituent encoders are in all-zeros state, and the encoders return to the all-zeros state after having encoded the entire input. This leads to a much better decoding performance. In WCDMA, the interleaver sizes are allowed to be $40 \leq N \leq 5114$ in order to adapt to different block lengths, and prime interleavers, which can be efficiently generated, are used [62]. In CDMA2000, N must be one of the twelve specific values between 378 and 20730 bits, and a similar turbo encoder structure is adopted. CDMA2000 also uses tail bits to bring back the encoders to the all-zeros state.

Both convolutional and turbo encoders are trellis encoders that map a long input sequence to a coded stream. The coded stream of convolutional coding can be viewed as the convolution of the input stream with an encoder polynomial. For both encoders, the constraint length K corresponds to the encoding delay in bits, and the ratio of the number of input bits to the number of output bits is the code rate.

Example 15.12: Consider the conventional rate-1/3 turbo coder with RSC encoders given by Fig. 15.11b. Let the interleaver to be $\prod = (3, 5, 6, 0, 2, 4, 1)$. The input to the turbo encoder is $a = (1101010)$. Determine the encoded bit sequence. Convert the turbo encoder to rate 1/2, and give the encoded bit sequence.

The interleaver first maps the input sequence a to a', $a' = (1101001)$. Assuming both encoders are initialized to state 0, we have

$$b^{(1)} = (1001100), \quad b^{(2)} = (1001110).$$

The encoder output is (111100000111011101000). The rate 1/2 turbo encoder can be realized by alternately deleting parity bits as (11100011011000).

15.8.2 Turbo decoder

The optimum ML decoding for the turbo code is too complex to implement. Instead, an iterative symbol probability-based decoding algorithm is employed. The reliability of individual transmitted or information symbols is calculated, rather than decoding the whole sequence. The most popular symbol probability decoding algorithm is the *a posteriori* probability (APP) algorithm, also known as the *BCJR algorithm* or the *forward-backward algorithm* [5].

The decoder can be broken up into several simple decoders, and soft information is exchanged between them. For the classical turbo code, decoding uses two soft-in/soft-out (SISO) decoders, also called *probabilistic decoders*. Each SISO decoder outputs the various bits as well as its confidence in the decision, that is, the *a posteriori* probability. The MAP rule is given by

$$\hat{b}_i = \max_{b=\{0,1\}} P\left[b_i = b \middle| \mathbf{y}^{(0)}, \mathbf{y}^{(1)}, \mathbf{y}^{(2)}\right], \qquad (15.83)$$

where $\mathbf{y}^{(0)}$ is the received systematic bit sequence, and $\mathbf{y}^{(1)}$ and $\mathbf{y}^{(2)}$ are the received parity sequences corresponding to the two constituent encoders.

Log-likelihood ratios

For binary codes, the confidence is described by the log-likelihood ratio (LLR), which is defined as

$$L(b_i) = \ln\left[\frac{\Pr(b_i = +1 | \mathbf{y})}{\Pr(b_i = -1 | \mathbf{y})}\right] = \ln\left[\frac{\Pr(b_i = +1)}{1 - \Pr(b_i = +1)}\middle| \mathbf{y}\right] \qquad (15.84)$$

where $\mathbf{y}$ is the received sequence. The sign of $L(b_i)$ can be used for hard decision, while the magnitude represents the confidence. A larger difference between $\Pr(b_i = +1 | \mathbf{y})$ and $\Pr(b_i = -1 | \mathbf{y})$ leads to a larger magnitude of $L(b_i)$, while the confidence of a decision is low with $L(b_i) = 0$ when the probabilities are equally likely.

From (15.84), the probability of a correct decision $\hat{b} = b_i \in \{0, 1\}$ is given as

$$\Pr\left(\hat{b} = b_i\right) = \frac{e^{\left|L(\hat{b})\right|}}{1 + e^{\left|L(\hat{b})\right|}}. \qquad (15.85)$$

The definition of LLR can be extended to nonbinary turbo codes, such as the double-binary turbo code used in WiMAX and DVB-RCS (Return Channel via Satellite).

The extrinsic information related to b_i is defined as the difference between the input and output LLRs associated with b_i

$$L_e(b_i) = L_{\text{out}}(b_i) - L_{\text{in}}(b_i). \qquad (15.86)$$

It is a reliable measure for each component decoder's estimate of the transmitted information symbol, based on the received component parity sequence only. The received systematic sequence can be used by each component decoder directly. The component decoders only exchange extrinsic information.

Applying Bayes' rule on (15.84), the a-posteriori LLR is obtained as

$$L(b_i) = L_{e,i}^{(2)} + L_{e,i}^{(1)} + L_s, \qquad (15.87)$$

where $L_{e,i}^{(2)}$ and $L_{e,i}^{(1)}$ are the extrinsic information contributed by the second and first decoders, respectively, and L_s is the a-posteriori LLR of the systematic bits

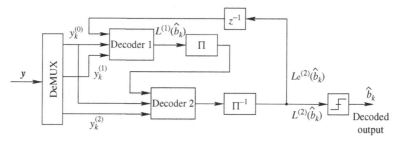

$$L_s = \ln \frac{P\left(y_i^{(0)}|b_i = 1\right)}{P\left(y_i^{(0)}|b_i = 0\right)}. \tag{15.88}$$

Based on the LLR, the estimate $\hat{b}_k$ of data bit b_k is derived

$$\hat{b}_k = \text{sign}\left(L(b_k)\right). \tag{15.89}$$

Decoding algorithms

The structure of the classical turbo decoder is shown in Fig. 15.15. Each SISO decoder processes its own input constituent codes and passes the extrinsic information to the other SISO decoder. The interleavers are identical to the interleavers in the turbo encoder so that the sequence at each decoder is properly aligned. Turbo decoding can be based on a message-passing or belief propagation algorithm.

The *a posteriori* probabilities or LLRs in the SISO decoder can be generated or approximated by using the MAP algorithm, which is a modified BCJR algorithm [5], or by using the soft output Viterbi algorithm (SOVA) [34].

Unlike the Viterbi algorithm whose decision is hard, the decisions of the BCJR are soft. While the Viterbi algorithm yields the most likely sequence as the codeword, the BCJR algorithm produces the most likely symbol along with its reliability at each time. The MAP algorithm outperforms the Viterbi algorithm by quite a margin at low E_b/N_0 and high BERs. The BCJR algorithm can be used to estimate *a posteriori* LLRs in systems represented by a trellis diagram; thus it can be used for the decoding of convolutional and linear block codes and for channel equalization.

Due to the overwhelming complexity of the BCJR algorithm, many reduced complexity variations have been developed. These include the M-BCJR and T-BCJR algorithms [26], reduced-complexity BCJR algorithms for turbo equalization [22], and the reduced-state BCJR (RS-BCJR) algorithm [17]. These algorithms are based on search over a reduced number of states on the trellis [26], or on a full search on a reduced-state trellis [17], or on search over a reduced number of paths on the trellis [22]. The BCJR algorithm has also been extended to trellis rate-distortion data compression [3].

In general, the SOVA is less complex than the MAP algorithm, but at a cost of performance. The SOVA is also used in decoding convolutional codes. For small memory, the SOVA is roughly half as complex as the log-MAP algorithm [60].

The decoding process using the MAP algorithm is given as follows:

- Set $L_e = 0$; decoder 1 decodes the received signal like a convolutional decoder, but gives soft output.
- From the soft output of decoder 1, L_e for decoder 2 is calculated.
- L_e for decoder 2 is fed back to decoder 1, resulting a refinement of the data estimate.
- This is then used as *a priori* information for a second iteration of code 1.
- This procedure continues until the specified number of iterations is reached or the convergence is achieved.

The BER performance improves as the number of iterations increases.

The main drawbacks of the turbo code are the relatively high decoding complexity and an increase in latency due to large interleaver size, and a performance floor at high SNR.

Discussion

Both Viterbi and turbo decoders are based on trellis processing of the received sample sequence. Unlike the Viterbi decoder, the turbo decoder makes multiple decoding passes over the received samples. The number of decoding passes for the turbo decoder is determined by SNR, and in 3G standards the maximum number of decoding passes is typically 6 to 10. Turbo decoder quantization usually employs 4-bit quantization.

TI's C64x DSPs integrate Viterbi and turbo co-processors that provide ASIC-like performance. In [73], a unified Viterbi/turbo decoder that shares some function blocks between the two decoding techniques is designed for 3GPP standards; this decoder operates in either the convolutional or the turbo decoding mode, and it can switch between the two modes or different turbo block sizes very rapidly. A high-speed radix-4 log-MAP turbo decoder targeted for HSDPA-MIMO mobile terminals has also been designed in [73]. The radix-4 architecture increases the throughput using parallelism in the trellis computations.

15.8.3 MAP algorithm

The turbo decoder may be implemented using the MAP algorithm. The MAP algorithm achieves the soft decision decoding by making two passes through the trellis, one pass in the forward direction and the other in the backward direction. For this reason, the MAP and its variants are sometimes called the *forward-backward algorithm*.

Let k be the time index, $\alpha_k(m)$ be the probability of the prior observations y up to time $k - 1$ with the decoder ending in state m at time k, $\beta_{k+1}(m)$ be the probability of future observations y at time k with the decoder starting at state m at time $k + 1$, $\gamma_i(y_k, m', m)$, for the data bit $i = 0, 1$, be the probability of transition at time k from state m' to state m for each branch, that is, the branch metric.

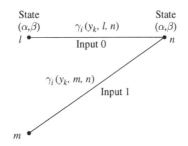

Figure 15.16

Trellis diagram for calculations of α, β, γ.

The trellis diagram for the calculation of α, β, γ is shown in Fig. 15.16. Each current state combines two paths from the previous states, one for input 1 and the other for input 0. There are the following relations [1, 10, 47]:

$$\alpha_k(m) = \frac{\sum_{m'=0}^{M-1} \sum_{i=0}^{1} \gamma_i \left(y_k, m', m\right) \alpha_{k-1}\left(m'\right)}{\sum_{m=0}^{M-1} \sum_{m'=0}^{M-1} \sum_{i=0}^{1} \gamma_i \left(y_k, m', m\right) \alpha_{k-1}\left(m'\right)}, \tag{15.90}$$

$$\beta_k(m) = \frac{\sum_{m'=0}^{M-1} \sum_{i=0}^{1} \gamma_i \left(y_{k+1}, m', m\right) \beta_{k+1}\left(m'\right)}{\sum_{m=0}^{M-1} \sum_{m'=0}^{M-1} \sum_{i=0}^{1} \gamma_i \left(y_{k+1}, m', m\right) \beta_{k+1}\left(m'\right)}, \tag{15.91}$$

where M is the number of decoder states and $\gamma_i\left(y_k, m', m\right)$ is dependent on the transmission channel; for the AWGN channel we have

$$\gamma_i\left(y_k, m', m\right) = p\left(y_k^{(u)} | b_k = i, m', m\right) \cdot p\left(b_k = i | m', m\right) \cdot p\left(m | m'\right), \quad u = 0, 1, \dots \tag{15.92}$$

with

$$p\left(y_k^{(0)} | b_k = i, m', m\right) = \frac{1}{\sqrt{2\pi\sigma^2}} e^{-\frac{1}{2\sigma^2}\left[y_k^{(0)} - a_k(i)\right]^2}, \tag{15.93}$$

$$p\left(y_k^{(u)} | b_k = i, m', m\right) = \frac{1}{\sqrt{2\pi\sigma^2}} e^{-\frac{1}{2\sigma^2}\left[y_k^{(u)} - b_k^{(u)}(i, m', m)\right]^2}, \quad u = 1, \dots, \tag{15.94}$$

$a_k(i)$, $b_k^{(u)}(i)$, $u = 1, \dots$, being the signal and parity bits on the trellis.

Example 15.13: Consider the RSC given by Fig. 15.11b, with input sequence $a = (1011)$. The encoder starts at state 0. Assuming that antipodal bits are transmitted over the AWGN channel with channel SNR 3 dB, use hard decoder and the MAP decoder to decode the received sequence.

Assuming a unit average signal power, from the channel SNR $= 2\,\text{dB} = 10^{0.2}$, we have the noise variance $\sigma^2 = 10^{-0.2}$.

The trellis of the RSC encoder is shown in Fig. 15.17. The encoded sequence is thus given as (11011010), or in antipodal bits $(+1 +1 -1 +1 +1 -1 +1 -1)$.

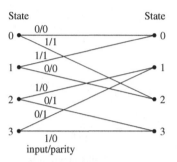

State   State

input/parity

Figure 15.17

Trellis diagram of the RSC encoder.

For a random run, the received sequence is

$$y = (1.1022, 1.5215, -1.9276, 0.6341, 0.7915, -1.9636, -0.0481, -0.2603)$$

If hard decision is employed, the sequence is (11011000), there is one bit in error.

Now we turn to soft decision using the MAP algorithm. The decoding is also based on the trellis. Let the probabilities of the signals for inputs 0 and 1 be p_s^0, p_s^1, and those of the parities for inputs 0 and 1 be p_p^0, p_p^1. The probabilities for the signal and parity are assumed to be independent. At time k, the probability of transition is given by

$$\gamma_i\left(y_k, m, m'\right) = p_s^i p_p^i, \quad i = 0, 1. \tag{15.95}$$

We now calculate $\alpha_k(m)$. At time $k = 0$, since the encoder starts at state 0, we have

$$[\alpha_0(0), \alpha_0(1), \alpha_0(2), \alpha_0(3)] = [1, 0, 0, 0].$$

At time $k = 1$,

$$\alpha_1(0) = \alpha_0(0)\gamma_0(y_1, 0, 0) + \alpha_0(1)\gamma_1(y_1, 1, 0) = \gamma_0(y_1, 0, 0) = p_s^0 p_p^0,$$

where

$$p_s^0 = \frac{1}{\sqrt{2\pi\sigma^2}} e^{-\frac{1}{2\sigma^2}\left(y_1^{(0)} - a_1\right)^2}, \quad p_p^0 = \frac{1}{\sqrt{2\pi\sigma^2}} e^{-\frac{1}{2\sigma^2}\left(y_1^{(1)} - b_1^{(1)}\right)^2}.$$

Note that $y_1^{(0)} = 1.1022$, $y_1^{(1)} = 1.5215$. From trellis, $a_1 = -1$, $b_1^{(1)} = -1$, we have $p_s^0 = 0.0108$, $p_p^0 = 0.0023$, $\gamma_0(y_1, 0, 0) = 2.4868 \times 10^{-5}$, and $\alpha_1(0) = 2.4868 \times 10^{-5}$.

$$\alpha_1(1) = \alpha_0(2)\gamma_1(y_1, 2, 1) + \alpha_0(3)\gamma_0(y_1, 3, 1) = 0,$$

$$\alpha_1(2) = \alpha_0(0)\gamma_1(y_1, 0, 2) + \alpha_0(1)\gamma_0(y_1, 1, 2) = \alpha_0(0)\gamma_1(y_1, 0, 2) = 0.1017,$$

$$\alpha_3(1) = 0.$$

The normalized values of $\alpha_k(m)$ are given by

$$\alpha_1(0) = 0.0002, \quad \alpha_1(1) = 0, \quad \alpha_1(2) = 0.9998, \quad \alpha_1(0) = 0.$$

Repeating the same procedure for $k = 2, 3, 4$, we get

state	0	1	2	3
α_0	1 (1)	0 (0)	0 (0)	0 (0)
α_1	0.0002 (0)	0 (0)	0.9998 (1)	0 (0)
α_2	0.0000 (0)	0.0003 (0)	0.0000 (0)	0.9997 (1)
α_3	0.0000 (0)	0.0002 (0)	0.0000 (0)	0.9998 (1)
α_4	0.0000 (0)	0.3379 (0)	0.0001 (0)	0.6620 (1)

The results in parentheses are for the case of no noise.

Following the same procedure, by starting from β_4 to β_0, we have, at time $k = 4$,

$$[\beta_4(0), \beta_4(1), \beta_4(2), \beta_4(3)] = [0, 0, 0, 1],$$

and we get the values of β to be

state	0	1	2	3
β_0	0.9980 (1)	0.0013 (0)	0.0004 (0)	0.0003 (0)
β_1	0.0001 (0)	0.9597 (0)	0.0000 (1)	0.0401 (0)
β_2	0.0010 (0)	0.0002 (0)	0.0398 (0)	0.9591 (1)
β_3	0 (0)	0.3379 (0)	0.0952 (0)	0.6621 (1)
β_4	0 (0)	0 (0)	0 (0)	1.000 (1)

The posterior input bit at time k is decided by computing the sum of all the probabilities for $b_k = 1$ conditioned on y, $\Pr(b_k = 1|y)$, and the sum of all the probabilities for $b_k = 0$ conditioned to y, denoted $\Pr(b_k = 0|y)$. From the trellis diagram shown in Fig. 15.18, we can extract these branches with input 1 for the calculation of $\Pr(b_k = 1|y)$.

Thus,

$$\Pr(b_k = 1|y) = \alpha_k(0)\gamma_1(y_k, 0, 2)\beta_{k+1}(2) + \alpha_k(1)\gamma_1(y_k, 1, 0)\beta_{k+1}(0)$$
$$+ \alpha_k(2)\gamma_1(y_k, 2, 1)\beta_{k+1}(1) + \alpha_k(3)\gamma_1(y_k, 3, 3)\beta_{k+1}(3),$$

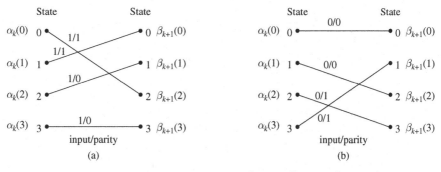

Trellis diagram for calculations of (a) $\Pr(b_k = 1|y)$, (b) $\Pr(b_k = 0|y)$.

$$\Pr\left(b_k = 0 | \mathbf{y}\right) = \alpha_k(0)\gamma_0(y_k, 0, 0)\beta_{k+1}(0) + \alpha_k(1)\gamma_0(y_k, 1, 2)\beta_{k+1}(2)$$
$$+ \alpha_k(2)\gamma_0(y_k, 2, 3)\beta_{k+1}(3) + \alpha_k(3)\gamma_0(y_k, 3, 1)\beta_{k+1}(1).$$

At $k = 1$, we have

$$\Pr\left(b_k = 1 | \mathbf{y}\right) = 2.3277 \times 10^{-6}, \quad \Pr\left(b_k = 0 | \mathbf{y}\right) = 3.2416 \times 10^{-9},$$

where the γ_i's are available when calculating $\alpha_k(m)$'s. The normalized values are

$$\Pr\left(b_k = 1 | \mathbf{y}\right) = 0.99861, \quad \Pr\left(b_k = 0 | \mathbf{y}\right) = 0.00139.$$

Thus, the decision is $\hat{b}_0 = 1$. Repeating the procedure for $k = 2, 3, 4$, we have

k	$\Pr\left(b_k = 1 | \mathbf{y}\right)$	$\Pr\left(b_k = 0 | \mathbf{y}\right)$	$\hat{b}_k$
1	0.99861	0.00139	1
2	7.0485×10^{-8}	1.00000	0
3	0.99992	8.2247×10^{-5}	1
4	0.99999	1.2387×10^{-5}	1

The MAP decoder generates the same output as the input sequence.

The quantization of channel values is considered in [18]. For an AWGN channel, even with four bits of quantization, the result based on quantized data is practically equal to the MAP algorithm operating on infinite precision data. The decoder's internal variables, $\alpha_k(m)$'s and $\beta_k(m)$'s, can also be quantized; however, these internal variables may not present a major storage problem, if sliding window schemes are used. The quantization of extrinsic information passed between two component decoders in a binary turbo decoder is considered in [42]. It is found that decoding with single-bit quantization introduces a performance loss of only about 0.6 dB, compared to infinite precision turbo decoding.

Log-MAP

The MAP algorithm is usually implemented in the logarithmic domain on a specific hardware.

The logarithm of a sum can be calculated by

$$\ln\left(e^{x_1} + e^{x_2}\right) = \max[x_1, x_2] + \ln\left(1 + e^{-|x_1 - x_2|}\right). \tag{15.96}$$

That is, it can be calculated as the maximum of x_1 and x_2, plus a correction term. The correction term depends on the difference between x_1 and x_2, and has a small range for quantization. An LUT can be used to avoid the computation of exponential and logarithmic functions, and a simple LUT for 3-bit quantization is given in Table 15.2. This version of the BCJR algorithm is known as *log-MAP* [61].

Table 15.2. Look-up table for the correction term in (15.96) for 3-bit quantization.								
$\lvert x_1 - x_2 \rvert$	0.00	0.25	0.50	0.75	1.00	1.25	1.50	1.75	> 2.00
$\ln\left(1 + e^{-\lvert x_1 - x_2 \rvert}\right)$	0.69	0.58	0.47	0.39	0.31	0.25	0.20	0.16	0.00

The max function (15.96) can be extended to k sum-terms in the logarithm function

$$\ln\left(\sum_{i=1}^{k} e^{x_i}\right) = \max(x_i) + \ln\left(\sum_{i=1}^{k} e^{x_i - \max(x_i)}\right). \tag{15.97}$$

This equation is useful for calculating the LLR for nonbinary constellations.

A static LUT introduces error in the LLR estimation due to its limited size and resolution. A dynamic LUT, which has partial resolutions in separate decision regions and is updated at each iteration of decoding, is described in [39]. By using dynamic LUTs of sufficient length, the LLR values obtained are close to their natural logarithmic counterparts. The dynamic-LUT architecture effectively eliminates the error floor that static LUT has.

Max-log-MAP

A suboptimal but lower-complexity max-log-MAP algorithm is implemented as an approximation of the optimum log-MAP algorithm by neglecting the correction terms in (15.96). The performance of the max-log-MAP degrades slightly, but it is still better than that of the SOVA. The hard decisions of the max-log-MAP output equal those of the Viterbi algorithm [60].

The max-log-MAP algorithm can be implemented using a pair of Viterbi algorithms, one for the forward direction and the other for the backward direction [77]. The complexity of the log-MAP is about twice that of the SOVA; however, the log-MAP is more suited to parallel processing [60].

15.8.4 Analysis of the turbo code

The BER performance of the turbo code has a distinct turbo cliff region and an error floor region. The turbo code has an impressive performance at low SNR (E_b/N_0), but at high SNR the performance is not very remarkable; this can be seen from the BER performance: An error floor (i.e. a change in the slope of the error rate curve) has been introduced at high SNR, due to the relatively small free distance of the turbo code [6, 62]. The excellent performance of the turbo code at low SNR is due to the sparse distance spectrum that results from a pseudorandom interleaver. The free distance as well as the distance spectrum can be calculated using the algorithm described in [28].

Distance spectrum analysis of the turbo code

The BER performance of a turbo code with ML decoding is also upper bounded by the union bound for the convolutional code, as given by (15.74) [62]

$$P_b \leq \sum_{d=d_{\text{free}}}^{\infty} \frac{N_d \tilde{w}_d}{N} Q\left(\sqrt{d \frac{2R_c E_b}{N_0}}\right). \tag{15.98}$$

Although they have the same form, for the turbo code $N_d \tilde{w}_d \ll N$ for low-weight codewords due to the pseudorandom interleaver. After the distance spectrum is obtained, this bound can be determined.

For moderate and high SNRs, the free distance term in the union bound dominates. In this case, the asymptotic performance of the convolutional and turbo codes with ML decoding is given by

$$P_b \approx \frac{N_{\text{free}} \tilde{w}_{\text{free}}}{N} Q\left(\sqrt{d_{\text{free}} \frac{2R_c E_b}{N_0}}\right), \tag{15.99}$$

where N_{free} and $\tilde{w}_{\text{free}}$ correspond to N_d and $\tilde{w}_d$ for $d = d_{\text{free}}$. This is the free distance asymptote of the code, which can be used to characterize the error floor of turbo codes. Increasing the interleaver length N while preserving d_{free} and N_{free} will lower the asymptote without changing its slope. The influence of N is illustrated in Fig. 15.19 for the same turbo code at a given number of iterations. For a fixed N, the error floor can be modified by increasing d_{free} while keeping N_{free} unchanged; this increases the slope of the asymptote.

The interleaver is most important for the turbo code. It should be selected as a pseudorandom interleaver. For a $(37, 21, 65536)$ turbo code with pseudorandom interleaver, it is found that $N_{\text{free}} = 3$, $d_{\text{free}} = 6$, and $\tilde{w}_{\text{free}} = 2$ [62]. The effective multiplicity $N_{\text{free}}/N = 3/65536$ is very small. When the interleaver is selected as a rectangular interleaver, the result is totally different. For a $(37, 21, 14400)$ turbo code with a 120×120 rectangular interleaver, it is found that $d_{\text{free}} = 12$ with a multiplicity of $N_{\text{free}} = 28,900$

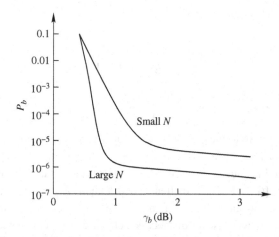

Figure 15.19   The influence of the interleaver size on the performance of a turbo code.

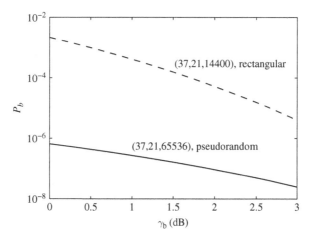

Figure 15.20   The free distance asymptotes for turbo code with different interleavers.

and $\tilde{w}_{\text{free}} = 4$ [62]. Thus, the effective multiplicity $N_{\text{free}}/N = 28900/14400 \approx 2$, which is much larger than that for the pseudorandom interleaver. This large multiplicity of d_{free} for a rectangular interleaver leads to relatively poor performance. According to the analysis given in [62]: for the rectangular interleaver, N_{free} is of the order of N, and increasing N does not lead to an increase in d_{free}; thus N_{free}/N does not change significantly and an increase in N does not lead to a sufficient lowering of the error floor.

Example 15.14: The free distance asymptotes for the $(37, 21, 65536)$ and $(37, 21, 14400)$ turbo codes discussed above can be obtained from (15.99). They are plotted in Fig. 15.20. These asymptotes correspond to the error floor for medium and large SNR. For the $(37, 21, 65536)$ turbo code, the contribution of the free distance term dominates the BER for SNR as low as 0.75 dB [62].

In the low SNR region, the contribution of other distances d in addition to d_{free} must be considered. In this case, the distance spectrum is referred to as *sparse or spectral thin*. For a convolutional code, higher distance spectral lines have an overall BER contribution greater than that of the free distance term for small SNR. The free distance asymptote dominates the BER performance only for large SNR, and this is due to the rapid increase of the multiplicity for large d. Such a distance spectrum is referred to as *spectrally dense*.

Weight enumeration analysis of the turbo code

For convolutional or turbo codes, the distance spectrum information can be rewritten in the form of the IOWEF

$$A(W, X) = \sum_{d=d_{\text{free}}}^{\infty} \sum_{w} A_{w,d} W^w X^d, \qquad (15.100)$$

where $A_{w,d}$ is the number of codewords of weight d caused by information sequences of weight w, and W and X are dummy variables. For convolutional codes, the IOWEF can be calculated via the transfer function method. However, for turbo codes the transfer function is hard to obtain due to the use of an interleaver. An approximate weight enumeration function (WEF) can be obtained by using a probabilistic uniform interleaver model. The corresponding bound on the BER of the turbo code with uniform interleavers is derived as [7, 62]

$$P_b \leq \sum_{w=w_{\min}}^{N} \frac{w}{N} W^w A_w(Z) \big|_{W=Z=e^{-R_c E_b/N_0}}, \tag{15.101}$$

where $w_{\min}$ is the minimum weight of the information sequence that generates a code-word in the terminated convolutional component code, and $A_w(Z)$ is the conditional WEF (CWEF) for the overall turbo code, Z being a dummy variable.

For the turbo code with uniform interleaving, the overall CWEF can be approximated by [7, 62]

$$A_w(Z) \approx \frac{w!}{\left(n_{\max_w}!\right)^2} N^{2n_{\max_w} - w} \left(A_w^{n_{\max_w}}(Z)\right)^2, \tag{15.102}$$

where $n_{\max_w}$ is the maximum number of error events due to a weight w information sequence that can be positioned in an interleaver of length N.

Substituting this approximation into (15.101), the bound obtained for P_b can be used for developing some design rules for turbo codes. These rules are given as follows [7, 62]:

- **Rule 1**. Select component encoders with feedback.
- **Rule 2**. Select the feedback polynomial as a primitive polynomial.
- **Rule 3**. Select an interleaver with appropriate spreading factors.

For rule 3, a rectangular interleaver has too much regularity, leading to a good spreading factor, but yielding poor overall performance due to large multiplicity. On the other hand, a pseudorandom interleaver has a thin distance spectrum but has a poor spreading factor. The s-random interleaver achieves a desired spreading property and its random-like index points avoid large multiplicity; thus, it yields an excellent performance in the error floor region, since its free distance is larger than that of the pseudorandom interleaver. Also, rules 2 and 3 are based on the approximations that are only valid for high SNR and large N.

EXIT analysis of the turbo code

The behavior of turbo codes at low SNR, or in the turbo cliff region, can be analyzed by the extrinsic information transfer (EXIT) analysis [72]. The component decoder is treated as a nonlinear LLR transfer function that transforms the input extrinsic LLR (mutual *a priori* information) $\bar{I}_a$ into an output extrinsic LLR $\bar{I}_e$.

The EXIT analysis takes into consideration the iterative nature of the decoding process. It makes use of the iterative decoding algorithm. The method is also used for prediction of the turbo cliff for SCCCs [71].

15.9 Serially concatenated convolutional codes

The SCCC with iterative decoding is an alternative to the PCCC. The SCCC encoder has the same block diagram as that of the conventional concatenated code, as shown in Fig. 15.12. The SCCC uses a recursive or nonrecursive convolutional outer code with rate $R_o = k/m$, encoder memory K_o, and generator matrix $G_o(D)$, and a recursive convolutional inner code with rate $R_i = m/n$, encoder memory K_i, and generator matrix $G_i(D)$. The overall code rate is $R = R_o R_i = k/n$ if the effect of trellis termination is negligible. Unlike the conventional concatenated code, for the SCCC the interleaver that separates the inner and outer encoders is mandatory. The interleave length N is usually divisible by m. In the SCCC, the interleaver is used for spectral thinning, as opposed to the spectrally dense case of the convolutional code. It also introduces randomness necessary for iterative decoding.

The BICM resembles the SCCC code, and iterative decoding can be applied to approximate the *a priori* probability [16, 46]. Uniform soft-decision feedback iterative decoding [46] provides a tradeoff between hard-decision and soft-decision feedback iterative decoding schemes in terms of performance gain and complexity.

Without the interleaver, the concatenated code simply becomes a convolutional code with an overall code rate $R_c = R_o R_i$ and overall generator matrix $G(D) = G_o(D)G_i(D)$. This is called a *cascaded convolutional code*, and can be decoded by using Viterbi decoding. This cascade will yield a reasonably large free distance, which increases with the component code complexity. But the free distance is not maximal.

SCCC versus PCCC

The PCCC or turbo code is an inherently unequal error protecting code [67]. Occasional bad interleaver mappings make a few bit positions affected by the dominant error events, leading to small d_{free} and hence a BER floor [67]. On the contrary, for the convolutional code all bit positions in the input sequence are subject to the same error events, leading to equal likelihood of error. Analysis and simulation for the BER performance show that the SCCC has a distinct advantage over the PCCC with random interleavers for large N and SNR [8, 62] in the AWGN channel. For the PCCC, the interleaving gain is fixed at N^{-1}, while for the SCCC it is $N^{-d_{\text{free}}^o/2}$, where d_{free}^o is the free distance of the outer encoder. Due to its large free distance, the SCCC does not manifest an error floor at high SNR. However, at very low SNR, the SCCC is inferior to the PCCC. The same conclusion is obtained in the presence of ISI [45].

Product code

The product code is a well-known serial concatenated block code [47]. Two minor component codes encode a block of identical, but permuted, information symbols. The product code employs the regular rectangular block interleaver. Its powerful potential can be achieved by using the turbo decoding principle [62]. For this reason, the product code is also known as a *block turbo code*. Like the SCCC, the product code does not suffer from the problem of an error floor as the turbo code does, and no special interleaver design is required. The product code with iterative decoding is now used as optional FEC in IEEE 802.16e.

15.9.1 Design of the SCCC

Like the PCCC, analysis of the SCCC can be based on the IOWEF of the code. A bound on the BER probability of the SCCC is derived in [8, 62], based on which some design rules for the SCCC are given as follows.

- **Rule 1**. The inner encoder should be a recursive convolutional encoder (i.e., with feedback) so that a gain is achieved by increasing the interleaving size N.
- **Rule 2**. The effective free distance, d_{eff}, of the inner encoder should be maximized by selecting a primitive feedback polynomial.
- **Rule 3**. The feedback polynomial of the inner encoder should have a factor of $1 + D$.
- **Rule 4**. The outer encoder should have the free distance, d_{free}^o, as large as possible to achieve a large interleaving gain.

The first two design rules are also used for selecting component encoders for the PCCC. Thus, a component encoder optimized for the turbo code is also suitable for using as the encoder of the inner code in the SCCC. For rule 4, a good nonrecursive convolutional code is suitable. A nonrecursive convolutional code has a slight advantage in BER performance for large SNR, and should be preferred for the outer encoder. Rule 3 is in conflict with rule 2, since the feedback polynomial constructed using rule 3 will not be primitive. As with the turbo code, the use of a primitive or nonprimitive feedback polynomial leads to a tradeoff between the performance at low and high SNRs.

Analysis and simulation show that if both the inner and outer convolutional encoders have no feedback, the concatenated code has a free distance $d_{\text{free}} = d_{\text{free}}^o d_{\text{free}}^i$ [62]. The interleaver introduces a significant improvement in free distance compared to the cascaded convolutional code case. However, there is no interleaving gain associated with the free distance multiplicity, leading to a free distance asymptote that is not associated with N. Nonetheless, for large N the pseudorandom interleaver has a spectral thinning effect, yielding an improved performance at moderate SNR.

15.9.2 Decoding of the SCCC

The SCCC can be decoded in a manner very similar to that of the PCCC. The interleavers decouple the concatenated codes so that the extrinsic informations obtained by the two decoders are mutually independent. Exchanging extrinsic information between the two SISO decoders may improve the overall performance. The block diagram of the iterative SCCC decoder is shown in Fig. 15.21. The two SISO decoders are the same as those used in the iterative decoder of the turbo code. A difference from the turbo decoder is that the outer SISO decoder of the SCCC has no access to channel observations, and can only rely on information received from the inner decoder.

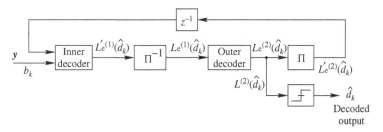

Figure 15.21   Iterative decoder for the SCCC code.

For each input symbol b_k, the inner decoder receives the channel LLR and the extrinsic information from the outer decoder and calculates a new LLR as

$$L_k^{(1)} = \ln \frac{\Pr\left(b_k' = 1 | y\right)}{\Pr\left(b_k' = 0 | y\right)} = \ln \frac{\Pr\left(y | b_k' = 1\right)}{\Pr\left(y | b_k' = 0\right)} + \ln \frac{\Pr\left(b_k' = 1\right)}{\Pr\left(b_k' = 0\right)}$$

$$= L_{e,k}^{(1)'} + L_k^{(1)'}. \tag{15.103}$$

Only $L_{e,k}^{(1)'}$ is passed to the outer decoder.

Likewise, the outer decoder output

$$L_k^{(2)} = \ln \frac{\Pr\left(b_k' = 1 | L_k^{(1)}\right)}{\Pr\left(b_k' = 0 | L_k^{(1)}\right)} = \ln \frac{\Pr\left(L_k^{(1)} | b_k' = 1\right)}{\Pr\left(L_k^{(1)} | b_k' = 0\right)} + \ln \frac{\Pr\left(b_k' = 1\right)}{\Pr\left(b_k' = 0\right)}$$

$$= L_{e,k}^{(2)} + L_{e,k}^{(1)'}. \tag{15.104}$$

Only $L_{e,k}^{(2)}$ is passed to the first decoder. The decoder performs a few iterations, and a decoding decision is then made on the symbol.

15.10 Low-density parity-check codes

The LDPC code [27], developed by Gallager in 1960, is another error-correcting code that approaches the Shannon limit. The LDPC code and an iterative decoding algorithm, which involves calculation of probabilistic information that is passed from one iteration to another, were first introduced by Gallager in [27], but they were ignored until the invention of turbo codes in 1993. The LDPC code was rediscovered by MacKay and Neal [49, 50]. The LDPC code can be constructed for any rate.

15.10.1 LDPC code: a linear block code

The LDPC code is a linear block code whose parity check matrix **H** is primarily composed of zeros. LDPC codes can be categorized into regular and irregular LDPC codes. The original Gallager codes are regular binary LDPC codes. The LDPC encoder uses the parity check matrix $\mathbf{H}$ rather than the generator matrix $\mathbf{G}$. The size of $\mathbf{H}$ is usually very large, but the density of nonzero elements is very low. By suitably defining $\mathbf{H}$, decoding of the LDPC code is made easy.

When the parity-check matrix $\mathbf{H}$, $(n - k) \times n$, has the same number p of ones in each column and the same number q of ones in each row, the code is a regular (p, q) LDPC code of length n, or denoted as an (n, p, q) LDPC code. Thus, each information bit is involved with p parity checks, and each parity-check bit is involved with q information bits. The fraction of ones in the parity-check matrix is $\frac{(n-k)q}{(n-k)n} = q/n$, which is very small for large block length n. For a regular code, we have $(n - k)q = np$, thus $p < q$. If all rows are linearly independent, the code rate is $(q - p)/q$, otherwise it is k/n. Typically, $p \geq 3$. When $\mathbf{H}$ has multiple column weights and multiple row weights, we get an irregular LDPC code.

When $p \geq 3$, there is at least one LDPC code whose minimum distance $d_{\min}$ grows linearly with the block length n [27]; thus a longer code length yields a better coding gain. Most regular LDPC codes are constructed with p and q on the order of 3 or 4.

A $(n = 20, p = 3, q = 4)$ code constructed by Gallager [27] is given as

$$
\begin{bmatrix}
1 & 1 & 1 & 1 & 0 & 0 & 0 & 0 & 0 & 0 & 0 & 0 & 0 & 0 & 0 & 0 & 0 & 0 & 0 & 0 \\
0 & 0 & 0 & 0 & 1 & 1 & 1 & 1 & 0 & 0 & 0 & 0 & 0 & 0 & 0 & 0 & 0 & 0 & 0 & 0 \\
0 & 0 & 0 & 0 & 0 & 0 & 0 & 0 & 1 & 1 & 1 & 1 & 0 & 0 & 0 & 0 & 0 & 0 & 0 & 0 \\
0 & 0 & 0 & 0 & 0 & 0 & 0 & 0 & 0 & 0 & 0 & 0 & 1 & 1 & 1 & 1 & 0 & 0 & 0 & 0 \\
0 & 0 & 0 & 0 & 0 & 0 & 0 & 0 & 0 & 0 & 0 & 0 & 0 & 0 & 0 & 0 & 1 & 1 & 1 & 1 \\
1 & 0 & 0 & 0 & 1 & 0 & 0 & 0 & 1 & 0 & 0 & 0 & 1 & 0 & 0 & 0 & 0 & 0 & 0 & 0 \\
0 & 1 & 0 & 0 & 0 & 1 & 0 & 0 & 0 & 1 & 0 & 0 & 0 & 0 & 0 & 0 & 1 & 0 & 0 & 0 \\
0 & 0 & 1 & 0 & 0 & 0 & 1 & 0 & 0 & 0 & 0 & 0 & 0 & 1 & 0 & 0 & 0 & 1 & 0 & 0 \\
0 & 0 & 0 & 1 & 0 & 0 & 0 & 0 & 0 & 0 & 1 & 0 & 0 & 0 & 1 & 0 & 0 & 0 & 1 & 0 \\
0 & 0 & 0 & 0 & 0 & 0 & 0 & 1 & 0 & 0 & 0 & 1 & 0 & 0 & 0 & 1 & 0 & 0 & 0 & 1 \\
1 & 0 & 0 & 0 & 0 & 1 & 0 & 0 & 0 & 0 & 0 & 1 & 0 & 0 & 0 & 0 & 0 & 1 & 0 & 0 \\
0 & 1 & 0 & 0 & 0 & 0 & 1 & 0 & 0 & 0 & 1 & 0 & 0 & 0 & 0 & 1 & 0 & 0 & 0 & 0 \\
0 & 0 & 1 & 0 & 0 & 0 & 0 & 1 & 0 & 0 & 0 & 0 & 1 & 0 & 0 & 0 & 0 & 0 & 1 & 0 \\
0 & 0 & 0 & 1 & 0 & 0 & 0 & 0 & 1 & 0 & 0 & 0 & 0 & 1 & 0 & 0 & 1 & 0 & 0 & 0 \\
0 & 0 & 0 & 0 & 1 & 0 & 0 & 0 & 0 & 1 & 0 & 0 & 0 & 0 & 1 & 0 & 0 & 0 & 0 & 1 \\
\end{bmatrix}.
$$

The matrix is composed of p submatrices, each has a single 1 in each column. The first submatrix contains all its 1s in a descending order: the ith row contains all 1s in columns $(i - 1)q + 1$ to iq. The lower submatrices are column permutations of the first submatrix. Analysis based on all codes constructed in this way shows that for fixed p, the error probability of the optimum decoder decreases exponentially for sufficiently low noise and sufficiently long block length [27]. The typical minimum distance increases linearly with the block length [27].

The irregular LDPC code achieves better performance than the turbo code and the regular LDPC code. The regular LDPC code can utilize memory more efficiently and this simplifies the implementation of the LDPC coder. A method for construction of the parity check matrix for the irregular LDPC code was proposed in [48].

The LDPC code can be decoded using the iterative soft-decision decoding algorithm based on the factor graph. It is constructed in a random manner, and the complexity of the decoding algorithm is linear in the block length of the code. Like RCPC codes, rate-compatible puncturing LDPCs provide different code rates [32, 75]. For channels with memory in the form of burst erasures, structured LDPC codes are designed in [38].

LDPC code versus turbo code

The LDPC code has better block error performance compared to the turbo code [58], and has no error floor, whereas error floors are commonly exhibited in the turbo code. In a sequal, the LDPC code becomes a serious competitor to the turbo code. In fact, the irregular LDPC code has achieved the closest proximity to capacity [62]. The irregular repeat accumulate codes are a class of irregular LDPC codes with linear encoding complexity; they have a performance that is superior to that of turbo codes of comparable complexity and as good as the best known irregular LDPC codes [37].

The LDPC code tends to have a higher encoding complexity than the turbo code, but has a lower decoding complexity than the turbo code. The LDPC code is extremely good in terms of the BER performance for large code block length. This large block length, however, introduces a significant decoding delay. The turbo code also has good BER performance, but the error floor occurs at relatively high BERs. The turbo code is more suitable for medium block size or constraint lengths.

The LDPC code has been adopted in the DVB-S2 standard, which was approved as the replacement of DVB-S by ETSI in 2005. More importantly, the turbo code is a patented technique, while the LDPC patents have expired. There are some hardware implementations of the LDPC codes such as Flarion Technologies' programmable LDPC decoder, and the LDPC codes for optical networking by Lucent Technologies and Agere. The LDPC code is specified as an optional channel coding scheme in IEEE 802.16e, 802.11n, and 802.20.

Representation using Tanner graphs

The LDPC code can be visualized by using the Tanner graph [69]. The Tanner graph is a bipartite graph associated with $\mathbf{H}$, where the nodes are divided into two disjoint classes with edges only between nodes in different classes. The information bits are placed on certain variable nodes, and the values for the remaining variables are determined by satisfying all the parity check constraints. Variable nodes correspond to the columns of $\mathbf{H}$, while constraint nodes correspond to the rows of $\mathbf{H}$. There is an edge (connection) between constraint node i and variable node j, if $h_{ij} = 1$. The number of edges incident upon a node is called the *degree of the node*. Thus, the bipartite graph of a (p, q) LDPC code has n variable nodes of degree p and m check nodes of degree q. A graph without cycles is said to be *acyclic* and is called a *tree*. The length of the shortest cycle in a graph is defined as the *girth* of the graph.

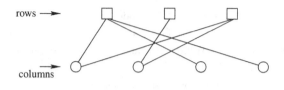

rows ⟶

columns ⟶

Figure 15.22 A Tanner graph.

Example 15.15: Consider

$$H = \begin{bmatrix} 1 & 0 & 1 & 1 \\ 0 & 1 & 0 & 0 \\ 1 & 1 & 0 & 1 \end{bmatrix}.$$

The Tanner graph is shown in Fig. 15.22, where circles denote variable nodes and squares denote constraint nodes. This graph is acyclic.

15.10.2 LDPC encoder and decoder

LDPC encoder

After the selection of the parity check matrix $\mathbf{H}$, we need to calculate the generator matrix $\mathbf{G}$. This can be obtained by using Gaussian elimination and column reordering to transform $\mathbf{H}$ into $\tilde{\mathbf{H}} = \left(-\mathbf{P}^T | \mathbf{I}\right)$. Accordingly, the generator matrix can be selected as $\mathbf{G} = (\mathbf{I} | \mathbf{P})$. For an (n,k) LPDC code, if $\mathbf{H}$ is espressed as

$$\mathbf{H} = \begin{bmatrix} \mathbf{A}_1^T & \mathbf{A}_2^T \end{bmatrix}, \tag{15.105}$$

where $\mathbf{A}_1$ is a $k \times (n-k)$ matrix and $\mathbf{A}_2$ is an $(n-k) \times (n-k)$ matrix, then

$$\mathbf{G} = \begin{bmatrix} \mathbf{I}_{k \times k} & \mathbf{A}_1 \mathbf{A}_2^{-1} \end{bmatrix}. \tag{15.106}$$

The generated $\mathbf{G}$ is usually not sparse. The information is encoded as

$$c = m\mathbf{G}. \tag{15.107}$$

Encoding by matrix multiplication has a complexity of $O\left(n^2\right)$, whereas the decoding complexity of the LDPC code is linear with n. Some examples of LDPC codes with encoding complexity of $O(n)$ are the irregular repeat accumulate codes [37], and two rate-compatible LDPC codes [41, 79]. In [59], an encoding algorithm at $O(n)$ is proposed for systematic codes. Encoding can also be implemented via message-passing [62].

LDPC decoder

Soft information from the observed signals is connected to the variable nodes. The probability of a variable node being in a certain state is then determined for the given soft information. Decoding can be done by a procedure of belief or probability propagation (also known as *message-passing*) on the Tanner graph [43]. Based on the Tanner graph, decoding can be performed at each node in parallel processing. Belief or probability propagation in the Bayesian network was first proposed for use in acyclic graphs [55] for modeling human reasoning, but it is applicable in any probabilistic inference problem such as in error control decoding, where the probabilistic value of transmitted information symbols needs to be inferred from noisy received symbols.

The sum-product algorithm, first proposed in [27], is a belief propagation algorithm for LDPC decoding. It implements probability calculations at each node based on local distributions, and passes a message on to connecting nodes in a factor graph. The sum-product algorithm has three steps: initialization, horizontal step and vertical step. It can operate on conditional probability measures or LLR. The use of LLR can reduce the computational complexity substantially. The min-sum algorithm approximates the vertical step using minimum operation rather than summation and hyperbolic functions, leading a further reduction in complexity with a slight performance degradation [2]. The optimality of the sum-product algorithm on acyclic (cycle-free) graphs was proved in [69]. A detailed exposition of the sum-product algorithm can be found in [50].

The sum-product algorithm is suboptimal in general. For the LDPC code, the calculations for each bit node are independent of all the other bit nodes, and the calculations for each check node are also independent of all the other check nodes. Thus, all the check nodes and the bit nodes can be potentially implemented in parallel [44]. The sum-product algorithm has a complexity lower than the MAP or the log-MAP algorithm for turbo decoding [44].

The MAP algorithm for standard turbo decoding can be closely approximated by belief propagation on a graph with cycles [53, 62]. For the turbo code, the MAP algorithm requires the serial calculation of all of the state metrics, and this limits its parallel implementation. The optimal decoder for BPSK signals in the AWGN channel is the MAP decoder that computes the LLR and makes a decision by comparing the LLR to the threshold zero.

There are also hard-decision decoding algorithms for the LDPC code, such as the majority-logic decoding [47, 51] and iteratively bit-flipping decoding [27]. The flipping algorithm has a lower complexity but with a reduced performance when compared to the sum-product algorithm. It is more suitable for very-high-speed scenarios such as optical networking. Recent research in the LDPC code has lead to coding systems with lower complexity, yet matching or outperforming the turbo code.

Recent hardware LDPC decoders are capable of throughputs of hundreds of Mbits/s. In [40], the FPGA-based LDPC decoder implementation is able to perform 15 decoding iterations with the (1944, 972) code at around 300 Mbits/s. The flexible decoder architecture for irregular LDPC codes supports twelve combinations of code lengths and code rates based on the IEEE 802.11n standard. All the codes correspond to a block-structured parity check matrix, in which the sub-blocks are either a shifted identity matrix or a zero matrix.

15.11 Adaptive modulation and coding

The time-varying wireless channel condition leads to a time-varying system capacity. To achieve the optimum performance, adaptivity in data rate, power control, coding, bandwidth, antenna, and protocols is necessary. In future generation wireless communication systems, link adaptation such as AMC is necessary.

AMC corresponds to data rate adaptation. It changes the modulation and coding format according to the channel condition. Users close to the BS can be assigned a modulation with a high code rate, while users close to the cell boundary can be assigned a modulation with a lower code rate. However, AMC is sensitive to channel measurement error and delay.

AMC has been deployed in many wireless standards such as GSM, IS-95, wireless LANs, HSDPA, WCDMA and WiMAX, and will also be deployed in 4G standards. It achieves a robust and spectrally efficient transmission over time-varying channels. The transmitter needs to have the CSI before it can adapt its transmission scheme according to the channel condition. The CSI is usually estimated at the receiver, and is then fed back to the transmitter. For a multicarrier system such as OFDM, modulation and/or coding can be different for each subcarrier.

The power gain or the received SNR at time i is estimated, this information is then used to adapt the modulation and coding parameters such as the data rate $R(i)$, transmit power $P(i)$, and coding parameters $C(i)$. These parameters are adjusted according to the received SNR at time i.

Some adaptation techniques are described below [31].

Variable-rate modulation

In variable-rate modulation, the data rate $R(\gamma)$ is adjusted according to γ. This is done either by fixing the symbol rate $R_s = 1/T$ but adjusting the modulation schemes or constellation sizes, or by fixing the modulation scheme but adjusting R_s. Varying symbol rate R_s leads to a varying signal bandwidth, and this is not practical for bandwidth allocation. Rather, with R_s fixed, changing the constellation size or modulation type is used in current systems such as EDGE, HSDPA, and IEEE 802.11a/g. For implementation of variable-rate modulation, the modulation parameters are fixed over a block or frame of symbols.

After the information bit is channel encoded, one of the K M-ary modulation schemes can be selected at the transmitter, based on the short-term channel quality such as SNR measured at the receiver. This channel quality is fed back to the transmitter. According to the level of channel quality, a suitable modulation mode is selected. This is illustrated in Fig. 15.23.

Variable coding

Different channel codes are used for different channel conditions. A stronger error correcting code is used when γ is small, and a weaker code or no coding is used for large γ. This

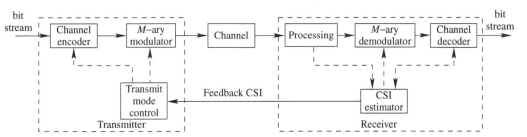

Figure 15.23 Model for adaptive modulation and coding. Power control can also included in the transmitter after the channel modulator module.

technique requires a constant channel over the block length of the code. It is particularly useful for fixed modulation due to the PAPR constraint.

The RCPC code also achieves different code rates $R_c = k/n$ by not transmitting certain coded bits. RCPC codes can be used for AMC. The code rate change for RCPC codes can be implemented frame by frame or even bit by bit. The RCPC code is used in EDGE.

The instantaneous BER can also be adapted subject to an average BER constraint $\overline{P}_b$. The BER is typically adapted along with some other form of adaptation such as data rate. There are also various hybrids of these schemes. The variable rate and power techniques for M-ary modulations have been described in [31]. The objective is to maintain a fixed instantaneous BER for each symbol while maximizing the average data rate by adapting the transmit power and constellation size. Implementation of adaptive coding is also illustrated in Fig. 15.23.

AMC combines adaptive modulation and adaptive coding, and adjusts the data rate according to the channel quality. On poor channels, lower data rates can be achieved by using a small constellation and a robust (low-rate) error-correcting code, while on good channels the higher data rates can be achieved with large constellations and a high-rate error correcting code. The normalized throughput is given by

$$T = (1 - \text{BLER})R_c \log_2 M \quad \text{(bits/s/Hz)} \tag{15.108}$$

where BLER is the block-error rate, $R_c \leq 1$, and M is the constellation size.

Variable-power transmission

The variable-power technique is used to compensate for SNR variation due to fading. It aims to achieve a constant received SNR, or equivalently a fixed BEP. The channel inversion scheme cannot realize any target P_b for Rayleigh fading, because $E[1/\gamma] = \infty$. The truncated channel inversion scheme inverts the fading above a given cutoff γ_0.

Variable frame length

For packet radio, frame length adaptation is also very important [29]. With a large frame length, the goodput of the link can drop to zero if every packet is corrupted. Thus the radio

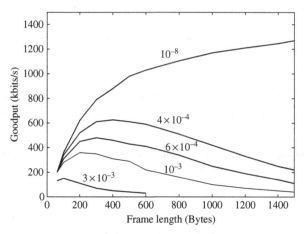

Figure 15.24 Goodput versus frame length for different BERs [15]. Redrawn from Fig.5a in [15]. ©1999, IEEE.

will continue transmitting or retransmitting and thereby wasting battery energy. Reducing frame length can increase the goodput and thus the energy efficiency due to the reduction of retransmission. However, the relative overhead of the frame header also increases. For a given BER, we can find an optimum frame length that maximizes the goodput or minimizes the transmit power in a fading channel [15]. In [15], a goodput versus frame length for different BERs is given in Fig. 15.24, which was obtained from measurements of a real peer-to-peer wireless link. An adaptive radio was designed in [15] that adapted the frame length, error control such as FEC/ARQ and SACK, processing gain, and equalization to different channel conditions, while minimizing battery energy. These adaptations can be achieved in MAC layer protocol.

Variable processing gain

In shared bands such as for the CDMA technique, the signal-to-interference ratio (SIR) can be less than 0 dB. The processing gain can be controlled at the link layer. Given a fixed chip rate (baseband bandwidth), if the SIR is greater than a threshold, then the product of the processing gain and the data rate (user throughput) is constant and equals the occupied bandwidth, that is, increasing the processing gain reduces the data rate. For a very low SIR, a large processing gain can be used. For the fixed chip rate, the transmitted power keeps constant, thereby a higher processing gain yields a higher energy per information bit. For an adaptive-processing-gain DSSS radio with a fixed chip rate of 2 Mchips/s [15], the result is shown in Fig. 15.25.

For packet radios, changes in frame size, error control, and processing gain used in each frame must be passed by the sender node to the receiver so that it can decode the packets. This may lead to a high signaling overhead. A better solution is to make each packet self-describing, and in this case, no synchronization between the sender and receiver nodes is necessary.

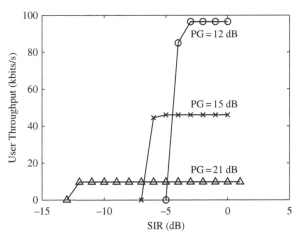

Figure 15.25 Throughput versus SIR for different processing gains [15]. Redrawn from Fig.12a in [15].
©1999, IEEE

15.12 ARQ and hybrid-ARQ

HARQ is an important topic for 3G/4G wireless systems. An introduction of ARQ and HARQ is given in this section.

ARQ

ARQ is typically used for packet transmissions in computer networking. For wireless applications, it can be used in WCDMA or WiMAX data services, or wireless networking. ARQ-enabled transmission requires each transmitted packet to be acknowledged by the receiver. Unacknowledged packets are assumed to be lost and are retransmitted. ARQ has three basic types: stop-and-wait ARQ, go-back-N ARQ, and selective-repeat ARQ.

Stop-and-go ARQ is the simplest ARQ. The transmitter waits for an acknowledgement from the receiver. If a positive acknowledgment (ACK) is received, the next message is transmitted; if a negative acknowledgment (NACK) is received, the message is retransmitted. Packet retransmission continues until a positive acknowledgment is received. The scheme is not efficient, since the channel stays idle between transmissions. Stop-and-go ARQ is very effective in the presence of high error rates. The throughput is very poor, and is given by [64]

$$\eta = \frac{K(1 - P_B)}{L_B + R_s \Delta T} \quad \text{(bits/symbol)}, \tag{15.109}$$

where K is the number of bits per block, L_B is the block length in symbols, P_B is the block error probability, R_s, in symbols/s, is the signaling speed over the channel, and ΔT is the overall round-trip delay.

In case of low error rate, go-back-N ARQ is more efficient. Groups of N packets are transmitted, and each group requires only one acknowledgment. When one or more packets of a group have been received incorrectly, the last N packets are retransmitted. This method eliminates the waiting period to receive acknowledgment for every packet in stop-and-go ARQ. The throughput for go-back-N ARQ is given by [64]

$$\eta = \frac{K(1 - P_B)}{L_B + R_s \Delta T P_B} \quad \text{(bits/symbol)}. \tag{15.110}$$

Selective-repeat ARQ further improves the efficiency of go-back-N ARQ, by retransmitting only those packets that have not been received correctly. The throughput of selective-repeat ARQ is given by [64]

$$\eta = \frac{K}{L_B}(1 - P_B) \quad \text{(bits/symbol)}. \tag{15.111}$$

The performance of an ARQ system is measured by its reliability and its throughput efficiency. The reliability is measured by the probability of undetected errors, $P(E)$.

For a linear code, let the probabilities of a received codeword containing no error, a detectable error pattern, and an undetectable error pattern be P_c, P_d, and P_e, respectively; of course, $P_c + P_d + P_e = 1$. A received codeword is accepted only when it contains no error or an undetectable error pattern, thus the probability of the receiver committing an error is given by

$$P(E) = \frac{P_e}{P_c + P_e}, \tag{15.112}$$

where P_c is determined by the channel. For a BSC with a transition probability p, $P_c = (1 - p)^n$ for a codeword of length n, and P_e can also be calculated for the corresponding code.

Example 15.16: A $(7, 4)$ Hamming code is used for error detection in an ARQ system. This code has $d_{\min} = 3$. There are $2^4 - 1 = 15$ specific error patterns, corresponding to nonzero codewords. Analysis of the weight structure of the code shows that there are seven codes with Hamming weight 3, seven with 4, and one with 7. The occurrence of these error patterns is undetectable, thus the probability of undetectable error is given by

$$P_e = 7p^3(1 - p)^4) + 7p^4(1 - p)^3 + p^7,$$

where p is the BSC error probability. The probability of correct transmission is

$$P_c = (1 - p)^7.$$

Accordingly, the probability of detectable error $P_d = 1 - P_e - P_c$.

The block error probability P_B of the Hamming code is the probability that two or more errors occur during transmission, as the Hamming code can only correct one error. P_B is calculated as 1.0 minus the probability of the occurrence of 0 or 1 error

$$P_B = 1.0 - (1 - p)^n - \sum_{i=1}^{n} p(1 - p)^{i-1}.$$

Hybrid-ARQ

The combination of ARQ and FEC leads to the HARQ scheme. This can eliminate the drawbacks of either scheme. HARQ uses a code that is designed for simultaneous error detection/correction. If the received codeword is detected in error, the receiver will determine whether to correct or request a retransmission according to the error-correction capability of the code. In HARQ, FEC is used to correct the error patterns that occur most frequently. When a less frequent error pattern is detected, a retransmission is requested. This achieves a simple decoder, a high reliability as well as a high throughput.

Unlike in conventional ARQ techniques, where all transmissions are decoded independently, in HARQ all transmissions associated with the same data block are jointly decoded to reduce the probability of decoding error. Compared with ARQ, HARQ reduces the control delay between transmissions by operating HARQ at the physical layer, and advocates efficient and complex signal combining strategies. An optimal combining scheme combines information from all transmissions, without discarding those packets containing errors. For ARQ, a simple one-bit feedback (ACK/NACK) can be sent to the transmitter, when the transmitter operates without CSI.

For type-I HARQ, often called *chase combining*, all retransmissions are identical to the first transmission. The receiver combines the current and all previous HARQ transmissions of the data block before decoding it. In noise-limited cases, MRC can be performed.

For type-II HARQ, also called *incremental redundancy*, the parity-check digits for error correction are transmitted to the receiver only when needed. Two linear codes are used: A high-rate code is for error detection only, while an invertible low-rate code is used for simultaneous error detection/correction. The invertible property of the low-rate code enables reconstruction of the message uniquely from the parity block, and only the parity-check digits are transmitted for error correction.

RCPC codes have an advantage with type-II HARQ protocols. A transmitter sends partial redundancies by puncturing a mother code based on the CSI. If the receiver fails to recover the message, it progressively requests additional redundancies which were previously punctured. For a given code rate, a punctured code usually has a poorer coding gain than the corresponding dedicated code, which is optimized for that rate. Thus, punctured bits must be chosen such that the performance gap between the dedicated and punctured codes is minimized. Likewise, by puncturing bits from a specific mother code, good rate-compatible LDPC codes of finite length are obtained, and they achieve a favorable complexity/performance tradeoff compared to dedicated LDPC codes [32, 41, 75]. Rate-compatible LDPC codes can also be constructed from a mother code by using puncturing and extending [79], or by combining rows of the lowest-rate parity-check matrix to produce the parity-check matrices for higher rates [13].

Both type-I and type-II HARQ schemes provide a significant benefit at low SNR, and type-II HARQ provides higher gain, most prominently for higher code rates [4]. At high SNR, the benefit from HARQ is not significant, as the FEC blocks can generally be decoded without error in the first transmission. Type-II HARQ leads to lower BER and block error rate (BLER) than type-I HARQ.

The selective-repeat type-I HARQ is supported in recent 3G standards such as IEEE 802.16e (WiMAX), 1xEV-DO, HSDPA, HSUPA and 3GPP LTE. These standards also optionally support type-II HARQ.

Problems

15.1 Using the CRC-CCITT polynomial, generate the 16-bit CRC code for a message consisting of two 1s followed by fourteen 0s by using long division.

15.2 Consider the (4,3) even-parity code with the 8 codewords (1 1 0 0), (0 1 1 0), (1 1 1 1), (1 0 0 1), (0 0 0 0), (0 0 1 1), (0 1 0 1), (1 0 1 0). Verify that the code is linear.

15.3 A (10,5) LDPC code has the following parity check matrix $\mathbf{H}$:

$$\begin{bmatrix} 1 & 1 & 0 & 0 & 1 & 1 & 1 & 1 & 0 & 0 \\ 1 & 0 & 1 & 1 & 0 & 1 & 0 & 1 & 0 & 1 \\ 0 & 1 & 0 & 1 & 1 & 0 & 0 & 1 & 1 & 1 \\ 1 & 0 & 1 & 0 & 1 & 0 & 1 & 0 & 1 & 1 \\ 0 & 1 & 1 & 1 & 0 & 1 & 1 & 0 & 1 & 0 \end{bmatrix}.$$

Determine the generator matrix $\mathbf{G}$.

15.4 For the (7, 3) code with generator matrix

$$\mathbf{G} = \begin{bmatrix} 1 & 0 & 0 & 1 & 0 & 1 & 1 \\ 0 & 1 & 0 & 1 & 1 & 1 & 0 \\ 0 & 0 & 1 & 0 & 1 & 1 & 1 \end{bmatrix},$$

determine the correctable error patterns and their syndromes.

15.5 A parity-check code has the parity-check matrix

$$\mathbf{H} = \begin{bmatrix} 1 & 1 & 0 & 1 & 1 & 0 & 0 \\ 1 & 1 & 1 & 0 & 0 & 1 & 0 \\ 0 & 1 & 1 & 1 & 0 & 0 & 1 \end{bmatrix}.$$

Find the generator matrix and all possible codewords.

15.6 Determine the parity check matrix for the (15, 11) Hamming code.

15.7 Show that the (5,2) linear code with nonzero codewords (0 1 1 0 1), (1 0 1 1 1), (1 1 0 1 0) is not cyclic.

15.8 Calculate the probabilities of correct decoding P_c, decoding error P_e, and decoding failure P_f for (a) a (4,3) single-parity-check code, and (b) a (4,1) repetition code.

Figure 15.26

Figure for Problem 15.12

15.9 A block code with parity-check matrix

$$\mathbf{H} = \begin{bmatrix} 0 & 0 & 0 & 1 & 1 & 1 & 1 \\ 0 & 1 & 1 & 0 & 0 & 1 & 1 \\ 1 & 0 & 1 & 0 & 1 & 0 & 1 \end{bmatrix}$$

is used for communication over a BSC with $0 < p < 0.5$.
(a) Find d_{min}.
(b) How many errors can the code correct?
(c) How many errors can be detected?
(d) For each syndrome, give the error pattern that corresponds to the error-correcting capability of the code.
(e) For the received message (0111011), find the codeword based on the maximum-likelihood decision.

15.10 Given a (31, 15) RS code, find the number of bits included in each symbol of this code. What is the minimal code distance? How many symbol errors can be corrected?

15.11 Represent the vector $\mathbf{y} = (11001101011)$ as a polynomial in D. (Hint: $\mathbf{y}(D) = y_0 + y_1 D + \ldots + y_{n-1} D^{n-1}$.)

15.12 An LFSR with high-order input is shown in Fig. 15.26. Determine the contents of the register for the input (1 1 0 1 1 0 0).

15.13 Divide the polynomial $m(D) = 1 + D^5$ by $g(D) = 1 + D + D^3$ to obtain both the quotient and the remainder.

15.14 The generator polynomials for a convolutional code are given as

$$g_1 = 1 + D + D^4, \quad g_2 = 1 + D + D^3 + D^4, \quad g_3 = 1 + D^2 + D^4.$$

(a) What is the constraint length of the code?
(b) Draw the shift-register encoder.
(c) Determine the state number of the trellis diagram of this code.
(d) For message $\mathbf{m}(D) = 1 + D + D^4 + D^5 + D^6 + D^{10}$, what is the encoder output?

15.15 Develop a flow chart of the Viterbi algorithm for software implementation.

15.16 A convolutional code is generated by $g_1 = [1, 0, 1]$, $g_2 = [1, 1, 1]$, $g_3 = [1, 1, 1]$.
(a) Draw the encoder. (b) Draw the state diagram. (c) Draw the trellis diagram for this code.
(d) When the convolutional code is transmitted over an AWGN channel with hard-decision decoding, find the transmitted sequence for the received sequence (101 001 011 110).

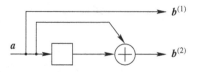

Figure 15.27

Figure for Problem 15.17.

15.17 Given a convolutional encoder shown in Fig. 15.27.
(a) Draw the trellis corresponding to four information digits.
(b) Find the number of codewords represented by the trellis in (a).
(c) When the received sequence is (11 01 10 01) over a BSC with $0 < p < 0.5$, decode the codeword using the Viterbi algorithm.

15.18 Write a program to implement the pseudorandom interleaver given by (15.82). Also, write a program to deinterleave the received bitstream.

15.19 Write a MATLAB program to implement the BCJR algorithm.

15.20 Consider a LDPC code defined as the null space of the parity-check matrix

$$
\mathbf{H} = \begin{bmatrix}
1 & 1 & 0 & 1 & 0 & 0 & 0 \\
0 & 1 & 1 & 0 & 1 & 0 & 0 \\
0 & 0 & 1 & 1 & 0 & 1 & 0 \\
0 & 0 & 0 & 1 & 1 & 0 & 1 \\
1 & 0 & 0 & 0 & 1 & 1 & 0 \\
0 & 1 & 0 & 0 & 0 & 1 & 1 \\
1 & 0 & 1 & 0 & 0 & 0 & 1
\end{bmatrix}.
$$

What is the minimum distance of this code? Plot the Tanner graph. Show that the girth of this Tanner graph is 6.

References

[1] M. A. Abu-Rgheff, *Introduction to CDMA Wireless Communications* (Oxford, UK: Academic Press, 2007).

[2] A. Amastasopoulos, A comparison between the sum-product and the min-sum iterative detection algorithms based on density evolution. In *Proc. IEEE GlobeCom*, Nov 2001, 1021–1025.

[3] J. B. Anderson, T. Eriksson & N. Goertz, On the BCJR algorithm for rate-distortion source coding. *IEEE Trans. Inf. Theory*, **53**:9 (2007), 3201–3207.

[4] J. G. Andrews, A. Ghosh & R. Muhamed, *Fundamentals of WiMAX: Understanding Broadband Wireless Networking* (Upper Saddle River, NJ: Prentice Hall, 2007).

[5] L. Bahl, J. Cocke, F. Jelinek & J. Raviv, Optimal decoding of linear codes for minimizing symbol error rate. *IEEE Trans. Inf. Theory*, **20**:2 (1974), 284–287.

[6] S. Benedetto & G. Montorsi, Unveiling turbo codes: some results on parallel concatenated coding schemes. *IEEE Trans. Inf. Theory*, **42**:2 (1996), 409–428.

[7] S. Benedetto & G. Montorsi, Design of parallel concatenated convolutional codes. *IEEE Trans. Commun.*, **44**:5 (1996), 591–600.

[8] S. Benedetto, D. Divsalar, G. Montorsi & F. Pollara, Serial concatenation of interleaved codes: Performance analysis, design, and iterative decoding. *IEEE Trans. Inf. Theory*, **44**:3 (1998), 909–926.

[9] E. R. Berlekamp, *Algebraic Coding Theory* (New York: McGraw-Hill, 1968).

[10] C. Berrou, A. Glavieux & P. Thitimajshima, Near Shannon limit error-correcting coding and decoding: turbo-codes. *Proc. IEEE ICC*, Geneva, Switzerland, May 1993, 1064–1070.

[11] C. Berrou & M. Jezequel, Non-binary convolutional codes for turbo coding. *Electron. Lett.*, **35**:1 (1999), 39–40.

[12] G. Caire, G. Taricco & E. Biglieri, Bit-interleaved coded modulation. *IEEE Trans. Inf. Theory*, **44**:3 (1998), 927–945.

[13] A. I. V. Casado, W.-Y. Weng, S. Valle & R. D. Wesel, Multiple-rate low-density parity-check codes with constant blocklength. *IEEE Trans. Commun.*, **57**:1 (2009), 75–83.

[14] R. T. Chien, Cyclic decoding procedure for the Bose-Chaudhuri-Hocquenghem codes. *IEEE Trans. Inf. Theory*, **10**:4 (1964), 357–363.

[15] C. Chien, M. B. Srivastava, R. Jain, P. Lettieri, V. Aggarwal & R. Sternowski, Adaptive radio for multimedia wireless links. *IEEE J. Sel. Areas Commun.*, **17**:5 (1999), 793–813.

[16] A. Chinapol & J. A. Ritcey, Design, analysis, and performance evaluation for BICM-ID with square QAM constellations in Rayleigh fading channels. *IEEE J. Sel. Areas Commun.*, **19**:5 (2001), 944–957.

[17] G. Colavolpe, G. Ferrari & R. Raheli, Reduced-state BCJR-type algorithms. *IEEE J. Sel. Areas Commun.*, **19**:5 (2001), 848–859.

[18] U. Dasgupta & C. N. Georghiades, Turbo decoding of quantized data. *IEEE Trans. Commun.*, **50**:1 (2002), 56–64.

[19] S. Dolinar & D. Divsalar, Weight distributions for turbo codes using random and nonrandom permutations. In *TDA Progress Report 42ÍC122*, Jet Propulsion Laboratory (JPL), CA, USA, Aug 1995, 56–65.

[20] P. Elias, Error-free coding. *IRE Trans. Inf. Theory*, **4**:4 (1954), 29–37.

[21] R. M. Fano, A heuristic discussion of probabilistic decoding. *IEEE Trans. Inf. Theory*, **9**:4 (1963), 64–74.

[22] D. Fertonani, A. Barbieri & G. Colavolpe, Reduced-complexity BCJR algorithm for turbo equalization. *IEEE Trans. Commun.*, **55**:12 (2007), 2279–2287.

[23] G. D. Forney, Jr., *Concatenated Codes* (Cambridge, MA: MIT Press, 1966).

[24] G. D. Forney, Jr., Coding and its application in space communications. *IEEE Spectrum*, **7** (1970), 47–58.

[25] G. D. Forney, Jr., Convolutional codes II: Maximum likelihood decoding. *Inf. Control*, **25**:3 (1974), 222–266.

[26] V. Franz & J. B. Anderson, Concatenated decoding with a reduced search BCJR algorithm. *IEEE J. Sel. Areas Commun.*, **16**:2 (1998), 186–195.

[27] R. Gallager, Low-density parity-check codes. *IRE Trans. Inf. Theory*, **8**:1 (1962), 21–28.

[28] R. Garello, P. Pierleni & S. Benedetto, Computing the free distance of turbo codes and serially concatenated codes with interleavers: Algorithms and applications. *IEEE J. Sel. Areas Commun.*, **19**:5 (1995), 800–812.

[29] S. Glisic, *Advanced Wireless Communications: 4G Technologies*, 2nd edn (Chichester, UK: Wiley-IEEE, 2007).

[30] M. J. E. Golay, Notes on digital coding. *Proc. IRE*, **37**:6 (1949), 657.

[31] A. Goldsmith, *Wireless Communications* (Cambridge, UK: Cambridge University Press, 2005).

[32] J. Ha, J. Kim, D. Klinc & S. W. McLaughlin, Rate-compatible punctured low-density parity-check codes with short block lengths. *IEEE Trans. Inf. Theory*, **52**:2 (2006), 728–738.

[33] J. Hagenauer, Rate compatible punctured convolutional codes and their applications. *IEEE Trans. Commun.*, **36**:4 (1988), 389–400.

[34] J. Hagenauer & P. Hoeher, A Viterbi algorithm with soft-decision outputs and its applications. In *Proc. IEEE GLOBECOM*, Dallas, TX, Nov 1989, 1680–1686.

[35] J. Hagenauer, E. Offer & L. Papke, Iterative decoding of binary block and convolutional codes. *IEEE Trans. Inf. Theory*, **42**:2 (1996), 429–445.

[36] R. W. Hamming, Error detecting and error correcting codes. *Bell Syst. Tech. J.*, **29**:2 (1950), 147–160.

[37] H. Jin, A. Khandekar & R. McEliece, Irregular repeat-accumulate codes. In *Proc. 2nd. Int. Symp. Turbo Codes Related Topics*, Brest, France, Sep. 2000, 1–8.

[38] S. J. Johnson, Burst erasure correcting LDPC codes. *IEEE Trans. Commun.*, **57**:3 (2009), 641–652.

[39] S. Kahveci, I. Kaya & K. Turk, Dynamic look-up-table-based maximum *a posteriori* probability algorithm. *Wireless Pers. Commun.*, **46**:3 (2008), 317–328.

[40] M. Karkooti, P. Radosavljevic & J. R. Cavallaro, Configurable, high throughput, irregular LDPC decoder architecture: tradeoff analysis and implementation. In *Proc. IEEE ASAP*, Steamboat Springs, CO, Sep 2006, 360–367.

[41] J. Kim, A. Ramamoorthy & S. W. McLaughlin, The design of efficiently-encodable rate-compatible LDPC codes. *IEEE Trans. Commun.*, **57**:2 (2009), 365–375.

[42] L. Krzymien & C. Schlegel, Turbo decoding with one bit extrinsic quantization. *IEEE Commun. Lett.*, **9**:8 (2005), 732–734.

[43] F. R. Kschischang, B. J. Frey & H. A. Loeliger, Factor graphs and the sum-product algorithm. *IEEE Trans. Inf. Theory*, **47**:2 (2001), 498–519.

[44] B. Levine, R. R. Taylor & H. Schmit, Implementation of near Shannon limit error-correcting codes using reconfigurable hardware. In *Proc. IEEE FCCM*, Napa Valley, CA, Apr 2000, 217–226.

[45] Y. Li & W. H. Mow, Iterative decoding of serially concatenated convolutional codes over multipath intersymbol-interference channels. In *Proc. IEEE ICC*, Vancouver, Canada, Jun 1999, **2**, 947–951.

[46] T. Li, W. H. Mow & K. B. Letaief, Low complexity iterative decoding for bit-interleaved coded modulation. *IEEE Trans. Wireless Commun.*, **5**:8 (2006), 1966–1970.

[47] J. Lin & D. J. Costello, Jr., *Error Control Coding: Fundamentals and Applications*, 2nd edn (Upper Saddle River, NJ: Pearson Prentice Hall, 2004).

[48] M. G. Luby, M. Mitzenmacher, M. A. Shokrollahi, and D. A. Spielman, Improved low-density parity-check codes using irregular graphs. *IEEE Trans. Inf. Theory*, **47**:2 (2001), 585–598.

[49] D. J. C. MacKay & R. M. Neal, Near Shannon limit performance of low density parity check codes. *Electron. Lett.*, **32**:18 (1996), 1645–1646.

[50] D. J. C. MacKay & R. M. Neal, Good error-correcting codes based on very sparse matrices. *IEEE Trans. Inf. Theory*, **45**:2 (1999), 399–432.

[51] J. L. Massey, *Threshold Decoding* (Cambridge, MA: MIT Press, 1978).

[52] J. L. Massey, Foundations and methods of channel coding. In *Proc. Int. Conf. Inf. Theory & Syst.*, NTG-Fachberichte, Berlin, Germany, Sep 1978, **65**, 148–157.

[53] R. J. McEliece, D. J. C. MacKay & J.-F. Cheng, Turbo decoding as an instance of Pearl's 'belief propagation' algorithm. *IEEE J. Sel. Areas Commun.*, **16**:2 (1998), 140–152.

[54] J. E. Meggitt, Error correcting codes and their implementation for data transmission systems. *IRE Trans. Inf. Theory*, **7**:4 (1961), 232–244.

[55] J. Pearl, Fusion, propagation, and structuring in belief networks. *Artif. Intell.*, **29**:3 (1986), 241–288.

[56] J. G. Proakis & M. Salehi, *Digital Communications*, 5th edn (New York: McGraw-Hill, 2008).

[57] R. M. Pyndiah, Near optimum decoding of product codes: block turbo codes. *IEEE Trans. Commun.*, **46**:6 (1998), 1003–1010.

[58] T. J. Richardson, M. A. Shokrollahi & R. L. Urbanke, Design of capacity-approaching irregular low-density parity-check codes. *IEEE Trans. Inf. Theory*, **47**:2 (2001), 619–637.

[59] T. J. Richardson & R. L. Urbanke, Efficient encoding of low-density parity-check codes. *IEEE Trans. Inf. Theory*, **47**:2 (2001), 638–656.

[60] P. Robertson, E. Villebrun & P. Hoeher, A comparison of optimal and sub-optimal MAP decoding algorithms operating in the log-domain. In *Proc. IEEE ICC*, Seattle, WA, Jun 1995, 1009–1013.

[61] P. Robertson, P. Hoeher & E. Villebrun, Optimal and sub-optimal maximum *a posteriori* algorithms suitable for turbo-decoding. *Euro. Trans. Telecommun.*, **8**:2 (1997), 119–125.

[62] C. B. Schlegel & L. C. Perez, *Trellis and Turbo Coding* (Piscataway, NJ: IEEE Press, 2004).

[63] C. E. Shannon, A mathematical theory of communication. *Bell Syst. Tech. J.*, **27** (1948), 379–423, 623–656.

[64] A. U. H. Sheikh, *Wireless Communications: Theory and Techniques* (Boston, MA: Kluwer, 2004).

[65] B. Sklar, A primer on turbo code concepts. *IEEE Commun. Mag.*, **35**:12 (1997), 94–102.

[66] B. Sklar, *Digital Communications: Fundamentals and Applications*, 2nd edn (Englewood Cliffs, NJ: Prentice Hall, 2001).

[67] G. L. Stuber, *Principles of Mobile Communication*, 2nd edn (Boston, MA: Kluwer, 2001).

[68] O. Y. Takeshita & D. J. Costello, Jr., New deterministic interleaver designs for turbo codes. *IEEE Trans. Inf. Theory*, **46**:6 (2000), 1988–2006.

[69] R. M. Tanner, A recursive approach to low complexity codes. *IEEE Trans. Inf. Theory*, **27**:5 (1981), 533–5547.

[70] H. Taub & D. L. Schillings, *Principles of Communications Systems*, 2nd edn (New York: McGraw-Hill, 1986).

[71] S. ten Brink, Rate one-half code for approaching the Shannon limit by 0.1 dB. *Electron. Lett.*, **36**:15 (2000), 1293–1294.

[72] S. ten Brink, Convergence behavior of iteratively decoded parallel concatenated codes. *IEEE Trans. Commun.*, **49**:10 (2001), 1727–1737.

[73] C. Thomas, M. A. Bickerstaff, L. M. Davis, T. Prokop, B. Widdup, G. Zhou, D. Garrett & C. Nicol, Integrated circuits for channel coding in 3G cellular mobile wireless systems. *IEEE Commun. Mag.*, 2003, 150–159.

[74] G. Ungerboeck, Channel coding with multi-level/phase signals. *IEEE Trans. Inf. Theory*, **28**:1 (1982), 55–67.

[75] B. N. Vellambi & F. Fekri, Finite-length rate-compatible LDPC codes: a novel puncturing scheme, *IEEE Trans. Commun.*, **57**:2 (2009), 297–301.

[76] A. J. Viterbi, Error bounds for convolutional codes and an asymptotically optimum decoding algorithm. *IEEE Trans. Inf. Theory*, **13**:2 (1967), 260–269.

[77] A. J. Viterbi, An intuitive justification and a simplified implementation of the MAP decoder for convolutional codes. *IEEE J. Sel. Areas Commun.*, **16**:2 (1998), 260–264.

[78] J. K. Wolf, Efficient maximum-likelihood decoding of linear block codes using a trellis. *IEEE Trans. Inf. Theory*, **24**:1 (1978), 76–80.

[79] M. R. Yazdani & A. H. Banihashemi, On construction of ratecompatible low-density parity-check codes. *IEEE Commun. Lett.*, **8**:3 (2004), 159–161.

16 Source coding I: speech and audio coding

16.1 Introduction

Source coding or data compression is used to remove redundancy in a message so as to maximize the storage and transmission of information. In Chapter 14, we have introduced Shannon's source-coding and rate-distortion theorems. We have also introduced lossless data compression based on the source-coding theorem. In Chapters 16 and 17, we will address speech/audio and image/video coding. Lossy data compression is obtained by quantizing the analog signals, and the performance of quantization is characterized by the rate-distortion bound. Source coding is the procedure used to convert an analog or digital signal into a bitstream; both quantization and noiseless data compression may be part of source coding.

16.1.1 Coding for analog sources

Source coding can be either lossy or lossless. For discrete sources, a lossless coding technique such as entropy coding is used. Huffman coding is a popular entropy coding scheme. Lossless coding uses more radio spectrum. For analog sources, lossy coding techniques are usually used.

PCM is a digital representation of an analog signal. The signal magnitude, sampled regularly at uniform intervals, is quantized to a series of symbols in a digital, usually binary code. This can be performed by using A/D converters. The demodulation of PCM signals can be performed by DACs. The PCM code is the original waveform for source coding. There are three approaches to source coding.

Waveform coding

The waveform coder uses fewer bits to encode the original PCM waveform. Increasing the bit rate will asymptotically lead to lossless transcoding of the original PCM waveform. A coding error signal can be defined as the difference between the original and decoded waveforms. Waveform coding can be either temporal or spectral waveform coding. These coders typically work at a bit rate of 16 kbits/s and above.

Temporal waveform coding is the most fundamental source coding technique. A temporal PCM waveform can be encoded by using the differential PCM (DPCM), adaptive PCM, delta modulation (DM), or continuous variable slope DM (CVSDM) technique.

Spectral waveform coding can be either sub-band coding or adaptive transform coding. In sub-band coding, the signal is divided into a few sub-bands and the temporal waveform in each sub-band is encoded separately. For speech signals, the low-frequency bands contain more energy and also the human ear is more sensitive to low-frequency bands; thus lower-band signals should be assigned more bits. The QMF [35] is generally used. The lower band is repeatedly subdivided by a factor of two, each subdivision using a pair of QMFs. The signal in each sub-band can be encoded by an adaptive PCM algorithm.

Transform coding can be treated as a kind of sub-band coding. In adaptive transform coding, the signal is sampled and subdivided into frames of a number of samples. Each frame is transformed into the frequency domain for coding and transmission. The Karhunen-Loève transform is optimum since it generates uncorrelated spectral values [10]. DFT and DCT also yield good performance. DCT achieves a performance very close to that of the Karhunen-Loève transform, and is generally used for speech and image coding [5]. DWT is a subset of sub-band decomposition, in which the transformed representation provides an intrinsic multiresolution data structure. All these transform coding techniques are used for speech and image coding. Due to the Gibbs phenomenon in linear filters, DCT and DWT introduce artifacts, called the *ringing effect*, at high compression factors (around 50).

Model-based coding

Model-based coders use an explicit source model with a small set of parameters. Such coders can achieve a very low rate transmission. The source is modeled as a linear filter; instead of transmitting the samples of the source waveform to the receiver, the filter parameters as well as an appropriate excitation signal are transmitted. This leads to a large compression ratio. Since the decoded waveform is not synchronized with the input waveform, the waveform error signal is not suitable for characterizing the performance of the model-based coder.

Linear predictive coding (LPC) and mixed excitation linear prediction (MELP) are two model-based coding techniques. The model-based speech coding techniques are generally called *vocoders (voice coders)*. For speech coding, the model-based coding approach generates a very low bit rate.

Hybrid coding

Hybrid coding combines the waveform and model-based coding techniques. The residual waveform, which is the prediction error in a predictive signal model, is transmitted using a waveform coder, while the model parameters are quantized and transmitted as side information. The code-excited linear prediction (CELP) vocoder and its many variants, including the residual excited linear prediction (RELP), the multipulse LPC, and the vector-sum excited linear prediction (VSELP) vocoders, are hybrid coding techniques. This class of vocoders dominates medium-bit-rate speech coding.

16.2 Quantization

Quantization is applied to waveform samples or model parameters. It can be scalar quantization or vector quantization (VQ). For scalar quantization, each of the multiple variables is quantized independently of one another. VQ is an extension of scalar quantization to multi-dimensions. VQ combines d variables and jointly quantizes them to a single value (codeword). At the decoder, each codeword is restored to a prototype vector.

16.2.1 Scalar quantization

Given a random variable x with a known pdf $p_x(x)$, we find a set of M quantization levels that minimizes the MSE distortion

$$D = E\left[(x - \hat{x})^2\right] = \int_{-\infty}^{\infty} p_x(x)\,(x - \hat{x})^2\,dx, \tag{16.1}$$

where $\hat{x}$ is one of the M quantization levels $\hat{x}_i$, and

$$(x - \hat{x})^2 = \min_i (x - \hat{x}_i)^2. \tag{16.2}$$

The optimum solution is the Lloyd-Max algorithm. The performance measure of the quantizer is SQNR.

The different quantization levels are collectively known as the *codebook*, and each quantization level is called a *codeword*. The codebook can be designed based on the Lloyd-Max algorithm by using a set of training data. For the Lloyd-Max quantizer, the codewords are the centroids over the decision intervals. Each parameter to be quantized is compared with a codebook where the codewords are stored, and the codeword closest to the parameter is selected as the quantized value. The scalar quantizer is illustrated in Fig. 16.1a. Scalar quantization has been treated in more detail for the uniform quantizer in Subsection 12.3.

Scalar quantization is generally used in speech coders standardized prior to 1994. It is used in LPC-based vocoder standards such as TIA IS54 VSELP, ETSI GSM RPE-LTP (regular pulse excitation with long-term prediction), and FS1015 LPC, as well as in waveform codecs such as PCM and adaptive differential PCM (ADPCM).

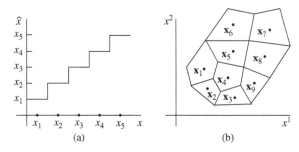

Figure 16.1 Scalar quantization and vector quantization. (a) Scalar quantizer (1-D). (b) Vector quantizer (2-D).

16.2.2 Vector quantization

The vector quantizer maps an input vector signal onto a finite set of reproduction vectors, collectively known as the *codebook*. For a multi-dimensional pdf $p(x)$, the MSE distortion is defined as

$$D = E\left[\|x - \hat{x}\|^2\right] = \int_{-\infty}^{\infty} (x - \hat{x})^T (x - \hat{x}) p(x)dx, \tag{16.3}$$

where $\hat{x}$ is one of the M quantization levels $\hat{x}_i$, and

$$\|x - \hat{x}\|^2 = \min_i \|x - \hat{x}_i\|^2. \tag{16.4}$$

The optimum solution can be derived in the same manner as that for the Lloyd-Max quantizer. The codewords are the centroids over the decision regions. VQ for the two-dimensional case is also illustrated in Fig. 16.1b.

When the joint pdf is obtained from a finite set of input vectors, the C-means algorithm is obtained as the optimal VQ algorithm, where C is the number of codewords. The Linde-Buzo-Gray (LBG) or *generalized Lloyd algorithm* is a batch implementation of C-mean. VQ is very powerful and it can produce results very close to the theoretical limit. Compared to scalar quantization, codebook generation for VQ is computationally very complex. Dozens of online C-mean and VQ algorithms that are based on competitive learning have been described in [10].

For a given distortion, the size of the codebook can be significantly reduced, compared to scalar quantization. This leads to a lower number of bits for coding. This advantage is obtained by exploiting the correlation among different components of the vectors. Therefore, a transformation of the vector that reduces this correlation decreases the advantage of VQ. In addition, VQ provides flexibility in partitioning high-dimensional space.

Shannon showed that VQ with an increasing d can achieve the rate-distortion bound for source coding [27]. VQ is now being used in most of the recent vocoders for quantizing the linear prediction coefficients and in image/video codecs.

Some VQ strategies

Multistage VQ of a vector x has K codebooks. In the encoder, at each stage, the input vector x is compared with

$$\hat{x} = \sum_{1}^{k} y_{i_k}^{(k)} \tag{16.5}$$

for $k = 1, 2, \ldots, K$, where $y_{i_k}^{(k)}$ is the i_kth codevector from the kth stage codebook. The principle is very similar to that of the SAR A/D converter. The output is an index set $\{i_1, i_2, \ldots, i_K\}$ that minimizes the distance between x and $\hat{x}$. Multistage VQ and its codebook design are discussed in [7].

A predictive VQ scheme can be used to eliminate the correlation between consecutive vectors that are to be quantized, as an extension of the DPCM in the scalar case. Other popular VQ methods are split VQ and conjugate VQ. Split VQ simply splits the vector into n subvectors, so that the search of each subvector is performed independently in its codebook. Split VQ has n codebooks. The conjugate VQ can be treated as a special two-stage VQ. The two codebooks are conjugate in some sense [7]. Conjugate VQ is effective to reduce the quantization distortion for a noisy channel.

16.3 Speech production and auditory systems

16.3.1 Speech production

Through speech, information is communicated from a speaker to a listener. Speech consists of acoustic sound waves produced by the human speech production system, which consists of the lungs, trachea (windpipe), larynx (organ of voice production), vocal tract, and nose. The vocal tract is comprised of the oral cavity from the larynx to the lips (the throat and mouth); the nose as well as the nasal passage is often called the *nasal tract*. The vocal tract takes on different lengths and cross-sections during speech activity. It has an average length of 17 cm in a typical adult male and is shorter for females, and the varying cross-section can be up to $20 \, \text{cm}^2$.

Speech signals are nonstationary, and they can be considered quasi-stationary over a very short period, typically 5–20 ms. Speech can be voiced (e.g., /a:/, /e/), unvoiced (e.g., /f/, /sh/), or mixed. Voiced speech is quasi-periodic in the time domain and has a harmonic structure in the frequency domain, while unvoiced speech is more like noise and with a broad bandwidth. The vocal cord does not participate in the voice production of unvoiced speech. The unvoiced sounds can be classified as *fricatives* (e.g., /th/), *plosives* (e.g., /p/, /t/), and *whispered* (e.g., /h/). The mixed sounds can be *voiced fricatives* (e.g., /z/) or *voiced plosives* (e.g., /b/). The energy of the voiced speech segments is generally higher than that of the unvoiced segments.

Spectrograms

A speech waveform consists of a sequence of sounds. A single Fourier transform of the entire waveform cannot capture its time-varying frequency content. The short-time Fourier transform is a Fourier transform of pieces of the waveform under a sliding window. The spectrogram is a graphical representation of the magnitude of the time-varying spectrum, and is given by

$$S(\omega, \tau) = |X(\omega, \tau)|^2, \tag{16.6}$$

where $X(\omega, \tau)$ is the short-time Fourier transform of the speech signal $x(n)$. There are two kinds of spectrograms: *narrowband* and *wideband*, arising from the length of the window. Narrowband spectrograms give good spectral resolution, while wideband spectrograms give good temporal resolution. A narrowband spectrogram is obtained by using a long window, typically with a duration of at least two pitch periods, while a wideband spectrogram uses a window with a duration less than a single pitch period.

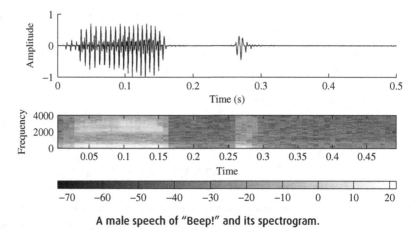

Figure 16.2

A male speech of "Beep!" and its spectrogram.

Example 16.1: A male speech of "Beep!" and its spectrogram are shown in Fig. 16.2. Here $f_s = 8000$ samples/s, the FFT length is 256. A Hanning window of 128 is employed. Peaks in the spectrum appear as dark red horizontal bands, which correspond to formant resonances.

Speech production modeling

Human speech production can be modeled by three separate sections, namely the source excitation, vocal-tract shaping, and the effect of speech radiation

$$S(z) = U(z)H(z)R(z), \tag{16.7}$$

where $U(z)$ denotes the voice waveform, $H(z)$ the dynamics of the vocal tract, and $R(z)$ the radiation effect. The vocal-tract transfer function can be modelled by an all-pole model

$$H(z) = \frac{1}{1 - \sum_{i=1}^{M} a_i z^{-i}}. \tag{16.8}$$

The source excitation can be $u(n) = \sum_{q=-\infty}^{\infty} \delta(n - qT)$, T being the peak time of the pulse, for the voiced case, and be a zero mean, unitary variance, uncorrelated noise for the unvoiced case.

Note that IIR filters are more popular for speech synthesis since the human auditory system is not sensitive to phase information. The vocal tract can be modeled by a series of uniform lossless acoustic tubes [9]. To approximate this vocal tract model using an all-pole synthesis filter, the delay should be at least twice the time for the sound travel along the tract. Given a vocal tract of 17 cm long, the sound speed of 340 m/s, and a sampling rate of 8 kHz, the corresponding filter order should be at least 8, which corresponds to 8 taps of 0.125 ms, or 1 ms. The sound propagation in the vocal tract during speech production is usually modeled by representing the vocal tract as a concatenation of short lossless uniform acoustic tubes.

Pitch and formants

Voiced speech is characterized by a set of resonant frequencies, which depend on the shape and physical dimensions of the vocal tract. The nominal center frequencies of the resonances are referred to as *formants* by speech scientists, since they form (or shape) the spectrum. The first three formants fall below 3000 Hz for adult speakers. For most vowels, the formant frequency is near 2600 Hz, and the first two formants can be used to distinguish vowels.

For voicing, the rate of vibration is called the *fundamental frequency of phonation*, also known as *pitch*. The pitch for men is typically in the range of 50–250 Hz, while for women and children the range is from 120–500 Hz [7, 9]. The period of pitch for a male is between 4 and 20 ms, while for a female it is 2 to 8 ms [7]. Everyone has a habitual pitch level, and the pitch is shifted up and down in speaking due to stress, intonation, and emotion. Pitch and formants are two important parameters for speech analysis, synthesis, and coding. The human ear is very sensitive to pitch estimation errors, but is relatively tolerant to errors in formant frequencies.

The basic unit for describing how speech conveys linguistic meaning is called a *phoneme*. Vowels are voiced speech sounds and are normally among the phonemes of largest amplitudes. Vowels can be distinguished by the location of the first three formant frequencies. An overview of the fundamentals of speech science is given in [9].

Most speech codecs are bandlimited to the conventional 300 to 3400 Hz telephone band, sampled at 8 kHz, and then converted to 16-bit linear PCM for further processing. All vocoders attempt to model the speech generation process as a dynamic physical system.

16.3.2 Psychoacoustics

Human hearing system

The auditory system is composed of the outer ear, middle ear, and inner ear. Sounds enter the eardrum through the auditory canal in the outer ear, resulting in vibrations of the eardrum. The eardrum is connected to the middle ear. The middle ear is composed of three interconnected bones, which transform the vibrations of the eardrum to the inner ear by the oval window. The inner ear has a cochlea, which is a coiled, snail-like tube and is filled with fluid. Vibrations of the eardrum result in a compression sound wave in the cochlear fluid via the oval window. The compression wave causes a vertical vibration of the basilar membrane. Different regions of the basilar membrane respond maximally to different frequencies. The cell bodies embedded within the membrane will cause a chemical reaction that causes a firing of short-duration electrical pulses in the nerve fibers. The electrical pulses finally reach the auditory processing region in the brain along the auditory nerve, and are perceived as sound.

The human ear can hear sounds in the range of 20 Hz to 20,000 Hz. In the Human auditory system, one sound may mask the perception of another sound. Masking is a prominent feature of human audio perception. Masking can be spectral or temporal. This can be used to reduce the bit rate in a compression scheme. The human ear is sensitive only to the magnitude spectrum of the speech, and not to the phase. Retaining some phase information, however, adds naturalness to the synthetic speech.

Sound intensity and loudness

The human hearing system can sense a sound pressure ranging from 10^{-5} to $100\,\text{Pa}$ (1 Pa $= 1\,\text{N/m}$). This range is usually defined as the sound pressure level (SPL)

$$\text{SPL} = 10\log_{10}\left(\frac{p}{p_0}\right)^2 \quad (\text{dB}), \tag{16.9}$$

where $p_0 = 20\,\mu\text{Pa}$ is roughly the sound pressure at the hearing threshold at frequencies around 2 kHz [3, 36]. The sound intensity I is proportional to p^2.

The hearing sensation is the loudness of the sound. The loudness of an audio signal depends on its duration, its temporal and spectral structure, as well as its intensity. A high frequency tone at a lower intensity level may have the same loudness as a low frequency tone at a higher intensity level. The loudness level of a test tone is defined as the loudness of the sound relative to the level of a 1-kHz sound tone. It is determined by adjusting the level of a 1-kHz tone until it sounds equally as loud as the test stimulus, and the level above the threshold of the 1-kHz tone is the loudness level of the test stimulus. The unit of loudness level in decibels is the *phon*.

The unit of loudness L, the *sone*, is defined as the loudness of a binaural 1-kHz tone at 40-dB sound pressure level above the hearing threshold. Above the hearing threshold, the loudness L and intensity I are related by [3]

$$L \propto I^{0.3}. \tag{16.10}$$

That is, the loudness is doubled when the intensity is increased tenfold.

Hearing threshold

The hearing threshold is the lowest sound level that can be heard at a given frequency. The hearing threshold for 20 year old subjects can be approximated by [31]

$$\text{SPL} = 3.64\left(\frac{f}{1000}\right)^{-0.8} - 6.5e^{-0.6\left(\frac{f}{1000}-3.3\right)^2} + 10^{-3}\left(\frac{f}{1000}\right)^4 \quad (\text{dB}). \tag{16.11}$$

This is plotted in Fig. 16.3. For frequency below 2 kHz, the threshold is independent of age. Above 2 kHz, the curve shifts upward as the age grows. Given the limit of human hearing, more hearing loss at higher frequency occurs with age.

This curve is extremely important for audio coding since frequency components in a signal that are below this level do not need to be transmitted. The quantization noise in frequency components cannot be perceived by the human hearing system, as long as it is below the hearing threshold.

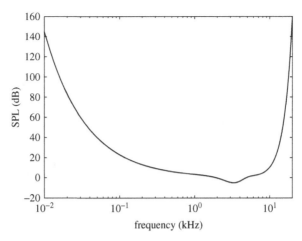

Figure 16.3

The hearing threshold for 20 year old subjects.

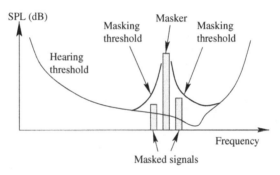

Figure 16.4

Frequency masking: A loud signal masks signals at nearby frequencies.

Masking phenomenon

Masking of soft sounds by louder ones is embodied in human hearing. Masking can be either frequency or temporal masking. Temporal masking is the dominant effect for sounds with transients, while frequency masking is dominant in steady state conditions.

Given a masker at a frequency, a masking threshold is generated for those frequencies at both sides of the masker. This is illustrated in Fig. 16.4. Thus, frequency masking temporarily raises the hearing threshold in the spectrum near the masker. These maskees are not perceived by the human hearing even if they are well above the hearing threshold. The masking threshold in a critical band is given in [16].

Temporal masking corresponds to masking that occurs prior to and after the presence of the masker. Pre-masking is unexpected since it occurs before the masker is switched on, while post-masking has a stronger effect and has a much longer duration. Premasking tends to last only about 5 ms, whereas postmasking extends anywhere from 50 to

300 ms, depending upon the strength and duration of the masker [36]. A good monograph on psychoacoustics is [36].

16.4 Speech/audio quality

The audio or visual quality of a coding system can be examined by subjective or objective methods. Subjective methods are based on listening and viewing tests, by comparing the original and processed speech/video data. Since speech and videos are perceived by the human auditory and visual systems, listening and viewing tests are very reliable in the evaluation of codecs. However, their cost is very high, it is time consuming, and normally they give little insight for the improvement of a coding algorithm.

Objective methods can quantify performance over a variety of scales, and can assist in the improvement of a coding algorithm, but its usefulness lies in its correlation with subjective testing. Objective measures provide a quantitative, repeatable, and accurate means of codec performance evaluation. These measures are only meaningful for comparing similar algorithms applied to the same original speech or image. They provide mathematical tractability. However, for speech and images, the human auditory and visual perception features are also widely used to compare the performance of compression algorithms, and thus are used for compression. The human auditory and visual systems depend on the frequency, and this makes sub-band coding popular. Usually, objective measures are used at the development stage of a speech or video coding algorithm, while subjective measures are used to verify the final design.

To evaluate the quality of speech/audio codecs, a test vector can be generated from the TIA acoustic database [32]. Both speech and noise files of the database have been preprocessed by a modified-IRS (intermediate reference system) filter to emulate the transfer function of the telephone network. They are sampled at 8 kHz and linear quantized in 16 bits per sample. The database assumes two types of background noise: street and car noise. Clean speech is mixed digitally with the same length of noise with different SNRs.

Noise cancellation may be required prior to speech/audio coding. For audio signals, due to the wide variation in the amplitude, median filter or other outlier elimination techniques [10], which are very effective for eliminating impulse noise in images, yield poor performance. The median filter may mistake a true signal as impulse noise. For speech signals, impulse noise removal can be performed by using linear prediction models. Active noise cancellation, which generates a controllable secondary source to compensate a noise in a certain spatial region, can also be applied [34].

16.4.1 Subjective quality measures

The most reliable and most popular performance measure is the mean opinion score (MOS), which is a subjective evaluation and is defined in ITU-T Rec. P.800. Listeners

Rating	Speech quality	distortion level
5	Excellent	Imperceptible
4	Good	Perceptible but not annoying
3	Fair	Perceptible and slightly annoying
2	Poor	Annoying but not objectionable
1	bad	Very annoying and objectionable

Table 16.1. MOS five-point scale.

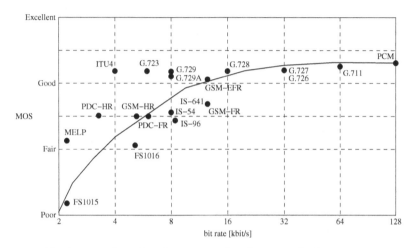

Figure 16.5   Subjective speech quality of some speech codecs. From [12] ©IEEE, 2001

are requested to group the quality of the encoded/decoded speech into five categories, that is *excellent*, *good*, *fair*, *poor*, or *bad*, with values ranging from 5 to 1. The MOS five-point scale is given in Table 16.1.

The MOS for the coder is the average of all the numeric ratings. If the variance of the MOS value is too large, the test is unreliable. A value between 4.0 to 4.5 is considered toll quality, such as for the ITU-T G.711 standard.

Other subjective measures include the diagnostic rhyme test (DRT) and diagnostic acceptability measure (DAM). DRT measures intelligibility, and DAM provides a more complete quality evaluation.

Subjective speech quality comparison of speech codecs

A comparison of the subjective speech quality of some of the speech codecs has been made in [12], and is illustrated in Fig. 16.5. These codecs are described in the following sections. In the figure, ITU4 denotes the future 4 kbits/s ITU codecs. The target performance for ITU4 is also shown. Note that all ITU codecs achieve an MOS of above 4.0.

16.4.2 Objective quality measures

Objective quality assessment is based on a mathematical comparison of the original and processed signals. The distortion introduced by the compression process can be characterized by the MSE or the mean absolute difference (MAD, or MAE, mean absolute error) between the original $\{s_n\}$ and reconstructed $\{\hat{s}_n\}$ sequences. The MSE is most widely used, and is defined by

$$\sigma_d^2 = \frac{1}{N} \sum_{n=1}^{N} \left(s_n - \hat{s}_n\right)^2, \tag{16.12}$$

where N is the length of the sequence.

SNR

SNR is a simple and common measure for evaluating the performance of a compression algorithm based on waveform coding. It is defined by

$$\begin{aligned} \text{SNR} &= 10 \log_{10} \frac{\sigma_s^2}{\sigma_d^2} \\ &= 10 \log_{10} \frac{\sum_{n=0}^{M-1} s^2(n)}{\sum_{n=0}^{M-1} \left[s(n) - \hat{s}(n)\right]^2} \quad \text{(dB)}, \end{aligned} \tag{16.13}$$

where $s(n)$ and $\hat{s}(n)$ are the original and decompressed signals, respectively. This measure is for long-term characterization of a codec, with very large M.

The SNR measure is unable to give equal weighting to high- and low-energy speech segments. Instead, its value is dominated by the SNR of the higher-energy voiced speech segments. The SNR measure is often used for waveform coding schemes such as the PCM and ADPCM.

SNR is a poor estimator for a broad range of speech distortions, since it is not related to any subjective attribute of speech quality and because the method weights all the time-domain errors in the waveform equally [9]. In addition, since the human auditory system is insensitive to phase distortion, encoders based on the speech spectrum focus only on the magnitude of the speech spectrum. As a consequence, although the decoded time-domain speech is perceived to be very similar by the listener, it can be quite different from the original, making the SNR measure meaningless.

Segmental SNR

Segmental SNR (SEGSNR) is a frame-based measure that characterizes the time-varying speech by averaging over a number of N-point speech segments

$$\text{SEGSNR} = \frac{1}{L} \sum_{i=0}^{L-1} \text{SNR}_i$$

$$= \frac{10}{L} \sum_{i=0}^{L-1} 10 \log_{10} \frac{\sum_{n=0}^{N-1} s^2(iN+n)}{\sum_{n=0}^{N-1} \left[s(iN+n) - \hat{s}(iN+n) \right]^2} \quad \text{(dB)}, \quad (16.14)$$

where L is the number of frames, each of length N.

Compared to the SNR measure, SEGSNR gives a fairer weighting to low-energy unvoiced segments by computing the geometric mean of the SNR values instead of the arithmetic mean. Thus, the SEGSNR measure penalizes more the coders with varying performance. It correlates better with subjective speech quality measures.

Problems can arise when using the SEGSNR measure if frames of silences are included, since this leads to large negative SEGSNR. The silent periods can be detected and then excluded from the SEGSNR calculation. Another method to eliminate the influence of silence is to set a lower threshold for all frames with lower SEGSNR. At the other extreme, frames with SEGSNR greater than 35 dB cannot be perceived by listeners as being very similar, and thus an upper threshold of 35 dB can be used to replace any larger SEGSNR values.

Like SNR, SEGSNR is meaningful only for waveform coders. Both SNR and SEGSNR are extremely sensitive to waveform alignments and phase distortions. For most vocoders, the original waveform cannot be preserved after synthesis. SNR and SEGSNR are thus not applicable, since they do not account for the perceptual properties of the human ear. For example, the 32 kbits/s ADPCM codec has a SEGSNR of about 28 dB, while the 13 kbits/s GSM RPE-LTP codec has a SEGSNR of about 16 dB, but both have a MOS of about 4.0 [12].

Other measures

Frequency-domain codecs can be best characterized in terms of spectral distortion between the original and processed speech signals. The articulation index (AI) measure splits the speech band into 20 sub-bands and computes the sub-band SNRs, and takes the average of the sub-band SNRs. The cepstral distance (CD) measure is highly correlated with subjective measures. It is defined in terms of the cepstral coefficients of the original and processed speech signals

$$\text{CD} = \left[\left(c_0^{\text{in}} - c_0^{\text{out}} \right)^2 + 2 \sum_{i=1}^{\infty} \left(c_i^{\text{in}} - c_i^{\text{out}} \right)^2 \right]^{1/2}, \quad (16.15)$$

where the cepstral coefficients of the input and output speech can be obtained from the LPC coefficients a_i [12].

Research on perceptual objective measurement has made much progress and perceptual objective measurement has become an important complement to the assessment of audio codecs. Perceptual objective measurement predicts the basic audio quality by incorporating psychoacoustic principles. The PEAQ (perceptual evaluation of audio quality) measurement is an objective measurement that is conducted in conjunction with formal listening tests, and is adopted by ITU as ITU-R BS.1387.

The PESQ (perceptual evaluation of speech quality) algorithm was standardized in 2001 by the ITU's Telecommunication Standardization Sector (ITU-T) as Rec. P.862, which is suitable for end-to-end measurement of telecommunication networks. PESQ demonstrates high superiority over other perceptual objective measures for various wireless, fixed networks and VoIP (voice over IP) codecs, under a variety of background noise and language conditions. For 22 known ITU-T benchmark tests, 69.2% and 91.3% of the prediction errors are within 0.25 and 0.50 MOS, respectively. The average correlation between PESQ and the subjective MOS is 0.935 [15]. PESQ is not suitable for evaluation of noise-reduction algorithms, since it is not designed to evaluate loudness loss.

PEAQ and PESQ are intrusive methods that use a listening-only model. Intrusive tests inject a reference signal into the network under test, so as to produce a MOS rating by comparing it with the signal that is being tested. Other factors related to two-way conversations, such as delay and echo are not considered. Thus, it is possible to have a high objective MOS, but the overall speech quality could still be poor.

Nonintrusive methods need no reference signal. They are not as accurate as their intrusive counterparts, but they are useful where the reference signals are inaccessible, such as in monitoring live network speech quality. Nonintrusive methods are based on an analysis of degraded speech signals, voice packet header information, or voice and network parameters. Nonintrusive methods are quite challenging because of the absence of a reference signal. ITU-T Rec. P.563, standardized in 2004, is the first ITU-T single-ended method for predicting the subjective quality in narrowband telephone networks. P.563 generates a subjective MOS. It reconstructs a pseudo reference, and this complicates implementation compared to PESQ, though they have similar application coverage. For 24 known ITU-T benchmark experiments, the average correlation between P.563 and subjective MOS scores was 0.88 [14].

Many subjective and objective measures are described in [9].

16.5 Speech coding

For a raw sound file, which is pure binary recordings of eight-bit input data, experiments show that simple Huffman coding produces more compression than LZSS dictionary coding in all the simulated cases [21]. This is because LZSS is effective when there are repeated strings of characters in the file, while Huffman coding is effective in case of overall frequency differences for individual sequences. In typical sound files, there are some overall frequency differences between the various codes, but not as many repeated strings.

When applying lossy compression to speech coding, a lossless stage usually follows. This is because lossy compression frequently smooths out the data, making it more suitable for lossless compression.

Major standard bodies for standardizing speech coders are ITU-T, the TIA of the ANSI, and the ETSI. For each standard, a reference source code, most commonly in the floating-point C language, is usually provided with the standard for the encoder and the decoder. The reference source code is very general, and can be modified for a specific platform. The developer still needs to transform it into fixed-point C code or DSP assembly code. A set of test vectors are also provided for verification of the implementation. Various speech coders used in 2G wireless standards and their resource requirements are listed in [17].

A classic tutorial on speech coding is given in [28], and two excellent texts on speech coding and speech standards are [7] and [12].

16.5.1 Logarithmic PCM coding

Uniform quantization used in PCM is undesirable when signal amplitudes are not uniformly distributed over the entire dynamic range of the signal. In this case, a nonuniform quantizer provides a better solution. The amplitude of a signal is first compressed using a nonlinear device, and the compressed signal is then subjected to a uniform quantizer. At the receiver, the inverse of the nonlinear element used in the transmitter is used to restore the signal.

The speech signal is band-limited to between 300 Hz and 3400 Hz. For digital telephony, analog speech is sampled by using the PCM format at a sampling frequency of 8 kHz. Due to the large dynamic range of the human voice, a resolution of 12 or 13 bits is required in a telephone conversation. Each sample is quantized into 12- or 13-bit linear PCM format, which is encoded by a logarithmic compression system, to 8 bits; this gives a 64 kbits/s digital signal known as *DS0 (Digital Signal 0)*. Depending on the logarithmic relation, we get either the μ-law (mu-law) PCM (North America and Japan) or the A-law PCM (Europe and most of the rest of the world). This nonlinear quantization method is called *companding (compressing-expanding)*. This system is described by ITU-T G.711. The MOS speech quality is 4.3 out of 5.

The A-law is defined by

$$F(x) = \begin{cases} \text{sgn}(x)\frac{1+\ln(A|x|)}{1+\ln A}, & 1/A < |x| < 1 \\ \text{sgn}(x)\frac{A|x|}{1+\ln A}, & 0 < |x| < 1/A \end{cases}, \tag{16.16}$$

where $x = \frac{v}{v_{\text{peak}}}$ is the normalized input, and sgn(x) is the signum function. A is commonly adopted as 87.6, and this yields a 24 dB improvement on SQNR over linear PCM for small signals ($|x| < 1/A$) and a relatively constant SQNR of 38 dB for large signals ($|x| > 1/A$). The method compresses 13-bit linear PCM into 8-bit companded PCM. The A-law is normally implemented into a 13-segment piecewise linear approximation. The 8-bit PCM specifies the location on the segment.

The μ-law is similar to the A-law. It slightly improves SQNR for voice signals compared to the A-law, but with a slightly reduced dynamic range. The μ-law is defined by

$$F(x) = \text{sgn}(x)\frac{\ln(1 + \mu|x|)}{\ln(1 + \mu)}, \tag{16.17}$$

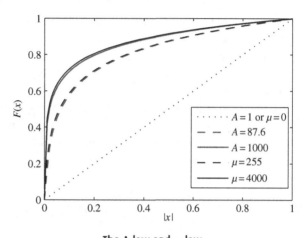

Figure 16.6

The A-law and μ-law.

where $\mu = 255$ is selected in G.711, resulting in about 24 dB improvement on SQNR over linear PCM. The method compresses 12-bit linear PCM into 8-bit companded PCM. The μ-law uses 15-segment piecewise linear approximation.

Example 16.2: The A-law and μ-law are plotted in fig. 16.6. For $\mu = 0$ or $A = 1$, there is no compression. The two laws are almost identical when $A = 87.6$ and $\mu = 255$. They are also very close when $A = 1000$ and $\mu = 4000$.

The logarithmic compression achieves a very desirable property: It ensures a constant SNR across the signal's dynamic range, regardless of the shape of the pdf of the signal [12]. That is, a large signal may have a large error, while a small signal must have a small error. The SNRs for the A-law and μ-law companders are, respectively, given by [12]

$$\text{SNR}_\mu = 6.02R + 4.77 - 20\log_{10}(\ln(1 + \mu)) \quad \text{(dB)}, \tag{16.18}$$

$$\text{SNR}_A = 6.02R + 4.77 - 20\log_{10}(1 + \ln A) \quad \text{(dB)}. \tag{16.19}$$

16.5.2 Linear prediction analysis and synthesis

LPC compresses speech by modeling the vocal tract that produces the speech. LPC is basically an auto-regressive model. The minimum-phase filter is selected to be an all-pole filter. This selection uses poles to shape its frequency response. These poles decide the frequency peaks, and human perception is more sensitive to spectral peaks than to valleys. The signal model can thus be written, in the z-transform domain, as

$$X(z) = \frac{U(z)}{A(z)}, \tag{16.20}$$

where $X(z)$ denotes the speech signal, $U(z)$ the excitation signal, and $1/A(z)$ the filter. The equivalent representation in the time domain is given by

$$x(n) = -\sum_{i=1}^{m} a_i x(n-1) + u(n), \tag{16.21}$$

where m is the order of the filter or predictor, and a_i, $i = 1, 2, \ldots, m$, are known as the *LPC coefficients*, and $a_0 = 1$. For speech sampled at 8 kHz, m is selected as 10.

For each speech frame, a_i's are estimated, and the prediction error signal is given by

$$e(n) = x(n) - \hat{x}(n) = \sum_{i=1}^{m} (\hat{a}_i - a_i) x(n-i) + u(n). \tag{16.22}$$

For uncorrelated excitation $u(n)$, the quadratic function of the prediction error signal is minimized when $\hat{a}_i = a_i$, $i = 1, 2, \ldots, m$, leading to $e(n) = u(n)$. For short-time stationary speech, for every frame with N samples, a window of $L \geq N$ samples is used, and there is overlap between windows. Minimizing the windowed squared error function yields a system of linear equations for $\hat{a}_i$.

Estimation of the LPC coefficients is based on the minimization of the MSE criterion

$$J = E\left[e^2(n)\right] = E\left[\left(x(n) - \sum_{i=1}^{m} a_i x(n-i)\right)^2\right]. \tag{16.23}$$

The optimal LPC coefficients can be found by setting $\frac{\partial J}{\partial a_k} = 0$, for $k = 1, 2, \ldots, m$. This leads to

$$\sum_{i=1}^{m} a_i R_x(i-k) = -R_x(k), \quad k = 1, 2, \ldots, m, \tag{16.24}$$

where the autocorrelations $R_x(i-k) = E[x(n-i)x(n-k)]$ and $R_x(k) = E[x(n)x(n-k)]$.

In matrix form, we obtain the normal equation

$$\mathbf{R}_x \mathbf{a} = -\mathbf{r}_x, \tag{16.25}$$

where

$$\mathbf{R}_x = \begin{pmatrix} R_x(0) & R_x(1) & \cdots & R_x(m-1) \\ R_x(1) & R_x(0) & \cdots & R_x(m-2) \\ \vdots & \vdots & \ddots & \vdots \\ R_x(m-1) & R_x(m-2) & \cdots & R_x(0) \end{pmatrix}, \tag{16.26}$$

$\mathbf{a} = (a_1, a_2, \ldots, a_m)^T$, and $\mathbf{r}_x = (R_x(1), R_x(2), \ldots, R_x(M))^T$.

The optimal LPC coefficient vector $\mathbf{a}$ can be solved from the normal equation (16.25) by finding the matrix inverse for $\mathbf{R}_x$. This linear equation can generally be solved by using

Gaussian elimination or similar linear algebra algorithms. By taking advantage of the special structure of $\mathbf{R}_x$, efficient algorithms such as the Levinson-Durbin algorithm can be used.

The Levinson-Durbin algorithm is an iterative-recursive process that finds the solution to the mth-order predictor from that of the $(m-1)$th-order predictor. For the derivation of the algorithm, the reflection coefficients (RCs) k_i are defined, which are an alternative representation of the LPC coefficients. A one-to-one correspondence exists between the set of RCs and the set of LPC coefficients [7]. The Levinson-Durbin algorithm is given in [7, 12]. In speech coding, estimation for a is performed for each frame.

The RCs k_i, $i = 1, \ldots, M$, can be derived from a_i, $i = 1, \ldots, M$. For $l = M, \ldots, 1$, [7]

$$k_l = -a_l^{(l)}, \tag{16.27}$$

$$a_i^{(l-1)} = \frac{a_i^{(l)} + k_l a_{l-i}^{(l)}}{1 - k_l^2}, \quad i = 1, \ldots, l-1, \tag{16.28}$$

where $a_i = a_i^{(M)}$.

The Levinson-Durbin algorithm is computationally very efficient. Also, it generates a set of RCs k_i, which can be used for the verification of the minimum-phase property of the prediction-error filter $A(z)$. The prediction-error filter $A(z)$ is a minimum-phase system if and only if all the associated RCs satisfy $|k_i| < 1$ [7]. When $A(z)$ is a minimum-phase system, all its zeros are within the unit circle in the z-plane. Thus, the inverse system $1/A(z)$ has all its poles inside the unit circle, and this guarantees the stability of the inverse system.

The Levinson-Durbin algorithm relies on the values of the LPCs, which may be over a large dynamic range and have no known bound. This may cause difficulties for fixed-point implementation, and careful planning is necessary to ensure all variables to be within the allowed range [6]. The Leroux-Gueguen algorithm provides a method to compute RCs from the autocorrelation values directly, and thus is suited for fixed-point implementation. The Levinson-Durbin algorithm is generally more popular than the Leroux-Gueguen algorithm [6, 7].

Quantization

Log area ratio (LAR)

The RCs are good candidates for quantization, since the stability of the quantized value can be verified from its magnitude. When $|k_i|$'s approach unity, they require more bits for quantization, since the spectral sensitivity function tends to be very large. Thus nonlinear quantization is desirable. The LAR is a nonlinear transform of RCs

$$f(k_i) = \log_{10}\left(\frac{1 + k_i}{1 - k_i}\right). \tag{16.29}$$

The LAR is subject to a uniform quantization.

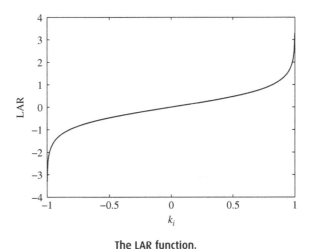

Figure 16.7   The LAR function.

Example 16.3: The LAR function is plotted in Fig. 16.7. Although LAR takes on value from $-\infty$ to ∞, from the figure, when $|k_i| = 0.99$, LAR $= 3.3$. Thus, the value of LAR is practically very limited. For $|k_i < 0.7|$, the LAR function is almost linear, the quantization performance in the RC domain is similar to that in the LAR domain using a uniform quantizer, and we can apply uniform quantizer in the RC domain. Typically, $|k_i| < 0.7$ for $i > 3$ [7]; in this case, high sensitivity for large $|k_i|$ is eliminated.

Each of the LPC coefficients can be assigned a different number of bits for scalar quantization to represent the quantized codebook index. For example, in IS-54 VSELP, the ten RCs are quantized using 6, 5, 5, 4, 4, 3, 3, 3, 3, 2 bits, respectively, yielding a total of 38 bits/frame.

Another nonlinear transformation is the inverse sine transformation $S_i = \sin^{-1}(k_i)$. The 5.6 kbits/s half-rate (HR) GSM codec implements the VQ of S_i.

Line spectral frequency (LSF)

LSF is another alternative representation of the LPC coefficients. LSFs are also known as *linear spectral pairs (LSPs)*, since zeros in the prediction-error filter $A(z)$ occur in complex conjugate pairs. Two polynomials $P(z)$ and $Q(z)$ are defined according to $A(z)$

$$P(z) = A(z)(1 + G(z)), \quad Q(z) = A(z)(1 - G(z)), \tag{16.30}$$

where

$$G(z) = z^{-(m+1)} \frac{A(z^{-1})}{A(z)}. \tag{16.31}$$

LSFs are defined as those values of frequency ω such that

$$P\left(e^{j\omega}\right) = 0 \quad \text{or} \quad Q\left(e^{j\omega}\right) = 0 \tag{16.32}$$

with $0 < \omega < \pi$. For real coefficients, the zeros occur in complex conjugate pairs. If $A(z)$ is minimum-phase, all the zeros of $P(z)$ and $Q(z)$ are on the unit circle, and they are interlaced with each other. During quantization, if the interlacing property is preserved, the stability of the synthesis filter is guaranteed. The LSF representation has a benefit that the LPC coefficients and RCs do not have, namely, modification to one LSF has a local effect on the PSD. The LSFs are suitable for quantization, and they are highly suitable for fixed-point representation since they are within the $(0, 1)$ interval.

The conversion between LSFs ω_i, $i = 1, \ldots, M$, and LPC coefficients a_i, $i = 1, \ldots, M$, are given in [7]. The LSFs are an efficient means to represent spectral envelopes, and become the dominant LPC representation for modern speech coders.

In FS1016 CELP, scalar quantization of the LSFs is applied, where quantization is implemented on $\cos\omega$, with a interval of $[-1, 1]$. LSF VQ is used in most state-of-the-art vocoders including 8 kbits/s G.729 ACELP (algebraic codebook excited linear prediction), dual-rate G.723.1 MP-MLQ (multipulse maximum likelihood quantization)/ACELP, FS MELP, GSM EFR (enhanced full rate) ACELP, and the 7.4 kbits/s IS-136 codec. Major techniques for LSF VQ are split VQ, multistage VQ, and predictive VQ. Split VQ is usually used, since its implementation complexity is low, but it has some performance degradation. Multistage VQ is used in FS MELP. In G.723.1 MP-MLQ/ACELP, predictive VQ with split structure is used. In G.729 ACELP, a combination of multistage VQ, split VQ, and PVQ-MA (predictive VQ with moving average) with switched predictor is employed. In ETSI GSM EFR ACELP, a combination of PVQ-MA and split VQ is used. In the IS-136 codec, split VQ is used.

Interpolation of LPC coefficients

For each frame, LPC coefficients are obtained by linear prediction analysis. Since each frame has a period of 20 to 30 ms, rapid changes in the LPC coefficients for adjacent frames can introduce undesirable transients in the synthesized speech signal. One method to deal with this is to subdivide the frame into subframes, and perform interpolation of the LPC coefficients between these subframes. Interpolation of the LPC coefficients is implemented in various standards such as TIA IS-54 VSELP, ETSI GSM RPE-LTP, and FS1016 CELP. Due to the stability problem, interpolation of the LPC coefficients is usually performed on the RC, LAR, or LSF parameters. The LSF provides the best interpolation performance [7].

Long-term linear prediction analysis

A typical tenth-order linear predictor is not capable of accurately modeling the voice signal with a pitch period of 50 samples. By increasing the prediction order to include one pitch period, the periodicity in the prediction error is substantially removed, leading to an increase in the prediction gain. However, this results in an excess bit rate as well as complexity.

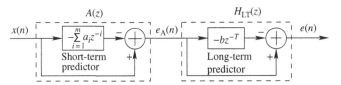

Figure 16.8 The cascade of a short-term prediction (STP)-error filter to a long-term prediction (LTP)-error filter.

The traditional solution is to cascade a short-term predictor with a long-term predictor, as shown in Fig. 16.8. The short-term predictor, as we discussed so far, typically has a relatively low prediction order m in the range of 8 to 12, to eliminate the correlation between nearby samples. The long-term prediction-error filter eliminates correlation between samples that are one pitch period apart.

The input to the long-term filter is the error generated from the short-term predictor and it produces the final output error $e(n)$. The long-term prediction-error filter has a transfer function

$$H_{LT}(z) = 1 + bz^{-T}, \qquad (16.33)$$

where T is the pitch period, and b is the long-term LPC coefficient, also known as *pitch gain*. The second term bz^{-T} is known as the *long-term predictor*.

The long-term linear prediction aims to minimize

$$J = E\left[e^2\right] = \frac{1}{N}\sum_{n=1}^{N}(e_A(n) + be_A(n - T))^2, \qquad (16.34)$$

where $e_A(n)$ is the error generated from the short-term predictor $A(z)$. The optimal pitch gain can be derived by setting $\frac{\partial J}{\partial b} = 0$, yielding

$$b = -\frac{\sum_n e_A(n)e_A(n - T)}{\sum_n e_A^2(n - T)}. \qquad (16.35)$$

The optimum LTP parameters can be obtained by computing J over all possible values of T, typically 20 to 147 samples, for a sampling rate of 8 kHz [12].

Linear prediction synthesis

The synthesis of the received speech signal is the inverse of the analysis process. It can be based on the LPC coefficients extracted from the original signal by using a synthesis filter

$$H(z) = \frac{1}{A(z)} = \frac{1}{1 + \sum_{i=1}^{M} a_i z^{-i}}. \qquad (16.36)$$

For the analysis process, as long as the prediction error is close to a white noise, the estimated LPC coefficients are satisfactory. At the synthesis stage, when the excitation is a white noise with unit variance, the synthesis filter will generate the synthesized speech.

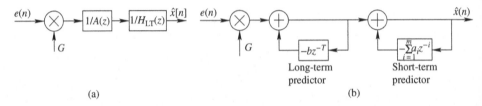

The cascade of a long-term synthesis filter and a short-term synthesis filter for speech production. $e(n)$ is a unit variance white noise. (a) The simplified model. (b) The expanded model.

Similarly, the long-term linear prediction model can also be taken into consideration. The long-term or *pitch synthesis filter* is given by

$$H_{LT}(z) = \frac{1}{1 + bz^{-T}}. \tag{16.37}$$

In this case, the short-term synthesis filter is known as the *formant synthesis filter*.

The overall speech synthesis (production) model is shown in Fig. 16.9. The gain G in Fig. 16.9 can be obtained by comparing the power level of the synthesized speech. It is given by [7]

$$G = \gamma \sqrt{R_x(0) + \sum_{i=1}^{m} a_i R_x(i)}, \tag{16.38}$$

where γ is a scaling constant, whose value depends on the type of window selected, typically ranging from 1 to 2.

We now discuss the stability of the two synthesis filters. As discussed earlier, for the short-term synthesis filter, as long as the RCs satisfy $|k_i| < 1$, $A(z)$ is a minimum-phase filter; then $1/A(z)$ has all its poles inside the unit circle of the z-plane, thus guaranteeing its stability. For the long-term synthesis filter, we can easily derive that in order to keep the filter stable the pitch gain must satisfy $|b| < 1$.

Prediction gain

The prediction gain of a predictor is defined by

$$PG = 10 \log_{10}\left(\frac{\sigma_x^2}{\sigma_e^2}\right) = 10 \log_{10}\left(\frac{R_x(0)}{R_e(0)}\right) = 10 \log_{10}\left(\frac{E\left[x^2(n)\right]}{E\left[e^2(n)\right]}\right) \quad \text{(dB)}. \tag{16.39}$$

Prediction gain can be defined for frame m, $PG(m)$. Similar to SEGSNR, the segmental prediction gain is the average of the prediction gains for many frames.

16.5.3 Predictive coding

Predictive coding eliminates the redundancy in the data by using values previously coded, and only codes the prediction error. This method is also known as *DPCM*. If the prediction error is not quantized, a lossless coding scheme using entropy encoding can be applied; if it is quantized, any further coding will yield lossy coding. At the decoder, the received error signal is added to the prediction and the signal is reconstructed.

Delta modulation

Delta modulation (DM) uses the difference between two successive PCM samples. The difference can usually be coded in fewer bits than the value of the samples themselves. In the simplest case, the difference is represented by only 1 bit denoting ± 1 PCM step. When the prediction error $d(n) = x(n) - \hat{x}(n)$ is not less than 0, the output $\hat{x}(n + 1)$ is increased by δ, where $\hat{x}(n)$ is the predicted value and $x(n)$ is the incoming data sequence; otherwise, it is decreased by δ

$$\hat{x}(n + 1) = \hat{x}(n) \pm \delta. \tag{16.40}$$

This method has the slope overload problem. That is, when the signal has a slope above $\pm \delta/T$, where T is the sampling period, the system is not capable of tracking the signal. The slope overload problem can be overcome by oversampling or increasing the step size. The former reduces the advantage of using delta modulation, while the latter causes the granularity problem; these problems are illustrated in Fig. 16.10.

DM is not suitable for speech signals, due to the wide variation in amplitudes of sounds. Adaptive delta modulation is used to alleviate the problems in delta modulation. The CVSDM technique is a particularly effective and popular technique.

DPCM and ADPCM

DPCM is an extension of the delta modulation, but uses a quantizer with p values. A block diagram of the DPCM encoder and decoder is shown in Fig. 16.11. The predictor for predicting the future value of a signal is of a high order, and is commonly implemented as an FIR filter. The error between the actual and predicted values are quantized. The predictor is implemented as the linear weighted sum of the previous samples (i.e., a transversal digital filter).

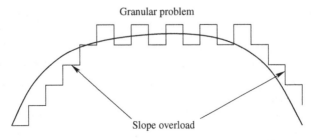

Figure 16.10   Illustration of the slope overload and granularity problems.

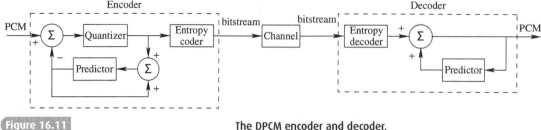

Figure 16.11   **The DPCM encoder and decoder.**

In time-domain waveform coding, the scalar quantization should be adaptive for optimal performance for nonstationary signals such as speech, which has a wide dynamic
range. In the adaptive PCM technique, two types of adaptive quantizers are used,
namely, the forward gain-adaptive quantizer and backward gain-adaptive quantizer. The
forward-adaptation method can accurately control the gain level of the input sequence,
but the gain must be sent to the decoder. In the backward-adaptation technique, the
gain is estimated from the quantizer output, thus the gain need not to be transmitted. However, a transmission error will lead to forward error propagation. When
the adaptation techniques are applied to the DPCM, the resulting system is called
ADPCM.

The forward-adaptation scheme introduces a delay, since the processing requires to
collect a sufficient number of samples. This delay may generate echo and annoying artifacts. Backward adaptation is preferred in those applications that have strict restriction on
delay. However, in the backward-adaptation scheme, the predictor coefficients are determined from the previously recovered speech, and the frequency of the LPC coefficient
update is limited by the codec complexity. The autocorrelation method and the Levinson-
Durbin algorithm can also be used to solve the filter coefficients in adaptive forward
predictive DPCM. Some backward-adaptation standards are ITU-T G.721, G.726, G.727,
and G.728.

G.721 defines the ADPCM scheme. A series of 8-bit nonuniform PCM samples (G.711)
is mapped onto a series of 4-bit ADPCM samples, reducing the data rate to 32 kbits/s. This
codec is popular due to its very low implementation complexity and its high speech quality.
G.721 ADPCM is used in several cordless telephone systems such as CT2, DECT, PHS,
and PACS.

G.726 and G.727 are two derivatives of G.721, and both can operate at transmission
rates of 16, 24, 32 and 40 kbits/s. G.721 has an attractive feature: its decoder can operate
without prior knowledge of the transmission rate. This feature is useful in packeted speech
transmission. G.727 is an embedded ADPCM codec, in which the decision levels of the
lower-rate codec constitute a subset of those of the higher-rate codecs. A codeword is
composed of core bits that must be decoded and enhancement bits that can be discarded by
the coder.

Time-domain waveform encoding techniques, such as PCM, DPCM, ADPCM, DM, and
adaptive DM, attempt to faithfully represent at the output the waveform from the source.
Scalar quantization is used for these time-main predictive coding techniques.

16.5.4 Frequency-domain waveform coding

Frequency-domain waveform coding filters a speech signal into a number of subbands, and separately encodes the signal in each sub-band. In sub-band coding, the signal is typically divided into four to eight sub-bands. Since the lower-frequency bands contain most of the spectral energy in voiced speech, more bits are assigned to the signals in the lower-frequency bands. The QMF bank can be used for subband coding [33]. The lower-frequency band can be repeatedly subdivided by a factor of two.

Transform coding is similar to sub-band coding. It is a block coding technique. A frame of samples is transformed into the spectral domain for coding. At the decoder, the spectral samples are subject to the inverse transform and the speech signal is synthesized. The transform from the time domain to the frequency domain aims at obtaining uncorrelated spectral samples. The Karhunen-Loève transform is optimal in the sense that it produces uncorrelated outputs. For a block of N samples, conventionally, the Karhunen-Loève transform requires a complexity of $O\left(N^4\right)$ operations [9]. The Karhunen-Loève transform can be effectively solved using the adaptive principal component analysis (PCA) method [10], which reduces the complexity to $O(Nn)$ operations, where n is the number of iterations. DCT is a suboptimal solution and is generally used. Transform coding can achieve toll-quality speech at 16 kbits/s, and communication-quality speech at 9.6 kbits/s.

Wideband speech covers the range of 50 Hz to 7 kHz, with a quality similar to that of an FM radio. The sampling rate is 16 kHz. G.722 is a standard for wideband speech, and is used in the H.323 video-conferencing standard. It separates the speech into two subbands and uses ADPCM to code each band. G.722.1 and G.722.2 are two wideband speech codecs. G.722.1 employs the filter bank method by using two QMFs, and it performs well for speech and audio, operating at 64 kbits/s. G.722.2 is the adaptive multi-rate wideband (AMR-WB) speech coder standardized by 3GPP. G722.2 is based on ACELP.

16.5.5 Voice activity detection

During the conversation of two persons, when one person speaks, the other person is silent. Statistics show that speech activity occupies only 3/8 of all the connection time [4]. In order to save radio spectrum, it is desirable to transmit speech frames over the air interface only when the person is actively speaking. This requires voice activity detection. The idea behind voice activity detection is that there is more energy in lower frequencies in the voiced speech than in the voiceless speech in the same band. Silence compression is the equivalent of RLE on data files.

Discontinuous transmission at the inactive speaker also leads to less power consumption and lower interference to other users. This is implemented by switching off speech transmission after several nonactive frames, and a silence descriptor (SID) frame is sent as a model for regeneration of a comfort noise at the receiver. At the encoder, the comfort noise generation algorithm extracts the noise parameters, namely, the residual energy level and the spectrum of the background noise, to form the silence descriptor packets. At the decoder, the noise information contained in the silence descriptor packets is used to generate the comfort noise. The voice activity detection algorithm is computationally very intensive. It employs both the energy and LPC analyses, with the latter requiring a great deal of computations.

In GSM, when the user speaks, the speech is encoded at 13 kbits/s. Otherwise, it is encoded at 500 bits/s. At the receiver, some *comfort noise* is generated when no signal is received. At the beginning of the nonactive period, a silence descriptor frame is transmitted, followed by interval transmission of silence descriptor frames, at least two frames per second, until the nonactive period is over. The ETSI AMR (adaptive multirate) codec utilizes discontinuous transmission, with voice activity detection and comfort noise generation to reduce bandwidth usage during the silence periods. Some earlier voice activity detection algorithms were implemented in G.729 Annex B (G.729B) and G.723.1 Annex A. There are some algorithms that provide better performance than that of G.729B [8, 18].

16.5.6 Linear predictive coding

Within each frame, the speech signal is a sample function of a stationary stochastic process. The Wold decomposition guarantees that a stationary stochastic process can be decomposed into the sum of a regular component $x_r(n)$ and a singular component $x_s(n)$

$$x(n) = x_r(n) + x_s(n). \tag{16.41}$$

The regular term is a filtered noise term that cannot be perfectly predicted by using a linear system, and the singular term is a sum of sinusoids that can be perfectly predicted by a linear system. The regular component can be represented as white noise passing through a linear filter. The linear filter is required to be causal and stable, and have a causal and stable inverse. Such a filter is a minimum-phase filter. This same filter is used for generating the singular component, which can be modeled by a periodic pulse train passing through the linear filter. The linear predictive vocoder is an important stochastic model for speech.

Vocoders are based on modeling the speech production system using an all-pole model of the vocal tract. The classical LPC vocoder is shown in Fig. 16.12. For voiced speech, the excitation is a periodic impulse train with a period that is equal to the pitch period of the speech. For unvoiced speech, it is a white noise sequence. The model parameters are the voicing status, gain, LPC coefficients, and pitch period. The voicing status uses one bit to denote whether the frame is voiced or unvoiced. The pitch period is estimated only for voiced frames. A encoder estimates the model parameters for every frame, quantizes and codes them, and then sends them to the receiver for speech reconstruction. The decoder is essentially the linear predictive model of speech production with parameters controlled by the bitstream.

In LPC, speech is classified as voiced and unvoiced. A frame with clear periodicity is treated as voiced, and a frame with noise-like appearance is treated as unvoiced. LPC requires a strict voice/unvoiced classification. The voiced speech has an energy concentration in the low-frequency band, since the pitch frequency is generally less than 500 Hz.

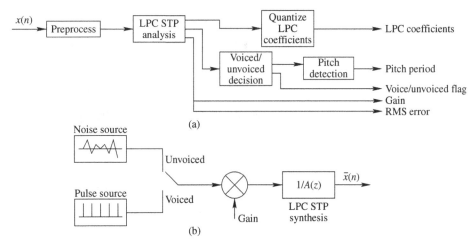

Figure 16.12   The traditional LPC vocoder.

The zero-crossing rate is also relatively low for voiced speech. These features can be used for voice detection.

Most vocoders assume the speech signal to be fairly constant over a time of approximately 20 ms. LPC is a popular vocoding method, and it typically achieves communication-quality speech at data rates ranging from 2.4 to 9.6 kbits/s. The FS1015 LPC (LPC-10) codec is a U.S. Government standard that uses 10 coefficients. LPC-10 achieves a bit rate of 2.4 kbits/s.

Speech quality in LPC can be improved by computing and encoding a residual error (as in DPCM), leading to a higher bit rate. This is implemented in the RELP vocoder, which provides communication-quality speech at about 9.6 kbits/s.

Example 16.4: Given the original speech of 6 seconds long, the synthesized waveform using the LPC algorithm is shown in Fig. 16.13. The frame length is 25 ms.

Very-low-bit-rate coding

There are a number of problems with the LPC technique [7]. First, transition frames cannot be simply classified as voiced and unvoiced, otherwise artifacts like buzzes and tonal noises will be generated. Secondly, synthesis of voiced and unvoiced speech using strictly periodic pulse train or strictly random noise is not a true representation of the real speech. Also, no phase information of the original speech is captured by the model.

The MELP coder overcomes some of the problems of the LPC technique such as the buzzy problem [20]. It utilizes a more complex and accurate speech production model. The excitation signal is a combination of a filtered periodic pulse sequence with a filtered noise sequence. The MELP coder classifies the speech into voiced, unvoiced, and jittery

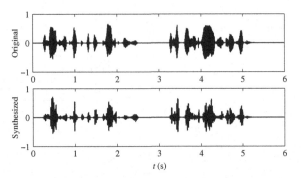

Figure 16.13   **The original speech and the LPC synthesized waveform.**

voiced. Due to the use of VQ, the MELP coder does not elevate the overall bit rate of LPC. The FS MELP coder operates at 2.4 kbits/s.

The zinc-basis function excitation (ZFE) waveform can efficiently model the LPC STP residual while reducing the buzziness of the synthesized speech. A prototype waveform interpolation (PWI) ZFE speech coder achieves a bit rate of 1.9 kbits/s [12]. The mixed-multiband excitation (MMBE) technique addresses the buzziness problem of the LPC coder by splitting the speech into multiple frequency bands on a frame basis. MMBE can be added to the LPC vocoder or the PWI-ZFE coder to yield a very-low-bit-rate voice compression, as low as 1.85 kbits/s. Sinusoidal transform coding, such as PWI and MMBE, is frequently used for bit rates of less than 4 kbits/s. The three voiced excitation methods, namely ZFE, MMBE, and sinusoidal transform coding, have been found to have similar performance [12].

16.5.7 Pitch period estimation

One of the most important parameters in speech analysis, synthesis, and coding is the pitch of the voiced speech. Estimation of pitch makes sense only for voiced frames, since unvoiced frames have random nature. Accurate pitch detection is one of the most difficult problems in speech analysis, especially when the Fourier component at the pitch frequency is missing, as in the case of telephone communications. For vocoders that require the estimation of pitch, an accurate estimation of the pitch period is of a prime concern, since the human ear is sensitive to pitch estimation errors. Checking for pitch doubling and halving together with pitch tracking are frequently performed for pitch detection.

The simplest method for pitch estimation is the *pitch picker* that picks the prominent peaks of the signal or its envelope. This method may find successive pitch periods, but the peak locations may be shifted by the formant frequencies, leading to a quavery voice in the synthesized speech signal. Popular pitch detection methods are the autocorrelation-based method and the cepstrum method. The dyadic wavelet transform is also used for voiced, unvoiced, and transient speech classification, and for pitch detection [12].

Autocorrelation-based method

Pitch can be estimated by using the short-time autocorrelation method. By calculating the autocorrelation values for the entire range of lag, the pitch period is estimated as the lag associated with the highest autocorrelation. The ACF is defined as

$$R(l, m) = \sum_{n=m-N+1}^{m} s(n)s(n - l), \qquad (16.42)$$

where N is the frame length, m is the time instant that the frame ends, and $l > 0$ is the time lag.

This method can find strong peaks at a period of T, and next strong peaks at $2T, 3T, \ldots$, where T is the pitch period. Since pitch associated with voice is below 500 Hz, a more accurate estimate for pitch is obtained by lowpass filtering the speech so as to eliminate the high-frequency components and noise.

Unfortunately, there are secondary maxima in the autocorrelation function that are related to the formant frequencies. These formant peaks may exceed the pitch-period maximum, during phoneme transitions, leading to large pitch estimation errors. A method frequently used to improve the autocorrelation method is oversampling, and this is known as the *fractional pitch period detection* [7]. Oversampling is employed in the FS1016 and GSM HR coders.

Magnitude difference function method

The magnitude difference function (MDF) method does not use multiplication. It is based on the expectation that $|s(n) - s(n - l)|$ is small for $l = 0, T, 2T, \ldots$. The magnitude difference function is defined by

$$\text{MDF}(l, m) = \sum_{n=m-N+1}^{m} |s(n) - s(n - l)|. \qquad (16.43)$$

Example 16.5: Given a speech of "Beep!", for frame 5, the autocorrelation and magnitude difference as functions of the lag are plotted in Fig. 16.14, for a frame length of 20 ms. For the autocorrelation-based method, the second largest correlation value is at delay $l = 56$. For the MDF case, the second smallest MDF also occurs at delay $l = 56$. This corresponds to a pitch period of 7 ms or a pitch frequency of 142.86 Hz. The results for the two methods are very close for frames 1 to 8, which are voiced frames.

Cepstrum method

The most effective method for suppressing the formants during pitch extraction is the cepstrum method. The cepstrum is defined as the Fourier transform of the logarithm of the power spectrum. In the cepstrum, the influence of the formants is almost completely removed from the pitch extraction.

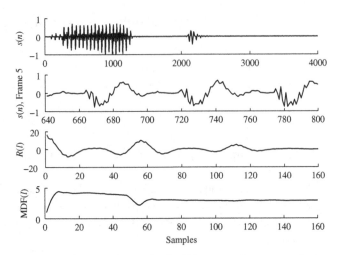

Figure 16.14   The autocorrelation and magnitude difference values versus the lag.

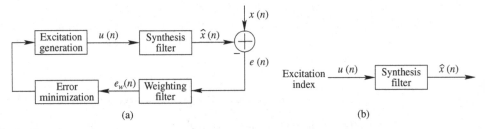

Figure 16.15   Speech coding based on analysis-by-synthesis. (a) The analysis-by-synthesis encoder. (b) The analysis-by-synthesis decoder.

Similarly to the autocorrelation method, if the peak cepstral exceeds a threshold, the speech is classified as voiced; otherwise, the zero-crossing counting is further performed. If the number of zero-crossings exceeds a threshold, the speech is classified as unvoiced; otherwise it is voiced. For voiced frames, the pitch period is the location of the peak cepstral value.

16.5.8 Analysis by synthesis

Most existing vocoders employ the principle of linear prediction based on analysis-by-synthesis [17]. Multipulse LPC is an analysis-by-synthesis method for synthesizing good-quality speech at 9600 bits/s. This data rate can be further reduced in CELP [26] by selecting the excitation sequence $e(n)$ from a codebook of zero-mean Gaussian sequences. CELP and its variants are widely used in modern speech codecs.

Analysis-by-synthesis can be in either open-loop or closed-loop form. The architecture of closed-loop analysis-by-synthesis is shown in Fig. 16.15. Encoding (analysis) is performed by perceptually optimizing the decoded (synthesized) signal in a closed loop.

Excitation generation is performed by multiplication of the excitation codebook and a gain. The closed-loop search procedure is repeated for all excitation codevectors stored in the codebook, and the best excitation sequence is encoded. Instead of minimizing the usual MSE term to generate the best waveform reproduction, the analysis-by-synthesis codec can minimize the perceptually weighted error $e_w(n)$. The same codebook is used at the decoder.

Multipulse excitation model

In predictive coding, it is not necessary to encode all prediction errors. Typically, by preserving a small percentage of the nonzero samples, the synthesized speech has little perceptual distortion. The preserved samples are encoded by their amplitudes and positions. This approach is called the *multipulse excitation (MPE) model* [2].

The regular pulse excitation (RPE) model is a popular MPE model. The prediction-error sequence is down-sampled to several subsequences, with the subsequence having the highest energy selected. The RPE codec is an open-loop excitation optimization assisted codec.

For the RPE codec, only the amplitude of the pulses and the subsequence number are encoded. A variant of RPE, namely RPE-LTP, is used in the GSM 6.10 full-rate (FR) codec standard at the source bit rate of 13 kbits/s. RPE-LTP is a hybrid method that tries to match in the magnitude spectrum domain using the LPC model, as well as in the time domain by approximating the original waveform. RPE-LTP preprocesses the sampled audio signal, applies STP analysis and filtering, and then LTP analysis and filtering. The remnant RPE, obtained by subtracting STP and LTP signal parts, are then evaluated. It uses a frame length of 20 ms.

RPE-LTP provides toll quality speech (MOS of 4.0). It has a higher robustness than the 32 kbits/s ADPCM codec, but has increased complexity and encoding delay. The predictor is implemented as a cascade of STP and LTP. A large amount of bits are allocated to the excitation signal due to the use of scalar quantization.

The MPE and RPE codecs can provide high quality speech at 10 kbits/s and higher, since scalar quantization is employed and a large amount of bits must be used for transmitting the positions and amplitudes of the excitation pulses. The GSM EFR ACELP coder, operating at 12.2 kbits/s, is designed to replace the GSM FR RPE-LTP coder. ACELP is a CELP type of algorithm.

Code-excited linear prediction

CELP is a descendant of the multipulse LPC. Like LPC, CELP utilizes a cascade of LTP and STP to avoid the voiced/unvoiced decision, and phase information is partially retained through a closed-loop analysis-by-synthesis method. In CELP, instead of using a codebook of pulse patterns, a codebook of excitation signals is used. Like RPE-LTP, CELP is also a hybrid method.

A speech signal is coded as a set of parameters such as LPC coefficients, gain coefficients, and codebook indices. In CELP, the synthesis filter is a cascade of an LTP filter and

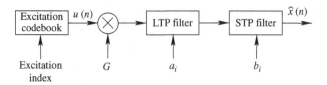

Figure 16.16

Speech synthesis of CELP.

a STP filter. The LTP filter is used for pitch synthesis, and the STP filter for generation of the spectral envelope (formants) of the speech signal. The LTP and STP filters are, respectively, used to model the long-term (fixed codebook, innovation) and short-term correlations (adaptive codebook, pitch) of the speech signal. Typically, the STP uses a 10th-order all-pole filter, and the LTP uses a 1st- to 3rd-order all-pole filter to extract those parameters. Coders usually operate on a block of speech samples for a duration of 20–40 ms. The speech synthesis part of the CELP model is shown in Fig. 16.16.

In the CELP decoder, the excitation is produced by summing the contributions from an adaptive codebook (pitch) and a fixed codebook (innovation). A synthesis filter of the all-pole model $1/A(z)$ is used to shape the excitation. For the generic CELP encoder, STP analysis is performed in each frame to yield the LPC coefficients, and the output STP error is then divided into subframes, which are further subject to LTP analysis. Excitation codebook search is the major part of the CELP coder.

For a typical CELP algorithm, the LPC coefficients, pitch period, pitch gain, codebook index, and codebook gain are transmitted. Unlike MPE and RPE codecs, VQ of model parameters is used for LPC coefficients for reducing the bit rate.

Signal Analysis

The prediction error or residual waveform, $e(n) = x(n) - \hat{x}(n)$, is transmitted with the source signal model. This residual waveform is also quantized for transmission

$$e^q(n) = e(n) + q(n), \tag{16.44}$$

where the quantization noise $q(n)$ is modeled by a white noise. The decoder reconstructs the signal sample as

$$x^{(d)}(n) = \hat{x}(n) + e^q(n) = -\sum_{i=1}^{m} a_i x^{(d)}(n-i) + e^q(n). \tag{16.45}$$

In the frequency domain, we have

$$A(z)X^{(d)}(z) = E(z) + Q(z). \tag{16.46}$$

For closed-loop prediction, the reconstructed sample for prediction at the encoder is the same as that in the decoder

$$e(n) = x(n) - \hat{x}(n) = x(n) + \sum_{i=1}^{m} a_i x^{(d)}(n-1). \tag{16.47}$$

As a result, the decoder output is derived as

$$x^{(d)}(n) = x(n) + q(n),$$ (16.48)

or in the frequency domain

$$X^{(d)}(z) = X(z) + Q(z).$$ (16.49)

Error weighting filter

The masking phenomenon in the human auditory system is employed in CELP. A perceptual noise-shaping filter is used to shape the noise-like speech spectrum so that it is inaudible. The weighting filter is implemented as

$$W(z) = \frac{A(z)}{A(z/\gamma)} = \frac{1 + \sum_{i=1}^{m} a_i z^{-i}}{1 + \sum_{i=1}^{m} a_i \gamma^i z^{-i}},$$ (16.50)

where γ is a constant in $[0, 1]$, typically in the range of 0.6 to 0.85.

The error weighting filter amplifies the error signal spectrum in nonformant regions, but deemphasizes the error signal spectrum in formant regions. This is because error in the regions of spectrum peaks is easily masked due to the high signal energy at these regions, while at spectrum valleys the signal has low energy and the error is prominent. Other forms of the error weighting filter are also used. For example, in G.728, the filter takes the form

$$W(z) = \frac{A\,(z/\gamma_1)}{A\,(z/\gamma_2)}$$ (16.51)

with $\gamma_1 = 0.9$ and $\gamma_2 = 0.6$.

Postfilter

A postfilter can be used to augment spectral prominence and to attenuate the components in the spectral valley. This alters the waveform, but reduces the audible noise in the reconstructed speech. Thus, it yields a further improvement in the subjective quality of the synthesized speech.

Postfiltering is used in G.721. Both adaptive short-term and long-term postfilters are employed in G.728. In G.728, the long-term postfilter is a comb filter that enhances the spectral peaks in the vicinity of the harmonics of the pitch frequency, while the adaptive short-term postfilter consists of a 10th-order pole-zero filter concatenated with a first-order single-zero filter. Note that the weighting filter is implemented only in the encoder and the post-filter is implemented only in the decoder.

16.5.9 CELP-based codecs

Along with its variants, such as ACELP, relaxed CELP (RCELP), low-delay CELP (LD-CELP) and VSELP, CELP is currently the most widely used speech coding algorithm. CELP scales well to both low and high bit rates. The U.S. federal standard FS1016 is the first CELP codec. It is a 4.8 kbits/s CELP coder that uses a standard CELP structure. The frame length is 30 ms, split into four 7.5 ms subframes.

CELP can achieve communications of toll quality speech at bit rates between 4.8 and 16 kbits/s. At 16 kbits/s, G.728 produces reconstructed speech that is indistinguishable from the 64 kbits/s G.711 PCM speech and is superior to the 32 kbits/s G.726 ADPCM speech, while at 4.8 kbits/s the FS1016 codec produces good communications-quality speech.

The PDC HR speech codec employs the pitch synchronous innovation CELP (PSI-CELP) principle [19]. It has a bit rate of 3.45 kbits/s, and a total channel-coded rate of 5.6 kbits/s. The frame length is 40 ms, which is divided into four 10 ms subframes. The PSI-CELP codec exploits the periodicity that occurs in voiced speech.

Speex is an open source/free software speech codec under the BSD license,[1] and is based on CELP. It is targeted at VoIP and has a wide range of bit rates available (from 2 to 44 kbits/s). Some of Speex's features are: narrowband (8 kHz), wideband (16 kHz), and ultra wideband (32 kHz) compression in the same bitstream; intensity stereo encoding; packet loss concealment; variable bit rate; voice activity detection and discontinuous transmission; fixed-point port; acoustic echo canceller; and noise suppression. Intensity stereo encoding, integration of multiple sampling rates in the same bitstream (embedded coding), and a variable-bit-rate mode are unique features in Speex. In the wideband mode, Speex uses a QMF to split the band in two.

Many mobile communication standards use codecs at different bit rates. For example, GSM uses FR/HR/EFR codecs. Other standards such as PDC and IS-136 also define HR, FR, EFR algorithms but may be at different bit rates.

Some important CELP variants are introduced below.

Vector-sum excited linear prediction

VSELP is a CELP coder with a particular codebook structure that has a reduced computational cost. In VSELP, excitation vectors from the stochastic codebook are obtained by a linear combination of a number of fixed basis vectors [11]. The TIA IS-54 VSELP standard is very similar to FS1016, except for the form and structure of the stochastic codebook. VSELP is implemented in TIA IS-54, operating at 7.95 kbits/s, in ETSI HR GSM 6.20 at 5.6 kbits/s, and in the PDC FR STD-27 codec at 6.7 kbits/s. The channel-code data rates are 13 kbits/s for IS-54 VSELP, 11.4 kbits/s for the HR GSM codec, and 11.2 kbits/s for the PDC FR codec. The additional data rate is utilized for channel coding and synchronization.

In the IS-54 VSELP coder, a total of 159 bits are allocated for each frame of 20 ms, leading to a bit rate of 7950 bits/s. The IS-54 coder outperforms FS1016. The HR GSM 6.20 and PDC FR STD-27 VSELP codecs are based on a structure similar to that of the IS-54 VSELP, and they were all originally developed by Gerson et al at Motorola Inc. They all use a frame length of 20 ms (four subframes), but use a different number of stochastic codebooks and/or basis vectors.

---

[1] http://www.speex.org

Algebraic codebook excited linear prediction

Excitation codebook search is the most computationally intensive among all CELP operations. ACELP uses simple algebraic rules such as addition and shift to create the excitation codevectors, thus the storage of the codebook is avoided. The original idea of ACELP was first introduced by Adoul et al [1], which was further refined and used in many standards, such as ITU-T G.723.1 MP-MLQ/ACELP, G.729 Conjugate Structure (CS)-ACELP, TIA IS-641 ACELP, ETSI GSM EFR ACELP, and ETSI AMR ACELP.

G.729 uses an algebraic excitation codebook and conjugate VQ for the involved gain. The coder operates at 8 kbits/s, with a delay of 25 ms. It has a frame length of 10 ms (with two subframes). G.729 gives synthesis speech quality equivalent to the 32-kbits/s G.726 ADPCM codec in error-free conditions, but significantly outperforms G.726 in case of channel errors. G.729 Annex A (G.729A) is a modification of G.729, but its bitstream is compatible with G.729, with approximately half the complexity (12 MIPS vs. 22 MIPS for full-duplex operation using fixed-point DSP). G.729A operates at 8 kbits/s and gives speech quality equivalent to G.729 in most conditions, with only a slight degradation in performance in case of background noise and in case of three tandems. G.729 is a most prominent low-delay speech codec.

The G.723.1 MP-MLQ/ACELP coder, standardized for very-low-bit-rate videophone, uses a 30 ms frame length. It operates at 5.3 kbits/s (using ACELP) and 6.3 kbits/s (using MP-MLQ excitation), with a coding delay of 67.5 ms and relatively low implementation complexity. The G.723.1 coder has a complexity of 14.6 MIPS for the 5.3 kbits/s mode and 16.0 MIPS for the 6.3 kbits/s mode. G.723.1 is amenable to voice-activity controlled discontinuous transmission and comfort noise injection. The G.723.1 speech codec and the H.263 video codec form the H.324 multimedia compression and transmission standard. Note that G.723.1 is different from the G.723 ADPCM standard. G.723.1 is one of the earliest codecs for VoIP applications, although today G.711, G.729, and G.722 are also supported in VoIP systems.

The IS-641 standard, which is used as the EFR IS-136 codec, was proposed to replace the 7.95 kbits/s IS-54 coder. IS-641 operates on a frame of 20 ms (four subframes), with a bit rate of 7.4 kbits/s and a channel coded bit rate of 13 kbits/s. The EFR IS-136 codec is very similar to the EFR GSM ACELP codec. IS-641 uses the same algebraic codebook as G.729.

The GSM EFR ACELP operates at 12.2 kbits/s. It was targeted to improve on the 13 kbits/s GSM FR coder (6.10 RPE-LTP), and it uses the same channel bit rate of 22.8 kbits/s as for the FR coder.

Adaptive multi-rate

Both channel and source statistics vary over time. AMR coding is a joint coding of the source and the channel. The source statistics vary very rapidly, while the CSI is updated much more slowly. For this reason, the source coder uses different number of source bits per frame, while the channel coder uses a varying number of bits for adequate channel protection, so that the total bit rate is kept constant.

ETSI AMR ACELP, also known as *network-controlled multimode* coder, operates at eight modes 12.2, 10.2, 7.95, 7.40, 6.70, 5.90, 5.15, and 4.75 kbits/s. There is also a silence descriptor mode operating at 1.8 kbits/s. The 12.2 kbits/s mode employs the GSM EFR coder, the 7.4 kbits/s mode is compatible with the IS-641 EFR, and the 6.7 kbits/s mode is compatible with the 6.7 kbits/s PDC FR.

AMR coding utilizes recursive systematic convolutional coding, which is very similar to Turbo coding, and thus has an error correction capability superior to convolutional coding used in GSM EFR. For GSM, AMR has two traffic channel modes, adaptive full-rate (AFS) at 22.8 kbits/s and adaptive half-rate (AHS) at 11.4 kbits/s. AFS has a MOS performance superior to GSM EFR, and AHS has a performance slightly worse, but close to that of GSM EFR.

AMR is now widely used in GSM and UMTS. Changing AMR adaptation in every frame is theoretically possible in WCDMA. AMR has also been considered for VoIP applications, where AMR adaptation is similar to WCDMA. For VoIP, rate adaptation is implemented so as to reduce network congestion.

Low-delay CELP

In PCM and ADPCM that encode the speech signal on a sample-by-sample basis, the delay is negligible. For CELP-based coders, a high compression ratio is achieved by processing the signal on a frame-by-frame basis. Conventional CELP codecs have a frame length of typically 20 to 30 ms, leading to a coding delay of at least 50 ms. The ITU-T G.728 LD-CELP codec, approved in 1991, was developed in [6]. G.728 operates at a bit rate of 16 kbits/s with a one-way coding delay of less than 2 ms.

G.728 is a very successful low-delay coder. The G.728 coder uses many techniques to achieve low delay [7]:

- The method reduces the frame length to 20 samples, which are further divided into four five-sample subframes.
- It uses recursive autocorrelation estimation for the calculation of the LPC coefficients, known as *external prediction*; LPC coefficients are estimated from the past frame due to the relatively short frames.
- It employs backward adaptive linear prediction, with the LPC coefficients being obtained from the synthetic speech. This avoids transmission of LPC coefficients.
- A short-term synthesis filter with a prediction order of 50 is employed, and avoids the error-sensitive long-term predictor.
- It employs backward excitation gain adaptation.

G.728 LD-CELP achieves a speech quality that is better than or equivalent to that of the 32 kbits/s G.726 ADPCM, while maintaining a high robustness against transmission errors. The G.728 coder has an implementation complexity of about 12.4 million operations (multiplys and adds) per second for the encoder, and 8.7 million operations per second for the decoder, leading to a full-duplex complexity of about 21 million operations per second [12]. This complexity is higher than that of many low bit-rate coders. The codec provides an average SEGSNR of 20.1 dB [12]. This coder has been extended to lower bit-rate codes, at 8, 6.5, and 4.67 kbits/s [6, 30].

Variable-bit-rate (VBR) CELP

The TIA IS-96 variable-bit-rate (VBR) CELP standard, also known as *Qualcomm CELP (QCELP)*, was developed by Qualcomm to increase the capacity of CDMA systems in 1994. IS-96 employs the conventional CELP structure. It is a source-controlled variable-bit-rate coder, where the control mechanism for bit rate depends on the background noise estimation and the signal energy. QCELP was later replaced by EVRC.

The IS-96 coder uses a frame length of 20 ms, and depending on speech activity the data rate is dynamically selected among one of four choices every frame, 8.55, 4, 2, and 0.8 kbits/s, known as *FR*, *HR*, *quarter-rate*, and *eighth-rate*, respectively. Depending on the bit rate, the frame is divided into different number of subframes, and the subframes can be either pitch subframe or codebook subframe. The encoder codes active speech frames at FR, and background noise and silence at lower rates. For a typical conversation, QCELP operates at an average bit rate of less than 4 kbits/s, but the speech quality is comparable to that of the 8 kbits/s IS-54 VSELP codec. In another version of QCELP, the FR is 13.2 kbits/s.

Enhanced variable rate codec

The enhanced variable rate codec (EVRC), developed in 1995, was used to replace QCELP in IS-95. EVRC is also used in CDMA2000. EVRC is practically a fixed-rate coder with a silence-coding mode [13]. It uses the RCELP codebook for representing the fixed codebook. EVRC provides better voice quality than the variable-rate QCELP, but with lower bit rates. It reduces the number of bits for representing the pitch and uses more bits for representing the excitation.

In EVRC, speech is sampled at 8 kHz with 16-bit samples. The length of each frame is 20 ms. The data rates for FR, HR, eighth rate are, respectively, 8.55 kbits/s, 4.0 kbits/s, and 0.8 kbits/s. The eighth rate is used for background noise. The average bit rate varies based on the network conditions. A quarter rate was later added into EVRC-B.

In addition to RCELP, EVRC-B uses the prototype pitch period (PPP) for the coding of stationary voice frames and noise excitation linear prediction (NELP) for the coding of unvoiced or noise frames. Using NELP and PPP coding provides EVRC-B with superior flexibility in rate assignment. EVRC-B is replacing EVRC in CDMA2000. A wideband extension to EVRC-B, EVRC-Wideband (EVRC-WB), is also available.

Selectable mode vocoder

The selectable mode vocoder (SMV) is used in CDMA2000 and as an upgrade of EVRC. However, SMV is also being replaced by EVRC-B for CDMA2000.

SMV is based on a variant of CELP called *extended CELP*. The operation mode of SMV is set by network operators. Rate adaptation in SMV is achieved as follows. For each frame (20ms) select the best rate (eighth rate, quarter rate, HR, or FR) based on the operation mode, according to the classification of input speech (silence, noise-like, unvoiced, onset, nonstationary voiced, or stationary voiced). This avoids multiple convolutional codes. Self-power control leads to better power management and longer battery life. SMV provides a MOS performance of 4.1 at FR.

16.5.10 Wideband speech coding

For most speech codecs, the speech signal is band-limited to 300 to 3400 Hz and sampled at 8 kHz. The filtering removes, on average, less than 1% of the energy of the speech signal. This filtered band is, however, important for maintaining an improved intelligibility and naturalness. Although the perceived quality of speech reproduced from wideband speech codecs is not significantly better, a better quality is desired in some applications.

SB-ADPCM

The ITU-T G.722 sub-band-split ADPCM (SB-ADPCM) codec operates at 64 kbits/s. G.722 band-limits the signal within 50 to 7000 Hz, and then samples at 16 kHz.

Sub-band splitting is performed by the aliasing-free QMF, which splits the 0–8000 Hz frequency band into 0–4000 Hz and 4000–8000 Hz bands, which are then sampled at 8 kHz by decimation. The lower band contains the major part of the signal energy, and thus is much more subjectively important than the upper band. The principle of embedded ADPCM coding that is used in G.727 is also used in G.722, so as to support 8 or 16 kbits/s data transmission.

G.722 has three operation modes: speech only at 64 kbits/s in Mode 1, 56 kbits/s speech plus 8 kbits/s data in Mode 2, and 48 kbits/s speech plus 16 kbits/s data in Mode 3. In Mode 1, the lower band is using 6 bits/sample ADPCM, accounting for 48 kbits/s, and the upper band is encoded using 2 bits/sample, or at 16 kbits/s. The coded bitstreams for the two bands are then multiplexed for transmission.

Transform coding

The G.722.1 wideband audio coding standard is a transform coding codec, which is designed to replace the 64 kbits/s G.722 SB-ADPCM codec. It has a frame length of 20 ms, sampled at 16 kbits/s. G.722.1 is based on the modulated lapped transform (MLT), which is a derivative of the DCT. MLT can be used where blocking effects cause severe signal distortion. Quantization of the transform coefficients employs a psychoacoustic model, and Huffman coding is then employed. G.722.1 generates output bit rates of 16, 24, or 32 kbits/s. The total delay (encoding plus decoding) is of the order of 60 ms.

Adaptive multi-rate–wideband

ETSI has standardized AMR-WB, by extending AMR to the 50–7000 Hz band and sampling at 16 kHz [13]. AMR-WB is also based on ACELP. AMR-WB is approved as a 3GPP and ITU-T G.722.2 wideband standard. AMR-WB is used in GSM and UMTS.

Like AMR, AMR-WB processes audio data at blocks of 20 ms with a 5 ms internal subframe structure. AMR-WB provides excellent speech quality and is robust to error-prone radio channels. Discontinuous transmission is applied in AMR-WB during silence and background noise only. Like AMR, it has 9 different data rates for rate adaptation, 6.6, 8.85, 12.65, 14.25, 15.85, 18.25, 19.85, 23.05, and 23.85 kbits/s. The lower rates are used for speech communications, and higher data rates are for music, bad background noise, or multiparty conversation. The speech quality of AMR-WB is equivalent to the 48 kbits/s mode of G.722 at 8.85 kbits/s, to 56 kbits/s mode of G.722 at 12.65 kbits/s, and to the 64 kbits/s mode of G.722 at 23.85 kbits/s.

3GPP has further extended AMR-WB to AMR-WB+, by using transform coding in addition to ACELP. It supports both mono and stereo audios, with a bit rate ranging from 5.2 to 48 kbits/s. AMR-WB+ is targeted for multimedia services. It provides backward compatibility with AMR-WB.

The variable multi-rate wideband (VMR-WB) speech codec was developed by 3GPP2 for CDMA2000. It is also capable of processing conventional narrow speech, and is interoperable with AMR-WB at certain bit rates. The VMR-WB core technology is based on AMR-WB, but differs in the definition of the operation modes. Multimode operation in VMR-WB was inspired by 3GPP2 SMV. VMR-WB is based on CELP coding including RCELP and ACELP.

Wideband CELP

CELP can be employed for wideband coding. The 16 kbits/s G.728 coder can be extended by using 16 kHz sampling, leading to a bit rate of 32 kbits/s. A better result can be achieved by paying special attention to the 4–7 kHz band. A LD-CELP based 32 kbits/s wideband codec was proposed in [23], and it is similar to G.728. It achieves a speech quality that is similar to that of the 64 kbits/s G.722 codec, but with a higher complexity. The lower and upper sub-bands are split by the QMF band-splitting scheme, as used in G.722.

There are also a number of full-band wideband ACELP coding schemes, including the backward-adaptive 32 kbits/s wideband ACELP and the forward-adaptive 9.6 kbits/s wideband ACELP, which were developed by the Sherbrooke University team [12].

16.6 Audio coding

Audio quality codecs usually cover the frequency range of 20 Hz to 20 kHz. In the consumer electronics applications, audio CD uses PCM format sampled at 44.1 kHz, each sample having a 16-bit resolution, and reaching a dynamic range (SNR) of 96 dB.

Lossless audio data compression is usually based on Huffman coding. The data compression ratio varies considerably with the audio waveforms, and is typically less than 3. The general-purpose LZW coding is not suitable for lossless audio compression, since the compressed files are typically 90% of the original file size [25]. Audio signals typically exhibit a degree of sample-to-sample correlation. This feature can be used to predict the value of the next audio sample, and LPC can be used to remove the redundancy. Entropy coding can then be applied.

For speech signals, the LPC vocoder models the vocal tract and works very well for low data rates. Audio and music signals have a production mechanism different from that of the human speech, and thus the CELP codecs are unable to properly code music signals. Also, the LPC vocoders do not function well when several speakers speak simultaneously, since the speech production model is derived from a single voice source; in this case, an audio codec can be used.

Perception of speech is different from nonspeech audio, since the human brain has a region in charge of speech processing. Speech coding seldom exploits psychoacoustics. The CELP coder uses simple psychoacoustics in the perceptual weighting filter and the postfilter. In contrast, audio coding relies heavily on psychoacoustic models. This may be due to the high complexity arising from pschoacoustic coding.

In perceptual coding of audio signals, the input signal is subdivided into frequency components, which are then encoded separately. Frequency-domain coding has the ability to code each component separately with appropriate accuracy depending on its spectral strength. Based on the power spectrum, excitation patterns can be computed for each component from empirical masking data, and by appropriate bit allocation the quantization noise can be made inaudible. Given a predetermined bit rate, the number of bits available for each block of frequency samples can be calculated. These sub-band amplitudes can be sorted, and bits allocated to the highest amplitude sub-bands until the bit pool is exhausted. Optimal bit allocation can be achieved by minimizing the average block error power [3]. A water-filling strategy for bit allocation is also popular. Frequency-domain coding is more popular for audio signals, since audio signals are highly tonal.

Some lossy and lossless audio codecs

The Dolby AC-3, also known as *Dolby Digital*, is similar in many aspects to MPEG-1 and MPEG-2 audio systems. They are lossy compression methods, and are widely used as the audio part in the DVD-Video, DVB, and North American HDTV standards. The sampling rate for AC-3 is 32, 44.1, and 48 kHz. Each channel is encoded with a bitstream ranging from 32 and 640 kbits/s. Dolby Digital supports the coding of up to six channels. DVD-Video provides multichannel audio, and the sampling rate for audio in DVD-Video is 48 kHz.

Super-audio CD uses the direct stream digital (DSD) technique to record 1-bit data at a sampling rate of 2,822.4 kHz, achieving a dynamic range of over 120 dB. DVD-Audio uses 16-, 20-, 24-bit linear PCM data, and it also uses the Meridian lossless packing (MLP) lossless compression technique of Dolby Laboratories to provide high-quality audio at sampling rates of up to 96 kHz for up to six channels or 192 kHz for two channels. The MLP typically achieves a compression ratio of 2:1. The dynamic range of DVD-Audio can reach 144 dB. DVD-Audio also supports the audio CD format and is compatible with DVD-Video. The MLP decoder reverses the encoding process and has a relatively low complexity: 27 MIPS for extracting a two-channel stream at 192 kHz or 40 MIPS for decoding six channels at 96 kHz.

In the following, we introduce the most popular MPEG Audio standards.

16.6.1 MPEG-1 and MPEG-2 Audio

MPEG is an ISO/IEC audio/video standardization project. The MPEG (Moving Pictures Experts Group) develops standards for moving pictures, associated audio and their combination. MPEG-1 (ISO/IEC 11172) achieves a total data rate of 1.5 Mbits/s, and MPEG-2 (ISO/IEC 13818) has a total data rate of 10 Mbits/s. MPEG-4 (ISO/IEC 14496) targets at a wide range of applications including wired, wireless, streaming, digital broadcasting, multimedia, and high quality audio/video. MPEG-7 (ISO/IEC 15938) addresses the description of multimedia content for multimedia database search. MPEG-21 (ISO/IEC N4318) addresses the many elements needed to build an infrastructure for the usage of multimedia content. All these MPEG standards standardize the bitstream and decoder specifications only, but not the encoder.

MPEG-1 Audio

MPEG-1 Audio supports one or two main channels, depending on the configuration and sampling frequencies of 32, 44.1, and 48 kHz. MPEG-1 Audio aims at providing a perceptually lossless coding scheme. The data rates vary between 32 and 224 kbits/s per channel, corresponding to compression ratios ranging from 2.7:1 to 24:1. MPEG-1 Audio maintains CD-like quality but with reduced data rate.

MPEG-1 Audio specifies three layers, which offer increasingly higher audio quality but slightly increased complexity. All the three layers cause lossy audio compression. Layer 1 is used in the digital compact cassette (DCC), Layer 2 in DAB and DVB, and Layer 3 for portable audio players and Internet. The MP3 file format is a modification of the MPEG Layer-3 format at lower sampling frequencies. MPEG-1 Layers 2 and 3 are also used for broadcasting in ITU-R BS.1115.

The basic structures of the MPEG Audio encoder and decoder are shown in Fig. 16.17. The input PCM signal is first mapped to the frequency domain, and bit allocation and coding of these frequency components are then performed. The input PCM is also fed to a psychoacoustic model whose output is used to determine the bit allocation. The bitstream formatting block interleaves the coded PCM signal with side information, and generates the encoded bitstream. The decoding process separates the quantized spectral components of the signal, and then reconstructs the time-domain representation of the signal.

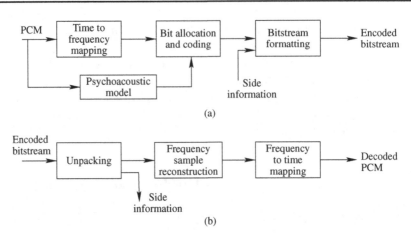

Figure 16.17 The basic structure of MPEG Audio encoder and decoder. (a) Encoder. (b) Decoder.

Layers 1 and 2 of MPEG-1 (and also MPEG-2) use $M = 32$ channel pseudo-QMF banks with the base filter $h(n)$ having 511 taps. The filter length N is 513, with the first and last coefficients being zero. Using (13.194), ϕ_k is selected as [3]

$$\phi_k = \frac{N - 1 - M}{2K}\left(k + \frac{1}{2}\right)\pi \qquad (16.52)$$

to ensure near neighbor alias cancellation. The prototype filter $h(n)$ has symmetrical coefficients, and is defined in ISO/IEC 11172-3. A uniform midtread quantizer is then used to quantize the frequency components. The psychoacoustic model is obtained based on the Hanning windowed DFT. Layers 1 and 2 exploit the MPEG psychoacoustic model 1.

For Layer 3, each of the 32 pseudo-QMF outputs is cascaded with an 18 frequency-line MDCT, generating a total of 576 frequency channels. Blocks of 36 sub-band samples are overlapped by 50%, multiplied by a sine window, and then processed by the MDCT transform. The MDCT outputs are further subject to a nonuniform midtread quantizer, and the MPEG psychoacoustic model 2 is used for quantization. The psychoacoustic model is also obtained by using a Hanning-windowed DFT. For Layer 3, Huffman coding is further applied to the quantization bits.

MPEG AAC and Dolby AC-3 are based on MDCT, and a Kaiser window is used in both the systems. In MPEG AAC, the Kaiser window with $\alpha = 4$ for steady state conditions, and $\alpha = 6$ for transients, is used; in Dolby AC-3, $\alpha = 5$.

MP3 is a popular audio file for sharing music over Internet. Typically, it achieves a compression ratio of 10:1. It achieves the best perceived audio quality at a specific, guaranteed bit rate. MP3 is a sub-band coding method. Each input sample enters a filter bank consisting of 32 polyphase sub-band filters. Each sub-band is encoded with a variable number of bits, which is determined by a psychoacoustic model based on the masking properties of the human ear [29]. A loud signal at one frequency makes the quantization noise at the neighboring frequencies inaudible, and thus fewer bits can be used to encode the neighboring sub-bands. The filter bank is followed by MDCT, and Huffman coding is finally applied to the quantized signal.

Introduction to perceptual audio coding, and MPEG Audio and Dolby AC-3 standards are given in [3].

MPEG-2 Audio

MPEG-2 is a multichannel extension of MPEG-1. MPEG-2 BC (Backwards Compatible) defines audio coding at half the MPEG-1 sampling frequencies. MPEG-2 AAC (Advanced Audio Coding) defines a higher-quality multichannel standard than is achievable with MPEG-1 extensions. AAC combines the coding efficiency of a high-resolution filter bank, prediction techniques, and Huffman coding to achieve very good quality audio at low data rates.

MPEG-2 AAC provides near-transparent subjective audio quality at a bit rate of 256 to 320 kbits/s for five channels and at 96 to 128 kbits/s for stereophonic signals. The MPEG-2 AAC coder is similar to the MPEG-1/2 Layer-3 coder since they both employ a switched filter bank, a nonuniform power-law quantizer, and Huffman coding.

16.6.2 MPEG-4 Audio

MPEG-4 specifies a set of tools, which are used as components in different coding algorithms. MPEG-4 Audio profiles define subsets of the MPEG-4 Audio functionalities for specific applications. MPEG-4 Audio provides a high coding efficiency, reaching lower data rates.

MPEG-4 has many audio coding tools. Major audio tools are speech coding, general audio coding, synthesized audio coding, and synthesized speech coding tools. The first two are used for natural speech/audio coding, while the last two are used for synthetic speech/audio coding. MPEG-4 Audio provides error-robustness tools that allow improved performance on error-prone transmission channels.

MPEG-4 speech coding

The speech coding tools support speech coding at bit rates from 2 up to 24 kbits/s using a CELP coder and a parametric coder known as *harmonic vector excitation coding (HVXC)*. HVXC allows the user to modify speed and pitch during playback.

MPEG-4 CELP is similar to that of a regular CELP codec, but is more flexible. It supports two different types of excitation generators: MPE and RPE. The MPEG-4 CELP MPE generator can operate in narrowband or wideband mode, at a sampling frequency of 8 kHz or 16 kHz. The bit rate can be adjusted in small steps by changing the frame length, the subframe length, and the number of pulses per subframe. The bit rate varies from 3.85 to 12.2 kbits/s for 8 kHz sampling, and from 10.9 to 23.8 kbits/s for 16 kHz sampling. The RPE generator is used for 16 kHz sampling, and it provides a bit rate ranging from 14.4 to 22.5 kbits/s.

MPEG-4 CELP not only provides a wide range of bit rates, but also supports scalable coding in the MPE mode. The full-rate stream is composed of a base-layer stream and one or more enhancement-layer streams. The enhancement layer leads to an increase in the bit rate or bandwidth scalability.

Silence compression is a special MPEG-4 CELP mode. It reduces the bit rate in frames with no or low voice activity. The basic idea is similar to that of HVXC coding. The decoder generates comfort noise for nonactive frames. Comfort noise is generated as a random excitation signal using the same excitation generator as for active frames, but with randomly generated control parameters such as pulse positions or amplitudes.

MPEG-4 HVXC coding allows speech coding at very low bit rates with reasonable quality: 2 and 4 kbits/s [22]. This is achieved by representing LPC residuals with harmonic coding for voiced segments and vector excitation coding (VXC) for unvoiced segments. HVXC coding is based on LPC coding. It has a high complexity, but has speed-control and pitch-modification functionalities during decoding.

MPEG-4 general audio coding

The MPEG-4 general audio coding tools can be either a time/frequency (T/F) coder or a parametric coder. Scalability is supported in both the coders. A parametric audio coder, called *harmonic and individual lines plus noise (HILN)*, can operate at bit rates down to about 4 kbits/s. MPEG-4 T/F is based on MPEG-2 AAC, and is capable of coding music and speech.

The MPEG-4 T/F coder defines a number of extensions to MPEG-2 AAC, in order to enhance the compression performance, such as perceptual noise substitution, LTP, and transform-domain weighted interleave vector quantization (TwinVQ). These enable the operation at extremely low bit rates, with very low delays and under error-prone transmission conditions. Bit-rate scalability is also provided in the T/F coder. When TwinVQ is used in conjunction with the LTP and perceptual noise substitution tools, the code can operate at bit rates between 6 and 16 kbits/s/ch; and this method is frequently used in the scalable configurations of the MPEG-4 T/F audio coder. Also, the coding quality scales with bit rate for all types of signals.

Parametric audio coding uses a signal representation that is based on parameterized models. This approach has long been used in musical analysis/synthesis, as well as in speech vocoders. Sinusoidal modeling has been used for analysis/synthesis of musical instruments, speech, and audio. There are also many other sound models such as harmonic models and noise models. Perceptual models are also considered in the parametric audio coder.

HILN audio coding [24] achieves very low bit rates, down to 4 kbits/s. HILN uses three types of components of the source models: individual sinusoids (described by their frequencies and amplitudes), a harmonic tone (described by its fundamental frequencies, amplitude, and spectral envelope of its partials), and a noise component (by its amplitude and spectral envelope). For the typical target bit rates of 6 to 16 kbits/s, only the parameters for a small number of components can be transmitted, and thus a perceptual model is employed to select those most important components. The parameter estimation can be based on LPC estimation. HILN can be combined with the MPEG-4 HVXC parametric speech coder to cover a wider range of signals and bit rates.

Problems

16.1 In an adaptive PCM speech coding system, the speech is sampled at 8 kHz, and each sample is represented by 8 bits. The quantizer step size is recomputed every 20 ms, and it is encoded using 4 bits. What is the bit rate of the speech coder? What is the SQNR of this system?

16.2 Consider a mobile communication system with the uplink frequency band between 800 MHz to 850 MHz and the downlink frequency band between 900 MHz to 950 MHz. Traffic channels occupy 80% the bandwidth. The system is required to support at least 2500 simultaneous calls using FDMA. The modulation scheme has a spectral efficiency of 1.68 bits/s/Hz, and rate 2/3 FEC codes are used for channel coding. Determine the maximum transmission bit rate of the speech coder.

16.3 Take a voiced portion of any speech waveform for pitch period estimation. Compute and plot the autocorrelation for a delay of $l = 20$ to 150 samples. Estimate the pitch period.

16.4 Clipping is an effective method to improve pitch period estimation by using the autocorrelation method. It reduces the influence of formant frequencies by eliminating low-amplitude samples. Define a clipping function

$$f(x) = \begin{cases} x + c, & x < -c \\ 0, & -c \leq x \leq c \\ x - c, & x > c \end{cases},$$

where c is a clipping limit. Solve Problem 16.3 by using the clipped signal as the input.

16.5 In order to estimate the pitch period, it is common to lowpass the speech signal at a frequency between 500 and 800 Hz before applying the autocorrelation method. Solve Problem 16.3 after lowpassing the speech signal.

16.6 Take a voice portion of any speech waveform. Solve Equation (16.25) for the LPC coefficients by using the Levinson-Durbin algorithm.

16.7 Given the all-pole filter $x(n) = u(n) - a_1 x(n-1) - a_2 x(n-2) - a_3 x(n-3)$ and all-zero filter $x(n) = u(n) + a_1 u(n-1) + a_2 u(n-2) + a_3 u(n-3)$, if $a_1 = -0.9$, $a_2 = 0.6$, and $a_3 = -0.5$, derive the RCs k_i. [Hint: The result is the same for both the filters.]

16.8 An autocorrelation coefficient can be defined to replace the autocorrelation function for pitch period estimation. Using the autocorrelation coefficient yields more precise estimations due to compensation for changing signal amplitudes. Apply this method for a segment of a voiced speech waveform.

16.9 For a voiced portion of a speech waveform, find the LPC coefficients. Using a prediction order of ten and a frame length of 240 samples, calculate the segmental prediction gain by averaging over many frames.

16.10 Train scalar quantizers of 2, 3, 4, 5, 6, and 7 bits with a segment of speech waveforms (say, 2000 samples) by using the Lloyd algorithm. Plot the quantizer transfer characteristic for each case. Plot the distortion sum as a function of the quantizer size. Test the performance of the quantizers using a segment of another speech waveform.

16.11 The pdf of speech samples is commonly accepted to be approximated by the Laplacian distribution

$$f(x) = \frac{1}{\sqrt{2}\sigma} e^{-\sqrt{2}|x|/\sigma}$$

Verify this by using a large amount of speech samples.

16.12 For input vector $x = (x_1, x_2)^T$, a VQ is designed. Each dimension of the input vector is quantized using an identical uniform scalar quantizer, and $0 \le x_i \le 1$. Plot the cells and codewords in the 2-D plane when the scalar quantizers use two bits each.

16.13 Implement voiced/unvoiced detection for LPC by using energy detection and ZCR detection. Test and compare the two methods on a segment of speech signals on a frame-by-frame basis.

16.14 The LPCs of a CELP encoder are given as $a_1 = -1.30$, $a_2 = 1.14$, $a_3 = -1.05$, $a_4 = 0.70$, $a_5 = -0.30$, $a_6 = 0.38$, $a_7 = -0.07$, $a_8 = 0.01$, $a_9 = 0.05$, $a_{10} = 0.06$. Plot the magnitude functions of the formant synthesis filter and the formant analysis filter. Plot the magnitude function of the perceptual weighting filter for different values of γ.

16.15 The G.729 coder uses the following filter to process the input speech

$$H(z) = \frac{0.46364 - 0.92725z^{-1} + 0.46364z^{-2}}{1 - 1.9059z^{-1} + 0.91140z^{-2}}.$$

Plot the magnitude response of the filter and explain its purpose.

16.16 The TIA/EIA IS96 defines the pseudorandom number generator as

$$x(n) = (521x(n-1) + 259)\mathrm{mod}\ 2^{16}.$$

Generate several sequences using different initial conditions. Verify their whiteness.

16.17 Download the codec Speex from http://www.speex.org. Compile speex under UNIX/Linux or Windows/Cygwin. The configuration option for compiling can be for different hardware platforms both floating-point and fixed-point architectures, including TI C5xxx and C6xxx, ADI Blackfin, x86, ARM etc. The code works on various operating systems including many Linux/UNIX variants, Symbian, and MacOS X. Learn the Speex Codec API (the libspeex library) and their control options.

Design a speech encoder for transmitter and a speech decoder for receiver. The speech encoder can adaptively change its encoding mode according to the channel quality which is measured at the receiver and is fed back from the receiver. At the receiver, the speech decoder is able to decode the speech. Write a C program for this purpose.

References

[1] J. P. Adoul, P. Mabilleau, M. Delprat & S. Morissette, Fast CELP coding based on algebraic codes. In *Proc IEEE ICASSP*, Dallas, TX, May 1987, 1957–1960.

[2] B. S. Atal & J. R. Remde, A new method of LPC excitation for producing natural-sounding speech at low bit rates. *IEEE ICASSP*, Paris, France, May 1982, 614–617.

[3] M. Bosi & R. E. Goldberg, *Introduction to Digital Audio Coding and Standards* (Boston, MA: Kluwer, 2003).

[4] P. T. Brady, A statistical analysis of on-off patterns in 16 conversations. *Bell Syst. Tech. J.*, **47**:1 (1968), 73–91.

[5] S. J. Campanella & G. S. Robinson, A comparison of orthogonal transformations for digital speech processing. *IEEE Trans. Commun.*, **19**:6 (1971), 1045–1049.

[6] J.-H. Chen, R. V. Cox, Y.-C. Lin, N. Jayant & M. J. Melchner, A low-delay CELP coder for CCITT 16 kb/s speech coding standard. *IEEE J. Sel. Areas Commun*, **10**:5 (1992), 830–847.

[7] W. C. Chu, *Speech Coding Algorithms: Foundation and Evolution of Standardized Coders* (Hoboken, NJ: Wiley, 2003).

[8] W. Chu, M. O. Ahmad & M. N. S. Swamy, Modified silence suppression algorithms and their performance tests. In *Proc. IEEE Midwest Symp. Circ. Syst.*, Cincinnati, OH, Aug 2005, **1**, 436–439.

[9] J. R. Deller, Jr., J. H. L. Hansen & J. G. Proakis, *Discrete-Time Processing of Speech Signals* (New York: Wiley-IEEE, 2000).

[10] K.-L. Du & M. N. S. Swamy, *Neural Networks in a Softcomputing Framework* (London: Springer, 2006).

[11] I. A. Gerson & M. A. Jasiuk, Vector sum excited linear prediction (VSELP) speech coding at 8kbps. In *Proc. ICASSP*, Albuquerque, NM, Apr 1990, **1**, 461–464

[12] L. Hanzo, F. C. A. Somerville & J. P. Woodard, *Voice Compression and Communications: Principles and Applications for Fixed and Wireless Channels* (New York: IEEE Press, 2001).

[13] K. Homayounfar, Rate adaptive speech coding for universal multimedia access. *IEEE Signal Process. Mag.*, **20**:2 (2003), 30–39.

[14] ITU-T, *Single-ended Method for Objective Speech Quality Assessment in Narrowband Telephony Applications*, ITU-T Rec. P.563, Geneva, May 2004.

[15] ITU-T, *Perceptual Evaluation of Speech Quality (PESQ): An Objective Method for End-to-end Speech Quality Assessment of Narrowband Telephone Networks and Speech Codecs*, ITU-T Rec. P.862, Geneva, Feb 2001.

[16] J. D. Johnston, Transform coding of audio signals using perceptual noise criteria. *IEEE J. Sel. Areas Commun.*, **6**:2 (1988), 314–323.

[17] Z. Kostic & S. Seetharaman, Digital signal processors in cellular radio communications. *IEEE Commun. Mag.*, **35**:12 (1997), 22–35.

[18] K. Li, M. N. S. Swamy & M. O. Ahmad, An improved voice activity detection using higher order statistics. *IEEE Trans. Speech Audio Process.*, **13**:5 (2005), 965–974.

[19] K. Mano, F. Moriya, S. Mild, H. Ohmuro, K. Ikeda & J. Ikedo. Design of a pitch synchronous innovation CELP coder for mobile communications. *IEEE J. Sel. Areas Commun.*, **13**:1 (1995), 31–41.

[20] A. V. McCree & T. P. Barnwell III, A mixed excitation LPC vocoder model for low bit rate speech coding. *IEEE Trans. Speech Audio Process.*, **3**:4 (1995), 242–250.

[21] M. Nelson, *The Data Compression Book* (New York: M&T Books, 1992).

[22] M. Nishiguchi, A. Inoue, Y. Maeda & J. Matsumoto, Parametric speech coding— HVXC at 2.0–4.0 kbps. In *Proc. IEEE Workshop Speech Coding*, Porvoo, Finland, Jun 1999, 84–86.

[23] E. Ordentlich & Y. Shoham, Low-delay code-excited linear-predictive coding of wideband speech at 32 kbps. In *Proc. IEEE ICASSP*, Toronto, Canada, May 1991, 9–12.

[24] H. Purnhagen & N. Meine, HILN: the MPEG-4 parametric audio coding tools. In *Proc. IEEE ISCAS*, Geneva, Switzerland, May 2000, 201–204.

[25] K. Sayood (ed), *Lossless Compression Handbook* (San Diego, CA: Academic Press, 2003).

[26] M. R. Schroeder & B. S. Atal, Code-excited linear prediction (CELP): high-quality speech at very low bit rates. In *Proc IEEE ICASSP*, Tampa, FL, Mar 1985, **3**, 937–940.

[27] C. E. Shannon, Coding theorems for a discrete source with a fidelity criterion. *Proc. IRE National Convention Record*, New York, Mar 1959, part 4, 142–163.

[28] A. S. Spanias, Speech coding: a tutorial review. *Proc. IEEE*, **82**:10 (1994), 1541– 1582.

[29] D. Stranneby & W. Walker, *Digital Signal Processing and Applications*, 2nd edn (London: Elsevier, 2004).

[30] H. Sun, C. Wang, M. O. Ahmad & M. N. S. Swamy, An 8 kb/s low delay CELP speech coder. In *Proc. 7th IEEE Int. Conf. Electron. Circ. Syst.*, Jounieh, Lebanon, Dec 2000, **1**, 286–289.

[31] E. Terhardt, Calculating virtual pitch. *Hearing Research*, **1**:2 (1979), 155–182.

[32] TIA/EIA, *TDMA Cellular/PCS–Radio Interface–Minimum Performance Standards for Discontinuous Transmission Operation of Mobile Stations*, ch16:IS-727, Jun 1998.

[33] J. M. Tribolet & R. E. Crochiere, Frequency domain coding of speech. *IEEE Trans. Acoust. Speech Signal Process.*, **27**:5 (1979), 512–530.

[34] W. Tong, E. I. Plotkin & M. N. S. Swamy, Active acoustic noise cancellation with audio signal enhancement based on an almost-symmetrical time-varying autoregressive-moving average model. *J. Acoust. Soc. Amer.*, **99**:6 (1996), 3528–3538.

[35] P. P. Vaidyanathan, Multirate digital filters, filter banks, polyphase networks, and applications: a tutorial. *Proc. IEEE*, **78**:10 (1990), 56–93.

[36] E. Zwicker & H. Fastl, *Psychoacoustics: Facts and Models* (Berlin: Springer-Verlag, 1990).

17 Source coding II: image and video coding

17.1 Introduction

A digital image is a rectangular array of picture elements (pixels), arranged in m rows and n columns. The resolution of the image is $m \times n$. Images can be categorized into bi-level, grayscale, and color images. A natural scene, such as a picture taken by a digital camera or obtained by using a scanner, is typically a continuous-tone image, where the colors vary continuously to the eye and there is a lot of noise in the picture. An artificial image, such as a graphical image, does not have the noise or blurring of a natural image. A cartoon-like image consists of uniform color in each area, but adjacent areas have different colors.

The features in each type of image can be exploited to achieve a better compression. For example, for the bi-level image, each pixel is represented by one bit. A pixel has a high probability of being the same as its neighboring pixels, and thus RLE is suitable for compressing such image. The image can be scanned column by column or in zigzag. For the grayscale image, each pixel is represented by n bits, and a pixel tends to be similar to its immediate neighbors but may be not identical, thus RLE is not suitable. By representing the image using a Gray code that differs in only one bit for two consecutive integers, a grayscale image can be separated into n bi-level images, and each can be compressed by using RLE. The similarity of neighboring pixels can be exploited by using a kind of prediction encoding.

A video is composed of many frames of image. It can be compressed by intraframe (spatial) as well as interframe (temporal) coding. Video compression is based on image compression. Image or video compression aims to remove the redundancy.

Lossless compression

Conventional lossless compression techniques that use Huffman coding or dictionary-based compression exploit the statistical variations in the frequency of individual symbols or the frequency of strings of symbols (i.e., phrases). For continuous-tone images, these lossless compression techniques cannot generate impressive compression. Since the pixels tend to be well spread over their entire range, their histograms based on frequency are not so spiky, and this limits the compression. When Huffman coding is used to encode a grayscale image, which typically has a set of 255 levels, the compression result is usually very disappointing, with a reduction of only about 1/2 to 1 bit/pixel [25].

Making use of the similarity of neighboring pixels, the difference between the real pixel value and its estimated value can be encoded with less bits. This is known as the

DPCM technique. A ensuing Huffman coding step can lead to a substantial improvement in compression ratio. However, this technique cannot generate good compression ratio for a scene that looks much like white noise, since it is unpredictable both spatially and temporally.

For a color image, a pixel consists of three color components, red, green and blue. A continuous-tone color image can thus be separated into three grayscale images, and each grayscale image can be compressed separately. For discrete-tone images, they have uniform regions, and this feature can be exploited to increase the compression ratio. Based on the similarity of neighboring pixels, many spatial prediction methods have been proposed.

Popular lossless image compression formats are PNG, GIF, and TIFF. They make use of the LZ77 variants or LZW coding. PNG was developed for improving and replacing GIF. It combines LZ77 and Huffman coding with prediction.

The JBIG standard (ITU-T Rec. T.82, ISO/IEC-11544) is a lossless image compression standard used for bi-level images. JBIG is used for compression of facsimile pages or scanned document compression. JBIG is based on prediction and context-based arithmetic coding. The lossless mode of the JBIG2 (ITU-T T.88, ISO/IEC 14492) standard generates files that are 2 to 4 times smaller than JBIG. JBIG-KIT is a free C program for JBIG encoder and decoder.[1]

Lossy compression

Image or video compression requires a compression ratio of 20:1 to 200:1. This makes conventional lossless coding techniques such as entropy coding and DPCM not very viable. Lossy compression using transform coding is the dominant image and video compression approach. In transform coding, the coding bit rate and picture quality are the two critical factors. The coding bit rate decides the bandwidth for transferring the images. Picture quality can be characterized by the MSE between the coded image or video and the original one. Both the reconstruction error and the data rate are controlled by the quantization parameters.

The JPEG and JPEG2000 formats are the most important image formats, and will be introduced in Sections 17.6 to 17.8. JBIG2 allows lossy compression of bilevel images [26]. Open-source JBIG2 encoder and decoder programs are available as jbig2enc and jbig2dec.[2,3] JBIG2, JPEG, and JPEG2000 compression techniques for images, as well as popular general-purpose compression techniques such as RLE and LZW, are all integrated into Portable Document Format (PDF), which is a de facto standard for documentation.

Image processing

In some situations, an image may be subject to special processing before compression. An image may require the following processing:

---

[1] http://www.cl.cam.ac.uk/$\sim$mgk25/jbigkit/
[2] http://www.imperialviolet.org/jbig2.html
[3] http://jbig2dec.sourceforge.net/

- *Noise filtering*: A corrupted image may require the removal of noise by a suitable filter. Averaging, Gaussian smoothing, and median filters are usually used for denoising.
- *Redundancy reduction*: to eliminate redundancy and unimportant information for a specific image analysis.
- *Importance enhancement*: such as edge enhancement for emphasizing the shapes in an image, image thresholding and edge detection for image segmentation.

Mathematical morphology is an important tool for image processing and analysis. Image dilation and image erosion are two basic morphology operations. Image restoration, segmentation, texture analysis, shape analysis, geometric transformations, and corner detection are all important topics of image processing. A good text on image processing and computer vision is by Sonka et al [31].

Denoising is one of the most important processing steps prior to image/video compression. Impulse noise can be modeled by using the Bernoulli-Gaussian process or the Poisson-Gaussian process, where the random time of occurrence of impulses is modeled by a binary Bernoulli process or a Poisson process, while the random amplitude of the impulse noise is modeled by a Gaussian process. Impulse noise is more detectable in the time domain than in the frequency domain. The median filter is traditionally used for the removal of impulse noise. The median filter is a nonlinear filter that takes the median of a set of numbers. The median is insensitive to an outlier, which is a sample with an unusually large value. The median filter is especially useful for image processing, since it removes noise and speckles but preserves the edges or stepwise discontinuities in the image. A number of techniques, which use an influence function to limit the influence of the outliers, are discussed in [10]. Other denoising techniques are Kalman filtering for impulsive noise [23], DWT filtering for correlated noise [15, 23], sampled function weighted order filtering for multiplicative noise [27], video denoising based on spatial filtering of the noisy wavelet coefficients [21], spatially adaptive wavelet-based methods for speckle noise [3, 5] and for AWGN noise [4], adaptive wavelet-based Bayesian denoising [22], denoising AWGN and speckle in videos based on spatial and temporal filtering [30], and despeckling using lossy compression [14].

17.2 Perception of human vision

17.2.1 Human visual system

The human visual system is an imaging system. Light enters the cornea, and the amount of light is controlled by the pupil in the iris. The light then passes through a crystalline lens and a vitreous body, which is filled with a transparent jelly known as *vitreous humor*, and is focused on the retina by the lens. The retina senses the light and then passes the signal to the brain via the optic nerve. The brain finally perceives the image. The eye is just a spatial sampling device.

In the retina, there are two types of discrete receptors called *rods* and *cones* from their shape. Rods dominate the periphery of the retina, while cones occupy a central area. Vision using the rods is monochromatic, has low resolution, but is effective at very low light level. On the other hand, using the cones leads to high resolution and color vision, but this also requires more light.

The cones on the retina occur in three different types, corresponding to different colors. The human vision is restricted to a light wavelength from 400 to 700 nanometers (nm), and the response is not uniform: It is very poor for blue lights, but it peaks in the area of green. This triple receptor characteristic of the eye enables to generate a range of colors by combining light sources having just three different wavelengths in different proportions. This is known as *additive color matching*. Weber's law states that manipulation of colors is a linear operation: Any color can be created by a linear combination of the three colors and such a combination is unique.

The eye is sensitive to light over a very large range of intensities: The upper end is about 10^{10} times the lower end [25]. However, at any instant, the eye adapts only to an average brightness level, and the eye can only perceive a small fraction of the total range. The contrast sensitivity of the eye is the smallest brightness difference that is visible. The contrast sensitivity is not constant, but increases in proportion to brightness. That is, a brightness change of one per cent is equally detectable for all brightness. Human vision is sensitive to small changes in luminance, but is relatively insensitive to changes in chrominance (color). Thus, for image compression, the chrominance can be compressed at a higher ratio.

The human visual system is not sensitive to high frequency components of spatial frequency. These high frequency components are usually treated as noise. The sensitivity of the eye to spatial frequency falls from the maximum, as the frequency increases from zero. This feature is usually used in image compression.

Masking occurs between periodic patterns with similar orientation and radial frequency, and between aperiodic patterns such as luminance borders, and an even greater masking effect occurs in textured regions. For videos, large-area flicker is less perceptible at rates of the order of 60 Hz. These perceptual models are exploited in most image, television, and video systems.

Visually lossless compression

Like speech/audio compression, the performance of image compression is evaluated by human perception, and thus these data can be slightly modified without affecting the perceived quality. Visually lossless compression is actually a kind of lossy compression, but the human viewer cannot detect a visual degradation. The strategy is to keep the difference between the original image and the reconstructed image below the threshold of perception. This principle is widely used in color space transformation and transform coding.

17.2.2 Color spaces

In computer image or video systems, a pixel of a color image can be represented as either RGB, YUV, or YIQ format. The RGB format represents the intensity values of red (R), green (G), and blue (B). The RGB color space is well suited to capture and display color images. RGB images are displayed on a color CRT (cathode ray tube) or an LCD (liquid crystal display) by separately illuminating the red, green and blue components of each pixel according to their intensities.

YCbCr, also known as *YUV*, is used in the European analog PAL (Phase Alternation Line) color TV system and international standards of digital video. The human visual system is less sensitive to color than to luminance (luma). In the RGB color space, the three components have the same resolution. The YUV format uses an independent intensity value (luminance, Y) plus two color (chrominance, U and V) values known as *hue* and *saturation*. The luminance component Y is a weighted average of R, G, and B

$$Y = k_r R + k_g G + k_b B, \tag{17.1}$$

with

$$k_b + k_r + k_g = 1. \tag{17.2}$$

ITU-R Rec. BT.601 defines $k_b = 0.114$ and $k_r = 0.299$. Accordingly [16]

$$Y = 0.299R + 0.587G + 0.114B, \tag{17.3}$$

$$C_b = 0.564(B - Y), \quad C_r = 0.713(R - Y), \tag{17.4}$$

where $0 \le R, G, B \le 1$ are normalized R, G, B values. The luma can be assigned a higher resolution than color.

Separation of luminance and color information is for the purpose of compatibility with old black and white TV systems. The Y, Cr, Cb components are further processed by using a DCT or wavelet transform.

There are also some other color spaces such as CMYK format for color printing and the HSV format for perceptual uniform color space.

Analog video systems

In addition to the PAL system, two other major analog color TV systems are the NTSC (National Television System Committee) system and the SECAM (SEquential Couleur Avec Memoire) system. The NTSC system is employed in America and Japan, the SECAM system is used in France, Eastern Europe, Russia, and the Middle East, and PAL is used in West Europe, China, India, and Australia.

PAL and SECAM use the 625/50 (625 scan lines and 50 fields/s) system, and NTSC uses the 525/60 system. All the three systems are interlaced with a 4:3 aspect ratio. For interlaced scanning, each frame is formed by two successive (odd lines and even lines) scanning passes (two fields); thus, the frame rate of the system is either 25 or 30 frames/s.

The three analog TV systems differ mainly in the way the luminance/chrominance components are calculated from the RGB component. They calculate luma by (17.3) for compatibility with the monochrome TV system, but use different coefficients for calculating the chroma components. NTSC uses YIQ, where I and Q represent the in-phase and quadrature axes on the same color space spanned by the U and V axes. SECAM uses $YD_b D_r$, where $D_b = R - Y$ and $D_r = B - Y$.

YCbCr sampling

The YCbCr data can be sampled by the 4:4:4, 4:2:2, or 4:2:0 format. The 4:4:4 sampling has the same resolution for Y, Cb and Cr, and has a sample of each component for every pixel. This provides no compression on the chrominance components. In 4:2:2 (2h1v) sampling, Cb and Cr have the same vertical resolution (1:1) as Y, but are downsampled at 2:1 horizontally. The 4:2:2 sampling is used for high-quality color reproduction; it compresses the image to 2/3 of its original size.

In the popular 4:2:0 (2h2v) sampling, Cb and Cr are downsampled at a ratio of 2:1 both horizontally and vertically, each having half the horizontal and half the vertical resolutions of Y. The 4:2:0 sampling requires half as many samples as 4:4:4 sampling.

In digital video, a pixel can be represented by either RGB or YUV format. Typically, a pixel can be represented by RGB format of 24 bits, where R, G, B each have 8 bits, or by YUV format of 16 bits, where Y uses 8 bits and U, V each use 4 bits. For video systems, the amount of data is excessive, and must be compressed for transmission and storage.

17.3 Quality of image and video coding

Like speech and audio coding, the quality of image and video coding can be evaluated by subjective or objective criteria. The subjective quality evaluation depends on the perception of the human visual system, as well as other factors like the viewer's state of mind and visual attention.

The U. Waterloo Repertoire is a suite of 32 test images for comparing different image compression algorithms.[4] MPEG has a wide library of video and audiovisual sequences for evaluation. The selection from this library is based on the criticality and the structure of the scene.

Subjective quality evaluation

The quality of image or video quality can also be subjectively evaluated by using the MOS measure. The test procedures for subjective quality evaluation are defined in ITU-R Rec. BT.500-11. According to ITU-R Rec. BT.500-10, the number of nonexpert viewers that participate in a test should be not less than 15, and a lower number for expert viewers. The viewers must be screened for visual acuity to detect color blindness or strong color vision deficiencies. At the beginning, the viewers are subject to a training phase. The tests are carried out based on ITU-R BT.500 and ITU-T P.910.

The commonly-used procedure from BT-500 is the double stimulus continuous quality scale (DSCQS) method. In a typical test session, a series of sequence pairs are shown to the viewer in order to grade each pair (original and impaired). The order of the two sequences is randomized during the test session.

---

[4] http://links.uwaterloo.ca/bragzone.base.html

Objective quality evaluation

The subjective evaluation method is costly and time-consuming. Objective quality evaluation methods are more widely employed during the period of image and video compression algorithm development. The most heavily employed measure is the peak signal-to-noise ratio (PSNR).

PSNR is defined as the ratio of the square of the highest possible signal value in the image/video to the MSE between the original and the impaired image or video frame

$$\text{PSNR} = 10\log_{10}\left(\frac{x_{\text{peak}}^2}{\sigma_d^2}\right)$$

$$= 10\log_{10}\frac{(2^n-1)^2}{\sigma_d^2} \quad (\text{dB}), \tag{17.5}$$

where $x_{\text{peak}} = 2^n - 1$ is the peak value of the signal, n is the number of bits per image sample, and the MSE σ_d^2 is the variance of the distortion.

The MAD measure is also quite frequently used for evaluating image compression algorithms

$$\text{MAD} = \frac{1}{N}\left|x - \hat{x}\right|. \tag{17.6}$$

The PSNR measure usually indicates the quality of an image or video frame. However, in some special cases, the change in the background or a local area can lead to very low PSNR, while the subjective quality is still very high as long as there is not much degradation in the area of interest. This is also true with MAD.

17.4 Predictive coding

For a source with memory, predictive coding known as *DPCM* is usually used to exploit the inherent dependence within the various events. DPCM is not in itself a compression method. Instead it is a technique for data decorrelation. Differential coding and adaptive coding are successful for speech coding, but their performance is very poor for image data. This difference arises from the fact that audio data tends to be repetitive. This repetitive nature is exploited in the LPC and ADPCM. In general, differential coding of images does not produce compression much better than the high-performance lossless algorithms.

Adaptive coding, usually used with differential coding, predicts the upcoming data (pixels) using previously seen data (pixels), and can adjust the coding scheme accordingly. The prediction MSE is significantly influenced by just the first three neighboring pixels, but less influenced by changes in pixels beyond that. The order of the predictors is the number of pixels used for prediction. Second- or third-order predictors are typically two-dimensional, that is, the predictor uses pixels that are displaced both horizontally and vertically from the pixel that is predicted. Given that the (i,j)th pixel is to be predicted, a second-order 2-D predictor can be defined by

$$\hat{x}(i,j) = 0.5x(i,j-1) + 0.5x(i-1,j) \tag{17.7}$$

and a third-order 2-D predictor by

$$\hat{x}(i,j) = 0.75x(i,j-1) - 0.5x(i-1,j-1) + 0.75x(i-1,j). \tag{17.8}$$

After using the predictive coding, the remanent error signal exhibits a very high spike at the location of zero, which is particularly suitable for entropy coding. Entropy coding is then employed to generate the compressed image data.

Although adaptive coding exhibits good performance, the compression is still far from being desirable. Adaptive coding combined with entropy coding is used in lossless compression in JPEG. Predictive coding is also used in lossy modes of JPEG and MPEG.

Example 17.1: Given an original image *Cynric.bmp* of size 240 × 320, (Fig. 17.1a) the outputs for the second- and third-order 2-D predictors, (17.7) and (17.8), are very similar. The predicted and error images obtained, when the second-order 2-D predictor is used, are also shown in Fig. 17.1. The histograms for the original image and the error images obtained by the two predictors are plotted in Fig. 17.2. The histograms of the two predictors are very similar, and they are very spiky and suitable for entropy coding.

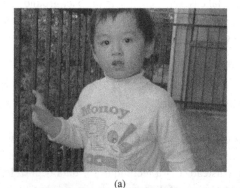

(a)

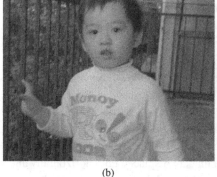 

(b)   (c)

Figure 17.1   Predictive coding. (a) The original image. (b) The predicted image. (c) The error image.

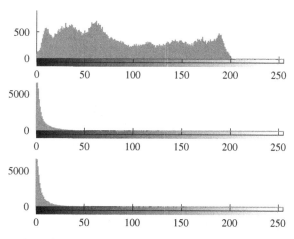

Figure 17.2 Predictive coding: the histograms of the original, predicted, and error images.

17.5 Transform-based image compression

In Chapter 13, we introduced the Laplace, z-, and Fourier transforms. DCT is a very popular transform used in image and video processing [1], and belongs to a family of sine and cosine transforms. There are also many other important transforms, such as the Karhunen-Loève transform and DWT.

The Karhunen-Loève transform [10] is an optimum transform that minimizes the overall MSE between the original and reconstructed images. The Karhunen-Loève transform is more complex than alternate suboptimum transforms such as the sine, cosine, or Fourier transform. A smoothly varying image area has an isotropic statistical dependency, which can be described by a first-order Markov process with a correlation coefficient of close to unity; under this condition, DCT has an image quality similar to the Karhunen-Loève transform, and is used in image or video compression standards.

For image compression, lossy compression based on DCT or DWT can generate remarkable compression, and both techniques are commonly employed in modern image and video compression.

Quantization of transform coefficients

The transform itself does not yield any compression. A block of 64 pixels will generate 64 DCT coefficients. Due to the orthogonality of the transform, the energy in both the pixel and transform domains are equal, and thus no compression is achieved. However, the significant part of the image energy is at the lower frequency components, while the majority of the higher frequency coefficients have little energy. Also, the human eye is not sensitive to higher frequencies. By discarding the higher frequencies or quantizing them using large step size, compression is achieved.

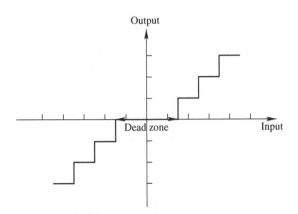

Figure 17.3   **The uniform quantizer with dead zone.**

The quantizers used in all standard image/video codecs are based on the uniform quantizer. Typically, the uniform quantizer is used for quantizing the dc coefficients, while the uniform quantizer with dead zone (UQ-DZ) is used for the ac coefficients. This primarily transforms more nonsignificant ac coefficients to zeros to increase the compression efficiency. The uniform quantizer is shown in Fig. 12.6, which can be classified as the midtread and midrise uniform quantizers. The midtread uniform quantizer is preferred for dc coefficient quantization, since zeros are generated for small dc coefficients. The UQ-DZ is shown in Fig. 17.3.

17.6 JPEG standard

The JPEG (Joint Photographic Experts Group) is a joint ISO/IEC and ITU-T standard (IS10918-1, ITU-T T.81) for compression of still images, and was completed in 1992.

17.6.1 Four modes of operation

The JPEG specifications define four modes of operation, namely

- *Baseline sequential encoding*: The image is encoded block-by-block from left to right, top to bottom.
- *Progressive encoding*: The image is encoded in such a manner that during decoding the reconstructed image is successively refined for each decoded block (known as *scan*).
- *Lossless encoding*: The decoded image is the same as the original image bit-by-bit.
- *Hierarchical encoding*: Multiple copies of the image are bundled together with each copy having a different resolution.

These modes rely on distinctly different technologies. The three lossy compression modes are all based on DCT, and the difference between them is the way in which the DCT

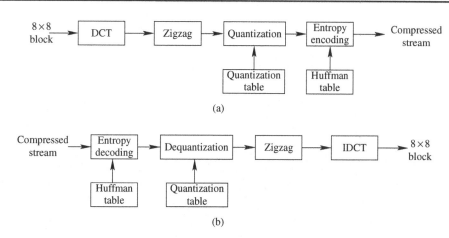

Figure 17.4

The JPEG system. (a) Encoder. (b) Decoder.

coefficients are transmitted. The lossless compression utilizes the predictive/adaptive coding and a Huffman code output stage.

The baseline sequential mode is the simplest DCT-based coding technique. It is sufficient for many applications, and is the default DCT-encoder for JPEG. The sequential encoding/decoding process is shown in Fig. 17.4. JPEG achieves image compression by discarding visually insignificant information such as the high frequency components.

The association between pixels tends to diminish rapidly. A pixel that is even 15 pixels away is of little use for the prediction of the current pixel [20]. For this reason, a DCT block of 64×64 might not compress much better than if we compress four 16×16 blocks, but DCT of a 64×64 block consumes considerably longer time. JPEG applies DCT to blocks of 8×8 image pixels, yielding 64 output frequency coefficients.

Progressive encoding is also desirable when the compressed image is transmitted and viewed in real time such as in a Web browser. JPEG uses DCT to transform the image into its spatial frequency components. In the progressive implementation of JPEG, the low-frequency components are compressed and transmitted first. These low-frequency components provide the general information of an image, while high-frequency components give details of the image. On coding a coefficient, a specified number of MSBs can be encoded first, and the other bits can be encoded in subsequent scans. This is known as *successive approximation* or *bit plane encoding*.

In the hierarchical encoding mode, an image is coded as a sequence of layers in a pyramid. At different layers of the pyramid, the resolution is different. Each lower layer provides prediction for the next upper layer. The lower-layer image is a downsampling version of the source image. Except for the top layer, for each layer, the difference between the source image and its lower reference layer is encoded, for the luminance and chrominance components, by using DCT or lossless procedure. Thus, the hierarchical mode also provides a progressive representation of an image, but is used in case multiresolution is required.

The lossless encoding calculates a predicted value for each pixel, and the difference between the pixel and its predicted value is encoded using Huffman coding.

The decoding procedure is exactly the inverse of the encoding process. JPEG achieves a compression ratio of 2:1 for lossless compression to 20:1 for lossy compression. JPEG works best on continuous-tone images, but performs poorly on bi-level images.

The lossy JPEG standard is a modification to the scheme proposed by Chen and Pratt [8]. DCT is a lossless transformation that does not perform compression, while the quantization of the DCT coefficients is lossy. Sample C code for demonstrating DCT compression and its expansion is given in [20], and free C source code of JPEG encoder is provided by the Independent JPEG Group.[5]

Note that the lossless mode of JPEG is different from the later JPEG-LS (ISO/IEC 14495-1:1999) standard for lossless coding of still images. JPEG-LS provides a low-complexity solution for lossless image coding with a compression efficiency of around 3 to 4. It is also based on adaptive prediction. JPEG-LS uses exponential-Golomb variable-length codes.

17.6.2 Quantization

In JPEG, the RGB image is first transformed into a luminance/chrominance color space to remove correlation. The pixels of each of the color components are then organized into 8×8 blocks. If the number of rows or columns are not a multiple of 8, the bottom row or the rightmost column are duplicated as many times as necessary. DCT is then applied to each 8×8 block, and extracts 64 frequency components.

The input image is first level shifted by 2^{P-1} from each pixel value, where P is the number of bits used for each pixel. DCT is then applied for each 8×8 block, and 64 DCT coefficients are obtained. Uniform midtread quantization is applied to quantize these coefficients, and the quantizer step sizes are given in the form of tables. The quantization tables for luminance and chrominance recommended by the JPEG standard are given in Table 17.1. These tables are obtained empirically from psychovisual thresholding for luminance and chrominance with 2:1 horizontal subsampling.

The step size generally increases as we move from the dc coefficients to the high-order coefficients. Since the human visual system is more acute to low frequency image information, the low frequency coefficients are assigned more bits, and the dc coefficient is assigned the largest number of bits. This creates more zeros for the high-frequency coefficients, and in the bottom right of the DCT coefficient array, there are some zeros. Compression mainly occurs at this stage. Thus, the quantization table $\mathbf{Q} = [Q(i,j)]$ can be alternatively implemented by

$$Q(i,j) = 1 + (i+j)R, \tag{17.9}$$

[5] http://www.ijg.org/

Table 17.1. The quantization tables giving the $Q(i,j)$ values, as recommended by JPEG.

luminance (grayscale)								chrominance (Cr and Cb)							
16	11	10	16	24	40	51	61	17	18	24	47	99	99	99	99
12	12	14	19	26	58	60	55	18	21	26	66	99	99	99	99
14	13	16	24	40	57	69	56	24	26	56	99	99	99	99	99
14	17	22	29	51	87	80	62	47	66	99	99	99	99	99	99
18	22	37	56	68	109	103	77	99	99	99	99	99	99	99	99
24	35	55	64	81	104	113	92	99	99	99	99	99	99	99	99
49	64	78	87	103	121	120	101	99	99	99	99	99	99	99	99
72	92	95	98	112	100	103	99	99	99	99	99	99	99	99	99

where R is a parameter provided by the user.

In image compression, each input vector is a two-dimensional $m \times n$ image block. Pattern matching occurs between the input vector and the constructed codebook. A set of RGB components can be quantized together.

In order to control the compression ratio, a quality factor q, in percentage value between 1 and 100 per cent, is used to scale the quantization matrix $\mathbf{Q}$. In the implementation provided by Independent JPEG Group,[6] the quantization matrix is scaled by a compression factor

$$\alpha = \begin{cases} \frac{50}{q}, & 1 \le q \le 50 \\ 2\left(1 - \frac{q}{100}\right), & 50 \le q \le 99 \end{cases}, \tag{17.10}$$

subject to the condition that the minimum element of $\alpha \mathbf{Q}$ is 1. The value $q = 100$ corresponds to lossless compression, and in this case all elements of $\alpha \mathbf{Q}$ are set as 1.

For each of the DCT coefficients, a label is obtained

$$l(i,j) = \left\lfloor \frac{S(i,j)}{Q(i,j)} + 0.5 \right\rfloor, \tag{17.11}$$

where $Q(i,j)$ is the (i,j)th element of the quantization table, $S(i,j)$ is the (i,j)th DCT coefficient, and $\lfloor x \rfloor$ takes the largest integer that is less than x. The reconstructed value of $S(i,j)$ is obtained by

$$\hat{S}(i,j) = l(i,j)Q(i,j). \tag{17.12}$$

The quantization error $\hat{S}(i,j) - S(i,j)$ is generally large for high-frequency coefficients due to the large step size.

As the midtread quantizers have a zero output level, they also function as thresholding and generate zeros for those DCT coefficients whose magnitude is less than half the step size. This typically leads to a long run of zeros.

---

[6] http://www.ijg.org/

17.6.3 Coding

Coding of the baseline encoder is implemented in two steps. The first step converts the quantized DCT coefficients into a set of symbols. Entropy coding is then applied to each symbol. JPEG specifies two alternative entropy coding methods, namely Huffman coding and arithmetic coding. For baseline coding, only Huffman coding is specified, while for all other modes of JPEG either Huffman or arithmetic coding can be used.

For dc and ac coefficients, as well as luminance and chrominance components, four different entropy tables may be used. Entropy tables are obtained from the environment.

Coding of dc coefficients

The top-left coefficient (or label) in the DCT coefficient matrix $\mathbf{S} = [S(i,j)]$ is the dc term, and it is the mean signal level of the block. This component changes slowly between adjacent blocks, and thus can be encoded using DPCM, that is, the dc label is coded as the difference from that of the previous block. The differences can be quite large, and a Huffman code for such a large alphabet can be inefficient.

JPEG resolves this by partitioning the possible values of the differences into categories. The size of the categories grows as a power of two: category 0 has one member, 0; category 1 has two members, -1 and 1, and category n has 2^n members, $-2^n, -2^n+1, \ldots, -2^{n-1}-1;$ $2^{n-1} + 1, \ldots, 2^n - 1, 2^n$, for $n \geq 2$. Thus, for a difference of 2^N possible values, only N bits are required for the Huffman code, and the elements within category n require extra n bits.

Coding of ac coefficients

The ac coefficients (or labels) are arranged in a zigzag sequence within the block, as shown in Fig. 17.5 in order to code all other coefficients in an ascending order of frequency. This leads to a zigzag scan of spatial frequency. The ac coefficients are also encoded in a manner similar to that for the dc coefficients. The possible values of the ac coefficients are

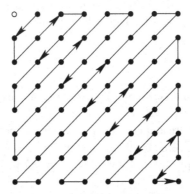

Figure 17.5   The zigzag scanning of the 8 × 8 DCT coefficients for an 8 × 8 image block.

partitioned into categories, but Huffman coding is applied to the ac coefficients themselves rather that to the differences, as in the dc coefficient case. RLE can be applied if a long run of zeros occurs, and if a long run of zeros occurs at the end of the block, an end-of-block (EOB) codeword is used instead. This yields a varying data rate. An increase in compression ratio can be achieved by increasing the step sizes. The higher the compression ratio, the more frequency components in a block turn to zero.

Given an input array, RLE generates a series of (*run*, *level*) pairs, where *run* indicates the number of zeros preceding a nonzero coefficient and *level* the nonzero coefficient. For example, given an input array $18, 0, 0, 0, -2, -8, 0, 0, 0, 0, 0, -1, -9$, the RLE output is $(0, 18), (3, -2), (0, -8), (5, -1), (0, -9)$. Each of these run-level pairs is then encoded as a separate symbol by the entropy encoder.

The nonzero ac labels are further bit-encoded using Huffman or arithmetic coding. The Huffman table is generated from the statistics of the image. Huffman coding is used in most JPEG implementations.

17.7 Wavelet-transform-based image coding

The DCT and other block-based transforms generate blocking artifacts in the image, and these artifacts are more visible at the block boundaries. LOT reduces the artifacts by overlapping adjacent blocks. The wavelet transform is a special type of LOT, and it can eliminate the blocking artifacts. The wavelet transform is used in JPEG2000. Coding of still images under MPEG-4 is also based on the wavelet transform.

17.7.1 Sub-band decomposition

In order to adapt the frequency response of the decomposed pictures to the human visual system, filters are employed for octave bands in sub-band coding. The wavelet transform is a special kind of sub-band coding. For image processing, the wavelet transform provides a multiresolution representation of an image signal.

By using subsampling in the rows and columns, an $N_r \times N_c$ image is decomposed into many subimages. If each row is filtered and downsampled, two $N_r \times \frac{N_c}{2}$ subimages are obtained. If we further filter and downsample the columns of the two subimages, we get four $\frac{N_r}{2} \times \frac{N_c}{2}$ subimages. The four subimages are, respectively, obtained by lowpass filtering the rows and columns (LL), lowpass filtering the rows and highpass filtering the columns (LH), highpass filtering the rows and lowpass filtering the columns (HL), and highpass filtering the rows and columns (HH). This procedure is illustrated in Fig. 17.6, where H_0 and H_1 are lowpass and highpass filters, respectively.

In fact, an original image can be divided into a variety of sub-band structures. For example, each of the four subimages in Fig. 17.6 can be further filtered and subsampled, in a recursive fashion. A popular decomposition is to further subdivide the LL subimage into

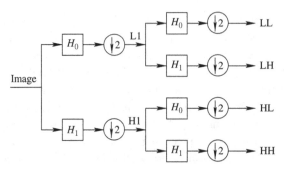

Figure 17.6

Sub-band decomposition of an image.

LL2	LH2	LH1	LH
HL2	HH2		
HL1		HH1	
HL		HH	

Figure 17.7

A popular structure for image sub-band decomposition.

four more subimages (LL1, LH1, HL1, HH1), and LL1 is further divided into four more subimages, resulting altogether 10 subimages. This is shown in Fig. 17.7.

17.7.2 Wavelet filter design

In sub-band coding, perfect reconstruction is achieved if (13.178) is satisfied. By adding some constraints on $H_0(z)$ and $H_1(z)$ such that the property of the wavelet transform is maintained, wavelet filters can be designed. The wavelet definition requires $H_0(z)$ and $H_1(z)$ to be continuous functions of z. It is desirable to have the largest possible number of continuous derivatives; this corresponds to having zeros at $z = -1$ for the wavelet filters. On the other hand, the wavelet filter is required to have linear phase response, since the phase carries information in the case of the image.

Thus, for a wavelet filter having at least n zeros at $z = -1$, we can write [13]

$$P(z) = H_0(z)H_1(-z) = \left(\frac{1+z^{-1}}{2}\right)^{2n} Q(z), \tag{17.13}$$

where $Q(z)$ has n unknown coefficients, and the notations $H_0(z)$, $H_1(z)$, and $P(z)$ are the same as those used in Section 13.11.1.

When $n = 2$ and $Q(z) = -1 + 4z^{-1} - z^{-2}$, we have [13]

$$P(z) = \frac{1}{16} \left(1 + z^{-1}\right)^4 \left(-1 + 4z^{-1} - z^{-2}\right). \tag{17.14}$$

By factorization, we have a (5,3) filter bank

$$H_0(z) = \frac{1}{8} \left(-1 + 2z^{-1} + 6z^{-2} + 2z^{-3} - z^{-4}\right), \tag{17.15}$$

$$H_1(-z) = \frac{1}{2} \left(1 + 2z^{-1} + z^{-2}\right). \tag{17.16}$$

The (5,3) wavelet filter pair is implemented using integer arithmetic, and the resulting transform is reversible, enabling lossless compression. The (5,3) integer filter pair is recommended for lossless image coding in JPEG2000. In JPEG, lossless coding is not possible if DCT is used, since the cosine elements of the transformation matrix can only be obtained by approximation. In order to preserve the image energy in the pixel and wavelet domains, the filter coefficients are normalized for unit gain. It should be noted that the paired lowpass and highpass filters are *biorthogonal* [13]. The highpass filter is given as

$$H_1(z) = \frac{1}{2} \left(1 - 2z^{-1} + z^{-2}\right). \tag{17.17}$$

$P(z)$ can also be factored differently to obtain a (4,4) filter bank

$$H_0(z) = \frac{1}{4} \left(1 + 3z^{-1} + 3z^{-2} + z^{-3}\right), \tag{17.18}$$

$$H_1(-z) = \frac{1}{4} \left(-1 + 3z^{-1} + 3z^{-2} - z^{-3}\right). \tag{17.19}$$

Both pairs give $P(z) - P(-z) = 2z^{-3}$.

When $n = 3$ and $Q(z) = 1 - 6z^{-1} + \frac{38}{3}z^{-2} - 6z^{-3} + z^{-4}$, $P(z)$ can be factorized into a (9,3) pair of Daubechies wavelet filters [9, 13]:

$$H_0(z) = 0.99436891104360 + 0.41984465132952 \left(z + z^{-1}\right)$$
$$- 0.17677669529665 \left(z^2 + z^{-2}\right) - 0.06629126073624 \left(z^3 + z^{-3}\right)$$
$$+ 0.03314563036812 \left(z^4 + z^{-4}\right), \tag{17.20}$$

$$H_1(z) = 0.70710678118655 - 0.35355339059327 \left(z + z^{-1}\right). \tag{17.21}$$

The (9,3) Daubechies wavelet filter pair is recommended for still image coding in MPEG-4.

Another pair of wavelet filters supported by JPEG2000 is the (9,7) floating-point wavelet. The (9,7) wavelet transform uses filters with floating-point impulse responses

$H_0(z)$ and $H_1(z)$ of length 9 and 7, as the lowpass and highpass analysis filters, respectively [13]:

$$H_0(z) = 0.85269865321930 + 0.37740268810913 \left(z + z^{-1}\right)$$
$$- 0.11062402748951 \left(z^2 + z^{-2}\right) - 0.02384929751586 \left(z^3 + z^{-3}\right)$$
$$+ 0.03782879857992 \left(z^4 + z^{-4}\right), \tag{17.22}$$

$$H_1(z) = -0.7884848720618 + 0.41809244072573 \left(z + z^{-1}\right)$$
$$0.04068975261660 \left(z^2 + z^{-2}\right) - 0.06453905013246 \left(z^3 + z^{-3}\right). \tag{17.23}$$

The QMF solution for the low- and highpass synthesis filters is given by

$$g_0(n) = -(-1)^n h_1(n), \quad g_1(n) = (-1)^n h_0(n). \tag{17.24}$$

This transform is known as the *irreversible transform*, and is used for high-performance lossy compression. The synthesis filter pair is related to the analysis filter pair by (13.176).

Nearly orthogonal filter banks with linear phase have been designed by optimization in [45], which makes a symmetric filter with a certain number of zeros at $z = -1$ to approximate the power complementary condition. Symmetric biorthogonal wavelets can be designed from a symmetric regular filter [44].

17.7.3 Coding of wavelet subimages

The lowest band of the wavelet subimages is a replica of the original image, but at a reduced size. When the number of wavelet decomposition levels is very high, as in JPEG2000, the correlation between the pixels in this subimage is very small. In this case, pixel-by-pixel coding can be used. In MPEG-4 still image coding, where the number of decomposition levels is not as high as that of JPEG2000, some correlation remains between pixels, and DPCM is used to reduce the correlation.

EZW and SPIHT

In Fig. 17.7, the subimages in each band are similar to its higher band subimages. A coefficient in a lower band corresponds to four coefficients in its immediate higher band. For example, a coefficient in HL2 corresponds 4 coefficients in HL1, and 16 coefficients in HL. This is also true for LH2 and HH2 bands. Thus, if a coefficient in a sub-band is zero, the corresponding coefficients in the higher bands are likely zero. This yields a tree of zeros, called *zero tree*. This method is very efficient for representing a very large number of zeros in the wavelet coefficients, since we need only specify the root of the zero tree.

The embedded zero-tree wavelet (EZW), or *embedded coding using zero-trees of wavelet coefficients*, algorithm [28] is a breakthrough in coding of the wavelet coefficients. For

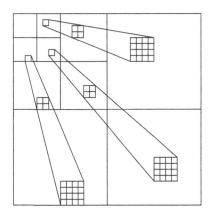

Figure 17.8   Illustration of the quad zero tree representation of the bands.

compression of higher bands, the wavelet coefficients are coded with a zero-tree structure using the EZW algorithm. It is observed that the nonsignificant coefficients from bands of the same orientation tend to be in the same locations, thus a zero-tree representation of the bands can be implemented. This is illustrated in Fig. 17.8 for a three-stage wavelet transform.

The EZW method and its variants are based on quantization by successive approximation, where the wavelet coefficients are represented by progressively decreasing quantization step sizes and the number of approximation passes depends on the specified quantization distortions. An adaptive arithmetic encoder is used to encode the generated symbols. In order to achieve spatial or SNR scalability, the algorithm scans the wavelet coefficients from the lowest to the highest frequency sub-bands (spatial scalability), or for each tree from the root to the leaves.

The set partitioning in hierarchical trees (SPIHT) algorithm [24] is a variant of the EZW algorithm. SPIHT outperforms the EZW algorithm, even without arithmetic coding. Both EZW and SPIHT inherently offer only SNR scalability. By modifying the bitstream, spatial scalability can be achieved. This, however, leads to the loss of SNR scalability. The zero-tree structure also yields error propagation through the sub-bands.

Wavelet-Based Codecs in Source Code

There is a wavelet transform coder construction kit coded in C, which can be used for testing the performance of wavelets.[7] The code is very modular, and it allows for simple replacements of individual components such as the quantizer, entropy coder, and wavelet filters. Although it does not use zerotrees, its performance is comparable to that of the EZW coder.

The SPIHT codec is also available in C and MATLAB code.[8] The SPIHT codec achieves good image quality and high PSNR, especially for color images. It is optimized

---

[7] http://www.geoffdavis.net/dartmouth/wavelet/wavelet.html
[8] http://www.cipr.rpi.edu/research/SPIHT/

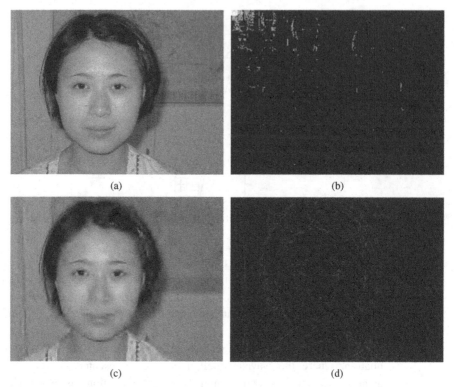

Figure 17.9   SPIHT coding at 0.1 bit/pixel. (a) The original image. (b) The compressed wavelet transform. (c) The reconstructed image. (d) The error image.

for progressive image transmission, and produces a fully embedded bitstream with error protection. It can code to exact bit rate or distortion. It can also be used for lossless compression.

Example 17.2: Given an original image *Huiyu.jpg* of size 240 × 320 and a compression rate of 0.1 bit/pixel, the result from SPIHT coding is displayed in Fig. 17.9. There are four images, namely the input, compressed wavelet transform, reconstructed, and error images. The PSNR is 29.7679 dB. The reconstructed pictures for two rates 0.25 and 0.5 bit/pixel are shown in Fig. 17.10 for comparison with the 0.1 bit/pixel case, with corresponding PSNRs 32.7689 and 35.6568 dB. It is shown that SPIHT coding achieves a very good image quality even at a very high compression ratio.

EBCOT

The deficiencies of EZW and SPIHT can be avoided if each sub-band is coded independently. Embedded block coding with optimized truncation (EBCOT) [33, 34] exploits

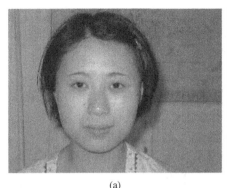 

(a)   (b)

Figure 17.10 **The reconstructed images of SPIHT coding: (a) at 0.25 bit/pixel. (b) at 0.5 bit/pixel.**

this idea. It further partitions sub-band samples into small blocks and codes each block independently. EBCOT offers an excellent compression performance together with a set of bitstream features, including resolution scalability, SNR scalability and a spatial random access capability. All features can exist/coexist within a single bitstream without a substantial sacrifice in compression efficiency. The EBCOT algorithm is adopted in JPEG2000.

The EBCOT algorithm uses a wavelet transform to generate the sub-band samples that are to be quantized and coded. Block coding is independently performed on nonoverlapping blocks within individual sub-bands. All blocks within a sub-band are required to have the same dimensions, except for those blocks on the right or lower boundaries.

EBCOT in Comparison with EZW and SPIHT

The embedded coding strategy used in EBCOT is similar to that used in EZW, but their data structures are different. EZW is capable of compressing an image with excellent rate times distortion performance, and the encoding rate can be precisely controlled with optimal performance for all rates. EZW utilizes the fact that some wavelet coefficients in different sub-bands represent the same location in the image. In EBCOT, since each block is coded independently, it is suitable for parallel hardware encoding and decoding. Unlike EZW and SPIHT that use the zero-tree structure, the similarities between sub-bands are not exploited in EBCOT, but the more efficient context-based arithmetic coding and the post-compression rate distortion optimization are used. EBCOT avoids the tree data structure, but uses a quadtree data structure.

An entropy-coded bitstream is extremely sensitive to channel errors. In EBCOT, each code block is independently encoded, and the individual quality packets can be independently decoded, and thus the effects of the channel errors are confined to a small area. In contrast, due to the zero-tree structure together with entropy coding, a single bit error may destroy a whole picture encoded by EZW or SPIHT.

EBCOT has low implementation complexity and excellent compression performance. Among EZW, SPIHT, and EBCOT, EZW provides the poorest compression performance, and the SPIHT and EBCOT have similar compression efficiency.

EBCOT Implemented in Three Stages

In EBCOT, the code block size is typically 32×32 or 64×64. The code block size can be freely selected, but the dimensions must be an integer power of two and the height cannot be less than four. EBCOT can be implemented in three stages, namely *bit plane quantization, binary arithmetic coding (tier 1 coding)*, and *bitstream organization (tier 2 coding)*.

At the quantization stage, all code blocks in a sub-band use the same quantizer, and successive approximation takes the form of bit plane quantization. The wavelet coefficients are first represented by their maximum precision and the process continues until the least precision; that is, they are encoded one bit at a time, starting with the MSB and terminating at the LSB. When the bitstream is truncated, some or all of the samples in a block may omit one or more LSBs. This is the same as using a UQ-DZ with a dead zone width of $\Delta_b 2^p$, where Δ_b is the uniform quantization stepsize for K-bit representation, and p is the number of LSBs truncated.

After the bitstream is generated, EBCOT identifies whether a coefficient should be coded based on the binary significant state and then codes it. The arithmetic coding of the bit plane, called *tier 1 coding*, generates one independent embedded bitstream for each code block. Tier 2 coding organizes the bitstreams for transmission. By signaling the ordering of the code bit plane pass, the SNR and spatial scalability is enabled.

In EBCOT, context-based adaptive binary arithmetic coding (CABAC) is used for entropy coding of each symbol, where the status of eight immediate neighbors is used for adaptation of the probability model. The entire bit plane is not encoded in one pass. Instead, each bit plane is encoded in three sub-bit-plane passes. This is called *fractional bit plane coding*. This creates an optimum bitstream after truncation, and the technique is known as *post compression rate distortion (PCRD) optimization*. The first pass of the fractional bit plane coding gives the largest reduction in the encoding distortion. The PCRD optimization makes efficient rate control possible.

Spatial and SNR Scalability

In EBCOT, spatial scalability is easily realized due to the octave sub-band decomposition. The bitstream is organized such that all the bit planes of the lower level sub-bands precede those of the higher sub-bands. For SNR scalability, a succession of images of the same spatial resolution is involved. The lowest layer that corresponds to the lowest quality image is called the *base layer*, and the parts of the bitstream for enhancing the image quality are called *enhancement layers*. This is also achieved by bit plane coding.

17.8 Wavelet-based image coding standards

17.8.1 JPEG2000 standard

JPEG2000 is a new joint ISO/IEC and ITU-T standard (IS 15444-1 and ITU-T T.800) for still image coding, and was completed in 2001. JPEG2000 is based on the two-dimensional DWT and EBCOT [33, 34]. JPEG may introduce blocky appearance for a large compression ratio, and JPEG2000 outperforms the older JPEG standard by approximately 2 dB of PSNR, for several images across all compression ratios. JPEG2000 achieves a substantially decreased distortion at low-bit rates. In JPEG2000, a bit rate of less than 0.25 bit per pixel is expected for grayscale images. JPEG2000 is capable of compressing both continuous tone and bilevel images.

JPEG2000 provides a signal decompression architecture for greater interoperability. It combines all four JPEG modes. Many types of scalability are supported by JPEG2000, and this is very desirable for transmission over slow communication links. Progressivity can be achieved in terms of quality, resolution, spatial location, and component. It also supports region of interest (ROI) compression. Since only the data required by the viewer needs to be transmitted, the effective compression ratio at the client is much greater than the actual compression ratio that is calculated from the file size at the server. JPEG2000 also includes error-correcting codes into its compressed stream to achieve error resilience for reliable image transmission in noisy channels.

The JPEG2000 encoder is based on the EBCOT algorithm. Some specific features of JPEG2000 are introduced here.

Tiling

In JPEG2000, the first step is to divide an image into nonoverlapping rectangular blocks of the same size called *tiles*, each being encoded independently. The tile size can be arbitrary, and it can be as large as the whole image or as small as a single pixel. Due to tiling, correlation between pixels in adjacent tiles is not exploited.

Color decorrelation

As in JPEG, dc level offsetting is applied to the RGB components within each tile. This is implemented by shifting all RGB components by 2^{B-1}, where B is the resolution of the color component.

For multicomponent images, two types of color decorrelation transforms, namely the *irreversible color transform* and the *reversible color transform*, are recommended. Irreversible color transform is used to generate the YCrCb components from RGB components, and this is described in Section 17.2.2. The irreversible transform is used only for lossy compression. The reversible color transform is an approximate version of the irreversible transform and generates only integer outputs

$$\begin{bmatrix} Y \\ U \\ V \end{bmatrix} = \begin{bmatrix} 0.25 & 0.25 & 0.25 \\ 1 & -1 & 0 \\ 0 & -1 & 1 \end{bmatrix} \begin{bmatrix} R \\ G \\ B \end{bmatrix}. \tag{17.25}$$

Although the color decorrelation of the reversible color transform is not as good as that of the irreversible color transform, the original RGB values can be recovered if YUV is losslessly coded.

In JPEG, the chrominance components are subsampled by 4:2:2 or 4:2:0 image formats. In JPEG2000, color subsampling is not performed by spatially subsampling a block, but by setting the wavelet coefficients of the highest LH, HL, and HH sub-bands to zero.

DWT and quantization

Instead of DCT, DWT has been applied to each tile. DWT is implemented by using the (9,7) or (5,3) wavelet filter bank. The wavelet coefficients generated are then quantized using a uniform quantizer with a dead zone. The quantization step size varies from sub-band to sub-band and from tile to tile. Exploiting the human visual system, the selection of quantization step size could be similar to the use of the quantization matrices recommended in JPEG. EBCOT uses successive approximation for quantizing the wavelet transform coefficients.

The basic quantizer step size Δ_b in the sub-band b is related to the dynamic range of the sub-band. It is represented by [13]

$$\Delta_b = 2^{R_b - \epsilon_b} \left(1 + \frac{\mu_b}{2^{11}} \right), \tag{17.26}$$

where R is the number of bits for representing the dynamic range of the sub-band, μ_b is a 11-bit mantissa, and ϵ_b is a 5-bit exponent. 2^{R_b} is greater than the magnitude of the largest coefficient in the sub-band b. A total of 2 bytes (μ_b and ϵ_b) is sent to the decoder. For the (5,3) filter bank, $\mu = 0$ and $\epsilon_b = R_b$, thus $\Delta_b = 1$.

Entropy coding and postprocessing

The indices of the quantized coefficients in each sub-band are encoded by using arithmetic coding. EBCOT is used for efficient coding of the indices, and the compressed bitstream is created. The compressed bitstream for each code block is then organized in a way similar to that in EBCOT.

17.8.2 MPEG-4 still image mode

MPEG-4 also supports still texture (image) coding, known as *MPEG-4 Visual Texture Coding (VTC)*. MPEG-4 VTC is very similar to JPEG2000. For coding of still images, DWT is used, and a high compression efficiency as well as spatial and SNR scalability is achieved [13]. MPEG-4 VTC does not provide a lossless functionality.

The Daubechies (9,3)-tap biorthogonal filter is used. The wavelet coefficients of the lowest band are DPCM-coded with a uniform quantizer. A layered structure, which is a variant of EZW, is used to code the higher band wavelet coefficients. The quantizer for each layer is uniform with a dead zone of twice the quantization step size of the layer.

The resulting bitstream is further encoded by an arithmetic encoder. A shape-adaptive DWT (SA-DWT) is used for compression of arbitrarily shaped textures. As in the case of the shape-adaptive DCT (SA-DCT) used in the MPEG-4 Video mode, the number of the transform coefficients obtained for SA-DWT is exactly the same as the number of pixels in the arbitrarily shaped region. When the object boundary is rectangular, SA-DWT reduces to the regular wavelet transform.

17.9 Comparison of image coding standards

17.9.1 Comparison of six popular standards

A comparison of JPEG2000 with JPEG-LS, MPEG-4 VTC, JPEG and PNG, and well established algorithms, such as SPIHT has been performed in [29], by using seven images from the official JPEG2000 test set. The comparison makes use of the software implementations:

- The JPEG2000 Verification Model (VM) 6.1 (ISO/IEC JTC1/SC29/WG1 N 1580).
- The MPEG-4 MoMuSys VM of Aug. 1999 (ISO/IEC JTC1/SC29/WG11 N 2805).
- The Independent JPEG Group JPEG implementation,[9] version 6b.
- The SPMG JPEG-LS implementation of the University of British Columbia,[10] version 2.2.
- The Lossless JPEG codec of Cornell University,[11] version 1.0.
- The libpng implementation of PNG,[12] version 1.0.3.
- The SPIHT codecs,[13] version 8.01.

For lossless compression, amongst JPEG2000, JPEG-LS, lossless JPEG (L-JPEG), PNG and SPIHT, JPEG-LS generally yields the best performance. JPEG2000 provides competitive compression ratios with the added benefit of scalability. The performance of SPIHT and PNG are very close to that of JPEG2000 on most images. The performance of lossless JPEG is poorer than the others. They can be ordered as

$$\text{L-JPEG} < \text{SPIHT, PNG} \lesssim \text{JPEG2000} < \text{JPEG-LS} \quad \text{(Lossless Compression)}$$

For lossy compression, amongst JPEG2000, JPEG, MPEG-4 VTC, and SPIHT, in case of nonprogressive bitstreams, JPEG2000 outperforms all other algorithms at all bit rates,

---

[9] http://www.ijg.org/
[10] http://spmg.ece.ubc.ca/
[11] ftp://ftp.cs.cornell.edu/pub/multimed
[12] ftp://ftp.uu.net/graphics/png
[13] http://www.cipr.rpi.edu/research/SPIHT/

and the performance of JPEG is the poorest at all rates. The advantage of JPEG2000 over the others becomes significant as the compression ratio increases. In case of SNR-scalable bitstreams across various bit rates, JPEG2000 outperforms all the other algorithms, and JPEG produces the worst performance. The performance of the various algorithms is also very close to their nonprogressive cases, except for JPEG. The order is

$$\text{JPEG} \ll \text{MPEG-4 VTC, SPIHT} < \text{JPEG2000} \quad \text{(Lossy Compression)}$$

As for the complexity, in case of lossless compression and decompression, JPEG-LS provides the fastest compression, JPEG2000 is considerably more complex, and the complexity of SPIHT is higher than that of JPEG2000. The complexity of L-JPEG is slightly higher than that of JPEG-LS, and the complexity of PNG is higher than that of L-JPEG. That is,

$$\text{JPEG-LS} \lesssim \text{L-JPEG} < \text{PNG} \ll \text{JPEG2000} < \text{SPIHT} \quad \text{(Complexity, Lossless)}$$

For nonprogressive coding, JPEG2000, MPEG-4 VTC and SPIHT are all significantly slower than JPEG, with MPEG-4 VTC significantly slower than the others. Similar behavior can be observed for the progressive case. That is,

$$\text{JPEG} \ll \text{JPEG2000, SPIHT} \ll \text{MPEG-4 VTC} \quad \text{(Complexity, Lossy)}$$

For error resilience, JPEG and JPEG-LS provide basic error resilience mechanisms, while JPEG2000 and MPEG-4 VTC offer stronger ones. PNG only provides support for error detection. The quality of reconstructed images is higher for JPEG2000 than for JPEG, for all encoding bit rates and error rates. That is,

$$\text{PNG} < \text{JPEG, JPEG-LS} \ll \text{JPEG2000, MPEG-4 VTC} \quad \text{(Error resilience)}$$

ROI coding is a novel functionality of JPEG2000. The JPEG2000 implementation used for evaluation was JJ2000,[14] version 4.0, contained in the JPEG2000 Part-5 FCD (ISO/IEC JTC 1/SC 29/WG 1, ISO/IEC FCD 15444-5: WG 1 N 2020, January 2000), since it provides better ROI support than the VM.

JPEG2000 provides more progressive types and orderings than MPEG-4 VTC. MPEG-4 VTC is the only evaluated standard, which is capable of coding individual objects of arbitrary shapes. Overall, JPEG2000 offers the richest set of features and provides superior rate distortion performance.

In summary, JPEG is suitable for lossy compression with the requirement of low complexity; JPEG-LS provides the best option for lossless compression, with the best compression efficiency at a low complexity; PNG is also a good solution for lossless compression. MPEG-4 VTC may be desirable when arbitrarily shaped objects are required to be coded. JPEG2000 provides the most flexible solution, providing good compression efficiency as well as many other features.

---

[14] http://jj2000.epfl.ch

17.9.2 DjVu and adaptive binary optimization (ABO)

Different compression methods are normally used for different types of images. For example, JBIG is used for bi-level images, and JPEG for continuous images. In some pictures, such as a scanned document, features of various types of pictures are contained in one image. Djvu from AT&T Laboratories is designed for this purpose.[15] DjVu is a free file format. It has been proposed as an alternative to the PDF format for documentation.

DjVu is a progressive compression method. DjVu typically achieves a compression factor as high as 1000 for scanned pages. For colored documents with both text and pictures, a DjVu file is typically 5 to 10 times smaller than the JPEG file with the same quality. DjVu files are about five times smaller than the PDF files for scanned documents.

DjVu decomposes the scanned document into three components, namely, *mask*, *foreground*, and *background*. The mask images contain the text and lines that are in bi-level form, and the foreground images contains their colors. The background is a continuous-tone image. Since each component has a different feature, they can be compressed at different resolutions. The background and foreground images are compressed using IW44, which is a method using integer wavelet, while the mask is compressed using an algorithm called JB2 (similar to JBIG2).

MatrixView's adaptive binary optimization (ABO) is a versatile and pervasive technique as it allows optimization of any form of digital content including images, videos, sound, and text.[16] By using repetition and correlation coding (RCC), ABO exploits the repetition and correlation found in the data. It generates lossless compression and native-embedded encryption of data. ABO is claimed to be able to achieve at least 300% more lossless compression than the levels reached by JPEG and JPEG2000 in the lossless mode. Also ABO has a much lower complexity than JPEG and JPEG2000, achieving higher encoding/decoding and transmission speeds.

17.10 Video data compression

Video is a succession of images. When each image is coded as an independent JPEG image, the process is called *motion JPEG*. Motion JPEG is very resilient to loss of information. Motion JPEG is popular for video capture, video editing, and security surveillance. It is most suitable for transmission over packet networks such as the Internet.

Video can be viewed as a sequence of correlated images. This temporal correlation is used by most video compression algorithms to remove redundancy. The 3-D DCT can be applied on an m-row, n-column, k-frame block. This method requires memories for k frames at both the encoder and the decoder to buffer the frames. This also limits its real-time applications, and k is typically selected between 2 and 4.

[15] http://www.djvu.org/
[16] http://www.matrixview.com

The standard video compression approach is based on motion compensation. The previous reconstructed frame is used to generate a prediction of the current frame, and only the difference between the prediction and the current frame, known as the *prediction error*, is encoded. The prediction operation must provide motion compensation to the motion of the objects in the frame. Currently, video codecs are standardized by MPEG and ITU-T. The structures of MPEG standards and ITU-T H-series standards are similar since many of the members have joined both the groups.

17.10.1 Frame format

ITU-R Rec. BT.601 defines a digital video format for broadcasting. It defines two interlaced systems: 525/60 and 625/50, which are based on the 4:2:2 YCbCr sampling. For storage applications, the source input format (SIF) was defined. SIF is based on a noninterlaced (or progressive) system with 4:2:0 YCbCr sampling. SIF has a luminance resolution of 352×240 at 30 frames/s or 352×288 at 25 frames/s, corresponding to the two ITU-R Rec. BT.601 systems. The quarter-SIF (QSIF) has half the resolution of SIF in both the row and column directions.

Common video compression standards define a number of video frame formats. The common intermediate format (CIF) is the basis for a popular set of formats. The CIF is defined for compatibility between the two ITU-R Rec. BT.601 systems. It is a progressive system, with 4:2:0 mid-sited sampling and a frame rate of 30 frames/s. The CIF has a luminance resolution of 352×288. Other four frame formats in the CIF family are 16CIF (1408×1152), 4CIF (704×576), quarter-CIF (QCIF, 176×144), and sub-QCIF (128×96). The selection of frame resolution depends on the application and transmission capacity. These frame formats are used for standard-definition television (SDTV), DVD (digital versatile disc) video, video conference, and mobile multimedia. For HDTV (high definition television), some other formats are also defined.

17.10.2 Frame types

MPEG defines three coded pictures: intra-picture (I), predicted-picture (P), and bidirectional predicted picture (B). I-frames are coded independently of the other frames, and act as the start for replay sequences. P-frames are used to predict motion from the previous reference frames, which can be the last I-frame or P-frame, whichever is closer to it. I- and P-frames are called *anchor frames*, because they are used as references in the coding of other frames using motion compensation.

B-frames are generated from a previous anchor frame and a future anchor frame by using motion compensation. The B-frame achieves a higher compression, and it is generated only after the future anchor frame has been generated. The use of B-frames can effectively compress the video, especially in case of a sudden change from one frame to the next. Use of B-frames increases the motion compensation efficiency, since occluded parts of moving objects can be compensated for from a future frame. Also, B-frames are not used for predicting any other frames, and thus can be discarded without affecting the pictures that follow. For this reason, B-frames can tolerate more error, and bit rate control can be performed by controlling the number of B-frames or the quantization of B-frames. However, the use of B-frames leads to a delay of several frames, which is equal to the number of B-frames between two anchor frames, for the encoder and the decoder. This limits its application to telecommunications.

Finally, there is also a fourth picture type called *D-frame*. It is intraframe-coded, where only the dc coefficients are retained. D-frames are intended for fast visible search modes, but not used in groups of pictures (GOPs). Of course, this fast search mode can also be supported by using the I-frames in a sequence.

17.10.3 Motion compensation

Motion compensation can effectively increase the compression efficiency of interframe coding. To perform motion compensation, motion estimation has to be performed first. Block-based motion compensation, first proposed by Jain and Jain in 1981 [17], is implemented in all the standard video codecs, and the technique is called the *block matching algorithm*. The motion estimation and compensation, sometimes referred to as *DPCM*, is used together with a DCT stage and an entropy encoder in all the major video standards such as H.261/3/4 and MPEG-1/2/4 Visual. Motion estimation and compensation is optional in H.261, but is an important integral part of all later MPEG and ITU-T H-series video codecs standards.

In a standard codec such as MPEG-1/2/4 or H.261/263/264, a block of 16×16 pixels, called a *macroblock*, is motion estimated and compensated. Motion estimation is performed only on the luminance parts of a frame. The motion vector is used for motion compensation of all the four luminance blocks in the macroblock, and is scaled for compensation of the chrominance blocks. After motion estimation, the residual block is encoded and transmitted, and the motion vector is also transmitted. Motion vectors for neighboring partitions are often highly correlated and each motion vector can thus be predicted from the motion vectors of nearby partitions, that is, the motion vectors can be DPCM coded and then entropy encoded.

Full-search method

In a typical block matching algorithm, the frame is first divided into $M \times M$ blocks. For each block, the algorithm searches the previous reconstructed frame for the $M \times M$ block that most closely matches it, and the distance between the two blocks is measured in pixels. When the distance is greater than a prespecified threshold, the block is encoded without using prediction; otherwise, a motion vector is included into the bitstream. The motion vector gives the relative location of the two blocks. For a maximum motion speed of w pixels per frame, the current block of pixels with the upper-left corner at (m, n) is matched against a block with upper-left corner at $(m + i, n + j)$, $-w \leq i,j \leq w$, in the previous frame. This is shown in Fig. 17.11.

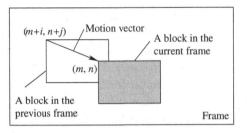

Figure 17.11   **Motion estimation based on the comparison of the current and previous frame.**

The motion vector corresponds to the best match, according to an MSE or MAD criterion

$$\text{MSE}(i,j) = \frac{1}{M^2} \sum_{m=1}^{M} \sum_{n=1}^{M} \left[p_0(m,n) - p_{-1}(m+i,n+j) \right]^2, \qquad (17.27)$$

$$\text{MAD}(i,j) = \frac{1}{M^2} \sum_{m=1}^{M} \sum_{n=1}^{M} \left| p_0(m,n) - p_{-1}(m+i,n+j) \right|, \qquad (17.28)$$

for $-w \le i,j \le w$, where $p_0(m,n)$ is a pixel of the current block at coordinate (m,n), and $p_{-1}(m+i,n+j)$ is a pixel of the corresponding block in the previous frame. When the $\frac{1}{M^2}$ term is dropped, the MSE criterion becomes the sum of squared differences or errors (SSD or SSE) criterion, and the MAD criterion reduces to the sum of absolute differences or errors (SAD or SAE) criterion. SAD has lower complexity, and is used in all the video codecs.

The block matching algorithm based on full search requires $(2w + 1)^2$ tests of the criterion. The search range $w = \pm 15$ pixels is sufficient for low-bit-rate applications, and H.263 uses a maximum displacement of ± 15 pixels. In a typical video sequence, motion vectors are concentrated around $(0,0)$. Full search thus starts at $(0,0)$ and proceeds in a spiral pattern around this location. Motion estimations can occupy 50–70 per cent of the overall complexity of the encoder if the full search method is used [13]. Thus, the full search method is only suitable for hardware implementation.

Motion estimation is conventionally based on the assumptions that the motion is translational and that the illumination is uniform. When these assumptions are not satisfied, such as in the case of the movement of human eyes, a geometrical transformation that converts a square block of $N \times N$ pixels into an irregular quadrilateral shape can be used, by using a bilinear transform [13]. This type of motion estimation/compensation is extremely complex, and numerous methods for simplifying the operations are available.

Example 17.3: In this example, we select the first 30 frames of the *coastguard* sequence in QCIF format at a frame rate of 8.33 frames/s (a frame skip of 3), and apply the block matching algorithm on the luma component only. Blocks of size 16×16 are used, and the maximum allowed motion displacement of ± 15 pixels in both directions is applied. Motion is estimated and compensated using original previous frames, while the motion

(a)

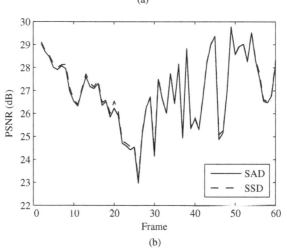

Frame

(b)

Figure 17.12 The block matching algorithm. (a) The 1st and 60th frames. (b) PSNRs of the reconstructed frames.

vectors are restricted to points inside the reference frame. Motion vectors can be encoded using the predictive coding followed by variable-length coding. Neglecting the difference between the original and matched blocks, the PSNR values using the SSD and SAD measures are plotted in Fig. 17.12. It is seen that the SSD criterion yields a slightly better PSNR performance than the SAD criterion does.

Fast block motion estimation

Instead of searching over all possible blocks within the search window, a number of fast block matching algorithms have been proposed, where the search is carried out over a subset of the blocks. Such algorithms are based on the assumption of unimodal error surface, namely that the block distortion measure increases monotonically as the search location moves away from the best-match location [17]; this may cause false estimations (a local minimum of distortion), for high motion speeds. Fast block matching algorithms, such as the two-dimensional logarithmic search [17] and the three-step search (TSS) method [18], typically have a complexity that grows with $\log_2 w$, where w is the search range [13]. They are more suitable for slow moving objects, as in video conferencing. The TSS requires $1 + 8\log_2 w$ tests, and is recommended by ITU-T for software-based H.261.

A fast block matching algorithm based on a feature-bit-plane matching scheme has been proposed in [42], whose accuracy is very close to that of the full-search method, but with a complexity of only one-tenth that of the latter. A block of pixels are requantized into two levels based on a given grey level T and full search is then applied on the feature bit plane. Also, the addition operation is replaced by the XOR operation.

In [11], a fast block matching algorithm, which achieves the same accuracy of the full-search algorithm, is proposed. The method is well suited for SIMD implementation. Parallel SIMD implementation of block motion estimation can substantially decrease the encoding time [12]. In [37], motion estimation based on pyramidal data structure has been implemented by exploiting the inter-level motion correlation as well as the intra-level motion correlation. Motion estimation based on multiplicationless Burt and Adelson's pyramids was implemented in [38].

Variable-size block motion compensation can effectively reduce the overhead of motion vectors, leading to very-low-bit-rate video coding [43]. In addition, global motion compensation is useful for increasing the coding efficiency at low bit rates for some video sequences that contain global motion arising from camera motion. This can significantly reduce the number of motion vectors.

Motion estimation with fractional pixel accuracy

Motion estimation can be implemented with fractional pixel accuracy, and it can be with half pixel precision, quarter pixel precision, or eighth pixel precision. In the search process of fractional pixel accuracy, normal block matching with integer pixel position is performed first. Then, the subpixel positions and their pixels are obtained by interpolation from the pixels of the macroblock in the previous frame.

In H.261, motion compensation with the full pixel precision is only optional. In MPEG-1 and H.263, motion compensation with only half pixel accuracy is used. MPEG-4 uses quarter pixel accuracy. H.263++, as an enhancement to H.263, also recommends quarter and eighth pixel precisions. H.264 uses quarter pixel accuracy as the default mode for lower complexity mode and eighth pixel accuracy for the higher complexity mode.

Wavelet-based motion compensation

Motion compensation in the wavelet transform domain, known as *multiresolution motion estimation*, was originally proposed in [41], and a number of subsequent multiresolution motion estimation algorithms have been based on this work [36, 39].

The multiresolution motion estimation method is similar to DCT-based sub-band coding. The wavelet coefficients of each sub-band (block) are extracted and compared to those of the reference sub-band (block). The motion vectors in the lowpass sub-band can be properly scaled and used as the final motion vectors for all the other sub-bands. It starts with estimating the motion vectors at the lowest resolution level. The motion vectors thus obtained are then used as predictions for the higher resolutions, where they are either accepted as final motion vectors or further refined. For the multiresolution motion estimation of video sequences, the 7/9 biorthogonal wavelet, one of the best wavelets for the coding of still images, is still a good wavelet for multiresolution motion estimation also [40].

Compared to the conventional spatial-domain motion estimation technique, the wavelet-based multiresolution motion estimation technique has the advantages of a substantial reduction in the computational load and ease of generating an embedded video bitstream. JPEG2000 is implemented by coding block by block, and it can be extended into a video format using multiresolution motion estimation.

17.10.4 Basic structure of video

Group of pictures (GOP)

The different frames of a video are organized in GOPs. The beginning of a GOP is a point for random access in the video sequence. Each GOP must be independently decoded, thus the first frame of a GOP must be an I-frame, or a B-frame that can be reconstructed from the following I-frame. For a closed GOP, all predictions take place within the GOP, while in an open GOP every frame boundary has predictions that cross it. A regular GOP has a fixed pattern of P- and B-frames between I-frames. Regular GOPs can be characterized by the GOP length N and the distance M between the anchor frames. GOPs are important for random access, fast forward, fast or normal reverse play of a video. A typical GOP pattern used in MPEG-1 is shown in Fig. 17.13, where all the three main frame types have the same size with 4:2:0 chroma format.

Macroblocks and blocks

The frames of the video sequence are divided into macroblocks of size 16×16. MPEG-1 uses a 4:2:0 structure for color information coding. This leads to a 16×16 luminance block, a 8×8 block for C_b, and a 8×8 block for C_r. The 16×16 luminance block is further divided into four 8×8 blocks, since 8×8 blocks are used for all DCT coding. The intracode of a macroblock in MPEG-1 is almost exactly the same as that for JPEG. In MPEG-2, the macroblock types may range from 4:2:0, 4:2:2, and 4:4:4 chroma formats.

Figure 17.13   **A typical GOP pattern used in MPEG-1.**

Slices

A group of macroblocks is called a *slice* (In H.261, it is called a *group of blocks (GOB)*). The reason for defining slices is to confine the channel error within the slices by resetting entropy coding. Slices can have different sizes within a frame, and the division of slices can be different in different frames. The organization of the slices introduces extra slice overhead. In order to optimize the slice structure, short slices can be used for macroblocks with significant energy (such as macroblocks in I-frames), and long slices for macroblocks with less significant energy (such as macroblocks in B-pictures). A slice has a raster scan structure.

17.10.5 Video encoder/decoder

The basic structure of the MPEG compression algorithm is very similar to that of ITU-T H.261. Implementation of MPEG starts from transforming analog RGB pictures into YUV pictures with Y, Cr, Cb components, which are then digitalized. YUV pictures can be sampled by the 4:2:0 or 4:2:2 format.

The interframe predictive coding is first applied. The difference between pixels in the current frame and their prediction is DCT-coded, quantized, and transmitted. The transmission bit rate is controlled by the predictor. A better predictor based on motion estimation leads to a smaller error signal, thus yielding a smaller bit rate. At the receiver, the decoded error signal of each pixel is added to the prediction value to reconstruct the frame.

DCT is applied to the difference image between the digitized image and the predicted image to extract spatial frequencies using 16×16 macroblocks. The DCT coefficients are then quantized with the high spatial frequencies in coarser steps. The implementation is very similar to that of JPEG. In MPEG-1 and H.261, zigzag scanning of the DCT coefficients is used, as in JPEG. In MPEG-2, the alternate scan is used, as shown in Fig. 17.14. Since the vertical correlation in the interlaced pictures is very small, an alternate scan generates better results when the interlaced related operation field prediction is used [13]. In H.263, for intracoding, in addition to zigzag scan, alternate horizontal and alternate vertical scans are also used. The alternate vertical scan is the same as the alternate scan mode in MPEG-2, as shown in Fig. 17.14, and the alternate horizontal scan is the transposed form of the alternate vertical scan.

In MPEG-1 the quantization matrix for intraframe coding is different for different formats. The quantization matrix for the SIF size ($352 \times 288 \times 25$ or 30) is derived from the vision contrast sensitivity curve for a nominal viewing distance [13], and it can be different for best perceptual results if the resolution or the viewing distance is changed. The default intraframe quantization table is shown in Table 17.2.

The quantization table for interframe coding uses the same weighting for all the frequency components. The reason is that high frequency interframe error does not necessarily mean high spatial frequency. It may arise from poor motion compensation or block boundary artifacts. In MPEG-1, all the 64 elements of the interframe quantization matrix are selected as 16.

Table 17.2. The intraframe quantization table used in MPEG-1.							
8	16	19	22	26	27	29	34
16	16	22	24	27	29	34	37
19	22	26	27	29	34	34	38
22	22	26	27	29	34	37	40
22	26	27	29	32	35	40	48
26	27	29	32	35	40	48	58
26	27	29	34	38	46	56	69
27	29	35	38	46	56	69	83

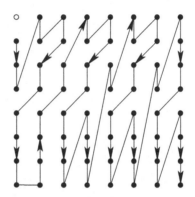

Figure 17.14 The alternate scanning of the 8 × 8 DCT coefficients for an 8 × 8 image block.

In H.261, all the pictures are interframe coded, thus only the interframe quantiza-tion matrix is defined. In MPEG-2, both linear and nonlinear quantization of the DCT coefficients are supported. In H.261 and MPEG-1, the DCT coefficients are coded, via a zigzag scanning process, as two-dimensional events (*run, index*), where *run* indi-cates the number of zero-valued coefficients in front of a nonzero coefficient in the zigzag scan, and *index* is the normalized magnitude of a nonzero coefficient. In H.263, the DCT coefficients are represented by a three-dimensional event (*last, run, index*), where the additional event *last* takes the value 0 or 1, corresponding to whether there are more nonzero coefficients in the block or it is the last nonzero coefficient in the block. Thus, it is intended to replace the end-of-block (EOB) code of H.261 and MPEG-1.

The procedure for MPEG-2 encoding is given in Fig. 17.15. The H261/3 and MPEG-1/4 encoders are also similar to that of MPEG-2.

The decoding process, as shown in Fig. 17.16, is the inverse of the encoding process. The decoder demultiplexes the received video bitstream, entropy-decodes it, dequantizes it, and then applies IDCT on it. The result is then added to an interframe prediction, and a final frame is obtained. The frame is then reordered, and the output pictures are obtained. Some control functions such as fast play, pause/resume, and reverse play are also implemented in the decoder.

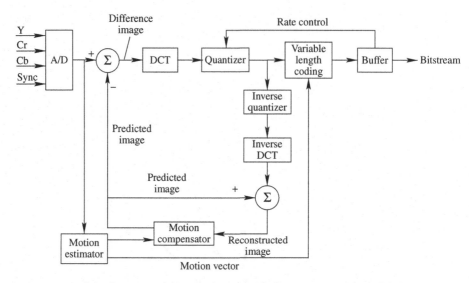

Figure 17.15   The MPEG-2 encoder. Reproduction with the permission of IEEE [2].

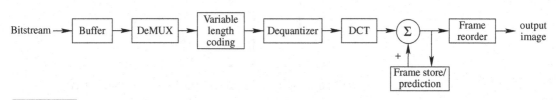

Figure 17.16   The MPEG-2 decoder.

17.10.6 Scalability

Scalability is supported in MPEG-2. Scalable video coding, also known as *layered video coding*, encodes the frames into the base layer and the enhancement layer. In case of network congestion, the enhancement layer data can be discarded.

Scalability provides a means of delivering better quality video, but at a cost of higher encoder complexity and bit rate overhead. Scalability finds wide applications in telecommunications, and broadcasting of TV and video over networks. MPEG-2 supports a variety of scalability [13]:

- *Data partitioning*. It is intended for use when two channels are available for transmission of a video bitstream. It divides the bitstream into two layers: The base layer contains the critical parts of the bitstream, such as the headers, motion vectors and some lower-order DCT coefficients; the enhancement layer contains other data such as the higher-order DCT coefficients. The former is transmitted in a better quality channel, while the latter is transmitted in a channel with poorer quality. Data partitioning is only used for P and B-frames, but not for I-frames, since in I-frames all DCT coefficients are important.

- *SNR scalability*. It generates two video layers of the same spatio-temporal resolution from a single video source. The basic layer by itself provides the basic video quality, and the enhanced layer, when added to the basic layer, can generates a higher quality video.
- *Spatial scalability*. Two spatial resolution video streams are generated from a single video. The base layer by itself provides basic spatial resolution, and the enhancement layer, when added to the spatially interpolated base layer, generates full spatial resolution of the input video. Spatial scalability is a most complex form of scalability, where each layer requires a complete encoder/decoder.
- *Temporal scalability*. It partitions video frames into layers, with the basic layer providing the basic temporal rate and the enhancement layer being coded with temporal prediction based on the base layer. The spatial resolution of the frame in each layer is the same. In MPEG-1 and MPEG-2, the I- and P-frames can be treated as the base layer, while the B-frames can be treated as the enhancement layer.
- *Hybrid scalability*. In MPEG-2, individual scalabilities can be combined to form a hybrid scalability. For each increment of scalability, an additional enhancement layer is added to the existing base and enhancement layers. Decoding of the upper enhancement layers requires the availability of the base and lower enhancement layers.

In H.263, data partitioning, temporal, SNR, spatial, and hybrid scalability are supported as optional modes. Unlike MPEG-2, the scalability in H.263 is not used for distribution purpose, but as a layered coding technique. By using unequal error protection to the base layer as well as error resilience techniques, the robustness of H.263 is further improved.

In H.263, the data partitioning mode is performed to put important data ahead of other data. This is to reduce the influence of channel errors on variable-length encoded data, since between two resynchronization markers, symbols that appear earlier in the bitstream are less influenced by errors than those that appear later, due to the cumulative impact of variable-length coding on the decoding of the subsequent data.

The scalability discussed so far is also applicable to MPEG-4. But due to the object-based coding, scalability of MPEG-4 can be different. In MPEG-4, SNR scalability is replaced by the fine granularity scalability (FGS). The basic layer is coded in a way similar to that of an SNR scalable coder, but the difference between the original DCT coefficients and the quantized coefficients are represented in bit planes, as opposed to being quantized with a finer quantizer step size as in the SNR scalable coder. In MPEG-4, object-based scalability is also supported, where scalability is applied to individual objects rather than the entire frame. The object-based temporal scalability is of particular interest. It increases the frame rate of an object such that the object has a smoother motion than the other objects in the frame.

17.10.7 Integer DCT transform

In earlier MPEG or H-26x standards, the 8×8 floating-point precision DCT transform was employed. In practical implementation, the floating-point number is implemented with a finite precision. Thus, a mismatch in DCT and IDCT arises. Approximated 8×8 DCT transforms can also be implemented to save computational complexity [6]. H.264 is a unique standard codec that uses the integer transform to replace the DCT transform.

H.264 specifies a 4×4 integer DCT as an approximation to the regular 4×4 DCT. In the one-dimension case, the integer transform of length 4, $\mathbf{T}_4$, and the DCT of length 4, $\mathbf{T}_{DCT}$, are given by

$$\mathbf{T}_4 = \begin{bmatrix} 1 & 1 & 1 & 1 \\ 2 & 1 & -1 & -2 \\ 1 & -1 & -1 & 1 \\ 1 & -2 & 2 & -1 \end{bmatrix}, \quad \mathbf{T}_{DCT} = \begin{bmatrix} 1 & 1 & 1 & 1 \\ 1.3 & 0.54 & -0.54 & -1.3 \\ 1 & -1 & -1 & 1 \\ 0.54 & -1.3 & 1.3 & -0.54 \end{bmatrix}.$$

(17.29)

Like the integer transform for color space transform, the integer DCT has an exact inverse transform, and there is no mismatch between the encoder and decoder. The use of smaller block size reduces the blocking artifacts. Also, the use of integer numbers significantly improves the computation speed, since the multiplication by two can be simply implemented by bit shift. Two-dimensional integer transform can be derived in the same way as 2-D DCT.

H.264 also gives an optional integer transform of length 8, which is also an approximation to the DCT of length 8. The elements in the transform matrix are not powers of two, and some multiplications are required. This optional mode is called *adaptive block transform*, since the block size can be selected as either 4×4, 4×8, 8×4, or 8×8, depending on the texture of the image. Adaptive block transform is implemented by applying

$$\mathbf{A} = \mathbf{T}_v \mathbf{B} \mathbf{T}_h^T \tag{17.30}$$

where $\mathbf{A}$ and $\mathbf{B}$ are the frequency and spatial blocks of the transform, and $\mathbf{T}_v$ and $\mathbf{T}_h$ can be either $\mathbf{T}_4$ or $\mathbf{T}_8$. $\mathbf{T}_8$ is given by

$$\mathbf{T}_8 = \begin{bmatrix} 13 & 13 & 13 & 13 & 13 & 13 & 13 & 13 \\ 19 & 15 & 9 & 3 & -3 & -9 & -15 & -19 \\ 17 & 7 & -7 & -17 & -17 & -7 & 7 & 17 \\ 9 & 3 & -19 & -15 & 15 & 19 & -3 & -9 \\ 13 & -13 & -13 & 13 & 13 & -13 & -13 & 13 \\ 15 & -19 & -3 & 9 & -9 & 3 & 19 & -15 \\ 7 & -17 & 17 & -7 & -7 & 17 & -17 & 7 \\ 3 & -9 & 15 & -19 & 19 & -15 & 9 & -3 \end{bmatrix}.$$

(17.31)

This transform can be implemented with an efficient fast algorithm given in [35].

17.10.8 Shape coding

Shape can be coded as either binary or grayscale shape. The binary shape is encoded using content-adaptive encoding, while the grayscale shape is encoded using the SA-DCT. SA-DCT is applied to 8×8 blocks that are located on the object boundary of an arbitrarily

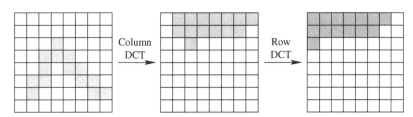

Figure 17.17
Operation of shape adaptive DCT.

shaped video object plane (VOP). SA-DCT applies one-dimensional DCT vertically and horizontally according to the number of active pixels in the row and column of the block, respectively. The size of each vertical DCT is the same as the number of active pixels in that column. After vertical DCTs are performed, the dc coefficients of all vertical DCTs are lined up in the first row, the first-order coefficients are lined up in the second row, and this rule applies for coefficients of all orders. Horizontal DCT is then applied to each row, and the coefficients from the horizontal DCTs are then lined up in rows. The final coefficients of SA-DCT are concentrated into the upper-left corner of the block. This is illustrated in Fig. 17.17.

In SA-DCT, the active (opaque) pixels in the boundary 8×8 are encoded without padding, leading to the number of DCT coefficients equal to the number of opaque pixels. At the encoder, shape adaptive IDCT (SA-IDCT) can be applied to reconstruct the pixels. The regular zigzag scan is modified according to the SA-DCT coefficients for RLE. For a block of 8×8 active pixels, SA-DCT reduces to the regular 8×8 DCT. All SA-DCT coefficients are quantized and coded in the same way as regular DCT coefficients by using the same quantizers and variable-length code tables.

SA-DCT is more complex than regular 2-D DCT, but it improves coding efficiency for boundary mainblocks. SA-DCT is used in MPEG-4.

17.10.9 Object-based coding and sprite coding

MPEG-4 relies on a content-based visual data representation of scenes. A scene is viewed as a composition of video objects (VOs) that have shape, motion, and texture. A VO is an area of the video scene that may occupy an arbitrary-shaped region for an arbitrary period. At a particular time, we have a VOP. The video frames are defined as layers of VOPs. The background is represented by VOP_0, and the objects of interest are represented as VOP_1 to VOP_n. The VOPs can be constructed beforehand or be defined by segmentation of the video frames. The VOP of an arbitrary shape is encapsulated into a bounding rectangle that has the minimum number of macroblocks.

In MPEG-4, the compression efficiency can be significantly improved by not coding the video background. The coding of the background does not need to be exact since viewers do not pay attention to the background. The background can be efficiently encoded by using a global motion model, which can be motion compensation with the spatial transform.

Global motion compensation is based on the transmission of a static sprite panorama, which is a still image describing the background over all frames in the sequence. The panorama sprite is transmitted to the receiver only once and for each following frame, only the camera parameters relevant to the background are transmitted to the decoder. The sprite can be encoded and transmitted as the first frame of a sequence or in a progressive manner. The moving foreground is transmitted separately as VOPs.

After defining the VOPs, the encoder then sends the VOPs, as well as the information of when and where each VOP will be displayed. At the decoder, the chosen VOPs are extracted to a single bitstream, and the output video composed for display. The encoder for the VOPs consists of a shape encoder and a traditional motion/texture encoder (as used in H.263).

MPEG-4 uses the motion-compensated DCT for color/texture coding, but augments this by an explicit compressed representation of the shape of arbitrary VOs [7]. The compression technology associated with the normative shape representation is block-based CABAC, where the symbol probability is calculated from the context of the neighboring pixels.

The object-based functionalities of MPEG-4 require segmentation algorithms for segmenting a video scene into objects. Image segmentation can be performed by using the watershed transform algorithm and region motion estimation [32].

17.10.10 Rate control

Rate control is necessary for maximizing the visual quality at a given target average bit rate by controlling the allocation of the bits. Rate control algorithms take into account the constraints of the decoder buffer. Various rate control schemes have been applied in MPEG-2 Test Model, MPEG-4 Verification Model (VM), and H.264/AVC Joint Model.

In MPEG-2, the TM5 (Test Model 5) rate control consists of three steps: target bit allocation, rate control, and adaptive quantization. Target bit allocation is the picture-level bit allocation. The bit estimate is based on a linear model between the bit budget and the reciprocal of the quantization step. The second step, rate control, is the macro-level bit allocation, which ensures that no buffer overflow and underflow occur. Based on the macro-level bit allocation, a quantizer step is implemented. Adaptive quantization modulates the quantization step from the previous step by increasing it for active areas and reducing it for smooth areas, that is, more bits are used in smooth areas and fewer bits for active areas. This is because the eye is not sensitive to the quantization noise in active areas, but is sensitive to quantization noise in smooth areas.

Unlike the linear model used in MPEG-2 TM5 rate control, MPEG-4 VM uses a quadratic model for rate control. The basic H.264/AVC rate control is similar to MPEG-4 VM rate control, and uses the quadratic rate distortion model. For H.264/AVC, the rate control algorithm consists of GOP-level rate control, picture-level rate control, and the optional basic unit level rate control. The GOP-level rate control calculates the total number of bits for the remaining pictures in the GOP, and also computes the initial quantization parameter of instantaneous decoding refresh (IDR) picture. The picture-level rate control has two stages: pre-encoding and post-encoding. Pre-encoding computes the quantization parameter for each frame, while post-encoding updates the parameter sets and adds the generated bits to the buffer. The basic unit is defined as a group of continuous macroblocks. The basic unit rate control is similar to picture-level rate control, and it is used to determine the quantization parameters for all basic units so that the total generated bits satisfy the picture target bits.

17.11 Introduction to video standards

The MPEG (Moving Pictures Experts Group, ISO/IEC JTC1/SC29/WG11) is a joint ISO/IEC group for developing audio-visual coding standards.[17] The MPEG standards involve the compression of digital images and sound, as well as their synchronization. MPEG is based on JPEG, but has additional features of motion compensation and variable length coding. MPEG uses block-based DCT and quantization, block-based motion estimation and compression, and lossless compression using entropy coding for motion vectors and the quantized DCT coefficients.

ITU-T also has developed its H-series video standards by its Video Coding Experts Group (VCEG, ITU-T SG16 Q.6).[18] Since members of VCEG also participated in MPEG, standards generated by the two groups are similar in many aspects. MPEG and VCEG also set up the Joint Video Team (JVT) for the development of H.264/MPEG-4 Part 10, which is entitled *Advanced Video Coding (AVC)*.

Currently, the wavelet technique has not been used in ITU or MPEG video standards. This is because the DCT-based block-coding technique can effectively make use of motion compensation. DWT-based image coding does not fit well with block-based motion compensation.

The ITU-T video standards are used for telecommunications, and a fast encoder and decoder may be required. In MPEG, the decoder must be fast for real-time operation, but the encoder can be slow. These standards typically specify the format for the bitstream and the decoder specifications, but not the encoder.

All MPEG standards give the syntax for combining audio, video, and data into a single bitstream. A packet structure is defined for multiplexing coded audio and video into one stream and keeping them synchronized. Data can also be multiplexed into the packet. MPEG provides reference software modules which are in C or C++ language.

ITU-T H.120, issued in 1984, is the first international standard for video coding. H.120 was targeted for videoconferencing applications at 1.544 Mbits/s and 2.048 Mbits/s. It uses the conditional replenishment or motion compensation technique for interframe coding and DPCM for intraframe encoding. H.120 was commercially unsuccessful. In the following, special features of other ITU-T and MPEG video standards are described.

---

[17] http://www.mpeg.org
[18] http://www.itu.int

H.261

The ITU-T H.261 standard was completed in 1990. It was intended for visual telephony or videoconference over the integrated services digital networks (ISDNs). It achieves a variable bit rate of $p \times 64$ kbits/s, where p is an integer between 1 and 30. H.261 is normally used for *head-and-shoulders* pictures with a CIF or QCIF spatial resolution. In H.261, both the encoder and the decoder must be fast enough for real-time operation.

In H.261, motion compensation is optional. H.261 uses only I-pictures and P-pictures, but not B-pictures. H.261 uses a single quantization coefficient instead of an 8×8 table of quantized coefficients. Coarse quantization of the coefficients causes loss of high frequency components. This artifact can be reduced by using a loop filter

$$h(x, y) = \frac{1}{16} \begin{bmatrix} 1 & 2 & 1 \\ 2 & 4 & 2 \\ 1 & 2 & 1 \end{bmatrix}. \tag{17.32}$$

The loop filter has a picture blurring effect, and should be activated only for macroblocks with motion. The loop filter is only defined in H.261. The loop filter is switched on only at bits rate lower than 6×64 kbits/s = 386 kbits/s. At a higher bit rate, it does not improve the subjective quality of a picture [13].

MPEG-1

MPEG-1 (IS11172-2), completed in 1991, was intended for intermediate data rate, with a bit rate of 1.5 Mbits/s. MPEG-1 is an extension of H.261, with many features being the same, but their bitstreams are incompatible. MPEG-1 is used to code CIF or QCIF spatial resolution. It is used for storing visual information on storage media such as CD-ROM. MPEG-1 is used in consumer applications such as Video CD (VCD).

MPEG-2/H.262

MPEG-2 (ISO/IEC 13812-2), also referred to as *H.262*, is a generic video codec, jointly developed by ISO/IEC and ITU-T. MPEG-2 was completed in 1994, and was intended for video transmission over the broadband ISDNs (B-ISBNs) using asynchronous transfer mode (ATM) transport that has a bit rate of at least 10 Mbits/s. MPEG-3 was intended for HDTV with a target bit rate of 40 Mbits/s, but was finally merged into MPEG-2. MPEG-2 can be used for a wide range of applications including DAB, SDTV (standard-definition television), HDTV broadcast, cable TV distribution, as well as consumer applications such as Super VCD (SVCD) and DVD-Video.

MPEG-2 defines two types of streams: the program stream and the transport stream. The program stream is similar to the MPEG-1 stream. The MPEG-2 decoder is required to be capable of decoding MPEG-1 bitstream. The program stream is intended for an error-free environment such as digital storage. The transport stream is intended for transmission, where channel errors may occur. The transport stream offers error resilience, and has the ability to assemble multiple program streams into a single bitstream. The basic data structure for both the program and transport streams is the packetized elementary stream (PES) packet.

In MPEG-1, frames are noninterlaced (progressive). But in MPEG-2, most frames are interlaced. The simple nonscalable mode of MPEG-2 is the direct extension of MPEG-1 that accommodates interlaced video coding. For interlaced video, a macroblock can be split into two 16×8 field blocks, and interpicture prediction can be performed between the fields, since they are close to each other. Field prediction is similar to frame prediction, except that the pixels of the target macroblock belong to the same field. MPEG-2 has the scalability function.

H.263

H.263, approved in 1996, is intended for video communications via telephony, Internet or mobile networks, with a much lower bit rate. H.263 is based on H.261, and includes the functionalites of H.261, with some improvements. H.263 has a target bit rate on the order of 24 kbits/s. H.263 provides error protection, robustness and resilience for wireless transmissions. H.263+ (approved in 1998) and H.263++ (approved in 2000) are enhancements to H.263. These enhancements are intended to develop video coding at bit rates less than 64 kbits/s and more specifically at less than 24 kbits/s. H.263+ and H.263++ can be used for real-time applications. H.26L is a further enhancement of H.263, which later evolved into H.264.

Like H.261 and MPEG-1, H.263 employs source images of the CIF family with 4:2:0 chroma format. The source encoder operates on noninterlaced pictures at approximately 29.97 frames per second. The pictures can be one of the five members of the CIF family. The picture layer of H.263 contains the picture header, the GOB header, various coding decisions on the macroblocks in a GOB, and the coded transform coefficients. This structure is also used in H.261, MPEG-1, and MPEG-2. H.263 differs from the other codecs by defining the type information called *PTYPE* in the header that gives information of the whole picture.

B-frames are important for low-bit-rate applications. At very low bit rates, the blocks of pixels are mainly made of low frequency DCT coefficients, and the blocking artifacts are obvious. In H.263, the overlapped block matching motion compensation is an optional advanced prediction mode, which has four motion vectors per macroblock, and this technique can reduce the blocking artifacts to a certain extent. H.263 recommends the deblocking of the picture by the block edge filter.

In normal mode of H.263, Huffman coding is employed. H.263 has an optional syntax-based binary arithmetic coding mode. Reversible variable-length codes are used to reduce the influence of the channel errors. H.263 also specifies the optional advanced intra/inter variable-length coding modes to improve encoding efficiency.

In H.263, the error correction code is a BCH (511,493). H.263 also implements error concealment, which replaces the damaged pixels with pixels from parts of the same frame (by interpolation) or previous frame (by controlling the motion vector) that have maximum resemblance. These are known as intraframe and interframe error concealment. Loss of packet is also concealed in H.263.

MPEG-4

MPEG-4 (ISO/IEC 14496), completed in 2000, is an open ISO standard for rich multimedia compression and interactivity. MPEG-4 defines all aspects of an end-to-end multimedia solution. MPEG-4 provides a framework for digital rights management and protection. It offers high quality images at data rates from below 64 kbits/s up to around 1,200 Mbits/s. MPEG-4 also supports synthetic images and still images.

MPEG-4 is organized as tools, profiles, and levels. It has system tools, visual tools, audio tools, and other tools. A tool is a subset of coding functions to support a specific feature. Major visual tools are video compression tools, robustness in error-prone environments, fine grain scalability, shape and alpha channel coding, face and body animation, coding of 2D meshes, and coding of 3D meshes. The video compression tools are video compression algorithms for bit rates between 5 kbits/s and 1 Gbits/s and for resolution from subQCIF to studio resolution ($4k \times 4k$ pixels). A profile is a set of VO types that a codec is targeted for handling, and it defines a subset of coding tools, since a VO is coded using one or more tools. A profile in MPEG-4 is similar to a set of annexes in H.263. Profiles are a mechanism for interoperability between codecs. Within a profile, there are levels that define constraints on the parameters of the bitstream.

The MPEG-4 Visual's Simple profile is for low-cost applications, such as video over mobile and Internet. Baseline H.263 was adopted as the core of this profile. There are three levels that define bit rates of 64, 128, and 384 kbits/s. It supports most optionalities (annexes) of H.263, and can decode a bitstream generated by the core H.263. In the Simple profile, to avoid the disadvantage of Huffman coding, pre-calculated Huffman-based coding is applied. A variable-length code table of codewords for its (*run, level, last*) combinations is pre-calculated, based on the probability distributions of generic video material. A total of 102 specific combinations of (*run, level, last*) have variable-length codes assigned to them. This method is also applied to variable-length coding of the differentially coded motion vectors.

In the Advanced Coding Efficiency profile, SA-DCT is applied to 8×8 blocks that are located on the boundary of an arbitrarily shaped VOP. Sprite coding is supported in MPEG-4 Main profile. MPEG-4 allows optionally the use of wavelets.

MPEG-4 and H.324 are multimedia standards compatible with the 3GPP standard, as their required processing power is affordable. Note that H.324 is composed of H.263 and the G.723.1 speech codec.

H.264/AVC

H.264/AVC, or MPEG-4 Part 10, is a joint ITU-T and ISO/IEC standard (ISO/IEC 14496-10), published in 2003 [19]. It achieves a significant improvement in compression efficiency over existing standards such as MPEG-2, and a substantial superiority of video quality over that of H.263 and MPEG-4 [13]. Compared with H.263+ and MPEG-4 Simple profile, H.264 allows up to 50 per cent in bit rate saving for a similar degree of encoder optimization [13]. H.264 is expected to overtake MPEG-2/H.262 in common use, and to be an important part of wireless multimedia services. It covers all common video applications ranging from mobile services, video-conferencing to IPTV, HDTV, and high-definition video storage.

H.264 provides functionality similar to that provided by H. 263+ and MPEG-4 Simple profile, but with much better compression and more reliable transmission. Some optional modes in H.263, such as high precision spatial accuracy for motion estimation, data partitioning, multiple previous reference pictures for prediction, and the deblocking filter in motion compensation, have become a part of the core of H.264. H.264 defines a set of three profiles. The Baseline profile supports intra-/inter-coding and entropy coding with CAVLC, and is targeted for videotelephony, videoconferencing, and wireless communications. The Main profile supports CABAC, and can be used for television broadcasting and video storage. The Extended profile provides improved error resilience, and is suitable for streaming media applications.

For application like video streaming, H.264 introduces a new type of picture, called *switching or secondary picture (S-picture)*, for transition from one bitstream to another. S-pictures are generated by the server at the time of switching, and are transmitted as the first frame after the switching. This effectively solves the picture drift problem of other techniques. In MPEG codecs, random access is based on the regular spaced I-frames. This is not viable in the low bit rate H.264. H.264 employs a method that creates S-pictures at the time of access.

H.264 is the unique standard that employs the integer transform instead of the floating-point DCT transform. All integer transforms in H.264 as well as their inverse transforms can be implemented using only shift and add operations in $(8 + b)$-bit arithmetic precision for b-bit input video. The transform coefficients are then quantized and entropy coded. Like H.263, H.264 uses a UQ-DZ and zigzag scanning of the quantized coefficients. In H.263 there are 32 quantization levels, while in H.264 there are 52 levels. In the Baseline profile, the quantized coefficients are entropy coded using CAVLC, and all other syntax elements are coded using fixed-length or exponential-Golomb variable-length codes. In the Main profile, a CABAC is used, which is up to 10% better than Huffman coding in terms of bit saving [13]. The decoder complexity of H.264 is about four times that of MPEG-2 and two times that of MPEG-4 Visual Simple profile.

Another important feature of H.264 is the separation of the video coding layer (VCL) from the network adaptation layer (NAL), where the VCL gives the format of the video content and specification of the NAL depends on the type of the transport network. The VCL is based on conventional block-based motion-compensated hybrid video coding, used in prior video coding standards such as MPEG-2 or H.263, but has a substantially higher degree of diversification, sophistication, and adaptability.

H.264 Baseline was targeted for royalty-free use. Microsoft's proprietary Windows Media format will also be a main contender to MPEG-4 and H.264. Source code of the H.264 reference model is available in C language.[19]

---

[19] http://iphome.hhi.de/suehring/tml/

MPEG-7

MPEG-7, called *multimedia content-based description standard*, specifies a standard set of descriptors for describing various types of multimedia information coded with the standard codecs. MPEG-7 does not specify how to extract features, nor how to search for them. Most fundamental descriptors are color descriptors, texture descriptors, shape descriptors, motion descriptors, and localization descriptors. Retrieval can be based on the features of each of these descriptors.

To describe audio-visual events, the visual features are first extracted as shape, size, texture, color, movement, and their positions in a frame; then the audio features are extracted as key, mood, tempo, tempo changes, and their positions in time domain. These low levels of abstraction can be performed automatically. Higher level abstraction requires human intervention to define the semantic relations between the lower level features. Information about the multimedia data, such as copyright information, is also required to be included.

All applications that make use of multimedia will benefit from MPEG-7. The features are indexed and are used for retrieval purpose. For example, the face recognition descriptor can be used for retrieval of face images from a video.

MPEG-21

MPEG-21 defines a multimedia framework that enables transparent and augmented uses of multimedia resources across various networks and devices, i.e., the delivery and consumption of video contents. It provides interfaces and protocols that enable creation, identification, description, management, protection, manipulation, search, access, storage, delivery and consumption of the contents, across the whole content-creation and consumption chain.

MPEG-21 provides solutions for universal multimedia access, based on digital item adaptation. A digital item is a structured digital object, and is a fundamental unit of distribution and transaction with the MPEG-21 framework. Video transcoding or scalable coding is used to maintain the interoperability between different video formats such as MPEG-2 and MPEG-4.

Problems

17.1 For a window of $2K + 1$ samples, the median filter is expressed by

$$\hat{y}(m) = \text{median}\{x(m - K), \ldots, x(m), \ldots, x(m + K)\}.$$

Select an image, add some impulse noise, and then denoise it by using the median filter.

17.2 Evaluate the transform matrix $\mathbf{C}_N$, defined by (13.145), for a 4×4 DCT. Calculate the dc component of the 4×4 block:

$$\begin{bmatrix} 1 & 1 & 5 & 5 \\ 2 & 4 & 12 & 8 \\ 3 & 9 & 15 & 12 \\ 5 & 11 & 22 & 8 \end{bmatrix}.$$

17.3 Evaluate Problem 17.1 using the integer transform, defined by (17.29). Find the difference between the transform outputs.

17.4 Given an input array: 3, 0, 0, −2, 5, 0, 0, 0, −14, 0, 0, 0, 7, ... Encode the sequence in RLE of the (run, level) pairs.

17.5 Many image processing operations are implemented using convolution of the image with a kernel. Implement the following operations on an image.

(a) Sharpening using the Laplacian kernel $\begin{bmatrix} -1 & -1 & -1 \\ -1 & -9 & -1 \\ -1 & -1 & -1 \end{bmatrix}$;

(b) Embossing using a directional derivative kernel $\begin{bmatrix} +2 & +1 & 0 \\ +1 & 0 & -1 \\ 0 & -1 & +2 \end{bmatrix}$;

(c) Edge extraction using the Sobel kernels $\begin{bmatrix} +1 & +2 & +1 \\ 0 & 0 & 0 \\ -1 & -2 & -1 \end{bmatrix}$ and $\begin{bmatrix} +1 & 0 & -1 \\ +2 & 0 & -2 \\ +1 & 0 & -1 \end{bmatrix}$

for vertical and horizontal derivatives. The edge corresponds to the gradient magnitude.

17.6 Select an image, and add "salt-and-pepper noise" to simulate an extreme case of locked or dead pixels in a camera. Apply a 3 × 3 median filter to remove the noise.

17.7 Implement a delta and variable-length coding method for compression of grayscale images. Compare the size of the compressed image to the theoretical limit defined by the image entropy.

17.8 Given seven eight-bit resolution luminance samples: 20, 34, 60, 243, 200, 49, 89. (a) Perform DPCM encoding. (b) Use the three-bit quantizer

$$y = \begin{cases} \pm 2, & |x| \leq 5 \\ \pm 5, & 5 < |x| \leq 12 \\ \pm 12, & 12 < |x| \leq 40 \\ \pm 40, & \text{else} \end{cases}$$

to quantize the DPCM data, and find the reconstructed samples. (c) Calculate the PSNR of the reconstructed samples.

17.9 Generate the DCT coefficients for an 8 × 8 pixel block (luminance) using random numbers between -50 and 800, and then apply quantization using Table 17.1. Find the quantization indices for the baseline JPEG with a quality factor of 50%.

17.10 Conduct a project to encode an image using JPEG with sequential coding and progressive coding. Compare the performance of the two modes.

17.11 Download the MATLAB toolbox jpegtool from `http://www.dms.auburn.edu/compression/download.html`. Install and test the program. Select any picture, experiment JPEG encoding by different choices of quantizer and block size to smooth the blocking artifacts. Compare the input and output images, and report the compression ratio.

17.12 Download the Independent JPEG Group JPEG software from `http://www.ijg.org/`. Read the manual. Write your own program to call the API functions.

17.13 Find the bit rate for the following video formats: (a) ITU-R BT.601/625, 4:2:2. (b) SIF/525, 4:2:0. (c) CIF. (d) SIF/525, 4:2:2. (e)subQCIF, 15 Hz.

17.14 An MPEG-2 video with its associated audio and channel coding has a bit rate of 6 Mbits/s. With 16QAM modulation, how many such videos can be transmitted over a 8-MHz channel in the UHF band? Assume that each modulated symbol occupies 1.25 MHz and each channel contains a 2 MHz guard band.

17.15 Download the H.264/AVC JM reference software from `http://iphome.hhi.de/suehring/tml/`. Install it. Learn the JM reference software manual. Experiment raw videos in RGB or YUV format by encoding/decoding with different options.

References

[1] N. Ahmed, T. Natarajan & K. R. Rao, Discrete cosine transform. *IEEE Trans. Comput.*, **23**:1 (1974), 90–93.

[2] B. Bhatt, D. Birks & D. Hermreck, Digital television: making it work. *IEEE Spectrum*, **34**:10 (1997), 19–28.

[3] M. I. H. Bhuiyan, M. O. Ahmad & M. N. S. Swamy, Spatially adaptive wavelet-based methods using the Cauchy prior for despeckling SAR images. *IEEE Trans. Circ. Syst. Video Tech.*, **17**:4 (2007), 500–507.

[4] M. I. H. Bhuiyan, M. O. Ahmad & M. N. S. Swamy, Wavelet-based image denoising with the normal inverse Gaussian prior and linear minimum mean squared error estimator. *IET Image Process.*, **2**:4 (2008), 203–217.

[5] M. I. H. Bhuiyan, M. O. Ahmad & M. N. S. Swamy, Spatially-adaptive thresholding in wavelet domain for despeckling of ultrasound images. *IET Image Process.*, **3**:3 (2009).

[6] S. Bouguezel, M. O. Ahmad & M. N. S. Swamy, Low-complexity 8×8 transform for image compression. *Electron. Lett.*, **44**:21 (2008), 1249–1250.

[7] N. Brady, MPEG-4 Standardized methods for the compression of arbitrarily shaped video objects. *IEEE Trans. Circ. Syst. Video Technol.*, **9**:8 (1999), 1170–1189.

[8] W.-H. Chen & W. K. Pratt, Scene adaptive coder. *IEEE Trans. Commun.*, **32**:3 (1984), 225–232.

[9] I. Daubechies, The wavelet transform, time frequency localization and signal analysis. *IEEE Trans. Inf. Theory*, **36**:5 (1990), 961–1005.

[10] K.-L. Du & M. N. S. Swamy, *Neural Networks in a Softcomputing Framework* (London: Springer, 2006).

[11] C. J. Duanmu, M. O. Ahmad & M. N. S. Swamy, A vector based fast block motion estimation algorithm for implementation on SIMD architectures. In *Proc. IEEE ISCAS*, Phoenix, AZ, May 2002, **4**, 337–340.

[12] C. J. Duanmu, M. O. Ahmad & M. N. S. Swamy, Fast block motion estimation with eight-bit partial sums using SIMD architecture. *IEEE Trans. Circ. Syst. Video Tech.*, **17**:8 (2007), 1041–1053.

[13] M. Ghanbari, *Standard Codecs: Image Compression to Advanced Video Coding* (London: IEE Press, 2003).

[14] N. Gupta, M. N. S. Swamy & E. I. Plotkin, Despeckling of medical ultrasound images using data and rate adaptive lossy compression. *IEEE Trans. Medical Imaging*, **24**:6 (2005), 743–754.

[15] N. Gupta, M. N. S. Swamy & E.I. Plotkin, Wavelet domain based video noise reduction using temporal DCT and hierarchically-adapted thresholding. *IET Image Process.*, **1**:1 (2007), 2–12.

[16] ITU-R, *Studio Encoding Parameters of Digital Television for Standard 4:3 and Wide-screen 16:9 Aspect Ratios*, ITU-R Rec. BT.601-5, Geneva, 1995.

[17] J. R. Jain & A. K. Jain, Displacement measurement and its application in interframe image coding. *IEEE Trans. Commun.*, **29**:12 (1981), 1799–1808.

[18] T. Koga, K. Iinuma, A. Hirano, Y. Iijima & T. Ishiguro, Motion-compensated inter-frame coding for video conferencing. In *Proc. National Telecommun. Conf. (NTC'81)*, New Orleans, LA, Dec 1981, G5.3.1–G5.3.5.

[19] A. Luthra, G. J. Sullivan & T. Wiegand, eds., *Special Issue on the H.264/AVC Video Coding Standard. IEEE Trans. Circ. Syst. Video Technol.*, **13**:7 (2003).

[20] M. Nelson, *The Data Compression Book* (New York: M&T Books, 1992).

[21] S. M. M. Rahman, M. N. S. Swamy & M. O. Ahmad, Video denoising based on inter-frame statistical modeling of wavelet coefficients. *IEEE Trans. Circ. Syst. Video Tech.*, **17**:2 (2007), 187–198.

[22] S. M. M. Rehman, M. O. Ahmad & M. N. S. Swamy, Bayesian wavelet-based image denoising using the Gauss-Hermite expansion. *IEEE Trans. Image Process.*, **17**:10 (2008), 1755–1771.

[23] K. D. Rao, M. N. S. Swamy & E. I. Plotkin, Adaptive filtering approaches for colour image and video restoration. *IEE Proc. Vis. Image Signal Process.*, **150**:3 (2003), 168–177.

[24] A. Said & W. A. Pearlman, A new, fast and efficient image codec based on set partitioning in hierarchical trees. *IEEE Trans. Circ. Syst. Video Technol.*, **6**:3 (1996), 243–250.

[25] K. Sayood, *Introduction to Data Compression*, 2nd edn (San Mateo, CA: Morgan Kaufmann, 2000).

[26] K. Sayood (ed), *Lossless Compression Handbook* (San Diego, CA: Academic Press, 2003).

[27] D. Sen, M. N. S. Swamy & M. O. Ahmad, Unbiased homomorphic system and its application in reducing multiplicative noise. *IEE Proc. Vis. Image Signal Process.*, **153**:5 (2006), 521–537.

[28] J. M. Shapiro, Embedded image coding using zero-trees of wavelet coefficients. *IEEE Trans. Signal Process.*, **41**:12 (1993), 3445–3462.

[29] D. Santa-Cruz, R. Grosbois & T. Ebrahimi, JPEG 2000 performance evaluation and assessment. *Signal Process.: Image Commun.*, **17**:1 (2002), 113–130.

[30] D. Sen, M. N. S. Swamy & M.O. Ahmad, Computationally fast techniques to reduce AWGN and speckle in Videos. *IEE Proc. Vis. Image Signal Process.*, **1**:4 (2007), 319–334.

[31] M. Sonka, V. Hlavac & R. Boyle, *Image Processing, Analysis, and Machine Vision*, 3rd edn (Toronto, Canada: Thomson Learning, 2008).

[32] C. K. Tan & M. Ghanbari, Using non-linear diffusion and motion information for video segmentation. In *Proc. IEEE ICIP*, New York, Sep 2002, **2**, 769–772.

[33] D. Taubman & A. Zakhor, Multirate 3-D subband coding of video. *IEEE Trans. Image Process.*, **3**:5 (1994), 572–588.

[34] D. Taubman, High performance scalable image compression with EBCOT. *IEEE Trans. Image Process.*, **9**:7 (2000), 1158–1170.

[35] M. Wien, Variable block-size transforms for H.264/AVC. *IEEE Trans. Circ. Syst. Video Technol.*, **13**:7 (2003), 604–613.

[36] J. Zan, M. O. Ahmad & M. N. S. Swamy, New techniques for multi-resolution motion estimation. *IEEE Trans. Circ. Syst. Video Tech.*, **12**:9 (2002), 793–802.

[37] J. Zan, M. O. Ahmad & M. N. S. Swamy, Pyramidal motion estimation techniques exploiting intra-level motion correlation. *IEEE Trans. Circ. Syst. II*, **50**:2 (2003), 83–93.

[38] J. Zan, M. O. Ahmad & M. N. S. Swamy, Multiplicationless Burt and Adelson's pyramids for motion estimation. *IEEE Trans. Circ. Syst. Video Tech.*, **14**:1 (2004), 136–141.

[39] J. Zan, M. O. Ahmad & M. N. S. Swamy, A multiresolution motion estimation technique with indexing. *IEEE Trans. Circ. Syst. Video Technol.*, **16**:2 (2006), 157–165.

[40] J. Zan, M. O. Ahmad & M. N. S. Swamy, Comparison of wavelets for multiresolution motion estimation. *IEEE Trans. Circ. Syst. Video Tech.*, **16**:3 (2006), 439–446.

[41] Y.-Q. Zhang & S. Zafar, Motion-compensated wavelet transform coding for color video compression. *IEEE Trans. Circ. Syst. Video Technol.*, **2**:3 (1992), 285–296.

[42] J. Zhang, M. O. Ahmad & M. N. S. Swamy, Feature-bit-plane matching technique for estimation of motion vectors. *Electron. Lett.*, **34**:11 (1998), 1090–1091.

[43] J. Zhang, M. O. Ahmad, and M. N. S. Swamy, Quadtree structured region-wise motion compensation for video compression. *IEEE Trans. Circ. Syst. Video Technol.*, **9**:5 (1999), 808–822.

[44] Y. Zhao & M. N. S. Swamy, Technique for designing biorthogonal wavelet filters with an application to image compression. *Electron. Lett.*, **35**:12 (1999), 1530–1532.

[45] Y. Zhao & M. N. S. Swamy, New technique for designing nearly orthogonal wavelet filter banks with linear phase. *IEE Proc.-Vis. Image Signal Process.*, **147**:6 (2000), 527–533.

18 Multiple antennas: smart antenna systems

18.1 Introduction

Wireless channels suffer from time-varying impairments such as multipath fading, interference, and noise. Diversity, such as time, frequency, space, polarization, or angle diversity, is typically used to mitigate these impairments. Diversity gain is achieved by receiving independent-fading replicas of the signal.

The multiple antenna system employs multiple antennas at either the transmitter or the receiver, and it can be either multiple-input single-output (MISO) for beamforming or transmit diversity at the transmitter, single-input multiple-output (SIMO) for diversity combining at the receiver, or MIMO, depending on the numbers of transmit and receive antennas. The MISO, SIMO, and MIMO channel models can be generated by using the angle-delay scattering function.

Multiple antenna systems are generally grouped as smart antenna systems and MIMO systems. A smart antenna system is a subsystem that contains multiple antennas; based on the spatial diversity and signal processing, it significantly increases the performance of wireless communication systems. Direction-finding and beamforming are the two most fundamental topics of smart antennas. Direction-finding is used to estimate the number of emitting sources and their DoAs, while beamforming is used to estimate the signal-of-interest (SOI) in the presence of interference.

A MIMO system consists of multiple antennas at both the transmitter and the receiver. They are typically used for transmit diversity and spatial multiplexing. Spatial multiplexing can maximize the system capacity by transmitting at each transmit antenna a different bitstream. MISO, SIMO, and MIMO can be collectively treated as MIMO, and thus the smart antenna system can be regarded as a special case of the MIMO system.

For implementation of multiple antenna systems, a major problem is the knowledge of the channel at the transmitter as well as the receiver. The situation may be such that CSI is available at both the receiver and the transmitter, CSI is available at the receiver alone, or no CSI is available. The performance can be improved if more CSI is available. Multiple antennas are difficult to implement on the MS due to the size and power restrictions.

18.1.1 The concept of smart antennas

Research on the topics of smart antennas started in the 1960s. Smart antennas are generally used for spatial filtering for interference reduction (SFIR) or SDMA. For SFIR, a BS steers its radiation pattern towards the desired MS, and this generates little interference to other users. SFIR is similar to sectorization. The average CCI can be reduced, and thus the reuse distance D can be decreased, leading to an increase in the capacity. SDMA achieves simultaneous communications with two or more users in the same cell that use the same frequency, time, and code. The BS creates many beams, each directed towards one user. Even if these beams are separated by an angle, the surrounding scatters may lead to multipath, which may arrive in the directions of other beams. In both the SFIR and SDMA cases, for an array of N_a elements, the maximum number of users or interfering sources it can support is N_a.

Based on the space diversity of an antenna array and digital signal processing techniques, a smart antenna system dynamically adjusts the beamforming weights so as to steer the main lobe of the beampattern toward the desired user on the transmit path. This considerably reduces interference to other users, when compared to the beampattern of a conventional omnidirectional antenna. This transmit beamforming function corresponds to a MISO channel. Transmit beamforming is also known as *transmit diversity combining*.

On the uplink, the BS nullifies the interference signals to maximize the SINR for the desired signal. Consequently, use of smart antennas can significantly reduce the overall interference in a cell and hence, improve the capacity of the system. This corresponds to a SIMO channel, and the function of the smart antenna system is known as *receive beamforming*. When there is only the desired signal and no interfering sources, this function is known as *diversity combining*.

Both types of smart antenna systems make use of the antenna diversity to improve the system performance. The beamforming function leads to an array gain, while diversity combining yields a diversity gain. The concept of smart antennas is illustrated by Fig. 18.1.

18.1.2 Smart antennas in mobile communications

Smart antenna technology can considerably reduce radiation hazards, and it becomes an effective method in controlling RF power levels. For a simulated network with four antennas, the capacity of a CDMA network with power control is increased by four to five times, compared with the same system using one antenna [1].

Smart antenna systems can increase the system capacity in a number of ways [29]. In a TDMA/FDMA system, they can be used to decrease the reuse distance by reducing the

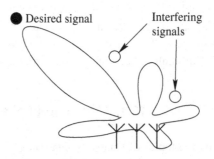

Figure 18.1   Schematic of a smart antenna system.

interference, thus decreasing the cluster size for frequency reuse. An alternative solution is to keep the cluster size unchanged, and increase the number of users within a cell. Multiple users are served on the same time/frequency slot, and they are distinguished by the BS using their different spatial signatures.

Smart antenna technology significantly improves the spectral efficiency and QoS. Smart antenna technology is not a core requirement for 3G mobile communications, due to the very expensive computation involved. Even though smart antenna technology is not mandatory in WCDMA and CDMA2000, many vendors have already introduced their smart antenna BSs. TD-SCDMA is the only 3G standard that specifies smart antenna technology using a uniform circular array with eight elements at the BS. Future-generation wireless systems may adopt smart antenna technology as an integral part of the system.

In [4], a smart-antenna based on a combined DoA and joint channel estimation scheme has been investigated for UTRA-TDD applying MUD at the uplink receiver. A user-specific training sequence is used for a combined DoA and joint channel estimation. The number of impinging DoAs is estimated according to a modified minimum description length criterion for centro-symmetric array configurations [42]. Smart antennas using uniform circular arrays have been examined for DoA estimation and beamforming in [26].

For IS-95 CDMA that has strict power control, due to the processing gain G, the power of a desired user, P_d, is related to the power of an interfering user, P_i, by

$$P_d = GP_i. \tag{18.1}$$

Application of an array of N_a elements can increase P_d by a factor of N_a. Assuming a threshold of SIR for deciding the number of admissible users in the cell, the adoption of an antenna array at the BS can increase the number of admissible users by a factor of N_a. Beamforming for narrowband CDMA BS receivers has been discussed in [38], where a comparison of the beam steering and the eigenfilter methods has also been made.

For 3G CDMA systems, a high-data rate user has a low spreading factor, and thus becomes a more severe interferer. In this case, the interference can be suppressed by placing a null in that direction. Downlink beamforming for WCDMA has been discussed in [30]. Within one logical cell with three sectors using standard panel antennas, user-specific and fixed beamforming modes can be used. User-specific beamforming generates individual beams to each MS without any restriction on the selection of the weight vectors. On the other hand, fixed beamforming synthesizes a finite set of beams at the BS and multiple MSs may receive signals transmitted under the same beam. Beamforming is not allowed on all channels; some channels only support fixed beamforming, while user-specific beamforming is optional.

18.2 Direction-finding

For a uniform linear array, as shown in Fig. 18.2, there is a time delay τ_n between the signal at element n and the reference element $n = 1$:

$$\tau_n = (n - 1)\frac{d}{c}\sin\theta, \tag{18.2}$$

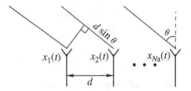

Figure 18.2

Time delays for a uniform linear array.

where d is the inter-element spacing, c is free-space speed of light, and θ is the angle of the incident signal, $-90° < \theta < 90°$. Thus

$$x_n(t) = x_1(t - \tau_n). \tag{18.3}$$

This model is also applicable for the case of the transmitter.

The phase differences for all the array elements are denoted by the steering vector

$$\boldsymbol{d} = \left(d_1, \ldots, d_{N_a}\right)^T, \tag{18.4}$$

where N_a is the number of the array elements and $d_i = e^{-j2\pi f \tau_i}$.

18.2.1 Pseudospectrums

The DoAs of incident signals are traditionally obtained by using a pseudospectrum method, which defines a function $P(\theta)$ that gives an indication of the DoAs at the maxima. DoAs can be easily obtained by using the Fourier transform of the signal vector $\boldsymbol{x}$. This produces the DoAs θ_i with a resolution of $2\pi/N_a$. The angular spectrum is given by

$$P(\theta) = \frac{\boldsymbol{d}^H(\theta)\mathbf{R}_{xx}\boldsymbol{d}(\theta)}{\boldsymbol{d}^H(\theta)\boldsymbol{d}(\theta)}, \tag{18.5}$$

where the superscript H denotes conjugate transpose, $\boldsymbol{d}(\theta)$ is the steering vector in the direction θ, and $\mathbf{R}_{xx} = E\left[\boldsymbol{x}\boldsymbol{x}^H\right]$. This spectrum, however, has too low a resolution.

Capon's method [6] is a minimum variance distortionless response (MVDR) technique, also known as the *ML technique*. After the weight is estimated by using the MVDR beamforming algorithm, the pseudospectrum is derived as

$$P_{\text{Capon}}(\theta) = \frac{1}{\boldsymbol{d}^H(\theta)\mathbf{R}_{xx}^{-1}\boldsymbol{d}(\theta)}. \tag{18.6}$$

This method introduces poor resolution when the sources are highly correlated, as in the case of the MVDR beamformer. The method provides a reasonable resolution, but suffers from bias. In [14], a pseudospectrum that has a form similar to that of Capon's pseudospectrum is obtained from a single snapshot of array measurement; the main lobes on the pseudospectrum correspond to the DoAs of the sources.

Some of the popular high-resolution direction-finding algorithms are MUSIC (MUltiple SIgnal Classifications) [35], ESPRIT (Estimation of Signal Parameters via Rotational Invariance Techniques) [32, 39], and the minimum-norm method. MUSIC is the most popular method and is applicable for arbitrary arrays, and is introduced in Section 18.2.2. The minimum-norm method is only suitable for a uniform linear array. Most high-resolution direction-finding algorithms including ML and MUSIC have an inherent, computationally expensive search procedure over d. ESPRIT eliminates this search procedure, but demands multiple identical arrays called *doublets*, which are separate arrays or two subarrays of one larger array. A thorough review of direction-finding methods is given in [24], and a good introduction to these algorithms is given in [25].

ESPRIT and MUSIC perform poorly in the presence of highly correlated multipath signals, which frequently occur in urban environments. In this case, the simple conventional beamformer spectrum can be used

$$P(\theta) = d^H(\theta)\mathbf{R}_{xx}d(\theta). \tag{18.7}$$

The maximum of $P(\theta)$ corresponds to the direction of the sources.

For an antenna array of N_a elements and N_s sources, the observed data vector is given by

$$x(t) = D(\theta)s(t) + n(t), \tag{18.8}$$

where $\mathbf{D} = \left[d_1|d_2|\ldots|d_{N_s}\right]$, $s(t) = \left(s_1(t), s_2(t), \ldots, s_{N_s}(t)\right)^T$, and n is an N_a-dimensional noise vector with variance σ^2.

The autocorrelation of $x(t)$ is thus given by

$$\mathbf{R}_{xx} = \mathbf{D}\mathbf{R}_{ss}\mathbf{D}^H + \sigma_n^2\mathbf{I}, \tag{18.9}$$

where $\mathbf{R}_{ss} = E\left[ss^H\right]$.

18.2.2 MUSIC

MUSIC is an eigenstructure method. It provides unbiased estimates of the number of signals, the DoAs, and the strengths of the signals. Given the number of sources N_s, the number of noise eigenvalues and eigenvectors is $N_a - N_s$. MUSIC exploits the eigenvector subspace, and thus is a subspace method.

By eigendecomposition of $\mathbf{R}_{xx}$, we have

$$\mathbf{R}_{xx}\mathbf{E} = \mathbf{E}\Lambda, \tag{18.10}$$

where $\Lambda = \text{diag}\left\{\lambda_1, \ldots, \lambda_{N_a}\right\}$, and $\mathbf{E} = \left[e_1|\ldots|e_{N_a}\right]$ is the matrix comprising all the eigenvectors. The first N_s significant eigenvalues correspond to the N_s sources.

The $N_a - N_s$ noise eigenvectors are given by

$$\mathbf{E}_N = \left[e_{N_s+1}|\ldots|e_{N_a}\right]. \tag{18.11}$$

For uncorrelated signals, the smallest eigenvalues are equal to the variance of the noise.

The noise subspace eigenvectors are orthogonal to the steering vector at the DoAs of the sources, thus at these DoAs

$$d^H(\theta)\mathbf{E}_N\mathbf{E}_N^H d(\theta) = 0. \tag{18.12}$$

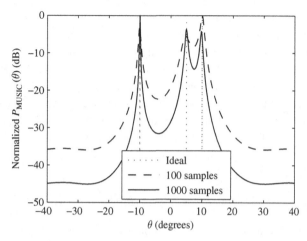

Figure 18.3   **Pseudospectrums of the MUSIC algorithm: for $N_a = 8$ and $N_s = 3$.**

Accordingly, the angular spectrum for the MUSIC algorithm can be defined by

$$P_{\text{MUSIC}}(\theta) = \frac{1}{d^H(\theta)\mathbf{E}_N\mathbf{E}_N^H d(\theta)}. \tag{18.13}$$

At the DoAs, there are peaks in the spectrum.

Example 18.1: For a uniform linear array of $N_a = 8$ elements with element spacing $d = \lambda/2$, given three uncorrelated sources with amplitudes $a_1 = a_2 = a_3 = 1$ impinging from directions $-10°$, $5°$ and $10°$, and $\sigma_n^2 = 0.1$, Figure 18.3 plots the angular spectrums of the MUSIC algorithm. The three curves in the figure correspond to the case when $\mathbf{R}_{xx}$ is obtained by the ideal signal and noise statistics, by averaging over 100 samples, and by averaging over 1000 samples, respectively.

MUSIC achieves very high resolution when the noise variance for all the elements are identical and the signals are completely uncorrelated. Otherwise, the resolution will diminish. MUSIC is applicable to an arbitrary array. The root-MUSIC algorithm developed for a uniform linear array in [3] reduces the complexity of MUSIC by finding the roots of a polynomial instead of plotting the pseudospectrum and searching for the peaks. For a uniform linear array, the same strategy has been applied to the min-norm algorithm, leading to the root-min-norm solution.

Most modulated communication signals have the cyclostationary property [16, 22]. Cyclic MUSIC is a direction-finding algorithm that exploits the cyclostationary property of the signal by replacing the correlation $\mathbf{R}_{xx}$ by the cyclic or conjugate cyclic correlation $\mathbf{R}_{xx}^\alpha$ and $\mathbf{R}_{xx^*}^\alpha$ [34]. Cyclic MUSIC provides a much better performance when the separation between two DoAs is very small, but performs worse when the signals are farther apart. Cyclic MUSIC is valid, independent of the number of signals and the nature of noise, as long as the number of signals having the specified cycle frequency is less than the number of array elements N_a. However, cyclic MUSIC requires information of the cyclic frequency.

DoA estimation under coherent signal conditions

The performance of MUSIC degrades severely in a coherent or highly correlated signal environment, as in the case of multipath propagation. For this purpose, various modifications to MUSIC have been proposed by modifying the covariance matrix through spatial smoothing [28]. Spatial smoothing imposes restrictions on the type and structure of the array. In [19], a uniform linear array with N_a antennas is divided into overlapping forward subarrays, and then the subarray covariance matrices are averaged. As long as the number of subarrays L is greater than or equal to the number of sources N_s ($L \geq N_s$), the modified covariance matrix of the signals obtained is always nonsingular regardless of the coherence of the signals [31]. The forward averaging spatial smoothing scheme can detect only $N_a/2$ coherent signals, as opposed to $M - 1$ noncoherent signals by using conventional MUSIC. By using a set of forward and conjugate backward subarrays simultaneously, up to $2N_a/3$ coherent signals can be detected.

18.3 Beamforming

Beamforming, also known as *spatial filtering* or *spatial diversity combining*, combines signals from an array of antennas to form an output. The signals corresponding to each antenna are multiplied by a complex weight and the output is formed for the receiver or the transmitter. For an optimal SNR performance, the receiver must always know the channel.

Assuming that the received signals at an antenna array are x_i, $i = 1, \ldots, N_a$, as shown in Fig. 18.4, the output of the beamformer is given by

$$y = \sum_{i}^{N_a} w_i^* x_i = \boldsymbol{w}^H \boldsymbol{x}, \tag{18.14}$$

where w_i is the beamforming weight at the ith antenna, $*$ denotes conjugation, and $\boldsymbol{w} = \left(w_1, w_2, \ldots, w_{N_a}\right)^T$.

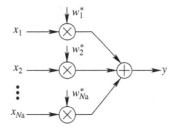

Figure 18.4   Beamforming for an antenna array of N_a elements.

In the SDR architecture, the beamforming block is the first block (for receiver) and the last block (for transmitter) in the baseband processing module. The calculated weight can be fed to the analog part of the circuit to adjust the gain at the analog part.

The performance of the beamformer is measured by the output SINR, defined by

$$\text{SINR} = \frac{w^H d R_{ss} d^H w}{w^H \mathbf{R}_I w}, \tag{18.15}$$

where the autocorrelation R_{ss} is the average power of the desired signal $s(t)$, $R_{ss} = E[s(t)s(t)]$, d is the steering vector of the desired signal, and $\mathbf{R}_I$ is the autocorrelation of the interference-plus-noise.

18.3.1 Blind source separation

Blind source separation consists of extracting the source signals from instantaneous mixtures of source signals. The source signals can be obtained by using a set of linear beamformers. The antenna measurements can be described by

$$x(k) = \mathbf{H}s(k) + n, \tag{18.16}$$

where $x(k)$ is a vector comprising the N_a antenna measurements at discrete time k, $s(k)$ is a vector comprising N_s source signals, $\mathbf{H}$ characterizes the channel, and n is AWGN.

A beamformer is used to find a weight matrix $\mathbf{W}$ so that the beamformer output $y(k)$ approximates $s(k)$

$$y(k) = \mathbf{W}^H x(k). \tag{18.17}$$

Note that the beamformer with weight matrix $\mathbf{W}$ is composed of a set of vector beamformers with weight vectors $w_1, \ldots, w_{N_s}$, that is, $\mathbf{W} = \left[w_1 | w_2 | \ldots | w_{N_s} \right]$. The pseudoinverse solution is $\mathbf{W}^H \mathbf{H} = \mathbf{I}$.

In a typical case, both $\mathbf{W}$ and $\mathbf{H}$ are unknown. This is a blind source separation problem [15]. In order to solve for the signal $s(k)$ based only on the output $x(k)$, the sources must be assumed to be independent and non-Gaussian. This is an independent component analysis (ICA) problem. A well-known ICA algorithm is the FastICA; many blind source separation algorithms are discussed in [15].

18.3.2 ZF, MRC, and Wiener beamformers

ZF beamformer

The beamforming problem can be characterized by the same equations as given by (18.16) and (18.17). The difference between beamforming and blind source separation lies in that the structure of the antenna array or signals can be used. When $\mathbf{H}$ is known, the pseudoinverse solution is the simplest

$$\mathbf{W} = \mathbf{H}^\dagger, \tag{18.18}$$

where $\mathbf{H}^\dagger$ is the pseudoinverse of $\mathbf{H}$. The pseudoinverse solution is also known as the *ZF beamformer*, since all the interfering sources are cancelled but the noise is not cancelled. A transmit ZF beamformer places nulls in the direction of the interfering sources, thus all the CCIs are eliminated and the ZF beamformer maximizes the SIR at the output.

MRC beamformer

When there are no interferers, the weights are just the MRC weights. In a MISO system, when the BS has the knowledge of the channel $\boldsymbol{h} = \left(h_1, h_2, \ldots, h_{N_a}\right)^T$ to its user, the transmit-MRC technique can be used as the beamforming algorithm

$$\boldsymbol{w}_{\mathrm{MRC}} = \frac{\boldsymbol{h}^H}{\sqrt{\|\boldsymbol{h}\|_F^2}}. \tag{18.19}$$

This is also applicable for receive-MRC in the case of a SIMO system.

The BER for receive-MRC is bounded by [40]

$$P_b \leq (1 + \gamma)^{-N_r}, \tag{18.20}$$

where γ is the SNR and N_r is the number of receive antennas. A receiver using ZF combining with N_r antennas and $N_s - 1$ interferers has the same performance as a receiver with $N_r - N_s + 1$ antennas and no interferences [40]. Thus, a ZF beamformer can null out $N_s - 1$ interferers with a diversity order of $N_r - N_s + 1$.

MMSE beamformer

When the objective is to minimize the difference between the output signal vector and the desired signal vector $s(k)$, for all k, in the MSE sense, we have

$$\mathbf{W} = \hat{\mathbf{R}}_{xx}^{-1} \mathbf{H}^H. \tag{18.21}$$

This is the linear MMSE or Wiener receiver.

The MMSE receiver maximizes the output SINR. Typically, there are many more interferers than the number of antennas, but only one or two are dominant interferers; in this case, the MMSE combiner provides a substantial increase in the performance. The MMSE combiner has the same performance as the MRC in the absence of any interference, but performs much better in the case of a strong interference, such as in interference-limited cellular systems.

18.3.3 Switched-beam antennas

Beamforming can be either based on switched-beam or adaptive beamforming systems. This subsection describes switched-beam beamforming systems, and adaptive antenna arrays are described in next section.

Switched-beam antenna arrays are an extension to the cellular sectorization scheme obtained by subdividing the sectors into multiple microsectors. Each microsector corresponds to a predetermined beam pattern. The antenna system switches between the beams so that the it always receives the strongest signal. The switched-beam approach is very simple, and only a single signal is selected and downconverted for baseband processing. Due to the fixed beams, it cannot null signals in an arbitrary direction. The Butler matrix [18] and the Blass matrix [5] are two common realizations of the fixed beams.

The Butler matrix is applied for an array of 2^n elements, performs a spatial FFT, and provides 2^n orthogonal beams, which are linearly independent combinations of the array element patterns. For a linear array, the Butler matrix produces beams that overlap at 3.9 dB below the maxima. Depending on the element patterns and the spacing, the array can cover 360°. The conventional Butler Matrix is realized using 3-dB 90° hybrids, crossovers, and phase shifters.

For a linear array of $N_a = 2^n$ elements, the array factor can be derived as

$$\text{AF}(\theta) = \frac{\sin\left(N_a \pi \frac{d}{\lambda} \sin\theta - l\pi\right)}{N_a \pi \frac{d}{\lambda} \sin\theta - l\pi}, \tag{18.22}$$

where $l = \pm\frac{1}{2}, \pm\frac{3}{2}, \ldots, \pm\frac{N_a-1}{2}$. If $d = \lambda/2$, the beams are evenly distributed over the full range of $-90° \leq \theta \leq 90°$.

The Butler matrix is easy to implement. The loss involved is very small. However, in a Butler matrix, the beamwidth and the beam angles vary with frequency, causing the beam to squint with frequency. For large matrices, the crossovers make interconnections complex.

Example 18.2: The array factor, given by (18.22), has been plotted in Fig. 18.5 for $N_a = 8$ and $d = \lambda/2$. It is seen that for different phase shifts (different l) the main lobes almost equally divide the angle span of θ. When $d > 0.5\lambda$, the beams span over an ever decreasing interval of angles. By controlling the phase shift l, one can switch from one beam to another.

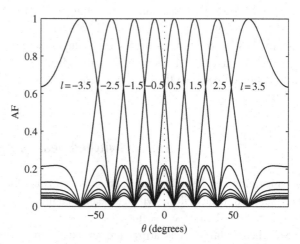

Figure 18.5   **Beams generated by using the Butler matrix approach: for $N_a = 8$.**

The Blass matrix forms beams using transmission lines and directional couplers via time delays, and thus is more suitable for broadband operation. It can be used for any number of elements. It is simple but has a low performance due to the loss arising from the resistive terminations. Although it has simpler interconnections compared to the Butler matrix since no crossover is involved, it requires more components, thus making it more expensive and heavier. Beam squinting with frequency does not arise.

18.4 Adaptive beamforming

Unlike switched-beam antenna arrays, adaptive antenna arrays can adaptively steer beams towards the desired signals and null the interfering signals. Various beamforming algorithms have been developed; they are derived by optimizing a certain criterion such as maximizing the output SNR or minimizing the output interference [12, 24]. A beamformer with N_a antenna elements can suppress up to $N_a - 1$ interferers, or achieve an increase in SNR by a factor of N_a. The beamformer can create $N_a - 1$ nulls in the directions of the interference sources, thus achieving a diversity gain of N_a.

An adaptive beamforming algorithm usually extracts an output, compares it with a known criterion or training sequence, and adjusts the beamforming weights based on the differences. Adjustment of the weights can be defined based on some optimum criterion, such as ML or MMSE.

18.4.1 DoA-based beamforming

The first class of beamforming algorithms is the spatial reference algorithms. The antenna array forms a beam that has a maximum in the main DoA of the desired signal and nulls in the DoAs of the multipath components of the interferers.

The approach typically works in three steps. It first uses a direction-finding algorithm such as MUSIC or ESPRIT to find the DoAs of all the signals. These DoAs are then associated with specific users. This identification problem is then solved using a training sequence or some user-specific properties. Finally, a beamforming algorithm such as the MVDR or the generalized sidelobe canceller algorithm is applied to extract the desired signal, or to form the beam for transmitting toward the desirable user.

No reference signal is required. Due to the limited number of degrees of freedom, this method requires the number of sources to be less than the number of antenna elements. This method is not applicable to small cell and non-LOS environments.

Beam steering

The simplest beamforming is the delay-and-sum beamformer which makes all weights equivalent in amplitude but the phases are selected to steer the main beam in the desired direction θ_0 [24]

$$w = \frac{1}{N_a} d\left(\theta_0\right). \tag{18.23}$$

This requires the knowledge of the desired signal DoA.

This method is suitable only when there is no dominant interference. The SNR at the output of the beamformer is defined by

$$\text{SNR} = \frac{w^H d \mathbf{R}_{ss} d^H w}{w^H \mathbf{R}_{nn} w} = \frac{N_a P}{\sigma_n^2}, \tag{18.24}$$

where P is the mean output power.

MVDR beamformer

MVDR is one of the most popular beamforming methods. The MVDR beamformer [7] minimizes the array output power subject to specific gain in the desired signal direction. This leads to the minimization of the contributions due to noise and interference

$$\min \sigma_{\text{MV}}^2 = w^H \mathbf{R}_{uu} w, \quad \text{subject to} \quad w^H d = 1, \tag{18.25}$$

where d is the steering vector of the desired signal, and $\mathbf{R}_{uu}$ is the autocorrelation of the interference-plus-noise signal:

$$\mathbf{R}_{uu} = \mathbf{R}_{ii} + \mathbf{R}_{nn}. \tag{18.26}$$

The method assumes that the desired and unwanted signals have zero mean.

By applying the Lagrange multiplier method, the optimum weight is derived as

$$w_{\text{MV}} = \frac{\mathbf{R}_{uu}^{-1} d}{d^H \mathbf{R}_{uu}^{-1} d}. \tag{18.27}$$

The MVDR method requires only the direction of the desired signal. It creates a mainlobe in the direction of the desired signal, while creating nulls in the directions of the interfering sources, thus maximizing the SIR.

When there is only one desired source and the noise, (18.27) reduces to

$$w_{\text{ML}} = \frac{\mathbf{R}_{nn}^{-1} d}{d^H \mathbf{R}_{nn}^{-1} d}. \tag{18.28}$$

This is the ML solution, and can be derived from the ML approach.

Example 18.3: Given four sources of equal amplitudes, impinging from $-50°$, $-20°$, $10°$, $40°$, respectively, and a uniform linear array of $N_a = 8$ antennas with a spacing of $\lambda/2$, $\sigma_n^2 = 0.1$. The source from $-20°$ is selected as the desired source. The beam patterns of the MVDR algorithm are shown in Fig. 18.6. In the ideal case, $\mathbf{R}_{xx}$ is diagonal, and there are deep nulls in the directions of the interfering sources. However, when $\mathbf{R}_{xx}$ is calculated over a finite number of samples, correlation is introduced and $\mathbf{R}_{xx}$ is not diagonal, and those nulls at the interfering directions become shallower or diminished, leading to a degradation of the MVDR algorithm. Figure 18.6 shows the degradation when $\mathbf{R}_{xx}$ is calculated over 100 and 500 samples.

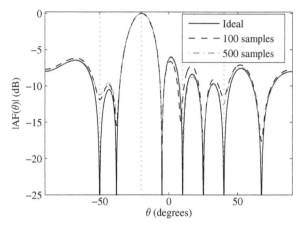

Figure 18.6   Beam patterns of the MVDR method: $N_a = 8$ and $d = \lambda/2$.

LCMV beamformer

The MVDR applies only one constraint. When there are multiple constraints given by

$$w^H \mathbf{D}_c = c, \qquad (18.29)$$

where $\mathbf{D}_c = [d(\theta_1)| \ldots |d(\theta_L)]$ is the spatial constraint matrix consisting of the steering vectors corresponding to the DoAs of the multipaths associated with the desired source, and c is a gain vector, which can be set to unity for EGC, or alternatively it can be optimized for MRC combining. θ_i's can also be specified as the DoAs of the interfering signals to null out the interference, and in this case, the corresponding components in c should be set to zero.

The optimal weight vector, referred to as the *linearly constrained minimum variance (LCMV) beamformer*, is given by [20]

$$w_{\mathrm{LCMV}} = \mathbf{R}_{uu}^{-1} \mathbf{D}_c \left(\mathbf{D}_c^H \mathbf{R}_{uu}^{-1} \mathbf{D}_c \right)^{-1} c. \qquad (18.30)$$

The LCMV beamformer can perfectly cancel up to $N_a - 1$ interferers.

Closed-form expressions for the outage probability of wireless systems with LCMV beamforming that cancels a number of dominant interferers using a uniform linear array have been derived in [27], for scenarios with a Rayleigh, Ricean, or Nakagami faded signal and Rayleigh faded interferers. The outage probability is shown to be very sensitive to the directions of the dominant interferers.

Beam-space processing

Most beamforming algorithms are based on element-space processing, where signals derived from each element are weighted and summed to produce the array output. Beam-space processing consists of first taking the array signals as input and generating a set of beams in different directions by using fixed weighting of the array signals. These patterns are then adaptively weighted and summed to produce the array output.

For an N_a-element array, beam-space processing produces a main beam steered in the direction of the desired signal and a set of up to $N_a - 1$ secondary (auxiliary) beams. The weighted output of the secondary beams is subtracted from the main beam to remove the interference, where the weights are adjusted to produce an estimate of the interference in the main beam. The beam-space processors have different names such as *Howells-Applebaum array*, *generalized sidelobe canceller*, *post-beamformer interference canceller*, and multiple-beam antennas [24]. These arrays are more useful when the number of sources $N_s - 1$ is less than the number of array elements N_a, $N_s < N_a$. In this case, only $N_s - 2$ weights are adjusted, as compared to N_a weights in the case of the element-space processor. When the number of beams is equal to N_a, these arrays are fully adaptive and have the same capabilities as those of the element-space array. In the absence of errors, both processing schemes produce identical results.

18.4.2 Training-based beamforming

Another class of methods is temporal reference beamforming, which requires a training sequence in the transmitted packets. The beamforming weights w are adapted by the criterion that the deviation of the combiner output from the training sequence is minimized. The criterion can be based on SIR, MMSE, or BER during the training period. This requires an adaptive algorithm such as the LMS, RLS, or the direct matrix inversion algorithm. This approach is computationally effective, but it consumes some spectrum. The major difficulty of this approach lies in the implementation of the prior carrier and signal recovery.

The most popular MMSE criterion is not optimal in terms of the bit error probability of the system. Adaptive beamforming algorithms have been derived by minimizing the BER cost function directly, such as the stochastic gradient [8], gradient Newton [10], and block-Shanno minimum BER [33] algorithms.

For noncoherent multipaths, both the DoA-based and training-based beamforming have the ability to overcome ISI, since the correlation of two different symbols of the same user will look like noise. However, for some systems such as the GSM system, the training sequence is not unique for each user; when there is CCI, errors may occur. For CDMA systems, since each user has a unique spreading code, training can extract the signal of each user. This feature is used in the least-squares despread-respread multitarget array (LS-DRMTA) algorithm [28], where the beamformer output is first despread by the desired user's code and then respread for the weight calculation.

MMSE beamformer

The MMSE beamformer is derived by minimizing the MSE between the desired or reference signal $d(t)$ and the output of the beamformer

$$\text{MSE} = E\left[e(t)e^*(t)\right], \tag{18.31}$$

where the error signal

$$e(t) = d(t) - \boldsymbol{w}^H \boldsymbol{x}(t). \tag{18.32}$$

This leads to the optimum Wiener solution

$$\boldsymbol{w}_{\text{opt}} = \mathbf{R}_{xx}^{-1} \boldsymbol{r}, \tag{18.33}$$

where

$$\boldsymbol{r} = E\left[d^*(t)\boldsymbol{x}(t)\right]. \tag{18.34}$$

When the interference is orthogonal to the desired signal, we have

$$\boldsymbol{w}_{\text{opt}} = K_{\text{MMSE}} \mathbf{R}_{nn}^{-1} \mathbf{D}_c, \tag{18.35}$$

where K_{MMSE} is a scaling factor. The MMSE beamformer requires the knowledge of the desired or reference signal.

Adaptive algorithms

When the DoAs change with time, adaptive algorithms are needed so that the beamforming weights can be updated for each sample or each block of samples. Various adaptation techniques can be applied to the optimization criteria defined for beamforming, such as those used in the MMSE beamformer and MVDR beamformer. Popular adaptation techniques are LMS and RLS [25]. Other adaptation techniques are conjugate gradient and Kalman filtering. These techniques are given in detail in [15].

Least mean squares

The MMSE beamformer requires the inversion of $\mathbf{R}_{xx}$. By applying gradient descent to the MSE criterion (18.31), we obtain the LMS algorithm

$$\begin{aligned}
\boldsymbol{w}(k+1) &= \boldsymbol{w}(k) - \frac{1}{2}\mu \frac{\partial \text{MSE}(\boldsymbol{w}(k))}{\partial \boldsymbol{w}} \\
&= \boldsymbol{w}(k) + \mu e^*(k)\boldsymbol{x}(k),
\end{aligned} \tag{18.36}$$

where μ is the step size. Note that in the second equality the expectation operation is replaced by the result for a single sample. The stability of the LMS algorithm is ensured when $0 \leq \mu \leq \frac{1}{2\lambda_{\text{max}}}$, where λ_{max} is the largest eigenvalue of $\mathbf{R}_{xx}$[15].

Direct matrix inversion

Direct matrix inversion, or *sample matrix inversion* is another method for obtaining the Wiener solution (18.33). It estimates $\mathbf{R}_{xx}$ and $\boldsymbol{r}$ based on the average of multiple samples, and then by applying matrix inversion. The estimate of $\mathbf{R}_{xx}$ may be updated for each new sample, leading to a new estimate of the weights. The convergence of the direct matrix inversion method is much faster than that for the LMS method, but the correlation matrix may be ill-conditioned and the matrix inversion requires $\frac{N^3}{2} + N^2$ complex multiplications [25]. The direct matrix inversion method can be applied for weight update in the case of two antenna elements.

Recursive least squares

The RLS method introduces a forgetting factor α when calculating $\mathbf{R}_{xx}(k)$ and $r(k)$. By using matrix inversion lemma, $\mathbf{R}_{xx}$ is derived as a recursive equation, thus the matrix inversion operation is avoided. The RLS algorithm is given as [25]

$$w(k) = w(k-1) + g(k)\left(d^*(k) - x^H(k)w(k-1)\right),\tag{18.37}$$

$$g(k) = \mathbf{R}_{xx}^{-1}(k)x(k),\tag{18.38}$$

$$\mathbf{R}_{xx}^{-1}(k) = \frac{1}{\alpha}\mathbf{R}_{xx}^{-1}(k-1) - \frac{\alpha^{-2}\mathbf{R}_{xx}^{-1}(k-1)x(k)x^H(k)\mathbf{R}_{xx}^{-1}(k-1)}{1 + \alpha^{-1}x^H(k)\mathbf{R}_{xx}^{-1}(k-1)x(k)},\tag{18.39}$$

where $0 \le \alpha \le 1$.

The RLS algorithm has a convergence speed that is an order of magnitude faster than that of the LMS algorithm, and the selection of step size is avoided. However, the complexity of RLS per sample is of the order of N^2, as opposed to the order of N for LMS. More detail on the RLS method is given in [15, 25]. The conjugate gradient method converges to the minimum of an error surface within N iterations for a rank-N matrix equation, and thus is faster than the RLS algorithm [9].

Example 18.4: Given a uniform linear array of 8 elements with spacing $d = \lambda/2$ and $\sigma_n^2 = 0.01$. The desired signal impinges at $20°$, with a source signal $s(k) = \sin(5\pi k/K)$. There are also two time-varying interfering sources at $-15°$ and $40°$, with $i_1(k) = 3\cos(4\pi k/K)$ and $i_2(k) = \exp(2k/K)\mathcal{N}(0, 1)$, respectively, where $\mathcal{N}(0, 1)$ is a normally distributed random variable with zero mean and unit variance.

Comparisons between the LMS and RLS algorithms are made in Figs. 18.7–18.10. For the LMS implementation, $\mu = \frac{1}{4*\lambda_{\max}}$. For the RLS implementation, $\alpha = 0.98$, and $\mathbf{R}_{xx}^{-1}(0) = 0.01\mathbf{I}$. From these figures, it is seen that the RLS can more rapidly track the

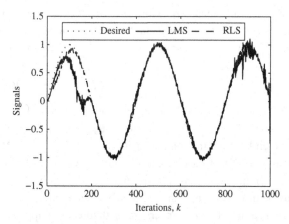

Figure 18.7

Acquisition and tracking of the desired signal.

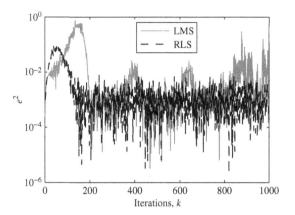

Figure 18.8 The square errors of the LMS and RLS methods.

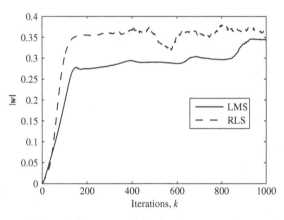

Figure 18.9 The magnitudes of the array weights for the LMS and RLS methods.

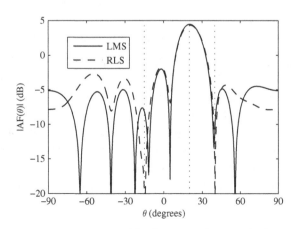

Figure 18.10 Beam patterns created by the LMS and RLS methods.

desired signal in a time-varying interfering environment. The training-based method does not require information about the source and interfering signals, and it automatically creates nulls in the directions of the interference and has a main lobe in the direction of the desired signal.

18.4.3 Blind beamforming

The third class of beamforming methods belongs to the blind approach based on the signal structure, and exploits the temporal and/or spectral properties of the received signals. Blind algorithms make no assumptions about the channel, and do not need any calibration of the antenna array. A training sequence is also not needed. The signal properties that are exploited can be the constant modulus and the cyclostationary properties. Blind beamforming algorithms have more convergence problems, compared to other methods.

The blind approach has the same drawbacks as blind equalization does, and thus limits its industrial application. Most blind algorithms assume that the channel is time-invariant during the period it collects data for determination of the signal statistics, but the sample collection time may be very long. Also, initialization of the estimate is critical for the convergence of the algorithm. For this reason, semi-blind algorithms that use very short training sequences to make the initial estimate can be used.

In a wireless communication environment, multipath propagation almost always takes place. This invalidates all eigendecomposition-based techniques and MVDR. In case of a non-Gaussian desired signal and directional Gaussian interferers, a cumulant-based blind method is used to first identify the steering vector of the desired signal and MVDR beamforming is then applied to remove Gaussian interference components [11]. The method behaves as the optimum beamformer that maximizes the SINR even in the presence of coherence (multipath).

Blind beamforming can also be implemented by using neural networks [12]. Although this method may lead to local minima on the MSE surface in some cases, it is generally robust and is suitable for analog implementation.

Constant modulus algorithms

The constant modulus algorithm has gained widespread popularity for blind source separation and blind equalization of communication signals. The constant modulus property is retained in most frequency- or phase-modulated signals, such as analog FM and PM, and digital FSK, PSK, MPSK signals. These signals have a constant amplitude $|s(k)|$, and information is carried in the phase. The constant modulus algorithm tries to find a beamformer $\boldsymbol{w}$ so that the output $\hat{s}(k) = \boldsymbol{w}^H \boldsymbol{x}(k)$ has the same normalized modulus for all k.

A simple implementation is minimizing a cost function, such as $E[||\hat{s}|^2 - 1|]$, by using the gradient-descent technique. This gives

$$\boldsymbol{w}(k+1) = \boldsymbol{w}(k) - \eta \boldsymbol{x}\boldsymbol{x}^H \boldsymbol{w}(k) \left(\left| \boldsymbol{w}^H(k)\boldsymbol{x} \right|^2 - 1 \right), \qquad (18.40)$$

where η is a small step size.

Since the frequency-selective channel destroys the constant modulus property of the signal, in order to use the constant modulus algorithm, the amplitude of the original signal needs to be restored or equalized before applying the constant modulus algorithm. The least-squares-based constant modulus algorithm (LS-CMA) is much faster than the conventional constant modulus algorithm [25].

The constant modulus algorithm is useful for eliminating correlated arrivals and is effective for constant modulated envelope signals. The algorithm, however, is not appropriate for the CDMA system because of the required power control [24].

Cyclostationary beamforming algorithms

Most of the man-made communication signals are cyclostationary and/or conjugate cyclostationary in nature [16, 22]. Based on this observation, a number of adaptive blind beamforming algorithms have been proposed in the literature [2, 13, 16, 41].

The two spectral self-coherence restoral (SCORE) algorithms, namely LS-SCORE and cross-SCORE, are the earliest successful cyclostationary beamforming algorithms [2]. As long as the cyclic frequency of a signal is available, the signal can always be extracted whatever be the number of interfering signals. The cyclic frequency is dependent on the carrier frequency and/or the symbol rate. The complexity of the algorithms for each sample is $O\left(N_a^3\right)$.

The cyclic adaptive beamforming (CAB) algorithm [41] is a simple and fast blind beamforming algorithm for low interfering environments, and it has a complexity of $O\left(N_a\right)$. Another cyclostationary beamforming algorithm with a complexity of $O\left(N_a\right)$ has been proposed in [13].

Five new cyclostationary algorithms, such as the adaptive cross-SCORE (ACS), adaptive CAB (ACAB), adaptive phase-SCORE (APS), maximally constrained autocorrelation (MCA), and constrained least-squares (CLS) algorithms, have been proposed in [16]. They all have a complexity of $O\left(N_a^2\right)$ complex multiplications. The ACS algorithm provides an excellent SINR performance in all interference environments, and is a practical one for wireless communications.

For implementation of cyclostationary algorithms, each signal is required to have a unique cycle frequency, which is a linear combination of its carrier frequency and its baud rate. The number of signals that can be extracted by this method is not limited by the spatial resolution provided by the antenna array, i.e., the number of array elements.

The IS-95 CDMA system is used for speech communications. In IS-95, all users occupy the same frequency and have the same baud rate. Thus, cyclostationary beamforming algorithms are not applicable. The 3G and future-generation mobile communications are targeted for mobile multimedia communications, where each user can typically be assigned a different data (baud) rate; thus, cyclostationary beamforming algorithms can be applied in these systems.

In conventional FDMA-based narrowband communication systems, each signal occupies a different frequency band. These signals are separated by using many narrowband bandpass filters, one for each signal. This leads to a considerable system cost in the BS. It is possible to use cyclostationary beamforming algorithms to extract the signal of each user, as long as the frequency of each signal is known to the beamformer; this can be accomplished by using a common pilot channel to broadcast each user's carrier frequency as well as other information.

18.5 Cyclostationary beamforming

18.5.1 Preliminaries on cyclostationarity

Man-made modulated signals are in general coupled with sinewave carriers, pulse trains, coding, repeating spreading, hopping sequences or cyclic prefixes, resulting in built-in periodicity. These modulated signals are characterized as having second-order cyclostationarity if their mean and autocorrelation display periodicity. For cyclostationary signals, nonoverlapping frequency bands are uncorrelated, and further, the inherent periodicity implies some spectral redundancy, which results in correlation between nonoverlapping spectral components separated by some multiple of the cycle frequencies [17, 21].

In the time domain, a second-order cyclostationary process is a random process for which the statistical properties (namely, the mean and autocorrelation) change periodically as functions of time, with a period T [17, 21]

$$m_x(t) = m_x(t+T) \quad \text{for all } t, \tag{18.41}$$

$$R_{xx}(t_1, t_2) = R_{xx}(t_1 + T, t_2 + T) \quad \text{for all } t_1, t_2. \tag{18.42}$$

Since $R_{xx}(t_1, t_2)$ is periodic, it has a Fourier-series representation. By denoting $t_1 = t + \frac{\tau}{2}$ and $t_2 = t - \frac{\tau}{2}$, we have a Fourier series of the form

$$R_{xx}\left(t + \frac{\tau}{2}, t - \frac{\tau}{2}\right) = \sum_\alpha R_{xx}^\alpha(\tau) e^{j2\pi\alpha t}, \tag{18.43}$$

where the Fourier coefficient $R_{xx}^\alpha(\tau)$ is called the *cyclic autocorrelation function* or *spectral correlation function*, and α is known as the *cycle frequency*. A communication signal may have cycle frequencies that are related to the carrier frequency, the symbol rate and its harmonics, the chip rate, guard period, the scrambling code period, and the channel coding scheme.

A scalar waveform $x(t)$ is said to be *spectrally self-coherent* (or *conjugate self-coherent*) at a frequency α, if the spectral correlation function or the cyclic (or cyclic conjugate) autocorrelation function, that is, the correlation between $x(t)$ and $x(t)$ shifted in frequency by α, is nonzero for some delay τ [2, 16]

$$R_{xx^{(*)}}^\alpha(\tau) = \left\langle x\left(t + \frac{\tau}{2}\right)[x^*]^{(*)}\left(t - \frac{\tau}{2}\right) e^{-j2\pi\alpha t}\right\rangle_\infty \neq 0, \tag{18.44}$$

where $*$ is the conjugate operator, the optional conjugation $(*)$ is applied in the conjugate self-coherence case, $<\cdot>_\infty$ denotes infinite time-averaging, and α is the cycle (or conjugate cycle) frequency. Note that $(x^*)^* = x$. When $\alpha = 0$, it corresponds to the conventional autocorrelation, as used in the energy detector. If a signal is cyclostationary with period T, then cyclic autocorrelation has a component at $\alpha = 1/T$. For stationary signals, $R^{\alpha}_{xx(*)} = 0$ for any $\alpha \neq 0$.

The spectral self-coherence (or conjugate self-coherence) coefficient $\rho^{\alpha}_{xx(*)}(\tau)$ is defined as [2, 16]

$$\rho^{\alpha}_{xx(*)}(\tau) = \frac{R^{\alpha}_{xx(*)}(\tau)}{R_{xx}(0)} \neq 0 \qquad (18.45)$$

at some value of τ.

The spectrum cyclic density (SCD) is the Fourier transform of the cyclic (or conjugate cyclic) autocorrelation

$$S^{\alpha}_{xx(*)}(f) = \int_{-\infty}^{\infty} R^{\alpha}_{xx(*)}(\tau)e^{-j2\pi f\tau}\,d\tau. \qquad (18.46)$$

Calculation of $S^{\alpha}_{xx(*)}(f)$ can be implemented as FFT plus spectral correlations. For a fixed number of samples N, the SCD can be estimated by

$$\hat{S}^{\alpha}_{xx(*)}(f) = \frac{1}{N}\frac{1}{T}\sum_{n=0}^{N-1} X_T\left(n, f + \frac{\alpha}{2}\right)[X_T^*]^{(*)}\left(n, f - \frac{\alpha}{2}\right), \qquad (18.47)$$

where $X_T(n,f)$ is T-point FFT of $x(t)$ around the nth sample

$$X_T(n,f) = \int_{n-\frac{T}{2}}^{n+\frac{T}{2}} x(u)e^{-j2\pi fu}\,du. \qquad (18.48)$$

For the received signal $x(t)$

$$x(t) = h(t)s(t) + w(t), \qquad (18.49)$$

where $s(t)$ is the user signal, $h(t)$ is the channel impulse response, and $w(t)$ is the additive noise. If the signal s and the noise w are uncorrelated, we have

$$S^{\alpha}_{xx(*)}(f) = H\left(f + \frac{\alpha}{2}\right)[H^*]^{(*)}\left(f - \frac{\alpha}{2}\right)S^{\alpha}_{ss(*)}(f) + S^{\alpha}_{ww(*)}(f), \qquad (18.50)$$

where $H(f) = \sum_{t=-\infty}^{\infty} h(t)e^{-j2\pi ft}$ is the spectrum of the channel. $S^{\alpha}_{xx(*)}(f)$ is a two-dimensional complex transform on a support set (f, α). The AWGN is assumed to be a wide-sense stationary process. Since it is not a cyclostationary process and the observation period approaches infinity, $S^{\alpha}_{ww(*)}(f) = 0$ for $\alpha \neq 0$. Thus, the spectral correlation function differentiates the noise energy from modulated signal energy.

The cyclic (or cyclic conjugate) cross-correlation function $R^{\alpha}_{sz(*)}(\tau)$ of two signals $s(t)$ and $z(t)$ is defined by

$$R^{\alpha}_{sz(*)}(\tau) = \left\langle s\left(t + \frac{\tau}{2}\right)\left[z^{(*)}\left(t - \frac{\tau}{2}\right)e^{j2\pi \alpha t}\right]^*\right\rangle_{\infty} \qquad (18.51)$$

and the spectral cross-correlation (or conjugate cross-correlation) coefficient $\rho^{\alpha}_{sz(*)}(\tau)$ of the two signals is defined by

$$\rho^{\alpha}_{sz(*)}(\tau) = \frac{R^{\alpha}_{sz(*)}(\tau)}{\sqrt{R_{ss}(0)R_{zz}(0)}}. \qquad (18.52)$$

For the signal vector x, the cyclic (or cyclic conjugate) autocorrelation function is defined by [2, 16]

$$\mathbf{R}_{xx^{(*)}}^{\alpha}(\tau) = \left\langle x\left(t+\frac{\tau}{2}\right)\left(x^H\right)^{(*)}\left(t-\frac{\tau}{2}\right)e^{-j2\pi\alpha t}\right\rangle_{\infty}. \tag{18.53}$$

Note that $\left(\mathbf{x}^H\right)^{(*)} = \mathbf{x}^T$, the superscript T denoting the transpose.

For most communication signals, the cyclic (or cyclic conjugate) autocorrelation or cross-correlation function becomes nonzero only at discrete periodic cyclic frequencies α [2, 22]. Theoretically, at cyclic (or cyclic conjugate) frequencies, the values of these functions are nonzero for any value of τ [2]. For digital signals, τ is usually taken as mT, where m is a non-negative integer and T is the sampling period. For many communication signals, the maximum self-coherence occurs at $\tau = 0$ [2, 16]. Hence $\tau = 0$ is usually chosen for implementation.

18.5.2 Summary of some algorithms

A reference signal $r(t)$ is defined as

$$r(t) = c^H u(t), \tag{18.54}$$

where c is a control vector and

$$u(t) = x^{(*)}(t-\tau)e^{j2\pi\alpha t}. \tag{18.55}$$

LS-SCORE is obtained by minimizing the difference between the beamformer output $y(t)$ and the reference signal $r(t)$ [2]

$$\min_{w,c}\left\langle|y(t) - r(t)|^2\right\rangle_T, \tag{18.56}$$

where $< \cdot >_T$ denotes time-averaging over the period T. By minimizing this difference subject to normalized w and c, that is, $w^H w = 1$ and $c^H c = 1$, the CLS algorithm is derived [16].

Cross-SCORE is obtained by maximizing the strength of the cross-correlation coefficient between $y(t)$ and $r(t)$ [2]

$$\max_{w,c}\left|\hat{\rho}_{yr}^{\alpha}(\tau)\right|^2. \tag{18.57}$$

CAB is derived by maximizing the strength of the cyclic cross-correlation function [41]

$$\max_{w,c}\left|\hat{R}_{yr}\right|^2, \tag{18.58}$$

subject to normalized w and c.

Auto-SCORE is obtained by maximizing the strength of the self-coherence coefficient of the beamformer output $y(t)$ [2]

$$\max_{w}\left|\hat{\rho}_{yy(*)}^{\alpha}(\tau)\right|. \tag{18.59}$$

Table 18.1. Comparison of the algorithms.

	Interference	CAB	LS-SCORE	ACAB	ACS	CLS	APS	MCA
Num. of complex multiplications		$2N_a$	$4.75N_a^2 + 2.25N_a$	$3N_a^2 + 2N_a$	$6.75N_a^2 + 4.25N_a$	$5.5N_a^2 + 3.5N_a$	$4.75N_a^2 + 4.25N_a$	$2N_a^2 + 2N_a$
Conjugate cyclostationary algorithms	Very strong	−	+	−	+++	−	N/A	N/A
	Strong	+	+	+	+++	−		
	Medium	+++	+	+++	++	++		
	Weak	++++	+	++++	++	++++		
Cyclostationary algorithms	Very strong	−	−	−	++	−	++	−
	Strong	+	+	+	+++	−	++	+
	Medium	++	+	++	++	+	++	++
	Weak	+++	+	++++	++	++++	++	+++

N/A denotes *not applicable*, N_a = Number of antenna elements. Taken from [16]. © 2008, Springer.

Phase-SCORE is a special case of auto-SCORE, and is a suboptimum algorithm that is derived for the cyclostationary signal only.

MCA is derived by maximizing the strength of the cyclic autocorrelation function of the beamformer output $y(t)$ [16]

$$\max_{\boldsymbol{w}} \left| \hat{R}^{\alpha}_{yy(*)}(\tau) \right|, \tag{18.60}$$

subject to the normalized weight $\boldsymbol{w}$. MCA derived in [16] is valid for the cyclostationary signal only.

Adaptive implementation of these algorithms can significantly reduce the complexity of these algorithms. The ACS, ACAB, and APS algorithms are derived from the same objective functions as those of the cross-SCORE, CAB, and phase-SCORE algorithms, respectively, but they are treated as different algorithms. ACS, ACAB, and APS solve an eigenvalue problem to find the dominant mode for all the samples by using the power method, while cross-SCORE and phase-SCORE solve an eigenvalue problem to find the dominant mode for each sample. The power method [37] is a simple method to find the dominant mode. For example, if the recursive computation of $\hat{\mathbf{R}}^{-1}_{xx}(k)$ is used, for cross-SCORE, the total number of complex multiplications at each sample is $\left(6.75N_a^2 + 4.25N_a\right) m$, m being the number of iterations needed to solve the eigenvalue to a specified accuracy. The value of m may be different at each sample for a specified iteration accuracy, which also limits a pipeline implementation using DSPs.

Extensive simulations for seven algorithms have been performed in [16], based on a complex signal environment comprising of twelve BPSK or 16QAM signals for different powers, carrier frequencies f_c, and baud rates f_{baud}. The SINR performances of the seven algorithms are compared in Table 18.1 [16]. In this table, "−" denotes that the SINR is sometimes below 5 dB, indicating that the performance of the beamformer is unacceptable, and each "+" represents 10 dB in SINR. ACS has the best SINR performance among all the seven algorithms.

18.5.3 ACS algorithm

The eigenvalue problem obtained for cross-SCORE is given by [2, 16]

$$\lambda w = \hat{\mathbf{R}}_{xx}^{-1}\hat{\mathbf{R}}_{xu}c, \tag{18.61}$$

$$\lambda c = \hat{\mathbf{R}}_{uu}^{-1}\hat{\mathbf{R}}_{xu}^{H} w, \tag{18.62}$$

where λ is a positive number.

The scalar λ in (18.61) and (18.62) only affects the gain of the beamformer, and it can be eliminated in adaptive implementation. The ACS algorithm is given by [16]

$$w(k) = \hat{\mathbf{R}}_{xx}^{-1}(k)\hat{\mathbf{R}}_{xu}(k)c(k-1), \tag{18.63}$$

$$w(k) = \frac{w(k)}{\|w(k)\|}, \tag{18.64}$$

$$c(k) = \hat{\mathbf{R}}_{uu}^{-1}(k)\hat{\mathbf{R}}_{ux}(k)w(k). \tag{18.65}$$

This is actually the power method for solving an eigenvalue problem [2, 37], and can very rapidly converge to the dominant mode of the eigenvalue problem in a rank-1 spectral self-coherence environment. For slow fading channels, we have

$$\hat{\mathbf{R}}_{xx}(k) = \frac{1}{k}\sum_{i=1}^{k}x(i)x^{H}(i)$$

$$= \frac{k-1}{k}\left[\hat{\mathbf{R}}_{xx}(k-1) + \frac{1}{k-1}x(k)x^{H}(k)\right], \tag{18.66}$$

$$\hat{\mathbf{R}}_{xu}(k) = \frac{k-1}{k}\left[\hat{\mathbf{R}}_{xu}(k-1) + \frac{1}{k-1}x(k)u^{H}(k)\right]. \tag{18.67}$$

Also,

$$\hat{\mathbf{R}}_{ux}(k) = \hat{\mathbf{R}}_{xu}^{H}(k), \quad \hat{\mathbf{R}}_{uu}^{-1}(k) = \left[\hat{\mathbf{R}}_{xx}^{-1}(k)\right]^{(*)}. \tag{18.68}$$

By using the matrix inverse lemma, from (18.66), we have

$$\hat{\mathbf{R}}_{xx}^{-1}(k) = \frac{k}{k-1}\left\{\hat{\mathbf{R}}_{xx}^{-1}(k-1) - \frac{\hat{\mathbf{R}}_{xx}^{-1}(k-1)x(k)x^{H}(k)\hat{\mathbf{R}}_{xx}^{-1}(k-1)}{(k-1)+x^{H}(k)\hat{\mathbf{R}}_{xx}^{-1}(k-1)x(k)}\right\}. \tag{18.69}$$

Example 18.5: We use a uniform linear array of $N_a = 4$ elements with a spacing of half a wavelength of the carrier 2 GHz. We choose the noise variance of the array $\sigma_{a,n}^2 = 0.1$, $\tau = 0$, the sampling rate $f_s = 48$ Msamples/s, and all the signals are assumed to be at the same noise level $\sigma_{s,n}^2 = 0.2$.

Table 18.2. Impinging signals.					
Signal	DoA	f_c, MHz	SNR, dB	f_s/f_{baud}	Modulation
A	20°	2025	15	10	QAM16
B	60°	2080	10	13	BPSK
C	−10°	1980	5	19	QAM16
D	−30°	1950	20	23	BPSK

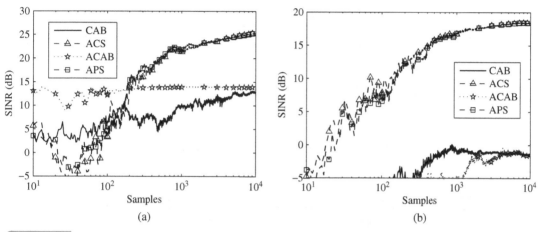

Figure 18.11 SINR performance. (a) Signal D as SOI (DoA=−30°, SNR=20 dB). (b) Signal C as SOI (DoA=−10°, SNR=5 dB).

Let us assume that there are four digital signals with a 100% cosine roll-off, impinging from different directions, as listed in Table 18.2, where f_c is the carrier frequency, f_{baud} is the baud rate, and DoA is the impinging angle. Simulation is based on the cyclostationarity of the signals, and the cyclostationary frequency is taken as $\alpha = f_{baud}$.

Four algorithms, namely, the CAB, ACS, ACAB, and APS, are considered in this example. Figure 18.11 shows the SINR performance of the algorithms. The beampatterns at the 10^4th sample of a random run are plotted in Fig. 18.12. We simulate for the cases of the strongest signal (signal D) and the weakest signal (signal C) as the SOI. The SINR performance is computed as an average over 20 independent runs, the average being carried in the sense of geometric mean.

For ACS, ACAB, and APS, $\mathbf{R}_{xx}^{-1}(0)$ is taken as $1/P_{max} \cdot \text{diag}(\boldsymbol{a})$, where $\boldsymbol{a}$ is an N_a vector with random elements between 0 and 1. $\mathbf{R}_{xu}(0)$ is taken as an $N_a \times N_a$ zero matrix.

The SINR performances of ACS and APS are very close to each other. ACS and APS always substantially outperform ACAB and CAB. ACAB and CAB fail to extract the SOI in the case of a strong interference. The beampatterns of ACS and APS almost coincide at the 10^4th sample. ACS and APS tend to create nulls in the directions of the interfering sources, while ACAB and CAB do not process this property.

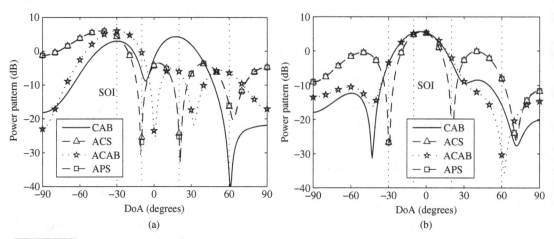

Figure 18.12 Beampattern at the 10^4th sample. (a) Signal D as SOI (DoA=−30°, SNR=20 dB). (b) Signal C as SOI (DoA=−10°, SNR=5 dB).

18.6 Wideband beamforming

Conventional narrowband beamformers are not suitable for wideband signals. For wideband signals, the phase at each antenna element for a specified DoA is a function of the frequency. For the same distance, a lower frequency signal component corresponds to a smaller phase shift. For wideband beam pattern synthesis, a simple method is to perform multiple independent narrowband designs over a range of frequencies. A more popular approach is to employ a tapped-delay-line structure, resulting in an array pattern that scales with temporal frequency. The frequency-invariant synthesis uses FIR filters to design an array pattern that is the same at all frequencies in a band of interest [36].

18.6.1 Tapped-delay-line structure

For wideband beamforming, a tapped-delay-line is normally used for each antenna element, and the structure of the beamformer is shown in Fig. 18.13. Each tapped-delay-line serves as an equalizer to make the phase response the same for different frequency components. This results in a large amount of coefficients, which are usually solved in the frequency domain by using FFT and IFFT. The length of the tapped-delay-lines depends on the bandwidth of the signals. As a result, wideband beamforming uses spatial as well as temporal filtering.

The output of the beamformer is given by

$$y(n) = \sum_{m=1}^{N_a} \sum_{k=0}^{K-1} w_{m,k} x_m(n-k), \qquad (18.70)$$

where $w_{m,k}$ is a weight.

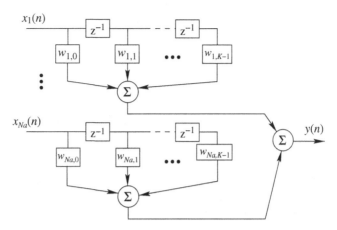

Figure 18.13
Tapped delay line structure of length K for an N_a-element beamformer.

For wideband or UWB antenna arrays, the antenna spacing d is usually determined by the highest frequency, f_h, of the input wave. For the uniform linear array, d can be given by

$$d = \frac{c}{2f_h}. \tag{18.71}$$

In order to cancel the directional interferences while having the specified response in the look direction, the beamformer can be formulated by minimizing the mean output power [24]

$$P(\boldsymbol{w}) = \boldsymbol{w}^T \mathbf{R} \boldsymbol{w}, \tag{18.72}$$

subject to

$$\mathbf{C}^T \boldsymbol{w} = \boldsymbol{F}, \tag{18.73}$$

where $\boldsymbol{w}$ is an MK-dimensional vector, the array correlation matrix $\mathbf{R}$ is an $MK \times MK$ real matrix whose elements represent the correlation between various tap outputs, $\boldsymbol{F}$ is a K-dimensional vector that specifies the frequency response in the look direction, and $\mathbf{C}$ is an $MK \times K$ constraint matrix. The solution is given by [24]

$$\hat{\boldsymbol{w}} = \mathbf{R}^{-1}\mathbf{C}(\mathbf{C}^T \mathbf{R}^{-1} \mathbf{C})^{-1} \boldsymbol{F}. \tag{18.74}$$

18.6.2 Pure delay-line wideband transmitter beamformer

One simple implementation is the pure delay-line wideband transmitter beamformer, where $w_{m,k} = 0$ or 1, and the tapped-delay-line degenerates to a time delay. The time delay at the mth antenna is set to be

$$T_m = T_0 + \tau_m. \tag{18.75}$$

To ensure positive T_m, one can select $T_0 \geq (N_a - 1)\,d/c$, d being the spacing of the uniform linear array. This beamformer uses no multiplication.

In the frequency domain, the filtering characteristic of the beamformer is given by [23]

$$H(f, \theta) = \frac{Y(f, \theta)}{X(f)} \tag{18.76}$$

$$= A_1(\theta) A_2 e^{-j2\pi f \alpha_0} e^{-j\pi f(N_a - 1)\frac{d}{c}(\sin\theta_0 - \sin\theta)} \tag{18.77}$$

$$\times \frac{\sin\left[\pi f N_a \frac{d}{c}(\sin\theta_0 - \sin\theta)\right]}{\sin\left[\pi f \frac{d}{c}(\sin\theta_0 - \sin\theta)\right]}. \tag{18.78}$$

where $A_1(\theta)$ is the angle-dependent gain of an antenna element, A_2 is attenuation due to propagation, θ_0 is the desired angle, and $\alpha_0 = T_0 + \tau_0$, τ_0 being the constant transmission delay of the first element. The directional patterns of the beamformer for different frequencies are derived from this equation.

Example 18.6: For an UWB signal with a center frequency of 7 GHz and a bandwidth of 6 GHz, given $\theta_0 = 30°$, we use an array of $N_a = 12$ elements and a perfect antenna ($A_1(\theta) = 1$), and $A_2 = 1$. The directional pattern, given by 18.78, is plotted in Fig. 18.14, for the cases of (a) $d = \frac{\lambda}{2f_h} = 1.5$ cm and (b) $d = \frac{\lambda}{2f_l} = 3.75$ cm. It is shown that for case (b), the grating lobes are very large and the inter-null beamwidths are reduced as compared to case (a). This confirms that d, given by (18.71), is reasonable. In this case, a perfect frequency-independence characteristic is obtained only along the desired angle, while the inter-null beamwidth and sidelobe characteristics depend on the frequency of operation.

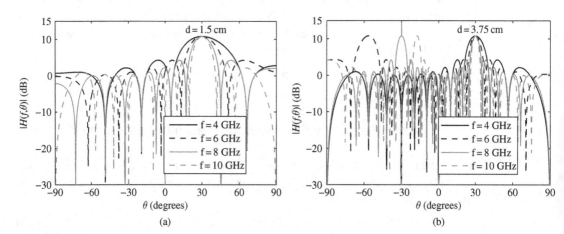

Figure 18.14 Directional patterns of a delay-line beamformer for different frequencies in UWB: (a) $d = \frac{\lambda}{2f_h} = 1.5$ cm. (b) $d = \frac{\lambda}{2f_l} = 3.75$ cm.

Problems

18.1 Determine the steering vector for a uniform linear array of N isotropic elements along the z-axis. The spacing of adjacent elements is d and the plane wave impinges from the direction (θ, ϕ).

18.2 Determine the steering vector for a uniform circular array of N isotropic elements on the x-y plane. The radius of the array circle is a and the plane wave impinges from the direction (θ, ϕ).

18.3 For an N-element uniform linear array, the elements are fed with currents of equal amplitude and a progressive phase lead of δ. The antenna spacing is d. Selecting the physical center of the array as the reference point,
(a) Verify that the array factor is given by $AF = \sin\left(\frac{N}{2}\psi\right) / \sin(\psi/2)$, $\psi = \beta d \cos\theta + \delta$.
(b) For $N = 8$ and $d = \lambda/2$, find the nulls of the array.
(c) To become a broadside array, that is, the main lobe in $\theta = 90°$, how should one select δ?
(d) To become an end-fire array with maximum lobe in $\theta = 0$, find δ.

18.4 According to the results of Problem 18.3, verify that the directivity of the an N-element uniform linear array is: (a) $D \approx 2(L/\lambda)$ for the broadside array and (b) $D \approx 4(L/\lambda)$ for the end-fire array. The array length $L = (N-1)d$. [Hint: $\int_{-\infty}^{\infty}(\sin x/x)^2 dx = \pi$.].

18.5 For a uniform planar array of 6×6 elements with a spacing of a half-wavelength in both directions, that is, $d_x = d_y = \lambda/2$, determine the DoA for a signal. Assume that the signal is from $\theta = 30°$ and $\phi = 90°$.

18.6 Derive the LCMV beamformer given by (18.30).

18.7 Write a MATLAB program to beamform a uniform linear array of 10 elements with a uniform $\lambda/2$ spacing toward the broadside ($\theta = 0$) by using the LMS algorithm. Assume that there are several impinging signals from different directions.

References

[1] 3GPP2, *C.S0002 Physical Layer Standard for CDMA 2000 Spread Spectrum Systems*, ver. 3.0, Jun 2001.
[2] B. G. Agee, S. V. Schell & W. A. Gardner, Spectral self-coherence restoral: a new approach to blind adaptive signal extraction using antenna arrays. *Proc. IEEE*, **78**:4 (1990), 753–766.
[3] A. Barabell, Improving the resolution of eigenstructure-based direction-finding algorithms. In *Proc. IEEE ICASSP*, Boston, MA, 1983, 336–339.
[4] J. J. Blanz, A. Papathanassiou, M. Haardt, I. Furio & P. W. Baier, Smart antennas for combined DOA and joint channel estimation in time-slotted CDMA mobile radio systems with joint detection. *IEEE Trans. Veh. Tech.*, **49**:2 (2000), 293–306.

[5] J. Blass, Multidirectional antenna: a new approach to stacked beams. In *IRE Int. Convention Record*, New York, Mar 1960, **8**, 48–C50.

[6] J. Capon, R. Greenfield & R. Kolker, Multidimensional maximum likelihood processing of a large aperture seismic array. *Proc. IEEE*, **55**:2 (1967), 192–211.

[7] J. Capon, High-resolution frequency-wavenumber spectrum analysis. *Proc. IEEE*, **57**:8 (1969), 1408–1418.

[8] S. Chen, N. N. Ahmad & L. Hanzo, Adaptive minimum bit-error rate beamforming. *IEEE Trans. Wireless Commun.*, **4**:2 (2005), 341–348.

[9] S. Choi & T. K. Sarkar, Adaptive antenna array utilizing the conjugate gradient method for multipath mobile communications. *Signal Process.*, **29**:3 (1992), 319–333.

[10] R. C. de Lamare & R. Sampaio-Neto, Adaptive multiuser receivers for DS-CDMA using minimum BER gradient-Newton algorithms. In *Proc. IEEE PIMRC*, Lisboa, Portugal, Sep 2002, **3**, 1290–1294.

[11] M. C. Dogan & J. M. Mendel, Cumulant-based blind optimum beamforming. *IEEE Trans. Aerospace Electron. Syst.*, **30**:3 (1994), 722–741.

[12] K.-L. Du, A. K. Y. Lai, K. K. M. Cheng & M. N. S. Swamy, Neural methods for antenna array signal processing: a review. *Signal Process.*, **82**:4 (2002), 547–561.

[13] K.-L. Du & M. N. S. Swamy, Simple practical cyclostationary beamforming algorithm. *IEE Proc. Vision Image Signal Process.*, **151**:3 (2004), 175–179.

[14] K.-L. Du & M. N. S. Swamy, A deterministic direction finding approach using a single snapshot of array measurement. In *Proc. IEEE Canadian Conf. Electr. Computer Eng.*, Saskatoon, Canada, May 2005, 1188–1193.

[15] K.-L. Du & M. N. S. Swamy, *Neural Networks in a Softcomputing Framework* (London: Springer, 2006).

[16] K.-L. Du & M. N. S. Swamy, A class of adaptive cyclostationary beamforming algorithms. *Circ. Syst. Signal Process.*, **27**:1 (2008), 35–63.

[17] K.-L. Du & W. H. Mow, *Exploiting Multiple Antennas for Spectrum Sensing in Cognitive Radio Networks*, U. S. Patent Application, 2009.

[18] J. Butler & R. Lowe, Beam-forming matrix simplifies design of electrically scanned antennas. *Electron. Design*, **9** (1961), 170–173.

[19] J. E. Evans, J. R. Johnson & D. F. Sun, *High Resolution Angular Spectrum Estimation Techniques for Terrain Scattering Analysis and Angles of Arrival Estimation in ATC Navigaton and Surveillance System*, MIT Lincoln Lab, Lexington, MA, Report 582, 1982.

[20] O. L. Frost, An algorithm for linearly constrained adaptive array processing. *Proc. IEEE*, **60**:8 (1972), 926–935.

[21] W. A. Gardner, *Statistical Spectral Analysis: An Nonprobabilistic Theory* (Englewood Cliffs, NJ: Prentice-Hall, 1987).

[22] W. A. Gardner, Exploitation of spectral redundancy in cyclostationary signals. *IEEE Signal Process. Mag.*, **8**:2 (1991), 14–36.

[23] M. Ghavami, L.B. Michael & R. Kohno, *Ultra Wideband: Signals and Systems in Communication Engineering*, 2nd edn (Chichester, UK: Wiley, 2007).

[24] L. C. Godara, Application of antenna arrays to mobile communications, Part II: beamforming and direction-of-arrival considerations. *Proc. IEEE*, **85**:8 (1997), 1195–1245.

[25] F. Gross, *Smart Antennas for Wireless Communications* (New York: McGraw-Hill, 2005).

[26] P. Ioannides & C. A. Balanis, Uniform circular arrays for smart antennas. *IEEE Anten. Propagat. Mag.*, **47**:4 (2005), 192–206.

[27] H. Li, Y.-D. Yao & J. Yu, Outage probabilities of wireless systems with LCMV beamforming. *IEEE Trans. Wireless Commun.*, **6**:10 (2007), 3515–3523.

[28] J. Liberti & T. Rappaport, *Smart Antenna for Wireless Communications* (Englewood Cliffs, NJ: Prentice Hall, 1999).

[29] A. F. Molisch, *Wireless Communications* (Chichester, UK: Wiley-IEEE, 2005).

[30] K. I. Pedersen, P. E. Mogensen & J. Ramiro-Moreno, Application and performance of downlink beamforming techniques in UMTS. *IEEE Commun. Mag.*, **41**:10 (2003), 134–143.

[31] S. U. Pillai & B. H. Kwon, Forward/backward spatial smoothing techniques for coherent signal identification. *IEEE Trans. Acoust. Speech Signal Process.*, **37**:1 (1989), 8–15.

[32] R. Roy & T. Kailath, ESPRIT – estimation of signal parameters via rotational invariance techniques. *IEEE Trans. Acoust. Speech Signal Process.*, **37**:7 (1989), 984–995.

[33] T. A. Samir, S. Elnoubi & A. Elnashar, Block-Shanno minimum bit error rate beamforming. *IEEE Trans. Veh. Tech.*, **57**:5 (2008), 2981–2990.

[34] S. V. Schell, Performance analysis of the cyclic MUSIC method of direction estimation for cyclostationary signals. *IEEE Trans. Signal Process.*, **42**:11 (1994), 3043–3050.

[35] R. Schmidt, Multiple emitter location and signal parameter estimation. *IEEE Trans. Anten. Propagat.*, **34**:2 (1986), 276–280.

[36] D. P. Scholnik & J. O. Coleman, Optimal array-pattern synthesis for wideband digital transmit arrays. *IEEE J. Sel. Topics Signal Process.*, **1**:4 (2007), 660–677.

[37] G. W. Stewart, *Introduction to Matrix Computations* (New York: Academic Press, 1973).

[38] J. S. Thompson, P. M. Grant & B. Mulgrew, Smart antenna arrays for CDMA systems. *IEEE Pers. Commun.*, **3**:5 (1996), 16–25.

[39] F. Wan, W.-P. Zhu & M. N. S. Swamy, A spatial extrapolation-based blind DOA estimation approach for closely spaced sources. *IEEE Trans. Aerospace Electron. Syst.*, **45** (2009).

[40] J. H. Winters, J. Salz & R. D. Gitlin, The impact of antenna diversity on the capacity of wireless communication systems. *IEEE Trans. Commun.*, **42**:2–4 (1994), 1740–1750.

[41] Q. Wu & K. M. Wong, Blind adaptive beamforming for cyclostationary signals. *IEEE Trans. Signal Process.*, **44**:11 (1996), 2757–2767.

[42] G. Xu, R. H. Roy & T. Kailath, Detection of number of sources via exploitation of the centro-symmetry property. *IEEE Trans. Signal Process.*, **42**:1 (1994), 102–112.

Multiple antennas: MIMO systems

19.1 Introduction

MIMO systems are wireless systems with multiple antenna elements at both ends of the link. MIMO systems can be used for beamforming, diversity combining, or spatial multiplexing. The first two applications are the same as for the smart antennas, while spatial multiplexing is the transmission of multiple data streams on multiple antennas in parallel, leading to a substantial increase in capacity. MIMO technology and turbo coding are the two most prominent recent breakthroughs in wireless communications. MIMO technology promises a significant increase in capacity.

MIMO systems have the ability to exploit, rather than combat, multipath propagation. The separability of the MIMO channel relies on the presence of rich multipath, which makes the channel spatially selective. Thus, MIMO effectively exploits multipath. In contrast, some smart antenna systems perform better in the LOS case, and their optimization criteria are based on the DoA/DoD. Although some smart antenna systems generate good results in the non-LOS channel, they mitigate multipath rather than exploit it.

The maximum spatial diversity obtained for a non-frequency-selective fading MIMO channel is proportional to the product of the numbers of receive and transmit antennas. In the uncorrelated Rayleigh fading channel, the MIMO channel capacity/throughput limit grows linearly with the number of transmit or receive antennas, whichever is smaller – i.e., $\min(N_t, N_r)$ [40]. According to the analysis and simulation performed in [40], MIMO can provide a spectral efficiency as high as 20–40 bits/s/Hz.

MIMO and OFDM are commonly thought to be the key techniques for next-generation wireless LAN and 4G mobile communications. MIMO-OFDM is used in IEEE 802.11n, IEEE 802.16m, and LTE.

19.2 MIMO system

19.2.1 MIMO system model

Narrowband MIMO systems with N_t transmit and N_r receive antennas have a channel that can be described by an $N_r \times N_t$ matrix $\mathbf{H} = [h_{ij}]$, where h_{ij} represents the channel transfer function from the jth transmitter to the ith receiver. $\mathbf{H}$ is modeled as a random matrix characterized by an uncorrelated or correlated Rayleigh fading channel, or an uncorrelated or correlated Ricean fading channel.

Assuming that each transmit antenna has an energy for each input symbol as E_s/N_t, E_s being the total energy transmitted from all antennas per input symbol, the received signal at antenna j is given by

$$y_j(k) = \sqrt{\frac{E_s}{N_t}} \sum_{i=1}^{N_t} h_{ji}x_i(k) + n_j(k), \tag{19.1}$$

where n_j's are the i.i.d. additive zero-mean circularly symmetric complex Gaussian (ZMC-SCG) variables with two-sided PSD $N_0/2$, and h_{ji} is the flat-fading channel gain from transmit antenna i to receive antenna j, h_{ji} being also a ZMCSCG variable in the non-LOS case.

In matrix form, we have

$$\mathbf{y}(k) = \sqrt{\frac{E_s}{N_t}} \mathbf{H}(k)\mathbf{x}(k) + \mathbf{n}(k), \tag{19.2}$$

where $\mathbf{y}$, an N_r-dimensional vector, corresponds to the output signals at the receiver antennas, $\mathbf{x}$ is an N_t-dimensional vector, whose element x_j denotes the signal transmitted from the jth transmitter, and $\mathbf{n}$ is the additive ZMCSCG noise vector with covariance matrix $\mathbf{R}_w$ at the receive antennas. The SNR at each receive antenna is E_s/N_0.

MIMO implementation relies on the rich scattering about the transmitter and receiver antennas. Insufficient scattering frequently occurs when the channel is approximately LOS, or when beamforming or directional antennas are used for interference reduction or long-range transmission.

For space-time codeword of block length N_{st}, N_{st} receive vector symbols in the codeword can be stacked together in a matrix form and be processed together. For the frequency-selective channel, the channel $\mathbf{H}$ can be represented as $\mathbf{H}(l)$, $l = 0, 1, \ldots, L-1$, where L is the maximum channel length; in this case, multiple continuous received vector samples can be stacked to solve $\mathbf{H}(l)$, $l = 0, 1, \ldots, L-1$.

19.2.2 Spatial correlation and MIMO channel model

For a MIMO system, the elements of $\mathbf{H}$ are usually assumed to be statistically independent of one another. This assumption is not always accurate, since correlation may exist due to the propagation environment (such as the presence of LOS component), the polarization of antenna elements, and the spacing between antenna elements. The fading correlation associated with $\mathbf{H}$ can be decomposed into two independent components [74]

$$\mathbf{H} = \mathbf{R}_r^{1/2} \mathbf{H}_w \mathbf{R}_t^{1/2}, \tag{19.3}$$

where $\mathbf{R}_r$ and $\mathbf{R}_t$ are called *receive correlation* and *transmit correlation* matrices, respectively, $\mathbf{H}_w$ is a matrix with independent Gaussian elements and unity variance, and the superscript $1/2$ denotes the Hermitian square root of a matrix. $\mathbf{R}_r$ determines the correlation between the rows of $\mathbf{H}$, independent of the transmit antennas. Similarly, $\mathbf{R}_t$ determines the covariance of the columns of **H**, independent of the receive antennas. This model is widely used in MIMO implementation, and it has been adopted by IEEE 802.11n and IEEE 802.20 as a MIMO channel model.

The correlation matrices $\mathbf{R}_r$ and $\mathbf{R}_t$ can be measured, or be computed by assuming the scattering distribution around the receive and transmit antennas. For uniform linear arrays at the transmitter and the receiver, the correlation matrices $\mathbf{R}_r$ and $\mathbf{R}_t$ can be calculated according to two different methods given in [87, 160]. From [160], we have

$$\mathbf{R}_r, \mathbf{R}_t = \begin{bmatrix} 1 & \rho & \rho^4 & \cdots & \rho^{(N-1)^2} \\ \rho & 1 & \rho & \ddots & \vdots \\ \rho^4 & \rho & 1 & \ddots & \rho^4 \\ \vdots & \ddots & \ddots & \ddots & \rho \\ \rho^{(N-1)^2} & \cdots & \rho^4 & \rho & 1 \end{bmatrix}, \tag{19.4}$$

where N is equal to N_r or N_t, corresponding to the receive or transmit antenna array, and ρ is the fading correlation between two adjacent receive or transmit antenna elements, which can be approximated by

$$\rho(d) \approx e^{-23\Delta^2 d^2}, \tag{19.5}$$

with Δ being the angular spread and d the inter-element distance. Note that for small $r(d)$, the higher-order terms are negligible and the correlation matrices take the form of triagonal matrices.

In practical cases, the degenerate channel phenomena called *keyholes* may arise, where the antenna elements at both the transmitter and the receiver have very low correlation, yet the channel matrix **H** has only a single degree of freedom, yielding a single mode of communication [6, 23, 43, 124]. This phenomenon is very similar to the case when rich-scattering transmit and receive antennas are separated by a screen with the wave passing through a keyhole. This model also applies for indoor propagation through hallways, narrow tunnels or waveguides. Relay channels in the amplify-and-forward mode can be treated as keyhole channels. Thus, low correlation is not a guarantee for achieving high capacity. For outdoor environments, roof edge diffraction is perceived as a keyhole by a vertical base array, whereas the keyhole effect may be avoided by employing a horizontally oriented transmitter array [23]. Instantaneous SNR and outage capacity distributions of spatially correlated keyhole MIMO channels has been investigated in [77].

A double-scattering MIMO channel model that includes both the fading correlation and rank deficiency was introduced in [43]. The multikeyhole channel is analyzed in [77]. For a large number of antennas, the capacity of a multikeyhole channel is a normally distributed sum of the capacities of single keyhole channels.

The correlation between antennas is typically not a problem for MIMO systems with well designed antennas. This is due to the fact that even for the worst case the correlation rarely exceeds 0.7, and this yields a degradation of less than 1 dB with MRC [62] and even less with the MMSE combiner. The capacity achievable with isotropic inputs is lowered by antenna correlation, while for nonisotropic inputs correlation may not be detrimental

[88]. For example, transmit correlation may be advantageous for small SNR and for $N_t > N_r$ [88].

The impact of channel correlation on the capacity of a MIMO system is negligible when the two-element array beamwidth, defined as $\frac{\lambda}{d\cos\phi}$, d being the inter-element distance and ϕ the mean DoA, is smaller than the angular spread of the incoming multipath signals [87]. Fully correlated fading destroys diversity gain, but array gain is retained. LOS component stabilizes the link, improving the SER performance, but it reduces the MIMO system capacity. These have been discussed in [103].

3GPP/3GPP2 have defined a cross-polarized channel model for MIMO systems [2]. The 3GPP cross-polarized channel model neglects the elevation spectrum. A composite channel impulse model for the cross-polarized channel that takes into account both the azimuth and elevation spectrums has been proposed in [121], based on which closed-form expressions for the spatial correlation have been derived and the impact of the various factors on the mutual information of the system has also been studied.

19.2.3 MIMO decoding

The optimum ML detector estimates x by

$$\hat{x} = \arg \min_{x \in \mathcal{B}^{N_t}} \left\| y - \sqrt{\frac{E_s}{N_t}} \mathbf{H} x \right\|, \tag{19.6}$$

where each element of x can select one of the M symbol constellations in $\mathcal{B}$. There are altogether M^{N_t} possible values of x. The optimum ML decoding has an exponential complexity.

For a group of N_{st} vectors, we have

$$\mathbf{Y} = \sqrt{\frac{\overline{E}_s}{N_t}} \mathbf{H} \mathbf{X} + \mathbf{N}, \tag{19.7}$$

where $\mathbf{X} = [x(0)|x(1)| \cdots |x(N_{st} - 1)]$ is an $N_t \times N_{st}$ matrix, and $\mathbf{Y}$ and $\mathbf{N}$ are defined in the same manner, but have dimension $N_r \times N_{st}$.

For (19.7), the ML decoding yields

$$\hat{\mathbf{X}} = \arg \min_{\mathbf{X} \in \mathcal{B}^{N_t \times N_{st}}} \left\| \mathbf{Y} - \sqrt{\frac{\overline{E}_s}{N_t}} \mathbf{H} \mathbf{X} \right\|_F^2, \tag{19.8}$$

where $\overline{E}_s$ is the average symbol energy over the block.

19.2.4 MIMO channel decomposition

The channel $\mathbf{H}$ can be decomposed by using SVD,

$$\mathbf{H} = \mathbf{U} \mathbf{\Sigma} \mathbf{V}^H, \tag{19.9}$$

where $\mathbf{U}$, Σ, and $\mathbf{V}$ are, respectively, of size $N_r \times N_r$, $N_r \times N_t$, and $N_t \times N_t$. $\mathbf{U}$ and $\mathbf{V}$ are unitary matrices, and $\Sigma = \text{diag}(\sigma_i)$. The rank of the channel matrix $\mathbf{H}$, r_H, must satisfy $r_H \leq \min(M_t, M_r)$. When $\mathbf{H}$ is of full rank, the channel is a rich-scattering environment, and

$$r_H = \min(M_t, M_r). \tag{19.10}$$

The squared Frobenius norm of $\mathbf{H}$ is the total power gain of the channel, and it is given by

$$\|\mathbf{H}\|_F^2 = \text{tr}\left(\mathbf{H}\mathbf{H}^H\right) = \sum_{i=1}^{N_r} \lambda_i, \tag{19.11}$$

where λ_i's are the eigenvalues of $\mathbf{H}\mathbf{H}^H$,

$$\lambda_i = \begin{cases} \sigma_i^2, & i = 1, 2, \ldots, r_H \\ 0, & i = r_H + 1, \ldots, N_r. \end{cases} \tag{19.12}$$

The value $\|\mathbf{H}\|_F^2$ is also a random variable. The statistics of $\|\mathbf{H}\|_F^2$ determine the diversity performance. For the i.i.d. (spatially white) channel, $\mathbf{H} = \mathbf{H}_w$, the pdf of $\|\mathbf{H}\|_F^2$ is a χ^2 distribution with $k = 2N_tN_r$ degrees of freedom [103],

$$p(x) = \frac{x^{N_tN_r-1}}{(N_tN_r - 1)!} e^{-x}, \quad x \geq 0. \tag{19.13}$$

The MIMO system is capable of achieving very high spectral efficiency and reliability in rich-scattering environments. However, when there exists spatial correlation arising from closely spaced antennas or limited angular spread in the MIMO channel, performance degrades significantly. A precoder can be used to enhance the capacity or performance of the MIMO system in a fading channel, when instantaneous or long-term CSI is available at the transmitter. In particular, a precoder using long-term CSI is favorable for correlated MIMO systems with low-rate feedback [10, 128].

19.2.5 Channel estimation

Like the single-antenna case, channel estimation for MIMO can be implemented based on training sequences or by using a blind technique. In the MIMO system, there are more channel parameters to be estimated. It is desirable to keep the training sequences from the multiple transmit antennas mutually orthogonal in some form (time, frequency, or code) to enhance estimation accuracy. The training sequences should typically have good auto- and cross-correlation properties. The number of required training samples should be $N_{tr} \geq N_t \times L$.

There are many transmission schemes over MISO or MIMO channels in the literature. These schemes rely on certain CSI available at the transmitter and/or receiver side. CSI at the receiver can be easily estimated using a training-based or blind technique. CSI at the transmitter can be obtained by feedback of the receiver's channel estimation based on the downlink training data (e.g., in TDD mode), or from the use of training or pilot data in the uplink (e.g. in FDD mode). Feedback usually causes some delay δ_D, and this delay should be insignificant compared to the coherent time T_c, $\delta_D \ll T_c$. In a fast changing channel, frequent channel estimation and feedback is required; in order to reduce the overhead on the reverse link, the slowly changing statistics or partial information of the channel can be fed back over the channel.

When full CSI is available at the transmitter, the transmitter can employ the eigenvector steering technique to approach the full capacity of the MIMO channel. The availability of CSI at the receiver is assumed by most multi-user MIMO methods.

The CSI is estimated by using training sequences. During the training phase, training sequences of L_t symbols long are transmitted from all the transmitting antennas, from which an estimate of the channel, $\hat{\mathbf{H}}$, is calculated. Data sequences are then transmitted and jointly detected.

Channel estimation methods

Coherent space-time decoding always requires knowledge of the MIMO channel as well as timing and frequency synchronization at the receiver. A timing offset introduces a pure delay convolutional channel; thus, timing synchronization can be lumped into channel estimation. Frequency synchronization requires separate implementation.

Preamble-based channel estimation is the most fundamental channel estimation method. A number of training or pilot symbols that are known to both the transmitter and the receiver are placed at the start of a frame. These symbols are known as the *preamble*. Assuming perfect frequency synchronization, for a frequency-flat MIMO channel, in the input-output relationship (19.7), the $N_t \times N_{st}$ matrix $\mathbf{X}$ is known at the receiver, and $\mathbf{Y}$ is also known at the receiver by measurement. Thus, the channel $\mathbf{H}$ can be estimated. In the same manner, channel estimation can be performed for frequency- or time-selective, single- and multi-carrier systems. LS and linear MMSE estimators can be obtained in a manner we have discussed in Chapter 9. Optimal training sequence design must consider the constraints of transmit power and the rate resources between training and information symbols.

The decision-directed channel estimation method begins with channel estimation using the preamble. Using the estimated channel, symbols are then decoded for a block within the coherent interval of the channel. These decoded symbols are used as the training symbols, and a refined channel estimation is obtained. The decision-directed method is valid for slowly varying channels, at least for high SNR. The semiblind channel estimation approach can improve spectrum efficiency by exploiting the signal properties as well as the preamble [150].

For channel estimation, perfect synchronization is assumed. Frame and frequency synchronization for single-antenna OFDM has been studied in Chapter 9. For the flat-fading MIMO channel, timing offset estimators have been proposed and their MSE performance studied in [156]. The CRB achieved by the ML estimator is inversely proportional to N_r, but does not depend on N_t [156].

19.2.6 CSI or partial CSI at the transmitter

Typical MIMO systems including the space-time coding and spatial multiplexing schemes are open-loop MIMO systems, where the transmit matrix $\mathbf{X}$ is designed independent of the channel conditions. Closed-loop MIMO systems make use of the adaptive nature of the channel, and the transmit matrix $\mathbf{X}$ is designed as a function of the channel conditions. In a coherent MIMO system, the receiver has an estimate of the forward link channel $\mathbf{H}$ for decoding purposes, and this CSI can be fed back to the transmitter. This leads to an improved error performance, better spectral efficiency, and simplified decoding [103]. The linear precoding technique is a closed-loop MIMO technique.

CSI at the transmitter can be obtained by either feedback from the receiver or channel estimation at the transmitter. Since channel estimation is always performed at the receiver, CSI feedback from the receiver is a simple solution.

Viability of CSI feedback

The viability of feedback is based on the reciprocity principle, which asserts that the forward and reverse channels are the same only if the used time, frequency, and antennas for both the channels are the same. Due to the duplexing operation, the two channels must have some difference in time, frequency, or antenna to reduce interference [103].

- In TDD mode, the duplexing time delay, δ_t, between the forward time slot and the backward time slot must satisfy $\delta_t \ll T_c$ so that the forward and reverse channels can be very close to each other.
- In FDD mode, the reverse channel can be approximated by the forward channel only when the duplexing frequency difference, δ_f, between the forward and reverse channels satisfies $\delta_f \ll B_c$. This usually cannot be satisfied in FDD, since δ_f is typically 5% of the carrier frequency f_c.
- In antenna division duplex (ADD) mode, the same frequency and time are used on the forward and reverse channels, but a different antenna on each channel is used. The duplexing location difference, δ_d, is defined as the separation between the forward antenna and the reverse antenna. Similarly, to ensure the forward and reverse channels to be close to each other, $\delta_d \ll D_c$. ADD is rarely used as a duplexing scheme, since there is not sufficient isolation between the forward and reverse links and the requirement $\delta_d \ll D_c$ may be difficult to meet when D_c is as small as half the carrier wavelength, $\lambda_c/2$.

Thus, among the various reciprocity methods, only pure TDD mode is suitable for channel estimation at the transmitter.

Partial CSI at the transmitter

Partial CSI at the transmitter is the more common situation for channel-adaptive transmission systems. The partial CSI is obtained by feedback from the receiver. Partial CSI contains typically a few bits fed back from the receiver to the transmitter. The partial CSI can take the form of a mean vector or a covariance matrix of the channel, or other channel parameters that do not change fast.

In TDD mode, the transmitter and the receiver can obtain the channel statistics directly and no feedback is necessary, since the forward and reverse links share the same physical channel. For finite-rate CSI, the system design must be based on how the MIMO channel is best characterized by a finite number of bits and what criterion is used for optimization. The first question requires a vector quantizer, while the second requires a criterion for the error probability, average capacity, or transmission rate [44].

Limited feedback is implemented by quantizing the channel vector (or the precoder) based on a predetermined codebook known at both the BS and the MS. Only the index is sent from the receiver to the transmitter. The channel matrix $\mathbf{H}$ is first transformed into an $N_t \times N_r$-dimensional complex vector by stacking the columns of $\mathbf{H}$, and a VQ algorithm is then applied to quantize this high-dimensional vector. VQ has been introduced in Section 16.2.2, and has been described in detail in [31]. The LBG or generalized Lloyd algorithm is a popular VQ algorithm. The minimum codebook size for quantized precoding has been discussed in [83].

Structured codebooks based on QAM and PSK sequences are used for limited feedback of beamforming weights in [113]. Such codebooks perform arbitrarily close to the random, numerically derived codebooks as $M \rightarrow \infty$, in terms of average SNR, BER and outage probability. The codebooks achieve the full diversity order. The QAM codebooks are used for quantizing the MRC vector, and the PSK codebooks are used for quantizing the EGC vector since PSK symbols have equal envelope. Since there is simple bits-to-symbol mapping for QAM and PSK constellations, no codebook storage is required at either the transmitter or receiver. These codebooks can be searched with complexity orders of magnitude smaller than an exhaustive search.

Quantized precoding for MIMO-OFDM with limited feedback has been adopted as an option in IEEE 802.16e, 3GPP LTE, 3GPP2 UWB, and IEEE 802.11n. A broad overview of limited feedback wireless communications is given in [84].

19.3 Capacity in i.i.d. slow fading channels

It is well-known that a real or complex zero-mean Gaussian random variable (or vector) has the entropy-maximizing property [27, 74, 141]. For two random vectors x and y, mutual information is defined by

$$I(y, x) = H(y) - H(y|x), \tag{19.14}$$

where the conditional entropy

$$H(x|y) = -E\left[\log_2(p(x|y))\right]. \tag{19.15}$$

For a linear complex model with channel $\mathbf{H}$

$$y = \mathbf{H}x + e, \tag{19.16}$$

where x has a zero mean and a covariance matrix $\mathbf{P}$, and e is a ZMCSCG vector with covariance matrix $\sigma_n^2 \mathbf{I}$. Thus, y has a zero mean and the covariance matrix $\mathbf{HPH}^H + \sigma_n^2 \mathbf{I}$. The mutual information is derived as [74]

$$I(y, x | \mathbf{H}) = \log_2 \det \left(\mathbf{I} + \frac{1}{\sigma_n^2} \mathbf{HPH}^H \right). \tag{19.17}$$

For a given channel $\mathbf{H}$, the Shannon capacity is given by the maximum mutual information between the received vectors $\{y_n\}$ and the transmitted vectors $\{x_n\}$. Only when x is zero-mean Gaussian, the mutual information reaches its maximum. Thus, the channel capacity is given as [74, 141]

$$C(\mathbf{H}) = B \log_2 \det \left(\mathbf{I} + \frac{1}{\sigma_n^2} \mathbf{HPH}^H \right), \tag{19.18}$$

where B is the bandwidth of the channel, and the total transmit power $P_{\text{tot}} = \text{tr}(\mathbf{P})$.

For a fading channel, the channel $\mathbf{H}$ itself is a random variable, and thus the channel capacity $C(\mathbf{H})$ is also a random variable. The ergodic capacity is the average capacity $E_{\mathbf{H}}[C(\mathbf{H})]$. When $\mathbf{H}$ is a random Gaussian matrix with i.i.d elements, the maximum ergodic capacity subject to a power constraint $\text{tr}(\mathbf{P}) \leq P$ is achieved at [141]

$$\mathbf{P} = \frac{P}{N_t} \mathbf{I}. \tag{19.19}$$

Thus, the average capacity is maximized when each antenna transmits an uncorrelated stream with the same average power. This result can be used when the transmitter does not have the CSI.

When full transmitter CSI and receiver CSI are available, the capacity reaches its maximum. The capacity of the MIMO system in a nonfading channel is given by [40]

$$C = \sum_{i=1}^{r_H} \log_2 \left(1 + \frac{P_i}{\sigma_n^2} \sigma_i^2 \right), \tag{19.20}$$

where r_H is the rank of $\mathbf{H}$, σ_i is the ith singular value of $\mathbf{H}$, P_i is the power allocated to the ith eigenmode, and σ_n^2 is the noise variance.

Assuming that all the transmit antennas are subject to a total power P, an equivalent expression for the capacity is given by [141]

$$C = \max_{\text{tr}(\mathbf{R}_{xx})=N_t} \log_2 \left[\det \left(\mathbf{I}_{N_r} + \frac{\overline{\gamma}}{N_t} \mathbf{HR}_{xx}\mathbf{H}^H \right) \right], \tag{19.21}$$

where $\mathbf{I}_{N_r}$ is the $N_r \times N_r$ identity matrix, $\overline{\gamma} = \overline{E}_s/N_0$ is the mean SNR of each receiver branch, and $\mathbf{R}_{xx} = E\left[xx^H \right]$ is the correlation matrix of the transmit data.

Distribution of power among the different antennas is dependent on the amount of CSI at the transmitter. Assuming that the receiver has perfect CSI, the capacity increases linearly with $\min(N_t, N_r, N_s)$, N_s being the number of scatters [40]. When N_s is sufficiently large, the capacity increases linearly with $\min(N_r, N_t)$, whether the transmitter has the CSI or not.

19.3.1 No CSI at the transmitter

When there is full CSI at the receiver but no CSI at the transmitter, the reasonable solution is to assign equal transmit power to all the transmitter antennas and use uncorrelated data streams [141]. In this case, $\mathbf{R}_{xx}$ in (19.21) is an identity matrix, and

$$
\begin{aligned}
C &= \log_2 \left[\det \left(\mathbf{I}_{N_r} + \frac{\bar{\gamma}}{N_t} \mathbf{H} \mathbf{H}^H \right) \right] \\
&= \sum_{i=1}^{r_H} \log_2 \left(1 + \frac{\bar{\gamma}}{N_t} \lambda_i \right).
\end{aligned}
\tag{19.22}
$$

Thus, the capacity of the MIMO channel is expressed as the sum of the capacities of r_H SISO channels, each having power gain λ_i.

Case 1. SIMO: $N_t = 1$ and $N_r > 1$

In this case, $\lambda = N_r$, and

$$
C_{\text{SIMO}} = \log_2 \left(1 + \bar{\gamma} N_r \right) > C_{\text{AWGN}},
\tag{19.23}
$$

where C_{AWGN} is the scalar channel capacity. The CSI at the transmitter does not affect the SIMO channel capacity.

Case 2. MISO: $N_t > 1$ and $N_r = 1$

In this case, $\lambda = N_t$, and

$$
C_{\text{MISO}} = \log_2 \left(1 + \bar{\gamma} \right) = C_{\text{AWGN}}.
\tag{19.24}
$$

The reason for $C_{\text{SIMO}} > C_{\text{MISO}}$ is that the transmitter in the MISO case cannot exploit the antenna array gain since it has no CSI.

Case 3. $N_r > 1$ and N_t is large

For $\|\mathbf{H}\|^2 = 1$, applying the strong law of large number leads to

$$
\frac{1}{N_t} \mathbf{H} \mathbf{H}^H \rightarrow N_r \mathbf{I}_{N_r}, \quad \text{almost surely (a.s.).}
\tag{19.25}
$$

Thus

$$
C \approx N_r \log_2 (1 + \bar{\gamma}) = N_r C_{\text{AWGN}}.
\tag{19.26}
$$

The capacity is N_r times the scalar channel capacity.

19.3.2 CSI known at the transmitter

When there is full CSI at both the transmitter and the receiver, the channel $\mathbf{H}$, decomposed using SVD, is given by (19.9), which for convenience is repeated below

$$\mathbf{H} = \mathbf{U}\Sigma\mathbf{V}^H. \tag{19.27}$$

Precoding can be applied to both the transmitter and the receiver, by pre-multiplying the signal vector to be transmitted, $\tilde{x}$, by $\mathbf{V}$ at the transmitter,

$$x = \mathbf{V}\tilde{x} \tag{19.28}$$

and left-multiplying the received signal vector y by $\mathbf{U}^H$,

$$\tilde{y} = \mathbf{U}^H y. \tag{19.29}$$

This is shown in Fig. 19.1.

In this way, the MIMO channel $\mathbf{H}$ is decomposed into r_H parallel scalar channels

$$\tilde{y}_i = \sqrt{\frac{E_s}{N_t}}\sqrt{\lambda_i}\tilde{s}_i + \tilde{n}_i, \quad i = 1, 2, \ldots, r_H, \tag{19.30}$$

where $\lambda_i = \sigma_i^2$, $\sigma_i = \Sigma_{i,i}$. For i.i.d. Rayleigh fading channels, $\mathbf{H}$ is usually of full rank, and thus $r_H = \min\{N_t, N_r\}$. The maximum capacity can be obtained by maximizing the sum of the individual scalar channel capacities subject to the total energy constraint:

$$C = \max_{\sum_{i=1}^{r_H} E_i = N_t} \sum_{i=1}^{r_H} \log_2\left(1 + \frac{\overline{\gamma}}{N_t}E_i\lambda_i\right) \tag{19.31}$$

subject to

$$\sum_{i=1}^{r_H} E_i = N_t, \tag{19.32}$$

where $E_i = E\left[|x_i|^2\right]$.

Optimal power allocation can be derived through the water-filling technique [27, 99, 119, 141]. The optimal capacity is achieved when the energy is allocated as

$$\begin{cases} \frac{N_0}{\sigma_i^2} + \frac{E_s}{N_t}E_i = \mu, & \frac{N_0}{\sigma_i^2} < \mu \\ E_i = 0, & \frac{N_0}{\sigma_i^2} \geq \mu \end{cases}$$

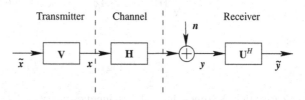

Precoder for MIMO when full CSI is known at both the transmitter and the receiver.

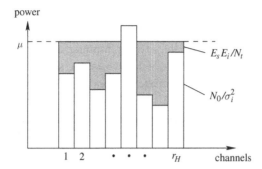

Figure 19.2   Illustration of the water-filling power-allocation technique.

or equivalently

$$E_i = \left(\frac{\mu N_t}{E_s} - \frac{N_t}{\overline{\gamma}\lambda_i} \right)^+, \tag{19.33}$$

where μ is chosen so as to satisfy the power constraint.

This can be visualized through a water-filling illustration, as shown in Fig. 19.2. Treating the noise power as sediment of a container, the available energy for each channel corresponds to the height from the surface to the sediment. Note that water-filling can achieve the optimum power distribution for Gaussian input signals, but for discrete symbols such as PSK or QAM symbols, the problem of optimum power distribution has not yet been solved [73].

In practice, the optimal power allocation is estimated by iteration [103]. At each iteration, the constant μ is obtained from the power constraint, and then the power allocated to each mode is calculated. If the power allocated to a mode is negative, this mode is dropped, and the power allocated to the other modes is recalculated. This process continues until the power allocated to each mode is nonnegative.

At low SNR, the water-filling algorithm (19.33) reduces to allocating all the power to the dominant eigenmode. Thus, the ergodic capacity is given by

$$\begin{aligned}
C &= E\left[\log_2 \left(1 + \overline{\gamma}\lambda_{\max} \right) \right] \\
&\approx \overline{\gamma} E\left[\lambda_{\max}\right] \log_2 e.
\end{aligned} \tag{19.34}$$

When $\mathbf{H} = \mathbf{H}_w$, we get from (19.34) the following result [99]

$$C \approx \overline{\gamma} \frac{r_H}{N} \left(\sqrt{N} + \sqrt{r_H} \right)^2 \log_2 e, \tag{19.35}$$

where $N = \max \{N_t, N_r\}$.

At high SNR, the water-filling solution approaches an equal power allocation to all the nonzero eigenmodes. The capacity for an equal power allocation has been discussed in Section 19.3.1. At high SNR, the capacity can be approximated by [99]

$$\overline{C} \approx N_1 \log_2 \left(\frac{\overline{\gamma}}{N_1} \right) + \frac{1}{\ln 2} \left(\sum_{k=1}^{N_1} \sum_{p=1}^{N_2-k} \frac{1}{p} - N_1 \gamma_E \right), \tag{19.36}$$

where $N_1 = \min(N_t, N_r)$, $N_2 = \max(N_t, N_r)$, and $\gamma_E \approx 0.57721566$ is Euler's constant.

19.3.3 Channel capacities for transmitter with versus without CSI

When the CSI is available at the transmitter, a higher capacity can be expected than the case when no CSI is available at the transmitter, since the transmitter with CSI can optimize its transmit covariance matrix for a given channel.

For a system with $N_r \geq N_t$, when $\text{SNR} = \frac{P}{\sigma_n^2} \to \infty$, the capacity for a transmitter with CSI approaches that for a transmitter without CSI, as the water-filling solution approaches equipower [12, 74]

$$\frac{C_{\text{CSI}}}{C_{\text{no-CSI}}} \to 1. \tag{19.37}$$

When $N_t > N_r$, CSI at the transmitter increases the capacity even at high SNR, with an incremental capacity gain [147]

$$G = \max\left\{N_r \log_2\left(\frac{N_t}{N_r}\right), 0\right\}. \tag{19.38}$$

This gain increases nearly linearly with N_r if $N_t > N_r$. This result is accurate for high SNR, but is optimistic at low SNR.

At low SNR, only the strongest channel eigenmode is used, yielding [147]

$$\frac{C_{\text{CSI}}}{C_{\text{no-CSI}}} \approx N_t \frac{\lambda_{\max}\left(\mathbf{HH}^H\right)}{\text{tr}\left(\mathbf{HH}^H\right)}, \tag{19.39}$$

where $\lambda_{\max}(\cdot)$ takes the maximum eigenvalue of the matrix within, and $\text{tr}(\cdot)$ is the trace of a matrix. Thus, the CSI at the transmitter is more important for low SNR. For example, for an MISO system with N_t transmit antennas and one receive antenna, $\mathbf{HH}^H$ is a scalar, and the right-hand side of (19.39) reduces to N_t, and we have for low SNR $\frac{C_{\text{CSI}}}{C_{\text{no-CSI}}} \approx N_t$.

For the i.i.d. Rayleigh fading channel, the capacity ratio at low SNR, in the limit of a large number of antennas ($N_t, N_r \to \infty$), is given by [147]

$$\frac{C_{\text{CSI}}}{C_{\text{no-CSI}}} \to \left(1 + \sqrt{\frac{N_t}{N_r}}\right)^2 > 1 \quad \text{(Rayleigh fading)}. \tag{19.40}$$

This ratio can be significant when $N_t > N_r$.

Capacity gain due to correlation CSI

The capacity gain due to statistical CSI at the transmitter is also given in [147]. Consider the correlation CSI at the transmitter with a known transmit antenna correlation $\mathbf{R}_t$ and $\mathbf{R}_r = \mathbf{I}$. Such CSI suggests transmission along the eigenvectors of $\mathbf{R}_t$, with a water-filling-type power allocation.

For $N_t \leq N_r$, if $\mathbf{R}_t$ is of full-rank, this capacity gain diminishes at high SNR. In this case, the incremental capacity gain at high SNR is given by [147]

$$G = \max\left\{r_t \log_2 \frac{N_t}{r_t}, 0\right\}, \tag{19.41}$$

where r_t is the rank of $\mathbf{R}_t$.

For $N_t > N_r$, full-rank correlation CSI at the transmitter still increases the capacity. When $\mathbf{R}_t$ is rank-deficient, the correlation CSI helps to increase the capacity at all SNRs for all antenna configurations.

If the receive correlation $\mathbf{R}_r \neq \mathbf{I}$, both the receive and transmit antenna correlations generally reduce the channel ergodic capacity at high SNR, compared to an i.i.d. channel [147]. A stronger channel correlation, measured by the condition number of the correlation matrix, yields a larger precoding gain from the correlation CSI. However, a strong correlation usually reduces the channel capacity [147].

At low SNR, transmit correlation helps to increase capacity [147]. At low SNR,

$$\frac{C_{\text{corr-CSI}}}{C_{\text{no-CSI}}} = \frac{N_t \lambda_{\max}(\mathbf{R}_t)}{\text{tr}(\mathbf{R}_t)}. \tag{19.42}$$

For the rank-one correlated Rayleigh fading channel, this ratio equals N_t.

19.4 Capacity in i.i.d. fast fading channels

For i.i.d. fast fading channels, the capacity $C(\mathbf{H})$ is a random variable, which is a function of the random channel $\mathbf{H}$. As a random variable, $C(\mathbf{H})$ can be characterized by its pdf. The outage capacity C_{out} and the ergodic capacity $\overline{C}$ are two measures of $C(\mathbf{H})$.

19.4.1 Outage and ergodic capacities

Here, we discuss the case of flat Rayleigh fading. The case of correlated Rayleigh fading channels has been considered in [66].

Outage capacity

The capacity outage probability P_{out} is the probability that the conditional channel capacity $C(\mathbf{H})$ drops below a certain rate R

$$P_{\text{out}} = \Pr(C(\mathbf{H}) < R). \tag{19.43}$$

This is the cdf of the random variable $C(\mathbf{H})$. An exact expression of P_{out} is derived in [44, 153]. The pdf of $C(\mathbf{H})$ can be approximated by that of a Gaussian random variable, with mean μ_C and variance σ_C^2. The mean and variance of the pdf of $C(\mathbf{H})$ increase with SNR.

If $N_1 = \min(N_t, N_r)$ is sufficiently large, $\mu_C \gg \sigma_C^2$, and channel hardening occurs; that is, the capacity behaves more like a deterministic quantity, for large N_1. Given an outage probability P_{out}, the outage rate can be correspondingly determined.

Ergodic capacity

The ergodic capacity is defined as

$$\overline{C} = E_{\mathbf{H}}\big[C(\mathbf{H})\big]. \tag{19.44}$$

For i.i.d. Rayleigh fading channels, $\mathbf{H}$ can be modeled as a spatial white matrix $\mathbf{H}_w$. The ergodic capacity is given by [141]

$$\begin{aligned}
\overline{C} = \int_0^\infty & \log_2\left(1 + \frac{\overline{\gamma}}{N_t}\gamma\right) \\
& \times \sum_{k=0}^{N_1-1} \frac{k!}{(k + N_2 - N_1)!} \left[L_k^{N_2-N_1}(\gamma)\right]^2 \gamma^{N_2-N_1} e^{-\gamma} d\gamma,
\end{aligned} \tag{19.45}$$

where

$$\begin{aligned}
L_k^n(\gamma) &= \frac{1}{k!} e^\gamma \gamma^{-n} \frac{d^k}{d\gamma^k}\left(e^{-\gamma}\gamma^{n+k}\right) \\
&= \sum_{l=0}^k (-1)^l \binom{k+n}{k-l} \frac{\gamma^l}{l!}
\end{aligned} \tag{19.46}$$

is the Laguerre polynomial of order k, $N_1 = \min(N_t, N_r)$, and $N_2 = \max(N_t, N_r)$.

The expression (19.45) can be numerically evaluated. Simulations show the following [44]:

- For a fixed $(N_t + N_r)$, the ergodic capacity is higher for a balanced N_t and N_r, that is, $N_1 = \min(N_t, N_r)$ as large as possible.
- For a fixed $N_1 = \min(N_t, N_r)$, it is always better to have $N_r > N_t$.

The first conclusion can be justified as $N_1 = \text{rank}(\mathbf{H})$ corresponds to the number of independent parallel channels between the transmitter and the receiver. The second conclusion is reasonable, since for a fixed N_t the total transmit power remains constant and more power can be collected by using more receive antennas.

Inserting (19.46) into (19.45), we also have [11, 124]

$$\begin{aligned}
\overline{C} = \log_2(e)\frac{N_1!}{(N_2-1)!} & e^{\frac{N_t}{\overline{\gamma}}} \sum_{l=0}^{N_1-1} \sum_{\mu=0}^{N_1} A_{l\mu} \frac{(-1)^{l+\mu}(l+\mu+N_2-N_1)!}{l!\,\mu!} \\
& \times \sum_{p=0}^{l+\mu+N_2-N_1} E_{p+1}\left(\frac{N_t}{\overline{\gamma}}\right),
\end{aligned} \tag{19.47}$$

where

$$\begin{aligned}
A_{l\mu} &= \binom{N_2-1}{N_1-1-l}\binom{N_2}{N_1-1-\mu} - \binom{N_2-1}{N_1-2-l}\binom{N_2}{N_1-\mu} \\
&= \binom{N_2-1}{N_1-1-l}\binom{N_2}{N_1-\mu}\left(\frac{N_1-\mu}{N_2-N_1+\mu+1} - \frac{N_1-l-1}{N_2-N_1+l+1}\right)
\end{aligned} \tag{19.48}$$

and

$$E_k(x) = \int_1^\infty e^{-xy} y^{-k} dy \tag{19.49}$$

is the exponential integral function of order k.

From (19.45) or (19.47), some insights are obtained for the following special cases.

Case 1. $N_t = N_r = 1$

This reduces to the single-antenna system transmitting over the flat Rayleigh fading channel. This has been treated in Section 14.6.1.

Case 2. $N_t > N_r = 1$

We have

$$\overline{C} = \frac{1}{(N_t - 1)!} \int_0^\infty \log_2 \left(1 + \frac{\overline{\gamma}}{N_t}\gamma\right) \gamma^{N_t-1} e^{-\gamma} d\gamma. \tag{19.50}$$

On the other hand, for $N_r = 1$, $\mathbf{H}$ becomes a row vector $\boldsymbol{h} = (h_1, \ldots, h_{N_t})^T$. From (19.21), we have

$$C(\boldsymbol{h}) = \log_2 \left(1 + \frac{\overline{\gamma}}{N_t} \sum_{i=1}^{N_t} |h_i|^2\right). \tag{19.51}$$

By the law of large numbers, for large N_t

$$\frac{1}{N_t} \sum_{i=1}^{N_t} |h_i|^2 \to 1. \tag{19.52}$$

Thus, for large N_t, we have

$$\overline{C} \approx \log_2(1 + \overline{\gamma}) = C_{\text{AWGN}}. \tag{19.53}$$

That is, for large N_t, the effect of fading diminishes asymptotically.

Example 19.1: For $N_r = 1$, the capacity versus N_t, given by (19.50), is plotted in Fig. 19.3, for different SNRs. It is seen that for $N_t \geq 4$, the ergodic capacity $\overline{C}$ is sufficiently close to C_{AWGN}. Thus, for the single receive antenna case, four transmit antennas are sufficient for achieving the ergodic capacity.

Case 3. $N_r > N_t = 1$

We have

$$\overline{C} = \frac{1}{(N_r - 1)!} \int_0^\infty \log_2 (1 + \overline{\gamma}\gamma) \gamma^{N_r-1} e^{-\gamma} d\gamma. \tag{19.54}$$

In a way similar to the derivation of (19.53), we have, for large N_r

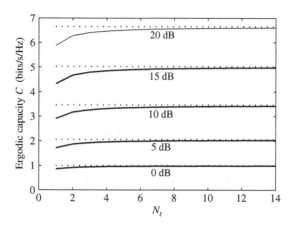

Figure 19.3 Ergodic capacity $\overline{C}$ for $N_t > 1$ and $N_r = 1$. The dotted lines correspond to the AWGN capacity.

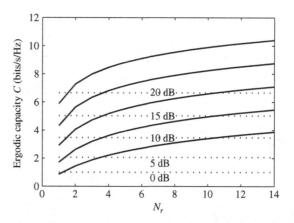

Figure 19.4 Ergodic capacity $\overline{C}$ for $N_t = 1$ and $N_r > 1$. The dotted lines correspond to the AWGN capacity.

$$\overline{C} \approx \log_2 \left(1 + N_r \overline{\gamma}\right) > C_{\text{AWGN}}. \tag{19.55}$$

Thus, by using multiple receiver antennas, it is possible to achieve a channel capacity beyond C_{AWGN}.

Example 19.2: For $N_t = 1$, the capacity versus N_r, given by (19.54), is plotted in Fig. 19.4, for different SNRs.

Case 4. $N_r \gg N_t > 1$

From (19.21), we have

$$\frac{1}{N_r} \mathbf{H} \mathbf{H}^H \rightarrow \mathbf{I}_{N_t} \tag{19.56}$$

and thus

$$\overline{C} \approx N_t \log_2 \left(1 + \frac{N_r}{N_t}\overline{\gamma}\right) > N_t C_{\text{AWGN}}. \tag{19.57}$$

Case 5. $N_t \gg N_r > 1$

From (19.21), we have

$$\frac{1}{N_t}\mathbf{HH}^H \to \mathbf{I}_{N_r} \tag{19.58}$$

and thus

$$\overline{C} \approx N_r \log_2 (1 + \overline{\gamma}) = N_r C_{\text{AWGN}}. \tag{19.59}$$

Case 6. $N_t = N_r > 1$

Capacity is calculated from (19.47) by setting $N_t = N_r = N_1 = N_2 = N$

$$\overline{C} = \int_0^\infty \log_2 \left(1 + \frac{\overline{\gamma}}{N}\gamma\right) \sum_{k=0}^{N-1} \left[L_k^0(\gamma)\right]^2 e^{-\gamma} d\gamma. \tag{19.60}$$

An approximation of the capacity for this case has been derived as [124]

$$\overline{C} \approx e^{1/\overline{\gamma}} \log_2(e)E_1 \left(\frac{1}{\rho}\right)$$
$$+ (N-1)\left[2\log_2 \left(1 + \sqrt{4\overline{\gamma}+1}\right) - \frac{\log_2(e)}{4\overline{\gamma}}\left(\sqrt{4\overline{\gamma}+1}-1\right)^2 - 2\right]. \tag{19.61}$$

Again for large N, applying the law of large numbers, we have

$$\frac{1}{N}\mathbf{HH}^H \to \mathbf{I}_N. \tag{19.62}$$

From (19.21), we have

$$\overline{C} \approx N \log_2 (1 + \overline{\gamma}) = N C_{\text{AWGN}}. \tag{19.63}$$

This shows that for large N, the ergodic capacity $\overline{C}$ increases linearly with N. This is the most powerful reason for deploying MIMO for high data rates.

Example 19.3: The ergodic capacity versus $N_t = N_r = N$, given by (19.60), is plotted in Fig. 19.5, for different SNRs.

In [124], a tight upper bound for the ergodic capacity of spatially correlated/double-scattering MIMO channels has also been obtained. A closed-form capacity formula for the keyhole MIMO channel has also been derived as $C(N_t, N_r) \leq \log_2 (1 + N_r\gamma)$.

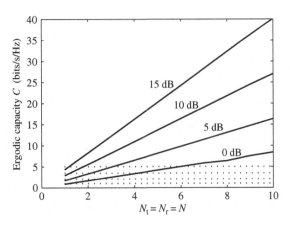

Figure 19.5 Ergodic capacity $\overline{C}$ for $N = N_t = N_r$. The dotted lines correspond to the AWGN capacity.

Case 7. At High SNR

At high SNR, the ergodic capacity of an i.i.d. flat fading MIMO channel can be approximated by [144]

$$\overline{C} \approx N_1 \log_2 \left(\frac{\overline{\gamma}}{N_t} \right) + \sum_{i=1}^{N_1} E\left[\log_2 \lambda \right]$$

$$= N_1 \log_2 \left(\frac{\overline{\gamma}}{N_t} \right) + \sum_{i=|N_t-N_r|+1}^{N_2} E\left[\log_2 \chi_{2i}^2 \right]. \tag{19.64}$$

Another approximation is given by [99, 101]

$$\overline{C} \approx N_1 \log_2 \left(\frac{\overline{\gamma}}{N_t} \right) + \frac{1}{\ln 2} \left(\sum_{k=1}^{N_1} \sum_{p=1}^{N_2-k} \frac{1}{p} - N_1 \gamma_E \right). \tag{19.65}$$

The capacity in the i.i.d. slow fading channels, C_{slow}, is always higher than the ergodic capacity in the fast fading case, C_{fast}, for given N_t, N_r, and SNR. As SNR increases, the gap decreases. At high SNR, the gap between C_{slow}, given by (19.36), and C_{fast}, given by (19.64) is

$$C_{\text{slow}} - C_{\text{fast}} = N_1 \log_2 \left(\frac{N_t}{N_1} \right), \tag{19.66}$$

which is close to zero for $N_r \geq N_t$, and is greater than zero for $N_r < N_t$.

Case 8. At Low SNR

When there is no CSI at the transmitter, equal power allocation is usually implemented at the transmit antennas. By using the approximation $\log_2(1 + x) \approx x \log_2 e$ for small x, we have

$$
\begin{aligned}
\overline{C} &= \sum_{i=1}^{N_1} E\left[\log_2\left(1 + \frac{\overline{\gamma}}{N_t}\lambda_i\right)\right] \\
&\approx \sum_{i=1}^{N_1} \frac{\overline{\gamma}}{N_t} E\left[\lambda_i\right]\log_2 e \\
&= \frac{\overline{\gamma}}{N_t}\sum_{i=1}^{N_1} E\left[\mathrm{tr}\left(\mathbf{HH}^H\right)\right]\log_2 e \\
&= N_t\overline{\gamma}\log_2 e.
\end{aligned}
\tag{19.67}
$$

Thus, at low SNR, a capacity or power gain of N_r over a single antenna system is obtained, since a receiver beamforming can be applied. Without CSI at the transmitter, multiple transmit antennas do not improve the capacity at low SNR, since transmit diversity cannot be achieved.

19.4.2 Capacity bounds

In the flat-fading MIMO channel, fading is independent on each channel, thus the channel matrix $\mathbf{H}$ is most probably of full rank, and the eigenvalues are close to one another. From (19.20), capacity increases linearly with the number of antenna elements. Thus, existence of heavy multipath, with many multipath components of similar strength, is very beneficial to the MIMO system. This is opposed to most conventional systems.

For a Rayleigh fading channel, $\left|h_{ij}\right|^2$ is chi-squared variate but normalized with $E\left[\left|h_{ij}\right|^2\right] = 1$. When there is perfect CSI at the receiver but no CSI at the transmitter, the capacity distribution has upper and lower bounds, for $N_t \geq N_r$, and these are given by [40]

$$
\sum_{k=N_t-N_r+1}^{N_t} \log_2\left(1 + \frac{\overline{\gamma}}{N_t}\chi_{2k}^2\right) \leq C \leq \sum_{k=1}^{N_t}\log_2\left(1 + \frac{\overline{\gamma}}{N_t}\chi_{2N_r}^2\right),
\tag{19.68}
$$

where χ_{2k}^2 is a chi-squared-distributed random variable with $2k$ degrees of freedom. The ergodic capacity is obtained by taking the mean over the χ^2 variables. Note that a χ^2-distributed random variable is related to a Gaussian random variable: given a Gaussian random variable X, then $Y = X^2$ has a chi-squared distribution. When X has zero mean, it is central chi-squared distributed, otherwise it is noncentral chi-squared distributed. Rayleigh and Rice an distributions can be derived from central and noncentral chi-squared distributions.

The lower bound in (19.68) is improved in [101] as

$$
\overline{C} \geq \sum_{k=1}^{N_1}\log_2\left(1 + \frac{\overline{\gamma}}{N_t}\exp\left(\sum_{p=1}^{N_2-k}\frac{1}{p} - \gamma_E\right)\right).
\tag{19.69}
$$

An improved lower bound at low SNR is derived as [101]

$$\overline{C} \geq \log_2 \left(1 + \frac{\overline{\gamma}}{N_t} \exp \left(\sum_{k=1}^{N_t N_r - 1} \frac{1}{k} - \gamma_E \right) \right). \tag{19.70}$$

In the case when CSI is available at both the transmitter and the receiver, if $N_t = N_r$, the capacity gain by water-filling is small compared with the equal-power distribution case. When $N_t > N_r$, water-filling can achieve a better gain. Since the transmitter has CSI, it can perform beamforming and steer the beam toward the receive array, and this can increase the SNR and thus, the capacity. Thus, a larger N_t leads to a larger capacity.

Another pair of MIMO capacity bounds for the Rayleigh fading channel is given in [127]. Assuming that no CSI is available at the transmitter and $N_t < N_r$. $\mathbf{H}$ can be modeled as spatial white matrix $\mathbf{H}_w$. Implementing QR decomposition [31], we have $\mathbf{H}_w = \mathbf{QR}$, where $\mathbf{Q}$ is a unitary matrix and $\mathbf{R}$ is an upper-triangular matrix. The elements of $\mathbf{R}$ above the main diagonal entries are statistically independent, each having a Gaussian distribution with zero mean and unit variance, while the magnitude squares of the diagonal elements, $|R_{ll}|^2$, are χ^2-distributed with $2(N_r - l + 1)$ degrees of freedom. After some manipulation, we have

$$C = \log_2 \left[\det \left(\mathbf{I} + \frac{P}{N_t} \mathbf{H} \mathbf{H}^H \right) \right] = \log_2 \left[\det \left(\mathbf{I} + \frac{P}{N_t} \mathbf{R} \mathbf{R}^H \right) \right]. \tag{19.71}$$

Finally, the lower and upper bounds can be obtained as [127]

$$\sum_{l=1}^{N_t} \log_2 \left(1 + \frac{p}{N_t} |R_{ll}|^2 \right) \leq C \leq \sum_{l=1}^{N_t} \log_2 \left(1 + \frac{p}{N_t} \left(|R_{ll}|^2 + \sum_{m=l+1}^{N_t} |R_{lm}|^2 \right) \right). \tag{19.72}$$

The capacity is lower-bounded by the sum of the capacities of N_t independent subchannels whose power gains are χ^2-distributed with degrees of freedom $2N_r, 2(N_r - 1), \ldots,$ $2(N_r - N_t + 1)$, and is upper-bounded by the sum of the capacities of N_t independent subchannels whose power gains are χ^2-distributed with degrees of freedom $2(N_r + N_t - 1)$, $2(N_r + N_t - 3), \ldots, 2(N_r - N_t + 1)$. The difference between the mean values of the upper and lower bounds is less than 1 bit/s/Hz per spatial dimension.

A dynamic model for CSI at the transmitter that takes into account channel temporal variation was constructed in [147]. The dynamic model covers smoothly from perfect CSI to statistical CSI at the transmitter. The capacity gains and the optimal input with dynamic CSI are analyzed asymptotically at low and high SNRs, and summarized in [148].

19.4.3 Ricean channels

The Ricean MIMO channel $\mathbf{H}_{\text{Ricean}}$ can be modeled as the sum of a constant $\mathbf{H}_{\text{LOS}}$ and a variable Rayleigh component caused by scattering, $\mathbf{H}_{\text{Rayleigh}}$ [103]

$$\mathbf{H}_{\text{Ricean}} = \sqrt{\frac{K_r}{K_r + 1}} e^{j\phi_0} \mathbf{H}_{\text{LOS}} + \sqrt{\frac{1}{K_r + 1}} \mathbf{H}_{\text{Rayleigh}}, \tag{19.73}$$

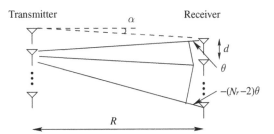

Figure 19.6   Geometry of a MIMO channel with transmit and receive linear antenna arrays.

where K_r is the Ricean factor, ϕ_0 is the phase shift of the signal due to propagation from a transmit antenna element to a receive antenna element, and $\mathbf{H}_{\text{Rayleigh}} = \mathbf{H}_w$.

Given a MIMO system with a uniform linear array of N_t transmit antennas and a uniform linear array of N_r receive antennas, as shown in Fig. 19.6, when the distance R between the two arrays are very large, the $\mathbf{H}_{\text{LOS}}$ matrix can be derived as

$$\mathbf{H}_{\text{LOS}} = \mathbf{d}_r \mathbf{d}_t^T = \begin{bmatrix} 1 & e^{j\theta} & \cdots & e^{j(N_t-1)\theta} \\ e^{-j\theta} & 1 & \cdots & \vdots \\ \vdots & \vdots & \ddots & e^{j\theta} \\ e^{-j(N_r-1)\theta} & e^{-j(N_r-2)\theta} & \cdots & 1 \end{bmatrix}, \tag{19.74}$$

where $\mathbf{d}_r$ and $\mathbf{d}_t$ are the transmit and receive steering vectors, and θ is the phase shift between two adjacent array elements. The second equality holds only when the transmit and receive antennas are almost parallel. It is easily seen that rank($\mathbf{H}_{\text{LOS}}$) = 1. When the angle α between the two arrays is very small and $R \gg d$, θ is very small and we get $\mathbf{H}_{\text{LOS}}$ with all the elements as unity, denoted $\mathbf{H}(1)$.

For the Ricean MIMO channel with i.i.d. Rayleigh part, the exact ergodic mutual information with equal power allocation has been given in [67]. The existence of a Ricean component reduces the multipath richness. For a given SNR, the Ricean channel has a capacity lower than that of the Rayleigh fading channel, but is higher than that of the AWGN channel; it approximates the capacity of the Rayleigh fading channel as $K_r \to 0$, and that of the AWGN channel as $K_r \to \infty$. However, for a given transmit power, the Ricean channel introduces a higher SNR, since there are no obstructions. For $N_t = N_r = 1$, the capacity increases monotonically with K_r. For the multiple-antenna case, depending on the singular values of $\mathbf{H}_{\text{LOS}}$, there are instances where the capacity is either improved or degraded by the LOS component [88]. The upper bound on ergodic capacity for correlated and well separated dual-polarized antennas in a Ricean channel is given in [98, 99].

19.5 Space-time coding

CSI at the transmitter can be avoided by applying space-time coding. Space-time coding is specially designed for use with multiple transmit antennas. Space-time codes introduce temporal and spatial correlation into signals that are transmitted by different antennas

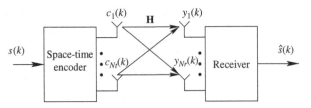

Figure 19.7   Space-time coding system.

to provide diversity and coding gain without sacrificing the bandwidth. Most space-time codes are designed for quasi-static channels, where the channel is constant over a block of N_{st} symbol times and the channel is unknown at the transmitter. A block of N_{st} symbols is encoded together.

Space-time coding is illustrated in Fig. 19.7. The information symbol $s(k)$ is encoded into N_t code symbols $c_1(k)$, $c_2(k)$, ..., $c_{N_t}(k)$, each code symbol being transmitted simultaneously from a different antenna. The N_t code symbols are encoded in such a way that both the coding gain and the diversity gain at the receiver are maximized.

Space-time coding increases redundancy over space and time, since each antenna transmits the same, but differently encoded signal. Space-time coding thus maximizes the diversity gain. Space-time coding techniques can be classified into space-time block codes (STBCs) and space-time trellis codes (STTCs), following the well-known convention in coding theory. For space-time diversity codes such as STTCs and STBCs, which extract full diversity order, the spatial rate $r_s \leq 1$.

Typical STBCs are linear codes over the field of complex numbers, which are symbols in two-dimensional modulation constellation points, while STTCs are not linear over complex numbers, since they are functions of the encoded bits. STTCs introduce memory and achieve a better performance than OSTBCs do, typically by about 2 dB. However, the decoding complexity of STTCs scales as $O\left(M^{\min\{N_t,N_r\}}\right)$, where M is the constellation size, as opposed to $O\left(\min\{N_t,N_r\}\right)$ for OSTBCs.

19.5.1 Performance analysis of space-time codes

The MIMO model can be rewritten as

$$y(k) = \mathbf{H}x(k) + n(k), \tag{19.75}$$

where $x(k) = \left(x_1(k), \ldots, x_{N_t}(k)\right)^T$ is the unnormalized signal vector. The channel coefficients are assumed to be constant during one encoded frame. For a group of N_{st} vectors, we have

$$\mathbf{Y} = \mathbf{HX} + \mathbf{N}, \tag{19.76}$$

where $\mathbf{X} = \left[x(0)\middle|x(1)\middle| \cdots \middle|x(N_{st} - 1)\right]$ is an $N_t \times N_{st}$ matrix, and $\mathbf{Y}$ and $\mathbf{N}$ are defined in a similar manner but have dimension $N_r \times N_{st}$.

Slow fading MIMO channels

For ML decoding and known CSI $\mathbf{H}$ at the receiver, the conditional pairwise error probability for two competing codewords $\mathbf{X}$ and $\tilde{\mathbf{X}}$ is given by

$$\Pr\left(\mathbf{X} \to \tilde{\mathbf{X}} | \mathbf{H}\right) = \frac{1}{2}\mathrm{erfc}\left(\sqrt{\frac{\left\|\mathbf{HX} - \mathbf{H}\tilde{\mathbf{X}}\right\|_F^2}{4\sigma_N^2}}\right)$$

$$= \frac{1}{2}\mathrm{erfc}\left(\sqrt{\left\|\mathbf{H}\left(\mathbf{B} - \tilde{\mathbf{B}}\right)\right\|_F^2 \frac{E_s}{4N_0}}\right), \tag{19.77}$$

where the last equality is obtained by normalizing the space-time codewords to

$$\mathbf{B} = \frac{\mathbf{X}}{\sqrt{E_s/T}}, \quad \tilde{\mathbf{B}} = \frac{\tilde{\mathbf{X}}}{\sqrt{E_s/T}} \tag{19.78}$$

and inserting $\sigma_n^2 = N_0/T$ for complex-valued signals, T being the symbol period.

After some manipulation, the following bound is obtained [73, 138]

$$\Pr\left(\mathbf{B} \to \tilde{\mathbf{B}}\right) = E_{\mathbf{H}}\left[\Pr\left(\mathbf{B} \to \tilde{\mathbf{B}} | \mathbf{H}\right)\right]$$

$$\leq \frac{1}{2}\left(\prod_{\nu=1}^{r} \frac{1}{1 + \lambda_\nu \frac{E_s}{4N_0}}\right)^{N_r}$$

$$< \frac{1}{2}\left[\frac{E_s}{4N_0}\left(\prod_{\nu=1}^{r} \lambda_\nu\right)^{1/r}\right]^{-rN_r}, \tag{19.79}$$

where

$$r = \mathrm{rank}\left(\Delta\right) = \mathrm{rank}\left(\mathbf{B} - \tilde{\mathbf{B}}\right), \tag{19.80}$$

$$\Delta = \left(\mathbf{B} - \tilde{\mathbf{B}}\right)\left(\mathbf{B} - \tilde{\mathbf{B}}\right)^H \in \mathcal{C}^{N_t \times N_t}, \tag{19.81}$$

and λ_ν is a nonzero eigenvalue of Δ. The last inequality is obtained by dropping the 1 in the denominator of the first inequality, at large SNR.

Rank and Determinant Criteria

The exponent rN_r in (19.79) is known as the *diversity gain*, G_d. In order to achieve the maximum diversity gain, the minimum rank r of all pairwise differences should be maximized

$$G_d = N_r \min_{\left(\mathbf{B}, \tilde{\mathbf{B}}\right)} \mathrm{rank}\left(\mathbf{B} - \tilde{\mathbf{B}}\right). \tag{19.82}$$

The coding gain leads to a horizontal shift of the error rate curve, and is given by

$$G_c = \min_{\left(\mathbf{B}, \tilde{\mathbf{B}}\right)} \left(\prod_{\nu=1}^{r} \lambda_\nu \right)^{1/r}. \tag{19.83}$$

For full rank $\mathbf{B} - \tilde{\mathbf{B}}$, i.e., $r = N_t$, we have $\prod_{\nu=1}^{r} \lambda_\nu = \det\left(\mathbf{B} - \tilde{\mathbf{B}}\right)$, and

$$G_c = \min_{\left(\mathbf{B}, \tilde{\mathbf{B}}\right)} \left(\det\left(\mathbf{B} - \tilde{\mathbf{B}}\right) \right)^{1/N_t}. \tag{19.84}$$

At high SNR, the pairwise error probability can be characterized by G_c and G_d. They affect the BER in different ways on a log-log scale, as shown in Fig. 5.2.

From (19.82) and (19.84), two criteria, namely the *rank criterion* and *determinant criterion*, can be obtained for space-time codeword design. The rank criterion optimizes the spatial diversity, while the determinant criterion optimizes the coding gain [138]. From the two criteria, space-time codes should be designed such that Δ has full rank for any $\mathbf{X} \neq \mathbf{X}' \in \mathcal{X}$, and such that the minimum value of $\det(\Delta)$ for any $\mathbf{X} \neq \mathbf{X}' \in \mathcal{X}$ is maximized. Note that the rank-determinant criterion is valid at high SNR.

Rank and Trace Criteria

At low SNR or when N_r is large, the pairwise error probability can be derived as a function of the trace of Δ [19]. The determinant criterion turns the trace criterion, and maximizes the minimum trace of the error matrix Δ over all pairs of codewords

$$d_e = \min_{\left(\mathbf{B}, \tilde{\mathbf{B}} | \mathbf{B} \neq \tilde{\mathbf{B}}\right)} \mathrm{tr}\{\Delta\}. \tag{19.85}$$

Fast fading MIMO channels

In i.i.d. fast Rayleigh fading channels, an upper bound of the average pairwise error probability is derived as [99, 138]

$$\Pr\left(\mathbf{X} \to \tilde{\mathbf{X}}\right) \leq \prod_{k=0}^{N_{st}-1} \left(1 + \frac{1}{4N_0} \|\mathbf{x}(k) - \tilde{\mathbf{x}}_k\|^2 \right)^{-N_r}$$

$$\leq \left(\frac{1}{4N_0} \right)^{-N_r L_{\mathbf{X}, \tilde{\mathbf{X}}}} \prod_{k=1}^{L_{\mathbf{X}, \tilde{\mathbf{X}}}} \|\mathbf{x}(k) - \tilde{\mathbf{x}}_k\|^{-2N_r}, \tag{19.86}$$

where $L_{\mathbf{X}, \tilde{\mathbf{X}}}$ is the effective length of the pair of codewords $\mathbf{X}$ and $\tilde{\mathbf{X}}$, and defined as the size of the set of time instants, $1 \leq k \leq N_{st}$, such that $\mathbf{x}(k) \neq \tilde{\mathbf{x}}(k)$. The second inequality is obtained for high SNR.

Distance-Product Criterion

At high SNR, the distance-product criterion is derived from (19.86):

- *Distance criterion.* The minimum effective length of the code over all pairs of codewords $\{X, \tilde{X}\}$ is maximized

$$L_{\min} = \min_{\{X, \tilde{X} | X \neq \tilde{X}\}} L_{\{X, \tilde{X}\}}. \tag{19.87}$$

- *Product criterion.* The minimum product distance of the code over all pairs of codewords is maximized

$$d_p = \min_{\{X, \tilde{X} | X \neq \tilde{X}\}} \prod_{k=1}^{L_{\min}} \|x(k) - \tilde{x}_k\|^2. \tag{19.88}$$

From (19.86), it is seen that the use of multiple antennas at the transmitter does not change the achievable diversity gain, but it increases the coding gain [138].

19.5.2 Orthogonal space-time block codes

The OSTBC [5, 97, 139] provides the full diversity order and a simple decoding process, but no beamforming gain or coding gain. The OSTBC uses orthogonal design (achieving orthogonal **H**) to separate signals from different transmit antennas, and achieves full diversity by using a simple linear ML decoder. The MIMO ML decoding process is decoupled into several SIMO ML decoding processes. The OSTBC introduces considerable loss in bandwidth efficiency, for $N_t > 2$ with complex constellations. The OSTBC is a linear space-time code.

The signal bits are first mapped onto symbols s_l. The space-time block encoder collects a block of N_s symbols, $s = (s_1, s_2, \ldots, s_{N_s})^T$, and encodes them as a sequence of N_x consecutive vectors $x(k) = (x_1(k), \ldots, x_{N_t}(k))^T$, $k = 0, 1, \ldots, N_x - 1$. The code rate is $R_c = N_s/N_x$. For the OSTBC, given two information blocks s and s', the error matrix Δ has full rank, achieving full diversity order.

The most popular OSTBC is the Alamouti code that is designed for two transmit antennas, $N_t = 2$ [5]. Transmit diversity using the Alamouti code achieves full diversity order of $2N_r$, and has been introduced in Section 5.3. The whole codeword is arranged in space and time as

$$X_2 = [x(2k) \,|\, x(2k+1)] = \frac{1}{\sqrt{2}} \begin{bmatrix} s_1 & -s_2^* \\ s_2 & s_1^* \end{bmatrix}, \tag{19.89}$$

where s_1 and s_2 are the two symbols in the blocks of $N_s = 2$, and $N_x = 2$. So $R_c = 1$. The factor $\frac{1}{\sqrt{2}}$ is used to normalize the average transmit power to E_s/T.

In WCDMA, the Alamouti code is implemented in a different form

$$X_2 = [x[2k] \,|\, x[2k+1]] = \frac{1}{\sqrt{2}} \begin{bmatrix} s_1 & s_2 \\ -s_2^* & s_1^* \end{bmatrix}. \tag{19.90}$$

Thus, the original symbols s_1 and s_2 are transmitted over the same antenna. The first antenna is used in the same way as in the single antenna case, and adding a second antenna does not influence the transmission of the first antenna.

The OSTBC [139] extends the Alamouti scheme to $N_t > 2$ based on orthogonality and unitarity principles, and it achieves the full diversity order of $N_t N_r$. For rate-1 real modulation, codes have been found for any number of transmit antennas, while a rate-1 complex modulation code only exists for $N_t = 2$. The series of rate-1 real modulation schemes can be extended to a series of rate-1/2 complex modulation schemes [139], and this doubles the delay. Thus, half-rate codes exist for an arbitrary number of transmit antennas [139]. Several orthogonal codes have been found with lower rates [139].

For a block of symbols $s = \left(s_1, s_2, \ldots, s_{N_s}\right)^T$, the OSTBC matrix $\mathbf{X}_o(s)$ is an $N_t \times N_x$ matrix, which must satisfy

$$\mathbf{X}_o \mathbf{X}_o^H = P \mathbf{I}_{N_t}, \tag{19.91}$$

where $P = \sum_{n=1}^{N_s} |s_n|^2$ is used to normalize the transmit power. In the generalized OSTBC, all the rows of $\mathbf{X}_o$ are also orthogonal but the norms are different. The generalized OSTBC enjoys full spatial diversity and very simple ML detection exactly as OSTBCs do [135].

According to (19.91), the case of $N_t = 2$ is given by (19.89) by dropping the factor $\frac{1}{\sqrt{2}}$. For $N_t = 3$ and 4, $\mathbf{X}_o$ takes the form

$$\mathbf{X}_o = \begin{bmatrix} s_1 & -s_2^* & -s_3^* & 0 \\ s_2 & s_1^* & 0 & -s_3^* \\ s_3 & 0 & s_1^* & s_2^* \end{bmatrix} \quad (N_t = 3), \tag{19.92}$$

$$\mathbf{X}_o = \begin{bmatrix} s_1 & -s_2^* & -s_3^* & 0 \\ s_2 & s_1^* & 0 & -s_3^* \\ s_3 & 0 & s_1^* & s_2^* \\ 0 & s_3 & -s_2 & s_1 \end{bmatrix} \quad (N_t = 4). \tag{19.93}$$

Since N_s symbols are transmitted over N_x time slots, the code rate of OSTBCs is

$$\eta_{\text{OSTBC}} = \frac{N_s}{N_x} \quad \text{(symbols per channel use)}. \tag{19.94}$$

Thus, for general complex constellations, we have $\eta_{\text{OSTBC}} = 1$ for $N_t = 2$, and $\eta_{\text{OSTBC}} = 3/4$ for $N_t = 3$ and 4. The OSTBC also exists for $N_t > 4$ but the code rate $\eta_{\text{OSTBC}} = 1/2$ [139]. For real constellations, $\eta_{\text{OSTBC}} = 1$ can be achieved for all $N_t \geq 2$.

The OSTBC can be decoded by linear ML decoding. ML decoding of s from $\mathbf{Y}$ can be transformed into N_s independent serial transmissions of its entry s_n over a flat channel with coefficient $\|\mathbf{H}\|_F^2$ and noise variance $\|\mathbf{H}\|_F^2 \sigma_n^2$ [116]. The effective SNR at the decision point is given by

$$\gamma = \frac{\|\mathbf{H}\|_F^2}{N_t} \bar{\gamma}. \tag{19.95}$$

Based on this and the modulation type, the BER performance of the OSTBC system can be calculated.

Some special STBCs

Quasi-orthogonal STBC

The quasi-orthogonal STBC (QO-STBC) trades diversity and decoding simplicity of the OSTBC for transmission rate enhancement [61]. The QO-STBC uses lower-dimensional OSTBCs as the building blocks of a higher-dimensional code. The major drawback of the QO-STBC is its low achievable diversity. Full diversity can be achieved by signal rotations which expand the constellation, thus leading to the rotated QO-STBC [122].

Modified OSTBC

A modified OSTBC for combating time selectivity has been proposed in [146]. It is systematically constructible for any number of antennas. The modified OSTBC has exactly the same complexity as the OSTBC, but is less subject to the performance degradation caused by symbol-by-symbol channel variation.

Perfect STBC

The perfect $(n \times n)$ STBCs [100] are a class of linear dispersion space-time codes that have full-rate, full-diversity, nonvanishing constant minimum determinant, uniform average transmitted energy per antenna, and good shaping for increasing spectral efficiency. These lead to optimality in terms of the diversity-multiplexing tradeoff, as well as excellent low SNR performance. These codes thus maximize the diversity and coding advantages for a MIMO system. Construction of perfect STBCs has been discussed for 2, 3, 4, and 6 antennas in [100]. In [34], perfect codes are constructed for any channel dimension, and the notion of perfect code is extended to the rectangular case.

Diversity embedded code

In transmitting over an unknown channel, a conservative strategy is to design for the worst channel for a given reliability (outage). Another strategy is an opportunistic one where information is embedded in such a manner as to deliver part of the information if the channel is bad, but more information when the channel is good. Diversity embedded codes are high-rate space-time codes that have a high-diversity code embedded within them. Diversity embedded codes [28, 29] allow multiple levels of reliabilities for different messages. This is desirable as wireless networks need to support applications with very different QoS requirements. Diversity embedded codes provide unequal error protection (UEP) with respect to the diversity metric suitable for fading channels [29]. They allow a form of communication where the high-rate code opportunistically takes advantage of good channel realizations while the embedded high-diversity code guarantees that at least part of the information is received reliably. In terms of unequal error protection, rate opportunism, and packet delay optimization, diversity-embedded codes are shown to have the potential to outperform the traditional single-layer codes in moderate SNR regimes [29]. Diversity-embedded codes provide a means of provisioning queues with different priorities (reliabilities). A HARQ scheme that uses feedback to schedule information over these prioritized queues has been proposed in [29]. Code designs [29] and fundamental limits of performance for such codes have been developed. In [33], these ideas are summarized in a unified framework.

Capacity and error probability of the OSTBC

The capacity achieved by the OSTBC, $C_{\text{OSTBC}}(\mathbf{H})$, is given by [116]

$$C_{\text{OSTBC}}(\mathbf{H}) = \eta_{\text{OSTBC}} \log_2 \left(1 + \frac{\overline{\gamma}}{N_t} \|\mathbf{H}\|_F^2 \right) \quad \text{(bits per channel use)}. \tag{19.96}$$

The difference from the channel capacity $C(\mathbf{H})$, given in (19.20), always satisfies [116]

$$\Delta C = C(\mathbf{H}) - C_{\text{OSTBC}}(\mathbf{H}) \geq 0. \tag{19.97}$$

The equality holds only when $N_t = 2$ and rank$(\mathbf{HH}^H) = 1$. Thus, the OSTBC with $N_r > 1$ will almost surely cause a capacity loss [44, 116]. Only in the case of the Alamouti code with $N_t = 2$ and a single receive antenna ($N_r = 1$) can the channel capacity be achieved.

The BER performance of OSTBCs in various MIMO channels has been studied. In [68], closed-form BER expressions of OSTBCs are derived for MPAM, MQAM, and MPSK constellations in a correlated Rayleigh MIMO channel, and the BER is also analyzed for correlated shadowed Ricean MIMO channels. The upper bounds on the BERs of OSTBCs in a correlated Rayleigh channel are given in [63]. Closed-form BER expressions of OSTBCs are derived for correlated Nakagami-fading channels in [37]. In [157], closed-form expressions for symbol error probabilities of OSTBCs that use MPSK and MQAM modulations over Ricean–Nakagami channels, are given. The symbol error and outage probabilities for OSTBCs in correlated Ricean fading have been studied in [96]. Closed-form exact BER expressions of the generalized OSTBCs for MQAM and MPSK constellations in both spatially white and correlated Rayleigh MIMO channels are derived in [69].

Closed-form expressions for the error probability of OSTBCs as well as diversity gains of keyhole MIMO channels are derived in [48]. The maximum spatial diversity gain of a keyhole MIMO channel is $\min(N_t, N_r)$ when $N_t \neq N_r$, and the achievable diversity gain is less than N_r but higher than $N_r - 1$ when $N_t = N_r$ [43, 48, 115]. For a 2×2 OSTBC system in keyhole MIMO channels, closed-form expressions have been derived for the average SEP for several modulation schemes, outage probability, amount of fading and ergodic capacity in [106], and used to study the performance of the system.

Applications of the OSTBC

Although orthogonal codes do not provide coding gain, their decoding simply requires some linear combinations of the received symbols, and they provide full diversity degree. The use of the OSTBC generally reduces the capacity of the MIMO channel. Thus, the OSTBC should be a good choice for improving the performance of a single antenna system by adding an extra transmission antenna, but it is argued that it is less attractive in terms of channel capacity for multiple antenna systems [74].

The Alamouti code has been adopted in WCDMA, CDMA2000, and IEEE 802.16d/e. The Alamouti code is known as *space-time transmit diversity (STTD)* in WCDMA. In CDMA2000, the space-time spreading (STS) scheme is used, where Alamouti-coded symbols are separated by using two orthogonal codes instead of two time slots.

The Alamouti code can be extended. When each of the two antennas has two polarizations, this leads to $N_t = 4$. This is possible for use in future wireless systems. In order to use the Alamouti code in case of $2N$ transmit antennas, one can group them into N two-antenna space-time encoders. Double-STTD (DSTTD), which consists of two parallel Alamouti STBC units and transmits with 4 transmit antennas, is adopted in 3GPP [1] and IEEE 802.11n. The Alamouti code scheme is generalized to the OSTBC that achieves a full diversity order with an arbitrary number of transmit antennas [139].

STBC, STBC design, and more discussion of multiple antenna techniques are given in detail in [74, 99].

19.5.3 Space-time trellis codes

The STTC is an extension of the conventional trellis code to the MIMO system [138]. The technique extends Ungerbock's TCM scheme to the spatial domain. Like the STBC, the STTC provides a full diversity order of $N_t N_r$, but it also provides a coding gain. The coding gain increases with the number of states as well as the number of receive antennas. This leads to an improved error performance, since the error performance depends on both the diversity order and coding gain. The STTC provides a coding gain by increasing the minimum distance between the code sequences.

Each STTC can be represented by a trellis. At time l, a vector $\boldsymbol{d}(l) = (d_1(l), \ldots, d_K(l))^T$ is fed into a linear shift register consisting of L_c blocks, each having K bits, L_c being the constraint length. The total length of the register is $L_c K$ bits. The major difference from the binary convolutional code is the way the register content is combined to generate the outputs for different antennas. The Viterbi algorithm can be used for decoding.

A simple STTC is illustrated in Fig. 19.8 for a vector modulation consisting of two QPSK symbols. Two transmit antennas are used, and the transmission achieves 2 bits/s/Hz. The four possible symbols $\{1, j, -1, -j\}$ are, respectively, labeled as $\{0, 1, 2, 3\}$. The corresponding space-time trellis has four incoming and four outgoing branches per state. As common practice, the space-time trellis is always initialized and ended at the zero state. At time t, a transition branch is selected according to the state of the encoder and the input bits. At time t, one symbol or two bits is transmitted at an antenna, depending on which state transition is made. The symbol transmitted at the other antenna is chosen according to the previous state, and the two antennas select their symbols for transmission based on the alternating state. A number of STTCs are described in the literature [44, 73, 138].

Design criteria for codes that are robust to the keyhole condition are given in [115]. It is reported that for the case of $N_t \leq N_r$, the design criteria in the keyhole condition is the same as that under rich scattering, but for $N_t > N_r$ the design criteria are indeed affected.

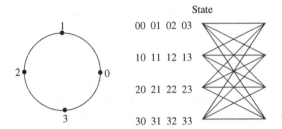

Figure 19.8   **Trellis of a four-state QPSK STTC for the delay diversity code.**

Delay diversity codes

The simplest coherent space-time code is the delay diversity code. The delay diversity code is capable of achieving the maximum possible diversity gain for any N_t and for any signal constellation. For a block $\boldsymbol{s} = \left(s_1, s_2, \ldots, s_{N_s}\right)^T \in \mathcal{C}^{N_s \times 1}$, a simplest delay scheme of one symbol per antenna corresponds to the space-time code matrix

$$
\mathbf{X}_{\mathrm{DD}} =
\begin{bmatrix}
s_1 & s_2 & \cdots & s_{N_s} & 0 & \cdots & 0 \\
0 & s_1 & s_2 & \cdots & s_{N_s} & \cdots & 0 \\
\vdots & \vdots & \ddots & \ddots & \ddots & \ddots & \vdots \\
0 & \cdots & 0 & s_1 & s_2 & \cdots & s_{N_s}
\end{bmatrix}
\in \mathcal{C}^{N_t \times (N_s + N_t - 1)},
\tag{19.98}
$$

where the zeros correspond to no data transmissions. The time span of the space-time code matrix is $N_x = N_s + N_t - 1$, leading to a code rate of $\eta = N_s / (N_s + N_t - 1)$ symbols per channel use. For any two different symbol blocks $\boldsymbol{s} \neq \boldsymbol{s}'$, the error matrix Δ always has full rank N_t. Decoding can be based on the Viterbi algorithm. The delay diversity code can be treated as a special case of STTCs.

Viterbi decoding of STTCs

The only difference between a space-time trellis and a trellis for the single-antenna case is that each branch of a space-time trellis has N_t labels instead of one. The Viterbi algorithm can be applied by choosing the branch metric as

$$
m(n) = \sum_{j=1}^{N_r} \left| y_j(n) - \sum_{i=1}^{N_t} \sqrt{\frac{\gamma}{N_t}} h_{ji} x_i(n) \right|^2.
\tag{19.99}
$$

For N_x symbols, the Viterbi algorithm finds the shortest path that minimizes $\sum_{n=1}^{N_x} m(n)$. At any time instant, the number of branches to be computed is $N_{\mathrm{state}} \times 2^{R_{\mathrm{st}}}$, where the number of states $N_{\mathrm{state}} = M$ and the transmission rate $R_{\mathrm{st}} = \eta_{\mathrm{st}} \log_2 M$ for M-ary modulation, with $\eta_{\mathrm{st}} = N_s / N_x$ as the code rate, when N_s symbols are transmitted in N_x time slots.

The STTC requires a multidimensional Viterbi algorithm for decoding. The complexity for decoding increases exponentially with the number of states, which is a major obstacle for its application in practical systems. A number of STTCs have been compared by simulation in [73]. It is concluded that the coding gain of the STTC promised by the determinant criterion cannot be achieved in general. The coding gain is visible only for high diversity degrees [73].

19.5.4 Differential space-time coding

Motivated by differential modulation and detection described in Section 7.11, differential space-time coding has also been proposed in [57, 60]. The space-time code matrix constructed must be unitary, and thus this differential space-time scheme is often called *differential unitary space-time coding*.

Differential space-time coded systems do not require the knowledge of the MIMO channel. By using ML decoding, the error performance of differential unitary space-time decoding is 3 dB inferior to its coherent counterpart. Design of the group of unitary space-time code matrices follows the same criteria as the design of the coherent space-time coding, that is, to maximize both the diversity and coding gains. Design of unitary space-time matrices can be implemented by using the Cayley transform [44].

A differential version of the Alamouti code is also available [140]. It retains most of the advantages of the Alamouti code, but does not require any channel knowledge for decoding. It achieves a higher transmission rate at a lower decoding complexity, when compared to the differential unitary space-time code for $N_t = 2$. The differential Alamouti code has been extended to the differential OSTBC in [41].

19.6 Spatial multiplexing

Unlike space-time coding, spatial multiplexing aims at maximizing the data rate. Spatial multiplexing is also referred to as *information MIMO*, while space-time coding is referred to as *diversity MIMO*. Spatial multiplexing multiplexes multiple spatial channels to send as many independent data as possible over different antennas for a given error rate. The antennas transmit parallel data streams termed *layers*, and spatial multiplexing is also called the *layered space-time scheme*. Spatial multiplexing with no coding can be treated as a space-time code with spatial rate $r_s = N_t$ and diversity order N_r. In general, in order to implement spatial multiplexing, it is required that $N_r \geq N_t$. The MIMO receiver for spatial multiplexing has to deal with multistream interference.

Generally, the space-time coding technique can provide a better QoS due to antenna diversity; on the other hand, spatial multiplexing can provide a high bandwidth efficiency, but due to lack of antenna diversity the poor link-level error performance may decrease the throughput, especially at low SNR. One compromise between diversity and data rate is to use linear space-time precoding/decoding with N_t greater than the number of data substreams [114]. In order to achieve spectral efficiency that is similar to that of spatial multiplexing techniques, space-time codes need to use higher modulation orders.

The linear dispersion code is a unified description of space-time coding and spatial multiplexing transmission [54]. The method enables a tradeoff between diversity and multiplexing gain [56]. The linear dispersion code can be efficiently decoded using a sphere decoder, which has a polynomial decoding complexity for a wide range of SNR. Compared to the OSTBC, the linear dispersion code achieves a higher rate, but at the cost of signal constellation expansion and not guaranteeing the maximum diversity gain.

Recently, spatial modulation has been proposed as an alternative MIMO transmission scheme to space-time coding and spatial multiplexing [91]. In the spatial modulation scheme, a block of information bits is jointly encoded by symbol constellation and the transmit antenna number, and only one transmit antenna is active at any instant. At the receiver, MRC is used to retrieve the transmitted block of information bits. Spatial modulation increases the overall spectral efficiency by $\log_2 N_t$, as compared to a linear increase for the V-BLAST system. For the same spectral efficiency of MIMO-OFDM systems, spatial modulation results in a substantial reduction in receiver complexity as compared to V-BLAST and nearly the same receiver complexity as the Alamouti scheme. Spatial modulation achieves better BER performance, as compared with the other two techniques [91].

19.6.1 Layered space-time receiver structures

The layered space-time schemes, known as the *Bell Laboratories layered space-time (BLAST) schemes*, were developed by Foschini and coworkers at Bell Labs. These are the first practical schemes to achieve transmission rates above one symbol per channel use. Coded BLAST can approach the fundamental limit provided by the ergodic capacity of the MIMO channel. The BLAST schemes do not incur any mutual information loss relative to the capacity of the MIMO channel [40].

Horizontal BLAST

Horizontal encoding BLAST (H-BLAST) [39] is the simplest layered space-time receiver structure. The data stream is first converted into N_t parallel substreams, each being encoded separately (temporal coding, interleaving, and symbol mapping), and then submitted to a different transmit antenna. The arrangement of the layers is shown in Fig. 19.9.

The H-BLAST code matrix employs the same number of transmit antennas and layers $N_t = N_l$, and is given by

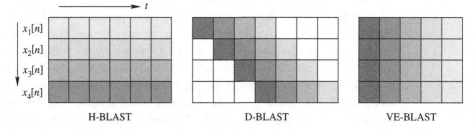

Figure 19.9 The layers of the H-BLAST, D-BLAST and V-BLAST schemes, for $N_t = 4$. Sub-blocks of the same gray level constitute a substream.

$$\mathbf{X} = \begin{bmatrix} x_1(1) & \cdots & x_1(N_x) \\ x_2(1) & \cdots & x_2(N_x) \\ \vdots & \ddots & \vdots \\ x_{N_t}(1) & \cdots & x_{N_t}(N_x) \end{bmatrix}, \tag{19.100}$$

where $x_l(n)$, $l = 1, 2, \ldots, N_t$, is the nth symbol of the lth layer. Each $x_l(n)$ can be either an uncoded symbol or an entry of codeword. The code rate for the uncoded case is given as

$$\eta_{\text{H-BLAST}} = N_t \quad \text{(symbols per channel use)}. \tag{19.101}$$

Joint ML decoding can be applied on each column, comprising of N_t symbols. This leads to a complexity that is exponential in N_t. This complexity can be further decreased by using the DFE decoder for MIMO, called the *nulling canceling decoder*. The maximum diversity order is at most N_r.

At the receiver side, these data streams are separated by keeping one data stream and suppressing all other data stream based on optimum combining. This requires $N_r \geq N_t$ to suppress all $N_t - 1$ interfering data streams. H-BLAST can at most achieve a diversity order of N_r. An array gain of N_r is also achievable.

The data streams are then detected in a serial order. The procedure is very similar to MUD: For different streams coming from different users, H-BLAST is actually an ordered serial (successive) interference cancellation (OSIC) receiver. To mitigate the error propagation problem, the receiver should first decode the streams in a descending order of SINR. In addition, the H-BLAST receiver has been shown to be equivalent to a ZF or MMSE generalized DFE, depending on the criterion used for interference reduction [45].

Diagonal BLAST

H-BLAST [39] does not provide diversity: The first substream dominates the performance at high SNR. The diagonal BLAST (D-BLAST) scheme is a better solution. D-BLAST is similar to H-BLAST, the only difference being that the generated N_t substreams are subjected to a stream interleaver. Each substream is subdivided into many sub-blocks, and these sub-blocks are transmitted by different antennas according to a round-robin schedule by padding some antennas with zeros. Decoding is performed substream by substream: Each decoded sub-block is subtracted from the signals at the other antenna elements, and thus the quality of the residual signal is enhanced. The layers of D-BLAST are shown in Fig. 19.9

The space–time code matrix of D-BLAST is given by

$$\mathbf{X} = \begin{bmatrix} x_1(1) & x_2(1) & \cdots & x_{N_l}(1) & & \\ & x_1(2) & x_2(2) & \cdots & x_{N_l}(2) & \\ & & \ddots & & & \ddots \\ & & & x_1(N_t) & x_2(N_t) & \cdots & x_{N_l}(N_t) \end{bmatrix}, \tag{19.102}$$

where $x_l(n)$ is the nth symbol of the lth layer transmitted by the nth antenna, N_l is the number of layers, and the blank spaces are filled with zeros, representing no transmission. Each $x_l(n)$ can be an uncoded symbol or a codeword entry. This space-time code matrix resembles that of the delay diversity code matrix, instead of repeating the same symbol per delay diversity layer.

From the space-time code matrix, D-BLAST transmits N_l layers carrying N_t symbols per layer over $N_l + N_t - 1$ time slots; thus the code rate of uncoded D-BLAST is given by

$$\eta_{D-BLAST} = \frac{N_t N_l}{N_l + N_t - 1} \quad \text{(symbols per channel use).} \tag{19.103}$$

This code rate is N_t times that of the delay diversity code. As N_l grows, the code rate approaches N_t symbols per channel use.

ML decoding of D-BLAST can collect a diversity order of $N_t N_r$, and the complexity is $O\left(M^{N_t N_l}\right)$ for M-ary modulation. Linear ZF or MMSE equalization can be used to reduce the complexity to cubic order, but at the expense of the error performance. A compromise is to use DFE. The ZF and MMSE versions of DFE, known as *SIC* and *PIC*, have been discussed for MUD of CDMA systems. The nulling canceling decoder requires $N_r \geq N_t$; it is computationally attractive, but only collects the receiver diversity N_r. The diversity order of D-BLAST is between N_r and $N_t N_r$.

The data stream is detected in the order of the antenna elements, as in the H-BLAST case. Since each data stream sees all the transmit antennas, it alternatively experiences good or bad SINR, as the amount of the contribution of the other streams having already been subtracted varies. Thus, each stream experiences full diversity order of $N_t N_r$. An array gain of N_r can be achieved.

D-BLAST achieves the lower bound of capacity given in (19.68) [7]. For $N_t = N_r$, the data streams alternatively see a channel with diversity order of 1, diversity order of 2, until diversity order of N_t. Note for a channel with diversity order k, its SNR has a pdf that is a chi-squared distribution with $2k$ degrees of freedom, χ^2_{2k}.

Vertical BLAST

In vertical encoding BLAST (V-BLAST) [46], the bitstream is first temporally encoded, interleaved, and symbol mapped. The resulting N_s symbols are then demultiplexed into N_t substreams, and these symbols are then transmitted over the antennas. V-BLAST can split each information bit across more than one antenna, and thus can achieve a diversity order of more than N_r; but, it requires joint decoding of the substream and thus, can be very complex. An array gain of N_r can be achieved. The layer arrangement of V-BLAST is shown in Fig. 19.9.

Detection of V-BLAST is simpler than that in the diagonal architecture. Detection and estimation of the transmitted symbols is performed on a vector-by-vector basis, and the detection can be performed using interference cancellation with SNR reordering.

19.6.2 Space-time receivers

Popular space-time receivers are the conventional MUD receivers such as the ZF, MMSE, optimum ML, and SIC receivers. For all these receivers, the number of receive antennas

N_r is assumed to be not less than the number of transmitted data streams N_t, $N_r \geq N_t$; otherwise, the channel is ill-conditioned and the data may not be correctly decoded. The effects of the ratio N_t/N_r on the spectral efficiency for the ZF and MMSE receivers are compared in [103] for the case of H-BLAST.

Linear ZF receiver

The ZF receiver is also known as *decorrelator* or *interference nulling* receiver. Given MIMO transmission

$$y = \mathbf{H}x + n. \tag{19.104}$$

A linear receiver is defined by

$$\tilde{y} = \mathbf{A}y = \mathbf{A}\mathbf{H}x + \mathbf{A}n. \tag{19.105}$$

The ZF receiver is obtained by setting $\mathbf{A}\mathbf{H} = \mathbf{I}$ in (19.105), that is, by eliminating all the nondiagonal elements of $\mathbf{A}\mathbf{H}$

$$\tilde{y} = \mathbf{H}^{\dagger}y, \tag{19.106}$$

where $\mathbf{H}^{\dagger}$ is the Moore-Penrose pseudoinverse of $\mathbf{H}$. This yields

$$\tilde{y} = x + \mathbf{H}^{\dagger}n, \tag{19.107}$$

which leads to noise enhancement. The ZF receiver has a diversity gain limited to $N_r - N_t + 1$ [99, 103]. The system is underdetermined if $N_t > N_r$.

At high SNR, the maximum achievable rate for the ZF receiver can be derived as [144]

$$R_{\text{ZF}} \approx N_1 \log_2\left(\frac{\bar{\gamma}}{N_t}\right) + N_t E\left[\log_2 \chi^2_{2(N_r - N_t + 1)}\right]. \tag{19.108}$$

Comparing (19.108) to the MIMO channel capacity given by (19.64), the first term is the same, thus the ZF receiver can fully harness the spatial degrees of freedom of the MIMO channel. However, the ZF receiver introduces a performance degradation in view of the second term [144].

In the multiuser system, scheduling can be exploited to overcome the drawback of noise enhancement of the ZF receiver by avoiding a poor channel [4, 55]. In this case, the performance of the ZF receiver approaches that of the optimal receiver, when K, the number of users, is large [55]. As K approaches infinity, the capacity achieved by the ZF receiver together with multiuser diversity scales identically compared to that of MIMO broadcast channels [4]. In [22], the sum-rate capacity of the multiuser MIMO system that employs spatial multiplexing and ZF receiver with a number of scheduling algorithms is given by closed-form expressions.

Linear MMSE receiver

The MMSE receiver minimizes the joint effects of the off-diagonal elements of $\mathbf{A}\mathbf{H}$ and of the filtered noise $\mathbf{A}n$:

$$\min E\left[\|\mathbf{A}y - x\|_F^2\right]. \tag{19.109}$$

This gives

$$\tilde{y} = \left(\mathbf{H}^H\mathbf{H} + \frac{N_0}{E_s}\mathbf{I}_{N_t}\right)^{-1}\mathbf{H}^H y, \tag{19.110}$$

where E_s is the average energy of each component of x (a transmitted symbol), and N_0 is the noise variance.

Linear receivers have low complexity, but the performance is lower than that of the more complex ML receiver and BLAST. The ZF receiver decomposes the MIMO channel into N_t parallel channels with enhanced additive noise. At low SNR, the MMSE receiver approximates a matched filter and outperforms the ZF receiver, while at high SNR it is close to the ZF receiver. For the simple H-BLAST, both ZF and MMSE realize a diversity order of $N_r - N_t + 1$ [103]. Linear receivers have a complexity that is cubic with the dimensions of the $\mathbf{H}$, $O\left(N^3\right)$, where $N = \max\left(N_r, N_t\right)$.

SIC receiver

The SIC receiver successively decodes the symbol streams, and subtracts the decoded symbol streams layer by layer. It can be ZF or MMSE-based. The complexity is of the fourth order of the dimension of $\mathbf{H}$, that is, $O\left(N^4\right)$. The complexity of the SIC receiver can be reduced to $O\left(N^3\right)$ by using the square-root algorithm proposed in [53]. SIC has a diversity order of $N_r - N_t + 1$. It suffers from error propagation.

In order to reduce the influence of error propagation, OSIC selects the stream with the highest SINR for peeling at the beginning of each stage. This is the method used in H-BLAST [39]. The complexity of OSIC is the same as that of SIC, that is, $O\left(N^4\right)$, plus the complexity of the ordering operation, which is $O\left(k^3\right)$ for a set of k elements. Due to the ordering operation, the diversity order collected by the OSIC receiver is greater than $N_r - N_t + 1$, but still lower than N_r. The OSIC receiver reduces the block error by a factor of about ten compared to a purely linear receiver, or equivalently, decreases the required SNR by about 4 dB [46]. By using OSIC, the V-BLAST prototypes have exhibited spectral efficiencies above 20 bits/s/Hz [46].

For spatial multiplexing with equal power allocation in i.i.d. fast Rayleigh fading MIMO channels, the ZF-SIC receiver achieves the ergodic MIMO capacity at asymptotically high SNR. At high SNR, the maximum achievable rate for the ZF-SIC receiver can be derived as[144]

$$R_{\text{ZF-SIC}} \approx N_1 \log_2\left(\frac{\overline{\gamma}}{N_t}\right) + \sum_{k=1}^{N_t} E\left[\log_2 \chi^2_{2(N_r-N_t+k)}\right]. \tag{19.111}$$

Comparing (19.111) to the MIMO channel capacity given by (19.64), it is seen that the achievable capacity of the ZF receiver is exactly equal to the channel capacity.

The MMSE-SIC receiver achieves the ergodic capacity of the MIMO channel for all SNRs [99, 144, 145]. Assuming that the SINR and the rate for stream k are SINR_k and $\log_2\left(1 + \text{SINR}_k\right)$ respectively, for any given $\mathbf{H}$, the sum rate is equal to the channel capacity

$$R_{\mathrm{MMSE-SIC}} = \sum_{k=1}^{N_t} \log_2 (1 + \mathrm{SINR}_k) = I(\boldsymbol{y}, \boldsymbol{x}|\mathbf{H}), \qquad (19.112)$$

where $I(\boldsymbol{y}, \boldsymbol{x}|\mathbf{H})$ is the mutual information of the channel $\mathbf{H}$, given by (19.17). Thus, the MMSE-SIC receiver is optimal.

The LLR-ordered SIC scheme is analyzed in multiuser multimode MIMO systems in which each user may choose between spatial multiplexing and beamforming [24]. The main idea is to detect and cancel the user signal in the order of LLR which provides a-posteriori information about the reliability of detection. LLR-ordered SIC provides 1–3 dB gain over the conventional SNR-ordered SIC in a multiuser MIMO system and the gain increases with the number of users [24]. However, for a channel that changes every L symbol intervals, the LLR-ordered SIC is about L times more complex than the SNR-ordered SIC.

Outage and error probability performance of the ordered ZF V-BLAST have been evaluated for i.i.d. Rayleigh fading channels in [85, 86]. The results are applicable to uncoded D-BLAST and the MUD system with SIC.

ML receiver

Spatial multiplexing with equal power allocation achieves the ergodic capacity of i.i.d. fast Rayleigh fading channels. The joint encoding of the data streams to be transmitted from different antenna elements, in conjunction with ML detection, is the optimum solution that approximates the capacity of a MIMO system. This method is only useful for a small number of antenna elements and a small modulation alphabet such as BPSK, and is complex for most practical cases. The ML receiver is optimal, and it achieves a diversity order of N_r. For a constellation of size M, ML decoding needs to search over M^{N_t} possible solutions.

Sphere decoding has been extended for near-ML decoding of MIMO systems [44, 94]. It can be used for decoding of space-time block and layered coded MIMO systems. The average decoding complexity is a polynomial function of N_t when $N_r \geq N_t$. The ZF or nulling canceling solution can be used for calculating the initial radius for searching. Sphere decoding may fail to return the ML solution in polynomial time for very large N_t, or large constellation size, or low SNR, or $N_r < N_t$ [44].

A number of improved versions of the sphere decoding algorithm, with improved error performance and/or reduced complexity compared to that of the basic sphere decoding algorithm, are available [44, 79, 93, 94]. List sphere decoding can generate bit-level soft information for use in capacity-approaching iterative MIMO decoders, and has been applied to turbo-coded V-BLAST systems [59]. A soft-sphere decoding algorithm for demodulating a MIMO system with $N_t = N_r = 4$, which achieves a transmission rate of 38.8 Mbits/s over a 5-MHz band, has been implemented for use in HSDPA [42].

K-best Schnorr-Euchner (KSE) decoding is a low complexity MIMO decoding scheme that approaches near-ML performance for MIMO detection [51]. KSE decoders have been implemented for 4×4 16QAM MIMO detection in CMOS technology, achieving a decoding throughput of up to 53.3 Mbits/s for the hard-output decoder, and more than 100 Mbits/s for the soft-output decoder.

Remarks

Practical systems employ coded BLAST structure, which breaks the demodulation process into multiple, separate, low-complexity pieces. The H-BLAST architecture is a popular spatial multiplexing scheme for its simple implementation. All the above receivers can be used for H-BLAST.

In the H-BLAST structure, if the MMSE receiver is used, the N_t streams are first separated and then decoded independently. If the OSIC receiver is used, the streams are separated layer by layer and at each layer they are ordered, and then the MMSE algorithm and optimal decoding are applied. OSIC significantly outperforms the MMSE receiver, but with a much higher complexity.

19.6.3 Spatial precoding

The BLAST structure, as a capacity-achieving technique, is a layered receiver-centric technique for spatial multiplexing. Despite its satisfactory performance in laboratories [46], the BLAST structure is not practical in cellular environments, since it depends highly on high SNR for joint decoding of the various streams, or else catastrophic error propagation will occur. Another capacity-achieving approach to spatial multiplexing is the transmitter-centric spatial precoding technique, which allows multiple data stream to be transmitted to different users.

The spatial precoding technique is motivated by the concept known as *writing on dirty paper* [26]. The principle of dirty paper coding (DPC) states that the effect of the interference can be canceled by proper coding at the transmitter and the resulting capacity is the same as that of a channel without interference, but with the same power constraint. Specifically, assume that a received signal is given by

$$y = s + i + n, \tag{19.113}$$

where s is the transmitted signal, i is the interference known deterministically to the transmitter, and n is AWGN. The capacity of the system is the same as if there were no interference, no matter how strong the interference is and whether it is known to the receiver.

In case of spatial precoding for the multi-user MIMO channel, the CSI and the transmitted signals are known at the transmitter, thus the transmitter knows how a signal designed for one user interferes with the other users and compensates for this.

SVD precoding

From Section 19.2.4, parallel decomposition of a channel can be defined by defining a linear transformation on the channel input x and output y, known as *transmit precoding* and *receiver shaping*. Linear precoding serves as a beamformer with one or more beams with per-beam power allocation.

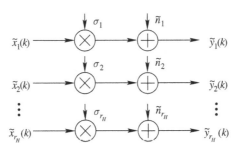

Figure 19.10   The MIMO channel is decomposed into r_H parallel scalar Gaussian channels through SVD.

In case of transmit precoding, as shown in Fig. 19.1, the antenna output is given by $x = V\tilde{x}$, where $\tilde{x}$ is the input vector, while at receiver shaping the channel output is given by $\tilde{y} = U^H y$, where y is the measurement at the receive antennas. Transmit precoding and receiver shaping transform the MIMO channel into r_H parallel scalar channels with input $\tilde{x}$ and output $\tilde{y}$

$$\tilde{y} = \Sigma \tilde{x} + \tilde{n}, \tag{19.114}$$

where $\tilde{n} = U^H n$. Since U is a unitary matrix, n and $\tilde{n}$ are identically distributed AWGN vectors. The r_H parallel scalar channels are illustrated in Fig. 19.10.

From (19.114), the MIMO channel is decomposed into r_H parallel Gaussian channels. Each channel has a power of σ_i^2. The r_H parallel channels do not interfere with one another. By sending independent data over each of the parallel channels, MIMO can support a capacity increase by a factor R_H. However, the capacity of each of the channels is dependent on its gain σ_i.

The complexity for SVD of the channel H is at $O\left(N_r N_t^2\right)$ if $N_r \geq N_t$, and a relatively accurate channel matrix is fed back. SVD precoding does not introduce noise enhancement, as does the open-loop linear detectors. Nonetheless, the closed-loop spatial multiplexing can achieve high performance, but at much lower decoding complexity than the ML detector in open-loop spatial multiplexing.

Linear precoding

Linear precoding based on ZF beamforming achieves a large fraction of dirty paper coding capacity while exhibiting reduced complexity. A ZF beamforming strategy can achieve the same asymptotic sum capacity as that of dirty paper coding, as the number of users goes to infinity. Based on the user grouping, the round-robin and proportional-fair scheduling schemes provide this asymptotic result [159].

More generally, linear precoding at the transmitter and postcoding at the receiver can be jointly designed by optimizing such criteria as the information capacity, BER, received SNR, or detection MSE [132]. The joint precoding/postcoding method can always decompose the MIMO channel into a set of parallel subchannels. A linear precoder that optimizes the sum capacity is given in [133]. A transceiver architecture based on ZF beamforming and linear receiver combining was proposed in [143]. The receiver combining and quantization for CSI feedback are jointly designed in order to maximize the expected SINR for each user. A design methodology for generating tree-structure codebooks tailored for arbitrary spatial correlation statistics is proposed in [143].

The linear precoder in fact decouples the input signal into orthogonal signal modes in the form of eigen-beams, each being allocated a power that is based on the CSI at the transmitter. In the case of perfect CSI, the precoded orthogonal spatial modes match the channel eigen-directions, and there is no interference between these signal streams. With partial CSI, precoder design must reduce the interference among signals sent on these beams. Exploiting the CSI available at the transmitter – perfect CSI, correlation CSI, mean CSI, and mean plus correlation CSI – precoder design can be based on such criteria as the ergodic capacity, error exponent, pairwise error probability, and detection MSE [147]. Optimal power allocation can be implemented by water-filling, based on the design criterion.

Precoding has been incorporated in IEEE 802.16e as a closed-loop MIMO scheme. An open-loop precoding proposal based on the reciprocity principle is also being considered by IEEE 802.11n. 3GPP uses a closed-loop beamforming technique, and precoding is being considered in its HSDPA. For precoding implementation, channel sounding may be used for channel estimation.

Precoding exploits the CSI at the transmitter to add an array gain. For perfect CSI at the transmitter, a diversity gain can also be delivered. In addition, precoding allows parallel channel transmissions, which reduce receiver complexity for high spatial rates. Linear space-time precoding schemes for downlink transmission in a MISO CDMA system have been treated in [110]. Spatial Tomlinson-Harashima precoding is a nonlinear pre-equalization technique for the MIMO system [38].

19.6.4 Other closed-loop MIMO schemes

Orthogonalized spatial multiplexing achieves orthogonality between transmitted symbols by applying phase rotation at the transmitter [70, 76]. It is a closed-loop MIMO system. By applying rotation operations to the transmitted symbols, orthogonalized spatial multiplexing allows a simple symbol decodable ML decoder at the receiver. It requires only one phase feedback value for the transmitter. Orthogonalized spatial multiplexing exhibits a performance that is almost the same as that of the optimum closed-loop system, but significantly reduces the processing complexity at both the transmitter and the receiver as well as the feedback overhead does.

The closed-loop V-BLAST scheme has been proposed for the MIMO-OFDM block-fading channel in [162]. It is the V-BLAST scheme with per-antenna-based power and rate feedback. The receiver jointly optimizes the power and rate assignments for all transmit antennas, and then returns them to the transmitter via a low-rate feedback channel. The power and rate optimization minimizes the total transmit power for support of an aggregate transmission rate during each fading block. The per-antenna-based power and rate control can be readily modified to combine with linear MIMO transmit precoding as an efficient and capacity-approaching partial-CSI-feedback scheme. The closed-loop

V-BLAST approaches closely the MIMO-OFDM channel capacity assuming perfect CSI at both the transmitter and the receiver [162].

TAS/MRC

Another closed-loop MIMO scheme that combines single transmit antenna selection (TAS) and receiver MRC is referred to as the *TAS/MRC* scheme. A single transmit antenna, which maximizes the total received signal power at the receiver, is selected for transmission at any time, and all the other transmit antennas are inactive. This requires a low-rate feedback channel. The error performance of the TAS/MRC scheme has been analyzed in [18, 20, 21, 112].

The average SNR gain of the TAS/MRC in slow Rayleigh fading is quantified and compared with those of receiver MRC and STBCs in [20]. The TAS/MRC scheme outperforms some more complex STBCs of the same spectral efficiency. The diversity order is $N_t N_r$ at high SNR, and channel estimation errors based on pilot symbols have no impact on the diversity order [20]. In [18], the performance of the TAS/MRC scheme is investigated for flat Nakagami-m fading channels with arbitrary real-valued m. The outage probability, the asymptotic BER expressions for BPSK modulation, and the exact SER expressions for MPSK and MQAM have been derived. The asymptotic SER expressions reveal a diversity order of $mN_t N_r$.

The TAS system with MRC or EGC in Nakagami fading is analyzed in [112], and exact expressions for the SNR statistics as well as the average BER and SER of several modulation techniques are derived.

The performance of multiuser diversity in multiuser point-to-multipoint TAS/MRC systems over slow, flat Rayleigh fading channels has been analyzed in [21]. With ideal CSI feedback, tight closed-form expressions of outage capacity and average SEP for the multiuser TAS/MRC system have been derived. The outage capacity increases with the increase of mean of effective average SNR achieved from K users and decreases with the increase of variance of the effective SNR. In order to obtain a higher outage capacity, it is necessary to have $N_r \le N_t$. From the SEP performance, the multiuser TAS/MRC system achieves a diversity order of approximately $KN_t N_r - KN_t + 1$, or approximately $KN_t N_r$ for large N_r. In the multiuser TAS/MRC system, users can be equivalent to virtual transmit antennas, which is the source of the multiuser diversity.

19.6.5 Beamspace MIMO

Parasitic arrays such as switched parasitic arrays and the electronically steerable passive array radiator (ESPAR) antenna [136] are alternatives to antenna arrays, but they require only one RF branch. They provide a compact, cost-effective solution for beamforming [107, 136], diversity [117, 120], and MIMO STBC [154] applications. Hence, they make it possible to integrate multiple-antenna technology into small wireless devices.

Beamspace MIMO achieves capacity gain in rich-scattering environments by using a single RF front end and a parasitic array capable of changing its radiation pattern on each symbol period [65]. Diverse symbol streams are mapped onto orthogonal bases (radiation patterns) defined in the beamspace domain of the transmitting array far-field region. The diverse symbol streams are then simultaneously transmitted towards different angles of departure. In rich-scattering environments, these symbol streams experience multipath fading in a manner similar to the conventional MIMO transmission. Using a traditional MIMO receiver it is then feasible to retrieve the transmitted information at a rate that is equal to that of comparable conventional MIMO systems. Beamspace MIMO can be implemented in both STBC or spatial multiplexing mode [65].

19.7 Diversity, beamforming, versus spatial multiplexing

19.7.1 Diversity, beamforming, and spatial multiplexing gains

The MIMO technology can be used for spatial multiplexing, diversity and/or beamforming, but all these goals at full scale cannot be achieved at the same time. These goals are contradictory to one another. When a MIMO system is used in a LOS environment, the achievable beamforming gain is $N_t N_r$, with N_t at the transmitter and N_r at the receiver; however, there is no diversity gain since there is no fading. In a heavy scattering environment, the diversity order is $N_t N_r$, but the maximum beamforming gain is upper-limited by $\left(\sqrt{N_t} + \sqrt{N_r}\right)^2$ [92]. In addition, the transmit diversity and beamforming are two complementary techniques. The first provides diversity with no array gain, while the second provides array gain with no diversity.

The gain of spatial multiplexing can be defined as [163]

$$G_{\mathrm{sm}} = \lim_{\overline{\gamma} \to \infty} \frac{C_{\mathrm{out}}(\overline{\gamma})}{\log_2 \overline{\gamma}}, \tag{19.115}$$

where C_{out} is the outage capacity in bits/s/Hz.

Similarly, the diversity gain is defined by (5.3). The diversity gain G_d can be obtained by approximating P_e or P_{out} at high SNR

$$P_e \approx \gamma^{-G_d} \quad \text{or} \quad P_{\mathrm{out}} \approx \gamma^{-G_d}. \tag{19.116}$$

Scalar Rayleigh channel

For the scalar slow fading Rayleigh channel, the diversity-multiplexing tradeoff for the PAM, QAM, and the channel itself are given by [144]

$$G_d = 1 - 2G_{\mathrm{sm}}, \quad G_{\mathrm{sm}} \in \left[0, \frac{1}{2}\right] \quad \text{(PAM)}, \tag{19.117}$$

$$G_d = 1 - G_{\mathrm{sm}}, \quad G_{\mathrm{sm}} \in [0, 1] \quad \text{(QAM)}, \tag{19.118}$$

$$G_d = 1 - G_{\mathrm{sm}}, \quad G_{\mathrm{sm}} \in [0, 1] \quad \text{(channel)}. \tag{19.119}$$

Thus, the uncoded QAM scheme trades off diversity and multiplexing gains optimally.

For example, (19.119) can be derived as follows. For the slow Rayleigh channel, the outage probability at $R = C_{out} = G_{sm} \log_2 \gamma$ is

$$
\begin{aligned}
P_{out} &= \Pr \left(\log_2 \left(1 + |h|^2 \gamma \right) < G_{sm} \log_2 \gamma \right) \\
&= \Pr \left(|h|^2 < \frac{\gamma^{G_{sm}} - 1}{\gamma} \right) \\
&\approx \gamma^{-(1-G_{sm})}
\end{aligned}
\tag{19.120}
$$

at high SNR. The last approximation is obtained by using $\Pr \left(|h|^2 < \epsilon \right) \approx \epsilon$ for small ϵ in Rayleigh fading. Thus, the diversity-multiplexing tradeoff (19.119) is obtained.

ISI channel

From a diversity perspective, ISI can be exploited by averaging the fluctuations in the channel gains across the different signal paths. The optimal diversity-multiplexing tradeoff curve for the ISI channel is [50]

$$
G_d^* = L(1 - G_{sm}) \quad \text{(ISI channel)}, \tag{19.121}
$$

where L is the number of taps corresponding to the ISI channel.

Analysis shows that the MLSE and linear DFE equalizers achieve the optimal diversity-multiplexing tradeoff without coding, but the ZF and MMSE equalizers do not [50]. However, if each transmission block is ended with a period of silence lasting the coherence time of the channel, both the ZF and MMSE equalizers achieve the optimal diversity-multiplexing tradeoff; thus, at high SNR, a simple precoding strategy can facilitate reliable communication with low complexity [50]. This avoids the use of the complex MLSE or linear DFE receiver. Besides, the OFDM system utilizes a prefix to obtain the optimal diversity-multiplexing tradeoff as the block length tends to infinity, and it also requires an additional linear encoding architecture at the transmitter.

The diversity-multiplexing tradeoff for a V-BLAST OSIC receiver with ZF or MMSE processing at each stage is analyzed in [161]. It is verified that under general settings the optimal ordering rule for a V-BLAST SIC receiver will not improve its performance regarding diversity-multiplexing tradeoff in point-to-point channels. Particularly, when the rates of data streams are fixed, the diversity order is not improved by user ordering.

Parallel and MISO Rayleigh channel

The optimal diversity-multiplexing tradeoff for the parallel channel with L diversity branches or the MISO channel with $N_t = L$ can be derived as [144]

$$
G_d = L(1 - G_{sm}), \quad G_{sm} \in [0, 1] \quad \text{(channel)}. \tag{19.122}
$$

When transmitting the same QAM symbol over the L subchannels, the tradeoff is given by [144]

$$
G_d = L(1 - L G_{sm}), \quad G_{sm} \in \left[0, \frac{1}{L} \right] \quad \text{(QAM, repetition)}. \tag{19.123}
$$

MIMO Rayleigh channel

Based on the definitions of G_{sm} and G_d, a tradeoff between G_{sm} and G_d has been proposed in [163]. For a MIMO system with i.i.d. Rayleigh fading channels, if the space-time code matrix has a time span $N_x \geq N_t + N_r - 1$, the optimal diversity achieved over all schemes for a given G_{sm} is given by [163]

$$G_d(G_{sm}) = (N_t - G_{sm})(N_r - G_{sm}) \quad \text{(channel)} \qquad (19.124)$$

for $G_{sm} = 0, 1, \ldots, \min(N_t, N_r)$. The maximum diversity order $N_t N_r$ and the maximum rate $\min(N_t, N_r)$ cannot be achieved simultaneously [163]. Note that at $G_{sm,max} = \min(N_t, N_r)$, we have $G_{d,min} = 0$. This is unreasonable since no channel provides zero diversity. This problem arises from the definition of the metric G_{sm}, which has no physical interpretation over finite SNR. At the other extreme, zero spatial multiplexing gain is achieved when G_d reaches it maximum $N_t N_r$.

The diversity-multiplexing tradeoff for many space-time coding/spatial multiplexing structures with different receivers (e.g. ZF, MMSE, ML, SIC) are described in [99]. The diversity-multiplexing tradeoff and outage performance for Ricean MIMO channels is analyzed in [126]. The diversity-multiplexing tradeoff characteristics of Rayleigh and Ricean channels are shown to be identical. In a high SNR regime, the outage probability versus SNR curve for a Ricean channel is a shifted version of that for the corresponding Rayleigh channel.

Diversity of the space-time-frequency selective fading channel

The diversity of a space-time-frequency selective fading channel is determined by the codeword dimensions and coherent parameters. Given a codeword duration T, the available diversity can be up to T/T_c; for a bandwidth B, the independent frequency diversity branches can be up to B/B_c. For space diversity, the number of antennas N_t and N_r and their topology determine the diversity orders to be N_t/P_t and N_r/P_r, P_t and P_r being, respectively, the packing factors of the receive and transmit arrays, where the packing factor is the number of coherent distances occupied by at least one antenna. To achieve the maximum space diversity, the antenna arrays at both the receiver and the transmitter should be spaced D_c apart. The maximum available diversity is thus given by [103, 118]

$$G_d = \frac{T}{T_c} \cdot \frac{B}{B_c} \cdot \frac{N_t}{P_t} \cdot \frac{N_r}{P_r}. \qquad (19.125)$$

19.7.2 Error probabilities for MIMO systems

SIMO

For a SIMO channel, assuming a rich-scattering environment and the separation between the antennas at the receiver to be greater than the coherent distance D_c, the average probability of symbol error in Rayleigh fading for MRC is given by [103]

$$\overline{P}_e \leq \overline{N}_e \left(1 + \frac{\gamma d_{\min}^2}{4}\right)^{-N_r} \approx \overline{N}_e \left(\frac{\gamma d_{\min}^2}{4}\right)^{-N_r}, \tag{19.126}$$

where $\overline{N}_e$ and $d_{\min}$ are the number of nearest neighbors and the minimum distance of separation of the constellation, respectively. The approximation on the RHS is for high SNR γ. The method achieves full diversity gain N_r and array gain N_r, but may have limitations when it is deployed in MSs.

MISO with no CSI at the transmitter

For a MISO or MIMO channel, transmit antenna diversity can be used. When the channel is unknown to the transmitter, the Alamouti scheme [5] can be used for a two transmit antenna MISO or MIMO system; in this case, at high SNR, $\overline{P}_e$ is upper-bounded by [103]

$$\overline{P}_e \leq \overline{N}_e \left(\frac{\gamma d_{\min}^2}{8}\right)^{-2}. \tag{19.127}$$

This achieves a full diversity order of 2. Space-time spreading is similar to the Alamouti scheme [58], and is used in CDMA2000. Due to no CSI available at the transmitter, array gain is not allowed at the transmitter. From (19.126), if $N_r = 2$, for the same P_e, SNR for transmit diversity is twice, or 3 dB above, that for receive diversity, since the array gain of 2 is achieved in the receive diversity scheme.

MISO with CSI at the transmitter

For a MISO with known channel at the transmitter, the signal at the transmitter can be first combined by a beamforming weight w so that the SNR is maximized at the receiver. The beamforming weight vector w has the same value as the weight vector for MRC diversity combining, and so this scheme is known as *transmit-MRC* [81]. In a rich-scattering environment, the average probability of symbol errors at high SNR is upper-bounded by

$$\overline{P}_e \leq \overline{N}_e \left(\frac{\gamma d_{\min}^2}{4}\right)^{-N_t}. \tag{19.128}$$

The MISO system using transmit-MRC has the same error performance as the SIMO system using receive-MRC.

MIMO with no CSI at the transmitter

For MIMO with no CSI at the transmitter, the Alamouti scheme can be used. For a two-transmit two-receive MIMO, the average symbol error probability at high SNR is upper-bounded by [103]

$$\overline{P}_e \leq \overline{N}_e \left(\frac{\gamma d_{\min}^2}{8}\right)^{-4}. \tag{19.129}$$

Table 19.1. Array gain and diversity order for multiple antenna systems.

Configuration	CSI at transmitter	Array gain	Diversity order
SIMO	yes or no	N_r	N_r
MISO, Alamouti/OSTBC	no	1	N_t
MISO, transmit-MRC	yes	N_t	N_t
MIMO, Alamouti/OSTBC	no	N_r	$N_t N_r$
MIMO, Dominant eigenmode	yes	$E[\lambda_{\max}]$	$N_r N_t$

Thus, a full diversity order of $N_t N_r = 4$ is achieved. Only receive array gain of $N_r = 2$ is achieved. The Alamouti scheme can be used for a MIMO system with two transmit antennas and any N_r receive antennas; this leads to a diversity order of $2N_r$ and an array gain of N_r.

MIMO with CSI at the transmitter

When CSI is known at the transmitter, spatial diversity of the MIMO system can be extracted through the dominant-eigenmode transmission technique. The array gain in this case is given by $E[\lambda_{\max}]$, where $\lambda_{\max} = \sigma_{\max}^2$ is the maximum eigenvalue of $\mathbf{HH}^H$. In the rich-scattering MIMO, [103]

$$\max(N_t, N_r) \leq \lambda_{\max} \leq N_t N_r. \tag{19.130}$$

Thus, the array gain is greater than or equal to that for no CSI at the transmitter.

Accordingly, by approximating the ML detection error in the Q-function by using the Chernoff bound, $Q(x) \leq e^{-x^2/2}$, which is a closed bound for large x, the SER at high SNR can be derived as [103]

$$\overline{N}_e \left(\frac{\gamma d_{\min}^2}{4} \right)^{-N_t N_r} \leq \overline{P}_e \leq \overline{N}_e \left(\frac{\gamma d_{\min}^2}{4 \min(N_t, N_r)} \right)^{-N_t N_r}. \tag{19.131}$$

In summary, the diversity order and the array gain that are achievable for different multiple antenna systems are summarized in Table 19.1.

Example 19.4: A comparison of SER for the above schemes is given in Fig. 19.11, for different combinations of (N_t, N_r), where BPSK modulation is employed. From the figure, it is seen that the transmit-MRC and the receive-MRC achieve the same result for $N_t = N_r$, and that the transmit-MRC achieves a gain of 3 dB compared to the Alamouti scheme for transmit diversity. For MIMO, although the Alamouti scheme and the dominant-eigenmode transmission achieve the same SER upper bound when $N_t = N_r = 2$, the dominant-eigenmode transmission achieves an array gain that is greater than or equal to that for the Alamouti scheme.

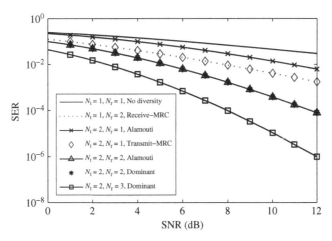

Figure 19.11   Comparison of various diversity schemes.

In addition, a MIMO system in a double-scattering channel with N_s effective scatters has been shown to achieve a diversity order of $N_t N_s N_r / \max(N_t, N_s, N_r)$ by using OSTBCs [125]. This implies that the full spatial order diversity of $N_t N_r$ can be achieved when $N_s \geq \max\{N_t, N_r\}$.

MIMO with imperfect CSI at the transmitter

In [9], error performance has been analyzed for a MIMO ZF receiver with power adaptation at the transmitter in the presence of both imperfect CSI and feedback delay over an i.i.d. Rayleigh flat-fading channel. The power allocation strategy is to maximize a weighted sum of the SNRs of all the substreams subject to a total transmit power constraint. Closed-form approximate upper bounds on BER are derived for MPSK and MQAM. These upper bounds are functions of the channel estimation error, the feedback delay, the constellation size, and the receive diversity order. The presence of channel estimation error and/or feedback delay cause an error floor, as the noise variance approaches zero.

Under the assumptions of equal power allocation at the transmitter and Gaussian distributed CCI, exact closed-form expressions for the BER of the SVD-based MIMO system using MPSK and MQAM with optimal detection in the presence of imperfect CSI at the transmitter are given in [8]. Similarly, imperfect CSI introduces CCI, and hence an error floor of BER.

19.7.3 MIMO beamforming

When multiple antennas are used to obtain beamforming or diversity gain instead of capacity gain, the same symbol, which is weighted by a complex scale factor, is sent over each transmit antenna, thereby leading to an input covariance matrix with unit rank. This is known as *MIMO beamforming* or *transmit beamforming*. By transmitting in the direction of the eigenvector corresponding to the largest eigenvalue of the channel, the output SNR after MRC at the receiver is maximized. The full diversity gain provided by a MIMO channel can be achieved by transmit beamforming and receive combining.

The beamforming strategy corresponds to transmitter precoding and receiver shaping with $\mathbf{V} = v$ and $\mathbf{U} = u$, where u and v are the normalized transmit and receive weight vectors, $\|u\| = \|v\| = 1$. The received signal is given by [47]

$$y = u^H \mathbf{H} v x + u^H n. \tag{19.132}$$

When $\mathbf{H}$ is known, u and v can be selected as the first columns of $\mathbf{U}$ and $\mathbf{V}$, which correspond to the maximum singular value of $\mathbf{H}$, $\sigma_1 = \sigma_{\max}$. In this case, the capacity corresponds to that of a scalar channel with channel power gain $\sigma_{\max}^2$

$$C = B \log_2 \left(1 + \sigma_{\max}^2 P / \sigma_n^2\right). \tag{19.133}$$

In this case, the array gain of beamforming is between $\max(M_t, M_r)$ and $M_r M_t$, and the diversity gain is $M_t M_r$ [47].

The performance of BER and outage probability of transmit beamforming is analyzed in [49] for uncoded binary transmission over MIMO channels for flat Rayleigh fading channels. In [90], closed-form expressions have been derived for the outage probability and ergodic capacity of transmit-beamforming MIMO MRC systems under Rayleigh fading.

Implementation of transmit beamforming requires CSI at the transmitter. One solution is quantized beamforming. Instead of sending the quantized CSI to the transmitter, the receiver quantizes the beamforming vector using a fixed codebook available at both the transmitter and the receiver. The quantized index is then sent to the transmitter. Systematic codebook design has been performed for the uncorrelated Rayleigh fading channels in [83, 95] and for correlated Rayleigh fading channels in [108].

WCDMA contains explicit support for closed-loop transmit diversity. The CSI is estimated from the two common pilot channels (CPICHs), and is then fed back to the transmitter to control the beamforming weights at different transmitting antennas. Even a crude feedback signaling can be extremely useful in improving the downlink performance. When CSI is not available at the transmitter, the Alamouti or STBC scheme can be used to obtain full diversity gain and array gain.

19.8 MIMO for frequency- or time-selective fading channels

For frequency-selective fading MIMO channels, ISI occurs. Space-time channel equalizers can be used, but the complexity is too high. A more popular method is to use the OFDM technique to convert the frequency-selective fading MIMO channel into many flat-fading channels. This leads to a parallel set of narrowband MIMO channels.

The capacity of a frequency-selective fading MIMO channel can be calculated by dividing the band into multiple narrow, frequency-flat subchannels, and then summing the capacities of the frequency-flat subchannels. This has been discussed in Section 14.6.3.

19.8.1 MIMO-SC

For a MIMO-SC (MIMO single-carrier) system with a frequency-selective but time-flat fading channel, the channel input-output relationship is given by

$$y(k) = \sqrt{\frac{E_s}{N_t}} \sum_{l=0}^{L} \mathbf{H}(l) x(k-l) + n(k), \tag{19.134}$$

where L is the channel order, $y(k) \in \mathcal{C}^{N_r}$, $x(k) \in \mathcal{C}^{N_t}$, and $w(k) \in \mathcal{C}^{N_r}$ is an AWGN vector, and for each tap l, $\mathbf{H}(l) = [h_{ij}(l)] \in \mathcal{C}^{N_r \times N_t}$. When $L = 0$, it reduces to the flat MIMO model.

For a group of N_x vectors, we have

$$\mathbf{X} = [x(0) | x(1) | \ldots x(N_x - 1)] \in \mathcal{C}^{N_t \times N_x}. \tag{19.135}$$

Define a group of $N_t \times (N_x + L)$ matrices

$$\left. \begin{aligned} \mathbf{X}^{(0)} &= [\mathbf{X}, \mathbf{0}_{N_t \times L}] \\ \mathbf{X}^{(1)} &= [\mathbf{0}_{N_t \times 1}, \mathbf{X}, \mathbf{0}_{N_t \times (L-1)}] \\ &\vdots \\ \mathbf{X}^{(L)} &= [\mathbf{0}_{N_t \times L}, \mathbf{X}] \end{aligned} \right\}. \tag{19.136}$$

The received space-time matrix is given by

$$\mathbf{Y} = [y(0) | \cdots | y(N_x + L)] = \sqrt{\frac{E_s}{N_t}} \sum_{l=0}^{L} \mathbf{H}(l) \mathbf{X}^{(l)} + \mathbf{N}, \tag{19.137}$$

where $\mathbf{N}$ is obtained in the same manner as $\mathbf{Y}$.

Equation (19.137) can be rewritten as

$$\mathbf{Y} = \sqrt{\frac{E_s}{N_t}} \mathbf{H}_{eq} \mathbf{X}_{eq} + \mathbf{N}, \tag{19.138}$$

where

$$\mathbf{H}_{eq} = [\mathbf{H}(0) \cdots \mathbf{H}(L)], \quad \mathbf{X}_{eq} = \begin{bmatrix} \mathbf{X}^{(0)} \\ \mathbf{X}^{(1)} \\ \vdots \\ \mathbf{X}^{(L)} \end{bmatrix}. \tag{19.139}$$

Thus, the space-time code matrix $\mathbf{X}_{eq}$ can be viewed as being transmitted from $N_t(L + 1)$ antennas over a flat fading channel. In rich-scattering environments, the maximum diversity achievable has been proved to be $N_t N_r(L + 1)$, for a suitably designed space-time code matrix $\mathbf{X}$ [44]. The space-time code should be carefully designed to achieve the maximum diversity. The Alamouti scheme has been extended to the MIMO-SC case with $N_t = 2$ [78].

SC-FDE systems equipped with multiple receive antennas have been discussed in [25] using either the LMS or RLS algorithm. Two hybrid time-frequency DFE receivers for MIMO-SC systems, namely, the MIMO-DFE and the layered space-time (LST) DFE (LST-DFE), have been introduced in [64]. LST-DFE generally outperforms the MIMO-DFE.

A comparison with different MIMO-OFDM receivers shows that the MIMO-SC receivers are superior to the MIMO-OFDM receivers in terms of BER and throughput, but at a higher complexity [64].

MIMO-CDMA

MIMO can be overlaid on the spread-spectrum model, yielding MIMO-SS (MIMO spread spectrum) modulation or MIMO-CDMA. MIMO-SS for diversity transmission now becomes part of the WCDMA standard, where the BS supports Alamouti scheme-based transmit diversity [58]. A MIMO space-division, code-division multiple-access (MIMO-SCDMA) scheme, where each user is distinguished jointly by its spreading code signature and its unique spatial signature, has been investigated in [158].

For a single-user MIMO channel, coordination among all the transmitters or receivers is not a problem. In the multi-user channel, there is no coordination between users. In the uplink, the BS needs to separate the signals from different users. This can be done by using a rake receiver or by using MUD. In the downlink, the BS can synchronize the data transmission for all the users, and the MS can use MUD to overcome MAI. A cheaper solution is to mitigate MAI at the transmitter by beamforming, based on CSI available at the transmitter. The rake receiver is also generalized as the ST-rake or a 2-D rake for the SIMO system, and the transmit analog of the ST-rake is a ST pre-rake [35].

Popular space-time processing receivers include the ST-MMSE and ST-ML receivers. The ST-ML receiver mitigates CCI and ISI, with the best performance, but it has a high complexity and also requires the knowledge of the interfering channels. The ST-ML provides a space diversity order of N_r and a multipath diversity order of L, and thus a total diversity order of N_rL. The ST-MMSE receiver has a diversity order that is slightly less than N_rL due to the loss arising from CCI.

MIMO-UWB

The UWB applications are typically in a rich-scattering indoor environment, which is especially suitable for MIMO implementation. The GHz center frequency of the UWB signal enables a compact antenna array. Thus, the combination of UWB and MIMO provides a viable and inexpensive solution to very high data rate for future-generation wireless PANs.

Space-time-coded pulsed UWB systems with TH-BPPM, TH-BPSK, and DS-BPSK are described in [129, 130]. Simulation results show that a DS space-time system can support the same number of users, but with a lower BER than that of the TH system [130]. The block diagram of a MIMO-CDMA or a multiuser MIMO-UWB system is illustrated in Fig. 19.12.

19.8.2 MIMO-OFDM

A frequency-selective MIMO channel has to use an ML or a suboptimal equalization, which has a complexity that grows exponentially with the product of the bandwidth

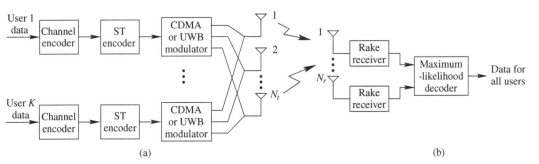

Figure 19.12 Block diagram of an $N_t \times N_r$ MIMO-CDMA or multiuser MIMO-UWB system. (a) Transmitter. (b) Receiver.

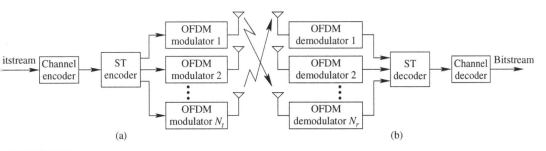

Figure 19.13 Block diagram of an $N_t \times N_r$ MIMO-OFDM system. (a) Transmitter. (b) Receiver.

and the delay spread. OFDM modulation can transform a frequency-selective fading channel of bandwidth B into N orthogonal flat fading channels, which can be efficiently calculated using IFFT. MIMO-OFDM is based on the same idea: it converts a frequency-selective MIMO channel into multiple flat fading MIMO channels using the OFDM technique. MIMO-OFDM is preferred over MIMO-SC in recent wireless communication standards, such as HSPA+, 3GPP LTE, 3GPP2 UMB, IEEE 802.16e/m, and IEEE 802.11n.

As in MC-CDMA, where CDMA is overlaid on OFDM, MIMO can also easily be overlaid on OFDM for diversity transmission or spatial multiplexing [15]. Unlike the single antenna OFDM case, the MIMO delay-spread channel, in general, provides both a higher diversity gain and a higher multiplexing gain than the MIMO flat-fading channel does [15]. The bitstream is first demultiplexed into a few substreams. Each of the substreams is then further OFDM-modulated and transmitted from an antenna. This procedure is reversed at the receiver. The resulting MIMO-OFDM system is shown in Fig. 19.13. A MIMO-OFDM receiver must perform time synchronization, frequency offset estimation and correction, and parameter estimation, by using a preamble consisting of one or more training sequences. These have been discussed in detail in [134].

MIMO-OFDM can be implemented as space-time coded OFDM (ST-OFDM), space-frequency coded OFDM (SF-OFDM), and space-time-frequency coded OFDM (STF-OFDM). ST-OFDM enables only a space diversity of $N_t N_r$, while SF-OFDM and

STF-OFDM can usually achieve the maximum diversity order of $(L+1)N_tN_r$, L being the number of multipaths [44, 130]. For this reason, the SF-OFDM is the most popular coding technique in MIMO-OFDM. Note that, for SF-OFDM, the maximum achievable diversity is derived as $G_d = \min((L+1)N_tN_r, NN_r) = (L+1)N_tN_r$, where N is the number of OFDM subcarriers and typically $N > (L+1)N_t$ [130].

If space-time coding is employed in OFDM, the STBC is implemented over two OFDM symbols. Since the OFDM symbol duration is quite long, the channel should be constant over the subsequent OFDM symbols. For the space-frequency block code (SFBC) [14], adjacent subcarriers are coded over in an identical way as for the STBC. This assumes the adjacent subcarriers to have the same amplitude and phase, which is approximately true in practice. SFBC is used in 3GPP LTE.

For the full MIMO-OFDM system, the LDPC-based space-time coding significantly outperforms the STTC in system performance and outperforms the turbo-code-based space-time coding scheme by exhibiting a lower receiver complexity and a more flexible scalability [89]. The optimal power and subchannel allocation for MIMO-OFDMA and MIMO-MC-CDMA has been discussed in [82].

For the MIMO MB-OFDM UWB scheme, due to the band hopping, K OFDM symbols in each STF codeword are sent over different sub-bands. The maximum achievable diversity is K times the case without the band hopping [130]

$$G_d = \min(KLN_tN_r, KNN_r). \tag{19.140}$$

An improved V-BLAST receiver which takes the decision errors into account was introduced in [75], based on the MMSE criterion. An iterative detection and decoding scheme for coded, layered space-time architectures in MIMO-OFDM systems is presented in [75]. The iterative scheme combined with the improved V-BLAST performs almost as well as the near-optimal turbo-MIMO approach does, while providing tremendous savings in complexity.

Blind channel estimation algorithms have been proposed for estimating MIMO-OFDM channels using the cyclostationary statistics [16] or the constant-modulus property [80]. In [123], a subspace-based blind channel estimation technique that exploits virtual carriers with no or insufficient cyclic prefix has been proposed for MIMO-OFDM. Training-based semiblind algorithms for MIMO-OFDM channel estimation can be found in [149, 151]. In [44], a hopping-pilot-based method and a kurtosis-based blind method are described for frequency-offset estimation of MIMO-OFDM.

AMC for MIMO-OFDM systems in a slow fading channel was investigated in [137]. Based on the EXIT analysis, an accurate packet error-rate prediction with channel estimation errors is given. The link adaptation algorithm is based on searching and selecting the best modulation and coding scheme. The performance of the proposed link adaptation algorithm was evaluated for the IEEE 802.11n MIMO-OFDM system, and good system throughput is achieved compared to the SNR-based algorithm.

Based on the IEEE 802.11a standard, an FPGA prototype for a full 4-stream MIMO-OFDM transceiver is implemented in [52]. It is capable of transmitting 216 Mbits/s in 20-MHz bandwidth, which is four times the peak data rate of 54 Mbits/s for IEEE 802.11a.

In [13], an efficient MMSE iterative receiver for 4×4 MIMO-OFDM systems has been integrated in a FPGA testbench.

19.8.3 MIMO for time-selective channels

In case of time-varying MIMO channels, the space-time Doppler (STDO) code is designed to provide joint STDO diversity. Based on the time-frequency duality, the space-time code for the time-selective or doubly-selective MIMO channel can be obtained by mapping from the corresponding space-time code designed for the frequency-selective MIMO channel [44].

In the deterministic basis expansion model, the doubly-selective fading channel is given by [44]

$$h(n; l) = \sum_{q=0}^{Q} h_q(l) e^{j\omega_q n}, \quad l = 0, 1, \ldots, L, \tag{19.141}$$

where $\omega_q = 2\pi \left(q - \frac{Q}{2} \right) / N$, during each block of N symbols. For the time-selective, frequency-flat fading channel, $L = 0$.

For the time-selective fading channel with $Q + 1$ bases, if the channel coefficients h_q are complex Gaussian distributed, the maximum diversity order is $Q + 1$. For the doubly-selective channel with Gaussian distributed channel coefficients $h_q(l)$, the maximum diversity order is $(L + 1)(Q + 1)$. STDO code design is described in detail in [44].

19.9 Space-time processing

We first describe the space-time channel model. Assume that there is an $N_t \times N_r$ MIMO system. For transmit antenna i, the N_r-dimensional vector representing the complex baseband outputs of the receive array is

$$r_i(t) = \sum_l s_{il} h_i(t - lT) + n_i(t), \tag{19.142}$$

where s_{il} is the complex-valued data symbol transmitted at time lT from the transmit antenna i, $h_i(t) = \left(h_{1i}, \ldots, h_{N_r i} \right)^T$ is the impulse response from the transmit antenna i to the receive array, and $n_i(t)$ is a vector representing the additive noise. $h_i(t)$ is also called the *spatial-temporal signature* of the radio link from the transmitter i to the receive array.

For N_t transmitters

$$r(t) = \sum_{i=1}^{N_t} r_i = \sum_l \mathbf{H}(t - lT) s_l + n(t), \tag{19.143}$$

where $\mathbf{H}(t) = \left[h_1(t) | h_1(t) | \cdots | h_{N_t}(t) \right]$ is the $N_r \times N_t$ channel impulse response matrix, $s_l = \left(s_{1l}, \ldots, s_{N_t l} \right)^T$, and $n_i(t)$ is the corresponding AWGN vector.

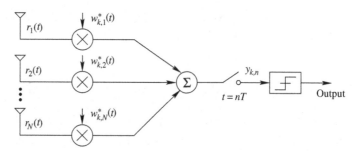

Architecture of a general linear space-time receiver.

19.9.1 Linear space-time processing model

Space-time processing combines the merits of both the spatial and temporal diversities. Temporal processing uses temporal taps for ISI mitigation, but it cannot provide a gain against CCI or thermal noise. Spatial processing uses spatial taps (antennas) for CCI mitigation, and can also eliminate MAI and noise. Space-time processing can use the spatial diversity to suppress CCI signals with DoAs different from that of a desired user, and use temporal diversity to remove CCI with a DoA which is the same as that of the desired signal.

A general linear space-time processor has an architecture as shown in Fig. 19.14. The space-time processor first processes the receive-antenna measurements $r(t)$ by convolving with linear filters represented by a vector impulse response of $w_k(t) = \left(w_{k1}(t), \ldots, w_{kN_r}(t)\right)^T$. The output of the combiner for user k, $y_k(t)$, is sampled, and decision making is performed to obtain the original symbols. The sampled output for user k at time $t = nT$ is given by

$$y_{k,n} = \int_{-\infty}^{\infty} w_k(\tau)^H r(nT - \tau) d\tau, \tag{19.144}$$

where $r(t) = \left(r_1, r_2, \ldots, r_{N_r}\right)^T$,

$$r_j = \sum_{i=0}^{N_t} s_i(t) * h_{ji}(t) + n_j(t), \tag{19.145}$$

where $s_i(t)$ is the complex data symbol transmitted at time t from antenna $i, i = 1, 2, \ldots, N_r$, and $n_j(t)$ is the additive noise on receive antenna j.

19.9.2 ZF and MMSE receivers

ZF receiver

To obtain the original symbols, $y_{k,n}$ should contain no CCI and ISI. This is the ZF criterion. Based on the ZF criterion, a relation between $w_k(t)$ and $\mathbf{H}(t)$ is established

$$\int_{-\infty}^{\infty} \boldsymbol{w}_k^H(\tau)\boldsymbol{H}(nT - \tau)d\tau = \begin{cases} \boldsymbol{e}_k, & n = 0 \\ \boldsymbol{0}, & n \neq 0 \end{cases}, \tag{19.146}$$

where $\boldsymbol{e}_k$ is a vector with the kth entry as 1 and all other entries as 0, and $\boldsymbol{0}$ is an all-zeros vector. By using the Fourier transform, the relation in the frequency domain can be established as a set of linear equations. This method can completely suppress ISI and CCI up to the limits imposed by the bandwidth, the number of antenna elements, and the number of interfering signals. The ZF receiver suffers from noise enhancement and adaptation problems [36].

MMSE receiver

Space-time processing can be based on the ST-MLSE or ST-MMSE method, depending on its optimization criterion. For the TDMA system, ST-MMSE is more suitable for eliminating CCI, while ST-MLSE is well suited when ISI is more severe than CCI.

The MMSE criterion is suitable for adaptation and evaluation at each step. The MSE for user k is given by

$$J_{\text{MSE}} = E\left[|y_{k,n} - s_{k,n}|^2\right], \tag{19.147}$$

where $s_{k,n}$ is the complex data symbol of user k at $t = nT$.

Practical linear equalizers based on the transversal filter that follows fixed antialiasing filters operate at a sampling rate that equals the symbol rate or a multiple of the symbol rate. For N_r antennas and user k, the equalizer with $2M + 1$ taps per antenna element is given by

$$\boldsymbol{w}_k = \begin{bmatrix} \boldsymbol{w}_{k,-M} \\ \vdots \\ \boldsymbol{w}_{k,M} \end{bmatrix}. \tag{19.148}$$

Thus, $\boldsymbol{w}_k$ is a $(2M + 1)N_r$-dimensional vector.

For a sampling rate of $1/\Delta$, the linear space-time combiner output for user k is given by

$$y_k(iT) = \boldsymbol{w}_k^H \tilde{\boldsymbol{r}}_i, \tag{19.149}$$

where the channel output sample vector

$$\tilde{\boldsymbol{r}}_i = \begin{bmatrix} \boldsymbol{r}(iT + M\Delta) \\ \vdots \\ \boldsymbol{r}(iT - M\Delta) \end{bmatrix} \tag{19.150}$$

is a $(2M + 1)N_r$-dimensional vector.

The MMSE solution for user k is given by [36]

$$\boldsymbol{w}_{k,\text{MMSE}} = \mathbf{R}^{-1}\boldsymbol{v}, \tag{19.151}$$

where

$$\mathbf{R} = E\left[\tilde{\boldsymbol{r}}_i \tilde{\boldsymbol{r}}_i^H\right], \quad \boldsymbol{v} = E\left[\tilde{\boldsymbol{r}}_i s_{k,i}^*\right]. \tag{19.152}$$

The MMSE is given by

$$J_{\text{MMSE}} = 1 - v^H R^{-1} v. \tag{19.153}$$

At the MMSE, the SINR is maximized

$$\text{SINR}_{\text{max}} = \frac{1 - J_{\text{MMSE}}}{J_{\text{MMSE}}}. \tag{19.154}$$

A simple and useful Chernoff-type upper bound on the symbol error rate is given by [36, 155]

$$P_s \leq e^{-\frac{1}{J_{\text{MMSE}}}}, \tag{19.155}$$

where J_{MMSE}, given by (19.153), applies for unit data symbol variance.

19.10 Space-time processing for CDMA systems

Combination of the MUD technique with antenna arrays represents the direction for future-generation CDMA mobile communications. Space-time MUD (ST-MUD) algorithms for CDMA systems have been proposed in a number of papers [30, 32, 102, 111, 152]. A combination of an adaptive antenna array in the spatial domain and an adaptive interference canceller in the temporal domain are also given in [71, 72]. In [17], some algorithms that combine multiantenna beamforming and temporal processing have been reviewed. A practical space-time receiver for CDMA is the 2-D rake receiver that uses the training-based MMSE beamforming for each path, followed by a conventional rake receiver. Blind adaptive algorithms for ST-MUD have also been given in [102, 111, 152].

MMSE-based ST-MUD can significantly increase the capacity of the CDMA system. The method is based on the MMSE between the data stream and the linear combiner output. Several MMSE-based ST-MUD algorithms for DS-CDMA are available [30, 32, 102]; these algorithms have been proved to be equivalent, and thus their theoretical BER performances are the same. Due to different implementation, their performances are actually different. Most remarkably, the ST-MUD algorithm proposed in [30, 32] does not require explicit estimation of channel and signaling information. This avoids any channel estimation error, and the method is thus more robust and provides a performance better than that provided by other ST-MUD algorithms in adaptive implementation. Adaptation of the ST-MUD algorithms is implemented by using training sequences.

19.10.1 Signal model

For a DS-CDMA system with K cochannel users, signals are spread by a random binary spreading sequence of length N and the short code is used, that is, $G_p = N$. There are $L_k, k = 1, \ldots, K$, resolvable multipath components for the received signal of the kth user.

The receiver at the BS is equipped with N_r antenna elements. The received signal at the pth element during a given symbol period can be modeled as

$$\boldsymbol{r}_p = \mathbf{SD}_p\mathbf{A}\boldsymbol{b} + \boldsymbol{n}_p, \qquad (19.156)$$

where $\boldsymbol{r}_p$ is a complex N-dimensional vector, $\mathbf{S}$ is an N-by-$\left(\sum_{k=1}^{K} L_k\right)$ spreading code matrix, $\mathbf{D}_p$ is a $\left(\sum_{k=1}^{K} L_k\right)$-by-$K$ steering matrix for all the users at the pth element, $\mathbf{A} = \mathrm{diag}\,(a_1,\dots,a_K)$ is a real $K \times K$ diagonal matrix of the amplitudes, $\boldsymbol{b} = (b1,\dots,b_K)^T$ is a complex K-dimensional vector of the input data symbols at a given instant t, and $\boldsymbol{n}_p$ is a zero-mean complex Gaussian noise vector with power spectral density σ^2.

The spreading code matrix $\mathbf{S}$ is given by

$$\mathbf{S} = \begin{bmatrix} s_{1,1},\dots,s_{1,L_1}; s_{2,1},\dots,s_{2,L_2};\dots,s_{K,1},\dots,s_{K,L_K} \end{bmatrix}. \qquad (19.157)$$

Each user is assigned a unique normalized spreading code (signature waveform) s_k, an N-dimensional vector, $k = 1,\dots,K$, and $s_{k,l}$, $l = 1,\dots,L_k$, is the delay of s_k corresponding to the lth multipath. Thus, the Euclidean norm

$$\|s_k\| = \|s_{k,l}\| = 1, \quad \text{for all } k,l. \qquad (19.158)$$

The normalized signature waveform of the kth user can be written as

$$s_k(t) = \frac{1}{\sqrt{N}} \sum_{n=0}^{N-1} c_k(n) u_{T_c}\,(t - nT_c), \quad 0 \le t \le T, \qquad (19.159)$$

where the symbol interval $T = NT_c$, T_c is the chip interval, $\{c_k(n)|n = 0,1,N-1\}$ is the spreading sequence for user k, u_{T_c} is a unit-height rectangular waveform of duration T_c. Mathematically,

$$\int_0^T s_k^2(t)\mathrm{d}t = 1. \qquad (19.160)$$

Note that s_k is the discrete-time vector version of $s_k(t)$.

The steering matrix for all the users at the pth element, $\mathbf{D}_p$, is given by

$$\mathbf{D}_p = \mathrm{blockdiag}\left(\begin{bmatrix} d_{p,1,1} \\ \vdots \\ d_{p,1,L_1} \end{bmatrix}, \cdots, \begin{bmatrix} d_{p,K,1} \\ \vdots \\ d_{p,K,L_K} \end{bmatrix}\right), \qquad (19.161)$$

where $d_{p,k,l}$ is the steering response of the kth user to the pth element via the lth multipath component.

For all the array elements, we have

$$\boldsymbol{r} = \tilde{\mathbf{S}}\mathbf{D}\mathbf{A}\boldsymbol{b} + \boldsymbol{n}, \qquad (19.162)$$

where $\boldsymbol{r} = \left(\boldsymbol{r}_1^H,\dots,\boldsymbol{r}_{N_r}^H\right)^H$ is an $N_r N$-dimensional vector,

$$\mathbf{D} = \begin{bmatrix} \mathbf{D}_1^H \,|\, \mathbf{D}_2^H \,|\, \cdots \,|\, \mathbf{D}_{N_r}^H \end{bmatrix}^H \qquad (19.163)$$

is a $\left(\sum_{k=1}^{K} L_k\right) N_r$-by-$K$ steering matrix,

$$\tilde{\mathbf{S}} = \mathbf{I}_{N_r \times N_r} \otimes \mathbf{S} = \text{blockdiag}(\mathbf{S}, \cdots, \mathbf{S})_{N_r \times N_r} \tag{19.164}$$

is an $N_r N$-by-$\left(\sum_{k=1}^{K} L_k\right) N_r$ matrix, $\otimes$ being the Kronecker product, and $\boldsymbol{n} = \left(\boldsymbol{n}_1^H \ldots \boldsymbol{n}_{N_r}^H\right)^H$ is an $N_r N$-dimensional vector.

19.10.2 Space-time detection algorithms

The spreading codes have good mutual orthogonality and autocorrelation properties. By using these properties, we define a sufficient statistic for further processing

$$\boldsymbol{x} = \tilde{\mathbf{S}}^T \boldsymbol{r} = \tilde{\mathbf{S}}^T \tilde{\mathbf{S}} \mathbf{D} \mathbf{A} \boldsymbol{b} + \boldsymbol{n}_x, \tag{19.165}$$

where $\boldsymbol{x} = \left(\boldsymbol{x}_1^H, \boldsymbol{x}_2^H, \ldots, \boldsymbol{x}_{N_r}^H\right)^H$ is a $\left(\sum_{k=1}^{K} L_k\right) N_r$ vector, $\boldsymbol{n}_x = \tilde{\mathbf{S}}^T \boldsymbol{n}$ is the corresponding noise term, and

$$\tilde{\mathbf{S}}^T \tilde{\mathbf{S}} = \text{blockdiag}\left(\mathbf{S}^T \mathbf{S}, \cdots, \mathbf{S}^T \mathbf{S}\right)_{N_r \times N_r} \tag{19.166}$$

is a $\left(\sum_{k=1}^{K} L_k\right) N_r$-by-$\left(\sum_{k=1}^{K} L_k\right) N_r$ matrix, which is close to a diagonal matrix.

Space-time matched filter

The single-user space-time matched filter (ST-MF) is given as [102]

$$\boldsymbol{y} = \mathbf{D}^H \boldsymbol{x} = \mathbf{M} \mathbf{A} \boldsymbol{b} + \boldsymbol{n}_y$$
$$= \mathbf{Q}_y \boldsymbol{b} + \boldsymbol{n}_y, \tag{19.167}$$

where $\boldsymbol{n}_y = \mathbf{D}^H \boldsymbol{n}_x$ is the noise vector, $\boldsymbol{y}$ is a K vector, and $\mathbf{M}$, a K-by-K matrix, is the space-time correlation matrix

$$\mathbf{M} = \mathbf{D}^H \tilde{\mathbf{S}}^T \tilde{\mathbf{S}} \mathbf{D} = \sum_{p=1}^{N_r} \mathbf{D}_p^H \mathbf{S}^T \mathbf{S} \mathbf{D}_p. \tag{19.168}$$

Under strict power control, $\mathbf{M}$ is close to a diagonal matrix, with positive diagonal elements [30]

$$\mathbf{M} \simeq \text{diag}\left(\sum_{p=1}^{N_r} \sum_{l=1}^{L_1} |d_{p,1,l}|^2, \cdots, \sum_{p=1}^{N_r} \sum_{l=1}^{L_K} |d_{p,K,l}|^2\right), \tag{19.169}$$

where the nondiagonal elements of $\mathbf{M}$ are quantities that are small relative to $\|s_{k,l}\|^2 = 1$. Thus,

$$\boldsymbol{y} \simeq \left(\left[a_1 \sum_{p=1}^{P} \sum_{l=1}^{L_1} |d_{p,1,l}|^2\right] b_1, \cdots, \left[a_K \sum_{p=1}^{P} \sum_{l=1}^{L_K} |d_{p,K,l}|^2\right] b_K\right)^T. \tag{19.170}$$

In this case, the ST-MF detector gives a satisfactory result.

The output of the ST-MF detector can be demodulated according to the signal constellation. For BPSK signals, the demodulated symbols are given by

$$\hat{b} = \text{sgn}(\Re(y)), \tag{19.171}$$

where sgn($\cdot$) is the signum function. MAI is not accounted for by the ST-MF detector.

ST-MUD algorithm I

Both x in (19.165) and the ST-MF output y in (19.167) can be used as sufficient statistics for MUD. When applying a linear transformation to y and then minimizing the MSE between the resulting vector and the data vector b, we have an MMSE-based ST-MUD that satisfies [102]

$$\mathbf{U} = \arg\min_{\mathbf{U}} \left\{ E\left[\left\| \mathbf{U}^H y - b \right\|^2 \right] \right\}, \tag{19.172}$$

where $E[\cdot]$ is the expectation operator. The standard Wiener solution is given as [102]

$$\mathbf{U}^H = \mathbf{AM} \left(\mathbf{MA}^2\mathbf{M} + \sigma^2\mathbf{M} \right)^{-1}, \tag{19.173}$$

where $\mathbf{U}^H$ is a $K \times K$ matrix. We denote it as ST-MUD-I.

The MUD output is given as

$$z = \mathbf{U}^H y = \left(\mathbf{U}^H \mathbf{MA} \right) b + \mathbf{U}^H n_y. \tag{19.174}$$

ST-MUD algorithm II

When x is used to synthesize the multiuser detector, the MMSE-based ST-MUD is defined by [102]

$$\mathbf{V} = \arg\min_{\mathbf{V}} \left\{ E\left[\left\| \mathbf{V}^H x - b \right\|^2 \right] \right\}. \tag{19.175}$$

Likewise, we have [102],

$$\mathbf{V}^H = \mathbf{AD}^H \left(\tilde{\mathbf{S}}^T \tilde{\mathbf{S}} \mathbf{DA}^2 \mathbf{D}^H + \sigma^2\mathbf{I} \right)^{-1}, \tag{19.176}$$

where $\mathbf{V}^H$ is a K-by-$\left(\sum_{k=1}^K L_k \right) N_r$ matrix and $\mathbf{I}$ is the $\left(\sum_{k=1}^K L_k \right) N_r$-by-$\left(\sum_{k=1}^K L_k \right) N_r$ identity matrix. We identify it as ST-MUD-II.

The MUD output for demodulating the data bits of the K users is given as

$$z = \mathbf{V}^H x = \left(\mathbf{V}^H \tilde{\mathbf{S}}^T \tilde{\mathbf{S}} \mathbf{DA} \right) b + \mathbf{V}^H n_x. \tag{19.177}$$

ST-MUD algorithm III

If we use r to synthesize the multiuser detector, the MMSE-based ST-MUD is defined by [30, 32]

$$\mathbf{W} = \arg\min_{\mathbf{W}} \left\{ E\left[\left\| \mathbf{W}^H r - b \right\|^2 \right] \right\}. \tag{19.178}$$

The standard Wiener solution is derived as

$$\mathbf{W}^H = E\left[\boldsymbol{b}\boldsymbol{r}^H\right]\left(E\left[\boldsymbol{r}\boldsymbol{r}^H\right]\right)^{-1}$$
$$= \mathbf{A}\mathbf{D}^H\tilde{\mathbf{S}}^T\left(\tilde{\mathbf{S}}\mathbf{D}\mathbf{A}^2\mathbf{D}^H\tilde{\mathbf{S}}^T + \sigma^2\mathbf{I}\right)^{-1}, \tag{19.179}$$

where $\mathbf{W}^H$ is a K-by-PN matrix, and $\mathbf{I}$ is the PN-by-PN identity matrix. We denote this as ST-MUD-III.

The MUD output is given as

$$\boldsymbol{z} = \mathbf{W}^H\boldsymbol{r} = \left(\mathbf{W}^H\tilde{\mathbf{S}}\mathbf{D}\mathbf{A}\right)\boldsymbol{b} + \mathbf{W}^H\boldsymbol{n}. \tag{19.180}$$

Architectures for ST-MUD-I/II/III

The architectures of the three ST-MUD algorithms are depicted in Fig. 19.15. Note that between $\boldsymbol{r}_p$ and $\boldsymbol{x}_p$, there are a bank of matched filters, each having a time delay τ_{k,l_k}, $k = 1, \ldots, K$, $l_k = 1, \ldots, L_k$. The correlation $< x, y > = \int_0^T x(t)y(t)\mathrm{d}t$, or denotes the inner product for vectors.

The MUD output $\boldsymbol{z}$ for ST-MUD-I/II/III, given by (19.174), (19.177), and (19.180), can be expressed as

$$\boldsymbol{z} = \mathbf{Q}_z\boldsymbol{b} + \boldsymbol{n}_z, \tag{19.181}$$

where $\mathbf{Q}_z$ is the $K \times K$ transfer matrix between the transmitted data $\boldsymbol{b}$ and the detected data $\boldsymbol{z}$, and $\boldsymbol{n}_z$ is the noise.

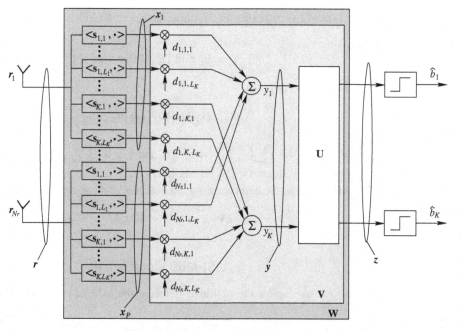

Figure 19.15   Architecture of the ST-MUD algorithms.

Symbol estimate can be made from z

$$\hat{b} = \mathrm{dec}(z) = [\mathrm{dec}\,(z_1), \cdots, \mathrm{dec}\,(z_K)]^T, \tag{19.182}$$

that is, $\hat{b}$ contains the alphabet symbols closest to the entries of z. For BPSK signals, bit estimate is given by a hard limiter

$$\hat{b} = \mathrm{sgn}(\Re(z)). \tag{19.183}$$

ST-MUD-III does not require any information on the signaling waves $\tilde{S}$ and the channel D. ST-MUD-II does not require channel information but needs to know the signaling waves $\tilde{S}$. ST-MUD-I requires both the signaling waves $\tilde{S}$ and the channel D. Although in CDMA systems, the signaling waveforms of the users, S, are usually available at the receiver, estimation of multipaths and multipath delays is still difficult. ST-MUD-III provides an attractive way to circumvent such limitations. Explicit estimation of the channel and the signaling waves is not necessary, since it is implicit in the training sequences or the properties of signals. Thus, ST-MUD-III has a BER performance better than that of ST-MUD-I/II in practical implementation, since it avoids the computation load as well as the error arising from estimating the signaling waves and the channel. Numerical results have shown that ST-MUD-III provides a BER performance superior to that of ST-MUD-I/II and ST-MF [32].

Equivalence of ST-MUD-I/II/III

The ST-MUD-I/II/III algorithms are equivalent to one another, for both asynchronous and synchronous CDMA systems, and there exist relations [30]

$$V^H = U^H D^H, \tag{19.184}$$

$$W^H = V^H \tilde{S}^T. \tag{19.185}$$

The equivalence of the three algorithms is given in [30]. The three ST-MUD algorithms differ in the size of the linear weight matrices.

Since the three ST-MUD algorithms are essentially equivalent, Q_z and n_z in (19.181) are the same

$$Q_z = W^H \tilde{S} D A = V^H \tilde{S}^T \tilde{S} D A = U^H M A, \tag{19.186}$$

$$n_z = W^H n = V^H \tilde{S}^T n = U^H D^H S^T n. \tag{19.187}$$

Theoretical BER performance for ST-MUD-I/II/III

The theoretical BER performance is the same for ST-MUD-I/II/III. For BPSK signals, the BER performance for user k can be approximated by $Q(\sqrt{\gamma_k})$, where γ_k is the received SNR of user k.

Conditioned on the users' bits, the covariance of z is

$$\mathrm{cov}(z) = \sigma^2 W^H W = \sigma^2 U^H \tilde{S}^T \tilde{S} U = \sigma^2 V^H M V, \tag{19.188}$$

which is equivalent to the covariance of the noise cov (n_z), where σ^2 is the variance of n. The BER of user k is thus given by [30, 102]

$$P_k = Q\left(\sqrt{\frac{z_k^2}{[\text{cov}\,(n_z)]_{k,k}}}\right) = Q\left(\frac{|\mathbf{V}^H\mathbf{M}\mathbf{A}b|_k}{\sigma\sqrt{[\mathbf{V}^H\mathbf{M}\mathbf{V}]_{k,k}}}\right), \tag{19.189}$$

where $[\cdot]_{k,k}$ and $[\cdot]_k$ denote the kth diagonal element of a square matrix and the kth element of a vector, respectively.

For the conventional ST-MF detector, $\mathbf{V}$ in (19.189) reduces to the identity matrix $\mathbf{I}$; thus, its BER performance is given by

$$P_k = Q\left(\frac{|\mathbf{M}\mathbf{A}b|_k}{\sigma\sqrt{\mathbf{M}_{k,k}}}\right). \tag{19.190}$$

The single-user lower bound (SULB) for the BER performance of any MUD algorithm is obtained by eliminating MAI, i.e., $\mathbf{Q}_y$ and $\mathbf{M}$ are diagonal matrices. Thus

$$P_k = Q\left(\frac{\mathbf{A}_k\sqrt{\mathbf{M}_{k,k}}}{\sigma}\right). \tag{19.191}$$

The theoretical BER performances for ST-MF, MMSE-based ST-MUD, and SULB have been given, and simulated for correlated and uncorrelated models of the channel in [30, 102]. ST-MUD significantly outperforms ST-MF for all the scenarios of multipath separation. The theoretical BER performance of ST-MUD is much closer to the lower bound than that of ST-MF. The use of antenna arrays has proved to be an effective measure to improve the BER performance of ST-MF and ST-MUD. Antenna gain, diversity gain, and processing gain contribute to the BER performance of ST-MUD, making it significantly superior to that of ST-MF under different scenarios of multipath azimuthal separations. Thus, ST-MUD significantly enhances the system capacity.

19.10.3 Adaptive implementation of ST-MUD

The stochastic gradient method is usually applied in adaptive implementation of algorithms, since it is simple and fast. In the case of MUD, typical values for K, L_k, $k = 1, \ldots, K$, P, and N in a cellular environment may result in a large number of coefficients in $\mathbf{U}^H$, $\mathbf{V}^H$, and $\mathbf{W}^H$. ST-MUD can be obtained adaptively by using the MMSE criterion (19.175), (19.172) or (19.178)

$$\mathbf{J} = \arg\min_{\mathbf{J}} F(\mathbf{J}) = E\left[\|z(i) - b(i)\|^2\right], \tag{19.192}$$

where $\mathbf{J} = \mathbf{U}$, $\mathbf{V}$, and $\mathbf{W}$,

$$z(i) = \mathbf{J}^H(i)g(i) \tag{19.193}$$

is the K-by-1 combiner output at the ith iteration, and $g(i)$ corresponds to x, y, and r, respectively.

The corresponding gradient-based algorithm is given as

$$\mathbf{J}(i+1) = \mathbf{J}(i) - \mu(i+1)\mathbf{g}(i)[z(i) - \boldsymbol{b}(i)]^H, \tag{19.194}$$

where μ is the step size. The problem with the gradient-based method is the difficulty in selecting a suitable step size [31]. Usually, μ is selected as a small positive constant. Actually, when μ is too large, $\mathbf{J}$ may be erratic and the algorithm unstable. On the contrary, when μ is too small, convergence is so slow that the algorithm cannot satisfy the real-time requirement. A fast converging algorithm is usually obtained by forcing the a-posteriori error to be zero after each iteration, that is,

$$\mathbf{J}^H(i+1)\mathbf{g}(i) = \boldsymbol{b}(i). \tag{19.195}$$

From this, a variable step size is obtained as

$$\mu(i+1) = \frac{1}{\|\mathbf{g}(i)\|^2}. \tag{19.196}$$

Thus, in the training mode, a normalized LMS algorithm is given as

$$\mathbf{J}(i+1) = \mathbf{J}(i) - \frac{\mathbf{g}(i)}{\|\mathbf{g}(i)\|^2}[z(i) - \boldsymbol{b}(i)]^H. \tag{19.197}$$

After training, algorithm (19.197) can be switched to decision-directed mode and continue adapting $\mathbf{J}(i)$

$$\mathbf{J}(i+1) = \mathbf{J}(i) - \frac{\mathbf{g}(i)}{\|\mathbf{g}(i)\|^2}\left[\boldsymbol{b}(i) - \hat{z}(i)\right]^H, \tag{19.198}$$

where $\hat{\boldsymbol{b}}(i) = \text{dec}(z(i))$ contains the alphabet symbols closest to the entries of $z(i)$. If the step size is suitably selected, both the algorithms given by (19.194) and (19.197) will converge to the vicinity of the MMSE solution.

The sizes of $\mathbf{U}^H$, $\mathbf{V}^H$, and $\mathbf{W}^H$ are $K \times K$, $K \times \left(\sum_{k=1}^K L_k\right) N_r$, and $K \times N_r N$, respectively; thus, the less the information required before applying an ST-MUD algorithm, the more the parameters required to be estimated, and accordingly the computational complexity is also increased. The computational as well as the storage complexity is $O\left(K^2\right)$ for ST-MUD-I, $O\left(KN_r \sum_{k=1}^K L_k\right)$ for ST-MUD-II, and $O\left(KN_r N\right)$ for ST-MUD-III.

Example 19.5: We conduct simulations of the adaptive implementations of the ST-MF and the three ST-MUD algorithms. A uniform linear array with an antenna spacing of half the carrier wavelength is selected. The lth resolvable multipath of the kth user impinge from direction $\theta_{k,l}$, thus the channel coefficient $d_{p,k,l} = q_{k,l}e^{j\pi(p-1)\sin\theta_{k,l}}$, where $q_{k,l}$ is the complex channel coefficient of this component and is in the Rayleigh distribution.

For asynchronous CDMA systems, the multipaths of a user signal are assumed to have random phases. To simplify the simulation, we assume that the multipath delays are of integer chip intervals, and the multipath spread of any user signal is limited to at most one symbol interval; in this case, an equalizer is not necessary. For adaptive implementation, knowledge of some of the channel parameters, including the fading coefficients, multipath

DoAs, spreading codes, and array response parameters may be not required, depending on the specific algorithm. We still need to assume knowledge of the timing of the users.

Gold codes of length $N = 63$ are used as the PN sequences. Assume that the interference power from other cells is 30% of the in-cell interference power. Thus, the variance σ^2, the total power of the interference, is $0.3(K - 1)a_0^2/N$, where a_0 is the nominal amplitude of a user signal.

Let us assume the number of active users $K = 20$ and the number of elements $N_r = 3$. All the signals are spread BPSK signals. The signal of user k has L_k Rayleigh-fading paths, where L_k is a random integer from 1 to 7. The direction $\theta_{k,l}$ is randomly distributed in $(75°, 75°)$. For loose power control, different users are subject to the same total average fading, $E\left[\sum_{l=1}^{L_k} \|q_{k,l}\|^2\right] = 1$. This cannot eliminate the instantaneous fluctuation of the total power of any of the users. For strict power control, different users are assumed to have the same total power, $\sum_{l=1}^{L_k} \|q_{k,l}\|^2 = 1$, at any given instant, thus the near-far effect is totally removed.

Given a random run of the three adaptive ST-MUD algorithms, the evolutions of the rms error between the MUD outputs and the input data symbols for asynchronous, strict or loose power control, CDMA systems are similar. The adaptive ST-MUD algorithms are run in the training mode for 200 iterations, and are then switched to the decision-directed mode for 2000 iterations. The power of each user is randomly selected within the range of ± 5 dB of the nominal power. As a typical illustration of the error evolutions, a random run of the algorithms for an asynchronous CDMA system using loose power control is given in Fig. 19.16.

For a random run, the amplitudes of the elements of matrices Q_z and Q_y are shown in Fig. 19.17. It is seen that Q_z and Q_y are close to diagonal matrices, where the diagonal elements correspond to the eigenvalues of the matrices. The eigenvalue spread of Q_z or Q_y is a measure of the output MAI. From Fig. 19.17, we find that the three ST-MUD algorithms achieve similar eigenvalue spreads, which are very close to unity. For

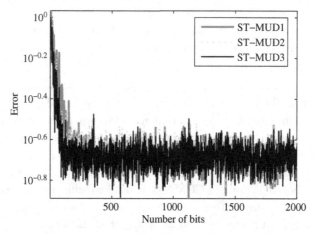

Figure 19.16 Error performance of the ST-MUD algorithms for an asynchronous CDMA system using loose power control: $N_r = 3$, $K = 20$, and a training sequence of 200 bits.

| | Spread of the diagonals of $\mathbf{Q}$: $\frac{\max\{\text{diag}(\mathbf{Q})\}}{\min\{\text{diag}(\mathbf{Q})\}}$ | | | |
	$\mathbf{Q}_z$, ST-MUD1	$\mathbf{Q}_z$, ST-MUD2	$\mathbf{Q}_z$, ST-MUD3	$\mathbf{Q}_y$, ST-MF
Syn. strict	1.1077	1.0881	1.0765	1.3206
Asyn. strict	1.0986	1.0843	1.0748	1.2852
Syn. loose	1.0837	1.0743	1.0624	3.3630
Asyn. loose	1.0736	1.0783	1.0633	2.7982

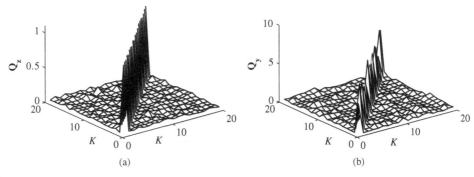

(a)   (b)

Figure 19.17 Eigenvalue spread of $\mathbf{Q}_y$ and $\mathbf{Q}_z$ for asynchronous CDMA system using loose/strict power control, from a random run: $N_r = 3$, $K = 20$, $L_k = 1, \ldots, 7$, and a training sequence of 200 bits.

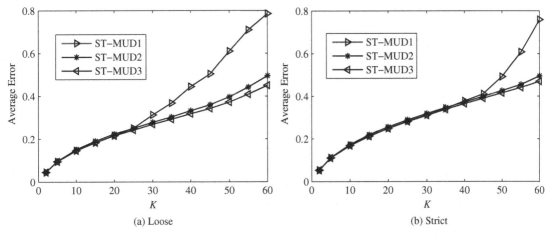

(a) Loose   (b) Strict

Figure 19.18 Average error versus K for the ST-MUD algorithms: uniform DoAs, $N_r = 3$, $L_k = 1, \ldots, 7$, and a training sequence of 200 bits.

ST-MF, the eigenvalue spread is relatively large and it is sensitive to the method of power control.

Monte Carlo simulations are used to analyze the BER performance. The DoAs of all the users' multipaths are randomly distributed in the 150° sector. Given the number of users K, the number of runs is taken as $2500/K$, which is used to guarantee the resolution for BER calculation to be 1.0×10^{-7}. For a training sequence of 200 bits, the average error in the rms sense for estimating each bit versus K is shown in Fig. 19.18, and the average BER P_b versus K is shown Fig. 19.19.

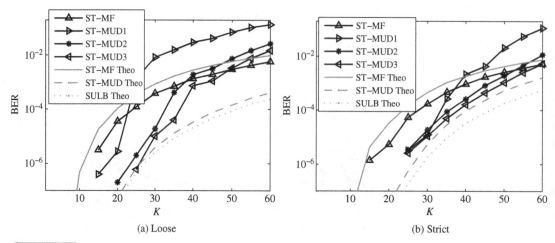

Figure 19.19 BER versus K for the ST-MF and ST-MUD algorithms: uniform random DoAs, $N_r = 3$, $L_k = 1, \ldots, 7$, and a training sequence of 200 bits.

From Fig. 19.18 it is seen that ST-MUD-I performs significantly worse than ST-MUD-II and ST-MUD-III in terms of the estimation error, while the performance of ST-MUD-III is slightly better than that of ST-MUD-II. Accordingly, the BER performance of ST-MUD-I is significantly worse than that of ST-MUD-II/III, and ST-MUD-III has a BER performance better than that of the ST-MUD-II.

From Fig. 19.19, it is noted that the BER performance of ST-MF is better than that of ST-MUD-I, when loose power control is applied. This is because ST-MF does not need any training sequence, while the performance of the ST-MUD algorithms is dependent on the length of the training sequences. A training sequence of 200 bits is not sufficient to estimate the algorithmic parameters for ST-MUD-I, but it is still sufficient for ST-MUD-II/III for loose/strict power control, asynchronous CDMA systems. By increasing the length of the training sequence, the BER performance of ST-MUDs can be significantly improved. For example, when the length of the training sequences is increased to 500 bits, the average error versus K is significantly reduced for ST-MUD-I, but slightly reduced for ST-MUD-II/III. The BER performance is shown in Fig. 19.20. In this case, the BER performance of ST-MUD-I is better than that of ST-MF.

Now, if we limit the number of users to $K = 20$ and assume the interfering power to be $0.01x(K-1)a_0^2/N$, where x is variable between 661.587 and 2.634, then the SNR for each user changes between -3 and 21 dB. The simulation result for asynchronous CDMAs under loose and strict power controls are shown in Fig. 19.21. From the figure, we see that when $x < 59$, the performance of the ST-MUD algorithms are significantly better than that of ST-MF.

Based on all the simulations, ST-MUD-III has a BER performance that is slightly better than that of ST-MUD-II, and both provide a BER performance much better than that of ST-MUD-I. ST-MUD-II/III typically performs better than ST-MF. ST-MUD-I is more sensitive to the length of training sequences: With a short training sequence it may provide a BER performance worse than that of ST-MF. ST-MUD-I approximates the performance

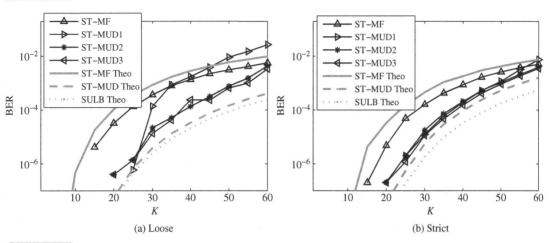

Figure 19.20 BER versus *K* for the ST-MF and ST-MUD algorithms: uniform random DoAs, $N_r = 3$, $L_k = 1, \ldots, 7$, and a training sequence of 500 bits.

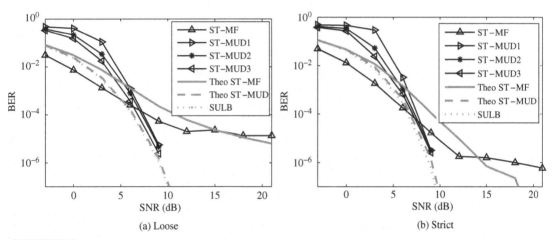

Figure 19.21 BER versus SNR for the ST-MF and ST-MUD algorithms: uniform random DoAs, $N_r = 3$, $K = 20$, $L_k = 1, \ldots, 7$, and a training sequence of 500 bits.

of ST-MUD-II/III, only when the training sequence is sufficiently long and strict power control is applied. The theoretical BER for ST-MUD is very close to that of the single-user lower bound (SULB).

19.11 MIMO in wireless standards

MIMO promises high spectral efficiencies to wireless systems, and as high as 20 to 40 bit/s/Hz has been reported in [46]. In contrast, current cellular systems typically operate with a spectral efficiency of 0.05 to 2 bits/s/Hz. MIMO has been implemented in WCDMA, CDMA2000, IEEE 802.16e and IEEE 802.11n. It will be the enabling technique for 4G.

In the WCDMA standard released in 1999, two-antenna open-loop and closed-loop transmit diversities have been specified at the BS. The open-loop transmit diversity is called *STTD*, which is based on the Alamouti code. The closed-loop transmit diversity makes use of feedback from the MS, and it is similar to transmit beamforming. The weights are so selected that the signals from the two antennas arrive coherently at the receiver. Thus, the scheme provides a two-fold diversity as well as 3 dB beamforming gain. 3GPP also extends the specification of diversity MIMO to the case of more than two transmit antennas, and considers spatial multiplexing MIMO for WCDMA/HSDPA in Release 6.

Within 3GPP technology, in Release 7, 2×1 and 4×2 MIMO configurations are used, whereas in Release 8, 2×2 and 4×4 configurations are used. HSDPA supports two-antenna open-loop and closed-loop transmit diversities. The open-loop scheme typically provides no additional gain for HSPDA at low to moderate moving speeds for MSs, if proportional fairness scheduling is employed [109]. However, it does provide a capacity gain of 10% to 20%, when a round-robin scheduler is employed, or when the MSs are moving very fast so that the proportional fairness scheduler cannot track the fast fading radio channel [109]. The closed-loop scheme provides higher gain, but requires an associated dedicated transport channel. Many MIMO algorithms for HSDPA have been proposed within 3GPP [3]. These algorithms employ either a space-time coding or a linear precoding scheme.

In an environment of small angular spread, downlink beamforming yields an array gain which dominates diversity gain, and transmit beamforming improves the downlink capacity in WCDMA and HSDPA more significantly than transmit diversity. A capacity gain of 150% can be achieved by forming a grid of fixed beams for an antenna array of four elements [104]. Multiple scrambling codes can be introduced in each BS to assign each beam with a different code. This increases the code space. The beamforming introduces a capacity gain over single-antenna transmission for HSDPA of 2.5 for four antennas and 4.7 for eight antennas, respectively [105]. In addition, beamforming also helps to improve the uplink performance. However, transmit beamforming loses its good performance when the angle spread is large, since the beams cannot be accurately steered to users any more. On the other hand, receive beamforming is not necessarily the best solution for the uplink due to the lack of diversity gain [142]. For this reason, STBCs are adopted by 3GPP as an open-loop STTD method.

CDMA2000 provides two kinds of transmit diversity: STS and orthogonal transmit diversity (OTD). STS is an extension of the Alamouti code [5], obtained by multiplying two Walsh codes. STS achieves diversity at the symbol level. OTD transmits alternative bits over spatially separated or orthogonally polarized antennas, each antenna using a different Walsh code. OTD achieves diversity in the Viterbi decoding path metrics, but has no symbol-level diversity. For a single antenna receiver, the Viterbi path metrics have large variations for slow speeds, but are less variable due to interleaving for high speeds. With OTD, interleaving across antennas allows for diversity in the path metrics even at low speeds. However, the diversity achieved with OTD is not as high as that achieved by STS [131]. The diversity gain achieved by OTD depends on the decoding process, but for STS this is not the case.

IEEE 802.16e (WiMAX) supports transmit diversity using the Alamouti code, beam-forming at both the uplink and the downlink, and spatial multiplexing. IEEE 802.16e also defines an optional transmit diversity mode, known as the *frequency-hopping diversity code*, which uses two antennas to encode in the space and frequency domains. In the frequency-hopping diversity code, the first antenna transmits the OFDM symbols without any encoding, as in a single-antenna transmission; the second antenna transmits the OFDM symbol by encoding over two consecutive subchannels using the Alamouti matrix. Unlike transmit diversity and transmit beamforming, spatial multiplexing is employed only under good SINR conditions. Both open-loop and closed-loop spatial multiplexing are defined in IEEE 802.16e. Antenna selection for the closed-loop MIMO is also defined in IEEE 802.16e. As an optional scheme for MIMO channel feedback mechanism in IEEE 802.16e, the BS can instruct the MS to perform channel sounding over allowed subcarriers. This method is most bandwidth intensive, but it provides the BS with the most accurate downlink channel estimate. The 4×4 MIMO configuration is also supported.

A 2×2 MIMO scheme is the baseline configuration for LTE downlink. Configurations with 4 antennas are also being considered. Different MIMO modes are envisaged, the selection between spatial multiplexing and transmit diversity depends on the channel condition. Precoding on the transmitter side is used to support spatial multiplexing. Transmit diversity is used when the selected number of streams (rank) is one. Cyclic delay diversity can be used in conjunction with spatial multiplexing in LTE.

IEEE 802.11n defines transmitter beamforming, transmitter diversity using STBC, DSTTD, and spatial multiplexing. Several MIMO configurations, including 2×2, 2×3, 3×3, and 4×4 are defined in IEEE 802.11n. IEEE 802.11n uses a 802.11a legacy packet, but its preamble also includes MIMO training sequences, followed by payload. MIMO is also being considered for IEEE 802.20 and IEEE 802.22. For example, the 625k-MC mode of IEEE 802.20 supports adaptive antenna arrays.

Problems

19.1 Consider the flat fading MIMO channel

$$y = \mathbf{H}x + w,$$

where $y \in \mathcal{R}^n$, the channel $\mathbf{H}$ is of dimension $n \times m$, and the input x is m-dimensional Gaussian $\mathcal{N}(\bar{x}, \mathbf{R}_x)$. Derive the MMSE (Wiener) estimate of $\hat{x}$. Prove the following orthogonality principle:

$$\operatorname{cov}\left((x - \hat{x}), y\right) = 0.$$

19.2 Given a deterministic 3×4 MIMO channel

$$\mathbf{H} = \begin{bmatrix} 0.4 + 0.6j & j & 2 & -0.8 + j \\ -0.8 & 0.4 + 0.2j & 1.5 - 0.6j & 0.9 \\ 0.6j & -0.7 & -0.1 + 1.1j & 0.6 \end{bmatrix},$$

(a) Find all the nonzero singular values of $\mathbf{H}$ and the eigenvalues of $\mathbf{H}^H\mathbf{H}$.

(b) Give an equivalent representation with parallel channels.

(c) When the transmitter allocates all its power to the strongest channel, this is known as transmitter beamforming [103]. Find the beamforming weights $\mathbf{V}$ for the antennas.

(d) Plot the channel capacity obtained by equal power allocation, beamforming, and optimal water-filling.

19.3 Assuming that the same channel defined in Problem 19.2 is given and that no CSI is available at the transmitter, implement Monte Carlo simulations to estimate the achievable information capacity for: (a) BPSK input and (b) QPSK input.

19.4 For a MISO slow fading channel, prove that the transmit-diversity scheme radiates energy in an isotropic manner if, and only if, the signals transmitted are uncorrelated with the same power. Show that the Alamouti code scheme radiates energy isotropically.

19.5 Show that the capacity of an $N_1 \times N_2$ MIMO channel with channel matrix $\mathbf{H}$ is the same as that of an $N_2 \times N_1$ MIMO channel with channel matrix $\mathbf{H}^*$ and the same power constraint.

19.6 Find the capacity and optimal power allocation for the MIMO channel defined in Problem 19.2. Assume that $P/\sigma_n^2 = 20$ dB and $B = 1$ Hz.

19.7 Consider a flat fading channel [144]

$$y(i) = h(i)x(i) + n(i), \quad i = 0, 1, \ldots,$$

where $h(i)$ is Gaussian with zero mean and unit variance, $x \in \{-a, a\}$, and $n(i)$ is AWGN with zero mean and unit variance. The signal is encoded as $\mathbf{x}_A = (a, 0)^T$ for a and $\mathbf{x}_B = (0, a)^T$ for $-a$. Detection of x is based on the block output

$$\mathbf{y} = \begin{pmatrix} y(0) \\ y(1) \end{pmatrix} = \begin{pmatrix} h(i)x(0) \\ h(i+1)x(1) \end{pmatrix} + \begin{pmatrix} n(i) \\ n(i+1) \end{pmatrix}.$$

(a) Derive $\Pr(y(0)|\mathbf{x}_A)$, $\Pr(y(1)|\mathbf{x}_A)$, $\Pr(y(0)|\mathbf{x}_B)$, and $\Pr(y(1)|\mathbf{x}_B)$.

(b) Verify the LLR function

$$\Lambda(\mathbf{y}) = \ln\left(\frac{p(\mathbf{y}|\mathbf{x}_A)}{p(\mathbf{y}|\mathbf{x}_B)}\right) = \frac{(y^2(0) - y^2(1))a^2}{1 + a^2}.$$

(c) Give the ML detection rule. (Hint: $\mathbf{x}_A$ when $\Lambda(\mathbf{y}) \geq 0$, and $\mathbf{x}_B$ when $\Lambda(\mathbf{y}) < 0$.)

(d) Derive the error probability P_e of the ML detection.

19.8 Find the outage capacity for a 3×3 MIMO channel at 5% outage for $P/\sigma_n^2 = 10$ dB and $B = 1$ MHz.

19.9 Find the diversity orders of the following MIMO systems over full scattering channels with quasi-static fading:

(a) $N_t = 2$, $N_r = 1$, number of delays $L = 2$, Alamouti scheme with ML decoding.

(b) $N_t = 3$, $N_r = 2$, $L = 2$, VBLAST scheme with optimal decoding.

19.10 For a $2 \times N_r$ MIMO system over an ergodic Rayleigh fading channel, the transmit correlation matrix is $\mathbf{R}_t = \begin{bmatrix} 1 & r \\ r & 1 \end{bmatrix}$. Plot the ergodic channel capacity for $r = 0$, 0.5, 0.75, 0.9 for the following two cases when there is no CSI at the transmitter:
(a) $N_r = 2$, the receive correlation $\mathbf{R}_r = \mathbf{R}_t$.
(b) $N_r = 1$.

19.11 Derive the ergodic channel capacity for the keyhole channel as a function of $N_t = N_r = n$. Plot the relation for different SNRs.

19.12 Given the linear dispersion code

$$\begin{bmatrix} x_1 & x_2 & x_3 & x_4 \\ -x_2^* & x_1^* & -x_4^* & x_3^* \\ x_8 & -x_7 & -x_6 & x_5 \\ -x_7^* & -x_8^* & x_5^* & x_6^* \end{bmatrix},$$

(a) Determine the rate of the code in symbols per channel use.
(b) Find the diversity orders offered by the code by using the rank criterion.

19.13 Design a MIMO-OFDM system to provide a raw data rate of 40 Mbits/s over a bandwidth of 10 MHz. Assume that $N_t = 2$, multipath spread $\tau_{max} = 0.5$ ms, and the Doppler spread is $f_D = 5$ Hz. Specify the following design parameters: the number of carriers, OFDM symbol duration, the length of the cyclic prefix, and the modulation scheme.

References

[1] 3GPP, *Double-STTD Scheme for HSDPA Systems with Four Transmit Antennas: Link Level Simulation Results*, 3GPP TSG RAN WG1 document, TSGR1#21(01)-0701, Jun 2001.
[2] 3GPP, *Spatial Channel Model for Multiple Input Multiple Output MIMO Simulations*, TR 25.996, Ver. 6.1.0, Sep 2003.
[3] 3GPP, *Multiple-input Multiple-output in UTRA*, TR 25.876 Ver. 1.5.1, May 2004.
[4] M. Airy, R. W. Heath, Jr. & S. Shakkottai, Multi-user diversity for the multiple antenna broadcast channel with linear receivers: asymptotic analysis. In *Proc. Asilomar Conf. Signals Syst. Computers*, Pacific Grove, CA, Nov 2004, **1**, 886–890.
[5] S. M. Alamouti, A simple transmit diversity technique for wireless communications. *IEEE J. Sel. Areas Commun.*, **16**:8 (1998), 1451–1458.
[6] P. Almers, F. Tufvensson & A. F. Molisch, Measurements of keyhole effect in a wireless multiple-input multiple-output (MIMO) channel. *IEEE Commun. Lett.*, **7**:8 (2003), 373–375.
[7] S. L. Ariyavisitakul, Turbo space-time processing to improve wireless channel capacity. *IEEE Trans. Commun.*, **48**:8 (2000), 1347–1359

[8] E. K. S. Au & W. H. Mow Exact bit error rate for SVD-based MIMO systems with channel estimation errors. In *Proc. IEEE ISIT*, Seattle, USA, Jul 2006, 2289–2293.

[9] E. K. S. Au, C. Wang, S. Sfar, R. D. Murch, W. H. Mow, V. K. N. Lau, R. S. Cheng & K. B. Letaief, Error probability for MIMO zero-forcing receiver with adaptive power allocation in the presence of imperfect channel state information. *IEEE Trans. Wireless Comun.*, **6**:4 (2007), 1523–1529.

[10] H. R. Bahrami, T. Le-Ngoc, A. M. N. Nasrabadi & S. H. Jamali, Precoder design based on correlation matrices for MIMO systems communications. In *Proc. IEEE ICC*, Seoul, Korea, May 2005, **3**, 2001–2005.

[11] E. Biglieri & G. Taricco, Fundamentals of MIMO channel capacity. In H. Bolcskei, D. Gesbert, C.B. Papadias & A.-J. van der Veen, eds., *Space-Time Wireless Systems: From Array Processing to MIMO Communications* (Cambridge, UK: Cambridge University Press, 2006), pp. 66–86.

[12] D. W. Bliss, K. W. Forsythe, A. O. Hero, III & A. F. Yegulalp, Environmental issues for MIMO capacities. *IEEE Trans. Signal Process.*, **50**:9 (2002), 2128–2142.

[13] L. Boher, R. Rabineau & M. Helard, FPGA implementation of an iterative receiver for MIMO-OFDM systems. *IEEE J. Sel. Areas Commun.*, **26**:6 (2008), 857–866.

[14] H. Bolcskei & A. J. Paulraj, Space-frequency coded broadband OFDM systems. In *Proc. IEEE WCNC*, Chicago, IL, Sep 2000, 1–6.

[15] H. Bolcskei, D. Gesbert & A. J. Paulraj, On the capacity of OFDM based spatial multiplexing systems. *IEEE Trans. Commun.*, **50**:2 (2002), 225–234.

[16] H. Bolcskei, R. W. Heath, Jr. & A. J. Paulraj, Blind channel identification and equalization in OFDM-based multiantenna systems. *IEEE Trans. Signal Process.*, **50**:1 (2002), 96–108.

[17] A. O. Boukalov & S.-G. Haggman, System aspects of smart-antenna technology in cellular wireless communications – an overview. *IEEE Trans. Microwave Theory Tech.*, **48**:6 (2000), 919–929.

[18] Z. Chen, Z. Chi, Y. Li & B. Vucetic, Error performance of maximal-ratio combining with transmit antenna selection in flat Nakagami-m fading channels. *IEEE Trans. Wireless Commun.*, **8**:1 (2009), 424–431.

[19] Z. Chen, J. Yuan & B. Vucetic, Improved space-time trellis coded modulation scheme slow Rayleigh fading channels. *Electron. Lett.*, **37**:7 (2001), 440–441.

[20] Z. Chen, J. Yuan & B. Vucetic, Analysis of transmit antenna selection/maximal-ratio combining in Rayleigh fading channels. *IEEE Trans. Veh. Tech.*, **54**:4 (2005), 1312–1321.

[21] X. Zhang, Z. Lv & W. Wang, Performance analysis of multiuser diversity in MIMO systems with antenna selection. *IEEE Trans. Wireless Commun.*, **7**:1 (2008), 15–21.

[22] C.-J. Chen & L.-C. Wang, Performance analysis of scheduling in multiuser MIMO systems with zero-forcing receivers. *IEEE J. Sel. Areas Commun.*, **25**:7 (2007), 1435–1445.

[23] D. Chizhik, G. J. Foschini, M. J. Gans & R. A. Valenzuela, Keyhole, correlations, and capacities of multielement transmit and receive antennas. *IEEE Trans. Wireless Commun.*, **1**:2 (2002), 361–368.

[24] Y. J. Chun & S. W. Kim, Log-likelihood-ratio ordered successive interference cancellation in multi-user, multi-mode MIMO systems. *IEEE Commun. Lett.*, **12**:11 (2008), 837–839.

[25] M. V. Clark, Adaptive frequency-domain equalization and diversity combining for broadband wireless communications. *IEEE J. Sel. Areas Commun.*, **16**:8 (1998), 1385–1395.

[26] M. H. N. Costa, Writing on dirty paper. *IEEE Trans. Inf. Theory*, **29**:3 (1983), 439–441.

[27] T. M. Cover & J. A. Thomas, *Elements of Information Theory* (Hoboken, NJ: Wiley, 2006).

[28] S. N. Diggavi, N. Al-Dhahir & A. R. Calderbank, Diversity embedded space-time codes. In *Proc. IEEE Globecom*, San Francisco, CA, Dec 2003, 1909–1914.

[29] S. N. Diggavi, A. R. Calderbank, S. Dusad & N. Al-Dhahir, Diversity embedded space-time codes. *IEEE Trans. Inf. Theory*, **54**:1 (2008), 33–50.

[30] K.-L. Du & M. N. S. Swamy, Performance of multiuser detection schemes for CDMA systems using antenna arrays. In *Proc. World Wireless Congress (WWC'04)*, San Francisco, CA, May 2004, 433–438.

[31] K.-L. Du & M. N. S. Swamy, *Neural Networks in a Softcomputing Framework* (London: Springer, 2006).

[32] K.-L. Du & M. N. S. Swamy, An adaptive space-time multiuser detection algorithm for CDMA systems. In *Proc. IEEE WiCOM*, Wuhan, China, Sep 2006, 1–5.

[33] S. Dusad, S. N. Diggavi, N. Al-Dhahir & A. R. Calderbank, Diversity embedded codes: theory and practice. *IEEE J. Sel. Topics Signal Process.*, **2**:2 (2008), 202–219.

[34] P. Elia, B. A. Sethuraman & P. V. Kumar, Perfect space-time codes for any number of antennas. *IEEE Trans. Inf. Theory*, **53**:11 (2007), 3853–3868.

[35] R. Esmailzadeh & M. Nakagawa, Prerake diversity combination for direct sequence spread spectrum mobile communication systems. *IEICE Trans. Comm.*, **E76-B**:8 (1993), 1008–1015.

[36] D. Falconer, Spatial-temporal signal processing for broadband wireless systems. In M. Shafi, S. Ogose & T. Hattori, eds, *Wireless Communications in the 21st Century* (New York: IEEE Press, 2002).

[37] G. Fememias, BER performance of linear STBC from orthogonal design over MIMO correlated Nakagami-m channels. *IEEE Trans. Veh. Technol.*, **53**:2 (2004), 307–317.

[38] R. F. H. Fischer, C. Windpassinger, A. Lampe & J. B. Huber, MIMO precoding for decentralized receivers. In *Proc. IEEE ISIT*, Lausanne, Switzerland, Jun–Jul 2002, 496.

[39] G. J. Foschini & M. J. Gans, Layered space-time architecture for wireless communication in a fading environment when using multiple antennas. *Bell Labs Tech. J.*, **1**:2 (1996), 41–59.

[40] G. J. Foschini & M. J. Gans, On limits of wireless communications in a fading environment when using multiple antennas. *Wireless Pers. Commun.*, **6**:3 (1998), 311–335.

[41] G. Ganesan & P. Stoica, Differential modulation using space-time block codes. *IEEE Signal Process. Lett.*, **9**:2 (2002), 57–60.

[42] D. Garrett, L. Davis., S. ten Brink, B. Hochwald & G. Knagge, Silicon complexity for maximum likelihood MIMO detection using spherical decoding. *IEEE J. Solid-State Circ.*, **39**:9 (2004), 1544–1552.

[43] D. Gesbert, H. Bolcskei, D. A. Gore & A. J. Paulraj, Outdoor MIMO wireless channels: models and performance prediction. *IEEE Trans. Commun.*, **50**:12 (2002), 1926–1934.

[44] G. B. Giannakis, Z. Liu, X. Ma & S. Zhou, *Space-Time Coding for Broadband Wireless Communications* (Hoboken, NJ: Wiley, 2007).

[45] G. Ginis & J. Cioffi, On the relation between V-BLAST and the GDFE. *IEEE Commun. Lett.*, **15**:9 (2001), 364–366.

[46] G. D. Golden, G. J. Foschini, R. A. Valenzuela & P. W. Wolbiansky, Detection algorithm and initial laboratory results using the V-BLAST space-time communication architecture. *Electron. Lett.*, **35**:1 (1999), 14–16.

[47] A. Goldsmith, *Wireless Communications* (Cambridge, UK: Cambridge University Press, 2005).

[48] Y. Gong & K. B. Letaief, On the error probability of orthogonal space-time block codes over keyhole MIMO channels. *IEEE Trans. Wireless Commun.*, **6**:9 (2007), 3402–3409.

[49] A. J. Grant, Performance Analysis of Transmit Beamforming. *IEEE Trans. Commun.*, **53**:4 (2005), 738–744.

[50] L. H. Grokop & D. N. C. Tse, Diversity-multiplexing tradeoff in ISI channels. *IEEE Trans. Inf. Theory*, **55**:1 (2009), 109–135.

[51] Z. Guo & P. Nilsson, Algorithm and implementation of the K-best sphere decoding for MIMO detection. *IEEE J. Sel. Areas Commun.*, **24**:3 (2006), 491–503.

[52] S. Haene, D. Perels & A. Burg, A real-time 4-stream MIMO-OFDM transceiver: system design, FPGA implementation, and characterization. *IEEE J. Sel. Areas Commun.*, **26**:6 (2008), 877–889.

[53] B. Hassibi, An efficient square-root algorithm for BLAST. In *Proc. IEEE ICASSP*, Istanbul, Turkey, Jun 2000, **2**, 737–740.

[54] B. Hassibi & B. Hochwald, High rate codes that are linear in space and time. *IEEE Trans. Inf. Theory*, **48**:7 (2002), 1804–1824.

[55] R. W. Heath, Jr., M. Airy & A. J. Paulraj, Multiuser diversity for MIMO wireless systems with linear receivers. In *Proc. Asilomar Conf. Signals Syst. Computers*, Pacific Grove, CA, Nov 2001, **2**, 1194–1199.

[56] R. W. Heath, Jr. & A. J. Paulraj, Linear dispersion codes for MIMO systems based on frame theory. *IEEE Trans. Signal Process.*, **50**:10 (2002), 2429–2441.

[57] B. Hochwald & W. Sweldens, Differential unitary space-time modulation. *IEEE Trans. Commun.*, **48**:12 (2000), 2040–2052.

[58] B. Hochwald, T. L. Marzetta & C. B. Papadias. A transmitter diversity scheme for wideband CDMA systems based on space-time spreading. *IEEE J. Sel. Areas Commun.*, **19**:1 (2001), 48–60.

[59] B. M. Hochwald & S. ten Brink, Achieving near-capacity on a multiple-antenna channel. *IEEE Trans. Commun.*, **51**:3 (2003), 389–399.

[60] B. L. Hughes, Differential space-time modulation. *IEEE Trans. Inf. Theory*, **46**:7 (2000), 2567–2578.

[61] H. Jafarkhani, A quasi-orthogonal space-time block code. *IEEE Commun. Lett.*, **49**:1 (2001), 1–4.

[62] W. C. Jakes, Jr., ed., *Microwave Mobile Communications* (New York: Wiley, 1974).

[63] E. A. Jorswieck & A. Sezgin, Impact of spatial correlation on the performance of orthogonal space-time block codes. *IEEE Commun. Lett.*, **8**:1 (2004), 21–23.

[64] R. Kalbasi, D. D. Falconer, A. H. Banihashemi & R. Dinis, A comparison of frequency-domain block MIMO transmission systems. *IEEE Trans. Veh. Tech.*, **58**:1 (2009), 165–175.

[65] A. Kalis, A. G. Kanatas & C. B. Papadias, A novel approach to MIMO transmission using a single RF front end. *IEEE J. Sel. Areas Commun.*, **26**:6 (2008), 972–980.

[66] M. Kang & M. S. Alouini, Capacity of correlated MIMO Rayleigh channels. *IEEE Trans. Wireless Commun.*, **5**:1 (2006), 143–155.

[67] M. Kang & M. S. Alouini, Capacity of MIMO Rician channels. *IEEE Trans. Wireless Commun.*, **5**:1 (2006), 112–122.

[68] I.-M. Kim, Exact BER analysis of OSTBCs in spatially correlated MIMO channels. *IEEE Trans. Commun.*, **54**:8 (2006), 1365–1373.

[69] I.-M. Kim & D. Kim, Closed-form exact BER and optimization of generalized orthogonal STBCs. *IEEE Trans. Wireless Commun.*, **7**:9 (2008), 3323–3328

[70] Y.-T. Kim, H. Lee, S. Park & I. Lee, Optimal precoding for orthogonalized spatial multiplexing in closed-loop MIMO systems. *IEEE J. Sel. Areas Commun.*, **26**:8 (2008), 1556–1566.

[71] R. Kohno, H. Imai, M. Hatori & S. Pasupathy, Combination of an adaptive array antenna and a canceller of interference for direct-sequence spread-spectrum multiple-access system. *IEEE J. Sel. Areas Commun.*, **8**:4 (1990), 675–682.

[72] R. Kohno, Interference cancellation and multiuser detection. In M. Shafi, S. Ogose & T. Hattori, eds., *Wireless Communications in the 21st Century* (New York: IEEE Press, 2002).

[73] V. Kuhn, *Wireless Communications over MIMO Channels: Applications to CDMA and Multiple Antenna Systems* (Chichester, UK: Wiley, 2006).

[74] E. G. Larsson & P. Stoica, *Space-Time Block Coding for Wireless Communications* (Cambridge, UK: Cambridge University Press, 2003).

[75] H. Lee, B. Lee & I. Lee, Iterative detection and decoding with an improved V-BLAST for MIMO-OFDM systems. *IEEE J. Sel. Areas Commun.*, **24**:3 (2006), 504–513.

[76] H. Lee, S. Park & I. Lee, Orthogonalized spatial multiplexing for closed-loop MIMO systems. *IEEE Trans. Commun.*, **55**:5 (2007), 1044–1052.

[77] G. Levin & S. Loyka, On the outage capacity distribution of correlated keyhole MIMO channels. *IEEE Trans. Inf. Theory*, **54**:7 (2008), 3232–3245.

[78] E. Lindskog & A. J. Paulraj, A transmit diversity scheme for channels with intersymbol interference. In *Proc. IEEE ICC*, New Orleans, LA, Jun 2000, **1**, 307–311.

[79] C. Ling, W. H. Mow, K. H. Li & A. C. Kot, Multiple-antenna differential lattice decoding. *IEEE J. Sel. Areas Commun.*, **23**:9 (2005), 1821–1829.

[80] Z. Liu, G. B. Giannakis, S. Barbarossa & A. Scaglione, Transmit-antennae space-time block coding for generalized OFDM in the presence of unknown multipath. *IEEE J. Sel. Areas Commun.*, **19**:7 (2001), 1352–1364.

[81] T. Lo, Maximal ratio transmission. *IEEE Trans. Comm.*, **47**:10 (1999), 1458–1461.

[82] E. S. Lo, P. W. C. Chan, V. K. N. Lau, R. S. Cheng, K. B. Letaief, R. D. Murch & W. H. Mow, Adaptive resource allocation and capacity comparison of downlink multiuser MIMO-MC-CDMA and MIMO-OFDMA. *IEEE Trans. Wireless Commun.*, **6**:3 (2007), 1083–1093.

[83] D. J. Love, R. W. Heath, Jr. & T. Strohmer, Grassmannian beamforming for multiple-input multiple-output wireless systems. *IEEE Trans. Inf. Theory*, **49**:10 (2003), 2735–2747.

[84] D. J. Love, R. W. Heath Jr, V. K. N. Lau, D. Gesbert, B. D. Rao & M. Andrews, An overview of limited feedback in wireless communication systems. *IEEE Sel. Areas Commun.*, **26**:8 (2008), 1341–1365.

[85] S. Loyka & F. Gagnon, Performance analysis of the V-BLAST algorithm: an analytical approach. *IEEE Trans. Wireless Commun.*, **3**:4 (2004), 1326–1337.

[86] S. Loyka & F. Gagnon, On outage and error rate analysis of the ordered V-BLAST. *IEEE Trans. Wireless Commun.*, **7**:10 (2008), 3679–3685.

[87] S. Loyka & G. Tsoulos, Estimating MIMO system performance using the correlation matrix approach. *IEEE Commun. Lett.*, **6**:1 (2002), 19–21.

[88] A. Lozano, A. M. Tulino & S. Verdu, Multiantenna capacity: myths and realities. In H. Bolcskei, D. Gesbert, C. B. Papadias & A.-J. van der Veen, eds., *Space-Time Wireless Systems: From Array Processing to MIMO Communications* (Cambridge, UK: Cambridge University Press, 2006), pp. 87–107.

[89] B. Lu, X. Wang & K. Narayanan, LDPC-based space-time coded OFDM systems over correlated fading channels: performance analysis and receiver design. *IEEE Trans. Commun.*, **50**:1 (2002), 74–88.

[90] A. Maaref & S. Aissa, Closed-form expressions for the outage and ergodic Shannon capacity of MIMO MRC systems. *IEEE Trans. Commun.*, **53**:7 (2005), 1092–1095.

[91] R. Y. Mesleh, H. Haas, S. Sinanovic, C. W. Ahn & S. Yun, Spatial modulation. *IEEE Trans. Veh. Tech.*, **57**:4 (2008), 2228–2241

[92] A. F. Molisch, *Wireless Communications* (Chichester, UK: Wiley-IEEE, 2005).

[93] W. H. Mow, Maximum likelihood sequence estimation from the lattice viewpoint. *IEEE Trans. Inf. Theory*, **40**:5 (1994), 1591–1600.

[94] W. H. Mow, Universal lattice decoding: principle and recent advances. *Wireless Commun. Mobile Comput.*, **3**:5 (2003), 553–569.

[95] K. K. Mukkavilli, A. Sabharwal, E. Erkip & B. Aazhang, On beamforming with finite rate feedback in multiple antenna systems. *IEEE Trans. Inf. Theory*, **49**:10 (2003), 2562–2579.

[96] R. U. Nabar, H. Bolcskei & A. J. Paulraj, Diversity and outage performance in space-time block coded Ricean MIMO channel. *IEEE Trans. Wireless Commun.*, **4**:5 (2005), 2519–2532.

[97] A. Naguib, N. Seshadri & A. R. Calderbank, Space-time coding and signal processing for high data rate wireless communications. *IEEE Signal Process. Mag.*, **17**:3 (2000), 76–92.

[98] C. Oestges, Mutual information of non-Kronecker structured dual-polarized 2×2 channels. *IEEE Trans. Veh. Tech.*, **56**:1 (2006), 410–413.

[99] C. Oestges & B. Cerckx, *MIMO Wireless Communications: From Real-World Propagation to Space-Time Code Design* (Oxford, UK: Academic Press, 2007).

[100] F. Oggier, G. Rekaya, J.-C. Belfiore & E. Viterbo, Perfect space-time block codes. *IEEE Trans. Inf. Theory*, **52**:9 (2006), 3885–3902.

[101] O. Oyman, R. U. Nabar, H. Boelcskei & A. J. Paulraj, Characterizing the statistical properties of mutual information in MIMO channels. *IEEE Trans. Signal Process.*, **51**:11 (2003), 2784–2795.

[102] C. B. Papadias & H. Huang, Linear space-time multiuser detection for multipath CDMA channels. *IEEE J. Sel. Areas Commun.*, **19**:2 (2001), 254–265.

[103] A. Paulraj, R. Nabar & D. Gore, *Introduction to Space–Time Wireless Communications* (Cambridge, UK: Cambridge University Press, 2003).

[104] K. I. Pedersen, P. E. Mogensen & J. Ramiro-Moreno, Application and performance of downlink beamforming techniques in UMTS. *IEEE Commun. Mag.*, **41**:10 (2003), 134–143.

[105] K. I. Pedersen & P. E. Mogensen, Performance of WCDMA HSDPA in a beamforming environment under code constraints. In *Proc. IEEE VTC*, Orlando, FL, Oct 2003, **2**, 995–999.

[106] K. Peppas & A. Maras, Performance evaluation of space-time block codes over keyhole Weibull fading channels. *Wireless Pers. Commun.*, **46**:4 (2008), 385–395.

[107] S. L. Preston, D. V. Thiel, J. W. Lu, S. G. O'Keefe & T. S. Bird. Electronic beam steering using switched parasitic patch elements. *Electron. Lett.*, **33**:1 (1997), 7–8.

[108] V. Raghavan, R. W. Heath, Jr. & A. M. Sayeed, Systematic codebook designs for quantized beamforming in correlated MIMO channels. *IEEE J. Sel. Areas Commun.*, **25**:7 (2007), 1298–1310.

[109] J. Ramiro-Moreno, K. I. Pedersen & P. E. Mogensen, Network performance of transmit and receive antenna diversity in HSPDA under different packet scheduling strategies. In *Proc. IEEE VTC*, Jeju, Korea, Apr 2003, **2**, 1454–1458.

[110] S. M. Razavizadeh, A. K. Khandani, V. T. Vakili & W. Tong, Space-time precoding for downlink transmission in multiple antenna CDMA systems. *IEEE Trans. Veh. Tech.*, **56**:5 (2007), 2590–2602.

[111] D. Reynolds, X. Wang & H. V. Poor, Blind adaptive space–time multiuser detection with multiple transmitter and receiver antennas. *IEEE Trans. Signal Process.*, **50**:6 (2002), 1261–1276.

[112] J. M. Romero-Jerez & A. J. Goldsmith, Performance of multichannel reception with transmit antenna selection in arbitrarily distributed Nagakami fading channels. *IEEE Trans. Wireless Commun.*, **8**:4 (2009), 2006–2013.

[113] D. J. Ryan, I. V. L. Clarkson, I. B. Collings, D. Guo & M. L. Honig, QAM and PSK codebooks for limited feedback MIMO beamforming. *IEEE Trans. Commun.*, **57**:4 (2009), 1184–1196.

[114] H. Sampath, P. Stoica & A. Paulraj, Generalized linear precoder and decoder design for MIMO channel using the weighted MMSE criterion. *IEEE Trans. Commun.*, **49**:12 (2001), 2198–2206.

[115] S. Sanayei, A. Hedayat & A. Nosratinia, Space time codes in keyhole channels: analysis and design. *IEEE Trans. Wireless Commun.*, **6**:6 (2007), 2006–2011.

[116] S. Sandhu & A. Paulraj, Space-time block codes: a capacity perspective. *IEEE Commun. Lett.*, **4**:12 (2000), 384–386.

[117] T. Sawaya, K. Iigusa, M. Taromaru & T. Ohira, Reactance diversity: proof-of-concept experiments in an indoor multipath-fading environment with a 5-GHz prototype planar ESPAR antenna. In *Proc. IEEE CCNC*, Las Vegas, NV, Jan 2004, 678–680.

[118] A. Sayeed & V. Veeravalli, The essential degrees of freedom in space-time fading channels. In *Proc. IEEE PIMRC*, Lisbon, Portugal, Sep 2002, **4**, 1512–1516.

[119] C. B. Schlegel & L. C. Perez, *Trellis and Turbo Coding* (Piscataway, NJ: IEEE Press, 2004).

[120] N. L. Scott, O.-L. Taylor & R. G. Vaughan, Diversity gain from a single-port adaptive antenna using switched parasitic elements illustrated with a wire and monopole prototype. *IEEE Trans. Anten. Propagat.*, **47**:6 (1999), 1066–1070.

[121] M. Shafi, M. Zhang, A. L. Moustakas, P. J. Smith, A. F. Molisch, F. Tufvesson & S. H. Simon, Polarized MIMO channels in 3-D: models, measurements and mutual information. *IEEE J. Sel. Areas Commun.*, **24**:3 (2006), 514–527.

[122] N. Sharma & C. B. Papadias, Improved quasi-orthogonal codes through constellation rotation. *IEEE Trans. Commun.*, **51**:3 (2003), 332–335.

[123] C. Shin, R. W. Heath, Jr. & E. J. Powers, Blind channel estimation for MIMO-OFDM systems. *IEEE Trans. Veh. Tech.*, **56**:2 (2007), 670–685.

[124] H. Shin & J. H. Lee, Capacity of multiple-antenna fading channels: spatial fading correlation, double scattering and keyhole. *IEEE Trans. Inf. Theory*, **49**:10 (2003), 2636–2647.

[125] H. Shin & M. Z. Win, MIMO diversity in the presence of double scattering. *IEEE Trans. Inf. Theory*, **54**:7 (2008), 2976–2996.

[126] W.-Y. Shin, S.-Y. Chung & Y. H. Lee, Diversity-multiplexing tradeoff and outage performance for Rician MIMO channels. *IEEE Trans. Inf. Theory*, **54**:3 (2008), 1186–1196.

[127] D.-S. Shiu, G. J. Foschini, M. J. Gans & J. M. Kahn, Fading correlation and its effect on the capacity of multielement antenna system. *IEEE Trans. Commnun.*, **48**:3 (2000), 502–513.

[128] O. Simeone, Y. Bar-Ness & U. Spagnolini, Linear and nonlinear preequal-ization/equalization for MIMO systems with long-term channel state infor-mation at the transmitter. *IEEE Trans. Wireless Commun.*, **3**:2 (2004), 373–377.

[129] W. P. Siriwongpairat, M. Olfat & K. J. R. Liu, On the performance evaluation of the TH and DS UWB MIMO systems. In *Proc. IEEE WCNC*, Atlanta, GA, Mar 2004, **3**, 1800–1805.

[130] W. P. Siriwongpairat & K. J. R. Liu, *Ultra-Wideband Communications Systems: Multiband OFDM Approach* (Piscataway, NJ: Wiley-IEEE, 2008).

[131] R. A. Soni & R. M. Buehrer, On the performance of open loop transmit diver-sity techniques for IS-2000 systems: a comparative study. *IEEE Trans. Wireless Commun.*, **3**:5 (2004), 1602–1615.

[132] Q. H. Spencer, J. W. Wallace, C. B. Peel, T. Svantesson, A. L. Swindlehurst, H. Lee & A. Gumalla, Performance of multi-user spatial multiplexing with measured chan-nel data. In G. Tsoulos, ed., *MIMO System Technology for Wireless Communications* (Boca Raton, FL: CRC Press, 2006), pp. 175–205.

[133] M. Stojnic, H. Vikalo & B. Hassibi, Rate maximization in multi-antenna broadcast channels with linear preprocessing. In *Proc. IEEE Globecom*, Dallas, TX, Nov–Dec 2004, 3957–3961

[134] G. L. Stuber, J. R. Barry, S. W. McLaughlin, Y. G. Li, M. A. Ingram & T. G. Pratt, Broadband MIMO-OFDM wireless communications. *Proc. IEEE*, **92**:2 (2004), 271–294.

[135] W. Su & X.-G. Xia, Two generalized complex orthogonal spacetime block codes of rates 7/11 and 3/5 for 5 and 6 transmit antennas. *IEEE Trans. Inf. Theory*, **49**:1 (2003), 313–316.

[136] C. Sun, A. Hirata, T. Ohira & N. C. Karmakar, Fast Beamforming of electroni-cally steerable parasitic array radiator antennas: theory and experiment. *IEEE Trans. Anten. Propagat.*, **52**:7 (2004), 1819–1832.

[137] P. H. Tan, Y. Wu & S. Sun, Link adaptation based on adaptive modulation and cod-ing for multiple-antenna OFDM system. *IEEE J. Sel. Areas Commun.*, **26**:8 (2008), 1599–1606.

[138] V. Tarokh, N. Seshadri & R. A. Calderbank, Space-time codes for high data rate wireless communication: Performance criterion and code construction. *IEEE Trans. Inf. Theory*, **44**:2 (1998), 744–765.

[139] V. Tarokh, H. Jafarkhani & A. R. Calderbank, Space-time block codes from orthogonal designs. *IEEE Trans. Inf. Theory*, **45**:5 (1999), 1456–1467.

[140] V. Tarokh & H. Jafarkhani, A differential detection scheme for transmit diversity. *IEEE J. Sel. Areas Commun.*, **18**:7 (2000), 1169–1174.

[141] I. E. Telatar, Capacity of multi-antenna Gaussian channels. *Euro. Trans. Telecom.*, **10**:6 (1999), 585–595.

[142] E. Tiirola & J. Ylitalo, Comparison of beam-forming and diversity techniques in terms of UTRA FDD uplink capacity. In *Proc. Nordic Radio Symp.*, Oulu, Finland, Aug 2004.

[143] M. Trivellato, F. Boccardi & H. Huang, On transceiver design and channel quantization for downlink multiuser MIMO systems with limited feedback. *IEEE J. Sel. Areas Commun.*, **26**:8 (2008), 1494–1504.

[144] D. Tse & P. Viswanath, *Fundamentals of Wireless Communications* (Cambridge, UK: Cambridge University Press, 2005).

[145] M. K. Varanasi & T. Guess, Optimum decision feedback multiuser equalization with successive decoding achieves the total capacity of the Gaussian multiple-access channel. In *Proc. Asilomar Conf. Signals, Systems, Computers*, Montery, CA, Nov 1997, **2**, 1405–1409.

[146] G. P. Villardi, G. T. F. De Abreu, & R. Kohno, Modified orthogonal space-time block codes for time-selective fading channels. *IEEE Trans. Veh. Tech.*, **57**:6 (2008), 3921–3927.

[147] M. Vu, Precoding design. In E. Biglieri, R. Calderbank, A. Constantinides, A. Goldsmith, A. Paulraj & H. V. Poor, eds., *MIMO Wireless Communications* (Cambridge, UK: Cambridge University Press, 2007), pp. 88–139.

[148] M. Vu & A. Paulraj, On the capacity of MIMO wireless channels with dynamic CSIT. *IEEE J. Sel. Areas Commun.*, **25**:7 (2007), 1269–1283.

[149] F. Wan, W.-P. Zhu & M. N. S. Swamy, A semiblind channel estimation approach for MIMO-OFDM systems. *IEEE Trans. Signal Process.*, **56**:7 (2008), 2821–2834.

[150] F. Wan, W.-P. Zhu & M. N. S. Swamy, A signal-perturbation-free whitening-rotation-based approach for MIMO channel estimation. *IEEE Trans. Signal Process.*, **57**:8 (2009), 3154–3166.

[151] F. Wan, W.-P. Zhu & M. N. S. Swamy, Frequency-domain semi-blind channel estimation for MIMO-OFDM systems. Submitted to *IEEE Trans. Wireless Commun.*

[152] X. Wang & H. V. Poor, Space-time multiuser detection in multipath CDMA channels. *IEEE Trans. Signal Process.*, **47**:9 (1999), 2356–2374.

[153] Z. Wang & G. B. Giannakis, Outage mutual information of space-time MIMO channels. *IEEE Trans. Inf. Theory*, **50**:4 (2004), 657–662.

[154] M. Wennstrom & T. Svantesson, An antenna solution for MIMO channels: The switched parasitic antenna. In *Proc. IEEE PIMRC*, San Diego, CA, Sep 2001, **1**, A-159–A-163.

[155] J. H. Winters, J. Salz & R. D. Gitlin, The impact of antenna diversity on the capacity of wireless communication systems. *IEEE Trans. Commun.*, **42**:2–4 (1994), 1740–1750.

[156] Y. C. Wu, S.-C. Chan & E. Serpedin, Symbol-timing estimation in space-time coding systems based on orthogonal training sequences. *IEEE Trans. Wireless Commun.*, **4**:2 (2005), 603–613

[157] F. Xu, D.-W. Yue, F. C. M. Lau & Q. F Zhou, Closed-form expressions for symbol error probability of orthogonal space-time block codes over Rician-Nakagami channels. *IET Commun.*, **1**:4 (2007), 655–661.

[158] L.-L. Yang, MIMO-assisted space-code-division multiple-access: linear detectors and performance over multipath fading channels. *IEEE J. Sel. Areas Commun.*, **24**:1 (2006), 121–131.

[159] T. Yoo & A. Goldsmith, On the optimality of multiantenna broadcast scheduling using zero-forcing beamforming. *IEEE J. Sel. Areas Commun.*, **24**:3 (2006), 528–541.

[160] A. van Zelst & J. S. Hammerschmidt, A single coefficient spatial correlation models for multiple-input multiple-output (MIMO) radio channels. In *Proc. URSI 27th General Assembly*, Maastricht, Netherlands, 2002, 657–660.

[161] H. Zhang, H. Dai & B. L. Hughes, Analysis on the diversity-multiplexing tradeoff for ordered MIMO SIC receivers. *IEEE Trans. Commun.*, **57**:1 (2009), 125–133.

[162] R. Zhang, Y.-C. Liang, R. Narasimhan & J. M. Cioffi, Approaching MIMO-OFDM capacity with per-antenna power and rate feedback. *IEEE J. Sel. Areas Commun.*, **25**:7 (2007), 1284–1297.

[163] L. Zheng & D. L. C. Tse, Diversity and multiplexing: a fundamental tradeoff in multiple-antenna channels. *IEEE Trans. Inf. Theory*, **49**:5 (2003), 1073–1096.

Ultra wideband communications

20.1 Introduction

UWB technology, also known as *impulse radio*, was first used to transmit Morse codes by Marconi in 1900 through the transatlantic telegraph. Modern UWB technology has been used for radar and communications since the 1960s. Like CDMA systems, early UWB systems were designed for military covert radar and communications. The early applications of UWB technology were primarily related to radar, driven by the fine-ranging resolution that comes with large bandwidth. UWB technology for wireless communications was pioneered by Scholtz [48]. With the intent of operating UWB in an unlicensed mode that overlaps licensed bands, the FCC issued rules under the FCC Rules and Regulations Part 15 for UWB operation in February 2002.

The FCC defined a UWB transmitter as "an intentional radiator that, at any point in time, has a fractional bandwidth equal to or greater than 0.20, or has a UWB bandwidth equal to or greater than 500 MHz, regardless of the fractional bandwidth". "The UWB bandwidth is the frequency band bounded by the points that are 10 dB below the highest radiated emission, as based on the complete transmission system including the antenna."

According to the FCC regulations, the transmitter sends pulses with a bandwidth of at least 500 MHz that is within the band 3.1 to 10.6 GHz, for output power densities below -41.25 dBm/MHz. The FCC Part 15 limit of 500 μV/m at 3 meters is equivalent to an effective isotropic radiated power (EIRP) of -41.25 dBm/MHz. The FCC spectrum masks for indoor and outdoor commercial UWB systems are plotted in Fig. 20.1. In Europe, the ETSI TG31 defined its UWB spectrum emission mask specification based on the FCC regulation. Many other countries proposed their UWB frequency allocations based on the FCC regulation.

Features of UWB technology

Due to its wideband nature, UWB is suitable for short-range ground- and tree-penetrating imaging. Some advantages of UWB communications are

- *Extremely simple radio, and thus low cost.* This arises from the carrier-free nature of signal transmission: the UWB signal spans a wide frequency spectrum, and RF mixing circuitry is not required. Thus, the entire UWB transceiver can be implemented as a single CMOS chip, enabling small size, low cost, and low power.
- *High data rates.* The wide bandwidth enables very high capacity at short distances.

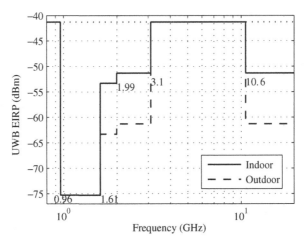

Figure 20.1   FCC spectrum masks for indoor and outdoor commercial UWB systems.

- *Multipath immunity.* The wide bandwidth provides frequency diversity to overcome multipath fading. The fine delay resolution property makes UWB radio a viable candidate for communications in dense multipath environments.
- *Large processing gain in presence of narrowband interference.*
- *Easy implementation of location.* The fine time resolution provides this capability at the centimeter level.
- *Long battery life.*

UWB communications are mainly in rich-scattering indoor environment, which is desired for MIMO implementation. The GHz center frequency of UWB radio relaxes the requirements on the antenna spacing. As a result, UWB-MIMO, the combination of UWB and MIMO technology is a viable solution for achieving very high data-rate, short-range wireless communication.

The difficulties in implementation are multipath energy capture, ISI, and the need for high-sampling-rate ADCs. The difficulty in multipath energy capture is due to the extremely low PSD of the UWB signals. This low PSD allows them to share the spectrum with existing RF devices. If a rake receiver is used to collect the multipath energy, a large number of fingers is needed due to the large number of multipath components.

DSSS has a slightly better BER performance than UWB for the same number of users, given the same frequency bandwidth constraint [22]. However, signal processing for CDMA is most difficult due to the very short chip period, and UWB offers a much cheaper solution. For the UWB bandwidth of several gigahertz, a solution using DSSS is impossible.

Detection and avoidance

UWB shares the frequency bands with other standards such as the WiMAX (3.4–3.8 GHz) and IEEE 802.11g (5 GHz). The detection and avoidance (DAA) technique is used for solving the interference issues. The UWB transmitter first detects the presence of another active device as well as its likelihood of interference, and avoids that specific band. This allows the UWB system to operate across a continuous range of spectrum. The detection is based on FFT. After a potential interfering signal is detected, a number of techniques can be used to reduce interference – transmit power control, frequency notching, and more advanced techniques. Transmit power control transmits the lowest possible power for reception.

Applications of UWB

UWB technology can be viewed as a kind of spread spectrum technology. Initial research on UWB radio is based on time-hopping spread spectrum (THSS). THSS can be implemented as a multiple access technology. Time-hopping UWB (TH-UWB) is the traditional UWB scheme, and it is an example of *impulse radio UWB* or *pulsed UWB*. TH-UWB has strict restriction on synchronization. TH-UWB can be implemented in analog or digital form. Impulse radio represents each symbol by one or a sequence of short pulses. DS-UWB is another pulsed UWB scheme that is based on DS-CDMA. Another popular UWB scheme is OFDM-UWB, which is based on OFDM modulation.

Pulsed UWB is a form of bandpass communications. Due to its broad bandwidth, UWB enables both high-data-rate short-range wireless PAN connectivity and longer-range, low-data-rate applications such as sensor, positioning, and identification networks. In addition to its applications in wireless PANs, UWB technology can be used for wireless LANs, wireless sensor networks, radar imaging, vehicular radar systems, location and tracking. FCC has specified a spectral mask for UWB systems for each of these applications. It can also be delivered over wire lines and cables without affecting the existing services, but needs double the bandwidth. UWB technology has now aroused worldwide interest.

Due to the characteristics of low power, high data rate, and limited range, UWB is positioned for the market of high-speed wireless PANs. UWB is now targeted at in-home multimedia applications, such as IEEE 802.15.3a. IEEE 802.15.3a is an initial implementation of UWB with a short range (less than 10 m) and a data rate above 100 Mbits/s. The DS-UWB and MB-OFDM schemes are two major proposals for IEEE 802.15.3a. In its spectrum band, UWB may interfere with, or be interfered with by, wireless LANs at the 5-GHz band. Interference from the 5-GHz wireless LAN band can be removed by using a notch filter or using multiband. In addition, DS-UWB is used in the baseline IEEE 802.15.4a standard.

UWB is an enabling technology for evolution to high-speed Wireless USB, Wireless 1394, Bluetooth 3.0, and ECMA-368/369. Wireless USB is the wireless evolution of the most widespread wired USB (Universal Serial Bus) 2.0 standard, and Wireless 1394 is the wireless evolution of IEEE 1394. Wireless USB is being developed by the USB Implementer's Forum and Wireless 1394 by the 1394 Trade Association. ECMA-368/369 (ISO/IEC 26907/26908) is specified by ECMA (European Computer Manufacturers Association) International, an ISO-accredited international standards body, for high-speed and short-range wireless PANs. All these standards employ the MB-OFDM UWB technology standardized by the WiMedia Alliance, and utilize all or part of the spectrum between 3.1 and 10.6 GHz with peripheral device interconnections at up to 480 Mbits/s. ECMA is also developing its TC32-TG20 standard for wireless PAN, which targets at 2 to 10 Gbits/s data transport at the 60-GHz band.

UWB is believed to have its most wide perspective for interfacing consumer electronics and computer peripherals. Both the WiMedia Alliance and the UWB Forum try to support Certified Wireless USB, wireless streaming video, Wireless 1394, and Bluetooth over UWB in their standard activities. UWB is also very promising for use in wireless BAN due to the very low power consumption.

20.2 UWB indoor channel

The power spectrums of a narrowband signal, a DS-CDMA signal, and a UWB signal are compared in Fig. 20.2. For a given transmit power, the UWB has the lowest PSD.

The IEEE 802.15.3a Task Group recommended a channel model in November 2002 [15], which is basically a modified version of the Saleh-Valenzuela model. Model parameters corresponding to several ranges are provided for both LOS and NLOS scenarios. The IEEE 802.15.3a Task Group developed four UWB indoor channel models to support its evaluation of the proposed UWB standards [34]. The IEEE 802.15.3a UWB model could provide reasonably accurate simulation results for UWB channels channels in vehicular environments (within a vehicle and outdoors in proximity to the vehicle) [42].

The UWB radio channel around the human body in a typical indoor environment has been measured and modeled in [16]. The body area channel consists of an initial cluster of components diffracting around the body and subsequent clusters of components reflecting from surrounding objects. Components diffracting around the body are well described by a high path loss exponent and correlated lognormal variables. Subsequent clusters have a more complex structure that can be described by a modified Saleh-Valenzuela model. Based on the measurements, a simple statistical channel model is given in [16].

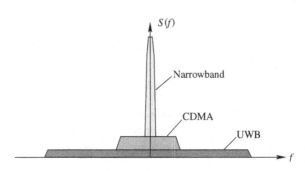

Figure 20.2   Power spectrums of narrowband, CDMA, and UWB signals.

Large-scale channel model

In indoor environments, measurements show that shadowing dominates the channel. At a given distance d, the received signal has a path loss that is log-normally distributed about the mean [23]

$$PL(d) = PL_{fs} + 10\alpha \log_{10}\left(\frac{d}{d_0}\right) + X_\sigma \quad (dB), \tag{20.1}$$

where d_0 is the reference distance, α is the exponent of path loss, X_σ is a zero-mean log-normally distributed random variable with standard deviation σ dB, and PL_{fs} is the free-space path loss.

For UWB communications, the frequency-dependence of the propagation loss may be significant, and this can be considered in the UWB path loss model by multiplying the distance-dependent $PL_{fs}(d)$ term by a factor that is associated with frequency $PL_f(f)$ [52]. In the channel model adopted in IEEE 802.15.3a, PL_{fs} is calculated by using the center frequency $f_c = \sqrt{f_H f_L}$, where f_H and f_L are obtained at the -10 dB edges of the spectrum; X_σ (dB) has a zero-mean normal distribution with $\sigma = 3$. Some results for the in-home channel such as path loss parameters, the number of multipath components, mean excess delay, and rms delay spread are given in [21].

Measurements show that some LOS environments provide better path loss than free space (that is, $\alpha < 2$). This is due to the additional power collected via reflections in indoor channels with an LOS path, and the received signal power is greater than that in free space [5]. This phenomenon has also been demonstrated for narrowband channels [41].

Small-scale fading model

The small-scale fading of the UWB channel is very different from that of the narrowband channel. The two major differences are [52]

- The number of multipath components received at the receiver within the period of an UWB waveform is very small. Thus, channel fading is not as severe as that in the narrow-band channel, and a large number of resolvable multipath components can be observed at the receiver.
- Due to the fine time resolution, the time of arrival of the multipath components will not be continuous, but is in clusters. This is different from the continuous case for the narrowband channel.

IEEE 802.15.3a Task Group considered three main indoor channel models for UWB systems [15]: the tapped-delay-line fading model [7], the Δ-K model [24], and the Saleh-Valenzuela model [47]. The tapped-delay-line fading model allows frequency selectivity of the UWB channel to be taken into consideration, but the clustering characteristic of the UWB channel is not reflected. This clustering property is considered in the Δ-K and Saleh-Valenzuela models. The Saleh-Valenzuela model was introduced in Section 3.1.7.

Table 20.1. Channel model parameters adopted in IEEE 802.15.3a.

Parameters	CM1	CM2	CM3	CM4
Λ (1/ns)	0.0233	0.4	0.0667	0.0667
λ (1/ns)	2.5	0.5	2.1	2.1
Γ	7.1	5.5	14	24
γ	4.3	6.7	7.9	12

Standard UWB channel model

Based on the log-normal path loss distribution and the Saleh-Valenzuela fading channel model, four sets of channel model (CM) parameters for different measurement environments are adopted in IEEE 802.15.3a, namely CM1, CM2, CM3, and CM4, for 3.1–10.6 GHz. CM1 and CM2 are, respectively, for the LOS and non-LOS scenarios with the distance between the transmitter and receiver being less than 4 m. CM3 is for a non-LOS scenario for a distance of 4 to 10 m. CM4 is defined for an environment with a strong delay dispersion, which has a delay spread of 25 ns. The model parameters for the four standard models are listed in Table 20.1 [8, 52]. IEEE 802.15.3c uses the 57–64 GHz unlicenced band, and channel parameters for some environments are given in [59].

A low-frequency channel model for the band of 300–800 MHz was proposed in [7], based on a measurement campaign performed in a typical office environment using baseband 1-ns pulses. It characterizes the shape of the PDP using a tapped-delay-line model. The low-frequency model has been accepted in IEEE 802.15.4a for performance evaluation of UWB systems operating below 1 GHz for indoor office-type environments [35]. A modified Saleh-Valenzuela channel model for low-data-rate UWB communications has also been formulated in IEEE 802.15.4a, and channel parameters are specified for CM-1 through CM-9, corresponding to indoor residential (LOS and non-LOS), indoor office (LOS and non-LOS), industrial (LOS and non-LOS), outdoor (LOS and non-LOS), and open outdoor (non-LOS) environments [35].

IEEE 802.15.4a also specifies a channel model for UWB BAN at 2 to 6 GHz [35]. As the model is based on FDTD simulations with 2 GHz bandwidth, it is not applicable to systems with larger bandwidth. FDTD simulations indicate that in the 2-6 GHz range, no energy is penetrating through the body.

The high-frequency (i.e., 3.1–10.6 GHz) and low-frequency channel models differ mainly in the arrival statistics and the amplitude distributions of multipath components [8]. The high-frequency model is sparse, i.e., there are resolvable delay bins that do not carry significant power. The amplitude distribution is lognormal for the high-frequency model while it is Nakagami-m for the low-frequency model. The variance of the lognormal distribution is assumed to be independent of delay, whereas the m-parameter of the Nakagami distribution decreases with delay. The pdf of the high-frequency model is generally non-monotonic and sparse, and the direct path is not necessarily the strongest one.

20.3 UWB capacity

For link budget analysis, the narrowband Friis transmission formula is not applicable, since it applies only to the power in continuous-wave (CW) sinusoidal signals and does not account for pulse distortion effects at either the transmit or receive antenna or even the type of waveform. Pulse radio transmission differentiates from a narrowband continuous-wave system in the distortion introduced by practical transmit and receive antennas. However, the narrowband model can be used to approximate the path loss for a UWB system [22].

Given the 7.5 GHz UWB band, the operating bandwidth can be estimated by link budget analysis. Assuming that the transmitted PSD is flat, the received signal power P_{RX} is given by

$$P_{RX} = -41.25 + 10 \log_{10} (f_U - f_L) - 20 \log_{10} \left(\frac{4\pi \bar{f}}{c} \right) - 10\alpha \log_{10} d \quad \text{(dBm)}, \quad (20.2)$$

where f_L and f_U are the lower and upper frequencies in the UWB band as per the FCC rules, $\bar{f}$ is the geometric mean of f_L and f_U, d is the distance in meters, α is the exponent of the path loss, and $c = 3 \times 10^8$ m/s is the speed of light. According to this equation, given $f_L = 3.1$ GHz and a distance of 10 m, the received power increases as f_U increases.

The SNR at the receiver can be determined as

$$\gamma = P_{RX} - L - N_0 - 6 \quad \text{(dB)}, \quad (20.3)$$

where L is an additional implementation loss, $N_0 = -114$ dBm/MHz is the standard thermal noise PSD, and a noise figure of 6 dB is applied.

The Shannon capacity for a single user in the AWGN channel is given by

$$C(d) = B \log_2(1 + \gamma) \quad \text{(bits/s)}. \quad (20.4)$$

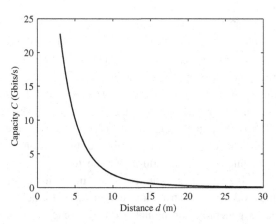

Figure 20.3   **UWB capacity versus distance.**

Example 20.1: The capacity of the UWB system, given by (20.4), is shown in Fig. 20.3, for $f_L = 3.1\,\text{GHz}$, $f_U = 10.6\,\text{GHz}$, the path loss exponent $\alpha = 3$ and $L = 35\,\text{dB}$.

20.4 Pulsed UWB

UWB has the attractive features of the time-domain nature of signal transmission. The UWB system can be designed as a single-band or multiband system, with each sub-band greater than 500 MHz. A single sub-band can occupy up to 7,500 MHz UWB spectrum. These schemes generate different performance tradeoffs and design challenges. Early single-band UWB systems generate simple, short pulses with wide spectral occupation. Multiband UWB systems use high-order modulation on constrained bandwidth to enable channelization.

The single-band approach is commonly treated as *pulsed* or *carrier-free communications*. Information is directly modulated into a sequence of impulse-like waveforms that occupy a bandwidth of up to 7.5 GHz. The single-band approach is the traditional method for UWB implementation.

20.4.1 Pulse shape

A typical pulse shape for UWB is the Gaussian doublet. This waveform can be easily generated from a square pulse by exploiting pulse shaping and filtering effects of the transmit and receive antennas [22]. The fast switching on and off at nanosecond order or shorter for square pulse generation actually generates a Gaussian-like pulse shape that approximates a Gaussian function

$$G(x) = \frac{1}{\sqrt{2\pi\sigma^2}} e^{-\frac{x^2}{\sqrt{2\sigma^2}}}. \tag{20.5}$$

This waveform is known as a *Gaussian pulse*. The transmit antenna has the highpass filtering effect of a derivative operation, and the transmitted waveform is known as a *Gaussian monocycle*. The receive antenna has the same effect as a derivative operation, and thus the received second-derivative of the Gaussian pulse is known as a *Gaussian doublet*. These pulses are illustrated in Fig. 20.4. In addition, free-space propagation produces a lowpass filtering or integration effect on the transmitted UWB signal [22].

The basic Gaussian pulse is represented by

$$y_{GP}(t) = K_1 e^{-\left(\frac{t}{\tau}\right)^2}, \tag{20.6}$$

where τ is the time-scaling factor and K_1 is a constant. Derivative of the equation is similar to highpass filtering the Gaussian pulse.

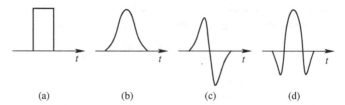

Figure 20.4 Pulse generation for the TH-UWB system. (a) Square pulse. (b) Gaussian pulse. (c) Gaussian monocycle. (d) Gaussian doublet.

Taking the first- and second-order derivatives of the Gaussian pulse, we have the Gaussian monocycle and the Gaussian doublet

$$y_{\text{GM}}(t) = K_2 \frac{-2t}{\tau^2} e^{-(\frac{t}{\tau})^2}, \tag{20.7}$$

$$y_{\text{GD}}(t) = K_3 \frac{-2}{\tau^2} \left(1 - \frac{2t^2}{\tau^2}\right) e^{-(\frac{t}{\tau})^2}, \tag{20.8}$$

where K_2 and K_3 are constants.

The constants K_1, K_2, and K_3 can be determined by the energy of the transmitted pulse, E, that is,

$$E = \int_{-\infty}^{\infty} y^2(t) dt, \tag{20.9}$$

where y corresponds to y_{GP}, y_{GM}, or y_{GD}. Accordingly, [22]

$$K_1 = \sqrt{\frac{E}{\tau\sqrt{\pi/2}}}, \quad K_2 = \sqrt{\frac{\tau E}{\sqrt{\pi/2}}}, \quad K_3 = \tau\sqrt{\frac{\tau E}{3\sqrt{\pi/2}}}. \tag{20.10}$$

The frequency spectrums of these pulses are, respectively, obtained by applying the Fourier transform to these pulses

$$Y_{\text{GP}} = K_1 \tau \sqrt{\pi} e^{-(\pi\tau f)^2}, \tag{20.11}$$

$$Y_{\text{GM}} = K_2 \tau \sqrt{\pi} (j2\pi f) e^{-(\pi\tau f)^2}, \tag{20.12}$$

$$Y_{\text{GD}} = K_3 \tau \sqrt{\pi} (j2\pi f)^2 e^{-(\pi\tau f)^2}. \tag{20.13}$$

Example 20.2: The waveforms of the pulses given by (20.6)–(20.8) and their frequency spectrums for $\tau = 0.1$ ns and $E = 1$ are shown in Fig. 20.5. From the figure, we notice that the 3-dB bandwidth of each waveform is around 3 GHz. The Gaussian doublet has the best 3-dB bandwidth, which is 3.68 GHz. These UWB waveforms are almost uniformly distributed in a wide frequency spectrum, and thus they are similar to noise.

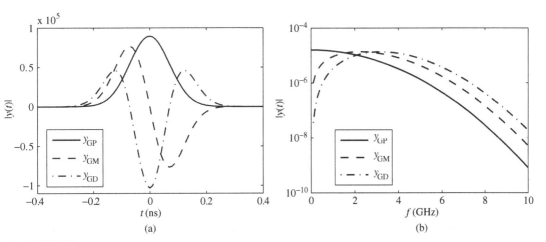

Figure 20.5 Gaussian pulse, monocycle, and doublet. (a) Time domain. (b) Frequency domain.

A UWB waveform may not cover the specified band. For example, the Gaussian pulse covers the frequency band from zero upwards, and the dc component is undesirable. A UWB waveform can be moved to a specified band with center frequency f_c by multiplying it by a sinusoid $\cos(2\pi f_c t)$. This method also eliminates the dc component in the Gaussian pulse.

Other useful UWB waveforms are orthogonal Hermite pulses and orthogonal prolate spheroidal wave functions (PSWFs) [22]. A set of PSWF pulses can be generated using a single-source signal, and they are suitable for M-ary communications; PSWF pulses of different orders are orthogonal to one another, and they have the same pulse width and the same bandwidth. In [30], the Rayleigh monocycle, Laplacian monocycle, cubic monocycle are also proposed as UWB waveforms. In order for the UWB waveforms to comply with the FCC's spectrum mask, both filtering and up-conversion have to be performed.

Generally, the spectrum bandwidth is the inverse of the pulse duration in the time domain, which is determined by the time scale factor τ in the case of Gaussian waveforms. UWB pulses can be generated by using avalanche transistors, tunnel diodes, and step-recovery diodes [54].

20.4.2 Modulation and multiple access for pulsed UWB

Modulation

The modulation technique used in the pulsed UWB system is usually selected from PAM, PPM, OOK, or biphase modulation. These modulation techniques are illustrated in Fig. 20.6. Biphase modulation is usually treated as BPSK.

In binary PPM signaling, '1' and '0' are represented by a pulse without any delay and by a pulse with a delay relative to the time reference, respectively. For the M-ary PPM system, the set of time shifts for representing different symbols is specified. The orthogonal PPM

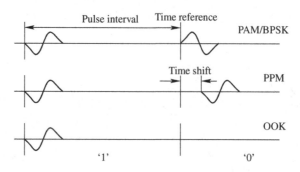

Figure 20.6

Modulation techniques for pulsed UWB.

scheme uses UWB pulse shape that is orthogonal to its time-shifted version. These time shifts depend on the UWB pulse.

PAM modulates the bits using different amplitudes, OOK represents bits by the presence of a pulse, while biphase modulation denotes the bits using the pulse polarity. Orthogonal pulse modulation requires special pulse shapes that are orthogonal to one another.

PPM and biphase modulation are the most common techniques. Orthogonal pulse modulation is useful for multiple access. PAM and OOK are rarely used. For PAM, a signal with smaller amplitude is more susceptible to noise interference than its larger amplitude counterpart, and also more power is required to transmit the higher amplitude pulse. For OOK, the presence of multipaths makes the determination of the absence of a pulse very difficult. Also, OOK, like biphase modulation, cannot be extended to M-ary modulation.

Biphase modulation and binary PPM have similar performance, but OOK is inferior in error performance. However, the OOK signal can be detected by a simple energy detection scheme. On the other hand, PAM inherently provides smoother PSD than OOK and PPM, both generating discrete spectrum lines, which may become severe interference to existing narrowband radios.

Multiple access

Duty-cycled DSSS systems achieve the wideband capacity as long as the number of independently faded resolvable paths increases sublinearly with the bandwidth, whereas duty-cycled PPM systems can achieve this capacity only if the number of paths increases sublogarithmically [40]. The difference arises from the fact that DSSS is spectrally more efficient than PPM and hence allows more bursty transmission.

For pulsed UWB, TH and DS spreading are two good multiple-access techniques. TH-UWB [48] utilizes low-duty-cycle pulses, where the time spreading between the pulses is used to provide time multiplexing of users. Typically, the pulses in TH-UWB have a duty cycle of about 1%. DS-UWB [14] utilizes high-duty-cycle pulses, whose polarities follow PN sequences.

Both TH- and DS-UWB techniques assign each user with a different PN sequence to encode the pulses in either position (for PPM) or polarity (for biphase modulation). Each bit is represented by using N_c pulses. Channelization is based on the codes. The spreading leads to a processing gain G_p, which increases the effective SNR as the UWB signal is processed by a correlating receiver based on matched filtering. For TH-UWB, the processing gain is given by [22]

$$G_p = \frac{N_c}{T_p/T_f}, \tag{20.14}$$

where T_p/T_f is the duty cycle, T_p is the period of the pulse and T_f is the frame period.

TH-UWB works in a way very similar to PPM. It is actually a kind of PPM where a code sequence determines the times for the transmitter to key on and off. The major difference between PPM and TH-UWB lies in the fact that the former employs pulse position patterns to represent the data symbols, whereas the latter denotes a code sequence which is further used for decoding the data information.

The TH and DS multiple-access schemes have been compared for single-antenna systems in [53]. It is shown that the TH-UWB system is suitable for analysis, but DS-UWB is more suitable for single-carrier UWB communications [52, 53]. THSS is not as popular as DSSS and FHSS, and this is due to the implementation difficulty in generating narrow impulses at nanosecond scale and providing very good timing accuracy.

Due to the technical difficulties associated with carrier sensing in UWB systems, CSMA is not effective for random access MAC. Slotted ALOHA may be used for UWB, but its inefficiency makes it unsuitable for high-speed data transmission. A channel-partition MAC such as hybrid TMDA/CDMA is better suited for high-data-rate UWB networks. For control signaling, random access schemes such as slotted ALOHA can be used.

UWB technology is capable of accurate positioning, and this enhances the UWB MAC performance as routing and power control can be simplified.

20.4.3 Time-hopping and direct-sequence UWB signals

A single pulse does not represent much information. Data needs to be modulated onto a sequence of pulses called a *pulse train*. A pulse train $s(t)$ can be written as

$$s(t) = \sum_{k=-\infty}^{\infty} p\left(t - kT_f\right), \tag{20.15}$$

where T_f is the pulse repetition or frame period, and $p(t)$ is the pulse shape. The resulting spectrum contains spectrum lines at frequencies $\frac{k}{T_f}$, $k = 0, 1, 2, \ldots$. These spectrum lines are called *comb lines*. Increasing the pulse rate $\frac{1}{T_f}$ leads to an increase in the amplitude of the frequency spectrum.

The pulsed-UWB system depends on the pulse train, since it is modulated in the time domain. Spectral lines occur for both PPM and time hopping. The spectral lines either violate the regulations, or require a power backoff. The powers of the comb lines can be reduced by dithering, which changes the regular pulse time of each pulse by a small random offset. When the offset is cyclic according to a known PN code $\{c(k)|k = 0, 1, \ldots, N_c - 1\}$, we get a TH-UWB system, where TH is also used for multiple access purpose. The frame interval T_f is divided into N_c segments of period T_c, and $N_c T_s \leq T_f$.

Randomization of the polarity of the transmitted symbols or pulses eliminates the spectral lines [36]. The use of long TH sequences makes synchronization more difficult. Randomizing the polarity of each transmitted pulse without changing the pulse positions can solve this disadvantage. A polarity randomization sequence is used for this purpose. The symbol waveform is the combination of a TH sequence and a polarity randomization sequence.

TH-UWB signals

The TH pulse train is written as

$$x(t) = \sum_{k=-\infty}^{\infty} p\left(t - kT_f - c(k)T_c\right). \tag{20.16}$$

In a synchronous network, orthogonal TH sequences satisfy $c_u(k) \neq c_{u'}(k)$ for any two users u and u' to avoid collision.

The TH pulse train can be combined with PAM, PSK, or PPM modulation to generate a TH-UWB signal. With MPAM modulation, we have the TH-UWB signal

$$x(t) = \sum_{k=-\infty}^{\infty} a_m(k)p\left(t - kT_f - c(k)T_c\right), \tag{20.17}$$

where $a_m(k)$, $m \in \{0, 1, \ldots, M - 1\}$, is the amplitude of the pulse carrying information $m(k)$. For BPSK modulation, $a_m(k) = \pm 1$. With M-ary PPM modulation, the information is carried by the positions of the pulses

$$x(t) = \sum_{k=-\infty}^{\infty} p\left(t - kT_f - c(k)T_c - m(k)T_d\right), \tag{20.18}$$

where T_d is the modulation delay for PPM, and $MT_d \leq T_f$.

TH impulse radio can be treated as a DS-CDMA system, where the spreading sequence has a large number of 0s and a small number of 1s. In comparison, the conventional DS-CDMA system uses spreading sequences that have an almost equal number of $+1$s or -1s. Due to this similarity, many results for DS-CDMA such as MUD can be applied to TH-UWB.

The processing gain of TH-UWB is very high. This offers an excellent multipath immunity since the high time resolution allows the separation and combination of all multipath components coherently at the receiver. The high processing gain ensures a low PSD, reducing interference to other existing systems.

DS-UWB signals

In DS-UWB with BPSK modulation, the binary symbol $b(k) \in \{-1, 1\}$ to be transmitted over the kth frame interval is spread by a sequence of N_c pulses, $\{c(n_c)p\left(t - kT_f - n_c T_c\right) | n_c = 0, 1, \ldots, N_c - 1\}$, where the polarities of the pulses are determined by the spreading sequence $\{c(n_c) \in \{-1, 1\}|n_c = 0, 1, N_c - 1\}$, which is unique for each user.

The signal of DS-UWB with BPSK modulation can be written as [14]

$$x(t) = \frac{1}{\sqrt{N_c}} \sum_{n=-\infty}^{\infty} b(n) \sum_{n_c=0}^{N_c-1} c(n_c) p(t - nT_f - n_c T_c), \qquad (20.19)$$

where the hop period T_c is greater than the pulse period T_p

As in TH-UWB, modulation techniques such as PAM, OOK, and PSM can be used in DS-UWB. DS-UWB performs in a way similar to the DS-CDMA system. In the DS-UWB system, the ternary code is desirable, where the chips may take on a zero value, that is, a pulse may be absent. Ternary codes have more different codes available, leading to better code correlation properties.

The Ipatov ternary sequence has perfect periodic autocorrelation [26]: The periodic autocorrelation function is a train of equally spaced, equal amplitude impulses. Unlike the m-sequence that exists only for length $2^m - 1$, the Ipatov ternary sequence exists for many more lengths: 7, 13, 21, 31, 57, 63, 73, 91, 127, 133, 183, 273, 307, 341, 364, 381, 511, $\cdots$. The lengths are typically of the form $N^2 - N + 1$, but not all numbers of this type are represented. The Ipatov ternary sequence can be generated from cyclic difference sets [25].

In the DS-UWB proposal by UWB Forum for IEEE 802.15.3a, the modulation scheme is either BPSK or 4-ary biorthogonal keying (4-BOK). The ternary code is used in DS-UWB for acquisition.

A large family of UWB signals, including TH and DS signaling with pulse position, interval, and amplitude modulations, can be modeled by shot noise processes [43]. The exact power spectrum can be evaluated using shot noise spectral theory. Different features of the signal model contribute clearly and separately to the resulting spectral expressions.

20.4.4 Pulsed-UWB transceivers

Pulsed-UWB uses baseband pulse shapes having rise and fall times at the sub-nanosecond level. Such pulses have a broadband spectrum, ranging from near-dc to several gigahertz. The pulse shape is applied directly to a transmit antenna, without using upconversion. Spectral shaping can be performed by tuning the shape of the ultra-short duration pulse, or by tuning the loading parameters of the transmit antenna. Pulsed-UWB transmitters that are compatible with CMOS and other IC technology are described in [54].

The most attractive feature of pulsed-UWB technology is its low complexity and low cost. The simplest TH-UWB transmitter may just contain a pulse generator, a timing circuit, and an antenna. The low complexity enables its inclusion in consumer electronics. For M-ary systems, a bit-to-symbol mapper is also required, and the symbols rather than bits are mapped to pulses. Precise timing circuitry is required, and is especially restricted for PPM. Pulses may be amplified before transmitting. FEC schemes may also be included in the system.

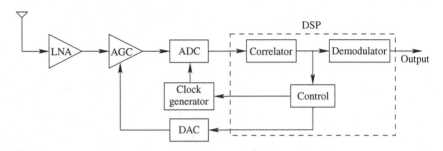

Figure 20.7   **Block diagram of a digital pulsed-UWB receiver.**

The block diagram of a general digital pulsed-UWB receiver is shown in Fig. 20.7. The correlator and control circuitry is used for acquisition and tracking of the required pulses among the other signals. The correlation receiver, also known as *matched filter*, is the optimal receiving technique. The correlator multiplies the received pulse by a template waveform, and then integrates the output over the duration of the pulse to generate a single dc voltage, from which the relative time positions are estimated for PPM. The template waveform is generated by using the pulse generator. For a multipath indoor channel, the matched filter receiver becomes the rake receiver.

The DS-UWB receiver has a structure similar to that of the general pulsed-UWB structure shown in Fig. 20.7, except that a code correlator, followed by Viterbi decoder and descrambler, is implemented in the demodulator block. The channel correlator (matched filter) and the code correlator are two linear operations, and their order can be swapped. This results in a modification of the other components.

Since the noise signal has zero correlation after pulse integration, the UWB receiver can acquire, track, and demodulate UWB signals that are significantly below the noise floor. The demodulator performs pulse-to-symbol mapping, and a symbol-to-bit mapper may be required for M-ary systems. The matched filter is often implemented as a rake receiver for multipath demodulation. A bandpass filter is usually used ahead of the LNA to remove any out-of-band interference.

Intel's pulsed-UWB proposal for IEEE 802.15.3a considers a bandwidth of 7.26 GHz. Its PPM transceiver has direct RF sampling and a 20-GHz 1-bit ADC with 1,280 parallel digital matched filters. Tunable notch filters are used for pulse shaping at the transmitter [2].

XtremeSpectrum (now a part of Freescale Semiconductor) proposed the DS-UWB system for IEEE 802.15.3a in May 2003 [56]. The Ipatov ternary sequence is used in the proposal. The DS-UWB transmitter first scrambles the transmit data to remove any correlation between adjacent bits. These scrambled bits are then coded with a convolutional code. The coded bits are then mapped into symbols, which are fed into the pulse generator to generates chips. If a chip is positive, a pulse is transmitted; for a zero chip, no pulse is transmitted; for a negative chip, a negative pulse is transmitted. The pulse has a spectral shape close to the desired final shape. The generated pulse train is subject to a bandpass filter to fine-tune the spectral shape, and is then passed to the antenna.

Pulsed UWB such as TH-UWB and DS-UWB spreads each information bit across the whole bandwidth in use, thus high attenuation at some frequencies is not very detrimental.

The high processing gain also helps to alleviate this problem, and it enables great immunity to distortion and noise. The noise can be a large quantization noise, which in turn allows a relatively low precision of the receiver's ADC. DS-UWB has a lot in common with TH-UWB.

A comprehensive introduction of various UWB receiver architectures designed for different modulation formats and signaling schemes is given in [2].

20.4.5 Challenges for pulsed UWB systems

The pulsed UWB signal can be detected by using conventional receivers. The rake receiver and the matched filter receiver require channel estimation. Differential detection [9], transmitted reference (TR) receivers [9, 17, 44], and differential TR receivers [9] do not need channel estimation. A matched filter is equivalent to a rake receiver with infinite fingers and perfect channel estimation. For differential detection systems, the symbols are differentially encoded, and each received data pulse is correlated with the previous one. Each pulse serves as a template for the next. Whatever the receiver architecture, a synchronization circuit must provide accurate information on the arrival times of the incoming pulses. This poses a serious challenge to UWB radios.

Single-band pulsed UWB has an inflexible spectrum mask. It also requires very wideband circuitry that introduces difficulties in implementation, especially in RF-CMOS implementation, and high sampling rates in DACs and ADCs. As with DS-CDMA, a strong interferer may block the wanted signal. The single-band system is more sensitive to ISI. Multiband UWB is a better choice. Multiband UWB can be implemented as a pulsed multiband or an MB-OFDM UWB. The multiband system needs an FH strategy for interference avoidance.

For a given data rate, the thermal noise of the single-band system, which is calculated by $N = -174 + 10 \log_{10} B$ dBm with B being the bandwidth, is several dB higher than that of the multiband system. The signal power is increased by the same amount, and thus the SNR is the same. In the single-band system, this has to be performed by using notch filters which, however, increase the noise figure of the receiver. The notch filter adds cost to the unit, and is not easily tunable. The multiband system can adaptively select sub-bands to provide good interference rejection, and thus is easier to implement.

An optimum UWB receiver is a matched filter plus the optimum MLSE detector (Viterbi algorithm). When implementing matched filtering, it is most difficult to generate exact pulses at the exact time at the receiver. A train of wider rectangular pulses coded with the same PN sequence is usually used as the template signal. A suboptimum receiver uses a generalized rake receiver structure, plus suboptimum equalizers such as zero-forcing equalizers. For example, the DS-UWB proposal of IEEE 802.15.3a adopts a rake receiver structure plus MMSE equalizer.

There are critical requirements on the ADC for the single-band pulsed-UWB system. The ADC must have a sampling rate of a few gigasamples/s. For a mimimal bandwidth of 500 MHz, the ADC must support a sampling rate at gigahertz level. As the received UWB signals are commonly immersed in noise or interference, an ADC with a few bits is sufficient [37]. However, the ADC needs a resolution greater than 4 bits to resolve signals from narrowband radios [46]. A typical solution is an interleaved flash ADC [13], where each channel is based on a flash converter; such an ADC consumes power at a scale that increases exponentially with bit precision. precision. MAXIM-Dallas provides the MAX108 chip, which is an 8-bit ADC with a sampling rate of 1.5 Gsamples/s, and the MAX19693, which is a 12-bit, 4.0 Gsamples/s DAC.

Synchronization and timing are extremely important for almost all of the pulsed-UWB receiver techniques (with the exception of the threshold detector receiver). Particularly, with synchronization error, correlation-based receivers will have a dramatic drop in the SNR at the receiver output. Most synchronizers are based on the ML principle that requires a clean template of the received pulse. However, the clean template is not available when the multipath channel is unknown. Dirty template based algorithms exploit the unique maximum of the cross-correlation of dirty templates extracted from the received waveform [58]. Relying on symbol-rate samples and integrate-and-dump operations, the algorithms ensure rapid synchronization by collecting multipath energy without knowledge of the channel. Compared to the nondata-aided mode, the data-aided mode expedites the synchronization and also enables timing in a multiuser environment. Dirty template synchronization algorithms have been developed for multiaccess PAM-UWB systems [58] and multiaccess PPM-UWB systems [57].

For implementation of single-band UWB, transmitter pulse shaping may have to be performed by purely analog filtering, since the computational complexity is daunting for using digital filtering. However, analog filtering reduces the available accuracy and repeatability of the transmitted pulse shape.

An additional constraint on the input impedance is placed on the LNA. It is in general difficult to design the LNA to minimize the effective noise figure subject to the input impedance requirement across the entire frequency band. Recent UWB LNAs that are based on CMOS technology can provide about 10–15 dB power gain with power consumption of around 10 mW over the specified UWB bandwidths [46].

The performance of the UWB antenna is also important for the UWB system. The antenna induces dispersion, which is often due to frequency dependence of material parameters such as permittivity. The antenna is a major source of distortion for time-domain pulses. Pulse distortion can be compensated by equalization.

20.4.6 Rake receivers

The rake receiver is a popular coherent receiver for pulsed UWB systems. The rake receiver needs to collect the rich multipath diversity: It needs to estimate the arrival time, the amplitude, and the phase of each multipath component. The detection performance critically depends on accurate estimation of channel tap coefficients and time delays. Synchronization becomes extremely difficult. The total number of multipath components for a UWB channel is significantly greater than that for a narrowband channel in a typical indoor environment, due to the large bandwidth of UWB. In an indoor environment, each pulse generates hundreds of resolvable multipath echoes, and this requires a large number of fingers to capture a significant part of the signal energy [33], causing a formidable complexity.

The very large bandwidth also leads to the high power consumption of an ADC which, following the Nyquist theorem, must operate at a rate of several Gsamples/s. It also causes the distortion that each path undergoes due to the frequency selectivity of the propagation environment. The channel estimation can be data-aided or non-data-aided by using an ML approach [33].

In typical UWB scenarios, the available number of MPCs at the receiver is often more than 100 [7]. Typically, 20 or more rake fingers are necessary to get a satisfactory performance. This places strong restriction on the equalizer and the rake receiver, since the complexity of a rake receiver increases exponentially with the number of multipath components. A reduced complexity rake receiver can be used by [7, 8]:

- *Maximum selection.* Select the L instantaneously strongest paths of L_1 resolvable paths.
- *Partial selection.* Select the first L nonzero arriving paths.
- *Threshold selection.* Select the first L paths the magnitudes of the gains of which are greater than a threshold.

This reduction in implementation complexity is achieved at the price of performance loss. For dense channels, the simpler partial selection rake scheme is almost as good as the maximum selection rake scheme in terms of bit error probability, even for a small number of fingers. In sparse channels, however, the maximum selection rake outperforms the partial selection rake significantly [8].

A pilot-channel-assisted LLR selective combining (PCA-LLR-SC) scheme for UWB rake receivers was proposed in [11] for long-range low-rate applications targeted by the IEEE 802.15.4a. The pilot and data channels are constructed using quadrature sinusoidal bursts that have the same Gaussian envelope. The system parameters are optimized through jointly minimizing the channel estimation mean square error and maximizing the receiver output SNR. The PCA-LLR-SC scheme is capable of providing robust low-rate UWB communications in fast-fading multipath channels and in the presence of multi-user interference.

20.4.7 Transmitted-reference receivers

TR system model

The TR system was originally proposed for spread-spectrum communications [9, 17, 44], but it did not gain popularity, since accurate channel estimation can be achieved by using rake receivers [17]. The TR receiver is a noncoherent receiver, and channel estimation is not required. It is regaining popularity in UWB communications, as accurate channel estimation is very challenging in this case.

In the TR scheme, a reference pulse, which is known to the receiver, is sent prior to each data pulse for channel sounding. This is based on the assumption of a constant channel over the symbol duration. The channel response to the former is exploited as a template for the latter. At the receiver, the received reference pulse is used as signal template for the data demodulation using correlation. Using an autocorrelation receiver, the total energy of all the multipath components is gathered to detect the signal. TR receivers collect full multipath diversity without channel estimation, and are robust to small timing offsets. A single sample may be sufficient to detect one data symbol. The mechanism can be improved by averaging the channel responses from several reference pulses [9, 17]. TR systems are robust to pulse distortions.

The TR receiver is a correlation receiver employing template signals, which are a noisy version of the channel response, followed by thresholding. The estimated symbol is

$$\hat{a}_j = \text{sgn}\left(r_j^T \hat{h}_j\right), \quad j = N_t, \ldots, N_s - 1, \tag{20.20}$$

where $\hat{h}_j$ and r_j are, respectively, the template signal and the vector of received samples for the jth data symbol.

Training-based TR systems construct their template signal by combining only the received signal corresponding to the training symbols

$$\hat{h}_{\text{TB}} = \hat{h}_j = \sqrt{\frac{N_f}{E_t N_t}} \sum_{i=0}^{N_t-1} a_i r_i, \tag{20.21}$$

where E_i is the energy of the ith symbol, $\{a_i\}$ are the information symbols taking values ± 1 with equal probability, N_f is the number of frames per symbol, N_t is the number of training symbols, and N_s is the number of symbols.

The TR receiver has to face noise enhancement. Since both the reference and data pulses experience the same noisy channel, the correlating operation over the full frame or symbol period accumulates (and colors) the noise. To improve the tradeoff between energy capture and noise accumulation, oversampling can be applied to take multiple samples per frame by speeding up the integrate and dump operation. Sparse channel responses with large delay spreads that are typical of UWB signals necessitate a long correlator integration time, leading to the collection of more noise energy. The TR scheme suffers from MAI and a data rate loss. Transmission of the reference symbol causes a power penalty of 3 dB. Large timing errors need to be compensated prior to transmit reference detection. A statistical analysis of the narrowband interference to the TR scheme has been made in [1]. The interference signals can be efficiently suppressed by using linear LS and MMSE combiners [1].

The TR scheme is often considered a low-data-rate scheme because of implicit assumptions that the pulse spacing D in a doublet should be longer than the channel length T_h to prevent inter-pulse interference, and the frame period T_f should be chosen such that there is no inter-frame interference, leading to $T_f > 3T_h$ [12]. A TR scheme that uses integrate and dump with oversampling was proposed in [12] to enable higher data rates in a multiuser context. Multiple reference delays are used to further improve the system performance similar to the role of multiple antennas in communication systems.

The performance of the TR scheme in the UWB channel has been derived in [39], where the fading of each resolvable multipath component is characterized by Nakagami distribution. Based on the derivation of exact and approximated expressions for the pdf of the output SNR, closed-form expressions for average SNR, amount of fading, outage probability, and average BEP of the receiver are given. In [29], the SEP analysis of multiuser TH-UWB systems employing M-ary PPM and a correlation receiver in the presence of MAI and timing jitter is performed by using the Gaussian quadrature rules (GQR) method.

Two timing recovery algorithms for UWB communications have been proposed in [6]. The algorithms exploit the samples of the received signal to estimate frame timing and symbol timing. Channel estimation comes out as a by-product and can be used for coherent matched filter detection. Both algorithms require sampling rates in the order of the inverse of the pulse duration. In [19], a semiblind ML-based synchronization scheme for recovering both symbol and frame timing has been developed. The proposed algorithm performs joint timing and channel estimation, and the impact of a low resolution ADC is shown to be marginal.

A balanced TR signaling scheme that combines the dual pulse with M-ary orthogonal encoding is designed in [27]. This approach mitigates both the inter-pulse interference and the MAI by increasing the frame time without sacrificing the data rate. A closed-form expression for the averaged SINR is derived for such systems in a realistic and standard UWB channel in [28].

Signals designed for both coherent and TR receivers

The TR signal can be demodulated with a coherent receiver, by simply throwing away the reference pulses. This, however, causes a 3-dB signal energy penalty compared to a signal that is designed for coherent receivers only. On the other hand, signals designed for coherent receivers cannot be demodulated by a TR receiver. A hybrid modulation method proposed in [60] enables coherent rake receivers and TR receivers in the same wireless network to simultaneously receive the signal. The key idea is to embed information into the reference pulse, without modifying the phase relationship between the reference pulse and the data pulse, since this phase relationship is critical for the TR receiver operation. This makes sure that the energy in the reference pulse is not wasted for the coherent receiver, and recovers the 3-dB loss by the normal TR receiver. Furthermore, we let the information in the reference pulse depend on the previous information symbol. The proposed signaling scheme is applicable not only to pulsed UWB systems, but also to narrowband or conventional spread spectrum systems.

Generalized TR receivers

To reduce this power penalty due to the reference symbol, a block of symbols that contains fewer reference symbols than data symbols can be transmitted, giving the generalized TR receiver [10, 18]. ML- and generalized likelihood ratio test (GLRT)-based receivers for UWB systems operating in a dense multipath channel exploiting training signals are derived in [18]. Performance gains of 3 and 4 dB for the ML and GLRT schemes in comparison to the training-based receiver have been observed. A 4-bit ADC in combination with an optimal choice for the AGC gain can provide a performance close to that of an infinite resolution ADC for the classical training-based receiver and the GLRT receiver, whereas an ADC with less than 3 bits introduces a substantial performance degradation

[20]. In [10], three simple generalized TR receivers are derived by optimally combining the channel estimates using the reference symbols, the data decisions and the data symbols. These receivers outperform the receivers proposed in [18].

20.5 Multiband UWB

The traditional TH-UWB scheme faces a number of difficulties for implementation. On the other hand, due to the technical maturity of DS-CDMA and OFDM, the DS-UWB and MB-OFDM UWB schemes are more popular. For UWB communications, DS-UWB is preferred for low-data-rate systems, while MB-OFDM UWB is preferred for high-data-rate systems. For example, IEEE 802.15.4a employs the DS-UWB scheme; IEEE 802.15.3a has an MB-OFDM UWB proposal and a DS-UWB proposal, but favors the MB-OFDM scheme. For location purpose, short pulsed UWB is preferred. PulsON is a UWB chipset developed by Time Domain Corporation to enable precise location services.

In the UWB system, there is ISI caused by multipath delays. For DS-UWB, inter-path interference may be caused by multipath components received within the observation window of a particular symbol. There is a high probability of collision for TH-UWB.

Multiband UWB enables digital filtering for transmitter pulse shaping, although the computation load is still very high. Information can be processed over a much smaller bandwidth, reducing the complexity of design and the power consumption, and improving spectral flexibility and worldwide compliance. Multiband UWB also allows the transmit power in each sub-band to be independently managed. The multiband approach can deal with both in-band and out-of-band interference rejection.

Multiband schemes for IEEE 802.15.3a

IEEE 802.15.3a employs a multiband UWB scheme for spectrum flexibility and easy implementation. It can be either a pulsed multiband DS-UWB scheme or an MB-OFDM scheme.

The IEEE 802.15.3a Task Group consolidated 26 UWB PHY specifications into two proposals: MB-OFDM UWB, supported by the WiMedia Alliance, and DS-UWB, supported by the UWB Forum. Due to the confliction between the two different industrial alliances, IEEE 802.15.3a project authorization request (PAR) was withdrawn on January 19, 2006.

Pulsed multiband UWB transmission is actually composed of many single-band UWB transmissions. The receiver structures for single-band UWB pulses are also applicable to these single-band UWB transmissions on the sub-bands. The pulsed multiband approach has a small design complexity and reduced power consumption, but it has the difficulties of collecting significant multipath energy using a single RF chain, sensitivity of group delay variations introduced by the analog front-end components, and stringent frequency-switching time requirements.

MB-OFDM UWB [3] uses OFDM symbols to convey the information on each of the sub-bands and interleaves them across all the sub-bands. MB-OFDM UWB avoids the disadvantages of pulsed multiband UWB: it has the ability to capture multipath energy using a single RF chain, it is insensitive to group delay variations, and has a relaxed frequency-switching time requirement. Compared to the pulsed multiband approach, the MB-OFDM approach has a slightly more complex transmitter due to the inverse DFT operation, and a higher PAPR.

A dual-antenna UWB CMOS transceiver for the MB-OFDM proposal of IEEE 802.15.3a was developed by Tzero Technologies [49]. XtremeSpectrum, which is now a part of Freescale Semiconductor, developed the XS100 TRINITY chipset for the DS-UWB proposal of IEEE 802.15.3a. Many other vendors including Intel also produce their standard-based UWB solutions. In addition, Pulse~LINK provides its proprietary CWave solution, which achieves data rates up to 1 Gbits/s over coax and wireless networks from the same chipset based on continuous pulse UWB technology.

20.5.1 Modulation of pulsed multiband UWB

For pulsed multiband UWB modulation, the UWB pulse must be shaped to the specified bandwidth of each sub-band. The pulses must be modulated to each band with different center frequencies. Multiple UWB pulses on different sub-bands can be transmitted at the same time. Modulation to sub-band i with carrier f_i can be achieved by mixing a sinusoidal signal

$$s_i(t) = \sqrt{2} y_{GP}(t) \cos(2\pi f_i t). \tag{20.22}$$

The Fourier transform is given by

$$S_i(f) = \frac{1}{\sqrt{2}} \left[Y_{GP}(f - f_i) + T_{GP}(f + f_i) \right]. \tag{20.23}$$

By removing the unwanted band, we get the modulated pulsed UWB signal. When signals at many different bands are transmitted simultaneously, we have

$$s(t) = \sum_{i=1}^{N_B} s_i(t) = \sqrt{2} y_{GP}(t) \sum_{i=1}^{N_B} \cos(2\pi f_i t), \tag{20.24}$$

where N_B is the number of bands. The resulting spectrum is spread over the full frequency band. When a band is interfered by other sources, that band is just not used. This leads to a null in the spectrum.

20.5.2 MB-OFDM UWB

MB-OFDM is now the leading technology for use in high-speed wireless UWB communications. It is supported by WiMedia and ECMA. In the ECMA standard, the 7.5 GHz UWB spectrum is divided into 14 sub-bands, each having a bandwidth of 528 MHz. This is also applicable for the MB-OFDM proposal of IEEE 802.15.3a. For each sub-band of 528 MHz, OFDM is applied with 128 subcarriers. Among the 128 subcarriers, 100 are used for data transmission, 12 as pilot subcarriers, 10 as guard subcarriers, and 6 are unused by setting them to zero. The subcarrier bandwidth in each sub-band is $\Delta f = 528/128 = 4.125$ MHz, and the FFT period is $1/\Delta = 242.42$ ns. The guard interval is 9.5 ns, and the prefix length is 60.6 ns. This yields an OFDM symbol length of $242.42 + 9.5 + 60.6 = 312.5$ ns. QPSK is used to modulate the symbols to be transmitted at each subcarrier. The use of QPSK is due to the limitation of the transmitter power, which must be lower than -41.3 dBm/MHz.

The MB-OFDM proposal for IEEE 802.15.3a suggests the use of only three sub-bands below the 5-GHz band as a mandatory mode. This mandatory mode supports eight different data rates ranging from 53.3 to 480 Mbits/s, which are obtained by using different channel coding rate, frequency spreading gain, or time spreading gain. Depending on the data rate, QPSK or dual-carrier modulation (DCM) constellation is used. The BER of the MB-OFDM UWB system is analyzed and simulated in [52]. Multiple access is achieved by using different preambles and time-frequency codes (associated with sub-bands) for different users [3]. For zero-padded MB-OFDM UWB systems, preamble-based low complexity algorithms for synchronization, channel estimation and equalization have been presented in [32].

For simple UWB transceiver, the MB-OFDM proposal for IEEE 802.15.3a assumes that there is no bit loading and that power is equally distributed across subcarriers within each sub-band. A sub-band assignment and power allocation algorithm is given for power-controlled channel allocation [51, 52]. Based on the IEEE 802.15.3a UWB channel models, performance analysis of MB-OFDM UWB systems is performed in [31], with considerations of imperfect frequency and timing synchronizations and the effect of ISI.

The MB-OFDM solution is very similar to the conventional wireless OFDM scheme, but with many measures that reduce the implementation complexity. A time-frequency code is used to transmit different OFDM symbols on different bands. This provides frequency diversity, but also enables multiple access between piconets operating in the same vicinity [4]. OFDM is used to modulate the information in each sub-band so as to capture the multipath energy efficiently. The smaller bandwidth of each sub-band eases the requirement on the sampling rate of the ADC.

The channel characteristics in the conventional OFDM system and in the MB-OFDM UWB one are quite different. Channels in the conventional OFDM system are less dispersive than those in the MB-OFDM UWB system, since the latter has a much larger bandwidth. Multipath channel coefficients in the conventional OFDM system are Rayleigh distributed, while those in the MB-OFDM UWB system are independent log-normally distributed [15].

Antenna dispersion faced by pulsed-UWB is not a problem for OFDM-based UWB. In the MB-OFDM UWB system, the linearity of the phase response over the entire UWB band is not important as long as the bandwidth of each subchannel is sufficiently small. Also, OFDM is normally used in multipath environments, where the multipath delay spread is much higher than any time delays created by the antennas.

For the purpose of detection and avoidance (DAA), at the MB-OFDM transmitter, the most common technique for frequency notching is to zero out tones that overlap the interfering bands at the FFT stage to achieve notches in the transmit spectrum. There is no increase in the transmitter complexity. At the receiver, tones carrying no information show a deep frequency notch. To achieve a desired notch depth, multiple tones have to be zeroed out. A deeper notch can be obtained by inserting data-specific dummy tones that are subject to a constraint on the amount of energy on either side of the interfering band; this method sacrifices fewer tones in order to achieve a notch-specific depth [4].

Pulsed-OFDM is an enhancement to an MB-OFDM system by combining some of the benefits of the pulsed UWB systems with those of the OFDM UWB systems [45]. It has a performance superior or comparable to MB-OFDM in multipath fading channels, and also has intrinsic low-complexity and power consumption advantages compared with MB-OFDM.

In [50], the performance of the MB-OFDM MIMO UWB system is analyzed in case of Nakagami frequency-selective fading channels. The maximum achievable diversity is shown to be the product of the numbers of transmit (N_t) and receive (N_r) antennas, the number of multipath components L, and the number of jointly encoded OFDM symbols [50]. In [55], a general STFC MB-OFDM UWB model is proposed, and is analyzed in case of the log-normally distributed multipath channel coefficients. The maximum achievable diversity order is found to be the product of N_t, N_r, and the FFT size.

Impact of timing jitter

Nonideal sampling clocks in OFDM-based UWB systems introduce random timing jitter, which results in ICI. For OFDM-based UWB systems, the ICI for a system with white timing jitter is bounded by [38]

$$\frac{\pi^2 (B\sigma_J)^2}{3} \left[1 - \frac{3\pi^2 (B\sigma_J)^2}{10} \right] \leq \frac{P_J}{\sigma_s^2} \leq \frac{\pi^2}{3} (B\sigma_J)^2, \qquad (20.25)$$

where P_J is the average ICI power caused by timing jitter, σ_s^2 is the variance of the symbol transmitted on each subcarrier, B is the bandwidth of a sub-band of at least of 500 MHz, and σ_J^2 is the variance of the timing jitter. σ_J is typically 10 to 150 ps.

For correlated jitters, the ICI power is in general subcarrier-dependent, but the average ICI power over all subcarriers is given by [38]

$$\frac{\overline{P_J}}{\sigma_s^2} = \frac{1}{3}\pi^2 (B\sigma_J)^2 \qquad (20.26)$$

which is independent of the correlation coefficients of the timing jitters. Therefore, the correlation between timing jitters does not change the overall ICI power.

A universal upper bound on the ICI power due to colored timing jitter is given in [38]. The impact of timing jitter can be reduced by oversampling. In case of uncorrelated jitters, oversampling by M can reduce the ICI power by a factor of M [38].

Problems

20.1 Using (20.24), plot the spectrum when six different frequencies, 4, 5, 6, 7, 8, 9 GHz, are utilized for modulating Gaussian pulses, where $\tau = 0.5$ ns. Also, plot the spectrum if the frequency of 5 GHz is not used. In order to create a deep null at 5 GHz, one can select a proper value of τ and divide the UWB band into more sub-bands. Show this by illustration.

20.2 Consider a UWB signal $s(t) = \sum_{i=-\infty}^{\infty} a_k p(t - kT_f)$, where $p(t)$ is the Gaussian pulse given by (20.6), the frame period T_f is larger than the pulse width, and a_k is uniformly randomly distributed on $\{-1, +1\}$. Calculate the average PSD. For $\tau = 0.4$ ns, $T_f = 100$ ns, calculate the maximum value of K_1 in (20.6) such that the average PSD of $s(t)$ is less than -41.25 dBm/Hz.

20.3 The average ICI power for an OFDM-based UWB system is given by (20.25) for white timing jitters and by (20.26) for correlated jitters. Assume that σ_J is 80 ps, and $B = 528$ MHz. Plot the average ICI power as a function of $B\sigma_J$.

References

[1] Y. D. Alemseged & K. Witrisal, Modeling and mitigation of narrowband interference for transmitted-reference UWB systems. *IEEE J. Sel. Topics Signal Process.*, **1**:3 (2007), 456–469.

[2] A. Annamalai, S. Muthuswamy, D. Sweeney, R. M. Buehrer, J. Ibrahim, C. R. Anderson & D. S. Ha, Receiver design principles. In J. H. Reed, ed., *An Introduction to Ultra Wideband Communication Systems* (Upper Saddle River, NJ: Prentice Hall, 2005), 253–377.

[3] A. Batra et al, *Multi-band OFDM Physical Layer Proposal for IEEE 802.15 Task Group 3a*, IEEE P802.15-03/268r2, Nov 2003.

[4] J. Balakrishnan & A. Batra, Multiband OFDM. In R. Aiello & A. Batra, eds., *Ultra Wideband Systems* (Oxford, UK: Elsevier, 2006), 211–247.

[5] R. M. Buehrer, Channel Modeling. In J. H. Reed, ed., *An Introduction to Ultra Wideband Communication Systems* (Upper Saddle River, NJ: Prentice Hall PTR, 2005).

[6] C. Carbonelli & U. Mengali, Synchronization algorithms for UWB signals. *IEEE Trans. Commun.*, **54**:2 (2006), 329–338.

[7] D. Cassioli, M. Z. Win & A. F. Molisch, The ultra-wide bandwidth indoor channel: from statistical model to simulations. *IEEE J. Sel. Areas Commun.*, **20**:6 (2002), 1247–1257.

[8] D. Cassioli, M. Z. Win, F. Vatalaro & A. F. Molisch, Low complexity rake receivers in ultra-wideband channels. *IEEE Trans. Wireless Commun.*, **6**:4 (2007), 2007.

[9] Y.-L. Chao & R. Scholtz, Optimal and suboptimal receivers for ultra-wideband transmitted reference systems. In *Proc. IEEE GLOBECOM*, San Francisco, CA, Dec 2003, **6**, 759–763.

[10] Y. Chen & N. C. Beaulieu, Improved receivers for generalized UWB transmitted reference systems. *IEEE Trans. Wireless Commun.*, **7**:2 (2008), 500–504.

[11] X. Chu, R. Murch, J. Liu & M. Ghavami, Selective rake combining for low-rate ultra-wideband communications. *IEEE Trans. Commun.*, **56**:8 (2008), 1313–1323.

[12] Q. H. Dang & A.-J. van der Veen, A decorrelating multiuser receiver for transmit-reference UWB systems. *IEEE J. Sel. Topics Signal Process.*, **1**:3 (2007), 431–442.

[13] W. Ellersick, C. K. K. Yang, W. Horowitz & W. Dally, GAD: a 12GS/s CMOS 4-bit A/D converter for an equalized multi-level link. In *Proc. IEEE Symp. VLSI Circ.*, Kyoto, Japan, Jun 1999, 49–52.

[14] J. R. Foerster, The performance of a direct-sequence spread ultrawideband system in the presence of multipath, narrowband interference, and multiuser interference. In *Proc. IEEE UWBST*, Baltimore, MD, May 2002, 87–91.

[15] J. R. Foerster, *Channel Modeling Sub-committee Final Report*, IEEE P802.15-02/490r1-SG3a, IEEE P802.15 Working Group for WPAN, Oct 2005.

[16] A. Fort, J. Ryckaert, C. Desset, P. De Doncker, P. Wambacq & L. Van Biesen, Ultra-wideband channel model for communication around the human body. *IEEE J. Sel. Areas Commun.*, **24**:4 (2006), 927–933.

[17] S. Franz & U. Mitra, On optimal data detection for UWB transmitted reference systems. In *Proc. IEEE Globecom*, San Francisco, CA, Dec 2003, **2**, 744–748.

[18] S. Franz & U. Mitra, Generalized UWB transmitted reference systems. *IEEE J. Sel. Areas Commun.*, **24**:4 (2006), 780–786.

[19] S. Franz, C. Carbonelli & U. Mitra, Joint semi-blind channel and timing estimation for generalized UWB transmitted reference systems. *IEEE Trans. Wireless Commun.*, **6**:1 (2007), 180–191.

[20] S. Franz & U. Mitra, Quantized UWB transmitted reference systems. *IEEE Trans. Wireless Commun.*, **6**:7 (2007), 2540–2550.

[21] S. S. Ghassemzadeh, R. Jana, C. W. Rice, W. Turin & V. Tarokh, A statistical path loss model for in-home UWB channels. In *Proc. IEEE UWBST*, Baltimore, MD, May 2002, 71–74.

[22] M. Ghavami, L.B. Michael & R. Kohno, *Ultra Wideband: Signals and Systems in Communication Engineering*, 2nd edn (Chichester, UK: Wiley, 2007).

[23] S. Glisic, *Advanced Wireless Communications: 4G Technologies*, 2nd edn (Chichester, UK: Wiley-IEEE, 2007).

[24] H. Hashemi, Impulse response modeling of indoor radio propagation channels. *IEEE J. Sel. Areas Commun.*, **11**:7 (1993), 967–978.

[25] T. Hoholdt & J. Justesen, Ternary sequences with perfect periodic autocorrelation. *IEEE Trans. Inf. Theory*, **29**:4 (1983), 597–600.

[26] V. P. Ipatov, Ternary sequences with ideal autocorrelation properties. *Radio Eng. Electron. Phys.*, **24** (1979), 75–79.

[27] D. I. Kim & T. Jia, M-ary orthogonal coded/balanced ultra-wideband transmitted-reference systems in multipath. *IEEE Trans. Commun.*, **56**:1 (2008), 102–111.

[28] D. I. Kim, Multiuser performance of M-ary orthogonal coded/balanced UWB transmitted-reference systems. *IEEE Trans. Commun.*, **57**:4 (2009), 1013–1024.

[29] N. V. Kokkalis, P. T. Mathiopoulos, G. K. Karagiannidis, and C. S. Koukourlis, Performance analysis of M-ary PPM TH-UWB systems in the presence of MUI and timing jitter. *IEEE J. Sel. Areas Commun.*, **24**:4 (2006), 822–828.

[30] C. J. Le Martret & G. B. Giannakis, All-digital PAM impulse radio for multiple-access through frequency-selective multipath. In *Proc. IEEE Globecom*, San Francisco, CA, Nov–Dec 2000, **1**, 77–81.

[31] H.-Q. Lai, W. P. Siriwongpairat & K. J. R. Liu, Performance analysis of multiband OFDM UWB systems with imperfect synchronization and intersymbol interference. *IEEE J. Sel. Topics Signal Process.*, **1**:3 (2007), 521–534.

[32] Y. Li, H. Minn & R. M. A. P. Rajatheva, Synchronization, channel estimation, and equalization in MB-OFDM systems. *IEEE Trans. Wireless Commun.*, **7**:11 (2008), 4341–4352.

[33] V. Lottici, A. N. D'Andrea & U. Mengali, Channel estimation for ultra-wideband communications. *IEEE J. Sel. Areas Commun.*, **20**:12 (2002), 1638–1645.

[34] A. F. Molisch, J. R. Foerster & M. Pendergrass, Channel models for ultrawideband personal area networks. *IEEE Wireless Commun.*, **10**:6 (2003), 14–21.

[35] A. F. Molisch *et al.*, *Status of Models for UWB Propagation Channel*, IEEE 802.15.4a Channel Model (Final Report), 2005, IEEE doc: IEEE 802.15-04-0662-02-004a.

[36] Y.-P. Nakache & A. F. Molisch, Spectral shaping of UWB signals for time-hopping impulse radio. *IEEE J. Sel. Areas Commun.*, **24**:4 (2006), 738–744.

[37] P. P. Newaskar, R. Blazquez & A. P. Chandrakasan, A/D precision requirements for an ultra-wideband radio receiver. In *Proc IEEE SIPS*, San Diego, CA, Oct 2002, 270–275.

[38] U. Onunkwo, Y. G. Li & A. Swami, Effect of timing jitter on OFDM-based UWB systems. *IEEE J. Sel. Areas Commun.*, **24**:4 (2006), 787–793.

[39] M. Pausini & G. J. M. Janssen, Performance analysis of UWB autocorrelation receivers over Nakagami-fading channels. *IEEE J. Sel. Topics Signal Process.*, **1**:3 (2007), 443–455.

[40] D. Porrat, D. N. C. Tse & S. Nacu, Channel uncertainty in ultra-wideband communication systems. *IEEE Trans. Inf. Theory*, **53**:1 (2007), 194–208.

[41] T. S. Rappaport, *Wireless Communications: Principles & Practice*, 2nd edn (Upper Saddle River, NJ: Prentice Hall, 2002).

[42] P. C. Richardson, W. Xiang & W. Stark, Modeling of ultra-wideband channels within vehicles. *IEEE J. Sel. Areas Commun.*, **24**:4 (2006), 906–912.

[43] A. Ridolfi & M. Z. Win, Ultrawide bandwidth signals as shot noise: a unifying approach. *IEEE J. Sel. Areas Commun.*, **24**:4 (2006), 899–905.

[44] C. K. Rushforth, Transmitted-reference techniques for random or unknown channels. *IEEE Trans. Inf. Theory*, **10**:1 (1964), 39–42.

[45] E. Saberinia, J. Tang, A. H. Tewfik & K. K. Parhi, Pulsed-OFDM modulation for ultrawideband communications. *IEEE Trans. Veh. Tech.*, **58**:2 (2009), 720–726.

[46] Z. Sahinoglu, S. Gezici & I. Guvenc, *Ultra-wideband Positioning Systems* (Cambridge, UK: Cambridge University Press, 2008).

[47] A. A. M. Saleh & R. A. Valenzuela, A statistical model for indoor multipath propagation. *IEEE J. Sel. Areas Commun.*, **5**:2 (1987), 128C-137.

[48] R. A. Scholtz, Multiple access with time-hopping impulse modulation. In *Proc. IEEE MILCOM*, Boston, MA, Oct 1993, 447–450.

[49] I. Sever, S. Lo, S.-P. Ma, P. Jang, A. Zou, C. Arnott, K. Ghatak, A. Schwartz, L. Huynh & T. Nguyen, A dual-antenna phase-array ultra-wideband CMOS transceiver. *IEEE Commun. Mag.*, **44**:8 (2006), 102–110.

[50] W. P. Siriwongpairat, W. Su, M. Olfat & K. J. R. Liu, Multiband-OFDM MIMO coding framework for UWB communication systems. *IEEE Trans. Signal Process.*, **54**:1 (2006), 214–224.

[51] W. P. Siriwongpairat, Z. Han & K. J. R. Liu, Power controlled channel allocation for multiuser multiband UWB systems. *IEEE Trans. Wireless Commun.*, **6**:2 (2007), 583–592.

[52] W. P. Siriwongpairat & K. J. R. Liu, *Ultra-Wideband Communications Systems: Multiband OFDM Approach* (Piscataway, NJ: Wiley-IEEE, 2008).

[53] V. S. Somayazulu, Multiple access performance in UWB systems using time hopping vs. direct sequence spreading. In *Proc. IEEE WCNC*, Mar 2002, **2**, 522–525.

[54] D. Sweeney, D. S. Ha, A. Annamalai & S. Muthuswamy, Transmitter design. In J. H. Reed, ed., *An Introduction to Ultra Wideband Communication Systems* (Upper Saddle River, NJ: Prentice Hall, 2005), pp. 213–251.

[55] L. C. Tran & A. Mertins, Space-time-frequency code implementation in MB-OFDM UWB communications: design criteria and performance. *IEEE Trans. Wireless Commun.*, **8**:2 (2009), 701–713.

[56] M. Welborn, M. M. Laughlin & R. Kohno, *DS-CDMA Proposal for IEEE 802.15.3a*, IEEE P802.15-03/334r3-TG3a, IEEE P802.15 Working Group for Wireless Personal Area Networks (WPANs), May 2003.

[57] L. Yang, Timing PPM-UWB signals in ad hoc multiaccess. *IEEE J. Sel. Areas Commun.*, **24**:4 (2006), 794–800.

[58] L. Yang & G. B. Giannakis, Timing ultra-wideband signals with dirty templates. *IEEE Trans. Commun.*, **53**:11 (2005), 1952–1963.

[59] S. -K. Yong, *TG3c Channel Modeling Sub-committee Final Report*, IEEE WPAN TG3, IEEE 15-07-0584-00-003c, Mar 2007.

[60] S. Zhao, P. Orlik, A. F. Molisch, H. Liu & J. Zhang, Hybrid ultrawideband modulations compatible for both coherent and transmit-reference receivers. *IEEE Trans. Wireless Commun.*, **6**:7 (2007), 2551–2559.

Cognitive radios

21.1 Conception of software-defined radio

Conception of software-defined radio (SDR) started in the early 1990s, and has now become a core technology for future-generation wireless communications. In 1997, the U.S. DoD recommended replacing its 200 families of radio systems with a single family of SDRs in the programmable modular communications system (PMCS) guideline document [45]. An architecture outlined in this document includes a list of radio functions, hardware and software component categories, and design rules [45]. The ultimate objective of SDR is to configure a radio platform like a freely programmable computer so that it can adapt to any typical air interface by using an appropriate programming interface. SDR is targeted to implement all kinds of air interfaces and signal processing functions using software in one device. It is the basis of the 3G and 4G wireless communications.

Proliferation of wireless standards has created the dramatic need for an MS architecture that supports multiband, multimode, and multistandard low-power radio communications and wireless networking. SDR has become the best solution. By using a unified hardware platform, the user needs only to download software of a radio and run it, and immediately shift to a new radio standard for a different environment. The download of the software can be over the air or via a smart card. For example, several wireless LAN standards, including IEEE 802.11, IEEE 802.15, Bluetooth, and HomeRF, use the 2.4 GHz ISM band, and they can be implemented in one SDR system [25].

The programmable chips used in SDR are typically general-purpose DSPs or FPGAs. The internal blocks of SDR hardware must also have adequate calibration and programming capacities so as to adapt for optimum performance. Tunable RF preselect filters are an enabling technique, but are still being developed by using the MEMS technology.

SDR is versatile, based on a simple, unified hardware platform. The downloadable software packages are wireless standards. The versatility of signal processing algorithms enables high frequency usage and more extensive services to users. SDR can even adaptively change the appropriate access scheme according to user's QoS such as transmission rate, tolerable BER or delay. Based on SDR, it is easy to realize a multimode, multiband MS. There are also solutions using adaptive RF circuits, where some components such as LNAs and VCOs have adaptivity in performance.

Multiple-antenna techniques are usually incorporated into SDRs. They can increase system capacity. In a wireless system, the use of multiple antennas makes the MAC protocol more complicated. MAC protocol must cope with impairments arising from shadowing, multipath delay spread, and CCI. Training fields are needed in a radio burst for channel estimation.

In the literature, there are many software realizations for various radio systems such as GSM [44] and DS-CDMA indoor system [4, 37]. The software realization of the GSM BS implements each GSM function block between the RF signal and the source speech signal in software. A unified SDR receiver is described in [24]. A SDR testbed for a 2×2 MIMO space-time-coded OFDM wireless LAN is described in [48].

21.2 Hardware/software architecture of software-defined radio

Hardware architecture

For SDRs, the architecture is featured by wideband antenna, wideband A/D and D/A conversions, and digital baseband signal processing modules. For multimode SDR, a reconfigurable baseband digital signal processing stage and a multiband or broadband analog RF stage are necessary.

Ideally, the A/D and D/A conversions are applied directly at radio frequency. This requires ADCs of an extremely wide band, and thus is not practical. More practical SDRs have an RF circuit sitting between the antenna and the digital part, as shown in Fig. 21.1. The RF circuit contains filters, LNA/power amplifier, as well as IF modules. Signals are amplified after each filtering. A/D and D/A conversions are applied on the IF signals rather than on baseband signals. DSPs are used for digital processing, and down-conversion can be done based on an NCO. This scheme can be called a *low-IF* architecture. SDR exercises strict restriction on the ADC: It must have a sampling frequency of at least the Nyquist frequency of the radio signal and have sufficient resolution.

For the current CDMA-based 3G systems, the baseband processing functions are divided into chip-rate processing functions, such as the path searcher, access detection and rake receiver, and symbol-rate processing functions. On the transmit path, the baseband processing module receives data from the link layer. The data is spread and modulated, multiplexed, and then sent to the DAC. On the receive path, the data from the ADC is subject to demultiplexing, channel selection, despreading, demodulation, channel decoding, and then is passed to the link layer.

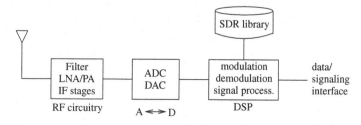

Figure 21.1   SDR architecture

For wideband wireless communications, various function blocks for baseband signal processing may consume many millions of multiply-accumulate (MMAC) operations that may be equivalent to the processing power of dozens of the most recent DSPs. For example, even a BS of the WCDMA system requires at least hundreds of thousands of MIPS, and today the best DSPs can only provide ten thousand MIPS. Multiple DSPs can be connected in series or parallel to provide more MIPS. Thus, using multiple DSPs in the design may be very costly. For this reason, SDR is more easily deployed in handsets.

FPGA-based hardware implementation would be much cheaper since only one single FPGA device can provide a computational capability of dozens of such DSPs. FPGAs now are the more realistic choice than the DSP-based solution.

In CDMA-based SDR, the chip-rate functions are typically much more MIPS-intensive than the symbol-rate functions. A desirable solution is to partition the chip-rate and symbol-rate functions so that chip-rate processing is mainly assigned to a reconfigurable processor such as an FPGA, while symbol-rate processing is mainly assigned to a DSP. Processing at symbol level can also be performed in microcontroller units (MCUs). Higher-layer protocols are handled in MCUs. For the non-CDMA based narrowband 2G systems, the entire baseband processing can be implemented in only one DSP.

For different standards, the sampled data must be decimated by a different scale so as to adapt the digitalization rate with the symbol rate of that particular standard. For this purpose, one can use a cascade of multistage decimation filters that progressively remove the out-of-band signal. Finally, a postfilter is used to precisely delimit the channel of interest and attenuate the interference.

Today, vendors such as ADI and National Semiconductors supply high-performance ADCs that are suitable for SDR. Vendors such as TI and ADI supply high-performance DSPs for wideband radio communications. Xilinx and Alterra are the two major vendors for FPGAs.

Software architecture

A SDR system is a system that uses a wideband hardware platform and different software components for different radios. Object-oriented design is ideal since each radio standard can be encapsulated into one class, and the code can be reused.

The software for the digital subsystem is divided into DSP software, FPGA software, and MCU software. There are some organizations dedicated to software standardization for SDR, such as the United States Joint Tactical Radio System (JTRS) Joint Program Office and the SDR Forum. The architectures proposed by both organizations are object-oriented, and described by using Unified Modeling Language (UML) notation. Each of the architectures defines an operating environment and its services and interfaces. They employ a POSIX-based operating system; noncore components are abstracted away from the hardware, and are all connected by using common object request broker architecture (CORBA). The core framework, as part of the operating environment, is the core set of open application layer interfaces and services, which provide an abstraction of the underlying software and hardware layers for application software design.

The Object Management Group (OMG) defines the open specifications Unified Modeling Language (UML), CORBA and Interface Definition Language (IDL). These specifications are popular in the definition of the software architecture of SDR. The OMG also has a special interest group (SIG) for SDR.

SpectruCell, developed by Advanced Communications Technologies, is a SDR platform for application development. It is targeted to the 3G BS as well as other wireless systems. SpectruCell provides application developers with a platform of hardware and middleware for implementing 2G and 3G air interfaces. It is broadly compliant with the Software-Defined Radio Forum or JTRS specifications. AdapDev, developed by Mercury Computer, is another SDR platform compliant with the JTRS specification. They are suitable for the development of SDR applications. Alcatel and Eurocom have also proposed their 3G SDR testbeds.

21.3 Conception of cognitive radio

Many licensed frequency bands are actually unused. Meanwhile, some frequency bands are overcrowded due to high-speed data services. For the frequency band up to 100 GHz, at any instant, only 5 to 10% of the spectrum is being used. In 2002, the FCC Spectrum Policy Task Force proposed spectrum policy reform, which allows secondary users to utilize those unused frequency bands licensed to the primary users. The FCC introduces the proposal of dynamic spectrum licensing, and cognitive radio is the enabling technology for this purpose. Cognitive radios will drive next-generation wireless networks.

The concept of cognitive radio was also introduced by Mitola [31]. A cognitive radio extends the functionality of SDR by adapting intelligently to its environment. Cognitive radio, also known as *smart radio* and *spectrum-agile radio*, can dynamically change its spectrum to an unused band to avoid interference by sensing its environment. A cognitive radio can make decisions on the network, frequency selection, modulation, coding, and all other parameters for flexible spectrum use according to its environment. The use of the idle frequency spectrum is purely on an opportunity-driven basis. Cognitive radio can use any idle spectrum, but must stop using it the instant the primary user of the spectrum begins to use it.

The FCC permits unlicensed devices to temporarily borrow spectrum from licensed holders as long as no excess interference is sensed by the primary user. The unlicensed devices are required to identify unused frequency bands before transmission by using dynamic frequency selection and incumbent profile detection. Basic cognitive radio techniques, such as dynamic frequency selection and transmit power control, have been used in some unlicensed devices.

Cognitive cycle

Cognitive radio transforms radio nodes from blind execution of predefined protocols to radio-domain-aware intelligent agents that provide a means to deliver the services that

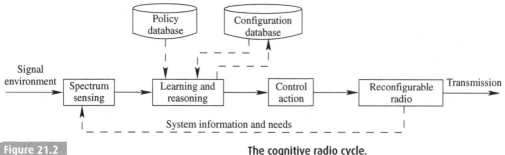

Figure 21.2 **The cognitive radio cycle.**

a user desires. To be successful, these radios must passively learn user preferences and program themselves. *Cognitive capability* and *reconfigurability* are two major characteristics of cognitive radio.

The cognitive cycle consists of three major components in its transceiver [22, 31]: (a) *RF or spectrum sensing*, such as radio-scene analysis and channel identification, at the receiving module, (b) *cognition/management*, such as dynamic spectrum management, routing, and QoS provisioning, at the transmitting module, and (c) *control action*, such as transmit power control, AMC, and rate control, at the transmitting module. The receiver that communicates with the cognitive radio needs also to feed back the spectrum information and quantized channel capacity to the transmitter. The cognitive cycle is illustrated in Fig. 21.2

The spectrum sensing functionality in cognitive radio systems can be divided into two subtasks: occupancy sensing and identity sensing. Spectrum sensing is addressed as a cross-layer design problem, where the physical layer implements sensing, cognition, and adaptation, whereas the MAC layer implements cooperation [12].

21.3.1 Topics in cognitive radio

Modulation and primary user recognition

For radio signal analysis, modulation recognition and bitstream analysis are applied to identify whether an alarm corresponds to a primary user, or a secondary user, or noise (false alarm). *Modulation recognition* is an intermediate step between signal detection and demodulation. Knowledge of the types of service operating on a channel can assist in the decision of jumping channels in a way that minimizes overhead to the cognitive radio and its impact on the primary users of the spectrum. A cognitive radio should be able to recognize other cognitive radios on the link channel to prevent them sensing one another as primary users and jumping channels.

Modulation recognition can be likelihood-based or feature-based [9]. The former is based on the likelihood function of the received signal and the decision is made by comparing the likelihood ratio against a threshold. The likelihood-based algorithms are optimal in the Bayesian sense: they minimize the probability of false classification, but suffer from high computational complexity. The feature-based approach makes decisions based on the observed values of typically, several features. In general, modulation recognition is a challenging task, especially in a noncooperative environment, where no prior knowledge of the incoming signal is available.

A number of features can be extracted to identify the type of the primary users from the received data using signal processing techniques. These features can be frequency-domain features (e.g., bandwidth, center frequency, single carrier versus multi-carrier), or time-domain features (e.g., the variance of the centered normalized signal amplitude, the variance of the zero-crossing interval, maximum duration of a signal, multiple-access technique, duplexing technique, frame duration, spreading codes or hopping patterns), or its statistical properties (e.g., moments, cumulants, cyclic cumulants). Some or all of the features can be extracted from the received data and used for classification. The primary users are identified by using the a-priori information about their transmission parameters. The classifier can be a neural network classifier [10]. The classifier can be trained off-line, but the recognition process must be performed online on the incoming signal at an affordable complexity.

Spectrum management

Spectrum management functions address four major challenges: *spectrum sensing*, *spectrum decision*, *spectrum sharing*, and *spectrum mobility*. Spectrum mobility allows a cognitive radio to exchange its frequency of operation in a dynamic manner by allowing the cognitive radios to operate in the best available frequency band seamlessly. Spectrum sharing deals with fair spectrum scheduling, which is a MAC functionality. A cognitive radio needs to maintain an up-to-date list of available channels within a band. The channel usage database can also be used to avoid the occupied licensed channels, and a secondary user estimates its position and checks a database to find out which channels are vacant in its vicinity.

Erasure correction codes

A secondary user selects a set of subchannels from the primary user band to establish a secondary user link that adapts itself in accordance with the primary user spectral activity on that band. The secondary user is required to vacate a subchannel as soon as a primary user becomes active on that subchannel. This causes a secondary user to lose packets on that subchannel. In order to compensate for this loss, a class of erasure correction codes called *LT (Luby transform) codes* or *Fountain codes* [30] can be used for packet-based channels with erasures before transmitting secondary user packets on these subchannels. This provides packet-level protection at the transport layer or higher, augmenting the bit-level protection that may be provided by the MAC and physical layers.

Spectrum awareness

Spectrum awareness over an entire operating bandwidth of interest is a basic and crucial task of a cognitive radio. Cognitive radio sensitivity is required to outperform the primary user receiver by a large margin in order to prevent the hidden terminal problem. This makes spectrum sensing, or spectrum awareness, a very challenging research problem. For cognitive radios, a cross-layer design approach is desirable since the process delay is required to be small.

Spectrum awareness can be either passive or active awareness. In passive awareness, the spectrum use pattern is obtained not by sensing by the secondary system itself, but by negotiating with primary users, or from a server or database. The passive awareness approach results in considerable signaling for distribution of frequency information. In active awareness, the secondary user actively senses the frequency spectrum to obtain the spectrum use pattern. Active spectrum awareness can be performed in a noncooperative or cooperative manner. Cooperation helps to cope with the hidden terminal problem.

Active spectrum awareness can be either reactive or proactive, depending on how white spaces (unused frequency bands) are searched. Reactive schemes operate on an on-demand basis, where a cognitive radio starts to sense the spectrum only when it has data to send. Proactive schemes, on the other hand, minimize the delay of secondary users by finding an idle band through maintaining a list of licensed bands currently available for opportunistic access through periodic spectrum sensing.

Cognitive radio as relay

When the direct channel on the primary link to the destination is weak compared to the channel from the primary transmitter to the cognitive radio, the cognitive radio can act as a relay. Packets from the primary link may not be received correctly at the destination, but they might be decoded successfully at the cognitive radio. The latter can queue and forward these packets to the intended receiver. In this case, packet relaying by the cognitive radio helps to empty the queue of the primary transmitter, thus also creating transmitting opportunities for the cognitive radio. According to the cognitive principle, the primary transmitter is unaware of the presence of a secondary node. Relaying by the cognitive radio has been analyzed in an information-theoretic framework in [6]. In a scenario with two single-user links, a primary and a cognitive link, the maximum stable throughput of the cognitive link is derived in [40] for a fixed throughput of the primary link. When the cognitive radio is allowed to act as a transparent relay for the primary link, a stable throughput of the cognitive link with relaying is also derived in [40]. The benefits of relaying depend highly on the topology of the network.

21.3.2 Cognitive radio in wireless standards

Cognitive radio will be a component of many future radio systems and networks. Many radios have to include an ability to coexist with other radios using different protocols in the same bands, and cognitive radio is a solution to coexistence. IEEE 802.19 creates rules for fostering coexistence within IEEE 802 standards operating in unlicensed bands. The

Software-Defined Radio Forum, ITU-R, and the IEEE are involved in the standardization of cognitive radio.

IEEE SCC41 (Standards Coordinating Committee 41),[1] Dynamic Spectrum Access Networks (DySPAN), formerly known as the IEEE 1900 Standards Group, was initially established in 2005, and reorganized to the current form in March 2007. IEEE SCC41 focuses on dynamic spectrum access among 3G/4G, WiFi, and WiMAX networks.

The CSMA used in IEEE 802.3 standard is somewhat similar to cognitive radio: both systems sense before transmission. Cognitive radio also bears some resemblance to dynamic channel selection/allocation. Frequency-hopping techniques used in GSM and Bluetooth are kinds of cognitive radios for interference avoidance.

The IEEE 802.11h standard, which was approved in September 2004, has incorporated dynamic frequency selection and transmit power control for IEEE 802.11a to solve interference with satellites and radar using the same 5 GHz frequency band. IEEE 802.22 (Wi-TV) is now being developed based on cognitive radio. WiMAX is also trying to reuse the UHF/VHF band based on cognitive radio. Spectrum sensing using energy detection is chosen in IEEE 802.11y for shared IEEE 802.11 operation with other users in the 3.6 GHz band. IEEE 802.16h is a provision for cognitive radios to avoid interference with other WiMAX devices.

The FCC also approved UWB as another solution for improving spectrum efficiency in 2002. For communications, the FCC allows UWB to overlay the band 3.1–10.6 GHz with existing spectrum users by limiting the transmit power to a sufficiently low level so that it does not impact the existing spectrum users.

21.4 Spectrum sensing

21.4.1 Secondary user-based local spectrum sensing

Approach to active spectrum sensing

The most effective way for spectrum sensing is to detect the primary users that are receiving data within the secondary user communication range. A wireless device typically senses the presence of a primary user by using one of the three active spectrum sensing techniques: *energy detection*, *matched filtering*, and *feature detection* [12].

- When there is no prior information of the primary users, energy detection using FFT is a simple, general approach. This method is valid only if the energy detected is above a certain threshold. The energy detector technique has some drawbacks such as difficulty in setting the threshold due to the changing background noise and interference level, incapability of differentiating between modulated signals, noise and interference, and the method is invalid for spread-spectrum signals. The performance of the method also

---

[1] http://www.scc41.org/

degrades in a fading environment and when the secondary user needs to cooperate with other nodes for spectrum sensing.

- A matched filter is optimum for signal detection in communication systems. This requires demodulation of a primary user signal. Thus, cognitive radio is required to have a-priori knowledge of a primary user signal at both physical and MAC layers (e.g., modulation type and order, pulse shaping, packet format), and then it implements coherent demodulation by performing timing and carrier synchronization, and even channel equalization.

- Feature detection, most typically cyclostationary feature detection, improves the detection performance by exploiting standards specifics. A signal with strong cyclostationary properties can be detected at very low SNR. Cyclostationary feature detection has the highest complexity among the three techniques.

The biggest challenge for spectrum sensing is in developing blind sensing techniques that are capable of detecting very weak primary user signals while being sufficiently fast and low cost to implement.

As energy detection has far lower complexity than feature detection does, feature detection is performed only after a predefined number of consecutive energy detection alarms are issued [26]. This effectively avoids feature detection in case of false alarms.

A fourth approach, cyclostationary beamforming-based spectrum sensing, is proposed in [12]. The idea is to use the ACS algorithm [11] to extract the desired signal at a unique cycle frequency from the measurement of an antenna array. Based on the spectrum analysis of the extracted signal, one can judge whether a specified channel is occupied by a primary user or a secondary user, or it is vacant. The complexity of this method runs between those of the energy detection and cyclic spectrum analysis methods. In contrast, energy detection can only detect whether the channel is occupied, but it cannot identify whether the received signal is a primary user or a secondary user signal. The energy detector is also unable to identify the existence of a user signal that is buried in noise, but cyclostationarity-based techniques can.

In order to avoid interference from the cognitive radio network itself, a cognitive radio usually does not access the spectrum during spectrum detection; this period is called the *quiet period*. In most existing spectrum sensing techniques, in-band sensing performs measurement during the specified quiet periods. There are some in-band sensing techniques without a quiet period proposed for OFDM-based IEEE 802.22. Another in-band sensing approach utilizes the complementary adjacent OFDM symbols in the preambles or in the pilots to eliminate the effect of channel and self-signal of the network, and to perform power detection [29]. These methods are customized for OFDM signals by using the training sequence. The cyclostationary beamforming-based spectrum sensing approach also eliminates the need for quiet periods [12]. In [12], a secondary user can measure while it is simultaneously transmitting, as long as its transmitted signal does not have the same cycle frequency; the method is simple, and is suitable for all kinds of communication signals, compared to existing in-band sensing techniques.

Occupancy: binary hypothesis test

Detection of whether the spectrum is currently occupied by a primary user can be stated by a binary hypothesis test. A null hypothesis $\mathcal{H}_0$ corresponds to the absence of a signal, and a hypothesis $\mathcal{H}_1$ corresponds to the presence of a signal. The signal received at the cognitive radio is given by

$$x(t) = \begin{cases} w(t), & \mathcal{H}_0 \\ hs(t) + w(t), & \mathcal{H}_1 \end{cases}, \tag{21.1}$$

where $s(t)$ is the primary user signal, h is the amplitude gain of the channel, and $w(t)$ is AWGN with zero mean and variance σ_n^2. Given a decision statistic d, the probabilities of detection P_d, false alarm P_{fa}, and missed alarm P_m are, respectively, given by

$$P_d = E\left[\Pr\left(\mathcal{H}_1|\mathcal{H}_1\right)\right] = \Pr\left(d > d_{\text{th}}|\mathcal{H}_1\right), \tag{21.2}$$

$$P_{\text{fa}} = E\left[\Pr\left(\mathcal{H}_1|\mathcal{H}_0\right)\right] = \Pr\left(d > d_{\text{th}}|\mathcal{H}_0\right), \tag{21.3}$$

$$P_m = E\left[\Pr\left(\mathcal{H}_0|\mathcal{H}_1\right)\right] = 1 - P_d, \tag{21.4}$$

where d_{th} is a threshold level, which can be calculated from a specified P_{fa} [3]. Naturally, P_{fa} is independent of the SNR, since under the $\mathcal{H}_0$ hypothesis there is no primary user signal present. Conversely, P_d is dependent on the SNR of the received signal as well as the channel conditions. These probabilities are illustrated in Fig. 21.3.

From the incumbent protection point of view, a higher P_{fa} is more tolerable than a lower P_d. In IEEE 802.22, $P_d = 0.9$ is chosen at a SNR of -20 dB. Note that if the primary user requires 100% protection in its frequency band, it is not allowed for the secondary usage in that frequency band.

For energy detection, a popular decision statistic is defined by

$$d = \sum_{n=1}^{N} |x(n)|^2. \tag{21.5}$$

Under hypothesis $\mathcal{H}_0$, the test statistic d is a random variable whose pdf $p_0(x)$ is a χ^2 distribution with $2N$ degrees of freedom for the complex-valued case, or with N degrees of freedom for the real-valued case.

The minimum required number N of samples is derived as a function of P_{fa}, P_d, and SNR γ [3]. In case of the energy detector, at low SNR ($\gamma \ll 1$), N scales as $O\left(1/\gamma^2\right)$; P_{fa}, P_d,

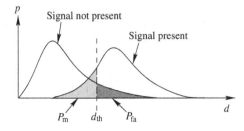

Figure 21.3   Probabilities of detections, missed alarms and false alarms.

P_m are given as functions of the energy threshold and the SNR [12, 28]. Complementary receiver operating characteristic (ROC) curves (P_m versus P_{fa}) of the energy detection method can thus be plotted.

A matched filter requires $O(1/\gamma)$ samples to meet a predetermined probability of error (P_e) constraint [3]. When the transmitter transmits random Gaussian noise, the optimal detector becomes an energy detector. The optimal detector for a zero-mean constellation is proved to behave like an energy detector in the limit of low SNR, and thus zero-mean signals are difficult to detect at low SNR, even with knowledge of the modulation scheme [38]. In this case, $O\left(1/\gamma^2\right)$ samples are required to meet a P_e constraint. This number of samples is significant and it imposes a limit on the frequency agility of a cognitive radio: Short duration gaps cannot be exploited in band usage.

When exploiting the cyclostationary property, spectrum sensing can be formulated as hypotheses testing on the received signal $x(t)$: Under hypothesis $\mathcal{H}_0$, $x(t)$ is (wide-sense) stationary and the band is treated as vacant; under hypothesis $\mathcal{H}_1$, $x(t)$ is cyclostationary and the band is judged occupied.

Wideband spectrum sensing is conventionally performed by using a tunable narrowband bandpass filter to sense one narrowband at a time. Multiband joint detection detects the signal energy levels over multiple frequency bands at a time, by jointly optimizing a bank of multiple narrowband detectors, in order to improve the opportunistic throughput capacity of cognitive radios and reduce their interference to primary users [36]. By applying a wavelet transform of the PSD function of the received wideband signal, the singularities of the PSD, which correspond to the discontinuities between sub-bands, can be located and thus the unoccupied sub-bands can be identified [43]. This method, however, does not work for spread spectrum signals.

The average P_d for energy-detector-based spectrum sensing over Rayleigh, Nakagami, and Ricean fading channels is derived in closed form in [7]. The performances of EGC, selection combining, and switch combining are investigated for energy detector-based spectrum sensing under Rayleigh fading. The EGC approach provides the highest gain in terms of P_m over the no-diversity case.

21.4.2 Cooperative spectrum sensing

Local spectrum sensing cannot avoid the hidden terminal problem. In wireless LAN, this problem can be solved by a four-way handshake using Request-To-Send/Clear-To-Send. However, it is impossible to implement the handshake in cognitive radio networks, as the primary users will not add additional resources for handshake. The spectrum sensing performance can be improved, and the hidden terminal problem can be avoided if cooperative spectrum sensing is applied. Cooperative sensing allows to mitigate the multipath fading and shadowing effects, and improves the detection probability. It provides significantly higher spectrum capacity gains than local sensing. This, however, introduces complexity, traffic overhead, and the need for a control channel.

An alternative technique is external sensing, where an external agent performs sensing and broadcasts the channel occupancy information to the cognitive radios. This overcomes the hidden primary user problem as well as the uncertainty due to shadowing and fading. Further, as the cognitive radios do not spend time for sensing and thus do not need sensing ability, spectrum efficiency is increased. The hidden terminal problem can be solved by employing a sufficient number of external sensing devices in the cognitive radio networks [21]. External sensing is among the spectrum-sensing methods proposed for the IEEE 802.22 standard.

In cooperative spectrum sensing, information from different cognitive radios is combined to make a decision on the presence or absence of the primary user. Cooperative spectrum sensing can be in centralized or distributed form. In general, centralized cooperative sensing is performed as follows. Each cognitive radio makes a binary decision based on its local measurement, and then forwards one bit of the decision to the common receiver, where all one-bit decisions are fused according to an OR logic or according to the voting rule. An alternative scheme is that each cognitive radio just forwards its measurement to the common receiver. A hard decision approach can perform almost as well as a soft decision one in terms of detection performance, but it needs a low bandwidth control channel [28]. To reduce the overhead due to sending the decisions, censoring can be applied by not sending the uncertain decisions. When the common receiver makes a final decision on the K binary decisions using an OR rule, we have [28]

$$P_{fa} = \Pr(\mathcal{H}_1|\mathcal{H}_0) = 1 - \Pr(\mathcal{H}_0|\mathcal{H}_0) = 1 - \prod_{i=1}^{K} (1 - P_{fa,i}), \qquad (21.6)$$

$$P_m = \Pr(\mathcal{H}_0|\mathcal{H}_1) = \prod_{i=1}^{K} P_{m,i}, \qquad (21.7)$$

where $P_{fa,i}$ and $P_{m,i}$ denote the false alarm and miss probabilities of the ith cognitive radio in its local spectrum sensing, respectively. If each cognitive radio achieves the same $P_{fa,0}$ and $P_{m,0}$, we have $P_m = P_{m,0}^K$; in this case, K is treated as the *sensing diversity order*, which is provided by the space diversity of the multiple cognitive radios.

In the case of distributed sensing, cognitive nodes share information among one another, but they make their own decisions as to which part of the spectrum they can use. When a cognitive radio is far from a primary user, cooperative spectrum sensing allows two secondary users to cooperate by treating the secondary user that is close to the primary user as a relay. This achieves a diversity gain arising from the relay protocol. One of the cognitive radios acts as a relay for the other, resulting in lower outage probabilities. This effectively combats shadowing and the hidden terminal problem. When two cognitive radios are in close proximity, they can be used as a virtual antenna array; that is, the measurements of the two cognitive radios are exchanged, and the two cognitive radios then jointly transmit using the Alamouti space-time code to combat fading [39].

Multiuser diversity is a form of selection diversity in which the user with the highest SNR is chosen as the transmission link. Multiuser diversity can be exploited in cooperative spectrum sensing to relay the sensing decision of each cognitive radio to the common receiver. This helps to reduce the reporting error probability [28].

21.5 Spectrum sensing using cyclostationary property

The cycle (or conjugate cycle) frequencies of conventional modulated signals are listed in [1]. Spread-spectrum and OFDM signals are also important, as they are used in 3G and 4G wireless systems, respectively. For spread-spectrum signals with a spreading code or hopping sequence of period N, the spectrum cyclic density (SCD) function exhibits cycle frequencies at $\alpha = \frac{k}{NT}$, for integer k and data symbol period T [17]. In [16], a general class of continuous-time DSSS signals is shown to exhibit cyclostationarity at cycle frequencies that are sums or differences of multiples of the symbol rate and the chip rate, that is, $\alpha = \frac{m}{T_0} + \frac{n}{T_c}$, where T_0 is the bit interval, T_c is the chip period, and m, n are integers. The discrete-time signal obtained by uniformly sampling a continuous-time long-code DSSS signal exhibits cyclostationarity, provided that at least two samples per chip are taken. When $\alpha \neq 0$, the degree of cyclostatioarity is very weak.

OFDM signals may be represented as a composite of N statistically independent sub-channel QAM signals. Without a cyclic prefix, subcarrier orthogonality destroys the cyclostationarity of the individual QAM signals. The inclusion of a cyclic prefix causes a loss of subcarrier orthogonality, but allows inherent QAM signal features to be detected [46]. Cyclic-prefix OFDM signaling is cyclostationary with the OFDM symbol period T, and thus exhibits nonconjugate cyclostationarity at cycle frequencies $\alpha = k/T$, $k = 0, \pm 1, \ldots$, and potentially other frequencies depending on the coding scheme, where $T = T_s + T_{cp}$, T_s being the length of the useful symbol data and T_{cp} the length of the cyclic prefix [23, 35, 46]. The cyclic autocorrelation function for $\alpha = k/T$ peaks at $\tau = \pm T_s$ [35].

21.5.1 Spectrum-cyclic-analysis-based spectrum sensing

The SCD function allows for modulation recognition based on the pattern of the cycle (or conjugate cycle) frequencies. The performance of the feature detector also depends on how much energy a feature contains, and different modulation schemes have features with different energy patterns. Cyclostationary feature detection estimates the peaks on the SCD function. The SCD function can be calculated by combining FFT and spectral correlation, as shown in Fig. 21.4.

The sufficient statistic can be defined for single or multiple cycle frequencies by

$$Z(N) = \sum_{\alpha} \int_{-\frac{f_s}{2}}^{\frac{f_s}{2}} S_{xx}^\alpha(f)^* \hat{S}_{xx}^\alpha(f) df, \qquad (21.8)$$

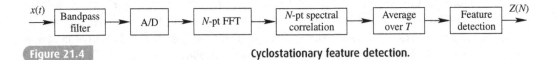

Figure 21.4   Cyclostationary feature detection.

where f_s is the sampling frequency and $S_{xx}^{\alpha}(f)$ is the SCD function. If $Z(N)$ is above a threshold, a signal is judged to be existing. Cyclostationary feature detection has a computational complexity of $O\left(N^2 + \frac{N}{2}\log_2 N\right)$ for a specified α, which is much larger than that of energy detection using FFT, namely, $O\left(\frac{N}{2}\log_2 N\right)$.

The detection of multiple cycle frequencies requires the calculation of a large fraction of the SCD function over a wide range of cyclic frequencies, followed by finding the peaks. This results in an intensive computational complexity and typically long observation times. The cycle frequency is required to have a much finer resolution than the spectral resolution, $\Delta\alpha \ll \Delta f \approx \frac{1}{T}$ [17]. This provides an improved detector performance and facilitates dichotomizing among the primary user and secondary user signals and different waveforms used.

Cyclostationary signatures provide an effective mechanism for overcoming the complexity of the SCD-based approach while maintaining the key advantages of spectrum cyclic analysis. By manipulating the IFFT frequency bins used for OFDM signal generation, correlation patterns can be artificially introduced in the power spectrum of that signal to form a cyclostationary signature, which is then used to aid the detection and classification of the OFDM-based signals in dynamic spectrum access networks [41, 42].

Cyclostationary feature detection is very attractive, as it does not require coherency and synchronization. It can also distinguish various signal properties and can work at low SNR. Spectrum cyclic analysis allows the detection and classification of signals with a performance approaching those of optimal coherent schemes [18], while maintaining the generality of other noncoherent approaches.

An OFDM radio is easy to integrate with spectrum-cyclic-analysis-based spectrum sensing, since both of them use FFT cores. When there are non-contiguous holes after the determination of the spectrum occupancy, the OFDM radio is the natural solution. This spectrum sensing approach has been widely used, and it has been also used for IEEE 802.22 cognitive radio.

21.5.2 Cyclostationary beamforming-based spectrum sensing

We have described signal cyclostationarity and cyclostationary beamforming in Section 18.5. In [12], spectrum sensing based on cyclostationary beamforming has been proposed. The approach exploits an antenna array, which can be installed on a fixed agent or on cognitive radios, to detect the occupancy of all the primary user and secondary user channels. If an antenna array is installed on a fixed agent, it is an external sensing method; the occupancy information is then broadcast to all the secondary users by using a control channel.

For a typical cognitive radio application, modulation parameters such as the carrier frequencies, data rates, and bandwidths of the possible channels for primary user signals are defined in a standard and are known by all the cognitive radios. The cycle frequencies for each primary user or secondary user mode on each channel are known and usually are designed to be unique, if the channel is occupied by a primary user or a cognitive radio.

Each class of cognitive radios may also know such modulation parameters of the other classes. The carrier frequencies of transmission channels are typically specified as

$$f_{c,n} = f_0 + \left(n - \frac{1}{2}\right) B, \quad n = 1, 2, \ldots, N, \tag{21.9}$$

where f_0 is the start frequency of the band, B is the bandwidth of each channel, and N specifies the number of channels. This is also applicable for OFDM signals. One or more cochannel signals may be present on each channel, as in the spread-spectrun communication scenarios.

A cognitive radio is required to scan the channels to find a vacant channel before accessing the network. A cognitive radio can test all the channels for the presence of transmission. In order to differentiate between primary user and secondary user signal types, their respective cycle or conjugate cycle frequencies must be selected to be different. Cyclostationary beamforming-based spectrum sensing first uses the ACS algorithm [11] to extract a signal of a specified α from the antenna array measurement. Based on the spectrum of the extracted signal, one can judge whether a channel is occupied by a primary user, a secondary user, or is vacant.

Cyclostationary beamforming-based spectrum sensing is advantageous over the spectrum cyclic analysis-based detection approach, since the latter has a much higher complexity. Search for the peaks on the SCD function is also computationally complex. The cyclostationary beamforming-based technique, however, has a complexity that is higher than that of the energy detector. At low SNR, the minimum number of required samples for reliable sensing is significantly less than that required for the energy detector.

The ACS algorithm has a performance superior to that of several other cyclostationary beamforming algorithms [11]. From the experiments given in [12], the ACS algorithm generally converges at around 2000 samples when the SIR γ is -20 dB or less. The number of samples required, N, should scale at between $O(1/\gamma)$ and $O(1/\gamma^2)$. That is, the minimum number of required samples is less than that required for the energy detector.

Example 21.1:

To examine the performance of the cyclostationary beamforming-based spectrum-sensing algorithm, we perform a few simulations in this example. We assume that the entire spectrum is divided into a fixed number of six channels at carrier frequencies 2001.2, 2002.2, ..., 2010.2 MHz. A uniform linear array with $n = 8$ antenna elements is used. The spacing between adjacent elements is half the wavelength at the carrier frequency of 2 GHz. The standard deviation of the noise at the array is $\sigma_n^2 = 0.1$. The signal environment for benchmarking is shown in Table 21.1: There are 5 primary user signals representing a primary service, and 2 secondary user signals representing different cognitive radio modes. The sampling rate f_s at the receiver is chosen to be 20 Msps, and all signals are at the same noise power level $\sigma_s^2 = 0.01$, and the signal power can be obtained from $SNR = \frac{P_s}{\sigma_s^2}$. The signals are BPSK or 16QAM signals, which are raised-cosine filtered with a roll-off factor of 0.22.

Table 21.1. Signals for benchmarking.

Signal	Carrier (MHz)	$f_{baud}/(f_s/20)$	Modulation	DoA	SNR (dB)	Primary/secondary user
A	2001.2	1/5	BPSK	35°	15	Primary user
B	2002.2	1/5	BPSK	20°	25	Primary user
C	2003.2	1/5	BPSK	−10°	30	Primary user
D	2006.2	1/5	BPSK	−35°	10	Primary user
E	2008.2	1/5	BPSK	60°	20	Primary user
F	2005.2	1/3	16QAM	-40°	15	Secondary user
G	2009.2	1/11	16QAM	15°	5	Secondary user

For these modulations, the maximum self-coherence occurs at $\tau = 0$ [1], and we select $\tau = 0$ here. The BPSK signal has conjugate cycle frequencies at $\alpha = \pm 2f_c + mf_{baud}$, $m = 0, \pm 1, \ldots$, and has cycle frequencies at $\alpha = mf_{baud}$, $m = \pm 1, \pm 2, \ldots$ [1, 11]. In a multiple signal environment, we needs to design the signals so that their selected cycle frequencies should have a least common multiple that is as large as possible [11]. The 16QAM signal has only nonconjugate cyclostationarity with cycle frequencies at $\alpha = mf_{baud}$, $m = \pm 1, \pm 2, \ldots$. The ACS algorithm can be implemented equally well for both nonconjugate and conjugate cyclostationary signals, depending on the properties of the desired signals.

The assignment of carrier frequencies and the selection of (conjugate) cycle frequencies must be very carefully done and some guidelines are given in [11].

We now implement spectrum sensing using Monte Carlo simulation. After applying the conjugate cyclostationarity-based ACS at $\alpha = 2f_c + f_{baud}$ at f_c, if there is a spectrum sidelobe at the channel, it may be either a primary user or a secondary user. We can test whether it is a secondary user mode by testing all the possible secondary user data rates using the cyclostationarity-based ACS algorithm until a spectrum mainlobe at the channel occurs; otherwise, it is not a secondary user and can be judged as a primary user signal or a hole. The threshold for spectrum occupancy can be lowered to 3 times the mean power in the band. The result is shown in Fig. 21.5. For this example, we assume that the possible secondary user signals are of type 16QAM with baud rates $\frac{1}{3}f_s/20$, $\frac{1}{7}f_s/20$, and $\frac{1}{11}f_s/20$.

Energy detection is a simple and popular technique for spectrum sensing. This method cannot discriminate between primary user and secondary user signals. It cannot detect a signal if it is buried in noise. When we set the detection threshold as 6 times the mean power in each channel, the result is shown in Fig. 21.6. Note that for the energy detection approach only one antenna is used and σ_a^2 is the noise power at the single antenna. Compared to the result obtained by the proposed approach (see Fig. 21.5), for a similar correct detection probability of channel occupancy (primary user plus secondary user), the probability of correct detection of spectrum holes is relatively lower in the energy detection approach; this corresponds to lower spectrum efficiency. By decreasing the detection threshold, the

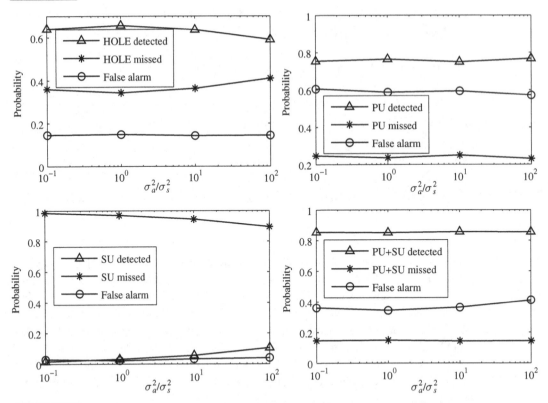

Figure 21.5 Detection probabilities at different noise levels: Conjugate cyclostationarity-based ACS at $\alpha = 2f_c + f_{baud}$ is used to extract the primary user signals, cyclostationarity-based ACS at $\alpha = f_{baud}$ to extract the secondary user signals. σ_a^2/σ_s^2 is used to characterize the SNR.

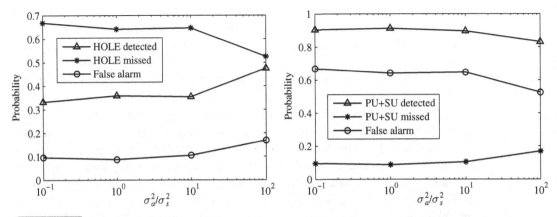

Figure 21.6 Detection probabilities of channel occupancy at different noise levels: energy detection.

probability of correct detection of occupied channels is made more close to 100%; this, however, further reduces the probability of correct detection of spectrum holes, and hence the spectrum efficiency.

The beamforming-based approach has the capability of correctly identifying primary user and secondary user signals. In Fig. 21.5, the probability of correctly identifying the secondary user signal is quite far from 100%; this is because one of the two secondary users (signal G) is too weak. Another advantage is that the beamforming-based method can extract and sense all the cochannel signals on the same channel (f_c), as long as they have different α values or cyclostationarity properties. This property can be applied to spread-spectrum signals, but it requires each signal to have a different data rate or coding scheme.

21.6 Dynamic spectrum access

Spectrum sensing and dynamic spectrum access are two critical aspects for cognitive radio networks. Network protocols in general exhibit some form of implicit cooperation, which leads to fairness and good performance for the entire network. Dynamic spectrum access of the secondary users can be formulated as a constraint optimization problem, which can be solved by using water-filling or game theory.

21.6.1 Water-filling for dynamic spectrum access

Water-filling is the traditional approach to dynamic spectrum access. In [8], a near optimal scheme with linear complexity has been designed for jointly allocating channels and power levels among cognitive radios. The proposed iterative water-filling scheme maximizes the sum capacity subject to individual constraints on users' power budgets. In [13], a more general formulation for dynamic spectrum access is given, and an iterative water-filling algorithm is derived.

In this section, we describe a dynamic spectrum access solution proposed in [13]. The dynamic spectrum access problem is formulated as maximizing the sum channel capacity while satisfying the power budgets of individual cognitive radios as well as the SINR constraints on both the secondary and primary users.

Consider N available channels and K contending cognitive radios in a cognitive radio network. We can maximize the sum capacity of the network while each cognitive radio has the constraints on BER, power, and channel usage, that is,

$$c_{n,k}, P_{n,k} = \arg\max C = \sum_{k=1}^{K} C_k = \sum_{k=1}^{K}\sum_{n=1}^{N} c_{n,k} B \log_2\left(1 + \frac{P_{n,k} g_{n,k}}{N_0 B}\right)$$

$$= \sum_{k=1}^{K}\sum_{n=1}^{N} c_{n,k} B \log_2\left(1 + P_{n,k}\bar{g}_{n,k}\right) \tag{21.10}$$

subject to

$$\sum_{k=1}^{K} c_{n,k} = 1, \quad \text{for all } n, \tag{21.11}$$

$$c_{n,k} \in \{0,1\}, \quad \text{for all } n,k, \tag{21.12}$$

$$\sum_{n=1}^{N} P_{n,k} \le P_{k,\max}, \quad \text{for all } n, \tag{21.13}$$

$$P_{n,k} \ge 0, \quad \text{for all } n,k, \tag{21.14}$$

where n runs from 1 to N, k runs from 1 to K, $c_{n,k} \in \{0,1\}$ indicates the assignment of channel n to cognitive radio k, $g_{n,k}$ is the channel power gain on channel n for cognitive radio k, $\overline{g}_{n,k} = \frac{g_{n,k}}{N_0 B}$, N_0 is the one-sided noise PSD, B is the bandwidth of each channel, and $P_{k,\max}$ is the maximum power of cognitive radio k. Equation (21.11) states that each channel can only be assigned to one cognitive radio, and (21.13) specifies the maximum power of each cognitive radio. In (21.10), C can be replaced by its normalized version $\overline{C} = C/B$.

Since each channel admits only one cognitive radio, the SINR of each cognitive radio k on channel n should be constrained by a threshold

$$\gamma_{n,k} = \overline{g}_{n,k} P_{n,k} \ge \gamma_{\text{th},k}. \tag{21.15}$$

It can be written as

$$P_{n,k} \ge \frac{\gamma_{\text{th},k}}{\overline{g}_{n,k}} = P_{n,k,\min}. \tag{21.16}$$

In a conservative design, the system allows one or more cognitive radios to coexist with a primary user, but the received SINR constraint at each primary user must be ensured. If the BS for the primary users transmits with power P_T at a distance of $d_{B,i}$ from primary user i and there is one cognitive radio on the same channel n, the received SINR at primary user i should be above a threshold [8]

$$\gamma_{\text{PU},i} = \frac{P_T d_{B,i}^{-r}}{P_{n,k} d_{k,i}^{-r} + N_0' B} \ge \gamma_0, \tag{21.17}$$

where $P_{n,k}$ is the transmit power of cognitive radio k on channel n, $d_{k,i}$ is the distance between cognitive radio k and primary user i, and r is the exponent of propagation loss. This constraint is equivalent to

$$P_{n,k} \le \frac{P_T}{\gamma_0} \left(\frac{d_{B,i}}{d_{k,i}} \right)^{-r} - \frac{N_0' B}{d_{k,i}^{-r}} = P_{n,k,\max}, \tag{21.18}$$

where $N_0' B$ is the noise power of a primary user.

Solving using Karush-Kuhn-Tucker theorem

The above problem is a mixed-integer nonlinear programming problem. The discreteness of $c_{n,k}$ endows the problem with analytical and algorithmic intractabilities. In order to solve it using nonlinear programming, one can convert the discrete variable $c_{n,k} \in \{0,1\}$ into a continuous variable. In [8], $c_{n,k}$ is treated as a continuous variable in the interval $[0,1]$, and its final result is quantized to 0 or 1. This result is suboptimal, since $c_{n,k}$'s usually stay in the middle part of the interval. In [13], $c_{n,k}$ in (21.10) is changed to $c_{n,k}^m$, where $m \geq 1$ is a fuzzifier. By selecting a large m, the optimization process will force one of the $c_{n,k}$'s, $k = 1, \ldots, K$, approximates unity, while all the other $c_{n,k}$'s approach zero, since otherwise $c_{n,k}^m$'s will be very small for all k.

With primary user protection, the problem can be reformulated as

$$c_{n,k}, P_{n,k} = \arg\max \overline{C} = \sum_{k=1}^{K} \sum_{n=1}^{N} \left(c_{n,k}\right)^m \log_2 \left(1 + P_{n,k}\overline{g}_{n,k}\right), \tag{21.19}$$

subject to

$$\sum_{k=1}^{K} c_{n,k} = 1, \quad \text{for all } n, \tag{21.20}$$

$$c_{n,k}(1 - c_{n,k}) = 0, \quad \text{for all } n, k, \tag{21.21}$$

$$\sum_{n=1}^{N} P_{n,k} \leq P_{k,\max}, \quad \text{for all } n, \tag{21.22}$$

$$P_{n,k} \geq c_{n,k} P_{n,k,\min}, \quad \text{for all } n, k, \tag{21.23}$$

$$P_{n,k} \leq c_{n,k} P_{n,k,\max}, \quad \text{for all } n, k, \tag{21.24}$$

where $c_{n,k}$ in (21.23) identifies that there is no SINR constraint for $P_{n,k}$ if channel n is not assigned to cognitive radio k.

This is a constrained convex optimization problem. To solve this problem using the Karush-Kuhn-Tucker theorem, the Lagrangian is defined as

$$L(P_{n,k}, c_{n,k}, \lambda_k, \beta_n, \mu_{n,k}) = \sum_{k=1}^{K} \sum_{n=1}^{N} \left(c_{n,k}\right)^m \log_2 \left(1 + P_{n,k}\overline{g}_{n,k}\right)$$

$$+ \sum_{n=1}^{N} \beta_n \left(\sum_{k=1}^{K} c_{n,k} - 1\right) + \sum_{n=1}^{N} \sum_{k=1}^{K} \beta'_{n,k} c_{n,k}(1 - c_{n,k})$$

$$+ \sum_{k=1}^{K} \lambda_k \left(P_{k,\max} - \sum_{n=1}^{N} P_{n,k}\right) + \sum_{n=1}^{N} \sum_{k=1}^{K} \mu_{n,k} \left(P_{n,k} - c_{n,k} P_{n,k,\min}\right)$$

$$+ \sum_{n=1}^{N} \sum_{k=1}^{K} \mu'_{n,k} \left(c_{n,k} P_{n,k,\max} - P_{n,k}\right). \tag{21.25}$$

The Karush-Kuhn-Tucker conditions are derived as

$$\mu_{n,k} \geq 0, \quad \mu'_{n,k} \geq 0, \quad \lambda_k \geq 0, \tag{21.26}$$

$$\frac{\partial L}{\partial P_{n,k}} = 0 \rightarrow (c_{n,k})^m \frac{1}{\ln 2} \frac{\overline{g}_{n,k}}{(1 + \overline{g}_{n,k} P_{n,k})} - \lambda_k + \mu_{n,k} - \mu'_{n,k} = 0, \tag{21.27}$$

$$\frac{\partial L}{\partial c_{n,k}} = 0 \rightarrow m(c_{n,k})^{m-1} \log_2(1 + \bar{g}_{n,k} P_{n,k}) + \beta_n$$
$$+ \beta'_{n,k}(1 - 2c_{n,k}) - \mu_{n,k} P_{n,k,\min} + \mu'_{n,k} P_{n,k,\max} = 0, \qquad (21.28)$$

$$\sum_{k=1}^{K} \lambda_k \left(P_{k,\max} - \sum_{n=1}^{N} P_{n,k} \right) = 0, \qquad (21.29)$$

$$\sum_{n=1}^{N} \sum_{k=1}^{K} \mu_{n,k} \left(P_{n,k} - c_{n,k} P_{n,k,\min} \right) = 0, \qquad (21.30)$$

$$\sum_{n=1}^{N} \sum_{k=1}^{K} \mu'_{n,k} \left(c_{n,k} P_{n,k,\max} - P_{n,k} \right) = 0. \qquad (21.31)$$

Since there are too many parameters to solve from the Karush-Kuhn-Tucker conditions, it is difficult to obtain the optimum solution. Iterative water-filling can be employed.

From (21.27), we can derive $P_{n,k}$ as a water-filling solution

$$P_{n,k} = (c_{n,k})^m \frac{1}{\lambda'_k \ln 2} - \frac{1}{\bar{g}_{n,k}}$$
$$= \begin{cases} P^0_{n,k} = \frac{1}{\lambda'_k \ln 2} - \frac{1}{\bar{g}_{n,k}}, & \text{if } \lambda'_k < \frac{\bar{g}_{n,k}}{\ln 2}, c_{n,k} = 1, \\ 0, & \text{otherwise} \end{cases} \qquad (21.32)$$

where $\lambda'_k = \lambda_k - \mu_{n,k} + \mu'_{n,k}$.

Substituting (21.32) into (21.22), and assuming λ_k is constant relative to n (that is, assuming $\mu_{n,k} - \mu'_{n,k} = $ constant relative to n), we have

$$\lambda'_k = \frac{|\mathcal{N}_k| \frac{1}{\ln 2}}{P_{k,\max} + \sum_{n \in \mathcal{N}_k} \frac{1}{\bar{g}_{n,k}}}, \qquad (21.33)$$

where $|\cdot|$ denotes the cardinal of the set within, and $\mathcal{N}_k$ is the set of channels that are assigned to user k, that is, $c_{n,k} = 1$, for all $n \in \mathcal{N}_k$.

We now consider the power constraints (21.23) and (21.24). After estimating the power by (21.32), we then force the estimated power to satisfy the power constraints

$$P_{n,k} = \max \left(\min \left(P^0_{n,k}, P_{n,k,\max} \right), P_{n,k,\min} \right), \quad \text{if } c_{n,k} = 1. \qquad (21.34)$$

Accordingly, λ'_k must be updated by replacing $P_{k,\max}$ in the denominator of (21.33) by $\min(P_{k,\max}, \sum_{n \in |\mathcal{N}_k|} P_{n,k,\max})$.

Iterative water-filling is implemented at a centralized node of the network. The node collects all the channel gains $g_{n,k}$'s, calculates the channel and power allocation, and then broadcasts the result to the cognitive radios. The algorithm assigns each channel to a user with a maximal capacity for that channel, subject to the power constraints. Given a channel, if the power constraints cannot be satisfied for any of the users, the channel is not assigned so as to save the power of the user batteries. An iterative water-filling algorithm for the above joint power and channel allocation scheme is given in detail in [13], and the algorithm converges to the final solution within a few iterations.

The complexity of the algorithm is $O(NKm)$, where m is the number of iterations. This algorithm is of a general-purpose nature, and additional constraints can be easily inserted into the algorithm. Simulation results demonstrate that there is a substantial capacity gap between the nonconservative and conservative designs.

Example 21.2:

We now give an example to demonstrate the performance of the iterative water-filling algorithm. All the cognitive radios are assumed to have the same maximum power $P_{k,\max} = 200$ mwatt, $i = 1, \ldots, K$. Set the SINR threshold for a cognitive radio $\gamma_{\text{th},k} = 5$ dB, the noise power at a cognitive radio $N_0 B = 0.1$ dBm. For the constraint (21.18), the transmit power at the BS of the primary user network $P_T = 3$ watts, the noise power at a primary user $N_0' B = 0.2$ dBm, the distance from the ith primary user to its BS $d_{B,i} = 20$ m, the distance from the kth cognitive radio to the ith primary user $d_{k,i} = 5$ m, the propagation exponent $r = 2.5$, and the SINR threshold for the primary user $\gamma_0 = 5$ dB.

To begin with, we assume that $N = 6$ channels are assigned to $K = 4$ cognitive radios. Assume that the gain vectors g_n, $n = 1, \ldots, N$, are generated by $g_n = (n/N)^3 a$, a being a random vector with entries uniformly distributed in $(0, 1)$. For a random run with nonconservative design, we have

$$
[g_{n,k}] = \begin{bmatrix}
0.0030 & 0.0011 & 0.0033 & 0.0045 \\
0.0247 & 0.0021 & 0.0199 & 0.0323 \\
0.1126 & 0.0879 & 0.0700 & 0.0423 \\
0.2542 & 0.2867 & 0.1447 & 0.1557 \\
0.3486 & 0.1662 & 0.1218 & 0.0088 \\
0.6945 & 0.3422 & 0.3643 & 0.1634
\end{bmatrix}.
$$

The algorithm generates the channel assignment $\mathbf{C} = [c_{n,k}]$ and the corresponding power allocation $\mathbf{P}$ as

$$
[c_{n,k}] = \begin{bmatrix}
0 & 0 & 1 & 0 \\
0 & 0 & 0 & 1 \\
0 & 1 & 0 & 0 \\
0 & 1 & 0 & 0 \\
1 & 0 & 0 & 0 \\
1 & 0 & 0 & 0
\end{bmatrix}, \quad
\mathbf{P} = \begin{bmatrix}
0 & 0 & 0.2000 & 0 \\
0 & 0 & 0 & 0.2000 \\
0 & 0.0996 & 0 & 0 \\
0 & 0.1004 & 0 & 0 \\
0.0999 & 0 & 0 & 0 \\
0.1001 & 0 & 0 & 0
\end{bmatrix}
$$

and $P_{n,k,\max} = P_{k,\max} = 0.2$. The sum capacity is 41.4935 bits/s/Hz.

When the constraint for conservative design is applied,

$$
[c_{n,k}] = \begin{bmatrix}
0 & 0 & 0 & 0 \\
0 & 0 & 0 & 0 \\
1 & 0 & 0 & 0 \\
0 & 1 & 0 & 0 \\
1 & 0 & 0 & 0 \\
1 & 0 & 0 & 0
\end{bmatrix}, \quad
\mathbf{P} = \begin{bmatrix}
0 & 0 & 0 & 0 \\
0 & 0 & 0 & 0 \\
0.0180 & 0 & 0 & 0 \\
0 & 0.0185 & 0 & 0 \\
0.0185 & 0 & 0 & 0 \\
0.0185 & 0 & 0 & 0
\end{bmatrix}
$$

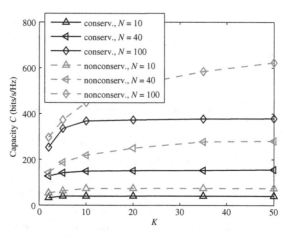

Figure 21.7   The capacity for different values of N and K, with conservative design.

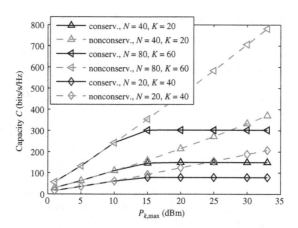

Figure 21.8   The sum capacity versus $P_{k,\max}$.

and $P_{n,k,\max} = 0.0185 \ll P_{k,\max} = 0.2$. The sum capacity is 23.2102 bits/s/Hz. Note that the first channel is not assigned in the conservative design due to poor channel quality.

The convergence of the iterative water-filling algorithm is very fast, usually needing 2 to 4 iterations for the algorithm to converge. For different values of N and K as well as conservative and nonconservative designs, the sum capacity as a function of K is plotted in Fig. 21.7. For different N and K values, the sum capacity as a function of $P_{k,\max}$ is plotted in Fig. 21.8. It is seen that when $P_{k,\max} > P_{n,k,\max}$, the sum capacity tends to saturate rapidly for conservative design, while it increases at a logarithmic rate for nonconservative design. For this example, it is seen that the conservative design substantially reduces the sum capacity.

21.6.2 Basic game theory

Game theory resolves situations in which people's interests conflict. Here we describe basic game theory using the notations given in [32, 50]. A learning automaton is an object that can choose from a finite number of actions according to its random environment. The learning automaton is capable of improving its performance with time while operating in an unknown environment. In a multiple-automata game, as illustrated in Fig. 21.9, N automata (players), A_1, A_2, ..., A_N, participate in a game. Each automaton A_i is represented by a 4-tuple $\{S_i, r_i, T_i, \boldsymbol{p}_i\}$, where S_i is a finite set of actions or pure strategies, with a cardinality m_i, $r_i \in [0, 1]$ is the random payoff to player i and the result of each play is a random payoff to each player, $\boldsymbol{p}_i(k) = \left(p_{i1}(k), \dots, p_{im_i}(k)\right)^T$ is the action choice probability distribution vector of player i at time k, $p_{il}(k)$ being the probability with which player i chooses the lth pure strategy, and T_i is the stochastic learning algorithm for updating $\boldsymbol{p}_i(k)$

$$\boldsymbol{p}_i(k+1) = T_i \left(\boldsymbol{p}_i(k), a_i(k), r_i(k) \right), \tag{21.35}$$

$a_i \in S_i$ being the action selected by player i.

The expected payoff or utility function of player i is defined as

$$d^i (a_1, \dots, a_N) = E\left[r_i | \text{player } j \text{ chooses action } a_j, a_j \in S_j, j = 1, \dots, N \right]. \tag{21.36}$$

The objective of each player is to maximize its expected payoff.

The learning algorithm for updating the action probabilities $\boldsymbol{p}_i(k)$ can be a reinforcement learning scheme, such as the linear reward-penalty ($L_{\text{R-P}}$) scheme [32]. The learning automata A_i's are updated as follows.

Algorithm 21.1 *(Learning Automata)*

1. Initialization.

    Choose $a_i \in S_i$, $i = 1, \dots, N$, at random, with random $\boldsymbol{p}_i(0)$.

2. `do while`

    $k = k + 1$.

    `for` $i = 1$ `to` N *(for all the N players)*

        Apply a_i, and observe the response r_i from the environment.

        Update $\boldsymbol{p}_i(k)$ by (21.35).

    `end for`

  `until` *a stopping criterion is reached or $\boldsymbol{p}_i$ converges.*

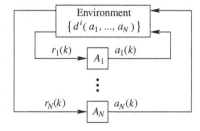

Figure 21.9   The automata game block diagram.

Given the probabilistic strategy for player i, p_i, the expected payoff for player i is

$$g^i(p_1, \ldots, p_N) = E\left[d^i | \text{player } j \text{ employs strategy } p_j, j = 1, \ldots, N\right]$$

$$= \sum_{j_1, \ldots, j_N} d^i(j_1, \ldots, j_N) \prod_{s=1}^N p_{sj_s}. \tag{21.37}$$

In order to map a multiple-agent problem, such as a wireless network, to a game, we need to define three components of a game: players, strategy, and utility function. The players are the nodes in the network. The strategy is the action related to the functionality being investigated, and the utility function is a performance metric.

Nash equilibrium

The N-tuple of strategies $(p_1^0, \ldots, p_N^0)$ is said to be a *Nash equilibrium*, if for each i, $i = 1, \ldots, N$, we have

$$g^i\left(p_1^0, \ldots, p_{i-1}^0, p_i^0, p_{i+1}^0, \ldots, p_N^0\right)$$

$$\geq g^i\left(p_1^0, \ldots, p_{i-1}^0, p_i, p_{i+1}^0, \ldots, p_N^0\right), \text{ for any } p_i \in [0,1]^{m_i}. \tag{21.38}$$

The vector p_i is a pure strategy if it is a unit probability vector, which has only one component being unity and all other components being zeros. A Nash equilibrium is said to be in *pure* strategies, if each p_i^0 is a unit probability vector. The components of p_i^0 have a sum of unity and are in one-to-one correspondence with the pure strategies of player i. If more than one pure strategy is assigned positive probabilities, p_i^0 is a mixed strategy, and the corresponding Nash equilibrium is a nondegenerated mixed Nash equilibrium. In general, each p_i^0 may be a mixed strategy and the Nash equilibrium defined by (21.38) is in mixed strategies. It is well known that every finite strategic game has a mixed-strategy Nash equilibrium [15, 32]. Correlated equilibrium is Nash equilibrium of an expanded game with a correlating device. Like Nash equilibria, correlated equilibria also exist in finite games.

The central concept in noncooperative game theory is an equilibrium under certain conditions, such as common knowledge of rationality. A powerful way of finding Nash equilibria of games is to eliminate dominated strategies. Given two strategies p_i and p_i' for player i in a normal form game, p_i' is said to be *strictly (or weakly) dominated* by p_i if, for every choice of strategies of the other players, player i's payoff from choosing p_i is *strictly greater than (or at least as great as)* player i's payoff from choosing p_i'. A practical method to compute the Nash equilibrium would be to use the sequential play where each player maximizes its own utility function sequentially while other players' strategies are fixed. A game can be solved by using an iterative process of *elimination for strictly dominated strategies* for all players.

A stochastic learning-based distributed solution is simple and efficient. It is used to learn p_i's by considering the history of play. The learning automata algorithm needs less information and control signaling to operate than the sequential play, but it converges much more slowly than the sequential play.

In an N-player environment, there are a lot of Pareto solutions: there is no better solution in which all players simultaneously do better. Bargaining theory can be used for finding a good Pareto solution. Through cooperation, another outcome, which is better than the noncooperative outcome (called the *threatpoint*), is obtained. For solving the cooperation by using a noncooperative game theory, the most commonly used incentives are pricing and reputation.

In a cooperative game, players cooperate by forming a coalition, described by a single payoff. The share of the payoff received by all players in the coalition is a payoff vector.

Potential games

A potential game is such a game that any change in the utility function of any player due to a unilateral deviation by that player is correspondingly reflected in a global function, referred to as the *potential function*. The existence of a potential function is desirable, since it makes analysis very easy and it maximizes a global utility by only trying to maximize a player's own utility. Exact and ordinal potential games possess a useful convergence property: players of the game are guaranteed to converge to a Nash equilibrium by playing their best response.

Repeated games

Many wireless systems co-exist with the same set of competing systems over a long period of time. This scenario can be modeled as a repeated (or dynamic) game, where the players play multiple rounds, remembering their past experience in the choice of the power allocation for the next round. At the end of each stage, all the players can observe the outcome of the stage game and can use the complete history of the play to decide on the future action. A strategy in the repeated game is a complete plan of action. For the repeated game, sequences of strategy profiles that form a Nash equilibrium in the stage game form a Nash equilibrium in the dynamic game. The dynamic game also allows for a much richer set of Nash equilibria. The players can agree through a standardization process to operate in any Nash equilibrium of the dynamic game. This provides more flexibility in obtaining a fair and efficient resource allocation. By considering a repeated game, cooperation decisions are made based on the long-term payoff of each cooperating partner. Specifically, the overall utility function is chosen to be the weighted sum of the individual utility from current cooperation period and over the next $K - 1$ cooperation periods.

Auction games

The auction game belongs to a special class of game with incomplete information. In an auction game, a principal (auctioneer) conditions his auctions on some information that is privately known by the other players, called *agents (bidders)*. According to an explicit set of rules, the principle determines resource allocation and prices on the basis of bids from the agents.

For example, in the pricing game for routing in the MANET, the source can be viewed as the principal, who attempts to buy the forwarding services from the candidate nodes that forward routes. The possible forwarding nodes are the bidders who compete with each other for serving the source node, by which they may gain extra payments for future use. In order to maximize their own interests, the selfish forwarding nodes will not reveal their actual forwarding costs. They compete for the forwarding request by eliciting their willingness of the payments in the form of bids. Thus, the sender is able to lower its forwarding payment by the competition among the candidate routing nodes based on the auction rules.

21.6.3 Four persona models

Selfish behavior by players may lead to a Nash equilibrium that is socially undesirable and is a suboptimal equilibrium. Therefore, it is imperative to make the network robust to selfish behavior, by punishing selfish behavior, or by credit-exchange-based and reputation-based incentive mechanisms.

The conventional representation of the human actor, the so-called *rational actor model* or *Homo economicus model*, is adopted in game theory. The players choose their actions independently, but their choice impacts on all the players in the network. The players are *rational*: they act in their best interest by maximizing their own utility. In a noncooperative game, players act selfishly such that their actions maximize their own utilities. Cooperation of such users can be implemented by using coalitional game theory or by designing cooperative protocols based on a noncooperative game, for which the selfish utility of each player is modified by a pricing function to enable cooperation. For coalitional games, players cooperate by forming coalitions, each coalition having a single payoff. Rational learning does not converge to Nash equilibrium in general [20]. No-regret learning is able to learn and converge to mixed strategy equilibria in games, and also to converge to a pure strategy Nash equilibrium if it exists for a given game [20]. The no-regret learning formulation is particularly useful to accommodate selfish users. The method replaces explicit opponent modeling with an implicit regret matrix, θ_n^l, for player l at time n.

In the rational actor model, people behave like self-interested, outcome-oriented actors with anonymous, nonstrategic interactions. Here, we discuss a common pool resource game. For a group of n players, each player i is expected to invest so as to control the probability p_i of damaging the system. The more player i invests, the lower p_i is expect to be. Each player is selfish, and tries to maximize its own utility, but does not consider the overall condition of the system. In this case, the payoff for player i can be defined as [19]

$$u_i(p_i) = p_i \left(1 - \frac{1}{n} \sum_{i=1}^{n} p_i \right) - cp_i^2, \tag{21.39}$$

where $c > 0$ is a cost for not investing. If no player invests, u_i is close to zero. This is an example of the "tragedy of the commons". On the other hand, if most players invest a sufficient amount, then the payoff of player i is nearly p_i. The unique Nash equilibrium in this game is $p_i^* = \frac{1}{1+2c+\frac{1}{n}}$, and [19]

$$u_i = \frac{c + \frac{1}{n}}{(1 + 2c + \frac{1}{n})^2}. \tag{21.40}$$

Three other persona models are usually considered in game theory in case of strategic interactions [19]. The *Homo reciprocans* model pays importance on reciprocity in strategic interaction. The persona *Homo reciprocans* exhibits strong reciprocity, which tends to cooperate and share with others similarly disposed, to punish those who violate cooperative norms, even at personal cost and even with no future rewards from so doing. *Homo reciprocans* is an altruist in the sense that he improves the welfare of a group of unrelated individuals at the expense of his personal welfare. In contrast, the conventional *Homo economicus* cooperates and punishes only if it is in his long-term interest. The persona *Homo egualis* may be willing to reduce his own payoff to increase the equality in the group when on top, but is displeased and exhibits a strong urge to reduce inequality when on the bottom. The persona *Homo parochius* divides the world into insiders and outsiders, and values insiders more highly than outsiders, and partially suppresses personal goals in favor of the goals of the group of insiders.

The *Homo egualis* society, as a distributed model, can be modeled by the utility function of player i in an n-player game [19]

$$u_i = x_i - \frac{\alpha_i}{n-1} \sum_{x_j > x_i} (x_j - x_i) - \frac{\beta_i}{n-1} \sum_{x_j < x_i} (x_i - x_j), \tag{21.41}$$

where $x_i, \ldots, x_j$ are the payoffs for the players, and $0 \le \beta_i < \alpha_i \le 1$. *Homo egualis* exhibits a weak urge to increase inequality when doing better than the others and a strong urge to reduce inequality when doing worse than the others, that is, the *Homo egualis* agents have the inequality aversion property, which is especially desirable for dealing with fairness.

21.6.4 Game-theoretic models for dynamic resources allocation

The most popular tool for analyzing wireless networks is game theory. A cognitive radio network can be regarded as a chaotic game, trying to find the equilibrium of conflicting actions that meet the nodes' operational goals across different network layers in a specific environment. From game theory, if the cognitive radios do not cooperate, there is a unique Nash equilibrium. At this point, the achievable rate is bounded by a constant, regardless of the available transmit power [27].

Game theory is especially useful for dynamic resource allocation, routing, cross-layer protocol design of a wireless network, including cognitive networks and wireless ad hoc/sensor networks. In a game-theoretic framework, the players of the game are the network nodes, such as cognitive radios or sensor nodes, their actions are their choices of transmission parameters (e.g., transmission powers, access probability, backoff interval, relaying nodes), and their utilities are their defined performance measures, such as their generic QoS metrics, which can be a combination of throughput, delay, energy, interference-related measures (e.g., SIR or SINR, or equivalent BER requirement), etc.

Cooperation in wireless networks can improve the performance by exploiting some form of multiuser diversity. In a cooperative spectrum sensing strategy, all cognitive radios collaborate by sharing their decisions regarding spectrum occupancy of the primary users. Cooperative detection can be exploited to relax the requirements on the detectors of the individual cognitive radios or to improve the agility of the entire network.

Game theory has been applied to a wireless network at the physical layer (for distributed power control and waveform adaptation), link layer (for channel assignment), and network layer (for packet forwarding, throughput management).

A game-theoretic model to obtain the optimal pricing for dynamic spectrum sharing in cognitive radio networks has been proposed in [34], where multiple primary services compete to offer spectrum access opportunities to the secondary service and the firms adjust their prices dynamically to gain the highest profit. In [2], a pricing policy was introduced for voice services in a wireless LAN environment considering both QoS performance and users' willingness-to-pay.

In [50], stochastic learning-based techniques were exploited to discover the equilibrium solution. Let λ and μ represent the arrival and service rates of a cognitive radio type, respectively. Each cognitive radio only contends for the spectrum with probability p_i, so the actual traffic load to the system can be approximated by $p_i\lambda_i$. To obtain the optimum p_i, from the Markov model for general traffic load, all the λ_i's and μ_i's are required and this, however, is impractical in a real access scenario. A more practical scheme is to learn p_i with only local information.

In [33], distributed adaptive channel allocation is described with the assumption that the cognitive radios can measure the local interference temperature on different frequencies and can adjust it by optimizing the information transmission rate. Targeting at a BER or an equivalent SIR requirement, the channel allocation problem is modeled as a potential game which converges to a deterministic Nash equilibrium point. No-regret learning is applied with cooperation on the potential game.

Pricing and resource allocation are closely related, as a service provider wants to maximize its revenue and the user desires to maximize his satisfaction in terms of QoS performance and price. A joint power/channel allocation scheme that uses a distributed pricing strategy has been proposed in [47]. The proposed price-based iterative waterfilling algorithm enables cognitive radios to reach a good Nash equilibrium. The algorithm can be implemented distributively, with cognitive radios repeatedly negotiating their best transmission powers and spectrum.

The auction mechanism can be used for resource allocation in wireless networks. In [5], a bandwidth allocation algorithm has been proposed for wireless ad hoc networks, in which flows are bidding for resources. It is a distributed, QoS-aware, price-based allocation algorithm that allocates bandwidth to flows based on information that is locally available. In [51], a sensor network is constructed with a mix of mobile and static sensors to achieve a balance between sensor coverage and sensor cost. Bidding protocols are used to guide the movement of mobile sensors to increase the coverage. In the protocols, static sensors detect coverage holes locally by using Voronoi diagrams and bid mobile sensors to move.

Fairness and efficiency

Efficiency and fairness are the main goals of a spectrum etiquette. An important metric for dynamic spectrum access is the average airtime per cognitive radio system. Airtime is the ratio of allocation time per cognitive radio to the reference time. Fairness can be achieved by cooperation. For perfect fairness and maximal airtime for each cognitive radio, there is an access probability vector that no cognitive radio system can do better in terms of airtime share without harming the other cognitive radio systems [49, 50]. When fairness is not considered, the most spectrum-efficient access is that all users compete for the spectrum greedily; this may, however, always block some cognitive radios from spectrum access.

In practice, the traffic loads of different cognitive radio systems and the operator's policy for the purpose of revenue must also be considered. This requires the definition of weighted fairness, that is, the airtime t_{air} for each cognitive radio over the product of the priority parameter θ_i and the traffic load λ_i, that is, $\frac{t_{air}}{\theta_i \lambda_i}$, is the same for all the cognitive radios. Efficiency is achieved if each cognitive radio is subject to the weighted fairness. The airtime can be replaced by the average cumulative on-spectrum time per cognitive radio t_{onair}, and the payoff $x_i = \frac{t_{onair}}{\theta_i \lambda_i}$, where λ_i can be estimated by historical usage records of cognitive radio i.

In a noncooperative strategy, an access technique inspired by the *Homo egualis* model can be used for achieving fairness in spectrum access. Each cognitive radio can update its access probability p_i by [50]

$$p_i = \max\left(0, \min\left(1, p_i + \frac{\alpha_i}{n-1} \sum_{x_j \geq x_i} \left(\frac{x_j - x_i}{x_j}\right) - \frac{\beta_i}{n-1} \sum_{x_j < x_i} \left(\frac{x_j - x_i}{x_j}\right)\right)\right),$$
$$\text{for all } j \neq i, \qquad (21.42)$$

where initially $p_i = 1$, and n is the number of cognitive radios. The average cumulative on-spectrum times of the cognitive radios that use the same spectrum block are needed. This can be done by book-keeping the busy time of the required spectrum based on periodical spectrum scanning and cognitive radio detection technologies. Thus, each cognitive radio can access the spectrum based only on its own book-keeping and its own local measurement.

The blocking probability of each cognitive radio is another important metric, and it is determined by all the state probabilities π_j and the traffic loads for different cognitive radios λ_j [49]. By modeling the arrival traffic as a Poisson's process, all the cognitive radios have almost the same airtime share and blocking probability by using this access scheme; this is fairness [49, 50].

QoS and interference temperature constraints

When revenue is treated as the primary concern, we may also consider the QoS and interference temperature constraints. For secondary spectrum sharing among spread spectrum users, the interference temperature constraint decides the SINR (or equivalently, BER) of each user and it is translated to a total received power threshold P_{th} at the measuring point. The QoS constraint defines different priorities or throughputs, which are related to the payments of the secondary users. Assuming a_i to be the priority parameter for link i, the operator problem can be formulated as maximization of the network revenue, $\max \sum_i a_i$, subject to the constraints on SIR, interference temperature, and transmission power of the ith link that transmits. The priority parameter a_i increases with the price c_i. In particular, one can select $a_i = c_i^{\alpha}$, $0 \le \alpha \le 1$; that is, a small α allows more active secondary links, while a large α guarantees services to users that pay higher price.

In this game, each cognitive radio maximizes its utility function u_i by its choice of being active or not. The system will reach an operating point where the network revenue is maximized while satisfying QoS and interference temperature constraints. A stochastic learning algorithm for the operator problem has been given in [50], where the probabilities are updated using the linear reward-inaction (L_{R-I}) reinforcement learning. It is composed of two phases. The coordination process controls the optimal set of active secondary links that access the spectrum, and the power control phase ensures the minimum target link SIRs given the set of active links.

When multiple cognitive radios access (or overlay on) the same channel (e.g., in a MC-CDMA system), one needs to minimize the total power consumption for all the K users while satisfying the data rate (or SINR) requirement of each user. Assuming that P_i and $\gamma_{\text{th},i}$, $i = 1, \ldots, K$, are the transmit power and the corresponding SINR threshold for cognitive radio i, respectively, the received SINR at cognitive radio i should satisfy the data rate constraint

$$\gamma_i = \frac{\alpha_{ii} P_i}{\sum_{j \neq i} \alpha_{ji} P_j + N_i} \ge \gamma_{\text{th},i}, \tag{21.43}$$

where α_{ji} is the channel gain between the jth transmitter and the ith receiver, and N_i is the noise power at the ith receiver. A simple iterative distributed power control algorithm to solve (21.43) subject to the power constraint $P_i \le P_{i,\max}$, which achieves the optimal power allocation, is given by [14]

$$P_i(k+1) = \frac{\gamma_{\text{th},i}}{\gamma_i(k)} P_i(k). \tag{21.44}$$

Problems

21.1 Briefly describe the salient features of MCUs, DSPs, FPGAs, and ASICs. Give the trade-offs in using these chips.

21.2 Assume a cognitive radio system that needs to identify BPSK, QPSK, and GMSK modulation formats. Which parameters can effectively differentiate them? Design a classifier for this purpose.

21.3 Define a noncooperative game to solve the DSA problem described in Section 21.6.1. Give your utility function.

21.4 For Problem 21.3, solve the defined game. Compare the sum capacity with that obtained in Example 21.2.

References

[1] B. G. Agee, S. V. Schell & W. A. Gardner, Spectral self-coherence restoral: A new approach to blind adaptive signal extraction using antenna arrays. *Proc. IEEE*, **78**:4 (1990), 753–766.

[2] L. Badia, S. Merlin, A. Zanella & M. Zorzi, Pricing VoWLAN services through a micro-economic framework. *IEEE Wireless Commun.*, **13**:1 (2006), 6–13.

[3] D. Cabric & R. Brodersen, Robust spectrum sensing techniques for cognitive radio networks. In F. H. P. Fitzek & M. D. Katz, eds., *Cognitive Wireless Networks* (Dordrecht, The Netherlands: Springer, 2007), 373–394.

[4] K. Chapman, P. Hardy, A. Miller & M. George, *CDMA Matched Filter Implementation in Virtex Devices*, Xilinx XAPP212 (v1.0), 2000.

[5] C. Curescu & S. Nadjm-Tehrani, A bidding algorithm for optimized utility-based resource allocation in ad hoc networks. *IEEE Trans. Mobile Comput.*, **7**:12 (2008), 1397–1414.

[6] N. Devroye, P. Mitran & V. Tarokh, Achievable rates in cognitive radio. *IEEE Trans. Inf. Theory*, **52**:5 (2006), 1813–1827.

[7] F. F. Digham, M.-S. Alouini & M. K. Simon, On the energy detection of unknown signals over fading channels. In *Proc. IEEE ICC*, Seattle, WA, May 2003, **5**, 3575–3579.

[8] F. F. Digham, Joint power and channel allocation for cognitive radios. In *Proc. IEEE WCNC*, Las Vegas, NV, Mar–Apr 2008, 882–887.

[9] O. A. Dobre, A. Abdi, Y. Bar-Ness & W. Su, Survey of automatic modulation classification techniques: classical approaches and new trends. *IET Commun.*, **1**:2 (2007), 137–156.

[10] K.-L. Du & M. N. S. Swamy, *Neural Networks in a Softcomputing Framework* (London: Springer, 2006).

[11] K.-L. Du & M. N. S. Swamy, A class of adaptive cyclostationary beamforming algorithms. *Circ. Syst. Signal Process.*, **27**:1 (2008), 35–63.

[12] K.-L. Du & W. H. Mow, *Exploiting Multiple Antennas for Spectrum Sensing in Cognitive Radio Networks*, U. S. Patent Application, 2009.

[13] K. -L. Du, M.N. S. Swamy & Q. Ni, A dynamic spectrum access scheme for cognitive radio networks. In *Proc. IEEE CCECE*, St. John's, Canada, May 2009.

[14] G. J. Foschini & Z. Miljanic, A simple distributed autonomous power control algorithm and its convergence. *IEEE Trans. Veh. Tech.*, **42**:4 (1993), 641–646.

[15] D. Fudenberg & J. Tirole, *Game Theory* (Cambridge, MA: The MIT Press, 1992).

[16] T. Fusco, L. Izzo, A. Napolitano & M. Tanda, On the second-order cyclostationarity properties of long-code DS-SS signals. *IEEE Trans. Commun.*, **54**:10 (2006), 1741–1746.

[17] W. A. Gardner, *Statistical Spectral Analysis: An Nonprobabilistic Theory* (Englewood Cliffs, NJ: Prentice-Hall, 1987).

[18] W. A. Gardner, Signal interception: a unifying theoretical framework for feature detection. *IEEE Trans. Commun.*, **36**:8 (1988), 897–906.

[19] H. Gintis, *Game theory Evolving: A Problem-Centered Introduction to Modeling Strategic Behavior* (Princeton, NJ: Princeton University Press, 2000).

[20] A. Greenwald & A. Jafari, A general class of no-regret algorithms and game-theoretic equilibria. In *Proc. 16th Ann. Conf. Learning Theory (COLT)*, Washington, DC, Aug 2003, 1–11.

[21] Z. Han & H. Jiang, Replacement of spectrum sensing and avoidance of hidden terminal for cognitive radio. In *Proc. IEEE WCNC*, Las Vegas, NV, Mar–Apr 2008, 1448–1452.

[22] S. Haykin, Cognitive radio: brain-empowered wireless communications. *IEEE J. Sel. Areas Commun.*, **23**:12 (2005), 201–220.

[23] R. W. Heath, Jr. & G. B. Giannakis, Exploiting input cyclostationarity for blind channel identification in OFDM systems. *IEEE Trans. Signal Process.*, **47**:3 (1999), 848–856.

[24] X. H. Huang, K.-L. Du, A. K. Y. Lai & K. K. M. Cheng, A unified software radio architecture. In *Proc. IEEE SPAWC*, Taoyuan, Taiwan, Mar 2001, 330–333.

[25] A. Jamin, P. Mahonen & Z. Shelby, Software radio implementability of wireless LANs. In E. Del Re, ed., *Software Radio: Technologies and Services* (London: Springer, 2001).

[26] W. S. Jeon, D. G. Jeong, J. A. Han, G. Ko & M. S. Song, An efficient quiet period management scheme for cognitive radio systems. *IEEE Trans. Wireless Commun.*, **7**:2 (2008), 505–509.

[27] E. G. Larsson & E. A. Jorswieck, The MISO interference channel: competition versus collaboration. In *Proc. Allerton Conf. Commun., Control, and Computing*, Monticello, AR, Sep 2007.

[28] K. B. Letaief & W. Zhang, Cooperative spectrum sensing. In E. Hossain & V. Bhargava, eds., *Cognitive Wireless Communication Networks* (Berlin: Springer, 2007), pp. 115–138.

[29] T. Li, W. H. Mow, V. K. N. Lau, M. Siu, R. S. Cheng & R. D. Murch, Robust joint interference detection and decoding for OFDM-based cognitive radio systems with unknown interference. *IEEE J. Sel. Areas Commun.*, **25**:3 (2007), 566–575.

[30] M. Luby, LT codes. In *Proc. IEEE FOCS*, Vancouver, Canada, Nov 2002, 271–282.

[31] J. Mitola III & G. Q. Maguire, Jr., Cognitive radio: making software radios more personal. *IEEE Pers. Commun.*, Aug 1999, 13–18.

[32] K. Narendra & M. A. L. Thathachar, *Learning Automata: An Introduction* (Englewood cliffs, NJ: Prentice Hall, 1989).

[33] N. Nie & C. Comaniciu, Adaptive channel allocation spectrum etiquette for cognitive radio networks. In *Proc. IEEE DySPAN*, Baltimore, MD, Nov 2005, 269–278.

[34] D. Niyato & E. Hossain, Competitive pricing for spectrum sharing in cognitive radio networks: Dynamic game, inefficiency of Nash equilibrium, and collusion. *IEEE J. Sel. Areas Commun.*, **26**:1 (2008), 192–202.

[35] M. Oner & F. Jondral, Air interface recognition for a software radio system exploiting cyclostationarity. In *Proc. IEEE PIMRC*, Barcelona, Spain, Sep 2004, **3**, 1947–1951.

[36] Z. Quan, S. Cui, A. H. Sayed & H. V. Poor, Wideband spectrum sensing in cognitive radio networks. In *Proc. IEEE ICC*, Beijing, China, May 2008, 901–906.

[37] X. Reves, A. Gelonch & F. Casadevall, Software radio implementation of a DS-CDMA indoor subsystem based on FPGA devices. In *Proc. IEEE PIMRC*, San Diego, CA, Sep–Oct 2001, **1**, D-86–D-90.

[38] A. Sahai, N. Hoven & R. Tandra, Some fundamental limits on cognitive radio. In *Proc. 42nd Allerton Conf. Commun. Contr. and Comput.* Monticello, IL, Oct 2004., 1–11.

[39] A. Sendonaris, E. Erkip & B. Aazhang, User cooperation diversity – part I: system description; part II: implementation aspects and performance analysis. *IEEE Trans. Commun.*, **51**:11 (2003), 1927–1948.

[40] O. Simeone, Y. Bar-Ness & U. Spagnolini, Stable throughput of cognitive radios with and without relaying capability. *IEEE Trans. Commun.*, **55**:12 (2007), 2351–2360.

[41] P. D. Sutton, K. E. Nolan & L. E. Doyle, Cyclostationary signatures for rendezvous in OFDM-based dynamic spectrum access networks. In *Proc. IEEE DySPAN*, Dublin, Ireland, Apr 2007, 220–231.

[42] P. D. Sutton, K. E. Nolan, and L. E. Doyle, Cyclostationary signatures in practical cognitive radio applications. *IEEE J. Sel. Areas Commun.*, **26**:1 (2008), 13–24.

[43] Z. Tian and G. B. Giannakis, A wavelet approach to wideband spectrum sensing for cognitive radios. In *Proc. IEEE 1st Int. Conf. Cognitive Radio Oriented Wireless Networks and Commun.*, Greece, Jun 2006, 1–5.

[44] T. Turlett, H. J. Bentzen & D. Tennenhouse, Toward the software realization of a GSM base station. *IEEE J. Sel. Areas Commun.*, **17**:4 (1999), 603–612.

[45] US DoD, *Programmable Modular Communications System (PMCS) Guidance Document*, US Department of Defense, Washington DC, Jul 31, 1997.

[46] D. Z. Vucic, M. M. Obradovic & D. M. Obradovic, Spectral correlation of OFDM signals related to their PLC applications. In *Proc. 6th Int. Symp. Power-Line Comm. and Its Appl. (ISPLC)*, Athens, Greece, Mar 2002, 1–4.

[47] F. Wang, M. Krunz & S. Cui, Price-based spectrum management in cognitive radio networks. *IEEE J. Sel. Topics Signal Process.*, **2**:1 (2008), 74–87.

[48] W. Xiang, T. Pratt & X. Wang, A software radio testbed for two-transmitter two-receiver space-time coding OFDM wireless LAN. *IEEE Commun. Mag.*, **42**:6 (2004), S20–S28.

[49] Y. Xing, R. Chandramouli, S. Mangold & S. Shankar N, Dynamic spectrum access in open spectrum wireless networks. *IEEE J. Sel. Areas Commun.*, **24**:3 (2006), 626–637.

[50] Y. Xing, H. Kushwaha, K. P. Subbalakshmi & R. Chandramouli, Codes and games for dynamic spectrum access. In H. Arslan, ed., *Cognitive Radio, Software Defined Radio, and Adaptive Wireless Systems* (Berlin: Springer, 2007), pp. 161–187.

[51] G. G. Wang, G. Cao, P. Berman & T. F. La Porta, Bidding protocols for deploying mobile sensors. *IEEE Trans. Mobile Comput.*, **6**:5 (2007), 515–528.

Wireless ad hoc and sensor networks

22.1 Introduction

A wireless ad hoc network is an autonomous, self-organized, distributed, peer-to-peer network of fixed, nomadic, or mobile users that communicate over bandwidth-constrained wireless links. There is no preexisting infrastructure. Wireless ad hoc networks can be relay, mesh, or star networks comprising special cases. When the nodes are connected in mesh topology, the network is also known as a *wireless mesh network*. It is known as a *mobile ad hoc network (MANET)*, when the nodes are mobile.

A wireless ad hoc network has a mutlihop relaying of packets, as shown in Fig. 22.1. It can be easily and rapidly deployed, and expensive infrastructures can be avoided. Without an infrastructure, the nodes handle the necessary control and networking tasks by themselves, generally by distributed control. Typical applications are wireless PANs for emergency operations, civilian, and military use. The wireless ad hoc network is playing an increasing role in wireless networks, and wireless ad hoc networking mode has been or is being standardized in most IEEE families of wireless networks.

MANETs and wireless sensor networks (WSNs) are the two major types of wireless ad hoc networks. They both are distributed, multi-hop systems. A MANET is an autonomous collection of mobile routers (and associated hosts) connected by wireless links. Each node is an information appliance, such as a personal digital assistant (PDA), equipped with a radio transceiver. The nodes are fully mobile. A WSN is a special kind of wireless ad hoc network with distributed sensing and processing capability. There may be densely distributed sensors that are equipped with low-power wireless transceivers, memory, and batteries. In both MANETs and WSNs the nodes act both as hosts and as routers. WSNs are distinguished from MANETs by processing the data gathered by sensors. The goal of WSNs is the detection/estimation of some events of interest, and not just communication. MANETs employ a node-centric communication model.

22.1.1 Wireless sensor networks

WSNs are an emerging technology that has the potential to revolutionize many aspects of our life. A WSN consists of a number of self-powered nodes with built-in computing power and sensors for acquiring data from the environment, as well as wireless communication capability. Many WSNs require little power and could potentially be deployed for a number of years. Attaching sensors to RFID tags or wireless nodes adds awareness to

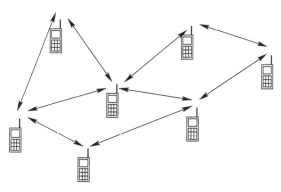

Figure 22.1   Wireless ad hoc network.

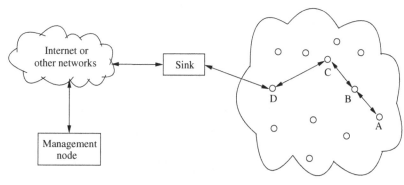

Figure 22.2   WSN illustration.

ubiquitous networks. This awareness means that the network can detect and respond to the environment, without human interaction.

A WSN is illustrated in Fig. 22.2. Sensed information at the nodes is relayed to a sink node (BS) by using multihop communication. The sink node has gateway functions to link to external networks such as the Internet. Typical receiver sensitivities are between −85 and −110 dBm.

Unlike a general ad hoc network, the WSN has the features of application-specificity, environment interaction, simplicity and scalability, energy-efficiency, self configurability, QoS, and data-centricity. More specifically, [5, 55]

- The number of sensor nodes in a WSN can be huge.
- Sensor nodes are densely deployed.
- Sensor nodes are prone to failures, or may die.
- The topology of a sensor network changes very frequently.
- Sensor nodes mainly communicate by broadcast, whereas most ad hoc networks are based on point-to-point communications.
- Information exchange between end-to-end nodes is rare in WSNs.

- Sensor nodes are limited in resources, such as power, computational resources, and bandwidth. A sensor node (or mote) is powered by a battery, which it is often unfeasible to recharge. Thus, energy efficiency is the primary concern in WSN design. Due to power restriction, computation is preferred over the network, rather than at the node.
- Each node has its own identifier, but may not have global identification (ID) because of the large amount of overhead and large number of sensors.

WSNs are similar to the MANETs, but have several major differences: WSNs involve little or low mobility, and they have much tighter computation and communication constraints.

Some application examples of WSNs are given below.

- Environmental monitoring: habitat monitoring, integrated biology, structural monitoring, forest wildfire monitoring, flood detection, volcano monitoring, disaster detection, monitoring of drinking water, monitoring of air pollution, smart environments, climate and weather forecasting.
- Healthcare: telemonitoring of human physiological data, tracking and monitoring of doctors and patients, machine health monitoring, remote virus monitoring.
- Military: battlefield surveillance, reconnaissance of enemy forces, monitoring and handling of nuclear, biological and chemical attacks, asset monitoring and management.
- Home applications: intrusion detection, home automation, high-security smart homes, building comfort, climate control in large buildings, remote metering.
- Interactive and control: pursuer-evader, industrial automation.
- Warehouse management: tracking of goods in retailer stores, tracking of articles in warehouse.
- Commercial applications: monitoring of material fatigue, monitoring of product quality, bridge monitoring, vehicle-based traffic monitoring.
- Precision agriculture: crop and livestock management, precise control of fertilizer concentrations.
- Public security and surveillance: tracking, identifications, and personalization.
- Scientific exploration: exploration of outer space and deep oceans.

In addition, mobile WSNs that use robots to carry sensors are also used for hazard processing. A mobile sensor network is a multi-robot system.

IEEE 802.15.4 (ZigBee) and 802.15.4a are most promising for WSNs. IEEE 1451 is the standard for smart sensor networks. IEEE 1451 makes it easier for different manufacturers to develop smart sensors and to interface those devices to networks. IEEE 802.15.4 (ZigBee) is especially suited to WSNs. It uses CSMA/CA and optional time-slotting protocol. While IEEE 802.15.1 (Bluetooth) supports moderate duty cycle with short battery life, but with a very high QoS and low latency, ZigBee supports very low duty cycle with an optional low latency and long battery life. However, Bluetooth Piconet supports a maximum number of eight nodes, whereas ZigBee supports up to 254 nodes in its network.

22.2 Routing

Multihop communication in a WSN can effectively overcome shadowing and path loss effects. Routing of wireless ad hoc networks faces severe challenges, such as securing broadcast wireless communication in an untrusted environment as well as finding the route itself. Routing is more difficult in MANETs due to the dynamic network topology. Enforcing collaboration is an important aspect in designing a secure and reliable wireless ad hoc network.

The traditional routing problems of unicast, multicast, anycast, and convergecast routing exist in wireless ad hoc networks for various purposes. Anycast refers to the case where a message is sent to an object name that has potentially multiple instantiations in the network, and any of these will do. Convergecast describes the notion of collecting data from several sources at a central point.

Flooding is a simple approach in which all nodes other than the target node forward a transmitted packet at least once. Flooding guarantees deliverability even if there are malicious nodes, as long as at least one adversary-free path exits. A node sends the received data or the management packets to its neighbors by broadcasting, unless a maximum number of hops for the packets is reached or the destination is gained. Flooding is not energy-/resource-aware, and it also introduces overlapping of measuring region and implosion of data.

In the gossiping protocol [25], each node forwards the incoming data/packets to a randomly selected neighbor node (including the neighbor sending to it). Gossiping can save energy, and solve the implosion problem, but it still cannot avoid overlapping. Gossiping causes delays in propagation of data through the nodes. Alternatively, location-based routing protocols utilize positional information to make routing decisions.

While flooding and gossiping are simple approaches that combine routing and data transmission, they cause huge network traffic. Many other routing protocols divide communication into routing operations (route discovery and route maintenance) and data communication. Once a route is selected, multiple data packets can be transmitted along the same route during the route lifetime. This reduces the traffic overhead, since only the nodes on the route retransmit the packet. Routing protocols that take this two-step strategy are classified as proactive, reactive, and hybrid.

Proactive routing protocols, also known as *table-driven routing* protocols, utilize tables to determine the next hop to reach the required destination. Each node maintains current routing information regarding the connectivity of every node to all other nodes that participate in the network. Every node has information on the network topology by propagating periodic updates. All nodes are able to make immediate decisions on the forwarding of a specific packet, thus the session establishment time is greatly reduced. This causes a constant amount of signaling traffic in the network for keeping routes up-to-date. Early proactive routing schemes were based on distance-vector routing (DVR) protocols that exploit the distributed Bellman-Ford (DBF) algorithm for computing. The shortest path in a weighted graph represents the network. Destination-sequenced distance-vector (DSDV) routing [53] is a DBF-algorithm-based routing protocol for ad hoc networks.

Optimized-link-state routing (OLSR) [32] is a proactive MANET routing protocol that is based on the link-state algorithm.

Reactive routing is a source-initiated on-demand approach. A route is created only when the source node requires a route to a specific destination. Reactive route discovery is usually based on a query/reply exchange, where flooding is used to reach the desired destination. Once the route discovery is completed and a path is established, the buffered data packets are sent. An established route is maintained depending on whether it is required or not. Reactive routing avoids the potential wastage of channel capacity and energy. However, a discovery delay is incurred at the time of transmission. The dynamic-source-routing (DSR) protocol [36] and ad hoc on-demand distance-vector (AODV) routing [54] are two well-known reactive routing protocols, and they are, respectively, standardized for routing Internet packets over MANETs and other wireless ad-hoc networks in IETF RFCs 4728 and 3561. AODV tries to improve DSR by maintaining routing tables at the nodes, so that the data packets do not contain the source-destination path.

An example of hybrid routing protocols is the zone-routing protocol (ZRP) [24]. The network is divided into zones, where every zone is a r-hop neighborhood of a node. The intra-zone routing protocol is a proactive routing protocol, while the inter-zone routing protocol is a reactive routing protocol.

QoS guarantees are required by most multimedia and other applications. Due to unreliable wireless channel, node mobility, lack of centralized control, channel contention, and limited node resources, QoS guarantees in MANETs are most challenging. The QoS routing protocol is required for any QoS solution since it selects the nodes, if any, to serve the requirements of the applications. Consequently, it plays a crucial role in data session admission control. In this case, both routes and QoS states are required to be discovered. Hybrid route discovery/state discovery schemes are feasible. One solution is to discover the routes proactively, but the QoS state is only sought for a QoS-constrained data session. Another solution is the QoS state discovery following the proactive/reactive route discovery.

Routing protocols for MANETs can also be classified into topological routing and geographic routing. A hybrid routing protocol can take advantage of both schemes by exploiting topological information, geographic information, and hierarchy information at the same time.

22.3 Security

22.3.1 Security problems

Security for wired and wireless communications provides the functions of:

- *Confidentiality*. Data is only revealed to the intended user. Confidentiality refers to data privacy, which is typically achieved via cryptographic mechanisms.

- *Integrity*. Data cannot be modified during the transmission. Integrity protects against data tampering, typically achieved through message authentication codes or by one-way hash functions.
- *Authentication*. An entity must pass the identification before getting served. Authentication can be considered a special integrity class, known as *origin integrity*.
- *Authorization*. An entity must get authorization before it takes action.
- *Access control*. Ensures that only authorized actions can be performed.
- *Nonrepudiation*. Prevents an entity from denying its actions. There is safeguard to prevent a node from denying that it signed a given message.
- *Availability*. Ensures that authorized actions can take place. For example, routes retuned by routing protocols must remain functional.

Security also needs to protect the privacy of users. The security problems for wireless ad hoc networks are basically the same as those for other wireless networks.

Security is implemented on multiple layers. At the link layer, strong encryption should be used to prevent over-the-air eavesdropping, and access control is applied to prevent unauthorized users from using the precious wireless channels. Authentication and authorization can be implemented by using AAA (authentication, authorization, and accounting) protocols and some other schemes such as IPsec and firewalls, at the network layer in the IP stack. At the transport layer, TLS (Transport Layer Security) is employed by using the certificate architecture. At the application layer, a number of schemes can be selected, such as digital signature and certificates, to provide both privacy and authentication.

In addition to providing confidentiality and data integrity between source and destination, cryptographic mechanisms are an essential requirement for MANET routing operations. Secure routing schemes rely on authentication (digital signatures or keyed MACs) to ensure only trusted insiders make up the routes.

Attacks in wireless ad hoc networks

Two kinds of security attacks on a wireless ad hoc network may take place: active attacks and passive attacks. An active attack is a deliberate disruption of network activity. Some examples of active attacks are:

- *Denial-of-service attack*. An attacker excludes legitimate users from network services by flooding, or by occupying valuable network resources, or destroying configuration information.
- *Wormhole attack*. An attacker connects two distant wormhole nodes using a private wormhole link. Due to tunneling, routing between the two regions tends to use the shortcut wormhole link, and the two wormhole nodes may drop packets, or selectively forward packets to avoid detection, or spy on the packets going through.
- *Blackhole or sinkhole attack*. This is suction of packets towards a malicious node by advertising itself as having the shortest path to all nodes in the network. The attacker then drops all the received packets, or performs selective forwarding, or alters the data passing through it, or can monitor and analyze the traffic.

- *Grayhole attack.* An attacker drops all data packets but lets control messages route through it. This makes the detection of the attack more difficult.
- *Man-in-the-middle attack.* The attacker impersonates the receiver with respect to the sender, and the sender with respect to the receiver.
- *Sybil attack.* Launched by a malicious node that illegally acquires multiple identities.
- *Blackmail attack.* The attacker causes false identification of a legitimate node as a malicious node. An attacker may blackmail a good node and tell other nodes in the network to add that node to their blacklists, thus avoiding the victim node in future routes.
- *Routing table poisoning.* A malicious node sends false routing updates, resulting in suboptimal routing, network congestion, or network partition.
- *Misrouting attack.* A malicious node sends a data packet to the wrong destination by modifying the final destination address of the data packet or by forwarding a data packet to the wrong next hop in the route to the destination.
- *Sleep deprivation.* An attacker prevents victim nodes from sleeping by bombarding them with legitimate requests, or making requests to the victims only as often as necessary to keep them awake. This makes the nodes soon run out of energy.
- *Detour attack.* An attacker adds some virtual nodes into a route during the route discovery phase so as to divert the traffic to other routes that appear to be shorter and might contain malicious nodes which could create other attacks. This attack is specific to source routing protocols.
- *Rushing attack.* A malicious node attempts to tamper with RouteRequest packets, modify the node list, and hurry its packet to the next node.

A passive attack can be a selfish node attack or eavesdropping. It may drop packets so as to prioritize its own traffic or conserve its energy, and this does not cause any intentional damage to the network. In order to stimulate cooperation between nodes, collaboration schemes are proposed based on credit, reputation, game theory, and other measures. Countermeasures to many security problems are discussed in [15].

Key management has remained a challenging issue in wireless ad hoc networks. This is especially true for WSNs due to the constraints of node resources. Before a network can exchange data securely, encryption keys must be established among nodes. Key distribution refers to the distribution of multiple keys among the nodes. Key management includes the processes of key setup, the initial distribution of keys, and key revocation.

Intrusion detection attempts to identify those systems or users who are trying to break into and misuse a system without authorization and those who have authorization, but are abusing their privileges. Intrusion detection can be formulated as a pattern classification problem, in which observed activities are classified as normal or intrusive.

22.3.2 Encryption

Encryption changes the plaintext (a stream or block of data to be protected) into ciphertext by using an encryption key, to combat eavesdropping. The receiver can decrypt the ciphertext to extract the plaintext by using the same or a different key, and the two cases are known as *symmetric key encryption* and *asymmetric key encryption*.

The theoretical framework for symmetric key-based cryptography was also established by Shannon in 1949 [59]. At the transmission end, there are two sources, namely, a message source and a key source. The key source is transmitted to the receiving end by some means, which are supposed to be not interceptible. The message source is enciphered by using the key, and the resulting ciphertext is transmitted to the receiving end by possibly interceptible means. The receiving end deciphers the received ciphertext to recover the message. The encipher and decipher algorithms may be known to the public.

WEP (Wired Equivalent Privacy) is a symmetrical key encryption algorithm. A block of plaintext is bitwise XORed with a pseudorandom key sequence of equal length. The key is generated by the WEP algorithm. WEP is used in the IEEE 802.11 family of standards. WEP has some security problems, and in October 2002, the Wi-Fi Alliance announced WPA (Wi-Fi Protected Access) to supercede WEP. WPA is designed to work with existing 802.11-based products and has forward compatibility with IEEE 802.11i. It addresses all the known shortcomings of WEP.

The AES (Advanced Encryption Standard) algorithm, which became effective as a U.S. government standard in May, 2002 to replace its predecessor, the DES (Data Encryption Standard), which became a U.S. government standard in 1976. AES is a symmetric key, block-ciphering encryption system. It is specified as a link-layer encryption method. It operates on a data block of 128 bits, which is organized in a 4×4 array of bytes called a *state*. The size of the encryption key can be 128, 192, or 256 bits long. It has strong cryptographic properties and is easy to implement in hardware or software, and thus has gained widespread adoption. Triple DES (3DES) is a block cipher formed from a DES cipher by using it three times.

Symmetric key encryption has the problem of secure key distribution over the network. Asymmetric key encryption solves this problem by using two keys, a public key, which is disclosed to the public, and a private key, which is kept secret. When a ciphertext is encrypted using one key, it can be decrypted only by the other key. Both keys are simultaneously generated by using the RSA (Rivest-Shamir-Adleman) algorithm [56]. For example, user A can send data to user B by encrypting the data using the public key of user B, and only user B can decrypt the data using his private key. This approach is widely used on the Internet for authentication, nonrepudiation and message integrity, and digital certificates. A Pretty Good Privacy (PGP) version of RSA is available in the public domain for noncommercial use on the Internet. Key management is a difficult problem in multihop communications, such as for wireless ad hoc or sensor networks, since most centralized secure protocols cannot be directly applied in distributed wireless networks.

AES is used for both authentication and encryption of the secure payload in Wi-Fi and WiMedia. For Wi-Fi, the security features are enhanced by IEEE 802.11i, which relies on the AES. HiperLAN/2 uses the DES or 3DES algorithm to secure the data. In IEEE 802.16 family, the DES, 3DES, AES, and RSA algorithms are used for the encryption services in the MAC security sublayer.

A hash function maps a long message to a fixed-length bit string, called the *digest* of the message. Hash functions are extensively used in cryptographic protocols such as in digital signature protocols. To prevent attack, a message digest is required to be at least 128 bits long. Construction of secure hash functions can be based on the Merkle-Damgard (MD) algorithm [57]. Some practical hash function families are MD, SHA, and RIPEMD.

Due to multihop wireless communications and physical exposure of the nodes, the security of wireless ad hoc networks is much more challenging. The nodes may belong to different authorities, and this introduces the cooperation of nodes and trust management. Attacks may be external and internal, and the internal attacks are not easy to prevent.

22.4 Technical overview for wireless ad hoc networks

As MANET and WSN nodes are energy-constrained devices, power consumption is a major consideration for a wireless ad hoc network. Power consumption is divided into two parts: the idle mode and the transmit/receive mode. Power-aware routing is targeted to maximize the lifetime of the network. It is formulated as an NP-complete problem.

Standards for wireless ad hoc networks

Ad hoc or mesh mode is now being supported by more and more IEEE 802 standards, including 802.11, 802.15, 802.16, and 802.20. The IEEE 802.11s ESS (extended service set) Mesh Task Group was formed in May 2004 to address the need for wireless mesh in wireless LANs. The standard, which is now underway, promises to be a highly interoperable wireless LAN mesh standard for wireless ad hoc networks.

The IEEE 802.16a standard incorporates the mesh mode in addition to the point-to-multiple-points (PMP) mode defined in the baseline IEEE 802.16. The IEEE 802.16a mesh mode targets only fixed broadband applications, and is not compatible with the existing PMP mode. The IEEE 802.16 working group established the Mobile Multihop Relay (MMR) study group in July 2005 to study the possibility for extending the PMP mode to support MSs by using multihop relaying techniques. Two mechanisms for scheduling data transmission in mesh mode are defined in IEEE 802.16: centralized and distributed scheduling. In centralized scheduling, the BS works like a clusterhead and arranges user nodes to share the channel in different time slots. In distributed scheduling, every node competes for channel access using a pseudo-random election algorithm based on the scheduling information of its two-hop neighbors. Data subframes are then allocated using a request-grant-confirm three-way handshaking protocol.

The IEEE 802.15.5 task group, initiated in May 2004, is currently working to provide a framework for interoperable, stable, and scalable wireless mesh topologies for wireless PAN devices. ZigBee (IEEE 802.15.4) is now a market-ready wireless mesh standard.

Capacity regions

Capacity regions for wireless ad hoc networks have been studied in [60]. These regions describe the set of achievable rate combinations between all source–destination pairs in the network under various transmission strategies, such as variable-rate transmission, single-hop or multihop routing, power control, and SIC. Numerical results indicate that multihop routing, the ability for concurrent transmissions, and SIC significantly increase the capacity of ad hoc and multihop cellular networks. On the other hand, gains from power control are significant only when variable-rate transmission is not used. Also, time-varying flat-fading and node mobility actually improve the capacity. Finally, multihop routing greatly improves the performance of energy-constraint networks.

Flat or hierarchical architecture

A multihop ad hoc network can have a flat or hierarchical architecture. A flat structure of a large number of nodes leads to a number of challenges in terms of network organization, gathering of the information, throughput, routing, and energy management; that is, a flat structure leads to low scalability and a complex network-wide coordination.

Hierarchical architectures can solve the scalability problem. Clustering enables in-network data aggregation: Nodes transmit their information to their clusterheads, and the clusterheads aggregate the received information and forward it to the destination. A clustering scheme can be either identifier-based clustering, topology-based clustering, or energy-based clustering. Periodic re-clustering is applied to select nodes with higher residual energy as clusterheads. The clustering hierarchy is illustrated in Fig. 22.3.

LEACH (low-energy adaptive clustering hierarchy) [26] is a clustering scheme that uses randomized rotation of the clusterheads and the corresponding clusters to distribute the work load evenly among nodes in the network. In LEACH, relay is not used, and each clusterhead directly reports to the sink. The chain scheme further improves the energy efficiency in such a way that each node transmits only to its closest neighbor. In a tree-based network, nodes are organized into a tree, and data aggregation is performed at intermediate nodes along the tree until it reaches the root node.

Coverage

One fundamental issue in wireless ad hoc networks is the coverage problem. Coverage is the spatial sensing range of a node. It has to be coordinated among nodes to avoid

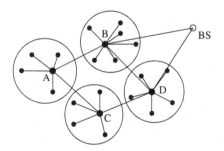

Figure 22.3   A clustering hierarchy. Nodes A, B, C, D are clustering heads. Nodes in each cluster transmit their data to their cluster heads. Each cluster head transmit its aggregated data to the destination using other cluster heads as relays.

redundancy, by considering communication distance and other characteristics of sensing tasks. Sensor node placement and dispatch are two important deployment problems in WSNs [63].

The target field may have coverage holes, that is, areas not covered by any node, due to random spatial deployment, presence of obstructions, or node failures. Routing holes, areas devoid of any nodes, may also occur in the deployed topology. A wireless ad hoc network may fail if some of the nodes cannot sense or relay the data.

Adversaries may deliberately damage wireless ad hoc networks by setting malicious nodes to jam the communication, yielding jamming holes, or to overwhelm regions in the ad hoc network by denial-of-service attacks to hinder their operation, which is normally based on trust [2]. Wireless ad hoc networks are highly susceptible to denial-of-service attacks due to their inherent limited resources coupled with use of insecure wireless channels [2].

QoS for wireless ad hoc networks

QoS support in ad hoc networks involves QoS model, QoS resource reservation signaling, QoS routing, and QoS MAC. A QoS mode specifies an architecture. QoS signaling coordinates the behavior of QoS routing, QoS MAC, and other components. The QoS routing process [11] searches for a path with enough resources but does not reserve resources, and thus, resources can be assured when QoS signaling needs to reserve resources. All upper-layer QoS components are dependent on and coordinate with the QoS MAC protocol. Trustworthiness-based QoS routing is a secure routing protocol with QoS support, which includes secure route discovery, secure route setup, and trustworthiness-based QoS routing metrics [68].

QoS for wireless ad hoc networks can be application-specific QoS or network QoS. For application-specific QoS (such as in WSNs), QoS parameters can be selected as coverage, exposure, measurement errors, and optimum number of active sensors. For network QoS (such as MANET), we consider how the underlying communication network can deliver the QoS-constrained data while efficiently utilizing network resources.

Congestion control

Congestion not only causes packet loss, but also leads to excessive energy consumption. In addition, congestion control is necessary to improve fairness and provide better QoS in case of wireless multimedia networks [65].

Two types of congestion could occur in wireless ad hoc networks. Node-level congestion, caused by buffer overflow in the node, is common in conventional networks. For wireless networks that are based on CSMA-like protocols, link-level congestion may arise: Collisions occur when multiple nodes access the channel at the same time. Congestion can result in packet loss, increased delay, and decrease of both the link utilization and the overall throughput. Retransmissions consume additional energy, which is critical to WSNs. Congestion has a direct impact on energy-efficiency and QoS.

Congestion control can be implemented through network resource management and traffic control. The first approach increases network resource when congestion occurs. In a wireless network, power control and multiple radio interfaces can increase the bandwidth and weaken congestion. Traffic control adjusts traffic at source nodes or intermediate nodes for congestion control. This approach is helpful in saving network resource. Traffic control can be on an end-to-end or on a hop-by-hop basis.

Network capacity

With only fixed rate point-to-point communications, finite bandwidth and large power, the transport capacity of planar networks scales like $\sqrt{nA}$, where n is the node density and A is the area occupied by the network, and the per-node throughput capacity is a decreasing function of n [20]. This result has been extended to three dimensions in [19]. Under minimal conditions on the attenuation, and for networks with constant n, the rate per communication pair in a wireless ad hoc network is shown to tend to zero as the number of users gets large [43]. For an UWB wireless ad hoc network with a power constraint and the explicit use of link adaptation, the uniform throughput per node is shown to increase as a function of n [49], and this is in contrast to the decreasing per-node throughput given in [20]. This capacity advantage justifies the promise of UWB technology for wireless ad-hoc networks.

The transmission capacity of a wireless ad hoc network is defined as the maximum spatial intensity of successful transmissions below a specified outage probability. Upper and lower bounds for the transmission capacity of wireless ad hoc networks with SIC receivers are developed in [64], for both perfect and imperfect SIC. Any imperfections in the interference cancellation rapidly degrade its usefulness. Only a few, often just one, interfering nodes need to be canceled in order to get the majority of the available performance gain.

22.5 Technical overview for wireless sensor networks

In additions to the common features of wireless ad hoc networks, WSNs have their own specific features. Many routing, power management, and data dissemination protocols have been designed for WSNs to support various unique requirements and constraints to make WSNs practically useful and operable. Power management is of critical importance. It is targeted at the twofold goal: minimizing the total energy consumption of all nodes in the network (to extend the lifetime of each node) and achieving a homogeneous consumption of energy throughout the network (to maximize the lifetime of the network).

For WSN design, energy-efficiency is treated as the top priority. The design should also be scalable for large networks. Latency and bandwidth efficiency are of secondary concern, while per-node fairness is usually not considered.

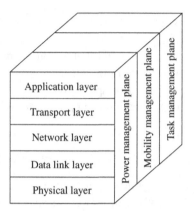

Figure 22.4   **The WSN protocol stack.**

Protocol stack

The architecture of protocol stack [5] at the sink and sensor nodes is shown in Fig. 22.4. The protocol stack integrates energy-aware routing, data aggregation, power efficiency, and cooperation of sensor nodes. This protocol stack is made up of five layers (physical, data link, network, transport, and application layers), and three planes (power management, mobility management, and task management planes).

The physical layer addresses the needs of robust modulation, transmission and reception. The management of time and time-related operations in WSNs is essential to timing events and network synchronization operations. The 915 MHz ISM band has been suggested for WSNs. Due to the hardware, power, and cost constraints, binary modulation and UWB technique are attractive for baseband transmission.

The data link layer is responsible for the multiplexing of data streams, data frame detection, medium access and error control. It ensures reliable point-to-point and point-to-multipoint connections in a communication network. It can be decomposed into MAC and LLC sublayers. Collision avoidance is the basic task of all MAC protocols. The MAC protocol for a multihop self-organizing WSN must achieve two goals, namely, the creation of the network infrastructure and the fair and efficient sharing of communications resources between sensor nodes. For error control, ARQ is limited by the retransmission energy cost and overhead in multihop WSNs, and simple FEC is preferred.

The transport layer is necessary when the sensor network is accessed via Internet or other external networks. The transport layer helps to maintain the flow of data if the application requires it. TCP can be used at the sink for interface to the Internet, but special transport layer protocol can be used between the sink node and the sensor nodes. A UDP-type protocol can be used between the sink and sensor nodes, since each sensor node is restricted by resources, and is not based on global addressing.

The network layer is in charge of routing the data supplied by the transport layer. The network layer shares some commonalities with that for ad hoc networking, but has new features to meet the more stringent requirements on scalability, energy efficiency and data-centricness. Finally, the application layer is dependent on the sensing tasks, such as a sensor management protocol.

Flow control through a WSN can be separated into two problems: routing and congestion control. Routing can be addressed as a discrete optimization problem with optimality criteria defined in terms of distance, energy consumed, time delay through the network, and bandwidth maximization. It is similar to an NP-hard traveling salesman problem as it aims to select data routes (hops) with minimal cost between the wireless nodes forming a graph. Congestion control finds and regulates the optimal flow rates between the network nodes subject to network capacity constraints. Priority-based Congestion Control Protocol (PCCP) is an upstream congestion control protocol for WSNs [62]. Based on the congestion degree and node priority index, PCCP utilizes a cross-layer optimization and imposes a hop-by-hop approach to control congestion.

The power management plane manages the power consumption among the three operations (sensing, computation, and wireless communications). The mobility management plane detects and registers the movement/mobility of sensor nodes as a network control primitive. The task management plane balances and schedules the events' sensing and detecting tasks from a specific area by cooperation of sensor nodes.

As WSNs are highly task-specific, and most often stand-alone implementations, they do not need to follow the strictly layered design, and a revolutionary approach to cross-layer design can be applied.

MAC protocols

The MAC determines when and how a node can access the medium and send its data. Design of MAC protocols in WSNs needs to consider collision avoidance, energy efficiency, scalability in node density, latency, fairness, throughput, and bandwidth utilization. Although many MAC protocols and algorithms are available for wireless ad hoc networks, they are not well suited to WSNs.

Energy is the primary concern for MAC design for WSNs. Collision is a first source of energy waste. Idle listening and overhearing are two other sources of energy waste. Overhearing occurs when a node receives packets that are destined to other nodes. Another major source is control packet overhead. Two techniques that have been explored for MAC to enable multihop routing while powering down radios for power conservation are:

- *Adaptive duty-cycling.* The set of nodes whose radios are powered down is carefully chosen such that a network backbone is continually maintained, while radios of nonbackbone nodes can be put to sleep. To balance the load, the active subset of nodes are adaptively cycled, based on parameters such as available energy and radio coverage.
- *Wakeup on demand.* This technique uses nodes with multiple radios: A low-power radio is used exclusively to wake up the high power radio, when the need arises. This is much like a paging channel in cellular networks. Such a technique is especially useful when bandwidths and data rates are higher, and this warrants the use of multiple radios.

Sensor-MAC (S-MAC) is a well-known MAC for WSNs [66]. The main goal of S-MAC is to reduce energy consumption caused by idle listening, collisions, overhearing, and control overhead. Since most of the time a wireless node is in idle listening, S-MAC turns off the node's transceiver periodically. This introduces some latency. Fairness among nodes can be tolerable as all the nodes are expected to serve the same application. Collision avoidance in S-MAC is similar to the distributed coordinated function (DCF) for 802.11 ad hoc mode.

ZigBee (IEEE 802.15.4) also has a MAC that is suited to WSNs. Receiver sensitivities for ZigBee are -85 dBm for 2.4 GHz and -92 dBm for 868/915 MHz. ZigBee allows up to 254 nodes. In ZigBee, the MAC layer defines two types of nodes: reduced function devices and full function devices. Full function devices are equipped with a full set of MAC layer functions, which enables them to act as a network coordinator or a network end-device.

Operating systems and databases

A sensor node is an embedded system, and an embedded operating system can be used. Middleware is often used to bridge the gap between the operating system and the application. This eases the development of distributed applications. Due to resource constraints, unreliability of wireless networks, and diversity, middleware for WSN presents a number of new challenges.

TinyOS was the first operating system specifically designed for WSNs [27]. TinyDB is an inquiry processing system for WSNs that operates on TinyOS by using a data-centric approach [46]. Both TinyOS and TinyDB were initially developed at the University of California, Berkeley, in cooperation with Intel Research. TinyOS is now being developed by a consortium, the TinyOS Alliance, and is open source, making it easy for developers to customize it as required.[1] It has the largest user base and is the benchmark for other operating systems. TinyOS is an event-driven operating system designed for sensor nodes that have very limited resources. The core requires 400 bytes of code and data memory, combined. A series of components is supplied to program conveniently, to easily acquire, and to process data acquired by sensors.

In a data-centric approach, each node keeps its data, and nodes execute retrieval and aggregation, with on-demand based operation to deliver the data to external applications. TinyDB supports a data aggregation function via SQL query, which supports selection, projection, determining sampling rate, group aggregation, user-defined aggregation, event trigger, lifetime query, setting storing point and simple joining.

Time synchronization

Time synchronization is necessary to provide temporal coordination among all the nodes engaged in a collaborative and distributed interaction. This can be achieved by sending timing messages to the target sensors. Approaches to time synchronization can be categorized as sender-receiver synchronization, receiver-receiver synchronization, and

---

[1] http://www.tinyos.net/

receive-only synchronization [52]. Sender-receiver synchronization is based on two-way message exchanges between a pair of nodes. In receiver-receiver synchronization, the nodes receive a beacon packet from a common sender, and then compare among them (excluding the common sender) their relative clock offset based on their receiving times of the beacon packet. Receive-only synchronization minimizes the use of timing messages for the purpose of saving energy. Many time synchronization protocols are introduced in [52].

Location

To be context-aware, location tracking is a major concern in WSNs. Localization is necessary for assigning geographic coordinates to each node in the WSN. This is necessary for monitoring the roaming path of a moving object and determining from which location a measurement came. Location discovery in WSNs poses significant design challenges. Because of constraints in size and cost, it is impractical to use GPS receivers at the sensor nodes. Moreover, WSNs may be deployed in regions where satellite signals may not be available.

Approaches to WSN location typically deploy a few known nodes called *beacons* or *anchors*, which are aware of their own locations. Due to the cost constraint of the sensor nodes, localization utilizing RSSI measurement is practical [71]. Many IEEE 802.x wireless standards, e.g. IEEE 802.15.4, support RSSI measurement to evaluate link quality. Tracking multiple interacting targets is more challenging due to the curse of dimensionality.

Security

The security of large, densely deployed WSNs requires efficient key distribution and management mechanisms. Security requirements in WSNs are similar to those for ad hoc networks. In addition, WSNs have two specific requirements:

- *Survivability*: ability to provide a minimum level of service in the presence of power exhaustion, failures or attacks;
- *Degradation of security services*: ability to change security level according to resource availability.

The security problems may be physical or logical in nature. Physical security problems can be caused by damaged or stolen nodes, inaccurate measurement, jamming, battery exhaustion attacks, as well as malicious nodes. Logical security problems can be due to eavesdropping, injection, attacks to protocols from external or internal nodes, as well as identity and instruction integrity.

TinySec is the first fully implemented, minimal protocol for link-layer cryptography in WSNs [37]. It is feasible to implement acceptable cryptographic protection for WSNs entirely in software. TinySec is a research platform that is easily extensible and has been incorporated into higher-level protocols.

Battery technologies

Battery technologies are important for WSNs. Sensor node lifetime is mainly dependent on the battery lifetime. There are three common battery technologies for WSNs — alkaline, lithium, and nickel metal hydride. The AA alkaline battery provides a cheap, high capacity energy source, but it has a wide voltage range and a large physical size. In addition, lifetime beyond five years is not possible due to battery self-discharge. Lithium batteries provide a very compact power source. They also provide a constant voltage supply that decays little as the battery is drained. Unlike alkaline batteries, lithium batteries are able to operate at temperatures down to −40°C. Nickel metal hydride batteries are easily rechargeable, but they have a significant decrease in energy density. If there is no energy-harvesting source, a nonrechargeable battery is a good choice since it has higher energy density. Some designs try to harvest energy from the ambient energy in the form of electromagnetic radiation, heat, or mechanical energy [4]. Energy harvesting techniques are now becoming mature, and both TI and Intel have solar energy harvesting solutions for powering WSNs.

22.6 Data aggregation and routing for WSNs

22.6.1 Data aggregation

Data or in-network aggregation is tightly coupled with how data is gathered at the sensor nodes as well as how packets are routed through the network, and has a significant impact on overall network efficiency. Sensor nodes may generate significant redundant data. Similar packets from a single or multiple nodes can be aggregated (fused) at intermediate nodes with the objective of reducing the number of transmissions and improving the detection performance, thereby reducing resource consumption and increasing network lifetime. Data aggregation can take two forms. The first one leads to a size reduction by combining and compressing data coming from different nodes and then sending the reduced data to save node resources. The second form merges packets from different sources into one packet without data processing to reduce the overhead: These packets from different sources may carry different physical quantities. Data fusion requires the transmission of data and control messages. Data aggregation can be based on a certain aggregation function (e.g., duplicate suppression, minima, maxima, and average). This technique has been used in a number of routing protocols. Data aggregation is performed at a sink, and it may require some form of synchronization among nodes.

Sensor data collection can be event-driven or demand-driven. In the event-driven case, when one or more sensor nodes have detected an event, the nodes report the data to the monitoring station. In the demand-driven case, the monitoring station enquires the sensor nodes, and the sensor nodes send their data in response to an explicit request. For example, a damage monitoring system is event-driven, whereas an inventory control system is demand-driven. A hybrid system can be both event- and demand-driven.

In flat networks, all sensor nodes play the same role and are equipped with approximately the same battery power. Data aggregation is accomplished by data-centric routing, and the sink usually transmits a query message to the sensors. A flat network causes excessive communication and computation burdens at the sink node. This will deplete the battery power at the sink node rapidly, causing a breakdown of the network. Hierarchical data aggregation performs better in view of scalability and energy efficiency. This approach conducts data fusion at some nodes to reduce the traffic to the sink.

The LEACH protocol [26] organizes the sensor nodes into clusters for data fusion. The sensor nodes transmit data to a designated node (clusterhead), where data aggregation is performed. The clusterhead in each cluster transmits the fused data from several sensors in its cluster to the sink.

Sensor network with mobile access (SENMA) [47] is a hierarchical architecture in which sensors are orchestrated by a few powerful mobile access points. By allowing sensors to propagate data directly to mobile access points over multiaccess channels, SENMA relieves sensors from energy-consuming network functions; this improves energy efficiency over the multihop ad hoc architecture by orders of magnitude. Mobile access points retrieve the data from the sensors and deliver it to a remote control center. In SENMA, sensors are driven by mobile access points; the presence of a strong beacon from the mobile access point significantly simplifies timing recovery and synchronization.

The common static sink strategy does not scale with the network size and increases the network congestion. Moreover, it may limit the network lifetime as the one-hop neighbors of the sink are the bottleneck of the network; in this sense, mobile sinks are more energy-effective, thus enhancing the network lifetime. Also, sink mobility makes security attack more difficult, and it may improve the network connectivity. There are several data dissemination protocols with mobile sinks available [22].

22.6.2 Routing

Data aggregation requires a different forwarding paradigm than classical routing. Classical routing protocols typically forward data along the shortest path to the destination in terms of some specified metric. In order to aggregate data to minimize energy expenditure, nodes should route packets based on packet content and choose the next hop in order to promote in-network aggregation. This type of data forwarding is often referred to as *data-centric routing*.

Many routing, power management, and data dissemination protocols are designed for WSNs, with energy-awareness as a primary concern. Routing protocols in WSNs may depend on the application and network architecture. Routing in WSNs is very challenging as WSNs are inherently distinguished from other wireless networks like MANETs or cellular networks.

Geographic routing and data-centric routing are present in WSNs. Geographic routing uses a region rather than a node identifier as the target of a packet; any node within the region can be accepted as a destination node and can receive and process a message. Geographic routing is important for WSNs when requesting sensor data from a region (e.g., request temperature in a room). Data-centric routing is a core abstraction of WSNs, and it finds routes from multiple sources to a single destination, which allows in-network data aggregation.

In addition to the basic routing techniques, such as flooding and gossiping, for general wireless ad hoc networks, other most well-known routing protocols for WSNs are SPIN [40], directed diffusion [31], and rumor routing [8].

SPIN

SPIN (Sensor Protocols for Information via Negotiation) [40] is a family of adaptive protocols (e.g., SPIN-1 and SPIN-2) for WSNs. It is a data-centric routing scheme. SPIN avoids the drawbacks of flooding by utilizing meta-data negotiation and resource-adaptive algorithms. Energy efficiency is achieved by sending meta-data (processed sensor data). Also, each node has its own resource manager that keeps track of energy resource consumption and is polled by the nodes before data transmission to extend its lifetime. Topological changes are localized, since each node needs to know only its single-hop neighbors. In SPIN, sensors advertise the availability of data, allowing interested nodes to query that data. This data negotiation mechanism eliminates the redundant data transmission. However, SPIN does not establish any path for data transmission, and thus, it does not guarantee the delivery of data.

Directed diffusion

Directed diffusion [31] is a popular data aggregation/data-centric routing paradigm for WSNs. All data generated by sensor nodes are characterized by attribute-value pairs. The data coming from different sources are combined en route to minimize the number of transmissions. All sensor nodes are application-aware, which allows energy savings by selecting empirically good paths, and by caching and processing data in the network.

At the beginning, the sink sends out its interest or task description to all nodes. The interest entry contains a timestamp field and several gradient fields. Each node then stores the interest entry in its cache. As the interest is propagated throughout the WSN, the gradients from the source back to the sink are set up. When the source has data satisfying the interest, it sends the data along the interest's gradient path, and at the same time the sink refreshes and reinforces the interest when it starts to receive data from the source.

In directed diffusion, the sink queries the sensor nodes if a specific data is available by flooding some tasks. Since it is data-centric, all communication is neighbor-to-neighbor with no need for node addressing. The on-demand nature as well as no need for maintaining global network topology makes direct diffusion highly energy-efficient. As it is based on a query-driven data delivery model, directed diffusion is not a good choice as a routing protocol for applications such as environmental monitoring, where continuous data delivery to the sink is required.

REEP (Reliable and Energy Efficient Protocol) is a data-centric, energy-aware routing protocol that is based on directed diffusion [70]. REEP is an interactive on-demand protocol, in which path establishment can be done based on the choice of any user or an application. Each node maintains an energy threshold and participates in path setup with adequate energy. REEP uses the request priority queue for loop prevention and alternate path setup in case of failed path, without invoking periodic flooding. The performance of REEP is superior to directed diffusion [70].

Rumor routing

Directed diffusion uses initial flooding to discover good paths between the sources and sinks. If only a small amount of data is requested from the nodes, the use of flooding is unnecessary. Rumor routing [8] is a variation of directed diffusion and is mainly intended for applications where geographic routing is not feasible, because a coordinate system is not available or the phenomenon of interest is not geographically correlated. The idea is to route the queries to the nodes that have observed a particular event to retrieve information about the occurring events. Rumor routing maintains only one path between the source and destination.

Routing for flat and hierarchical architectures

For each class of architecture, the routing protocols are different. Flat routing protocols are similar to the conventional multihop ad hoc routing protocols, where each sensor node determines its parent node(s) to forward data packets. Using this approach, all the nodes can reach the BS irrespective of their position.

For hierarchical (tree-based or cluster-based) architecture, the nodes are organized in hierarchical form. Aggregation trees (clusters) have the robustness problems, since each node has to send the partial result of its aggregation to a single parent or clusterhead. Many routing protocols are based on clustering. Such algorithms work well in relatively static networks. An alternative solution is that each node can send the data to its (possibly) multiple neighbors, and hence, data may flow from the sources to the sinks along multiple paths and aggregation may be performed by each node.

22.7 Relay, user cooperation, and MIMO relay networks

Relay is closely related with routing. It is a fundamental technique for wireless ad hoc networks. Relay is also being considered for cellular coverage enhancement through efforts like IEEE 802.16j (Multihop Relay Specification for 802.16) to incorporate relay capabilities in mobile WiMAX (IEEE 802.16e). IEEE 802.16j will be fully compatible with IEEE 802.16e mobile and subscriber stations, but a BS specific to 802.16j will be required for relays to operate. Multihop communications is one of the enabling techniques for 4G, and it is being standardized for IEEE 802.16m and LTE-Advanced.

In wireless ad hoc networks, cooperation among nodes can be exploited to improve system performance. The fundamental ideas behind cooperation can be found in the literature on the relay channel [13]. The cooperative relaying transmission results in cooperative diversity, and it is definitely an option for the future-generation wireless networks.

22.7.1 Relay

Basic relay schemes are amplify-and-forward (AF) , decode-and-forward (DF) , and compress-and-forward (CF) protocols [42]. In the AF scheme, the receiver receives repeated codes from two separate transmitters, except that the relay transmitter amplifies its own receiver noise. The destination can decode its received signal by using a diversity combining technique. In the DF scheme, the relay may decode the entire source codeword or perform symbol-by-symbol decoding, forward the data possibly using a different code, and let the destination perform full decoding, depending on the resources available at the relay. Again, the destination can perform diversity combining. In the CF scheme, the relay forwards the quantized/compressed/estimated version of its observations.

The transmitter cooperative channel model is illustrated in Fig. 22.5, where S stands for source, R for relay, and D for destination. In the receiver cooperative case, the relay is close to the receiver.

DF assumes that the source-relay channel is outage-free so that the relay always retrieves the source packet correctly. However, practical wireless channels experience fading from time to time, and this results in a drastic performance degradation. DF is limited by direct transmission between the source and relay. When the channel between the source and the relay, $h_{s,r}$, is very poor, the source simply continues its transmission to the destination, by using repetition or more powerful codes. Selection relaying schemes can be used to select among several relay or repetition coding schemes so as to adapt to channel measurements between the cooperating nodes [42]. A single bit of feedback from destination to relay to indicate the success/failure of the source transmission to the destination allows DF to achieve the full diversity available in the single-antenna relay channel [42].

AF can better resolve the problem of poor source-relay channel, but sampling, amplifying, and retransmitting have certain complexity; in addition, the destination may receive transmissions that are too noisy from the source and the relay due to the use of analog signals. In the low SNR and low outage probability regime, AF performs very poorly, whereas the bursty AF protocol is optimal and achieves the outage capacity of the network [6]. Moreover, bursty AF can achieve this performance without a-priori channel knowledge at the receivers. In contrast, DF is strictly suboptimal in this regime.

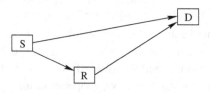

Figure 22.5   The transmitter cooperative channel model.

CF can be a better solution to fading, but DF generally outperforms CF when the source-relay channel is in good condition [39]. DF approaches capacity when the relay is near the transmitter, whereas CF is close to optimum when the relay is near the receiver [39]. Therefore, in [51], DF is used in transmitter cooperation, while CF is used in receiver cooperation.

Wyner-Ziv cooperation [28] is a practical CF scheme. It exploits Wyner-Ziv codes in wireless user cooperation to help combat inter-user outage. The encoder of a Wyner-Ziv system usually consists of two parts: quantizer and index encoder. The relay, after performing channel decoding, stores soft reliability information. The direct source copy from the source to the destination is viewed as the decoder side information, and the relay can borrow ideas from Wyner-Ziv coding to process and transmit the soft reliability information. Wyner-Ziv cooperation is most useful when the noisy copy retrieved at the relay has a high correlation with the original source. In [21], the SISO decoder at the relay decodes the received signal and generates various a-posteriori probabilities, and then forwards the soft information to the destination. At the destination, the soft information is used as extrinsic information for iterative decoding. Compared to AF or DF with MRC, the use of soft value as a-priori information affects the decoding behavior and improves the performance with less overhead [21], since in MRC a large amount of information is exchanged in order to combine the paths. Similar soft relaying schemes are given in [9, 16].

For high data rates, incremental relaying protocols can be used. Limited feedback, which indicates the success/failure of the direct transmission, is transmitted from the destination [42]. This feedback information can be broadcast to both the source and the relay. If the source-destination channel is sufficiently good, the feedback indicates that direct transmission is successful and the relay does nothing; otherwise, the relay forwards what it received from the source. In the latter case, the destination will combine the two transmissions. This results in gains over AF by increasing the rate for good source-relay conditions [42]. Incremental relaying protocols can be viewed as extensions of incremental redundancy, or HARQ. Performance for various cooperative diversity schemes can be accordingly derived [17].

For the DF strategy, a low-complexity coherent demodulator at the destination has been derived in [61]. The coherent demodulator is in the form of a weighted combiner, termed *cooperative MRC*. The weights are selected adaptively to account for the quality of both the source-relay-destination and source-destination links. The coherent demodulator achieves the maximum possible diversity, regardless of the underlying constellation. Its error performance tightly bounds that of ML demodulation with DF relaying. Multihop, multibranch cooperative diversity is also considered. The cooperative MRC can achieve almost the same performance as the optimum ML detector does, but with a much lower complexity [61].

For a fixed rate of transmission, increased power savings can be achieved by using the feedback for power control. Only a few bits of feedback are sufficient to achieve most of the gains of the optimal power control with CSI at the transmitter. The finite-rate feedback results are given for the low-complexity, full-diversity AF protocol in [3]. When no CSI is available at the transmitters, transmitting with equal power at the source and relay is close to optimal, especially for relays positioned close to the source [3]. The outage probability for AF with one-bit feedback has at least a fourth-order diversity [3]. To obtain large performance improvements over constant power transmission, it is imperative to have feedback for CSI to allow for temporal power control [3]. Maximizing the diversity order over a fading relay channel by using DF with quantized CSI feedback has been investigated in [38].

Opportunity-driven multiple access (ODMA) is a misnomer as it is not a true multiple access, but a relaying protocol. ODMA breaks the path into smaller hops, and makes use of other MS in the cell to relay the signal. The optimum routing tries to achieve the minimum total path loss for the transmission. It is shown in [23] that a relaying system with distributed intelligence can exhibit an average reduction of 21 dB in transmission power. Multihop relaying routing protocols have been investigated for CDMA air interface in conventional cellular scenarios [23]. The ODMA protocol utilizes the path loss between the terminals as the metric to determine the routing [1, 23]. ODMA was once proposed for UTRA-TDD [1], but was finally dropped due to implementation problems.

In addition, joint source-channel coding for cooperative relay networks is shown to significantly improve the performance compared to the conventional scheme of source coding followed by cooperative channel coding [18].

22.7.2 User cooperation

Multiple antennas can provide spatial diversity. However, it is not easy to implement this in a mobile unit due to size restriction. Another form of spatial diversity can be achieved by the cooperation of the in-cell users. When two users have their data to send, this is not a simple relay problem. The user cooperative diversity improves the achievable rate region and reduces the outage probability, and a user with more fading benefits most from the cooperation [17, 58].

Two-user cooperation model

Assuming that two users, denoted U1 and U2, have their own data to send. They cooperate with a target to send this information to the receiver with an increased rate. The receiver can be a BS or another MS in an ad hoc network, and here we denote it as BS to differentiate it from the two partners. Each MS receives an attenuated version of the partner's transmitted signal and combines it with its own data to construct its transmit signal. The channel model is illustrated in Fig. 22.6. User cooperation is a method of transmit diversity for mobile users. Each user has information of his own to send rather than simply act as the other user's relay.

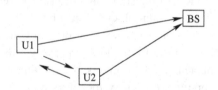

Figure 22.6   The cooperative channel model.

Information-theoretic analysis shows that the achievable rate region with user cooperation is always larger than the noncooperative region, implying that there is always a cooperative strategy in which both users benefit from cooperation [58]. The MS with the better channel can help the other MS achieve some acceptable level of performance while sacrificing only a small fraction of its own data rate. A simple, modified MRC detector, referred to as λ-*MRC*, is given in [58].

The channel is described as [58]

$$Y_0(t) = h_{10}X_1(t) + h_{20}X_2(t) + n_0(t), \tag{22.1}$$

$$Y_1(t) = h_{21}X_2(t) + n_1(t), \tag{22.2}$$

$$Y_2(t) = h_{12}X_1(t) + n_2(t), \tag{22.3}$$

where $Y_0(t)$, $Y_1(t)$, and $Y_2(t)$ are the received baseband signals at the BS, user 1 and user 2, respectively, during one symbol period, X_i, $i = 1, 2$, is the signal transmitted by user i, n_i's, $i = 0, 1, 2$, are the zero-mean white complex Gaussian noises with variance σ_i^2 at the BS, user 1, and user 2, respectively, and the fading coefficients h_{ij}'s are zero-mean complex Gaussian random variables with Rayleigh distributed variance, which remain constant over at least one symbol period. To eliminate the direct effect of X_i on Y_i, $i = 1, 2$, it is necessary to isolate the transmitted signal from the received one. This may be achieved by using two separate channels or two colocated antennas. The channels h_{ij} can be estimated by the corresponding receiver, and $h_{21} = h_{12}$ due to the reciprocity of the channel.

In the two-user cooperative system, each node acts as a data source as well as a relay, and each node operates under overall (source + relay) power and bandwidth constraints. The Alamouti code is used to achieve transmit diversity. This is a distributed space-time coding scheme. A source first broadcasts to its destination and all potential relays. The relays then forward the information to the destination. The system can provide a diversity order of 2, and it can even provide multiplexing gain.

All the two-user orthogonal cooperative diversity protocols, including AF, selection relaying, and incremental relaying, achieve full diversity of 2, though at a loss of spectral efficiency due to half-duplex operation and possibly at the cost of additional receiver hardware [42].

Two-user coded cooperation

In the two-user cooperation, typically, if an error is detected within the received packet, a user transmits his own data to BS instead of relaying the other user's data. Transmit diversity cannot be obtained. In the two-user coded cooperation scheme [33], each user tries to transmit incremental redundancy for his partner. Whenever that is not possible, the users automatically revert back to a noncooperative mode. Coded cooperation is managed automatically through code design, and feedback between users is not needed. The codeword of each user is partitioned into two portions: One partition is transmitted by the user, and the other by the partner. By allowing different code rates and partitions, coded cooperation provides a great degree of flexibility to adapt to channel conditions. RCPC, turbo [72] and LDPC [10, 29] codes can be used for coded cooperation.

In the two-user coded-cooperation scheme [30, 33], each block is encoded with an FEC code, so that for an overall rate R, we have N total coded symbols allocated for each source block. The two users cooperate by dividing the transmission of their coded source blocks into two successive time segments, called *frames*. In the first frame, each user transmits a rate-R_1 codeword ($R_1 > R$) with N_1 symbols. This itself is a valid codeword which can be decoded to obtain the original information. If the user successfully decodes the partner's first-frame transmission, the user computes and transmits N_2 additional parity symbols for the partner's data in the second frame, where $N_1 + N_2 = N$. These additional parities are selected such that they can be combined with the first-frame codeword to produce a more powerful rate-R codeword. If the user does not successfully decode the partner, N_2 additional parity bits for his own data are transmitted. Each user always transmits a total of N bits per source block over the two frames, and the users only transmit in their own orthogonal multiple-access channels.

The two users act independently in the second frame, with no knowledge of whether their own first frame was correctly decoded by their partner. In other words, no feedback is assumed between the cooperating partners. As a result, there are four possible cooperative cases for the transmission of the second frame based on the first frame decoding results at the two users, and different protocols are designed. Outage expressions confirm that full diversity is achieved by coded cooperation [30]. An alternative cooperative method is given in [16]: If the CRC decoder detects error within the received packet, a user transmits a soft decision symbol which is obtained from the decoded data. Thus, transmit diversity always can be achieved.

Coded cooperation achieves full diversity when both users cooperate and gives impressive gains in BER for the case of slow fading. Space-time cooperation aims to capture space-time diversity in fast fading by applying space-time coding principles to coded cooperation. When a source initiates transmission to its destination, many relays potentially receive the transmission. Those MSs that can fully decode the transmission utilize a space-time code to cooperatively relay to the destination. Repetition-based and space-time-coded cooperative diversities in nonergodic, multipath fading have been analyzed in terms of outage probability in [41]. Both the algorithms can provide full spatial diversity in the number of cooperating MSs, and the effective coding or SNR gain/loss can be characterized as a function of the interterminal average SNRs. Space-time-coded cooperation offers higher diversity order and higher spectral efficiencies than repetition-based algorithms.

22.7.3 MIMO relay networks

Virtual MIMO schemes can be used to improve communication performance of wireless ad hoc networks. Virtual MIMO is a network-based approach. Different nodes in the network can act as elements of an antenna array. As the array elements are not physically connected, a large amount of information must be sent to the combining nodes. In this approach, multiple individual single-antenna nodes cooperate for energy-efficient communications. Examples of virtual MIMO schemes are an Alamouti-encoded scheme for single-hop transmissions [58], a STBC-encoded scheme without perfect synchronization [44], and a clustered topology-based, time-division, DF multirelay, space-time coded MIMO channel for multihop transmissions [12].

Clustered relaying

A wireless network with fading and a single source-destination pair is considered in [7]. The information reaches the destination via multiple hops through a sequence of layers of single-antenna relays. Each layer of relays can be treated as virtual MIMO. At high SNR, AF is optimal in terms of degrees of freedom, as it achieves the degrees of freedom equal to that of a point-to-point MIMO system. Hence, lack of coordination in relay nodes does not reduce the achievable degrees of freedom. The performance of this AF strategy degrades with increasing network size. This phenomenon is analyzed by finding the tradeoffs between network size, rate, and diversity.

A multihop WSN with nodes grouped in cooperative clusters that exploits transmit and receive cooperation among cluster nodes is proposed in [14]. Multihop transmission is carried out by concatenating single cluster-to-cluster hops, where every cluster-to-cluster link is a cooperative distributed MIMO channel. Transmit diversity is exploited through a DF scheme using two time slots: the intracluster slot for data sharing within the cluster and the intercluster slot for transmission between clusters. At the receiver side, a distributed reception protocol is devised based on selection diversity. The multihop cooperative WSN is designed for minimum end-to-end outage probability by deriving the optimum time and power allocated on the intracluster and intercluster slots of every single hop, subject to a per-link energy constraint. The scheme achieves a diversity equal to that of the equivalent MIMO system. The cluster-based strategy limits cooperation inside the clusters, and this reduces synchronization and resource management complexity.

A cluster-based cooperative strategy using incremental redundancy cooperative coding for slow Rayleigh fading is proposed in [45]. The collaborative cluster consists of $M - 1$ nodes between the sender and the destination. The transmitted message is encoded using a mother code which is partitioned into M blocks, each assigned to one of the M transmission slots. In the first slot, the sender broadcasts its information by transmitting the first block, and its helpers attempt to decode this message. In the remaining slots, each of the next $M - 1$ blocks is sent either through a helper which has successfully decoded the message or directly by the sender where a dynamic schedule is based on the ACK-based feedback from the cluster. An average frame-error rate upper bound and its asymptotic version are derived as a function of the average fading channel SNRs and the code threshold. Based on the asymptotic bound, diversity, coding, and transmission energy gains in both the high and moderate SNR regimes have been investigated for three different scenarios: transmitter clustering, receiver clustering, and cluster hopping [45].

Distributed STBCs

Distributed space-time coding achieves cooperative diversity without CSI at the relays. Using this scheme, antennas of the distributive relays work as transmit antennas of the sender and generate a space-time code at the receiver. It achieves the maximal diversity when the transmit power is infinitely large. OSTBCs are particularly suitable for transmission in the network setting using distributed space-time coding, where each node transmits a different column of the OSTBC matrix [41]. Distributed space-time coding achieves higher diversity than selection DF using the same orthogonal designs, when there is more than one relay [35].

Distributed STBCs are designed for wireless networks that have a large set of single-antenna relay nodes, but only a small, a-priori unknown subset of nodes is active at any given time [67]. The signal transmitted by an active relay node is the product of an information-carrying code matrix and a unique node signature vector of length N_c. This approach allows convenient exploitation of existing coherent, differential, and non-coherent STBCs originally designed for N_c co-located antennas, and accordingly allows for low-complexity coherent, differential, and noncoherent detection. Existing STBCs designed for $N_c > 2$ co-located antennas are favorable choices for the code matrix, guaranteeing a diversity order of $d = \min\{N_s, N_c\}$ in case of N_s active nodes. The performance loss entailed by the distributed implementation for $N_s > N_c$ has been analytically characterized.

Transmitter and receiver cooperations

In [50], capacity improvement from transmitter and receiver cooperations is investigated in a two-transmitter, two-receiver network with phase fading and full CSI available at all nodes. The transmitters cooperate by first exchanging messages over an orthogonal transmitter cooperation channel, then encoding jointly with dirty-paper coding. The receivers cooperate by using Wyner-Ziv CF over an analogous orthogonal receiver cooperation channel. Transmitter cooperation outperforms receiver cooperation and improves capacity over noncooperative transmission under most operating conditions when the cooperation channel is strong. However, a weak cooperation channel limits the transmitter cooperation rate; in this case, receiver cooperation is more advantageous. Transmitter-and-receiver cooperation, i.e., a scheme that uses transmitter cooperation as well as receiver cooperation, offers sizable additional capacity gain over transmitter-only cooperation at low SNR, whereas at high SNR transmitter cooperation alone captures most of the cooperative capacity improvement.

Under quasi-static channels, when all the nodes have equal average transmit power along with full CSI, transmitter cooperation outperforms receiver cooperation, whereas the opposite is true when power is optimally allocated among the cooperating nodes but CSI is only available at the receiver [51]. When the system is under optimal power allocation with full CSI, DF transmitter cooperation outperforms CF receiver cooperation in terms of capacity. Similar conclusions follow under Rayleigh fading in the high-SNR regime. In a large cluster of M cooperating nodes, transmitter cooperation without CSI at the transmitter, or receiver cooperation under equal power allocation, provides no capacity gain in a static channel, and at most a constant capacity gain that fails to grow with M in a fading channel.

A virtual MIMO architecture with V-BLAST receiver has been proposed for WSNs in [34]. The scheme does not require transmitter-side cooperation. The energy and delay efficiencies are derived for networks with both single- and multiple-antenna data gathering nodes.

In [48], a group of receivers collaborating to decode a message that none of the receivers can individually decode is considered. Improved least-reliable bits (I-LRB) collaborative decoding is applied for user cooperation in bandwidth-limited scenarios. Collaborative decoding utilizes reliability information and information about competing paths in SISO decoders. The cooperating nodes iterate between a process of information exchange and decoding. The approach offers a significant performance advantage over a constrained-overhead, incremental form of MRC. Collaborative decoding may be considered a CF scheme.

Diversity-multiplexing tradeoff

A general multiple-antenna network with multiple sources, multiple destinations, and multiple relays is considered in [69] in terms of the diversity-multiplexing tradeoff. In case of a full-duplex relay, while DF achieves optimal diversity-multiplexing tradeoff when each of the nodes has one antenna, it may not maintain its good performance when the degrees of freedom in the direct link are increased, whereas CF continues to perform optimally. For a half-duplex relay, CF achieves optimal diversity-multiplexing tradeoff as well. For a system with multiple relays, each node with a single antenna, even under the ideal assumption of full-duplex relays and a clustered network, this virtual MIMO system can never fully mimic a real MIMO diversity-multiplexing tradeoff.

Problems

22.1 For a source and destination, show how the multi-hop network provides a significant energy saving over a single-hop network for the same distance. [Hint: The N-hop transmission has a power advantage over a single-hop transmission: $\eta = \frac{P_{N-hop}}{NP_{1-hop}} = N^{\gamma-1}$, where γ is the propagation exponent.]

22.2 WSNs use multiple sensors. Increasing the sensor density by a factor of n, show the SNR is improved by $10\log_{10} n$ dB.

22.3 For a power metric d^γ with the propagation exponent $\gamma \geq 2$ and d is the distance between the sender and the receiver, show that relay traffic along an intermediate collinear node is always a better choice. When the power metric is $d^\gamma + a$, with constant $a > 0$, is the proposition true?

22.4 Consider a geographical region of 100 square meters. Assume that N nodes are distributed in this region according to a uniformly random distribution and that each node can communicate for a distance of R meters. Find the average number of nodes $E[N]$ as a function of R, for $1 \leq R \leq 10$, if the network is fully connected. Note that $E[N]$ can be calculated by averaging.

References

[1] 3GPP, *Physical layer items not for inclusion in Release'99*, Tech. Report TR 25.833 V1.1.0, ETSI, Apr 2000.

[2] N. Ahmed, S. S. Kanhere & S. Jha, The holes problem in wireless sensor networks: a survey. *ACM Mobile Comput. & Commun. Rev.*, **1**:2 (2005), 1–14.

[3] N. Ahmed, M. A. A. Khojastepour, A. Sabharwal & B. Aazhang, Outage minimization with limited feedback for the fading relay channel. *IEEE Trans. Commun.*, **54**:4 (2006), 659–669.

[4] Y. Ammar, A. Buhrig, M. Marzencki, B. Charlot, S. Basrour, K. Matou & M. Renaudin, Wireless sensor network node with asynchronous architecture and vibration harvesting micro power generator. In *Proc. Joint sOc-EUSAI Conf.*, Grenoble, Oct 2005, 1–6.

[5] I. F. Akyildiz, W. Su, Y. Sankarasubramaniam & E. Cayirci, A survey on sensor networks. *IEEE Commun. Mag.*, **40**:8 (2002), 102–114.

[6] A. S. Avestimehr & D. N. C. Tse, Outage capacity of the fading relay channel in the low-SNR regime. *IEEE Trans. Inf. Theory*, **53**:4 (2007), 1401–1415.

[7] S. Borade, L. Zheng & R. Gallager, Amplify-and-forward in wireless relay networks: rate, diversity, and network size. *IEEE Trans. Inf. Theory*, **53**:10 (2007), 3302–3318.

[8] D. Braginsk & D. Estrin, Rumor routing algorithm for sensor networks. In *Proc. ACM WSNA*, Atlanta, GA, Sep 2002, 22–31.

[9] T. Bui & J. Yuan, A decode and forward cooperation scheme with soft relaying in wireless communication. In *Proc. IEEE SPAWC*, Helsinki, Finland, Jun 2007, 1–5.

[10] A. Chakrabarti, A. de Baynast, A. Sabharwal & B. Aazhang, Low density parity check codes for the relay channel. *IEEE J. Sel. Areas Commun.*, **25**:2 (2007), 280–291.

[11] S. Chakrabarti & A. Mishra, QoS issues in ad hoc wireless networks. *IEEE Commun. Mag.*, **39**:2 (2001), 142–148.

[12] A. D. Coso, S. Savazzi, U. Spagnolini & C. Ibars, Virtual MIMO channels in cooperative multi-hop wireless sensor networks. In *Proc. 40th Ann. Conf. Inf. Sci. Syst. (CISS)*, Princeton, NJ, Mar 2006, 75–80.

[13] T. M. Cover & A. A. El Gamal, Capacity theorems for the relay channel. *IEEE Trans. Inf. Theory*, **25** :5 (1979), 572–584.

[14] A. del Coso, U. Spagnolini & C. Ibars, Cooperative distributed MIMO channels in wireless sensor networks. *IEEE J. Sel. Areas Commun.*, **25**:2 (2007), 402–414.

[15] R. Falk, C.-T. Huang, F. Kohlmayer & A.-F. Sui, Security in wireless mesh networks. In Y. Zhang, J. Luo & H. Hu, eds., *Wireless Mesh Networking* (Boca Raton, FL: Auerbach, 2007), pp. 183–223.

[16] Y. Fukuyama, O. Takyu, K. Adachi & M. Nakagawa, Relay method of scheduling soft decision symbol based on the result of error detecting code in cooperative communication. *IEICE Trans. Fundamentals*, **E90-A**:11 (2007), 2404–2412.

[17] S. Glisic, *Advanced Wireless Communications: 4G Technologies*, 2nd edn (Chichester, UK: Wiley-IEEE, 2007).

[18] D. Gunduz & E. Erkip, Source and channel coding for cooperative relaying. *IEEE Trans. Inf. Theory*, **53**:10 (2007), 3454–3475.

[19] P. Gupta & P. R. Kumar, Internets in the sky: the capacity of three dimensional wireless networks. *Commun. Inf. Syst.*, **1**:1 (2001), 33–50.

[20] P. Gupta & P. R. Kumar, The capacity of wireless networks. *IEEE Trans. Inf. Theory*, **46**:2 (2000), 388–404.

[21] Y. Hairej, A. Darmawan, H. Morikawa, Cooperative Diversity using Soft Decision and Distributed Decoding. In *Proc. 16th IST Mobile Wireless Commun. Summit*, Budapest, Hungary, Jul 2007, 1–5.

[22] E. B. Hamida & G. Chelius, Strategies for data dissemination to mobile sinks in wireless sensor networks. *IEEE Wireless Commun.*, **15**:6 (2008), 31–37.

[23] T. J. Harrold & A. R. Nix, Intelligent relaying for future personal communications systems. In *IEE Colloquium on Capacity & Range Enhancement Techniques for 3G*, London, UK, Feb 2000, 9/1–9/5.

[24] Z. J. Haas, The routing algorithm for the reconfigurable wireless networks. In *Proc. IEEE ICUPC*, San Diego, CA, Oct 1997, **2**, 562–566.

[25] S. M. Hedetniemi & A. Liestman, A survey of gossiping and broadcasting in communication networks. *Networks*, **18**:4 (1988), 319–349.

[26] W. R. Heinzelman, A. Chandrakasan & H. Balakrishnan, Energy-efficient communication protocol for wireless microsensor networks. In *Proc. IEEE HICSS*, Maui, HI, Jan 2000, 3005–3014.

[27] J. Hill, R. Szewczyk, A. Woo, S. Hollar, D. Culler & K. Pister, System architecture directions for networked sensors. *ACM SIGPLAN Notices*, **35**:11 (2000), 93–104.

[28] R. Hu & J. T. Li, Practical compress-forward in user cooperation: Wyner-Ziv cooperation. In *Proc. IEEE ISIT*, Seattle, WA, Jul 2006, 489–493.

[29] J. Hu & T. M. Duman, Low density parity check codes over wireless relay channels. *IEEE Trans. Wireless Commun.*, **6**:9 (2007), 3384–3394.

[30] T. E. Hunter, S. Sanayei & A. Nosratinia, Outage analysis of coded cooperation. *IEEE Trans. Inf. Theory*, **52**:2 (2006), 375–391.

[31] C. Intanagonwiwat, R. Govindan & D. Estrin, Direct diffusion: A scalable and robust communication paradigm for sensor networks. In *Proc. ACM MobiCom*, Boston, MA, Aug 2000, 56–67.

[32] P. Jacquet, P. Muhlethaler, T. Clausen, A. Laouiti, A. Qayyum & L. Viennot, Optimized link state routing protocol for ad hoc networks. In *Proc. IEEE INMIC*, Lahore, Pakistan, Dec 2001, 62–68.

[33] M. Janani, A. Hedayat, T. E. Hunter & A. Nosratinia, Coded cooperation in wireless communications: space-time transmission and iterative decoding. *IEEE Trans. Signal Process.*, **52**:2 (2004), 362–371.

[34] S. K. Jayaweera, V-BLAST-based virtual MIMO for distributed wireless sensor networks. *IEEE Trans. Commun.*, **55**:10 (2007), 1867–1872.

[35] Y. Jing and H. Jafarkhani, Using orthogonal and quasi-orthogonal designs in wireless relay networks. *IEEE Trans. Inf. Theory*, **53**:11 (2007), 4106–4118.

[36] D. B. Johnson & D. A. Maltz, Dynamic source routing in ad hoc wireless networks. *Mobile Comput.*, **353** (1996), 153–181

[37] C. Karlof, N. Sastry & D. Wanger, TinySec: a link layer security architecture for wireless sensor networks. In *Proc. ACM SenSys*, Baltimore, MD, Nov 2004, 162–175.

[38] T. T. Kim, G. Caire & M. Skoglund, Decode-and-forward Relaying with quantized channel state feedback: an outage exponent analysis. *IEEE Trans. Inf. Theory*, **54**:10 (2008), 4548–4564.

[39] G. Kramer, M. Gastpar & P. Gupta, Cooperative strategies and capacity theorems for relay networks. *IEEE Trans. Inf. Theory*, **51**:9 (2005), 3037–3063.

[40] J. Kulik, W. R. Heinzelman & H. Balakrishnan, Negotiation base protocols for disseminating information in wireless sensor networks. *Wireless Netw.*, **8** (2002), 169–185.

[41] J. N. Laneman & G. W. Wornell, Distributed space–time-coded protocols for exploiting cooperative diversity in wireless networks. *IEEE Trans. Inf. Theory*, **49**:10 (2003), 2415–2425.

[42] J. N. Laneman, D. N. C. Tse & G. W. Wornell, Cooperative diversity in wireless networks: efficient protocols and outage behavior. *IEEE Trans. Inf. Theory*, **50**:12 (2004), 3062–3080.

[43] O. Leveque & I. E. Telatar, Information-theoretic upper bounds on the capacity of large extended ad hoc wireless networks. *IEEE Trans. Inf. Theory*, **51**:3 (2005), 858–865.

[44] X. Li, M. Chen & W. Liu, Application of STBC-encoded cooperative transmissions in wireless sensor networks. *IEEE Signal Process. Lett.*, **12**:2 (2005), 134–137.

[45] R. Liu, P. Spasojevic & E. Soljanin, Incremental redundancy cooperative coding for wireless networks: cooperative diversity, coding, and transmission energy gains. *IEEE Trans. Inf. Theory*, **54**:3 (2008), 1207–1224.

[46] S. R. Madden, M. J. Franklin, J. M. Hellerstein & W. Hong, TinyDB: an acquisitional query processing system for sensor networks. *ACM Trans. Database Syst.*, **30**:1 (2005), 122–173.

[47] G. Mergen, Q. Zhao & L. Tong, Sensor networks with mobile access: energy and capacity considerations. *IEEE Trans. Commun.*, **54**:11 (2006), 2033–2044.

[48] A. 'Nayagam, J. M. Shea & T. F. Wong, Collaborative decoding in bandwidth-constrained environments. *IEEE J. Sel. Areas Commun.*, **25**:2 (2007), 434–446.

[49] R. Negi & A. Rajeswaran, Capacity of ultra wide band wireless ad hoc networks. *IEEE Trans. Wireless Commun.*, **6**:10 (2007), 3816–3824.

[50] C. T. K. Ng, N. Jindal, A. J. Goldsmith & U. Mitra, Capacity gain from two-transmitter and two-receiver cooperation. *IEEE Trans. Inf. Theory*, **53**:10 (2007), 3822–3827.

[51] C. T. K. Ng & A. J. Goldsmith, The impact of CSI and power allocation on relay channel capacity and cooperation strategies. *IEEE Trans. Wireless Commun.*, **7**:12 (2008), 5380–5389.

[52] K.-L. Noh, Y.-C. Wu, K. Qaraqe & E. Serpedin, Time synchronization for wireless sensor networks. In M. Ibnkahla, ed., *Adaptive Signal Processing in Wireless Communications* (Boca Raton, FL: CRC Press, 2009), pp. 373–410.

[53] C. E. Perkins & P. Bhagwat, Highly dynamic destination-sequenced distance-vector routing (DSDV) for mobile computers. In *Proc. ACM SIGCOMM*, London, UK, Aug 1994, 234–44.

[54] C. E. Perkins & E. M. Royer, Ad-hoc on-demand distance vector routing. In *Proc IEEE WMCSA*, New Orleans, LA, Feb 1999, 90–100

[55] C. E. Perkins, ed., *Ad Hoc Networking* (Reading, MA: Addison-Wesley, 2000).

[56] R. Rivest, A. Shamir & L. Adleman, A method for obtaining digital signatures and public-key cryptosystems. *Commun. ACM*, **21**:2 (1978), 120–126.

[57] P. Sarkar, Overview of cryptographic primitives for secure communication. In N. Sklavos, X. Zhang, eds., *Wireless Security and Cryptography* (Boca Raton, FL: CRC Press, 2007).

[58] A. Sendonaris, E. Erkip & B. Aazhang, User cooperation diversity – part I: system description; part II: implementation aspects and performance analysis. *IEEE Trans. Commun.*, **51**:11 (2003), 1927–1948.

[59] C. E. Shannon, Communication theory of secrecy systems. *Bell Syst. Tech. J.*, **28** (1949), 656–715.

[60] S. Toumpis & A. J. Goldsmith, Capacity regions for wireless ad hoc networks. *IEEE Trans. Wireless Commun.*, **2**:4 (2003), 736–748.

[61] T. Wang, A. Cano, G. B. Giannakis & J. N. Laneman, High-performance cooperative demodulation with decode-and-forward relays. *IEEE Trans. Commun.*, **55**:7 (2007), 1427–1438.

[62] C. Wang, B. Li, K. Sohraby, M. Daneshmand & Y. Hu, Upstream congestion control in wireless sensor networks through cross-layer optimization. *IEEE J. Sel. Areas Commun.*, **25**:4 (2007), 786–795.

[63] Y.-C. Wang, C.-C. Hu & Y.-C. Tseng, Efficient placement and dispatch of sensors in a wireless sensor network. *IEEE Trans. Mobile Comput.*, **7**:2 (2008), 262–274.

[64] S. P. Weber, J. G. Andrews, X. Yang & G. de Veciana, Transmission capacity of wireless ad hoc networks with successive interference cancellation. *IEEE Trans. Inf. Theory*, **53**:8 (2007), 2799–2814.

[65] G. Xie, M. N. S. Swamy & M. O. Ahmad, Joint optimal multipath routing and rate control for multidescription coded video streaming in ad hoc networks. *IEEE Trans. Multimedia*, **10**:8 (2008), 1687–1697.

[66] W. Ye, J. Heidemann & D. Estrin, An energy-efficient MAC protocol for wireless sensor networks. In *Proc. InfoCom*, New York, Jun 2002, **3**, 3–12.

[67] S. Yiu, R. Schober & L. Lampe, Distributed space-time block coding. *IEEE Trans. Commun.*, **54**:7 (2006), 1195–1206.

[68] M. Yu & K. K. Leung, A trustworthiness-based QoS routing protocol for wireless ad hoc networks. *IEEE Trans. Wireless Commun.*, **8**:4 (2009), 1888–1898.

[69] M. Yuksel & E. Erkip, Multiple-antenna cooperative wireless systems: a diversity-multiplexing tradeoff perspective. *IEEE Trans. Inf. Theory*, **53**:10 (2007), 3371–3393.

[70] F. Zabin, S. Misra, I. Woungang, H. F. Rashvand, N.-W. Ma & M. Ahsan Ali, REEP: data-centric, energy-efficient and reliable routing protocol for wireless sensor networks. *IET Commun.*, **2**:8 (2008), 995–1008

[71] R. Zemek, D. Anzai, S. Hara, K. Yanagihara & K. Kitayama, RSSI-based localization without a prior knowledge of channel model parameters. *Int J. Wireless Inf. Netw.*, **15**:3/4 (2008), 128–136.

[72] Z. Zhang & T. M. Duman, Capacity approaching turbo coding and iterative decoding for relay channels. *IEEE Trans. Commun.*, **53**:11 (2005), 1895–1905.

Computation of probabilities that involves a Gaussian process requires the calculation of the integral of the Gaussian pdf

$$p(x) = \frac{1}{\sigma\sqrt{2\pi}}e^{-\frac{(x-m)^2}{2\sigma^2}}. \tag{A.4}$$

The probability of a Gaussian random variable x exceeding x_0 is given by

$$\Pr(x \geq x_0) = \int_{x_0}^{\infty} p(x)dx. \tag{A.5}$$

Substituting $y = \frac{x-m}{\sigma}$, we have

$$\Pr\left(y > \frac{x_0 - m}{\sigma}\right) = \int_{\frac{x_0-m}{\sigma}}^{\infty} \frac{1}{\sqrt{2\pi}}e^{-y^2/2}dy, \tag{A.6}$$

where the kernel of the integral is the normalized Gaussian pdf with zero mean and unit variance.

The Q-function is defined as

$$Q(x) = \int_{x}^{\infty} \frac{1}{\sqrt{2\pi}}e^{-y^2/2}dy, \tag{A.7}$$

which is the complementary Gaussian cdf $Pr(y \geq x)$ for the pdf of the zero-mean, unit-variance Gaussian random variable. There is no closed-form solution for $Q(x)$. Some analysis results for the Q-function are listed here [1]:

$$Q(x) + Q(-x) = 1, \tag{A.8}$$

$$Q(x) > \frac{1}{x\sqrt{2\pi}}\left(1 - \frac{1}{x^2}\right)e^{-x^2/2}, \quad x > 1, \tag{A.9}$$

$$Q(x) < \frac{1}{x\sqrt{2\pi}}e^{-x^2/2}, \quad x > 0, \tag{A.10}$$

$$Q(x) \leq \frac{1}{2}e^{-x^2/2}, \quad x \geq 0, \tag{A.11}$$

$$Q(x) \leq e^{-x^2/2}, \quad x \geq 0, \tag{A.12}$$

$$Q(x) \leq \frac{1}{2}e^{-x\sqrt{2/\pi}}, \quad x \geq 0, \tag{A.13}$$

$$Q(x) = \frac{1}{\pi} \int_0^{\pi/2} e^{-x^2/(2\sin^2\theta)}, \quad x \geq 0, \tag{A.14}$$

$$Q(x) = \frac{1}{2} - \frac{1}{\sqrt{2\pi}} \left(x - \frac{x^3}{2} + \frac{x^5}{2 \cdot 4} - \frac{x^7}{2 \cdot 4 \cdot 8} + \cdots \right), \tag{A.15}$$

$$Q(x) = \frac{e^{-x^2/2}}{x\sqrt{2\pi}} \left(1 - \frac{1}{x^2} + \frac{1 \cdot 3}{x^4} - \frac{1 \cdot 3 \cdot 5}{x^6} + \cdots \right), \tag{A.16}$$

where (A.9) and (A.10) are very tight lower and upper bounds, obtained from (A.16). Equation (A.12) is the well-known Chernoff bound.

The error function erf(x) and the complementary error function erfc(x) are sometimes used. They are defined as

$$\mathrm{erfc}(x) = \frac{2}{\sqrt{\pi}} \int_z^\infty e^{-x^2} dx, \quad \mathrm{erfc}(x) = \frac{2}{\sqrt{\pi}} \int_0^z e^{-x^2} dx, \tag{A.17}$$

$$\mathrm{erf}(x) = 1 - \mathrm{erfc}(x). \tag{A.18}$$

They relate to the Q-function by

$$Q(x) = \frac{1}{2}\mathrm{erfc}\left(\frac{x}{\sqrt{2}} \right), \tag{A.19}$$

$$\mathrm{erfc}(x) = 2Q\left(x\sqrt{2} \right). \tag{A.20}$$

MATLAB provides the `erfc` and `erf` functions for numerical evaluation of these functions.

Reference

[1] S. Verdu, *Multiuser Detection* (Cambridge, UK: Cambridge University Press, 1998).

Appendix B Wirtinger calculus

The optimization of system parameters depends on the definition of certain criterion functions. An analytical optimum solution can be derived by setting the derivatives with respective to the adjustable system parameters to zero. In digital communications, signals and systems are typically represented in complex form. The Wirtinger calculus is defined for deriving the derivative of the criteria with respect to the complex parameters.

The complex derivative of a complex function $f(z)$, $f'(z) = \frac{df(z)}{dz}$, is a very fundamental notion in complex analysis. If f' exists in a region $\mathcal{X} \subset \mathcal{C}$, $f(z)$ is said to be *analytic* or *holomorphic* in $\mathcal{X}$. In order for $f(z)$ to be holomorphic,

$$f(z) = u(x, y) + jv(x, y), \quad z = x + jy, \tag{B.21}$$

the Cauchy-Riemann equations must be satisfied

$$\frac{\partial u(x, y)}{\partial x} = \frac{\partial v(x, y)}{\partial y}, \quad \frac{\partial v(x, y)}{\partial x} = -\frac{\partial u(x, y)}{\partial y}. \tag{B.22}$$

Then, $f'(z)$ can be expressed by [1]

$$\frac{df(z)}{dz} = \frac{\partial u(x, y)}{\partial x} + j\frac{\partial v(x, y)}{\partial x}. \tag{B.23}$$

In digital communications, we face the problem of optimization of a real function with respect to complex parameters. Optimization based on complex cost functions makes no sense, since no ordering operation is defined for complex numbers. The real function $f(z)$ is not holomorphic, unless it is a real constant.

The optimization of a real function of a complex variable, $f(z) = u(x, y)$, can be implemented as optimizing $u(x, y)$ with respect to two real variables x and y. The Wirtinger calculus is based on this idea, but gives a compact notation.

Definition (Wirtinger Calculus):

Given a (complex) function $f(z)$ of a complex variable $z = x + jy \in \mathcal{C}$, $x, y \in \mathcal{R}$, its derivatives with respect to z and z^* are defined, respectively, as

$$\frac{\partial f}{\partial z} = \frac{1}{2}\left(\frac{\partial f}{\partial x} - j\frac{\partial f}{\partial y}\right), \quad \frac{\partial f}{\partial z^*} = \frac{1}{2}\left(\frac{\partial f}{\partial x} + j\frac{\partial f}{\partial y}\right). \tag{B.24}$$

Based on the above definition, we have the following important results:

$$f(z) = cz \implies \frac{\partial f}{\partial z} = c, \quad \frac{\partial f}{\partial z^*} = 0, \tag{B.25}$$

$$f(z) = cz^* \implies \frac{\partial f}{\partial z} = 0, \quad \frac{\partial f}{\partial z^*} = c, \tag{B.26}$$

$$f(z) = zz^* \implies \frac{\partial f}{\partial z} = z^*, \quad \frac{\partial f}{\partial z^*} = z. \tag{B.27}$$

These results can be very easily verified. For example, for $f(z) = zz^*$, $\frac{\partial f}{\partial z^*}$ can be derived as

$$\frac{\partial f}{\partial z^*} = \frac{\partial}{\partial z^*}(zz^*) = \frac{1}{2}\left(\frac{\partial(x^2+y^2)}{\partial x} + j\frac{\partial(x^2+y^2)}{\partial y}\right) = \frac{1}{2}(2x + j2y) = z. \tag{B.28}$$

Differentiation using the Wirtinger calculus is similar to that with real functions with real variables. Note that z^* is treated as constant when differentiating with respect to z and vice versa. It can be verified that the sum, product, quotient, and chain rules for the derivatives of a real function still hold for Wirtinger calculus differentiation. For the product rule, given $f(z) = f_1(z)f_2(z)$, we have

$$\frac{\partial}{\partial z}\left[f_1(z)f_2(z)\right] = \frac{\partial f_1(z)}{\partial z}f_2(z) + f_1(z)\frac{\partial f_2(z)}{\partial z}. \tag{B.29}$$

For the chain rule, given $f(z) = h(g(z))$, $g(z) \in \mathcal{C}$, we have [1]

$$\frac{\partial f(z)}{\partial z} = \left.\frac{\partial h(w)}{\partial w}\right|_{w=g(z)} \cdot \frac{\partial g(z)}{\partial z} + \left.\frac{\partial h(w)}{\partial w^*}\right|_{w=g(z)} \cdot \frac{\partial g^*(z)}{\partial z}, \tag{B.30}$$

$$\frac{\partial f(z)}{\partial z^*} = \left.\frac{\partial h(w)}{\partial w}\right|_{w=g(z)} \cdot \frac{\partial g(z)}{\partial z^*} + \left.\frac{\partial h(w)}{\partial w^*}\right|_{w=g(z)} \cdot \frac{\partial g^*(z)}{\partial z^*}. \tag{B.31}$$

Unlike the complex derivative of a complex function, the Wirtinger derivative exists for all functions. For holomorphic function $f(z)$, the Wirtinger derivative with respect to z agrees with the ordinary derivative of a complex function.

For multiple complex variable systems, $z = (z_1, z_2, \cdots, z_n)^T \in \mathcal{C}^n$, and we have $f(z) \in \mathcal{R}$. The gradient can be defined as

$$\frac{\partial f}{\partial z} = \begin{bmatrix} \frac{\partial f}{\partial z_1} \\ \frac{\partial f}{\partial z_2} \\ \vdots \\ \frac{\partial f}{\partial z_n} \end{bmatrix}, \quad \frac{\partial f}{\partial z^*} = \begin{bmatrix} \frac{\partial f}{\partial z_1^*} \\ \frac{\partial f}{\partial z_2^*} \\ \vdots \\ \frac{\partial f}{\partial z_n^*} \end{bmatrix}. \tag{B.32}$$

At the optimum, these gradients are equal to zero vector. The following Wirtinger derivatives are important:

$$f(z) = c^T z = z^T c \implies \frac{\partial f(z)}{\partial z} = c, \quad \frac{\partial f(z)}{\partial z^*} = 0, \tag{B.33}$$

$$f(z) = c^T z^* = z^H c \implies \frac{\partial f(z)}{\partial z} = 0, \quad \frac{\partial f(z)}{\partial z^*} = c, \tag{B.34}$$

$$f(z) = z^H \mathbf{M} z = z^T \mathbf{M} z^* \implies \frac{\partial f(z)}{\partial z} = \mathbf{M}^T z^*, \quad \frac{\partial f(z)}{\partial z^*} = \mathbf{M} z, \tag{B.35}$$

where $c = (c_1, c_2, \cdots, c_n)^T \in \mathcal{C}^n$.

The gradient of a real function that has complex variables is usually defined as

$$\nabla = \begin{pmatrix} \frac{\partial}{\partial x_1} + j \frac{\partial}{\partial y_1} \\ \frac{\partial}{\partial x_2} + j \frac{\partial}{\partial y_2} \\ \vdots \\ \frac{\partial}{\partial x_n} + j \frac{\partial}{\partial y_n} \end{pmatrix}. \tag{B.36}$$

Thus,

$$\nabla f(z) = 2 \frac{\partial f(z)}{\partial z^*}, \tag{B.37}$$

$$(\nabla f(z))^* = 2 \frac{\partial f(z)}{\partial z}. \tag{B.38}$$

The gradient representation is not elegant, since it introduces a factor of 2. The Wirtinger Calculus is more elegant since it has clearer arithmetic rules.

Reference

[1] R. F. H. Fischer, C. Windpassinger, A. Lampe & J. B. Huber, MIMO precoding for decentralized receivers. In *Proc. IEEE ISIT*, Lausanne, Switzerland, Jun-Jul 2002, 496.

Index

Printed in the United States
By Bookmasters